PROGRESS IN BRAIN RESEARCH

VOLUME 146

NGF AND RELATED MOLECULES IN HEALTH AND DISEASE

PROGRESS IN BRAIN RESEARCH

VOLUME 146

NGF AND RELATED MOLECULES IN HEALTH AND DISEASE

EDITED BY

LUIGI ALOE

*Institute of Neurobiology and Molecular Medicine, National Research Council (CNR), Via K. Marx 15/43,
00137 Rome, Italy*

LAURA CALZÀ

*Department of Veterinary Morphophysiology and Animal Production DIMORFIPA, University of Bologna,
Via Tolara di Sopra 50, 40064 Ozzano Emilia, Bologna, Italy*

ELSEVIER

AMSTERDAM – BOSTON   HEIDELBERG – LONDON – NEW YORK – OXFORD
PARIS – SAN DIEGO – SAN FRANCISCO – SINGAPORE – SYDNEY – TOKYO
2004

ELSEVIER B.V.
Sara Burgerhartstraat 25
P.O. Box 211, 1000 AE
Amsterdam, The Netherlands

ELSEVIER Inc.
525 B Street, Suite 1900
San Diego, CA 92101-4495
USA

ELSEVIER Ltd
The Boulevard, Langford Lane
Kidlington, Oxford OX5 1GB
UK

ELSEVIER Ltd
84 Theobalds Road
London WC1X 8RR
UK

First edition 2004

Library of Congress Cataloging in Publication Data
A catalog record is available from the Library of Congress.

British Library Cataloguing in Publication Data
A catalogue record is available from the British Library.

ISBN: 0-444-51472-4 (volume)
ISBN: 0-444-80104-9 (series)
ISSN: 0079-6123

⊗ The paper used in this publication meets the requirements of ANSI/NISO Z39.48-1992 (Permanence of Paper).
Printed in The Netherlands.

List of Contributors

J. Alberch, Department of Cell Biology and Pathology, Medical School, IDIBAPS, University of Barcelona, Casanova 143, E-08036 Barcelona, Spain

E. Alleva, Behavioral Pathophysiology Section, Istituto Superior di Sanità (ISS), Viale Regina Elena 299, I-00161 Rome, Italy

L. Aloe, Institute of Neurobiology and Molecular Medicine, National Research Council (CNR) Viale Marx 15/43, I-00137 Rome, Italy

H.H. Althaus, Max-Planck Institute for Experimental Medicine, RU Neural Regeneration, H.-Reinstr. 3, D-37075 Göttingen, Germany

M. Anafi, Prescient NeuroPharma Inc., Laboratories of Protein Chemistry, Molecular Biology and Cell Biology, 96 Skyway Avenue, Toronto, ON M9W 4Y9, Canada

P. Anand, Peripheral Neuropathy Unit, Imperial College London, Department of Neurology, Hammersmith Hospital, Du Cane Road, London W12 0NN, UK

F. Angelucci, Institute of Neurology, Catholic University, Largo Gemelli 8, I-00168, Rome, Italy

A. Antonelli, Institute of Neurobiology and Molecular Medicine, National Research Council (CNR), Via K. Marx 15/43, 00137 Rome, Italy

H. Arien-Zakay, Department of Pharmacology and Experimental Therapeutics, School of Pharmacy, Faculty of Medicine, The Hebrew University of Jerusalem, P.O. Box 12065, Jerusalem 91120, Israel

P.A. Barker, Montreal Neurological Institute, McGill University, Montreal, QC H3A 2B4, Canada

P.F. Bartlett, Queensland Brain Institute, The University of Queensland, Brisbane, 4067 Qld, Australia

P.L. Belichenko, Department of Neurology and Neurological Sciences, Program in Neuroscience, Stanford University, Stanford, CA 94305, USA

A. Blesch, Department of Neurosciences-0626, University of California, San Diego, San Diego, La Jolla, CA 92093, USA

B. Blits, The Miami Project to Cure Paralysis, University of Miami School of Medicine, Miami, FL, USA

G.J. Boer, Netherlands Institute for Brain Research, Department of Neuroregeneration, Meibergdreef 33, 1105 AZ Amsterdam, The Netherlands

V.A. Botchkarev, Department of Dermatology, Boston University School of Medicine, Boston, MA, USA

N.V. Botchkareva, The Gillette Company, Needham, MA, USA

A. Braun, Immunology and Allergology, Fraunhofer Institute of Toxicology and Experimental Medicine, Nikolai-Fuchs-Str. 1, D-30625 Hannover, Germany

R.L. Burch, Department of Pharmacology and Physiology, University of Rochester School of Medicine and Dentistry, 601 Elmwood Avenue, Rochester, NY 14642, USA

L. Calzà, Department of Veterinary Morphophysiology and Animal Production, (DIMORFIPA), Università di Bologna, Via Tolara di Sopra 50, Ozzano dell'Emilia, I-40064 Bologna, Italy

J.M. Canals, Department of Cell Biology and Pathology, Medical School, IDIBAPS, University of Barcelona, Casanova 143, E-08036 Barcelona, Spain

G. Carmignoto, CNR Institute of Neuroscience and Department of Experimental Biomedical Sciences, Viale G. Colombo 3, 35121 Padova, Italy

B.D. Carter, Department of Biochemistry, Center for Molecular Neuroscience, Vanderbilt University Medical School, 655 Light Hall, Nashville, TN 37232, USA

T. Cataudella, Department of Pharmacological Sciences and Center of Excellence on Neurodegenerative Diseases, University of Milan, Via Balzaretti 9, I-20133 Milan, Italy

E. Cattaneo, Department of Pharmacological Sciences and Center of Excellence on Neurodegenerative Diseases, University of Milan, Via Balzaretti 9, I-20133 Milan, Italy

A. Chalazonitis, Department of Anatomy and Cell Biology, Columbia University, College of Physicians and Surgeons, 630 W. 168th Street, New York, NY 10032, USA

G.N. Chaldakov, Department of Forensic Medicine, Division of Cell Biology, Medical University, Varna, Bulgaria

J.W. Commissiong, Laboratory of Cell Biology, Prescient NeuroPharma Inc., 96 Skyway Avenue, Toronto, ON M9W 4Y9, Canada

L. Conti, Department of Pharmacological Sciences and Center of Excellence on Neurodegenerative Diseases, University of Milan, Via Balzaretti 9, I-20133 Milan, Italy

J.D. Cooper, Department of Neurology and Neurological Sciences, Program in Neuroscience, Stanford University, Stanford, CA 94305, USA

E. Coppola, NCI-FDRC, Neural Development Group, Frederick, MD 21702, USA

M.D. Coughlin, Department of Psychiatry and Behavioural Neurosciences, McMaster University, 1200 Main Street West, Hamilton, ON L8N 3Z5, Canada

E.J. Coulson, Queensland Brain Institute, The University of Queensland, Brisbane, 4072 Qld, Australia

R.J. Crowder, Department of Pathology, Washington University School of Medicine, St. Louis, MO 63110, USA

N. D'Ambrosi, Fondazione Santa Lucia, Via Ardeatina 354, I-00179 Rome and, Department of Neuroscience, University of Rome Tor Vergata, Rome, Italy

N. De Sordi, Department of Veterinary Morphophysiology and Animal Production, (DIMORFIPA), Università di Bologna, Via Tolara di Sopra 50, Ozzano dell'Emilia, I-40064 Bologna, Italy

J.-D. Delcroix, Department of Neurology and Neurological Sciences, Program in Neuroscience, Stanford University, Stanford, CA 94305, USA

G. D'Intino, Department of Veterinary Morphophysiology and Animal Production, (DIMORFIPA), Università di Bologna, Via Tolara di Sopra 50, Ozzano dell'Emilia, I-40064 Bologna, Italy

Y. Durocher, Biotechnology Research Institute, National Research Council of Canada, Animal Cell Technology Group, Montreal, QC, Canada

M. Fahnestock, Department of Psychiatry and Behavioural Neurosciences, McMaster University, 1200 Main Street West, Hamilton, ON L8N 3Z5, Canada

M. Fernandez, Department of Veterinary Morphophysiology and Animal Production, (DIMORFIPA), Università di Bologna, Via Tolara di Sopra 50, Ozzano dell'Emilia, I-40064 Bologna, Italy

M. Fiore, Institute of Neurobiology and Molecular Medicine, National Research Council (CNR), Via K. Marx 15/43, 00137 Rome, Italy

N. Francia, Behavioral Pathophysiology Section, Istituto Superior di Sanità (ISS), Viale Regina Elena 299, I-00161 Rome, Italy

R.S. Freeman, Department of Pharmacology and Physiology, University of Rochester School of Medicine and Dentistry, 601 Elmwood Avenue, Rochester, NY 14642, USA

V. Freund, Institut National de la Santé et de la Recherche Médicale, INSERM Unité 425, Neuroimmunopharmacologie Pulmonaire, Université Louis Pasteur-Strasburg I, F-67401 Illkirch Cedex, France

B. Fritzsch, Creighton University, Department of Biomedical Sciences, Omaha, NE 68178, USA

N. Frossard, Institut National de la Santé et de la Recherche Médicale, INSERM Unité 425, Neuroimmunopharmacologie Pulmonaire, Université Louis Pasteur-Strasburg I, F-67401 Illkirch Cedex, France

C.P. Genain, Neuroimmunology Laboratory, Room C-440, Department of Neurology, University of California, 505 Parnassus Avenue, San Francisco, CA 94143-0435, USA

J.J. Gentry, Department of Biochemistry, The Center for Molecular Neuroscience, Vanderbilt University Medical School, 655 Light Hall, Nashville, TN 37232, USA

P.I . Ghenev, Department of General and Clinical Pathology, Medical University, Varna, Bulgaria

L. Giardino, Pathophysiology Center for Nervous System, Hesperia Hospital, I-41100 Modena, Italy

A. Giuliani, Department of Veterinary Morphophysiology and Animal Production, (DIMORFIPA), Università di Bologna, Via Tolara di Sopra 50, Ozzano dell'Emilia, I-40064 Bologna, Italy

S. Gobbo, CNR Institute of Neuroscience and Department of Experimental Biomedical Sciences, Viale G. Colombo 3, 35121 Padova, Italy

M. Grim, Institute of Anatomy, First Faculty of Medicine, Charles University Prague, Prague, Czech Republic

W.T.J. Hendriks, Netherlands Institute for Brain Research, Department of Neuroregeneration, Meibergdreef 33, 1105 AZ Amsterdam, The Netherlands

C.L. Howe, Department of Neurology and Neurological Sciences, Program in Neuroscience, Stanford University, Stanford, CA 94305, USA

M.G. Hristova, Department of Forensic Medicine, Division of Cell Biology, Medical University, Varna, Bulgaria

K. Kawamoto, Laboratory of Clinical Immunology, Department of Veterinary Clinic, Faculty of Agriculture, Tokyo University of Agriculture and Technology, Tokyo, Japan

J. Kelly, Institute for Biological Studies, National Research Council of Canada, Ottawa, ON, Canada

S. Kerzel, Department of Clinical Chemistry and Molecular Diagnostics, Central Laboratory, Hospital of the Philipps University, Marburg University, Baldingerstr, D-35033 Marburg, Germany

T.J. Kilpatrick, The Walter and Eliza Hall Institute for Medical Research, Parkville, Vic., Australia

C.-F. Lai, Department of Neurology and Neurological Sciences, Program in Neuroscience, Stanford University, Stanford, CA 94305, USA

P. Lazarovici, Department of Pharmacology and Experimental Therapeutics, School of Pharmacy, Faculty of Medicine, The Hebrew University of Jerusalem, P.O. Box 12065, Jerusalem 91120, Israel

S. Lecht, Department of Pharmacology and Experimental Therapeutics, School of Pharmacy, Faculty of Medicine, The Hebrew University of Jerusalem, P.O. Box 12065, Jerusalem 91120, Israel

R. Levi-Montalcini, Institute of Neurobiology and Molecular Medicine, CNR, Viale Marx 15, I-00137 Rome, Italy

D.J. Lomb, Department of Pharmacology and Physiology, University of Rochester School of Medicine and Dentistry, 601 Elmwood Avenue, Rochester, NY 14642, USA

L. Lossi, Department of Veterinary Morphophysiology and Rita Levi-Montalcini Center for Brain Repair, Via Leonardo da Vinci 44, 10095 Grugliasco, Turin, Italy

B. Lu, Section on Neural Development and Plasticity, NICHD, NIH Building 49 Room 6A80, 49 Convent Drive, MSC 4480, Bethesda, MD 20892-4480, USA

L. Manni, Institute of Neurobiology and Molecular Medicine, National Research Council (CNR), Via K. Marx 15/43, 00137 Rome, Italy

M. Manservigi, Department of Veterinary Morphophysiology and Animal Production, (DIMORFIPA), Università di Bologna, Via Tolara di Sopra 50, Ozzano dell'Emilia, I-40064 Bologna, Italy

A.A. Mathé, Department of Pharmacology, Karolinska Institute, SE-17177, Stockholm, Sweden

H. Matsuda, Laboratory of Clinical Immunology, Department of Veterinary Clinic, Faculty of Agriculture, Tokyo University of Agriculture and Technology, Tokyo, Japan

A. Merighi, Department of Veterinary Morphophysiology and Rita Levi-Montalcini Center for Brain Repair, Via Leonardo da Vinci 44, 10095 Grugliasco, Turin, Italy

W.C. Mobley, Department of Neurology and Neurological Sciences, Program in Neuroscience, Stanford University, Stanford, CA 94305, USA

M.K. Moore, Prescient NeuroPharma Inc., Laboratories of Protein Chemistry, Molecular Biology and Cell Biology, 96 Skyway Avenue, Toronto, ON M9W 4Y9, Canada

S. Morley, Queensland Brain Institute, The University of Queensland, Brisbane, 4067 Qld, Australia

B. Murra, Fondazione Santa Lucia, Via Ardeatina 354, I-00179 Rome, Italy

A. Nakagawara, Division of Biochemistry, Chiba Cancer Center Research Institute, 662-2 Nitona, Chuoh-ku, Chiba 260-8717, Japan

C. Nassenstein, Immunology and Allergology, Fraunhofer Institute of Toxicology and Experimental Medicine, Nikolai-Fuchs-Str. 1, D-30625 Hannover, Germany

W.A. Nockher, Department of Clinical Chemistry and Molecular Diagnostics, Central Laboratory, Hospital of the Philipps University, Baldingerstr, D-35033 Marburg, Germany

R. Paus, Department of Dermatology, University Hospital Hamburg-Eppendorf, University of Hamburg, Martinistr. 52, D-20246 Hamburg, Germany

A. Peaire, Prescient NeuroPharma Inc., Laboratories of Protein Chemistry, Molecular Biology and Cell Biology, 96 Skyway Avenue, Toronto, ON M9W 4Y9, Canada

E. Pérez-Navarro, Department of Cell Biology and Pathology, Medical School, IDIBAPS, University of Barcelona, Casanova 143, E-08036 Barcelona, Spain

E.M.J. Peters, Biomedical Research Center, Charité, Campus Virchow Hospital, Humboldt University Berlin, Berlin, Germany

P.S. Petrova, Prescient NeuroPharma Inc., Laboratories of Protein Chemistry, Molecular Biology and Cell Biology, 96 Skyway Avenue, Toronto, ON M9W 4Y9, Canada

J. Pevsner, Department of Neurology, Kennedy Krieger Institute and Department of Neuroscience, Johns Hopkins School of Medicine, Baltimore, MD, USA

S. Pirondi, Department of Veterinary Morphophysiology and Animal Production, (DIMORFIPA), Università di Bologna, Via Tolara di Sopra 50, Ozzano dell'Emilia, I-40064 Bologna, Italy

A. Raibekas, Prescient NeuroPharma Inc., Laboratories of Protein Chemistry, Molecular Biology and Cell Biology, 96 Skyway Avenue, Toronto, ON M9W 4Y9, Canada

S.K. Raychaudhuri, Psoriasis Research Institute and Stanford University School of Medicine, 510 Ashton Avenue, Palo Alto, CA 94306, USA

S.P. Raychaudhuri, Psoriasis Research Institute and Stanford University School of Medicine, 510 Ashton Avenue, Palo Alto, CA 94306, USA

L.F. Reichardt, Howard Hughes Medical Institute, University of California, Neuroscience Unit, San Francisco, CA 94143-0724, USA

K. Reid, The Walter and Eliza Hall Institute for Medical Research, Parkville, Vic., Australia

H. Renz, Department of Clinical Chemistry and Molecular Diagnostics, Central Laboratory, Hospital of the Philipps University, Baldingerstr, D-35033 Marburg, Germany

M.J. Ruitenberg, Netherlands Institute for Brain Research, Department of Neuroregeneration, Meibergdreef 33, 1105 AZ Amsterdam, The Netherlands

A. Salehi, Department of Neurology and Neurological Sciences, Program in Neuroscience, Stanford University, Stanford, CA 94305, USA

C. Salio, Department of Veterinary Morphophysiology and Rita Levi-Montalcini Center for Brain Repair, Via Leonardo da Vinci 44, 10095 Grugliasco, Turin, Italy

D. Santucci, Behavioral Pathophysiology Section, Istituto Superiore di Sanità (ISS), Viale Regina Elena 299, I-00161 Rome, Italy

M.C. Schoell, Department of Pharmacology and Physiology, University of Rochester School of Medicine and Dentistry, 601 Elmwood Avenue, Rochester, NY 14642, USA

S. Sephanova, Department of Pharmacology and Experimental Therapeutics, School of Pharmacy, Faculty of Medicine, The Hebrew University of Jerusalem, P.O. Box 12065, Jerusalem 91120, Israel

K.M. Shipham, The Walter and Eliza Hall Institute for Medical Research, Parkville, Vic., Australia

V. Shridhar, Department of Experimental Pathology, Division of Laboratory Medicine, Mayo Clinic, Rochester, MN, USA

M. Sieber-Blum, Department of Cell Biology, Neurobiology and Anatomy, Medical College of Wisconsin, 8701 Watertown Plank Road, Milwaukee, WI 53226, USA

D.I. Smith, Department of Experimental Pathology, Division of Laboratory Medicine, Mayo Clinic, Rochester, MN, USA

I.S. Stankulov, Department of Forensic Medicine, Division of Cell Biology, Medical University, Varna, Bulgaria

J.A. Straub, Department of Pharmacology and Physiology, University of Rochester School of Medicine and Dentistry, 601 Elmwood Avenue, Rochester, NY 14642, USA

V. Szeder, Department of Cell Biology, Neurobiology and Anatomy, Medical College of Wisconsin, 8701 Watertown Plank Road, Milwaukee, WI 53226, USA and Institute of Anatomy, First Faculty of Medicine, Charles University Prague, Prague, Czech Republic

R. Tabakman, Department of Pharmacology and Experimental Therapeutics, School of Pharmacy, Faculty of Medicine, The Hebrew University of Jerusalem, P.O. Box 12065, Jerusalem 91120, Israel

L. Tessarolo, NCI-FDRC, Neural Development Group, Frederick, MD 21702, USA

M.H. Tuszynski, Department of Neurosciences-0626, University of California, San Diego, San Diego, La Jolla, CA 92093, USA

F. Vacca, Fondazione Santa Lucia, Via Ardeatine 354, I-00179 Rome, Italy and Department of Human Physiology and Pharmacology, University of Rome 'La Sapienza', Rome, Italy

J. Valletta, Department of Neurology and Neurological Sciences, Program in Neuroscience, Stanford University, Stanford, CA 94305, USA

A.M. Vergnano, Department of Veterinary Morphophysiology and Rita Levi-Montalcini Center for Brain Repair, Via Leonardo da Vinci 44, 10095 Grugliasco, Turin, Italy

J. Verhaagen, Netherlands Institute for Brain Research, Department of Neuroregeneration, Meibergdreef 33, 1105 AZ Amsterdam, The Netherlands

N. Vigo, Prescient NeuroPharma Inc., Laboratories of Protein Chemistry, Molecular Biology and Cell Biology, 96 Skyway Avenue, Toronto, ON M9W 4Y9, Canada

P. Villoslada, Neuroimmunology Laboratory, Department of Neurology, University of Navarra, Spain

C. Volonté, Fondazione Santa Lucia, Via Ardeatine 354, I-00179 Rome, Italy and Institute of Neurobiology and Molecular Medicine, CNR, Rome, Italy

C. Wu, Department of Neurology and Neurological Sciences, Program in Neuroscience, Stanford University, Stanford, CA 94305, USA

L. Xie, Department of Pharmacology and Physiology, University of Rochester School of Medicine and Dentistry, 601 Elmwood Avenue, Rochester, NY 14642, USA

G. Yu, Department of Psychiatry and Behavioural Neurosciences, McMaster University, 1200 Main Street West, Hamilton, ON L8N 3Z5, Canada

M. Zonta, CNR Institute of Neuroscience and Department of Experimental Biomedical Sciences, Viale G. Colombo 3, 35121 Padova, Italy

Preface

"Ideas won't keep: something must be done about them."
Alfred North Whitehead (1861–1947)

Growth factors are signaling proteins which play a crucial role in cell proliferation, differentiation and survival, and their functions are currently intensively studied in biomedical research. The paradigmatic growth factor nerve growth factor (NGF) has increasingly been demonstrated to be essential for the development and maintenance of specific populations of neurons in the peripheral and central nervous system. Originally identified as neurite outgrowth-stimulating factor, NGF is now known to also affect a number of non-neural cells, including cells of the immune system in a variety of ways, acting, most probably, as an immunomodulator of the 'cross-talk' between neurons and immunocytes.

NGF was the first discovered member of the family of neurotrophins, which also include brain-derived neurotrophic factor (BDNF), neurotrophin-3 (NT-3), NT-4/5, NT-6 and NT-7. All these neurotrophins share significant structural homology and sometimes overlapping functions mediated by two classes of cell surface receptors. The low-affinity neurotrophin receptor p75 (p75^{NTR}) binds all known neurotrophins with similar affinity, while the high-affinity receptor belongs to the superfamily of tyrosine protein kinase (Trk) including TrkA-C receptors that are different for the various neurotrophins. The discovery of the NGF first and of other related molecules later generates an intense investigation in a wide range of basic and clinical studies. This scientific research growth process resulted in the production of an enormous body of knowledge, not only in the nervous system, but more recently also in the pathogenesis of neuroimmune and (auto)immune inflammatory diseases. Also, through the use of recombinant technologies, other neurotrophic factors were later identified showing a wide range of effects on a variety of cells. Molecules with these activities include ciliary neurotrophic factor, glial cell line-derived neurotrophic factor, basic fibroblast growth factor, insulin-like growth factor, leukemia inhibitory factor, bone morphogenetic proteins and transforming growth factor-beta, to list a part of them only. In the last few years, these growth factors has attracted unparallel and a renewed interest in determining whether some of them could be useful potential therapeutic agents for human diseases.

Recent technological advance promotes the development of strategies using growth factors in general and neurotrophic molecules in particular in preventing the progressive neuronal loss, maintaining neuronal connections and function, and inducing additional regenerative benefits such as stimulation of neurotransmitter turnover and axonal sprouting. Up to date, several therapeutic strategies to deliver growth factors in animal models and in human diseases have been explored. These include recombinant human neurotrophin application, direct gene transfer using (non-)viral vectors, the implantation of ex vivo genetically engineered cells secreting neurotrophic factors, and the grafting of neural stem progenitor cells. Clinical trials have been attempted and others are currently in progress to evaluate whether neurotrophic factors can prevent or protect against neural cell degeneration in the peripheral and central nervous system. The spectrum of diseases targeted by such a neurotrophin-based therapy includes, for example, Parkinson's disease, Huntington's disease, Alzheimer's disease, the multiple sclerosis-like model of experimental allergic encephalomyelitis, diabetic neuropathies, and spinal cord injury. In addition, diseases featured by immune, inflammatory and (fibro)proliferative reactions, such as corneal and skin ulcers, rheumatoid arthritis, systemic sclerosis, tumors, bronchial asthma, atherosclerotic cardiovascular disease and even hair growth disorders, appeared to be of potential interest for clinical evaluations.

The book includes 33 manuscripts, which are divided into eight sections. We have tried to organize the sections following a functional scheme to guide readers through the current knowledge about NGF and related molecules. The articles discuss new concepts and findings regarding neurobiology, immunobiology and vascular biology of neurotrophic factors, particularly neurotrophins, in keeping with its mission to update these fast growing fields of biomedical research. The major goal of these articles is therefore to disseminate the latest important leads about NGF and related molecules. The present book is a useful guide for biomedical research in the field of neurotrophobiology. The editors hope that the data and hypotheses presented in the book will foster a tight interaction between basic scientists and clinicians and will convey to the readers some of the excitement that enlivens the current achievements about neurotrophic factor findings. And, in effect, sharing the importance of the concept that *ideas won't keep* unless scientists *do something about them*.

Luigi Aloe
Laura Calzà
Editors

Acknowledgments

The articles of this volume are part of scientific contributions presented and discussed at The 7th International Conference on NGF and Related Molecules held in Modena, Italy from 15 to 19 May, 2002. For other numerous aspects of neurotrophic factor biology including molecular and mechanistic advances, readers are also referred to recent reviews on neurotrophic factors and to the Meeting report published in *EMBO Journal* 2002; 3: 1029–1034.

This volume is dedicated to the Nobel Laureate Professor Rita Levi-Montalcini, who discovered the first cell growth factor and opened up the field of trophic factors in the biomedical science. Her presence at the Modena NGF2002 Conference and her article on 'The nerve growth factor and the neuroscience chess board' (this issue) represent her vital participation in this growing field of neurotrophin research.

We owe a debt of gratitude to all the participants of the Conference who provided the opportunity for exchanging results and sharing ideas about ongoing studies in NGF and related molecules. We believe, as stated by Albert Einstein, that *"if you have an apple and I have an apple, and we exchange apples, each of us will have an apple. However, if you have one idea and I have another idea, and we exchange ideas, each of us will have two ideas"*.

We acknowledge the contribution of Dr. George Chaldakov from the Medical University, Varna, Bulgaria on a sabbatical visit and collaboration with Dr. Luigi Aloe at the Institute of Neurobiology and Molecular Medicine, CNR, Rome, Italy for his inspiring enthusiasm and numerous daily scientific and intellectual discussions. Luigi Aloe and Laura Calzà are particularly grateful to Dr. Luciana Giardino for her constant scientific and intellectual contribution during the preparation and execution of the NGF Conference. We also like to acknowledge the qualified competence of Angela Rizzi of the Communication Ways who provided a graceful environmental support for the organization and execution of the social events of the NGF Conference.

We also gratefully acknowledge the financial contribution by Associazione Levi-Montalcini, associazione di promozione sociale, Torino, Italy and by Fondazione CARISBO, Bologna, Italy for the preparation of this volume.

Photograph taken at the 7th International Conference on NGF and Related Molecules held in Modena, Italy from 15-19 May, 2002.
From left to right: Rita Levi-Montalcini, Laura Calzà and Luigi Aloe

Contents

Growth factors and cell signaling

Progress in Brain Research, Vol. 146
ISSN 0079-6123

CHAPTER 1

Trafficking the NGF signal: implications for normal and degenerating neurons

Jean-Dominique Delcroix, Janice Valletta, Chenbiao Wu, Charles L. Howe,
Chun-Fai Lai, John D. Cooper, Pavel V. Belichenko, Ahmad Salehi and
William C. Mobley*

Department of Neurology and Neurological Sciences and of Pediatrics and the Program in Neuroscience, Stanford University, Stanford, CA 94305, USA

Abstract: Nerve growth factor (NGF) activates TrkA to trigger signaling events that promote the survival, differentiation and maintenance of neurons. The mechanism(s) that controls the retrograde transport of the NGF signal from axon terminals to neuron cell bodies is not known. The 'signaling endosome' hypothesis stipulates that NGF, TrkA and signaling proteins are retrogradely transported on endocytic vesicles. Here, we provide evidence for the existence of signaling endosomes. Following NGF treatment, clathrin-coated vesicles (CCVs) contain NGF bound to TrkA together with activated signaling proteins of the Ras/pErk1/2 pathway. NGF signals from isolated CCVs through the Erk1/2 pathway. Early endosomes appear to represent a second type of signaling endosomes. We found that NGF induced a sustained activation of Rap1, a small monomeric GTP-binding protein of the Ras family, and that this activation occurred in early endosomes that contain key elements of Rap1/pErk1/2 pathway. We discuss the possibility that the failure of retrograde NGF signaling in a mouse model of Down syndrome (Ts65Dn) may be due to the failure to retrograde transport signaling endosomes. It is important to define further the significance of signaling endosomes in the biology of both normal and degenerating neurons.

Keywords: nerve growth factor; signal transduction; endosome; Down syndrome; Alzheimer's disease

Introduction

Elucidating the cellular mechanisms that build, maintain and modify synaptic contacts is required for understanding normal and abnormal nervous system function. Neurotrophic factors (NTFs) are small proteins that act through specific receptors to guide the development and maintenance of neuronal connections. Current studies aimed at understanding how NTFs signal to accomplish these effects are focused on defining how the NTF signal generated in

receptors on the terminals of axons are sent to cell bodies (Sofroniew et al., 2001).

Neurotrophins, neurotrophin receptors and signals

The neurotrophins (NTs) are a family of NTFs whose actions influence neurons of both the peripheral nervous system (PNS) and central nervous system (CNS) (Huang and Reichardt, 2001; Sofroniew et al., 2001). Four members of the NT family are present in mammals: nerve growth factor (NGF), brain-derived neurotrophic factor (BDNF), NT-3 and NT-4. Highly homologous, NTs are synthesized as precursors (31–35 kD) that are processed to mature forms (13.2–15.2 kDa) which form non-covalently

---

*Correspondence to: W.C. Mobley, Department of Neurology and Neurological Sciences and of Pediatrics and the Program in Neuroscience, Stanford University, Stanford, CA 94305, USA. Tel.: +1-650-498-5858; E-mail: ngfv1@stanford.edu

DOI: 10.1016/S0079-6123(03)46001-9

bound homodimers (Sofroniew et al., 2001; Roux and Barker, 2002). Following the discovery of NGF by Levi-Montalcini and Hamburger more than 50 years ago (Levi-Montalcini and Hamburger, 1951), this prototypic NT became the focus of extensive study (Huang and Reichardt, 2001; Sofroniew et al., 2001; Roux and Barker, 2002). Each of the NTs binds to p75 neurotrophin receptor (p75^{NTR}), a transmembrane glycoprotein of the TNF receptor family. Signaling through p75^{NTR} is complex (Huang and Reichardt, 2001; Sofroniew et al., 2001; Roux and Barker, 2002). This receptor modulates Trk-mediated signaling induced by NTs; p75^{NTR} also signals independently of Trk and, in some cases, signals in the absence of NT binding. p75^{NTR} plays an important role in regulating survival. The complexity of p75^{NTR} signaling is illustrated by the fact that both proapoptotic and antiapoptotic effects have been recorded. Although there is much to learn about the signaling mechanisms of p75^{NTR}, activation of the JNK pathway contributes significantly to proapoptotic effects while signaling through NF-κB pathway promotes survival (Sofroniew et al., 2001; Roux and Barker, 2002).

In addition, NTs bind to a member(s) of the tyrosine kinase (Trk) family of receptors. TrkA preferentially binds NGF, TrkB preferentially binds BDNF and NT-4, and TrkC preferentially binds NT-3 (Patapoutian and Reichardt, 2001; Sofroniew et al., 2001; Roux and Barker, 2002). In the absence of p75^{NTR}, NT-3 also activates TrkA and TrkB. NGF signaling through TrkA elicits the classical neurotrophic actions ascribed to NGF (Loeb et al., 1991; Huang and Reichardt, 2001; Sofroniew et al., 2001). Binding of NGF causes receptor dimerization, autophosphorylation of tyrosine residues within the activation loop (Y670, Y674, Y675), followed by phosphorylation of seven additional tyrosine residues in the intracellular domain (ICD) of the receptor (Cunningham et al., 1997; Roux and Barker, 2002). pY490 and pY785 in activated TrkA (pTrkA) receptors are docking sites for adaptor proteins that link the receptor to downstream signaling pathways, including those mediated by Ras, Rap1, PI3K and PLC-γ (Stephens et al., 1994; Atwal et al., 2000). Several mitogen-activated protein kinase (MAPK) pathways are induced by TrkA activation. The pY490 site binds Shc to recruit the signaling proteins

of the Ras/Erk1/2 pathway: Grb2/SOS/Ras/c-Raf/ Mek1,2 and Erk1/2 (Sofroniew et al., 2001). Frs2, a scaffolding protein (Kao et al., 2001), and possibly Gab2 (Gu et al., 1998; Liu et al., 2001), also binds to this site to recruit signaling proteins of the Rap1/ Erk1/2 pathway, a pathway that plays an important role in differentiation: CrkL/C3G (CrK Sh3 guanine nucleotide exchange factor)/Rap1/B-Raf/ Mek1,2 and Erk1/2 (Meakin et al., 1999; Kao et al., 2001; Wu et al., 2001). Recent studies show that Erk5 is also activated by NT signaling (Kamakura et al., 1999; Watson et al., 2001) through a pathway that includes Ras, MEKK3 and MEK5 (Kamakura et al., 1999; Pearson et al., 2001). NGF also activates p38 MAPK (Morooka and Nishida, 1998; Xing et al., 1998; York et al., 1998; Klesse and Parada, 1999; Wu et al., 2001) and through this pathway regulates neurite outgrowth in PC12 cells (Morooka and Nishida, 1998). The adaptor complex that links pTrkA to p38 MAPK probably includes one or more of the MEKKs, and MKK3 and 6 (Pearson et al., 2001).

The PI3K pathway plays an important role in NT-mediated survival signaling (Klesse and Parada, 1999). Both Ras-dependent and Ras-independent pathways are engaged. The latter involves binding to pTrkA of a complex containing Shc, Grb2 and Gab-1; this complex then binds the regulatory (p85) and catalytic (p110) subunits of PI3K and brings them to membranes where PI3K-mediated phosphorylation of phosphoinositides (producing PI-3,4-P2 and PI-3,4,5-P3) allows for binding of 3-phosphoinositide-dependent kinase (PDK1) which phosphorylates and activates Akt (Stephens et al., 1998; Sofroniew et al., 2001). Activated Akt regulates the activity of a number of substrates (Sofroniew et al., 2001; Roux and Barker, 2002). The Ras-dependent pathway involves direct interaction of Ras with PI3K (Klesse and Parada, 1998, 1999; Mazzoni et al., 1999).

NGF also signals through TrkA to activate the PLC-γ pathway. pY785 serves as the docking site for PLC-γ (Vetter et al., 1991). Phosphorylation activates PLC-γ, resulting in the production of IP3 and DAG (Sofroniew et al., 2001; Roux and Barker, 2002) and the activation of classical, novel and atypical forms of PKC (protein kinase C) (Sofroniew et al., 2001). Through activation of PLC-γ, NGF mediates aspects of both neuronal differentiation and survival (Wooten et al., 2000); in part these actions may be mediated through effects on the activity of the Ras/Erk1/2 pathway (Marais et al., 1998).

Studies of NGF signaling have extensively used PC12 cells. Because of technical challenges, there is much less data for signaling in neurons (Klesse and Parada, 1999; Atwal et al., 2000). Nevertheless, studies on peripheral neurons show that the Erk1/2, Erk5, PI3K, p38 and PLC-γ signaling pathways are all represented and activated by NGF (Vogelbaum et al., 1998; Klesse and Parada, 1999; Ganju et al., 2001; Howe et al., 2001; Watson et al., 2001; Dodge et al., 2002; Ji et al., 2002; Delcroix et al., 2003).

NT retrograde signaling

An important challenge for understanding the biology of NTs is explaining how the NT-mediated activation of signaling pathways results in specific, well-coordinated biological responses. Among the most interesting questions is how NT signals are moved from the axon terminal to the neuron cell body. It has long been known that the targets of neuronal innervation play a vital role in regulating the survival, differentiation and phenotypic maintenance of innervating NT-responsive neurons. The discovery of NGF (Levi-Montalcini and Hamburger, 1951, 1953; Levi-Montalcini, 1987) provided an important clue to the mechanism; a soluble 'nerve growth factor' produced in the target dramatically influenced specific populations of neurons (Huang and Reichardt, 2001; Sofroniew et al., 2001). It is now well established that NGF and other NTs are produced and released in target tissues to activate specific receptors on the distal axons of innervating neurons and that the signals thus produced are retrogradely transported to the cell bodies to regulate cytosolic and nuclear events important for survival and differentiation (Huang and Reichardt, 2001; Sofroniew et al., 2001). However, there is little insight into how the signal is moved retrogradely. We and others (Ehlers et al., 1995; Grimes et al., 1996, 1997; Bhattacharyya et al., 1997, 2002; Riccio et al., 1997; Tsui-Pierchala and Ginty, 1999; Kuruvilla et al., 2000; Zhang et al., 2000; Howe et al., 2001; Watson et al., 1999, 2001; Jullien et al., 2002) have hypothesized that the NGF signal is transmitted via endocytosis of complexes containing NGF bound to pTrkA followed by retrograde transport of the 'signaling

endosome' thus formed. Herein, we will review the studies that have tested the signaling endosomes hypothesis.

Results of recent studies

Signaling endosomes derived from clathrin-coated membranes

An important test of the 'signaling endosome' hypothesis has been the isolation of signaling endosomes from NGF-treated cells. We began these studies by focusing on the possibility that clathrin-coated membranes would be a source of signaling endosomes. This was suggested by the fact that these membranes are used to internalize the activated receptor tyrosine kinases of other 'growth factors' (Sorkin and von Zastrow, 2002) and by studies in which TrkA was shown to colocalize with the clathrin heavy chain in PC12 cells treated with NGF, but not with vehicle (Beattie et al., 1996; Grimes et al., 1996). Indeed, in earlier studies we noted an apparent increase in the extent of clathrin coating of surface membranes in response to NGF treatment (Beattie et al., 2000), results that confirmed and extended earlier observations for an effect of NGF treatment on the coating of surface membranes in PC12 cells and sympathetic neurons (Connolly et al., 1981). Importantly, the effect was shown to be due to activation of TrkA (Beattie et al., 2000). To explore the idea that clathrin-coated membranes could serve as a source of signaling endosomes, we carried out studies of NGF signaling in PC12 cells. The original report of these studies, and the detailed methods in support of them, have been published (Howe et al., 2001).

NGF recruits clathrin to membranes

First, we confirmed that there was an increase in the association of clathrin with membranes in NGF-treated PC12 cells. Two protocols were used. In one, we enriched for surface membranes; in the second, we enriched for internal membranes (Howe et al., 2001). In both cases, NGF treatment resulted in a prompt increase in the amount of membrane-associated clathrin. NGF treatment for 2 min increased the amount of membrane-associated clathrin heavy chain (CHC) in the plasma membrane-enriched preparation by 58% ($\pm 5.6\%$; $n=7$; $p<0.001$ vs untreated) and in the internal membrane fraction by 145% ($\pm 10.1\%$; $n=3$; $p<0.001$ vs untreated) (Howe et al., 2001). These findings are evidence that NGF elicits the redistribution of clathrin to both surface and internal membranes.

Activated TrkA is associated with clathrin

Increased clathrin coating of membranes following NGF treatment suggested that the activated TrkA

receptor may be found in these membranes. Indeed, it also raised the possibility that activated TrkA would be found in complex with CHC and the clathrin adaptor protein AP2. To test if activated TrkA receptors were present in clathrin-coated membranes, we carried out immunoprecipitations with an antibody to tyrosine phosphorylated TrkA. We found CHC and AP2 in a complex with activated TrkA and showed that the amount increased markedly following NGF treatment (Fig. 1A). The association with CHC peaked at 2 min and returned to baseline levels by 15 min (Howe et al., 2001). These findings are

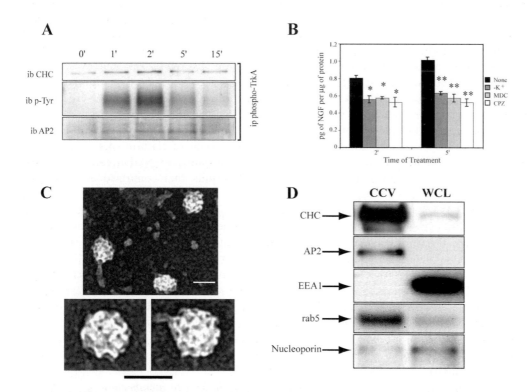

Fig. 1. NGF Induced the Formation of Complexes Containing Activated TrkA, CHC and AP2: (A) PC12 cells were treated at 37°C with NGF (2 nM) for the times indicated, or were left untreated (0'). Cell lysates were immunoprecipitated with an antibody directed against tyrosine phosphorylated TrkA (p-Y490). The western blot of the immuno-precipitates was probed for CHC, phosphotyrosine, or AP2. There was a robust increase in complexes containing activated TrkA, CHC, and AP2 that peaked at 2 min; (B) Inhibitors of clathrin-mediated endocytosis suppressed ^{125}I-NGF internalization. PC12 cells were subjected to hypotonic shock and potassium depletion (-K$^+$), or incubated with either 100 μM chlorpromazine (CPZ), 50 μM monodansylcadaverine (MDC), or no additive (None) for 15 min at 37°C. In the continued presence of the inhibitors, cells were incubated with ^{125}I-NGF (2 nM) for the times indicated. Values shown are expressed as pg of NGF internalized per μg of protein ($n=3$ for all conditions; * $p<0.05$ vs no inhibitor, ** $p<0.001$ vs no inhibitor). (C) Electron microscopic analysis of the CCV fraction showing typical CCV profiles. Scale bar in upper micrograph of panel (C) is 65 nm. Scale bar at bottom of panel (C) is 65 nm and refers to the two micrographs of individual CCVs shown. (D) Biochemical characterization of the CCV fraction. Relative levels of CHC, AP2, EEA1, rab5 and nucleoporin in the CCV fraction vs an equivalent amount of protein from WCL (reprinted, with permission, as modified from Howe et al., 2001).

consistent with the presence of activated TrkA in the clathrin-coated membranes of NGF-treated cells. In that NGF did not induce redistribution of clathrin in cells expressing kinase-defective TrkA (Beattie et al., 2000), our findings suggest that TrkA activation was responsible for the recruitment of CHC and AP2 to TrkA in NGF-treated cells. We conclude that NGF signaling results in the formation of complexes in which activated TrkA is present together with molecules that mediate endocytosis through the clathrin pathway.

NGF is internalized through the clathrin pathway

To test the prediction that complexes containing NGF bound to activated TrkA are internalized via the clathrin pathway, we asked whether ^{125}I-NGF internalization would be affected by several inhibitors of clathrin-mediated endocytosis: potassium depletion (Larkin et al., 1983), monodansylcadaverine (MDC) (Davies et al., 1984), and chlorpromazine (CPZ) (Sofer and Futerman, 1995). Hypotonic shock followed by potassium depletion (-K), or preincubation with either 100 μM CPZ or 50 μM MDC, blunted internalization of NGF (Fig. 1B). At 2 min, there was a statistically significant decrease in internalization with -K ($p = 0.047$), MDC ($p = 0.0075$) or CPZ ($p = 0.025$). Likewise, following 5 min of ^{125}I-NGF exposure, PC12 cells internalized significantly less NGF with -K ($p = 0.0003$), MDC ($p = 0.00054$) or CPZ ($p = 0.0041$). Furthermore, while NGF internalization increased significantly between 2 and 5 min in the absence of inhibitors ($p = 0.00068$ $2'$ vs $5'$), there was no significant difference in the amount of ^{125}I-NGF internalized between 2 and 5 min under conditions of -K or in the presence of MDC or CPZ ($p = 0.73$ $2'$ vs $5'$ -K; $p = 0.77$ $2'$ vs $5'$ MDC; $p = 0.99$ $2'$ vs $5'$ CPZ) (Howe et al., 2001). Importantly, at the concentrations used, neither MDC nor CPZ nor -K decreased TrkA activation in response to NGF treatment (data not shown). We also observed that NGF internalization in DRG (dorsal root ganglia) neurons was largely dependent upon endocytosis through the clathrin pathway (data not shown) (Howe et al., 2001). We conclude that NGF internalization is dependent, at least in part, upon the clathrin pathway.

Preparing a fraction enriched in CCVs

The increase in clathrin-coated membranes that followed NGF treatment suggested that NGF could signal to increase the number of CCVs. Isolation of CCVs generated a fraction that contained electron-dense particles approximately 65 nm in diameter that had the structure characteristic of CCVs (Fig. 1C) (Howe et al., 2001). Biochemical analysis showed that the CCV fraction was enriched 71.4-fold in CHC ($n = 3$; $p = 0.00011$) as compared to an equivalent amount of protein from whole cell lysates (WCL) (Fig. 1D). A similar enrichment was observed for AP2 in the CCV fraction. We found that both CCVs and the WCL contained Rab5, a GTPase that is associated with endocytic vesicles. Notably, Rab5 was enriched 17.7-fold (± 1.1-fold; $n = 2$; $p = 0.03$) in CCVs relative to the WCL (Howe et al., 2001). This finding is consistent with the role Rab5 plays in the biogenesis of CCVs and sorting endosomes (Woodman, 2000).

The extent to which the CCV-enriched fraction was contaminated with other membranes was evaluated. Compared to CCVs, WCL was enriched 19.1-fold ($n = 3$; $p < 0.001$) in nucleoporin, a specific marker of nuclear membranes (Fig. 1D). To assay plasma membrane contamination, we surface-iodinated PC12 cells and isolated CCVs. By comparing the labeled NGF present in CCVs isolated from warmed and unwarmed cells, we estimated the upper limit of plasma membrane contamination of the CCV fraction to be 10% (Howe et al., 2001). Next we tested for the presence of non-CCV endosomal membranes. While a substantial amount of the early endosomal marker EEA1 was detected in WCL, little if any of this marker was found in the CCV fraction (Fig. 1D), indicating that the CCV fraction was not detectably contaminated with early endosomes (Howe et al., 2001). Taken together, these data provide compelling evidence that the fraction isolated in these experiments was highly enriched in markers of the clathrin pathway and that it corresponds to a markedly enriched preparation of CCVs.

Consistent with the finding that NGF increased the association of CHC with internal membranes, as indicated above, we observed a significant increase in CCVs following NGF treatment. To examine NGF effects on CCVs, PC12 cells were incubated with

8

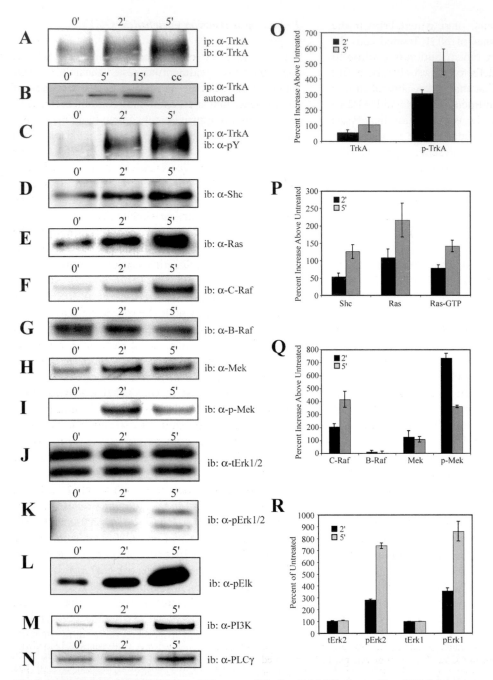

Fig. 2. CCVs Contained NGF, TrkA and Elements of the Ras-Erk1/2 Pathway: (A and O) TrkA was present in CCVs. PC12 cells were treated at 37°C with NGF (2 nM) for the times indicated, or were not treated (0′). CCVs isolated from equal numbers of cells were lysed and immunoprecipitated for TrkA. The data are displayed in A and quantified in O (B) NGF was bound to TrkA in CCVs. PC12 cells were incubated with ^{125}I-NGF (2 nM) for 1 h at 4°C, then warmed for the times indicated in the continued presence of NGF. CCV lysates were immunoprecipitated with a TrkA antibody. NGF binding to TrkA in CCVs increased markedly with warming and was specific because there was no signal when unlabeled NGF was present in excess (500 ×) during the incubation ('cc'). (C through M) The samples were produced in the same way as for Fig. 2A. (C and O) Activated TrkA was present in CCVs from NGF-treated cells. TrkA immunoprecipitates from CCVs were immunoblotted with an antibody to phosphotyrosine. (D and P) Shc was present in CCVs

NGF (2 nM) for 1 h at 4°C, and then warmed in the continued presence of NGF for 0, 2, 5 or 15 min. The amount of total protein in CCV fractions obtained from equal numbers of cells was measured. Following 2 min of warming, CCV protein was 166% of that in unwarmed cells, reflecting a 66% increase in CCVs ($p < 0.001$). At 5 min, there was a 53% increase ($p < 0.001$). The increases induced by NGF were transient, because by 15 min the amount of CCV protein had returned almost to the level found in untreated cells. Importantly, warming in the presence of the vehicle control did not substantially contribute to the formation of CCVs (5.4% increase ± 1.2%; $n = 3$; $p > 0.05$).

NGF was bound to TrkA in CCVs

To determine whether or not NGF was contained in CCVs, we isolated CCVs from PC12 cells incubated with ^{125}I-NGF. The amount of NGF in CCVs increased markedly with time of warming; there was a 3.78-fold increase relative to CCV protein between 2 and 15 min ($n = 4$; $p < 0.005$). Internalization was specific for NGF, since competition with a 1000-fold excess of unlabeled NGF resulted in nearly complete inhibition of radiolabeled NGF uptake (data not shown). CPZ and MDC each inhibited the internalization of ^{125}I-NGF into the CCV fraction that occurred upon warming (not shown) (Howe et al., 2001). We asked to what extent NGF internalization via CCVs was dependent on TrkA activation. First, we tested the effect of K252a, a relatively specific inhibitor of TrkA, on NGF internalization into cells. In the presence of K252a, NGF internalization decreased to 51% (± 11%; $n = 3$; $p < 0.05$) of the untreated value. Thus, in PC12 cells about 50% of

NGF internalization requires TrkA activation. Importantly, K252a reduced the internalization of NGF into CCVs to 10% (± 6%; $n = 3$; $p < 0.02$) of the untreated value, indicating that the vast majority of clathrin-mediated NGF internalization is dependent on TrkA activation (Howe et al., 2001).

We next determined whether NGF-TrkA complexes were present in CCVs. To begin, we examined TrkA in CCVs. TrkA was present in CCVs and the amount increased following NGF treatment (Fig. 2A, O). Following 2 min of NGF treatment, the amount was 157% ($n = 4$; $p < 0.0002$) of that in CCVs from untreated cells, reflecting a 57% increase. At 5 min, the increase was 108% ($n = 4$; $p < 0.0001$) above untreated. To clarify whether these changes were due to the increase in CCV production observed with NGF treatment, the amount of TrkA in CCVs from treated cells, as percent of the untreated control, was divided by the amount of CCV protein in treated cells, as percent of the untreated control. This gave a ratio for the extent of enrichment with treatment. A value significantly greater than one would indicate that the increase was larger than could be accounted for by the increase in CCVs. For TrkA, neither the ratio at 2 min (0.95; $p > 0.50$) nor that at 5 min (1.36; $p > 0.10$) was significantly greater than one (Howe et al., 2001). These data suggest that the increase in TrkA in CCVs was due to the increase in the number of CCVs.

Next, we asked whether or not the NGF in CCVs was bound to TrkA. CCVs isolated from cells warmed in the presence of ^{125}I-NGF were lysed and immunoprecipitated with an antibody against TrkA. The TrkA immunoprecipitate was submitted to SDS-PAGE and auto-radiography. Figure 2B shows a radiolabeled band corresponding to the monomer weight of NGF. In one experiment, bands were

and was increased with NGF treatment. (E and P) Ras was present in CCVs. CCVs from cells treated with NGF contained more Ras. (F and Q) C-Raf was present in CCVs. The amount was markedly increased by NGF treatment. (G and Q) B-Raf was present in CCVs, but did not increase with NGF treatment. (H and Q) Mek1/2 was found in the CCV fraction, and was modestly increased by NGF treatment. (I and Q) Activated Mek1/2 were present at extremely low levels in CCVs isolated from untreated cells, but were markedly increased in CCVs from NGF-treated cells. (M) PI-3 kinase was present in the CCV fraction, and was increased by NGF treatment. (N) Likewise, PLCγ was increased following NGF treatment. (J and R) Erk1/2 were found in CCVs. There was no increase in total Erk1/2 in CCVs following NGF treatment. (B and R) However, the same blot reprobed with anti-p-Erk1/2 (i.e., Erk1 and Erk2) showed a dramatic increase in activated Erk in CCVs isolated from NGF-treated cells at both 2 and 5 min. (L) CCVs from NGF-treated cells transmitted a signal in a cell-free assay. PC12 cells were treated at 37°C with NGF (2 nM) for the times indicated, or were not treated (0′). Following NGF treatment, CCVs exhibited a marked increase in the ability to phosphorylate the Elk-GST substrate (reprinted, with permission, as modified from Howe et al., 2001).

excised from the gel and counted. Compared to the unwarmed sample, the amount of ^{125}I-NGF was 296% at 5 min, and 562% at 15 min. These studies show that NGF is bound to TrkA in CCVs. They also demonstrate that the amount of NGF bound to TrkA in CCVs was out of proportion to the increase in CCV production, suggesting that TrkA receptors with bound NGF are enriched in the CCV fraction.

Given NGF binding to TrkA, it was expected that TrkA receptors in CCVs from NGF-treated cells would be activated. To confirm this, we tested for activated TrkA in CCVs. There was very little tyrosine phosphorylated TrkA in CCVs isolated from untreated cells. In contrast, activated TrkA was present in the CCVs of treated cells at both 2 and 5 min (Fig. 2C, O). Following treatment there was a 307% increase in pTrkA at 2 min ($n = 3$; $p = 0.004$) and a 510% increase at 5 min ($n = 3$; $p = 0.004$). The levels of phosphorylated TrkA were dramatically increased with respect to CCV protein at both 2 and 5 min (Howe et al., 2001). We conclude that NGF induces the formation of CCVs that are enriched in activated TrkA, and that NGF remains bound to activated TrkA in this organelle.

NGF signals from CCVs

To assess the ability of activated TrkA receptors in CCVs to signal, we asked whether proteins known to participate in the Ras-Erk1/2 signaling pathway were associated with CCVs. Several of these proteins were present in CCVs from untreated cells (Fig. 2D–M). Shc was present, as was Ras, Mek1/2 and Erk1/2. However, we detected only a small amount of C-Raf and very little of the activated forms of Ras, Mek1/2 or Erk1/2. NGF treatment resulted in marked changes in several of the signaling proteins present within CCVs. First, we noted that NGF treatment caused an increase in the amount of the 66 kD form of Shc associated with CCVs. At 5 min, the increase in Shc significantly exceeded that for CCV protein ($n = 3$; $p < 0.05$). Ras was also contained in CCVs (Fig. 2E,P). While some Ras was present in the CCVs of untreated cells, the amount was significantly increased by NGF treatment at both 2 and 5 min, and at 5 min exceeded the increase in CCVs

($n = 5$; $p < 0.005$). The increases followed a time-course similar to that for Shc. To determine whether activated Ras was present, we affinity precipitated the affinity GTP-bound form of Ras, using a GST-fusion protein consisting of the Ras-binding domain of Raf-1. We observed increases in Ras-GTP at both 2 and 5 min of NGF treatment (Fig. 2P). By 5 min, the amount of activated Ras was significantly enriched with respect to CCV protein ($n = 3$; $p < 0.05$). While there was little C-Raf in CCVs from untreated cells, the amount was markedly increased following NGF treatment, and was significantly in excess of CCV protein ($n = 3$ and $p < 0.001$ for both 2 and 5 min) (Fig. 2F, Q). In contrast, while B-Raf was present, the levels did not increase with NGF treatment (Fig. 2G, Q). Mek1/2 were also increased at both 2 and 5 min of NGF treatment, but they were not enriched with respect to CCV protein (Fig. 2H, Q). However, at both time-points pMek1/2 were markedly increased and were significantly enriched relative to CCV protein ($n = 3$ for both; $p < 0.0001$ for both) (Fig. 2I, Q).

Next, we asked whether Erk1/2 were associated with CCVs. They were present in both treated and untreated cells, and the amount of these proteins was not increased by NGF treatment (Fig. 2J, R). Indeed, relative to CCV protein levels, the amounts of Erk1 and Erk2 were decreased. Of note, however, activated Erk1/2 (i.e., pErk1 and pErk2) were present essentially only in CCVs from NGF-treated cells, and the degree of phosphorylation increased between 2 and 5 min (Fig. 2K, R). The pattern was highly reminiscent of that seen for the levels of Shc, Ras and C-Raf. Phosphorylated Erk1/2 were both significantly enriched in CCVs following NGF treatment ($n = 5$ and $p < 0.001$ for both pErk1 and pErk2 at both 2 and 5 min) (Howe et al., 2001). These studies provide evidence for the presence of components of the Ra/Erk1/2 pathway in CCVs from untreated cells and document significant changes in response to NGF treatment. The findings indicate that NGF signaling results in the increasing presence within CCVs of activated TrkA, Shc, Ras, activated Ras, C-Raf and activated Mek1/2, and suggest that these changes are linked to the activation of Erk1/2.

To test the signaling potential of CCVs, we examined the capacity of isolated CCVs to phosphorylate Elk, a well-known substrate of Erk.

We used as substrate an Elk-GST fusion protein that includes the Elk domain normally phosphorylated by Erk1/2 in vivo. Consistent with our findings for activated Erk1/2 in CCVs, there was a marked increase in activity associated with the CCVs isolated from cells treated with NGF (Fig. 2L). The increase in activity on Elk was significantly greater than the increase in CCV protein at both 2 and 5 min ($n = 3$; $p < 0.05$) (Howe et al., 2001). We conclude that CCVs can serve as a platform for propagating the NGF signal through the Ras/Erk1/2 pathway.

NGF recruits components of other signaling pathways to CCVs

NGF signals through a diverse set of pathways (Sofroniew et al., 2001). We entertained the possibility that protein components of non-MAPK pathways might be present in the CCV fraction. We found that the p85 subunit of PI-3 kinase and that of PLC-γ were both present in CCVs, and that their levels increased with NGF treatment. The levels of PI3-K were quantified. Relative to the untreated controls, the levels of PI3-K were 376% ($\pm 20.7\%$; $n = 3$; $p = 0.002$) at 2 min and 589% ($+ 72.0\%$; $n = 3$; $p = 0.0002$) at 5 min of NGF treatment. These increases were far greater than those for CCV protein (Howe et al., 2001). These findings are consistent with the view that a number of signaling pathways are present in CCVs and suggest that CCVs may be used to support NGF signaling from them.

Early endosomes serve as signaling endosomes

The isolation and the characterization of vesicles containing proteins of the Ras/Erk1/2 pathway suggested that CCV could be used as a platform for NGF signaling. An important question is whether or not CCVs could be used to move NGF signals from axon terminals to neuron cell bodies. Two findings suggested that CCVs would not prove to be the retrograde carrier. The first is that these are short-lived intermediates in the endocytic pathway (Sorkin and von Zastrow, 2002). The second consideration focuses on the temporal nature of the signals generated by signaling endosomes. It is possible to construct several models for the signaling endosome.

In one, signaling endosomes would continue to signal during retrograde transit. Under this model, the entire neuron (i.e., axon terminal, axon and cell body) would be informed by activation of signaling proteins present on the cytosolic surface of signaling endosomes. This model is attractive, but it poses the need for persistent activation of signaling pathways. Because the distance that endosomes must travel from the axon terminal to the cell body can be considerable, and because the speed of the retrograde transport is only about 2–20 mm/h (Grafstein and Forman, 1980), for this model to apply signaling would need to persist for many hours. A requirement for continued NGF signaling is inconsistent with the finding that NGF-induced activation of the Ras/Erk1/2 pathway is transient (Wu et al., 2001). Interestingly, long-lived activation of the Erk1/2 pathway has been shown to be necessary for the induction of the genetic and other cellular programs that promote neuronal differentiation of PC12 cells, events that may reflect those induced by the NGF retrograde signal (Sofroniew et al., 2001). The search for a pathway that could lead to the prolonged activation of Erk1/2 following NGF treatment led to the discovery of the Rap1/Erk1/2 pathway (York et al., 1998). Rap1 is a member of the Ras family (Wu et al., 2001). There is evidence that activation of Rap1 causes sustained activation of Erk1/2 (York et al., 1998), a finding that suggests that the Rap1 pathway could play a role in NGF retrograde signaling.

NGF signals through the Rap1/Erk1/2 pathway

To pursue the possibility that the Rap1/Erk1/2 pathway might serve to carry the NGF retrograde signal, we carried out studies in PC12 cells to examine the activation of this pathway following NGF treatment. In contrast to earlier studies, we examined NGF signaling under conditions in which the endogenous Rap1 signaling pathway was examined. The original report of these studies, and the detailed methods to support them, have been published (Wu et al., 2001). NGF, but not EGF, induced prolonged activation of Erk1/2 (Wu et al., 2001). To ask whether the activation of endogenous Ras could account for the sustained activation of Erk1/2 elicited by NGF, we examined Ras activation via

12

an established method (Herrmann et al., 1995). The method takes advantage of specific binding of a C-RafRBD/GST fusion protein to the activated form of Ras (i.e., Ras^{GTP}). EGF-induced Ras activation was detected as early as 10 s, reached its maximal activation at 2 min, decreased to a low level at 5 min and was detected only weakly thereafter (Fig. 3B). NGF-induced Ras activation was maximal at 2 min and was no longer detected after 5 min (Wu et al., 2001). Thus, the transient activation of endogenous Ras is insufficient to explain the prolonged activation of Erk1/2 seen in NGF-treated cells.

We next investigated whether endogenous Rap1 was activated by NGF. Similar to the assay used for activated Ras, a RalGDS-RBD/GST fusion protein was used to analyze Rap1 activation. No significant activation of Rap1 was detected after EGF treatment (Fig. 3C). In contrast, NGF treatment elicited a marked increase in $Rap1^{GTP}$. Moreover, unlike the transient pattern seen for Ras activation, NGF-induced Rap1 activation was persistent. While Rap1 was weakly activated at early time points (2–10 min), there was a marked increase at 30 min and the activation continued through at least 180 min (Wu et al., 2001). The increase at 30 min and thereafter was not due to newly synthesized Rap1, since we did not observe an increase in Rap1 protein by immunoblotting, and because a protein synthesis inhibitor had no influence on the effect of NGF treatment (data not shown).

To investigate further the NGF effect on Rap1, we tested 6-24 cells, a PC12 cell line that overexpresses TrkA (Stephens et al., 1994). 6-24 cells were treated with NGF (50 ng/ml). Activated Rap1 was assayed as described above (Fig. 3E). The results showed that NGF treatment induced long-term activation of Rap1 following a time course similar to that described above. Since Rap1 was more robustly activated in cells that overexpress TrkA, the possibility was raised that NGF signals through TrkA to induce Rap1 activation. To show that TrkA was required for NGF-induced persistent Rap1 activation, serum-starved PC12 cells were either pretreated with a vehicle or with 0.5 μM of K252a prior to NGF treatment. Pretreatment with K252a abolished NGF-induced Rap1 activation (Fig. 3D) (Wu et al., 2001). Thus, we conclude that NGF acts through TrkA to induce persistent activation of Rap1

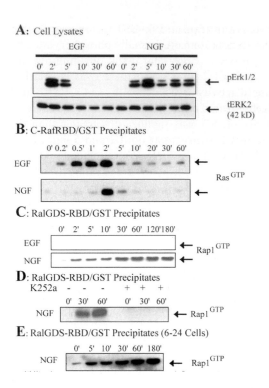

Fig. 3. Analysis of endogenous Ras and Rap1 activation in EGF- and NGF-treated cells. An equal number of serum-starved PC12 cells were treated with either EGF (50 ng/ml) or NGF (50 ng/ml) for the time intervals indicated, or the vehicle control (0 min). (A) The cells were rinsed and lysed in RIPA buffer. 40 μg proteins from each cell lysates were separated on SDS-PAGE and analyzed using immunoblotting. The blot was probed with a rabbit antibody that specifically recognizes activated (i.e., phosphorylated) Erk1/2 (top panel). The blot was then reprobed with a mouse antibody against Erk2. (B) Ras^{GTP} was precipitated with the C-RafRBD/GST fusion protein. The fusion protein was prebound to glutathione-agarose beads prior to incubation with cell lysates. Following incubation, the beads were washed and boiled in SDS-PAGE sample buffer. The samples were separated on 12.5% SDS-PAGE and proteins were transferred onto PVDF membrane. The membrane was probed with a mouse antibody to Ras. (C) The amounts of $Rap1^{GTP}$ following EGF- and NGF-treatment were assayed as in (B), except that the RalGDS-RBD/GST fusion proteins were used to precipitate $Rap1^{GTP}$. The resulting blot was probed with a mouse antibody to Rap1. (D) Cells were pretreated with either vehicle (−) or 0.5 μM K252a (+) for 15 min at 37°C prior to treatment with NGF (50 ng/ml) for the indicated times or not treated (0′). Activated Rap1 was assayed as in (C). (E) Serum-starved 6-24 cells were treated with NGF (50 ng/ml) for the time intervals indicated or the vehicle control (0′). The amount of $Rap1^{GTP}$ was assayed as in (C). All blots were visualized using SuperSignal and the results shown are representative of at least three independent experiments (reprinted, with permission, as modified from Wu et al., 2001).

and that this effect correlates temporally with sustained activation of Erk1/2.

Defining the components of the TrkA/Rap1/Erk1/2 pathway

To investigate how NGF induced persistent activation of Rap1, we examined the effect of NGF treatment on signaling proteins that regulate the activity of Rap1. The results for NGF were compared to those for EGF, since EGF failed to activate Rap1. C3G is a guanine nucleotide exchange factor that specifically interacts with and activates Rap1 by catalyzing the conversion of Rap1 from the inactive (Rap1^{GDP}) to the active form (Rap1^{GTP}) (Gotoh et al., 1995). Since tyrosine phosphorylation has been suggested to be required for C3G activity (Ichiba et al., 1999), we evaluated the tyrosine phosphorylation state of C3G. Serum-starved PC12 cells were treated with either EGF or NGF for time intervals ranging from 0 to 30 min and C3G was immunoprecipitated from cell lysates. C3G was constitutively tyrosine-phosphorylated in PC12 cells, as revealed by blotting with 4G10, an antibody to phosphotyrosine. EGF treatment induced rapid dephosphorylation of C3G and caused rapid association of C3G with a number of tyrosine-phosphorylated proteins whose estimated molecular masses ranged from 95 to 170 kDa (Wu et al., 2001). This observation suggested that tyrosine phosphorylation may regulate the ability of C3G to interact with other proteins. In contrast to the findings for EGF, NGF treatment did not produce an appreciable change in tyrosine-phosphorylated C3G. Nor was C3G shown to become associated with novel tyrosine-phosphorylated proteins in NGF-treated cells (Wu et al., 2001). To investigate further the presence of tyrosine phosphorylated C3G following EGF or NGF treatment, tyrosine phosphorylated proteins were immunoprecipitated with anti-phosphotyrosine antibodies (4G10) from cells that were treated with either EGF or NGF for time intervals ranging from 0 to 30 min. The precipitates were separated on SDS-PAGE and immunoblotted with anti C3G antibodies. Our results confirmed that C3G became transiently and markedly dephosphorylated in EGF-treated cells. No other bands corresponding to tyrosine

phosphorylated C3G were evident in EGF-treated cells at 2 min. In NGF-treated cells, the level of tyrosine phosphorylated C3G was somewhat decreased relative to the control, but this phospho-protein was readily evident at all time points (Wu et al., 2001).

In several cell lines, C3G has been shown to form complexes with Crk in response to growth factor stimulation (Gotoh et al., 1995; Boussiotis et al., 1997; Okada and Pessin, 1997). We asked which of the two predominant Crk isoforms, CrkII or CrkL, was present in the C3G immunoprecipitates. The 40 kDa CrkII was not detected in the complex by immunoblotting (data not shown). The 36 kDa CrkL was found to be constitutively associated with C3G in serum-starved PC12 cells. EGF treatment resulted in a marked decrease in the amount of associated CrkL. The decrease was apparent by 2 min and persisted through 30 min. NGF treatment did not result in a decrease in associated CrkL in complex with C3G (Wu et al., 2001). These results demonstrated clear differences for EGF and NGF signaling on tyrosine phosphorylation of C3G and on proteins found in complex with C3G.

CrkL has been shown to play an important role in regulating Rap1 activation and in suppressing Ras-dependent signaling in T lymphocytes (Boussiotis et al., 1997). When cotransfected with C3G, CrkL was effective in enhancing Rap1 activation in PC12 cells (York et al., 1998). These findings together with our data for the continued association of C3G and CrkL following NGF treatment pointed to the possibility that endogenous CrkL might regulate NGF signaling to Rap1. To examine a role for CrkL, serum-starved PC12 cells were treated with either EGF or NGF. CrkL was immunoprecipitated and analyzed using SDS-PAGE and immunoblotting. A short exposure of the blots probed with 4G10 showed that several tyrosine phosphorylated proteins were associated with CrkL in unstimulated cells (Fig. 4A) (Wu et al., 2001). There were at least three such prominent proteins with apparent molecular masses of 130, 95 and 68 kDa. None of these proteins were detected after 5 min of EGF treatment. Instead, a new protein with a molecular mass of 120 kD appeared in the CrkL immunoprecipitates in EGF treated cells. At 10 min, the tyrosine phosphorylated proteins seen in the quiescent state were again detected. A longer exposure (Fig. 4B) revealed additional differences between

14

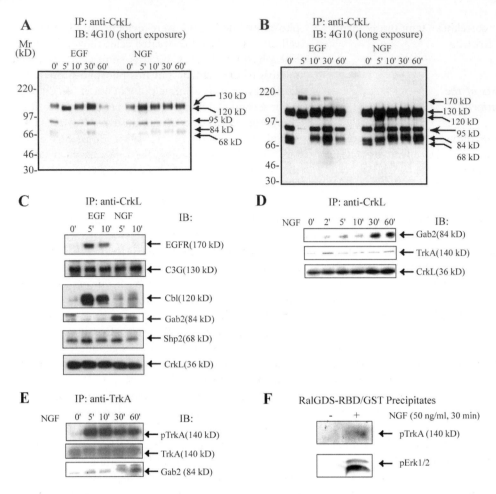

Fig. 4. Different CrkL-containing complexes were induced by NGF and EGF treatment. An equal number of serum starved-PC12 cells were treated with either EGF (50 ng/ml) or NGF (50 ng/ml) for the time intervals indicated, or the vehicle control (0′). The cells were rinsed, lysed and immunoprecipitated with antibodies to CrkL. The immunoprecipitates were separated on 7.5% SDS-PAGE and analyzed by immunoblotting with the antibodies indicated. (A) A short exposure of a blot probed with 4G10 to detect tyrosine phosphorylated proteins. (B) A longer exposure of (A). EGF treatment rapidly changed the proteins associated with CrkL. NGF recruited an 84 kD protein to CrkL. (C) The CrkL immunoprecipitate was probed with antibodies to a number of signaling proteins. After EGF treatment, a complex containing C3G/CrkL/Shp2/Cbl/EGFR was formed. In NGF-treated cells, a complex was formed that contained C3G/CrkL/Shp2/Gab2/TrkA; (D) CrkL was immunoprecipitated as in (C), except that cells were treated with NGF (50 ng/ml) for the time intervals indicated or the vehicle control (0′). The immunoprecipitates were separated on SDS-PAGE and the blot was probed with antibodies to Gab2, TrkA and CrkL. In NGF-treated cells, Gab2 and TrkA were both present in complex with CrkL for the duration of the experiment. (E) Cells were treated with NGF (50 ng/ml) for the time intervals indicated or the vehicle control (0′) and TrkA was immunoprecipitated with a mouse antibody to TrkA (MCTrks). The immunoprecipitates were separated on 10% SDS-PAGE and the blot was probed with the following antibodies: pTrkA(pY490), total TrkA (MCTrks) and Gab2. All blots were visualized using SuperSignal and the results shown are representative of at least three independent experiments. (F) An equal number of serum-starved PC12 cells were treated with NGF (50 ng/ml) or the vehicle control for 30 min. After lysing cells in a buffer containing NP40 plus glycerol (Wu et al., 2001), Rap1^{GTP} was precipitated as described. The resulting precipitates were washed and separated using a 7.5% SDS-PAGE and transferred onto PVDF membrane. The blot was probed with antibodies to either p-TrkA (pY490) or p-Erk1/2. All blots were visualized using SuperSignal and the results shown are representative of at least three independent experiments (reprinted, with permission, as modified from Wu et al., 2001).

EGF and NGF treatments. EGF induced transient association of CrkL with a 170 kD tyrosine-phosphorylated protein. In contrast, NGF treatment induced the appearance of a tyrosine-phosphorylated 84 kD protein that persisted in the CrkL complexes through 60 min (Wu et al., 2001). We conclude that CrkL forms complexes with different proteins in response to EGF and NGF treatment.

To identify the tyrosine-phosphorylated proteins that were either constitutively or transiently associated with CrkL, we probed the blots for a number of known signaling proteins. Our results revealed the following: (1) the 170 kDa tyrosine-phosphorylated protein that became associated with CrkL only after EGF treatment was the EGF receptor; (2) the 130 kD constitutively tyrosine-phosphorylated protein was C3G; (3) the 120 kD tyrosine-phosphorylated protein that transiently associated with CrkL in response to EGF was the adapter protein Cbl; (4) the tyrosine-phosphorylated 84 kD protein that became associated with CrkL in NGF-treated cells was the adapter protein Gab2; and (5) the 68 kDa constitutively tyrosine-phosphorylated protein was the tyrosine phosphatase Shp2 (Wu et al., 2001) (Fig. 4C). Notably, the association of Gab2 with the CrkL-containing complex was quite persistent. Though the maximal level of Gab2 was seen at 30 min following NGF treatment, a significant amount of Gab2 was still present in the CrkL immunocomplex at 60 min (Fig. 4D). When we reprobed the blots with a mouse anti-TrkA antibody, our results showed that TrkA was also prominently recruited to the CrkL complex at 2–5 min (Wu et al., 2001). Although the amount of TrkA decreased at later times, it was still present in the complex at 60 min (Fig. 4D). To examine whether NGF induced the association of Gab2 with TrkA, TrkA was immunoprecipitated from PC12 cells. The blot was first probed with anti-pTrkA (pY490) antibodies. The results showed that TrkA was activated in these cells even at 60 min following NGF treatment. While the total amount of TrkA did not appear to change, Gab2 was increasingly associated with TrkA between 5 and 60 min (Fig. 4E) (Wu et al., 2001).

In summary, our finding showed that proteins known to regulate Rap1 activity are found in markedly different complexes in cells treated with EGF and NGF. CrkL constitutively formed a complex with tyrosine-phosphorylated C3G and Shp2. This complex was rapidly changed by EGF treatment; C3G was dephosphorylated and a protein complex that included C3G/CrkL/Shp2/Cbl and tyrosine-phosphorylated EGFR was transiently formed. In contrast, C3G continued to be tyrosine-phosphorylated in NGF-treated cells and the C3G/CrkL/Shp2 complex became stably associated with Gab2 and TrkA. These results raised the possibility that NGF treatment caused formation of a complex in which activated TrkA is linked to activated Rap1.

We carried out studies to define whether activated TrkA was associated with activated Rap1. Serum-starved PC12 cells were treated with NGF (50 ng/ml) for 30 min, a condition under which Rap1 is nearly maximally activated. Cells were rinsed and lysed in a buffer containing NP-40 and glycerol. The resulting supernatants from both control and NGF-treated samples were incubated with the RalGDS-RBD/GST fusion protein to precipitate Rap1^{GTP} as described above. The precipitates were analyzed using SDS-PAGE and immunoblotting with an antibody that specifically recognizes the activated form of TrkA (pY490). Activated TrkA was detected only in the NGF-treated cells (Fig. 4F). Importantly, activated Erk1/2 was also present in the precipitates of NGF-treated cells, but not control cells (Wu et al., 2001). We conclude that activated Rap1 is associated, either directly or indirectly, with activated TrkA and activated Erk1/2 in NGF-treated cells.

Rap1 signaling from early endosomes

We next undertook studies to define the subcellular localization of activated Rap1 in NGF treated cells. Several observations suggested that intracellular membranes, including endosomes, would be involved in Rap1 signaling. First, Rap1 has been shown to reside primarily in endosomes, secretory granules and the Golgi apparatus (Kim et al., 1990; Beranger et al., 1991). Second, by 30 min of NGF treatment most surface TrkA receptors have been internalized (Grimes et al., 1996); this event coincides with the maximal activation of Rap1 seen in the current study. We used confocal microscopy to investigate the subcellular localization of Rap1 and several components of the Rap1 signaling complex by determining whether or not they colocalized with EEA1, a marker for early endosomes (Mu et al., 1995).

We first examined the distribution of TrkA, Rap1 and activated Erk1/2. We found that Rap1 was present in early endosomes in both vehicle and NGF-treated PC12 cells. In vehicle-treated cells, punctate EEA1 and Rap1 immunostaining decorated structures of various size that were widely distributed in the cytoplasm; partial colocalization of these markers was seen throughout the cytoplasm (Fig. 5C, D) (Wu et al., 2001). After NGF treatment, the immunostaining patterns for Rap1 and EEA1 were quite similar. There was increased colocalization of Rap1 and EEA1 with many bright puncta now clustered near the nucleus (Fig. 5A) (Wu et al., 2001). Punctate TrkA staining was widely distributed in the cytoplasm in vehicle-treated cells (Fig. 5A). There was little colocalization of TrkA and Rap1. Following NGF treatment, TrkA became concentrated in the perinuclear region and displayed significant colocalization with Rap1 (Fig. 5B) (Wu et al., 2001). Next, we stained PC12 cells for activated Erk1/2. As expected, there was no signal for activated Erk1/2 in vehicle-treated cells. However, in NGF-treated cells, activated Erk1/2 were seen in the perinuclear region where they were partially colocalized with EEA1 (Fig. 5F) (Wu et al., 2001). Taken together, these results are evidence that key components of the Rap1 signaling complex are present in early endosomes.

To define further the subcellular localization of Rap1, we used a step gradient density system of OptiPrep (5% : 10% : 15% : 20% : 25%) to fractionate PC12 cell homogenates. Membrane fractions were collected and analyzed using SDS-PAGE and immunoblotting. An earlier study using similar gradients revealed that plasma membrane markers were present in low density fractions and that internal membranes, such as early endosomes, were found in heavy density fractions (Sheff et al., 1999). Consistent with these findings, we showed that plasma membrane, as marked by EGFR, was detected in fractions 1 and 2 (Fig. 6A). The endosomal markers EEA1 and Rab5B were enriched in fractions 3 and 4. There were striking differences for the fractionation of Ras and Rap1 in untreated cells. While Ras was detected in lighter fractions, Rap1

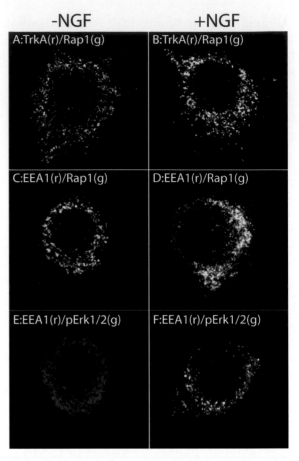

Fig. 5. Analysis of subcellular localization of components of the Rap1 signaling pathway by indirect immunofluorescence. PC12 cells were cultured for 24–48 h on cover glasses coated with matrigel. Cells were serum-starved and were treated with NGF (50 ng/ml) (B, D, F) or the vehicle control (A, C, E) for 30 min. Cells were then rinsed, fixed with 100% methanol, and processed for indirect immunofluorescence as described (Wu et al., 2001). Primary antibodies were: a rabbit antibody to Rap1 (1/250) and a mouse antibody to TrkA (1/400) (A, B); a rabbit antibody to Rap1 (1/250) and a mouse antibody to EEA1 (1/120) (C, D); a rabbit antibody to pErk1/2 (1/120) and a mouse antibody to EEA1 (1/120) (E, F). Alexa 568 goat anti-mouse IgG conjugates and Alexa 488 goat anti-rabbit IgG conjugates were used to visualize the primary antibodies. (g): green fluorescence signal; (r): red fluorescence signal. Colocalization of antigens is denoted by the yellow fluorescence signal. The scale bar is 10 μm (reprinted, with permission, as modified from Wu et al., 2001).

was present only in heavier fractions. The presence of Ras in fractions 1 and 2 is consistent with its known plasma membrane localization (Fig. 6A) (Leevers et al., 1994). Cofractionation of Rap1 with

A: OptiPrep Gradients

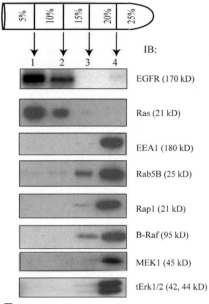

B: Fractions of OptiPrep Gradients

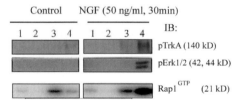

Fig. 6. Activation of Rap1 in endosomal fractions. (A) Post-nuclear supernatants generated from an equal number of serum-starved PC12 cells were fractionated by centrifugation using a step-density gradient system as described (Wu et al., 2001). Four membrane fractions were collected as indicated. Total proteins from each fraction were precipitated in 7% TCA and washed in acetone. The pellets were then dried and boiled in SDS-PAGE loading buffer. The proteins were analyzed using SDS-PAGE and immunoblotting. The blot was independently probed with antibodies to EGFR, Ras, EEA1, Rab5B, Rap1, B-Raf, Mek1 and Erk1/2. The molecular mass corresponding to each antigen is shown in parenthesis. (B) In the upper two panels, membrane fractions were collected from serum-starved PC12 cells that were either treated with NGF (50 ng/ml) or the vehicle control for 30 min. Total proteins from each fraction were precipitated, washed and analyzed using SDS-PAGE and immunoblotting as in (A). Activated TrkA and activated Erk1/2 were detected using specific antibodies. In the lowest panel, to detect activated Rap1, gradient fractions were collected and lysed as indicated in the text and the amount of Rap1^{GTP} was assayed as described in Fig. 3. All blots were visualized using SuperSignal and the results shown are representative of at least three independent experiments.

EEA1 and Rab5B is further evidence that Rap1 is localized in intracellular membranes including early endosomes (Wu et al., 2001). It is noteworthy that B-Raf, Mek1 and Erk1/2 were also detected in fractions 3 and 4, as were Gab2, C3G and CrkL (data not shown) (Wu et al., 2001).

To determine whether or not NGF induced Rap1 activation in endosome-enriched fractions, serum-starved PC12 cells were treated either with NGF (50 ng/ml) or the vehicle for 30 min. Step gradient density fractionation was then performed. The fractionation of Rap1 was unaffected by NGF treatment (data not shown). The membrane fractions collected from the gradient were adjusted to the lysis buffer used to examine the complex containing activated Rap1 together with pTrkA and pErk1/2 (Wu et al., 2001). The clear lysates were then incubated with RalGDS-RBD/GST to precipitate Rap1^{GTP}. The precipitates were analyzed using SDS-PAGE and immunoblotting. A basal level of Rap1^{GTP} was detected in the control sample (Fig. 6B). NGF treatment resulted in a marked increase in the amount of Rap1^{GTP} in fraction 4 (Wu et al., 2001). In a parallel experiment, the total proteins from each gradient fraction were precipitated with TCA (trichloroacetic acid) prior to analysis using SDS-PAGE and immunoblotting. We found that activated TrkA was detected only in membrane fractions isolated from NGF-treated cells. Significantly, activated TrkA was present only in the fractions enriched for Rap1 (Fig. 6B). Similar results were also obtained for activated Erk1/2. In preliminary studies, we found that NGF treatment resulted in increased levels of Gab2, C3G, CrkL, B-Raf, Mek1 and Erk1/2 in fraction 4 (data not shown) (Wu et al., 2001). Our findings point to the existence of complexes containing activated TrkA together with activated Rap1 and activated Erk1/2 in intracellular membranes in NGF-treated cells.

Together with the results from immunostaining studies, these findings indicate that Rap1 signaling through the Erk1/2 pathway can originate from early endosomes. In parallel with our studies on CCVs, preliminary studies examining Rap1-containing endosomes isolated from NGF treated cells have shown that these fractions can also catalyze the phophorylation of the Elk/GST fusion protein (C. Wu, W.C. Mobley, unpublished observations). These observations are evidence that NGF signaling through the Rap1/Erk1/2 pathway in early endosomes could be used to transmit the NGF retrograde signal.

Exploring a role for signaling endosomes in vivo

A critical test of the signaling endosome hypothesis is to show that physiologically significant signaling events can arise from them. Supporting the view that endosomes do carry significant NGF signals, we found that failed retrograde transport of NGF in the Ts65Dn mouse model of Down syndrome (DS) was linked to the degeneration of basal forebrain cholinergic neurons (BFCNs). The studies to support this conclusion, and the detailed methods that support it, have been published (Cooper et al., 2001). Ts65Dn is a mouse model of human Down syndrome (i.e., trisomy 21) in which there are three copies of many of the mouse orthologues of the genes present on distal human chromosome 21 (Reeves et al., 1995). We found degeneration of BFCNs in this model, a

phenotype that replicates findings in aged individuals with DS and in patients with Alzheimer's disease (AD). Using immunostaing for p75^{NTR} to mark BFCNs, there was a difference in both the size and number of these neurons in the medial septal nucleus in aged Ts65Dn mice (number: 2N = 1914 ± 123, Ts65Dn = 1463 ± 63. $p = 0.028$, $n = 5$; size: 2N = 139.5 ± 1.5 μm^2, Ts65Dn = 118.8 ± 4.5 μm^2. $p = 0.009$, $n = 5$). Interestingly, these changes were age-related as there was no difference in either parameter at age 6 months. The differences between Ts65Dn and 2N mice in number and size resulted, in part, from a failure of BFCNs in Ts65Dn mice to undergo the increases seen with aging in 2N mice (Cooper et al., 2001). Since NGF plays an important role in maintaining the phenotype of BFCNs (Sofroniew et al., 2001), we studied the possibility that failure of NGF signaling could be responsible for BFCN degeneration. First, we examined expression of the NGF gene. There was no difference in the amount of NGF mRNA in hippocampus, the target of degenerating BFCNs (unpublished observations). As determined by ELISA, there was also no decrease

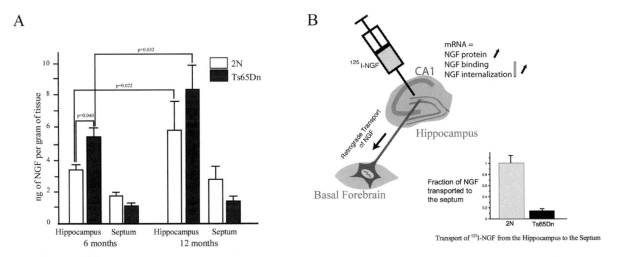

Fig. 7 NGF retrograde signaling in the mouse BFCN-hippocampal system: (A) Septal and hippocampal NGF levels in Ts65Dn and 2N mice. ELISA measurements of NGF (ng NGF per gram tissue wet weight; mean ± SEM) showed significant increases in the hippocampus of both 2N and Ts65Dn mice between 6 and 12 months of age. NGF levels were significantly higher in the hippocampus of Ts65Dn (black) vs 2N (white) mice at 6 months of age (2N = 3.4 ± 0.3; Ts65Dn = 5.5 ± 0.5; $p = 0.04$; $n = 5$), but were not significantly different at 12 months of age (2N = 5.9 ± 1.6; Ts65Dn = 8.7 ± 1.9; $p = 0.15$; $n = 4$). There was a trend toward a reduction in the level of NGF in the septal region of Ts65Dn mice at 6 months of age (2N = 1.7 ± 0.2; Ts65Dn = 1.0 ± 0.2; $p = 0.18$; $n = 5$) and at 12 months of age (2N = 3.0 ± 0.2; Ts65Dn = 1.4 ± 0.3; $p = 0.25$; $n = 4$), but these changes did not reach statistical significance. (B) Each of the steps needed for NGF signaling are represented schematically. Studies in the Ts65Dn mouse showed no reduction in the amount of mRNA^{NGF}, NGF protein, or NGF receptor binding and internalization nor was there a failure to respond to NGF when it was provided to cell bodies. However, there was a dramatic decrease in the retrograde transport of NGF.

TABLE 1

NGF binding and internalization in hippocampal synaptosomes from 2N and Ts65Dn mice

Method[a]	2N	Ts65Dn	p Values[b]
Surface bound before warming	31 ± 3.3	53 ± 2.6	< 0.005
Internalization during warming	11 ± 1.5	14 ± 1.6	< 0.05
Surface bound after warming	26 ± 0.6	39 ± 0.3	< 0.05

[a]All results are expressed as pg of NGF per μg synaptosomes.
[b]One-tailed t-test. $N = 3$ for all measurement.

in the amount of NGF present in the hippocampus (Fig. 7). Indeed, we discovered that the hippocampus of Ts65Dn mice contained more NGF than the hippocampus of control animals. Next, we examined NGF binding and internalization. Hippocampal synaptosomes were prepared from Ts65Dn mice and control mice. Remarkably, both binding and internalization were significantly increased in Ts65Dn synaptosomes (Table 1). These findings were evidence against a role for either NGF production or NGF-mediated receptor activation and internalization as a cause of BFCN degeneration. They left open the possibility that failed retrograde transport of NGF, and of an NGF signal that would be carried in endosomes containing NGF, might still contribute. We gathered evidence that this was the case. ^{125}I-NGF was injected into the hippocampus of Ts65Dn and 2N controls to monitor the retrograde transport of NGF. While there was robust transport of NGF in 2N mice, little, if any, transport was detected in Ts65Dn mice (Fig. 7B). To test whether or not failed transport of NGF was linked to degeneration of BFCNs, we conceived of a method to bypass the defect by directly treating BFCN cell bodies with NGF. We reasoned that if the retrograde transport defect was critical for the degenerative phenotypes, treatment with NGF in this way might reverse them. We bypassed retrograde transport by delivering NGF for 2 weeks to the lateral ventricle of aged Ts65Dn and 2N mice and then examined the size and number of BFCNs. NGF treatment resulted in significant increases in both the size and number of BFCNs, leading to their normalization relative to 2N mice (number: Ts65Dn, vehicle $= 1612 \pm 106$, $n = 8$, NGF $= 2004 \pm 119$, $n = 8$, $p = 0.027$; size: Ts65Dn, vehicle $= 104.7 \pm 4.1$ μm^2, $n = 8$, NGF $= 131.7 \pm 4.0$ μm^2, $n = 8$, $p = 0.002$) (Cooper et al., 2001).

These results draw a link between the degeneration of BFCNs and the lack of retrograde NGF transport. In view of the emerging story for the structure and function of signaling endosomes, including the presence within them of NGF bound to its receptors, these findings point to the possibility that failed retrograde transport of NGF marks the failed retrograde transport of endosomes from which NGF signals.

Discussion

Our studies show the existence of at least two classes of endosomes in which NGF-pTrkA complex is present together with an extended array of signaling proteins. In CCVs, we found activated components of the Ras/Erk1/2 pathway as well as p85 subunit of PI3K and PLC-γ. In early endosomes, we found elements of the Rap1/Erk1/2 pathway. While existing data are inadequate to define the physiological role played by these organelles, observations in a mouse model of BFCN degeneration point to the possibility that failed retrograde transport of signaling endosomes may contribute to pathogenesis. Taken together, the findings are evidence that endosomes could be used as platforms from which NGF signals are communicated.

Support for the 'signaling endosome' has come from a number of laboratories (Ehlers et al., 1995; Grimes et al., 1996, 1997; Riccio et al., 1997; Bhattacharyya et al., 1997, 2002; Tsui-Pierchala and Ginty, 1999; Watson et al., 1999, 2001; Kuruvilla et al., 2000; Zhang et al., 2000; Howe et al., 2001; Sofroniew et al., 2001; Jullien et al., 2002). Studies in vitro have shown that:

(1) NGF at axon terminals must be internalized for the signal to be propagated to CREB (cyclic AMP response element binding protein) (Riccio et al., 1997);
(2) NGF induces endocytosis of TrkA through both clathrin membrane-mediated and non clathrin-mediated uptake (Howe et al., 2001; Shao et al., 2002);
(3) retrograde NGF signals are decreased by inhibiting endocytosis (Watson et al., 2001);

(4) NGF is bound to pTrkA in endosomes (Grimes et al., 1996, 1997; Howe et al., 2001);

(5) NGF-pTrkA complexes are found in CCVs together with activated components of the Ras/Erk1/2 pathway as well as PI3K and PLC-γ (Howe et al., 2001);

(6) NGF signals from CCVs to Elk-1, a downstream target of Erk1/2 (Howe et al., 2001);

(7) NGF induces the formation of complexes containing pTrkA and activated components of the Rap1/Erk1/2 pathway in early endosomes (Mochizuki et al., 2001; Wu et al., 2001);

(8) retrogradely transported early endosomes contain NGF-pTrkA together with activated components of the Rap1/Erk1/2 and PI3K pathways (J.D. Delcroix and W.C. Mobley, personal communication) and may contain Erk5 (Watson et al., 2001);

(9) NGF-pTrkA and BDNF-pTrkB complexes are moved retrogradely in axons to cell bodies (Tsui-Pierchala and Ginty, 1999; Watson et al., 1999; Kuruvilla et al., 2000);

(10) retrograde transport is blocked by disrupting microtubules (Watson et al., 1999); and

(11) retrogradely transported pTrk is required for CREB phosphorylation (Riccio et al., 1997; Watson et al., 1999), c-fos induction (Watson et al., 1999) and Akt phosphorylation (Kuruvilla et al., 2000).

In vivo observations are consistent in showing that NGF, pTrk and signaling proteins of the Erk1/2 pathway are retrogradely transported in sciatic nerve (Ehlers et al., 1995; Bhattacharyya et al., 2002; Johnson, 2002). Taken together, the evidence is compelling that signaling endosomes serve as a platform for NGF and NT signaling.

While the evidence for signaling endosomes is strong, Campenot and colleagues have provided data pointing to an alternative mechanism for retrograde signaling (MacInnis and Campenot, 2002). Using compartmented cultures in which NGF was supplied only to distal axons, they found that NGF linked to beads, and therefore presumably incapable of undergoing endocytosis, supported the survival of developing sympathetic neurons. In contrast to earlier observations (Watson et al., 1999; Kuruvilla et al.,

2000), Campenot has shown in compartmented cultures that activation of TrkA in cell bodies is not required for phosphorylation of Akt or CREB that follows NGF-mediated activation of TrkA on distal axons (R. Campenot, personal communication). Whatever the explanation for the discrepancies in findings, these data point to the importance of defining what role the endosomes play in retrograde signaling.

Signaling endosomes and neurodegeneration

While it is likely that important target-derived messages are communicated through signaling endosomes to enhance these parameters, there is no proof for this. It is in the context of defining a role for signaling endosomes that our findings on the mouse model of DS may provide additional insights. Under the signaling endosomes hypothesis, the failed retrograde transport of NGF would mark the failed retrograde transport of the NGF signal derived in the target. Our studies showed that the defect was due, at least in part, to an event that followed NGF internalization and that failed transport was correlated with the degeneration of BFCNs, a neuronal population dependent on NGF for survival and differentiation. Importantly, key degenerative features were reversed by an intervention that bypassed the defect in transport. These findings are consistent with the view that retrograde transport of signaling endosomes that carry NGF are needed for the normal function of BFCNs. However, additional experiments are needed to prove the point and to extend the analysis to fully define the defects in NGF signaling. We have considered the possibility that conditions other than DS impact the interaction of BFCNs with their targets. Retrograde trophic relationships may also be influenced by degenerative disorders, a topic that has received relatively little attention. In the context of our studies on the mouse model of DS, the cellular neuropathology of AD is extremely interesting. It is remarkable that, as in the mouse model, in the AD brain NGF levels in cortex and hippocampus are increased, while those in the basal forebrain are decreased (Scott et al., 1995). The parallel in the findings for NGF levels raises the possibility that failed retrograde transport of

NGF may also characterize AD. The mechanism(s) responsible is unknown, but several possibilities have been suggested for blocking the trafficking of NGF and/or NGF signaling endosomes (Salehi et al., 2003). Both extracellular and intracellular defects can be envisioned. Among the most interesting are: (1) failure to properly traffic signaling endosomes due to disruption of microtubules or blockade of axons due to neurofibrillary tangles; (2) competition between signaling endosomes and intracellular protein aggregates for dynein-mediated transport; and (3) delivery of signaling endosomes to lysosomes before the signal can be delivered to neuron cell bodies. It will be important to explore further these suggestions and to test the relevance of failed retrograde signaling to neurodegeneration.

Abbreviations

AD	Alzheimer's disease
CCVs	Clathrin-coated vesicles
CNS	Central nervous system
MEK	MAPK kinase
MEKK	MAPK kinase kinase
NGF	Nerve growth factor
NTs	Neurotrophins
PNS	Peripheral nervous system

References

Atwal, J.K., Massie, B., Miller, F.D. and Kaplan, D.R. (2000) The TrkB-Shc site signals neuronal survival and local axon growth via MEK and PI3-kinase. Neuron, 27: 265–277.

Beattie, E.C., Howe, C.L., Wilde, A., Brodsky, F.M. and Mobley, W.C. (2000) NGF signals through TrkA to increase clathrin at the plasma membrane and enhance clathrin-mediated membrane trafficking. J. Neurosci., 20: 7325–7333.

Beranger, F., Goud, B., Tavitian, A. and de Gunzburg, J. (1991) Association of the Ras-antagonistic Rap1/Krev-1 proteins with the Golgi complex. Proc. Natl. Acad. Sci. USA, 88: 1606–1610.

Bhattacharyya, A., Watson, F.L., Bradlee, T.A., Pomeroy, S.L., Stiles, C.D. and Segal, R.A. (1997) Trk receptors function as rapid retrograde signal carriers in the adult nervous system. J. Neurosci., 17: 7007–7016.

Bhattacharyya, A., Watson, F.L., Pomeroy, S.L., Zhang, Y.Z., Stiles, C.D. and Segal, R.A. (2002) High-resolution imaging

demonstrates dynein-based vesicular transport of activated Trk receptors. J. Neurobiol., 51: 302–312.

Boussiotis, V.A., Freeman, G.J., Berezovskaya, A., Barber, D.L. and Nadler, L.M. (1997) Maintenance of human T cell anergy: blocking of IL-2 gene transcription by activated Rap1. Science, 278: 124–128.

Connolly, J.L., Green, S.A. and Greene, L.A. (1981) Pit formation and rapid changes in surface morphology of sympathetic neurons in response to nerve growth factor. J. Cell Biol., 90: 176–180.

Cooper, J.D., Salehi, A., Delcroix, J.D., Howe, C.L., Belichenko, P.V., Chua-Couzens, J., Kilbridge, J.F., Carlson, E.J., Epstein, C.J. and Mobley, W.C. (2001) Failed retrograde transport of NGF in a mouse model of Down's syndrome: reversal of cholinergic neurodegenerative phenotypes following NGF infusion. Proc. Natl. Acad. Sci. USA, 98: 10439–10444.

Cunningham, M.E., Stephens, R.M., Kaplan, D.R. and Greene, L.A. (1997) Autophosphorylation of activation loop tyrosines regulates signaling by the TRK nerve growth factor receptor. J. Biol. Chem., 272: 10957–10967.

Davies, P.J., Cornwell, M.M., Johnson, J.D., Reggianni, A., Myers, M. and Murtaugh, M.P. (1984) Studies on the effects of dansylcadaverine and related compounds on receptor-mediated endocytosis in cultured cells. Diabetes Care, 7(Suppl. 1): 35–41.

Dodge, M.E., Rahimtula, M. and Mearow, K.M. (2002) Factors contributing to neurotrophin-independent survival of adult sensory neurons. Brain Res., 953: 144–156.

Ehlers, M.D., Kaplan, D.R., Price, D.L. and Koliatsos, V.E. (1995) NGF-stimulated retrograde transport of trkA in the mammalian nervous system. J. Cell Biol., 130: 149–156.

Ganju, P., Davis, A., Patel, S., Nunez, X. and Fox, A. (2001) p38 stress-activated protein kinase inhibitor reverses bradykinin B(1) receptor-mediated component of inflammatory hyperalgesia. Eur. J. Pharmacol., 421: 191–199.

Gotoh, T., Hattori, S., Nakamura, S., Kitayama, H., Noda, M., Takai, Y., Kaibuchi, K., Matsui, H., Hatase, O., Takahashi, H., et al. (1995) Identification of Rap1 as a target for the Crk SH3 domain-binding guanine nucleotide-releasing factor C3G. Mol. Cell. Biol., 15: 6746–6753.

Grafstein, B. and Forman, D.S. (1980) Intracellular transport in neurons. Physiol. Rev., 60: 1167–1283.

Grimes, M.L., Beattie, E. and Mobley, W.C. (1997) A signaling organelle containing the nerve growth factor-activated receptor tyrosine kinase, TrkA. Proc. Natl. Acad. Sci. USA, 94: 9909–9914.

Grimes, M.L., Zhou, J., Beattie, E.C., Yuen, E.C., Hall, D.E., Valletta, J.S., Topp, K.S., LaVail, J.H., Bunnett, N.W. and Mobley, W.C. (1996) Endocytosis of activated TrkA: evidence that nerve growth factor induces formation of signaling endosomes. J. Neurosci., 16: 7950–7964.

Gu, H., Pratt, J.C., Burakoff, S.J. and Neel, B.G. (1998) Cloning of p97/Gab2, the major SHP2-binding protein in hematopoietic cells, reveals a novel pathway for cytokine-induced gene activation. Mol. Cell., 2: 729–740.

Herrmann, C., Martin, G.A. and Wittinghofer, A. (1995) Quantitative analysis of the complex between p21ras and the Ras-binding domain of the human Raf-1 protein kinase. J. Biol. Chem., 270: 2901–2905.

Howe, C.L., Valletta, J.S., Rusnak, A.S. and Mobley, W.C. (2001) NGF signaling from clathrin-coated vesicles: evidence that signaling endosomes serve as a platform for the Ras-MAPK pathway. Neuron, 32: 801–814.

Huang, E.J. and Reichardt, L.F. (2001) Neurotrophins: roles in neuronal development and function. Annu. Rev. Neurosci., 2: 677–736.

Ichiba, T., Hashimoto, Y., Nakaya, M., Kuraishi, Y., Tanaka, S., Kurata, T., Mochizuki, N. and Matsuda, M. (1999) Activation of C3G guanine nucleotide exchange factor for Rap1 by phosphorylation of tyrosine 504. J. Biol. Chem., 274: 14376–14381.

Ji, R.R., Samad, T.A., Jin, S.X., Schmoll, R. and Woolf, C.J. (2002) p38 MAPK activation by NGF in primary sensory neurons after inflammation increases TRPV1 levels and maintains heat hyperalgesia. Neuron, 36: 57–68.

Johnson, G. (2002) Signal transduction. Scaffolding proteins–more than meets the eye. Science, 295: 1249–1250.

Jullien, J., Guili, V., Reichardt, L.F. and Rudkin, B.B. (2002) Molecular kinetics of nerve growth factor receptor trafficking and activation. J. Biol. Chem., 277: 38700–38708.

Kamakura, S., Moriguchi, T. and Nishida, E. (1999) Activation of the protein kinase ERK5/BMK1 by receptor tyrosine kinases. Identification and characterization of a signaling pathway to the nucleus. J. Biol. Chem., 274: 26563–26571.

Kao, S., Jaiswal, R.K., Kolch, W. and Landreth, G.E. (2001) Identification of the mechanisms regulating the differential activation of the mapk cascade by epidermal growth factor and nerve growth factor in PC12 cells. J. Biol. Chem., 276: 18169–18177.

Kim, S., Mizoguchi, A., Kikuchi, A. and Takai, Y. (1990) Tissue and subcellular distributions of the smg-21/rap1/Krev-1 proteins which are partly distinct from those of c-ras p21s. Mol. Cell. Biol., 10: 2645–2652.

Klesse, L.J. and Parada, L.F. (1998) p21 ras and phosphatidy-linositol-3 kinase are required for survival of wild-type and NF1 mutant sensory neurons. J. Neurosci., 18: 10420–10428.

Klesse, L.J. and Parada, L.F. (1999) Trks: signal transduction and intracellular pathways. Microsc. Res. Tech., 45: 210–216.

Kuruvilla, R., Ye, H. and Ginty, D.D. (2000) Spatially and functionally distinct roles of the PI3-K effector pathway during NGF signaling in sympathetic neurons. Neuron, 27: 499–512.

Larkin, J.M., Brown, M.S., Goldstein, J.L. and Anderson, R.G. (1983) Depletion of intracellular potassium arrests coated pit formation and receptor-mediated endocytosis in fibroblasts. Cell, 33: 273–285.

Leevers, S.J., Paterson, H.F. and Marshall, C.J. (1994) Requirement for Ras in Raf activation is overcome by targeting Raf to the plasma membrane. Nature, 369: 411–414.

Levi-Montalcini, R. (1987) The nerve growth factor 35 years later. Science, 237: 1154–1162.

Levi-Montalcini, R. and Hamburger, V. (1951) Selective growth stimulating effects of mouse sarcoma on the sensory and sympathetic nervous system of chick embryo. J. Exp. Zool., 116: 321–361.

Levi-Montalcini, R. and Hamburger, V. (1953) A diffusible agent of mouse sarcoma, producing hyperplasia of sympathetic ganglia and hyperneurotization of viscera in the chick embryo. J. Exp. Zool., 123: 233–287.

Liu, Y., Jenkins, B., Shin, J.L. and Rohrschneider, L.R. (2001) Scaffolding protein Gab2 mediates differentiation signaling downstream of Fms receptor tyrosine kinase. Mol. Cell Biol., 21: 3047–3056.

Loeb, D.M., Maragos, J., Martin-Zanca, D., Chao, M.V., Parada, L.F. and Greene, L.A. (1991) The trk proto-oncogene rescues NGF responsiveness in mutant NGF-nonresponsive PC12 cell lines. Cell, 66: 961–966.

MacInnis, B.L. and Campenot, R.B. (2002) Retrograde support of neuronal survival without retrograde transport of nerve growth factor. Science, 295: 1536–1539.

Marais, R., Light, Y., Mason, C., Paterson, H., Olson, M.F. and Marshall, C.J. (1998) Requirement of Ras-GTP-Raf complexes for activation of Raf-1 by protein kinase C. Science, 280: 109–112.

Mazzoni, I.E., Said, F.A., Aloyz, R., Miller, F.D. and Kaplan, D. (1999) Ras regulates sympathetic neuron survival by suppressing the p53-mediated cell death pathway. J. Neurosci., 19: 9716–9727.

Meakin, S.O., MacDonald, J.I., Gryz, E.A., Kubu, C.J. and Verdi, J.M. (1999) The signaling adapter FRS-2 competes with Shc for binding to the nerve growth factor receptor TrkA. A model for discriminating proliferation and differentiation. J. Biol. Chem., 274: 9861–9870.

Mochizuki, N., Yamashita, S., Kurokawa, K., Ohba, Y., Nagai, T., Miyawaki, A. and Matsuda, M. (2001) Spatio-temporal images of growth-factor-induced activation of Ras and Rap1. Nature, 411: 1065–1068.

Morooka, T. and Nishida, E. (1998) Requirement of p38 mitogen-activated protein kinase for neuronal differentiation in PC12 cells. J. Biol. Chem., 273: 24285–24288.

Mu, F.T., Callaghan, J.M., Steele-Mortimer, O., Stenmark, H., Parton, R.G., Campbell, P.L., McCluskey, J., Yeo, J.P., Tock, E.P. and Toh, B.H. (1995) EEA1, an early endosome-associated protein. EEA1 is a conserved alpha-helical peripheral membrane protein flanked by cysteine 'fingers' and contains a calmodulin-binding IQ motif. J. Biol. Chem., 270: 13503–13511.

Okada, S. and Pessin, J.E. (1997) Insulin and epidermal growth factor stimulate a conformational change in Rap1 and dissociation of the CrkII-C3G complex. J. Biol. Chem., 272: 28179–28182.

Patapoutian, A. and Reichardt, L.F. (2001) Trk receptors: mediators of neurotrophin action. Curr. Opin. Neurobiol., 11: 272–280.

Pearson, G., Robinson, F., Beers Gibson, T., Xu, B.E., Karandikar, M., Berman, K. and Cobb, M.H. (2001) Mitogen-activated protein (MAP) kinase pathways: regulation and physiological functions. Endocr. Rev., 22: 153–183.

Reeves, R.H., Irving, N.G., Moran, T.H., Wohn, A., Kitt, C., Sisodia, S.S., Schmidt, C., Bronson, R.T. and Davisson, M.T. (1995) A mouse model for Down syndrome exhibits learning and behaviour deficits. Nat. Genet., 11: 177–184.

Riccio, A., Pierchala, B.A., Ciarallo, C.L. and Ginty, D.D. (1997) An NGF-TrkA-mediated retrograde signal to transcription factor CREB in sympathetic neurons. Science, 277: 1097–1100.

Roux, P.P. and Barker, P.A. (2002) Neurotrophin signaling through the p75 neurotrophin receptor. Prog. Neurobiol., 67: 203–233.

Salehi, A., Delcroix, J.-D. and Mobley, W.C. (2003) Traffic at the intersection of neurotrophic factor signaling and neurodegeneration. Trends Neurosc., 26: 73–80.

Scott, S.A., Mufson, E.J., Weingarten, J.A., Skau, K.A. and Crutcher, K.A. (1995) Nerve growth factor in Alzheimer disease: increased levels throught the brain coupled with declines in nucleus basalis. J. Neurosci., 15: 6213–6221.

Shao, Y., Akmentin, W., Toledo-Aral, J.J., Rosenbaum, J., Valdez, G., Cabot, J.B., Hilbush, B.S. and Halegoua, S. (2002) Pincher, a pinocytic chaperone for nerve growth factor/TrkA signaling endosomes. J. Cell Biol., 157: 679–691.

Sheff, D.R., Daro, E.A., Hull, M. and Mellman, I. (1999) The receptor recycling pathway contains two distinct populations of early endosomes with different sorting functions. J. Cell Biol., 145: 123–139.

Sofer, A. and Futerman, A.H. (1995) Cationic amphiphilic drugs inhibit the internalization of cholera toxin to the Golgi apparatus and the subsequent elevation of cyclic AMP. J. Biol. Chem., 270: 12117–12122.

Sofroniew, M.V., Howe, C.L. and Mobley, W.C. (2001) Nerve growth factor signaling, neuroprotection, and neural repair. Annu. Rev. Neurosci., 24: 1217–1281.

Stephens, L., Anderson, K., Stokoe, D., Erdjument-Bromage, H., Painter, G.F., Holmes, A.B., Gaffney, P.R., Reese, C.B., McCormick, F., Tempst, P., et al. (1998) Protein kinase B kinases that mediate phosphatidylinositol 3,4,5-trisphosphate-dependent activation of protein kinase B. Science, 279: 710–714.

Sorkin, A. and von Zastrow, M. (2002) Signal transduction and endocytosis: Close encounter of many kinds. Nature rev. Mol. Cell. Biol., 3: 600–614.

Stephens, R.M., Loeb, D.M., Copeland, T.D., Pawson, T., Greene, L.A. and Kaplan, D.R. (1994) Trk receptors use redundant signal transduction pathways involving SHC and PLC-gamma 1 to mediate NGF responses. Neuron, 12: 691–705.

Tsui-Pierchala, B.A. and Ginty, D.D. (1999) Characterization of an NGF-P-TrkA retrograde-signaling complex and age-dependent regulation of TrkA phosphorylation in sympathetic neurons. J. Neurosci., 19: 8207–8218.

Vetter, M.L., Martin-Zanca, D., Parada, L.F., Bishop, J.M. and Kaplan, D.R. (1991) Nerve growth factor rapidly stimulates tyrosine phosphorylation of phospholipase C-gamma 1 by a kinase activity associated with the product of the trk protooncogene. Proc. Natl. Acad. Sci. USA, 88: 5650–5654.

Vogelbaum, M.A., Tong, J.X. and Rich, K.M. (1998) Developmental regulation of apoptosis in dorsal root ganglion neurons. J. Neurosci., 18: 8928–8935.

Watson, F.L., Heerssen, H.M., Bhattacharyya, A., Klesse, L., Lin, M.Z. and Segal, R.A. (2001) Neurotrophins use the Erk5 pathway to mediate a retrograde survival response. Nat. Neurosci., 4: 981–988.

Watson, F.L., Heerssen, H.M., Moheban, D.B., Lin, M.Z., Sauvageot, C.M., Bhattacharyya, A., Pomeroy, S.L. and Segal, R.A. (1999) Rapid nuclear responses to target-derived neurotrophins require retrograde transport of ligand-receptor complex. J. Neurosci., 19: 7889–7900.

Woodman, P.G. (2000) Biogenesis of the sorting endosome: the role of Rab5. Traffic, 1: 695–701.

Wooten, M.W., Seibenhener, M.L., Neidigh, K.B. and Vandenplas, M.L. (2000) Mapping of atypical protein kinase C within the nerve growth factor signaling cascade: relationship to differentiation and survival of PC12 cells. Mol. Cell. Biol., 20: 4494–4504

Wu, C., Lai, C.F. and Mobley, W.C. (2001) Nerve growth factor activates persistent Rap1 signaling in endosomes. J. Neurosci., 21: 5406–5416.

Xing, J., Kornhauser, J.M., Xia, Z., Thiele, E.A. and Greenberg, M.E. (1998) Nerve growth factor activates extracellular signal-regulated kinase and p38 mitogen-activated protein kinase pathways to stimulate CREB serine 133 phosphorylation. Mol. Cell. Biol., 18: 1946–1955.

York, R.D., Yao, H., Dillon, T., Ellig, C.L., Eckert, S.P., McCleskey, E.W. and Stork, P.J. (1998) Rap1 mediates sustained MAP kinase activation induced by nerve growth factor. Nature, 392: 622–626.

Zhang, Y., Moheban, D.B., Conway, B.R., Bhattacharyya, A. and Segal, R.A. (2000) Cell surface Trk receptors mediate NGF-induced survival while internalized receptors regulate NGF-induced differentiation. J. Neurosci., 20: 5671–5678.

Progress in Brain Research, Vol. 146
ISSN 0079-6123

CHAPTER 2

The p75 neurotrophin receptor: multiple interactors and numerous functions

Jennifer J. Gentry[1], Philip A. Barker[2] and Bruce D. Carter[1,]*

[1]*The Center for Molecular Neuroscience and Department of Biochemistry, Vanderbilt University Medical School,
Nashville, TN, USA*
[2]*Montreal Neurological Institute, McGill University, Montreal, H3A 2B4 QC, Canada*

Abstract: The neurotrophin receptor p75 ($p75^{NTR}$), is involved in a diverse array of cellular responses, including apoptosis, neurite outgrowth and myelination. Stimulation of $p75^{NTR}$ with neurotrophin can activate multiple downstream signals, including the small GTP binding protein Rac, the transcription factor NF-κB and the stress activated kinase, JNK. How these signals are generated and regulated to produce a specific cellular effect has yet to be fully elucidated. A number of proteins have recently been shown to interact with the intracellular domain of $p75^{NTR}$. Here, we review these $p75^{NTR}$ interacting factors and the current evidence as to how they contribute to the functional effects of $p75^{NTR}$ activation.

Keywords: NGF; $p75^{NTR}$; NRIF; NRAGE; SC-1; NADE; apoptosis; TRAF

Introduction

The neurotrophins nerve growth factor (NGF), brain-derived neurotrophic factor (BDNF), neurotrophin-3 (NT-3), and NT-4/5 regulate neuronal survival and differentiation during the development of the vertebrate nervous system by binding to the tyrosine kinase (Trk) family of receptors, which includes TrkA, TrkB and TrkC, and to the p75 neurotrophin receptor ($p75^{NTR}$), a member of the tumor necrosis factor (TNF) receptor family (reviewed by Huang and Reichardt, 2001). The Trks are tyrosine kinase receptors that bind selectively to particular neurotrophins and promote the survival of specific populations of neurons during development. In contrast, all of the neurotrophins

bind with similar affinity to $p75^{NTR}$ (Fig. 1). When co-expressed, the two receptors form a high affinity binding complex that provides specificity in neurotrophin binding and enhances Trk tyrosine kinase signaling and subsequent trophic effects. It is now well established that neurotrophin binding to $p75^{NTR}$ in the absence of Trk signaling can activate a signaling cascade that results in apoptosis in specific cell types (reviewed by Roux and Barker, 2002) (Fig. 2). In addition, $p75^{NTR}$ signaling has recently been shown to affect myelination (Chan et al., 2001; Cosgaya et al., 2002) and neurite outgrowth (Wang et al., 2002; Wong et al., 2002). These Trk-independent functions of $p75^{NTR}$ have been suggested to be involved in the development of the nervous system and in a cellular response to injury. In an effort to understand how these very diverse signals are generated and regulated, attention has recently focused on the molecular mechanisms by which $p75^{NTR}$ elicits these responses. In this review we will discuss the numerous proteins that bind to the intracellular domain of the receptor and attempt

*Correspondence to: B.D. Carter, Department of Biochemistry, Center for Molecular Neuroscience; 655 Light Hall; Vanderbilt University Medical School, Nashville, TN 37232, USA.
Tel.: + 1-615-936-3041; Fax: + 1-615-343-0704;
E-mail: bruce.carter@vanderbilt.edu

DOI: 10.1016/S0079-6123(03)46002-0

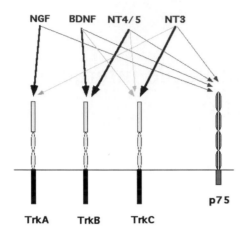

Fig. 1. The neurotrophins and their receptor binding specificity. While each of the neurotrophins binds with specificity to the Trk receptors, all of the neurotrophins bind with similar affinity to the p75 neurotrophin receptor.

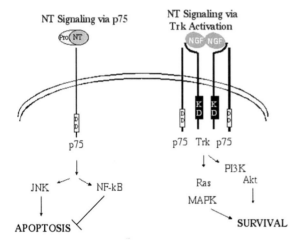

Fig. 2. Signaling through the neurotrophin receptors. Neurotrophin activation of the p75^{NTR} in the absence of Trk activation initiates a signal transduction cascade that can induce apoptosis. Neurotrophin binding to its cognate Trk receptor in the presence of p75^{NTR} leads to activation of the Trk activity, which promotes survival and differentiation.

to put them into the context of the known signal transduction pathways activated by p75^{NTR}.

Signals mediated by neurotrophin binding to p75^{NTR}

P75^{NTR} activates an apoptotic signal

During mammalian development, approximately 50% of the neurons that are generated undergo

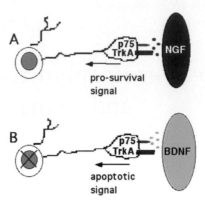

Fig. 3. Neurotrophin regulation of neuronal survival. (A) A developing neuron sends projections to an appropriate target tissue expressing NGF, leading to activation of pro-survival signals by TrkA and p75^{NTR}. (B) A developing neuron sends projections to an inappropriate target tissue expressing BDNF. BDNF cannot activate the Trk receptor and instead binds to p75^{NTR} and activates a pro-apoptotic signal.

programmed cell death (Oppenheim, 1991). While the advantages of this method of development have yet to be fully determined, it is clear that without the controlled elimination of massive numbers of immature neurons, a functional nervous system cannot be created. This is well exemplified by animals deficient in caspase-3, which display an embryonic lethal phenotype due to massive exencephaly (Kuida et al., 1998). To create the complex architecture of the vertebrate nervous system, developing neurons are dependent on the positive regulatory signals from neurotrophins secreted by target tissues. Neurons that fail to reach their target are eliminated through an apoptotic program that is thought to be initiated by the lack of Trk signaling. However, recent studies have suggested that some neurons may reach an inappropriate target, either spatially or temporally, and are eliminated through selective activation of the p75^{NTR} (Fig. 3). For example, Miller and colleagues demonstrated that there is hyperinnervation of the pineal gland by sympathetic neurons in *bdnf*−/− mice (Kohn et al., 1999). These neuronal projections reach this target at a time in which it expresses BDNF, since sympathetic neurons only express TrkA, which is selective for NGF, and p75^{NTR}, they suggested that normal pruning occurs through BDNF-dependent activation of p75^{NTR}.

Analysis of transgenic mice lacking the p75[NTR] has revealed a role for this receptor in the elimination of a number of neurons during development. Frade et al. demonstrated decreased apoptosis in the embryonic retina (Frade et al., 1996) and in the spinal cord (Frade and Barde, 1999) of *p75−/−* mice relative to their wild type littermates. Corresponding increases in the number of neurons in the basal forebrain (Naumann et al., 2002) and superior cervical ganglia (Bamji et al., 1998) were also reported in these animals. In addition, while *p75[NTR]* knockout (KO) animals have fewer sensory neurons in the dorsal root ganglia, presumably due to attenuated Trk signaling, sensory neurons in the trigeminal ganglion of the null mice show decreased TUNEL staining during the period of programmed cell death (Agerman et al., 2000). Neurotrophin binding to p75[NTR] has also been shown to induce apoptosis in cultured Schwann cells (Syroid et al., 2000; Khursigara et al., 2001), oligodendrocytes (Casaccia-Bonnefil et al., 1996), motor neurons of embryonic spinal cord (Frade and Barde, 1999; Sedel et al., 1999), hippocampal neurons (Friedman, 2000), neuroblastoma cells (Bunone et al., 1997) and smooth muscle cells (Wang et al., 2000). Various injuries to the nervous system also activate a p75[NTR]-dependent apoptotic program in, for example, in motor neurons following lesion of the facial motor nerve (Ferri et al., 1998), hippocampal neurons after the induction of seizures (Roux et al., 1999; Troy et al., 2002), Schwann cells after axotomy (Syroid et al., 2000), cortico-spinal neurons after axotomy (Giehl et al., 2001) and in oligodendrocytes after spinal cord injury (Beattie et al., 2002). In addition, increased p75[NTR] expression has been observed in patients with multiple sclerosis (Dowling et al., 1999), ALS (Lowry et al., 2001), and Alzheimer's disease (Salehi et al., 2000a; Mufson et al., 2002), leading to speculation that p75[NTR] may be involved in these neurodegenerative diseases. It is important to note that the physiological actions of p75[NTR] may not be restricted to the neural system; recent in vivo and in vitro data indicate a role for p75[NTR] in apoptosis within nonneuronal tissues such as the vasculature (Wang et al., 2000).

The mechanism by which p75[NTR]-mediated apoptosis is activated has not been fully elucidated; however, the stress activated MAP kinase, c-jun N-terminal kinase, JNK, has been shown to play an essential role (Yoon et al., 1998). Of the three isoforms of JNK, JNK3 is expressed exclusively in the nervous system and recent evidence indicates that this JNK isoform may be required for p75[NTR]-mediated cell death (Harrington et al., 2002). The targets of this kinase include the transcription factors c-jun, ATF-2 and p53. P75[NTR]-mediated apoptosis was shown to be independent of c-jun (Palmada et al., 2002) while blocking p53 prevented a death signal (Aloyz et al., 1998).

Execution of apoptosis requires activation of caspase molecules. These proteins exist as a prozymogen whose proteolytic activity is induced through cleavage. The resulting caspase cleavage cascade ultimately results in laddering of the DNA and cell death. However, unlike many other members of the TNF receptor family, p75[NTR] does not appear to activate the receptor-proximal enzyme, caspase-8, although caspases-1, -3, -6, and -9 have all been implicated in p75[NTR]-mediated cell death (Gu et al., 1999; Agerman et al., 2000; Troy et al., 2002). The mechanisms of caspase activation induced by neurotrophin binding to p75[NTR] and the specific role of the JNK signaling pathway in this process have yet to be determined.

p75[NTR] activates a pro-survival signal

The first recognized function of neurotrophin binding to p75, was to facilitate the pro-survival signals generated by the Trk receptor (reviewed by Roux and Barker, 2002). However, recent findings have demonstrated that, in specific contexts, p75[NTR] can also promote survival independent of the Trks. For example, neurotrophin binding to p75[NTR] has been shown to block the death of neuroblastoma cells (Oh et al., 1993; Cortazzo et al., 1996) and sensory neurons (Longo et al., 1997) deprived of trophic support, and hippocampal neurons following glutamate receptor activation (Bui et al., 2002). In addition, through studies of the *p75−/−* mice, it was shown that this receptor is required for survival of subplate neurons in the developing cortex (DeFreitas et al., 2001). The p75[NTR] null mice also exhibit a dramatic loss in sensory neurons (Lee et al., 1992), likely due to reduced signaling from a p75-Trk complex. However, given that selective activation of p75^{NTR} can promote sensory neuron survival in vitro (Longo et al., 1997), it is also possible that a portion of the neurons die due to loss of p75^{NTR} pro-survival signaling.

The finding that p75^{NTR} can initiate opposing pathways was not entirely surprising given that other members of the TNF receptor family also elicit both pro-death and pro-survival signals. One of the best characterized anti-apoptotic signals generated by receptors in this family is the activation of the transcription factor, NF-κB (reviewed by Karin and Lin, 2002). NF-κB is composed of homo- and heterodimers of the subunits p65, p50, p52, cRel, and RelB. The prototypic NF-κB molecule consists of a p65/p50 heterodimer, which is normally sequestered in the cytoplasm through binding to the inhibitory protein IκB. In response to various stimuli, IκB is phosphorylated by the upstream IKK complex, leading to ubiquitination and degradation of the inhibitor. Loss of IκB exposes the nuclear localization sequence of the NF-κB subunits, thereby allowing the dimer to translocate into the nucleus where it regulates transcription of target genes. Among the genes induced by NF-κB are several that promote survival, including members of the inhibitor of apoptosis family, IAP, and Bcl-2 homologs (reviewed by Pahl, 1999).

Activation of NF-κB by neurotrophin binding to p75^{NTR} functions as a pro-survival signal in a Schwannoma cell line (Gentry et al., 2000) and primary Schwann cells (Khursigara et al., 2001) which both express p75^{NTR} but not TrkA receptors. Neurotrophin treatment also induced a pro-survival signal via NF-κB in trigeminal (Hamanoue et al., 1999) and hippocampal neurons (Culmsee et al., 2002). In PC12 cells (Taglialatela et al., 1997; Foehr et al., 2000; Wooten et al., 2001) and sympathetic neurons (Maggirwar et al., 1998), NGF-dependent induction of NF-κB is required for survival; however, each of these cells coexpress TrkA and p75^{NTR} and it is not clear if both receptors must be activated to stimulate this transcription factor.

Another pro-survival signal that has been reported to emanate from the p75^{NTR} receptor is activation of the phosphatidyl inositol 3-kinase (PI3K)-Akt pathway (Roux et al., 2001; Bui et al., 2002). Akt, also known as protein kinase B, is known to prevent apoptosis through phosphorylation-dependent inactivation of caspase 9, the pro-death protein Bad and members of the Forkhead family of transcription factors (del Peso et al., 1997; Cardone et al., 1998; Brunet et al., 1999). This pathway was shown to be activated by p75^{NTR} in hippocampal neurons and to serve a protective function following excitotoxic injury (Bui et al., 2002). Interestingly, Akt can also mediate activation of NF-κB (Beraud et al., 1999), thus these two signals may lie on the same pathway.

New roles for p75^{NTR}

The biological actions of p75^{NTR} have recently expanded to include modulation of myelin formation and neurite outgrowth. Activation of p75^{NTR} by BDNF in Schwann cells enhanced the formation of myelin in Schwann cell-sensory neuron cocultures and in sciatic nerves whereas blocking the action of endogenous BDNF was inhibitory (Chan et al., 2001). These findings appear physiologically relevant since mice lacking the p75^{NTR} gene show deficient myelin formation (Cosgaya et al., 2002). Although the mechanisms by which p75^{NTR} regulates myelination are not known, it is notable that NF-κB activity is required for differentiation of Schwann cells into a myelinating phenotype (Nickols et al., 2003).

Signals from p75^{NTR} not only influence myelin formation, but also affect the ability of neurons to interact with myelin, at least in the CNS. Neurotrophin binding to p75^{NTR} was shown to activate Rac (Harrington et al., 2002) and inhibit Rho, resulting in neurite outgrowth (Yamashita et al., 1999). In surprising contrast, p75^{NTR} has been found to physically interact with the GPI-linked Nogo receptor and may be a signaling component of a receptor complex that activates Rho, induces growth cone collapse and inhibits growth (Wang et al., 2002; Wong et al., 2002). Nogo receptor ligands such as myelin-associated glycoprotein (MAG), oligodendrocyte myelin glycoprotein (OMGP) and Nogo provide potent stop-growth signals to the growth cone and are a major impediment to spontaneous regeneration of injured central neurons (Woolf and Bloechlinger, 2002). Understanding how p75^{NTR} differentially regulates small GTPases to affect neurite growth has emerged as a critically important question in the field.

During development, p75^{NTR} and Trk receptors collaborate to promote survival in appropriate target tissues yet in neurons lacking trophic support, p75^{NTR} facilitates cell death. It is interesting to speculate whether a similar functional p75^{NTR} interaction may function during neurite outgrowth. For example, p75^{NTR} and Trk may collaborate to promote axonal outgrowth in response to appropriate neurotrophin input, but in the presence of a repulsive Nogo receptor binding cue, p75^{NTR} may cause growth cone collapse and antagonize outgrowth. It is conceivable therefore that the delicate balance of p75^{NTR} signaling within a neuron tightly regulates both survival and neurite growth.

Upstream signaling modulators

p75^{NTR} activates a complex signaling network that can promote or prevent survival and neuritogenesis. To understand the ultimate effect of p75^{NTR} on a cell, it is essential to understand how downstream signals are generated by identifying each of the components of the signal transduction pathway. It is very likely that there is differential expression or activation of such upstream factors in various cell types under different physiological contexts. A great deal of effort has gone into elucidating receptor-proximal signal

transducers and there is now a growing list of intracellular p75^{NTR} interacting factors (Fig. 4).

The p75^{NTR} is the founding member of the TNF receptor superfamily. This family is characterized by a cysteine rich region in the extracellular domain and includes receptors that are critical for proper immune function (Baud and Karin, 2001). A number of receptors within this family can induce apoptosis through recruitment of adaptor proteins to an intracellular death domain, and subsequent activation of a caspase cascade (Nagata, 1997). The intracellular domain of p75^{NTR} also has a putative death domain and can initiate apoptosis through activation of caspases (Gu et al., 1999; Agerman et al., 2000; Troy et al., 2002).

While p75^{NTR} can activate downstream signals common to many TNF receptors, such as NF-κB and JNK, upstream signaling seems to diverge from other death domain containing receptors. Other pro-apoptotic receptors of the TNF family use adaptor proteins to activate caspase-8, but p75^{NTR} mediated apoptosis is caspase-8 independent (Gu et al., 1999). In addition, all other receptors of the family bind to ligands of the TNF-α family. These ligands are trimers and signaling occurs through a trimeric receptor complex, which recruits trimeric adaptor proteins. The receptor stoichiometry in the p75^{NTR} signaling complex has not been determined, but it has long been assumed that ligand dependent signaling occurs through receptor

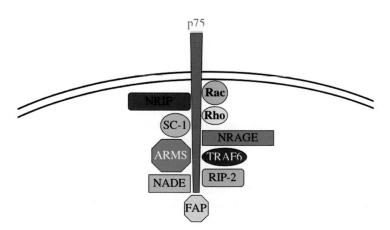

Fig. 4. p75^{NTR} and its interacting factors. TRAF6, RIP-2, and FAP-1 are associated with activation of NF-κB. NRIF, SC-1, NADE, NRAGE, and Rac are associated with cell cycle arrest and apoptosis. ARMS may mediate cross talk between p75^{NTR} and Trk receptors. RhoA inhibition by p75^{NTR} may facilitate axon outgrowth.

dimerization, since the mature neurotrophins form stable dimers.

The neurotrophins, like many peptide ligands, are initially translated in a pro-form and are processed through proteolytic cleavage to generate the mature trophic factor (Berger and Shooter, 1977). The pro-forms of the neurotrophins have recently been shown to bind to p75^{NTR} with higher affinity than the mature neurotrophins (Lee et al., 2001a). These pro-neurotrophins do not bind or activate Trk receptors (Delsite and Djakiew, 1999; Lee et al., 2001a; Mowla et al., 2001) but potently induce p75^{NTR}-dependent apoptosis in sympathetic neurons (Lee et al., 2001a) and oligodendrocytes (Beattie et al., 2002). Recent findings have shown that oligodendrocyte cell death that occurs following spinal cord injury in mice requires activation of p75^{NTR} by pro-NGF (Beattie et al., 2002), suggesting that at least for injury-induced apoptosis, the pro-neurotrophins may be the primary activating ligand for p75^{NTR}. It will be very interesting to identify the oligomeric structure of the pro-neurotrophins and determine whether these ligands activate the p75^{NTR} through a trimeric receptor complex, similar to other members of the TNF receptor family.

TNF receptor family signaling molecules

One group of homologous proteins that functions to mediate signaling through a number of TNF receptors is the TNF receptor-associated factor (TRAF) family, which currently consists of TRAF1 through TRAF6 in mammals (Bradley and Pober, 2001). These proteins are characterized by a carboxyl terminal TRAF domain which is further subdivided into a highly conserved TRAF-N domain that mediates homo and hetero-oligomerization, and the TRAF-C domain that determines the specificity of receptor binding and contributes to TRAF oligomerization. TRAF proteins contain an amino terminal activation domain, which in TRAF 2–6, consists of RING and zinc fingers. Of the six TRAF proteins, TRAF 2, 5 and 6, are capable of activating NF-κB and JNK, which has made these proteins attractive candidate molecules in p75^{NTR} signal transduction. TRAF6 is expressed in the nervous system and has been shown to associate with p75^{NTR} (Khursigara

et al., 1999; Ye et al., 1999; Foehr et al., 2000). When 293T cells were transfected with both p75^{NTR} and TRAF6, the proteins can be coimmunoprecipitated after stimulation with NGF (Khursigara et al., 1999). The association required the juxtamembrane sequence of p75^{NTR} and was transient. TRAF6 and p75^{NTR} could be coimmunoprecipitated from cells within 1 min of NGF stimulation but the association was lost within 20 min, suggesting that TRAF6 may dissociate following its activation. Expression of dominant negative TRAF6 blocked NGF induced nuclear translocation of the p65 subunit of NF-κB in Schwann cells and taken together, available data indicate that TRAF6 is recruited to the p75^{NTR} receptor upon NGF stimulation then released to modulate downstream signals, particularly NF-κB.

In addition to TRAF6, other members of the TRAF family have been suggested to associate with p75^{NTR} (Ye et al., 1999); however, this study relied on ectopic expression of both p75^{NTR} and TRAF proteins in cell lines to examine association through coimmunoprecipitation. The associations were not shown to be direct and since TRAF proteins can form heterologous complexes, the possibility that endogenous TRAFs could link other family members to the p75^{NTR} cannot be ruled out. Using expression of the intracellular domain of p75^{NTR} (p75^{ICD}) to activate NF-κB in 293cells, these authors found that TRAF2, like TRAF6, enhanced activation of NF-κB, while TRAF4 blocked activation through p75^{ICD}. This data indicate that TRAF4 may function to oppose NF-κB activation making the cell susceptible to p75^{NTR} mediated cell death. While these studies have provided some interesting clues as to the possible role of the TRAF proteins in p75^{NTR} signaling, it will be important to examine the association of endogenous proteins.

A new group of TRAF domain containing proteins, the TD-encompassing factor (TEF) proteins, has also been identified (Zapata et al., 2001). One of these proteins, MUL has been shown to be mutated in the developmental autosomal recessive disorder Mulibrey Nanism (Avela et al., 2000). MUL interacts with the p75^{NTR} as well as all the TRAF proteins and blocks TRAF6 mediated NF-κB activation (Zapata et al., 2001). These studies provide evidence that p75^{NTR} may use multiple members of the TRAF family and TRAF related proteins to regulate signal transduction.

Receptor interacting protein-2 (RIP-2) is another molecule involved in TNF receptor family signaling that has been implicated in p75^{NTR} function. RIP-2 is a serine/threonine kinase that can both activate NF-κB, and induce apoptosis (McCarthy et al., 1998). RIP-2 can recruit and activate caspase-1 to initiate apoptosis through its caspase recruitment domain (CARD) (Thome et al., 1998). Activation of caspase-1 can be blocked by RIP-2 competitively binding to the pro-survival molecules, IAP-1 (inhibitor of apoptosis 1), or COP, (CARD only protein) (McCarthy et al., 1998; Lee et al., 2001b). Like activation of caspase-1, RIP-2 induction of NF-κB requires the CARD domain of RIP-2 (McCarthy et al., 1998). The TRAF proteins 1,5, and 6 form complexes with RIP-2 and expression of dominant negative TRAF5 or 6 abrogates RIP-2 activation of NF-κB. RIP-2 has been shown to associate with the death domain of the p75^{NTR} receptor through the CARD domain of RIP-2 (Khursigara et al., 2001). The association was direct and was detected transiently after NGF stimulation in a time course similar to that observed for the association between TRAF6 and p75^{NTR}. Coexpression of RIP-2 and

p75^{NTR} resulted in activation of NF-κB and was enhanced in the presence of NGF. The authors used rat primary Schwann cells to further examine the role of RIP-2 in p75^{NTR} mediated signaling. Initially after plating, the Schwann cells responded to NGF stimulation by activating NF-κB and no apoptosis was detected; however, after a week or more in culture, NF-κB was no longer activated and NGF now caused cell death. In seeking an explanation for this signaling switch, Khursigara et al. (1999, 2001) noted that the early cultures expressed high levels of RIP-2 while in older cultures this signaling protein was barely detectable. In addition, expression of dominant negative RIP-2 in early Schwann cell cultures blocked NGF activation of NF-κB and the cells underwent NGF mediated apoptosis. In late cultures, transfection of wild type RIP-2 restored the ability of p75^{NTR} to activate NF-κB and the cells no longer died by apoptosis. These data provide compelling evidence that RIP-2 expression regulates a switch between the cellular outcomes of survival and death in response to p75^{NTR} activation.

At present, the data indicate that both RIP-2 and TRAF6 are important upstream signals in the activation of NF-κB by p75^{NTR} in Schwann cells (Fig. 5). Further studies will be needed to determine if

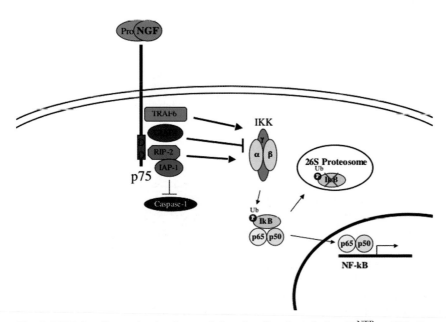

Fig. 5. TNF signaling molecules that modulate signaling through the p75^{NTR} to NF-κB.

TRAF6 functions upsteam of RIP-2, or if the proteins coordinately activate NF-κB.

GTP binding proteins

Two small GTP binding proteins, RhoA and Rac, have been implicated as receptor proximal modulators in p75^{NTR} signaling. A yeast two-hybrid screen with the intracellular domain of p75^{NTR} identified the actin remodeling protein, RhoA, as an interacting protein (Yamashita et al., 1999). P75^{NTR} associated with and activated RhoA, while NGF addition led to GTP hydrolysis and RhoA inactivation. Since RhoA regulates cytoskeletal organization and the formation of stress fibers, the interaction with p75^{NTR} suggested that neurotrophin binding to this receptor could regulate neurite outgrowth. Using ciliary ganglion, which express p75^{NTR} but not TrkA, the authors demonstrated that NGF increased neurite length, as did inactivation of RhoA, while constitutive activation of RhoA blocked neurite elongation by NGF. Moreover, p75^{NTR} KO animals were shown to have defects in neurite outgrowth and Schwann cell migration (Yamashita et al., 1999; Bentley and Lee, 2000). Hence, a model emerges in which unliganded p75^{NTR} bound to active GTP-RhoA, keeps actin fibers in place and prevents cellular extensions; however, upon NGF binding to the receptor RhoA is inactivated by GTP hydrolysis, leading to cytoskeletal rearrangement and neurite outgrowth.

Another GTPase, Rac-1, has been shown to be an upstream signal in p75^{NTR} mediated apoptosis of oligodendrocytes (Harrington et al., 2002). Although binding of Rac-1 to p75^{NTR} has not been demonstrated, NGF binding to p75^{NTR} in oligodendrocytes activated Rac-1 and a dominant negative Rac-1 blocked both p75^{NTR}-mediated JNK activation and induction of apoptosis. Based on their results, the authors suggested that Rac-1 is required upstream of JNK for NGF induced cell death. Interestingly, Rac and Rho often function antagonstically, with Rho preventing the sprouting of axons and Rac facilitating it (reviewed by da Silva and Dotti, 2002). Thus, depending on the cellular context, neurotrophin binding to p75^{NTR} can generate a neuritogenic signal or an apoptotic signal through the Rac and Rho GTP binding proteins.

Novel p75^{NTR} interacting factors

Yeast two-hybrid screening using the p75^{NTR} intracellular domain has identified several novel proteins including NRAGE, NADE, NRIF, and SC-1 (Fig. 6). NRAGE (for neurotrophin receptor-interacting MAGE homologue) is an 86 kD member of the MAGE family, all of which contain a 200 amino acid MAGE homology domain (Salehi et al., 2000b). It has been known for some time that MAGE proteins are often overexpressed in cancer where they are sometimes processed to peptides in the cytosol and presented as MHC-associated tumor-specific antigens, but the cellular function of this family of proteins has remained essentially unknown. The interaction of NRAGE with p75^{NTR} has been validated by coimmunoprecipitation of p75^{NTR} and NRAGE in PC12nnr5 cells (Salehi et al., 2000b) and PC12 cells (Tcherpakov et al., 2002). p75^{NTR}-mediated apoptosis is enhanced in MAH cells which overexpress NRAGE and in some cells, NRAGE overexpression acts as a potent activator of a JNK cascade that induces cytochrome C release, caspase-9 activation and cell death. Interestingly, NRAGE overexpression disrupts the physical interaction of p75^{NTR} with TrkA, but when TrkA is coexpressed with NRAGE in MAH cells, NGF mediated cell death was blocked (Salehi et al., 2000b). This indicates that TrkA signaling overrides apoptotic signaling through p75^{NTR}, even in the absence of its association with p75^{NTR}.

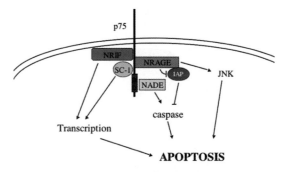

Fig. 6. Pro-apoptotic p75 binding proteins.

Two other MAGE family members (MAGE-H1 and Necdin) have recently been identified as p75^{NTR} binding proteins (Tcherpakov et al., 2002) and NRAGE has been identified as a pro-apoptotic binding partner of Unc5a1, which like p75^{NTR} has an intracellular death domain (Williams et al., 2003). These findings suggest that the NRAGE-p75^{NTR} interaction is only one of a growing network of receptor-MAGE protein interactions necessary for apoptosis.

Subsequent to its identification, NRAGE was pulled out of a separate yeast two-hybrid screen using the avian IAP protein ITA as bait (Jordan et al., 2001). NRAGE can coimmunoprecipitate with both ITA and a mammalian IAP family member, XIAP. NRAGE has been shown to enhance apoptosis mediated by IL-3 withdrawal and to prevent Bcl-2 protection of this cell death, potentially by sequestering XIAP. Although the effects of this interaction have not been examined in p75^{NTR} signaling, it is possible that NRAGE contributes to cell death in part by binding and sequestering XIAP.

Two potential transcription factors, SC-1 and NRIF, have also been identified through their interaction with p75^{NTR} (Chittka and Chao, 1999; Casademunt et al., 1999). Both NRIF and SC-1 contain Kruppel type zinc fingers in the carboxyl terminus and a potential PEST regulatory domain in the amino terminus. In addition to the Kruppel, or C$_2$H$_2$-type zinc fingers, SC-1 contains a PR, or positive regulatory, domain recently identified in a number of transcription factors. SC-1 binds to the juxtamembrane sequence of p75^{NTR} and translocates to the nucleus in response to NGF when cotransfected with p75^{NTR} in COS cells. SC-1 message was detected primarily in the brain of adult animals and may function to block cell cycle progression as its expression dramatically reduces BrdU incorporation (Chittka and Chao, 1999).

NRIF contains two amino terminal motifs found in many transcription factors of the C$_2$H$_2$-type zinc finger family, KRAB domains, which are thought to function as transcriptional repressor modules, as well as a SCAN domain, which mediates transcription factor oligomerization (Casademunt et al., 1999). Consistent with its role as a transcription factor, NRIF has been shown in vitro, to bind specifically to a predicted NRIF binding sequence of DNA (Gentry et al., 2001). In addition, KRAB domains have been shown to associate with KAP-1, an adaptor protein that recruits transcriptional repressors (Friedman et al., 1996). NRIF message is present in all cell types examined and has been shown to increase apoptosis and block BrdU incorporation when over expressed in 293T cells (Benzel et al., 2001). Analysis of *nrif*−/− mice revealed a role for NRIF in p75^{NTR}- mediated programmed cell death as these animals have a 50% decrease in apoptosis of early retinal cells, which are known to undergo p75^{NTR} dependent programmed cell death (Casademunt et al., 1999). Additional evidence linking NRIF to the p75^{NTR} cell death pathway was demonstrated by the complete loss of apoptosis induced specifically through activation of this receptor in sympathetic neurons isolated from *nrif* null mice (Linggi et al., 2002).

Although only one NRIF gene has been identified in humans, a second NRIF gene has now been cloned in mice (Benzel et al., 2001). The NRIF2 protein is 85% identical to NRIF1, and contains the same amino acid homology domains. NRIF1 and NRIF2 are most divergent within the p75^{NTR} binding domain yet NRIF2 also binds to the receptor. NRIF1 and NRIF2 can form heterodimers and expression of NRIF2, like NRIF1, blocks BrdU incorporation when over expressed in 293T cells. Although these results suggest functional redundancy between the two gene products, there must be distinct differences since the *nrif1*−/− mice have deficiencies in p75^{NTR} signaling. Further analysis is necessary to sort out the physiological functions of NRIF1 and 2.

Recently, NRIF1 was reported to interact with another p75^{NTR} interactor, TRAF6. This interaction led to perinuclear and subnuclear localization of NRIF and enhanced TRAF6 activation of JNK (Gentry et al., 2001). Thus, binding to p75^{NTR} may facilitate the interaction between these factors and thereby increase TRAF6 signaling while allowing for translocation of NRIF to the nucleus to regulate transcription. How TRAF6 allows nuclear translocation of NRIF is not clear; however, the ability of p75^{NTR} to bind potential transcription factors is particularly intriguing in light of a recent report that this receptor is cleaved by the membrane protease, γ-secretase, and the intracellular domain translocates to the nucleus (Schechterson et al., 2002). Together, these results suggest the existence of a

34

p75^{ICD}-NRIF complex that TRAF6 may shuttle into the nucleus.

Another novel protein isolated from a yeast two hybrid screen and implicated in the cell death pathway activated by p75^{NTR} is NADE, or p75^{NTR}-associated cell death executor (Mukai et al., 2000). The only amino acid homology domains identified within NADE sequence are a functional nuclear export sequence (NES) and two potential sites for ubiquitination. In vitro translated NADE binds to the death domain of p75^{NTR} and pull down assays indicate that the NES of NADE is required for this interaction. Endogenous p75^{NTR} and NADE were coimmunoprecipitated in PC12 cells and the neuroblastoma cell line SK-N-MC. In 293T cells expressing p75^{NTR} and NADE, NGF treatment led to association between the proteins and to proteolytic activation of caspases-2 and -3, followed by apoptosis. NGF induced cell death was also observed in NADE transfected PC12 cells, which express p75^{NTR} and TrkA, indicating that, unlike NRAGE, over expression of NADE may be sufficient to block survival signals through TrkA. A role for endogenous NADE in p75-mediated apoptosis of oligodendrocytes has been suggested based on the increase of mRNA^{NADE} following NGF treatment and because ability of a NADE fragment (residues 81–124) to block the induction of cell death (Mukai et al., 2002).

The mechanisms by which NADE functions have yet to be revealed; however the presence of the NES leads one to speculate that it may act in the nucleus. Given the numerous proteins found to associate with the intracellular domain of p75^{NTR} that appear to have a role in the nucleus (described above), the potential localization of the p75^{ICD} in the nucleus (Schechterson et al., 2002), and the requirement for protein synthesis in p75^{NTR} mediated apoptosis (Palmada et al., 2002), it is likely that this receptor activates a transcription-dependent cell death pathway.

Connections between p75^{NTR} and Trk receptors

The complex signaling mechanisms initiated by neurotrophin, binding to p75^{NTR} very likely can modulate and are influenced by Trk signaling since most cell types coexpress p75^{NTR} with a Trk receptor. In addition, Trk and p75^{NTR} can be coimmunoprecipitated (Bibel et al., 1999) and both receptors are required to form a high affinity complex for neurotrophin binding (Hempstead et al., 1991). A number of studies have clearly demonstrated that the presence of p75^{NTR} enables maximum Trk signaling (reviewed by Roux and Barker, 2002); however, less is known about the influence of Trk on p75's downstream effects. It has been shown that coactivation of Trk receptors and p75^{NTR} will block p75^{NTR} induced apoptosis (Yoon et al., 1998). This has been suggested to occur through Trk-dependent Ras activation, which can block the pro-apoptotic p75^{NTR} activated kinase, JNK (Kaplan and Miller, 2000). Trk activation of PI3K and the subsequent activation of the pro-survival molecule Akt may also counteract pro-apoptotic signals coming from the p75^{NTR}; for example, Akt can bind JIP and block its ability to scaffold kinases in the JNK pathway which normally leads to enhanced JNK signaling (Kim et al., 2002). Another way p75^{NTR} independent signaling could be modulated by Trk activation is through recruitment of p75^{NTR} to the activated Trk complex. Adaptor molecules that connect the two receptors may alter the conformation of p75^{NTR} or change the stoichiometry of the receptor complex, preventing binding of signaling molecules to the receptor, thus blocking specific pathways. Two such adaptor molecules have now been identified which could connect the p75^{NTR} and Trk receptors. The protein ankyrin repeat-rich membrane spanning (ARMS) was isolated from a yeast two-hybrid screen using the p75^{NTR} intracellular domain as bait (Kong et al., 2001). ARMS is a large protein, over 200 kD, containing four potential transmembrane domains, N-terminal ankyrin repeats, a SAM domain, and a potential PDZ binding domain at the extreme carboxyl terminus. mRNA^{ARMS} was localized to a number of neuronal populations that maintain synaptic plasticity through adulthood including hippocampal neurons, motor neurons of the spinal cord, cerebellar Purkinje cells, and neurons within the olfactory bulb. Sensory neurons of the DRG also express ARMS into adulthood and in sympathetic neurons it colocalizes with TrkA. ARMS coimmunoprecipitates with both p75^{NTR} and Trk receptors and is tyrosine phosphorylated in response to neurotrophin activation of Trks. These data lead to speculation that ARMS may link signaling between p75^{NTR} and Trk, although the functional effects of any connection remain to be examined.

A second adaptor protein, p62, may indirectly link the p75^{NTR} and Trk receptors. P62 has been shown to bind and alter the localization of the atypical protein kinase C (aPKC) in response to NGF (Samuels et al., 2001). These aPKC's, unlike other PKC isoforms, do not require either DAG or Ca^{++} for activity and are thought to function as upstream activators of NF-κB (Diaz-Meco et al., 1994). In PC12 cells, endogenous p62 has been shown to bind to both TrkA and the p75^{NTR} binding protein, TRAF6; and to facilitate the activation of NF-κB by NGF (Wooten et al., 2001). The interaction of TrkA with p62 was also reported to be essential for TrkA internalization into endosomes (Geetha and Wooten, 2003) thus, this bridging protein may facilitate synergistic signaling from the two receptors to promote survival and differentiation.

Conclusion

Accumulating evidence indicates that the p75^{NTR} regulates cell survival during development and in response to injury within the nervous system. In addition, p75^{NTR} signaling has been shown to play a role in myelination and neurite outgrowth. To understand the mechanisms by which this receptor functions, it is essential to elucidate the molecules activated by the receptor and determine how these signals mediate specific effects. Akt and NF-κB activation and JNK signaling represent opposing signals generated by p75^{NTR} but the molecular components and regulatory mechanisms of the pathways and their specific gene targets have yet to be fully elucidated. The rapidly expanding list of p75^{NTR} interacting factors illustrates the complexity of signal transduction through this receptor. It is likely that the expression or activation of specific pathways is temporally and spatially regulated and induced by specific cues and that the presence or absence of various interactors is necessary for the formation of signaling complexes necessary for specific signaling events. The sum of these signals in a given context

ultimately determines the p75^{NTR} response. While there is very little data yet on how these interactors may function together, the best evidence for formation of a receptor complex comes from examination of TRAF6. This protein binds not only to the p75^{NTR} and other members of the TRAF family, but also to the potential transcription factor, NRIF and the adaptor protein p62. Whether TRAF6 facilitates translocation of a p75^{ICD} complex with multiple transcription factors remains an open question.

In conclusion, the p75^{NTR}, once referred to as a 'nonsignaling receptor' (Schlessinger et al., 1995) has become a hotbed of activity. Given the many interacting proteins, the complex signaling pathways and the growing list of p75^{NTR} ligands, one can only wonder what surprises this receptor has in store for us next.

Abbreviations

aPKC	atypical protein kinase C
ARMS	ankyrin repeat-rich spanning
BDNF	brain-derived neurotrophic factor
CARD	caspase recruitment domain
CD40	cluster of differentiation 40
COP	CARD only protein
DRG	dorsal root ganglion
HP	heterochromatin protein
IAP	inhibitor of apoptosis
IκB	inhibitor of NF-κB
IL-1R	interleukin-1 receptor
IRAK	IL-1 receptor associated kinase
ITA	inhibitor of T cell apoptosis
JNK	c-Jun N-Terminal kinase
KAP-1	KRAB-associated protein-1
KRAB	Kruppel-associated box
MAG	myelin-associated glycoprotein
MAGE	melanoma-associated antigen
MAP Kinase	mitogen-associated protein kinase
MUL	muscle-liver-brain-eye nanism
NADE	p75-associated cell death executor
NES	nuclear export sequence
NF-κB	nuclear factor-kappa B
NGF	nerve growth factor
NIK	NF-kappaB-inducing kinase

NRAGE	neurotrophin receptor-interacting MAGE homologue
NRIF	neurotrophin receptor interacting factor
NT-3	neurotrophin-3
NT-4/5	neurotrophin-4/5
Omgp	oligodendrocyte myelin glycoprotein
p75^{NTR}	p75 neurotrophin receptor
PR Domain	positive regulatory domain
RANK	receptor activator of NF-κB
RIP-2	receptor interacting protein-2
SCAN	SRE-ZBP, Ct-fin-51, AW-1, number 18 cDNA
TEF	TRAF domain encompassing factor
TLR	Toll-like receptor
TNF	tumor necrosis factor
TRAF	TNF receptor associated factor
Trk	tropomyosin related kinase
TUNEL	TdT-mediated dUTP nick end labeling
XIAP	X-linked inhibitor of apoptosis

References

Agerman, K., Baudet, C., Fundin, B., Willson, C. and Ernfors, P. (2000) Attenuation of a caspase-3 dependent cell death in NT4- and p75- deficient embryonic sensory neurons. Mol. Cell. Neurosci., 16: 258–268.

Aloyz, R.S., Bamji, S.X., Pozniak, C.D., Toma, J.G., Atwal, J., Kaplan, D.R. and Miller, F.D. (1998) p53 is essential for developmental neuron death as regulated by the TrkA and p75 neurotrophin receptors. J. Cel. Biol., 143: 1691–1703.

Avela, K., Lipsanen-Nyman, M., Idanheimo, N., Seemanova, E., Rosengren, S., Makela, T.P., Perheentupa, J., Chapelle, A.D. and Lehesjoki, A.E. (2000) Gene encoding a new RING-B-box-Coiled-coil protein is mutated in mulibrey nanism. Nat. Genet., 25: 298–301.

Bamji, S.X., Majdan, A.E., Pozniak, C.D., Belliveau, D.J., Aloyz, R., Kohn, J., Causing, C.G. and Miller, F.D. (1998) The p75 neurotrophin receptor mediates neuronal apoptosis and is essential for naturally occurring sympathetic neuron death. J. Cell. Biol., 140: 911–923.

Baud, V. and Karin, M. (2001) Signal transduction by tumor necrosis factor and its relatives. Trends Cell. Biol., 11: 372–377.

Beattie, M.S., Harrington, A.W., Lee, R., Kim, J.Y., Boyce, S.L., Longo, F.M., Bresnahan, J.C., Hempstead, B.L. and Yoon, S.O. (2002) ProNGF induces p75-mediated death of oligodendrocytes following spinal cord injury. Neuron., 36: 375–386.

Bentley, C.A. and Lee, K.F. (2000) p75 is important for axon growth and schwann cell migration during development. J. Neurosci., 20: 7706–7715.

Benzel, I., Barde, Y.A. and Casademunt, E. (2001) Strain-specific complementation between NRIF1 and NRIF2, two zinc finger proteins sharing structural and biochemical properties. Gene, 281: 19–30.

Beraud, C., Henzel, W.J. and Baeuerle, P.A. (1999) Involvement of regulatory and catalytic subunits of phosphoinositide 3-kinase in NF-kappaB activation. Proc. Natl. Acad. Sci. USA, 96: 429–434.

Berger, E.A. and Shooter, E.M. (1977) Evidence for pro-beta-nerve growth factor, a biosynthetic precursor to beta-nerve growth factor. Proc. Natl. Acad. Sci. USA, 74: 3647–3651.

Bibel, M., Hoppe, E. and Barde, Y.A. (1999) Biochemical and functional interactions between the neurotrophin receptors trk and p75NTR. EMBO J., 18: 616–622.

Bradley, J.R. and Pober, J.S. (2001) Tumor necrosis factor receptor-associated factors (TRAFs). Oncogene., 20: 6482–6491.

Brunet, A., Bonni, A., Zigmond, M.J., Lin, M.Z., Juo, P., Hu, L.S., Anderson, M.J., Arden, K.C., Blenis, J. and Greenberg, M.E. (1999) Akt promotes cell survival by phosphorylating and inhibiting a Forkhead transcription factor. Cell., 96: 857–868.

Bui, N.T., Konig, H.G., Culmsee, C., Bauerbach, E., Poppe, M., Krieglstein, J. and Prehn, J.H. (2002) p75 neurotrophin receptor is required for constitutive and NGF-induced survival signalling in PC12 cells and rat hippocampal neurones. J. Neurochem., 81: 594–605.

Bunone, G., Mariotti, A., Compagni, A., Morandi, E. and Della Valle, G. (1997) Induction of apoptosis by p75 neurotrophin receptor in human neuroblastoma cells. Oncogene., 14: 1463–1470.

Cardone, M.H., Roy, N., Stennicke, H.R., Salvesen, G.S., Franke, T.F., Stanbridge, E., Frisch, S. and Reed, J.C. (1998) Regulation of cell death protease caspase-9 by phosphorylation. Science., 282: 1318–1321.

Casaccia-Bonnefil, P., Carter, B.D., Dobrowsky, R.T. and Chao, M.V. (1996) Death of oligodendrocytes mediated by the interaction of nerve growth factor with its receptor p75. Nature, 383: 716–719.

Casademunt, E., Carter, B.D., Benzel, I., Frade, J.M., Dechant, G. and Barde, Y.A. (1999) The zinc finger protein NRIF interacts with the neurotrophin receptor p75(NTR) and participates in programmed cell death. Embo. J., 18: 6050–6061.

Chan, J.R., Cosgaya, J.M., Wu, Y.J. and Shooter, E.M. (2001) Neurotrophins are key mediators of the myelination program in the peripheral nervous system. Proc. Natl. Acad. Sci. USA, 98: 14661–14668.

Chittka, A. and Chao, M.V. (1999) Identification of a zinc finger protein whose subcellular distribution is regulated by serum and nerve growth factor. Proc. Natl. Acad. Sci. USA, 96: 10705–10710.

Cortazzo, M.H., Kassis, E.S., Sproul, K.A. and Schor, N.F. (1996) Nerve growth factor (NGF)-mediated protection of neural crest cells from antimitotic agent-induced apoptosis: the role of the low-affinity NGF receptor. J. Neurosci., 16: 3895–3899.

Cosgaya, J.M., Chan, J.R. and Shooter, E.M. (2002) The neurotrophin receptor p75NTR as a positive modulator of myelination. Science., 298: 1245–1248.

Culmsee, C., Gerling, N., Lehmann, M., Nikolova-Karakashian, M., Prehn, J.H., Mattson, M.P. and Krieglstein, J. (2002) Nerve growth factor survival signaling in cultured hippocampal neurons is mediated through TrkA and requires the common neurotrophin receptor P75. Neuroscience, 115: 1089–1108.

da Silva, J.S. and Dotti, C.G. (2002) Breaking the neuronal sphere: regulation of the actin cytoskeleton in neuritogenesis. Nat. Rev. Neurosci., 3: 694–704.

DeFreitas, M.F., McQuillen, P.S. and Shatz, C.J. (2001) A novel p75NTR signaling pathway promotes survival, not death, of immunopurified neocortical subplate neurons. J. Neurosci., 21: 5121–5129.

del Peso, L., Gonzalez-Garcia, M., Page, C., Herrera, R. and Nunez, G. (1997) Interleukin-3-induced phosphorylation of BAD through the protein kinase Akt. Science, 278: 687–689.

Delsite, R. and Djakiew, D. (1999) Characterization of nerve growth factor precursor protein expression by human prostate stromal cells: a role in selective neurotrophin stimulation of prostate epithelial cell growth. Prostate, 41: 39–48.

Diaz-Meco, M.T., Dominguez, I., Sanz, L., Dent, P., Lozano, J., Municio, M.M., Berra, E., Hay, R.T., Sturgill, T.W. and Moscat, J. (1994) zeta PKC induces phosphorylation and inactivation of I kappa B-alpha in vitro. Embo. J., 13. 2842–2848.

Dowling, P., Ming, X., Raval, S., Husar, W., Casaccia-Bonnefil, P., Chao, M., Cook, S. and Blumberg, B. (1999) Up-regulated p75NTR neurotrophin receptor on glial cells in MS plaques. Neurology, 53: 1676–1682.

Ferri, C.C., Moore, F.A. and Bisby, M.A. (1998) Effects of facial nerve injury on mouse motoneurons lacking the p75 low-affinity neurotrophin receptor. J. Neurobiol., 34: 1–9.

Foehr, E.D., Lin, X., O'Mahony, A., Geleziunas, R., Bradshaw, R.A. and Greene, W.C. (2000) NF-kappa B signaling promotes both cell survival and neurite process formation in nerve growth factor-stimulated PC12 cells. J. Neurosci., 20: 7556–7563.

Frade, J.M. and Barde, Y.A. (1999) Genetic evidence for cell death mediated by nerve growth factor and the neurotrophin receptor p75 in the developing mouse retina and spinal cord. Development., 126: 683–690.

Frade, J.M., Rodriguez-Tebar, A. and Barde, Y.A. (1996) Induction of cell death by endogenous nerve growth factor through its p75 receptor. Nature, 383: 166–168.

Friedman, J.R., Fredericks, W.J., Jensen, D.E., Speicher, D.W., Huang, X.P., Neilson, E.G. and Rauscher, F.J., III (1996) KAP-1, a novel corepressor for the highly conserved KRAB repression domain. Genes Dev., 10: 2067–2078.

Friedman, W.J. (2000) Neurotrophins induce death of hippocampal neurons via the p75 receptor. J. Neurosci., 20: 6340–6346.

Geetha, T. and Wooten, M.W. (2003) Association of the atypical protein kinase C-interacting protein p62/ZIP with nerve growth factor receptor TrkA regulates receptor trafficking and Erk5 signaling. J. Biol. Chem., 278: 4730–4739.

Gentry, J.J., Casaccia-Bonnefil, P. and Carter, B.D. (2000) Nerve growth factor activation of nuclear factor kappaB through its p75 receptor is an anti-apoptotic signal in RN22 schwannoma cells. J. Biol. Chem., 275: 7558–7565.

Gentry, J.J., Rutkoski, N., Linggi, M.S., Musiek, E.S., Emeson, R. and Carter, B.D. (2001) A functional interaction between the p75 associated proteins NRIF and TRAF6. Abstr. 695.14, 31st Annual Mtg, Soc. Neurosci., San Diego, CA.

Giehl, K.M., Rohrig, S., Bonatz, H., Gutjahr, M., Leiner, B., Bartke, I., Yan, Q., Reichardt, L.F., Backus, C., Welcher, A.A., Dethleffsen, K., Mestres, P., Meyer, M. (2001) Endogenous brain-derived neurotrophic factor and neurotrophin-3 antagonistically regulate survival of axotomized corticospinal neurons in vivo. J. Neurosci., 21: 3492–3502.

Gu, C., Casaccia-Bonnefil, P., Srinivasan, A. and Chao, M.V. (1999) Oligodendrocyte apoptosis mediated by caspase activation. J. Neurosci., 19: 3043–3049.

Hamanoue, M., Middleton, G., Wyatt, S., Jaffray, E., Hay, R.T. and Davies, A.M. (1999) p75-mediated NF-kappaB activation enhances the survival response of developing sensory neurons to nerve growth factor. Mol. Cell. Neurosci., 14: 28–40.

Harrington, A.W., Kim, J.Y. and Yoon, S.O. (2002) Activation of Rac GTPase by p75 is necessary for c-jun N-terminal kinase-mediated apoptosis. J. Neurosci., 22: 156–166.

Hempstead, B.L., Martin-Zanca, D., Kaplan, D.R., Parada, L.F. and Chao, M.V. (1991) High-affinity NGF binding requires coexpression of the trk proto-oncogene and the low-affinity NGF receptor. Nature, 350: 678–683.

Huang, E.J. and Reichardt, L.F. (2001) Neurotrophins: roles in neuronal development and function. Annu. Rev. Neurosci., 24: 677–736.

Jordan, B.W., Dinev, D., LeMellay, V., Troppmair, J., Gotz, R., Wixler, L., Sendtner, M., Ludwig, S. and Rapp, U.R. (2001) Neurotrophin receptor-interacting mage homologue is an inducible inhibitor of apoptosis protein-

38

interacting protein that augments cell death. J. Biol. Chem., 276: 39985–39989.

Kaplan, D.R. and Miller, F.D. (2000) Neurotrophin signal transduction in the nervous system. Curr. Opin. Neurobiol., 10: 381–391.

Karin, M. and Lin, A. (2002) NF-kappaB at the crossroads of life and death. Nat. Immunol., 3: 221–227.

Khursigara, G., Bertin, J., Yano, H., Moffett, H., DiStefano, P.S. and Chao, M.V. (2001) A prosurvival function for the p75 receptor death domain mediated via the caspase recruitment domain receptor-interacting protein 2. J. Neurosci., 21: 5854–5863.

Khursigara, G., Orlinick, J.R. and Chao, M.V. (1999) Association of the p75 neurotrophin receptor with TRAF6. J. Biol. Chem., 274: 2597–2600.

Kim, A.H., Yano, H., Cho, H., Meyer, D., Monks, B., Margolis, B., Birnbaum, M.J. and Chao, M.V. (2002) Akt1 regulates a JNK scaffold during excitotoxic apoptosis. Neuron, 35: 697–709.

Kohn, J., Aloyz, R.S., Toma, J.G., Haak-Frendscho, M. and Miller, F.D. (1999) Functionally antagonistic interactions between the TrkA and p75 neurotrophin receptors regulate sympathetic neuron growth and target innervation. J. Neurosci., 19: 5393–5408.

Kong, H., Boulter, J., Weber, J.L., Lai, C. and Chao, M.V. (2001) An evolutionarily conserved transmembrane protein that is a novel downstream target of neurotrophin and ephrin receptors. J. Neurosci., 21: 176–185.

Kuida, K., Haydar, T.F., Kuan, C.Y., Gu, Y., Taya, C., Karasuyama, H., Su, M.S., Rakic, P. and Flavell, R.A. (1998) Reduced apoptosis and cytochrome c-mediated caspase activation in mice lacking caspase 9. Cell., 94: 325–337.

Lee, K.F., Li, E., Huber, L.J., Landis, S.C., Sharpe, A.H., Chao, M.V. and Jaenisch, R. (1992) Targeted mutation of the gene encoding the low affinity NGF receptor p75 leads to deficits in the peripheral sensory nervous system. Cell, 69: 737–749.

Lee, R., Kermani, P., Teng, K.K. and Hempstead, B.L. (2001a) Regulation of cell survival by secreted proneurotrophins. Science., 294: 1945–1948.

Lee, S.H., Stehlik, C. and Reed, J.C. (2001b) Cop, a caspase recruitment domain-containing protein and inhibitor of caspase-1 activation processing. J. Biol. Chem., 276: 34495–34500.

Linggi, M.S., Burke, T.L., Gentry, J.J. and Carter, B.D. (2002) NRIF, a proapoptotic protein is required for p75-mediated cell death. Abstr. #426.15, 32nd Annual Mtg. Soc. Neurosci., Orlando, FL.

Longo, F.M., Manthorpe, M., Xie, Y.M. and Varon, S. (1997) Synthetic NGF peptide derivatives prevent neuronal death via a p75 receptor-dependent mechanism. J. Neurosci. Res., 48: 1–17.

Lowry, K.S., Murray, S.S., McLean, C.A., Talman, P., Mathers, S., Lopes, E.C. and Cheema, S.S. (2001) A potential role for the p75 low-affinity neurotrophin receptor in spinal motor neuron degeneration in murine and human amyotrophic lateral sclerosis. Amyotroph. Lateral Scler. Other Motor Neuron Disord., 2: 127–134.

Maggirwar, S.B., Sarmiere, P.D., Dewhurst, S. and Freeman, R.S. (1998) Nerve growth factor-dependent activation of NF-kappaB contributes to survival of sympathetic neurons. J. Neurosci., 18: 10356–10365.

McCarthy, J.V., Ni, J. and Dixit, V.M. (1998) RIP2 is a novel NF-kappaB-activating and cell death-inducing kinase. J. Biol. Chem., 273: 16968–16975.

Mowla, S.J., Farhadi, H.F., Pareek, S., Atwal, J.K., Morris, S.J., Seidah, N.G. and Murphy, R.A. (2001) Biosynthesis and post-translational processing of the precursor to brain-derived neurotrophic factor. J. Biol. Chem., 276: 12660–12666.

Mufson, E.J., Ma, S.Y., Dills, J., Cochran, E.J., Leurgans, S., Wuu, J., Bennett, D.A., Jaffar, S., Gilmor, M.L., Levey, A.I. and Kordower, J.H. (2002) Loss of basal forebrain P75(NTR) immunoreactivity in subjects with mild cognitive impairment and Alzheimer's disease. J. Comp. Neurol., 443: 136–153.

Mukai, J., Hachiya, T., Shoji-Hoshino, S., Kimura, M.T., Nadano, D., Suvanto, P., Hanaoka, T., Li, Y., Irie, S., Greene, L.A. and Sato, T.A. (2000) NADE, a p75NTR-associated cell death executor, is involved in signal transduction mediated by the common neurotrophin receptor p75NTR. J. Biol. Chem., 275: 17566–17570.

Mukai, J., Shoji, S., Kimura, M.T., Okubo, S., Sano, H., Suvanto, P., Li, Y., Irie, S. and Sato, T.A. (2002) Structure-function analysis of NADE: identification of regions that mediate nerve growth factor-induced apoptosis. J. Biol. Chem., 277: 13973–13982.

Nagata, S. (1997) Apoptosis by death factor. Cell, 88: 355–365.

Naumann, T., Casademunt, E., Hollerbach, E., Hofmann, J., Dechant, G., Frotscher, M. and Barde, Y.A. (2002) Complete deletion of the neurotrophin receptor p75NTR leads to long-lasting increases in the number of basal forebrain cholinergic neurons. J. Neurosci., 22: 2409–2418.

Nickols, J.C., Valentine, W., Kanwal, S. and Carter, B.D. (2003) Activation of the transcription factor NF-kappaB in Schwann cells is required for peripheral myelin formation. Nat. Neurosci., 6: 161–167.

Oh, J., Zhong, L.T., Yang, J., Bitler, C.M., Butcher, L.L., Bredesen, D.E. and Rabizadeh, S. (1993) Induction of apoptosis by the low-affinity NGF receptor. Science, 261: 345–348.

Oppenheim, R.W. (1991) Cell death during development of the nervous system. Annu. Rev. Neurosci., 14: 453–501.

Pahl, H.L. (1999) Activators and target genes of Rel/NF-kappaB transcription factors. Oncogene, 18: 6853–6866.

Palmada, M., Kanwal, S., Rutkoski, N.J., Gufstafson-Brown, C., Johnson, R.S., Wisdom, R. and Carter, B.D. (2002) c-jun is essential for sympathetic neuronal death induced by NGF withdrawal but not by p75 activation. J. Cell. Biol., 158: 453–461.

Roux, P.P. and Barker, P.A. (2002) Neurotrophin signaling through the p75 neurotrophin receptor. Prog. Neurobiol., 67: 203–233.

Roux, P.P., Bhakar, A.L., Kennedy, T.E. and Barker, P.A. (2001) The p75 neurotrophin receptor activates Akt (protein kinase B) through a phosphatidylinositol 3-kinase-dependent pathway. J. Biol. Chem., 276: 23097–23104.

Roux, P.P., Colicos, M.A., Barker, P.A. and Kennedy, T.E. (1999) p75 neurotrophin receptor expression is induced in apoptotic neurons after seizure. J. Neurosci., 19: 6887–6896.

Salehi, A., Ocampo, M., Verhaagen, J. and Swaab, D.F. (2000a) P75 neurotrophin receptor in the nucleus basalis of meynert in relation to age, sex, and Alzheimer's disease. Exp. Neurol., 161: 245–258.

Salehi, A.H., Roux, P.P., Kubu, C.J., Zeindler, C., Bhakar, A., Tannis, L.L., Verdi, J.M. and Barker, P.A. (2000b) NRAGE, a novel MAGE protein, interacts with the p75 neurotrophin receptor and facilitates nerve growth factor-dependent apoptosis. Neuron, 27: 279–288.

Samuels, I.S., Seibenhener, M.L., Neidigh, K.B. and Wooten, M.W. (2001) Nerve growth factor stimulates the interaction of ZIP/p62 with atypical protein kinase C and targets endosomal localization: evidence for regulation of nerve growth factor-induced differentiation. J. Cell. Biochem., 82: 452–466.

Schechterson, L.C., Kanning, K.C., Hudson, M.P. and Bothwell, M. (2002) The neurotrophin receptor p75 is cleaved by regulated intramembranous proteolysis. Abstr. #822.10, 32nd Annual Mtg. Soc. Neurosci., Orlando, FL.

Schlessinger, J., Lax, I. and Lemmon, M. (1995) Regulation of growth factor activation by proteoglycans: what is the role of the low affinity receptors?. Cell, 83: 357–360.

Sedel, F., Bechade, C. and Triller, A. (1999) Nerve growth factor (NGF) induces motoneuron apoptosis in rat embryonic spinal cord in vitro. Eur. J. Neurosci., 11: 3904–3912.

Syroid, D.E., Maycox, P.J., Soilu-Hanninen, M., Petratos, S., Bucci, T., Burrola, P., Murray, S., Cheema, S., Lee, K.F., Lemke, G. and Kilpatrick, T.J. (2000) Induction of postnatal schwann cell death by the low-affinity neurotrophin receptor in vitro and after axotomy. J. Neurosci., 20: 5741–5747.

Taglialatela, G., Robinson, R. and Perez-Polo, J.R. (1997) Inhibition of nuclear factor kappa B (NFkappaB) activity induces nerve growth factor-resistant apoptosis in PC12 cells. J. Neurosci. Res., 47: 155–162.

Tcherpakov, M., Bronfman, F.C., Conticello, S.G., Vaskovsky, A., Levy, Z., Niinobe, M., Yoshikawa, K., Arenas, E. and Fainzilber, M. (2002) The p75 neurotrophin receptor interacts with multiple MAGE proteins. J. Biol. Chem., 277: 49101–49104.

Thome, M., Hofmann, K., Burns, K., Martinon, F., Bodmer, J.L., Mattmann, C. and Tschopp, J. (1998) Identification of CARDIAK, a RIP-like kinase that associates with caspase-1. Curr. Biol., 8: 885–888.

Troy, C.M., Friedman, J.E. and Friedman, W.J. (2002) Mechanisms of p75-mediated death of hippocampal neurons. Role of caspases. J. Biol. Chem., 277: 34295–34302.

Wang, K.C., Kim, J.A., Sivasankaran, R., Segal, R. and He, Z. (2002) P75 interacts with the Nogo receptor as a coreceptor for Nogo, MAG and OMgp. Nature, 420: 74–78.

Wang, S., Bray, P., McCaffrey, T., March, K., Hempstead, B.L. and Kraemer, R. (2000) p75(NTR) mediates neurotrophin-induced apoptosis of vascular smooth muscle cells. Am. J. Pathol., 157: 1247–1258.

Williams, M.E., Strickland, P., Watanabe, K. and Hinck, L. (2003) UNC5H1 induces apoptosis via its juxtamembrane domain through an interaction with NRAGE. J. Biol. Chem. [epub ahead of print].

Wong, S.T., Henley, J.R., Kanning, K.C., Huang, K.H., Bothwell, M. and Poo, M.M. (2002) A p75(NTR) and Nogo receptor complex mediates repulsive signaling by myelin-associated glycoprotein. Nat. Neurosci., 5: 1302–1308.

Woolf, C.J. and Bloechlinger, S. (2002) Neuroscience. It takes more than two to Nogo. Science, 297: 1132–1134.

Wooten, M.W., Seibenhener, M.L., Mamidipudi, V., Diaz-Meco, M.T., Barker, P.A. and Moscat, J. (2001) The atypical protein kinase C-interacting protein p62 is a scaffold for NF-kappaB activation by nerve growth factor. J. Biol. Chem., 276: 7709–7712.

Yamashita, T., Tucker, K.L. and Barde, Y.A. (1999) Neurotrophin binding to the p75 receptor modulates Rho activity and axonal outgrowth. Neuron, 24: 585–593.

Ye, X., Mehlen, P., Rabizadeh, S., VanArsdale, T., Zhang, H., Shin, H., Wang, J.J., Leo, E., Zapata, J., Hauser, C.A., Reed, J.C., Bredesen, D.E. (1999) TRAF family proteins interact with the common neurotrophin receptor and modulate apoptosis induction. J. Biol. Chem., 274: 30202–30208.

Yoon, S.O., Casaccia-Bonnefil, P., Carter, B. and Chao, M.V. (1998) Competitive signaling between TrkA and p75 nerve growth factor receptors determines cell survival. J. Neurosci., 18: 3273–3281.

Zapata, J.M., Pawlowski, K., Haas, E., Ware, C.F., Godzik, A. and Reed, J.C. (2001) A diverse family of proteins containing tumor necrosis factor receptor-associated factor domains. J. Biol. Chem., 276: 24242–24252.

Progress in Brain Research, Vol. 146
ISSN 0079-6123

CHAPTER 3

The role of neurotransmission and the Chopper domain in p75 neurotrophin receptor death signaling

E.J. Coulson[1],*, K. Reid[2], K.M. Shipham[2], S. Morley[1], T.J. Kilpatrick[2] and P.F. Bartlett[1]

[1]*Queensland Brain Institute, The University of Queensland, Brisbane, Qld, Australia*
[2]*The Walter and Eliza Hall Institute for Medical Research, Parkville, Vic., Australia*

Abstract: The role of p75 neurotrophin receptor ($p75^{NTR}$) in mediating cell death is now well charaterized, however, it is only recently that details of the death signaling pathway have become clearer. This review focuses on the importance of the juxtamembrane Chopper domain region of $p75^{NTR}$ in this process. Evidence supporting the involvement of K^+ efflux, the apoptosome (caspase-9, apoptosis activating factor-1, APAF-1, and Bcl-$_{xL}$), caspase-3, c-jun kinase, and p53 in the $p75^{NTR}$ cell death pathway is discussed and regulatory roles for the $p75^{NTR}$ ectodomain and death domain are proposed. The role of synaptic activity is also discussed, in particular the importance of neutrotransmitter-activated K^+ channels acting as the gatekeepers of cell survival decisions during development and in neurodegenerative conditions.

Keywords: p75 neurotrophin receptor; cell death pathways; K^+ channel; neurotransmission; neurotrophin; apoptosome; G-protein coupled receptors

Introduction

The apoptosis of neurons that occurs concurrently with synaptogenesis is one of the most important events in the developing nervous system. Neuronal apoptosis also occurs following neurotrauma and as a result of neurodegenerative disease, possibly by reactivating these same developmental cell death/ apoptosis pathways. The p75 neurotrophin receptor ($p75^{NTR}$) is expressed in neurons during development and is downregulated in most neuronal populations following the developmental period in which synaptogenesis and cell death occur. It is reexpressed after neuronal insult or degeneration at levels comparable to developmental expression (Gage et al., 1989; Mufson et al., 1992; Rende et al., 1993; Lee et al., 1995; Kokaia et al., 1998; Roux et al., 1999; Andsberg

et al., 2001; Giehl et al., 2001; Lowry et al., 2001b; Greferath et al., 2002). Since the discovery of the role of $p75^{NTR}$ in mediating cell death (Barrett and Bartlett, 1994), considerable evidence has emerged confirming the important role $p75^{NTR}$ plays in mediating these cell death processes (reviewed by Miller and Kapalan, 1998; Yoon et al., 1998; Casaccia-Bonnefil et al., 1999; Coulson et al., 1999b; Barrett, 2000; Roux and Barker, 2002).

A large number of compelling in vivo studies of both developmental, naturally occurring cell death and cell death triggered in the adult by an insult to the nervous system have demonstrated the essential role played by $p75^{NTR}$ (Lee et al., 1992; von Bartheld et al., 1994; Cheema et al., 1996; Frade et al., 1996; Majdan et al., 1997; Bamji et al., 1998; Davey and Davies, 1998; Frade and Barde, 1998; Roux et al., 1999; Soilu-Hänninen et al., 2000; Naumann et al., 2002). However, unlike other death pathways, such as Fas- or chemically-induced cell death, the effects of $p75^{NTR}$ are not easily replicated in vitro. Although

*Correspondence to: E.J. Coulson, Queensland Brain Institute, The University of Queensland, Brisbane, 4072 Qld, Australia. E-mail: e.coulson@uq.edu.au

DOI: 10.1016/S0079-6123(03)46003-2

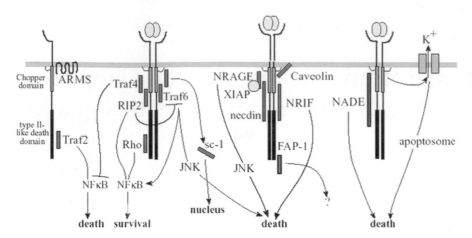

Fig. 1. Interaction of p75^{NTR} with the many described binding proteins. A large number of accessory proteins bind to the intracellular domain of p75^{NTR} and mediate a variety of effects, which can vary depending on experimental conditions. The majority of these proteins interact in the presence of neurotrophin ligand, and those that mediate cell death (or induce caspase activity) interact with the Chopper domain. Where known, any downstream mediators for the proteins are indicated.

many laboratories have found adequate ways to assay for downstream effects of p75^{NTR} signaling, the translation of techniques between research groups or the interpretation of results between cell types has not always been straightforward. Rather, the proposed downstream signaling pathway of p75^{NTR} has developed into a complex array of arrows pointing to outcomes (Fig. 1), many seemingly dependent on cell type, expression of Trk, p75^{NTR} expression level, developmental age, and other extraneous factors, such as the level of growth factors or the composition of the medium.

If p75^{NTR} is as important a regulator of cell death in the nervous system as has been asserted, why is it so difficult to promote in vitro? Neurons in their natural environment, with connections formed over many weeks and having interpreted thousands of environmental cues, will act differently from those placed in a minimalist culture system designed for manipulating individual cellular processes. Furthermore, axons and dendrites in vivo receive and interpret a vast array of trophic and synaptic signals, the subtle nuances of which are certainly lost in a crude culture system. We have hypothesized that the environment, in particular the presence of synaptic input, may actually be the key to regulating survival and death pathways during development and following neurotrauma or neurodegeneration. Indeed, because culture media are often designed to mimic the effects of synaptic activity in order to maintain neurons in culture (glutamate, forskolin and increased K$^+$ concentrations are regular additives to basal tissue culture media), the over-abundance of 'synaptic' signals might be an important contributing factor in the difficulty experienced in assaying p75^{NTR}-mediated cell death in culture.

This review attempts to formulate a model of the mechanisms by which p75^{NTR}-mediated cell death occurs. In doing so, we will explore how the regulation of synaptic activity might impinge on the p75^{NTR} death-signaling cascade. We will review the literature covering the regulators and downstream mediators of p75^{NTR}-mediated death, in particular arguing the importance of the Chopper domain. Finally, these data will be integrated into what we know in a broader context about the interactions between neurotrophin signaling and synaptic activity in an attempt to make sense of this enigmatic receptor-p75^{NTR}.

Factors controlling neuronal cell death

Trophic support is unquestionably required for neuronal survival and p75^{NTR} can only mediate cell death when Trk receptor signaling is deficient (Rabizadeh et al., 1993; Casaccia-Bonnefil et al.,

1996; Frade et al., 1996; Bamji et al., 1998; Frade and Barde, 1998; Coulson et al., 1999a; Lee et al., 2001; Beattie et al., 2002). Two key mechanisms might regulate p75^{NTR}-mediated cell death by neurotrophin signaling through Trk receptors. In the presence of neurotrophins, p75^{NTR} can cooperate with trk to assist in promoting survival signaling. In addition, activation of Trk phosphorylation cascades activate survival-promoting proteins and pathways that directly compete with p75^{NTR}-mediated death pathways. It is well recognized that cell excitation is also a crucial factor in neuronal survival but the effects of this are less well characterized (see Linden, 1994; Sherrard and Bower, 1998). We propose that, like Trk signaling, synaptic activity can also override the p75^{NTR} cell death signaling pathway.

Despite the increasing realization that both apoptotic signaling cascades and neurotrophic signals play an important role in modulating synaptic plasticity (Mattson and Duan, 1999; Schinder and Poo, 2000), the molecular interactions between death pathways and neuronal activity have not been extensively described. Emerging data suggest that, in addition to their roles in neurodegenerative processes, apoptotic proteases, caspases and calpains, play a significant part in the regulation of synaptic plasticity (Chan and Mattson, 1999). However, the extent to which crosstalk exists between signaling components higher up these pathways remains unclear. It has been widely hypothesized that synaptic input assists in controlling membrane potential and that neurons are kept alive by membrane depolarization, induced by raising the level of either extracellular K$^+$ or intracellular Ca^{2+} (see Sherrard and Bower, 1998; Mennerick and Zorumski, 2000; Yu et al., 2001). Indeed, there is overwhelming evidence that reduced levels of intracellular K$^+$ can mediate apoptosis (see below), and that ion channels involved in modulating synaptic activity can also regulate K$^+$ efflux.

We have recently found that one form of p75^{NTR}-mediated cell death, induced solely by the membrane-localized Chopper domain of p75^{NTR}, instigates an early and significant outward flux of K$^+$ from neurons. We have not only predicted that this provides an important contribution to propagating the death signal arising from p75^{NTR} but have gone on to establish an interaction with a K$^+$ channel

that is capable of regulating both cell death and synaptic activity (see below).

The Chopper domain of p75^{NTR}

Previously, we demonstrated that dorsal root ganglion (DRG) neurons during the period of naturally occurring cell death were susceptible to p75^{NTR}-mediated death pathways (Barrett and Bartlett, 1994) and that overexpression of p75^{NTR} promoted cell death of DRG neurons maintained in culture with cytokine support (Coulson et al., 1999a). In particular, the juxtamembrane region of p75^{NTR} (or the 'Chopper domain') was both sufficient and necessary to induce cell death in this context (Coulson et al., 1999a, 2000). The Chopper domain was also able to act constitutively as a death inducer in a variety of neuronal and non-neuronal cells, both in the presence and in the absence of neurotrophins (Coulson et al., 2000). Moreover, a form of the Chopper domain that lacks any transmembrane localization signals did not itself induce cell death and was capable, presumably by a dominant-negative mechanism, of preventing cell death induced by overexpression of either membrane-linked Chopper or p75^{NTR} constructs. A similar result was obtained in experiments involving nerve growth factor (NGF)-induced cell death in Schwann cells and, importantly, developmental cell death in the chick retina (Coulson et al., 2000), which has previously been shown to be mediated by p75^{NTR} (Frade et al., 1996). These data strongly suggest that the Chopper domain is the critical death-inducing portion of the p75^{NTR} molecule.

Binding proteins that interact with Chopper

An increasingly large number of proteins have been found to interact with the intracellular domain of p75^{NTR} (Fig. 1), including the neurotrophin receptor interacting factor (NRIF) (Casademunt et al., 1999), the zinc finger protein sc-1 (Chittka and Chao, 1999), neurotrophin receptor associating death effector (NADE) (Mukai et al., 2000), MAGE family members (Salehi et al., 2000; Tcherpakov et al., 2002), Fas-associated protein-1 (FAP-1) (Irie et al., 1999), caveolin (Bilderback et al., 1997, 1999), ankyrin repeat-rich membrane spanning proteins (ARMS) (Kong et al., 2001), and the GTPase family member Rho (Yamashita et al., 1999). These are the subjects of other papers in this issue (see also Roux and Barker, 2002). In addition, a number of proteins, tumor necrosis factor receptor (TNFR) associating factor (TRAF) (Khursigara et al., 1999; Ye et al., 1999), receptor interacting protein-2 (RIP2) (Khursigara et al., 2001), c-jun activating kinase (JNK) (Ladiwala et al., 1998; Yoon et al., 1998; Kaplan and Miller, 2000), Bcl-2 family members (Coulson et al., 1999b), GTPase family member Rac (Harrington et al., 2002), and p53 family members (Aloyz et al., 1998) can influence and participate in the signaling pathway of p75^{NTR}, even in the absence of a direct association with p75^{NTR}. The functional complexity of p75^{NTR} signaling has been indicated by the range of effects that these interacting proteins have on cells, either promoting cell death (TRAF proteins, NRIF, NADE, NRAGE, Rac) or mediating other cellular processes (sc-1, caveolin, GTPases, TRAF proteins, RhoA, ARMS, other MAGE proteins). Most of these proteins do not contain well characterized functional or catalytic domains to assist in piecing together their downstream signaling pathways; rather, functional outcomes without identification of intermediate steps have been documented, highlighting the fact that additional work in these areas is required.

Interestingly, the vast majority of the identified *death*-inducing p75^{NTR}-interacting proteins can interact with the Chopper domain (Fig. 1). Using this domain as bait in a yeast two-hybrid screen of DRG-expressed cDNAs, we isolated the p75^{NTR}-interacting protein Bex3/NADE (Brown and Kay, 1999), which was also found to interact with the death domain region of p75^{NTR} (Mukai et al., 2000). NADE appears to induce cortical neuronal death associated with zinc toxicity and the coincident upregulation of p75^{NTR} expression (Park et al., 2000). Surprisingly, we found no cooperation of NADE with either Chopper or p75^{NTR}-mediated cell death in DRG neurons (Reid et al., 2002).

NRIF also binds equally well to both the Chopper domain and the region with homology to the Fas and TNF death domains (Casademunt et al., 1999). However, despite strong evidence that when NRIF is genetically absent the resulting number of neurons in vivo is higher, the underlying pathway remains unknown.

The MAGE protein family member, NRAGE, also interacts preferentially with the Chopper domain (Salehi et al., 2000); and the cascade promoted by NRAGE fits well with what we know about p75^{NTR}-induced death pathways (see Gentry et al. in chapter 2 of this title, and below).

The death-inducing TRAF 4 protein, which presumably interacts via an accessory protein as no interaction has been observed in yeast two-hybrid screens (see Coulson et al., 1999a), binds preferentially to the Chopper domain (Ye et al., 1999), as does TRAF 6 (Khursigara et al., 1999; Ye et al., 1999). The possible signaling pathways downstream of TRAFs 4 and 6 have not been fully explored, although they may promote cell death by antagonizing cell survival signals such as NFκB (see below).

While the source of the underlying signal is not known, activation of the GTPase Rac by NGF-induced p75^{NTR}-mediated death signals is an intriguing finding (Xia et al., 1995). Rac appears to have sustained activity, including activating JNK3, following NGF but not brain-derived neurotrophic factor (BDNF) treatment (Xia et al., 1995). The role of JNK is elaborated below, but it is also possible that Rac might mediate Chopper/p75^{NTR}-mediated cell death by regulating other signaling pathways such as the phosphatidylinositol 4,5-bisphosphate (PIP2) pathway (Ren and Schwartz, 1998), which can control ion channel activity.

In addition to these death-inducing proteins, there are other proteins (including the MAGE family member necdin, caveolin, sc-1 and ARMS) that also interact with the Chopper domain (Bilderback et al., 1997; Chittka and Chao, 1999; Kong et al., 2001; Tcherpakov et al., 2002) but which have no known death-related effects.

How does Chopper/p75^{NTR} kill?

Overall, the evidence outlined above supports the original observation that the Chopper domain is pivotal in mediating death signals initiated through p75^{NTR}. However, are the death signals originating from the truncated form of the membrane-attached Chopper domain equivalent to those initiated by 'physiological' p75^{NTR}-activated cell death? Current knowledge regarding the components of the death cascade, from its endpoint back upstream to the p75^{NTR} binding proteins, is summarized in the following section.

Components of the death pathway

Caspases

The end point of an apoptotic cell death cascade is the activation of one or more cysteinyl aspartate-specific proteinases or caspases. Caspases mediate highly specific cleavage events in dying cells which collectively manifest the apoptotic phenotype of DNA fragmentation and the orderly dismantling of the cell (Nicholson and Thornberry, 1997; Thornberry and Lazebnik, 1998). There are more than a dozen apoptosis-promoting caspase proteins, each with its own panel of downstream targets. These are broadly defined as either upstream activator caspases or downstream effector caspases (Nicholson and Thornberry, 1997; Thornberry et al., 1997; Chan and Mattson, 1999). The best characterized upstream apoptotic caspases are caspase-8 (which is activated directly by the TNFR death domain binding protein, TRADD, and then cleaves and activates the downstream effector caspase-3) and caspase-9, the caspase most often activated by stress, irradiation and chemical insults in the so called 'intrinsic' or mitochondrial pathway, which in turn also activates caspase-3 (Thornberry and Lazebnik, 1998). In addition, a number of naturally occurring inhibitors or regulators of caspases (such as crmA) have been characterized and, based on their active-site preference, 'specific' peptide inhibitors have been produced and used to define which caspases may be activated following a particular death stimulus (Nicholson and Thornberry, 1997; Thornberry and Lazebnik, 1998; Ekert et al., 1999).

p75^{NTR}-induced cell death can be inhibited by crmA, synthetic caspase inhibitors or antisense downregulation targeted against caspases-1, -3, -6, -7 and -9 (Coulson et al., 1999a; Gu et al., 1999; Soilu-Hänninen et al., 1999; Mennerick and Zorumski, 2000; Troy et al., 2002). Chopper-mediated death also appears to be dependent on

caspases-9, -7 and -3 (Coulson et al., 2000). The effect of inhibiting these particular caspases is highly consistent between different cell types and various ways of stimulating p75^{NTR}-mediated cell death. In addition, when overexpressed together with p75^{NTR}, NADE induces capsase-3 activity (Mukai et al., 2000; Park et al., 2000) and NRAGE induces caspases-3, -9 and -7 (Salehi et al., 2002). In contrast, caspase 8 does not appear to play an important role in p75^{NTR}-mediated cell death, as it is not activated in the NGF/p75^{NTR}-mediated death of oligodendrocytes, despite its expression (Gu et al., 1999), and is not required for p75^{NTR} to mediate neurotrophin-induced death in hippocampal neurons (Troy et al., 2002). Caspase-8 peptide inhibitors also have no significant effect on Chopper-induced cell death (Coulson and Bartlett, unpublished data). Therefore, it is most probable that p75^{NTR} kills via a caspase-9/-3 pathway, rather than via the TNFR-activated death domain-caspase 8 pathway; this is consistent with the lack of apoptotic function of the p75^{NTR} death domain when chimerically attached to the Fas receptor ectodomain and activated with Fas (Kong et al., 1999). The role of caspases-6 and -7 in the downstream pathway is unclear, although Chopper-induced death can activate calpain (Coulson et al., 2000), which in turn activates caspases-7 and -3 (Chan and Mattson, 1999; Ruiz-Vela et al., 1999). The involvement of calpain in the p75^{NTR} pathway has not been investigated.

The apoptosome

Genetic analysis of apoptosis in the nematode *Caenorhabditis elegans* has revealed that the intrinsic cell death complex is composed of three core interacting components. An equivalent ternary complex was found in mammalian cells, involving apoptotic-protease-activating factor-1 (APAF-1), caspase-9, and Bcl-$_{xL}$, an anti-apoptotic member of the Bcl-2 family (Pan et al., 1998). Caspase-9 and APAF-1 bind to each other in the presence of dATP and cytochrome c released from the mitochondria, an event that leads to caspase-9 activation. Activated caspase-9 in turn cleaves and activates caspase-3 (Li et al., 1997). This complex has become known as the apoptosome.

Activation of the caspase-9/-3 pathway by p75^{NTR} would predict the upstream involvement of APAF-1,

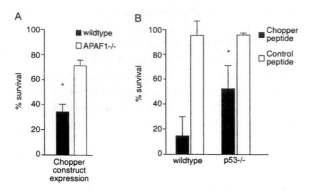

Fig. 2. APAF-1 and p53 are required for p75^{NTR}-mediated cell death. (A) Survival of DRG neurons isolated from E13 wild type and APAF-1 deficient littermates 24 h after microinjection with membrane-linked Chopper construct, as previously described (Coulson et al., 1999a). Chopper expression induced significant cell death in wild type neurons, whereas neurons deficient in APAF-1 were completely protected. *$p < 0.05$ $n = 3 \pm$ sd. (B) Survival of NGF-cultured DRG neurons isolated from newborn wild type and p53 deficient littermates, 20 h after a 2 h treatment with Chopper peptides, as previously described (Coulson et al., 2000). Death-inducing Chopper peptides induced significant cell death in wild type ($+/+$) neurons compared with control peptides. Significantly less death was induced in neurons deficient in p53 (p53$-/-$) *$p < 0.05$ $n = 3 \pm$ sd.

an hypothesis supported by the attenuation of NGF-induced p75^{NTR}-mediated death of hippocampal neurons by downregulation of APAF-1 with anti-sense oligonucleotides (Troy et al., 2002). In support of this, we find that the death of sensory neurons induced by overexpression of constructs encoding either Chopper or an ectodomain-deficient p75^{NTR} is prevented by genetic deletion of APAF-1 (Fig. 2A). This further demonstrates the similarity of the Chopper- and p75^{NTR}-mediated death pathways and supports the concept that the 'intrinsic' caspase-9/-3 death pathway is central to the cascade initiated by p75^{NTR}.

Proteins that regulate the 'intrinsic' cell death pathway such as the caspase-9 interacting inhibitor of apoptosis protein (IAP) and its antagonist DIABLO (Cain et al., 2001) are thus implicated as regulators of the p75^{NTR} death pathway, and there is indeed emerging evidence to support their role in the cascade. The neurotrophin-induced death of hippocampal neurons is mediated by DIABLO and IAP levels (Troy et al., 2002) and NRAGE has been found

to bind to XIAP and augment cell death (Jordan et al., 2001), although whether it does so by concurrently binding to the Chopper domain is still unknown.

Bax, p53 and the JNK pathway

A number of other factors known to influence survival following growth factor withdrawal have been implicated in p75^{NTR}-mediated cell death. However, although growth factor withdrawal appears to be influenced by p75^{NTR} signaling (Barrett and Bartlett 1994), the dominant-negative form of Chopper fails to prevent growth-factor-withdrawal-induced cell death to any extent (Coulson et al., 2000), indicating that these two pathways might actually operate independently.

p75^{NTR}- and NRAGE-mediated death, unlike that induced by growth factor withdrawal, are also not prevented by Bcl-2, the well characterized modulator of the intrinsic mitochondrial death pathway (Coulson et al., 1999a; Soilu-Hänninen et al., 1999; Jordan et al., 2001), although p75^{NTR}-mediated cell death can be prevented by increased expression of the Bcl-2 homologue Bcl-$_{xL}$ (Coulson et al., 1999a), an inhibitory component of the apoptosome (Pan et al., 1998).

The apoptotic protein Bax- and caspase-3-dependent cell death pathway activates JNK, is inhibited by Bcl-2 and Bcl-$_{xL}$, and is active in postmitotic neurons (Xiang et al., 1998; Cregan et al., 1999). While Bax appears to be essential for mediating growth-factor-withdrawal-induced cell death (Easton et al., 1997), it is not known whether it is essential for p75^{NTR}-mediated death. However, given that mice lacking the *Bax* gene do not appear to have a developmental increase in neuron number equivalent to that of mice harboring mutations in genes that are within the p75^{NTR} pathway, it is possible that Bax is not a key component.

p53 is also a common downstream component of signaling cascades that converge on the JNK pathway. This pathway is active at a similar time to p75^{NTR} in sympathetic neurons during development and plays a significant role in growth-factor-withdrawal-induced neuronal death (Aloyz et al., 1998; Pozniak et al., 2000). Chopper-mediated cell death in sensory neurons is diminished but not abolished by genetic deletion of p53 (Fig. 2B). One interpretation of these results is that the p53/Bax pathway is not essential for p75^{NTR}-mediated death but is, nonetheless, activated in a feedback loop and promotes the progress of the apoptotic process.

There are a large number of reports that p75^{NTR}-mediated cell death induces JNK activity (Xia et al., 1995; Casaccia-Bonnefil et al., 1996; Bamji et al., 1998; Ladiwala et al., 1998; Yoon et al., 1998; Roux et al., 2001; Harrington et al., 2002). CEP1347, an inhibitor of MLK which is upstream of JNK, inhibits NGF-induced p75^{NTR}-mediated oligodendrocyte apoptosis (Xia et al., 1995; Yoon et al., 1998), but not killing mediated by overexpression of NRAGE (Salehi et al., 2002). However, dominant-negative JNK constructs are effective in reducing both NRAGE- and p75^{NTR}-induced death (Harrington et al., 2002; Salehi et al., 2002). Surprisingly, gene deletion of c-jun, which prevents growth-factor-withdrawal-induced cell death, does not affect the ability of p75^{NTR} to kill (Palmada et al., 2002). Thus the observed activation of JNK by p75^{NTR}, although strongly implicated, may not always be essential for the induction of apoptosis (Xia et al., 1995), as is the case with TNF-induced apoptosis (Liu et al., 1996). It is also possible that a number of divergent JNK-like pathways are initiated by p75^{NTR} depending on the circumstances (Fanger et al., 1997).

These data suggest that the primary death-mediating pathways underlying growth-factor-withdrawal-induced cell death can contribute to, but are non-essential mechanisms for, promotion of p75^{NTR} induced cell death. Despite this, it is likely that these two independent pathways converge downstream, possibly at the point of activation of the apoptosome.

Potassium channels as mediators of apoptosis

One well characterized contribution to, and possible requirement for, apoptosome-mediated cell death is lowered intracellular K$^+$, since assembly of the active complex is suppressed by normal intracellular concentrations of K$^+$ (Cain et al., 2001). Studies suggest that the latter safeguards the cell against inappropriate formation of the apoptosome complex.

Activation of this complex requires either the extensive release of cytochrome c (to overcome the inhibitory effects of normal intracellular concentrations of K$^+$) or the rapid and extensive loss of intracellular K$^+$ (Cain et al., 2001).

It has been widely reported that the efflux of K$^+$ is a very early apoptotic response in neurons (Yu et al., 1997, 1999; McLaughlin et al., 2001; Yu et al., 2001; Xia et al., 2002), lymphocytes (Hughes et al., 1997; Bortner et al., 2001; Vu et al., 2001) and other cell types (Maeno et al., 2000; Reeves et al., 2002). It is a prerequisite for cell volume loss and is coincident with the appearance of annexin V on the extracellular lipid membrane. It also promotes activation of the apoptosome and subsequent cleavage of pro-caspase 3 to its active form, is associated with activation of JNK phosphorylation pathways, and promotes activation of endonucleases leading to DNA laddering (Hughes et al., 1997; Dallaporta et al., 1998; Wang et al., 1999; Maeno et al., 2000; Cain et al., 2001; Vu et al., 2001; Reeves et al., 2002). In addition, in the context of experimentally induced cell death, retaining normal levels of intracellular K$^+$ by inhibition of K$^+$ efflux, either by pharmacological agents (Yu et al., 1997; Wang et al., 1999; Nietsch et al., 2000; Xia et al., 2002) or by raising the K$^+$ gradient with high extracellular K$^+$ concentrations (Hughes et al., 1997; Yu et al., 1997, 1999), has been shown to block the progress of cell death signals. Interestingly, in situations involving p75^{NTR}-activated death signals, raised extracellular K$^+$ levels are not protective (Courtney et al., 1997; Bamji et al., 1998).

In nonneuronal cells, depending on the apoptotic stimulus, the proposed perpetrators of observed K$^+$ flux encompass a variety of K$^+$ channels, TEA, Ba$^+$ and/or quinine sensitive channels, the Na^{2+}/K$^+$ ATP pump, and voltage-gated K$^+$ channels (Deutsch and Chen, 1993; Avdonin et al., 1998; Wang et al., 1999; Nadeau et al., 2000; Nietsch et al., 2000; Xia et al., 2002). This suggests that, while activation of a channel to promote K$^+$ efflux is a widespread mechanism in pathways signaling cell death, there is a high level of variation in the channels used by apoptosis signaling pathways.

We have recent data that G-protein activated inwardly rectifying potassium (GIRK) channels contribute to both Chopper- and p75^{NTR}-mediated cell death, but not for growth-factor-withdrawal-mediated death. Rather, delayed rectifier (IK) and M-type K^+ flux has been implicated in the latter process (Yu et al., 1997; Xia et al., 2002).

GIRK channels

The inwardly rectifying channels, of which the GIRK channels are a subset (Kir3.0), have an increasing profile in mediating neuronal cell death in vivo. The four identified mammalian GIRK channels form both homo- and hetero-dimeric channel units and each is expressed widely in the nervous system in developing and adult postmitotic neurons (Liao et al., 1996; Chen et al., 1997; Karschin and Karschin, 1997; Signorini et al., 1997; Jelacic et al., 2000; Wickman et al., 2000). GIRK channels play a significant role in controlling resting membrane potential and they modulate excitatory synaptic input in neurons and cardiomyocytes by lowering the resting membrane potential to limit action potential firing (Kitamura et al., 2000; Kobrinsky et al., 2000; Cho et al., 2001; Takigawa and Alzheimer, 2002). GIRK4-deficient mice show impaired spatial learning and memory (Wickman et al., 2000), as well as blunted heart rate responses in cardiomyocytes (Wickman et al., 2000). Mice that harbor a natural point mutation in GIRK2 (*weaver*), or that have a genetic deletion of GIRK2, display a diverse range of neurological defects due to chronic synaptic silencing, including loss of dopaminergic neurons, cerebellar granule cell death and spontaneous seizures (Signorini et al., 1997). In addition, overexpression of Kir1.1 (ROMK1) causes drastic neuronal death due to K^+ loss and synaptic silencing of hippocampal neurons (Nadeau et al., 2000). Since caspase inhibitors can protect dying *weaver* neurons, misregulation of GIRK channels in apoptotic cascades is strongly implicated. This is supported by recent data showing that the degree of caspase activation by naturally occurring p75[NTR]-induced cell death in the embryonic chick retina can be significantly inhibited by blocking endogenous K^+ channel activity with GIRK-channel-specific bee venom.

For the most part, GIRK channels are activated via G-protein β/γ subunits, which bind and stabilize a PIP2-GIRK channel interaction and thus pore opening (Huang et al., 1998; Zhang et al., 1999; Kobrinsky et al., 2000; Lei et al., 2000; Fernandez-Fernandez et al., 2001). However, they may also be activated by the direct action of ethanol (Lewohl et al., 1999; Kobrinsky et al., 2000), by pertussis-insensitive G_q (Stevens et al., 1999), directly by cAMP (Mullner et al., 2000), pH (Schulte and Fakler, 2000), or by intracellular Na^+ (Ho and Murrell-Lagnado, 1999). When regulated by G-protein activation, only minor variations in intracellular K^+ levels tend to occur so as to set an appropriate resting membrane potential. The level of PIP2 controls this; channel inactivation occurs due to rapid PIP2 hydrolysis by phospholipase C (PLC) (Cho et al., 2001; Kobrinsky et al., 2000).

We have evidence that p75[NTR] activation of K^+ channels does not involve $G_{i/o}$-protein activity but does require the PIP2 binding domain for cell death to occur. We have postulated that, unlike neurotransmitter-activated channel opening, p75[NTR]-mediated opening is sustained, resulting in the considerable K^+ efflux observed (Coulson et al., 2003) and thus apoptosome induction. The GTPase Rac, which is an upstream regulator of PIP2 (Ren and Schwartz, 1998; Kost et al., 1999), binds to and activates PIP5K (the kinase that synthesizes PIP2 (see Ren and Schwartz, 1998), and can promote rapid PIP2 production (Hartwig et al., 1995). Since Rac is a downstream component of the p75[NTR]-mediated death pathway, and is induced in a sustained way by p75[NTR] (Harrington et al., 2002), Rac-stimulated PIP2 synthesis is a direct mechanism by which GIRK channel opening could be prolonged. In addition, in the absence of trk phosphorylation—a condition required for p75[NTR]-mediated death signaling—PLC activity (Segal et al., 1996) and thus hydrolysis of PIP2 (Kobrinsky et al., 2000) would be reduced.

A unique pathway

These results demonstrate that the Chopper- and p75[NTR]-mediated cell death cascades are indistinguishable from one other, strongly supporting the relevance of studying the Chopper domain as a tool for understanding the p75[NTR]-mediated death pathway. While p75[NTR]/Chopper-induced activation of the apoptosome by lowered intracellular K^+

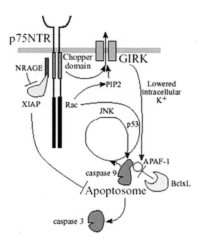

Fig. 3. Summary of what is known about the downstream death-signaling cascade. p75^{NTR} activates intracellular K$^+$ efflux through activation of GIRK channels. Despite identification of a wide variety of interacting proteins, none are known to mediate this activity; however Rac may contribute to the pathway by promoting PIP2 synthesis, and thus sustain channel opening, as well as by activating JNK. Loss of intracellular K$^+$ triggers activation of the apoptosome, and caspase-9 and -3 activity, which is known to be necessary for p75^{NTR}-mediated death. Apoptosome activity is in turn controlled by the anti-apoptotic IAP proteins, one of which, XIAP, is bound to and functionally antagonized by NRAGE. In addition, JNK and p53 participate in the signaling cascade, possibly in a feedback loop promoted by apoptosome activity.

(assisted by the JNK and p53 pathways) is strongly supported as the probable death cascade (Fig. 3), there is no direct signaling messenger linking these molecules, despite the identification of half a dozen death-inducing p75^{NTR}-interacting proteins. The mechanism by which p75^{NTR} activates GIRK channels resulting in lowered intracellular K$^+$ is not via G-protein activation, a process that can actually override the death signal, but it may be promoted by p75^{NTR}-induced, Rac-mediated PIP2 synthesis. These and other lipid-localized signaling molecules (Dobrowsky et al., 1994, 1995) remain likely candidates and are sure to undergo close scrutiny in the future.

Although both p75^{NTR}- and growth-factor-withdrawal-mediated cell death activate K$^+$ channels and promote death via the apoptosome, the pathways have different upstream signals, including those that activate the appropriate K$^+$ channel. Thus they are independent. The p75^{NTR}/Chopper pathway also

displays significant differences from the Fas/death domain-induced cell death pathway, indicating that p75^{NTR} mediates a unique and important signaling cascade.

The role of the death domain in regulating cell death

If the Chopper domain of p75^{NTR} is the key to signaling cell death, what alternate role does the large remaining portion of the intracellular domain play? There are two possibilities: (i) that the tail of the TNFR-like death domain acts to regulate other signaling process functions not related to cell death or survival (e.g., cell cycle control, neurite outgrowth and migration, or myelination); and (ii) that the non-Chopper portion of the intracellular domain acts to regulate Chopper-initiated death signals, including promoting competitive cell survival signals. There is evidence to support both hypotheses, and they are certainly not mutually exclusive.

In the first instance, it is known that many of the proteins that interact with the TNFR-like death domain region of p75^{NTR} have survival promoting functions, including the activation of NFκB (Carter et al., 1996; Ladiwala et al., 1998; Hamanoue et al., 1999; Ye et al., 1999; Khursigara et al., 2001). This aspect of p75^{NTR} signaling is reviewed extensively in this issue (see Carter and/or Barker herein, and elsewhere, see Barrett, 2000; Kaplan and Miller, 2000; Roux and Barker, 2002). However, promotion of survival as an alternative signaling pathway is unlikely to counter death pathways effectively, and thus is unlikely to represent the complete process.

If a second role of the death domain is to thwart induction of death signaling, then cell death should be promoted in its absence. This in fact is the case: removal of the death domain, with retention of the Chopper domain as either the full ectodomain- or ectodomain-deleted p75^{NTR} expression constructs, results in an increased ability to mediate death when overexpressed in sensory neurons as compared with constructs retaining the entire intracellular domain (Coulson et al., 1999a, 2000). The mechanism by which this occurs is speculated on below but remains unknown.

Tertiary structure

One interesting aspect of a number of the p75^{NTR} binding partners (MAGE protein necdin, NADE and NRIF) is their ability to interact with both the death domain region and the Chopper domain, despite possibly spanning over 60 amino acids of the intracellular domain of p75^{NTR} (Casademunt et al., 1999; Mukai et al., 2000; Tcherpakov et al., 2002). One structural characteristic of the p75^{NTR} intracellular domain, highlighted by Carlos and colleagues (Liepinsh et al., 1997), is the flexible linker joining the 'unstructured' juxtamembrane/Chopper domain to the alpha-helical and globular death domain region. It is possible that the intracellular domain can undergo a dramatic tertiary structure movement whereby the globular death domain folds back towards the membrane domain.

The interaction of Ras family member RhoA with the distal end of the p75^{NTR} death domain may support the fold-back model of p75^{NTR} signaling. Ras family members, including RhoA and Rac, contain a conserved sequence, the CAAX box. Here the protein is posttranslationally modified, including reversible attachment of a lipid moiety, so called palmitoylation/myristilation, which localizes the protein to the plasma membrane (Adamson et al., 1992).

p75^{NTR} has also been shown to be palmitoylated, (Barker et al., 1994) and the Chopper domain requires not only membrane attachment for its ability to induce death but also the juxtamembrane cysteine residue that is palmitoylated (Coulson et al., 2000). Presumably this modification localizes p75^{NTR} in an area of the membrane (lipid rafts) where other palmitoylated signal proteins are concentrated and promotes signaling (Harder and Simons, 1997; Brown and London, 1998).

If the membrane-localized Rho proteins were to interact with the distal portion of p75^{NTR}, flexing of the death domain toward the membrane would certainly facilitate the process. This folded structure of the p75^{NTR} intracellular domain might be one configuration that allows survival-promoting binding proteins to compete successfully with death-inducing and similarly palmitoylated NADE and MAGE proteins (Barker and Salehi, 2002) for docking space, thus precluding the activation of death signaling.

Alternatively, binding of death-inducing proteins such as NRIF and NADE to the folded p75^{NTR} intracellular domain may cause the death domain to become sequestered at the membrane and prevent binding of nonpalmitoylated survival-promoting proteins that require free access to a less hydrophilic and cytoplasmic environment for docking.

Binding partners compete

Further evidence for competitive signaling between proteins interacting with either the Chopper domain or the death domain is presented by Chao and colleagues (Khursigara et al., 2001), who found that RIP2 binds to the death domain region of p75^{NTR} and activates NFκB in a protective/survival role. This interaction could be antagonized by TRAF6, which interacts (albeit indirectly) within the Chopper domain. In the context of NGF-induced p75^{NTR} signaling in oligodendrocytes, it activates JNK and promotes cell death (Khursigara et al., 2001). Thus the competitive nature of TRAF6 and RIP2 binding to p75^{NTR} results in the promotion of alternative death or survival pathways, respectively.

In other studies conducted in the absence of NGF, when TRAF6 promotes activation of NFκB (Khursigara et al., 1999; Ye et al., 1999), competitive binding and subsequent signaling of TRAF4 from the Chopper domain was sufficient to override survival signals and promote the death of transfected HEK293 cells (Ye et al., 1999).

These results illustrate that the death domain of p75^{NTR} can play not only a survival-signaling role but may also be involved in competitive interactions at the protein-binding level between p75^{NTR}-mediated survival and death signaling cascades. This supports the hypothesis that the TNFR-like death domain of p75^{NTR} acts to regulate and to modulate death signals initiated at the Chopper domain rather than promoting cell death itself.

Cleavage of p75^{NTR}

Ectodomain cleavage

Results obtained by studying the Chopper domain also led us to question the role of the ectodomain of

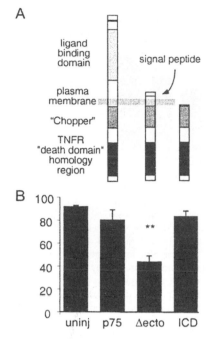

Fig. 4. Opposing roles of the ectodomain and the transmembrane domain of p75^{NTR} in promoting cell death. (a) Diagram of p75^{NTR}-derived expression construct-encoded proteins. (b) Survival of NGF-cultured neurons isolated from newborn mice 16 h after microinjection, as previously described (Coulson et al., 1999a), of the various expression constructs. In the presence of NGF, full-length p75^{NTR} induces no significant cell death compared with uninjected cells (uninj). Removal of the ectodomain (Δecto) promotes the ability of p75^{NTR} to induce cell death, even in the presence of NGF, whereas no cell death is induced by the construct with both the ecto- and transmembrane domains removed and encoding just the intracellular domain (ICD). $**p < 0.01$ $n = 3 \pm$ sd.

p75^{NTR} in regulating death signaling, given that removal of the ectodomain promotes cell death (Fig. 4; Coulson et al., 2000). We hypothesized that p75^{NTR} might be functionally regulated by proteolytic cleavage in a similar way to other type-I transmembrane proteins (Arribas et al., 1996). Supporting this hypothesis was considerable work predating the discovery of the cell death role of p75^{NTR}, which demonstrated that the extracellular domain of endogenous p75^{NTR} was released in vitro and in vivo from Schwann cells by metalloprotease cleavage (DiStefano and Johnson, 1988; Barker et al., 1991; DiStefano et al., 1991, 1993). Importantly, it also showed that cleavage was developmentally regulated (DiStefano et al., 1991), was upregulated

in times of neural injury and degeneration (DiStefano et al., 1993), and, furthermore, that inhibition of metalloprotease activity by 8-hydroxyquinine (8-OH) promoted repair and regeneration after sciatic nerve lesion (DiStefano et al., 1993). These data are consistent with our hypothesis that cleavage of the ectodomain would promote death signaling and reinforces the concept that an ectodomain-free (Chopper-like) p75^{NTR} protein is a physiological form.

What has yet to be identified are the regulatory factors that play a role in determining when p75^{NTR} is cleaved, and the nature of the actual metalloprotease responsible for the cleavage. Although we find that 8-OH does reduce ectodomain cleavage as well as promote Schwann cell survival (Fig. 5) and that NGF has no additional effects in this situation, it has yet to be conclusively determined whether such cleavage is a physiological death-activating event and, if so, whether it is indeed independent of neurotrophin binding as has been reported (DiStefano and Johnson, 1988). It is possible that another condition as well as neurotrophin binding may need to be met for cleavage and promotion of cell death, or perhaps that the pro-neurotrophins might act to promote death (Lee et al., 2001; Beattie et al., 2002) by enhancing the processing of p75^{NTR}, unlike the mature neurotrophins.

Intracellular domain cleavage

We also observed, using an overexpressed, truncated form of p75^{NTR}, that prevention of membrane attachment of the intracellular domain reversed the death-inducing capabilities of this domain (Fig. 4; Coulson et al., 2000). Again, there are a number of transmembrane proteins, e.g., the Alzheimer's disease amyloid protein precursor (APP), Notch and ErbB4, that are not only cleaved to release their ectodomains by metalloproteases but are also subsequently processed by presenilin-like proteases (Zhang et al., 2001) to release the intracellular domain that then participates in further signal transduction actions, including transcriptional regulation (De Strooper et al., 1999; Wilson et al., 1999; Lee et al., 2002b). Together with our colleagues (this issue), we (Shipham, Coulson, Bartlett and

A

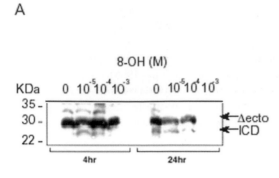

B

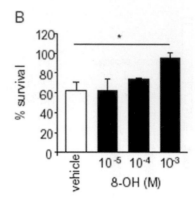

Fig. 5. Inhibition of metalloprotease activity with 8-hydroxyquinine (8-OH) inhibits cleavage of endogenous p75^{NTR} and promotes Schwann cell survival. Rat Schwann cells were cultured in basal medium, as previously described (Soilu-Hänninen et al., 1999), and were treated with increasing doses of 8-OH as previously described (DiStefano et al., 1993). (A) After 24 h in 8-OH, there was a dose-dependent reduction in the intensity of p75^{NTR} immunoreactive protein fragments by Western blot using anti-human p75^{NTR} antibodies (Promega). p75^{NTR} fragments (arrowed) are at equivalent molecular weights to proteins expressed by Δecto and ICD constructs (depicted in Fig. 6). (B) Duplicate cultures in Terasaki plates were also treated with 8-OH, and assayed for viability by MTT reaction and cell counts. In the untreated, basal condition, 40% of Schwann cells die over 24 h, possibly due to endogenous NGF. In the presence of 10^{-3} M 8-OH, the same concentration that shows maximal inhibition of protein fragmentation and promotes sciatic nerve repair (DiStefano et al., 1993), basal cell death was significantly reduced. *$p < 0.05$ $n = 6 \pm$ sem.

Kilpatrick, unpublished data; Kanning et al., 2003; Jung et al., 2003) have observed, in both Schwann cells (Fig. 5) and the PC12 neuroblastoma cell line, an endogenous intracellular immunoreactive protein fragment of equivalent size to that expressed from the cDNA construct expressing only the intracellular domain (ICD; Fig. 4), as well as one which corresponds to p75^{NTR} that lacks the ectodomain (Δecto; Fig. 4). The functional significance of these cleaved forms is predicated by the phenotypes of the cells that express construct forms of them, but remains to be fully investigated.

Interaction of synaptic activity and neurotrophin signaling controls neuronal survival

During development, growth factors such as the neurotrophins are generally believed to be present at the axonal target in limited quantities, resulting in a competitive environment and thus the death of many neurons (over 50% of some populations (Oppenheim, 1991)). This developmental cell death only occurs once synaptic connections are made, as inhibition of synapse formation can block developmental cell death (Verhage et al., 2000), Additionally, once synaptic connections are sought, many neurons

become dependent on synaptic transmission for survival, indicating that activity also plays a trophic role (Terrado et al., 2001; Petreanu and Alvarez-Buylla, 2002). Similar to neurotrophin dependence, there appears to be a window of activity dependence in some neuronal populations (Mennerick and Zorumski, 2000) that diminishes as the neurons mature, but also has the potential to be reactivated in situations of neurodegeneration.

Crosstalk between neurotrophins and neurotransmission

Removal of afferent input, mimicked by blocking action potentials (Fawcett et al., 1984; Lipton, 1986; Catsicas et al., 1992) or inhibiting excitatory cell receptors (Gould et al., 1994), results in neuronal death. Indeed, alternations in physiological patterns of activity render the neurons susceptible to death (Mennerick and Zorumski, 2000). It is well established that neurons deprived of synaptic input during development or in adulthood can be rescued from death by the increased supply of growth factors. Notably, it is in these same disease or injury situations that p75^{NTR}-mediated cell death is often observed. For example, cell death in vivo as

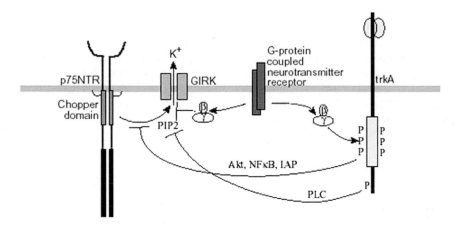

Fig. 6. Model of GIRK channels as the control point for cell survival decisions. The interaction of neurotrophin receptor signaling with neurotransmitter receptor signaling at the GIRK channel explains many experimental data showing that an increase in one can often compensate for a lack of activity in the other. GIRK channels are traditionally opened by G-protein β/γ subunits and PIP2 following neurotransmitter receptor activity. In the absence of neurotransmission, activated p75[NTR] stimulates sustained GIRK channel opening, initiating a cell death cascade. Stimulation of G-protein-coupled receptors can override p75[NTR]-mediated channel opening, while G-protein activity can also activate Trk phosphorylation. Trk signaling is known to activate survival signals (Akt, NFκB and inhibitors of apoptosis protein [IAPs]), which in turn inhibit p75[NTR] death signaling. In addition, phospholipase C (PLC) activation by Trk promotes PIP2 hydrolysis and thus may contribute to GIRK channel inactivation and inhibition of p75[NTR]-mediated cell death.

promoted by axotomy or experimentally induced encephalomyelitis can be ameliorated by the application of either cytokines, such as leukaemia inhibitory factor (Cheema et al., 1994; Butzkueven et al., 2002), neurotrophins (Sendtner et al., 1992a,b; Lowry and Cheema, 2000; Giehl et al., 2001), or downregulation of p75[NTR] by antisense oligonucleotides (Cheema et al., 1996; Soilu-Hänninen et al., 2000; Lowry et al., 2001a). Indeed, there is such a wealth of data demonstrating the efficacy of neurotrophic factors in preventing neuronal death in animal models that their possible clinical application remains an active area of research (Semkova and Krieglstein, 1999).

The converse is also true: in the absence of neurotrophins, increased afferent innervation increases the number of neurons surviving in vivo (Linden, 1994; Sherrard and Bower, 1998) and increased activity of neurotransmitter receptors (Lachica et al., 1998; Belluardo et al., 2000; Pereira et al., 2001) or the downstream derivative cAMP (Hanson et al., 1998) can compensate for the lack of growth factor signaling. However, there has been little direct evidence to determine whether the observed trophic support provided by neurotransmission is due to activity itself or to activity-promoted release, or

regulation of, the neurotrophins and their receptor-mediated signaling pathways.

In addition to receptor tyrosine kinase-mediated survival cascades (Kaplan and Miller, 2000), neurotrophins can promote and modify synaptic transmission (see Schinder and Poo, 2000; Blum et al., 2002; Yang et al., 2002). Synaptic activity can also regulate expression levels of neurotrophins and further facilitate growth factor signaling, suggesting there is synergism between these two cascades (Schinder and Poo, 2000).

A 'calcium set point' hypothesis has been formulated to explain the relationship between activity and survival, proposing that synaptic activity regulates the intracellular concentration of Ca^{2+}, which when low can promote apoptosis that is overcome by the presence of neurotrophins (Mennerick and Zorumski, 2000). This was supported by the ability of depolarizing concentrations of external K^+, which is also thought to affect internal Ca^{2+} levels, to protect against some forms of cell death. Both p75[NTR] and TrkA receptors have been shown to potentiate Ca^{2+} flux (Jiang et al., 1999) and intracellular Ca^{2+} is important for Chopper-mediated cell death (Coulson et al., 2003). Despite this and other evidence that developmental changes in Ca^{2+} are likely to be a contributing factor in susceptibility to neuronal death (Mennerick and Zorumski, 2000), this hypothesis cannot always explain experimental results as neither NGF treatment nor NGF withdrawal normally cause changes in Ca^{2+} (Franklin et al., 1995). Rather, the results often point more strongly to the regulation of K^+ being the key factor in the regulation of apoptotic cascades (see above), which in turn may fit better with the ability of neurotrophins to compensate for the loss of synaptic activity.

GIRK channels as cell survival gatekeepers

Our recent observation that GIRK channels appear important in relaying the $p75^{NTR}$ death cascade, additionally suggests that these channels might act in a dual role, controlling both synaptic activity and cell death. This is the first evidence that control of cell survival decisions is indeed due to direct molecular interplay between the neurotrophic factor receptors and components of synaptic transmission.

We have proposed a model (Fig. 6) based on our recent data in which, in the absence or deregulation of neurotransmitter receptor activity, activated $p75^{NTR}$ stimulates GIRK channel opening, depleting intracellular K^+ and activating the apoptosome death cascade. More importantly, in the presence of active G-proteins, either by stimulated neurotransmission or by provision of $\beta\gamma$ subunits, traditional GIRK channel activation could override an activated $p75^{NTR}$ death pathway and promote survival. This model explains the fact that the exogenous provision of nicotine, acetylcholine or GABA agonists is neuroprotective (Belluardo et al., 2000; Pereira et al., 2001). On another level, synaptic activity in the form of adenosine or adenylate cyclase activity can, in the absence of neurotrophin, activate the kinase domain of trk and promote survival signaling (Lee et al., 2002a). Activation of the $G_{i/o}$-coupled adenosine receptor could, of course, also directly inhibit any death-signaling role of $p75^{NTR}$/GIRK channels in this context.

Signals initiated by growth factor signaling through trks are, by and large, survival-promoting cascades. Activation and upregulation of proteins such as NFκB, Bcl-$_{xL}$ Akt and IAPs by trk signaling can in turn have direct anti-apoptotic functions, including regulation of the $p75^{NTR}$ death-signaling pathway (Kaplan and Miller, 2000). Such signals can also be mediated directly through $p75^{NTR}$ alone (Carter et al., 1996; Khursigara et al., 1999; Roux et al., 2001) or in conjunction with the Trk receptor in its traditional 'low-affinity' role (Berg et al., 1991). As hypothesized above, survival signaling through $p75^{NTR}$ may also occur at the intracellular domain in a way that directly competes with death signals initiated through $p75^{NTR}$. Any of these mechanisms could explain how neurotrophins actively override $p75^{NTR}$-mediated cell death permitted by the lack of synaptic transmission.

Recently, another link between neurotransmission and neurotrophins has been described. BDNF, like NGF (Lockhart et al., 1997; Malcangio et al., 1997), can promote activation of neurotransmission through muscarinic receptors in efferent cardiomyocytes (Rutherford et al., 1997; Yang et al., 2002). Indeed, Birren and colleagues have demonstrated that $p75^{NTR}$ is required for this cascade (Yang et al., 2002), which appears to be distinct from the death-signaling pathway evoked by BDNF in sympathetic neurons (Bamji et al., 1998). This finding further demonstrates that there is an interconnection in both synaptic- and $p75^{NTR}$-mediated cascades.

Summary

The data presented here demonstrate the existence of crosstalk between synaptic activity and neurotrophic signals. Already it has been shown that neurotransmitters signal through Trk receptors (Lee et al., 2002b), that neurotrophin receptors can mediate neurotransmission (Yang et al., 2002), and that death signaling through $p75^{NTR}$ requires the absence of neurotransmission and recruits a regulatory component of synaptic activity to act in the death pathway. It appears that the role of $p75^{NTR}$ in assisting Trk-mediated survival signaling may be surpassed by its fundamental role of stimulating signals that are directly interlinked with neurotransmission.

The fact that $p75^{NTR}$ signaling is so significantly affected by synaptic input means that the $p75^{NTR}$ research community must take care when designing and interpreting experiments aimed at characterizing and understanding the regulation of p75^{NTR} signaling. It is possible that one reason that p75^{NTR}-mediated cell death has often proved difficult to promote in culture is that inadvertently inhibitory pathways have been activated, or an appropriate environment for p75^{NTR} activation and cell death activity has not been adequately reconstituted. While culture systems have proved useful tools to piece together much of the p75^{NTR} death pathway, it will require more sophisticated methods, which acknowledge that synaptic activity can impinge on the system, to tease out the more subtle controls and regulatory processes involved in both activating and transmitting death signaling through p75^{NTR}.

Conclusion

There is an increasing body of evidence, as outlined above, that an interplay exists between neurotrophin signaling and signals resulting from synaptic activity. One of these links is the K$^+$ channel, which not only regulates synaptic activity but is also required for and can regulate the ability of p75^{NTR} to signal cell death. The discovery that a K$^+$ channel is a component of the p75^{NTR} death pathway assists in cementing the most likely steps in the cascade, these being the loss of intracellular K$^+$, leading to activation of the apoptosome and JNK, caspase activation and cellular dismantling. Frustratingly, the identification of binding proteins has not yet helped to reveal all the components of the pathway; however, there are some tantalizing leads to follow in this area. A strong case has been put forward that the Chopper domain is responsible for propagating the p75^{NTR} death cascade, given that to date the downstream signals emanating from Chopper correspond overwhelmingly with those promoting cell death and stemming from the full p75^{NTR} intracellular domain. Various proposals for the functionality of the remainder of the p75^{NTR} molecule have also been suggested, including cleavage within both the ectodomain and the transmembrane/intracellular domain as regulatory events or possibly fundamental steps required for p75^{NTR} activity. These and other p75^{NTR}-regulating processes are sure to provide the focus of considerable research in the near future.

Acknowledgments

Supported by the Motor Neurone Disease Research Institute of Australia, the National Health and Medical Research Council of Australia and the Multiple Sclerosis Society of Australia. We would like to thank R. Tweedale and I. H. Duncan for editorial assistance.

Abbreviations

APAF-1	apoptosis protease activating factor-1
ARMS	ankyrin repeat-rich membrane spanning protein
BDNF	brain-derived neurotrophic factor
DRG	dorsal root ganglia
FAP-1	Fas-associated protein-1
GIRK	G-protein activated inwardly rectifying K$^+$ channel
IAP	inhibitor of apoptosis
JNK	c-jun activating kinase
NADE	neutrophin receptor associating death effector
NGF	Nerve growth factor
NFκB	nuclear factor κB
NRAGE	neurotrophin receptor interacting MAGE protein
NRIF	neurotrophin receptor interacting factor
PIP2	phosphatidylinositol 4,5-bisphosphate
PLC	phospholipase C
p75^{NTR}	p75 neurotrophin receptor
RIP2	receptor interacting protein-2
TNFR	tumor necrosis factor receptor
TRAF	TNF receptor associating factor
XIAP	X-linked inhibitor of apoptosis

References

Adamson, P., Marshall, C.J., Hall, A. and Tilbrook, P.A. (1992) Post-translational modifications of p21rho proteins. J. Biol. Chem., 267: 20033–20038.

Aloyz, R.S., Bamji, S.X., Pozniak, C.D., Toma, J.G., Atwal, J., Kaplan, D.R. and Miller, F.D. (1998) p53 is essential for developmental neuron death as regulated by the TrkA and p75 neurotrophin receptors. J. Cell Biol., 143: 1691–1703.

Andsberg, G., Kokaia, Z. and Lindvall, O. (2001) Upregulation of p75 neurotrophin receptor after stroke in mice does not

56

contribute to differential vulnerability of striatal neurons. Exp. Neurol., 169: 351–363.

Arribas, J., Coodly, L., Vollmer, P., Kishimoto, T.K., Rose-John, S. and Massague, J. (1996) Diverse cell surface protein ectodomains are shed by a system sensitive to metalloprotease inhibitors. J. Biol. Chem., 271: 11376–11382.

Avdonin, V., Kasuya, J., Ciorba, M.A., Kaplan, B., Hoshi, T. and Iverson, L. (1998) Apoptotic proteins Reaper and Grim induce stable inactivation in voltage-gated K$^+$ channels. Proc. Natl. Acad. Sci. USA, 95: 11703–11708.

Bamji, S.X., Majdan, M., Pozniak, C.D., Belliveau, D.J., Aloyz, R., Kohn, J., Causing, C.G. and Miller, F.D. (1998) The p75 neurotrophin receptor mediates neuronal apoptosis and is essential for naturally occurring sympathetic neuron death. J. Cell Biol., 140: 911–923.

Barker, P.A., Barbee, G., Misko, T.P. and Shooter, E.M. (1994) The low affinity neurotrophin receptor, p75LNTR, is palmitoylated by thioester formation through cysteine 279. J. Biol. Chem., 269: 30645–30650.

Barker, P.A., Miller, F.D., Large, T.H. and Murphy, R.A. (1991) Generation of the truncated form of the nerve growth factor receptor by rat Schwann cells. Evidence for post-translational processing. J. Biol. Chem., 266: 19113–19119.

Barker, P.A. and Salehi, A. (2002) The MAGE proteins: emerging roles in cell cycle progression, apoptosis, and neurogenetic disease. J. Neurosci. Res., 67: 705–712.

Barrett, G.L. (2000) The p75 neurotrophin receptor and neuronal apoptosis. Prog. Neurobiol., 61: 205–229.

Barrett, G.L. and Bartlett, P.F. (1994) The p75 nerve growth factor receptor mediates survival or death depending on the stage of sensory neuron development. Proc. Natl. Acad. Sci. USA, 91: 6501–6505.

Beattie, M.S., Harrington, A.W., Lee, R., Kim, J.Y., Boyce, S.L., Longo, F.M., Bresnahan, J.C., Hempstead, B.L. and Yoon, S.O. (2002) ProNGF induces p75-mediated death of oligodendrocytes following spinal cord injury. Neuron, 36: 375–386.

Belluardo, N., Mudo, G., Blum, M. and Fuxe, K. (2000) Central nicotinic receptors, neurotrophic factors and neuroprotection. Behav. Brain Res., 113: 21–34.

Berg, M.M., Sternberg, D.W., Hempstead, B.L. and Chao, M.V. (1991) The low-affinity p75 nerve growth factor (NGF) receptor mediates NGF-induced tyrosine phosphorylation. Proc. Natl. Acad. Sci. USA, 88: 7106–7110.

Bilderback, T.R., Gazula, V.R., Lisanti, M.P. and Dobrowsky, R.T. (1999) Caveolin interacts with Trk A and p75(NTR) and regulates neurotrophin signaling pathways. J. Biol. Chem., 274: 257–263.

Bilderback, T.R., Grigsby, R.J. and Dobrowsky, R.T. (1997) Association of p75(NTR) with caveolin and localization of neurotrophin-induced sphingomyelin hydrolysis to caveolae. J. Biol. Chem., 272: 10922–10927.

Blum, R., Kafitz, K.W. and Konnerth, A. (2002) Neurotrophin-evoked depolarization requires the sodium channel Na(V)1.9. Nature, 419: 687–693.

Bortner, C.D., Gomez-Angelats, M. and Cidlowski, J.A. (2001) Plasma membrane depolarization without repolarization is an early molecular event in anti-Fas-induced apoptosis. J. Biol. Chem., 276: 4304–4314.

Brown, A.L. and Kay, G.F. (1999) Bex1, a gene with increased expression in parthenogenetic embryos, is a member of a novel gene family on the mouse X chromosome. Hum. Mol. Genet., 8: 611–619.

Brown, D.A. and London, E. (1998) Functions of lipid rafts in biological membranes. Annu. Rev. Cell. Dev. Biol., 14: 111–136.

Butzkueven, H., Zhang, J.G., Soilu-Hanninen, M., Hochrein, H., Chionh, F., Shipham, K.A., Emery, B., Turnley, A.M., Petratos, S., Ernst, M., Bartlett, P.F. and Kilpatrick, T.J. (2002) LIF receptor signaling limits immune-mediated demyelination by enhancing oligodendrocyte survival. Nat. Med., 8: 613–619.

Cain, K., Langlais, C., Sun, X.M., Brown, D.G. and Cohen, G.M. (2001) Physiological concentrations of K$^+$ inhibit cytochrome c-dependent formation of the apoptosome. J. Biol. Chem., 276: 41985–41990.

Carter, B.D., Kaltschmidt, C., Kaltschmidt, B., Offenhauser, N., Bohm-Matthaei, R., Baeuerle, P.A. and Barde, Y.A. (1996) Selective activation of NF-kappa B by nerve growth factor through the neurotrophin receptor p75. Science, 272: 542–545.

Casaccia-Bonnefil, P., Carter, B.D., Dobrowsky, R.T. and Chao, M.V. (1996) Death of oligodendrocytes mediated by the interaction of nerve growth factor with its receptor p75. Nature, 383: 716–719.

Casaccia-Bonnefil, P., Gu, C., Khursigara, G. and Chao, M.V. (1999) p75 neurotrophin receptor as a modulator of survival and death decisions. Microsc. Res. Tech., 45: 217–224.

Casademunt, E., Carter, B.D., Benzel, I., Frade, J.M., Dechant, G. and Barde, Y.A. (1999) The zinc finger protein NRIF interacts with the neurotrophin receptor p75(NTR) and participates in programmed cell death. EMBO J., 18: 6050–6061.

Catsicas, M., Pequignot, Y. and Clarke, P.G. (1992) Rapid onset of neuronal death induced by blockade of either axoplasmic transport or action potentials in afferent fibers during brain development. J. Neurosci., 12: 4642–4650.

Chan, S.L. and Mattson, M.P. (1999) Caspase and calpain substrates: roles in synaptic plasticity and cell death. J. Neurosci. Res., 58: 167–190.

Cheema, S.S., Barrett, G.L. and Bartlett, P.F. (1996) Reducing p75 nerve growth factor receptor levels using antisense oligonucleotides prevents the loss of axotomized sensory neurons in the dorsal root ganglia of newborn rats. J. Neurosc. Res., 46: 239–245.

Cheema, S.S., Richards, L., Murphy, M. and Bartlett, P.F. (1994) Leukemia inhibitory factor prevents the death of axotomised sensory neurons in the dorsal root ganglia of the neonatal rat. J. Neurosci. Res., 37: 213–218.

Chen, S.C., Ehrhard, P., Goldowitz, D. and Smeyne, R.J. (1997) Developmental expression of the GIRK family of inward rectifying potassium channels: implications for abnormalities in the weaver mutant mouse. Brain Res., 778: 251–264.

Chittka, A. and Chao, M.V. (1999) Identification of a zinc finger protein whose subcellular distribution is regulated by serum and nerve growth factor. Proc. Natl. Acad. Sci. USA, 96: 10705–10710.

Cho, H., Nam, G.B., Lee, S.H., Earm, Y.E. and Ho, W.K. (2001) Phosphatidylinositol 4,5-bisphosphate is acting as a signal molecule in alpha(1)-adrenergic pathway via the modulation of acetylcholine-activated $K(+)$ channels in mouse atrial myocytes. J. Biol. Chem., 276: 159–164.

Coulson, E.J., Reid, K., Baca, M., Shipham, K., Hulett, S.M., Kilpatrick, T.J. and Bartlett, P.F. (2000) Chopper, a new death domain of the p75 neurotrophin receptor which mediates rapid neuronal cell death. J. Biol. Chem., 275: 30537–30545.

Coulson, E.J., Reid, K., Barrett, G.L. and Bartlett, P.F. (1999a) p75NTR mediated neuronal death is promoted by Bcl-2 and prevented by Bcl-xL. J.f Biol. Chem., 274: 163787–163791.

Coulson, E.J., Reid, K. and Bartlett, P.F. (1999b) Signaling of neuronal cell death by the p75NTR neurotrophin receptor. Mol. Neurobiol., 20: 29–44.

Courtney, M.J., Akerman, K.E. and Coffey, E.T. (1997) Neurotrophins protect cultured cerebellar granule neurons against the early phase of cell death by a two-component mechanism. J. Neurosci., 17: 4201–4211.

Cregan, S.P., MacLaurin, J.G., Craig, C.G., Robertson, G.S., Nicholson, D.W., Park, D.S. and Slack, R.S. (1999) Bax-dependent caspase-3 activation is a key determinant in p53-induced apoptosis in neurons. J. Neurosci., 19: 7860–7869.

Dallaporta, B., Hirsch, T., Susin, S.A., Zamzami, N., Larochette, N., Brenner, C., Marzo, I. and Kroemer, G. (1998) Potassium leakage during the apoptotic degradation phase. J. Immunol., 160: 5605–5615.

Davey, F. and Davies, A.M. (1998) TrkB signalling inhibits p75-mediated apoptosis induced by nerve growth factor in embryonic proprioceptive neurons. Curr. Biol., 8: 915.

De Strooper, B., Annaert, W., Cupers, P., Saftig, P., Craessaerts, K., Mumm, J.S., Schroeter, E.H., Schrijvers, V., Wolfe, M.S., Ray, W.J., Goate, A. and Kopan, R. (1999) A presenilin-1-dependent gamma-secretase-like protease mediates release of Notch intracellular domain. Nature, 398: 518–522.

Deutsch, C. and Chen, L.Q. (1993) Heterologous expression of specific K^+ channels in T lymphocytes: functional consequences for volume regulation. Proc. Natl. Acad. Sci. USA, 90: 10036–10040.

DiStefano, P.S., Chelsea, D.M., Schick, C.M. and McKelvy, J.F. (1993) Involvement of a metalloprotease in low-affinity

nerve growth factor receptor truncation: inhibition of truncation in vitro and in vivo. J. Neurosci., 13: 2405–2414.

DiStefano, P.S., Clagett-Dame, M., Chelsea, D.M. and Loy, R. (1991) Developmental regulation of human truncated nerve growth factor receptor. Ann. Neurol., 29: 13–20.

DiStefano, P.S. and Johnson, E.M., Jr. (1988) Identification of a truncated form of the nerve growth factor receptor. Proc. Natl. Acad. Sci. USA, 85: 270–274.

Dobrowsky, R.T., Jenkins, G.M. and Hannun, Y.A. (1995) Neurotrophins induce sphingomyelin hydrolysis. Modulation by co-expression of p75NTR with Trk receptors. J. Biol. Chem., 270: 22,135–22,142.

Dobrowsky, R.T., Werner, M.H., Castellino, A.M., Chao, M.V. and Hannun, Y.A. (1994) Activation of the sphingomyelin cycle through the low-affinity neurotrophin receptor. Science, 265: 1596–1599.

Easton, R.M., Deckwerth, T.L., Parsadanian, A.S. and Johnson, E.M., Jr. (1997) Analysis of the mechanism of loss of trophic factor dependence associated with neuronal maturation: a phenotype indistinguishable from Bax deletion. J. Neurosci., 17: 9656–9666.

Ekert, P.G., Silke, J. and Vaux, D.L. (1999) Inhibition of apoptosis and clonogenic survival of cells expressing crmA variants: optimal caspase substrates are not necessarily optimal inhibitors. EMBO J., 18: 330–338.

Fanger, G.R., Gerwins, P., Widmann, C., Jarpe, M.B. and Johnson, G.L. (1997) MEKKs, GCKs, MLKs, PAKs, TAKs, and tpls: upstream regulators of the c-jun amino-terminal kinases? Curr. Opin. Genet. Dev., 7: 67–74.

Fawcett, J.W., O'Leary, D.D. and Cowan, W.M. (1984) Activity and the control of ganglion cell death in the rat retina. Proc. Natl. Acad. Sci. USA, 81: 5589–5593.

Fernandez-Fernandez, J.M., Abogadie, F.C., Milligan, G., Delmas, P. and Brown, D.A. (2001) Multiple pertussis toxin-sensitive G-proteins can couple receptors to GIRK channels in rat sympathetic neurons when expressed heterologously, but only native G(i)-proteins do so in situ. Eur. J. Neurosci., 14: 283–292.

Frade, J.M. and Barde, Y.A. (1998) Microglia derived nerve growth factor causes cell death in the developing retina. Neuron, 20: 35–41.

Frade, J.M., Rodriguez-Tebar, A. and Barde, Y.A. (1996) Induction of cell death by endogenous nerve growth factor through its p75 receptor. Nature, 383: 166–168.

Franklin, J.L., Sanz-Rodriguez, C., Juhasz, A., Deckwerth, T.L. and Johnson, E.M., Jr. (1995) Chronic depolarization prevents programmed death of sympathetic neurons in vitro but does not support growth: requirement for Ca^{2+} influx but not Trk activation. J. Neurosci., 15: 643–664.

Gage, F.H., Batchelor, P., Chen, K.S., Chin, D., Higgins, G.A., Koh, S., Deputy, S., Rosenberg, M.B., Fischer, W. and Bjorklund, A. (1989) NGF receptor reexpression and NGF-mediated cholinergic neuronal hypertrophy in the damaged adult neostriatum. Neuron, 2: 1177–1184.

Giehl, K.M., Rohrig, S., Bonatz, H., Gutjahr, M., Leiner, B., Bartke, I., Yan, Q., Reichardt, L.F., Backus, C., Welcher, A.A., Dethleffsen, K., Mestres, P., Meyer, M. (2001) Endogenous brain-derived neurotrophic factor and neurotrophin-3 antagonistically regulate survival of axotomized corticospinal neurons in vivo. J. Neurosci., 21: 3492–3502.

Gould, E., Cameron, H.A. and McEwen, B.S. (1994) Blockade of NMDA receptors increases cell death and birth in the developing rat dentate gyrus. J. Comp. Neurol., 340: 551–565.

Greferath, U., Mallard, C., Roufail, E., Rees, S.M., Barrett, G.L. and Bartlett, P.F. (2002) Expression of the p75 neurotrophin receptor by striatal cholinergic neurons following global ischemia in rats is associated with neuronal degeneration. Neurosci. Lett., 332: 57–60.

Gu, C., Casaccia-Bonnefil, P., Srinivasan, A. and Chao, M.V. (1999) Oligodendrocyte apoptosis mediated by caspase activation. J. Neurosci., 19: 3043–3049.

Hamanoue, M., Middleton, G., Wyatt, S., Jaffray, E., Hay, R.T. and Davies, A.M. (1999) p75-mediated NF-kappaB activation enhances the survival response of developing sensory neurons to nerve growth factor. Mol. Cell. Neurosci., 14: 28–40.

Hanson, M.G., Jr., Shen, S., Wiemelt, A.P., McMorris, F.A. and Barres, B.A. (1998) Cyclic AMP elevation is sufficient to promote the survival of spinal motor neurons in vitro. J. Neurosci., 18: 7361–7371.

Harder, T. and Simons, K. (1997) Caveolae, DIGs, and the dynamics of sphingolipid-cholesterol microdomains. Curr. Opin. Cell Biol., 9: 534–542.

Harrington, A.W., Kim, J.Y. and Yoon, S.O. (2002) Activation of Rac GTPase by p75 is necessary for c-jun N-terminal kinase-mediated apoptosis. J. Neurosci., 22: 156–166.

Hartwig, J.H., Bokoch, G.M., Carpenter, C.L., Janmey, P.A., Taylor, L.A., Toker, A. and Stossel, T.P. (1995) Thrombin receptor ligation and activated Rac uncap actin filament barbed ends through phosphoinositide synthesis in permeabilized human platelets. Cell, 82: 643–653.

Ho, I.H. and Murrell-Lagnado, R.D. (1999) Molecular mechanism for sodium-dependent activation of G protein-gated K$^+$ channels. J. Physiol., 520(Pt 3): 645–651.

Huang, C.L., Feng, S. and Hilgemann, D.W. (1998) Direct activation of inward rectifier potassium channels by PIP2 and its stabilization by Gbetagamma. Nature, 391: 803–806.

Hughes, F.M., Jr., Bortner, C.D., Purdy, G.D. and Cidlowski, J.A. (1997) Intracellular K$^+$ suppresses the activation of apoptosis in lymphocytes. J. Biol. Chem., 272: 30567–30576.

Irie, S., Hachiya, T., Rabizadeh, S., Maruyama, W., Mukai, J., Li, Y., Reed, J.C., Bredesen, D.E. and Sato, T.A. (1999) Functional interaction of Fas-associated phosphatase-1 (FAP-1) with p75(NTR) and their effect on NF-kappaB activation. FEBS Lett., 460: 191–198.

Jelacic, T.M., Kennedy, M.E., Wickman, K. and Clapham, D.E. (2000) Functional and biochemical evidence for G-protein-gated inwardly rectifying K$^+$ (GIRK) channels composed of GIRK2 and GIRK3. J. Biol. Chem., 275: 36211–36216.

Jiang, H., Takeda, K., Lazarovici, P., Katagiri, Y., Yu, Z.X., Dickens, G., Chabuk, A., Liu, X.W., Ferrans, V. and Guroff, G. (1999) Nerve growth factor (NGF)-induced calcium influx and intracellular calcium mobilization in 3T3 cells expressing NGF receptors. J. Biol. Chem., 274: 26209–26216.

Jordan, B.W., Dinev, D., LeMellay, V., Troppmair, J., Gotz, R., Wixler, L., Sendtner, M., Ludwig, S. and Rapp, U.R. (2001) Neurotrophin receptor-interacting mage homologue is an inducible inhibitor of apoptosis protein-interacting protein that augments cell death. J. Biol. Chem., 276: 39985–39989.

Jung, K.M., Tan, S., Landman, N., Petrova, K., Murray, S., Lewis, R., Kim, P.K., Kim, D.S., Ryu, S.H., Chao, M.V. and Kim, T.W. (2003) Regulated intramembrane proteolysis of the p75 neurotrophin receptor modulates its association with the TrKA receptor. J. Biol. Chem.

Kanning, K.C., Hudson, M., Amieux, P.S., Wiley, J.C., Bothwell, M. and Schecterson, L.C. (2003) Proteolytic processing of the p75 neurotrophin and two homologs generates C-terminal fragments with signaling capability. J. Neurosci., 23: 5425–5436.

Kaplan, D.R. and Miller, F.D. (2000) Neurotrophin signal transduction in the nervous system. Curr. Opin. Neurobiol., 10: 381–391.

Karschin, C. and Karschin, A. (1997) Ontogeny of gene expression of Kir channel subunits in the rat. Mol. Cell. Neurosci., 10: 131–148.

Khursigara, G., Bertin, J., Yano, H., Moffett, H., DiStefano, P.S. and Chao, M.V. (2001) A prosurvival function for the p75 receptor death domain mediated via the caspase recruitment domain receptor-interacting protein 2. J. Neurosci., 21: 5854–5863.

Khursigara, G., Orlinick, J.R. and Chao, M.V. (1999) Association of the p75NTR neurotrophin receptor with TRAF 6. J. Biol. Chem., 274: 2597–2600.

Kitamura, H., Yokoyama, M., Akita, H., Matsushita, K., Kurachi, Y. and Yamada, M. (2000) Tertiapin potently and selectively blocks muscarinic K(+) channels in rabbit cardiac myocytes. J. Pharmacol. Exp. Ther., 293: 196–205.

Kobrinsky, E., Mirshahi, T., Zhang, H., Jin, T. and Logothetis, D.E. (2000) Receptor-mediated hydrolysis of plasma membrane messenger PIP2 leads to K$^+$-current desensitization. Nat. Cell Biol., 2: 507–514.

Kokaia, Z., Andsberg, G., Martinez-Serrano, A. and Lindvall, O. (1998) Focal cerebral ischemia in rats induces expression of p75 neurotrophin receptor in resistant striatal cholinergic neurons. Neuroscience, 84: 1113–1125.

Kong, H., Boulter, J., Weber, J.L., Lai, C. and Chao, M.V. (2001) An evolutionarily conserved transmembrane protein that is a novel downstream target of neurotrophin and ephrin receptors. J. Neurosci., 21: 176–185.

Kong, H., Kim, A.H., Orlinick, J.R. and Chao, M.V. (1999) A comparison of the cytoplasmic domains of the Fas receptor and the p75 neurotrophin receptor. Cell Death Differ., 6: 1133–1142.

Kost, B., Lemichez, E., Spielhofer, P., Hong, Y., Tolias, K., Carpenter, C. and Chua, N.H. (1999) Rac homologues and compartmentalized phosphatidylinositol 4, 5-bisphosphate act in a common pathway to regulate polar pollen tube growth. J. Cell Biol., 145: 317–330.

Lachica, E.A., Kato, B.M., Lippe, W.R. and Rubel, E.W. (1998) Glutamatergic and GABAergic agonists increase $[Ca^{2+}]i$ in avian cochlear nucleus neurons. J. Neurobiol., 37: 321–337.

Ladiwala, U., Lachance, C., Simoneau, S.J., Bhakar, A., Barker, P.A. and Antel, J.P. (1998) p75 neurotrophin receptor expression on adult human oligodendrocytes: signaling without cell death in response to NGF. J. Neurosc., 18: 1297–1304.

Lee, F.S., Rajagopal, R., Kim, A.H., Chang, P.C. and Chao, M.V. (2002a) Activation of Trk neurotrophin receptor signaling by pituitary adenylate cyclase-activating polypeptides. J. Biol. Chem., 277: 9096–9102.

Lee, H.J., Jung, K.M., Huang, Y.Z., Bennett, L.B., Lee, J.S., Mei, L. and Kim, T.W. (2002b) Presenilin-dependent gamma-secretase-like intramembrane cleavage of ErbB4. J. Biol. Chem., 277: 6318–6323.

Lee, K.F., Li, E., Huber, L.J., Landis, S.C., Sharpe, A.H., Chao, M.V. and Jaenisch, R. (1992) Targeted mutation of the gene encoding the low affinity NGF receptor p75 leads to deficits in the peripheral sensory nervous system. Cell, 69: 737–749.

Lee, R., Kermani, P., Teng, K.K. and Hempstead, B.L. (2001) Regulation of cell survival by secreted proneurotrophins. Science, 294: 1945–1948.

Lee, T.H., Abe, K., Kogure, K. and Itoyama, Y. (1995) Expressions of nerve growth factor and p75 low affinity receptor after transient forebrain ischemia in gerbil hippocampal CA1 neurons. J. Neurosci. Res., 41: 684–695.

Lei, Q., Jones, M.B., Talley, E.M., Schrier, A.D., McIntire, W.E., Garrison, J.C. and Bayliss, D.A. (2000) Activation and inhibition of G protein-coupled inwardly rectifying potassium (Kir3) channels by G protein beta gamma subunits. Proc. Natl. Acad. Sci. USA, 97: 9771–9776.

Lewohl, J.M., Wilson, W.R., Mayfield, R.D., Brozowski, S.J., Morrisett, R.A. and Harris, R.A. (1999) G-protein-coupled inwardly rectifying potassium channels are targets of alcohol action. Nat. Neurosci., 2: 1084–1090.

Li, P., Nijhawan, D., Budihardjo, I., Srinivasula, S.M., Ahmad, M., Alnemri, E.S. and Wang, X. (1997) Cytochrome c and dATP-dependent formation of APAF-1/

caspase-9 complex initiates an apoptotic protease cascade. Cell, 91: 479–489.

Liao, Y.J., Jan, Y.N. and Jan, L.Y. (1996) Heteromultimerization of G-protein-gated inwardly rectifying K^+ channel proteins GIRK1 and GIRK2 and their altered expression in weaver brain. J. Neurosci., 16: 7137–7150.

Liepinsh, E., Ilag, L.L., Otting, G. and Ibanez, C.F. (1997) NMR structure of the death domain of the p75 neurotrophin receptor. EMBO J., 16: 4999–5005.

Linden, R. (1994) The survival of developing neurons: a review of afferent control. Neuroscience, 58: 671–682.

Lipton, S.A. (1986) Blockade of electrical activity promotes the death of mammalian retinal ganglion cells in culture. Proc. Natl. Acad. Sci. USA, 83: 9774–9778.

Liu, Z.G., Hsu, H., Goeddel, D.V. and Karin, M. (1996) Dissection of TNF receptor 1 effector functions: JNK activation is not linked to apoptosis while NF-kappaB activation prevents cell death. Cell, 87: 565–576.

Lockhart, S.T., Turrigiano, G.G. and Birren, S.J. (1997) Nerve growth factor modulates synaptic transmission between sympathetic neurons and cardiac myocytes. J. Neurosci., 17: 9573–9582.

Lowry, K.S. and Cheema, S.S. (2000) A comparison between antisense p75NTR oligonucleotides and neurotrophic factors in promoting the survival of postnatal sensory neurons in vitro. In vitro Cell Dev. Biol. Anim., 36: 520–526.

Lowry, K.S., Murray, S.S., Coulson, E.J., Epa, R., Bartlett, P.F., Barrett, G. and Cheema, S.S. (2001a) Systemic administration of antisense p75(NTR) oligodeoxynucleotides rescues axotomised spinal motor neurons. J. Neurosci. Res., 64: 11–17.

Lowry, K.S., Murray, S.S., McLean, C.A., Talman, P., Mathers, S., Lopes, E.C. and Cheema, S.S. (2001b) A potential role for the p75 low-affinity neurotrophin receptor in spinal motor neuron degeneration in murine and human amyotrophic lateral sclerosis. Amyotroph. Lateral Scler. Other Motor Neuron Disord., 2: 127–134.

Maeno, E., Ishizaki, Y., Kanaseki, T., Hazama, A. and Okada, Y. (2000) Normotonic cell shrinkage because of disordered volume regulation is an early prerequisite to apoptosis. Proc. Natl. Acad. Sci. USA, 97: 9487–9492.

Majdan, M., Lachance, C., Gloster, A., Aloyz, R., Zeindler, C., Bamji, S., Bhakar, A., Belliveau, D., Fawcett, J., Miller, F.D. and Barker, P.A. (1997) Transgenic mice expressing the intracellular domain of the p75 neurotrophin receptor undergo neuronal apoptosis. J. Neurosci., 17: 6988–6998.

Malcangio, M., Garrett, N.E., Cruwys, S. and Tomlinson, D.R. (1997) Nerve growth factor- and neurotrophin-3-induced changes in nociceptive threshold and the release of substance P from the rat isolated spinal cord. J. Neurosci., 17: 8459–8467.

Mattson, M.P. and Duan, W. (1999) 'Apoptotic' biochemical cascades in synaptic compartments: roles in adaptive plasticity and neurodegenerative disorders. J. Neurosci. Res., 58: 152–166.

McLaughlin, B., Pal, S., Tran, M.P., Parsons, A.A., Barone, F.C., Erhardt, J.A. and Aizenman, E. (2001) p38 activation is required upstream of potassium current enhancement and caspase cleavage in thiol oxidant-induced neuronal apoptosis. J. Neurosci., 21: 3303–3311.

Mennerick, S. and Zorumski, C.F. (2000) Neural activity and survival in the developing nervous system. Mol. Neurobiol., 22: 41–54.

Miller, F.D. and Kapalan, D.R. (1998) Life and death decisions: a biological role for the p75 neurotrophin receptor. Cell Death Differ., 5: 343–345.

Mufson, E.J., Brashers-Krug, T. and Kordower, J.H. (1992) p75 nerve growth factor receptor immunoreactivity in the human brainstem and spinal cord. Brain Res., 589: 115–123.

Mukai, J., Hachiya, T., Shoji-Hoshino, S., Kimura, M.T., Nadano, D., Suvanto, P., Hanaoka, T., Li, Y., Irie, S., Greene, L.A. and Sato, T.A. (2000) NADE, a p75NTR-associated cell death executor, is involved in signal transduction mediated by the common neurotrophin receptor p75NTR. J. Biol. Chem., 275: 17566–17570.

Mullner, C., Vorobiov, D., Bera, A.K., Uezono, Y., Yakubovich, D., Frohnwieser-Steinecker, B., Dascal, N. and Schreibmayer, W. (2000) Heterologous facilitation of G protein-activated K(+) channels by beta-adrenergic stimulation via cAMP-dependent protein kinase. J. Gen. Physiol., 115: 547–558.

Nadeau, H., McKinney, S., Anderson, D.J. and Lester, H.A. (2000) ROMK1 (Kir1.1) causes apoptosis and chronic silencing of hippocampal neurons. J. Neurophysiol., 84: 1062–1075.

Naumann, T., Casademunt, E., Hollerbach, E., Hofmann, J., Dechant, G., Frotscher, M. and Barde, Y.A. (2002) Complete deletion of the neurotrophin receptor p75NTR leads to long-lasting increases in the number of basal forebrain cholinergic neurons. J. Neurosci., 22: 2409–2418.

Nicholson, D.W. and Thornberry, N.A. (1997) Caspases: Killer proteases. Trends Biochem. Sci., 22: 299–306.

Nietsch, H.H., Roe, M.W., Fiekers, J.F., Moore, A.L. and Lidofsky, S.D. (2000) Activation of potassium and chloride channels by tumor necrosis factor alpha. Role in liver cell death. J. Biol. Chem., 275: 20556–20561.

Oppenheim, R.W. (1991) Cell death during development of the nervous system. Annu. Rev. Neurosci., 14: 453–501.

Palmada, M., Kanwal, S., Rutkoski, N.J., Gufstafson-Brown, C., Johnson, R.S., Wisdom, R. and Carter, B.D. (2002) c-jun is essential for sympathetic neuronal death induced by NGF withdrawal but not by p75 activation. J. Cell Biol., 158: 453–461.

Pan, G., O'Rourke, K. and Dixit, V.M. (1998) Caspase-9, Bcl-xL, and APAF-1 form a ternary complex. J. Biol. Chem., 273: 5841–5845.

Park, J.A., Lee, J.Y., Sato, T.A. and Koh, J.Y. (2000) Co-induction of p75NTR and p75NTR-associated death executor in neurons after zinc exposure in cortical culture or transient ischemia in the rat. J. Neurosci., 20: 9096–9103.

Pereira, S.P., Medina, S.V. and Araujo, E.G. (2001) Cholinergic activity modulates the survival of retinal ganglion cells in culture: the role of M1 muscarinic receptors. Int. J. Dev. Neurosci., 19: 559–567.

Petreanu, L. and Alvarez-Buylla, A. (2002) Maturation and death of adult-born olfactory bulb granule neurons: role of olfaction. J. Neurosci., 22: 6106–6113.

Pozniak, C.D., Radinovic, S., Yang, A., McKeon, F., Kaplan, D.R. and Miller, F.D. (2000) An anti-apoptotic role for the p53 family member, p73, during developmental neuron death. Science, 289: 304–306.

Rabizadeh, S., Oh, J., Zhong, L.T., Yang, J., Bitler, C.M., Butcher, L.L. and Bredesen, D.E. (1993) Induction of apoptosis by the low-affinity NGF receptor. Science, 261: 345–348.

Reeves, E.P., Lu, H., Jacobs, H.L., Messina, C.G., Bolsover, S., Gabella, G., Potma, E.O., Warley, A., Roes, J., Segal, A.W. (2002) Killing activity of neutrophils is mediated through activation of proteases by K+ flux. Nature, 416: 291–297.

Reid, K., Coulson, E.J., Faux, C. and Bartlett, P.F. (2002) Signaling cell death through the p75 neurotrophin receptor is not dependent on the level of NADE. Proc. Aust. Neurosci. Soc., 13: 65.

Ren, X.D. and Schwartz, M.A. (1998) Regulation of inositol lipid kinases by Rho and Rac. Curr. Opin. Genet. Dev., 8: 63–67.

Rende, M., Provenzano, C. and Tonali, P. (1993) Modulation of low-affinity nerve growth factor receptor in injured adult rat spinal cord motoneurons. J. Comp. Neurol., 338: 560–574.

Roux, P.P. and Barker, P. (2002) Neurotrophin signaling through the p75 neurotrophin receptor. Prog. Neurobiol., 67: 203–233.

Roux, P.P., Bhakar, A.L., Kennedy, T.E. and Barker, P.A. (2001) The p75 neurotrophin receptor activates Akt (protein kinase B) through a phosphatidylinositol 3-kinase-dependent pathway. J. Biol. Chem., 276: 23097–23104.

Roux, P.P., Colicos, M.A., Barker, P.A. and Kennedy, T.E. (1999) p75 neurotrophin receptor expression is induced in apoptotic neurons after seizure. J. Neurosci., 19: 6887–6896.

Ruiz-Vela, A., Gonzalez de Buitrago, G. and Martinez, A.C. (1999) Implication of calpain in caspase activation during B cell clonal deletion. EMBO J., 18: 4988–4998.

Rutherford, L.C., DeWan, A., Lauer, H.M. and Turrigiano, G.G. (1997) Brain-derived neurotrophic factor mediates the activity-dependent regulation of inhibition in neocortical cultures. J. Neurosci., 17: 4527–4535.

Salehi, A.H., Roux, P.P., Kubu, C.J., Zeindler, C., Bhakar, A., Tannis, L.L., Verdi, J.M. and Barker, P.A. (2000) NRAGE, a novel MAGE protein, interacts with the p75 neurotrophin receptor and facilitates nerve growth factor-dependent apoptosis. Neuron, 27: 279–288.

Salehi, A.H., Xanthoudakis, S. and Barker, P.A. (2002) NRAGE, a p75 neurotrophin receptor-interacting protein, induces caspase activation and cell death through a JNK-dependent mitochondrial pathway. J. Biol. Chem., 277: 48043–48050.

Schinder, A.F. and Poo, M. (2000) The neurotrophin hypothesis for synaptic plasticity. Trends Neurosci., 23: 639–645.

Schulte, U. and Fakler, B. (2000) Gating of inward-rectifier K^+ channels by intracellular pH. Eur. J. Biochem., 267: 5837–5841.

Segal, R.A., Bhattacharyya, A., Rua, L.A., Alberta, J.A., Stephens, R.M., Kaplan, D.R. and Stiles, C.D. (1996) Differential utilization of Trk autophosphorylation sites. J. Biol. Chem., 271: 20175–20181.

Semkova, I. and Krieglstein, J. (1999) Neuroprotection mediated via neurotrophic factors and induction of neurotrophic factors. Brain Res. Brain Res. Rev., 30: 176–188.

Sendtner, M., Holtmann, B., Kolbeck, R., Thoenen, H. and Barde, Y.A. (1992a) Brain-derived neurotrophic factor prevents the death of motoneurons in newborn rats after nerve section. Nature, 360: 757–759.

Sendtner, M., Schmalbruch, H., Stockli, K.A., Carroll, P., Kreutzberg, G.W. and Thoenen, H. (1992b) Ciliary neurotrophic factor prevents degeneration of motor neurons in mouse mutant progressive motor neuronopathy. Nature, 358: 502–504.

Sherrard, R.M. and Bower, A.J. (1998) Role of afferents in the development and cell survival of the vertebrate nervous system. Clin. Exp. Pharmacol. Physiol., 25: 487–495.

Signorini, S., Liao, Y.J., Duncan, S.A., Jan, L.Y. and Stoffel, M. (1997) Normal cerebellar development but susceptibility to seizures in mice lacking G protein-coupled, inwardly rectifying K^+ channel GIRK2. Proc. Natl. Acad. Sci. USA, 94: 923–927.

Soilu-Hänninen, M., Ekert, P., Bucci, T., Syroid, D., Bartlett, P.F. and Kilpatrick, T.J. (1999) Nerve growth factor signaling through p75 induces apoptosis in Schwann cells via a Bcl-2-independent pathway. J. Neurosci., 19: 4828–4838.

Soilu-Hänninen, M., Epa, R., Shipham, K., Butzkueven, H., Buci, T., Barrett, G., Bartlett, P. and Kilpatrick, T. (2000) Treatment of experimental autoimmune encephalomyelitis with antisense oligonucleotides against the low affinity neurotrophin receptor. J. Neurosci. Res., 69: 712–721.

Stevens, E.B., Shah, B.S., Pinnock, R.D. and Lee, K. (1999) Bombesin receptors inhibit G protein-coupled inwardly rectifying K^+ channels expressed in Xenopus oocytes through a protein kinase C- dependent pathway. Mol. Pharmacol., 55: 1020–1027.

Takigawa, T. and Alzheimer, C. (2002) Phasic and tonic attenuation of EPSPs by inward rectifier K^+ channels in rat hippocampal pyramidal cells. J. Physiol., 539: 67–75.

Tcherpakov, M., Bronfman, F.C., Conticello, S.G., Vaskovsky, A., Levy, Z., Niinobe, M., Yoshikawa, K., Arenas, E. and Fainzilber, M. (2002) The p75 neurotrophin receptor interacts with multiple MAGE proteins. J. Biol. Chem., 277: 49101–49104.

Terrado, J., Burgess, R.W., DeChiara, T., Yancopoulos, G., Sanes, J.R. and Kato, A.C. (2001) Motoneuron survival is enhanced in the absence of neuromuscular junction formation in embryos. J. Neurosci., 21: 3144–3150.

Thornberry, N.A. and Lazebnik, Y. (1998) Caspases: enemies within. Science, 281: 1312–1316.

Thornberry, N.A., Rano, T.A., Peterson, E.P., Rasper, D.M., Timkey, T., Garcia-Calvo, M., Houtzager, V.M., Nordstrom, P.A., Roy, S., Vaillancourt, J.P., Chapman, K.T. and Nicholson, D.W. (1997) A combinatorial approach defines specificities of members of the caspase family and granzyme B. Functional relationships established for key mediators of apoptosis. J. Biol. Chem., 272: 17907–17911.

Troy, C.M., Friedman, J.E. and Friedman, W.J. (2002) Mechanisms of p75-mediated death of hippocampal neurons. Role of caspases. J. Biol. Chem., 277: 34295–34302.

Verhage, M., Maia, A.S., Plomp, J.J., Brussaard, A.B., Heeroma, J.H., Vermeer, H., Toonen, R.F., Hammer, R.E., van den Berg, T.K., Missler, M., Geuze, H.J. and Sudhof, T.C. (2000) Synaptic assembly of the brain in the absence of neurotransmitter secretion. Science, 287: 864–869.

von Bartheld, C.S., Kinoshita, Y., Prevette, D., Yin, Q.W., Oppenheim, R.W. and Bothwell, M. (1994) Positive and negative effects of neurotrophins on the isthmo-optic nucleus in chick embryos. Neuron, 12: 639–654.

Vu, C.C., Bortner, C.D. and Cidlowski, J.A. (2001) Differential involvement of initiator caspases in apoptotic volume decrease and potassium efflux during Fas- and UV-induced cell death. J. Biol. Chem., 276: 37602–37611.

Wang, L., Xu, D., Dai, W. and Lu, L. (1999) An ultraviolet-activated K^+ channel mediates apoptosis of myeloblastic leukemia cells. J. Biol. Chem., 274: 3678–3685.

Wickman, K., Karschin, C., Karschin, A., Picciotto, M.R. and Clapham, D.E. (2000) Brain localization and behavioral impact of the G-protein-gated K^+ channel subunit GIRK4. J. Neurosci., 20: 5608–5615.

Wilson, C.A., Doms, R.W. and Lee, V.M. (1999) Intracellular APP processing and A beta production in Alzheimer disease. J. Neuropathol. Exp. Neurol., 58: 787–794.

Xia, S., Lampe, P.A., Deshmukh, M., Yang, A., Brown, B.S., Rothman, S.M., Johnson, E.M., Jr. and Yu, S.P. (2002) Multiple channel interactions explain the protection of sympathetic neurons from apoptosis induced by nerve growth factor deprivation. J. Neurosci., 22: 114–122.

Xia, Z., Dickens, M., Raingeaud, J., Davis, R.J. and Greenberg, M.E. (1995) Opposing effects of ERK and JNK-p38 MAP kinases on apoptosis. Science, 270: 1326–1331.

Xiang, H., Kinoshita, Y., Knudson, C.M., Korsmeyer, S.J., Schwartzkroin, P.A. and Morrison, R.S. (1998) Bax involvement in p53-mediated neuronal cell death. J. Neurosci., 18: 1363–1373.

Yamashita, T., Tucker, K.L. and Barde, Y.A. (1999) Neurotrophin binding to the p75 receptor modulates Rho activity and axonal outgrowth. Neuron, 24: 585–593.

Yang, B., Slonimsky, J.D. and Birren, S.J. (2002) A rapid switch in sympathetic neurotransmitter release properties mediated by the p75 receptor. Nat. Neurosci., 5: 539–545.

Ye, X., Mehlen, P., Rabizadeh, S., VanArsdale, T., Zhang, H., Shin, H., Wang, J.J., Leo, E., Zapata, J., Hauser, C.A., Reed, J.C. and Bredesen, D.E. (1999) TRAF family proteins interact with the common neurotrophin receptor and modulate apoptosis induction. J. Biol. Chem., 274: 30202–30208.

Yoon, S.O., Casaccia-Bonnefil, P., Carter, B. and Chao, M.V. (1998) Competitive signaling between trkA and p75 nerve growth factor receptors determines cell survival. J. Neurosci., 18: 3273–3281.

Yu, S.P., Canzoniero, L.M. and Choi, D.W. (2001) Ion homeostasis and apoptosis. Curr. Opin. Cell Biol., 13: 405–411.

Yu, S.P., Yeh, C., Strasser, U., Tian, M. and Choi, D.W. (1999) NMDA receptor-mediated K^+ efflux and neuronal apoptosis. Science, 284: 336–339.

Yu, S.P., Yeh, C.H., Sensi, S.L., Gwag, B.J., Canzoniero, L.M., Farhangrazi, Z.S., Ying, H.S., Tian, M., Dugan, L.L. and Choi, D.W. (1997) Mediation of neuronal apoptosis by enhancement of outward potassium current. Science, 278: 114–117.

Zhang, H., He, C., Yan, X., Mirshahi, T. and Logothetis, D.E. (1999) Activation of inwardly rectifying K^+ channels by distinct PtdIns(4,5)P2 interactions. Nat. Cell Biol., 1: 183–188.

Zhang, S.X., Guo, Y. and Boulianne, G.L. (2001) Identification of a novel family of putative methyltransferases that interact with human and *Drosophila* presenilins. Gene, 280: 135–144.

Progress in Brain Research, Vol. 146
ISSN 0079-6123

CHAPTER 4

The role of NT-3 signaling in Merkel cell development

Maya Sieber-Blum[1,*], Viktor Szeder[1,2] and Milos Grim[2]

[1]*Department of Cell Biology, Neurobiology and Anatomy, Medical College of Wisconsin, Milwaukee, WI 53226, USA*
[2]*Institute of Anatomy, First Faculty of Medicine, Charles University Prague, Prague, Czech Republic*

Abstract: Merkel cells originate from the neural crest. They are located in hairy and glabrous skin and have neuroendocrine characteristics. Together with A β afferents, Merkel cells form a slowly adapting mechanoreceptor, the Merkel nerve ending, which transduces steady skin indentation. Neurotrophin-3 (NT-3) plays important roles in neural crest cell development. We thus sought to determine whether neurotrophin signaling is essential for Merkel cell development in the whisker pad of the mouse. Our data indicate that at embryonic day 16.5 (E 16.5), NT-3 and its receptors, p75 neurotrophin receptor (p75^{NTR}) and tyrosine kinase receptor, TrkC are not expressed at detectable levels in Merkel cells. After a perinatal switch, however, Merkel cells in whiskers of newborn mice are immunoreactive for p75^{NTR}, TrkC and NT-3. Immunoreactivity of all three markers persists into adulthood. By contrast, innervating fibers are intensely p75^{NTR}-immunoreactive in E16.5 whiskers, but no TrkC immunoreactivity is detected. At birth, and at 6 weeks of age, afferent fibers are intensely immunoreactive for both p75^{NTR} and TrkC. In TrkC null whiskers, numerous Merkel cells are present at E16.5, and they are innervated. We draw three major conclusions from these observations: (i) NT-3 signaling through p75^{NTR} or TrkC is not required for the development and prenatal survival of either a major subset or of all Merkel cells, (ii) the postnatal survival of Merkel cells is supported by autocrine or paracrine NT-3, rather than by neuron-derived NT-3, and (iii) Merkel cell-derived NT-3 is not a chemoattractant for innervating Aβ fibers, but is likely to be involved in maintaining Merkel cell innervation postnatally.

Keywords: Merkel cell; mechanoreceptor; neural crest; TrkC; NT-3; p75^{NTR}

Introduction

Together with the terminals of sensory neurons Merkel cells form the Merkel nerve endings, slowly adapting cutaneous mechanoreceptors that are activated by steady skin indentation (Iggo and Muir, 1969; Halata et al., 2002). Both, avian (Halata et al., 1990; Grim and Halata, 2000a,b) and mammalian (Szeder et al., 2003a) Merkel cells are derived from

*Correspondence to: M. Sieber-Blum, Department of Cell Biology, Neurobiology and Anatomy, Medical College of Wisconsin, 8701 Watertown Plank Road, Milwaukee, WI 53226, USA.
Tel.: +1-414-456-8465; Fax: +1-414-456-6517;
E-mail: sieberbl@mcw.edu

DOI: 10.1016/S0079-6123(03)46004-4

the neural crest. Since NT-3 signaling through tyrosine kinase receptor, TrkC plays important roles during neural crest cell development, we have determined the time course of expression of NT-3 and its receptors, TrkC and p75 neurotrophin receptor (p75^{NTR}), and the effect of TrkC gene deletion on Merkel cell development. The mouse whisker pad is a suitable tissue for the study of Merkel cell development as the cells are present in defined locations. Within the whisker follicle, Merkel cells line the bulge area, which is a niche for keratinocyte stem cells (Oshima et al., 2001). At the opening of the hair follicle, Merkel cells are located in the rete ridge. Interfollicular Merkel cells are grouped into touch domes between hairs (Fig. 1; Halata et al., 2003).

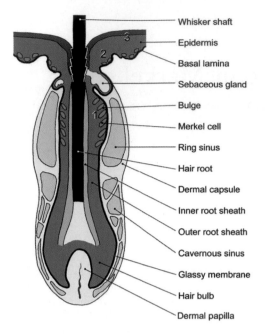

Whisker shaft

Epidermis

Basal lamina

Sebaceous gland

Bulge

Merkel cell

Ring sinus

Hair root

Dermal capsule

Inner root sheath

Outer root sheath

Cavernous sinus

Glassy membrane

Hair bulb

Dermal papilla

Fig. 1. Diagram of a whisker with locations of Merkel cells. Medial longitudinal view. The epidermis forms a pocket within the dermis, called the outer root sheet. It consists of a layer of stratified epithelium, which is thickened in the bulge area. The bulge contains keratinocyte stem cells. The dermal papilla, dermal sheet and blood sinuses are dermal derivatives. The Merkel cells are located in three characteristic places, in the outer root sheet (**1**) at the bulge region, in the rete ridge (**2**) at the opening of the follicle and in touch domes (**3**) in interfollicular areas. Merkel cells are innervated by Aβ afferents (not shown).

Structure and function of the Merkel cell

Merkel cells are located in the basal layer of the epithelium of the hair follicle in the thickened area (bulge) below the sebaceous gland. The bulge is a multilayered epithelial structure that contains keratinocyte stem cells, which in anagen phase migrate toward the dermal papillae to form new hair (Oshima et al., 2001). The bulge area is surrounded by the ring sinus. The entire follicle is surrounded by a basal membrane, the glassy membrane. Merkel cells are stacked obliquely to the glassy membrane. They pierce it with processes of about 3 μm and make contact with discoid nerve terminals (Halata et al., 2003). The myelinated axons have a diameter of about 5 μm, and lose their myelin sheath at the

basal lamina. Myelinated axons branch intensely so that each axon supplies up to 50 Merkel cells. The Merkel cell processes that face the nerve terminals contain typical large dense core granules (50–110 nm) (Iggo and Muir, 1969; Breathnach and Robins, 1970; Halata et al., 2002). The cytoplasmic membrane of Merkel cells is in close apposition to the membrane of an axonal terminal with areas of synaptic membrane specializations (Chen et al., 1973; Iggo and Findlater, 1984).

Merkel cells express low molecular-weight cytokeratins 8, 18, 19 and 20, but not the types of cytokeratin that are characteristic of fully differentiated keratinocytes. Antibodies against the low molecular-weight cytokeratins are used to identify Merkel cells light microscopically (Moll et al., 1984, 1986a; Saurat et al., 1984). Pan-neuronal markers, such as neuron-specific enolase, protein gene product 9.5 (PGP 9.5), synaptophysin and chromogranin A are also expressed by Merkel cells. Antibodies against these markers thus are used also to visualize Merkel cells (Zaccone, 1986; Gauweiler et al., 1988; Hartschuh and Weihe, 1988; Dalsgaard et al., 1989; Ramieri et al., 1992).

Recent reports indicate a direct involvement of the Merkel cell in the transduction of mechanical stimuli into action potentials in the afferent nerve fiber (Chan et al., 1996; Senok and Baumann, 1997; Tazaki and Suzuki, 1998; Fagan and Cahusac, 2001). Increases of intracellular calcium transients in Merkel cells have been observed during mechanical stimulation of Merkel nerve endings (Chan et al., 1996; Tazaki and Suzuki, 1998) and during swelling of Merkel cells in hyposmotic solution (reviewed by Halata et al., 2002). The increase in intracellular calcium transients were not observed in the absence of extracellular calcium or in the presence of the calcium channel blocker, amiloride. It is thus thought that mechanical stimulation causes an influx of calcium into Merkel cells, which in turn triggers further calcium release from intracellular stores (Senok and Baumann, 1997). Since free calcium is a requirement for synaptic transmitter release (Katz and Miledi, 1967) and because Merkel cells contain neurotransmitters, calcium-mediated release likely regulates the conversion of mechanic stimuli into action potentials. Furthermore, glutaminergic transmission (Fagan and Cahusac, 2001) appears to be essential for the slowly adapting responses during the static phase of mechanical stimuli (Ogawa, 1996; reviewed by Halata et al., 2002).

Ontogenetic origin of Merkel cells

Whether Merkel cells are of epidermal origin or whether they are derived from the neural crest has been the subject of a long-standing controversy in the literature (reviewed by Halata et al., 2003). Evidence for the epidermal origin of mammalian Merkel cells was based on their location in the basal layer of the epidermis (Lyne and Hollis, 1971; English et al., 1980), their expression of low molecular-weight cyto-keratins (Moll et al., 1984, 1986b), and the fact that Merkel cells developed in human fetal skin that was transplanted onto host nude mice (Moll et al., 1990). However, neither its location, nor its expression of a particular gene, or set of genes, necessarily defines the ontogenetic origin of a cell type. Moreover, in the fetal skin transplants, Merkel cell precursors and neural crest-derived glia were already present at the time of transplantation (Moore and Munger, 1989; Terenghi et al., 1993; Moll et al., 1996). In contrast, chick-quail transplantation experiments have provided solid evidence against the epidermal origin of avian Merkel cells (Halata et al., 1990; Grim and Halata, 2000a,b). We have used the double transgenic Wnt1-cre/R26R mouse (Friedrich and Soriano, 1991; Echelard et al., 1994; Danielian et al., 1998; Soriano, 1999; Chai et al., 2000; Jiang et al., 2002), in which all neural crest cells and their derivatives are marked permanently by their expression of beta-galactosidase to elucidate the developmental origin of Merkel cells. In Wnt1-cre/R26R double transgenic mice, Merkel cells express beta-galactosidase, providing evidence for the neural crest origin of mammalian Merkel cells (Szeder et al., 2003a).

NT-3 action in neural crest cell development

Analyses by in vitro colony assay (Sieber-Blum and Cohen, 1980; Sieber-Blum, 1989, 1999) have revealed fibroblast growth factor-2 (FGF-2) is a mitogen for quail neural crest stem cells in culture (Zhang et al., 1997). Proliferative descendants of neural crest stem cells become dependent on a neurotrophin. NT-3, brain-derived neurotrophic factor (BDNF) or nerve growth factor (NGF) are equally effective in supporting the survival of proliferating neural crest stem cells and their immediate descendants, suggesting that they act through the non-selective neuro-trophin receptor $p75^{NTR}$, which is expressed by neural crest cells during early migration (Zhang et al., 1997). Another function of NT-3 in neural crest cell development is to promote, in synergy with FGF-2 and transforming growth factor-beta 1 (TGF-beta1), expression of the norepinephrine transporter (NET) (Ren et al., 2001). The NET is expressed ubiquitously in avian and mammalian embryos in neuronal and non-neuronal tissues. Norepinephrine transport affects directly or indirectly the development of diverse neuronal and non-neuronal tissues (reviewed by Ren et al., 2003). Norepinephrine transport promotes the differentiation of progenitor cells into noradrenergic cells in both the neural crest (Zhang et al., 1997a) and the *locus ceruleus* (Sieber-Blum and Ren, 2000). In the melanogenic cell lineage, any neurotrophin of the NGF family of neurotrophins (NGF, BDNF or NT-3) synergizes with stem cell factor to support the survival of pigment cell pre-ccursors (Langtimm-Sedlak et al., 1996). TrkC is expressed by approximately 30% of cardiac neural crest cells during early in vivo migration and up to 50% during advanced migration (Youn et al., 2002). Deletion of the *TrkC* gene causes cardiac neural crest stem cells to become fate-restricted prematurely (Sieber-Blum and Zhang, 2002; Youn et al., 2002). This observation is reminiscent of the report by Farinas et al. (1996), who observed that in NT-3 deficient mice, neural crest-derived neuronal progenitors cease proliferation and differentiate prematurely. Overall, NT-3 plays several important roles in the survival, proliferation, diversification and differentiation of neural crest stem cells. Some functions are mediated by $p75^{NTR}$, others by TrkC.

Bound NT-3 activates TrkC kinase, leading to receptor dimerization and the recruitment and phosphorylation of signaling proteins that regulate neuronal proliferation, differentiation, and survival (Barbacid, 1994). *TrkC* null mice have a phenotype that is similar to that of NT-3-deficient mice, confirming that NT-3 acts predominantly through TrkC (Snider, 1994). Alternative splicing of the *TrkC*

66

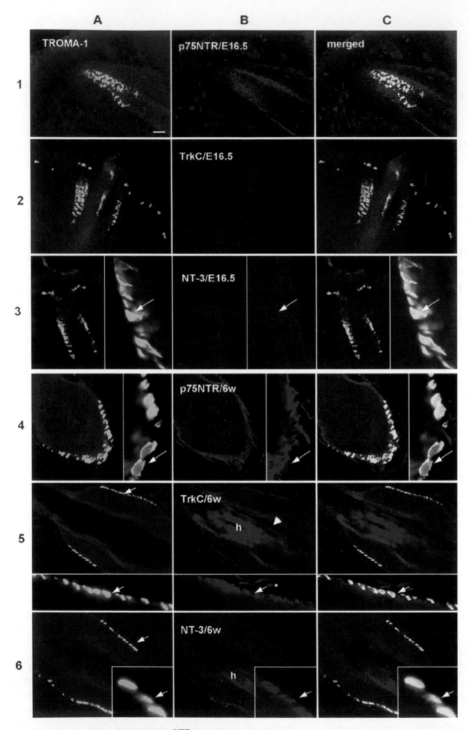

Fig. 2. Expression pattern of p75NTR, TrkC and NT-3 in the wild type whisker. Double indirect immunohistochemistry with TROMA-1 (**1A–6A**; Merkel cell marker; fluorescein fluorescence) and antibodies against p75NTR, TrkC and NT-3 (**1B–6B**; Texas red fluorescence) in longitudinal cryosections; **1C–6C**; merged images. (**1–3**), at E16.5. (**4–6**), at postnatal week 6. (**1A–C**) TROMA-1 and p75NTR antibodies. Merkel cells do not bind the p75NTR antibody. Innervating nerve fibers are intensely p75NTR immunoreactive.

gene can generate several types of truncated TrkC isoforms that lack the kinase domain (Tsoulfas et al., 1993; Valenzuela et al., 1993; Garner and Large, 1994). These noncatalytic receptors, which are well conserved among species, are thought to negatively modulate signaling by the catalytic receptor when both the catalytic and truncated isoforms are co-expressed (Palko et al., 1999). Both, catalytic and truncated TrkC receptors are absent in the TrkC null mouse line that was used in our present study (Tessarollo et al., 1997).

p75^{NTR}, TrkC and NT-3 expression during prenatal and postnatal Merkel cell development

p75^{NTR}

In E16.5 wild type mouse embryos, that is 2–3 days before they are born, Merkel cells do not express p75^{NTR} at detectable levels (Fig. 2(1A–C)). Merkel cells were identified with TROMA-1, an antibody against cytokeratin 8, which is a marker for prenatal and postnatal Merkel cells (Vielkind et al., 1995; Fig. 2(1A–6A)). By contrast, axons that innervate the whisker follicle are intensely p75^{NTR}-immunoreactive already at E16.5 (Fig. 2(1A–C)). p75^{NTR} staining on axons was confirmed by double immunostaining with antibodies against p75^{NTR} and against peripherin (Szeder et al., 2003b) (Table 1).

The p75^{NTR} expression pattern is identical in newborn (Szeder et al., 2003b) and in 6-week (6w) old mice (Fig. 2(4A–C)). In postnatal whiskers, axons are still intensely p75^{NTR}-immunoreactive. Moreover, all cells in the outer root sheet, including Merkel cells, are immunoreactive (Fig. 2(4A–C); arrows). Thus during perinatal stages, Merkel cells switch from being p75^{NTR}-immunonegative to showing intense p75^{NTR} immunoreactivity.

TABLE 1
Summary of p75^{NTR}, TrkC and NT-3 expression in wild type mice

	E16.5	Newborn	Six weeks
Merkel cells			
p75^{NTR}	−	+	+
TrkC	−	+	+
NT-3	±	+	+
Axons			
p75^{NTR}	+	+	+
TrkC	−	+	+
NT-3	−	−	−

E16.5 and Six weeks, this study. Newborn, from Szeder et al. (2003b).

TrkC

At E16.5, Merkel cells do not express TrkC (Fig. 2(2A–C)). In these images Merkel cells are visible in the bulge area of two whiskers, in the rete ridge and in interfollicular areas. Low immunofluorescence is attributable exclusively to background immunofluorescence. Interestingly, innervating axons do not express TrkC at detectable levels either.

In newborns (Szeder et al., 2003b) and at 6 weeks of age (Fig. 2(5A–C)), all cells in the outer root sheet, including Merkel cells, are TrkC immunoreactive. In addition, some cells in the inner root sheet appear TrkC-immunoreactive (Fig. 2(5B); arrow head). Interestingly, the inner layers of the outer root sheet in the bulge region are TrkC-negative. Innervating fibers in newborns (Szeder et al., 2003b) and at 6 weeks (Fig. 2(5B), asterisk) are TrkC-immunoreactive. The hair (h) shows unspecific fluorescence.

NT-3

At E 16.5, Merkel cells do not express NT-3, or at low levels only (Fig. 2(3A–C)). Non-Merkel cells in

(2A–C) TROMA-1 and TrkC immunoreactivity. Merkel cells and innervating fibers are TrkC-negative. (3A–C) TROMA-1 and NT-3 immunoreactivity. Left, low magnification; right, higher magnification at the bulge location. The inner layers of the outer root sheet are weakly immuno-reactive; the Merkel cells are either negative or they express NT-3 at low levels only. (4A–C) TROMA-1 and p75^{NTR} immunoreactivity at 6 weeks of age; oblique section. The Merkel cells (arrow) are p75^{NTR}-immunoreactive, as are the innervating afferents. (5A–C) TROMA-1 and TrkC immunoreactivity in the bulge area; upper panel, low magnification; lower panel, higher magnification of are marked with arrow in 5A, upper panel. Merkel cells (arrows) and innervating fibers (asterisk, 5B, lower panel) are TrkC-immunoreactive. In addition, cells in the inner root sheet are Trk C immunoreactive. The keratinized hair (h) binds antibody unspecifically. (6A–C) TROMA-1 and NT-3 immunoreactivity in the bulge region. All cells in the outer root sheet express NT-3, including the Merkel cells. Inserts; higher magnification of area marked by arrow in 6A. Bar (1A–6C), 25 μm.

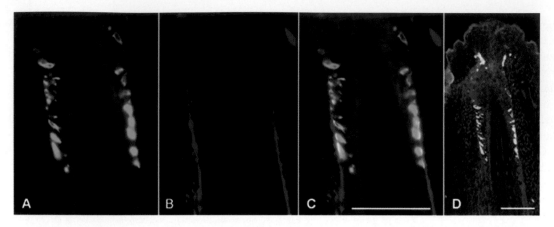

Fig. 3. Merkel cells in TrkC null mice. Double immuno-staining in longitudinal section through the middle of an E16.5 TrkC null whisker. (A) Numerous Merkel cells are located in the bulge region of the follicle and in the rete ridge. They are intensely TROMA-1 immunoreactive (fluorescein fluorescence). (B) Neuron-specific β-III tubulin-immunoreactive (Texas red fluorescence) innervating fibers are present. (C) Merged images. (D) Overview of whisker shown in (A–C); triple stain with TROMA-1 antibodies (fluorescein fluorescence), anti-p75^{NTR} antibodies (Texas red fluorescence) and DAPI nuclear stain (blue). Numerous Merkel cells are present in the outer root sheet of the bulge area and in the rete ridge. Bars, (A–C) 50 μm; (D), 50 μm.

the inner layers of the outer root sheet show low levels of NT-3 immunoreactivity (Fig. 2(3A–C)).

At birth (Szeder et al., 2003a) and at 6 weeks of age (Fig. 2(6A–C)), all cells in the outer root sheet, including Merkel cells are strongly NT-3-immuno-reactive. The hair (h) shows unspecific fluorescence. Double stain in three Merkel cells from the area marked in Fig. 2(6A) is shown at higher magnification (Fig. 2(6A–C), insert).

Merkel cell development in TrkC null mice

To determine whether a lack of TrkC signaling affects Merkel cell development, we have stained cryosections of TrkC null whisker pads with antibodies against cytokeratin 8 (TROMA-1). Numerous Merkel cells are present at E16.5 in the outer root sheet of the whisker follicle (Fig. 3) and they are innervated by neuron-specific beta-III tubulin-immunoreactive neurites (Fig. 3(B–D)), which are p75^{NTR} immunoreactive in wild type animals (Szeder et al., 2003b). A decrease in Merkel cell numbers appears unlikely. Quantification of Merkel cells in whiskers is difficult though, as their numbers change greatly with the orientation of the section through the whisker. Therefore, all Merkel cells within a whisker need to be scored, which was beyond the scope of our study.

Discussion

We have determined the temporal and spatial expression patterns of p75^{NTR}, TrkC and NT-3 in the murine whisker during Merkel cell development. There is a perinatal switch in Merkel cells from immunonegativity to detectable expression of p75^{NTR}, TrkC and NT-3 around birth that persists into adulthood. At E16.5, Merkel cells are not imunoreactive for p75^{NTR} and TrkC, and they express no NT-3, or low levels of NT-3 only. In contrast, innervating axons are strongly immuno-fluorescent for p75^{NTR} at E16.5, but they are TrkC-negative. At birth, and at 6 weeks of age, Merkel cells express all three markers, p75^{NTR}, TrkC and NT-3, and innervating axons are both p75^{NTR}-positive and TrkC-positive. In *TrkC* null whiskers, Merkel cells are present and they are innervated.

Three major conclusions can be drawn from these observations: (i) prenatal development and survival of Merkel cells are independent of NT-3-signaling through p75^{NTR} or TrkC, (ii) Merkel cell-derived NT-3 signaling through TrkC does not act as a chemoattractant for innervating fibers, and (iii) autocrine or paracrine NT-3, rather than neuron-derived NT-3, is likely to support the survival of Merkel cells and maintain their innervation postnatally.

Formation of Merkel cells does not require NT-3 signaling through p75^{NTR} or TrkC

Our p75^{NTR} data are in agreement with a report by Kinkelin et al. (1999), which showed that p75^{NTR} is required for the survival of postnatal Merkel cells in hairy skin. The postnatal loss of Merkel cells in backskin started at approximately 2 weeks of age and was completed at 2 months (Kinkelin et al. 1999). In contrast, Fundin et al. (1997) reported an increase of Merkel cells in the rete ridge, but not in the outer root sheet, in whiskers of 1–2-week-old p75^{NTR} null mice. The two groups were using the same p75^{NTR}-deficient mouse line, therefore the discrepancy is not based on genetical differences. It is conceivable that Merkel cells in different locations have distinct neurotrophin requirements, as has been shown for Merkel cells in different rugae in the mouse palate (Ichikawa et al., 2001). In summary, p75^{NTR} is not required for the prenatal development and survival of Merkel cells, but is important for their postnatal survival.

We find that Merkel cells do not express TrkC or p75^{NTR} until around birth. This observation is in agreement with the work of Fundin et al. (1997), who observed in TrkC kinase-deficient mice the onset of a decrease in innervation at around 1 week postnatally that was followed by a progressive loss of Merkel cells. Similarly, Airaksinen et al. (1996) reported the virtual absence of Merkel cells in 14-day-old NT-3-deficient mice. Conversely, Albers et al. (1996) found an increase in Merkel cell numbers in touch domes of 4–8-month-old mice that over-expressed NT-3 in the basal layer of the epidermis. These studies, as well as our data support the view that NT-3 signaling through TrkC or p75^{NTR} is not required for the formation of Merkel cells, but that it becomes essential for the survival of postnatal Merkel cells and for the integrity of their innervation.

The Rice group (Cronk et al., 2002) reports a total absence of Merkel cells in E14.5, E15.5 and P0 TrkC null mice. Since the same authors find Merkel cells in mice that have a nonfunctional TrkC kinase domain, but intact truncated receptors, they conclude that truncated TrkC receptors are essential for the formation of Merkel cells. These data and their interpretation are in conflict with our results, as we observe numerous Merkel cells in whiskers of E16.5 TrkC null mice. The reason for this discrepancy remains elusive so far. The mouse line used in both studies is identical (Tessarollo et al., 1997), the location observed within the whisker is identical, and the developmental stages studied are comparable. There are two differences, however, between the study by the Rice group and our own, the marker used and the orientation of the sections. Whereas Cronk et al. (2002) have used PGP 9.5 to identify Merkel cells, we have used antibodies that recognize cytokeratin 8. Both are established Merkel cell markers. The possibility that PGP 9.5 is not expressed in TrkC null mice can be excluded, because TrkC null nerve fibers are PGP 9.5-immunoreactive (Figs. 3I and 6E in Cronk et al., 2002). It is conceivable that in the oblique sections shown in the Rice study, the relatively small area occupied by follicular Merkel cells was missed. Alternatively, it may be possible that whiskers in different locations in the whisker pad have different growth factor requirements. In contrast to the Rice study, our data and reports from other investigators show that the development and prenatal survival either of a major subset or of all Merkel cells do not require NT-3 signaling through full length TrkC or through truncated TrkC.

Merkel cell-derived NT-3 does not act as a chemoattractant for Aβ afferents

Merkel cells start to express detectable levels of NT-3 around birth (Szeder et al., 2003b) and NT-3 expression is maintained into adulthood (this study). Schecterson and Bothwell (1992), Tessarollo et al. (1993) and Farinas et al. (1996) and Botchkarev et al. (1998) have described NT-3 immunoreactivity in postnatal whiskers as well. The mechanisms that regulate sensory innervation of Merkel cells in whiskers is not known. Our data provide evidence that excludes one potential mechanism, that of tropic action of Merkel cell-derived NT-3 to attract growth cones of Aβ fibers. At E16.5 innervation is completed, but Merkel cells express lower levels of NT-3, if any, than surrounding epidermal cells in the outer root sheet. Thus perinatal NT-3 expression in the outer root sheet is not Merkel cell-specific. Furthermore the onset and the levels of NT-3 expression are most likely too late and too low to exert any tropic or trophic actions. In addition, we show that at E16.5, the innervating fibers express $p75^{NTR}$, but not yet TrkC.

Merkel cells are not dependent on neuron-derived NT-3

English et al. (1980) and Nurse and Diamond (1984) have suggested that neuron-derived NT-3 released at nerve terminals may act to maintain Merkel cell survival. Based on our data, this possibility can be excluded. The main reason is that Merkel cells themselves are NT-3 immunoreactive. Moreover, Merkel cells develop even in the absence of NT-3 signaling through TrkC. At birth, Merkel cells express NT-3, TrkC and $p75^{NTR}$ (this study), whereas dorsal root ganglion neurons no longer express NT-3 after E16 (Schecterson and Bothwell, 1992). Postnatal survival of Merkel cells thus seems to be supported by autocrine or paracrine NT-3, rather than by neuron-derived NT-3.

In summary, our data and those of most other investigators indicate that NT-3 signaling through $p75^{NTR}$ or TrkC is not required for the prenatal development of Merkel cells, or for pathfinding and initial innervation by sensory Aβ fibers. Postnatally, however, autocrine or paracrine NT-3 may serve to support the survival of Merkel cells and maintain their innervation.

Acknowledgments

We thank Lino Tessarollo for providing TrkC null mice, Joan Ward and Eva Kluzáková for excellent technical assistance and Jan Kacvinsky for providing the artwork. The TROMA-1 antibody, developed by P. Brulet and R. Kemler, was purchased from the Developmental Studies Hybridoma Bank (NICHD; maintained by the University of Iowa). Supported by USPHS grant NS38500 (MSB) and postdoctoral fellowship 1F05-11111-02 (VS) from the National Institute of Neurological Disorders and Stroke, NIH, and by grants VZ 1111 00003-3G (MG) and LN 00A065 (MG) from the Ministry of Education of the Czech Republic.

Abbreviations

BDNF	brain-derived growth factor
FGF-2	fibroblast growth factor-2
NET	norepinephrine transporter
NGF	nerve growth factor
NT-3	neurotrophin-3
p75NTR	low-affinity neurotophin receptor
PGP 9.5	pan-neuronal protein gene product 9.5
TGF-beta1	transforming growth factor beta-1
TrkC	tyrosine kinase, high-affinity neurotrophin receptor
TROMA-1	antibody against cytokeratin-8; Merkel cell marker

References

Airaksinen, M.S., Koltzenburg, M., Lewin, G.R., Masu, Y., Helbig, C., Wolf, E., Brem, G., Toyka, K.V., Thoenen, H., Meyer, M. (1996) Specific subtypes of cutaneous mechanoreceptors require neurotrophin-3 following peripheral target innervation. Neuron, 16: 287–295.

Albers, K.M., Perrone, T.N., Goodness, T.P., Jones, M.E., Green, M.A. and Davis, B.M. (1996) Cutaneous overexpression of NT-3 increases sensory and sympathetic neuron number and enhances touch dome and hair follicle innervation. J. Cell Biol., 134: 487–497.

Barbacid, M. (1994) The Trk family of neurotrophin receptors. J. Neurobiol., 25: 1386–1403.

Botchkarev, V.A., Botchkareva, N.V., Albers, K.M., van der Veen, C., Lewin, G.R. and Paus, R. (1998) Neurotrophin-3 involvement in the regulation of hair follicle morphogenesis. J. Invest. Dermatol., 111: 279–285.

Breathnach, A.S. and Robins, J. (1970) Ultrastructural observation on Merkel cells in human fetal skin. J. Anat., 106: 411.

Chai, Y., Jiang, X., Ito, Y., Bringas, P., Jr., Han, J., Rowitch, D.H., Soriano, P., McMahon, A.P. and Sucov, H.M. (2000) Fate of the mammalian cranial neural crest during tooth and mandibular morphogenesis. Development, 127: 1671–1679.

Chan, E., Yung, W.H. and Baumann, K.I. (1996) Cytoplasmic Ca2 + concentrations in intact Merkel cells of an isolated, functioning rat sinus hair preparation. Exp. Brain Res., 108: 357–366.

Chen, S.Y., Gerson, S. and Meyer, J. (1973) The fusion of Merkel cell granules with a synapse-like structure. J. Invest. Dermatol., 61: 290–292.

Cronk, K.M., Wilkinson, G.A., Grimes, R., Wheeler, E.F., Jhaveri, S., Fundin, B.T., Silos-Santiago, I., Tessarollo, L.,

Reichardt, L.F., Rice, F. (2002) Diverse discrepancies of developing Merkel innervation on the trkA and both full-length and truncated isoforms of trkC. Development, 129: 3739–3750.

Dalsgaard, C.-J., Rydh, M. and Haegerstrand, A. (1989) Cutaneous innervation in man visualized with protein gene product 9.5 (PGP 9.5) antibodies. Histochemistry, 92: 385–389.

Danielian, P.S., Muccino, D., Rowitch, D.H., Michael, S.K. and McMahon, A.P. (1998) Modification of gene activity in mouse embryos in utero by a tamoxifen-inducible form of Cre recombinase. Curr. Biol., 8: 1323–1326.

Echelard, Y., Vassileva, G. and McMahon, A.P. (1994) Cis-acting regulatory sequences governing Wnt-1 expression in the developing mouse CNS. Development, 120: 2213–2224.

English, K.B., Burgess, P.R. and Kavka-Van Norman, D. (1980) Development of rat Merkel cells. J. Comp. Neurol., 194: 475–496.

Fagan, B.M. and Cahusac, P.M.B. (2001) Evidence for glutamate receptor mediated transmission at mechanoreceptors in the skin. Neuroreport, 12: 341–347.

Farinas, I., Yoshida, C.K., Backus, C. and Reichardt, L.F. (1996) Lack of neurotrophin-3 results in death of spinal neurons and premature differentiation of their precursors. Neuron, 17: 1065–1078.

Friedrich, G. and Soriano, P. (1991) Promoter traps in embryonic stem cells; a genetic screen to identify and mutate developmental genes in mice. Genes. Dev., 5: 1513–1523.

Fundin, B.T., Silos-Santiago, I., Ernfors, P., Fagan, A.M., Aldskogius, H., DeChiara, T.M., Phillips, H.S., Barbacid, M., Yancopoulos, G.D., Rice, F.L. (1997) Differential dependency of cutaneous mechanoreceptors on neurotrophins, trk receptors, and p76 LNGFR. Dev. Biol., 190: 94–116.

Garner, A.S. and Large, T.H. (1994) Isoforms of the avian trkC receptors: A novel kinase insertion dissociates transformation and process outgrowth from survival. Neuron, 13: 457–572.

Gauweiler, B., Weihe, E., Hartschuh, W. and Yanaihara, N. (1988) Presence and coexistence of chromogranin A and multiple neuropeptides in Merkel cells of mammalian oral mucosa. Neurosci. Lett., 89: 121–126.

Grim, M. and Halata, Z. (2000a) Developmental origin of avian Merkel cells. Anat. Embryol. (Berlin), 202: 401–410.

Grim, M. and Halata, Z. (2000b). Developmental origin of Merkel cells in birds. In: H. Suzuki and T. Ono (Eds.), Merkel Cells, Merkel Cell Carcinoma and Neurobiology of the Skin. Elsevier Science., Amsterdam,Tokyo, pp. 23–32.

Halata, Z., Grim, M. and Baumann, K.I. (2003) Friedrich Sigmund Merkel and his 'Merkel Cell', morphology, development and physiology-review and new results. Anat. Rec., 271A: 225–239.

Halata, Z., Grim, M. and Christ, B. (1990) Origin of spinal cord meninges, sheaths of peripheral nerves, and cutaneous receptors including Merkel cells. An experimental and ultrastructural study with avian chimeras. Anat. Embryol., 182: 529–537.

Hartschuh, W. and Weihe, E. (1988). Multiple messenger candidates and marker substances in the mammalian Merkel cell-axon complex: a light and electron microscopic immunohistochemical study. In: W. Hamann and A. Iggo (Eds.), Progress in Brain Research. Elsevier, pp. 181–187.

Ichikawa, H., Matsuo, S., Silos-Santiago, I., Jaequin, M.F. and Sugimoto, T. (2001) Developmental dependency of Merkel endings on trks in the palate. Brain Res. Mol. Brain Res., 88: 171–175.

Iggo, A. and Muir, A.R. (1969) The structure and function of a slowly adapting touch corpuscle in hairy skin. J. Physiol., 200: 763–796.

Iggo, A. and Findlater, G.S. (1984). A review of Merkel cell mechanisms. In: W. Hamann and A. Iggo (Eds.), Sensory Receptor Mechanisms. World Scientific Publ. Co., Singapore, pp. 117–131.

Jiang, X., Iseki, S., Maxson, R.E., Sucov, H.M. and Morriss-Kay, G.M. (2002) Tissue origins and interactions in the mammalian skull vault. Dev. Biol., 241: 106–116.

Katz, B. and Miledi, R. (1967) The timing of calcium action during neuromuscular transmission. J. Physiol., 189: 535–544.

Kinkelin, I., Stucky, S. and Koltzenburg, M. (1999) Postnatal loss of Merkel cells, but not of slowly adapting mechanoreceptors in mice lacking the neurotrophin receptor p75. J. Neurosci., 11: 3963–3969.

Langtimm-Sedlak, C.J., Schroeder, B., Saskowski, J.L., Carnahan, J.F. and Sieber-Blum, M. (1996) Multiple action of stem cell factor in neural crest cell differentiation in vitro. Dev. Biol., 174: 345–359.

Lyne, A.G. and Hollis, D.E. (1971) Merkel cells in sheep epidermis during fetal development. J. Ultrastruct. Res., 34: 464–472.

Moll, R., Moll, I. and Franke, W.W. (1984) Identification of Merkel cells in human skin by specific cytokeratin antibodies: changes of cell density and distribution in fetal and adult plantar epidermis Differentiation, 28: 136–154.

Moll, I., Moll, R. and Franke, W.W. (1986a) Formation of epidermal and dermal Merkel cells during human fetal skin development. J. Invest. Dermatol., 87: 779–787.

Moll, R., Osborn, M., Hartschuh, W., Moll, I., Mahrle, G. and Weber, K. (1986b) Variability of expression and arrangement of cytokeratin and neurofilaments in cutaneous neuroendocrine carcinomas (Merkel cell tumors): Immunocytochemical and biochemical analysis of twelve cases. Ultrastruct. Pathol., 10: 473–495.

Moll, I., Lane, A.T., Franke, W.W. and Moll, R. (1990) Intraepidermal formation of Merkel cells in xenografts of human fetal skin. J. Invest. Dermatol., 94: 359–364.

Moll, I., Ziegler, W. and Schmelz, M. (1996) Proliferative Merkel cells were not detected in human skin. Arch. Dermatol. Res., 288: 184–187.

Moore, S.J. and Munger, B.L. (1989) The early ontogeny of the afferent nerves and papillary ridges in human digital glabrous skin. Brain Res. Dev. Brain Res., 48: 119–141.

Nurse, C.A. and Diamond, J. (1984) A fluorescent microscopic study of the development of rat touch domes and their Merkel cells. Neuroscience, 11: 509–520.

Ogawa, H. (1996) The Merkel cell as a possible mechanoreceptor cell. Prog. Neurobiol., 49: 317–334.

Oshima, H., Rochat, A., Kedzia, C., Kobayashi, K. and Barrandon, Y. (2001) Morphogenesis and renewal of hair follicles from adult multipotent stem cells. Cell, 26: 233–245.

Palko, M.E., Coppola, V. and Tessarollo, L. (1999) Evidence for a role of truncated TrkC receptor isoforms mouse developments. J. Neurosci., 19: 775–782.

Ramieri, G., Panzica, G.C., Viglietti-Panzica, C., Modica, R., Springall, D.R. and Polak, J.M. (1992) Non-innervated Merkel cells and Merkel-neurite complexes in human oral mucosa revealed using antiserum to protein gene product 9.5. Arch. Oral Biol., 37: 263–269.

Ren, Z.G., Pörzgen, P.P., Zhang, J.M., Chen, X.R., Amara, S.G., Blakely, R.D. and Sieber-Blum, M. (2001) Autocrine regulation of norepinephrine transporter expression. Mol. Cell. Neurosci., 17: 539–550.

Ren, Z.G., Pörzgen, P.P., Youn, Y.-H. and Sieber-Blum, M. (2002) Ubiquitous embryonic expression of the norepinephrine transporter. Dev. Neurosci., in press.

Saurat, J.H., Didierjean, L., Skallii, O., Siegenthaler, G. and Gabbiani, G. (1984) The intermediate filament proteins of rabbit normal epidermal Merkel cells are cytokeratins. J. Invest. Dermatol., 83: 431–435.

Schecterson, L.C. and Bothwell, M. (1992) Novel roles for neurotrophins are suggested by BDNF and NT-3 mRNA expression in developing neurons. Neuron, 9: 449–463.

Senok, S.S. and Baumann, K.I. (1997) Functional evidence for calcium-induced calcium release in isolated rat vibrissal Merkel cell mechanoreceptors. J. Physiol., 500: 29–37.

Sieber-Blum, M. (1989) Commitment of neural crest cells to the sensory neuron lineage. Science, 243: 1608–1611.

Sieber-Blum, M. and Cohen, A.M. (1980) Clonal analysis of quail neural crest cells: they are pluripotent and differentiate in vitro in the absence of noncrest cells. Dev. Biol., 80: 96–106.

Sieber-Blum, M. and Ren, Z.G. (2000) Norepinephrine transporter expression and function in noradrenergic cell differentiation. Mol. Cell. Biochem., 212: 61–70.

Sieber-Blum, M. (1999). The neural crest colony assay: assessing molecular influences on development in culture. In: L.W. Haynes (Ed.), The Neuron in Tissue Culture. IBRO, John Wiley & Sons Ltd., pp. 5–22.

Sieber-Blum, M. and Zhang, Z. (2002) Neural crest and neural crest defects. Biomed. Rev., 13: 23–32.

Snider, W.D. (1994) Functions of the neurotrophins during nervous system development: What the knockouts are teaching us. Cell, 77: 627–638.

Soriano, P. (1999) Generalized lacZ expression with the ROSA26 Cre reporter strain. Nat. Genet., 21: 70–71.

Szeder, V., Grim, M., Halata, Z. and Sieber-Blum, M. (2003a) Neural crest origin of mammalian Merkel cells. Dev. Biol., 253: 258–263.

Szeder, V., Grim, M. and Sieber-Blum, M. (2003b). Role of neurotrophin-3 in mammalian Merkel cell development. Dev. Dyn., in press.

Tazaki, M. and Suzuki, T. (1998) Calcium inflow of hamster Merkel cells in response to hyposmotic stimulation indicate a stretch activated ion channel. Neurosci. Lett., 243: 69–72.

Terenghi, G., Sundaresan, M., Moscoso, G. and Polak, J.M. (1993) Neuropeptides and a neuronal marker in cutaneous innervation during human foetal development. J. Comp. Neurol., 328: 595–603.

Tessarollo, L., Tsoulfas, P., Martin-Zanca, D., Gilbert, D.J., Jenkins, N.A., Copeland, N.G. and Parada, L.F. (1993) trkc, a receptor for neurotrophin-3, is widely expressed in the developing nervous system and in non-neuronal tissues. Development, 118: 463–475.

Tessarollo, L., Tsoulfas, P., Donovan, M.J., Palko, M.E., Blair-Flynn, J., Hempstead, B.L. and Parada, L.F. (1997) Targeted deletion of all isoforms of the trkC gene suggests the use of alternate receptors by its ligand neurotrophin-3 in neuronal development and implicates trkC in normal cardiogenesis. Proc. Natl. Acad. Sci. USA, 94: 14776–14781.

Tsoulfas, P., Soppet, D., Escandon, E., Tessarollo, L., Mendoza-Ramirez, J.L., Rosenthal, A., Nikolics, K. and Parada, L.F. (1993) The rat trkC locus encodes multiple neurogenic receptors that exhibit differential response to neurotrophin-3 in PC12 cells. Neuron, 10: 975–990.

Valenzuela, D.M., Maisonpierre, P.C., Glass, D.J., Rojas, E., Nunez, L., Kong, Y., Gies, D.R., Stitt, T.N., Ip, N.Y., Yancopoulos, G.D. (1993) Alternative forms of rat TrkC with different functional capabilities. Neuron, 10: 963–974.

Vielkind, U., Sebzda, M.K., Gibson, I.R. and Hardy, M.H. (1995) Dynamics of Merkel cell patterns in developing hair follicles in the dorsal skin of mice, demonstrated by a monoclonal antibody to mouse keratin 8. Acta Anat., 152: 93–109.

Youn et al. (2003) Stem cell defect in the cardiac neural crest of TrkC hull mice. Mol. Cell. Neurosci., in press.

Zaccone, G. (1986) Neuron-specific enolase and serotonin in the Merkel cells of conger-eel (Conger conger) epidermis. An immunohistochemical study. Histochemistry, 85: 29–34.

Zhang, J.-M., Dix, J., Langtimm-Sedlak, C.J., Trusk, T., Schroeder, B., Hoffmann, R., Strosberg, A.D., Winslow, J.W. and Sieber-Blum, M. (1997) Neurotrophin-3 and Norepinephrine-mediated adrenergic differentiation and the inhibitory action of desipramine and cocaine. J. Neurobiol., 32: 262–280.

Zhang, J.M., Hoffmann, R.M. and Sieber-Blum, M. (1997) Mitogenic and anti-proliferative signals for neural crest cells and the neurogenic action of TGF-beta1. Dev. Biol., 208: 375–386.

Growth factors, progenitor cells and cell survival

Progress in Brain Research, Vol. 146
ISSN 0079-6123

CHAPTER 5

Stem cells and nervous tissue repair: from *in vitro* to *in vivo*

Laura Calzà[1,2,*], Mercedes Fernandez[1], Alessandro Giuliani[1], Stefania Pirondi[1], Giulia D'Intino[1], Marco Manservigi[1], Nadia De Sordi[1] and Luciana Giardino[1,2]

[1]*Department of Veterinary Morphophysiology and Animal Production (DIMORFIPA), University of Bologna, Bologna, Italy*
[2]*Pathophysiology Center for the Nervous System, Hesperia Hospital, Modena, Italy*

Abstract: Recent development in stem cell biology has indicated a new possible approach for the treatment of neurological diseases. However, in spite of tremendous hope generated, we are still on the way to understand if the use of stem cells to repair mature brain and spinal cord is a reliable possibility. In particular, we know very little on the in situ regulation of adult neural stem, and this also negatively impact on cell transplant possibilities. In this chapter we will discuss issues concerning the role and function of stem cells in neurological diseases, with regard to the impact of features of degenerating neurons and glial cells on in situ stem cells. Stem cell location and biology in the adult brain, brain host reaction to transplantation, neural stem cell reaction to experimental injuries and possibilities for exogenous regulation are the main topics discussed.

Keywords: in situ neural stem cells; neurodegenerative diseases; adult brain; NGF; thyroid hormone

Introduction

In the last years the association of such concepts as 'stem cell' and 'brain repair' has generated tremendous hope and great confusion not only in lay communities but in the scientific environment as well. On the one hand, important developments of stem cell biology, cell culture and cloning strategies have exposed a research area, that could lead to unexpected clinical applications, though leaving major questions still open. On the other hand, the extremely limited capacity of the adult nervous tissue for self-repair is forcing the need for new strategies of therapeutical intervention in neurodegenerative diseases. The idea to access a potentially unlimited source and to generate differentiated cells for repair purposes has enticed the neuroscience field. Mechanisms and regulatory molecules of neurogenesis and gliogenesis are under active investigation in embryonic and adult nervous tissue. In the field of neurotrophins, a new, fascinating and expanding area is represented by their possible role in the regulation of adult neurogenesis and gliogenesis.

In this chapter we will discuss issues concerning the role and function of stem cells in neurological diseases, with regard to the impact on features of degenerating neurons and glial cells on in situ stem cells. The successful use of stem-derived cell lines for brain and spinal cord repair relies in fact on the sound knowledge of stem cell biology, whose impressive progress over the past years and also on the parallel understanding of molecular signals necessary to maintain structure and function of mature neural networks in normal brain and how these are modified by neurological diseases.

*Correspondence to: L. Calzà, DIMORFIPA,Università di Bologna, Via Tolara di Sopra 50, I-40064 Ozzano dell'Emilia, Bologna Italy. Tel.: +39-051-792-950; Fax: +39-051-792-956; E-mail: lcalza@vet.unibo.it

DOI: 10.1016/S0079-6123(03)46005-6

Proliferating cells in the central nervous system of the adult rat

Stem cell biology has revealed how most tissues are generated during fetal life from pools of stem cells, which are also responsible for tissue regeneration and, in some instance, repair in adult life (Weissman et al., 2001). Stem cells are currently defined as single cells that are clonal precursors of further stem cells of the same type, or of a defined set of differentiated progeny (Gage, 2000; Weissman, 2000a,b; Weissman et al., 2001). In most tissues, stem cells are rare; only severe, restricted conditions guarantee their isolation (Reya et al., 2001; Weissman et al., 2001; Wurmser and Gage, 2002). It is presently debated whether organ-specific stem cells are lineage-restricted (stem cells in a specific tissue can generate all cell types of the same tissue) or whether 'trans-differentiation' or 'de-differentiation' occurs (any stem cell from either embryonic or adult tissues can generate any cell type under adequate stimulus) (Andersson et al., 2001; Morshead et al., 2002; Shih et al., 2002; Vescovi et al., 2002).

Precursor cells with 'stem cell-like' properties have been identified in the nervous tissue of adult vertebrates in the subventricular zone (SVZ) of the telencephalon (Alvarez-Buylla and Garcia-Verdugo, 2002), in the dentate gyrus of the hippocampus (Gould and Gross, 2002; Song et al., 2002) and possibly also in the spinal cord (Horner et al., 2000) and mesencephalon (Chichung et al., 2002). Cells derived from these areas, when cultured in presence of mitogens like epidermal growth factor (EGF) or basic-fibroblast growth factor (bFGF), generate 'neurospheres' of self-renewing, clustering cells (Fig. 1A–C). Upon mitogen withdrawal, these cells stop proliferating, attach to the bottom of the plate and differentiate into mature cell types; within

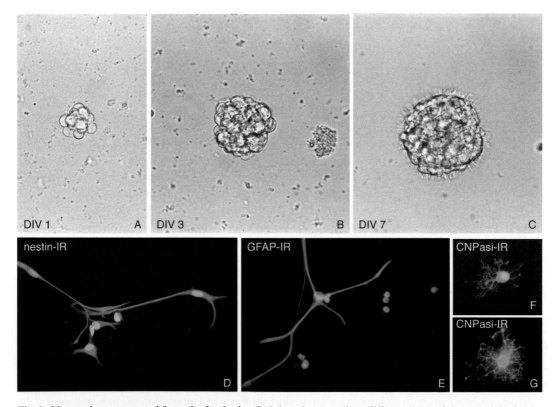

Fig. 1. Neurospheres generated from the forebrain of adult male rat, at three different times of the growth in the presence of epidermal growth factor (A–C). After mitogen withdrawal, precursors start to differentiate expressing markers for neuroectodermal cells (nestin, D), for astroglial cells (GFAP, E), oligodendrocytes (CNPase, F,G) and neurons. Nuclei are visualized by the nuclear dye Hoechst 33258.

8–12 days, the three main cell types of the mature nervous tissue, neurons, astrocytes and oligodendrocytes, can be identified in the culture by expression of selective antigens (Fig. 1D–G). As they can differentiate into all cell lines within the nervous tissue, these cells have been named 'neural stem cells' (Alvarez-Buylla and Lois, 1995; Gage et al., 1995; Weiss et al., 1996).

Labeling of proliferating cells in vivo has long limited the identification of 'stem cells' in the adult brain. This crucial step has now been superseded by labeling and identification of proliferating cells with cell-class specific markers. There are substances heavily incorporated by cycling cells, like ^{3}H-thymidine or the thymidine analog bromo-deoxyuridine (BrdU). However, as recently pointed out by several authors, BrdU marks DNA synthesis, not division: therefore more rigorous criteria should be applied to allow speculation on gliogenesis and neurogenesis in the adult brain (Cameron and McKay, 2001; Gould and Gross, 2002; Rakic, 2002). More recently, we and other authors have proposed immunolabeling of antigens normally expressed by cycling cells, like Ki67 (Fig. 1B, F) (Calzà et al., 1998; Giardino et al., 2000; Kee et al., 2002) or PCNA (Redmond et al., 1996); this procedure provides a useful tool for double labeling experiments, though it does not solve all specificity issues.

Lining the wall of the lateral ventricle in the adult mammalian brain, lies a large germinal zone, named subventricular zone (SVZ) (Luskin, 1993; Morshead et al., 1994), which has been extensively characterized in rodents (Doetsch and Alvarez-Buylla, 1996; Lois et al., 1996; Doetsch et al., 1997; Alvarez-Buylla and Garcia-Verdugo, 2002). Proliferating cells are allocated within tubes of glial cells (astrocytes) forming a migratory stream, which extends from the SVZ to the olfactory bulb (Fig. 2A–E). It has been calculated that thousands of young neurons migrate into the olfactory bulb every day (Lois and Alvarez-Buylla, 1994), but only a fraction of these cells survive to complete their differentiation. Interaction between newly generated neurons and astrocytes seems to be important not only to define or enhance migration of precursors, but also to support migrating cells. Newborn neurons can be incorporated into olfactory bulb circuits differentiating into granule cells (Petreanu and Alvarez-Buylla, 2001) and interneurons in the accessory olfactory bulb (Bonfanti et al., 1997).

Extensive neurogenesis has been described also in the dentate gyrus of the hippocampus (Altman and Das, 1965). Newly formed cells display passive membrane properties, action potentials and functional synaptic inputs similar to those found in mature dentate granule cells (Van Praag et al., 2002). It has been estimated that 9000 new cells are formed every day (Cameron and McKay, 2001); part of these immature cells turn into fully differentiated granule cells (Kempermann et al., 2003) capable to send axonal projections, containing synaptic vesicle proteins, to field CA3 (Markakis and Gage, 1999). Neurogenesis in the hippocampus has been demonstrated also in primates (Kornack and Rakic, 1999; Tonchev et al., 2003) and humans (Eriksson et al., 1998). The potential functional implications of neurogenesis in the hippocampus during adult life are impressive. Indeed, the hippocampus is thought to act as 'the gateway for memory' and a close relationship has been demonstrated between hippocampal neurogenesis and hippocampal behaviors (Kempermann et al., 2002). Neurogenesis in the hippocampus, in terms of proliferation and incorporation of newly formed neurons in existing circuits, is in fact highly regulated by different exogenous stimuli. Aging itself may regulate the balance between proliferation, leading to integration into existing circuitry, and proliferation followed by apoptosis (Kuhn et al., 1996); the same may also experience (Kempermann et al., 1998; Van Praag et al., 1999), hormones and growth factors (Cameron and McKay, 1999; Aberg et al., 2000), drugs (Brezun and Daszuta, 1999), and abnormal activation like during prolonged seizures (Parent et al., 2002).

Molecules involved in the regulation of the fate of adult 'stem cells' are poorly known. According to in vitro and neuroembryology studies, soluble molecules, including growth factors, and extracellular matrix proteins take part in the regulation of precursor fate, i.e., proliferation, migration and differentiation. Endogenous EGF and bFGF possibly participate in the regulation of neural and glial progenitors also in the mature nervous tissue (Xian and Zhou, 1999); in fact, EGF receptors are present in nestin-positive cells within the SVZ area (Fig. 2F–H). Nestin is an intermediate filament expressed by neuroectodermal cells. The embryonic form of the neuronal cell-adhesion molecule

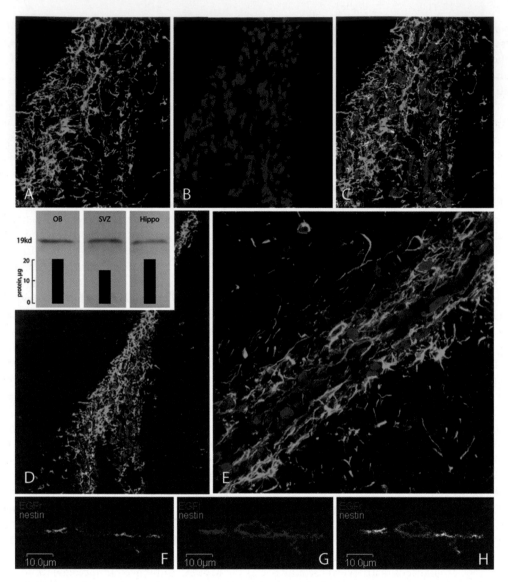

Fig. 2. The panels illustrate the population of neural stem cells localized in the subventricular zone in the wall of the lateral ventricles in the forebrain of adult rats, as visualized by using specific immunohistochemical markers and confocal laser scan microscopy. A large numbers of cells expressing the nuclear proliferating antigen Ki67 (red staining, B–E) are encompassed in the migratory stream to the olfactory bulb, circumscribed by astroglial cells (green staining, A, C–E), as observed in coronal (A–C, high magnification; D, low magnification) and sagittal (E, high magnification) sections. These cells also express nestin (F), an intermediate filament protein that is a marker for neuroectodermal cell, and receptors for mitogens like epidermal growth factor (G). These two proteins coexist in most of the proliferating cells in the SVZ (H). The blots included illustrate a possible approach to quantify proliferative activity in the brain, by using Western blot analysis of the Ki67 protein.

(polysialic acid-N-CAM) is also expressed by precursor cells in the SVZ and in the dentate gyrus, and can act as complex regulatory signal for maturation of specific cell populations from the SVZ. In fact, it has been shown that myelination, e.g., maturation of oligodendrocytes, and downregulation of PSA-NCAM expression by axons (Charles et al., 2000) coincide. Moreover, expression of PSA-NCAM in precursors seems to coincide with restriction to a glial fate (Ben-Hur et al., 1998).

We have also described expression of the low-affinity neurotrophin receptor p75 (p75^{NTR}) in the SVZ of adult rats (Calzà et al., 1998; Giuliani et al., 2002). This receptor was found in numerous dividing Ki-67 and BrDU-IR cells, thus suggesting a possible association with cell cycle regulation also in this system. Indeed, the interaction of this receptor with molecules involved in the regulation of cell cycle has been recently demonstrated (Frade, 2000a). During development, p75^{NTR} is present in the SVZ in mouse brain, where its expression parallels the expression of NRAGE (Kendall et al., 2002), p75^{NTR} intracellular binding protein regulating cell cycle progression, apoptosis and neurogenic disorders (Salehi et al., 2000; Barker et al., 2002). p75^{NTR} receptor is expressed in mitotically active neural crest cells (Stemple and Anderson, 1992; You et al., 1997), where it may influence differentiated neuron identity (Rifkin et al., 2000). Although other molecules may be implicated in p75^{NTR}-mediated cellular actions, converging evidence suggests that nerve growth factor (NGF) itself may be involved in the cell cycle regulation of neural stem cells: NGF is heavily incorporated by cells of the SVZ in neonatal rats (Calzà et al., 1998), and regulates proliferation and differentiation of neuronal stem cells previously exposed to basic fibroblast growth factor (Cattaneo and McKay, 1990). Moreover, young retinal neurons, cultured in vitro and treated with NGF, overexpress cyclin B and reenter the cell cycle (Frade, 2000b). Finally, NGF mediates protection of neural crest cells from apoptosis induced by antimitotic agent (Cortazzo et al., 1996).

Stem cell for transplantation

The extremely limited capability of the central nervous system to replace lost cells and reestablish long-distance projections suggested that brain and spinal cord would mainly benefit from transplantation therapies (Bjorklund and Lindvall, 2000); neural transplants have been proposed in animal research more than two decades ago (reviewed by Brundin et al., 1999) and have been introduced as a treatment for Parkinson's disease in the last 10 years (reviewed by Bakay et al., 1999). Parkinson's disease and the related animal models, characterized by a defect of a single transmitter, the dopamine, in a well defined brain area, the striatum, appeared to be the candidate condition for transplantation therapy, as compared to other neurological diseases. In both animal and clinical studies, different sources of potential dopaminergic cells have been used for transplant, including adrenal medulla, embryonic and fetal tissues. In all cases, postmitotic cells where used and implanted in the striatum. The general principle was therefore to provide the damaged brain with dopamine-releasing cells in the area where dopamine acts. Transplanted cells survive into the host's striatum; innervation efficacy depends from denervation model, cellular composition of the graft, concomitant delivery of trophic factors, etc.

However, in spite of impressive experimental results in animals and the early enthusiastic reports from uncontrolled clinical trials, the results of the first double-blind placebo-controlled trial of embryonic neural tissue transplantation in Parkinson disease, (Freed et al., 2001) have not been exciting. This NIH study, already initially argued against by European scientists for technical aspects including graft preparation strategies, immunotherapies and neurosurgical procedure (Widner, 1994), nonetheless has had the merit to revive the debate on limitations and perspectives of cell transplantation therapies in neurodegenerative diseases (Bjorklund and Lindvall, 2000; Brundin et al., 2001; Dunnett et al., 2001; Freed et al., 2001; Nikkhah, 2001; Robinson, 2003). A major issue, concerning the limited number of functionally active cells surviving transplantation and developing into functionally active synapses, suggests that postmitotic cells are not the best-suited source for transplantation. Moreover, logistic and ethical problems related to embryonic material collection (postmitotic cells cannot be expanded or stored longer that few days and cells from three to four embryos are needed for each transplant) make the large-scale use of neural fetal cells unreliable. Therefore, the idea to use neural stem cells for Parkinson disease has gained attractiveness, including the use of engineered multipotent stem cells to create a potentially unlimited source of neurons expressing a defined neurochemical phenotype (Arenas, 2002). Moreover this option may well be applied for other degenerative diseases, like stroke, dementia, and according to some author, demyelinating diseases.

Anyhow, the enthusiastic hope deriving from stem cell biology should be filtered through a clear mind, as pointed out in a recent review by Rossi and Cattaneo (2002). We believe that many crucial questions must be answered before considering the use of stem cells for brain and spine repair, including whether they are supposed to deliver missing substances, like neurotransmitters, growth or trophic factors, to reestablish local or long-distance connections, to remyelinate central axons or to replace extensive neural tissue loss with multiple neural phenotype. Moreover, neurobiology of the host tissue should also be outlined, considering whether it adapted to chronic disease over decades, or suffers an acute damage. Its reaction (integration or rejection) to transplanted cells must be considered, and also the presence of tissue inflammation and its positive or negative role with respect to disease neurobiology and transplantation should be taken into account. Unfortunately, unlike other tissues, brain and spine reconstruction does not simply follow introduction of new and healthy cells.

Why are endogenous stem cells ineffective in brain repair?

In vitro, neural stem cells from adult brain can generate all cell types if adequately stimulated. In vivo, after brain damage, they exhibit an extremely limited attitude, if any, to activate themselves, i.e., to increase proliferation rate, migrate to lesioned areas or differentiate into neurochemically-defined neurons. Based on this observation, the limitation can be reasonably attributed to local microenvironment rather than to intrinsic properties of 'adult stem cells' (see also Snyder and Park, 2002). Compelling evidences describe mature CNS as 'nonpermissive' for reconstruction of neural wiring. In particular, the mature CNS is an inhospitable milieu for axonal regrowth, which is on the contrary highly efficient in the hospitable milieu of the peripheral nervous system. Moreover, as experimental lesion further on alters the neurobiological context, a different molecular signaling on neural stem cells in lesioned as compared to intact brain cannot be excluded.

Endogenous neural stem cells seem to be unable to react in different experimental conditions involving lesion of selective neural population, e.g., by increasing proliferation rate. In 6-hydroxydopamine lesion, a widely used model for Parkinson disease, no spontaneous proliferation of forebrain stem cells was described (Fallon et al., 2000). Decreased rather than increased proliferation was described in SVZ (Fernandez et al., 2002) and in the hippocampus (Mohapel et al., 2002) after immunotargeting of cholinergic neurons obtained by injecting the saporin-conjugated [192]IgG-anti NGF low-affinity receptor. Moreover, proliferation and migration in the SVZ was disrupted by amyloid-beta peptide, as indicated by in vivo studies in APP-mutant mice and after intraventricular infusion of amyloid-beta peptide (Haughey et al., 2002). Though controversial (Bjorklund and Lindvall, 2000), targeted apoptosis has been reported on the contrary to induce a limited neurogenesis in the cerebral cortex of adult mouse (Magavi et al., 2000).

So far, spontaneous increase in DNA synthesis, cell proliferation and differentiation into mature phenotypes have been substantiated in stroke, mechanic brain and spinal cord injury and in experimental allergic encephalomyelitis (EAE). In ischemic lesions, a transient and regional specific increase in cell proliferation, as measured by BrdU uptake, has been shown in the SVZ (Jin et al., 2001; Zhang et al., 2001; Arvidsson et al., 2002), cerebral cortex (Zhang et al., 2001) and hippocampus, where newly-generated cells differentiate into neuron phenotype (Yagita et al., 2001; Arvidsson et al., 2002; Nakatomi et al., 2002). Traumatic brain (Chirumamilla et al., 2002) and spinal cord injury (Namiki and Tator, 1999) also induce cell proliferation in the SVZ and hippocampus of adult rat, producing not only new astrocytes for local scar building but also new cells in distant regions (Kernie et al., 2001). However, the efficiency of cell replacement is limited, as the great percentage of newly generated neuron dies shortly after generation, possibly due to the unfavorable environment (Arvidsson et al., 2002). This finding should also suggest that a similar fate is reserved to transplanted cells, which meet the same unfavorable environment than endogenous precursor, and further points out that understanding neurobiology of the host is mandatory also for successful transplantation protocols.

In demyelinating diseases too, neural stem cells and endogenous precursors are activated. Multiple sclerosis (MS) and EAE are degenerative diseases characterized by severe loss of oligodendrocytes and consequent demyelination, probably due to inflammatory and autoimmune components (Scolding et al., 1994). Oligodendrocyte precursors (OPCs) apt to differentiate into mature oligodendrocytes can be isolated from the adult brain, their presence being confirmed by NG2-immunoreactivity or probing for platelet-derived growth factor alpha-receptors (Levine et al., 2001). OPCs are present in both grey and white matter, where they comprise 5–8% of all the cells in the adult brain (Dawson et al., 2000). Are they able to repair demyelinating lesions? There is evidence of remyelination in different experimental conditions in the adult central nervous system (Levine et al., 2001). Previous results from our laboratory demonstrated that the proliferation rate and expression of nestin, a marker for neuroepithelial cells, are upregulated in the SVZ (Calzà et al., 1998) during EAE. Proliferating nuclei were visualized using both BrDU uptake and the antibody against cell cycle-associated antigen Ki67. A very high number of dividing cells was found in EAE animals, not only limited to SVZ, but also widespread in the white and grey matter in the brain and spinal cord and in between ependymal cells in the central canal. Ki67 and BrDU-positive cells were associated with inflammatory foci, but also located well apart from blood vessels and inflammatory cellular aggregates. Double labeling experiments revealed that only a few expressed OX42, a marker for macrophage/microglial lineage, or GFAP-positivity, associated with astrocytes. Moreover, cycling cells were also found among ependymal cells in the central canal; nestin-IR was found in the ependymal layer of the central canal in the spinal cord, in small, finely branched intraparenchymal cells and in reactive astrocytes. PSA-NCAM-IR cells, grouped in small aggregates, were also found in the spinal cord. These were newly-generated cells as revealed by double-labeling experiments using antibody against the proliferation-associated marker Ki67. Further research indicated that these cells evolve into myelinating oligodendrocytes (Calzà et al., 2002), as confirmed also by other investigators (Picard-Riera et al., 2002). However, although the involvement of gliogenic

areas like SVZ has been described in the EAE model (Calzà et al., 1998; Picard-Riera et al., 2002), local oligodendrocyte precursors are likely involved also in remyelination; in this context, migration of newly-generated cells might represent an accessory phenomenon.

On the contrary, although a significant number of oligodendrocyte precursor cells is found in early lesions in MS tissue (Bruck et al., 1994; Scolding et al., 1998), remyelination attempts in early plaques are not followed by repair (Perry, 1998) in chronic lesions, where oligodendrocyte precursors are in a relatively quiescent state (Wolswijk, 1998). These observations are as yet unexplained.

Although stroke, brain and spinal cord traumatic injuries and demyelinating diseases are different conditions, in which replacement involves different cell types (e.g., neurons or oligodendrocytes), nonetheless they share common features supposed to participate in the process of neural stem cells/precursors activation: inflammation and breakdown of the blood–brain barrier. Focal cerebral ischemia elicits a marked inflammatory response involving early recruitment of granulocytes and delayed infiltration of the ischemic area by T cells and macrophages (Stoll et al., 1998). In EAE and MS, multiple confluent foci of inflammation including mononuclear cells appear in many areas of the brain and spinal cord immediately after the acute phase of the disease. Mononuclear and polymorphonuclear, also including iNOS-positive cells, are found in perivascular and intraparenchymal infiltrates in a later phase (Calzà et al., 1997).

In the adult brain, inflammation is commonly considered detrimental for repair, mainly because it is followed by active phagocytosis and scar formation, which mechanically hamper cell migration or axonal growth. However, immune response and inflammation play a pivotal role in maintenance and repair of tissues other than the CNS; in fact, growing evidence indicates that under certain circumstances no repair is possible when the immune response is not coupled to some inflammation, also in the CNS (Schwartz et al., 1999; Feurstein and Wang, 2001; Warrington et al., 2001). This seems to occur in white matter injuries, including demyelinating diseases (Diemel et al., 1998), in which remyelination is impaired by substantial reduction of the inflammatory response

(Triarhou and Herndon, 1986) and by macrophages depletion (Kotter et al., 2001); in spinal cord injuries, where activated macrophages are beneficial for axon regrowth (Rapalino et al., 1998); in stroke, where it is still unclear whether the detrimental effects of inflammation outweigh neuroprotective mechanisms or vice versa (Stoll et al., 1998). Macrophages seem to act as key cells in helping tissue healing, also in the nervous system, possibly due to phagocytosis of debris; it has been suggested that a restricted inflammatory reaction in the CNS and, thus, a reduced migration of macrophages in the CNS, may represent a key impediment in axonal regeneration in the CNS (Lazarov-Spiegler et al., 1998).

Is there a link between inflammation, activated macrophages and neural stem cells? In other tissues, macrophages contribution to regeneration has been associated with the production of signaling molecules involved in cellular events of repair (DiPietro, 1995). Thus, also in the CNS, activated macrophages might as well synthesize and release factors affecting neural stem cell or progenitors. For example, activated macrophages express TGF-beta1 (Hinks and Franklin, 1999), which promotes oligodendrocyte progenitor differentiation (Franklin and Hinks, 1999). We have reported that a massive macrophages infiltration occurs in brain and spinal cord parenchyma during EAE (Calzà et al., 1997), a condition in which both neural stem cells in the SVZ (Calzà et al., 1998) and oligodendrocyte precursors are activated (Calzà et al., 2002). Interestingly, the conditioned medium from stimulated microglia or from a monocyte/macrophage cell line promotes differentiation of cholinergic neurons in the basal forebrain and synergizes with NGF in the developing forebrain (Jonakait et al., 2000).

Thus, in the damaged brain or spinal cord a highly complex cellular net between neurons, glial, inflammatory cells, stem cells and progenitors is established, thus creating an extracellular molecular milieu different from normal brain's. It is indeed in such a micro-environment that transplanted cells meet.

In vivo induction of stem cell: can we force it toward repair?

As described, the role of endogenous precursors in brain repair is extremely limited, due to the intrinsic properties of neural stem cells, to the unfavorable microenvironment of the adult brain, and to the injury itself, that seldom allows effective neurogenesis and gliogenesis. Manipulation of endogenous neural stem cells through exogenous administration of active substance has been rarely attempted, partly due to the limited knowledge of endogenous signaling molecules. The injection of both bFGF and EGF into the lateral ventricle of normal animals, expands SVZ progenitor population, leading most prominently to neurons after bFGF stimulation and to astrocytes after EGF stimulation in the olfactory bulb (Kuhn et al., 1997); b-FGF and EGF, alone or in combination, greatly increase proliferation rate around the fourth ventricle and central canal of the spinal cord, being the new cells destined to a gliogenic fate (Martens et al., 2002). This supports the view that lineage of endogenous neural stem cells develops according to the final location in the mature CNS. Also the angiogenic protein vascular endothelial growth factor stimulates neurogenesis in vivo (Jin et al., 2002). These findings offer a new perspective on the reciprocal regulation of histological components in the mature CNS, and in particular on the possible association between precursors migration in vivo and blood vessels (Palmer et al., 2000). They further support the view that conditions affecting the blood–brain barrier integrity, like inflammation, may induce microenvironmental conditions favoring neural stem cells activation. However, delivery of mitogens into the ventricular system in the adult brain is not inconsequential. For example, EGF intraventricular infusion causes local tumor formation in the majority of treated animals (Fig. 3A–D), in the form of massive intraventricular cystic formations with strong nestin-positivity (Fig. 3E).

Lately, Fallon and colleagues (2000) report that brain infusion of transforming growth factor-alpha (TGF-α) in animals with a selective lesion of the dopaminergic nigrostriatal system induces massive proliferation of forebrain stem cells, followed by migration of both glial and neural progenitors toward injection side. This is followed by increase in the number of differentiated neurons in the striatum and by improvement of the behavioral test for activation of dopaminergic nigrostriatal pathway. Two findings of this study are important to our

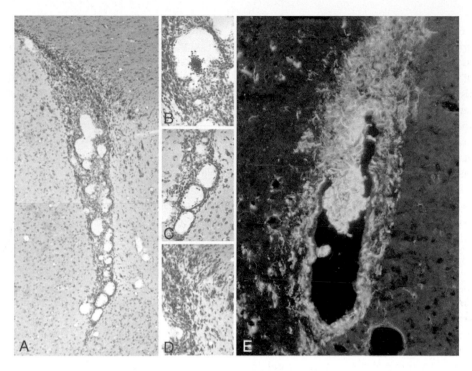

Fig. 3. Chronic delivery of epidermal growth factor into the ventricular system in the adult brain causes local tumor formation in the majority of treated animals (A–D), in the form of massive intraventricular cystic formations with strong nestin-positivity (E).

purpose: the fact that lesion itself is not sufficient to 'activate' endogenous precursors and that TGF-α is effective in lesioned animals only. This further supports the hypothesis that sensitivity of endogenous neural stem cells to microenvironmental signals only occurs during a temporal window, opened either by experimental or spontaneous pathological conditions and/or by administration of exogenous molecules.

Neurotrophins are essential regulators of nervous tissue development and of brain and spinal cord building. NGF is the essential growth factor for differentiation and survival of sympathetic, sensory and central cholinergic neurons. A new, fascinating and expanding area in the field of neurotrophins is represented by their role in the regulation of adult neurogenesis and gliogenesis. The multiple roles of neurotrophins as mediators in cell cycle regulation and differentiation during development point them out as likely candidates for physiological regulation of neural stem cell proliferation and differentiation in adult brain and as possible targets for exogenous regulation of such processes in brain repair as well

(Calzà et al., 2003a). The presence of p75[NTR] in the SVZ also suggests a possible regulatory role of neurotrophins. In fact, infusion of brain-derived neurotrophic factor (BDNF) into the lateral ventricle leads to proliferation of cells, which express neuron-specific proteins and expand from the SVZ into striatum, septum, thalamus and hypothalamus (Pencea et al., 2001). Moreover, when C17.2 cells, natural producers of NGF, BDNF and glial cell line-derived neurotrophic factor (GDNF), are genetically modified to overexpress neurotrophin-3, they also increase expression of NGF and BDNF, but not GDNF (Blesch et al., 2002).

We are exploring the possibility to use 'molecular cocktails' including neurotrophins, particularly NGF, to influence in vivo proliferation, migration and phenotype lineage of stem cells in experimentally induced pathological conditions. The immunotoxin [192]IgG-saporine induces a progressive and severe lesion of the cholinergic system in the basal forebrain, as indicated by the impairment of behavioral performance in the water maze test and by the reduction of choline acetyltransferase (ChAT) activity and of cholinergic innervation in the cortex and hippocampus. Using this paradigm, we implanted lesioned animals with a device for chronic icv delivery of EGF/bFGF and NGF and we concurrently administered retinoic acid per os. These combined treatments increased proliferation rate in SVZ in lesioned and unlesioned animals and slightly, but significantly, raise ChAT activity in the hippocampus (Fernandez et al., 2002; Calzà et al., 2003b).

Also in relation to remyelination exogenous substances may play an important role in the induction of precursors. In MS, a considerable amount of remyelination occurs; however, myelin is inappropriately thin for the corresponding axon and internodes are shorter, so that resultant remyelination is histopathologically and functionally inadequate (Franklin, 2002a; Stangel and Hartung, 2002). Increasing evidence indicates that, under certain environmental influences, cells committed to the oligodendrocyte lineage develop and migrate into the brain parenchyma to form differentiated myelinating oligodendrocytes. As this process is also present in experimental demyelinating lesions, remyelinating strategies are under active investigation in MS research. Neurotrophins, like NGF (Villoslada et al., 2000), mitogens, like bFGF and PDGF, insulin-like growth factor I (Stangel and Hartung, 2002) and tumor necrosis factor-alpha (TNF-α) (Arnett et al., 2001) have been administered to promote endogenous remyelination. However, the very low permeability of the blood–brain barrier to these proteins represent a limiting factor for the application of such strategies in human disease, as it imposes invasive delivery procedures.

Cell transplant of oligodendrocyte progenitors, Schwann cells, olfactory ensheathing cells has been also attempted, (Franklin, 2002b), with some improvement of remyelination processes. However, practical clinical application in MS therapy is unlikely considering the multiple interspersed plaques of demyelination all over the rostrocaudal extension of the brain and spinal cord and the consequent necessity to multiple transplant (in the order of tens). Moreover, effective myelination results from a complex interplay between axon and oligodendrocyte. In fact, electrical activity within the axon not only is involved in regulating differentiation and survival of oligodendrocyte precursors during

development (Barres and Raff, 1994) but also plays a key role in the induction of myelination (Demerens et al., 1996). MS also comprehends axonal pathology, which may contribute to further impairment of myelination (Ferguson et al., 1997; Bitsch et al., 2000).

In our opinion, future therapeutic strategies for MS should address remyelination improvement through administration of exogenous substances. We have investigated the capability of thyroid hormone to enhance the myelinating potential of OPCs in EAE. Thyroid hormone is known to induce more oligodendrocytes to form from multipotent neural stem cells (Rogister et al., 1999) and to influence oligodendrocyte development and maturation (Baas et al., 1997). The cellular action of the hormone is mediated by receptors of the family of nuclear receptors, which are transcription factors capable of repressing or inducing transcription of target genes (Zhang and Lazar, 2000). There are two thyroid hormone receptor genes (alpha and beta), each generating two isoforms (alpha1, alpha2, beta1, beta2) by alternative splicing. OPC cells express alpha isoforms, whereas expression of beta isoforms is confined to differentiated oligodendrocytes (Rodriguez-Pena, 1999). Early in development, thyroid hormone functions as an instructive agent, triggering exit from cell cycle (Durand and Raff, 2000). In postmitotic oligodendrocytes, it increases morphological and functional maturation by stimulating expression of various genes, such as myelin-oligodendrocyte glycoprotein, myelin-basic protein (MBP) and glutamine synthase. Moreover, a 120-bp enhancer element region, in the gene encoding for the intermediate filament protein nestin in progenitor cells, contains putative binding sites for nuclear hormone receptor (thyroid hormone). This suggests that nuclear receptors for thyroid hormone play a role in regulating the progenitor marker protein nestin (Lothian et al., 1999). Finally, a magnetic resonance spectroscopy study has shown that thyroxin therapy can reverse abnormal myelination in congenital hypothyroidism even when thyroxin therapy was initiated beyond the age limit when abnormalities in myelinogenesis are considered irreversible (Jagannathan et al., 1998). We have treated EAE animals with three administrations of thyroid hormone in the advanced stage of the disease, when a severe disorganization of myelin sheet was observed (Calzà et al., 2002; Giardino et al., 2002); thyroid hormone treatment reduced the number of proliferating cells in SVZ and spinal cord, and favored OPCs differentiation in EAE rats; the expression of markers for undifferentiated precursors (nestin) actually decreased. Thyroid hormone treatment induced onset of O4-positive cells and up-regulation of A2B5-IR and mRNA for platelet-derived growth factor receptor (PDGFR). PDGF is a powerful inductor of oligodendrocyte lineage from stem cells. More importantly, myelin basic protein (MBP), which decreases in EAE animals, is significantly upregulated in thyroid hormone treated EAE animals (Calzà et al., 2002); double labeling with BrdU and the precursor oligodendrocyte marker NG2 reveals that newly generated oligodendrocytes take part in the process. Moreover, the clinical course of the disease improves in T4-treated animals (Giardino et al., 2002).

Nerve growth factor and demyelinating diseases: beyond myelin repair

In recent years, different groups have produced convincing data supporting NGF role in remyelination. We and others (Bracci-Laudiero et al., 1992; Calzà et al., 1997) have described multiple molecular alterations of the NGF system in the CNS during EAE. For example, a widespread expression of TrkA and p75^{NGFR} can be observed in CNS white matter in cells sharing morphological similarity with astrocytes and/or microglial cells, in oligodendrocytes and in a proportion of mononuclear cells and macrophages comprising the inflammatory cellular infiltrates; such widespread expression correlates with severe alterations in NGF content in different brain areas (Calzà et al., 1997). Administration of human recombinant-NGF was shown to delay the onset of clinical EAE in marmoset, preventing the full development of inflammation and demyelination (Villoslada et al., 2000). We have also shown that treatments known to increase endogenous synthesis of NGF and to regulate oligodendrocytes maturation, as thyroid hormone administration, may improve the clinical course of EAE (Giardino et al., 2002).

NGF can improve the clinical course of EAE (Ransohoff and Trebst, 2000) acting on different cell types, including inflammatory and immune cells, glial cells and neurons and can regulate inflammatory processes and the immune response also in neural tissue (Villoslada et al., 2000). It may protect oligodendrocytes from injury and death induced by TNF-α as indicated by in vitro studies (Takano et al., 2000). Moreover, a possible direct effect of NGF on oligodendrocytes in myelin formation is suggested by in vitro findings showing that NGF induces proliferation of oligodendrocytes isolated from adult pig brain (Althaus et al., 1992); it also activates trkA-mediated intracellular pathways leading to genomic effect (Althaus et al., 1997), possibly participating in myelin formation (Althaus et al., 2001). Finally, NGF can be neuroprotective during EAE. In fact, axonal damage and neural loss have been described in MS (Ferguson et al., 1997; Evangelou et al., 2000a,b) and EAE (Kornek et al., 2000) and altered axonal function may also impair remyelination (Demerens et al., 1996).

Conclusions

In vivo neural stem cell biology is still largely unknown. We do not know, as yet, the endogenous factors regulating proliferation, migration and differentiation of neural stem cells, nor do we understand the balance between intrinsic properties of neural stem cells and microenvironmental signals regulating these processes during adulthood; the impact of pathological conditions is also by far undetermined. Although therapies based on stem cells, including cell transplant, certainly are a great opportunity and a future frontier for neurological diseases, in vivo regulation of neural stem cells and progenitors cannot be disregarded. Knowledge of the molecular mechanisms keeping the balance of endogenous stem cells and of mechanisms hindering repair in the damaged brain and regeneration of the central axons is fundamental for the successful application of transplantation strategies.

Enthusiastic hope for stem cell therapy in neurological diseases should be tempered by clear statements about the complexity of the goal and the many black holes scattered along the neurobiology of the candidate pathological conditions. According to Bjorklund and Lindvall (2000), 'we would do well to learn the lesson from the troubled path of gene therapy field: not to promise too much too early', also for stem cells therapy in neurological diseases.

Acknowledgments

This paper was supported by Fondazione Cassa di Risparmio, Bologna (CARISBO) and Telethon (grant 1336). Giulia D'Intino is receiving a Hesperia Hospital PhD fellowship. The English editing of the manuscript by Dr. Bianca Rossini is gratefully acknowledged.

Abbreviations

BDNF	brain-derived neurotrophic factor
bFGF	basic fibroblast growth factor
BrdU	bromo-deoxyuridine
ChAT	choline acetyl transferase
CNS	central nervous system
EAE	experimental allergic encephalomyelitis
EGF	epidermal growth factor
GDNF	glial cell line-derived neurotrophic factor
GFAP	glial fibrillary acid protein
MBP	myelin basic protein
MS	multiple sclerosis
iNOS	inducible-nitric oxide synthase
NGF	nerve growth factor
OPCs	oligodendrocyte precursor cells
PDGF	platelet-derived growth factor
p75[NTR]	neurotrophin low-affinity receptor
PSA-NCAM	polysialic acid neuronal adhesion molecule
SVZ	subventricular zone
T4	thyroxine
TGF	transforming growth factor
TNF	tumor necrosis factor
TrkA	receptor tyrosine kinase A

References

Aberg, M.A.I., Aberg, N.D., Hedbacker, H., Oscarsson, J. and Eriksson, P.S. (2000) Peripheral infusion of IGF-I selectively induces neurogenesis in the adult rat hippocampus. J. Neurosci., 20: 2896–2903.

Althaus, H.H., Kloppner, S., Schmidt-Schultz, T. and Schwartz, P. (1992) Nerve growth factor induces proliferation and enhances fiber regeneration in oligodendrocytes isolated from adult pig brain. Neurosci. Lett., 135: 219–223.

Althaus, H.H., Hempel, R., Kloppner, S., Engel, J., Schmidt-Schultz, T., Kruska, L. and Heumann, R. (1997) Nerve growth factor signal transduction in mature pig oligodendrocytes. J. Neurosci. Res., 50: 729–742.

Althaus, H.H., Mursch, K. and Kloppner, S. (2001) Differential response of mature TrkA/p75(NTR) expressing human and pig oligodendrocytes: aging, does it matter? Microsc. Res. Tech., 52: 689–699.

Altman, J. and Das, G.D. (1965) Autoradiographic and histological evidence of postnatal neurogenesis in rats. J. Comp. Neurol., 124: 319–335.

Alvarez-Buylla, A. and Lois, C. (1995) Neuronal stem cells in the brain of adult vertebrates. Stem Cells, 13: 263–272.

Alvarez-Buylla, A. and Garcia-Verdugo, J.M. (2002) Neurogenesis in adult subventricular zone. J. Neurosci., 22: 629–634.

Andersson, D.J., Gage, F.H. and Weissman, I.L. (2001) Can stem cells cross lineage boundaries? Nat. Med., 7: 393–395.

Arenas, E. (2002) Stem cells in the treatment of Parkinson's disease. Brain Res. Bull., 57: 795–808.

Arnett, H.A., Mason, J., Marino, M., Suzuki, K., Matsushima, G.K. and Ting, J.P.-Y. (2001) TNF-α promotes proliferation of oligodendrocyte progenitors and remyelination. Nat. Neurosci., 4: 1116–1122.

Arvidsson, A., Collin, T., Kirik, D., Kokaia, Z. and Lindvall, O. (2002) Neuronal replacement from endogenous precursors in the adult brain after stroke. Nat. Med., 8: 963–970.

Baas, D., Bourbeau, D., Sarlieve, L.L., Ittel, M.E., Dussault, J.H. and Puymirat, J. (1997) Oligodendrocyte maturation and progenitor cell proliferation are independently regulated by thyroid hormone. Glia, 19: 324–332.

Bakay, R.A.E., Kordower, J.H. and Starr, P.A. (1999). Restorative surgical therapies for Parkinson's disease. In: M.H. Tuszynski and J. Kordower (Eds.), CNS Regeneration. Basic Science and Clinical Advances. Academic Press, London, pp. 389–417.

Barker, P.A., Hussain, N.K. and McPherson, P.S. (2002) Retrograde signaling by the neurotrophins follows a well-worn trk. Trends Neurosci., 25: 379–381.

Barres, B.A. and Raff, M.C. (1994) Control of oligodendrocyte number in the developing rat optic nerve. Neuron, 12: 935–942.

Ben-Hur, T., Rogister, B., Murray, K., Rougon, G. and Dubois-Dalcq, M. (1998) Growth and fate of PSA-NCAM + precursors of the postnatal brain. J. Neurosci., 18: 5777–5788.

Bitsch, A., Schuchardt, J., Bunkowski, S., Kuhlmann, T. and Bruck, W. (2000) Acute axonal injury in multiple sclerosis. Correlation with demyelination and inflammation. Brain, 123: 1174–1183.

Bjorklund, A. and Lindvall, O. (2000) Cell replacement therapies for central nervous system disorders. Nat. Neurosci., 3: 537–544.

Blesch, A., Lu, P. and Tuszynski, M.H. (2002) Neurotrophic factors, gene therapy, and neural stem cells for spinal cord repair. Brain Res. Bull., 57: 833–838.

Bonfanti, L., Peretto, P., Merighi, A. and Fasolo, A. (1997) Newly-generated cells from the rostral migratory stream in the accessory olfactory bulb of the adult rat. Neuroscience, 81: 489–502.

Bracci-Laudiero, L., Aloe, L., Levi-Montalcini, R., Buttinelli, C., Schilter, D., Gillessen, S. and Otten, U. (1992) Multiple sclerosis patients express increased levels of beta-nerve growth factor in cerebrospinal fluid. Neurosci. Lett., 147: 9–12.

Brezun, J.M. and Daszuta, A. (1999) Depletion in serotonin decreases neurogenesis in the dentate gyrus and the subventricular zone of adult rats. Neuroscience, 89: 999–1002.

Bruck, W., Schmied, M., Suchanek, G., Bruck, Y., Breitschopf, H., Poser, S., Piddlesden, S. and Lassmann, H. (1994) Oligodendrocytes in the early course of multiple sclerosis. Ann. Neurol., 35: 65–73.

Brundin, P., Emgard, M. and Mundt-Petersen, U. (1999). Graft of embryonic dopamine neurons in rodent models of Parkinson's disease. In: M.H. Tuszynski and J. Kordower (Eds.), CNS Regeneration. Basic Science and Clinical Advances. Academic Press, London, pp. 299–320.

Brundin, P., Dunnett, A., Bjorklund, A. and Nikkhah, G. (2001) Transplanted dopaminergic neurons: more or less? Nat. Med., 7: 512–513.

Calzà, L., Giardino, L., Pozza, M., Micera, A. and Aloe, L. (1997) Time-course changes of nerve growth factor, corticotropin-releasing hormone, and nitric oxide synthase isoforms and their possible role in the development of inflammatory response in experimental allergic encephalomyelitis. Proc. Natl. Acad. Sci. USA, 94: 3368–3373.

Calzà, L., Giardino, L., Pozza, M., Bettelli, C., Micera, A. and Aloe, L. (1998) Proliferation and phenotype regulation in the subventricular zone during experimental allergic encephalomyelitis: in vivo evidence of a role for nerve growth factor. Proc. Natl. Acad. Sci. USA, 95: 3209–3214.

Calzà, L., Fernandez, M., Giuliani, A., Aloe, L. and Giardino, L. (2002) Thyroid hormone activates oligodendrocyte precursors and increases a myelin-forming protein and NGF content in the spinal cord during experimental allergic encephalomyelitis. Proc. Natl. Acad. Sci. USA, 99: 3258–3263.

Calzà, L., Fernandez, M., Giuliani, A., Pirondi, S., D'Intino, G. and Giardino, L. (2003a) Nerve growth factor in the central nervous system: more than neuron survival. Arch. Ital. Biol., 141: 93–102.

Calzà, L., Giuliani, A., Fernandez, M., Pirondi, S., D'Intimo, F., Aloe, L. and Giardino, L. (2003b) Neural stem cells and cholinerpic neurons: regulation by immunolesion and treatment with mitogens, retinoic acid, and nerve growth factor. Proc. Natl. Acad. Sci., 100: 7325–7330.

Cameron, H.A. and McKay, R.D. (1999) Restoring production of hippocampal neurons in old age. Nat. Neurosci., 2: 894–897.

Cameron, H.A. and McKay, R.D. (2001) Adult neurogenesis produces a large pool of new granule cells in the dentate gyrus. J. Comp. Neurol., 435: 406–417.

Cattaneo, E. and McKay, R. (1990) Proliferation and differentiation of neuronal stem cells regulated by nerve growth factor. Nature, 347: 762–765.

Charles, P., Hernandez, M.P., Stankoff, B., Aigrot, M.S., Colin, C., Rougon, G., Zalc, B. and Lubetzki, C. (2000) Negative regulation of central nervous system myelination by polysialylated-neural cell adhesion molecule. Proc. Natl. Acad. Sci. USA, 97: 7585–7590.

Chichung, D., Dziewczapolski, G., Willhoite, A.R., Kaspar, B.K., Shults, C.W. and Gage, F.H. (2002) The adult substantia nigra contains progenitor cells with neurogenic potential. J. Neurosci., 22: 6639–6649.

Chirumamilla, S., Sun, D., Bullock, M.R. and Colello, R.J. (2002) Traumatic brain injury induced cell proliferation in the adult mammalian central nervous system. J. Neurotrauma, 19: 693–703.

Cortazzo, M.H., Kassis, E.S., Sproul, K.A. and Schor, N.F. (1996) Nerve growth factor (NGF)-mediated protection of neural crest cells from antimitotic agent-induced apoptosis: the role of the low-affinity NGF receptor. J. Neurosci., 16: 3895–3899.

Dawson, M.R., Levine, J.M. and Reynolds, R. (2000) NG2-expressing cells in the central nervous system: are they oligodendroglial progenitors? J. Neurosci. Res., 61: 471–479.

Demerens, C., Stankoff, B., Logak, M., Anglade, P., Allinquant, B., Couraud, F., Zalc, B. and Lubetzki, C. (1996) Induction of myelination in the central nervous system by electrical activity. Proc. Natl. Acad. Sci. USA, 93: 9887–9892.

Diemel, L.T., Copelman, C.A. and Cuzner, M.L. (1998) Macrophages in CNS remyelination: friend or foe? Neurochem. Res., 23: 341–347.

DiPietro, L.A. (1995) Wound healing: the role of the macrophage and other immune cells. Shock, 4: 233–240.

Doetsch, F. and Alvarez-Buylla, A. (1996) Network of tangential pathways for neuronal migration in adult mammalian brain. Proc. Natl. Acad. Sci. USA, 93: 14895–14900.

Doetsch, F., Garcia-Verdugo, J.M. and Alvarez-Buylla, A. (1997) Cellular composition and three-dimensional organization of the subventricular germinal zone in the adult mammalian brain. J. Neurosci., 17: 5046–5061.

Dunnett, S.B., Bjorklund, A. and Lindvall, O. (2001) Cell therapy in Parkinson's disease — stop or go? Nat. Neurosci. Rev., 2: 365–369.

Durand, B. and Raff, M. (2000) A cell-intrinsic timer that operates during oligodendrocyte development. BioEssay, 22: 64–71.

Eriksson, P.S., Perfilieva, E., Bjork-Eriksson, T., Alborn, A.-M., Nordborg, C., Peterson, D.A. and Gage, F.H. (1998) Neurogenesis in the adult human hippocampus. Nat. Med., 4: 1313–1317.

Evangelou, N., Konz, D., Esiri, M.M., Smith, S., Palace, J. and Matthews, P.M. (2000a) Regional axonal loss in the corpus callosum correlates with cerebral white matter lesion volume and distribution in multiple sclerosis. Brain, 123: 1845–1849.

Evangelou, N., Esiri, M.M., Smith, S., Palace, J. and Matthews, P.M. (2000b) Quantitative pathological evidence for axonal loss in normal appearing white matter in multiple sclerosis. Ann. Neurol., 47: 391–395.

Fallon, J., Reid, S., Kinyamu, R., Opole, I., Baratta, J., Korc, M., Endo, T.L., Duong, A., Nguyen, G., Karkehabadhi, M., Twardzik, D. and Loughlin, S. (2000) In vivo induction of massive proliferation, directed migration, and differentiation of neural cells in the adult mammalian brain. Proc. Natl. Acad. Sci. USA, 97: 14686–14691.

Ferguson, B., Matyszak, M.K., Esiri, M.M. and Perry, V.H. (1997) Axonal damage in acute multiple sclerosis lesions. Brain, 120: 393–399.

Fernandez, M., Giuliani, A., Giardino, L. and Calzà, L. (2002) In vivo strategies for stem cells regulation in the adult brain: a chance for cholinergic neurons? Soc. Neurosci. Abstr., 483.14.

Feurstein, G.Z. and Wang, X. (2001) Inflammation and stroke: benefits without harm? Arch. Neurol., 58: 672–674.

Frade, J.M. (2000a) NRAGE and the cycling side of the neurotrophin receptor p75. Trends Neurosci., 23: 591–592.

Frade, J.M. (2000b) Unscheduled re-entry into the cell cycle induced by NGF precedes cell death in nascent retinal neurons. J. Cell Sci., 113: 1139–1148.

Franklin, R.J.M. and Hinks, G.L. (1999) Understanding CNS remyelination: clues from developmental and regeneration biology. J. Neurosci. Res., 58: 207–213.

Franklin, R.J.M. (2002a) Why does remyelination fail in multiple sclerosis? Nat. Neurosci. Rev., 3: 705–714.

Franklin, R.J.M. (2002b) Remyelination of the demyelinated CNS: the case for and against transplantation of central, peripheral and olfactory glia. Brain Res. Bull., 57: 827–832.

Freed, C.R., Greene, P.E., Breeze, R.E., Tsai, W.Y., DuMouchel, W., Kao, R., Dillon, S., Winfield, H., Culver, S., Trojanowski, J.Q., Eidelberg, D. and Fahn, S. (2001) Transplantation of embryonic dopamine neurons for severe Parkinsons's disease. N. Engl. J. Med., 344: 710–719.

Gage, F.H. (2000) Mammalian neural stem cells. Science, 287: 1433–1438.

Gage, F.H., Coates, P.W., Palmer, T.D., Kuhn, H.G., Fisher, L.J., Suhonen, J.O., Peterson, D.A., Suhr, S.T. and Ray, J. (1995) Survival and differentiation of adult neuronal progenitor cells transplanted to the adult brain. Proc. Natl. Acad. Sci. USA, 92: 11879–11883.

Giardino, L., Bettelli, C. and Calzà, L. (2000) In vivo regulation of precursor cells in the subventricular zone of adult rat brain by thyroid hormone and retinoids. Neurosci. Lett., 295: 17–20.

Giardino, L., Fernandez, M., Giuliani, A., Pirondi, S. and Calzà, L. (2002) Thyroid hormone improves clinical course of experimental allergic encephalomyelitis and stimulates oligo-dendrocyte precursor maturation. Soc. Neurosci. Abstr., 799.15.

Giuliani, A., D'Intino, G., Fernandez, M., Zanni, M., Giardino, L. and Calzà, L. (2002) p75-Immunoreactivity in the subventricular zone in adult male rats. Soc. Neurosci. Abstr., 127.7.

Gould, E. and Gross, C.G. (2002) Neurogenesis in adult mammals: some progress and problems. J. Neurosci., 22: 619–623.

Haughey, N.J., Liu, D., Nath, A., Borchard, A.C. and Mattson, M.P. (2002) Disruption of neurogenesis in the subventricular zone of adult mice, and in human cortical neuronal precursor cells in culture, by amyloid beta-peptide: implications for the pathogenesis of Alzheimer's disease. Neuromol. Med., 1: 125–1135.

Hinks, G.L. and Franklin, R.J.M. (1999) Distinctive patterns of PDGF-A, FGF-2, IGF-I and TGF-beta-1 gene expression during remyelination of experimentally-induced spinal cord demyelination. Mol. Cell. Neurosci., 14: 153–168.

Horner, P.J., Power, A.E., Kempermann, G., Kuhn, H.G., Palmer, T.D., Winkler, J., Thal, L.J. and Gage, F.H. (2000) Proliferation and differentiation of progenitor cells through-out the intact adult rat spinal cord. J. Neurosci., 20: 2218–2228.

Jagannathan, N.R., Tandon, N., Raghunathan, P. and Kochupillai, N. (1998) Reversal of abnormalities of myelina-tion by thyroxine therapy in congenital hypothyroidism: localized in vivo proton magnetic resonance spectroscopy (MRS) study. Dev. Brain Res., 109: 179–186.

Jin, K., Minami, M., Lan, J.Q., Ou Mao, X., Batteur, S., Simon, R.P. and Greenberg, D.A. (2001) Neurogenesis in dentate subgranular zone and rostral subventricular zone after focal cerebral ischemia in the rat. Proc. Natl. Acad. Sci. USA, 98: 4710–4715.

Jin, K., Zhu, Y., Sun, Y., Mao, X.O., Xie, L. and Greenberg, D.A. (2002) Vascular endothelial growth factor (VEGF) stimulates neurogenesis in vitro and in vivo. Proc. Natl. Acad. Sci. USA, 99: 11946–11950.

Jonakait, G.M., Wen, Y., Wan, Y. and Ni, L. (2000) Macrophage cell-conditioned medium promotes cholinergic differentiation of undifferentiated progenitors and synergizes with nerve growth factor action in the developing basal forebrain. Exp. Neurol., 161: 285–296.

Kee, N., Sivalingam, S., Boonstra, R. and Wojtowicz, J.M. (2002) The utility of Ki-67 and BrdU as proliferative markers of adult neurogenesis. J. Neurosci. Meth., 115: 97–105.

Kempermann, G., Kuhn, H.G. and Gage, F.H. (1998) Experience-induced neurogenesis in the senescent dentate gyrus. J. Neurosci., 18: 3206–3212.

Kempermann, G., Gast, D. and Gage, F.H. (2002) Neuroplasticity in old age: sustained fivefold induction of hippocampal neurogenesis by long-term environmental enrichment. Ann. Neurol., 52: 135–143.

Kempermann, G., Gast, D., Kronenberg, G., Yamaguchi, M. and Gage, F.H. (2003) Early determination and long-term persistence of adult-generated new neurons in the hippocampus of mice. Development, 130: 391–399.

Kendall, S.E., Goldhawk, D.E., Kubu, C., Barker, P.A. and Verdi, J.M. (2002) Expression analysis of a novel p75^{NTR} signaling protein, which regulates cell cycle progression and apoptosis. Mech. Dev., 117: 187–200.

Kernie, S.G., Erwin, T.M. and Parada, L.F. (2001) Brain remodeling due to neuronal and astrocytic proliferation after controlled cortical injury in mice. J. Neurosci. Res., 66: 317–326.

Kornack, D.R. and Rakic, P. (1999) Continuation of neurogenesis in the hippocampus of the adult macaque monkey. Proc. Natl. Acad. Sci. USA, 96: 5768–5773.

Kornek, B., Storch, M.K., Weissert, R., Wallstroem, E., Stefferl, A., Olsson, T., Linington, C., Schmidbauer, M and Lassmann, H. (2000) Multiple sclerosis and chronic autoimmune encephalomyelitis: a comparative quantitative study of axonal injury in active, inactive, and remyelinated lesions. Am. J. Pathol., 157: 267–276.

Kotter, M.R., Setzu, A., Sim, F.J., Van Rooijen, N. and Franklin, J.M. (2001) Macrophage depletion impairs oligodendrocyte remyelination following lysolecithin-induced demyelination. Glia, 35: 204–212.

Kuhn, H.G., Dickinson-Anson, H. and Gage, F.H. (1996) Neurogenesis in the dentate gyrus of the adult rat: age-related decrease of neuronal progenitor proliferation. J. Neurosci., 16: 2027–2033.

Kuhn, H.G., Winkler, J., Kempermann, G., Thal, L.J. and Gage, F.H. (1997) Epidermal growth factor and fibroblast growth factor-2 have different effects on neural progenitors in the adult rat brain. J. Neurosci., 17: 5820–5829.

Lazarov-Spiegler, O., Rapalino, O., Agranov, G. and Schwartz, M. (1998) Restricted inflammatory reaction in the CNS: a key impediment to axonal regeneration? Mol. Med. Today, 4: 337–342.

Levine, J.M., Reynolds, R. and Fawcett, J.W. (2001) The oligodendrocyte precursor cell in health and disease. Trends Neurosci., 24: 39–47.

Lois, C. and Alvarez-Buylla, A. (1994) Long-distance neuronal migration in the adult mammalian brain. Science, 264: 1145–1148.

Lois, C., Garcia-Verdugo, J.-M. and Alvarez-Buylla, A. (1996) Chain migration of neuronal precursors. Science, 271: 978–981.

Lothian, C., Prakash, N., Lendahl, U. and Wahlstrom, G.M. (1999) Identification of both general and region-specific embryonic CNS enhancer elements in the nestin promoter. Exp. Cell Res., 248: 509–519.

Luskin, M.B. (1993) Restricted proliferation and migration of postnatally generated neurons derived from the forebrain subventricular zone. Neuron, 11: 173–189.

Magavi, S.S., Leavitt, B.R. and Macklis, J.D. (2000) Induction of neurogenesis in the neocortex of adult mice. Nature, 405: 951–955.

Markakis, E.A. and Gage, F.H. (1999) Adult-generated neurons in the dentate gyrus send axonal projections to field CA3 and are surrounded by synaptic vesicles. J. Comp. Neurol., 406: 449–460.

Martens, D.J., Seaberg, R.M. and van der Kooy, D. (2002) In vivo infusions of exogenous growth factors into the fourth ventricle of the adult mouse brain increase the proliferation of neural progenitors around the fourth ventricle and the central canal of the spinal cord. Eur. J. Neurosci., 16: 1045–1057.

Mohapel, P., Leanza, G. and Lindvall, O. (2002) Alterations in forebrain acetylcholine influence hippocampal neurogenesis in the adult rodent. Soc. Neurosci. Abstr., 23.9.

Morshead, C.M., Reynolds, B.A., Craig, C.G., McBurney, M.W., Staines, W.A., Morassutti, D., Weiss, S. and Van der Kooy, D. (1994) Neural stem cells in the adult mammalian forebrain: a relatively quiescent subpopulation of subependymal cells. Neuron, 13: 1071–1082.

Morshead, C.M., Van der Kooy, D. and Iscove, N.N. (2002) Reply to 'Hematopoietic potential of neural stem cells'. Nat. Med., 8: 536–537.

Nakatomi, H., Kuriu, T., Okabe, S., Yamamoto, S., Kawahara, N., Tamura, A., Kirino, T. and Nakafuku, M. (2002) Regeneration of hippocampal pyramidal neurons after ischemic brain injury by recruitment of endogenous neural progenitors. Cell, 110: 429–441.

Namiki, J. and Tator, C.H. (1999) Cell proliferation and nestin expression in the ependyma of the adult rat spinal cord after injury. J. Neuropathol. Exp. Neurol., 58: 489–498.

Nikkhah, G. (2001) Neural transplantation therapy for Parkinson's disease: potential and pitfalls. Brain Res. Bull., 56: 509.

Palmer, T.D., Willhoite, A.R. and Gage, F.H. (2000) Vascular niche for adult hippocampal neurogenesis. J. Comp. Neurol., 425: 479–494.

Parent, J.M., Valentin, V.V. and Lowenstein, D.H. (2002) Prolonged seizures increase proliferating neuroblasts in the adult rat subventricular zone-olfactory bulb pathway. J. Neurosci., 22: 3174–3188.

Pencea, V., Bingaman, K.D., Wiegand, S.J. and Luskin, M.B. (2001) Infusion of brain-derived neurotrophic factor into the lateral ventricle of the adult rat leads to new neurons in the parenchyma of the striatum, septum, thalamus, and hypothalamus. J. Neurosci., 21: 6706–6717.

Perry, V.H. (1998) Reluctant remyelination: the missing precursors. Brain, 121: 2219–2220.

Petreanu, L.T. and Alvarez-Buylla, A. (2001) Granule cell replacement in the olfactory bulb of adult anosmic mice. Soc. Neurosci. Abstr., 140: 15.

Picard-Riera, N., Decker, L., Delarasse, C., Goude, K., Nait-Oumesmar, B., Liblau, R., Pham-Dinh, D. and Baron-Van Evercooren, A. (2002) Experimental autoimmune encephalomyelitis mobilizes neural progenitors from the subventricular zone to undergo oligodendrogenesis in adult mice. Proc. Natl. Acad. Sci. USA, 99: 13211–13216.

Rakic, P. (2002) Adult neurogenesis in mammals: an identity crisis. J. Neurosci., 22: 614–618.

Ransohoff, R.M. and Trebst, C. (2000) Surprising pleiotropy of nerve growth factor in the treatment of experimental autoimmune encephalomyelitis. J. Exp. Med., 191: 1625–1629.

Rapalino, O., Lazarov-Spiegler, O., Agranov, E., Velan, G.J., Yoles, E., Fraidakis, M., Solomon, A., Gepstein, R., Katz, A., Belkin, M., Hadani, M. and Schwartz, M. (1998) Implantation of stimulated homologous macrophages results in partial recovery of paraplegic rats. Nat. Med., 4: 814–821.

Redmond, L., Hockfield, S. and Morabito, M.A. (1996) The divergent homeobox gene PBX1 is expressed in the postnatal subventricular zone and interneurons of the olfactory bulb. J. Neurosci., 16: 2972–2982.

Reya, T., Morrison, S.J., Clarke, M.F. and Weissman, I.L. (2001) Stem cell, cancer, and cancer stem cells. Nature, 414: 105–111.

Rifkin, J.T., Todd, V.J., Anderson, L.W. and Lefcort, F. (2000) Dynamic expression of neurotrophin receptors during sensory neuron genesis and differentiation. Dev. Biol., 227: 465–480.

Robinson, R. (2003) Another set back for fetal transplants for Parkinson's disease. Lancet Neurol., 2: 69.

Rodriguez-Pena, A. (1999) Oligodendrocyte development and thyroid hormone. J. Neurobiol., 40: 497–512.

Rogister, B., Ben-Hur, T. and Dubois-Dalcq, M. (1999) From neural stem cells to myelinating oligodendrocytes. Mol. Cell Neurosci., 14: 287–300.

Rossi, F. and Cattaneo, E. (2002) Neural stem cell therapy for neurological diseases: dreams and reality. Nat. Neurosci. Rev., 3: 401–409.

Salehi, A., Zeindler, C., Kubu, C.J., Bhaker, A., Verdi, J.M. and Barker, P.A. (2000) NRAGE, a novel MAGE protein, interacts with p75 neurotrophin receptor and facilitates nerve growth factor dependent apoptosis. Neuron, 27: 279–288.

Schwartz, M., Moalem, G., Leibowitz-Amit, R. and Cohen, I.R. (1999) Innate and adaptive immune responses can be beneficial for CNS repair. Trends Neurosci., 22: 295–299.

Scolding, N.J., Zajicek, J.P., Wood, N. and Compston, D.A. (1994) The pathogenesis of demyelinating disease. Prog. Neurobiol., 43: 143–173.

Scolding, N.J., Franklin, R., Stevens, S., Heldin, C.H., Compston, A. and Newcombe, J. (1998) Oligodendrocyte progenitors are present in the normal adult human CNS and in the lesions of multiple sclerosis. Brain, 121: 2221–2228.

Shih, C.-C., Mamelak, A., LeBon, T. and Forman, S.J. (2002) Hematopoietic potential of neural stem cells. Nat. Med., 8: 535–536.

Snyder, E.Y. and Park, K.I. (2002) Limitations in brain repair. Nat. Med., 8: 928–930.

Song, H.J., Stevens, C.F. and Gage, F.H. (2002) Neural stem cells from adult hippocampus develop essential properties of functional CNS neurons. Nat. Neurosci., 5: 438–445.

Stangel, M. and Hartung, H.-P. (2002) Remyelinating strategies for the treatment of multiple sclerosis. Prog. Neurobiol., 68: 361–376.

Stemple, D.L. and Anderson, D.J. (1992) Isolation of a stem cell for neurons and glia from the mammalian neural crest. Cell, 71: 973–985.

Stoll, G., Jander, S. and Schroeter, M. (1998) Inflammation and glial responses in ischemic brain lesions. Prog. Neurobiol., 56: 149–171.

Takano, R., Hisahara, S., Namikawa, K., Kiyama, H., Okano, H. and Miura, M. (2000) Nerve growth factor protects oligodendrocytes from tumor necrosis factor-alpha-induced injury through Akt-mediated signaling mechanisms. J. Biol. Chem., 275: 16360–16365.

Tonchev, A.B., Yamashima, T., Zhao, L. and Okano, H. (2003) Differential proliferative response in the postischemic hippocampus, temporal cortex, and olfactory bulb of young adult macaque monkeys. Glia, 42: 209–224.

Triarhou, L.C. and Herndon, R.M. (1986) The effect of dexamethasone on l-α-lysophosphatidyl choline/lysolecithin-induced demyelination. Arch. Neurol., 43: 121–125.

Van Praag, H., Kempermann, G. and Gage, F.H. (1999) Running increases cell proliferation and neurogenesis in the adult mouse dentate gyrus. Nat. Neurosci., 2: 266–270.

Van Praag, H., Schinder, A., Christie, B.R., Toni, N., Palmer, T.D. and Gage, F.H. (2002) Functional neurogenesis in the adult hippocampus. Nature, 415: 1030–1034.

Vescovi, A.L., Rietze, R., Magli, M.C. and Bjornson, C. (2002) Hematopoietic potential of neural stem cells. Nat. Med., 8: 535.

Villoslada, P., Hauser, S.L., Bartke, I., Unger, J., Heald, N., Rosenberg, D., Cheung, S.W., Mobley, W.C., Fisher, S. and Genain, C.P. (2000) Human nerve growth factor protects common marmosets against autoimmune encephalomyelitis by switching the balance of T helper cell type 1 and 2 cytokines within the central nervous system. J. Exp. Med., 191: 1799–1806.

Warrington, A.E., Bieber, A.J., Ciric, B., Van Keulen, V., Pease, L.R., Mitsunaga, Y., Paz Soldan, M.M. and

Rodrıguez, M. (2001) Immunoglobulin-mediated CNS repair. J. Allergy Clin. Immunol., 108: S121–S125.

Weiss, S., Dunne, C., Hewson, J., Wohl, C., Wheatley, M., Peterson, A.C. and Reynolds, B.A. (1996) Multipotent CNS stem cells are present in the adult mammalian spinal cord and ventricular neuraxis. J. Neurosci., 16: 7599–7609.

Weissman, I.L. (2000a) Stem cells: units of development, units of regeneration, and units in evolution. Cell, 100: 157–168.

Weissman, I.L. (2000b) Translating stem and progenitor cell biology to the clinic: barriers and opportunities. Science, 287: 1442–1446.

Weissman, I.L., Anderson, D.J. and Gage, F. (2001) Stem and progenitor cells: origins, phenotypes, lineage committments, and transdifferentiations. Annu. Rev. Cell Dev. Biol., 17: 387–403.

Widner, H. (1994) NIH neural transplantation funding. Science, 263: 737.

Wolswijk, G. (1998) Oligodendrocyte survival, loss and birth in lesions of chronic-stage multiple sclerosis. Brain, 123: 105–115.

Wurmser, A.E. and Gage, F.H. (2002) Cell fusion causes confusion. Nature, 416: 485–487.

Yagita, Y., Kitagawa, K., Ohtsuki, T., Takasawa, K.-I., Miyata, T., Okano, H., Hori, M. and Matsumoto, M. (2001) Neurogenesis by progenitor cells in the ischemic adult rat hippocampus. Stroke, 32: 1890–1896.

You, S., Petrov, T., Chung, P.H. and Gordon, T. (1997) The expression of the low affinity nerve growth factor receptor in long-term denervated Schwann cells. Glia, 20: 87–100.

Xian, C.J. and Zhou, X.F. (1999) Roles of transforming growth factor-alpha and related molecules in the nervous system. Mol. Neurobiol., 20: 157–183.

Zhang, J. and Lazar, M.A. (2000) The mechanism of action of thyroid hormones. Annu. Rev. Physiol., 62: 439–466.

Zhang, R.L., Zhang, Z.G., Zhang, L. and Chopp, M. (2001) Proliferation and differentiation of progenitor cells in the cortex and the subventricular zone in the adult rat after focal cerebral ischemia. Neuroscience, 105: 33–41.

Progress in Brain Research, Vol. 146
ISSN 0079-6123

CHAPTER 6

Pathways of survival induced by NGF and extracellular ATP after growth factor deprivation

Nadia D'Ambrosi[1,2], Barbara Murra[1], Fabrizio Vacca[1,3] and Cinzia Volonté[1,4,*]

[1]*Fondazione Santa Lucia, Rome, Italy*
[2]*University of Rome Tor Vergata, Department of Neuroscience, Rome, Italy*
[3]*University of Rome 'La Sapienza', Department of Human Physiology and Pharmacology, Rome, Italy*
[4]*Institute of Neurobiology and Molecular Medicine, CNR, Rome, Italy*

Abstract: In a previous work we demonstrated that extracellular adenosine-5′-triphosphate (ATP), acting on P2 receptors, exerts neuritogenic and trophic effects on the phaeochromocytoma PC12 cell line. These actions are comparable to those sustained by nerve growth factor (NGF) and involve several overlapping pathways. In this work, we describe some of the mechanisms recruited by ATP and NGF in maintaining PC12 cell survival after serum deprivation. We show that both ATP and NGF upregulate the expression of the stress-induced heat shock protein HSP70 and HSP90, whilst glucose-response protein GRP75 and GRP78 are not affected. In parallel with NGF, ATP prevents the cleavage and activation of caspase-2 and inhibits the release of cytochrome c from mitochondria into the cytoplasm. Finally, neither NGF, nor ATP directly modulate the expression of P2 receptors in the induction of cell survival. Our data contribute to dissect the biological mechanisms activated by extracellular purines exerting trophic actions and to establish that survival and neurite outgrowth lie on different mechanistic pathways.

Keywords: P2 receptor; PC12 cells; caspase-2; cytochrome c; heat shock proteins

Introduction

The biological mechanisms behind differentiation, proliferation or cell death can be concurrently investigated by using a cellular model system such as the rat phaeochromocytoma PC12 cells (Greene and Tischler, 1976). Although the PC12 cellular model can only partially mimic the whole body complexity of an intact organism, it allows experimental control under defined culture conditions and especially a more direct investigation of the several overlapping pathways propagating a biological process. These cells share neural crest origin with sympathetic neurons and thus retain neuronal features, such as nerve growth factor (NGF)-dependency for survival

and differentiation in defined serum-free conditions (Greene, 1978). In the absence of NGF and in the presence of serum, they instead maintain their proliferation properties. Finally, when PC12 cells are deprived of trophic support, they undergo rapid apoptotic death. The addition of NGF hampers such death (Batistatou and Greene, 1991; Rukenstein et al., 1991) and, moreover, induces sympathetic-like differentiation. During development, many cell populations are acutely dependent for their survival on target-derived trophic factors and neurons that fail to find the appropriate targets and sources of neurotrophic factors will eventually die (Pettmann and Henderson, 1998). Sympathetic neurons, for example, require NGF for survival during late embryogenesis and early postnatal period. Therefore, cell death constitutes an integral part of the plasticity of the nervous system (Oppenheim, 1991), also playing an important role in the removal

*Correspondence to: C. Volonté, Fondazione Santa Lucia, Via Ardeatina, 354, I-00179 Rome, Italy. Tel.: +39-06-5150-1557; Fax: +39-06-5150-1556; E-mail: cinzia@in.rm.cnr.it

DOI: 10.1016/S0079-6123(03)46006-8

of damaged cells after neuronal injury and diseases (Mattson, 2000).

Cell death induced by trophic factor deprivation is driven by multiple mechanisms, comprising caspase-2 (Troy et al., 1996, 1997; Stefanis et al., 1998; Troy et al., 2001), caspase-3, caspase-9 activities (Park et al., 1996; Haviv et al., 1997; Green, 1998; Yoshimura et al., 1998), as well as release of cytochrome c from mitochondria (Yang et al., 1997; Kim et al., 1999). Indeed, interruption of caspase activation and/or cytochrome c release is a potential mechanism for the prevention of cell death. Recent studies suggest that also heat shock proteins (HSPs) display a crucial role in determining cell survival/death, following various cellular stresses (Jolly and Morimoto, 2000). HSPs can either positively or negatively modulate the pathways of cell death, acting at numerous sites in the apoptotic machinery (Maloney and Workman, 2002) and their pro-survival action involves the inhibition of caspase-3 (Pandey et al., 2000; Saleh et al., 2000; Beere and Green, 2001) and the activation of JNK (Mosser et al., 2000).

Several different survival agents have been identified, including serum, insulin-like growth factor-1, neurotrophins, and NMDA (Levi-Montalcini, 1987; D'Mello et al., 1993; Koh et al., 1995; Chao, 2003). Moreover, recent findings have demonstrated a direct role exerted by extracellular nucleotides in the development and maintenance of the nervous system (Rathbone et al., 1999). In particular, purines may induce neurite outgrowth during de novo neuritogenesis (Gysbers and Rathbone, 1992; D'Ambrosi et al., 2000), regeneration (D'Ambrosi et al., 2001), proliferation (Rathbone et al., 1999; Sanches et al., 2002), and survival after serum or growth factor deprivation (Fujita et al., 2000; D'Ambrosi et al., 2001). These actions are driven by both P1-adenosine and P2-ATP receptors. These last comprise fast, ionotropic channels (P2X$_{1-8}$, Khakh, 2001) and G protein-coupled, metabotropic receptors (P2Y$_{1,2,4,6,11,12,13,14}$) (Abbracchio et al., 2003).

In a previous work, we demonstrated that extracellular ATP and some of its slowly hydrolysable analogues induce survival in PC12 cells deprived of growth factors, in a way comparable to NGF. The trophic action exerted by ATP occurs through P2 receptors, since several P2 receptor

antagonists (but not P1 antagonists or inhibitors of ectonucleotidases) are able to prevent or attenuate it (D'Ambrosi et al., 2001). Nevertheless, little is still known about the mechanisms pertaining to survival, not only under physiological conditions and normal development, but also in aging and in the pathogenesis of many diseases. Since the most common markers of cell death, necrosis and apoptosis (Liou et al., 2003), mostly contribute to these conditions, the understanding of their molecular intermediates has assumed an increasingly important role in cell biology and medicine. Here we establish that the survival action exerted by extracellular ATP involves selective modulation of protein synthesis, protease activities and cytochrome c release.

Cell survival by ATP and NGF: induction of HSPs but not of GRPs

Molecular chaperones maintain the appropriate folding and conformation of proteins and are crucial in regulating the balance between protein synthesis and degradation. They play a key role in many cellular stress responses, such as apoptotic stimuli (Jolly and Morimoto, 2000). The expression of members of the HSP family, such as HSP70 and HSP90, is upregulated by serum, mitogens or growth factors (Helmbrecht et al., 2000), whilst it is generally downregulated when apoptosis is committed (Maloney and Workman, 2002). High temperatures, cerebral ischemia, seizures and other insults are also known to induce the expression of HSP (Sharp et al., 1999; Beere and Green, 2001). Such induction has been often associated with a protective effect against a subsequent, more severe insult. This has been directly demonstrated by overexpression of HSP70, the major inducible HSP found in all living cells, preserving fibroblasts (Angelidis et al., 1991) or neurons (Mailhos et al., 1994) from cell death. Nevertheless, its overproduction can ameliorate some, but not all, types of apoptosis-related cell death. With our study, we show that the addition of 1% serum, or 50 ng/ml NGF, or 100 μM ATP to serum-deprived PC12 cells increases the intracellular amount of HSP70 and HSP90 (Fig. 1). In particular, with respect to control cells kept in serum-free conditions, 1% serum, NGF or ATP raises the levels of HSP70 of about

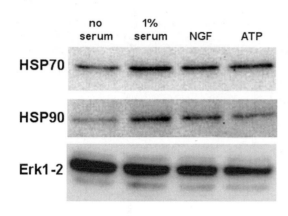

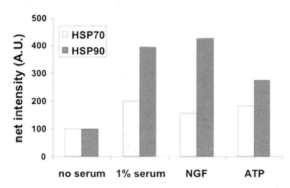

Fig. 1. Extracellular ATP upregulates the expression of HSP70 and HSP90 proteins. PC12 cells cultured on collagen-coated dishes in RPMI 1640 medium supplemented with 10% heat inactivated horse serum (HS) and 5% fetal calf serum (FCS) were serum-deprived by several cycles of centrifugation/resuspension in serum-free medium and replated without or with 1% serum, 50 ng/ml NGF or 100 μM ATP. After 24 h, the cells were lysed with ice-cold RIPA buffer (PBS, 1% Nonidet P-40 (NP-40), 0.5% sodium deoxycholate, 0.1% sodium dodecyl sulfate (SDS), in the presence of 1 mM phenylmethylsulphonyl fluoride (PMSF) and 10 μg/ml leupeptin), kept on ice for 30 min and then centrifuged for 10 min, at 4°C. Supernatants were collected and assayed for protein content by the method of Bradford (Bradford, 1976). Equal amount of protein (30 μg) from each sample was subjected to SDS-PAGE on 10% polyacrylamide gels, to western blotting and immunoreactions with anti-HSP90, HSP70 (1 : 200) (Santa Cruz Biotechnology, Santa Cruz, CA, USA) and Erk1-2 (1 : 500) (Calbiochem, Germany) antibodies. Band detection and quantification was performed by enhanced chemiluminescence (ECL) with horseradish peroxidase-coupled secondary antibodies, using Kodak Image Station 440CF with Kodak 1D 3.5 software. Data represent the net intensity of HSP90 and HSP70, expressed as arbitrary units (A.U.) and normalized with Erk1-2.

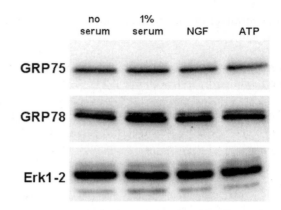

Fig. 2. GRP proteins are not modulated by addition of serum, NGF or ATP. PC12 cells were serum-deprived and replated either with 1% serum, 50 ng/ml NGF or 100 μM ATP. After 24 h, the cells were lysed and total protein collected in SDS-PAGE sample buffer. Equal amount of protein (30 μg) from each sample was subjected to western blotting and immunoreactions with anti-GRP75, GRP78 (1 : 200) (Santa Cruz Biotechnology, Santa Cruz, CA, USA) and Erk1-2 antibodies. Images were obtained by ECL, as described in the legend of Fig. 1.

99%, 56% and 83%, respectively, and the levels of HSP90 of 295%, 326% and 176% (Fig. 1). Therefore, it emerges that also in this specific case, HSP70 and HSP90 can be considered pro-survival markers.

Although HSP70 and HSP90 are produced in most brain cells, the expression of other genes encoding HSP as well as glucose-regulated proteins (GRP), sensors of intracellular glucose concentration (Lee et al., 1999), is modulated only in specific cells, or organelles, under selective physiological or pathological conditions. In ischemic brain (Lee, 1992), for instance, low levels of glucose/oxygen generally induce GRP75 and GRP78, in order to transport proteins, respectively within mitochondria and endoplasmic reticulum. Our present data show that, differently from HSP70/90, serum starvation, 50 ng/ml NGF, and 100 μM ATP up to 24 h does not affect GRP75 and GRP78 protein expression in PC12 cells (Fig. 2). Therefore, GRP75 and GRP78 do not seem to be involved in the apoptotic machinery of serum deprivation. It is conceivable that differential inhibition of selective apoptotic cascades (HSP and not GRP, for instance) might depend on the subcellular localization of the different targets (Marks and Berg, 1999). Cell- or organelle-specific responses that become differently modulated therefore provide additional details on the type and intracellular localization of the injury caused in PC12 cells by lack of serum.

Activation of caspase-2 and release of cytochrome c are inhibited by extracellular ATP

The activation of caspases is an additional signal transduction pathway that is regulated by serum deprivation. Caspase-2 is a member of ICE-like proteases that, after activation, induces apoptosis by triggering the caspase cascade (Cohen, 1997; Slee et al., 1999; Van de Craen et al., 1999). It was previously demonstrated that caspase-2 is processed and activated in PC12 cells after withdrawal of trophic support (Troy et al., 1996, 1997) and that the addition of NGF prevents the formation of the activated form (Stefanis et al., 1998). Now, we show here that 50 ng/ml NGF, or 1% serum, or 100 μM extracellular ATP reduces by 50% the formation of the cleaved (20 kD) activated form of caspase-2 in 24 h, as demonstrated by western blot and immunoreactions (Fig. 3A). Moreover, extracellular ATP inhibits by 30% the activation of caspase-2, as directly measured by in vitro activity assay (Fig. 3B). Both processing and direct activity of caspase-2 in PC12 cells is therefore increased under serum-free conditions and inhibited in the presence of either serum, NGF or extracellular ATP, sustaining cell survival. This would indicate that cell death induced by serum starvation oppositely regulates caspase-2 expression/activity and HSP70 or HSP90 expression.

It is becoming increasingly clear that, during the early phases of apoptosis, mitochondria undergo a permeability transition characterized by breakdown of the membrane potential. As a consequence, several mitochondrial proteins are released into the cytoplasm, including cytochrome c, which activates the caspase cascade by interacting with cytoplasmic factors (Yang et al., 1997). We show here that this occurs also in PC12 cells under serum-free conditions. In already 3 h, serum starvation causes a significant release (up to 4-fold) of cytochrome c into the cytoplasm. Again, the addition of either 50 ng/ml NGF or extracellular 100 μM ATP almost restores the basal levels obtained in the presence of serum (Fig. 4). Therefore, extracellular ATP and NGF similarly regulate caspase-2 and cytochrome c, during rescue of cell death.

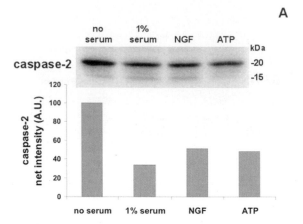

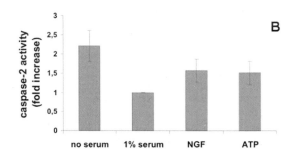

Fig. 3. Extracellular ATP prevents caspase-2 cleavage and activation induced by serum withdrawal. PC12 cells were serum-deprived and replated either with 1% serum, 50 ng/ml NGF or 100 μM ATP. After 24 h, the cells were lysed at 4°C in 25 mM HEPES, pH 7.5, 5 mM EDTA, 1 mM EGTA, 5 mM MgCl$_2$, 10 mM DTT for 30 min and the lysate was sonicated. (A) Equal amount of total protein (30 μg) was subjected to SDS-PAGE and western blot with anti-caspase-2 antibody (1 : 200) (Santa Cruz Biotechnology, Santa Cruz, CA, USA). Images were obtained by ECL, as described in the legend of Fig. 1. (B) 100 μg of total protein was incubated for 2 h at 37°C in reaction buffer (25 mM HEPES, pH 7.5, 40% sucrose, 0.4% Triton, 10 mM DTT), in the presence of 50 μM of the caspase-2 chromogenic substrate Ac-VDVAD-pN (Alexis Corporation, Lausen, Switzerland). Caspase-2 activity was quantified in a microtiter plate at 405 nm and net absorbancies produced by each sample were expressed as fold activation and referred to 1% serum (used as a control). Data represent means ± S.E.M. from three independent experiments.

ATP and NGF induce cell survival independently by the modulation of P2 receptors

It is well established that NGF-evoked neurite outgrowth and survival lie on different mechanistic pathways (Klesse et al., 1999; Kaplan and Miller, 2000)

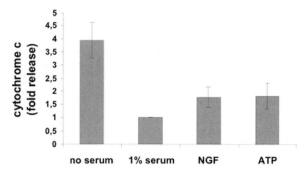

Fig. 4. Release of cytochrome c after serum deprivation is prevented by ATP. PC12 cells were serum-deprived and replated either with 1% serum, 50 ng/ml NGF or 100 μM ATP. After 3 h, the cells were lysed in a sucrose-containing lysis buffer (25 mM sucrose, 1 mM EDTA, 10 mM Tris pH 7.5, 1 mM PMSF, 10 μg/ml leupeptin) and kept on ice for 30 min. The lysate was centrifuged at $500 \times g$ for 10 min at 4°C, the supernatant was centrifuged again at $10000 \times g$ for 10 min, and the final supernatant containing the cytosolic fraction (15 μg) was loaded onto SDS-PAGE on 7% polyacrylamide gels and subjected to western blotting and immunoreactions with anti-cytochrome c antibody (1:200) (Santa Cruz Biotechnology, Santa Cruz, CA, USA). After ECL, band densitometry was performed, expressed as fold of cytochrome c release, with respect to 1% serum. Data represent means ± S.E.M. from three independent experiments.

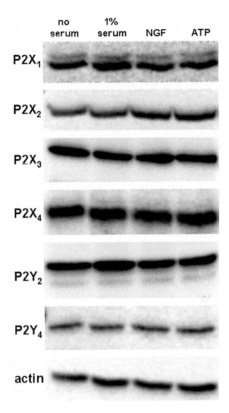

Fig. 5. After serum deprivation, P2 receptor expression is neither modulated by NGF nor by ATP. PC12 cells were serum-deprived and replated without or with 1% serum, 50 ng/ml NGF or 100 μM ATP. After 6 h, the cells were lysed and total protein collected in SDS-PAGE sample buffer. Equal amount of protein (30 μg) from each sample was subjected to western blotting and immunoreactions in the presence of anti-P2X$_1$, -P2X$_2$, -P2X$_4$, -P2Y$_2$ and -P2Y$_4$ (Alomone Labs, Jerusalem, Israel) (1:200), anti-P2X$_3$ (Neuromics, Minneapolis, MN, USA) (1:2000) and monoclonal anti-β-actin clone AC-15 (Sigma-Aldrich, MI, Italy) (1:5000). Images were obtained by ECL, using Kodak Image Station 440.

and that extracellular ATP can sustain both survival and neuritogenesis, with the distinction that whereas ATP and other agonists of P2 receptors are sufficient to sustain cell viability, they are only subsidiary to promote neuritogenesis (D'Ambrosi et al., 2001). We have also reported that, under neurite regenerating conditions, extracellular ATP, as well as NGF, transiently upregulates the protein expression of several P2 receptor subtypes, mainly P2X$_2$, P2X$_3$, P2X$_4$ and P2Y$_2$, but not P2Y$_4$ (D'Ambrosi et al., 2001). We observe here that P2X$_1$, P2X$_2$, P2X$_3$, P2X$_4$, P2Y$_2$ and P2Y$_4$ receptors are modulated neither by the apoptotic insult provided by serum deprivation, nor by the addition of 1% serum, 50 ng/ml NGF, 100 μM ATP. This occurs following 1 h treatment (not shown), 4–6 h exposure (Fig. 5) or 24 h treatment (not shown). These results support the notion that neuritogenic and survival actions are not overlapping in their downstream signal transduction pathway(s), but that extracellular ATP and NGF do share either common mechanisms or a potential interplay in the signals triggered on their target cells. Consistently, cell death mediated by lack of trophic

factor is generally accompanied by the presence of extracellular ATP coming from cellular outflow induced by membrane permeability loss, and from degradation of nucleic acids of dying cells. This release of ATP could then replenish a temporary gap in the supply of trophic agents.

Conclusions

We have described here some of the mechanisms employed by extracellular ATP (or by NGF) in being permissive to survival of PC12 cells temporarily deprived or insufficiently supplied with serum factors. In spite of the always new advances made in understanding the different pathways involved in cell death, an ideal pharmacological treatment should target different biological intermediates, since compounds blocking the initiating steps may not be as effective at halting the progression of cell death, as those directed towards the propagating steps. Under this light, extracellular ATP seems to apparently respond to both requirements, by being capable of influencing, directly or indirectly, several distinct biological markers.

Acknowledgments

The research presented was supported by Progetto Coordinato CNR Agenzia 2000. F.Vacca is the recipient of a PhD fellowship from the University of Rome 'La Sapienza', Dottorato di Ricerca in 'Neurofisiologia: Basi neurali di funzioni cognitive superiori'. We thank Dr. Delio Mercanti for kindly providing NGF.

Abbreviations

ATP	adenosine-5'-triphosphate
GRP	glucose-response protein
HSP	heat shock protein
JNK	c-Jun NH$_2$-terminal kinase
NGF	nerve growth factor
PBS	phosphate buffered saline
PAGE	polyacrylamide gel electrophoresis
SDS	sodium dodecyl sulphate

References

Abbracchio, M.P., Boeynaems, J.M., Barnard, E.A., Boyer, J.L., Kennedy, C., Miras-Portugal, M.T., King, B.F., Gachet, C., Jacobson, K.A., Weisman, G.A. and Burnstock, G. (2003) Characterization of the UDP-glucose receptor (re-named here the P2Y (14) receptor) adds diversity to the P2Y receptor family. Trends Pharmacol. Sci., 24: 52–55.

Angelidis, C.E., Lazaridis, I. and Pagoulatos, G., N. (1991) Constitutive expression of heat-shock protein 70 in mammalian cells confers thermoresistance. Eur. J. Biochem., 199: 35–39.

Batistatou, A. and Greene, L.A. (1991) Aurintricarboxylic acid rescues PC12 cells and sympathetic neurons from cell death caused by nerve growth factor deprivation: correlation with suppression of endonuclease activity. J. Cell Biol., 115: 461–471.

Beere, H.M. and Green, D.R. (2001) Stress management—heat shock protein-70 and the regulation of apoptosis. Trends Cell Biol., 11: 6–10.

Bradford, M.M. (1976) A rapid and sensitive method for the quantitation of microgram quantities of protein utilizing the principle of protein-dye binding. Anal. Biochem., 72:2 48–54.

Chao, M.V. (2003) Neurotrophins and their receptors: A convergence point for many signalling pathways. Nat. Rev. Neurosci., 4: 299–309.

Cohen, G.M. (1997) Caspases: the executioners of apoptosis. Biochem. J., 326: 1–16.

D'Ambrosi, N., Cavaliere, F., Merlo, D., Milazzo, L., Mercanti, D. and Volonté, C. (2000) Antagonists of P2 receptor prevent NGF-dependent neuritogenesis in PC12 cells. Neuropharmacology, 39: 1083–1094.

D'Ambrosi, N., Murra, B., Cavaliere, F., Amadio, S., Bernardi, G., Burnstock, G. and Volonté, C. (2001) Interaction between ATP and nerve growth factor signalling in the survival and neuritic outgrowth from PC12 cells. Neuroscience, 108: 527–534.

D'Mello, S.R., Galli, C., Ciotti, T. and Calissano, P. (1993) Induction of apoptosis in cerebellar granule neurons by low potassium: inhibition of death by insulin-like growth factor I and cAMP. Proc. Natl. Acad. Sci. USA, 90: 10989–10993.

Fujita, N., Kakimi, M., Ikeda, Y., Hiramoto, T. and Suzuki, K. (2000) Extracellular ATP inhibits starvation-induced apoptosis via P2X2 receptors in differentiated rat pheochromocytoma PC12 cells. Life Sci., 66: 1849–1859.

Green, D.R. (1998) Apoptotic pathways: the roads to ruin. Cell, 94: 695–698.

Greene, L.A. (1978) Nerve growth factor prevents the death and stimulates the neuronal differentiation of clonal PC12 pheochromocytoma cells in serum-free medium. J. Cell Biol., 78: 747–755.

Greene, L.A. and Tischler, A.S. (1976) Establishment of a noradrenergic clonal line of rat adrenal pheochromocytoma cells which respond to nerve growth factor. Proc. Natl. Acad. Sci. USA, 73: 2424–2428.

Gysbers, J.W. and Rathbone, M.P. (1992) Guanosine enhances NGF-stimulated neurite outgrowth in PC12 cells. Neuroreport, 3: 997–1000.

Haviv, R., Lindenboim, L., Li, H., Yuan, J. and Stein, R. (1997) Need for caspases in apoptosis of trophic factor-deprived PC12 cells. J. Neurosci. Res., 5: 69–80.

Helmbrecht, K., Zeise, E. and Rensing, L. (2000) Chaperones in cell cycle regulation and mitogenic signal transduction: a review. Cell Prolif., 33: 341–365.

Jolly, C. and Morimoto, R.I. (2000) Role of the heat shock response and molecular chaperones in oncogenesis and cell death. J. Natl. Cancer Inst., 92: 1564–1572.

Kaplan, D.R. and Miller, F.D. (2000) Neurotrophin signal transduction in the nervous system. Curr. Opin. Neurobiol., 10: 381–391.

Khakh, B.S. (2001) Molecular physiology of P2X receptors and ATP signalling at synapses. Nat. Rev. Neurosci., 2: 165–174.

Kim, Y.M., Chung, H.T., Kim, S.S., Han, J.A., Yoo, Y.M., Kim, K.M., Lee, G.H., Yun, H.Y., Green, A., Li, J., Simmons, R.L. and Billiar, T.R. (1999) Nitric oxide protects PC12 cells from serum deprivation-induced apoptosis by cGMP-dependent inhibition of caspase signaling. J. Neurosci., 19: 6740–6747.

Klesse, L.J., Meyers, K.A., Marshall, C.J. and Parada, L.F. (1999) Nerve growth factor induces survival and differentiation through two distinct signaling cascades in PC12 cells. Oncogene, 18: 2055–2068.

Koh, J.Y., Gwag, B.J., Lobner, D. and Choi, D.W. (1995) Potentiated necrosis of cultured cortical neurons by neurotrophins. Science, 268: 573–575.

Lee, A.S. (1992) Mammalian stress response: induction of the glucose-regulated protein family. Curr. Opin. Cell Biol., 4: 267–273.

Lee, J., Bruce-Keller A.J., Kruman, Y., Chan, S.L. and Mattson, M.P. (1999) 2-Deoxy-D-glucose protects hippocampal neurons against excitotoxic and oxidative injury: evidence for the involvement of stress proteins. J. Neurosci Res., 57: 48–61.

Levi-Montalcini, R. (1987) The nerve growth factor 35 years later. Science, 237: 1154–1162.

Liou, A., K.F., Clark, R.S., Henshall, D.C., Yin, X., M. and Chen, J. (2002) To die or not to die for neurons in ischemia, traumatic brain injury and epilepsy: a review on the stress-activated signaling pathways and apoptotic pathways. Prog. Neurobiol., 69: 103–142.

Mailhos, C., Howard, M.K. and Latchman, D.S. (1994) Heat shock proteins hsp90 and hsp70 protect neuronal cells from thermal stress but not from programmed cell death. J. Neurochem., 63: 1787–1795.

Maloney, A. and Workman, P. (2002) HSP90 as a new therapeutic target for cancer therapy: the story unfolds. Expert Opin. Biol. Ther., 2: 3–24.

Marks, N. and Berg, M., J. (1999) Recent advances on neuronal caspases in development and neurodegeneration. Neurochem. Int., 1999 35: 195–220.

Mattson, M.P. (2000) Apoptosis in neurodegenerative disorders. Nat. Rev. Mol. Cell Biol., 1: 120–129.

Mosser, D.D., Caron, A.W., Bourget, L., Meriin, A.B., Sherman, M.Y., Morimoto, R.I. and Massie, B. (2000) The chaperone function of hsp70 is required for protection against stress-induced apoptosis. Mol. Cell Biol., 20: 7146–7159.

Oppenheim, R.W. (1991) Cell death during development of the nervous system. Annu. Rev. Neurosci., 14: 453–501.

Pandey, P., Saleh, A., Nakazawa, A., Kumar, S., Srinivasula, S.M., Kumar, V., Weichselbaum, R., Nalin, C., Alnemri, E.S., Kufe, D. and Kharbanda, S. (2000) Negative regulation of cytochrome c-mediated oligomerization of Apaf-1 and activation of procaspase-9 by heat shock protein 90. EMBO J., 19: 4310–4322.

Park, D.S., Stefanis, L., Yan, C.Y., Farinelli, S.E. and Greene, L.A. (1996) Ordering the cell death pathway. Differential effects of BCL2, an interleukin-1-converting enzyme family protease inhibitor, and other survival agents on JNK activation in serum/nerve growth factor-deprived PC12 cells. J. Biol. Chem., 271: 21898–21905.

Pettmann, B. and Henderson, C.E. (1998) Neuronal cell death. Neuron, 20: 633–647.

Rathbone, M.P., Middlemiss, P.J., Gysbers, J.W., Andrew, C., Herman, M.A., Reed, J.K., Ciccarelli, R., Di Iorio, P. and Caciagli, F. (1999) Trophic effects of purines in neurons and glial cells. Prog. Neurobiol., 59: 663–690.

Rukenstein, A., Rydel, R.E. and Greene, L.A. (1991) Multiple agents rescue PC12 cells from serum-free cell death by translation- and transcription-independent mechanisms. J. Neurosci., 11: 2552–2563.

Saleh, A., Srinivasula, S.M., Balkir, L., Robbins, P.D. and Alnemri, E.S. (2000) Negative regulation of the Apaf-1 apoptosome by Hsp70. Nat. Cell Biol., 2: 476–483.

Sanches, G., de Alencar, L.S. and Ventura, A.L. (2002) ATP induces proliferation of retinal cells in culture via activation of PKC and extracellular signal-regulated kinase cascade. Int. J. Dev. Neurosci., 20: 21–27.

Sharp, F.R., Massa, S.M. and Swanson, R.A. (1999) Heat-shock protein protection. Trends Neurosci., 22: 97–99.

Slee, E.A., Adrain, C. and Martin, S.J. (1999) Serial killers: ordering caspase activation events in apoptosis. Cell Death Differ., 6: 1067–1074.

Stefanis, L., Troy, C.M., Qi, H., Shelanski, M.L. and Greene, L.A. (1998) Caspase-2 (Nedd-2) processing and death of trophic factor-deprived PC12 cells and sympathetic neurons occur independently of caspase-3 (CPP32)-like activity. J. Neurosci., 18: 9204–9215.

Troy, C.M., Stefanis, L., Prochiantz, A., Greene, L.A. and Shelanski, M.L. (1996) The contrasting roles of ICE family proteases and interleukin-1beta in apoptosis induced by trophic factor withdrawal and by copper/zinc superoxide dismutase down-regulation. Proc. Natl. Acad. Sci. USA, 93: 5635–5640.

Troy, C.M., Stefanis, L., Greene, L.A. and Shelanski, M.L. (1997) Nedd2 is required for apoptosis after trophic factor withdrawal, but not superoxide dismutase (SOD1) down-regulation, in sympathetic neurons and PC12 cells. J. Neurosci., 17: 1911–1918.

Troy, C.M., Rabacchi, S.A., Hohl, J.B., Angelastro, J.M., Greene, L.A. and Shelanski, M.L. (2001) Death in the balance: alternative participation of the caspase-2 and -9 pathways in neuronal death induced by nerve growth factor deprivation. J. Neurosci., 21: 5007–5016.

Van de Craen, M., Declercq, W., Van den Brande, I., Fiers, W. and Vandenabeele, P. (1999) The proteolytic procaspase activation network: an in vitro analysis. Cell Death Differ., 6: 1117–1124.

Yang, J., Liu, X., Balla, K., Kim, C.N., Ibrado, A.M., Cai, J., Peng, T.I., Jones, D.P. and Wang, X. (1997) Prevention of apoptosis by Bcl-2: release of cytochrome c from mitochondria blocked. Science, 275: 1129–1132.

Yoshimura, S., Banno, Y., Nakashima, S., Takenaka, K., Sakai, H., Nishimura, Y., Sakai, N., Shimizu, S., Educhi, Y., Tsujimoto, Y. and Nozawa, Y. (1998) Ceramide formation leads to caspase-3 activation during hypoxic PC12 cell death. Inhibitory effects of Bcl-2 on ceramide formation and caspase-3 activation. J. Biol. Chem., 273: 6921–6927.

Progress in Brain Research, Vol. 146
ISSN 0079-6123

CHAPTER 7

ProNGF: a neurotrophic or an apoptotic molecule?

Margaret Fahnestock*, Guanhua Yu and Michael D. Coughlin

Department of Psychiatry and Behavioral Neurosciences, McMaster University, Hamilton, ON, Canada

Abstract: Nerve growth factor (NGF) acts on various classes of central and peripheral neurons to promote cell survival, stimulate neurite outgrowth and modulate differentiation. NGF is synthesized as a precursor, proNGF, which undergoes processing to generate mature NGF. It has been assumed, based on studies in the mouse submandibular gland, that NGF in vivo is largely mature NGF, and that mature NGF accounts for the molecule's biological activity. However, recently we have shown that proNGF is abundant in central nervous system tissues whereas mature NGF is undetectable, suggesting that proNGF may have a function distinct from its role as a precursor. A recent report that proNGF has apoptotic activity contrasts with other data demonstrating that proNGF has neurotrophic activity. This chapter will review the structure and processing of NGF and what is known about the biological activity of proNGF. Possible reasons for the discrepancies in recent reports are discussed.

Keywords: neurotrophin; precursor; TrkA; p75^{NTR}; neurite outgrowth; survival

Introduction

In the half a century since the discovery of nerve growth factor (NGF) by Rita Levi-Montalcini and Viktor Hamburger (1953), this molecule has never ceased to be full of surprises. From the initial finding that NGF is produced in snake venom and in inexplicably high quantities in the male mouse submandibular gland (Cohen, 1959, 1960) to the most recent reports discussed at NGF2002, the field has seen the unfolding story of NGF filled with twists and turns, and above all, controversy. The present state of affairs with respect to the NGF precursor, proNGF, is no exception.

Biosynthesis of NGF

The structure, biosynthesis, and biological activity of NGF have been studied extensively in the mouse

*Correspondence to: M. Fahnestock, Department of Psychiatry and Behavioral Neurosciences, McMaster University, 1200 Main Street West, Hamilton, ON L8N 3Z5, Canada. Tel.: +1-905-525-9140 (Ext. 23344); Fax: +1-905-522-8804; E-mail: fahnest@mcmaster.ca

DOI: 10.1016/S0079-6123(03)46007-X

submandibular gland because of its extraordinarily high concentration in this tissue (Cohen, 1960; Levi-Montalcini and Angeletti, 1968; Fahnestock, 1991). NGF is encoded by a single gene that is over 45 kilobases in length (Ullrich et al., 1983). Two separate promoters and a total of four exons are alternatively spliced to yield two major and two minor transcripts with the coding sequence at the 3' end in exon four (Edwards et al., 1986; Selby et al., 1987; Racke et al., 1996). NGF protein is translated from the two major alternatively spliced transcripts to produce 34 and 27 kD prepro species, with translation initiation sites at amino acid positions −187 and −121, respectively (Fig. 1). Removal of the signal sequence in the endoplasmic reticulum reduces these translation products to proNGF species of 32 and 25kD (Darling et al., 1983; Ullrich et al., 1983; Edwards et al., 1986, 1988b; Selby et al., 1987).

ProNGF contains three potential glycosylation sites, two in the prosegment and one in the mature sequence (Fig. 1). It has been demonstrated that a 43kD form of NGF expressed by transfected BSC40 cells is N-glycosylated in the prosegment, whereas the glycosylation site in the mature sequence does not

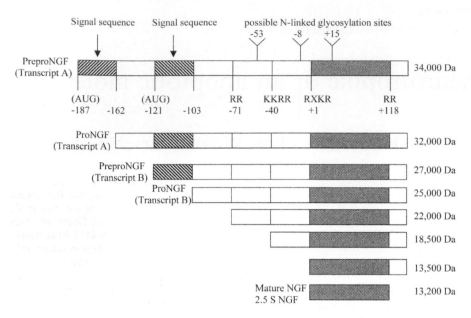

Fig. 1. Likely intermediates in NGF biosynthesis. Translation initiation sites at −187 and −121 are marked by vertical lines, as are potential cleavage sites at −71, −40 and +1. 'Y' represents potential glycosylation sites.

appear to be used (Seidah et al., 1996), except in a very minor proportion of molecules (Murphy et al., 1989). N-glycosylation of proNGF contributes to efficient protein expression (Suter et al., 1991) and may be important for exit of proNGF from the endoplasmic reticulum (Seidah et al., 1996). High-molecular-weight glycosylated NGF precursors, mostly of molecular weight 53 kD, have been identified both in vitro and in vivo (Edwards et al., 1988b; Lakshmanan et al., 1988; Bresnahan et al., 1990; Reinshagen et al., 2000; Yiangou et al., 2002). However, many tissues contain the unglycosylated proNGF (Chen et al., 1997; Fahnestock et al., 2001).

ProNGF undergoes further posttranslational processing at both amino- and carboxyl-terminal ends to generate the mature, biologically active product of 13.2 kD (Darling et al., 1983). Several intermediate sizes of NGF precursors are detectable in mouse and rat tissues, including a 27–29 kD species (Darling et al., 1983; Dicou et al., 1986) probably resulting from translation of the shorter transcript B identified by Edwards et al. (1986); a stable intermediate of 22–24 kD (Berger and Shooter, 1977; Dicou, 1992; Chen et al., 1997); and an 18–19 kD processing intermediate (Darling et al.,

1983; Dicou, 1989) (Fig. 1). The functions of the precursors and their intermediates are poorly understood.

Carboxyl-terminal processing of murine NGF occurs between a pair of arginine residues, whereas processing of intermediate and amino-terminal sites of NGF occurs at four-residue basic sites. Furin, and to a lesser extent other prohormone convertases, are able to process proNGF at its amino-terminus to produce the mature form of NGF, also known as βNGF (Bresnahan et al., 1990; Seidah et al., 1996). The kallikrein γNGF, which is found in a complex with betaNGF in the mouse submandibular gland, has been shown to process proNGF at both amino- and carboxyl-termini to produce intermediate and mature forms of NGF; following processing of the C-terminus of NGF by γNGF, the γNGF remains bound to βNGF at the carboxyl-terminal arginine (Greene et al., 1968; Mobley et al., 1976; Edwards et al., 1988a; Dicou, 1989; Jongstra-Bilen et al., 1989). It is not clear whether carboxyl-terminal processing occurs in species or tissues other than the mouse submandibular gland. Recently, both plasmin and matrix metalloprotease-7 (MMP-7) have also been shown to process the NGF precursor (Lee et al., 2001). Plasmin cleavage of proNGF produces the mature form, whereas MMP-7 cleavage results in a 17 kD intermediate that may be the same as the 18–19 kD intermediate identified previously (Darling et al., 1983; Dicou, 1989).

Secretion of proNGF

For many years, the literature has accepted that NGF biosynthesis in the mouse submandibular gland is representative of NGF biosynthesis in other species and tissues. Human NGF is highly homologous to murine NGF, and many of the same proteolytic cleavage sites and glycosylation consensus sites in the mouse sequence are conserved in human NGF (Ullrich et al., 1983). Neither NGF precursors nor intermediates have been reported in saliva, suggesting that processing occurs prior to secretion in the submandibular gland. Cell expression studies also support the model that NGF is largely processed into the mature form prior to secretion (Seidah et al., 1996). However, secretion of unprocessed proneurotrophins does occur from transfected mammalian cells, including neurons (Mowla et al., 1999), and from baculovirus-infected insect cells (Fig. 2). ProNGF, not NGF, is secreted from prostate cells, spermatids, and hair follicles (Chen et al., 1997; Delsite and Djakiew, 1999; Yardley et al., 2000). Primary sympathetic neurons reportedly secrete proNGF too (Smith et al., 2002). Here we show (Fig. 2) that primary rat cortical astrocytes also secrete proNGF. Therefore, in tissues other than the submandibular gland, NGF is likely secreted as proNGF, or at least as a mixture of proNGF and NGF.

ProNGF in tissues

Enzyme-linked immunosorbent assays (ELISA) for NGF are widely used and were thought to measure the mature 13.2 kD NGF protein in most tissues. It is now known that NGF in mouse, rat and human brain exists as 32 kD proNGF, with little or no mature NGF present (Fahnestock et al., 2001), suggesting that what was previously measured as NGF was actually proNGF. This is in contrast to the mouse submandibular gland, where large amounts of

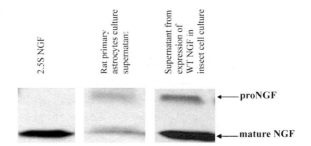

Fig. 2. ProNGF and NGF are secreted into conditioned medium of rat cortical astrocytes and baculovirus-infected insect cells in culture. Left: 2.5S NGF was isolated from mouse submandibular gland as previously described (Mobley et al., 1997; Petrides and Shooter, 1986). Middle: Single cells were isolated from cerebral cortex of newborn (0–24 h) Wistar rats as described in Hertz et al. (1989). After 14 days in culture, cells were shaken for 16 h at 200 rpm to remove oligodendrocytes. The purity of the astrocytes thus obtained was greater than 98% as estimated by indirect immunocytochemistry using antibody to glial fibrillary acidic protein. Cells were switched to serum-free DMEM. Medium was harvested after 24 h and concentrated 75-fold before analysis by Western blotting. Right: Recombinant wild-type NGF baculovirus was used to infect Sf9 insect cells in Sf-900 II serum-free medium (Invitrogen Life Technologies, Burlington, ON, Canada) as described in Van der Zee et al. (1995). Medium was harvested after three days. Western blotting was carried out as described in Fahnestock et al. (2001).

mature NGF are detectable, with lesser amounts of proNGF also found. The presence of proNGF and absence of mature NGF in central nervous system tissue is not due to rapid and selective degradation of NGF compared to proNGF, or to technical considerations such as antibody specificity, lack of sensitivity of the assay, or loss of NGF during tissue preparation. Others have noted that NGF is found as a precursor in tissues such as thyroid gland, retina, prostate, hair follicle, skin, colon, and dorsal root ganglia (Dicou et al., 1986; Chakrabarti et al., 1990; Delsite and Djakiew, 1999; Reinshagen et al., 2000; Yardley et al., 2000; Yiangou et al., 2002). Many of these tissues contain no detectable mature NGF.

It could be argued that mature NGF is taken up and transported so rapidly following secretion that it is not detectable by immunological assays, whereas proNGF is more long-lived. Basal forebrain cholinergic neurons projecting to cortex and hippocampus obtain NGF by retrograde transport from these targets and can efficiently internalize and transport NGF. In the Alzheimer's diseased brain, however, these cholinergic neurons degenerate and can no longer transport NGF as efficiently, leading to an increase in untransported NGF in the target tissue. If the mature form of NGF were the biologically active form, one would expect mature NGF to accumulate in the Alzheimer's brain. The NGF that accumulates in cortex and hippocampus is proNGF, with no mature NGF detectable (Fahnestock et al., 2001). This argues against the role of proNGF solely as a precursor to the biologically active, mature NGF, and suggests that proNGF is the form that is active in brain.

Physical characteristics of proNGF

ProNGF most likely is a dimer in solution. Rattenholl et al. (2001a) used a series of experiments including analytical ultracentrifugation, glutaraldehyde crosslinking followed by SDS-PAGE, and gel filtration to show that recombinant human proNGF exists as a dimer. We mutated a single amino acid at the mouse proNGF cleavage site (arginine to glycine at amino acid −1) to produce a cleavage-resistant recombinant proNGF, which we call proNGF (R-1G) (Yu et al., 2002; Fahnestock et al., submitted). We used gel filtration in the presence of 4 M urea to remove any associated proteins (mature NGF dissociates at 8M urea), and, in accordance with the data of Rattenholl et al. (2001a,b), we find that recombinant proNGF (R-1G) migrates as a dimer (Fig. 3). It has, furthermore, been suggested that proNGF dimers may associate into higher order structures for binding to p75 neurotrophin receptor (p75^{NTR}) (Ibanez, 2002).

NGF is a basic molecule, but proNGF is less basic. Isoelectric focusing of mature NGF gives an isoelectric point of 9.3, whereas proNGF (R-1G) has an isoelectric point of 8.1 (Fig. 4).

Biological activity of proNGF

That the NGF precursor has little or no biological activity has been widely assumed. Edwards et al. (1988a) demonstrated that the neurite outgrowth activity of proNGF on chick dorsal root ganglion

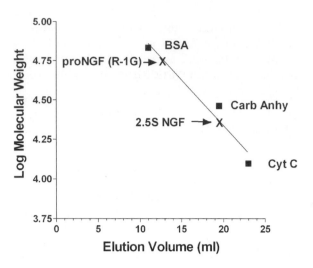

Fig. 3. ProNGF is a dimer in solution. Insect cell conditioned medium expressing a cleavage-resistant proNGF [proNGF (R-1G)] was concentrated forty-fold and chromatographed on a Sephadex G-75 column (Amersham Biosciences, Baie d'Urfé, Quebec) in 25 mM sodium phosphate pH 5.6–0.1 M NaCl–4 M urea. The column was pre-calibrated with bovine serum albumin (BSA, $M_r = 68,000$), carbonic anhydrase (Carb Anhyd, $M_r = 29,000$), and cytochrome C (Cyt C, $M_r = 12,500$). The calculated M_r of 2.5S NGF on this column is approximately 24,000. The calculated M_r of proNGF (R-1G) is approximately 60,000.

(DRG) neurons increased 10- to 20-fold after processing to mature NGF with trypsin. The role of the pro segment was thought to be to aid in folding of mature NGF (Suter et al., 1991; Rattenholl et al., 2001a). In contrast, a number of other investigators over the years have demonstrated that either the full-length proNGF or intermediate forms exhibit activity. Saboori and Young (1986) demonstrated that the 32 kD proNGF molecule isolated from mouse salivary gland promotes nerve growth prior to processing by γNGF. Reduced forms of the 53 kD NGF precursor bound on nitrocellulose membranes promoted neurite outgrowth and survival activity of PC12 cells (Lakshmanan et al., 1989). Suter et al. (1991) showed that a proNGF with a deletion including the processing site for conversion to mature NGF exhibited 50% of wild type neurite outgrowth activity on PC12 cells. Ibanez et al. (1992) constructed and expressed partially processed NGF precursors that exhibited neurite outgrowth and survival activity. The biological activity of full-length proNGF isolated from rat round spermatids was

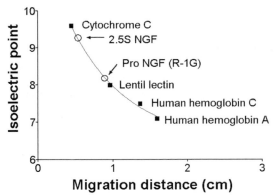

Fig. 4. Isoelectric focusing of NGF and proNGF. Isoelectric focusing in pH 3–10 gels and staining of protein standards (Bio-Rad Laboratories, Mississauga, ON, Canada) were carried out according to the manufacturer's instructions. Affinity-purified, cleavage-resistant proNGF [proNGF (R-1G)] from insect cell conditioned medium and 2.5S NGF purified from mouse submandibular glands (Mobley et al., 1976; Petrides and Shooter, 1986) were visualized using an antibody to NGF (Cedarlane, Hornby, ON) that recognizes both 2.5S NGF and proNGF (Fahnestock et al., 2001), and ECL (Amersham Biosciences). Isoelectric points of proteins used for standard curve: Cytochrome C, pI 9.6; Lentil lectin, pI 8.0; Human hemoglobin C, pI 7.5; Human hemoglobin A, pI 7.1. Calculated isoelectric point of 2.5S NGF is 9.3, and that of proNGF (R-1G) is approximately 8.1.

demonstrated in neurite outgrowth activity on PC12 cells and survival activity on Sertoli cells, whereas the 22 kD NGF intermediate exhibited less robust bioactivity (Chen et al., 1997). Using survival of DRG neurons as an assay, Rattenholl et al. (2001a) demonstrated that recombinant proNGF was equally as active as NGF. Most recently, our laboratory showed that a cleavage-resistant proNGF exhibits neurite outgrowth activity on both mouse sympathetic cervical ganglion (SCG) neurons and PC12 cells. The specific activity of proNGF in stimulating neurite outgrowth (based on the EC_{50}) is approximately five-fold less than that of mature NGF. In addition, this proNGF promotes survival of SCG neurons in culture and exhibits TrkA receptor binding and activation activity slightly less than to that of mature NGF (Yu et al., 2002; Fahnestock et al., submitted). These studies suggest that at least some forms of proNGF may indeed be neurotrophic.

In contrast, Lee et al. (2001) has recently demonstrated that a cleavage-resistant form of proNGF promotes apoptosis in primary SCG neurons and smooth muscle cells. So is proNGF apoptotic or neurotrophic?

ProNGF receptor binding

The answer may lie with the NGF receptors. NGF binds to two types of transmembrane receptors, a receptor tyrosine kinase, TrkA (Barbacid, 1995), and $p75^{NTR}$, a member of the tumor necrosis factor (TNF) receptor family (Friedman and Greene, 1999). TrkA mediates survival and neurite outgrowth by signaling through the PI3 kinase-Akt and Ras-MAP kinase pathways (Kaplan and Miller, 2000). When both receptors are present, $p75^{NTR}$ increases NGF binding affinity to TrkA, increases ligand discrimination, and enhances neurite outgrowth and survival by activating NF-kB and Rho signaling (Davies et al., 1993; Mahadeo et al., 1994; Bibel et al., 1999; Yamashita et al., 1999; Hamanoue et al., 1999). When $p75^{NTR}$ is present in the absence of TrkA, however, it can mediate cell death by signaling through the ceramide, p53, and c-Jun N-terminal kinase (JNK) pathways (Aloyz et al., 1998; Friedman, 2000; Brann et al., 2002). It follows, then, that a ligand that binds only to $p75^{NTR}$ is apoptotic, even in the presence of TrkA.

Mature NGF binds to TrkA with high affinity ($K_d \sim 10^{-11}$ M), whereas it binds to $p75^{NTR}$ with low affinity ($K_d \sim 10^{-9}$ M). Lee et al. (2001) demonstrated that their cleavage-resistant proNGF was able to bind to the low-affinity $p75^{NTR}$ receptor with enhanced affinity (10^{-10} M) but could not bind with high affinity to TrkA. Our cleavage-resistant proNGF, in contrast, binds to TrkA in cross-linking studies and activates TrkA phosphorylation at only slightly higher concentrations than NGF, suggesting it retains high-affinity binding to TrkA (Yu et al., 2002; Fahnestock et al., submitted). As expected of a molecule that binds TrkA, our proNGF exhibits neurotrophic activity.

Supporting the apoptotic model of proNGF, the amino-terminus of mature NGF has been implicated in binding to TrkA (Kahle et al., 1992; Ibanez et al., 1993; Woo et al., 1995; Kullander et al., 1997; Wiesmann et al., 1999), suggesting that unprocessed proNGF retaining an extended amino-terminus should be sterically hindered from binding to this receptor. Supporting the neurotrophic model, NGF loop regions have been implicated in the interaction between NGF and TrkA (Ibanez, 1995, 1998; Kullander et al., 1997) and in NGF binding to p75^{NTR} (Ibanez et al., 1992; Ryden and Ibanez, 1997). The loop regions might be expected to retain receptor-binding activity even in the presence of an amino-terminal pro extension. Circular dichroism suggests that the structure of mature NGF is largely maintained in the presence of the pro segment (Rattenholl et al., 2001a).

Differences between recombinant, cleavage-resistant proNGFs

Our cleavage-resistant proNGF differs from that of Lee et al. (2001) in several key structural features and in the expression and purification systems used. First, our proNGF contains a single amino acid substitution, a R-to-G substitution, at the -1 position. This substitution was designed to perturb the molecule as little as possible by substituting a small amino acid, glycine, for an arginine residue that forms part of the tetrabasic cleavage site. ProNGF from Lee et al. (2001) contains four separate amino acid substitutions: K-R to A-A at residues −1 and −2, and R-R to A-A at residues 118–119. The double alanine substitution at residues −1 and −2 was designed, like our single substitution, to eliminate the proNGF cleavage site. The double alanine substitution at residues 118–119 was designed to eliminate a carboxyl-terminal cleavage site that could release a polyhistidine tag. Our proNGF does not carry this polyhistidine tag and therefore we have no need for the carboxyl-terminal mutations. It is possible that the additional amino acid substitutions in the proNGF from Lee et al. (2001) and the fact that they are alanines rather than glycines, may unfold the molecule slightly more than a single substitution or change the structure from the native form. There is some precedent for polyhistidine tags disrupting protein structure. There is some precedent for polyhistidine tags disrupting protein structure. Polyhistidine tags have been shown to interfere with protein refolding and stabilization, and with receptor-ligand binding and biological activity (Ramage

et al., 2002; Ledent et al., 1997; Lawrence et al., 2001). This may be particularly true if the amino- and carboxyl-terminal ends interact. The pro segment of NGF has been shown to be important for proper folding of the molecule (Suter et al., 1991; Rattenholl et al., 2001a,b), whereas the carboxyl-terminus is important for stability and biological activity (Drinkwater et al., 1993; Kruttgen et al., 1997). Furthermore, the carboxyl-terminus of NT-3 is a key domain for interactions of NT-3 with p75^{NTR} (Urfer et al., 1994), suggesting this might also be true for NGF. Lee et al. (2001) have demonstrated no adverse effects of their polyhistidine tag on mature NGF expression or function. However, it is possible that a carboxyl-terminal polyhistidine tag might change both inter- and intramolecular interactions of proNGF in the presence of the pro domain.

It is interesting that the nickel columns commonly used to purify histidine-tagged proteins can cause oxidation and promote proteolysis by contaminating metalloproteases (Ramage et al., 2002). ProNGF from Lee et al. (2001) is extremely susceptible to cleavage by a variety of proteases including plasmin and MMP-7, a matrix metalloprotease, whereas native proNGF is also stable in tissue, as shown by its detection in a variety of sources including postmortem human brain (Fahnestock et al., 2001). It is possible that the multiple amino acid substitutions, the polyhistidine tag or the nickel column purification promote destabilization of the proNGF from Lee et al. (2001), whereas our single amino acid substitution and lack of polyhistidine tag may promote a more stable structure.

Finally, there are differences in expression systems used to produce the proNGF proteins. Lee et al. (2001) expressed proNGF in a mammalian expression system using 293 cells, whereas our proNGF is expressed in a baculovirus/insect cell system. The reasons we use baculovirus are that expressed proteins are produced in large amounts in serum-free medium for easy purification, and there are no endogenous neurotrophic factors produced by insect cells. Although 293 cells produce extremely low amounts of endogenous trophic factors, it has been demonstrated that neurotrophin subunits can freely recombine with each other (Treanor et al., 1995), which could produce artifacts. On the other hand, it has been suggested that there may be a chaperone present in insect cell supernatants that could interact with proNGF to produce artifactual biological activity. However, we partially purified proNGF by size exclusion chromatography in the presence of urea, and this eight-fold purified material exhibits similar neurite outgrowth activity as unpurified material (Yu et al., 2002; Fahnestock et al., submitted). A contaminating molecule would have to be the same molecular weight as proNGF and stay bound to it during size exclusion chromatography in the presence of urea to be a candidate, and so this alternative is unlikely.

Is endogenous proNGF neurotrophic or apoptotic?

The cleavage-resistant proNGF molecules discussed above are mutated recombinant proteins. The characteristics and physiological role of endogenous proNGF are still a puzzle. Mature NGF is effective in low amounts at promoting cell survival and neurite outgrowth via TrkA, yet the amounts required for activation of p75^{NTR} cell death pathways are at least an order of magnitude greater (Friedman, 2000). ProNGF, if its in vivo role is apoptotic, is a high-affinity ligand for p75^{NTR} that could promote cell death via this receptor at much lower doses than mature NGF. This suggests that the balance between cell survival and cell death could depend upon the relative amounts of proNGF and mature NGF in tissues (Ibanez, 2002). Yet, as we have seen, most tissues examined, with the exception of the mouse submandibular gland, contain primarily, if not exclusively, proNGF.

Cleavage-resistant proNGF is neurotrophic in vitro, as we have shown. If endogenous proNGF is similarly neurotrophic, then what is its role relative to that of mature NGF? We propose that proNGF is responsible for the normal neurotrophic activity in most tissues, but that injury increases the processing of proNGF to NGF and thereby provides a rapid and local supply of the more neurotrophically active mature NGF. In support of this hypothesis, both sciatic nerve transection and kainic acid-induced seizures upregulate prohormone convertases that can process neurotrophin precursors (Meyer et al., 1996; Marcinkiewicz et al., 1999).

Much work needs to be done to clarify the differences between the biological characteristics and activities of proNGF and NGF. It is not clear whether proNGF and NGF are both internalized by cells or whether they are internalized at the same rate. It is not known whether they are both retrogradely transported at the same rate, or if NGF-induced signal transduction pathways are activated similarly by proNGF. Although several studies have reported that proNGF has neurotrophic activity similar to that of mature NGF (Chen et al., 1997; Rattenholl et al., 2001a), we and others have found its activity to be anywhere from two to twenty-fold less potent than that of mature NGF (Edwards et al., 1988a; Suter et al., 1991; Yu et al., 2002; Fahnestock et al., submitted). Quantification of proNGF is problematic because of the lack of a pure standard, and therefore these numbers must be interpreted with some caution. The use of modified proNGF molecules facilitates purification of intact proNGF but also confounds interpretation of results. Purification of native, full-length proNGF to homogeneity and its further characterization is an important next step in the study of this surprising and controversial molecule.

Summary

The nature of the activity exhibited by full-length, native proNGF is still controversial. It is not clear which, if either, of the mutated proNGF molecules now under study are representative of the proNGF found in vivo. Structural and bioactivity comparisons of native proNGF with both noncleavable proNGFs, and comparisons of the elements that distinguish our proNGF from the proNGF constructed by Lee et al. (2001), will answer some of these questions. Further work will be required, for example, on retrograde transport and on regulation following injury, to elucidate the physiological role of proNGF and the function of proNGF processing in vivo.

Acknowledgments

We thank Sean Coughlin for technical contributions, and Dr. Jack Diamond for helpful discussions. This work was supported by grants from the Ontario Neurotrauma Foundation to M. Fahnestock and G. Yu, and by an award from the Scottish Rite Charitable Foundation of Canada to M. Fahnestock. M. Fahnestock's participation at the NGF2002 meeting was made possible by a grant from the Canadian Institutes for Health Research to R. Racine and M. Fahnestock.

Abbreviations

DRG	dorsal root ganglion
ELISA	enzyme-linked immunosorbent assay
JNK	c-Jun N-terminal kinase
MMP-7	matrix metalloprotease-7
NGF	nerve growth factor
p75^{NTR}	p75 neurotrophin receptor
SCG	sympathetic cervical ganglion
SDS–PAGE	sodium dodecyl sulfate- polyacrylamide gel electrophoresis
TNF	tumor necrosis factor

References

Aloyz, R.S., Bamji, S.X., Pozniak, C.D., Toma, J.G., Atwal, J., Kaplan, D.R. and Miller, F.D. (1998) p53 is essential for developmental neuron death as regulated by the TrkA and p75 neurotrophin receptors. J. Cell Biol., 143: 1691–1703.

Barbacid, M. (1995) Neurotrophic factors and their receptors. Curr. Opin. Cell Biol., 7: 148–155.

Berger, E.A. and Shooter, E.M. (1977) Evidence for pro-beta-nerve growth factor, a biosynthetic precursor to beta-nerve growth factor. Proc. Natl. Acad. Sci. USA, 74: 3647–3651.

Bibel, M., Hoppe, E. and Barde, Y.A. (1999) Biochemical and functional interactions between the neurotrophin receptors trk and p75NTR. EMBO J, 18: 616–622.

Brann, A.B., Tcherpakov, M., Williams, I.M., Futerman, A.H. and Fainzilber, M. (2002) Nerve growth factor-induced p75-mediated death of cultured hippocampal neurons is age-dependent and transduced through ceramide generated by neutral sphingomyelinase. J. Biol. Chem., 277: 9812–9818.

Bresnahan, P.A., Leduc, R., Thomas, L., Thorner, J., Gibson, H.L., Brake, A.J., Barr, P.J. and Thomas, G. (1990) Human fur gene encodes a yeast KEX2-like endoprotease that cleaves pro-β-NGF in vivo. J. Cell Biol., 111: 2851–2859.

Chakrabarti, S., Sima, A.A., Lee, J., Brachet, P. and Dicou, E. (1990) Nerve growth factor (NGF), proNGF and NGF

receptor-like immunoreactivity in BB rat retina. Brain Res., 523: 11–15.

Chen, Y., Dicou, E. and Djakiew, D. (1997) Characterization of nerve growth factor precursor expression in rat spermatids and the trophic effects of nerve growth factor in the maintenance of Sertoli cell viability. Mol. Cell. Endocrinol, 127: 129–136.

Cohen, S. (1959) Purification and metabolic effects of a nerve growth-promoting protein from snake venom. J. Biol. Chem., 234: 1129–1137.

Cohen, S. (1960) Purification of a nerve-growth promoting protein from the mouse salivary gland and its neuro-cytotoxic antiserum. Proc. Natl. Acad. Sci. USA, 46: 302–311.

Darling, T.L., Petrides, P.E., Beguin, P., Frey, P., Shooter, E.M., Selby, M. and Rutter, W.J. (1993) The biosynthesis and processing of proteins in the mouse 7S nerve growth factor complex. Cold Spring Harb. Symp. Quant. Biol., 48 Pt 1: 427–434.

Davies, A.M., Lee, K.F. and Jaenisch, R. (1993) p75-deficient trigeminal sensory neurons have an altered response to NGF but not to other neurotrophins. Neuron, 11: 565–574.

Delsite, R. and Djakiew, D. (1999) Characterization of nerve growth factor precursor protein expression by human prostate stromal cells: a role in selective neurotrophin stimulation of prostate epithelial cell growth. Prostate, 41: 39–48.

Dicou, E. (1989) Interaction of antibodies to synthetic peptides of proNGF with in vitro synthesized NGF precursors. FEBS Lett., 255: 215–218.

Dicou, E. (1992) Nerve growth factor precursors in the rat thyroid and hippocampus. Mol. Brain Res., 14: 136–138.

Dicou, E., Lee, J. and Brachet, P. (1986) Synthesis of nerve growth factor mRNA and precursor protein in the thyroid and parathyroid glands of the rat. Proc. Natl. Acad. Sci. USA, 83: 7084–7088.

Drinkwater, C.C., Barker, P.A., Suter, U. and Shooter, E.M. (1993) The carboxyl terminus of nerve growth factor is required for biological activity. J. Biol. Chem., 268: 23202–23207.

Edwards, R.H., Selby, M.J. and Rutter, W.J. (1986) Differential RNA splicing predicts two distinct nerve growth factor precursors. Nature, 319: 784–787.

Edwards, R.H., Selby, M.J., Garcia, P.D. and Rutter, W.J. (1988a) Processing of the native nerve growth factor precursor to form biologically active nerve growth factor. J. Biol. Chem., 263: 6810–6815.

Edwards, R.H., Selby, M.J., Mobley, W.C., Weinrich, S.L., Hruby, D.E. and Rutter, W.J. (1988b) Processing and secretion of nerve growth factor: Expression in mammalian cells with a vaccinia virus vector. Mol. Cell. Biol., 8: 2456–2464.

Fahnestock, M. (1991) Structure and biosynthesis of nerve growth factor. Curr. Top. Microbiol. Immunol., 165: 1–26.

Fahnestock, M., Michalski, B., Xu, B. and Coughlin, M.D. (2001) The precursor pro-nerve growth factor is the predominant form of nerve growth factor in brain and is increased in Alzheimer's disease. Mol. Cell. Neurosci., 18: 210–220.

Fahnestock, M., Yu, G., Michalski, B., Mathew, S., Colquhoun, A., Ross, G.M., and Coughlin, M.D. The nerve growth factor precursor proNGF exhibits neurotrophic activity qulatatively similar to that of mature nerve growth factor. Submitted.

Friedman, W.J. (2000) Neurotrophins induce death of hippocampal neurons via the p75 receptor. J. Neurosci., 20: 6340–6346.

Friedman, W.J. and Greene, L.A. (1999) Neurotrophin signaling via Trks and p75. Exp. Cell Res., 253: 131–142.

Greene, L.A., Shooter, E.M. and Varon, S. (1968) Enzymatic activities of mouse nerve growth factor and its subunits. Proc. Natl. Acad. Sci. USA, 60: 1383–1388.

Hamanoue, M., Middleton, G., Wyatt, S., Jaffray, E., Hay, R.T. and Davies, A.M. (1999) p75-mediated NF-kappaB activation enhances the survival response of developing sensory neurons to nerve growth factor. Mol. Cell. Neurosci., 14: 28–40.

Hertz, L., Juurlink, B.H.J., Hertz, E., Fosmark, H. and Schousboe, A. (1989). Preparation of primary cultures of mouse (rat) astrocytes. In: A. Shahar (Ed.), A Dissection and Tissue Culture Manual of the Nervous System. Alan R. Liss, Inc, New York, pp. 105–108, Chapter 21.

Ibanez, C.F. (1995) Neurotrophic factors: from structure-function studies to designing effective therapeutics. Trends Biotechnol., 13: 217–227.

Ibanez, C.F. (1998) Emerging themes in structural biology of neurotrophic factors. Trends Neurosci., 21: 438–444.

Ibanez, C.F. (2002) Jekyll-Hyde neurotrophins: the story of proNGF. Trends Neurosci., 25: 284–286.

Ibanez, C.F., Ebendal, T., Barbany, G., Murray-Rust, J., Blundell, T.L. and Persson, H. (1992) Disruption of the low affinity receptor-binding site in NGF allows neuronal survival and differentiation by binding to the trk gene product. Cell, 69: 329–341.

Ibanez, C.F., Ilag, L.L., Murray-Rust, J. and Persson, H. (1993) An extended surface of binding to Trk tyrosine kinase receptors in NGF and BDNF allows the engineering of a multifunctional pan-neurotrophin. EMBO J., 12: 2281–2293.

Jongstra-Bilen, J., Coblentz, L. and Shooter, E.M. (1989) The in vitro processing of the NGF precursors by the gamma-subunit of the 7S NGF complex. Mol. Brain Res., 5: 159–169.

Kahle, P., Burton, L.E., Schmelzer, C.H. and Hertel, C. (1992) The amino terminus of nerve growth factor is involved in the interaction with the receptor tyrosine kinase p140trkA. J. Biol. Chem., 267: 22707–22710.

Kaplan, D.R. and Miller, F.D. (2000) Neurotrophin signal transduction in the nervous system. Curr. Opin. Neurobiol., 10: 381–391.

Krutten, A., Heymach, J.V., Jr., Kahle, P.J. and Shooter, E.M. (1997) The role of the nerve growth factor carboxyl terminus in receptor binding and conformational stability. J. Biol. Chem., 272: 29222–29228.

Kullander, K., Kaplan, D. and Ebendal, T. (1997) Two restricted sites on the surface of the nerve growth factor molecule independently determine specific TrkA receptor binding and activation. J. Biol. Chem., 272: 9300–9307.

Lakshmanan, J., Burns, C. and Smith, R.A. (1988) Molecular forms of nerve growth factor in mouse submaxillary glands. Biochem. Biophys. Res. Commun., 152: 1008–1014.

Lakshmanan, J., Beattie, G.M., Hayek, A., Burns, C. and Fisher, D.A. (1989) Biological actions of 53 kDa nerve growth factor as studied by a blot and culture technique. Neurosci Lett., 99: 263–267.

Lawrence, D., et al. (2001) Differential hepatocyte toxicity of recombinant Apo2L/TRAIL versions. Nat. Med., 7: 383–385.

Ledent, P., Duez, C., Vanhove, M., Lejeune, A., Fonze, E., Charlier, P., Rhazi-Filali, F., Thamm, I., Guillaume, G., Samyn, B., Devreese, B., Van Beeumen, J., Lamotte-Brasseur, J., Frere, J.M. (1997) Unexpected influence of a C-terminal-fused His-tag on the processing of an enzyme and on the kinetic and folding parameters. FEBS Lett., 413: 194–196.

Lee, R., Kermani, P., Teng, K.K. and Hempstead, B.L. (2001) Regulation of cell survival by secreted proneurotrophins. Science, 294: 1945–1948.

Levi-Montalcini, R. and Hamburger, V. (1953) A diffusible agent of mouse sarcoma producing hyperplasia of sympathetic ganglia and hyperneurotization of viscera in the chick embryo. J. Exp. Zool., 123: 233–288.

Levi-Montalcini, R. and Angeletti, P.U. (1968) Nerve growth factor. Physiol. Rev., 48: 534–569.

Mahadeo, D., Kaplan, L., Chao, M.V. and Hempstead, B.L. (1994) High affinity nerve growth factor binding displays a faster rate of association than p140trk binding. Implications for multi-subunit polypeptide receptors. J. Biol. Chem., 269: 6884–6891.

Marcinkiewicz, M., Marcinkiewicz, J., Chen, A., Leclaire, F., Chretien, M. and Richardson, P. (1999) Nerve growth factor and proprotein convertases furin and PC7 in transected sciatic nerves and in nerve segments cultured in conditioned media: their presence in Schwann cells, macrophages, and smooth muscle cells. J. Comp. Neurol., 403: 471–485.

Meyer, A., Chretien, P., Massicotte, G., Sargent, C., Chretien, M. and Marcinkiewicz, M. (1996) Kainic acid increases the expression of the prohormone convertases furin and PC1 in the mouse hippocampus. Brain Res., 732: 121–132.

Mobley, W.C., Schenker, A. and Shooter, E.M. (1976) Characterization and isolation of proteolytically modified nerve growth factor. Biochemistry, 15: 5543–5551.

Mowla, S.J., Pareek, S., Farhadi, H.F., Petrecca, K., Fawcett, J.P., Seidah, N.G., Morris, S.J., Sossin, W.S. and Murphy, R.A. (1999) Differential sorting of nerve growth factor and brain-derived neurotrophic factor in hippocampal neurons. J. Neurosci., 19: 2069–2080.

Murphy, R.A., Chlumecky, V., Smillie, L.B., Carpenter, M., Nattriss, M., Anderson, J.K., Rhodes, J.A., Barker, P.A., Siminoski, K., Campenot, R.B., et al. (1989) Isolation and characterization of a glycosylated form of beta nerve growth factor in mouse submandibular glands. J. Biol. Chem., 264: 12502–12509.

Petrides, P.E. and Shooter, E.M. (1986) Rapid isolation of the 7S-nerve growth factor complex and its subunits from murine submaxillary glands and saliva. J. Neurochem., 46: 721–725.

Racke, M.M., Mason, P.J., Johnson, M.P., Brankamp, R.G. and Linnik, M.D. (1996) Demonstration of a second pharmacologically active promoter region in the NGF gene that induces transcription at exon 3. Mol. Brain Res., 41: 192–199.

Ramage, P., Hemmig, R., Mathis, B., Cowan-Jacob, S.W., Rondeau, J.M., Kallen, J., Blommers, M.J.J., Zurini, M. and Rüdisser, S. (2002) Snags with tags: Some observations made with (His)$_6$-tagged proteins. Life Sci. News, 11: 1–4. (Also available at Amersham Biosciences website, http://www.lsn-online.com).

Rattenholl, A., Lilie, H., Grossmann, A., Stern, A., Schwarz, E. and Rudolph, R. (2001a) The pro-sequence facilitates folding of human nerve growth factor from Escherichia coli inclusion bodies. Eur. J. Biochem., 268: 3296–3303.

Rattenholl, A., Ruoppolo, M., Flagiello, A., Monti, M., Vinci, F., Marino, G., Lilie, H., Schwarz, E. and Rudolph, R. (2001b) Pro-sequence assisted folding and disulfide bond formation of human nerve growth factor. J. Mol. Biol., 305: 523–533.

Reinshagen, M., Geerling, I., Eysselein, V.E., Adler, G., Huff, K.R., Moore, G.P. and Lakshmanan, J. (2000) Commercial recombinant human beta-nerve growth factor and adult rat dorsal root ganglia contain an identical molecular species of nerve growth factor prohormone. J. Neurochem., 74: 2127–2133.

Ryden, M. and Ibanez, C.F. (1997) A second determinant of binding to the p75 neurotrophin receptor revealed by alanine-scanning mutagenesis of a conserved loop in nerve growth factor. J. Biol. Chem., 272: 33085–33091.

Saboori, A.M. and Young, M. (1986) Nerve growth factor: biosynthetic products of the mouse salivary glands. Characterization of stable high molecular weight and 32,000-dalton nerve growth factors. Biochemistry, 25: 5565–5571.

Seidah, N.G., Benjannet, S., Pareek, S., Savaria, D., Hamelin, J., Goulet, B., Laliberté, J., Lazure, C., Chrétien, M., Murphy, R.A. (1996) Cellular processing of the nerve growth factor precursor by the mammalian pro-protein convertases. Biochem. J., 314: 951–960.

Selby, M.J., Edwards, R.H., Sharp, F. and Rutter, W.J. (1987) Mouse nerve growth factor gene: Structure and expression. Mol. Cell Biol., 7: 3057–3064.

Smith, P.G., Pedchenko, T., Krizsan-Agbas, D. and Hasan, W. (2002) Sympathetic postganglionic neurons synthesize and secrete pro-nerve growth factor protein in culture. Soc. Neurosci. Abst., 332.3.

Suter, U., Heymach, J.V., Jr. and Shooter, E.M. (1991) Two conserved domains in the NGF propeptide are necessary and sufficient for the biosynthesis of correctly processed and biologically active NGF. EMBO J., 10: 2395–2400.

Treanor, J.J., Schmelzer, C., Knusel, B., Winslow, J.W., Shelton, D.L., Hefti, F., Nikolics, K. and Burton, L.E. (1995) Heterodimeric neurotrophins induce phosphorylation of Trk receptors and promote neuronal differentiation in PC12 cells. J. Biol. Chem., 270: 23104–23110.

Ullrich, A., Gray, A., Berman, C. and Dull, T.J. (1983) Human beta-nerve growth factor gene sequence highly homologous to that of mouse. Nature, 303: 821–825.

Urfer, R., Tsoulfas, P., Soppet, D., Escandon, E., Parada, L.F. and Presta, L.G. (1994) The binding epitopes of neurotrophin-3 to its receptors trKC and gp75 and the design of a multifunctional human neurotrophin. EMBO J., 13: 5896–5909.

Van der Zee, C.E.E.M., Rashid, K., Le, K., Moore, K.A., Stanisz, J., Diamond, J., Racine, R.J. and Fahnestock, M. (1995) Intraventricular administration of antibodies to nerve growth factor retards kindling and blocks mossy fiber sprouting in adult rats. J. Neurosci., 15: 5316–5323.

Wiesmann, C., Ultsch, M.H., Bass, S.H. and de Vos, A.M. (1999) Crystal structure of nerve growth factor in complex with the ligand-binding domain of the TrkA receptor. Nature, 401: 184–188.

Woo, S.B., Timm, D.E. and Neet, K.E. (1995) Alteration of NH2-terminal residues of nerve growth factor affects activity and Trk binding without affecting stability or conformation. J. Biol. Chem., 270: 6278–6285.

Yamashita, T., Tucker, K.L. and Barde, Y.A. (1999) Neurotrophin binding to the p75 receptor modulates Rho activity and axonal outgrowth. Neuron, 24: 585 593.

Yardley, G., Relf, B., Lakshmanan, J., Reinshagen, M. and Moore, GP. (2000) Expression of nerve growth factor mRNA and its translation products in the anagen hair follicle. Exp. Dermatol., 9: 283–289.

Yiangou, Y., Facer, P., Sinicropi, D.V., Boucher, T.J., Bennett, D.L., McMahon, S.B. and Anand, P. (2002) Molecular forms of NGF in human and rat neuropathic tissues: decreased NGF precursor-like immunoreactivity in human diabetic skin. J. Peripher. Nerv. Syst., 7: 190–197.

Yu, G., Michalski, B., Mathew, S., Coughlin, M.D. and Fahnestock, M. The nerve growth factor precursor exhibits biological activity similar to that of mature NGF. Soc. Neurosci. Abstr., 332.2.

Progress in Brain Research, Vol. 146
ISSN 0079-6123

CHAPTER 8

NGF deprivation-induced gene expression: after ten years, where do we stand?

Robert S. Freeman*, Robert L. Burch, Robert J. Crowder, David J. Lomb, Matthew C. Schoell, Jennifer A. Straub and Liang Xie

Department of Pharmacology and Physiology, University of Rochester School of Medicine and Dentistry, Rochester, NY 14642, USA

Abstract: Nerve growth factor (NGF) is required for the survival of developing sympathetic and sensory neurons. In the absence of NGF, these neurons undergo protein synthesis-dependent apoptosis. Ten years have gone by since the first reports of specific genes being upregulated during NGF deprivation-induced cell death. Over the last decade, a few additional genes (*DP5, Bim, SM-20*) have been added to a list that began with *cyclin D1* and *c-jun*. In this chapter, we discuss the evidence that these genes act as regulators of neuronal cell death. We also suggest a hypothesis for how one gene, *SM-20*, may function to suppress a self-protection mechanism in NGF-deprived neurons.

Keywords: apoptosis; c-jun; cyclin D; Bim; hypoxia-inducible factor; SM-20; sympathetic neuron

Introduction

Ever since Rita Levi-Montalcini and Viktor Hamburger began their collaboration that led to the discovery of nerve growth factor (NGF) (historically reviewed by Cowan, 2001), an enormous effort has gone into understanding where, when, and how neurotrophic factors function in the vertebrate nervous system. Observations and predictions made during the three decades after the purification of NGF formed the basis of a *neurotrophic factor hypothesis*. Succinctly put, the neurotrophic factor hypothesis postulates that innervating nerve terminals compete for limiting quantities of survival-promoting neurotrophic factors that are produced by and released from cells in the target field. In the

simplest case, those neurons that successfully compete survive while the unsuccessful neurons die. Competition for survival-promoting neurotrophic factors is clearly an important mechanism for matching the size of a neuronal population to its target field, especially in the peripheral nervous system, where the survival of an entire population of neurons may depend on a single neurotrophic factor. It is now equally clear, however, that other factors participate in regulating the survival of developing neurons and that the role of neurotrophic factors is not limited to cell survival (Yuen et al., 1996).

The classic immunosympathectomy experiments of Levi-Montalcini and Booker (1960) provided early support for the neurotrophic factor hypothesis. In just a few days, the sympathetic ganglia in mice administered anti-NGF antibodies were reduced to one-sixth their normal size due to massive neuronal cell death. The results were the first demonstration of the importance of endogenous NGF for the development of the nervous system. Over the next 30 years, a requirement for NGF in the survival of developing sympathetic and sensory neurons was

*Correspondence to: R.S. Freeman, Department of Pharmacology and Physiology, University of Rochester School of Medicine and Dentisry, 601 Elmwood Avenue, Rochester, NY 14642, USA. Tel.: +1-585-273-4893; Fax: +1-585-273-2652; E-mail: Robert_Freeman@urmc.rochester.edu
[1]Present address: Department of Pathology, Washington University School of Medicine, St. Louis, MO 63110, USA.

DOI: 10.1016/S0079-6123(03)46008-1

firmly established. By the end of this period, a family of NGF-related neurotrophic factors, the neurotrophins, including NGF, brain-derived neurotrophic factor (BDNF), neurotrophin-3 (NT-3), and NT-4/5, had been discovered and their physiological significance and therapeutic potential had become apparent (Yuen and Mobley, 1996; Lewin and Barde, 1996).

NGF influences nervous system development and function in many ways other than promoting cell survival. For example, it can promote differentiation, regulate neurotransmitter expression and release, facilitate axon guidance and synapse formation, and modulate synaptic activity (reviewed by Huang and Reichardt, 2001). NGF exerts its effects by acting on cell surface receptors comprised of the tyrosine kinase A (TrkA) receptor and the common neurotrophin receptor p75 (p75[NTR]). Binding of NGF stimulates TrkA protein kinase activity and, in ways not fully understood, alters protein–protein interactions involving p75[NTR]. Activation of NGF receptors initiates an intracellular signal that is transduced by an increasingly complex network of biochemical events. The precise signal that is generated is shaped by the relative levels of TrkA and p75[NTR] and by other factors including the type of cell, its stage in development, and its extracellular environment (Sofroniew et al., 2001). Whereas TrkA is essential for the survival promoting action of NGF on sympathetic neurons, the role of p75[NTR] remains unclear due to its ability to mediate both cell survival and cell death (Barrett, 2000).

Over the last 15 years, two basic strategies have been used to study how NGF promotes cell survival and, conversely, why NGF-deprivation results in cell death. Investigators have either stimulated cells with NGF or taken it away and then examined the morphological and biochemical consequences. These experiments have revealed NGF-dependent signal transduction pathways that enhance cell survival and antagonize cell death mechanisms as well as opposing pathways that actively promote cell death in response to NGF withdrawal (Kaplan and Miller, 2000). In this chapter, we review some of the progress that has been made towards answering the question, why do developing neurons die when deprived of NGF? Emphasis is placed on the role of gene expression in the death of sympathetic neurons deprived of NGF.

NGF deprivation-induced gene expression and the emergence of a cell death hypothesis

Dissociated sympathetic neurons isolated from the superior cervical ganglia (SCG) of late term or newborn rats and mice are commonly used for analyzing the cellular and biochemical effects of NGF deprivation. This model has its roots in the demonstration by Levi-Montalcini and Angeletti (1963) that sensory and sympathetic neurons require NGF for survival in vitro. That this and other early cell culture systems (Greene, 1978) were not more widely exploited for studying the intracellular mechanisms regulating cell death and survival appears to have been a reflection of the general indifference with which the cell death process was viewed at the time. To be sure, the significance of cell death for the development of the nervous system was firmly established (Oppenheim, 1991), but the concept of cell death as an active, self-directed and gene-regulated process was still in its infancy. Because trophic factors such as NGF could promote cell growth and enhance certain metabolic functions, it was not unreasonable to assume that cell death as a result of insufficient NGF occurred because cellular metabolic capabilities declined to levels incompatible with life.

This hypothesis was put to the test by Martin et al. (1988) using dissociated rat sympathetic neurons maintained in vitro by NGF. These investigators reasoned that if NGF promoted survival primarily by sustaining cellular metabolic activities above a critical threshold, then the death that occurred after its removal might be hastened by drugs that inhibit macromolecule biosynthesis. On the other hand, if cell death was an active process requiring gene expression, then these same inhibitors might slow down or even block the death. Consistent with the latter hypothesis, RNA and protein synthesis inhibitors substantially slowed the death of NGF-deprived sympathetic neurons. RNA and protein synthesis inhibitors were found to delay death of other neurons deprived of different trophic factors (Scott and Davies, 1990) and, importantly, of sensory and motor neurons in vivo (Oppenheim et al., 1990). Although these observations did not prove the existence of cell death-promoting genes (Ratan et al., 1994), they sparked a search for genes expressed in dying neurons.

The discovery of cell-autonomous and genetically programmed cell death in *Caenorhabditis elegans* helped feed the prediction that analogous (and perhaps homologous) cell death genes would function in higher animals including humans (Ellis et al., 1991). Consistent with this prediction, several groups reported on the increased expression of various genes including *c-fos*, *c-myc*, and *p53* during cell death in regressing tissues such as prostate and mammary gland (reviewed by Freeman et al., 1993). However, in most cases it was unclear whether these genes were expressed in the cell that were dying and, if so, whether their expression actually contributed to cell death. Perhaps the most compelling evidence for an intrinsic cell death program in mammals was the discovery that the Bcl-2 protein was capable of protecting cells from a variety of death stimuli (Vaux et al., 1988). Bcl-2 was to become the first example of a human protein that functions mainly as an intracellular inhibitor of cell death. The ability of Bcl-2 to prevent death, first shown in hematopoietic cells, was extended a few years later to NGF-deprived sympathetic neurons (Garcia et al., 1992).

We now know that cell death after NGF withdrawal is carried out by specific biochemical pathways that culminate in the activation of cysteine proteases known as caspases, and that one function of NGF is to actively suppress these pathways. However, inhibiting caspase activity alone does not completely suppress NGF deprivation-induced death (Deshmukh et al., 2000). This is because NGF withdrawal does more than unleash pro-apoptotic pathways; it also produces a repressed metabolic state indicated by a decline in glucose uptake and in RNA and protein synthesis. When such a state persists long enough (e.g., in caspase-inhibited cells deprived of NGF), neurons undergo bioenergetic failure resulting in caspase-independent cell death (Chang et al., 2002). Thus, sympathetic neurons depend on NGF for survival because of its ability to block pro-apoptotic pathways and its ability to maintain a proper metabolic state.

Before discussing specific genes that are upregulated in NGF-deprived neurons, a brief review of the biochemical mechanisms that regulate the death of NGF-deprived neurons is necessary (see also Deshmukh and Johnson, 1997). As mentioned above, the death of sympathetic neurons in vitro is

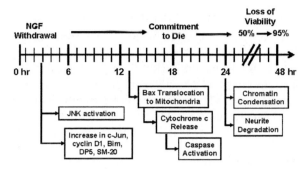

Fig. 1. Events during cell death in NGF-deprived sympathetic neurons. Studies over the last ten years have led to a model for the cell death that occurs after NGF withdrawal. JNK activity increases relatively early after removal of NGF resulting in increased expression of AP-1 transactivated genes. Some of these genes function to increase the activity of pro-apoptotic Bcl-2 family proteins, resulting in release of cytochrome c from the mitochondria to the cytosol. Cytochrome c release and other less well characterized events stimulate caspase activity, which accelerates the rate of death in NGF-deprived neurons.

stimulated by removing NGF from cultures of dissociated cells (Fig. 1). The first cell deaths occur about 18–24 h after NGF removal. Between 24 and 30 h, roughly half of the neurons will have died, with almost all of the neurons dying by 48 h. Neurons deprived of NGF show several features of apoptosis. Their cell bodies shrink and plasma membranes bleb, their neurites disintegrate, the chromatin undergoes marked condensation, and DNA is cleaved at internucleosomal sites (Deckwerth and Johnson, 1993; Edwards and Tolkovsky, 1994). With the possible exception of cell shrinkage, apoptotic morphology is not evident until about 15 h after NGF removal. That cell shrinkage is a relatively early event after NGF withdrawal may reflect loss of the trophic or 'nourishing' aspects of NGF action. A decrease in phosphatidylinositol 3-kinase activity and an increase in the activity of c-Jun N-terminal kinases (JNKs) after NGF withdrawal are two of the factors that contribute to cell atrophy and reduced metabolic activity (Tsui-Pierchala et al., 2000; Harris et al., 2002).

Despite the early decline in protein synthesis and glucose uptake in NGF-deprived neurons, mitochondrial function and cellular ATP levels are maintained at roughly normal levels until late in the cell death process. A marked increase in JNK activity occurs within 3–5 h that is paralleled by an increase in the level of phosphorylated c-Jun in neuronal nuclei (Ham et al., 1995; Virdee et al., 1997; Eilers et al., 1998). While overall mRNA levels decrease, the message levels of a small number of genes actually increase (see below). Pro-apoptotic members of the Bcl-2 family, most notably Bax and Bim, move from predominantly cytosolic pools to the mitochondrial membrane in NGF-deprived neurons (Putcha et al., 1999; Putcha et al., 2001; Whitfield et al., 2001). Translocation of Bax and Bim to mitochondria is closely associated with the release of cytochrome c and SMAC/DIABLO from the mitochondrial intermembrane space into the cytoplasm (Putcha et al., 1999, 2001; Deshmukh et al., 2002). In fact, Bax function is required for release of cytochrome c into the cytosol and loss of Bax completely suppresses NGF deprivation-induced death of sympathetic neurons (Deckwerth et al., 1996; Deshmukh and Johnson, 1998). As in other examples of apoptotic cell death, these events culminate in the activation of caspases (Deshmukh et al., 1996; McCarthy et al., 1997). Caspase inhibitors block cell death without affecting the kinetics of cytochrome c release indicating that caspases act at a relatively late step in this cell death paradigm (Deshmukh and Johnson, 1998; Neame et al., 1998). Caspase-2, caspase-3 and caspase-9 have all been implicated in death caused by NGF deprivation (Troy et al., 1997; Deshmukh et al., 2000; Troy et al., 2001).

As mentioned above, the ability of RNA and protein synthesis inhibitors to block neuronal death initiated a search for genes that are upregulated in neurons undergoing apoptosis. The first of these to by identified in trophic factor-deprived neurons included *cyclin D1*, *c-jun*, *c-fos*, *fosB*, *c-myb*, *mkp-1* and *NGFI-A* (Estus et al., 1994; Freeman et al., 1994; Ham et al., 1995). The proteins encoded by *c-jun*, *c-fos*, *fosB*, *c-myb* and *NGFI-A* are DNA binding transcription factors. *Cyclin D1* encodes a regulatory subunit for cyclin-dependent protein kinases (CDKs) 4 and 6 while the product of the *mkp-1* gene is a serine/threonine phosphatase that regulates the activity of MAP kinases. *Cyclin D1*, *c-myb c-jun*, and *mkp-1* expression increases within 5 h and peaks around 15 h after NGF withdrawal. In comparison, the increase in NGFI-A, c-fos, and fosB mRNA levels

are delayed, occurring mostly between 10 and 20 h after NGF removal. Recently, several additional genes have been found to be upregulated after NGF withdrawal including *DP5* (Imaizumi et al., 1997), *Bim* (Putcha et al., 2001; Whitfield et al., 2001), and *SM-20* (Lipscomb et al., 1999). In the rest of this chapter, we discuss the evidence that these genes have a role in the death of NGF-deprived neurons.

c-Jun and the JNK pathway

Jun family proteins (c-Jun, JunB, JunD) together with members of the Fos, ATF and Maf subfamilies of basic region-leucine zipper proteins comprise the dimeric transcription factor known as activator protein-1 (AP-1) (Shaulian and Karin, 2001). Numerous stimuli can activate AP-1 including growth factors, cytokines, hormones, neurotransmitters and a variety of chemical, biological and physical stresses. Of the various forms of AP-1, the best characterized are those comprised of c-Jun homodimers and c-Jun/Fos heterodimers. Regulation of AP-1 activity is complex and involves *jun* and *fos* transcription and mRNA stabilization, stabilization and phosphorylation of Jun and Fos proteins, and interactions between AP-1 and other proteins that modulate its activity.

The demonstration that c-jun and c-fos mRNA levels are upregulated after NGF withdrawal was an early indication that AP-1 might be involved in neuronal cell death. Other indications are an increase in AP-1 transactivation activity after NGF withdrawal (Ham et al., 1995) and the increased expression of two endogenous AP-1 regulated genes, *transin* and *collagenase* (Estus et al., 1994). Phosphorylation of c-Jun by JNKs enhances the transactivation potential of AP-1 and, as predicted, JNK activity increases and phosphorylated c-Jun accumulates in the nuclei of NGF-deprived neurons (Virdee et al., 1997; Eilers et al., 1998). More direct evidence for a role for AP-1 in neuronal death came from experiments in which c-Jun neutralizing antibodies (Estus et al., 1994) or a dominant negative c-Jun expression vector (Ham et al., 1995) were microinjected into sympathetic neurons. In both cases, the number of neurons undergoing cell death after NGF withdrawal decreased significantly.

Studies from several laboratories have established an important role for JNK activity in neuronal cell death. Inhibitors of JNK activation, such as the mixed-lineage protein kinase (MLK) inhibitor CEP-1347 (Maroney et al., 1999) and the JNK binding domain of JNK-interacting protein 1 (Eilers et al., 2001; Harding et al., 2001), efficiently block phosphorylation of c-Jun and prevent neuronal death after NGF withdrawal. Release of cytochrome c into the cytosol is also blocked by inhibiting JNK as are the decreases in metabolic activity and cell size that normally occur in NGF-deprived neurons (Harris et al., 2002), suggesting that the JNK pathway may be important for suppressing both the trophic and survival-promoting functions of NGF. Upstream regulators of JNK and AP-1 activation have recently been identified, mostly by overexpressing activated and dominant negative forms of candidate proteins in sympathetic neurons. These studies suggest a pathway involving Cdc42/Rac1, MLKs, mitogen activated protein kinase kinases 4 and 7, JNKs, and c-Jun/AP-1 (Bazenet et al., 1998; Mota et al., 2001; Xu et al., 2001). Finally, a recent study suggests that the multidomain scaffolding protein POSH might act to bring several JNK pathway proteins together in order to activate JNK and stimulate c-Jun phosphorylation (Xu et al., 2003).

Mice deficient in c-Jun and each of the three JNKs have been created using homologous recombination. Mice lacking functional c-Jun die early in embryogenesis, before the onset of programmed cell death in NGF-dependent neurons. To circumvent this problem, Palmada et al. (2002) created mice in which the c-jun coding exon was flanked by loxP sites. Sympathetic neurons from these mice were treated with a recombinant adenovirus expressing Cre recombinase to effectively delete the c-jun gene in the infected cells. Elimination of c-Jun in these neurons prevented cell death after NGF withdrawal, although it did not affect death induced by activation of p75^{NTR}.

Animals deficient in JNK1 and JNK2 lack any obvious neuronal phenotype. In contrast, JNK3-deficient mice were found to have decreased neuronal apoptosis after injection with kainic acid, suggesting that JNK3 may contribute to some forms of neuronal death (Yang et al., 1997). By comparing cultured sympathetic neurons from JNK3 −/− and JNK3 +/−

mice, Bruckner et al. (2001) found that disruption of JNK3 resulted in a reduction in the phosphorylation of c-Jun and a delay in cell death after NGF withdrawal. The lack of complete inhibition of cell death and c-Jun phosphorylation in JNK3 −/− neurons may have been due to redundant or compensatory mechanisms involving JNK1 and JNK2. To date, only a few genes regulated by AP-1 have been implicated in neuronal cell death (reviewed by Shaulian and Karin, 2002). One of these is the gene encoding Fas ligand (FasL), which was reported to contribute to cell death in NGF-deprived PC12 cells and in cerebellar granule neurons (Le Niculescu et al., 1999). The mRNA levels of FasL have been shown to increase slightly in sympathetic neurons deprived of NGF (Putcha et al., 2002). However, sympathetic neurons containing loss of function mutations in either Fas or FasL undergo NGF deprivation-induced death at the same rate as wild type neurons (Putcha et al., 2002). Therefore, the FasL pathway does not appear to be an essential target of c-Jun during NGF withdrawal-induced cell death.

Cyclin D1 and cyclin-dependent kinases

As cells traverse the cell cycle they round up and become less adherent, their chromatin condenses, their nuclear lamina disassemble, and their plasma membranes ruffle and invaginate. These changes are reversible and facilitate DNA replication and division of the genetic and cellular material into two daughter cells. During apoptosis, cells become rounded and less adherent, their chromatin condenses and the nuclear lamina breaks down. The plasma membranes of apoptotic cells bleb and invaginate, sometimes pinching off pieces called apoptotic bodies. The gross similarities in the morphologies of dividing cells and dying cells helped give rise to suggestions that the two processes might share overlapping mechanisms (Heintz, 1993). In fact, the ties between cell cycle control and cell death are more than just skin deep. The ability of numerous protooncogenes and cell growth regulators to promote cell death, along with overwhelming evidence that perturbations to the normal cell cycle can trigger apoptosis, have bolstered the hypothesis that cell death is initiated by conflicting growth regulatory signals (Freeman, 1998).

Cell division is driven by the formation of active complexes between cyclins and CDKs at progressive stages of the cell cycle, from G1-phase through the G2/M transition (Sherr, 1996). Progress through G1 and entry into S phase is regulated by the accumulation of cyclin D proteins that bind to and activate CDK4 and CDK6. Interestingly, the increase in cyclin D1 protein during G1-phase is at least partly due to transactivation of the *cyclin D1* gene by AP-1 (Shaulian and Karin, 2001). Accumulation of cyclins results in formation of cyclin–CDK complexes and the displacement of cyclin-dependent kinase inhibitors (CKIs) from the CDKs. A major target of the G1-phase cyclin D/CDK complexes is the retinoblastoma protein (Rb). In resting cells, underphosphorylated Rb binds the E2F family of transcription factors, thereby preventing E2F from transactivating genes critical for DNA replication (Harbour and Dean, 2000). CDK-dependent phosphorylation of Rb reduces its affinity for E2F proteins. Therefore, an important consequence of CDK activation during G1-phase is the release of E2F transcription factors from Rb resulting in activation as well as derepression of E2F transactivated genes.

It has been almost ten years since cyclin D1 mRNA levels were shown to increase in sympathetic neurons undergoing NGF withdrawal-induced apoptosis. Since then, an increase in cyclin D1 mRNA and/or protein has been reported in PC12 cells deprived of NGF (Davis et al., 1997), in potassium-deprived cerebellar granule cells (Padmanabhan et al., 1999; Sakai et al., 1999), in various neurons after treatment with kainic acid (Liu et al., 1996; Giardina et al., 1998; Ino and Chiba, 2001), and in brain and spinal cord neurons subjected to ischemia (Guegan et al., 1997; Li et al., 1997a; Kuroiwa et al., 1998; Sakurai et al., 2000; Katchanov et al., 2001). In most of these studies the significance of elevated cyclin D1 for cell death was not investigated and in certain models of focal or global ischemia, increased cyclin D1 was only detected in nonneuronal cells or in apparently healthy neurons (Wiessner et al., 1996; Li et al., 1997b; Small et al., 2001).

Several lines of evidence implicate cyclin/CDK involvement in neuronal death (Liu and Greene, 2001). For example, pharmacological inhibitors of CDKs such as flavopiridol, olomoucine and roscovitine block NGF deprivation-induced death and DNA damage-induced death of sympathetic neurons and PC12 cells, as well as potassium deprivation-induced death of cerebellar granule neurons (Park et al., 1996, 1997b; Padmanabhan et al., 1999). Pharmacological CDK inhibitors also protect cortical neurons from oxygen–glucose deprivation in vitro and in vivo (Osuga et al., 2000; Katchanov et al., 2001). Recombinant Sindbis virus vectors expressing the CKIs p16, p21 and p27 or dominant-negative forms of CDK4 and CDK6 have been reported to inhibit neuronal cell death caused by NGF withdrawal, DNA damage, and βamyloid (Park et al., 1997a, 1998; Giovanni et al., 1999). Additional evidence for the involvement of a cyclin/CDK pathway in neuronal cell death comes from demonstrations of increased Rb phosphorylation and E2F-1 protein levels in cerebellar granule cells undergoing death evoked by low potassium (Padmanabhan et al., 1999; O'Hare et al., 2000).

Despite reports of cyclin D1 upregulation and CDK involvement in various types of neuronal death, a requirement for cyclin D1 in neuronal death has not been demonstrated. While cyclin D1 overexpression can induce cell death in certain nonneuronal cells and in N1E-115 neuroblastoma cells (Kranenburg et al., 1996), we found that overexpression of cyclin D1 in sympathetic neurons fails to promote cell death in the presence of NGF and does not significantly affect the rate of cell death after NGF withdrawal (Fig. 2A, B). In experiments analogous to those performed by Park et al. (1997a), we tested the effects of expressing CKIs (p16 and p21) and dominant-negative CDK4 on the survival of NGF-deprived sympathetic neurons. In our experiments, however, none of the CDK inhibitors affected the extent of NGF deprivation-induced death (Fig. 2C). The reason for the discrepancy in our results may be related to the different methods used for gene delivery (i.e., microinjection vs recombinant Sindbis virus), differences in the expression levels of the introduced genes, differences in the culture conditions employed, or differences in the way viability was assessed. In at least one other model of trophic factor deprivation, induction of p21$^{CIP1/WAF1}$ in PC12 cells did not inhibit death caused by serum or NGF withdrawal (Erhardt and Pittman, 1998).

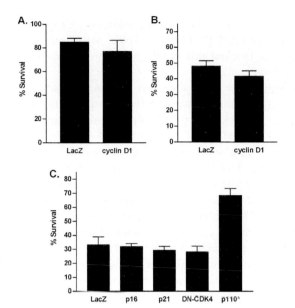

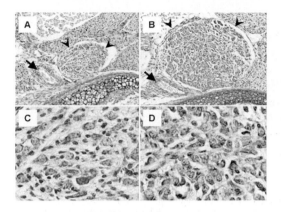

Fig. 3. Loss of cyclin D1 function negatively impacts the development of the SCG in mice. Images of the SCG (stained with cresyl violet) in a *cyclin D1* −/− mouse (A and C) and a wild type littermate (B and D) observed under a 10 × (A and B) or 40 × (C and D) objective. The sections shown were taken from approximately the same cross-sectional plane through the animal. Arrowheads in A and B point to the SCG; arrows indicate the position of the carotid artery. The mice used for this study (strain B6,129-*Ccnd1^{tm1}*) were obtained from The Jackson Laboratory (Bar Harbor, Maine).

Fig. 2. Effects of expressing cyclin D1, CKIs, and dominant-negative CDK4 on the survival of NGF-dependent neurons. SCG neurons from embryonic day-21 rats were dissociated and plated in medium containing NGF. On the fifth day after plating, neurons were microinjected with plasmid expression vectors for the indicated proteins and either maintained in the presence of NGF or deprived of NGF as described by Lipscomb et al. (1999). Survival was assessed after staining the cells with the DNA-binding dye Hoechst 33,342. Healthy neurons had diffusely-stained and decondensed chromatin with an intact nuclear membrane whereas the chromatin in unhealthy and dying cells was condensed or undetectable. Percent survival corresponds to the fraction of injected cells that were scored as healthy. (A) Neurons injected with beta-galactosidase (LacZ) or cyclin D1 vectors were maintained in the presence of NGF for 36 or 72 h (only the 36 h data are shown) before being scored for survival. The survival of LacZ- and cyclin D1-injected neurons (at either time point) was not significantly different (mean ± SEM, $n = 3$). (B) Neurons injected with LacZ or cyclin D1 vectors were deprived of NGF for 36 h and then scored for survival. In each case, NGF deprivation resulted in a similar amount of cell death (mean ± SEM, $n = 3$). (C) Neurons were injected with vectors for LacZ, p16, p21, dominant-negative CDK4 (containing a point mutation in the catalytic domain) and a constitutively active form of phosphatidylinositol 3-kinase (p110*) (Crowder and Freeman, 1998) and then deprived of NGF for 48 h. Only p110* provided significant protection from cell death (mean ± SEM, $n = 4$).

With the goal of assessing the importance of *endogenous* cyclin D1 in NGF deprivation-induced death, we obtained mice in which the *cyclin D1* gene was disrupted by homologous recombination

(Sicinski et al., 1995). Cyclin D1-deficient mice are viable at birth but grow at a reduced rate and only occasionally reach adulthood. Those that survive display an abnormal leg-clasping reflex, and have severe retinal hypoplasia. Adult *cyclin D1* −/− females exhibit reduced proliferation in their mammary glands during pregnancy, rendering them unable to lactate (Fantl et al., 1995; Sicinski et al., 1995). *Cyclin D1* +/− mice were mated to generate homozygous, heterozygous and wild type littermates. The SCG from each animal were dissected and cultured separately. Surprisingly, during the dissections we were unable to locate the SCG in a subset of mice from each litter. Genotyping revealed these mice to be *cyclin D1* −/−. Analysis of the offspring from multiple matings involving different *cyclin D1* +/− parents indicated that the SCG were anatomically displaced, shrunken, or altogether absent in 100% of the *cyclin D1* −/− animals. Microscopic analysis of sections stained with cresyl violet revealed that the SCG of newborn and postnatal day 1–4 *cyclin D1* −/− mice are significantly reduced in size (Fig. 3). By sectioning through the entire length of the ganglion, we were able to conclude that only its size, and not its anatomical location, was affected by the absence of cyclin D1. Estimates of neuronal number ($n = 4$)

indicate that the SCG from newborn *cyclin D1* −/− mice have less than 10% the number of neurons found in wild type and heterozygous littermates.

The reduced number of SCG neurons in cyclin D1-deficient mice could be due to decreased proliferation, migration, and/or differentiation of sympathetic precursors or increased cell death. Given the well-documented role of cyclin D1 in cell cycle control, and the fact that the decrease in cell number was evident at birth (before the onset of programmed cell death in the SCG), it seems most likely that the reduced number of neurons is due to a defect in the proliferation of sympathetic neuroblasts during embryogenesis. Similar conclusions were reached concerning the hypoplasia in the retinas of *cyclin D1* −/− mice and in the cerebellum of *cyclin D2* −/− mice (Sicinski et al., 1995; Huard et al., 1999). Regardless of the exact mechanism, the phenotype of these mice suggests that cyclin D1 plays an essential role in the embryonic development of sympathetic ganglia. It remains to be determined when the effects of loss of cyclin D1 are first manifested and what their consequences are for the innervation of sympathetic targets.

BH3-only proteins: DP5 and Bim

The *DP5* gene was identified from a differential screen of mRNAs expressed in NGF-deprived vs NGF-maintained sympathetic neurons (Imaizumi et al., 1997). *DP5* expression increases within 5 h after NGF withdrawal and reaches maximal levels by 10–15 h. Blocking NGF deprivation-induced death with cycloheximide, membrane depolarizing concentrations of potassium, or cpt-cAMP inhibited the induction of DP5 mRNA expression. Overexpression of DP5 was sufficient to promote cell death in the presence of NGF and this was partially reversed by coexpression with Bcl-2. DP5 mRNA levels have since been reported to be elevated in the spinal neurons of amyotrophic lateral sclerosis patients and in rat retinal ganglion cells after optic nerve transection (Shinoe et al., 2001; Wakabayashi et al., 2002). Upregulation of *DP5* during neuronal death may involve the JNK/AP-1 pathway. *DP5* expression increases in cerebellar granule neurons undergoing potassium deprivation-induced apoptosis and

treatment with the MLK inhibitor CEP-1347, which completely blocked JNK activation and partially blocked cell death, partially blocked DP5 induction

Around the same time that DP5 was discovered, its human ortholog, harakiri (Hrk), was identified in a yeast 2-hybrid screen based on its ability to interact with Bcl-2 and Bcl-X_L (Inohara et al., 1997). These results along with the identification of a Bcl-2 homology 3 (BH3) domain in the middle of the DP5/Hrk protein, immediately suggested a mechanism for DP5-induced cell death involving Bcl-2 family proteins. DP5 protein levels were found to increase in cortical neurons exposed to βamyloid peptide and at least some of the DP5 was detected in a complex with Bcl-X_L (Imaizumi et al., 1999). Moreover, although DP5 overexpression can promote neuronal death, it fails to do so in neurons from Bax −/− mice (Harris and Johnson, 2001). Together these studies suggest that upregulation of *DP5* during neuronal death may be partially regulated by the JNK/AP-1 pathway and that DP5-induced cell death requires Bax.

In *C. elegans*, the BH3-only protein Egl-1 is essential for initiating the programmed death of all somatic cells during development. Developmentally programmed cell death in mammals is decidedly more complex as reflected by the presence of at least 10 BH3-only proteins. Studies on mice lacking individual BH3-only proteins suggest that multiple BH3-only proteins may have evolved as a way of controlling the initiation of cell death in different cell types and in response to different apoptotic stimuli (Puthalakath and Strasser, 2002). In line with the increased complexity of cell death in mammals, a second BH3-only protein (Bim) was recently found to be upregulated in sympathetic neurons deprived of NGF (Putcha et al., 2001; Whitfield et al., 2001). As with DP5, the increase in Bim mRNA levels is partially dependent on JNK activity and it can be reversed by adding back NGF (Harris and Johnson, 2001; Whitfield et al., 2001). In addition to repressing *Bim* transcription, NGF may actively suppress the pro-apoptotic activity of Bim by stimulating its phosphorylation through a MEK/ERK-dependent pathway (Biswas and Greene, 2002). It is not known how phosphorylation regulates Bim but it may be involved in its sequestration in the cytosol in NGF-maintained cells. Dephosphorylation of Bim in the absence of survival factors could be one of the events that triggers its realignment with mitochondria during cell death.

Both Bim and DP5 can bind pro-survival Bcl-2 proteins suggesting that their increase during cell death may help promote the shift from Bax complexes that contain a relatively large proportion of pro-survival Bcl-2 proteins to complexes containing mostly Bax. Such a shift would be predicted to lead to cytochrome c release and caspase activation. In support of this hypothesis, death induced by over-expressing Bim is largely reduced in sympathetic neurons from *Bax* −/− mice. Unlike Bax, however, Bim is not absolutely required for NGF withdrawal-induced death of sympathetic neurons. Cell death in trophic factor-deprived neurons from *Bim* −/− mice is delayed by 12–24 h compared to neurons from *Bim* +/+ mice. The lack of an apparent requirement for Bim could be because DP5 performs a similar function during cell death or because DP5 can partially compensate for the loss of Bim in Bim-deficient neurons. An analysis of cell death in neurons lacking DP5, or both DP5 and Bim, should help clarify this issue.

SM-20 and prolyl hydroxylation

Using a differential display approach, we found that expression of the *SM-20* gene increases in sympathetic neurons deprived of NGF (Lipscomb et al., 1999). SM-20 mRNA levels increase within 5 h after removal of NGF and peak between 10 and 15 h. SM-20 protein levels are also increased within 15 h of NGF deprivation. Overexpression of SM-20 is sufficient to promote cell death in NGF-maintained sympathetic neurons (Lipscomb et al., 1999, 2001) and in PC12 cells stably expressing an inducible SM-20 transgene (Straub et al., 2003). The death is caspase-dependent and, at least in SM-20/PC12 cells, accompanied by an increase in cytosolic cytochrome c. Interestingly, induction of SM-20 in PC12 cells also leads to a substantial increase in the amount of cytochrome c present in whole cell lysates and in mitochondria-enriched subcellular fractions (Straub et al., 2003). The mechanism underlying the SM-20-induced increase in cytochrome c is not known

although it does not appear to involve an increase in cytochrome c mRNA levels.

SM-20 was originally identified as a growth factor inducible gene in vascular smooth muscle cells (Wax et al., 1994). In vivo, SM-20 mRNA and protein levels were found to increase in smooth muscle cells after vascular wall injury (Wax et al., 1996). Based on these findings, a function for SM-20 in muscle cell proliferation was proposed. Other reports, however, imply that SM-20 might function in growth arrest, differentiation and cell death. For example, SM-20 mRNA levels increase markedly over 48 h in muscle cell lines cultured in reduced serum, a procedure used to promote differentiation but that can also lead to significant cell death (Moschella et al., 1999). Expression of *SM-20* was also reported to increase in fibroblasts upon activation of a temperature-sensitive form of p53. Activation of p53 in these cells triggers both growth arrest and cell death and expression of *SM-20* in cells lacking functional p53 appears to recapitulate these effects (Madden et al., 1996).

Recently, SM-20-related sequences were identified in the rat, mouse and human genomes (Dupuy et al., 2000; Taylor, 2001). This new gene family has been dubbed the EGLN (*egl*-nine) family and currently consists of three genes in mammals (EGLN 1–3) and a single gene in both *C. elegans* (*egl-9*) and *Drosophila*. Mutations at the *C. elegans egl-9* locus were originally associated with a defective egg-laying phenotype 20 years ago, but the *egl-9* gene was not characterized (Trent et al., 1983). After discovering that loss-of-function mutations in *egl-9* conferred resistance to a paralytic toxin produced by *Pseudomonas*, Darby et al. (1999) identified the *egl-9* gene and putative protein coding sequences. Interestingly, the paralytic toxin has been identified as cyanide, a product of the *Pseudomonas* hydrogen cyanide synthase enzyme (Gallagher and Manoil, 2001). SM-20 is approximately 50% identical to the predicted Egl-9 protein over their aligned sequences. Egl-9 also contains a long N-terminal extension that is not conserved in SM-20 but that is homologous to a region in the N-terminus of mammalian EGLN1, suggesting that EGLN1 is the true ortholog of Egl-9. The greatest sequence identity among EGLN family members lies in their C-termini, which are 70% identical over a stretch of 126 amino acids (Fig. 4).

120

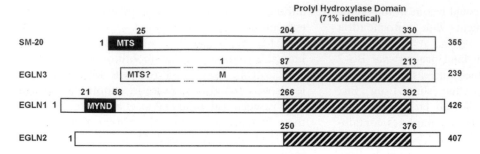

Fig. 4. Structural domains in mammalian EGLN proteins. EGLN3 is the likely human ortholog of rat SM-20 even though the start of the predicted EGLN3 coding sequence corresponds to methionine-117 in SM-20 (Taylor, 2001). Despite the reported absence of EGLN3 sequences corresponding to the mitochondrial targeting sequence in SM-20, SM-20-specific antibodies detect a protein in human muscle cells that is approximately the same molecular mass (≈ 40 kD) as SM-20 and shows a similar punctate cytoplasmic distribution (Wax et al., 1996). These observations raise the possibility that the reported EGLN3 sequence is incomplete. *M*, the first ATG codon in the human EGLN3 open reading frame predicted from expressed sequence tags. *MYND*, MYND-type zinc finger domain in EGLN1 (also present in *C. elegans* Egl-9). Amino acid numbers for rat SM-20 and human EGLN 1–3 are indicated.

A biochemical function for the EGLN protein family was recently elucidated. Using secondary structure prediction software as a tool to mine sequence databases, Aravind and Koonin (2001) identified Egl-9 and human EGLN3 (the human ortholog of rat SM-20) as potential members of the iron- and 2-oxoglutarate-dependent dioxygenase superfamily. Within this superfamily, the predicted Egl-9 structure was suggested to be most similar to mammalian prolyl-4-hydroxylase. About the same time, several laboratories were looking for candidate proteins that could catalyze the hydroxylation of proline residues in the transcription factor hypoxia-inducible factor (HIF)-1α. The hydroxylation of conserved prolines in HIF-1α was shown a few months earlier to mediate its oxygen-dependent destabilization, but the enzyme that catalyzed the hydroxylation remained unknown (Ivan et al., 2001; Jaakkola et al., 2001; Yu et al., 2001). Proline hydroxylation destabilizes HIF-1α because it increases its affinity for the von Hippel-Lindau protein, the substrate recognition subunit for an E3 ubiquitin ligase (Kaelin, 2002). Using secondary structure predictions, *C. elegans* genetics, and in vitro hydroxylation assays, Epstein et al. (2001) demonstrated that endogenous Egl-9 regulates the stability of *C. elegans* HIF and that mammalian EGLN proteins can hydroxylate HIF-1α in vitro. All three EGLN proteins (also called 'HIF prolyl hydroxylases' (HPH) and 'prolyl hydroxylase domain' (PHD)

proteins) have now been shown to be capable of destabilizing and inactivating HIF-1α in intact cells through the hydroxylation of two prolines, Pro-402 and Pro-564 (Bruick and McKnight, 2001; Huang et al., 2002). It remains to be determined which mammalian EGLN proteins target HIF-1α in vivo and whether other substrates for these enzymes exist.

HIF is clearly important in the physiologic response to oxygen deprivation through its ability to regulate the expression of genes involved in glycolysis, angiogenesis and erythropoiesis (Maxwell and Ratcliffe, 2002). But does HIF function as part of a neuroprotective pathway within neurons? Evidence supporting such a role for HIF comes from a study linking activation of HIF-1 with the ability of iron chelators to rescue cortical neurons from oxidative stress-induced death (Zaman et al., 1999). These same iron chelators were previously shown to protect sympathetic neurons from NGF deprivation-induced death (Farinelli and Greene, 1996). Because iron chelators stabilize HIF-1α by blocking the activity of SM-20/EGLN prolyl hydroxylases (which require iron in their active site), we suggest that the neuroprotective effects of these compounds may depend on inactivation of SM-20 and/or other EGLN proteins. Consistent with this possibility, we have found that expression of a stabilized form of HIF-1α (in which the prolines that undergo hydroxylation were mutated) significantly delays death caused by NGF withdrawal (L. Xie, unpublished data).

A working but still largely untested hypothesis for SM-20 involvement in trophic factor deprivation-induced death assumes that NGF withdrawal triggers self-protection mechanisms in addition to pro-apoptotic mechanisms. The self-protection mechanisms, which may include activation of HIF, could ensure that transient NGF deprivation does not irreversibly commit cells to apoptosis. After prolonged NGF withdrawal, however, sufficient SM-20 protein could accumulate to catalyze the hydroxylation and destabilization of HIF-1α. Thus, upregulation of *SM-20* after NGF withdrawal could be a means of silencing a HIF-mediated self-protection pathway. Currently there are several gaps in this hypothesis. For example, it is not yet known if HIF is activated after NGF withdrawal. If it is, does HIF contribute to self-protection and the ability of NGF-deprived neurons to be rescued by later re-addition of NGF? Does protection by HIF require its activity as a transcription factor and, if so, what are its target genes? Is the activity of HIF regulated by SM-20 in NGF-deprived neurons? Finally, is SM-20 function essential for cell death after NGF withdrawal?

Summary

The mechanisms that control cell survival and death in NGF-dependent neurons are tightly regulated. The requirement for new gene expression may help ensure that cell death mechanisms are not prematurely or inappropriately activated when neurons are transiently deprived of trophic support. The past ten years of searching for genes induced in trophic factor-deprived neurons has been fruitful on several accounts. The upregulation of Bim and DP5, which contributes directly to the activation of proapoptotic pathways, has validated initial speculation that the expression of cell death genes is induced by NGF withdrawal. The upregulation of c-Jun and cyclin D1 during NGF deprivation-induced death spearheaded studies that led to the recent consideration of CDKs and JNKs as potential therapeutic targets for stroke (Irving and Bamford, 2002; O'Hare et al., 2002). The expression of SM-20 in dying neurons and the recent discovery of its prolyl hydroxylase activity may add yet another layer of regulation to the control of cell death in NGF-deprived neurons. Although the studies highlighted in this chapter have provided many new insights, they have raised at least as many new questions—enough to keep cell death researchers busy for the next ten years.

Acknowledgments

Research conducted in the author's laboratory is funded by National Institute of Health (NIH) grants NS34400 and NS42224. We also acknowledge support from NIH predoctoral training grant DE07202 (M. Schoell) and predoctoral fellowship HG00183 (R. Burch). The dominant-negative CDK4 plasmid was a gift from Dr. Steve Dowdy.

Abbreviations

AP-1	activator protein-1
BH3	Bcl-2 homology 3
CDK	cyclin-dependent protein kinase
EGLN	*egl*-nine
FasL	Fas ligand
HIF	hypoxia-inducible factor
Hrk	harakiri protein
JNK	c-Jun N-terminal protein kinase
MLK	mixed lineage protein kinase
NGF	nerve growth factor
Rb	retinoblastoma protein
SCG	superior cervical ganglion/ganglia

References

Aravind, L. and Koonin, E.V. (2001) The DNA-repair protein AlkB, EGL-9, and leprecan define new families of 2-oxoglutarate- and iron-dependent dioxygenases. Genome Biol., 2: 7.1–7.8.

Barrett, G.L. (2000) The p75 neurotrophin receptor and neuronal apoptosis. Prog. Neurobiol., 61: 205–229.

Bazenet, C.E., Mota, M.A. and Rubin, L.L. (1998) The small GTP-binding protein Cdc42 is required for nerve growth factor withdrawal-induced neuronal death. Proc. Natl. Acad. Sci. USA, 95: 3984–3989.

Biswas, S.C. and Greene, L.A. (2002) Nerve growth factor (NGF) down-regulates the Bcl-2 homology 3 (BH3) domain-only protein Bim and suppresses its proapoptotic activity by phosphorylation. J. Biol. Chem., 277: 49511–49516.

Bruckner, S.R., Tammariello, S.P., Kuan, C.Y., Flavell, R.A., Rakic, P. and Estus, S. (2001) JNK3 contributes to c-Jun activation and apoptosis but not oxidative stress in nerve growth factor-deprived sympathetic neurons. J. Neurochem., 78: 298–303.

Bruick, R.K. and McKnight, S.L. (2001) A conserved family of prolyl-4-hydroxylases that modify HIF. Science, 294: 1337–1340.

Chang, L.K., Putcha, G.V., Deshmukh, M. and Johnson, E.M., Jr. (2002) Mitochondrial involvement in the point of no return in neuronal apoptosis. Biochimie., 84: 223–231.

Cowan, W.M. (2001) Viktor Hamburger and Rita Levi-Montalcini: the path to the discovery of nerve growth factor. Annu. Rev. Neurosci., 24: 551–600.

Crowder, R.J. and Freeman, R.S. (1998) Phosphatidylinositol 3-kinase and Akt protein kinase are necessary and sufficient for the survival of nerve growth factor-dependent sympathetic neurons. J. Neurosci., 18: 2933–2943.

Darby, C., Cosma, C.L., Thomas, J.H. and Manoil, C. (1999) Lethal paralysis of Caenorhabditis elegans by Pseudomonas aeruginosa. Proc. Natl. Acad. Sci. USA, 96: 15202–15207.

Davis, P.K., Dudek, S.M. and Johnson, G.V. (1997) Select alterations in protein kinases and phosphatases during apoptosis of differentiated PC12 cells. J. Neurochem., 68: 2338–2347.

Deckwerth, T.L. and Johnson, E.M., Jr. (1993) Temporal analysis of events associated with programmed cell death (apoptosis) of sympathetic neurons deprived of nerve growth factor. J. Cell Biol., 123: 1207–1222.

Deckwerth, T.L., Elliott, J.L., Knudson, C.M., Johnson, E.M., Jr., Snider, W.D. and Korsmeyer, S.J. (1996) BAX is required for neuronal death after trophic factor deprivation and during development. Neuron, 17: 401–411.

Deshmukh, M. and Johnson, E.M., Jr. (1997) Programmed cell death in neurons: focus on the pathway of nerve growth factor deprivation-induced death of sympathetic neurons. Mol. Pharmacol., 51: 897–906.

Deshmukh, M. and Johnson, E.M., Jr. (1998) Evidence of a novel event during neuronal death: development of competence-to-die in response to cytoplasmic cytochrome c. Neuron, 21: 695–705.

Deshmukh, M., Vasilakos, J., Deckwerth, T.L., Lampe, P.A., Shivers, B.D. and Johnson, E.M., Jr. (1996) Genetic and metabolic status of NGF-deprived sympathetic neurons saved by an inhibitor of ICE family proteases. J. Cell Biol., 135: 1341–1354.

Deshmukh, M., Kuida, K. and Johnson, E.M., Jr. (2000) Caspase inhibition extends the commitment to neuronal death beyond cytochrome c release to the point of mitochondrial depolarization. J. Cell Biol., 150: 131–143.

Deshmukh, M., Du, C., Wang, X. and Johnson, E.M., Jr. (2002) Exogenous smac induces competence and permits caspase activation in sympathetic neurons. J. Neurosci., 22: 8018–8027.

Dupuy, D., Aubert, I., Duperat, V.G., Petit, J., Taine, L., Stef, M., Bloch, B. and Arveiler, B. (2000) Mapping, characterization, and expression analysis of the SM-20 human homologue, c1orf12, and identification of a novel related gene, SCAND2. Genomics, 69: 348–354.

Edwards, S.N. and Tolkovsky, A.M. (1994) Characterization of apoptosis in cultured rat sympathetic neurons after nerve growth factor withdrawal. J. Cell Biol., 124: 537–546.

Eilers, A., Whitfield, J., Babij, C., Rubin, L.L. and Ham, J. (1998) Role of the Jun kinase pathway in the regulation of c-Jun expression and apoptosis in sympathetic neurons. J. Neurosci., 18: 1713–1724.

Eilers, A., Whitfield, J., Shah, B., Spadoni, C., Desmond, H. and Ham, J. (2001) Direct inhibition of c-Jun N-terminal kinase in sympathetic neurones prevents c-jun promoter activation and NGF withdrawal-induced death. J. Neurochem., 76: 1439–1454.

Ellis, R.E., Yuan, J.Y. and Horvitz, H.R. (1991) Mechanisms and functions of cell death. Annu. Rev. Cell Biol., 7: 663–698.

Epstein, A.C., Gleadle, J.M., McNeill, L.A., Hewitson, K.S., O'Rourke, J., Mole, D.R., Mukherji, M., Metzen, E., Wilson, M.I., Dhanda, A., Tian, Y.M., Masson, N., Hamilton, D.L., Jaakkola, P., Barstead, R., Hodgkin, J., Maxwell, P.H., Pugh, C.W., Schofield, C.J. and Ratcliffe, P.J. (2001) C. elegans EGL-9 and mammalian homologs define a family of dioxygenases that regulate HIF by prolyl hydroxylation. Cell, 107: 43–54.

Erhardt, J.A. and Pittman, R.N. (1998) p21WAF1 induces permanent growth arrest and enhances differentiation, but does not alter apoptosis in PC12 cells. Oncogene, 16: 443–451.

Estus, S., Zaks, W.J., Freeman, R.S., Gruda, M., Bravo, R. and Johnson, E.M., Jr. (1994) Altered gene expression in neurons during programmed cell death: identification of c-jun as necessary for neuronal apoptosis. J. Cell Biol., 127: t-27.

Fantl, V., Stamp, G., Andrews, A., Rosewell, I. and Dickson, C. (1995) Mice lacking cyclin D1 are small and show defects in eye and mammary gland development. Genes Dev., 9: 2364–2372.

Farinelli, S.E. and Greene, L.A. (1996) Cell cycle blockers mimosine, ciclopirox, and deferoxamine prevent the death of PC12 cells and postmitotic sympathetic neurons after removal of trophic support. J. Neurosci., 16: 1150–1162.

Freeman, R.S. (1998). The cell cycle and neuronal cell death. In: V.E. Koliatsos and R.R. Ratan (Eds.), Cell Death and Diseases of the Nervous System. Humana Press, Totowa, New Jersey, pp. 103–119.

Freeman, R.S., Estus, S., Horigome, K. and Johnson, E.M. (1993) Cell death genes in invertebrates and (maybe) vertebrates. Curr. Opin. Neurobiol., 3: 25–31.

Freeman, R.S., Estus, S. and Johnson, E.M., Jr. (1994) Analysis of cell cycle-related gene expression in postmitotic neurons: selective induction of Cyclin D1 during programmed cell death. Neuron, 12: 343–355.

Gallagher, L.A. and Manoil, C. (2001) Pseudomonas aeruginosa PAO1 kills Caenorhabditis elegans by cyanide poisoning. J. Bacteriol., 183: 6207–6214.

Garcia, I., Martinou, I., Tsujimoto, Y. and Martinou, J.C. (1992) Prevention of programmed cell death of sympathetic neurons by the bcl-2 proto-oncogene. Science, 258: 302–304.

Giardina, S.F., Cheung, N.S., Reid, M.T. and Beart, P.M. (1998) Kainate-induced apoptosis in cultured murine cerebellar granule cells elevates expression of the cell cycle gene cyclin D1. J. Neurochem., 71: 1325–1328.

Giovanni, A., Wirtz-Brugger, F., Keramaris, E., Slack, R. and Park, D.S. (1999) Involvement of cell cycle elements, cyclin-dependent kinases, pRb, and E2F/DP, in β-amyloid-induced neuronal death. J. Biol. Chem., 274: 19011 19016.

Greene, L.A. (1978) Nerve growth factor prevents the death and stimulates the neuronal differentiation of clonal PC12 pheochromocytoma cells in serum-free medium. J. Cell Biol., 78: 747–755.

Guegan, C., Levy, V., David, J.P., Ajchenbaum-Cymbalista, F. and Sola, B. (1997) c-Jun and cyclin D1 proteins as mediators of neuronal death after a focal ischaemic insult. Neuroreport, 8: 1003–1007.

Ham, J., Babij, C., Whitfield, J., Pfarr, C.M., Lallemand, D., Yaniv, M. and Rubin, L.L. (1995) A c-Jun dominant negative mutant protects sympathetic neurons against programmed cell death. Neuron, 14: 927–939.

Harbour, J.W. and Dean, D.C. (2000) The Rb/E2F pathway: expanding roles and emerging paradigms. Genes Dev., 14: 2393–2409.

Harding, T.C., Xue, L., Bienemann, A., Haywood, D., Dickens, M., Tolkovsky, A.M. and Uney, J.B. (2001) Inhibition of JNK by overexpression of the JNK binding domain of JIP-1 prevents apoptosis in sympathetic neurons. J. Biol. Chem., 276: 4531–4534.

Harris, C.A. and Johnson, E.M., Jr. (2001) BH3-only Bcl-2 family members are coordinately regulated by the JNK pathway and require Bax to induce apoptosis in neurons. J. Biol. Chem., 276: 37754–37760.

Harris, C.A., Deshmukh, M., Tsui-Pierchala, B., Maroney, A.C. and Johnson, E.M., Jr. (2002) Inhibition of the c-Jun N-terminal kinase signaling pathway by the mixed lineage kinase inhibitor CEP-1347 (KT7515) preserves metabolism and growth of trophic factor-deprived neurons. J. Neurosci, 22: 103–113.

Heintz, N. (1993) Cell death and the cell cycle: a relationship between transformation and neurodegeneration? Trends Biochem. Sci., 18: 157–159.

Huang, E.J. and Reichardt, L.F. (2001) Neurotrophins: roles in neuronal development and function. Annu. Rev. Neurosci., 24: 677–736.

Huang, J., Zhao, Q., Mooney, S.M. and Lee, F.S. (2002) Sequence determinants in hypoxia-inducible factor-1α for hydroxylation by the prolyl hydroxylases PHD1, PHD2, and PHD3. J. Biol. Chem., 277: 39792–39800.

Huard, J.M., Forster, C.C., Carter, M.L., Sicinski, P. and Ross, M.E. (1999) Cerebellar histogenesis is disturbed in mice lacking cyclin D2. Development, 126: 1927–1935.

Imaizumi, K., Tsuda, M., Imai, Y., Wanaka, A., Takagi, T. and Tohyama, M. (1997) Molecular cloning of a novel polypeptide, DP5, induced during programmed neuronal death. J. Biol. Chem., 272: 18842–18848.

Imaizumi, K., Morihara, T., Mori, Y., Katayama, T., Tsuda, M., Furuyama, T., Wanaka, A., Takeda, M. and Tohyama, M. (1999) The cell death-promoting gene DP5, which interacts with the BCL2 family, is induced during neuronal apoptosis following exposure to amyloid beta protein. J. Biol. Chem., 274: 7975–7981.

Ino, H. and Chiba, T. (2001) Cyclin-dependent kinase 4 and cyclin D1 are required for excitotoxin-induced neuronal cell death in vivo. J. Neurosci., 21: 6086–6094.

Inohara, N., Ding, L., Chen, S. and Nunez, G. (1997) Harakiri, a novel regulator of cell death, encodes a protein that activates apoptosis and interacts selectively with survival-promoting proteins Bcl-2 and Bcl-X$_{(L)}$. EMBO J, 16: 1686–1694.

Irving, E.A. and Bamford, M. (2002) Role of mitogen and stress-activated kinases in ischemic injury. J. Cereb. Blood Flow Metab., 22: 631–647.

Ivan, M., Kondo, K., Yang, H., Kim, W., Valiando, J., Ohh, M., Salic, A., Asara, J.M., Lane, W.S., Kaelin, W.G., Jr. (2001) HIFα targeted for VHL-mediated destruction by proline hydroxylation: implications for O$_2$ sensing. Science, 292: 464–468.

Jaakkola, P., Mole, D.R., Tian, Y.M., Wilson, M.I., Gielbert, J., Gaskell, S.J., Kriegsheim, A., Hebestreit, H.F., Mukherji, M., Schofield, C.J., Maxwell, P.H., Pugh, C.W., Ratcliffe, P.J. (2001) Targeting of HIF-α to the von Hippel-Lindau ubiquitylation complex by O$_2$-regulated prolyl hydroxylation. Science, 292: 468–472.

Kaelin, W.G., Jr. (2002) How oxygen makes its presence felt. Genes Dev., 16: 1441–1445.

Kaplan, D.R. and Miller, F.D. (2000) Neurotrophin signal transduction in the nervous system. Curr. Opin. Neurobiol., 10: 381–391.

Katchanov, J., Harms, C., Gertz, K., Hauck, L., Waeber, C., Hirt, L., Priller, J., von Harsdorf, R., Bruck, W., Hortnagl, H., Dirnagl, U., Bhide, P.G., Endres, M. (2001) Mild cerebral ischemia induces loss of cyclin-dependent kinase inhibitors and activation of cell cycle machinery before delayed neuronal cell death. J. Neurosci., 21: 5045–5053.

Kranenburg, O., van der Eb, A.J. and Zantema, A. (1996) Cyclin D1 is an essential mediator of apoptotic neuronal cell death. EMBO J., 15: 46–54.

Kuroiwa, S., Katai, N., Shibuki, H., Kurokawa, T., Umihira, J., Nikaido, T., Kametani, K. and Yoshimura, N. (1998) Expression of cell cycle-related genes in dying cells in retinal ischemic injury. Invest. Ophthalmol. Vis. Sci., 39: 610–617.

Le Niculescu, H., Bonfoco, E., Kasuya, Y., Claret, F.X., Green, D.R. and Karin, M. (1999) Withdrawal of survival factors results in activation of the JNK pathway in neuronal cells leading to Fas ligand induction and cell death. Mol. Cell. Biol., 19: 751–763.

Levi-Montalcini, R. and Angeletti, P.U. (1963) Essential role of the nerve growth factor in the survival and maintenance of dissociated sensory and sympathetic embryonic nerve cells in vitro. Dev. Biol., 7: 653–659.

Levi-Montalcini, R. and Booker, B. (1960) Destruction of the sympathetic ganglia in mammals by an antiserum to a nerve growth protein. Proc. Natl. Acad. Sci. USA, 46: 384–391.

Lewin, G.R. and Barde, Y.A. (1996) Physiology of the neurotrophins. Annu. Rev. Neurosci., 19: 289–317.

Li, Y., Chopp, M. and Powers, C. (1997a) Granule cell apoptosis and protein expression in hippocampal dentate gyrus after forebrain ischemia in the rat. Can. J. Neurol. Sci., 150: 93–102.

Li, Y., Chopp, M., Powers, C. and Jiang, N. (1997b) Immunoreactivity of cyclin D1/cdk4 in neurons and oligodendrocytes after focal cerebral ischemia in rat. J. Cereb. Blood Flow Metab., 17: 846–856.

Lipscomb, E.A., Sarmiere, P.D., Crowder, R.J. and Freeman, R.S. (1999) Expression of the SM-20 gene promotes death in nerve growth factor-dependent sympathetic neurons. J. Neurochem., 73: 429–432.

Lipscomb, E.A., Sarmiere, P.D. and Freeman, R.S. (2001) SM-20 is a novel mitochondrial protein that causes caspase-dependent cell death in nerve growth factor-dependent neurons. J. Biol. Chem., 276: 5085–5092.

Liu, D.X. and Greene, L.A. (2001) Neuronal apoptosis at the G1/S cell cycle checkpoint. Cell Tissue Res., 305: 217–228.

Liu, W., Bi, X., Tocco, G., Baudry, M. and Schreiber, S.S. (1996) Increased expression of cyclin D1 in the adult rat brain following kainic acid treatment. Neuroreport, 7: 2785–2789.

Madden, S.L., Galella, E.A., Riley, D., Bertelsen, A.H. and Beaudry, G.A. (1996) Induction of cell growth regulatory genes by p53. Cancer Res., 56: 5384–5390.

Maroney, A.C., Finn, J.P., Bozyczko-Coyne, D., O'Kane, T.M., Neff, N.T., Tolkovsky, A.M., Park, D.S., Yan, C.Y., Troy, C.M. and Greene, L.A. (1999) CEP-1347 (KT7515), an inhibitor of JNK activation, rescues sympathetic neurons and neuronally differentiated PC12 cells from death evoked by three distinct insults. J. Neurochem., 73: 1901–1912.

Martin, D.P., Schmidt, R.E., DiStefano, P.S., Lowry, O.H., Carter, J.G. and Johnson, E.M., Jr. (1988) Inhibitors of protein synthesis and RNA synthesis prevent neuronal death caused by nerve growth factor deprivation. J. Cell Biol., 106: 829–844.

Maxwell, P.H. and Ratcliffe, P.J. (2002) Oxygen sensors and angiogenesis. Sem. Cell Dev. Biol., 13: 29–37.

McCarthy, M.J., Rubin, L.L. and Philpott, K.L. (1997) Involvement of caspases in sympathetic neuron apoptosis. J. Cell Sci., 110: 2165–2173.

Moschella, M.C., Menzies, K., Tsao, L., Lieb, M.A., Kohtz, J.D., Kohtz, D.S. and Taubman, M.B. (1999) SM-20 is a novel growth factor-responsive gene regulated during skeletal muscle development and differentiation. Gene Expr., 8: 59–66.

Mota, M., Reeder, M., Chernoff, J. and Bazenet, C.E. (2001) Evidence for a role of mixed lineage kinases in neuronal apoptosis. J. Neurosci., 21: 4949–4957.

Neame, S.J., Rubin, L.L. and Philpott, K.L. (1998) Blocking cytochrome c activity within intact neurons inhibits apoptosis. J. Cell Biol., 142: 1583–1593.

O'Hare, M., Wang, F. and Park, D.S. (2002) Cyclin-dependent kinases as potential targets to improve stroke outcome. Pharmacol. Ther., 93: 135–143.

O'Hare, M.J., Hou, S.T., Morris, E.J., Cregan, S.P., Xu, Q., Slack, R.S. and Park, D.S. (2000) Induction and modulation of cerebellar granule neuron death by E2F-1. J. Biol. Chem., 275: 25358–25364.

Oppenheim, R.W. (1991) Cell death during development of the nervous system. Annu. Rev. Neurosci., 14: 453–501.

Oppenheim, R.W., Prevette, D., Tytell, M. and Homma, S. (1990) Naturally occurring and induced neuronal death in the chick embryo in vivo requires protein and RNA synthesis: evidence for the role of cell death genes. Dev. Biol., 138: 104–113.

Osuga, H., Osuga, S., Wang, F., Fetni, R., Hogan, M.J., Slack, R.S., Hakim, A.M., Ikeda, J.E. and Park, D.S. (2000) Cyclin-dependent kinases as a therapeutic target for stroke. Proc. Natl. Acad. Sci. USA, 97: 10254–10259.

Padmanabhan, J., Park, D.S., Greene, L.A. and Shelanski, M.L. (1999) Role of cell cycle regulatory proteins in cerebellar granule neuron apoptosis. J. Neurosci., 19: 8747–8756.

Palmada, M., Kanwal, S., Rutkoski, N.J., Gufstafson-Brown, C., Johnson, R.S., Wisdom, R. and Carter, B.D. (2002) c-jun is essential for sympathetic neuronal death induced by NGF withdrawal but not by p75 activation. J. Cell Biol., 158: 453–461.

Park, D.S., Farinelli, S.E. and Greene, L.A. (1996) Inhibitors of cyclin-dependent kinases promote survival of post-mitotic neuronally differentiated PC12 cells and sympathetic neurons. J. Biol. Chem., 271: 8161–8169.

Park, D.S., Levine, B., Ferrari, G. and Greene, L.A. (1997a) Cyclin dependent kinase inhibitors and dominant negative cyclin dependent kinase 4 and 6 promote survival of NGF-deprived sympathetic neurons. J. Neurosci., 17: 8975–8983.

Park, D.S., Morris, E.J., Greene, L.A. and Geller, H.M. (1997b) G1/S cell cycle blockers and inhibitors of cyclin-dependent kinases suppress camptothecin-induced neuronal apoptosis. J. Neurosci., 17: 1256–1270.

Park, D.S., Morris, E.J., Padmanabhan, J., Shelanski, M.L., Geller, H.M. and Greene, L.A. (1998) Cyclin-dependent kinases participate in death of neurons evoked by DNA-damaging agents. J. Cell Biol., 143: 457–467.

Putcha, G.V., Deshmukh, M. and Johnson, E.M., Jr. (1999) BAX translocation is a critical event in neuronal apoptosis: regulation by neuroprotectants, BCL-2, and caspases. J. Neurosci., 19: 7476–7485.

Putcha, G.V., Moulder, K.L., Golden, J.P., Bouillet, P., Adams, J.A., Strasser, A. and Johnson, E.M. (2001) Induction of BIM, a proapoptotic BH3-only BCL-2 family member, is critical for neuronal apoptosis. Neuron, 29: 615–628.

Putcha, G.V., Harris, C.A., Moulder, K.L., Easton, R.M., Thompson, C.B. and Johnson, E.M., Jr. (2002) Intrinsic and extrinsic pathway signaling during neuronal apoptosis: lessons from the analysis of mutant mice. J. Cell Biol., 157: 441–453.

Puthalakath, H. and Strasser, A. (2002) Keeping killers on a tight leash: transcriptional and post-translational control of the pro-apoptotic activity of BH3-only proteins. Cell Death Differ., 9: 505–512.

Ratan, R.R., Murphy, T.H. and Baraban, J.M. (1994) Macromolecular synthesis inhibitors prevent oxidative stress-induced apoptosis in embryonic cortical neurons by shunting cysteine from protein synthesis to glutathione. J. Neurosci., 14: 4385–4392.

Sakai, K., Suzuki, K., Tanaka, S. and Koike, T. (1999) Up-regulation of cyclin D1 occurs in apoptosis of immature but not mature cerebellar granule neurons in culture. J. Neurosci. Res., 58, 396–406.

Sakurai, M., Hayashi, T., Abe, K., Itoyama, Y., Tabayashi, K. and Rosenblum, W.I. (2000) Cyclin D1 and Cdk4 protein induction in motor neurons after transient spinal cord ischemia in rabbits. Stroke, 31: 200–207.

Scott, S.A. and Davies, A.M. (1990) Inhibition of protein synthesis prevents cell death in sensory and parasympathetic neurons deprived of neurotrophic factor in vitro. J. Neurobiol., 21: 630–638.

Shaulian, E. and Karin, M. (2001) AP-1 in cell proliferation and survival. Oncogene, 20: 2390–2400.

Shaulian, E. and Karin, M. (2002) AP-1 as a regulator of cell life and death. Nat. Cell Biol., 4: E131–E136.

Sherr, C.J. (1996) Cancer cell cycles. Science, 274: 1672–1677.

Shinoe, T., Wanaka, A., Nikaido, T., Kanazawa, K., Shimizu, J., Imaizumi, K. and Kanazawa, I. (2001) Upregulation of the pro-apoptotic BH3-only peptide harakiri in spinal neurons of amyotrophic lateral sclerosis patients. Neurosci. Lett., 313: 153–157.

Sicinski, P., Donaher, J.L., Parker, S.B., Li, T., Fazeli, A., Gardner, H., Haslam, S.Z., Bronson, R.T., Elledge, S.J., Weinberg, R.A. (1995) Cyclin D1 provides a link between development and oncogenesis in the retina and breast. Cell, 82: 621–630.

Small, D.L., Monette, R., Fournier, M.C., Zurakowski, B., Fiander, H. and Morley, P. (2001) Characterization of cyclin D1 expression in a rat global model of cerebral ischemia. Brain Res., 900: 26–37.

Sofroniew, M.V., Howe, C.L. and Mobley, W.C. (2001) Nerve growth factor signaling, neuroprotection, and neural repair. Annu. Rev. Neurosci., 24: 1217–1281.

Straub, J.A., Lipscomb, E.A., Yoshida, E.S. and Freeman, R.S. (2003) Induction of SM-20 in PC12 cells leads to increased cytochrome c levels, accumulation of cytochrome c in the cytosol, and caspase-dependent cell death. J. Neurochem., 85: 318–328.

Taylor, M.S. (2001) Characterization and comparative analysis of the EGLN gene family. Gene, 275: 125–132.

Trent, C., Tsuing, N. and Horvitz, H.R. (1983) Egg-laying defective mutants of the nematode Caenorhabditis elegans. Genetics, 104: 619–647.

Troy, C.M., Stefanis, L., Greene, L.A. and Shelanski, M.L. (1997) Nedd2 is required for apoptosis after trophic factor withdrawal, but not superoxide dismutase (SOD1) down-regulation, in sympathetic neurons and PC12 cells. J. Neurosci., 17: 1911–1918.

Troy, C.M., Rabacchi, S.A., Hohl, J.B., Angelastro, J.M., Greene, L.A. and Shelanski, M.L. (2001) Death in the balance: alternative participation of the caspase-2 and -9 pathways in neuronal death induced by nerve growth factor deprivation. J. Neurosci., 21: 5007–5016.

Tsui-Pierchala, B.A., Putcha, G.V. and Johnson, E.M., Jr. (2000) Phosphatidylinositol 3-kinase is required for the trophic, but not the survival-promoting, actions of NGF on sympathetic neurons. J. Neurosci., 20: 7228–7237.

Vaux, D.L., Cory, S. and Adams, J.M. (1988) Bcl-2 gene promotes haemopoietic cell survival and cooperates with c-myc to immortalize pre-B cells. Nature, 335: 440–442.

Virdee, K., Bannister, A.J., Hunt, S.P. and Tolkovsky, A.M. (1997) Comparison between the timing of JNK activation, c-Jun phosphorylation, and onset of death commitment in sympathetic neurones. J. Neurochem., 69: 550–561.

Wakabayashi, T., Kosaka, J. and Hommura, S. (2002) Up-regulation of Hrk, a regulator of cell death, in retinal ganglion cells of axotomized rat retina. Neurosci. Lett., 318: 77–80.

Wax, S.D., Rosenfield, C.L. and Taubman, M.B. (1994) Identification of a novel growth factor-responsive gene in vascular smooth muscle cells. J. Biol. Chem., 269: 13041–13047.

Wax, S.D., Tsao, L., Lieb, M.E., Fallon, J.T. and Taubman, M.B. (1996) SM-20 is a novel 40-kd protein whose expression in the arterial wall is restricted to smooth muscle. Lab. Invest., 74: 797–808.

Whitfield, J., Neame, S.J., Paquet, L., Bernard, O. and Ham, J. (2001) Dominant-negative c-Jun promotes neuronal survival by reducing BIM expression and inhibiting mitochondrial cytochrome c release. Neuron, 29: 629–643.

Wiessner, C., Brink, I., Lorenz, P., Neumann-Haefelin, T., Vogel, P. and Yamashita, K. (1996) Cyclin D1 messenger RNA is induced in microglia rather than neurons following transient forebrain ischaemia. Neuroscience, 72: 947–958.

Xu, Z., Maroney, A.C., Dobrzanski, P., Kukekov, N.V. and Greene, L.A. (2001) The MLK family mediates c-Jun N-terminal kinase activation in neuronal apoptosis. Mol. Cell. Biol., 21: 4713–4724.

Xu, Z., Kukekov, N.V. and Greene, L.A. (2003) POSH acts as a scaffold for a multiprotein complex that mediates JNK activation in apoptosis. EMBO J., 22: 252–261.

Yang, D.D., Kuan, C.Y., Whitmarsh, A.J., Rincon, M., Zheng, T.S., Davis, R.J., Rakic, P. and Flavell, R.A. (1997) Absence of excitotoxicity-induced apoptosis in the hippocampus of mice lacking the Jnk3 gene. Nature, 389: 865–870.

Yu, F., White, S.B., Zhao, Q. and Lee, F.S. (2001) HIF-1α binding to VHL is regulated by stimulus-sensitive proline hydroxylation. Proc. Natl. Acad. Sci. USA, 98: 9630–9635.

Yuen, E.C. and Mobley, W.C. (1996) Therapeutic potential of neurotrophic factors for neurological disorders. Ann. Neurol., 40: 346–354.

Yuen, E.C., Howe, C.L., Li, Y., Holtzman, D.M. and Mobley, W.C. (1996) Nerve growth factor and the neurotrophic factor hypothesis. Brain Dev., 18: 362–368.

Zaman, K., Ryu, H., Hall, D., O'Donovan, K., Lin, K.I., Miller, M.P., Marquis, J.C., Baraban, J.M., Semenza, G.L. and Ratan, R.R. (1999) Protection from oxidative stress-induced apoptosis in cortical neuronal cultures by iron chelators is associated with enhanced DNA binding of hypoxia-inducible factor-1 and ATF-1/CREB and increased expression of glycolytic enzymes, p21$^{(waf1/cip1)}$, and erythropoietin. J. Neurosci., 19: 9821–9830.

Progress in Brain Research, Vol. 146
ISSN 0079-6123

CHAPTER 9

Neural stem and progenitor cells: choosing the right Shc

Tiziana Cataudella, Luciano Conti and Elena Cattaneo*

Department of Pharmacological Sciences and Center of Excellence on Neurodegenerative Diseases, University of Milan, I-20133 Milan, Italy

Abstract: Neural stem cell (NSCs) are self-renewing, multipotent cells able to generate neurons, astrocytes and oligodendrocytes. Since their identification, these properties have made NSCs an attractive subject for therapeutic applications to the damaged brain. In this context, understanding the mechanisms and the molecules regulating their biological properties is important and it is focused to gain control over their proliferative and differentiative potential. Here we will discuss values and unsolved aspects of the system and the employment of potentially key molecular targets for proper control of NSCs fate.

Keywords: self-renewal; neuronal production; signaling proteins; progenitors

Introduction

Throughout the last decade there has been an increasingly enthusiastic interest in the cell biology of stem cells. These are widely considered as an invaluable potential tool for cell therapy approaches to a broad range of clinical conditions. According to its definition, the main physiologic function of a stem cell is to generate all of the differentiated cell types of the tissue in which it resides. Indeed, a stem cell is generally defined operationally as a cell that is: (i) multipotent, (ii) capable of self-renewal, and (iii) capable of generating a progeny that can functionally integrate into and repair the tissue of origin.

Regarding the neural stem cells (NCSs), i.e., those stem cells residing inside the nervous system, this implies that their progeny will include mature neurons, astrocytes and oligodendrocytes. Stem cell technology is particularly important for the central nervous system (CNS) since cell transplantation might help to overcome the intrinsic poor capability of the nervous tissue to replace elements lost in the course of injury or disease. Proper control over the differentiation pattern of brain stem cells may therefore eventually allow the treatment of a wide range of degenerative diseases characterized by neuronal or glial loss. However, relatively little is known about the molecular regulators and genetic cascades that control NSCs self-renewal and multipotency. In this article, we will review the recent advancements in the NSCs field aimed to improve the control of their proliferation or differentiation, aiming at an efficient and safe future clinical employment for cell replacement purposes.

Stem or progenitor cells: how to distinguish between the two?

Cells with stem-like properties, initially identified in the fetal, and more recently in the adult mammalian brain (Gage, 2000), can be grown in culture, displaying the potential to self renew and to generate

*Correspondence to: E. Cattaneo, Department of Pharmacological Sciences and Center of Excellence on Neurodegenerative Diseases, University of Milan, Via Balzaretti 9, I-20133 Milan, Italy. Tel.: +39-02-5031-8333; Fax: +39-02-5031-8284; E-mail: elena.cattaneo@unimi.it

DOI: 10.1016/S0079-6123(03)46009-3

the different cell types of the nervous system (McKay, 1997). During brain development, NSCs are localized in the epithelial layer of the germinal zone surrounding the ventricles (Temple, 2001). As brain maturation continues, postmitotic neurons migrate away from the ventricular zone, mainly guided by radially oriented glial processes, and the ventricular zone diminishes in size (Rao, 1999). In the adult brain, cells with similar stem-like properties also exist, mostly originating from two regions: the hippocampus and the subventricular zone (SVZ) of the lateral ventricles (Gage, 2002). Noteworthy, different studies indicate that NSCs from different fetal and adult brain areas are not identical, as demonstrated by different growth characteristics, trophic factor requirements, and specific patterns of differentiation (Temple, 2001). This is further demonstrated by in vitro experiments that revealed that NSCs differ in their potential according to the developmental stage at which they are isolated and to the site from where they were obtained (Temple, 2001). More generally, in vitro studies indicate that two types of neural stem-like cells have been isolated in multiple brain regions and appear to coexist. One type shows epidermal growth factor (EGF) responsiveness and can be expanded as floating cell aggregates, called neurospheres (Reynolds and Weiss, 1992), becoming fibroblast growth factor (FGF)-responsive with the in vitro passages (Vescovi et al., 1993; Represa et al., 2001). The second group has been shown to be FGF-dependent and can be propagated both as adherent cultures as well as neurospheres (Kalyani et al., 1997).

Importantly, all the above mentioned evidence indicating the existence of heterogeneous NSC populations may be the consequence of the lack of a neural stem-restricted marker for NSC prospective isolation procedure. This implies that it is currently very difficult to distinguish, but posteriorly and only following accurate clonal analysis, between a real NSC and a progenitor. Conceptually, these two populations differ for their differentiative capabilities. Indeed, while NSCs are multipotential, brain progenitor are generally considered to be more limited in their potential and able to produce only restricted phenotypes (McKay, 1997). Up to date, there are only few markers of putative NSCs (nestin, sox-1, musashi, AC133, PNA^{low}/HSA^{low}, Lex/ssea-1) that may be used with some degree of specificity being expressed by brain progenitors and not by other cell types (Rossi and Cattaneo, 2002).

In the absence of a clear in vivo assay to identify the NSCs, most authors use the ability to grow, in vitro, neurospheres which contain cells capable to differentiate as glia and neurons, as an operative definition for 'NSCs' (Rossi and Cattaneo, 2002). Nevertheless, this assay is not devoid of faults since neurospheres contain cells that are clearly not uniform in their differentiative stage and estimation of the number of *bona fide* stem cells contained in a preparation of cells dissociated from neurospheres varies as widely as for those contained in the fetal brain (Temple, 2001; Rossi and Cattaneo, 2002; Suslov et al., 2002). New strategies to allow efficient prospective isolation of stem cells from the brain are highly demanded to guarantee uniformity of results and lead to trustworthy conclusions. Increasing attempts in this direction are ongoing by combining sorting for cell size and antigenic properties (Rietze et al., 2001; Capela and Temple, 2002).

Transdifferentiation, transformation, fusion: deciphering the stem cells plasticity

Recent progress in stem cell research indicates that certain mammalian cells maintain a high degree of plasticity giving rise to multilineage cell differentiation. Indeed, some developmental peculiarities suggest that stem cells may be able to differentiate into cell types that are not of the same germ layers (Tajbakhsh et al., 1994). An early intriguing case is that of cultured neural stem cells derived from neurospheres of clonogenic origin which repopulate the hematopoietic system in sublethally irradiated allogenic host by cells (Bjornson et al. 1999). Similarly, in other studies, transplantation of adult bone marrow cells has generated a broad range of phenotypes, including muscle cells (Ferrari et al., 1998), liver cells (Petersen et al., 1999; Lagasse et al., 2000), brain cells (Brazelton et al., 2000; Mezey et al., 2000) and other (Krause et al., 2001). This seemed to indicate that the extracellular factors or cell-cell interaction might be sufficient for reprogramming putative somatic stem cells into a more pluripotent, embryonic stem cell (ES)-like, condition. This hypothesis was reinforced by the demonstration that when injected into blastocysts NSCs participate to the formation of most of the tissues of the mouse (Clarke et al., 2000). This hypothesis is challenged by results from various groups. Morshead and colleagues showed that, in their hands, putative NSCs extensively cultivated in vitro do not turn into hematopoietic cells with any appreciable frequence (Morshead et al., 2002). The same authors suggested that genetic instability of NSCs after long term in vitro expansion may explain the original results. They suggested that the claimed hematopoietic reconstitution could result as a consequence of transformation events or artefacts due to the in vitro procedures or methodological problems. While this study was criticized for the possible different number of stem cells in the culture and the presence of transformation events (Vescovi et al., 2002), it seems reasonable to conclude that transdifferentiation events, when present, appears to be a much rarer phenomenon than previously described and may be peculiar only to some cell clones. Because of that, the clinical relevance of such data is debated. Nevertheless, the possibility that even a few brain or blood stem cells can transdifferentiate remains of biological interest. However, no consistent demonstration of cell conversion followed the original claims, leaving the subject, at present, rather confused.

Following the report of Bjornson et al. (1999), we began a set of experiments aimed at assessing whether the reported ability of NSCs to transdifferentiate into blood was acquired due to in vitro expansion (and possible de-differentiation of the cells), or whether it was an intrinsic property of brain stem cells (Magrassi et al., 2003). To test this, we isolated fetal neural cells (fNC), from E10 embryos derived of transgenic mice expressing EGFP (at this stage the telencephalic vesicles do not have ingrowths of blood vessels and are highly enriched in NSCs and progenitors). The cells were dissociated and directly transplanted into sublethally irradiated C57Bl/6 as performed by Bjornson et al. (1999). Analysis of grafted animals at different time points did not reveal development of chimerism in the hematopoietic compartment even after very long survival times (16 months) after grafting. The same negative results were obtained by injecting the fNC into the tail vein or by transplanting them directly into the bone marrow cavity of sublethally irradiated mice. While

we could detect donor cells by PCR and Fluorescence Activated Cell Sorting (FACS) analyzes at the early time points, the same assays failed to detect the presence of circulating donor-derived fNC EGFP positive cell on a fraction of peripheral blood collected from the grafted animals at mid and at late time points (i.e., 30, 60, 120 and 495 days after grafting) (Magrassi et al., 2003). These results indicated that NSCs do not physiologically exhibit transdifferentiative capability and that, if present, this capability requires ex vivo expansion procedures that would possibly reprogramme the differentiative potential of the donor cells. This seems in agreement with recent results from Verfaille's group that has reported the isolation of a Multipotent Adult Progenitor cell (MAPc) able to extensively contribute to chimerae once injected in the early blastocyst and to repopulate all the adult tissues once transplanted into sublethally irradiated mice (Jiang et al., 2002). Interestingly, these cells can be established from adult murine mesenchymal cells only after at least 20 passages in vitro, reinforcing the idea that the in vitro expansion may somehow act to reprogram the cells toward an ES-like potential.

However, various studies now indicate that transdifferentiation of in vitro expanded stem cells may be explained by fusion of donor cells with host cells (Terada et al., 2002; Ying et al., 2002). While these studies were showing the possibility of cell fusion between somatic stem cells and cultured embryonic stem cells, more recent experiments now indicate the possibility that cell fusion may also occur among somatic tissues after in vivo transplantation (Vassilopoulos et al., 2003; Wang et al., 2003).

Future research is needed to establish the real ability of stem cells to transdifferentiate and the impact of this phenomena for therapeutical approaches.

Driving proliferation and differentiation events in neural cells through Shc(s) molecules

The identification of candidate molecular mechanisms able to modulate proliferation, differentiation and, possibly, plasticity of NCSs is of great interest in order to implement studies and trials for cell replacement therapies. To this regard, an interesting strategy may come from studies of signaling mechanisms downstream of growth factors receptors. Particularly, we pointed to the regulated expression and activity of Shc(s) adapter molecules, which couple signals from activated receptors to downstream effectors, as a potential mechanism to regulate division, survival and differentiation on stem cells in the brain. Shc(s) proteins indeed appear to play a role in the control of the proliferation and subsequent maturation of mitotically active neural stem/progenitor cells into postmitotic neurons (Cattaneo and Pelicci, 1998; Conti et al., 2001).

Up to date, three Shc(s) genes have been identified, named ShcA, ShcB/Sli and ShcC/Rai/N-Shc having a consistent homology (Pelicci et al., 1992, 1996; O'bryan et al., 1996; Luzi et al., 2000). These three Shc(s) molecules are characterized by the presence of phosphotyrosine regulatory residues and the PTB, CH1 (a proline rich domain) and SH2 domains in the presented order. Three isoforms are known for ShcA (of 66, 52 and 46 kD), two isoforms for ShcB (of 52 and 47 kD) and two for ShcC (of 54 and 69 kD). $p66^{ShcA}$ display a further N-terminal CH domain (CH2) that contains important regulatory serine residues. They share elevated homology in both the C terminus SH2 domain and the N terminus PTB domain, the most divergent sequence being in the proline and glycine rich CH1 (Collagene Homology 1) region. ShcA proteins have been extensively characterized and shown to be widely expressed outside the CNS. Their importance is indicated by (i) the early embryonic lethal phenotype of $p52^{ShcA}$ null mutation (Lai and Pawson, 2000), (ii) the impairment in thymocytes development in conditional $p52^{ShcA}$ knockout (Zhang et al., 2002), and (iii) by the increase in life span and resistance to stress stimuli in $p66^{ShcA}$ knockout animals (Migliaccio et al., 1999).

Despite the apparently constitutive presence of ShcA in extraneural tissues, ShcA expression and activity within the brain is tightly regulated during development and maximal in the embryonic day 10 neural tube. At later time points, ShcA remains confined to the germinal epithelium where mitotically active immature stem and progenitor cells are located. Instead, in the areas of the embryonic or postnatal brain where postmitotic neurons are present, $mRNA^{ShcA}$ is highly reduced. Similarly, the adult brain exhibited low ShcA expression, the main exception being the olfactory epithelium, which is a predominant area of active neurogenesis in the adult. These changes in the expression and activity of ShcA as a function of neuronal maturation were confirmed in vitro in differentiating neuronal cultures (Conti et al., 1997). Not only ShcA is present in actively dividing neurogenic areas, but is also susceptible of being activated. Indeed, in vivo immunoprecipitation of ShcA from the telencephalic vesicles of embryonic brains injected intraventricularly with mitogens like EGF revealed a higher phosphorylation of the $p52^{ShcA}$ isoform with respect to control animals (Conti et al., 1997). In treated samples, Grb2 coimmunoprecipitation was also observed, indicating that ShcA is not only present in the germinal epithelium, but it is also able to elicit a functional response with recruitment of downstream pathways.

The demonstration that ShcA availability is regulated during neurogenesis and becomes limited during NSCs maturation in vivo and in vitro led to the proposition that other Shc-like proteins may substitute for ShcA function in mature neurons (Cattaneo and Pelicci, 1998). Given the existence of two more recently identified Shc members, ShcB and ShcC, the latter being selectively expressed in the brain, we suggested that one or both of them could replace ShcA in mature neurons (Cattaneo and Pelicci, 1998). Analyzes of ShcC expression showed an opposite expression pattern with respect to ShcA, being absent in neural progenitors but present in early postmitotic neurons and reaching maximal levels in the adult brain where it is found localized only in neurons. Similar changes in ShcA and ShcC levels during neuronal maturation have been observed in several mammalian species (rat, mouse and human) (Conti et al., 2001). Notably, ShcC is found in neurons from various regions of the adult brain thus predicting a general role played by ShcC in these cells. Particularly, given the above described central roles of ShcA in signal transduction, ShcC appearance in differentiating NSCs has been hypothesized to serve different 'connector functions' compared with ShcA, allowing maturing cells to respond differently to environmental stimuli (Conti et al., 2001). To this regard, Lai and Pawson (2000) demonstrated the existence of a strict link between Shc levels and cell responsiveness. The authors showed that ShcA expression and activity are required in cells of the cardiovascular system to make them responsive to low concentrations of growth factors. Indeed, while a low concentration of growth factors is necessary to activate the MAPK pathway in mouse embryo fibroblasts (MEF), cells from ShcA knockout mice require a higher concentration of growth factors to activate the same signaling cascade. Transfection experiments in primary neural cells and in postmitotic neurons revealed that ShcC acts to promote neuronal differentiation and improve survival of these cells (Conti et al., 2001). It was also found that ShcC elicits these effects through a different kinetic of activation of downstream effector molecules with respect to ShcA. Indeed, ShcC elicits neuronal differentiation via prolonged stimulation of the MAPK (Conti et al., 2001; Pelicci et al., 2002). This behavior is reminiscent of that described in PC12 cells exposed to NGF, where persistent activation of MAPK is required for neuronal differentiation. On the contrary, ShcC-driven prosurvival effect occurs via recruitment of the PI3K-Akt pathway (Conti et al., 2001; Pelicci et al., 2002), as demonstrated by the fact that its pharmacological or molecular inhibition markedly abolishes this effect. To this respect, ShcC-induced Akt activation was found to cause phosphorylation (with inhibition) of Bad, a proapoptotic member of the Bcl2 family (Conti et al., 2001).

Single and double *ShcB/C* null mice have been recently described (Sakai et al., 2000). ShcB-deficient mice exhibit a loss of peptidergic and nonpeptidergic nociceptive sensory neurons. ShcC null mice appear not to show gross anatomical abnormalities. Noteworthy, mice lacking both ShcB and ShcC exhibit a significant additional loss of neurons within the superior cervical ganglia. This aspect may emphasize that the lack of phenotype in ShcC null mice could be due to a partial compensation by the other ShcB or other Shc members during development, thus masking ShcC real function in neural tissues. Further analyzes will be required to elucidate ShcC role in neuronal generation from stem cells .

Taken together these results unveil a new scenario within which physiological changes in the availability of ShcA and ShcC adaptors during brain development may act to modify neural stem/progenitor cell responsiveness as a function of the new and developing environment.

Conclusions

Fetal and adult NSCs are an important tool to be exploited for brain repair in neurodegenerative disease either through their transplantation or via their in situ activation in the brain. The understanding of the function of candidate genes responsible for proliferation and differentiation is essential to implement studies for cell replecement therapies; however, at present, the specification of NSC into the desired phenotypes is far from being efficiently controlled. The prominent activities of Shc(s) proteins in modulating cell responsiveness and the demonstration of their regulated expression at the transition from proliferation to differentiation in the brain, point at the Shc(s) signaling pathways as candidate targets for pharmacological modulation of stem cell division and differentiation in the brain.

Acknowledgments

The work of the authors described in this article is supported by grants from the Ministry of Research and University (MIUR, 2001055212-004), by a F.I.R.B. National Research Network on Neural Stem Cells coordinated by E.C. (MIUR, RBNE01YRA3-1), Telethon Onlus (#GP0215Y02), and Associazione Italiana Ricerca sul Cancro (AIRC, Italy) to E.C. and Telethon Onlus (#GGPO2457) to L.C.

Abbreviations

NSCs	neural stem cells
CNS	central nervous system
SVZ	subventricular zone
EGF	epidermal growth factor
FGF	fibroblast growth factor
EC	embryonic stem cell
fNC	fetal Neural stem and progenitor Cells
EGFP	enhanced green fluorescent protein
FACS	fluorescence activated cell sorting
MAPc	multipotent adult progenitor cell

Shc	Src Homologue and Collagene Homologue
PNA	peanut agglutinin
HAS	heat-stable antigen
LeX	Lewis X
SSEA-1	stage-specific embryonic antigen 1

References

Bjornson, C.R., Rietze, R.L., Reynolds, B.A., Magli, M.C. and Vescovi, al. (1999) Turning brain into blood: a hematopoietic fate adopted by adult neural stem cells in vivo. Science, 283: 534–537.

Brazelton, T.R., Rossi, F.M.V., Keshet, G.I. and Blau, H.M. (2000) From marrow to brain: expression of neural phenotypes in adult mice. Science, 290: 1775–1779.

Capela, A. and Temple, S. (2002) LeX/ssea-1 is expressed by adult mouse CNS stem cells, identifying them as non-ependymal. Neuron, 35: 865–875.

Cattaneo, E. and Pelicci, P.G. (1998) Emerging roles for Sh2/PTB-containing Shc adapter proteins in the developing mammalian brain. Trends Neurosci., 21: 476–481.

Clarke, D.L., Johansson, C.B., Wilbertz, J., Veress, B., Nilsson, E., Karlstrom, H., Lendahl, U. and Frisen, J. (2000) Generalized potential of adult neural stem cells. Science, 288: 1660–1663.

Conti, L., De Fraja, C., Gulisano, M., Migliaccio, E., Govoni, S. and Cattaneo, E. (1997) Expression and activation of SH2/PTB-containing ShcA adaptor protein reflects the pattern of neurogenesis in the mammalian brain. Proc. Natl. Acad. Sci. USA, 94: 8185–8190.

Conti, L., Sipione, S., Magrassi, L., Bonfanti, L., Rigamonti, D., Pettirossi, V., Peschanski, M., Haddad, B., Pelicci, P., Milanesi, G., Pelicci, G. and Cattaneo, E. (2001) Shc signaling in differentiating neural progenitor cells. Nat. Neurosci., 4: 579–586.

Ferrari, G., Cusella-De Angelis, G., Coletta, M., Paolucci, E., Stornaiuolo, A., Cossu, G. and Mavilio, F. (1998) Muscle regeneration by bone marrow-derived myogenic progenitors. Science, 279: 1528–1530.

Gage, F.H. (2000) Mammalian neural stem cells. Science, 287: 1433–1438.

Gage, F.H. (2002) Neurogenesis in the adult brain. J. Neurosci., 22: 612–613.

Jiang, Y., Jahagirdar, B.N., Reinhardt, R.L., Schwartz, R.E., Keene, C.D., Ortiz-Gonzalez, X.R., Reyes, M., Lenvik, T., Lund, T., Blackstad, M., Du, J., Aldrich, S., Lisberg, A., Low, W.C., Largaespada, D.A. and Verfaillie, C.M. (2002) Pluripotency of mesenchymal stem cells derived from adult marrow. Nature, 4418: 41–49.

Kalyani, A., Hobson, K. and Rao, M.S. (1997) Neuroepithelial stem cells from the embryonic spinal cord: isolation, characterization, and clonal analysis. Dev. Biol., 186: 202–223.

Krause, D.S., Theise, N.D., Collector, M.I., Henegariu, O., Hwang, S., Gardner, R., Neutzel, S. and Sharkis, S.J. (2001) Multi-organ, multi-lineage engrafment by a single bone marrow-derived stem cell. Cell, 105: 369–377.

Lagasse, E., Connors, H., Al-Dhalimy, M., Reitsma, M., Dohse, M., Osborne, L., Wang, X., Finegold, M., Weissman, I.L. and Grompe, M. (2000) Purified hematopoietic stem cells can differentiate into hepatocytes in vivo. Nat. Med., 6: 1229–1234.

Lai, K.M.V. and Pawson, T. (2000) The ShcA phosphotyrosine docking protein sensitizes cardiovascular signaling in the mouse embryo. Genes Dev., 14: 1132–1145.

Luzi, L., Confalonieri, S., DiFiore, P.P. and Pelicci, P.G. (2000) Evolution of Shc functions from nematode to human. Curr. Opin. Genet. Dev., 10: 668–674.

Magrassi, L., Castello, S., Ciardelli, L., Podesta', M., Gasparoni, A., Conti, L., Pezzotta, S., Frassoni, F. and Cattaneo, E. (2003) Freshly dissociated fetal neural stem/progenitor cells do not turn into blood. Mol. Cell. Neurosci., 22: 179–187.

McKay, R. (1997) Stem cells in the central nervous system. Science, 276: 66–71.

Mezey, E., Chandross, K.J., Harta, G., Maki, R.A. and McKercher, S.R. (2000) Turning blood into brain: cells bearing neuronal antigens generated in vivo fron bone marrow. Science, 290: 1779–1782.

Migliaccio, E., Giorgio, M., Mele, S., Pelicci, G., Reboldi, P., Panolfi, P.P., Lanfrancone, L. and Pelicci, P.G. (1999) The p66shc adaptor protein controls oxydative stress response and life span in mammals. Nature, 402: 309–313.

Morshead, C.M., Benveniste, P., Iscove, N.N. and Van der Kooy, D. (2002) Hematopoietic competence is a rare property of neural stem cells that may depend on genetic and epigenetic alterations. Nat. Med., 8: 268–273.

O'bryan, J.P., Songyang, Z., Cantley, L., Der, C.J. and Pawson, T. (1996) A mammalian adapter protein with conserved src homology 2 and phosphotyrosine binding domains is related to Shc and is specifically expressed in the brain. Proc. Natl. Acad. Sci. USA, 93: 2729–2734.

Pelicci, G., Lanfrancone, L., Grignani, F., McGlade, J., Cavallo, F., Forni, G., Nicoletti, I., Grignani, F., Pawson, T. and Pelicci, P.G. (1992) A novel trasforming protein (Shc) with an SH2 domain is implicated in mitogenic signal trasduction. Cell, 70: 93–104.

Pelicci, G., Dente, L., De Giuseppe, A., Verducci-Galletti, B., Giuli, S., Mele, S., Vetriani, C., Giorgio, M., Pandolfi, P.P., Cesareni, G. and Pelicci, P.G. (1996) A family of Shc related proteins with conserved PTB, CH1 and SH2 regions. Oncogene, 13: 633–641.

Pelicci, G., Troglio, F., Bodini, A., Melillo, R.M., Pettirossi, V., Coda, L., DeGiuseppe, A., Santoro, M. and Pelicci, P.G. (2002) The neuron-specific Rai (ShcC) adaptor protein inhibits apoptosis by coupling Ret to the Phosphoatidylinosiyol 3-kinase/AKT pathway. Mol. Cell. Biol., 22: 7351–7363.

Petersen, B.E., Bowen, W.C., Patrene, K.D., Mars, W.M., Sullivan, A.K., Murase, N., Boggs, S.S., Greenberger, J.S. and Goff, J.P. (1999) Bone marrow as a potential source of hepatic oval cells. Science, 284: 1168–1170.

Rao, M.S. (1999) Multipotent and restricted precursors in the central nervous system. Anat. Rec., 257: 137–148.

Represa, A., Shimazaki, T., Simmonds, M. and Weiss, S. (2001) EGF-responsive neural stem cells are a transient population in the developing mouse spinal cord. Eur. J. Neurosci., 14: 452–462.

Reynolds, B.A. and Weiss, S. (1992) Generation of neurons and astrocytes from isolated cells of the adult mammalian central nervous system. Science, 255: 1707–1710.

Rietze, R.L., Valcanis, H., Brooker, G.F., Thomas, T., Voss, A.K. and Bartlett, P.F. (2001) Purification of a pluripotent neural stem cell from the adult mouse brain. Nature, 412: 736–739.

Rossi, F. and Cattanco, E. (2002) Opinion: neural stem cell therapy for neurological diseases: dreams and reality. Nat. Rev. Neurosci., 3: 401–409.

Sakai, R., Henderson, J.T., O'Bryan, J.P., Elia, A.J., Saxton, T.M. and Pawson, T. (2000) The mammalian ShcB and ShcC phosphotyrosine docking proteins function in the maturation of sensory and sympathetic neurons. Neuron, 28: 819–833.

Suslov, O.N., Kukekov, V.G., Ignatova, T.N. and Steindler, D.A. (2002) Neural stem cell heterogeneity demonstrated by molecular phenotyping of clonal neurospheres. Proc. Natl. Acad. Sci. USA, 99: 14506–14511.

Tajbakhsh, S., Vivarelli, E., Cusella-De Angelis, G., Rocancourt, D., Buckingham, M. and Cossu, G. (1994) A population of myogenic cells derived from the mouse neural tube. Neuron, 13: 813–821.

Temple, S. (2001) The development of neural stem cells. Nature, 414: 112–117.

Terada, N., Hamazaki, T., Oka, M., Hoki, M., Mastalerz, D.M., Nakano, Y., Meyer, E.M., Morel, L., Petersen, B.E. and Scott, E.W. (2002) Bone marrow cells adopt the phenotype of other cells by spontaneous cell fusion. Nature, 416: 542–545.

Vassilopoulos, G., Wang, P.R. and Russell, D.W. (2003) Transplanted bone marrow regenerates liver by cell fusion. Nature, April 24, 422: 901–904.

Vescovi, A.L., Reynolds, B.A., Fraser, D.D. and Weiss, S. (1993) bFGF regulates the proliferative fate of unipotent (neuronal) and bipotent (neuronal/astroglial) EGF-generated CNS progenitor cells. Neuron, 11: 951–966.

Vescovi, A.L., Rietze, R., Magli, M.C. and Bjornson, C. (2002) Hematopoietic potential of neural stem cells. Nat. Med., 8: 535.

Wang, X., Willenbring, H., Akkari, Y., Torimaru, Y., Foster, M.Al-Dhalimy, M.Lagasse, E.Finegold, M.Olson, S. and Grompe, M. (2003) Cell fusion is the principal source of bone-marrow-derived hepatocytes. Nature, April 24, 422: 897–901.

Ying, Q.L., Nichols, J., Ewans, E.P. and Smith, A.G. (2002) Changing potency by spontaneous fusion. Nature, 416: 545–548.

Zhang, W., Camerini, V., Bender, T.P. and Ravichandran, K.S. (2002) A nonredundant role for the adapter protein Shc in thymic T cell development. Nat. Immunol., 3: 749–755.

Neurotrophins and CNS

Progress in Brain Research, Vol. 146
ISSN 0079-6123

CHAPTER 10

Acute and long-term synaptic modulation by neurotrophins

Bai Lu*

Section on Neural Development and Plasticity, NICHD, National Institute of Health, Bethesda, MD, USA

Abstract: While it has now been well accepted that neurotrophins play an important role in synapse development and plasticity, the specific effects of each neurotrophin on different populations of neurons at different developmental stages have just begun to be worked out. Moreover, the cellular and molecular mechanisms underlying the synaptic function of neurotrophins remain poorly understood. In general, synaptic effects of neurotrophins could be divided into two categories: *acute* effect on synaptic transmission and plasticity occurring within seconds or minutes after cells are exposed to a neurotrophin, and *long-term* effect on synaptic structures and function that takes days to accomplish. In this review I have considered the previous findings on neurotrophic regulation of synapses in view of these two categories. Acute and long-term effects of neurotrophins are reexamined in detail in three model systems: the neuromuscular junction, the hippocampus and the visual cortex. Potential molecular mechanisms that mediate the acute or long-term neurotrophic regulation are discussed. Efforts are made to understand the mechanistic differences between the two effects and their relationships. Further study of these mechanisms will help us better understand how neurotrophins can achieve diverse and synapse-specific modulation.

Keywords: neurotrophins; neuromuscular junction; hippocampus; visual cortex; synaptic transmission; synaptogenesis; synaptic plasticity

Introduction

Change in the efficacy of synaptic transmission, known as synaptic plasticity, is believed to be a general mechanism underlying many functions of the brain, particularly learning and memory. Two forms of synaptic plasticity are thought to mediate memories of different time courses: short-term changes in synaptic strength occurring within minutes and hours, and long-term modulation of the structure and function of synapses lasting over the course of days. A particularly well-studied example is the

mammalian hippocampus. A single high-frequency stimulation (HFS) of presynaptic fibers induces an increase in the efficacy of synaptic transmission that lasts for about 2–3 h and then returns to baseline. This is called early phase of long-term potentiation (E-LTP). In contrast, when HFS is applied repeatedly (3–4 times) at certain intervals, the synaptic potentiation can last for days (or as long as the recordings can be maintained). This is called long-lasting or late phase LTP (L-LTP). The L-LTP is thought to be the cellular basis for long-term memory (Abel et al., 1997; Miller et al., 2002; Villarreal et al., 2002) (reviewed by Kandel, 2001). Similar short- and long-term forms of synaptic plasticity have been observed in the sea slug *Aplysia*, a model system used by Kandel and others to study the molecular basis of learning and memory (Bailey et al., 1996). A single exposure of the neuromodulator serotonin facilitates

*Correspondence to: B. Lu, Section on Neural Development and Plasticity, NICHD, NIH, Building 49, Rm. 6A80, 49 Convent Dr., MSC4480, Bethesda, MD 20892-4480, USA. Tel.: +1-301-435-2970; Fax: +1-301-496-1777 (F); E-mail: bailu@mail.nih.gov

DOI: 10.1016/S0079-6123(03)46010-X

synaptic transmission at the sensory-motor synapses for a few hours. In contrast, continuous exposure to, or 4–5 repeated applications of, serotonin elicits a long-term facilitation that lasts for days. These forms of short- and long-term facilitation of synaptic transmission are believed to underlie short- and long-term sensitization, a simple form of learning and memory (Bailey et al., 1996; Kandel, 2001). Similarly, brief training of the fruit fly *Drosophila* results in short-term memory, while repeated, spaced training leads to long-term memory formation (Tully et al., 1994; Dubnau and Tully, 1998).

Several unique features set long-term modulation apart from short-term plasticity. First, the long-term form is dependent on new protein synthesis while short-term form is not (Frey et al., 1988; Tully et al., 1994; Bailey et al., 1996; Dubnau and Tully, 1998). Second, the transcription factor *cAMP response element binding* protein (CREB) has been repeatedly demonstrated to play a critical role in long-term, but not short-term, synaptic plasticity (Dash et al., 1990; Bourtchuladze et al., 1994; Davis et al., 1996; Bartsch et al., 1998; Silva et al., 1998). Third, long-term synaptic modulation often involves synaptic growth or structural alteration of synapses, in addition to changes in synaptic efficacy (Glanzman et al., 1990; Bailey and Kandel, 1993; Schacher et al., 1993; Schuster et al., 1996a,b). Moreover, the synaptic growth appears to be associated with endocytosis of cell surface molecules such as cell adhesion proteins (Bailey et al., 1992; Hu et al., 1993; Bailey et al., 1997). It has been proposed that repeated stimulation leads to a cAMP-dependent, sustained phosphorylation which in turn triggers the expression of several genes responsible for long-term structural and functional changes at synapses (Bailey et al., 1996; Dubnau and Tully, 1998; Silva et al., 1998). Serotonin has been identified as the neuromodulator that mediates both short- and long-term synaptic facilitation in *Aplysia* (Bailey et al., 1996).

The vertebrate counterparts of serotonin have not been established. Recent studies suggest that neurotrophins, traditionally thought to be involved in neuronal survival and differentiation, may serve as candidates for such a new class of neuromodulators that regulate synaptic transmission, development and plasticity in vertebrates (McAllister et al., 1999; Poo, 2001). Neurotrophins are a family of secretory proteins that consist of nerve growth factor (NGF), brain-derived neurotrophic factor (BDNF), neurotrophin-3 (NT-3) and NT-4/5 (Lewin and Barde, 1996; Huang and Reichardt, 2001). Two types of receptors have been identified to mediate neurotrophin signaling and function: receptor tyrosine kinases (Trk) and the p75 neurotrophin receptor (p75^{NTR}) (Barbacid, 1993; Lee et al., 2001). The first evidence for the synaptic function of neurotrophins was the demonstration that application of BDNF and NT-3 to cultured *Xenopus* neuromuscular synapses induces a rapid potentiation of synaptic transmission (Lohof et al., 1993). Similarly, neurotrophins have been shown to elicit rapid modulation of synapses in the central nervous system (CNS) both in culture (Kim et al., 1994; Lessmann et al., 1994; Levine et al., 1995) and in slices (Figurov et al., 1996; Patterson et al., 1996). Around the same time, the effects of long-term exposure of neurotrophins on synapses have been reported. For example, treatment of the *Xenopus* neuromuscular synapses with BDNF or NT-3 for 2–3 days results in not only a sustained increase in quantal size and a more reliable impulse-evoked synaptic transmission, but also changes in the structures of the presynaptic terminals (Wang et al., 1995). In the CNS, chronic exposure of the neurons in the visual cortex with neurotrophins exerts profound effects on dendritic growth in slices (McAllister et al., 1995) and on the development of ocular dominance columns in cat and rodents in vivo (Maffei et al., 1992; Cabeli et al., 1995). All these results strongly suggest that analogous to serotonin in *Aplysia*, synaptic modulation by neurotrophins may also elicit two temporally very different effects on vertebrate synapses: the *acute* modulation of synaptic transmission and plasticity that may be mediated by a cascade of protein phosphorylation events, and *long-term* regulation of the structure and function of synapses that may require new protein synthesis.

In this review I will try to put the studies of the effects of neurotrophins on vertebrate synapses into the conceptual framework of short-term (or acute) and long-term synaptic modulations that have often been used in the field of synaptic plasticity. Synaptic functions of neurotrophins have been most extensively studied in three model systems: the neuromuscular junction (NMJ), the hippocampus and the visual cortex. In all three systems, attempts will be made to interpret the published work in view of the two categories: the acute and long-term actions of neurotrophins. I will then discuss the potential molecular mechanisms that distinguish the acute and long-term effects. Readers are referred to some recent reviews for more detailed discussion of the synaptic functions of neurotrophins (Lu and Chow, 1999; McAllister et al., 1999; Schuman, 1999; Poo, 2001).

Neuromuscular junction

Because of its simplicity and accessibility, its clearly identifiable pre- and postsynaptic elements, and ease with which gene expression could be manipulated, the *Xenopus* NMJ has been an ideal model system to study both acute and long-term effects of neurotrophins. In fact, the first synaptic action of neurotrophins was demonstrated using this system (Lohof et al., 1993). In that study, application of either BDNF or NT-3 was shown to rapidly enhance synaptic transmission at the NMJ. The acute effect of the neurotrophins is due strictly to an enhancement of transmitter release probability at presynaptic sites (Lohof et al., 1993; Stoop and Poo, 1995). There are several mechanistic differences for the acute effect of these two neurotrophins. First, whereas a rise of intracellular Ca^{2+} ($[Ca^{2+}]i$) is required for both factors, the acute effect of BDNF is dependent on Ca^{2+} influx (Stoop and Poo, 1996) while that of NT3 on the release of Ca^{2+} from intracellular Ca^{2+} stores (He et al., 2000). Second, BDNF acts exclusively on the nerve terminals and not on neuronal soma (Stoop and Poo, 1995). In contrast, NT-3 may act on cell body to elicit an intracellular signal(s) that rapidly propagate along axon to the nerve terminal (Chang and Popov, 1999). Third, concomitant activation of PI3 kinase and an increase of $[Ca^{2+}]i$ appears to be both necessary and sufficient to mediate the NT-3-induced synaptic potentiation (Yang et al., 2001). However, the signal transduction mechanisms underlying the acute effect of BDNF remain largely unknown. Finally, protein synthesis inhibitor anisomycin or cycloheximide does not prevent the effects of NT3, suggesting that acute potentiation of synaptic transmission by neurotrophins is completely independent of protein synthesis (Chang and Popov, 1999). Whether protein synthesis is required for the acute potentiation of synaptic transmission induced by bath application of BDNF has not been firmly established. Interestingly, the acute effect of BDNF occurs preferentially to active synapses and requires cAMP as a gate (Boulanger and Poo, 1999a,b). The synthesis and/or secretion of neurotrophins from the postsynaptic muscle cells increase rapidly in response to presynaptic activity (Funakoshi et al., 1995; Wang and Poo, 1997; Xie et al., 1997). These results support the notion that neurotrophins may serve as retrograde signals that mediate activity-dependent potentiation and perhaps stabilization of active terminals during synaptic competition (Lu and Figurov, 1997; Wang and Poo, 1997; Xie et al., 1997).

Neurotrophins also elicit long-term regulatory effects at the NMJ. Treatment with BDNF or NT-3 (2–3 days) results in a sustained increase in quantal size as well as a more reliable impulse-evoked synaptic transmission (Wang et al., 1995; Liou and Fu, 1997; Liou et al., 1997). Moreover, BDNF or NT-3 enhances the expression of synaptic vesicle proteins such as synapsin-I and synaptophysin, and increases the number of synaptic varicosities in the presynaptic site (Wang et al., 1995). Muscle-derived NT-3 appears to control the quantal size and the excitability of the presynaptic motor neurons, and this effect requires synaptic activities mediated by postsynaptic ACh receptors (Liou and Fu, 1997; Nick and Ribera, 2000). These results suggest a positive feedback mechanism in which activity dependent presynaptic modulation is mediated by postsynaptically derived neurotrophins. In addition to the presynaptic effects, neurotrophin signaling through TrkB receptors appears to be involved in the regulation of the properties of ACh channels as well as the clustering of ACh receptor in the postsynaptic membranes (Fu et al., 1997; Wang and Poo, 1997; Gonzalez et al., 1999). Interestingly, in a triplet system in which a motor neuron innervates two muscle cells, one of which over-expresses NT4, the synaptic potentiation is restricted to the synapse made by the NT4-expressing cell without spreading onto the nearby synapse made by a muscle cell not expressing NT-4 (Wang et al., 1998). These results imply that the long-term effects of NT-4 at the

NMJ are highly localized at the site of NT-4 expression/secretion.

Hippocampus

In the CNS, the synaptic effects of neurotrophins have been extensively studied in the hippocampus. Experiments using cultured hippocampal neurons demonstrate that BDNF is capable of acutely potentiating excitatory synaptic transmission (Knipper et al., 1994; Lessmann et al., 1994; Levine et al., 1995; Takei et al., 1997). However, the effects are quite variable and may depend on the initial reliability and strength of the synapses and the nature of postsynaptic target cells (Lessmann and Heumann, 1998; Berninger et al., 1999; Schinder et al., 2000). Moreover, it has been difficult to reach consensus whether it is due to pre- or postsynaptic mechanisms (Lessmann and Heumann, 1998; Levine et al., 1998; Li et al., 1998a,b; Lin et al., 1998; Suen et al., 1998). These discrepancies may be due, at least in part, to the heterogeneity of the culture systems, which lack the appropriate afferent and efferent synaptic connections observed in vivo. In a remarkable set of experiments, puffing of BDNF onto CA1 neurons in hippocampal slices has been shown to elicit a cation channel-mediated depolarization as rapidly as glutamate (Kafitz et al., 1999; Blum et al., 2002). Similar rapid regulation of cation channels has been seen when BDNF is applied to pontine neurons, and this effect has been attributed to BDNF modulation of TRPC channels through PLC-γ (Li et al., 1999). An early report showed that exogenously applied BDNF can potentiate basal synaptic transmission at CA1 synapses through a presynaptic mechanism (Kang and Schuman, 1995), but these results seem to be difficult to reproduce by others (Figurov et al., 1996; Patterson et al., 1996; Tanaka et al., 1997; Frerking et al., 1998; Gottschalk et al., 1998).

More consistent results have been obtained regarding acute modulation of LTP in hippocampal slices. Application of exogenous BDNF facilitates tetanus-induced LTP at the CA1 synapses in neonatal hippocampal slices, in which the endogenous BDNF levels are low (Figurov et al., 1996). In contrast, inhibition of endogenous BDNF activity, either by the BDNF scavenger TrkB-IgG, or in BDNF gene knockout (KO) mice, reduces the magnitude of tetanus-induced LTP in adult hippocampus, in which the endogenous BDNF levels are high (Korte et al., 1995; Figurov et al., 1996; Patterson et al., 1996; Kang et al., 1997). The deficit in hippocampal LTP seen in the BDNF KO mice could be rescued either by acute application of recombinant BDNF or by virus mediate BDNF gene transfer (Korte et al., 1996; Patterson et al., 1996; Pozzo-Miller et al., 1999), suggesting an acute rather than long-term developmental modulation by BDNF. Elimination of TrkB receptor from the postsynaptic cells of the CA1 synapses using conditional knockout techniques confirms that BDNF modulation of LTP primarily through a presynaptic mechanism (Xu et al., 2000a). BDNF facilitates CA1 LTP by enhancing synaptic responses to high frequency, LTP-inducing tetanic stimulation (Figurov et al., 1996; Gottschalk et al., 1998; Pozzo-Miller et al., 1999). This effect requires activation of MAP kinase and PI3 kinase, and is insensitive to inhibitors of new protein synthesis (Gottschalk et al., 1999). Experiments using BDNF KO mice indicate that BDNF enhances high frequency synaptic transmission by increasing the number of readily releasable (docked) vesicles at CA1 synapses, possibly by regulating synaptic vesicle proteins (Pozzo-Miller et al., 1999). Consistent with this, BDNF induces a MAP kinase-dependent phosphorylation of synapsin-I, leading to an increase in availability of synaptic vesicles for release (Jovanovic et al., 2000). Thus, acute modulation of CA1 synapses by BDNF is achieved by protein synthesis-independent modifications of existing presynaptic proteins. At dentate synapses, a brief puff of BDNF evoked a rapid Ca^{2+} influx into the dendritic spines, and when paired with a weak presynaptic stimulation elicits a robust LTP (Kovalchuk et al., 2002). These results suggest a rapid and postsynaptic action of BDNF at the dentate synapses.

Less is known about the long-term effects of neurotrophins in the hippocampus. In dissociated cultures of hippocampal or cortical neurons, chronic application of BDNF results in an increase in different presynaptic (Takei et al., 1997) and postsynaptic (Narisawa-Saito et al., 1999; Takei et al., 2001; Narisawa-Saito et al., 2002) proteins, as well as complex effects on synaptic transmission (Rutherford et al., 1998; Vicario-Abejon et al., 1998;

Sherwood and Lo, 1999; Bolton et al., 2000). Again the variability of the results may reflect the heterogeneity of the culture systems. In neonatal hippocampus of TrkB KO (−/−) mice in vivo, there is a significant decrease in the density of axonal varicosities and synaptic contacts, in the number of synaptic vesicles, and in the levels of several synaptic proteins (Martinez et al., 1998). Conversely, long-term treatment of hippocampal slice cultures with BDNF markedly increases the frequency of miniature EPSPs, the number of docked vesicles, and the number of dendritic spines and synapses in the CA1 area (Tyler and Pozzo-Miller, 2001). This treatment also elicits a rapid increase in the expression of synaptophysin and synaptobrevin, but a delayed increase in that of synaptotagmin (Tartaglia et al., 2001). Interestingly, the BDNF-induced increase in synaptotagmin expression is mediated through a protein synthesis and cAMP-dependent mechanism (Tartaglia et al., 2001), characters reminiscent of late phase LTP (L-LTP) and long-term memory seen in many model systems.

The above results suggest that long-term synaptic modulation by BDNF may be involved in long-term memory. Indeed, tetanic stimulation used to induce L-LTP elicits an increase in hippocampal mRNA^{BDNF}, with a time course well correlated with L-LTP (Patterson et al., 1992; Castren et al., 1993; Dragunow et al., 1993; Kesslak et al., 1998; Morimoto et al., 1998). Inhibition of BDNF signaling, either by BDNF gene KO or by the BDNF scavenger TrkB-IgG, markedly attenuates L-LTP (Kang et al., 1997; Korte et al., 1998; Patterson et al., 2001). Moreover, application of TrkB-IgG 30–70 min after LTP induction reverses the already established LTP, and prevents the occurrence of L-LTP (Kang et al., 1997; Korte et al., 1998). Conceivably, long-term exposure to BDNF, similar to multiple tetanic stimuli, could induce the translation of certain proteins critical for the structural and functional changes that underlie L-LTP in the hippocampal synapses. Mice that lack TrkB receptors in their forebrain and hippocampus exhibit impairments in LTP as well as water maze test, pointing to an important role of BDNF signaling in learning and memory in vivo (Minichiello et al., 1999). Further characterization using knock-in techniques indicate that the TrkB mediates BDNF-induced modulation of L-LTP via recruitment of phospholipase C-γ (PLC-γ) and by subsequent phosphorylation of Ca^{2+}-calmodulin IV (CaMKIV) and CREB (Minichiello et al., 2002). While BDNF regulates both E-LTP and L-LTP, it is unclear whether it uses distinct mechanisms for these two forms of plasticity.

Visual cortex

The visual cortex is the third system in which synaptic actions of neurotrophins is studied quite extensively (Bonhoeffer, 1996; Berardi and Maffei, 1999; McAllister et al., 1999). For acute effects of neurotrophins on synaptic plasticity, the results from various laboratories are quite consistent. Similar to those observed in the hippocampus, BDNF (20 ng/ml) facilitates tetanus-induced LTP in layer III synapses of young adult rats (Akaneya et al., 1997; Huber et al., 1998; Jiang et al., 2001). Using BDNF heterozygous mice, Maffei and colleagues demonstrated that BDNF may selectively regulate LTP in the layer IV to layer III synapses but not in white matter to layer III synapses in the visual cortex (Bartoletti et al., 2002). Treatment of slices from the visual cortex with BDNF also blocks long-term depression (LTD) induced by low frequency stimulation, probably through a presynaptic mechanism (Akaneya et al., 1996; Huber et al., 1998; Kinoshita et al., 1999; Kumura et al., 2000). In a remarkable set of experiments, Tsumoto and his colleague have recently shown that the low frequency stimulation (1 Hz, 15 min) normally used to induce LTD in slices cannot induce LTD in the visual cortex in vivo (Jiang et al., 2003). However, the in vivo LTD can be induced after BDNF-TrkB signaling is inhibited by infusing TrkB-IgG or BDNF antibody into the visual cortex. An interesting implication of this work is that a synapse in the visual cortex may undergo LTP or LTD depending on the level of BDNF, which is controlled by different neuronal activity. The effects of BDNF on LTP and LTD induction have also been attributed to the significant enhancement of synaptic responses to repetitive stimulation (Huber et al., 1998). In addition to its role in synaptic plasticity, BDNF (and NGF) also seems to be able to potentiate basal synaptic transmission when applied at higher concentrations (200 ng/ml) (Akaneya et al., 1997;

Carmignoto et al., 1997), a phenomenon similar to that seen in the hippocampus (Kang and Schuman, 1995). Given that proforms of neurotrophins, proNGF and proBDNF, are now considered to be functional by interacting more effectively with $p75^{NTR}$ rather than Trk receptors (Lee et al., 2002), and many of the neurotrophin preparations used in the previous experiments may contain substantial amount of both proNGF and proBDNF (see Fahnestock et al. in this issue), it might be useful to reexamine the various effects on hippocampal and cortical synapses, using pure proforms and mature forms of neurotrophins.

An extensively characterized long-term effect of neurotrophins in the visual cortex is the regulation of dendritic and axonal growth (McAllister et al., 1999). Neurotrophin treatment of cultured slices derived from neonatal visual cortex elicits complex effects on the length and the complexity of both apical and basal dendrites of pyramidal neurons (McAllister et al., 1995). On basal dendrites, BDNF and NT3 have opposite effects (McAllister et al., 1997). Specific deletion of TrkB gene results in a marked retraction of the dendrites of pyramidal neurons (Xu et al., 2000b). Transfection of full length TrkB into visual cortex increases the proximal dendritic branching, while transfection of truncated form of TrkB, which serves as a dominant negative agent, promotes net elongation of distal dendrites (Yacoubian and Lo, 2000). BDNF also destabilize the dendritic spines (Horch et al., 1999). Interestingly, the effects of BDNF on dendritic growth can be blocked by inhibition of spontaneous neuronal activity, synaptic transmission and Ca^{2+} channels, suggesting an activity-dependent modulation (McAllister et al., 1996). While it takes about 36 h to see the full effect, BDNF acts as a spatially restricted signal that regulates dendritic branching locally near the site of BDNF secretion (Horch and Katz, 2002). In a different part of the visual system, BDNF has been shown to modulate activity-dependent arborization of retinal ganglion axons projected in the tectum (Cohen-Cory et al., 1996; Cohen-Cory and Lom, 1999; Lom and Cohen-Cory, 1999).

Another major long-term effect of neurotrophins in the visual cortex is the regulation of the ocular dominance plasticity (Bonhoeffer, 1996; Snider and Lichtman, 1996; Berardi and Maffei, 1999). This was first demonstrated by Maffei's group using 'monocular deprivation (MD)', one of the best models for experience-dependent plasticity. If vision is deprived in one eye during critical period, axonal inputs to the cortex from this eye become functionally disconnected, and dominance is shifted to the nondeprived eye. The development of ocular dominance reflects competition between afferent thalamic inputs driven by two eyes in an activity-dependent manner. Intraventricular infusion of NGF attenuates the electrophysiological and behavioral consequences of MD in rats (Domenici et al., 1991, 1992; Maffei et al., 1992; Fiorentini et al., 1995). Infusion of NGF antibodies interfere with the development of ocular dominance and prolong the critical period, suggesting a role of endogenous NGF (Berardi et al., 1994; Domenici et al., 1994). Further, exogenous supply of NGF also counters the effects of 'dark-rearing (DR)', another frequently used experimental paradigm known to arrest the development of visual cortex and prolong the critical period (Fagiolini et al., 1997; Pizzorusso et al., 1997). These experiments form the basis for a hypothesis that development of ocular dominance requires competition of axonal fibers from two eyes for NGF, which is produced in limited amount in the visual cortex. However, the NGF receptor TrkA is only marginally expressed in visual cortex, mostly reflecting the cholinergic afferents from basal forebrain (Schoups et al., 1995; Prakash et al., 1996; Rossi et al., 2002). On the other hand, TrkB, the receptor for BDNF and NT4, is highly expressed in the visual cortex during critical period (Allendoerfer et al., 1994; Cabeli et al., 1996). More importantly, the expression of BDNF, but not NGF, in the visual cortex is regulated by visual inputs (Castren et al., 1992; Schoups et al., 1995; Capsoni et al., 1999; Pollock et al., 2001), and monocular activity blockade elicits a striking decrease in $mRNA^{BDNF}$ and BDNF protein in the visual cortex corresponding to the deprived eye (Bozzi et al., 1995; Rossi et al., 1999; Lein and Shatz, 2000). These results suggest a potential role of TrkB ligands in visual cortex development. Indeed, infusion of BDNF or NT-4, but not NGF, into cat visual cortex blurs the segregation of eye-specific columns in cat, suggesting that too much supply of the TrkB ligand eliminates the need for competition between thalamic inputs representing two eyes (Cabeli et al., 1995).

Similar inhibition of ocular dominance column formation has been observed when endogenous TrkB ligands are scavenged by infusion of TrkB-IgG, but not by the NGF scavenger TrkA-IgG or NT-3 scavenger, TrkC-IgG (Cabeli et al., 1997). Injection of microbeads containing NT-4 also prevents shrinkage of LGN neurons induced by MD in ferrets (Riddle et al., 1995), and infusion of BDNF or NT-4 blocks the effects of MD on ocular dominance plasticity in cats (Gillespie et al., 2000; Hata et al., 2000). Thus, experimental data on neurotrophic regulation of ocular dominance plasticity are conflicting, with one set implicating NGF while the other BDNF or NT-4. It is possible that NGF is the preferred ligand that regulates ocular dominance in rodents (rat/mouse) (Lodovichi et al., 2000), while higher mammals (ferret/cat) use the TrkB ligands for the same purpose (Silver et al., 2001). Since proforms of neurotrophins are secreted and functional, it is also possible that activation of p75^{NTR} by pro-neurotrophins may mediate the regulation of ocular dominance plasticity. A number of recent studies have demonstrated a role of p75^{NTR} in the survival and function of subplate cells (DeFreitas et al., 2001; McQuillen et al., 2002), a transient population of neurons known to be involved in the development of ocular dominance (Ghosh and Shatz, 1992, 1994; Lein et al., 1999). One must demonstrate, however, that p75 binds preferentially to NGF in rats, but to TrkB ligands in cats.

Mechanistic differences between acute and long-term neurotrophic regulation

While overwhelming experimental data clearly indicate that neurotrophins elicit two very different forms of synaptic modulation, many important questions remain unanswered. What are the hallmarks for the acute and long-term effects other than their temporal differences? Are the acute and long-term neurotrophic regulations mediated fundamentally different mechanisms? Can acute effect be converted to long-term regulation? If so what are the mechanistic links between the acute and long-term effects? Mechanistic questions are often easier to address using simple model systems. Again the *Xenopus* neuromuscular synapses may be an ideal system to address these questions. Both acute and long-term treatments of the nerve–muscle coculture induce an enhancement of transmitter release at the neuromuscular synapses, as reflected by the increases in the frequency of spontaneous synaptic currents (SSCs) and the amplitude of the evoked synaptic currents (ESCs) (Lohof et al., 1993; Wang et al., 1995). However, only the long-term application of neurotrophins elicits changes in the properties of ACh receptors (Fu et al., 1997; Wang and Poo, 1997) and the clustering of ACh receptors in the postsynaptic membranes (Gonzalez et al., 1999). Moreover, the acute effects of NT-3 are transient, depending on the continuous presence of the factors. Synaptic efficacy returns to the baseline levels shortly after removal of NT-3 (Lohof et al., 1993). In contrast, chronic exposure of NT-3 elicited a profound change in synaptic efficacy that persists even after withdrawal of the neurotrophins (Wang et al., 1995). Most importantly, unlike the acute application, long-term treatment with BDNF or NT-3 induces remarkable morphological changes of presynaptic neurons. The expression of the synaptic vesicle proteins synapsin I and synpatophysin, as well as the number of synaptic varicosities, is markedly increased after 2–3 days treatment of BDNF or NT-3 (Wang et al., 1995). The recycling vesicle pool in spinal neurons, as reflected by FM-143 labeled fluorescence spots, is also enlarged (Je et al., 2002).

One of the fundamental differences in short- and long-term forms of activity-dependent synaptic modulation is the requirement of protein synthesis. This principle may be applied to the acute and long-term regulation of synapses by neurotrophins. The acute potentiation of synaptic transmission by NT-3 is insensitive to the protein synthesis inhibitor anisomycin or cycloheximide (Chang and Popov, 1999). Our recent experiments demonstrated that the long-term effects of NT-3 depend upon new protein synthesis (Je et al., 2002). We have used two independent methods to block protein translation. One is to inhibit mTOR, a kinase that phosphorylates and activates eIF4E, by bath application of the inhibitor rapamycin. The other is to express a coumermycin-inducible, double stranded RNA dependent protein kinase (PKR), which is known to inhibit translation by phosphorylating eIF2. In both cases, the physiological as well as the morphological effects induced by the long-term treatment with NT-3 was completely prevented. In contrast, the acute effect of NT-3 was not blocked by either rapamycin or activation of PKR. These results suggest that dependence on protein synthesis is a key mechanistic difference between the acute and long-term synaptic effects of NT-3. The situation for BDNF is a little complicated. In a recent study, Poo and colleagues showed that localized contact of a single BDNF-coated bead with the presynaptic axon resulted in a very localized increase in $[Ca^{2+}]i$ (restricted within 40 μm from the site of bead contact) as well as a very localized synaptic potentiation (the beads must be placed within 60 μm from the synapses) (Zhang and Poo, 2002). Moreover, the BDNF-bead could still enhance synaptic transmission at a neuromuscular synapse made by an isolated axon (cell body of the motor neuron severed), suggesting a local action of BDNF on presynaptic terminals. Remarkably, this type of synaptic potentiation could be blocked by a prolonged treatment (2 h) with, but not a short exposure of, the translation inhibitor anisomycin or cycloheximide. This is somewhat inconsistent with the idea that only long-term neurotrophic regulation requires protein synthesis because the effect of the BDNF beads was relatively fast. Curiously, prolonged treatment with anisomycin also blocked BDNF induced Ca^{2+} influx. Given that BDNF-induced synaptic potentiation requires Ca^{2+} influx (Stoop and Poo, 1996), it is therefore unclear whether anisomycin truly acted as a protein translation inhibitor or instead served as a blocker for Ca^{2+} influx. Some well-known protein synthesis inhibitors have been shown to block Ca^{2+} influx in CNS neurons (Linden, 1996).

In addition to the requirement for protein synthesis, the long-term synaptic modulation by NT-3 also differs from its acute effect in several ways. First, the long-term, but not the acute, synaptic modulation requires the endocytosis of NT-3-TrkC receptor complex (Zhou et al., 2001). The changes in synaptic transmission and synaptic growth induced by chronic treatment with NT-3 were completely blocked by presynaptic expression of dominant negative dynamin, which prevents endocytosis. Bead-conjugated NT-3 that is too large to be endocytosed could still induce acute effects, but

failed to elicit the long-term changes. Second, the acute and long-term neurotrophic modulation may use very different signaling pathways. The acute potentiation of synaptic transmission by NT-3 depends on the concomitant activation of PI-3 kinase and PLC-γ pathways in the presynaptic terminals (Yang et al., 2001). One of the major signaling events downstream of PLC-γ is the IP3-induced release of Ca^{2+} from intracellular Ca^{2+} stores, leading to the activation of CaMKII (He et al., 2000). The long-term effects of NT-3, however, do not require the PLC-γ-CaMKII signaling (Zhou et al., 2001). While both acute and long-term effects of NT-3 depend on activation of PI-3 kinase in the presynaptic neurons, only the long-term regulation requires Akt, a major downstream target of PI3 kinase. Inhibition of Akt by presynaptic expression of a dominant negative Akt prevented NT-3-induced long-term structural and functional changes, without affecting the acute potentiation of synaptic transmission by NT-3. Third, the activation of the transcription factor CREB appears to be critical for the NT-3-induced long-term changes at the neuromuscular synapses (Je et al., 2002). Application of NT-3 to the *Xenopus* nerve–muscle cocultures induced phosphorylation and activation of CREB. However, presynaptic expression of dominant negative CREB blocked long-term, but not acute, synaptic modulation by NT-3. Taken together, these results provide mechanistic insights into the relationships between acute and long-term changes induced by NT-3.

Abbreviations

BDNF	brain-derived neurotrophic factor
CREB	*c*AMP *r*esponse *e*lement *b*inding protein
E-LTP	early phase long-term potentiation
HFS	high-frequency stimulation
L-LTP	long-lasting or late phase long-term potentiation
LTP	long-term potentiation
NGF	nerve growth factor
NMJ	neuromuscular junction
NT-3	neurotrophin-3
NT-4/5	neurotrophin-4/5
p75[NTR]	p75 neurotrophin receptor
Trk	receptor tyrosine kinase

References

Abel, T., Nguyen, P.V., Barad, M., Deuel, T.A., Kandel, E.R. and Bourtchouladze, R. (1997) Genetic demonstration of a role for PKA in the late phase of LTP and in hippocampus-based long-term memory. Cell, 88: 615–626.

Akaneya, Y., Tsumoto, T. and Hatanaka, H. (1996) Brain-derived neurotrophic factor blocks long-term depression in rat visual cortex. J. Neurophysiol., 76: 4198–4201.

Akaneya, Y., Tsumoto, T., Kinoshita, S. and Hatanaka, H. (1997) Brain-derived neurotrophic factor enhances long-term potentiation in rat visual cortex. J. Neurosci., 17: 6707–6716.

Allendoerfer, K.L., Cabeli, R.J., Escandon, E., Kaplan, D.R., Nikolics, K. and Shatz, C.J. (1994) Regulation of neurotrophin receptors during the maturation of the mammalian visual system. J. Neurosci., 14: 1795–1811.

Bailey, C.H., Bartsch, D. and Kandel, E.R. (1996) Toward a molecular definition of long-term memory storage. Proc. Natl. Acad. Sci. USA, 93: 13445–13452.

Bailey, C.H., Chen, M., Keller, F. and Kandel, E.R. (1992) Serotonin-mediated endocytosis of apCAM: an early step of learning-related synaptic growth in Aplysia. Science, 256: 645–649.

Bailey, C.H., Kaang, B.K., Chen, M., Martin, K.C., Lim, C.S., Casadio, A. and Kandel, E.R. (1997) Mutation in the phosphorylation sites of MAP kinase blocks learning-related internalization of apCAM in Aplysia sensory neurons. Neuron, 18: 913–924.

Bailey, C.H. and Kandel, E.R. (1993) Structural changes accompanying memory storage. Annu. Rev. Physiol., 55: 397–426.

Barbacid, M. (1993) Nerve growth factor: a tale of two receptors. Oncogene, 8: 2033–2042.

Bartoletti, A., Cancedda, L., Reid, S.W., Tessarollo, L., Porciatti, V., Pizzorusso, T. and Maffei, L. (2002) Heterozygous knock-out mice for brain-derived neurotrophic factor show a pathway-specific impairment of long-term potentiation but normal critical period for monocular deprivation. J. Neurosci., 22: 10072–10077.

Bartsch, D., Casadio, A., Karl, K.A., Serodio, P. and Kandel, E.R. (1998) CREB1 encodes a nuclear activator, a repressor, and a cytoplasmic modulator that form a regulatory unit critical for long-term facilitation. Cell, 95: 211–223.

Berardi, N., Cellerino, A., Domenici, L., Fagiolini, M., Pizzorusso, T., Cattaneo, A. and Maffei, L. (1994) Monoclonal antibodies to nerve growth factor affect the postnatal development of the visual system. Proc. Natl. Acad. Sci. USA, 91: 684–688.

Berardi, N. and Maffei, L. (1999) From visual experience to visual function: roles of neurotrophins. J. Neurobiol., 41: 119–126.

Berninger, B., Schinder, A.F. and Poo, M.M. (1999) Synaptic reliability correlates with reduced susceptibility to synaptic potentiation by brain-derived neurotrophic factor. Learn. Mem., 6: 232–242.

Blum, R., Kafitz, K.W. and Konnerth, A. (2002) Neurotrophin-evoked depolarization requires the sodium channel Na(V)1.9. Nature, 419: 687–693.

Bolton, M.M., Pittman, A.J. and Lo, D.C. (2000) Brain-derived neurotrophic factor differentially regulates excitatory and inhibitory synaptic transmission in hippocampal cultures. J. Neurosci., 20: 3221–3232.

Bonhoeffer, T. (1996) Neurotrophins and activity-dependent development of the neocortex. Curr. Opin. Neurobiol., 6: 119–126.

Boulanger, L. and Poo, M.M. (1999a) Gating of BDNF-induced synaptic potentiation by cAMP. Science, 284: 1982–1984.

Boulanger, L. and Poo, M.M. (1999b) Presynaptic depolarization facilitates neurotrophin-induced synaptic potentiation. Nat. Neurosci., 2: 346–351.

Bourtchuladze, R., Frenguelli, B., Blendy, J., Cioffi, D., Schutz, G. and Silva, A.J. (1994) Deficient long-term memory in mice with a targeted mutation of the cAMP-responsive element-binding protein. Cell, 79: 59–68.

Bozzi, Y., Pizzorusso, T., Cremisi, F., Rossi, F.M., Barsacchi, G. and Maffei, L. (1995) Monocular deprivation decreases the expression of messenger RNA for brain-derived neurotrophic factor in the rat visual cortex. Neuroscience, 69: 1133–1144.

Cabeli, R.J., Allendoerfer, K.L., Radeke, M.J., Welcher, A.A., Feinstein, S.C. and Shatz, C.J. (1996) Changing patterns of expression and subcellular localization of TrkB in the developing visual system. J. Neurosci., 16: 7965–7980.

Cabeli, R.J., Horn, A. and Shatz, C.J. (1995) Inhibition of ocular dominance column formation by infusion of NT-4/5 or BDNF. Science, 267: 1662–1666.

Cabeli, R.J., Shelton, D.L., Segal, R.A. and Shatz, C.J. (1997) Blockade of endogenous ligands of trkB inhibits formation of ocular dominance columns. Neuron, 19: 63–76.

Capsoni, S., Tongiorgi, E., Cattaneo, A. and Domenici, L. (1999) Dark rearing blocks the developmental down-regulation of brain-derived neurotrophic factor messenger RNA expression in layers IV and V of the rat visual cortex. Neuroscience, 88: 393–403.

Carmignoto, G., Pizzorusso, T., Tia, S. and Vicini, S. (1997) Brain-derived neurotrophic factor and nerve growth factor potentiate excitatory synaptic transmission in the rat visual cortex. J. Physiol. (Lond), 498: 153–164.

Castren, E., Pitkanen, M., Sirvio, J., Parsadanian, A., Lindholm, D., Thoenen, H. and Riekkinen, P.J. (1993) The induction of LTP increases BDNF and NGF mRNA but decreases NT-3 mRNA in the dentate gyrus. Neuroreport, 4: 895–898.

Castren, E., Zafra, F., Thoenen, H. and Lindholm, D. (1992) Light regulates expression of brain-derived neurotrophic factor mRNA in the rat visual cortex. Proc. Natl. Acad. Sci. USA, 89: 9444–9448.

146

Chang, S. and Popov, S.V. (1999) Long-range signaling within growing neurites mediated by neurotrophin-3. Proc. Natl. Acad. Sci. USA, 96: 4095–4100.

Cohen-Cory, S., Escandon, E. and Fraser, S.E. (1996) The cellular patterns of BDNF and trkB expression suggest multiple roles for BDNF during Xenopus visual system development. Dev. Biol., 179: 102–115.

Cohen-Cory, S. and Lom, B. (1999) BDNF modulates, but does not mediate, activity-dependent branching and remodeling of optic axon arbors in vivo. J. Neurosci., 19: 9996–10003.

Dash, P.K., Hochner, B. and Kandel, E.R. (1990) Injection of the cAMP-responsive element into the nucleus of Aplysia sensory neurons blocks long-term facilitation. Nature, 345: 718–721.

Davis, G.W., Schuster, C.M. and Goodman, C.S. (1996) Genetic dissection of structural and functional components of synaptic plasticity. III. CREB is necessary for presynaptic functional plasticity. Neuron, 17: 669–679.

DeFreitas, M.F., McQuillen, P.S. and Shatz, C.J. (2001) A novel p75NTR signaling pathway promotes survival, not death, of immunopurified neocortical subplate neurons. J. Neurosci., 21: 5121–5129.

Domenici, L., Berardi, N., Carmignoto, G., Vantini, G. and Maffei, L. (1991) Nerve growth factor prevents the amblyopic effects of monocular deprivation. Proc. Natl. Acad. Sci. USA, 88: 8811–8815.

Domenici, L., Cellerino, A., Berardi, N., Cattaneo, A. and Maffei, L. (1994) Antibodies to nerve growth factor (NGF) prolong the sensitive period for monocular deprivation in the rat. Neuroreport, 5: 2041–2044.

Domenici, L., Parisi, V. and Maffei, L. (1992) Exogenous supply of nerve growth factor prevents the effects of strabismus in the rat. Neuroscience, 51: 19–24.

Dragunow, M., Beilharz, E., Mason, B., Lawlor, P., Abraham, W. and Gluckman, P. (1993) Brain-derived neurotrophic factor expression after long-term potentiation. Neurosci. Lett., 160: 232–236.

Dubnau, J. and Tully, T. (1998) Gene discovery in Drosophila: new insights for learning and memory. Annu. Rev. Neurosci., 21: 407–444.

Fagiolini, M., Pizzorusso, T., Porciatti, V., Cenni, M. and Maffei, L. (1997) Transplant of Schwann cells allows normal development of the visual cortex of dark-reared rats. Eur. J. Neurosci., 9: 102–112.

Figurov, A., Pozzo-Miller, L., Olafsson, P., Wang, T. and Lu, B. (1996) Regulation of synaptic responses to high-frequency stimulation and LTP by neurotrophins in the hippocampus. Nature, 381: 706–709.

Fiorentini, A., Berardi, N. and Maffei, L. (1995) Nerve growth factor preserves behavioral visual acuity in monocularly deprived kittens. Vis. Neurosci., 12: 51–55.

Frerking, M., Malenka, R.C. and Nicoll, R.A. (1998) Brain-derived neurotrophic factor (BDNF) modulates inhibitory, but not excitatory, transmission in the CA1 region of the hippocampus. J. Neurophysiol., 80: 3383–3386.

Frey, U., Krug, M., Reymann, K.G. and Matthies, H. (1988) Anisomycin, an inhibitor of protein synthesis, blocks late phases of LTP phenomena in the hippocampal CA1 region in vitro. Brain Res., 452: 57–65.

Fu, A.K., Ip, F.C., Lai, K.O., Tsim, K.W. and Ip, N.Y. (1997) Muscle-derived neurotrophin-3 increases the aggregation of acetylcholine receptors in neuron-muscle co-cultures. Neuroreport, 8: 3895–3900.

Funakoshi, H., Belluardo, N., Arenas, E., Yamamoto, Y., Casabona, A., Persson, H. and Ibanez, C.F. (1995) Muscle-derived neurotrophin-4 as an activity-dependent trophic signal for adult motor neurons. Science, 268: 1495–1499.

Ghosh, A. and Shatz, C.J. (1992) Involvement of subplate neurons in the formation of ocular dominance columns. Science, 255: 1441–1443.

Ghosh, A. and Shatz, C.J. (1994) Segregation of geniculocortical afferents during the critical period: a role for subplate neurons. J. Neurosci., 14: 3862–3880.

Gillespie, D.C., Crair, M.C. and Stryker, M.P. (2000) Neurotrophin-4/5 alters responses and blocks the effect of monocular deprivation in cat visual cortex during the critical period. J. Neurosci., 20: 9174–9186.

Glanzman, D.L., Kandel, E.R. and Schacher, S. (1990) Target-dependent structural changes accompanying long-term synaptic facilitation in Aplysia neurons. Science, 249: 799–802.

Gonzalez, M., Ruggiero, F.P., Chang, Q., Shi, Y.J., Rich, M.M., Kraner, S. and Balice-Gordon, R.J. (1999) Disruption of TrkB-mediated signaling induces disassembly of postsynaptic receptor clusters at neuromuscular junctions. Neuron, 24: 567–583.

Gottschalk, W., Pozzo-Miller, L.D., Figurov, A. and Lu, B. (1998) Presynaptic modulation of synaptic transmission and plasticity by brain-derived neurotrophic factor in the developing hippocampus. J. Neurosci., 18: 6830–6839.

Gottschalk, W.A., Jiang, H., Tartaglia, N., Feng, L., Figurov, A. and Lu, B. (1999) Signaling mechanisms mediating BDNF modulation of synaptic plasticity in the hippocampus. Learn. Mem., 6: 243–256.

Hata, Y., Ohshima, M., Ichisaka, S., Wakita, M., Fukuda, M. and Tsumoto, T. (2000) Brain-derived neurotrophic factor expands ocular dominance columns in visual cortex in monocularly deprived and nondeprived kittens but does not in adult cats. J. Neurosci., 20: RC57.

He, X., Yang, F., Xie, Z. and Lu, B. (2000) Intracellular Ca^{2+} and Ca^{2+}/Calmodulin-dependent kinase II mediate acute potentiation of neurotransmitter release by neurotrophin-3. J. Cell Biol., 149: 783–792.

Horch, H.W. and Katz, L.C. (2002) BDNF release from single cells elicits local dendritic growth in nearby neurons. Nat. Neurosci., 5: 1177–1184.

Horch, H.W., Kruttgen, A., Portbury, S.D. and Katz, L.C. (1999) Destabilization of cortical dendrites and spines by BDNF. Neuron, 23: 353–364.

Hu, Y., Barzilai, A., Chen, M., Bailey, C.H. and Kandel, E.R. (1993) 5-HT and cAMP induce the formation of coated pits and vesicles and increase the expression of clathrin light chain in sensory neurons of aplysia. Neuron, 10: 921–929.

Huang, E.J. and Reichardt, L.F. (2001) Neurotrophins: roles in neuronal development and function. Ann. Rev. Neurosci., 24: 677–736.

Huber, K.M., Sawtell, N.B. and Bear, M.F. (1998) Brain-derived neurotrophic factor alters the synaptic modification threshold in visual cortex. Neuropharmacology, 37: 571–579.

Je, H.-S., Zhou, J. and Lu, B. (2002) Neurotrophin-3 induced long-term synaptic modulation at Xenopus neuromuscular junction is dependent on protein synthesis and CREB. Soc. Neurosci. Abst., 28: 615–630.

Jiang, B., Akaneya, Y., Hata, Y. and Tsumoto, T. (2003) Long-term depression is not induced by low frequency stimulation in rat visual cortex in vivo: A possible preventing role of endogenous BDNF. J. Neurosci., 23: 3761.

Jiang, B., Akaneya, Y., Ohshima, M., Ichisaka, S., Hata, Y. and Tsumoto, T. (2001) Brain-derived neurotrophic factor induces long-lasting potentiation of synaptic transmission in visual cortex in vivo in young rats, but not in the adult. Eur. J. Neurosci., 14: 1219–1228.

Jovanovic, J.N., Czernik, A.J., Fienberg, A.A., Greengard, P. and Sihra, T.S. (2000) Synapsins as mediators of BDNF-enhanced neurotransmitter release. Nat. Neurosci., 3: 323–329.

Kafitz, K.W., Rose, C.R., Thoenen, H. and Konnerth, A. (1999) Neurotrophin-evoked rapid excitation through TrkB receptors. Nature, 401: 918–921.

Kandel, E.R. (2001) The molecular biology of memory storage: a dialogue between genes and synapses. Science, 294: 1030–1038.

Kang, H. and Schuman, E.M. (1995) Long-lasting neurotrophin-induced enhancement of synaptic transmission in the adult hippocampus. Science, 267: 1658–1662.

Kang, H., Welcher, A.A., Shelton, D. and Schuman, E.M. (1997) Neurotrophins and time: Different roles for TrkB signaling in hippocampal long-term potentiation. Neuron, 19: 653–664.

Kesslak, J.P., So, V., Choi, J., Cotman, C.W. and Gomez-Pinilla, F. (1998) Learning upregulates brain-derived neurotrophic factor messenger ribonucleic acid: a mechanism to facilitate encoding and circuit maintenance? Behav. Neurosci., 112: 1012–1019.

Kim, H.G., Wang, T., Olafsson, P. and Lu, B. (1994) Neurotrophin 3 potentiates neuronal activity and inhibits g-aminobutyrateric synaptic transmission in cortical neurons. Proc. Natl. Acad. Sci. USA, 91: 12341–12345.

Kinoshita, S., Yasuda, H., Taniguchi, N., Katoh-Semba, R., Hatanaka, H. and Tsumoto, T. (1999) Brain-derived neurotrophic factor prevents low-frequency inputs from inducing long-term depression in the developing visual cortex. J. Neurosci., 19: 2122–2130.

Knipper, M., da Penha Berzaghi, M., Blochl, A., Breer, H., Thoenen, H. and Lindholm, D. (1994) Positive feedback between acetylcholine and the neurotrophins nerve growth factor and brain-derived neurotrophic factor in the rat hippocampus. Eur. J. Neurosci., 6: 668 671.

Korte, M., Carroll, P., Wolf, E., Brem, G., Thoenen, H. and Bonhoeffer, T. (1995) Hippocampal long-term potentiation is impaired in mice lacking brain-derived neurotrophic factor. Proc. Natl. Acad. Sci. USA, 92: 8856–8860.

Korte, M., Kang, H., Bonhoeffer, T. and Schuman, E. (1998) A role for BDNF in the late-phase of hippocampal long-term potentiation. Neuropharmacology, 37: 553–559.

Korte, M., Staiger, V., Griesbeck, O., Thoenen, H. and Bonhoeffer, T. (1996) The involvement of brain-derived neurotrophic factor in hippocampal long-term potentiation revealed by gene targeting experiments. J. Physiol. (Paris), 90(3–4): 157–164.

Kovalchuk, Y., Hanse, E., Kafitz, K.W. and Konnerth, A. (2002) Postsynaptic induction of BDNF-mediated long-term potentiation. Science, 295: 1729–1734.

Kumura, E., Kimura, F., Taniguchi, N. and Tsumoto, T. (2000) Brain-derived neurotrophic factor blocks long-term depression in solitary neurones cultured from rat visual cortex. J. Physiol., 524: 195 204.

Lee, F.S., Kim, A.H., Khursigara, G. and Chao, M.V. (2001) The uniqueness of being a neurotrophin receptor. Curr. Opin. Neurobiol., 11: 281–286.

Lee, R., Kermani, P., Teng, K.K. and Hempstead, B.L. (2002) Regulation of cell survival by secreted proneurotrophins. Science, 294: 1945–1948.

Lein, E.S., Finney, E.M., McQuillen, P.S. and Shatz, C.J. (1999) Subplate neuron ablation alters neurotrophin expression and ocular dominance column formation. Proc. Natl. Acad. Sci. USA, 96: 13491–13495.

Lein, E.S. and Shatz, C.J. (2000) Rapid regulation of brain-derived neurotrophic factor mRNA within eye-specific circuits during ocular dominance column formation. J. Neurosci., 20: 1470–1483.

Lessmann, V., Gottmann, K. and Heumann, R. (1994) BDNF and NT-4/5 enhance glutamatergic synaptic transmission in cultured hippocampal neurons. Neuroreport, 6: 21–25.

Lessmann, V. and Heumann, R. (1998) Modulation of unitary glutamatergic synapses by neurotrophin-4/5 or brain-derived neurotrophic factor in hippocampal microcultures: presynaptic enhancement depends on pre-established paired-pulse facilitation. Neuroscience, 86: 399–413.

Levine, E.S., Crozier, R.A., Black, I.B. and Plummer, M.R. (1998) Brain-derived neurotrophic factor modulates hippocampal synaptic transmission by increasing N-methyl-d-aspartic acid receptor activity. Proc. Natl. Acad. Sci. USA, 95: 10235–10239.

Levine, E.S., Dreyfus, C.F., Black, I.B. and Plummer, M.R. (1995) Brain-derived neurotrophic factor rapidly enhances synaptic transmission in hippocampal neurons via postsynaptic tyrosine kinase receptors. Proc. Natl. Acad. Sci. USA, 92: 8074–8077.

Lewin, G.R. and Barde, Y.-A. (1996) Physiology of the neurotrophins. Annu. Rev. Neurosci., 19: 289–317.

Li, H.S., Xu, X.Z. and Montell, C. (1999) Activation of a TRPC3-dependent cation current through the neurotrophin BDNF. Neuron, 24: 261–273.

Li, Y.X., Xu, Y., Ju, D., Lester, H.A., Davidson, N. and Schuman, E.M. (1998a) Expression of a dominant negative TrkB receptor, T1, reveals a requirement for presynaptic signaling in BDNF-induced synaptic potentiation in cultured hippocampal neurons. Proc. Natl. Acad. Sci. USA, 95: 10,884–10,889.

Li, Y.X., Zhang, Y., Lester, H.A., Schuman, E.M. and Davidson, N. (1998b) Enhancement of neurotransmitter release induced by brain-derived neurotrophic factor in cultured hippocampal neurons. J. Neurosci., 18: 10,231–10,240.

Lin, S.Y., Wu, K., Levine, E.S., Mount, H.T., Suen, P.C. and Black, I.B. (1998) BDNF acutely increases tyrosine phosphorylation of the NMDA receptor subunit 2B in cortical and hippocampal postsynaptic densities. Brain Res. Mol. Brain Res., 55: 20–27.

Linden, D.J. (1996) A protein synthesis-dependent late phase of cerebellar long-term depression. Neuron, 17: 483–490.

Liou, J.C. and Fu, W.M. (1997) Regulation of quantal secretion from developing motoneurons by postsynaptic activity-dependent release of NT-3. J. Neurosci., 17: 2459–2468.

Liou, J.C., Yang, R.S. and Fu, W.M. (1997) Regulation of quantal secretion by neurotrophic factors at developing motoneurons in Xenopus cell cultures. J. Physiol. (Lond), 503: 129–139.

Lodovichi, C., Berardi, N., Pizzorusso, T. and Maffei, L. (2000) Effects of neurotrophins on cortical plasticity: same or different? Neuroscience, 20: 2155–2165.

Lohof, A.M., Ip, N.Y. and Poo, M.M. (1993) Potentiation of developing neuromuscular synapses by the neurotrophins NT-3 and BDNF. Nature, 363: 350–353.

Lom, B. and Cohen-Cory, S. (1999) Brain-derived neurotrophic factor differentially regulates retinal ganglion cell dendritic and axonal arborization in vivo. J. Neurosci., 19: 9928–9938.

Lu, B. and Chow, A. (1999) Neurotrophins and hippocampal synaptic plasticity. J. Neurosci. Res., 58: 76–87.

Lu, B. and Figurov, A. (1997) Role of neurotrophins in synapse development and plasticity. Rev. Neurosci., 8: 1–12.

Maffei, L., Berardi, N., Domenici, L., Parisi, V. and Pizzorusso, T. (1992) Nerve growth factor (NGF) prevents the shift in ocular dominance distribution of visual cortical neurons in monocularly deprived rats. J. Neurosci., 12: 4651–4662.

Martinez, A., Alcantara, S., Borrell, V., Del Rio, J.A., Blasi, J., Otal, R., Campos, N., Boronat, A., Barbacid, M., Silos-Santiago, I. and Soriano, E. (1998) TrkB and TrkC signaling are required for maturation and synaptogenesis of hippocampal connections. J. Neurosci., 18: 7336–7350.

McAllister, A.K., Katz, L.C. and Lo, D.C. (1996) Neurotrophin regulation of cortical dendritic growth requires activity. Neuron, 17: 1057–1064.

McAllister, A.K., Katz, L.C. and Lo, D.C. (1997) Opposing roles for endogenous BDNF and NT-3 in regulating cortical dendritic growth. Neuron, 18: 767–778.

McAllister, A.K., Lo, D.C. and Katz, L.C. (1995) Neurotrophins regulate dendritic growth in developing visual cortex. Neuron, 15: 791–803.

McAllister, A.M., Katz, L.C. and Lo, D.C. (1999) Neurotrophins and synaptic plasticity. Annu. Rev. Neurosci., 22: 295–318.

McQuillen, P.S., DeFreitas, M.F., Zada, G. and Shatz, C.J. (2002) A novel role for p75NTR in subplate growth cone complexity and visual thalamocortical innervation. J. Neurosci., 22: 3580–3593.

Miller, S., Yasuda, M., Coats, J.K., Jones, Y., Martone, M.E. and Mayford, M. (2002) Disruption of dendritic translation of CaMKIIalpha impairs stabilization of synaptic plasticity and memory consolidation. Neuron, 36: 507–519.

Minichiello, L., Calella, A.M., Medina, D.L., Bonhoeffer, T., Klein, R. and Korte, M. (2002) Mechanism of TrkB-mediated hippocampal long-term potentiation. Neuron, 36: 121–137.

Minichiello, L., Korte, M., Wolfer, D., Kuhn, R., Unsicker, K., Cestari, V., Rossi-Arnaud, C., Lipp, H.P., Bonhoeffer, T. and Klein, R. (1999) Essential role for TrkB receptors in hippocampus-mediated learning. Neuron, 24: 401–414.

Morimoto, K., Sato, K., Sato, S., Yamada, N. and Hayabara, T. (1998) Time-dependent changes in neurotrophic factor mRNA expression after kindling and long-term potentiation in rats. Brain Res. Bull., 45: 599–605.

Narisawa-Saito, M., Carnahan, J., Araki, K., Yamaguchi, T. and Nawa, H. (1999) Brain-derived neurotrophic factor regulates the expression of AMPA receptor proteins in neocortical neurons. Neuroscience, 88: 1009–1014.

Narisawa-Saito, M., Iwakura, Y., Kawamura, M., Araki, K., Kozaki, S., Takei, N. and Nawa, H. (2002) Brain-derived neurotrophic factor regulates surface expression of alpha-amino-3-hydroxy-5-methyl-4-isoxazoleproprionic acid receptors by enhancing the N-ethylmaleimide-sensitive factor/GluR2 interaction in developing neocortical neurons. J. Biol. Chem., 277: 40901–40910.

Nick, T.A. and Ribera, A.B. (2000) Synaptic activity modulates presynaptic excitability. Nat. Neurosci., 3: 142–149.

Patterson, S., Grover, L.M., Schwartzkroin, P.A. and Bothwell, M. (1992) Neurotrophin expression in rat hippocampal slices: A stimulus paradigm inducing LTP in CA1 evokes increases in BDNF and NT-3 mRNAs. Neuron, 9: 1081–1088.

Patterson, S.L., Abel, T., Deuel, T.A., Martin, K.C., Rose, J.C. and Kandel, E.R. (1996) Recombinant BDNF rescues deficits in basal synaptic transmission and hippocampal LTP in BDNF knockout mice. Neuron, 16: 1137–1145.

Patterson, S.L., Pittenger, C., Morozov, A., Martin, K.C., Scanlin, H., Drake, C. and Kandel, E.R. (2001) Some forms of cAMP-mediated long-lasting potentiation are associated with release of BDNF and nuclear translocation of phospho-MAP kinase. Neuron, 32: 123–140.

Pizzorusso, T., Porciatti, V., Tseng, J.L., Aebischer, P. and Maffei, L. (1997) Transplant of polymer-encapsulated cells genetically engineered to release nerve growth factor allows a normal functional development of the visual cortex in dark-reared rats. Neuroscience, 80: 307–311.

Pollock, G.S., Vernon, E., Forbes, M.E., Yan, Q., Ma, Y.T., Hsieh, T., Robichon, R., Frost, D.O. and Johnson, J.E. (2001) Effects of early visual experience and diurnal rhythms on BDNF mRNA and protein levels in the visual system, hippocampus, and cerebellum. J. Neurosci., 21: 3923–3931.

Poo, M.M. (2001) Neurotrophins as synaptic modulators. Nat. Rev. Neurosci., 2: 24–32.

Pozzo-Miller, L., Gottschalk, W.A., Zhang, L., McDermott, K., Du, J., Gopalakrishnan, R., Oho, C., Shen, Z. and Lu, B. (1999) Impairments in high frequency transmission, synaptic vesicle docking and synaptic protein distribution in the hippocampus of BDNF knockout mice. J. Neurosci., 19: 4972–4983.

Prakash, N., Cohen-Cory, S. and Frostig, R.D. (1996) Rapid and opposite effects of BDNF and NGF on the functional organization of the adult cortex in vivo. Nature, 381: 702–706.

Riddle, D.R., Lo, D.C. and Katz, L.C. (1995) NT-4-mediated rescue of lateral geniculate neurons from effects of monocular deprivation. Nature, 378: 189–191.

Rossi, F.M., Bozzi, Y., Pizzorusso, T. and Maffei, L. (1999) Monocular deprivation decreases brain-derived neurotrophic factor immunoreactivity in the rat visual cortex. Neuroscience, 90: 363–368.

Rossi, F.M., Sala, R. and Maffei, L. (2002) Expression of the nerve growth factor receptors TrkA and p75NTR in the visual cortex of the rat: development and regulation by the cholinergic input. J. Neurosci., 22: 912–919.

Rutherford, L.C., Nelson, S.B. and Turrigiano, G.G. (1998) BDNF has opposite effects on the quantal amplitude of pyramidal neuron and interneuron excitatory synapses. Neuron, 21: 521–530.

Schacher, S., Kandel, E.R. and Montarolo, P. (1993) cAMP and arachidonic acid simulate long-term structural and functional changes produced by neurotransmitters in Aplysia sensory neurons. Neuron, 10: 1079–1088.

Schinder, A.F., Berninger, B. and Poo, M. (2000) Postsynaptic target specificity of neurotrophin-induced presynaptic potentiation. Neuron, 25: 151–163.

Schoups, A.A., Elliott, R.C., Friedman, W.J. and Black, I.B. (1995) NGF and BDNF are differentially modulated by visual experience in the developing geniculocortical pathway. Brain Res. Dev. Brain Res., 86: 326–334.

Schuman, E.M. (1999) Neurotrophin regulation of synaptic transmission. Curr. Opin. Neurobiol., 9: 105–109.

Schuster, C.M., Davis, G.W., Fetter, R.D. and Goodman, C.S. (1996a) Genetic dissection of structural and functional components of synaptic plasticity. I. Fasciclin II controls synaptic stabilization and growth. Neuron, 17: 641–654.

Schuster, C.M., Davis, G.W., Fetter, R.D. and Goodman, C.S. (1996b) Genetic dissection of structural and functional components of synaptic plasticity. II. Fasciclin II controls presynaptic structural plasticity. Neuron, 17: 655–667.

Sherwood, N.T. and Lo, D.C. (1999) Long-term enhancement of central synaptic transmission by chronic brain-derived neurotrophic factor treatment. J. Neurosci., 19: 7025–7036.

Silva, A.J., Kogan, J.H., Frankland, P.W. and Kida, S. (1998) CREB and memory. Annu. Rev. Neurosci., 21: 127–148.

Silver, M.A., Fagiolini, M., Gillespie, D.C., Howe, C.L., Frank, M.G., Issa, N.P., Antonini, A. and Stryker, M.P. (2001) Infusion of nerve growth factor (NGF) into kitten visual cortex increases immunoreactivity for NGF, NGF receptors, and choline acetyltransferase in basal forebrain without affecting ocular dominance plasticity or column development. Neuroscience, 108: 569–585.

Snider, W.D. and Lichtman, J.W. (1996) Are neurotrophins synaptotrophins? Mol. Cell. Neurosci., 7: 433–442.

Stoop, R. and Poo, M.M. (1995) Potentiation of transmitter release by ciliary neurotrophic factor requires somatic signaling. Science, 267: 695–699.

Stoop, R. and Poo, M.M. (1996) Synaptic modulation by neurotrophic factors: differential and synergistic effects of brain-derived neurotrophic factor and ciliary neurotrophic factor. J. Neurosci., 16: 3256–3264.

Suen, P.C., Wu, K., Xu, J.L., Lin, S.Y., Levine, E.S. and Black, I.B. (1998) NMDA receptor subunits in the post-synaptic density of rat brain: expression and phosphorylation by endogenous protein kinases. Brain Res. Mol. Brain Res., 59: 215–228.

Takei, N., Sasaoka, K., Inoue, K., Takahashi, M., Endo, Y. and Hatanaka, H. (1997) Brain-derived neurotrophic factor increases the stimulation-evoked release of glutamate and the levels of exocytosis-associated proteins in cultured cortical neurons from embryonic rats. J. Neurochem., 68: 370–375.

Takei, N., Kawamura, M., Hara, K., Yonezawa, K. and Nawa, H. (2001) Brain-derived neurotrophic factor enhances neuronal translation by activating multiple initiation processes: comparison with the effects of insulin. J. Biol. Chem., 276: 42818–42825.

Tanaka, T., Saito, H. and Matsuki, N. (1997) Inhibition of GABAa synaptic responses by brain-derived neurotrophic factor (BDNF) in rat hippocampus. J. Neurosci., 17: 2959–2966.

Tartaglia, N., Du, J., Neale, E., Tyler, W.J., Pozzo-Miller, L. and Lu, B. (2001) Protein synthesis dependent and independent regulation of hippocampal synapses by brain-derivedneurotrophic factor. J. Biol. Chem., 276: 37585–37593.

Tully, T., Preat, T., Boynton, S.C. and Del Vecchio, M. (1994) Genetic dissection of consolidated memory in Drosophila. Cell, 79: 35–47.

Tyler, W.J. and Pozzo-Miller, L.D. (2001) BDNF enhances quantal neurotransmitter release and increases the number of docked vesicles at the active zones of hippocampal excitatory synapses. J. Neurosci., 21: 4249–4258.

Vicario-Abejon, C., Collin, C., McKay, R.D. and Segal, M. (1998) Neurotrophins induce formation of functional excitatory and inhibitory synapses between cultured hippocampal neurons. J. Neurosci., 18: 7256–7271.

Villarreal, D.M., Do, V., Haddad, E. and Derrick, B.E. (2002) NMDA receptor antagonists sustain LTP and spatial memory: active processes mediate LTP decay. Nat. Neurosci., 5: 48–52.

Wang, T., Xie, K.W. and Lu, B. (1995) Neurotrophins promote maturation of developing neuromuscular synapses. J. Neurosci., 15: 4796–4805.

Wang, X., Berninger, B. and Poo, M. (1998) Localized synaptic actions of neurotrophin-4. J. Neurosci., 18: 4985–4992.

Wang, X.H. and Poo, M.M. (1997) Potentiation of developing synapses by postsynaptic release of neurotrophin-4. Neuron, 19: 825–835.

Xie, K., Wang, T., Olafsson, P., Mizuno, K. and Lu, B. (1997) Activity-dependent expression of NT-3 in muscle cells in culture: implication in the development of neuromuscular junctions. J. Neurosci., 17: 2947–2958.

Xu, B., Gottschalk, W., Chow, A., Wilson, R.I., Schnell, E., Zang, K., Wang, D., Nicoll, R.A., Lu, B. and Reichardt, L.F. (2000a) The role of brain-derived neurotrophic factor receptors in the mature hippocampus: modulation of long-term potentiation through a presynaptic mechanism involving TrkB. J. Neurosci., 20: 6888–6897.

Xu, B., Zang, K., Ruff, N.L., Zhang, Y.A., McConnell, S.K., Stryker, M.P. and Reichardt, L.F. (2000b) Cortical degeneration in the absence of neurotrophin signaling: dendritic retraction and neuronal loss after removal of the receptor TrkB. Neuron, 26: 233–245.

Yacoubian, T.A. and Lo, D.C. (2000) Truncated and full-length TrkB receptors regulate distinct modes of dendritic growth. Nat. Neurosci., 3: 342–349.

Yang, F., He, X., Feng, L., Mizuno, K., Liu, X., Russell, J., Xiong, W. and Lu, B. (2001) PI3 kinase and IP3 are both necessary and sufficient to mediate NT3-induced synaptic potentiation. Nat. Neurosci., 4: 19–28.

Zhang, X. and Poo, M.M. (2002) Localized synaptic potentiation by BDNF requires local protein synthesis in the developing axon. Neuron, 36: 675–688.

Zhou, J., He, X., Wang, Y.-T., Xiong, W.C. and Lu, B. (2001) Role of receptor internalization and PI-3 Kinase in NT3-induced long-term synaptic potentiation at Xenopus neuromuscular junction. Soc. Neurosci. Abst., 27: 45.42.

Progress in Brain Research, Vol. 146
ISSN 0079-6123

CHAPTER 11

Neurotrophic factors and CNS disorders: findings in rodent models of depression and schizophrenia

Francesco Angelucci[1],*, Aleksander A. Mathé[2] and Luigi Aloe[3]

[1]*Institute of Neurology, Catholic University, Rome, Italy*
[2]*Department of Pharmacology, Karolinska Institute, Stockholm, Sweden*
[3]*Institute of Neurobiology and Molecular Medicine, CNR, Rome, Italy*

Abstract: Nerve growth factor (NGF) and brain-derived neurotrophic factor (BDNF) are proteins involved in neuronal survival and plasticity of dopaminergic, cholinergic and serotonergic neurons in the central nervous system (CNS). Loss of neurons in specific brain regions has been found in depression and schizophrenia, and this chapter summarizes the findings of altered neurotrophins in animal models of those two disorders under baseline condition and following antidepressive and antipsychotic treatments. In a model of depression (Flinders sensitive line/Flinders resistant line; FSL/FRL rats), increased NGF and BDNF concentrations were found in frontal cortex of female, and in occipital cortex of male 'depressed' FSL compared to FRL control rats. Using the same model, the effects of electroconvulsive stimuli (ECS) and chronic lithium treatment on brain NGF, BDNF and glial cell line-derived neurotrophic factors were investigated. ECS and lithium altered the brain concentrations of neurotrophic factors in the hippocampus, frontal cortex, occipital cortex and striatum. ECS mimic the effects of electroconvulsive therapy (ECT) that is an effective treatment for depression and also schizophrenia. Since NGF and BDNF may also be changed in the CNS of animal models of schizophrenia, we investigated whether treatment with antipsychotic drugs (haloperidol, risperidone, and olanzapine) affects the constitutive levels of NGF and BDNF in the CNS. Both typical and atypical antipsychotic drugs altered the regional brain levels of NGF and BDNF. Other studies also demonstrated that these drugs differentially altered neurotrophin mRNAs. Overall, these studies indicate that alteration of brain level of NGF and BDNF could constitute part of the biochemical alterations induced by antipsychotic drugs.

Keywords: NGF; BDNF; GDNF; electroconvulsive stimuli; lithium; Flinders Sensitive line; antipsychotic drugs

Introduction

Depression and schizophrenia are two major psychiatric disorders causing severe personal suffering and disability and imposing major strains on society. At the present time, the most convincing and apparently largest risk factor for schizophrenia is a genetic loading for the disorder. Additional risk

factors of moderate strength are represented by complications of pregnancy and birth, viral epidemics and other adverse circumstances (see Wright et al., 1995). In the last few years, a number of studies have consistently shown that a structural neuropathology exists in a substantial proportion of schizophrenic subjects. One of the best-confirmed structural findings in schizophrenia research indicates cytoarchitectural abnormalities in the entorhinal and anterior cingulate cortex. Indeed, experimental findings showed that interruption of cell proliferation during early corticogenesis induces marked damage to neocortical regions in the brain, such as the

*Correspondence to: F. Angelucci, Institute of Neurology, Catholic University, Largo Gemelli 8, I-00168, Rome, Italy. Tel.: +39-0328-6904589; E-mail: angeluccifrancesco@rm.unicatt.it

DOI: 10.1016/S0079-6123(03)46011-1

entorhinal cortex, anterior cingulate, and prefrontal regions. These reductions may also lead to an altered synthesis and release of neurotrophic factors. Studies in animal models indicated that entorhinal and cortical maldevelopment could lead to behavioral impairment that may be related to schizophrenic symptoms (Talamini et al., 1998; Arnold, 2000; Lewis and Levitt, 2002). Recent studies of neurodevelopmental mechanisms suggest that a combination of genetic susceptibility and environmental perturbations appears to be necessary for the expression of some psychotic disorders. Thus, altered synthesis and/or release of neurotrophic factors during specific developmental period may affect the pathological manifestation of these disorders (Shintani, 1999). Neurotrophic factors were first identified as regulators of neural growth and differentiation during development, but are now known to be potent regulators of plasticity and survival of adult neurons and glia. In collaboration with other groups we have recently been involved in these studies.

Neurotrophins and behavior

The first study on the possible role of NGF in behavioral response goes back to 1986 when it was demonstrated that aggressive behavior induced by 6–8 weeks of isolation in laboratory animals caused an increase in NGF in the bloodstream and brain (Aloe et al., 1986). This observation was later extended to other rodents and in human subjects experiencing anxiety induced by drug abstinence or parachute jumping. The results showed that the release of NGF was associated with anxiety-like behavior rather than just aggressiveness (Aloe et al., 1994, 1996).

Subsequent studies revealed that environmental changes (Aloe et al., 1990; Cirulli, 2001; Haddjiconstantinou et al., 2001), depression, and electroconvulsive therapy (ECT) also lead to a significant alteration of brain and blood NGF levels (Aloe et al., 2000). Moreover, Gall and Isackson (1989) discovered a link between NGF and convulsion by demonstrating that there is an increase of $mRNA^{NGF}$ in animal soon after a limbic seizure. Moreover, other studies revealed that pentylenetetriczole injected i.p. (Watson and Milbrandt, 1989; Mack et al., 1995), kainic acid injected into the lateral ventricle (Gall et al., 1991), bicuculline injected into the prepiriform cortex (Riva et al., 1992), NMDA injected into the ventricles or into the hippocampus (Gwag et al., 1993), and quinolonic acid injected into the left dorsal hippocampus (Rocamora et al., 1994) all lead to an increase of $mRNA^{NGF}$. Similar results were obtained following unilateral electrolytic lesions of the dentate gyrus hilum (Gall and Isackson, 1989; Isackson et al., 1991; Lauterborn et al., 1994), by implanting steel electrodes into the CA1-CA3 hippocampal areas (Ernfors et al., 1991; Bengzon et al., 1993), in the basal-lateral amygdala (Sato et al., 1996), and with low (50–70 mA) and high (150 mA) intensity electrostimuli (Follesa et al., 1994). These studies also showed that the increase of $mRNA^{NGF}$ occured within 30 min to 24 h after induction of convulsion. The brain areas where the increase of the messenger could be identified were the granular cells of the dentate gyrus, olfactory cortex and neocortex (layers II and III), pyriform cortex, pyramidal layer of the parietal cortex, CA1-CA3 regions of the amygdala, granular layer of the hippocampus, entorhinal cortex, most of the limbic areas, the cerebellum and the mid-brain.

To further investigate the relationship between antipsychotic effects and brain neurotrophin synthesis and release, the effects of halonperidol were studied in animal models and in humans. The results showed that administration of haloperidol, an antipsychotic, sedation-inducing drug, lowered NGF levels in the bloodstream and hypothalamus of adult mice (Alleva et al., 1996) and the circulating NGF levels in schizophrenic patients (Strange, 2001). These results are in line with the hypothesis that one of the effects of the antipsychotic drugs is to lower the constitutive presence of NGF in human plasma and in the brain as well. Whether the lower NGF plasma levels after haloperidol are due to a reduced synthesis of NGF or an enhanced uptake in the NGF-responsive cells in the peripheral or central nervous system (CNS) remains a matter for further investigations.

The rationale to study neurotrophins in CNS disorders is based on studies of animal models of altered brain neurogenesis which provided evidence that impaired cell migration in specific brain regions altered the distribution of brain protein and gene expression.

In the present review, we present data obtained using animal models which suggest that neurotrophic factors, particularly NGF and BDNF, might be implicated in mechanism(s) of action of antipsychotic and antidepressive treatment(s).

Depression and neurotrophins

Converging evidence from epidemiological studies points to three important factors contributing to the precipitation of depressive syndromes: Genetic factors, prenatal or early postnatal environmental factors, and life events (stress), such as grief, unemployment, relational breaks, death or withdrawal from drugs or abuse. Although these events are associated with an altered presence of neurotrophins in the blood and brain areas (Smith et al., 1995; Aloe et al., 1996; Shintani, 1999; Aloe et al., 2002) their role in the development of depression remains unknown.

In the past few years, results indicating that neurotrophins, particularly BDNF, could be involved in the pathophysiology of depression and stress-related affective disorders have been reported. Thus, chronic stress has been shown to cause atrophy and death of CA3 neurons in the hippocampus (Sapolsky, 1990; Magarinos et al., 1996). In addition, stress decreases the expression of BDNF in CA3 pyramidal and dentate gyrus granule cell layers in the hippocampus (Smith et al., 1995). Thus, downregulation of BDNF could contribute to the atrophy of CA3 neurons, or render these neurons more susceptible to other factors, such as adrenal glucocorticoids, that are induced in response to repeated stress. Lastly, brain imaging studies have reported that there is a volume reduction in the hippocampus of patients with depression or posttraumatic stress disorder (Sapolsky, 1990; Sheline et al., 1999; Bremner et al., 2000). Atrophy and/or decreased function of the hippocampus could explain the loss, observed in depressed patients, of the negative feedback control that this brain region

exerts on the hypothalamic-pituitary-adrenal axis (Young et al., 1990). Disturbance in the brain 5-hydroxytryptamine (5-HT, serotonin) system has been implicated in anxiety disorders, bulimia, chronic impulsive/aggressive behavior, and violent suicide (Arango et al., 2002) and these disorders are treated with compounds that augment 5-HT neurotransmission in the brain (Vaswani et al., 2003). It has recently been shown that BDNF influences the phenotype, structural plasticity, and survival of central serotonergic neurons (Siuciak et al., 1994). Moreover, pharmacological studies indicate that administration of BDNF has trophic effects on 5-HT neurons (Mamounas et al., 1995).

Upregulation of BDNF in response to antidepressant treatment could have similar behavioral effects, and could enhance 5-HT neurotransmitter function. These findings also indicate that there is a positive, reciprocal interaction between 5-HT and BDNF; chronic SSRI treatment increases levels of BDNF, and upregulated BDNF would be expected to increase 5-HT neuronal function. Supporting this hypothesis, chronic infusion of BDNF is reported to have antidepressant effects in the forced swim test and the learned helplessness paradigm (Siuciak et al., 1997). However, the role of this and other neurotrophins in the normal development and function of 5-TH neurons has not been elucidated.

Other studies have shown decreased levels of BDNF in patients with major depression (Karege et al., 2002) and an increased BDNF immunoreactivity in the hippocampus of patients treated with antidepressant drugs (Chen et al., 2001b). In addition to monoamines, several neuropeptides, including opioid peptides (Green et al., 1978; Emrich, 1984) and neuropeptide tyrosine (NPY) (Widerlöv et al., 1988; Wahlestedt et al., 1989), are thought to play a role in clinical depression or learned helplessness behavior in rats. Several recent papers have examined the effects of central administration of BDNF on neuropeptide distribution (Croll et al., 1994; Nawa et al., 1994). Therefore, the modulation of neuropeptide systems by BDNF, whether a direct or indirect effect, may contribute to the antidepressant-like effects of this protein. This raises the question as whether depression might be considered a neurodegenerative disorder (Altar, 1999).

Schizophrenia and neurotrophins

A number of studies published over the last few years suggest that the onset of schizophrenia in adulthood may be the consequence of early alterations during neurodevelopment, defined as neurodevelopmental encephalopathy (Fish, 1977; Weinberger and Lipska, 1995). According to this hypothesis, schizophrenia may be viewed as a general disorder of brain maturation and organization that appears phenotypically before the appearance of classical symptoms. Possible risk factors (in addition to genetic predisposition) could be operating in pregnancy and during the delivery (viral infections, nutritional disorders, alcohol or other drugs intake, obstetric complications) in the early stages of life as well as, later on, environmental stressful events (Jackob and Beckman, 1986; Sham et al., 1992; Spohr et al., 1993; Davis et al., 1995).

Recently, a hypothesis of schizophrenia as an alteration in neurodevelopment characterized by a disordered cell migration and a neuronal and glial disconnection has focused attention on the possible role played by neurotrophic factors, mainly NGF, BDNF, and NT-3.

NT-3 probably plays an important role in the immature CNS because it is a neurotrophin particularly present in those areas where the proliferation, differentiation and migration of neuronal precursors have origin. Moreover, mRNA^{NT-3} levels in the hippocampus of newborn animals are much higher than those of other neurotrophins in different brain regions (Maisonpierre et al., 1990). Studies in schizophrenic patients have demonstrated a significant presence of the A3 allele, which is located in the promoter region of the gene for the NT-3, compared with healthy controls. Genotype A3 was more frequent in schizophrenic patients in the homozygotic variant as well as in the heterozygotic variant and the risk of developing schizophrenia was equal to 2.4 in subjects where such polymorphism was present (Nanko et al., 1994). The authors of the study continued their research by demonstrating a high frequency of homo- and heterozigosis for the Glu-63 allele instead of Gly-63 in the AP-1 region of the gene for NT-3, in a selected sample of schizophrenic patients whose severe illness was due to early alterations in development was reported (Hattori and Nanko, 1995).

Findings that total brain tissues display a reduced presence of BDNF (Freedman et al., 1992, 1994) give additional support to the hypothesis of nuerotrophin involvement in schizophrenia. Further results from animal models of schizophrenia revealed that NGF synthesis and/or release, and most likely NGF receptor on target cells, are also altered in the CNS (Talamini et al., 1998).

NGF and BDNF in an animal model of depression

The observations that BDNF produces an antidepressant-like effect when injected into the brain (Siuciak et al., 1997) and that stress (Aloe et al., 1994) leads to changes of constitutive levels of neurotrophins (Ueyama et al., 1997) raise the possibility that alterations in neurotrophin production and/or function are instrumental in the pathophysiology of depression (Altar, 1999). To address this question, the brain levels of BDNF and NGF were measured in an animal model of depression, the Flinders Sensitive Line (FSL) rats, and their control counterparts, the Flinders Resistant Line (FRL). FSL animals had higher levels of both BDNF and NGF in the frontal cortex and occipital cortex, whereas in the hypothalamus only the levels of BDNF were affected (Angelucci et al., 2000a). These results are not in line with data obtained by others showing that the levels of neurotrophins are decreased in depression (Altar, 1999), but are consistent with the observations that antidepressant drugs increase mRNA^{BDNF} in the brain via 5-HT$_2$ and β-adrenoceptor (Duman et al., 1995) and that BDNF itself has an antidepressant effect (Siuciak et al., 1997). Specifically, an accumulation of neurotrophins in brain areas involved in cognition and volition, as well as states of hedonia/anhedonia, could reflect a decreased release/breakdown or a compensatory increase in neurotrophin synthesis in response to primary neurotransmitter and neuropeptides changes in the limbic system during depression (Duman et al., 1997). Thus the increase of NGF and BDNF might be involved in the pathophysiology of the disease or constitute a part of an adaptive response to it.

Since gender differences in neurotransmitters and neuropeptides both in the CNS and periphery have been described (De Vries, 1990), it was of interest to explore whether the same holds true for neurotrophins. Moreover, since women exhibit higher frequency of and greater susceptibility to depression (Weissman and Olfson, 1995; Maier et al., 1999), we investigated whether neurotrophins might be selectively changed in brains of FSL female rats. The results indicated that the male and female brains, irrespective of the strain, contain different levels of neurotrophins, at least in some brain regions (Angelucci et al., 2000a). Though the functional significance of these findings remains to be established, the data indicate that BDNF and NGF or their reported interaction with estrogens (Solum and Handa, 2002) may be involved in susceptibility to depression.

NGF and BDNF brain levels are altered by electroconvulsive stimuli in a rat model of depression

In order to further understand the machanisms of action of antidepressive treatments, e.g. possible effects of repeated electroconvulsive stimuli (ECS) on the brain regional concentrations of NGF and BDNF (Angelucci et al., 2003a) were investigated in the 'depressed' FSL animals. ECS increased NGF concentration in the hippocampus of FSL rats, but no changes were found in FRL or Sprague Dawley rats (Fig. 1). These data suggest that ECS selectively alter NGF in the hippocampus of 'depressed' rats. ECS also decreased NGF concentration in the frontal cortex of FSL rats. Whether the decrease in NGF synthesis is relevant to the pathophysiology of depression is not known. However, since a reduced functionality of medial prefrontal cortex has been observed in depressed patients (Drevets, 2000), the possibility exists that neurotrophin synthesis and release could also be altered in depression. Since striatal cholinergic neurons utilize NGF produced in the cortex, the increase in NGF levels observed after ECS in the striatum in both FSL and FRL rats suggests that ECS may alter NGF production and transport in the cortico-striatal pathway.

While ECS did not change BDNF concentration in the hippocampus of FSL and FRL rats (Fig. 2), it increased BDNF concentration in Sprague Dawley rats (Fig. 3). The reason for these different changes is not clear. In Sprague Dawley rats these data are in line with other studies showing that BDNF mRNA is increased by ECS (Duman and Vaidya, 1998) and BDNF can lead to the recovery of behavioral deficits in animal models of depression (Siuciak et al., 1997). Together these findings may indicate that in FSL and FRL rats ECS induces increased both BDNF gene expression and release metabolism, thus resulting in no apparent change in the protein level.

ECS alter NGF, BDNF, and GDNF brain levels in Sprague Dawley rats

To further characterize the role of neurotrophic factors in the action of ECS, we investigated the effects of repeated ECS on NGF, BDNF and GDNF brain concentrations in Sprague Dawley rats from which the two Flinders lines were originally bred (Angelucci et al., 2002). The results of these studies revealed that ECS increased NGF in the frontal cortex and BDNF in the hippocampus and striatum (Fig. 3). Thus, NGF concentrations changed only in the brain region where BDNF was unaffected, indicating a region specific action of ECS on these neurotrophins and no overlapping effects. These data are consistent with observations that acute and chronic ECS treatments lead to increased expression of mRNA^{NGF} (Follesa et al., 1994), mRNA^{BDNF} and BDNF receptor mRNA (Nibuya et al., 1995) in limbic brain regions.

ECS decreased GDNF in hippocampus and striatum (Angelucci et al., 2002). This effect is consistent with previous results showing that acute and chronic ECS enhance the expression of the GDNF receptor GFRα-1 and GFRα-2 mRNAs in the dentate gyrus of the rat hippocampus (Chen et al., 2001a). It is of interest that following ECS the levels of BDNF and GDNF, two growth factors affecting dopaminergic neurons, change in opposite direction in the hippocampus and striatum. One possible explanation is that although BDNF and GDNF and their receptors are present in the same ventral mesencephalic dopaminergic neurons (Engele, 1998), the two factors may have different signaling mechanisms, as reported in cultured ventral mesencephalic neurons (Feng et al., 1999a,b).

Effects of ECS and lithium on NGF in a rat model of depression

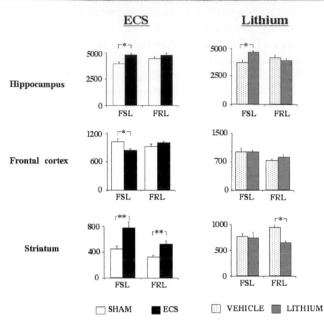

Fig. 1. Effects of electroconvulsive stimuli (ECS) and chronic lithium treatment on NGF brain concentration in FSL and FRL rats. Asterisks indicate significant between-group differences (*$p < 0.05$, **$p < 0.01$).

Effects of ECS and lithium on BDNF in a rat model of depression

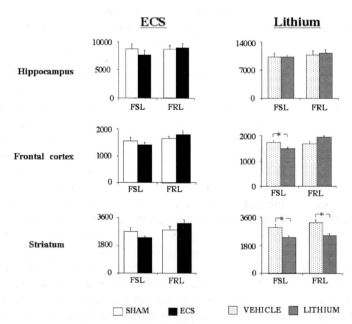

Fig. 2. Effects of electroconvulsive stimuli (ECS) and chronic lithium treatment on BDNF brain concentration in FSL and FRL rats. Asterisks indicate significant between-group differences (*$p < 0.05$).

Effects of ECS on NGF and BDNF in Sprague Dawley rats

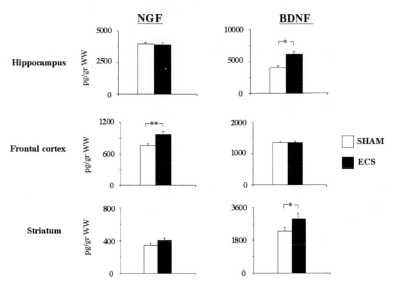

Fig. 3. Effects of electroconvulsive stimuli (ECS) on NGF and BDNF brain concentration in Sprague Dawley rats. Asterisks indicate significant between-group differences (*$p < 0.05$, **$p < 0.01$).

NGF, BDNF, and GDNF brain levels are altered by chronic lithium treatment in a rat model of depression

Lithium, probably the most effective mood stabilizer, has been shown to induce an increase in the expression of NGF and BDNF in the hippocampus and cortex of rats (Fukumoto et al., 2001; Hellweg et al., 2002). Using the FSL and the FRL rats, the effect of chronic lithium treatment on brain NGF, BDNF, and GDNF was studied. Lithium was administered as food supplementation during a 6–9 week period. The results demonstrate that chronic lithium treatment significantly altered the constitutive brain concentrations of NGF and BDNF (Figs. 1 and 2) and of GDNF (Angelucci et al., 2003b). In addition, the brain levels of these neurotrophic factors were differentially affected in FSL, compared to the FRL rats (Figs. 2 and 3). Indeed, in a recent paper (Fukumoto et al., 2001) it was reported that lithium and valproate treatment for 14 and 28 days increased the concentration of BDNF in the hippocampus, frontal cortex and temporal cortex of Wistar rats. These additional data demonstrated that lithium also alters the concentration of NGF and

GDNF, NGF was increased in the hippocampus of FSL rats (Fig. 1) while GDNF was decreased in the hippocampus of FRL rats (Angelucci et al., 2003b). In the striatum, lithium decreased BDNF in both strains (Fig. 2). In contrast to the study cited above (Fukumoto et al., 2001), it was observed a decrease in BDNF in the frontal cortex after lithium (Fig. 2). This apparent discrepancy could be due to the fact that healthy Wistar rats were used in that study but not 'depressed' FSL rats. Moreover, Fukumoto et al. reported an increase in BDNF after 14 days, but not at the end of the treatment, i.e., 28 days, while in the present study the treatment time was longer, i.e., 42 days. Although the significance of a decrease in BDNF is not clear, it is consistent with the notion that a dysfunction in medial prefrontal cortex areas may be implicated in phenomenon of anhedonia and perhaps pathogenesis of depressive symptoms (Drevets, 2000).

In the FRL rats, lithium increased GDNF concentrations in the frontal cortex, but decreased it in the hippocampus (Angelucci et al., 2003b). In the only other study that we could find (Fukumoto et al., 2001), no significant effects of lithium on GDNF were observed. The reasons for this discrepancy could be the same as mentioned above. Moreover, it is possible that GDNF changes are the result of a subsequent adaptation phenomena and not a primary cellular response to lithium.

Neurotrophin brain levels are altered following administration of antipsychotic drugs

Recent studies have raised the possibility that NGF and BDNF are abnormally regulated in the CNS of animal models of schizophrenia (Aloe et al., 2000; Fiore et al., 2000). These findings and the fact that ECT is also efficient in the treatment of schizophrenia suggest that antipsychotic drugs might alter neurotrophins. To test this hypothesis, the effects of chronic treatment with haloperidol, risperidone and olanzapine on NGF and BDNF concentrations in the rat brain were investigated. Haloperidol decreased NGF in the hippocampus striatum (Fig. 4). Likewise, risperidone reduced NGF level in the hippocampus (Fig. 4). In the same study (Angelucci et al., 2000c), choline acetyltransferase (ChAT) immunoreactivity in the neurons of the medial septum and Meynert basal nucleus was analyzed. Haloperidol and risperidone reduced the number of large-size ChAT-immunopositive neurons.

Since the hippocampus is involved in the regulation of learning and memory processes, the decrease in NGF hippocampal levels after treatment with haloperidol and risperidone raises the possibility of a functional correlation between low availability of neurotrophins and cognitive deficits observed in schizophrenic patients. This hypothesis is supported by the observation that the reduction of ChAT-immunorcactivity in the medial septum after haloperidol and risperidone is positively correlated with the decrease in NGF in the hippocampus and striatum (Fig. 4).

In the same study, haloperidol increased NPY in the occipital cortex, whereas risperidone increased NPY in the hypothalamus, occipital cortex, hippocampus and in a nonsignificant manner in the frontal cortex (Angelucci et al., 2000c). As mentioned in the 'Introduction', NGF can regulate NPY in the CNS. However, since antipsychotic drugs differently change NPY and NGF concentrations the data

seem to suggest that the mechanism by which haloperidol and risperidone alter NPY is not related to that of NGF. Haloperidol and risperidone decreased the basal levels of BDNF in frontal and occipital cortex and hippocampus. Moreover, haloperidol significantly decreased TrkB immunoreactivity in neurons of the hippocampus, substantia nigra and ventral tegmental area (Angelucci et al., 2000b). Risperidone also induced similar changes, though the differences were not statistically significant. The possibility that a higher risperidone dose would have resulted in some changes as seen after halosperidol cannot be excluded. The results also do not exclude the possibility that haloperidol and risperidone act on different target cells, or alternatively that these drugs influence BDNF-producing and/or responsive cells in a different manner. Speculatively, this could be due to the risperidone's lower D_2 antagonist potency combined with 5-HT_2 receptor-blocking properties. The atypical drug olanzapine also altered the brain concentrations of neurotrophins (Fig. 4). Chronic treatment with olanzapine decreased BDNF in the hippocampus (Angelucci et al., 2000b,c) implying existence of a common effect of antipsychotic drugs on neurotrophins. Olanzapine also reduced BDNF in the frontal cortex. Since olanzapine antagonizes 5-HT_{2A} receptors in the frontal cortex (Kapur et al., 1998; Bymaster et al., 1999), one possibility is that it could reduce mRNA^{BDNF} via this mechanism. This hypothesis is in line with the findings that activation of 5-HT_{2A} receptor in the cortex results in enhanced presynaptic release of glutamate, stimulation of AMPA receptors and subsequent increase in mRNA^{BDNF} synthesis (Vaidya et al., 1997).

In contrast to haloperidol and risperidone, olanzapine treatment increased NGF in the hippocampus (Fig. 4). Since treatment with olanzapine is associated with an anticholinergic action at the muscarinic receptor (Bymaster et al., 1999; Moore, 1999), it is possible that olanzapine may alter the synthesis and release of NGF in the target regions of basal forebrain cholinergic neurons. This hypothesis is supported by findings that ChAT-immunoreactivity, which is regulated by NGF, decreases in the septum and nucleus basalis of Meynert after haloperidol and risperidone administration (Angelucci et al., 2000c).

Effects of antipsychotic drugs on NGF and BDNF

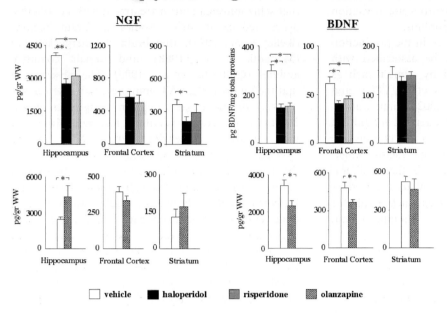

Fig. 4. Effects of antipsychotic drugs (haloperidol, risperidone, and olanzapine) treatment on NGF and BDNF concentration in rat brain. Asterisks indicate significant between-group differences (*$p < 0.05$, **$p < 0.01$).

Conclusions

ECS- and lithium-induced alterations in neurotrophin brain levels

The data presented here indicate that treatment with ECS leads to alteration in the synthesis of brain neurotrophins. It has been reported that ECS given to laboratory animals can affect neuronal morphology by increasing the sprouting of mossy fiber pathway and the number of granule cells in the hippocampus (Duman and Vaidya, 1998). These events can lead to establishment of new synaptic connections and nerve cell remodeling that may, in part, explain the clinical benefit of ECT (Sackeim et al., 1983). In animals, ECS can increase neurogenesis in adult rat hippocampus (Malberg et al., 2000) and, if also true in humans, contribute to the therapeutic actions of antidepressants. An enhanced availability of neurotrophic factors in specific brain regions might lead to longer neuronal survival, and consequently could be of relevance for the action of ECT (Duman et al., 1998; Jacobs et al., 2000). Moreover, since depressed patients also show memory deficits (Mitchell and Dening, 1996), it can

be hypothesized that ECT, as well as other therapeutic approaches, can reverse the atrophy of stress-vulnerable hippocampal neurons (Smith et al., 1995), most probably via regulation of neurotrophic factors (Duman and Vaidya, 1998). In our rat model of depression we observed opposite changes in NGF in frontal cortex and striatum after ECS. Since ECS increased NGF levels in the striatum, and NGF is transported from the cortex and utilized as trophic support for the cholinergic neurons it is possible that this increase is secondary to the reduced levels observed in the frontal cortex of FSL rats.

It has been previously demonstrated that lithium alters the brain levels of several neuropeptides, including NPY, neurokinin A (NKA), substance P (SP), and calcitonin gene-related peptide (CGRP) (Mathé et al., 1990; Mathé et al., 1994). Since NPY, NKA, SP and CGRP are, in part, regulated by neurotrophic factors the possibility arises that the action of lithium on neuropeptides is mediated, *inter alia*, by neurotrophic factors. Lithium increases the levels of neuroprotective proteins, such as Bcl-2 (Manji et al., 2000), and downregulates the proapoptotic gene Bax (Chen and Chuang, 1999) in the rat brain. As an increased synthesis of NGF in the hippocampus seems to be neuroprotective/antiapoptotic, it is conceivable that lithium exerts its neurotrophic/neuroprotective action through NGF and other neurotrophins.

It is of interest that an increase in the production of neurotrophic factors might be associated with increased neurogenesis induced by stimuli such as environmental enrichment (Ickes et al., 2000) and dietary restriction (Lee et al., 2002). Since antidepressant treatments increase neurogenesis in adult rat hippocampus (Malberg et al., 2000) it is conceivable that increased levels of neurotrophic factors might promote neurogenesis by increasing the proliferation of progenitor cells and/or by inducing their differentiation into neurons. Indeed, such a role for BDNF has been proposed by Fukumoto and coworkers (2001).

Neurotrophin level alterations associated with treatment with antipsychotic drugs

Disturbed neural development has been postulated as an important factor in the speculatively, pathophysiology of schizophrenia. Thus, speculatively, alterations in NGF and BDNF brain concentrations could in part explain the brain maldevelopment, which in postnatal and adult life can lead to the neuropathological manifestation of schizophrenic-like disorders. This hypothesis is supported by recent findings indicating that in schizophrenic patients BDNF protein is elevated in the anterior cingulate cortex and hippocampus (Takahashi et al., 2000; Durany et al., 2001) and mRNA^{BDNF} is reduced in the hippocampus and prefrontal cortex (Brouha et al., 1996; Kittell et al., 1997). Thus, the neurotrophin hypothesis proposes that alterations in neurotrophic factors expression could be responsible for neural maldevelopment and disturbed neural plasticity (Thome et al., 1998).

As shown in Fig. 4, haloperidol and risperidone reduced NGF and BDNF in the hippocampus, a brain region markedly affected in schizophrenia (Falkai and Bogerts, 1986; Weinberger, 1999). Since NGF and BDNF are known to play an important role in cognitive processes (Lo, 1995; Thoenen, 1995), it is possible that prolonged treatment with antipsychotic drugs might negatively influence

these processes. Indeed, recent studies indicate that schizophrenia and/or treatment with neuroleptic drugs reduce dendritic spines in brain regions (Kelley et al., 1997), alter brain cholinergic activity (Mahadik et al., 1988), and impair cognitive abilities (Gallhofer et al., 1996). The observation that the number of ChAT immunoreactive neurons is reduced in the basal forebrain (Angelucci et al., 2000c) is consistent with these findings. Moreover, the fact that haloperidol reduced NGF in the striatum suggests that this drug may interfere with the trophic effect of NGF on the septo-hippocampal pathway (Seiler and Schwab, 1984; Rylett and Williams, 1994). Previous studies showed that treatment with risperidone causes less severe deficit in patients' ability to perceive emotion, compared to haloperidol (Kee et al., 1998; Kern et al., 1998). However, whether the effect of risperidone on cognitive processes is linked to a lower efficacy of risperidone in reducing the brain level of neurotrophins remains to be demonstrated. The observation that in the frontal cortex and occipital cortex BDNF levels were reduced by antipsychotics suggests that the effects of chronic treatment with these drugs on cognitive processes may in part be mediated by neurotrophins.

Cumulatively, our data show that chronic antipsychotic treatment alters the levels of NGF and BDNF and their mRNAs in selected brain regions. Thus, it is possible that the different clinical profiles of antipsychotic drugs could be due to different effect of these drugs on neurotrophin distribution in the brain. These results suggest that evaluation of antipsychotic drugs should also include their possible effects on molecules, such as neurotrophins, involved in neuronal plasticity and neuroprotection.

Acknowledgments

Supported by the Swedish Medical Research Council grant 10414, the Italian National Research Council (CNR), fellowship bando n. 203.04.21, and the Karolinska Institutet. This study was supported in part by Fondazione CARISBO, Bologna, Italy and PF 2000 ISS (ICG 120/4RA 00-90) to Luigi Aloe.

Abbreviations

BDNF	brain-derived neurotrophic factor
CGRP	calcitonin gene-related peptide
ChAT	choline acetyltransferase
CNS	central nervous system
ECS	electroconvulsive stimuli
ECT	electroconvulsive treatment
FRL	Flinders resistant line
FSL	Flinders sensitive line
GDNF	glial cell line-derived neurotrophic factor
GFR	GDNF receptor
5-HT	5-hydroxytryptamine (serotonin)
NGF	nerve growth factor
NKA	neurokinin A
NPY	neuropeptide tyrosine
SP	substance P
TrkA	tyrosine kinase receptor A
TrkB	tyrosine kinase receptor B

References

Alleva, E., Della Seta, D., Cirulli, F. and Aloe, L. (1996) Haloperidol treatment decreases nerve growth factor levels in the hypothalamus of adult mice. Prog. Neuropsychopharmacol. Biol. Psychiatry, 20: 483–489.

Aloe, L., Alleva, E., Böhm, A. and Levi-Montalcini, R. (1986) Aggressive behavior induces release of nerve growth factor from mouse salivary gland into the bloodstream. Proc. Natl. Acad. Sci. USA, 83: 6184–6187.

Aloe, L., Alleva, E. and De Simone, R. (1990) Changes of NGF level in mouse hypothalamus following intermale aggressive behavior: biological and immunohistochemical evidence. Behav. Brain Res., 39: 53 61.

Aloe, L., Alleva, E. and Fiore, M. (2002) Stress and nerve growth factor: findings in animal models and humans. Pharmacol. Biochem. Behav., 73: 159–166.

Aloe, L., Bracci-Laudiero, L., Alleva, E., Lambiase, A., Micera, A. and Tirassa, P. (1994) Emotional stress induced by parachute jumping enhances blood nerve growth factor levels and the distribution of nerve growth factor receptors in lymphocytes. Proc. Nat. Acad. Sci. USA, 91: 10440–10444.

Aloe, L., Iannitelli, A., Angelucci, F., Bersani, G. and Fiore, M. (2000) Studies in animal models and humans suggesting a role of nerve growth factor in schizophrenia-like disorders. Behav. Pharmacol., 11: 235–242.

Aloe, L., Iannitelli, A., Bersani, G., Alleva, E., Agelucci, F., Maselli, P. and Manni, L. (1997) Haloperidol administration in humans lowers plasma nerve growth factor level: evidence

that sedation induces opposite effects to arousal. Neuropsychobiology, 36: 65–68.

Aloe, L., Tuveri, M.A., Guerra, G., Pinna, L., Tirassa, P., Micera, A. and Alleva, E. (1996) Changes in human plasma nerve growth factor level after chronic alcohol consumption and withdrawal. Alcohol. Clin. Exp. Res., 20: 462–465.

Altar, C.A. (1999) Neurotrophins and depression. Trends Pharmacol. Sci., 20: 59–61.

Angelucci, F., Aloe, L., Jiménez-Vasquez, P. and Mathé, A.A. (2000a) Mapping the differences in the brain concentration of brain-derived neurotrophic factor (BDNF) and nerve growth factor (NGF) in an animal model of depression. Neuroreport, 11: 1369–1373.

Angelucci, F., Mathé, A.A. and Aloe, L. (2000b) Brain-derived neurotrophic factor and tyrosine kinase receptor TrkB in rat brain are significantly altered after haloperidol and risperidone administration. J. Neurosci. Res., 60: 783–794.

Angelucci, F., Aloe, L., Gruber, S.H.M., Fiore, M. and Mathé, A.A. (2000c) Chronic antipsychotic treatment selectively alters nerve growth factor and neuropeptide Y immunoreactivity and the distribution of choline acetyl transferase in rat brain regions. Int. J Neuropsychopharmacol., 3: 13–25.

Angelucci, F., Aloe, L., Jiménez Vasquez, P. and Mathé, A.A. (2002) Electroconvulsive stimuli alter the regional concentrations of NGF, BDNF and GDNF in adult rat brain. J. Electr. Convul. Ther., 18: 138–143.

Angelucci, F., Aloe, L., Jiménez Vasquez, P. and Mathé, A.A. (2003a) Electroconvulsive stimuli alter nerve growth factor but not brain-derived neurotrophic factor concentrations in brains of a rat model of depression. Neuropeptides, 37: 51–56.

Angelucci, F., Aloe, L., Jiménez Vasquez, P. and Mathé, A.A. (2003b) Lithium treatment alters brain concentrations of nerve growth factor, brain-derived neurotrophic factor and glial cell line-derived neurotrophic factor in a rat model of depression. Int. J. Neuropsychopharmacol, in press.

Arango, V., Underwood, M.D. and Mann, J.J. (2002) Serotonin brain circuits involved in major depression and suicide. Prog. Brain Res., 136: 443–453.

Arnold, S.E. (2000) Cellular and molecular neuropathology of the parahippocampal region in schizophrenia. Ann. NY Acad. Sci., 911: 275–292.

Bengzon, J., Kokaia, Z., Ernfors, P., Kokaia, M., Leanza, G., Nilsson, O.G., Persson, H. and Lindvall, O. (1993) Regulation of neurotrophin and trkA, trkB and trkC tyrosine Kinase receptor messenger RNA expression in kindling. Neuroscience, 53: 433–446.

Bremner, J.D., Narayan, M., Anderson, E.R., Staib, L.H., Miller, H.L. and Charney, D.S. (2000) Hippocampal volume reduction in major depression. Am. J. Psychiatry, 157: 115–118.

Brouha, A.K., Shannon Weickert, C., Heyde, T.M., Herman, M.M., Murray, A.M., Bigelow, L.B., Weinberger, D.R. and Kleinmann, J.E. (1996) Reductions in brain derived neurotrophic factor mRNA in the hippocampus of patients with schizophrenia. Soc. Neurosci. Abstr., 22: 1680.

Bymaster, F., Perry, K.W., Nelson, D.L., Wong, D.T., Rasmussen, K., Moore, N.A. and Calligaro, D.O. (1999) Olanzapine: a basic science update. Br. J. Psychiatry, 37 (Suppl.): 36–40.

Chen, A.C., Eisch, A.J., Sakai, N., Takahashi, M., Nestler, E.J. and Duman, R.S. (2001a) Regulation of GFRalpha-1 and GFRalpha-2 mRNAs in rat brain by electroconvulsive seizure. Synapse, 39: 42–50.

Chen, B., Dowlatshahi, D., MacQueen, G.M., Wang, J.F. and Young, L.T. (2001b) Increased hippocampal BDNF immunoreactivity in subjects treated with antidepressant medication. Biol. Psychiatry, 50: 260–265.

Chen, R.W. and Chuang, D.M. (1999) Long term lithium treatment suppresses p53 and Bax expression but increases Bcl-2 expression. A prominent role in neuroprotection against excitotoxicity. J. Biol. Chem., 274: 6039–6042.

Cirulli, F. (2001) Role of environmental factors on brain development and nerve growth factor expression. Physiol. Behav., 73: 321–330.

Croll, S., Wiengand, S., Anderson, K., Lindsay, R. and Nawa, H. (1994) Regulation of neuropeptides in adult rat forebrain by the neurotrophins BDNF and NGF. Eur. J. Neurosci., 6: 1343–1353.

Davis, J.O., Phelps, J.A. and Bracha, H.S. (1995) Prenatal development of monozygotic twins and concordance for schizophrenia. Schizophr. Bull., 21: 357–366.

De Vries, G.J. (1990) Sex differences in neurotransmitter systems. J Neuroendocrinol., 2: 1–13.

Deutch, A.Y. (1993) Prefrontal cortical dopamine systems and the elaboration of functional corticostriatal circuits: implications for schizophrenia and Parkinson's disease. J. Neural Transm. Gen. Sect., 91: 197–221.

Drevets, W.C. (2000) Functional anatomical abnormalities in limbic and prefrontal cortical structures in major depression. Prog. Brain Res., 126: 413–431.

Duman, R.S., Heninger, G.R. and Nestler, E.J. (1997) A molecular and cellular theory of depression. Arch. Gen. Psychiatry, 54: 597–606.

Duman, R.S. and Vaidya, V.A. (1998) Molecular and cellular actions of chronic electroconvulsive seizures. J. Electr. Convul. Ther., 14: 181–193.

Duman, R.S., Vaidya, V.A., Nibuya, M., Morinobu, S. and Rydelek Fitzgerald, L. (1995) Stress, antidepressant treatments, and neurotrophic factors: Molecular and cellular mechanisms. Neuroscientist, 1: 351–360.

Durany, N., Michel, T., Zochling, R., Boissl, K.W., Cruz-Sanchez, F.F., Riederer, P. and Thome, J. (2001) Brain-derived neurotrophic factor and neurotrophin 3 in schizophrenic psychoses. Schizophr. Res., 52: 79–86.

Emrich, H.M. (1984) Endorphins in psychiatry. Psychiat. Dev., 2: 97–114.

Engele, J. (1998) Spatial and temporal growth factor influences on developing midbrain dopaminergic neurons. J. Neurosci. Res., 53: 405–414.

Ernfors, P., Bengzon, J., Kokaia, Z., Persson, H. and Lindvall, O. (1991) Increased levels of messenger RNAs for neurotrophic factors in the brain during kindling epileptogenesis. Neuron, 7: 165–176.

Falkai, P. and Bogerts, B. (1986) Cell loss in the hippocampus of schizophrenics. Eur. Arch. Psych. Neurol. Sci., 236: 154–161.

Feng, L., Wang, C.Y., Jiang, H., Oho, C., Dugich-Djordjevic, M., Mei, L. and Lu, B. (1999a) Differential signaling of glial cell line-derived neurotrophic factor and brain-derived neurotrophic factor in cultured ventral mesencephalic neurons. Neuroscience, 93: 265–273.

Feng, L., Wang, C.Y., Jiang, H., Oho, C., Mizuno, K., Dugich-Djordjevic, M. and Lu, B. (1999b) Differential effects of GDNF and BDNF on cultured ventral mesencephalic neurons. Brain Res. Mol. Brain Res., 66: 62–70.

Fiore, M., Korf, J., Angelucci, F., Talamini, L. and Aloe, L. (2000) Prenatal exposure to methylazoxymethanol acetate in the rat alters neurotrophin levels and behavior: considerations for neurodevelopmental diseases. Physiol. Behav., 71: 57–67.

Fish, B. (1977) Neurobiologic antecedentes of schizophrenia in children: evidence for an inherited, congenital neurointegrative defect. Arch. Gen. Psychiatry, 34: 1297–1313.

Follesa, P., Gale, K. and Mocchetti, I. (1994) Regional and temporal pattern of expression of nerve growth factor and basic fibroblast growth factor mRNA in rat brain following electroconvulsive shock. Exp. Neurol., 127: 37–44.

Freedman, R., Stromberg, I., Nordstrom, A.L., Seiger, A., Olson, L., Bygdeman, M., Wiesel, F.A., Granholm, A.C. and Hoffer, B.J. (1994) Neuronal development in embryonic brain tissue derived from schizophrenic women and grafted to animal hosts. Schizophr. Res., 13: 259–270.

Freedman, R., Stromberg, I., Seiger, A., Olson, L., Nordstrom, A.L., Wiesel, F.A., Bygdeman, M., Wetmore, C., Palmer, M.R. and Hoffer, B.J. (1992) Initial studies of embryonic transplants of human hippocampus and cerebral cortex derived from schizophrenic women. Biol. Psychiatry, 32: 1148–1163.

Fukumoto, T., Morinobu, S., Okamoto, Y., Kagaya, A. and Yamawaki, S. (2001) Chronic lithium treatment increases the expression of brain-derived neurotrophic factor in the rat brain. Psychopharmacology, 158: 100–106.

Gall, C.M. and Isackson, P.J. (1989) Limbic seizures increase neuronal production of messenger RNA for nerve growth factor. Science, 245: 758–761.

Gall, C.M., Murray, K. and Isackson, P.J. (1991) Kainic acid-induced seizures stimulate increased expression of nerve growth factor mRNA in rat hippocampus. Mol. Brain Res., 9: 113–123.

Gallhofer, B., Bauer, U., Lis, S., Krieger, S. and Gruppe, H. (1996) Cognitive dysfunction in schizophrenia: comparison of treatment with atypical antipsychotic agents and conventional neuroleptic drugs. Eur. Neuropsychopharmacol., 6(Suppl. 2): S13–S20.

Green, R., Peralta, E., Hong, J.S., Mao, C.C., Aterwill, C.K. and Costa, E. (1978) Alteration in GABA metabolism and met-enkephalin content in rat brain following repeated electroconvulsive shock therapy. J. Neurochem., 31: 607–611.

Gwag, B.J., Sessler, F.M., Waterhouse, B.D. and Springer, J.E. (1993) Regulation of nerve growth factor mRNA in the hippocampal formation: effects of N-Methil-D-Aspartate receptor activation. Exp. Neurol., 121: 160–171.

Haddjiconstantinou, M., McGuire, L., Duchemin, A.M., Laskowski, B., Kiecolto-Glaser, J. and Glaser, R. (2001) Changes in plasma nerve growth factor levels in older adult associated with chronic stress. J. Neuroimmunol., 116: 102–106.

Hattori, M. and Nanko, S. (1995) Association of neurotrophin-3 gene variant with severe forms of schizophrenia. Biochem. Bioph. Res. Co., 209: 513–518.

Hellweg, R., Lang, U.E., Nagel, M. and Baumgartner, A. (2002) Subchronic treatment with lithium increases nerve growth factor content in distinct brain regions of adult rats. Mol. Psychiatry, 7: 604–608.

Ickes, B.R., Pham, T.M., Albeck, D.S., Mohammed, A.H. and Granholm, A.C. (2000) Long-term environmental enrichment leads to regional increases in neurotrophin levels in rat brain. Exp. Neurol., 164: 45–52.

Isackson, P.J., Huntsman, M.M., Murray, K.D. and Gall, C.M. (1991) BDNF mRNA expression is increased in adult rat forebrain after limbic seizures: temporal patterns of induction distinct from NGF. Neuron, 6: 937–948.

Jackob, H. and Beckman, H. (1986) Prenatal developmental disturbances in the limbic allocortex in schizophrenics. J. Neural Transpl., 65: 303–326.

Jacobs, B.L., Praag, H. and Gage, F.H. (2000) Adult brain neurogenesis and psychiatry: a novel theory of depression. Mol. Psychiatry, 5: 262–269.

Kapur, S., Zipursky, R.B., Remington, G., Jones, C., DaSilva, J., Wilson, A.A. and Houle, S. (1998) 5-HT$_2$ and D$_2$ receptor occupancy of olanzapine in schizophrenia: a PET investigation. Am. J. Psychiatry, 155: 921–928.

Karler, R., Calder, L.D., Thai, L.H. and Bedingfield, J.B. (1994) A dopaminergic-glutamatergic basis for the action of amphetamine and cocaine. Brain Res., 658: 8–14.

Karege, F., Perret, G., Bondolfi, G., Schwald, M., Bertschy, G. and Aubry, J.M. (2002) Decreased serum brain-derived neurotrophic factor levels in major depressed patients. Psychiatry Res., 109: 143–148.

Kee, K.S., Kern, R.S., Marshall, B.D. and Green, M.F. (1998) Risperidone versus haloperidol for perception of emotion in treatment-resistant schizophrenia: preliminary findings. Schizophr. Res., 31: 159–165.

Kelley, J.J., Gao, X.M., Tamminga, C.A. and Roberts, R.C. (1997) The effect of chronic haloperidol treatment on dendritic spines in the striatum. Exp. Neurol., 146: 471–478.

Kern, R.S., Green, M.F., Marshall, B.D., Wirshing, W.C., Wirshing, D., McGurk, S., Marder, S.R. and Mintz, J. (1998) Risperidone versus haloperidol on reaction time, manual dexterity, and motor learning in treatment-resistant schizophrenia patients. Biol. Psychiatry, 44: 726–732.

Kittell, D.A., Shannon Weickert, C., Heyde, T.M., Herman, M.M. and Kleinmann, J.E. (1997) Reduction of BDNF mRNA in the prefrontal cortex of patients with schizophrenia, Am. College Neuropsychopharmacol., Abstract.

Lauterborn, J.C., Isackson, P.J. and Gall, C.M. (1994) Seizure-induced increases in NGF mRNA exhibit different time courses across forebrain regions and are biphasic in hippocampus. Exp. Neurol., 125: 22–40.

Lee, J., Duan, W. and Mattson, M. (2002) Evidence that brain-derived neurotrophic factor is required for basal neurogenesis and mediates, in part, the enhancement of neurogenesis by dietary restriction in the hippocampus of adult mice. J. Neurochem., 82: 1367–1375.

Lewis, D.A. and Levitt, P. (2002) Schizophrenia as a disorder of neurodevelopment. Annu. Rev. Neurosci., 25: 409–432.

Lo, D.C. (1995) Neurotrophic factors and synaptic plasticity. Neuron, 15: 979–981.

Mack, K.J., Yi, S., Chang, S., Millan, N. and Mack, P. (1995) NGFI-C expression is affected by physiological stimulation and seizures in the somatosensory cortex. Mol. Brain Res., 29: 140–146.

Magarinos, A.M., McEwen, B.S., Flugge, G. and Fuchs, E. (1996) Chronic psychosocial stress causes apical dendritic atrophy of hippocampal CA3 pyramidal neurons in subordinate tree shrews. J. Neurosci., 16: 3534–3540.

Mahadik, S.P., Laev, H., Korenovsky, A. and Karpiak, S.E. (1988) Haloperidol alters rat CNS cholinergic system: enzymatic and morphological analyses. Biol. Psychiatry, 24: 199–217.

Maier, W., Gansicke, M., Gater, R., Rezaki, M., Tiemens, B. and Urzua, R.F. (1999) Gender differences in the prevalence of depression: a survey in primary care. J. Affect. Disord., 53: 241–252.

Maisonpierre, P.C., Belluscio, L. and Friedman, B. (1990) NT3, BDNF and NGF in the developing rat nervous system: parallel as well as reciprocal patterns of expression. Neuron, 5: 501–509.

Malberg, J.E., Eisch, A.J., Nestler, E.J. and Duman, R.S. (2000) Chronic antidepressant treatment increases neurogenesis in adult rat hippocampus. J. Neurosci., 20: 9104–9110.

Mamounas, L.A., Blue, M.E., Siuciak, J.A. and Altar, C.A. (1995) Brain-derived neurotrophic factor promotes the survival and sprouting of serotonergic axons in rat brain. J. Neurosci., 15: 7929–7939.

Manji, H.K., Moore, G.J., Rajkowska, G. and Chen, G. (2000) Neuroplasticity and cellular resilience in mood disorders. Mol. Psychiatry, 5: 578–953.

Mathé, A.A., Jousisto-Hanson, J., Stenfors, C. and Theodorsson, E. (1990) Effect of lithium on tachykinins, calcitonin gene-related peptide, and neuropeptide Y in rat brain. J. Neurosci. Res., 26: 233–237.

Mathé, A.A., Norstedt Wikner, B., Stenfors, C. and Theodorsson, E. (1994) Effects of lithium on neuropeptide Y, neurokinin A and substance P in brain and peripheral tissues of rat. Lithium, 5: 241–247.

Mitchell, A.J. and Dening, T.R. (1996) Depression-related cognitive impairment: possibilities for its pharmacological treatment. J. Affect. Disord., 36: 79–87.

Moore, N.A. (1999) Olanzapine: preclinical pharmacology and recent findings. Br. J. Psychiatry, 37(Suppl.): 41–44.

Nanko, S., Hattori, M., Kuwata, S., Sasaki, T., Fukuda, R., Dai, X.Y., Yamaguchi, K., Shibata, Y. and Kazamatsuri, H. (1994) Neurotrophin-3 gene polymorphism associated with schizophrenia. Acta Psychiatr. Scand., 89: 390–392.

Nawa, H., Pelleymounter, M.A. and Carnahan, J. (1994) Intraventricular administration of BDNF increases neuropeptide expression in newborn rat brain. J. Neurosci., 14: 3751–3765.

Nibuya, M., Morinobu, S. and Duman, R. (1995) Regulation of BDNF and trkB mRNA in rat brain by chronic electroconvulsive seizure and antidepressant drug treatments. J. Neurosci., 15: 7539–7547.

Riva, M.A., Gale, K. and Mocchetti, I. (1992) Basic fibroblast growth factor mRNA increases in specific brain regions following convulsive seizures. Mol. Brain Res., 15: 311–318.

Rocamora, N., Massieu, L., Boddeke, H., Palacios, J.M. and Mengod, G. (1994) Differential regulation of the expression of nerve growth factor, brain-derived neurotrophic factor and neurotrophin-3 mRNAs in adult rat brain after intrahippocampal injection of quinolinic acid. Mol. Brain Res., 26: 89–98.

Rylett, R.J. and Williams, L.R. (1994) Role of neurotrophins in cholinergic-neurone function in the adult and aged CNS. Trends Neurosci., 17: 486–490.

Sackeim, H.A., Decina, P. and Prohovnik, I. (1983) Anticonvulsant and antidepressant properties of electroconvulsive therapy: a proposed mechanism of action. Biol. Psychiatry, 18: 1301–1310.

Sapolsky, R.M. (1990) Glucocorticoids, hippocampal damage and the glutamatergic synapse. Prog. Brain Res., 86: 13–23.

Sato, K., Kashihara, K., Morimoto, K. and Hayabara, T. (1996) Regional increases in brain-derived neurotrophic factor and nerve growth factor mRNAs during amygdaloid kindling, but not in acidic and basic fibroblast growth factor mRNAs. Epilepsia, 37: 6–14.

Seiler, M. and Schwab, M. (1984) Specific retrogade transport of nerve growth factor from neocortex to nucleus basalis in the rat. Brain Res., 266: 225–232.

Sham, P.C., O'Callaghan, E., Takey, N., Murray, G.K., Hare, E.H. and Murray, R.M. (1992) Schizophrenia following pre-natal exposure to influenza epidemics between 1939 and 1960. Br. J. Psychiatry, 160: 461–466.

Sheline, Y.I., Sanghavi, M., Mintun, M.A. and Gado, M.H. (1999) Depression duration but not age predicts hippocampal volume loss in medically healthy women with recurrent major depression. J. Neurosci., 19: 5034–5043.

Shintani, F. (1999) Cytokines and neurotrophins in psychiatric disorders. Biomed. Rev., 10: 69–73.

Siuciak, J.A., Altar, C.A., Wiegand, S.J. and Lindsay, R.M. (1994) Antinociceptive effect of brain-derived neurotrophic factor and neurotrophin-3. Brain Res., 633: 326–330.

Siuciak, J.A., Lewis, D.R., Wiegand, S.J. and Lindsay, R.M. (1997) Antidepressant-like effect of brain-derived neurotrophic factor (BDNF). Pharmacol. Biochem. Behav., 56: 131–137.

Smith, M.A., Makino, S., Kvetnansky, R. and Post, R.M. (1995) Effects of stress on neurotrophic factor expression in the rat brain. Ann. N.Y. Acad. Sci., 771: 234–239.

Solum, D.T. and Handa, R.J. (2002) Estrogen regulates the development of brain-derived neurotrophic factor mRNA and protein in the rat hippocampus. J. Neurosci., 22: 2650–2659.

Spohr, H.L., Williams, J. and Steinhausen, H.C. (1993) Prenatal alcohol exposure and long-term developmental consequences. Lancet, 341: 907–910.

Strange, P.G. (2001) Antipsychotic drugs: importance of dopamine receptors for mechanisms of therapeutic actions and side effects. Pharmacol. Rev., 53: 119–133.

Takahashi, M., Shirakawa, O., Toyooka, K., Kitamura, N., Hashimoto, T., Maeda, K., Koizumi, S., Wakabayashi, K., Takahashi, H., Someya, T. and Nawa, H. (2000) Abnormal expression of brain-derived neurotrophic factor and its receptor in the corticolimbic system of schizophrenic patients. Mol. Psychiatry, 5: 293–300.

Talamini, L.M., Koch, T., Ter Horst, G.J. and Korf, J. (1998) Methylazoxymethanol acetate-induced abnormalities in the entorhinal cortex of the rat; parallels with morphological findings in schizophrenia. Brain Res., 789: 293–306.

Thoenen, H. (1995) Neurotrophins and neuronal plasticity. Science, 270: 593–598.

Thome, J., Foley, P. and Riederer, P. (1998) Neurotrophic factors and the maldevelopmental hypothesis of schizophrenic psychoses. J. Neural Transm., 105: 85–100.

Ueyama, T., Kawai, Y., Nemoto, K., Sekimoto, M., Tone, S. and Senba, E. (1997) Immobilization stress reduced the expression of neurotrophins and their receptors in the rat brain. Neurosci. Res., 28: 103–110.

Vaidya, V.A., Marek, G.J., Aghajanian, G.K. and Duman, R.S. (1997) 5-HT$_{2A}$ receptor-mediated regulation of brain-derived neurotrophic factor mRNA in the hippocampus and the neocortex. J. Neurosci., 17: 2785–2795.

Vaswani, M., Linda, F.K. and Ramesh, S. (2003) Role of selective serotonin reuptake inhibitors in psychiatric disorders: a comprehensive review. Prog. Neuropsychopharmacol. Biol. Psychiatry, 27: 85–102.

Wahlestedt, C., Ekman, R. and Widerlöv, E. (1989) NPY and the central nervous system: Distribution and possible relationship to neurological and psychiatric disorders. Prog. Neuropsychopharmacol. Biol. Psychiatry, 13: 31–54.

Watson, M.A. and Milbrandt, J. (1989) The NGF-B gene, a transcriptionally inducible member of the steroid receptor gene superfamily: genomic structure and expression in rat brain after seizure induction. Mol. Cell Biol., 9: 4213–4219.

Weinberger, D.R. (1999) Cell biology of the hippocampal formation in schizophrenia. Biol. Psychiatry, 45: 395–402.

Weinberger, D.R. and Lipska, B.K. (1995) Cortical maldevelopment, schizophrenia, and antipsychotic drugs: a search for common ground. Schizophr. Res., 16: 87–100.

Weissman, M.M. and Olfson, M. (1995) Depression in women: implications for health care research. Science, 269: 799–801.

Widerlöv, E., Lindstrom, L.H., Wahlestedt, C. and Ekman, R. (1988) NPY and peptide YY as possible CSF markers for major depression and schizophrenia, respectively. J. Psychiatr. Res., 22: 69–79.

Wright, P., Takei, N., Rifkin, L. and Murray, R.M. (1995) Maternal influenza, obstetric complications, and schizophrenia. Am. J. Psychiatry, 152: 1714–1720.

Young, E.A., Spencer, R.L. and McEwen, B.S. (1990) Changes at multiple levels of the hypothalamo-pituitary adrenal axis following repeated electrically induced seizures. Psychoneuroendocrinology, 15: 165–172.

Progress in Brain Research, Vol. 146
ISSN 0079-6123

CHAPTER 12

Discovering novel phenotype-selective neurotrophic factors to treat neurodegenerative diseases

Penka S. Petrova[1], Andrei Raibekas[1], Jonathan Pevsner[2], Noel Vigo[1], Mordechai Anafi[1], Mary K. Moore[1], Amy Peaire[1], Viji Shridhar[3], David I. Smith[3], John Kelly[4], Yves Durocher[5] and John W. Commissiong[1],*

[1]*Prescient NeuroPharma Inc., Laboratories of Protein Chemistry, Molecular Biology and Cell Biology, Toronto, ON, Canada*
[2]*Department of Neurology, Kennedy Krieger Institute and Department of Neuroscience, Johns Hopkins School of Medicine, Baltimore, MD, USA*
[3]*Department of Experimental Pathology, Division of Laboratory Medicine, Mayo Clinic, Rochester, MN, USA*
[4]*Institute for Biological Studies, National Research Council of Canada, Ottawa, ON, Canada*
[5]*Biotechnology Research Institute, National Research Council of Canada, Animal Cell Technology Group, Montreal, QC, Canada*

Abstract: Astrocytes and neurons in the central nervous system (CNS) interact functionally to mediate processes as diverse as neuroprotection, neurogenesis and synaptogenesis. Moreover, the interaction can be homotypic, implying that astrocyte-derived secreted molecules affect their adjacent neurons optimally vs remote neurons. Astrocytes produce neurotrophic and extracellular matrix molecules that affect neuronal growth, development and survival, synaptic development, stabilization and functioning, and neurogenesis. This new knowledge offers the opportunity of developing astrocyte-derived, secreted proteins as a new class of therapeutics specifically to treat diseases of the CNS. However, primary astrocytes proliferate slowly in vitro, and when induced to immortalize by genetic manipulation, tend to lose their phenotype. These problems have limited the development of astrocytes as sources of potential drug candidates. We have successfully developed a method to induce spontaneous immortalization of astrocytes. Gene expression analysis, karyotyping and activity profiling data show that these spontaneously immortalized type-1 astrocyte cell lines retain the properties of their primary parents. The method is generic, such that cell lines can be prepared from any region of the CNS. To date, a library of 70 cell lines from four regions of the CNS: ventral mesencephalon, striatum, cerebral cortex and hippocampus, has been created. A phenotype-selective neurotrophic factor for dopaminergic neurons has been discovered from one of the cell lines (VMCL1). This mesencephalic astrocyte-derived neurotrophic factor (MANF) is a 20 kD, glycosylated, human secreted protein. Homologs of this protein have been identified in 16 other species including *C. elegans*. These new developments offer the opportunity of creating a library of astrocyte-derived molecules, and developing the ones with the best therapeutic indices for clinical use.

Keywords: Cell culture; cell lines; neuroprotection; Parkinson's disease; spontaneous immortalization; type-1 astrocytes

Introduction

The selective death of specific neuronal phenotypes that underlies the etiology of each of the four major neurodegenerative diseases: Parkinson's disease (PD), Alzheimer's disease (AD), Lou Gehrig's disease or amyotrophic lateral sclerosis (ALS) and Huntington's disease (HD), lead to debilitating clinical consequences, but is of profound scientific interest. Dopaminergic neurons in the zona compacta of the substantia nigra (SNc) die in PD (Fig. 1), while GABAergic neurons in the substantia nigra zona

*Correspondence to: J. Commissiong, Laboratory of Cell Biology, Prescient NeuroPharma Inc., 96 Skyway Avenue, Toronto, ON, M9W 4Y9, Canada. Tel.: +1-416-674-8047; Fax: +1-416-674-8060; E-mail: johnc@prescientneuropharma.com

DOI: 10.1016/S0079-6123(03)46012-3

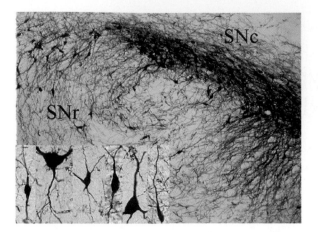

Fig. 1. Subphenotypes in the zona compacta of rat substantia nigra. Illustration of a normal zona compacta of the substantia nigra (SNc) of rat, immunostained to identify dopaminergic (TH+) neurons. Six morphological dopaminergic subphenotypes (inset, lower left) were identified in the SNc. Do different neurotrophic factors exert differing effects on these different subphenotypes?

reticulata (SNr) and dopaminergic neurons in the medial ventral tegmental region (VTA) survive (Hardman et al., 1996; McRitchie et al., 1997). These observations suggest that SNc dopaminergic neurons are more susceptible to death, compared with the adjacent SNr GABAergic or VTA dopaminergic neurons. All three types of neurons occupy the same small, ventral mesencephalic space.

The above results imply that the death of a dopaminergic neuron in the SNc in PD is a discrete, specific event. One possible mechanism is activity-dependent production of excessive reactive oxygen species (ROS) secondary to a high level of oxidative deamination (Gotz et al., 1990; Kristal et al., 2001; Li et al., 2001). It is unclear whether the metabolic profiles of the SNc and VTA dopaminergic neurons are so different to cause such opposing effects. Toxic, target-derived molecules could also be involved, given that the SNc and VTA dopaminergic neurons project to different targets, the neostriatum and nucleus accumbens respectively (Morgan et al., 1986; Anglade et al., 1995). However, no dopaminergic toxins have been identified in the striatum, although functional, injury-induced, dopaminergic neuroprotective activity has been reported in the striatum (Niijima et al., 1990; Maggio et al., 1997). Dopaminergic neurons in the Parkinsonian brain that express calbindin-D_{28K} appear to be less susceptible to neurodegeneration (Yamada et al., 1990). The significance of this observation to the general pathology in PD has been difficult to assess. A variety of other etiologies have been suggested for dopaminergic neuronal death in PD, including a complement-dependent toxin of serum origin, defects in cytochrome C oxidase in the SNc, functional defects in mitochondrial P450 enzymes (Defazio et al., 1994; Itoh et al., 1997; Riedl et al., 1998) and other less well-defined causes (Masalha et al., 1997; Plante-Bordeneuve et al., 1997; Yoritaka et al., 1997). None have been proven to date.

We are therefore faced with the task of treating and possibly curing a disease whose fundamental cause we do not understand. There are two essential requirements. The first is to arrest further neuronal death after diagnosis, and then protect the remaining dopaminergic neurons over the long-term. The second is to replace, at least in part, the neurons that died prior to diagnosis. It follows that neuroprotection combined with cell therapy is likely to be the most realistic and effective therapeutic approach to the treatment of PD. In recent years, we have acquired powerful tools to protect neurons in vitro. Indeed, neuroprotection, the ability to protect neurons from death in a consistent and robust way (still mainly in the laboratory), is one of the most powerful phenomena in modern neurobiology (Takeshima et al., 1996; Gozes, 2001). The challenge is to understand the cellular and molecular mechanisms that are the basis of this neuroprotection, and transfer the knowledge to the in vivo, clinical setting to treat the neurodegenerative diseases.

Three sources of neurotrophic factors are recognized: autocrine, in which the neuron elaborates self-protective molecules (Davies, 1996; Aliaga et al., 1998); target-derived, which is particularly important in developmental neurobiology (Hashino et al., 2001; Lotto et al., 2001; Von Bartheld and Johnson, 2001); and paracrine, in which the trophic molecules are derived from adjacent, mainly nonneuronal cells or type-1 astrocytes (Fig. 2A), (Takeshima et al., 1994; Albrecht et al., 2002).

The focus of this review will be on paracrine, astrocyte-derived, neuroprotective protein molecules. As therapeutics they tend to be less toxic, nonteratogenic and specific (Henderson, 1995; Jain, 1998).

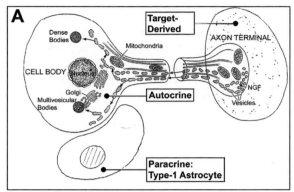

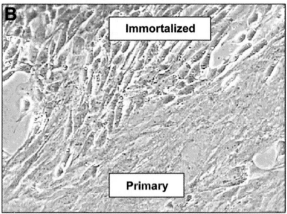

Fig. 2. Types of neurotrophic factors and spontaneous immortalization of astrocytes. (A) Neurotrophic factors may be autocrine, target-derived, or paracrine (modified from Purves and Litchman, 1985. Fig. 11, Chapter 7). The main source of paracrine neurotrophic factors in the CNS is type-1 astrocytes. (B) Spontaneous immortalization of type-1 astrocytes is signaled by a dramatic change in the morphology of the cells from broad and flat (primary) to small and spindle shaped (immortalized). Immortalized cells were marked, picked and subcloned to produce pure cell lines. The duration of the immortalization process is 45–65 days.

Increasing knowledge in the development peptide mimetics from neurotrophic factors will increase the value of these molecules as therapeutics to treat neurodegenerative diseases (Skaper and Walsh, 1998; Beglova et al., 2000; Xie and Longo, 2000).

In terms of a demonstrated, causal link between loss of neuroprotection in nigral dopaminergic neurons and degeneration of dopaminergic neurons leading to PD, the evidence is still indirect. Tooyama et al. (1994) have observed that the expression of basic fibroblast growth factor (bFGF) is diminished in dopaminergic neurons in Parkinsonian patients, but persists in the nigra of age-matched controls. bFGF acting independently, or indirectly as a mitogen for astrocytes which then secrete the neuroprotective molecules, is potently neuroprotective for dopaminergic neurons in vitro (Mena et al., 1995; Hou and Mytilineou, 1996). A reduction of hippocampal brain-derived neurotrophic factor (BDNF) has been shown to precede neuronal degeneration in the CA1 region (Yamasaki et al., 1998). Our current understanding is that the actions of neurotrophic factors are translated via complex molecular cascades (Rozengurt, 1995; Kahn et al., 1997), which therefore offer the opportunity of developing small molecules to promote those reactions that mediate neuroprotection, and inhibit cell death-promoting interactions. In terms of cell therapy, we have also acquired the capacity, although still limited, to expand dopaminergic neurons in vitro. Moreover, new methods of packaging cells optimally for use in cell therapy, and maintaining them at >90% viability for 24 h have recently been developed (Peaire et al., 2003). Within a few years, these new methods should allow us to produce enough dopaminergic neurons of high viability, for transfer to collaborative neurosurgical centers, to conduct a pilot clinical trial. The work presented in this chapter will explore the scientific basis and technologies associated with these new approaches to the treatment of PD. Our objective is to implement a workable strategy for the effective clinical treatment of PD.

Astrocytes and neuroprotection

There is a solid body of evidence demonstrating that secreted proteins present in conditioned medium (CM) prepared from ventral mesencephalic type-1 astrocytes are selectively neurotroprotective for dopaminergic neurons (O'Malley et al., 1992; Takeshima et al., 1994; Panchision et al., 1998; Petrova et al., 2003). There is also suggestive evidence that the astrocyte-neuron interaction can be homotypic. The term homotypic implies that molecules derived from ventral mesencephalic astrocytes are more effective in protecting ventral mesencephalic neurons compared with striatal or cerebral cortical neurons (O'Malley et al., 1994; Takeshima et al., 1994). It follows that neurotrophic factors to treat PD and HD should be derived from ventral mesencephalic and striatal astrocytes, respectively.

Most of the evidence in support of the above contention is derived from experiments utilizing primary astrocytes (Takeshima et al., 1994). However, primary astrocytes are an impractical source of molecules when used over an extended period of time. They are likely to be heterogeneous. They are difficult to maintain, and tend to expand slowly in culture. Quality control for inter-batch preparations is challenging. Errors, requiring an experiment to be repeated, pose practical problems. To be of value as a drug development tool, astrocytes must be immortalized, but with retention of their phenotype. Most methods of immortalization rely on the insertion of foreign DNA into the genome, generally resulting in a changed phenotype. The large tumor antigen from simian virus-40 (SV40-LTA) is the instrument of choice in many immortalization protocols (Azuma et al., 1996; Engele et al., 1996), although other methods have been used as well (Louis et al., 1992; Goletz et al., 1994; Seidman et al., 1997). We have successfully developed an alternative immortalization protocol that is based on the serial passage of proliferating astrocytes (Panchision et al., 1998). Gene expression analysis, karyotyping and biological activity profiling data suggest that cells that were immortalized by this new protocol retain their phenotype (Panchision et al., 1998). Immortalization is signaled by an abrupt change in the morphology of the cell from broad and flat (Fig. 2B, bottom), to small and spindle-shaped (Fig. 2B, top), and by localized expansion at the sites of immortalization in the culture dish. The putative, immortalized cells are then marked, picked and subcloned, and pure cell lines prepared. The method is generic. It can be used to develop cell lines from any region of the CNS. To date, we have prepared a library of seventy such rodent cell lines from four regions of the CNS: ventral mesencephalon (55), striatum (5), cerebral cortex (5) and hippocampus (5). Conditioned medium prepared from one of these cell lines, ventral mesencephalic cell line 1 (VMCL1), contained glial cell line-derived neurotrophic factor (GDNF) and transforming growth factor beta-2 (TGF-β2), both of which had been previously identified in the primary

parent cells by RT-PCR (unreported observation). An unidentified peak of activity was also present that was indicative of an unknown neuroprotective molecule. Standard chromatographic methods were used to separate the different biological activities. GDNF and TGF-β2 were identified by ELISA according to the manufacturer's instructions (Promega, Madison, WI).

One of the crucial elements in this search for novel neurotrophic factors is a robust, functional bioassay that is capable of identifying neurotrophic activity reliably during sequential purification of a factor. The assay utilizes a ventral mesencephalic cell culture that contains 20% of dopaminergic neurons (Fig. 3) (Takeshima et al., 1996). This high content of dopaminergic neurons provides a high, absolute number of dopaminergic neurons for analysis, permitting test compounds of similar neurotrophic potencies to be resolved reliably. As shown in Fig. 4B, in the absence of neuroprotection, almost all of the plated cells died after seven days (DIV7) in culture, in response to a serum-deprivation insult. However, when treated with astrocyte CM, neuronal survival was pronounced (Fig. 4C). We also know the neuronal phenotypic composition of the culture system as determined 24 h after plating (dopaminergic 20%, GABAergic 50%, serotonergic 5%, cholinergic 8%, bi-potential vimentin+ cells 5%, astrocytes (GFAP+) 0%, with 12% remaining unknown to date. Therefore, the relative specificity of a test CM or pure neurotrophic molecule for a given neuronal phenotype can be determined using this bioassay (Takeshima et al., 1996). It is evident that without a working knowledge of the phenotypic composition of a bioassay, it cannot be used to identify phenotype-selective neurotrophic factors.

In summary, we had decided on a combined strategy of neuroprotection and cell therapy; identified type-1 astrocytes as a source of neuroprotective molecules; built the concept of homotypic astrocyte-neuron interaction into our thinking and used it as a guide in selecting the regions of the CNS from which to develop cell lines; developed a method of preparing astrocyte cell lines that retain their phenotype; developed a novel bioassay of known phenotypic composition with a high content of dopaminergic neurons for use in identifying dopaminergic-selective neurotrophic factors. These initiatives are complementary

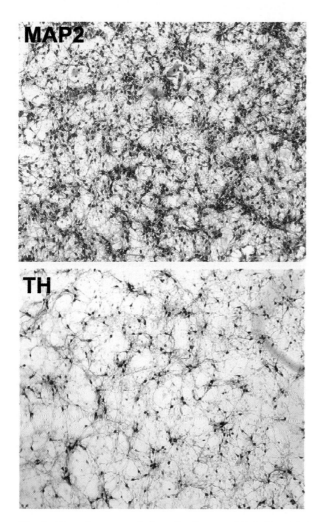

Fig. 3. Ventral mesencephalic cell culture at DIV1. Rat, E14 ventral mesencephalon (1.0 mm³) was microdissected, dispersed at a density of 1.0×10^6 cells/ml, and plated as 25 μl microisland droplets, in an area of 12.5 mm². This plating technique results in cells that are homogeneously distributed, at high density, at the center of the microisland (MAP2). The microdissection technique is used routinely, and results in cultures containing 20% of dopaminergic neurons (TH). The cells were immunostained using the Vector ABC method. MAP2+, × 10 Objective; TH+, × 10 Objective. Column factor: 1.0.

to other developments that were taking place in the field. Others have described progress in neurogenesis from adult stem cells and deriving dopaminergic neurons from progenitors (Studer et al., 1998; Song et al., 2002). Methods for the effective delivery of proteins to discrete sites in the brain are being developed (Kordower et al., 2000). The genes that are

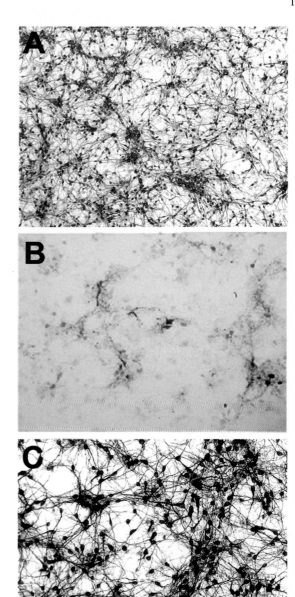

Fig. 4. Neuroprotection. The cultures were prepared and plated using the same methods explained in the legend of Fig. 3. The experiment was designed to illustrate the neuroprotective power of conditioned medium prepared from certain ventral mesencephalic type-1 astrocyte cell lines. (A) The center of the microisland culture (DIV1). At DIV7, and in the absence of neuroprotection, nearly all of the cells had died (B) When treated with 25% v/v of type-1 astrocyte conditioned medium, there was a dramatic rescue of the cells (C). (A), (B) and (C): MAP2 staining using Vector ABC method. A: × 10. B and C: × 20.

implicated in the small familial component of the Parkinsonian population have been cloned (Farrer et al., 1998; Shimura et al., 2000; Lee et al., 2002).

To date, it has been difficult to develop neurotrophic molecules deliberately as potential drug candidates targeted at specific diseases. Discovery has been mainly by serendipity. The sources have been diverse, and a wide variety of methods of testing biological activities have yielded conflicting and inconclusive results. The methods and principles being outlined in this report will likely change the field significantly, at least with regard to paracrine-derived neurotrophic factors targeted at the CNS.

Protein purification, sequencing, expression and testing

It took several decades (1948–1992) to discover the four members of the neurotrophin family: nerve growth factor (NGF), BDNF, neurotrophin-3 (NT-3) and NT-4/5 (Barde, 1994), but just five years (1993–1998) to discover the four members of the glial cell line-derived neurotrophic factor (GDNF) family ligands (GFLs): GDNF, neurturin (NRTN), artemin (ARTN) and persephin (PSPN) (Airaksinen and Saarma, 2002). Factors that are neuroprotective for dopaminergic neurons are illustrated in Table 1. Only deprenyl (selegiline) in this list is actually used clinically to treat PD. The clear implication is that factors to treat neurodegenerative diseases must be developed in a targeted way. We have attempted to pursue such a discovery strategy with MANF.

MANF is derived from a spontaneously immortalized VMCL1. The activity profile of VMCL1-CM demonstrated significant protection of dopaminergic neurons (TH+) in a dose-dependent manner, but not of the general neuronal population (MAP2+) (Panchision et al., 1998). Using classical protein purification and analysis, MANF was purified from 5 L VMCL1-CM, in five sequential chromatographic steps: affinity, gel filtration, weak interactions, ion exchange and high resolution gel filtration (Petrova et al., 2003). Our efforts were concentrated on a peak of activity lacking GDNF and TGF-β2, which were identified by ELISA. At each stage in the analysis, the chromatographic fractions were tested using the bioassay, usually over a 5-day period. The cultures

TABLE 1
Dopaminergic neurotrophic factors

Family	Examples	References
Neurotrophins	BDNF	Hyman et al., 1991
	NT-4/5	Altar et al., 1994
GDNF-Family ligands (GFLs)	GDNF	Lin et al., 1993
	NRTN	Kotzbauer et al., 1996
	ARTN	Baloh et al., 1998
	PSPN	Baloh et al., 1998
(New factor)	MANF	Petrova et al., 2003
TGF family	TGF-β1	Henrich-Noack et al., 1994
	TGF-β2	Krieglstein and Unsicker, 1994
	TGF-β3	Poulsen et al., 1994
	TGF-α	Alexi and Hefti, 1993
Cytokines	IL-2	Alonso et al., 1993
	IL-6	von Coelln et al., 1995
	IGF-II	Liu and Lauder, 1992
Mitogens	bFGF	Engele and Bohn, 1991
	EGF	Casper et al., 1991
	PDGF-AA	Giacobini et al., 1993
	PDGF-BB	Nikkhah et al., 1993
MAO B-I	Deprenyl	Roy and Bédard, 1993
	Rasagiline	Finberg et al., 1998
Miscellaneous	CNTF	Hagg and Varon, 1993
	S-100-β	Liu and Lauder, 1992
	VEGF	Silverman et al., 1999
	cAMP	Hartikka et al., 1992

cAMP: cyclic adenosine mono phosphate; ARNT: artemin; BDNF: brain-derived neurotrophic factor; CNTF: ciliary neurotrophic factor; EGF: epidermal growth factor; FGF: fibroblast growth factor; GDNF: glial cell line-derived neurotrophic factor; IGF: insulin-like growth factor; IL: interleukin; MANF: mesencephalic astrocyte-derived neurotrophic factor; MAO B-I: monoamine oxidase B inhibitors; NRTN: neurturin; NT-3: neurotrophin 3; PDGF: platelet-derived growth factor; PSPN: persephin; TGF: transforming growth factor; VEGF: vascular endothelial growth factor.

were treated at DIV0, 2 and 4, scored daily, fixed and stained on DIV5 to identify TH+ and MAP2+ cells. The live cultures ($n = 4$) were scored daily on a scale of 0–10, based on visual inspection.

At the last, high resolution gel filtration step, there were 40 fractions (Fig. 5). Biological activity was localized in fractions #18–24. The active fractions and inactive side fractions were pooled separately, and retested to verify the presence and absence respectively of biological activity. Further analysis of the samples is described in detail elsewhere (Petrova et al., 2003). The last analytical step was done using a

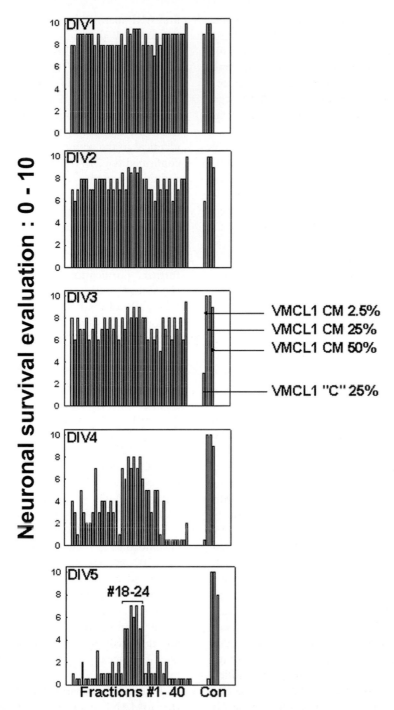

Fig. 5 Use of the bioassay to identify active fractions during purification of MANF. Neuroprotective activity was scored on a scale of 0–10, based on visualization of the live cells in culture, at 24 h intervals. Negative and positive controls were the unconditioned growth medium ('C') and VMCL1-CM (2.5, 25 and 50% v/v respectively). As shown, by DIV2, it was possible to begin to localize the active fractions. By DIV4/5 localization of the active fractions was readily apparent, as illustrated. Biological activity was localized in fractions #18–24 in the last purification step, using a BioSil 125 high resolution gel filtration column.

12% SDS–PAGE gel which yielded a single band in the expected 20 kD range. The gel was fixed (methanol/acetic acid/water in the ratio 50:10:40), washed (20, 10 and 10 min), stained (overnight, using GelCode Blue Stain Reagent at room temperature) and washed. The 20 kD band was excised and placed in a 1.5 ml microcentrifuge tube and stored on at −86°C until analyzed.

Prior to analysis, the gel band was destained (100 mM ammonium bicarbonate/30% acetonitrile, 30 to 60 min), washed (deionized water, 10, 10, 10 min), dehydrated (acetonitrile), rehydrated (50 mM ammonium bicarbonate) and digested (200 ng of modified trypsin, Promega, Madison, WI) overnight at 37°C. The digest was then transferred to an eppendorf tube, evaporated to dryness (Savant centrifugal evaporator), redissolved (5 μl 5% acetonitrile/0.1% trifluoroacetic acid (desalted, ZipTipTM C18 tips; Millipore, Bedford, MA), and reconstituted in 5 μl 75% methanol/0.5% acetic acid. Peptide analysis was done by nanoelectrospray ionization-tandem mass spectrometry (nESI-MS/MS) (Blackburn and Anderegg, 1997; Li et al., 2000). Two novel peptides were identified from the 20 kD band, ... DVTFSPATIE... and ...QIDLSTVDL. A search of the NCBI protein sequence database for matching proteins was done, as well as BLAST searches. The 20 kD protein was tentatively identified as a human arginine-rich protein or ARP, that contained translated sequences that were 100% homologous with the two peptides identified from nESI-MS/MS analysis. This human ARP cDNA had previously been cloned (Shridhar et al., 1996), but never expressed. The name arginine-rich protein is derived from the fact that 22 of the first 55 N-terminal residues in this transcript are arginines, a property not compatible with a secreted protein. A more detailed search of the public databases identified seventeen homologs of ARP, including those of rat, mouse, C. elegans and a second human transcript. All of these, including the second human transcript, lacked the Arg-rich, N-terminal sequence identified in the first human sequence. Additional analysis revealed that AA56 in ARP was methionine. It therefore seemed evident that the start methionine in the secreted rat protein (and the other homologs as well) responsible for the neuroprotective activity observed in the bioassay was the equivalent of the first ARP identified, minus the first 55 AAs. Further, the Signal-P program indicated two possible signal sequences: (1) The 15-AA sequence: M-56 to V-70 (MWATQGLAVRVALSV...), and (2) the 21-AA sequence: M-56 to A-76 (MWATQGLAVRV ALSVLPGSRA...). Edman microsequencing analysis of the secreted recombinant protein revealed a 5-AA N-terminal sequence of ...LRPGD, that is identical to the equivalent 5-AAs in ARP. This result confirmed the 21-AA signal sequence, and assigned a definitive sequence to the biologically active, secreted protein.

The original human ARP cDNA was therefore subcolned to delete the first 55-AAs, making it equivalent to the other homologs identified in the data base. The re-engineered cDNA was tagged for further purification (His-tag) and identification (HSV-tag) and ligated into the Novagen pTriEx vector. Proof reading PCR and the Novagen pTriEx vector system were used according to the Supplier's recommendations. The protein expressed lacked the R-rich, 55-AA N-terminal region of the original ARP molecule. It was therefore renamed mesencephalic astrocyte-derived neurotrophic factor (MANF) to reflect its origin from a mesencephalic astrocyte, and its neurotrophic activity. Tag-free MANF was expressed in E. coli, and tagged MANF in VMCL1, COS and HEK293 using standard expression protocols (Durocher et al., 2002). A larger scale expression of recombinant MANF (25 mg) in HEK293-EBNA cells was done as previously described (Durocher et al., 2002). MANF is encoded by a small 4.3 Kb gene with 4 exons, located on the short arm of human chromosome 3. Two domains in MANF: 39-AA and 109-AA respectively, and the 8 cysteines are conserved from C. elegans to man. The secondary structure is dominated by α-helices (47%) and random coils (37%). Studies to determine the localization of MANF in the brains of rat, monkey and man, as well as the receptor, signaling pathways and biologically active peptide mimetics are in progress.

Figure 6 demonstrates an SDS-PAGE gel of tag-free hrMANF expressed in E. coli. Unpurified and purified MANF are shown in lanes #2 and #3, respectively. The band in lane #3 was excised and analyzed by nESI-MS/MS. Eight peptides (box inset) were identified, all of which are derived from MANF.

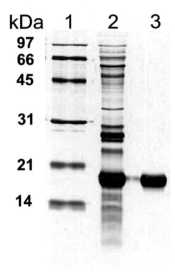

FYQDLK
INELMPK
KINELMPK
FYQDLKDR
QIDLSTVDLK
DVTFSPATIENELIK
IINEVSKPLAHHIPVEK
DRDVTFSPATIENELIK

Fig. 6. Highly purified tag-free MANF expressed in *E. coli*. Tag-free hrMANF expressed in *E. coli* (lane 2) was purified (lane 3) by high resolution gel filtration (BioSil 125), analyzed by SDS-PAGE and stained with coomassie blue. The single band indicated in lane 3 was destained, digested using papain, and the peptide digest analyzed by nESI-MS/MS. Eight peptides were identified, all of which were derived from hrMANF. No bacterial protein was identified. Lane 1: MW Standard.

There were no contaminating bacterial proteins. Tag-free MANF expressed in *E. coli* was shown to be biologically active (Petrova et al., 2003). It was used to quantitate MANF expressed in HEK293, as illustrated in Fig. 7. It is evident (Fig. 7A) that MANF expressed in HEK293 is glycosylated. Independent verification of glycosylation was obtained by the action of neuraminidase (Petrova et al., 2003). Polyclonal antibodies to MANF have been prepared (Fig. 8). At a dilution of 1:5000, 15.6 ng of MANF was easily identified (Fig. 8, lane #9). Experiments to identify MANF by immunocyto-chemistry in the human brain are in progress.

The first test of hrMANF expressed in VMCL1 was in the determination of the viability of ventral mesencephalic cells at DIV5, using the Live Cell/ Dead Cell Assay (Molecular Probes). Cell viability in the cultures treated with VMCL1-CM containing hrMANF was significantly increased (Fig. 9, #c, $p < 0.05$) vs positive controls (#b and #d). This was the first indication that hrMANF was biologically

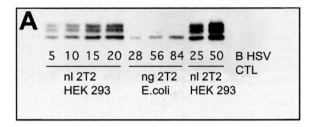

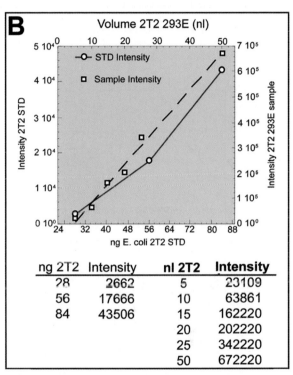

ng 2T2	Intensity	nl 2T2	Intensity
28	2662	5	23109
56	17666	10	63861
84	43506	15	162220
		20	202220
		25	342220
		50	672220

Fig. 7. Quantitation of hrMANF expressed in HEK293 cells. hrMANF expressed in HEK293 cells was quantitated using Western blot combined with densitometry. A standard curve was constructed (o) with 28, 56 and 84 ng of purified hrMANF expressed in *E. coli* (A: Center values, and B: [o]). Subsequently, secreted hrMANF in 5, 10, 15, 20, 25 and 50 nl of the HEK293 medium (A) was plotted as illustrated (B: [□]), and the equivalent quantities of hrMANF determined. Note the difference in units of density on the Y axis for pure hrMANF (left: ng 2T2) and unknown hrMANF (right: nl 2T2). 2T2 is the name assigned to hrMANF in this experiment.

active. The presence of hrMANF in the CM used in these experiments (Fig. 9) was independently verified by Western blot (Petrova et al., 2003). Glycosylated hrMANF expressed in HEK293 cells was tested in the bioassay, and the cultures stained by indirect, double immunofluorescence to identify TH and

176

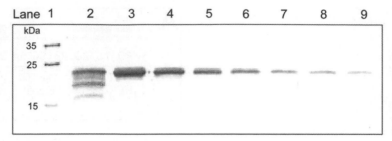

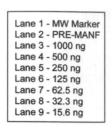

Lane 1 - MW Marker		
Lane 2 - PRE-MANF		
Lane 3 - 1000 ng		
Lane 4 - 500 ng		
Lane 5 - 250 ng		
Lane 6 - 125 ng		
Lane 7 - 62.5 ng		
Lane 8 - 32.3 ng		
Lane 9 - 15.6 ng		

Fig. 8. Sensitivity of anti-MANF polyclonal antibodies. Affinity-purified anti-MANF polyclonal antibodies diluted 1:5000 recognized 15.6 ng of hrMANF. Seven quantities of MANF (15.6–1000 ng, lanes #3–9) were all identified by the antibody. Lane #2 contains MANF prior to purification. Lane #1: MW Standard. The analysis was done by Western blot.

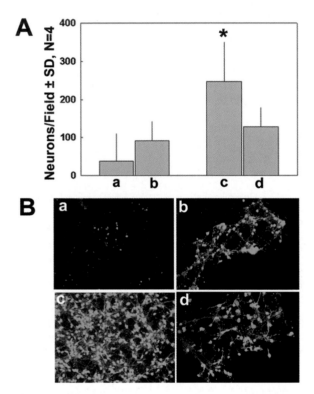

Fig. 9. hrMANF-induced increased viability of ventral mesencephalic neurons. (A) 25% v/v of the VMCL1-CM containing secreted hrMANF (verified independently by Western blot) caused a significant increase at DIV5 ($p < 0.05$ by SNK) in the viability of ventral mesencephalic cells (c) compared to the controls (a, b and d). Conditions a, b, and d are negative control (unconditioned growth medium), positive control (astrocyte conditioned medium) and CM from mock transfected VMCL1 cells respectively. * c is significantly different from a, b and d. ANOVA: $p < 0.01$. (B) Representative images from the four conditions illustrated in (A) above. There is a close correspondence between the number of cells/field indicated in the images (B), and the mean number of cells/field in the histograms (A).

MAP2 at DIV4 (Fig. 10). The data demonstrate that MANF at 0.05 and 50 ng/ml (Fig. 10A, #3 and #4, respectively), caused significant protection (P < 0.05) of dopaminergic neurons versus negative control. The MAP2 data suggest that MANF also increased general neuronal protection. However, we have suggested elsewhere that GABAergic and serotonergic neurons appear not to maintain the expression of GAD and tryptophan hydroxylase respectively in the presence of MANF, while the expression of TH is maintained in dopaminergic neurons (Petrova et al., 2003). This result represents a variation in mechanism of neuroprotection. The GABAergic and serotonergic neurons appear not to die. Instead, they appear to downregulate the enzymes that are unique to these neurons.

Understanding the problem

We are attempting to treat and possibly cure PD, but without a firm understanding of the etiology of the death of dopaminergic neurons that causes the disease (Ruberg et al., 1995; Plante-Bordeneuve et al., 1997; Hubble et al., 1998). As shown (Fig. 1, inset), six morphological subphenotypes of dopaminergic neurons can be identified in the SNc. Other evidence suggests that the expression of tyrosine hydroxilase (TH) mRNA in SNc dopaminergic neurons can also be binned into several levels of expression (Weiss-Wunder and Chesselet, 1991). The electrophysiological firing patterns of dopaminergic neurons are also heterogeneous (Grace and Bunney, 1984a, b). Heterogeneity of dopaminergic neurons has been identified even at the level of the dopamine

A

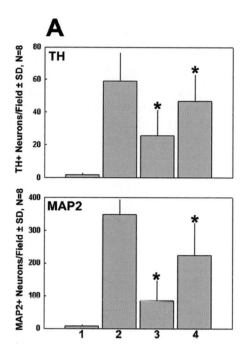

B

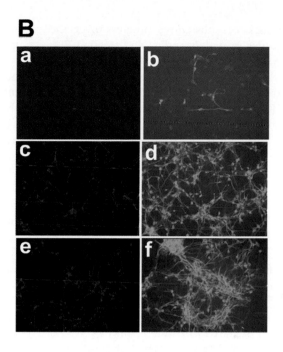

Fig. 10. hrMANF increases the survival of dopaminergic neurons in vitro. (A) Purified hrMANF expressed in HEK 293 was tested in the bioassay. At DIV4, hrMANF at 0.05 (#3) and 50 (#4) ng/ml increased the survival of dopaminergic neurons significantly ($p < 0.05$ by SNK) vs the negative control (#1: unconditioned growth medium). The positive control (#2) is 25% v/v astrocyte conditioned medium. See text for an explanation of the MAP2 data, with regard to the selectivity of MANF for the dopaminergic phenotype. * Significantly different, $p < 0.05$ vs negative control (#1). (B) Representative fluorescence images TH (left panels) and MAP2 (right panels) from the negative control (a and b, same field), positive control (c and d, same field) and hrMANF, 50 ng/ml (e and f, same field). Indirect immunofluorescence, TH (CY3): a, c and e; MAP2 (FITC): b, d and f. All fields × 20.

transporter (Burchett and Bannon, 1997; Katz et al., 1997). Dopaminergic neurons are also heterogeneous with regard to the expression of the calcium-binding protein calbindin-D_{28K} (Sanghera et al., 1994; Sanghera et al., 1995; Betarbet et al., 1997). We have usually assumed, but really do not know whether these neuronal sub-groups are homogeneous in terms of their physiological, neuroprotective requirements. Some of our recent evidence (unreported observation) suggests that CM from different astrocyte cell lines exert differential neuroprotective effects on different dopaminergic morphological subphenotypes.

These diverse results detailed above strongly suggest that dopaminergic neurons in the SNc may be a functionally heterogeneous population. If they are, especially in terms of their neuroprotective requirements, then the effective treatment of PD with a single neurotrophic factor might not be feasible. In that case, we will need to discover factors that are specific not to generic dopaminergic neurons, but to

subsets of dopaminergic neurons. If there is an equivalent functional heterogeneity in the ventral mesencephalic astrocyte population that matches the morphological heterogeneity of the nigral dopaminergic neurons, then the actual in vivo situation may be complex. At one extreme, and assuming a one to one match between astrocytes and dopaminergic neurons, at least six subphenotypes of ventral mesencephalic astrocytes may be implicated in the support of the six dopaminergic subtypes. Functional overlap may reduce this number to three. In that case a cocktail of at least two, and possibly three neurotrophic factors derived from ventral mesencephalic astrocytes would be needed to treat PD effectively. Although the scenario detailed above is somewhat speculative, and may seem unduly complex, it really is not, in terms of what is known about organizational complexity in other parts of the CNS, particularly the cerebellum (Eccles et al., 1967; Ito, 1997). Given the enormous effort, and lack of substantive progress over the last two decades in understanding and treating PD, we ought to begin to explore the difficult issues from a new perspective. The large number of molecules that are neuroprotective for dopaminergic neurons (Table 1), albeit in widely differing experimental paradigms, indirectly support the idea of multiple physiological neuroprotective molecules in vivo.

Patients with long-standing PD that have lost nearly 100% of their dopaminergic neurons will likely not benefit from neuroprotective drug therapy. Cell therapy will likely be the treatment of choice for this group of patients. Fortunately, substantial progress is being made in this area. The days of using minced pieces of nigral tissue in PD cell therapy should now be at end. We are transitioning into an era in which cells for cell therapy to treat PD will be prepared from stem cells or progenitors, packaged optimally, and prepared to retain viability of > 90% over a 24 h period. The phenotypic composition of the cell preparation should also be known, with dopaminergic neurons accounting for 25–50%, and astrocytes < 1.0%. These new ideas are presented in a recent publication (Peaire et al., 2003). When cell therapy is used to treat PD, it will be prudent to combine it with a neuroprotective strategy. The working assumption should be that exogenous dopaminergic neurons transplanted into the brain of a PD patient will die, as did their endogenous predecessors, if not protected. Significant progress has been made in delivering neuroprotective factors to localized regions of the primate CNS using lentiviral vectors, and the expression of the protein maintained over many months (Kordower et al., 2000; Galimi and Verma, 2002). The new Medtronics pump will likely prove to be clinically useful (Gardner et al., 1995). The issue of delivery remains an important topic in the application of protein therapeutics to treat neurodegenerative diseases. However, if there is a molecule with the right clinical indices that is strongly indicated for the treatment of PD, it is difficult to believe that an effective delivery mechanism will not be found. The effective movement of BDNF across the blood–brain barrier of rat has been described (Pardridge et al., 1994). Recent successes in the preparation of peptide mimetics that are more effective than their parent

precursors is also encouraging (Brenneman and Gozes, 1996).

The idea of astrocyte-derived secreted proteins as homotypic neuroprotective molecules may be an instance of a more general glia–neuron interaction in the nervous system. In a model system of tectal astrocytes and retinal ganglion cells (RGCs), astrocyte-derived molecules regulated the development, functioning and stabilization of synapses in vitro (Nagler et al., 2001; Ullian et al., 2001). In yet another paradigm, hippocampal astrocyctes were six-fold more effective in promoting neurogenesis from hippocampal adult stem cells, vs spinal cord astrocytes (Song et al., 2002). In two recent paradigms of differentiation of dopaminergic neurons, ventral mesencephalic glia were also involved (Wagner et al., 1999; Matsuura et al., 2001). Astrocyte-derived tumor necrosis factor alpha (TNFα) strengthened synaptic transmission in both dispersed and slice cultures, specifically by mediating increased expression of AMPA receptors in the postsynaptic neurons by an activity-dependent process (Beattie et al., 2002). Astrocytes can also modulate synaptic transmission at both inhibitory and excitatiory synapses (Kang et al., 1998; Newman and Zahs, 1998; Haydon, 2000). Similar results have been reported for the action of Schwann cells at the frog neuromuscular junction in the peripheral nervous system (PNS) (Robitaille, 1998). These diverse results of glia–neuron interactions indicate that we are beginning to acquire a clearer understanding of the functional and molecular interplay between glia and neurons in the CNS and PNS. These developments bode well for the development of drug candidates in the difficult search to find effective therapies to treat peripheral neuropathics and the major neurodegenerative diseases.

There has not yet been a dramatic breakthrough in the treatment of PD, but there are significant, modest accomplishments. The concept of homotypic, astrocyte-derived neurotrophic factors from the mesencephalon will likely prove to be important (Le Roux and Reh, 1995; Araque et al., 2001; Petrova et al., 2003). The use of lentiviral vectors to deliver factors to nondividing cells in the CNS over extended periods of several months is also a significant advance (Kordower et al., 2000). The preparation of dopaminergic neurons from progenitors initially, and eventually from stem cells for use in PD cell therapy, combined with methods to package the cells optimally and keep them viable over extended periods represents solid progress (Hynes and Rosenthal, 2000; Rodriguez-Pallares et al., 2001; Bjorklund et al., 2002; Peaire et al., 2003). There is an optimistic mood in the Parkinsonian research community. The challenge is to translate optimism into actual, effective therapy.

Abbreviations

cAMP	cyclic adenosine mono phosphate
ARTN	artemin
BDNF	brain-derived neurotrophic factor
CNS	central nervous system
CNTF	ciliary neurotrophic factor
EGF	epidermal growth factor
FGF	fibroblast growth factor
GDNF	glial cell line-derived neurotrophic factor
GFLs	GDNF family ligands
IGF	insulin-like growth factor
IL	interleukin
MANF	mesencephalic astrocyte-derived neurotrophic factor
MAO B-I	monoamine oxidase B inhibitors
NRTN	neurturin
NT-3	neurotrophin-3
PD	Parkinson's disease
PDGF	platelet-derived growth factor
PSPN	persephin
RGCs	retinal ganglion cells
SNc	substantia nigra zona compacta
SNr	substantia nigra zona reticulata
TGF	transforming growth factor
VEGF	vascular endothelial growth factor

References

Airaksinen, M.S. and Saarma, M. (2002) The gdnf family: signalling, biological functions and therapeutic value. Nat. Rev. Neurosci., 3: 383–394.

Albrecht, P.J., Dahl, J.P., Stoltzfus, O.K., Levenson, R. and Levison, S.W. (2002) Ciliary neurotrophic factor activates spinal cord astrocytes, stimulating their production and release of fibroblast growth factor-2, to increase motor neuron survival. Exp. Neurol., 173: 46–62.

Alexi, T. and Hefti, F. (1993) Trophic actions of transforming growth factor α on mesencephalic dopaminergic neurons developing in culture. Neuroscience, 55: 903–918.

Aliaga, E., Rage, F., Bustos, G. and Tapia-Aranchibia, L. (1998) BDNF gene transcripts in mesencephalic neurons and its differential regulation by NMDA. Neuroreport, 9: 1959–1962.

Alonso, R., Chaudieu, I., Diorio, J., Krishnamurthy, A., Quirion, R. and Boksa, P. (1993) Interleukin-2 modulates evoked release of [^{3}H]dopamine in rat cultured mesencephalic cells. J. Neurochem., 61: 1284–1290.

Altar, C.A., Boylan, C.B., Fritsche, M., Jackson, C., Hyman, C. and Lindsay, R.M. (1994) The neurotrophins NT-4/5 and BDNF augment serotonin, dopamine, and GABAergic systems during behaviorally effective infusions to the substantia nigra. Exp. Neurol., 130: 31–40.

Anglade, P., Tsuji, S., Javoy-Agid, F., Agid, Y. and Hirsch, E.C. (1995) Plasticity of nerve afferents to nigrostriatal neurons in Parkinson's disease. Ann. Neurol., 37: 265–272.

Araque, A., Carmignoto, G. and Haydon, P.G. (2001) Dynamic signaling between astrocytes and neurons. Annu. Rev. Physiol., 63: 795–813.

Azuma, M., Tamatani, T., Fukui, K., Yuki, T., Motegi, K. and Sato, M. (1996) Different signalling pathways involved in transforming growth factor-β_1-induced morphological change and type IV collagen synthesis in simian virus-40-immortalized normal human salivary gland duct and myoepithelial cell clones. Arch. Oral. Biol., 41: 413–424.

Baloh, R.H., Tansey, M.G., Lampe, P.A., Fahrner, T.J., Enomoto, H., Simburger, K.S., Leitner, M.L., Araki, T., Johnson, E.M., Milbrandt, J. (1998) Artemin, a novel member of the GDNF ligand family, supports peripheral and central neurons and signals through the GFRalpha3-RET receptor complex. Neuron, 21: 1291–1302.

Barde, Y.A. (1994) Neurotrophic factors: An evolutionary perspective. J. Neurobiol., 25: 1329–1333.

Beattie, E.C., Stellwagen, D., Morishita, W., Bresnahan, J.C., Ha, B.K., Von Zastrow, M., Beattie, M.S. and Malenka, R.C. (2002) Control of synaptic strength by glial TNFalpha. Science, 295: 2282–2285.

Beglova, N., Maliartchouk, S., Ekiel, I., Zaccaro, M.C., Saragovi, H.U. and Gehring, K. (2000) Design and solution structure of functional peptide mimetics of nerve growth factor. J. Med. Chem., 43: 3530–3540.

Betarbet, R., Turner, R., Chockkan, V., DeLong, M.R., Allers, K.A., Walters, J., Levey, A.I. and Greenamyre, J.T. (1997) Dopaminergic neurons intrinsic to the primate striatum. J. Neurosci., 17: 6761–6768.

Bjorklund, L.M., Sanchez-Pernaute, R., Chung, S., Andersson, T., Chen, I.Y., McNaught, K.S., Brownell, A.L., Jenkins, B.G., Wahlestedt, C., Kim, K.S., Isacson, O. (2002) Embryonic stem cells develop into functional dopaminergic neurons after transplantation in a Parkinson rat model. Proc. Natl. Acad. Sci. USA, 99: 2344–2349.

Blackburn, R.K. and Anderegg, R.J. (1997) Characterization of femtomole levels of proteins in solution using rapid proteolysis and nanoelectrospray ionization mass spectrometry. J. Am. Soc. Mass Spectrom., 8: 483–494.

Brenneman, D.E. and Gozes, I. (1996) A femtomolar-acting neuroprotective peptide. J. Clin. Invest., 97: 2299–2307.

Burchett, S.A. and Bannon, M.J. (1997) Serotonin, dopamine and norepinephrine transporter mRNAs: heterogeneity of distribution and response to'binge' cocaine administration. Mol. Brain Res., 49: 95–102.

Casper, D., Mytilineou, C. and Blum, M. (1991) EGF enhances the survival of dopamine neurons in rat embryonic mesencephalon primary cell culture. J. Neurosci. Res., 30: 372–381.

Davies, A.M. (1996) Paracrine and autocrine actions of neurotrophic factors. Neurochem. Res., 21: 749–753.

Defazio, G., Dal Toso, R., Benvegnù, D., Minozzi, M.C., Cananzi, A.R. and Leon, A. (1994) Parkinsonian serum carries complement-dependent toxicity for rat mesencephalic dopaminergic neurons in culture. Brain Res., 633: 206–212.

Durocher, Y., Perret, S. and Kamen, A. (2002) High-level and high-throughput recombinant protein production by transient transfection of suspension-growing human 293-EBNA1 cells. Nucleic Acids Res., 30: e9.

Eccles, J.C., Ito, M. and Szentagothai, J., Anonymous (1967) The Cerebellum as a Neuronal Machine. Springer-Verlag.

Engele, J. and Bohn, M.C. (1991) The neurotrophic effects of fibroblast growth factors on dopaminergic neurons in vitro are mediated by mesencephalic glia. J. Neurosci., 11: 3070–3078.

Engele, J., Rieck, H., Choi-Lundberg, D. and Bohn, M.C. (1996) Evidence for a novel neurotrophic factor for dopaminergic neurons secreted from mesencephalic glial cell lines. J. Neurosci. Res., 43: 576–586.

Farrer, M., Wavrant-De Vrieze, F., Crook, R., Boles, L., Perez-Tur, J., Hardy, J., Johnson, W.G., Steele, J., Maraganore, D., Gwinn, K. and Lynch, T. (1998) Low frequency of α-synuclein mutations in familial Parkinson's disease. Ann. Neurol., 43: 394–397.

Finberg, J.P.M., Takeshima, T., Johnston, J.M. and Commissiong, J.W. (1998) Increased survival of dopaminergic neurons by rasagiline, a monoamine oxidase B inhibitor. Neuroreport, 9: 703–707.

Galimi, F. and Verma, I.M. (2002) Opportunities for the use of lentiviral vectors in human gene therapy. Curr. Top. Microbiol. Immunol., 261: 245–254.

Gardner, B., Jamous, A., Teddy, P., Bergstrom, E., Wang, D., Ravichandran, G., Sutton, R. and Urquart, S. (1995) Intrathecal baclofen—a multicentre clinical comparison of the Medtronics Programmable, Cordis Secor and Constant Infusion Infusaid drug delivery systems. Paraplegia, 33: 551–554.

Giacobini, M.M.J., Almström, S., Funa, K. and Olson, L. (1993) Differential effects of platelet-derived growth factor isoforms on dopamine neurons in vivo: -BB supports cell survival, -AA enhances fiber formation. Neuroscience, 57: 923–929.

Goletz, T.J., Robetorye, S. and Pereira-Smith, O.M. (1994) Genetic analysis of indefinite division in human cells: Evidence for a common immortalizing mechanism in T and B lymphoid cell lines. Exp. Cell. Res., 215: 82–89.

Gotz, M.E., Freyberger, A. and Riederer, P. (1990) Oxidative stress: a role in the pathogenesis of Parkinson's disease. J. Neural. Transm. Suppl., 29: 241–249.

Gozes, I. (2001) Neuroprotective peptide drug delivery and development: potential new therapeutics. Trends. Neurosci., 24: 700–705.

Grace, A.A. and Bunney, B.S. (1984a) The control of firing pattern in nigral dopaamine neurons: burst firing. J. Neurosci., 4: 2877–2890.

Grace, A.A. and Bunney, B.S. (1984b) The control of firing pattern in nigral dopaminergic neurons: single spike firing. J. Neurosci., 4: 2866–2876.

Hagg, T. and Varon, S. (1993) Ciliary neurotrophic factor prevents degeneration of adult rat substantia nigra dopaminergic neurons in vivo. Proc. Natl. Acad. Sci. USA, 90: 6315–6319.

Hardman, C.D., McRitchie, D.A., Halliday, G.M., Cartwright, H.R. and Morris, J.G. (1996) Substantia nigra pars reticulata neurons in Parkinson's disease. Neurodegeneration, 5: 49–55.

Hartikka, J., Staufenbiel, M. and Lubbert, H. (1992) Cyclic AMP, but not basic FGF, increases the in vitro survival of mesencephalic dopaminergic neurons and protects them from MPP(+)-induced degeneration. J. Neurosci. Res., 32: 190–201.

Hashino, E., Shero, M., Junghans, D., Rohrer, H., Milbrandt, J. and Johnson, E.M.J. (2001) GDNF and neurturin are target-derived factors essential for cranial parasympathetic neuron development. Development, 128: 3773–3782.

Haydon, P.G. (2000) Neuroglial networks: neurons and glia talk to each other. Curr. Biol., 10: R712–R714.

Henderson, C.E. (1995) Neurotrophic factors as therapeutic agents in amyotrophic lateral sclerosis—Potential and pitfalls. Adv. Neurol., 68: 235–240.

Henrich-Noack, P., Prehn, J.H.M. and Krieglstein, J. (1994) Neuroprotective effects of TGF-beta1. J. Neural. Transm., 98(Suppl. 43): 33–45.

Hou, J.G.G. and Mytilineou, C. (1996) Secretion of GDNF by glial cells does not account for the neurotrophic effect of bFGF on dopamine neurons in vitro. Brain Res., 724: 145–148.

Hubble, J.P., Kurth, J.H., Glatt, S.L., Kurth, M.C., Schellenberg, G.D., Hassanein, R.E.S., Lieberman, A. and Koller, W.C. (1998) Gene–toxin interaction as a putative risk factor for Parkinson's disease with dementia. Neuroepidemiology, 17: 96–104.

Hyman, C., Hofer, M., Barde, Y.-A., Juhasz, M., Yancopoulos, G.D., Squinto, S.P. and Lindsay, R.M.

(1991) BDNF is a neurotrophic factor for dopaminergic neurons of the substantia nigra. Nature, 350: 230–232.

Hynes, M. and Rosenthal, A. (2000) Embryonic stem cells go dopaminergic. Neuron, 28: 11–14.

Ito, M. (1997) Cerebellar microcomplexes. Int. Rev. Neurobiol., 41: 475–487.

Itoh, K., Weis, S., Mehraein, P. and Müller-Höcker, J. (1997) Defects of cytochrome c oxidase in the substantia nigra of Parkinson's disease: An immunohistochemical and morphometric study. Mov. Disord., 12: 9–16.

Jain, R.K. (1998) The next frontier of molecular medicine: Delivery of therapeutics. Nat. Med., 4: 655–657.

Kahn, M.A., Huang, C.J., Caruso, A., Barresi, V., Nazarian, R., Condorelli, D.F. and De Vellis, J. (1997) Ciliary neurotrophic factor activates JAK/Stat signal transduction cascade and induces transcriptional expression of glial fibrillary acidic protein in glial cells. J. Neurochem., 68: 1413–1423.

Kang, J., Jiang, L., Goldman, S.A. and Nedergaard, M. (1998) Astrocyte-mediated potentiation of inhibitory synaptic transmission. Nat. Neurosci., 1: 683–692.

Katz, J.L., Newman, A.H. and Izenwasser, S. (1997) Relations between heterogeneity of dopamine transporter binding and function and the behavioral pharmacology of cocaine. Pharmacol. Biochem. Behav., 57: 505–512.

Kordower, J.H., Emborg, M.E., Bloch, J., Ma, S.Y., Chu, Y., Leventhal, L., McBride, J., Chen, E.Y., Palfi, S., Roitberg, B.Z., Brown, W.D., Holden, J.E., Pyzalski, R., Taylor, M.D., Carvey, P., Ling, Z., Trono, D., Hantraye, P., Deglon, N. and Aebischer, P. (2000) Neurodegeneration prevented by lentiviral vector delivery of GDNF in primate models of Parkinson's disease. Science, 290: 767–773.

Kotzbauer, P.T., Lampe, P.A., Heuckeroth, R.O., Golden, J.P., Creedon, D.J., Johnson, E.M., Jr. and Milbrandt, J. (1996) Neurturin, a relative of glial-cell-line-derived neurotrophic factor. Nature, 384: 467–470.

Krieglstein, K. and Unsicker, K. (1994) Transforming growth factor-β promotes survival of midbrain dopaminergic neurons and protects them against N-methyl- 4- phenylpyridinium ion toxicity. Neuroscience, 63: 1189–1196.

Kristal, B.S., Conway, A.D., Brown, A.M., Jain, J.C., Ulluci, P.A., Li, S.W. and Burke, W.J. (2001) Selective dopaminergic vulnerability: 3,4-dihydroxyphenylacetaldehyde targets mitochondria. Free Radic. Biol. Med., 30: 924–931.

Le Roux, P.D. and Reh, T.A. (1995) Astroglia demonstrate regional differences in their ability to maintain primary dendritic outgrowth from mouse cortical neurons in vitro. J. Neurobiol., 27: 97–112.

Lee, M.K., Stirling, W., Xu, Y., Xu, X., Qui, D., Mandir, A.S., Dawson, T.M., Copeland, N.G., Jenkins, N.A. and Price, D.L. (2002) Human alpha-synuclein-harboring familial Parkinson's disease-linked Ala-53- > Thr mutation causes neurodegenerative disease with alpha-synuclein aggregation

in transgenic mice. Proc. Natl. Acad. Sci. USA, 99: 8968–8973.

Li, J., Kelly, J.F., Chernushevich, I., Harrison, D.J. and Thibault, P. (2000) Separation and identification of peptides from gel-isolated membrane proteins using a microfabricated device for combined capillary electrophoresis/nanoelectrospray mass spectrometry. Anal. Chem., 72: 599–609.

Li, S.W., Lin, T.S., Minteer, S. and Burke, W.J. (2001) 3,4-Dihydroxyphenylacetaldehyde and hydrogen peroxide generate a hydroxyl radical: possible role in Parkinson's disease pathogenesis. Brain Res. Mol. Brain Res., 93: 1–7.

Lin, L.-F.H., Doherty, D.H., Lile, J.D., Bektesh, S. and Collins, F. (1993) GDNF: A glial cell-line derived neurotrophic factor for midbrain dopaminergic neurons. Science, 260: 1130–1132.

Liu, J.P. and Lauder, J.M. (1992) S-100β and insulin-like growth factor-II differentially regulate growth of developing serotonin and dopamine neurons in vitro. J. Neurosci. Res., 33: 248–256.

Lotto, R.B., Asavaritikrai, P., Vali, L. and Price, D.J. (2001) Target-derived neurotrophic factors regulate the death of developing forebrain neurons after a change in their trophic requirements. J. Neurosci., 21: 3904–3910.

Louis, J.C., Magal, E., Muir, D., Manthorpe, M. and Varon, S. (1992) CG-4, a new bipotential glial cell line from rat brain, is capable of differentiating in vitro into either mature oligodendrocytes or type-2 astrocytes. J. Neurosci. Res., 31: 193–204.

Maggio, R., Riva, M., Vaglini, F., Fornai, F., Racagni, G. and Corsini, G.U. (1997) Striatal increase of neurotrophic factors as a mechanism of nicotine protection in experimental parkinsonism. J. Neural. Trans., 104: 1113–1123.

Masalha, R., Herishanu, Y., Alfahel-Kakunda, A. and Silverman, W.F. (1997) Selective dopamine neurotoxicity by an industrial chemical: An environmental cause of Parkinson's disease? Brain Res., 774: 260–264.

Matsuura, N., Lie, D.C., Hoshimaru, M., Asahi, M., Hojo, M., Ishizaki, R., Hashimoto, N., Noji, S., Ohuchi, H., Yoshioka, H. and Gage, F.H. (2001) Sonic hedgehog facilitates dopamine differentiation in the presence of a mesencephalic glial cell line. J. Neurosci., 21: 4326–4335.

McRitchie, D.A., Cartwright, H.R. and Halliday, G.M. (1997) Specific A10 dopamiergic nuclei in the midbrain degenerate in Parkinson's disease. Exp. Neurol., 144: 202–213.

Mena, M.A., Casarejos, M.J., Gimenéz-Gallego, G. and Garcia De Yebenes, J. (1995) Fibroblast growth factors: Structure-activity on dopamine neurons in vitro. J. Neural. Transm. Park. Dis. Dement. Sect., 9: 1–14.

Morgan, S., Steiner, H., Rosenkranz, C. and Huston, J.P. (1986) Dissociation of crossed and incrossed nigrostriatal projections with respect to site of origin in the rat. Neuroscience, 17: 609–614.

Nagler, K., Mauch, D.H. and Pfrieger, F.W. (2001) Glia-derived signals induce synapse formation in neurones of the rat central nervous system. J. Physiol. (Lond.), 533: 665–679.

Newman, E.A. and Zahs, K.R. (1998) Modulation of neuronal activity by glial cells in the retina. J. Neurosci., 18: 4022–4028.

Niijima, K., Araki, M., Ogawa, M., Nagatsu, I., Sato, F., Kimura, H. and Yoshida, M. (1990) Enhanced survival of cultured dopamine neurons by treatment with soluble extracts from chemically deafferentiated striatum of adult rat brain. Brain Res., 528: 151–154.

Nikkhah, G., Odin, P., Smits, A., Tingström, A., Othberg, A., Brundin, P., Funa, K. and Lindvall, O. (1993) Platelet-derived growth factor promotes survival of rat and human mesencephalic dopaminergic neurons in culture. Exp. Brain Res., 92: 516–523.

O'Malley, E.K., Sieber, B.-A., Black, I.B. and Dreyfus, C.F. (1992) Mesencephalic type I astrocytes mediate the survival of substantia nigra dopaminergic neurons in culture. Brain Res., 582: 65–70.

O'Malley, E.K., Sieber, B.-A., Morrison, R.S., Black, I.B. and Dreyfus, C.F. (1994) Nigral Type I astrocytes release a soluble factor that increases dopaminergic neuron survival through mechanisms distinct from basic fibroblast growth factor. Brain Res., 647: 83–90.

Panchision, D.M., Martin-DeLeon, P.A., Takeshima, T., Johnston, J.M., Shimoda, K., Tsoulfas, P., McKay, R.D. and Commissiong, J.W. (1998) An immortalized, type-1 astrocyte of mesencephalic origin source of a dopaminergic neurotrophic factor. J. Mol. Neurosci., 11: 209–221.

Pardridge, W.M., Kang, Y.-S. and Buciak, J.L. (1994) Transport of human recombinant brain-derived neurotrophic factor (BDNF) through the rat blood–brain barrier in vivo using vector-mediated peptide drug delivery. Pharm. Res., 11: 738–746.

Peaire, A. E., Takeshima, T., Johnston, J. M., Isoe, K., Nakashima, K. and Commissiong, J. W (2003) Production of dopaminergic neurons for cell therapy in the treatment of Parkinson's disease. J. Neurosci. Meth., 124: 61–74.

Petrova, P.S., Raibekas, A., Pevsner, J., Vigo, N., Anafi., M., Moore, M. K., Peaire, A. E., Shridhar, V., Smith, D. I., Kelly, J., Durocher, Y. and Commissiong, J. W. (2003) MANF: A new mesencephalic, astrocyte-derived neurotrophic factor with selectivity for dopaminergic neurons, J. Mol. Neurosci., 20: 173–187.

Plante-Bordeneuve, V., Taussig, D., Thomas, F., Said, G., Wood, N.W., Marsden, C.D. and Harding, A.E. (1997) Evaluation of four candidate genes encoding proteins of the dopamine pathway in familial and sporadic Parkinson's disease: Evidence for association of a DRD2 allele. Neurology, 48: 1589–1593.

Poulsen, K.T., Armanini, M.P., Klein, R.D., Hynes, M.A., Phillips, H.S. and Rosenthal, A. (1994) TGFbeta2 and TGFbeta3 are potent survival factors for midbrain dopaminergic neurons. Neuron, 13: 1245–1252.

Purves, D. and Litchman, J.W. (1985). Principles of Neural Development. Sinauer Associates Inc., Sunderland, Mass.

Riedl, A.G., Watts, P.M., Jenner, P. and Marsden, C.D. (1998) P450 enzymes and Parkinson's disease: The story so far. Mov. Disord., 13: 212–220.

Robitaille, R. (1998) Modulation of synaptic efficacy and synaptic depression by glial cells at the frog neuromuscular junction. Neuron, 21: 847–855.

Rodriguez-Pallares, J., Rey, P., Soto-Otero, R. and Labandeira-Garcia, J.L. (2001) N-acetylcysteine enhances production of dopaminergic neurons from mesencephalic-derived precursor cells. Neuroreport, 12: 3935–3938.

Roy, E. and Bédard, P.J. (1993) Deprenyl increases survival of rat foetal nigral neurones in culture. Neuroreport, 4: 1183–1186.

Rozengurt, E. (1995) Convergent signalling in the action of integrins, neuropeptides, growth factors and oncogenes. Cancer Surv., 24: 81–96.

Ruberg, M., Scherman, D., Javoy-Agid, F. and Agid, Y. (1995) Dopamine denervation, age of onset, and Parkinson's disease. Neurology, 45: 392.

Sanghera, M.K., Manaye, K.F., Liang, C.-L., Iacopino, A.M., Bannon, M.J. and German, D.C. (1994) Low dopamine transporter mRNA levels in midbrain regions containing calbindin. Neuroreport, 5: 1641–1644.

Sanghera, M.K., Zamora, J.L. and German, D.C. (1995) Calbindin-D_{28k}-containing neurons in the human hypothalamus: Relationship to dopaminergic neurons. Neurodegeneration, 4: 375–381.

Seidman, K.J.N., Teng, A.L., Rosenkopf, R., Spilotro, P. and Weyhenmeyer, J.A. (1997) Isolation, cloning and characterization of a putative type-1 astrocyte cell line. Brain Res., 753: 18–26.

Shimura, H., Hattori, N., Kubo, S., Mizuno, Y., Asakawa, S., Minoshima, S., Shimizu, N., Iwai, K., Chiba, T., Tanaka, K. and Suzuki, T. (2000) Familial Parkinson disease gene product, parkin, is a ubiquitin-protein ligase. Nat. Genet., 25: 302–305.

Shridhar, V., Rivard, S., Shridhar, R., Mullins, C., Bostick, L., Sakr, W., Grignon, D., Miller, O.J. and Smith, D.I. (1996) A gene from human chromosomal band 3p21.1 encodes a highly conserved arginine-rich protein and is mutated in renal cell carcinomas. Oncogene, 12: 1931–1939.

Silverman, W.F., Krum, J.M., Mani, N. and Rosenstein, J.M. (1999) Vascular, glial and neuronal effects of vascular endothelial growth factor in mesencephalic explant cultures. Neuroscience, 90: 1529–1541.

Skaper, S.D. and Walsh, F.S. (1998) Neurotrophic molecules: strategies for designing effective therapeutic molecules in neurodegeneration. Mol. Cell Neurosci., 12: 179–193.

Song, H., Stevens, C.F. and Gage, F.H. (2002) Astroglia induce neurogenesis from adult stem cells. Nature, 417: 39–44.

Studer, L., Tabar, V. and McKay, R.D.G. (1998) Transplantation of expanded mesencephalic precursors leads to recovery in Parkinsonian rats. Nat. Neurosci., 1: 290–295.

Takeshima, T., Johnston, J.M. and Commissiong, J.W. (1994) Mesencephalic type 1 astrocytes rescue dopaminergic neurons from death induced by serum deprivation. J. Neurosci., 14: 4769–4779.

Takeshima, T., Shimoda, K., Johnston, J.M. and Commissiong, J.W. (1996) Standardized methods to bioassay neurotrophic factors for dopaminergic neurons. J. Neurosci. Meth., 67: 27–41.

Tooyama, I., McGeer, E.G., Kawamata, T., Kimura, H. and McGeer, P.L. (1994) Retention of basic fibroblast growth factor immunoreactivity in dopaminergic neurons of the substantia nigra during normal aging in humans contrasts with loss in Parkinson's disease. Brain Res., 656: 165–168.

Ullian, E.M., Sapperstein, S.K., Christopherson, K.S. and Barres, B.A. (2001) Control of synapse number by glia. Science, 291: 657–660.

Von Bartheld, C.S. and Johnson, J.E. (2001) Target-derived BDNF (brain-derived neurotrophic factor) is essential for the survival of developing neurons in the isthmo-optic nucleus. J. Comp. Neurol., 433: 550–564.

von Coelln, R., Unsicker, K. and Krieglstein, K. (1995) Screening of interleukins for survival-promoting effects on cultured mesencephalic dopaminergic neurons from embryonic rat brain. Brain Res. Dev. Brain Res., 89: 150–154.

Wagner, J., Akerud, P., Castro, D.S., Holm, P.C., Canals, J.M., Snyder, E.Y., Perlmann, T. and Arenas, E. (1999) Induction of midbrain dopaminergic phenotype in Nurr1-overexpressing neural stem cells by type 1 astrocytes. Nat. Biotech., 17: 653–659.

Weiss-Wunder, L.T. and Chesselet, M.-F. (1991) Subpopulations of mesencephalic dopaminergic neurons express different levels of tyrosine hydroxylase messenger RNA. J. Comp. Neurol., 303: 478–488.

Xie, Y. and Longo, F.M. (2000) Neurotrophin small-molecule mimetics. Prog. Brain Res., 128: 333–347.

Yamada, T., McGeer, P.L., Baimbridge, K.G. and McGeer, E.G. (1990) Relative sparing in Parkinson's disease of substantia nigra dopamine neurons containing calbindin-D_{28K}. Brain Res., 526: 303–307.

Yamasaki, Y., Shigeno, T., Furukawa, Y. and Furukawa, S. (1998) Reduction in brain-derived neurotrophic factor protein level in the hippocampal CA1 dendritic field precedes the delayed neuronal damage in the rat brain. J. Neurosci. Res., 53: 318–329.

Yoritaka, A., Hattori, N., Yoshino, H. and Mizuno, Y. (1997) Catechol-O-methyltransferase genotype and susceptibility to Parkinson's disease in Japan. J. Neural Trans., 104: 1313–1317.

Progress in Brain Research, Vol. 146
ISSN 0079-6123

CHAPTER 13

Neurobehavioral coping to altered gravity: endogenous responses of neurotrophins

Nadia Francia[1], Daniela Santucci[1,*], Luigi Aloe[2] and Enrico Alleva[1]

[1]*Behavioral Pathophysiology Section, Istituto Superiore di Sanità (ISS), Rome, Italy*
[2]*Institute of Neurobiology and Molecular Medicine, Consiglio Nazionale delle Ricerche (CNR), Rome, Italy*

Abstract: An altered gravitational environment represents a unique challenge for biological systems that have evolved against gravitational background. Ground-based and space research indicates that the developing nervous system is potentially affected by exposure to hyper/microgravity. With the construction of the orbiting International Space Station long-term research on the nervous system will be possible. With this perspective, we started ground-based studies to characterize mouse behavioral responses to rotation-induced 2 g hypergravity, using a custom-made centrifuge device. Brain levels of nerve growth factor (NGF) and brain-derived neurotrophic factor (BDNF) as well as NGF and BDNF expression and mast cell distribution in heart and lung, were evaluated and correlated with the changes in mouse behavior upon hypergravity exposure. Hypergravity strongly affected the spontaneous activity of the animals, selectively modifying mouse behavioral repertoire. Such changes were mainly related to variations in brain levels of NGF, while BDNF was slightly affected, thus confirming a role for these neurotrophins in neuronal plasticity underlying experience-induced neurobehavioral changes. Moreover, gender differences were observed in both behavioral and neurobiological responses to hypergravity. These results indicate that changes in the gravitational environment might represent a useful tool to investigate the neurobiological and behavioral responses to stressors and may provide insights into the mechanisms underlying development and plasticity of nervous system in brain, heart, and lung.

Keywords: BDNF; NGF; behavior; hypergravity; motion sickness; mouse

Space neuroscience: an emerging field

The first living creature to be launched into the space was the Soviet dog Laika aboard Sputnik 2 on November 1957, while the first human spaceflight was by Yuri Alekseyevich Gagarin aboard Vostok-1 on April 1961 (West, 2000). Physiological measurements were relatively crude in these early flights, but over the past 40 years the degree of methodological sophistication has greatly increased, culminating in the US Spacelab studies of the 1990s. The US Spacelab program has enabled high-caliber research

in microgravity. Eleven Spacelab missions were flown over a period of 15 years (from 1983 to 1998), and numerous experiments were performed involving research in different fields, such as cardiovascular/cardiopulmonary, musculoskeletal, endocrine, cell biology, and integrative physiology.

Neurolab (STS-90), launched on April 1998, was a dedicated 16-day Life Sciences mission that focused on neuroscience research (for more information, see http://lsda.jsc.nasa.gov/neurolab/neurolab.stm and 'online Neurolab mission brochure' website http://www.psu.edu/nasa/brochure/nl01.html). The Neurolab mission was a milestone in the space neuroscience history and with the assembly of a space station orbiting around the Earth, the International Space Station (ISS), long-duration space flights would have become common, opening new scientific perspectives

*Correspondence to: D. Santucci, Behavioral Pathophysiology Section, Istituto Superiore di Sanità (ISS) Viale Regina Elena, 299, I-00161 Rome, Italy. Tel.: + 39-06-4990-2039; Fax: + 39-06-495-7821; E-mail: santucci@iss.it

DOI: 10.1016/S0079-6123(03)46013-5

for the space neuroscience field. Moreover, the risks for astronauts associated to long-term permanence in the space led to the necessity of a better understanding of the effects induced by short- and/ or long-term exposure to altered gravitational environments.

When the orbit assembly sequence of ISS will be completed, important research into physiological alterations associated with long-term weightlessness exposure may be conducted. In this perspective, our research group started a series of ground-based experiments mainly focusing on setting up a mouse model to study the effects of exposure to hypergravitational fields. The ultimate purpose of our study was the smooth transfer of this animal model from ground-based settings to weightlessness aboard of the ISS.

A review of neurobehavioral effects induced by exposure to altered gravity

Experimental evidence from both space and ground-based research indicates that exposure to altered gravitational environment (weightlessness, micro- or hypergravity) induces changes in mammalian vestibular system, affecting postural and motor functions. Specifically, rats and hamsters data indicate that exposure to altered gravitational fields induces deficiencies in water immersion righting performance, a delay in surface righting response, difficulty to maintain the balance during the swim, a decrease in both swimming and walking speed, a postural attitude to increase the support surface at rest and motor hyperactivity (Sondag et al., 1996; Ronca and Alberts, 1997; Walton, 1998; Wubbels and de Jong, 2000; Bouët et al., 2003). Such disturbances have been related to altered gain in the gravity-sensitive portion of the vestibular system (Sondag et al., 1997; Fox et al., 1998; Wubbels and de Jong, 2000).

Most of these locomotor deficits are transient, normal behavior recovers after a few days are spent in normal Earth gravity (1 g). Nevertheless, when animals are exposed to altered gravity during 'critical phases' of their development, some anomalies in vestibular-induced behaviors may persist into adulthood. In rats these critical phases correspond to the period around birth (when the development of the

vestibular system, the descending pathways and postural reflex regulation occurs) and to the period around weaning (when the development of the complex adult locomotor patterns is completed; Clarac et al., 1998; Walton, 1998).

Giménez y Ribotta et al. (1998) observed a substantial delay of the development of the cortical monoaminergic projections to the spinal cord and an alteration in the organization and ultrastructure of these projections in young rats exposed to hypergravity from day 11 of gestation to postnatal days 15 (PND 15). These modifications continued to persist after 8 months of normal gravity. Moreover, rats exposed to rotational-generated hypergravity from gestational day 11 to PND 6 or PND 21 showed an altered cerebellar growth (Baer et al., 2000; Sajdel-Sulkowska et al., 2000), while rats flown in 1998 on the space shuttle (Neurolab) from PND 8 to PND 24 showed an abnormal development of extensor motor neurons (Inglis et al., 1999) and changes in the number and morphology of cortical synapses (DeFelipe et al., 2002). Finally, rats developed in a microgravity environment for 16 days (during the Neurolab project), as well as mice exposed to hypergravity, appeared to use different search strategies in the early phases of a spatial learning task in the absence of a clear deficit in the ability to use spatial information and to form memories of place (Temple et al., 1999, Mandillo et al., 2003).

In addition to postural, motor coordination and orientation problems, exposure to altered gravity has been extensively reported to also cause a motion sickness (MS) syndrome in mammals. MS has been described as an illness triggered by a 'sensory conflict' involving the vestibular system, occurring when sensory inputs regarding body position in space are contradictory or different from those predicted from experience (reviewed by Money, 1970; Yates et al., 1998). During space flights, astronauts may experience a related disorder, named space adaptation syndrome (Reschke et al., 1986; Homick et al., 1997), which may compromise the mission. In a large variety of animal species MS is associated with the occurrence of an emetic response. However, many species, including rodents and rabbits, lack the ability to vomit and thus are unable to exhibit the most obvious indicator of MS (Borison, 1989; Yates et al., 1998). Mitchell et al. (1977a,b) demonstrated that, after exposure to rotation-induced hypergravity, rats engaged in anti-nausea pica behavior (from the name of the bird *Pica pica*; the ingestion of significant amount of nonnutritive substances, such as kaolin, sawdust and wood). They investigated the effects of rotation on postrotational consumption of food, water and kaolin (a hydrated aluminum silicate with a clay-like taste) and observed a significant increase from baseline in postrotational kaolin consumption. Since kaolin consumption in species incapable of emesis corresponded to several of the principle characteristics of MS exhibited by species capable of emesis, pica behavior was validated as a quantitative index of MS.

From human and rodent studies it appears that susceptibility to MS and the vulnerability of the vestibular system upon exposure to rotational stimuli changes with age (Tyler and Bard, 1949; Money, 1970; McCaffrey and Graham, 1980; Chinn and Smith, 1993; Sondag et al., 1996, 1997; Wubbels and de Jong, 2000). In rats, effects of exposure to rotationally generated hypergravity have been reported to be particularly marked just after birth and again at weaning (Oyama and Platt, 1967; Baer et al., 2000; Sajdel-Sulkowska et al., 2000).

A mouse model for motion sickness: behavioral and neurobiological characterization

Initial aim of our research was to characterize MS in adult CD-1 mice exposed to rotation-induced hypergravity (Santucci et al., 2000). Since data on MS susceptibility related to sex are poor and controversial, and few data on sex differences in rodent pica behavior are presently available, mice of both sexes were studied. The animals were exposed to 1 h rotation-induced 2 g hypergravity using a custom-made centrifuge device (see below) and pica behavior was measured through kaolin daily intake. To finely characterize behavioral response to hypergravity stimulation, mouse spontaneous activity was monitored 1 h before, 1 h during, and 1 h after hypergravity exposure.

Hypergravity facility

Hypergravity condition was elicited by a centrifuge, a custom-made prototype designed and manufactured

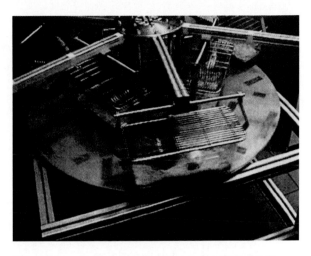

Fig. 1. The centrifuge facility. For a technical description see the text.

by Isolceram (I-00040 Rocca Priora, Italy) (Fig. 1). The centrifuge used during our hypergravity pilot experiments consisted of a turntable (radius = 50 cm) set in motion by a central rotor, the number of turns of which could be adjusted by a digital switch. The turntable could hold up to six home cages. During the experiment it was rotated at a constant rate: an angular velocity of 336°/s (56 rpm) producing a resultant linear acceleration of 2 g.

Successively, the centrifuge was modified in an apparatus consisting of six radial aluminum arms (50 cm long) mounted on a central rotor in the vertical axis of the centrifuge. Each arm was fitted with an adjustable bracket designed to hold a single cage of identical dimensions to the home cage (33 × 13 × 14 cm), such that cage hung and swung freely under the arm. The centrifuge was also equipped with a system of six wide-angle videocameras connected to a monitor and six videorecorders. This system allowed observing, as well as recording, the mouse spontaneous activity in each centrifuge cage. Because of the rotation, the centrifuge cage swung-out from the vertical direction at a constant angle, whose magnitude depended on the rotation speed. Since at a given rotation speed the distance along the arm of the bracket from the central rotor (corresponding to the axis of rotation) determined the force generated inside the centrifuge cage, during the experiment brackets were arranged at two levels: 21 and 45 cm from the central rotor so that rotation at a constant rate of 50 rpm produced a resultant linear acceleration of approximately 1 g (1.09 g) inside the inner cages and approximately 2 g (1.85 g) inside the outer cages. The mice were totally free-moving in the centrifuge cages, a fundamental condition for studying the effects of gravity changes on behavioral performance. Because of their own free movements during rotation, in addition to linear acceleration, mice were exposed to variable Coriolis forces, depending upon the speed and direction of animal motion within the centrifuge cages.

Behavioral observations

Acute exposure to 2 g hypergravity strongly affected the behavioral response of CD-1 adult mice (Fig. 2). Pica behavior aroused in mice, as evidenced by the increase in kaolin intake in the day immediately following the rotation. Moreover, decreases in spontaneous activity both during and immediately after the rotation was also observed, in agreement with data previously reported for rats (Eskin and Riccio, 1966; Ossenkopp and Frisken, 1982), suggesting that mice experienced MS as consequence of being exposed to hypergravity. Nevertheless, spontaneous activity recovered to levels similar to those observed in prerotation.

Of interest was the observation that hypergravity induced selective modifications of the mouse behavioral repertoire, almost totally suppressing specific behavioral endpoint, such as *rearing*, *digging* and *face washing*, while inducing *open eyes resting* behavior. This latter behavior showed a characteristic profile over time: substantially absent before rotation, it appeared after 5 min of rotation and rapidly increased, lasting to very high levels throughout all hypergravity exposure and slowly decreasing only after rotation. These observations suggested that in mice, in addition to pica behavior, *open eyes resting* might be a behavioral indication of the arousal of MS syndrome. Interestingly, although both males and females experienced MS, sex differences appeared to emerge in postrotational period where females constantly persisted in performing *open eyes resting*, while males already showed signs of recovery from MS syndrome 10 min after hypergravity stimulus cessation.

Data from the literature report sex differences in susceptibility to MS in Japanese house musk shrew (*Suncus murinus*) and humans (*Homo sapiens*) with females resulting, in general, more responsive than males to motion stimuli (Javid and Naylor, 1999; Dobie et al., 2001). Moreover, studies on humans indicate that although women show greater incidence in MS history, they did not differ from men in severity of symptoms (Jokerst et al., 1999; Park and Hu, 1999). An interesting aeromedical study on the role of women in military aviation underlines that, although men are on the average, larger, stronger, and more aerobically fit than women, there are large variations within each sex and a large overlap between the sexes. Gender differences disappear when allowance is made for size, strength, and fitness (Lyons, 1992).

Neurobiological observations: endogenous neurotrophins in the central nervous system

In order to correlate behavioral changes with neurobiological modifications, brain levels of NGF and BDNF were assessed 1 h after rotation. NGF and BDNF are well-studied polypeptide growth factors involved in the development and maintenance of specific peripheral and central populations of neuronal cells. In addition to their trophic function, NGF and BDNF act as modulators of synaptic plasticity, enhancing synaptic strength, altering spine density and promoting dendritic branching and axonal sprouting (reviewed by Thoenen, 1995; Lindsay, 1996). The importance of the role of NGF and BDNF on the neurohormonal pathways underlying the response to stress is increasingly recognized. In the last 20 years, our research group has been investigating the neurobehavioral responses to different stressing situations in rodent models, evidencing that stressful events may induce changes in the behavioral repertoire which are related to modifications in brain levels of neurotrophins, especially NGF (Alleva and Aloe, 1989; Aloe et al., 1994; Alleva and Santucci, 2001). Alleva and Aloe (1989) hypothesized that the rather rapid increase in the levels of brain NGF, which follows a stressful event, could allow some phenomena of brain plasticity to take place, as a result of appropriate

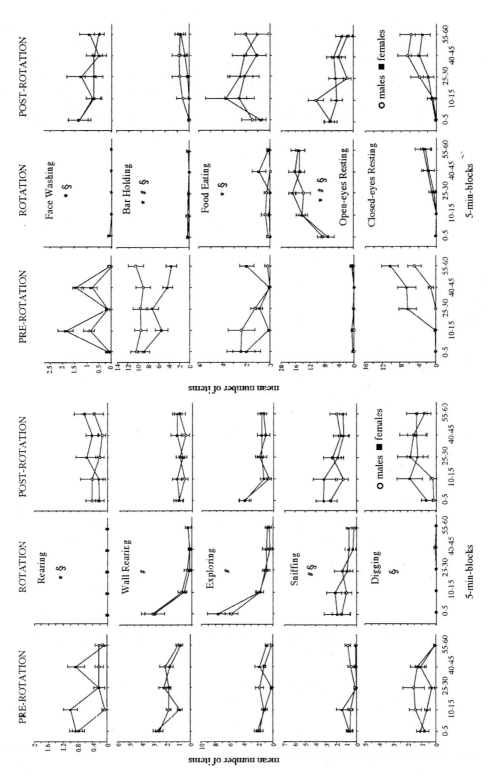

Fig. 2. Behavioral responses of adult male and female CD-1 mice occurring in 5-min blocks during 60 min pre-rotation and post-rotation periods. Data are means ± s.e. of eight subjects for final group. Significant comparisons ($p < 0.05$) of pre-rotational period vs rotational (*), or post-rotational (#) periods and rotational vs postand post-rotational periods (§) are reported. From Santucci et al. (2000).

(or inappropriate) coping strategies by way of modulating attentional and/or cognitive processes while controlling the stressor itself.

Mouse behavioral repertoire modifications observed under hypergravity environment were related to changes in brain levels of NGF and BDNF. Specifically, in male mice exposed to 1 h of rotation we found that the amount of NGF was dramatically increased in frontal cortex, hypothalamus and hippocampus of male mice 20 min after hypergravity exposure, while no changes were observed in brain levels of BDNF (Santucci et al., 2000) (Table 1).

Recent observations have evidenced that rotation also affects NGF, but not BDNF, brain levels of female mice (Fig. 3). In particular, as previously evidenced in rotated male, a tendency to increased NGF was also observed in the frontal cortex of rotated females ($F(1, 7) = 4.03$, $p = 0.085$). Differently from rotated males, in females no modification was noticed in the hippocampal area, while a strong reduction in the amount of NGF was observed in the cerebellum of rotated females ($F(1, 7) = 10.95$, $p = 0.013$). Finally, contrarily to what observed in males, in female mice rotation decreased NGF hypothalamic levels ($F(1, 7) = 11.85$, $p = 0.011$).

As previously mentioned, modifications in the levels of NGF synthesis in the hypothalamus as a neurobiological response to stressful events have been extensively described (Alleva and Aloe, 1989; Aloe et al., 1994; Alleva and Santucci, 2001). This may indicate that this neurotrophin plays a key role in the modulation of neuroendocrine activities, to maintain physiological homeostasis/homeodynamics

(Levi-Montalcini et al., 1990). Indeed, frontal cortex and hypothalamus are brain areas known to be involved in MS syndrome (Uno et al., 1997; Yates et al., 1998), while vestibulocerebellum (uvula/nodulus and flocculus) contributes to the vestibular reflexes which regulate gaze stability and postural control by modulating the central vestibular circuit (Bruce and Fritzsch, 1997; Uno et al., 2000, 2002). Our results strongly suggest an involvement of NGF in mouse neurobehavioral response to hypergravity, probably contributing to the modulation of vestibular pathways responsible for brain elaboration of vestibular inputs, ultimately inducing behavioral adaptive responses to the hypergravity stimulus. The gender differences in neurobehavioral responses encountered might mirror sex differences in strategies adopted to cope with changes in gravitational environment.

Effects of hypergravity on basal expression of NGF and BDNF in other systems

NGF and BDNF are also produced and released by cells of the visual system, where they play an important role in the maintenance, function and neuronal plasticity of retinal, cortical and geniculate cells during early development and in adult life (Castrén et al., 1992; McAllister et al. 1995; Cellerino and Maffei, 1996). Since neurons of the adult visual system undergo extensive and continual adaptation in response to environmental demands (Gu et al., 1998), the effects of changes due to gravitational forces on the basal levels of endogenous NGF and BDNF in

TABLE 1

Endogenous NGF and BDNF levels (expressed as pg/g of fresh tissue) in the CNS of unrotated and rotated male CD-1 mice

	NGF		BDNF	
	Unrotated	Rotated	Unrotated	Rotated
Olfactory Bulbs	8150 ± 980	6690 ± 768	$12,703 \pm 285$	$13,064 \pm 1308$
Frontal Cortex	1298 ± 190	2120 ± 125 [a]	$12,697 \pm 527$	$12,298 \pm 1095$
Parietal Cortex	3298 ± 688	2063 ± 115	$13,649 \pm 215$	$16,004 \pm 892$
Hippocampus	3752 ± 229	5034 ± 310 [a]	$13,082 \pm 1556$	$17,380 \pm 1849$
Hypothalamus	197 ± 30	612 ± 105 [a]	9471 ± 283	9312 ± 605
Cerebellum	1029 ± 230	1142 ± 141	6512 ± 923	5456 ± 559

From Santucci et al. (2000).
[a] $p < 0.05$.

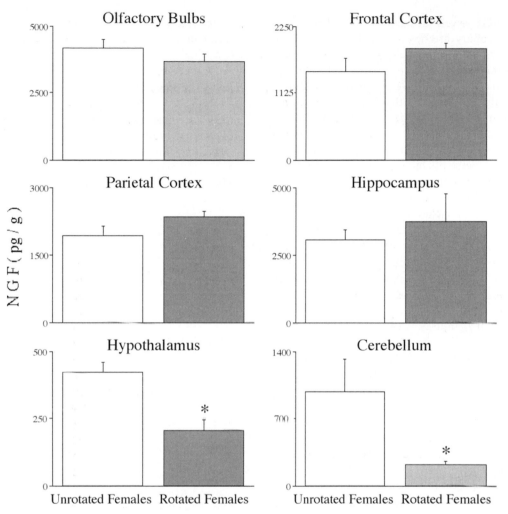

Fig. 3. Endogenous NGF levels (expressed as pg/g of fresh tissue) in the CNS of unrotated and rotated female CD-1 mice. * = p < 0.05.

the visual system were also investigated (Aloe et al. 2001). Results from cytochemical immunolocalization of neurotrophins in the visual system indicated that exposure to 1 h of hypergravity increased the expression of both NGF and BDNF in the visual cortex and in the geniculate bodies of adult mice, while reducing it in the retina. Moreover, a colocalization of immunoreactivity for NGF/BDNF, as well as for neuropeptide tyrosine (NPY) was also observed. In particular, the increase in both NGF and BDNF levels in the visual cortex and geniculate bodies were associated with higher NPY-immunoreactivity, while the decreased neurotrophins levels in the retina was associated with a downregulation of

NPY-immunopositivity. Since it has been reported that NGF and BDNF can synergically or separately modulate a variety of neurotransmitters and neuropeptides (Croll et al., 1994), these results suggested a functional link between NGF, BDNF and NPY in a dynamic remodeling response to hypergravity stimulus.

Today, there is accumulating evidence that NGF and BDNF are involved in the differentiation and activation of immune cells (Levi-Montalcini et al., 1990; Aloe et al., 1997; Simone et al., 1999). Furthermore, NGF and BDNF as well as other neurotrophins and their receptors are expressed in developing heart (reviewed by Hempstead, 2001), whereas NGF levels decreased in human athero-sclerotic coronary arteries (Chaldakov et al., 2001). Intriguingly, NGF exerts a protective role against post-ischemic dysfunction of sympathetic coronary innervation (Abe et al., 1997). Together these findings lead to the suggestion that the neurotrophins produced and released in the heart, and probably in the lung, could contribute to the maintenance of neural integrity in these organs. Mast cells, known to secrete a variety of biological active mediators, including neurotrophins, might also be involved in maintenance of neuronal integrity (Purcell and Atterwill, 1995; Tonchev et al., 1998). Since altered gravitational forces are known to influence the physiopathological activity of both heart and lungs (Portugalov et al., 1976; Tolkovsky, 1997; Beaumont et al., 1998; Fortrat et al., 1998), and since mast cells are source of and target to NGF (Aloe and Micera, 1999) and also implicated in respiratory and cardiovascular injuries (Aloe et al., 1997; Chaldakov et al., 2001), it is reasonable to hypothesize that hypergravity-induced stress-related activity could alter the local synthesis and release of NGF and BDNF. In our studies on CD-1 mouse, we found that exposure to short-term hypergravity induced significant variations in lung and heart neurotrophin levels and in mast-cell distribution when the heart and lungs of rotated mice were compared to nonrotated controls (Antonelli et al., 2002). The expression of neurotrophins' mRNA in both tissues was also significantly affected by hypergravity. And, mast cells increased in number in the proximity to heart blood vessels and in lung airway epithelium of rotated mice. This study indicates that hypergravity also influences cardiovas-cular and respiratory tissues and suggests that neurotrophins and mast cells are involved in the reaction to these environmental stimuli.

Acknowledgments

This study was supported by intramural ISS funding and financed by an Agenzia Spaziale Italiana (ASI) grant (ASI-ARS-I/R/123/01) to DS. The study involv-ing experimental animals was conducted in accor-dance with national and institutional guidelines for the protection of human subjects and animal welfare.

Abbreviations

NGF	nerve growth factor
BDNF	brain-derived neurotrophic factor
ISS	International Space Station
g	gravity
MS	motion sickness

References

Abe, T., Morgan, D.A. and Gutterman, D.D. (1997) Protective role of nerve growth factor against postischemic dysfunction of sympathetic coronary innervation. Circulation, 95: 213–220.

Alleva, E. and Aloe, L. (1989) Physiological roles of NGF in adult rodents: a biobehavioral perspective. Int. J. Comp. Psychol., 2: 147–163.

Alleva, E. and Santucci, D. (2001) Psychosocial vs. 'physical' stress situations in rodents and humans: role of neurotro-phins. Physiol. Behav., 73: 313–320.

Aloe, L., Tirassa, P. and Alleva, E. (1994) Cold water swimming stress alters NGF and low-affinity NGF receptor distribution in developing rat brain. Brain. Res. Bull., 33: 173–178.

Aloe, L., Bracci, L., Bonini, S. and Manni, L. (1997) The expanding role of nerve growth factor: from the neurotrophic activity to immunological diseases. Allergy, 52: 883–904.

Aloe, L. and Micera, A. (1999) Nerve growth factor: basic findings and clinical trials. In: L. Aloe and G.N. Chaldakov (Eds.), Nerve Growth Factor in Health and Disease. Biomedical Reviews, Vol. 10: The Bulgarian-American Center, Varna, pp. 3–14.

Aloe, L., Fiore, M., Santucci, D., Amendola, T., Antonelli, A., Francia, N., Corazzi, G. and Alleva, E. (2001) Effect of hypergravity on the mouse basal expression of NGF and BDNF in the retina, visual cortex and geniculate nucleus: correlative aspects with NPY immunoreactivity. Neurosci. Lett., 302: 29–32.

Antonelli, A., Santucci, D., Amendola, T., Triaca, V., Corazzi, G., Francia, N., Fiore, M., Alleva, E. and Aloe, L. (2002) Short-term hypergravity influences NGF and BDNF expression, and mast cell distribution in the lungs and heart of adult male mice. J. Gravit. Physiol., 9: 29–38.

Baer, L.A., Li, G.H., Sulkowsky, G.M., Ronca, A.E. and Wade, C.E. (2000) Early and late effects of perinatal hypergravity exposure on the developing CNS. Gravit. Space Biol. Bull., 14: 23.

Beaumont, M., Fodil, R., Isabey, D., Lofaso, F., Touchard, D., Harf, A. and Louis, B. (1998) Gravity effects on upper airway area and lung volumes during parabolic flight. J. Appl. Physiol., 84: 1639–1645.

Borison, H.L. (1989) Area postrema: chemoreceptor circumventricular organ of the medulla oblongata. Prog. Neurobiol., 32: 351–390.

Bouët, V., Gahéry, Y. and Lacour, M. (2003) Behavioural changes induced by early and long-term gravito-inertial force modification in the rat. Behav. Brain Res., 139: 97–104.

Bruce, L.L. and Fritzsch, B. (1997) The development of vestibular connections in rat embryos in microgravity. J. Gravit. Physiol., 4: 59–62.

Castren, E., Zafra, F., Thoenen, H. and Lindholm, D. (1992) Light regulates expression of brain-derived neurotrophic factor mRNA in rat visual cortex. Proc. Natl. Acad. Sci. USA., 89: 9444–9448.

Cellerino, A. and Maffei, L. (1996) The action of neurotrophins in the development and plasticity of the visual cortex. Prog. Neurobiol., 49: 53–71.

Chaldakov, G.N., Stankulov, I.S., Fiore, M., Ghenev, P.I. and Aloe, L. (2001) Nerve growth factor levels and mast cells distribution in human coronary atherosclerosis. Atherosclerosis, 159: 57–66.

Chinn, H.I. and Smith, P.K. (1993) Motion sickness. Pharmacol. Rev., 7: 33–82.

Clarac, F., Vinay, L., Cazalets, J.R., Fady, J.C. and Jamon, M. (1998) Role of gravity in the development of posture and locomotion in the neonatal rat. Brain Res. Rev., 28: 35–43.

Croll, S.D., Wiegand, S.J., Anderson, K.D., Lindsay, R.M. and Nawa, H. (1994) Regulation of neuropeptides in adult rat forebrain by the neurotrophins BDNF and NGF. Eur. J. Neurosci., 6: 1343–1353.

DeFelipe, J., Arellano, J.I., Merchan-Perez, A., Gonzalez-Albo, M.C., Walton, K. and Llinas, R. (2002) Spaceflight induces changes in the synaptic circuitry of the postnatal developing neocortex. Cereb. Cortex, 12: 883–891.

Dobie, T., McBride, D., Dobie, T., Jr. and May, J. (2001) The effects of age and sex on susceptibility to motion sickness. Aviat. Space Environ. Med., 72: 13–20.

Eskin, A. and Riccio, D. (1966) The effects of vestibular stimulation on spontaneous activity in the rat. Psychol. Rec., 16: 523–527.

Fortrat, J.-O., Somody, L. and Gharib, C. (1998) Autonomic control of cardiovascular dynamics during weightlessness. Brain Res. Rev., 28: 66–72.

Fox, A., Daunton, N.G. and Corcoran, M.L. (1998) Study of adaptation to altered gravity through systems analysis of motocontrol. Adv. Space Res., 22: 245–253.

Giménez y Ribotta, M., Sandillon, F. and Privat, A. (1998) Influence of hypergravity on the development of monoaminergic systems in the rat spinal cord. Dev. Brain Res., 111: 147–157.

Gu, Q., Liu, Y., Dyck, R.H., Booth, V. and Cynader, M.S. (1998) Effects of tetrodotoxin treatment in LGN on neuromodulatory receptor expression in developing visual cortex. Dev. Brain Res., 106: 93–99.

Hempstead, B. (2001) Role of neurotrophins in cardiovascular development. In: I. Mocchetti (Ed.), Neurobiology of the Neurotrophins. FP Graham Publishing Co., Johnson City, TN, pp. 141–148.

Homick, J.L., Reschke, M.F. and Miller, E.F. (1997) The effects of prolonged exposure to weightlessness on postural equilibrium, In: R.S. Johnston, L.F. Dietlein (Eds.), Biomedical Results from Skylab, Washington, DC: Scientific and Technical Information Office, NASA. pp. 104–112.

Inglis, E.M., Zuckerman, K.E., Roberts, M. and Kalb, R.G. (1999) Altered dendritic morphology in a population of motor neurons in rats reared in microgravity. Soc. Neurosci. Abstr., 91.99.

Javid, F.A. and Naylor, R.J. (1999) Variables of movement amplitude and frequency in the development of motion sickness in Suncus murinus. Pharmacol. Biochem. Behav., 64: 115–122.

Jokerst, M.D., Gatto, M., Fazio, R., Gianaros, P.J., Stern, R.M. and Koch, K.L. (1999) Effects of gender of subjects and experimenter on susceptibility to motion sickness. Aviat. Space Environ. Med., 70: 962–965.

Levi-Montalcini, R., Aloe, L. and Alleva, E. (1990) A role for nerve growth factor in nervous, endocrine and immune systems. Prog. NeuroEndocrinImmunol., 3: 1–10.

Lindsay, R.M. (1996) Role of neurotrophins and trk receptors in the development and maintenance of sensory neurons: an overview. Philos. Trans. R. Soc. Lond B. Biol. Sci., 351: 365–373.

Lyons, T.J. (1992) Women in the fast jet cockpit–aeromedical considerations. Aviat. Space Environ. Med., 63: 809–818.

Mandillo, S., Del Signore, A., Paggi, P., Francia, N., Santucci, D., Mele, A. and Oliverio, A. (2003) Effects of acute and repeated daily exposure to hypergravity on spatial learning in mice. Neurosci. Lett., 336: 147–150.

McAllister, A.K., Lo, D.C. and Katz, L.C. (1995) Neurotrophins regulate dendritic growth in developing visual cortex. Neuron, 15: 791–803.

McCaffrey, R.J. and Graham, G. (1980) Age-related differences for motion sickness in the rat. Exp. Aging Res., 6: 555–561.

Mitchell, D., Krusemark, M.L. and Hafner, D. (1977a) Pica: a species relevant behavioral assay of motion sickness in the rat. Physiol. Behav., 18: 125–130.

Mitchell, D., Laycock, J.D. and Stephens, W.F. (1977b) Motion sickness-induced pica in the rat. Am. J. Clin. Nutr., 30: 147–150.

Money, K.E. (1970) Motion sickness. Physiol. Rev., 50: 1–39.

Ossenkopp, K.P. and Frisken, N.L. (1982) Defecation as an index of motion-sickness in the rat. Physiol. Psychol., 10: 355–360.

Oyama, J. and Platt, W.T. (1967) Reproduction and growth of mice and rats under conditions of simulated increased gravity. Am. J. Physiol., 212: 164–166.

Park, A.H. and Hu, S. (1999) Gender differences in motion sickness history and susceptibility to optokinetic rotation-induced motion sickness. Aviat. Space Environ. Med., 70: 1077–1080.

Portugalov, V.V., Savina, E.A., Kaplansky, A.S., Yakovleva, G.I., Plakhuta-Plakutina, G.I., Pankova, A.S., Katunyan, P.I., Shubich, M.G. and Buvailo, A.S. (1976) Effect of space flight factors on the mammal: experimental-morphological study. Aviat. Space Environ. Med., 47: 813–816.

Purcell, W.M. and Atterwill, C.K. (1995) Mast cells in neuroimmune function: neurotoxicological and neuropharmacological perspectives. Neurochem. Res., 20: 521–532.

Reschke, M.F., Anderson, D.J. and Homick, J.L. (1986) Vestibulo-spinal modification as determined with the H-reflex during Spacelab-1 flight. Exp. Brain Res., 64: 367–379.

Ronca, A.E. and Alberts, J.R. (1997) Altered vestibular function in fetal and newborn rats gestated in space. J. Gravit. Physiol., 4: 63–66.

Sajdel-Sulkowska, E.M., Baer, L.A., Li, G.H., Sulkowsky, G.M., Koibuchi, N., Ronca, A.E. and Wade, C.E. (2000) The effect of hypergravity on cerebellar development may be mediated by altered thyroid status. Soc. Neurosci. Abstr., 115.111.

Santucci, D., Corazzi, G., Francia, N., Antonelli, A., Aloe, L. and Alleva, E. (2000) Neurobehavioural effects of hypergravity conditions in the adult mouse. Neuroreport, 11: 3353–3356.

Simone, M.D., De Santis, S., Vigneti, E., Papa, G., Amadori, S. and Aloe, L. (1999) Nerve growth factor: a survey of activity on immune and hematopoietic cells. Hematol. Oncol., 17: 1–10.

Sondag, H.N., de Jong, H.A. and Oosterveld, W.J. (1996) Altered behavior in hamsters by prolonged hypergravity. Acta Otolaryngol., 43: 289–294.

Sondag, H.N., de Jong, H.A. and Oosterveld, W.J. (1997) Altered behavior in hamsters conceived and born in hypergravity. Brain Res. Bull., 43: 289–294.

Temple, M.D., Kosik, K.S. and Steward, O. (1999) Spatial navigation and memory of place in animals that develop in microgravity. Soc. Neurosci. Abstr., 25(Pt 22): 1623.

Thoenen, H. (1995) Neurotrophins and neuronal plasticity. Science, 270: 593–598.

Tolkovsky, A. (1997) Neurotrophic factors in action-new dogs and new tricks. Trends Neurosci., 20: 1–3.

Tonchev, A.B., Valchanov, K.P., Sulaiman, O.A.R., Ghenev, P.I. and Chaldakov, G.N. (1998) A suggestive neurotrophic potential of mast cells in heart and submandibular glands of the rat. Biomed. Rev., 9: 143–145.

Tyler, D.B. and Bard, P. (1949) Motion sickness. Physiol. Rev., 29: 311–369.

Uno, Y., Horii, A., Uno, A., Fuse, Y., Fukushima, M., Doi, K. and Kubo, T. (2002) Quantitative changes in mRNA expression of glutamate receptors in the rat peripheral and central vestibular systems following hypergravity. J. Neurochem., 81: 1308–1317.

Uno, A., Takeda, N., Horii, A., Morita, M., Yamamoto, Y., Yamatodani, A. and Kubo, T. (1997) Histamine release from the hypothalamus induced by gravity change in rats and space motion sickness. Physiol. Behav., 61: 883–887.

Uno, A., Takeda, N., Kitahara, T., Sakata, Y., Yamatodani, A. and Kubo, T. (2000) Effects of vestibular cerebellum lesion on motion sickness in rats. Acta Otolaryngol., 120: 386–389.

Walton, K. (1998) Postnatal development under conditions of simulated weightlessness and space flight. Brain Res. Rev., 28: 25–34.

West, J.B. (2000) Physiology in microgravity. J. Appl. Physiol., 89: 379–384.

Wubbels, R.J. and de Jong, H.A. (2000) Vestibular-induced behavior of rats born and raised in hypergravity. Brain Res. Bull., 52: 349–356.

Yates, B.J., Miller, A.D. and Lucot, J.B. (1998) Physiological basis and pharmacology of motion sickness: an update. Brain Res. Bull., 47: 395–406.

Progress in Brain Research, Vol. 146
ISSN 0079-6123

CHAPTER 14

Neurotrophic factors in Huntington's disease

Jordi Alberch*, Esther Pérez-Navarro and Josep M. Canals

Department of Cell Biology and Pathology, Medical School, IDIBAPS, University of Barcelona, Barcelona, Spain

Abstract: Huntington's disease is a neurodegenerative disorder characterized by the selective loss of striatal neurons and, to a lesser extent, cortical neurons. The neurodegenerative process is caused by the mutation of huntingtin gene. Recent studies have established a link between mutant huntingtin, excitotoxicity and neurotrophic factors. Neurotrophic factors prevent cell death in degenerative processes but they can also enhance growth and function of neurons that are affected in Huntington's disease. The endogenous regulation of the expression of neurotrophic factors and their receptors in the striatum and its connections can be important to protect striatal cells and maintains basal ganglia connectivity. The administration of exogenous neurotrophic factors, in animal models of Huntington's disease, has been used to characterize the trophic requirements of striatal and cortical neurons. Neurotrophins, glial cell line-derived neurotrophic factor family members and ciliary neurotrophic factor have shown a potent neuroprotective effects on different neuronal populations of the striatum. Furthermore, they are also useful to maintain the integrity of the corticostriatal pathway. Thus, these neurotrophic factors may be suitable for the development of a neuroprotective therapy for neurodegenerative disorders of the basal ganglia.

Keywords: Huntington's disease; basal ganglia; striatum; neurotrophins; CNTF; GDNF; excitotoxicity; neuroprotection

Introduction

Neurodegenerative disorders entail the loss of specific neuronal populations. Furthermore, propagating mechanisms such as oxidative stress, free-radical activity, excitotoxicity, formation of intracellular aggregates, immunogenicity, mitochondrial dysfunction and apoptosis can also be common to many neurodegenerative diseases (Shoulson, 1998). Although these neurological disorders share some characteristics, they also have their own idiosyncrasy, which will require specific treatments. Therefore, understanding the mechanisms that lead to the death of these groups of neurons is critical for developing therapeutic strategies. However, none

of the drugs used to date has been successful. Thus, new therapeutic approaches are needed to prevent or slow the neurodegenerative processes.

Neurotrophic factors are naturally occurring proteins that promote survival of specific neuronal populations, stimulate their morphological differentiation and regulate neuronal gene expression through interaction with specific cellular receptors (Huang and Reichardt, 2001). Although neurotrophic factors have a key function during development, their presence is also required in the adult brain for maintaining neuronal function and phenotype (Sofroniew et al., 1990; Conner and Varon, 1996; Cooper et al., 1996). A number of studies have shown that neurotrophic factors have the potential to protect diseased or injured neurons from dying, to induce neuronal sprouting and to increase neuronal metabolism and function. These neuroprotective effects have been tested in several animal models of neurological disorders, such as toxic neuropathy

*Correspondence to: J. Alberch, Department of Cell Biology and Pathology, Medical School, IDIBAPS, University of Barcelona, Casanova 143, E-08036 Barcelona, Spain. Tel.: +34-93-403-5285; Fax: +34-93-402-1907; E-mail: alberch@medicina.ub.es

DOI: 10.1016/S0079-6123(03)46014-7

(Apfel, 1996), diabetic neuropathy (Apfel, 1999; Akkina et al., 2001), motor neuron disease (Mitsumoto et al., 1994; Corse et al., 1999), spinal cord trauma (Grill et al., 1997; Blesch and Tuszynski, 2002), stroke (Holtzman et al., 1996), Alzheimer's disease (Fischer et al., 1987; Hefti, 1994), Parkinson's disease (Akerud et al., 1999; Alexi et al., 2000; Kordower et al., 2000) and Huntington's disease (HD) (Emerich et al., 1997a; Alexi et al., 2000; Alberch et al., 2002). Although the lack of growth factors is not the origin of the neurodegenerative disorders, their potential therapeutic effects have been claimed because of their capacity to regulate cell death in specific neuronal populations that are more sensitive to neurodegeneration. However, neurotrophic factors do not cross the blood-brain barrier and high doses are required to obtain an effect in the brain when they are administered systemically. This explains the failure of clinical trials using neurotrophic factors for the treatment of various neurodegenerative disorders (Apfel, 2002; Thoenen and Sendtner, 2002). Therefore, to avoid indiscriminate administration and side effects, new delivery systems have been developed. Recent studies have used the transplantation of engineered cells to release neurotrophic factors or the injection of viral vectors mediating the overexpression of these factors, and local protective effects have been achieved (Kordower et al., 1999; Akerud et al., 2001; Alberch et al., 2002; Mittoux et al., 2002).

Some movement disorders are caused by a neurodegenerative process of the basal ganglia. These subcortical nuclei are mainly organized in the striatum (caudate nucleus and putamen) and globus pallidus (external and internal segment) connecting to subthalamic nucleus, substantia nigra (SN), thalamus and cerebral cortex (Gerfen, 1992). The anatomy of the basal ganglia connections suggests that these structures operate as part of recurrent circuits (loops), such as cortex–basal ganglia–thalamus–cortex or SN–basal ganglia–SN. These loops run through the striatum, which plays a pivotal role in the analysis and production of motor and cognitive functions (Graybiel, 2000). Neurodegenerative processes in the striatum cause important motor and cognitive dysfunctions, as observed in HD.

Here we discuss the role of neurotrophic factors in the basal ganglia and possible neuroprotective

strategies for the treatment of HD. Recent studies have characterized the neurotrophic factors that have neuroprotective effects on striatal neuronal populations using different models of HD. Neurotrophins [nerve growth factor (NGF), brain-derived neurotrophic factor (BDNF), neurotrophin-3 (NT-3), and NT-4/5], glial cell line-derived neurotrophic factor (GDNF) family and ciliary neurotrophic factor (CNTF) are the best candidates for the development of therapeutic strategies to protect striatal cells in HD.

Huntington's disease

HD is a dominantly inherited and progressive neurodegenerative disorder characterized by motor and cognitive dysfunctions (Martin and Gusella, 1986; Ross, 2002). The HD gene, huntingtin, was identified and located in the short arm of chromosome 4 and contains a polymorphic stretch of repeated CAG trinucleotides (The Huntington's Disease Collaborative Research Group, 1993). When the number of repeats exceeds 35, the gene encodes a version of huntingtin that leads to disease (McMurray, 2001). HD is found in all regions of the world and in all races, with a prevalence of 5–10 per 100,000 (Conneally, 1984; Harper, 1992). Motor disorders are a classical feature of the disease, with impairment of both voluntary and involuntary movements. The involuntary movements are choreatic with fast uncontrolled movements of limbs and trunk. Coordination of voluntary movement becomes progressively more difficult. Emotional disturbances usually accompany the impairment of cognitive functions, with depression and dementia. Personality changes, with irritability and apathy can also develop. The age of onset is usually between mid-30s and late-50s, although juvenile and late onset of HD occur (Greenamyre and Shoulson, 1994).

The most marked neuropathological features are the dramatic loss of neurons in the striatum, combined with a less-marked atrophy of the neocortex. The striatal degeneration is only observed in GABAergic medium-sized spiny neurons, which constitute 95% of all neurons in the striatum. They are projection neurons and are divided in two groups depending on the target area. The direct pathway projects to the substantia nigra (SN) pars reticulata and internal segment of the globus pallidus and contains GABA, substance P (SP) and dynorphin. On the other hand, the indirect pathway has enkephalin-containing GABAergic neurons and projects to the globus pallidus external segment (Gerfen, 1992; Graybiel, 2000). These neuronal populations are not uniformly affected in HD. Indeed, enkephalinergic neurons of the indirect pathway tend to die before those in the direct pathway (Reiner et al., 1988; Richfield et al., 1995). However, choline acetyltransferase (ChAT), GABA or somatostatin/NADPH diaphorase/neuropeptide-Y interneurons are spared (Dawbarn et al., 1985; Ferrante et al., 1985). In late stages of the disease, a loss of pyramidal neurons of layers II–III, V and VI of association and motor cortices is observed (Cudkowicz and Kowall, 1990; DiFiglia et al., 1997; MacDonald and Halliday, 2002). Recent studies give the corticostriatal pathway a critical role in striatal dysfunction and in the neurologic impairment (Laforet et al., 2001; MacDonald and Halliday, 2002).

The mechanisms of neurodegeneration in HD are unknown. Mutant huntingtin triggers a cascade of intracellular events that lead to cell death only in sensitive neuronal populations, although huntingtin gene is expressed in all cell types. However, the first goal is to understand the function of normal huntingtin. It is predominantly a cytoplasmic protein, and interacts with proteins associated with membrane vesicles or cytoskeleton, suggesting a role in endocytosis, intracellular trafficking, membrane recycling and retrograde and fast axonal transport (Gusella and MacDonald, 1998). Furthermore, huntingtin also interacts with transcriptional regulatory proteins (Petersen et al., 1999; Reddy et al., 1999). Huntingtin has an important role in normal development, since homozygous knockout (KO) mice die by embryonic day 7.5 (Duyao et al., 1995; Nasir et al., 1995). Recent studies also show that huntingtin has anti-apoptotic properties, acting downstream of Bcl-2 family members and upstream of caspase-3 activation (Rigamonti et al., 2000). Furthermore, normal huntingtin interacts with different proteins to block the apoptosis induced by caspase 8 (Gervais et al., 2002). Therefore, the loss of huntingtin function may activate several intracellular apoptotic pathways,

although there is some controversy as to whether mutant huntingtin leads to a gain or a loss of function (Cattaneo et al., 2001).

Many different experimental models of neurodegeneration may be relevant to HD. Excitotoxicity and impaired energy metabolism are involved in several neurodegenerative diseases (Coyle and Puttfarcken, 1993). However, these two processes are linked since impaired energy metabolism reduces the threshold for glutamate toxicity and can activate excitotoxic mechanisms (Beal et al., 1993). The use of quinolinic acid, an NMDA receptor agonist, or 3-nitropropionic acid, an inhibitor of succinate dehydrogenase, reproduces the pattern of striatal atrophy observed in HD (Beal et al., 1986; Brouillet et al., 1999).

A breakthrough in HD research was the development of transgenic mouse models. Transgenic mice expressing exon-1 of the human HD gene with 115 (R6/1) or 145 (R6/2) CAG repeats (Mangiarini et al., 1996) and mice expressing an yeast artificial chromosome containing the full-length human HD gene with 72 CAG repeats (YAC72) (Hodgson et al., 1999) develop a progressive neurological disorder. Furthermore, a conditional transgenic HD mouse was created using a tet-regulated system (Yamamoto et al., 2000). These animal models are excellent tools to study the mechanisms involved in the pathogenesis of HD and to develop therapeutic strategies.

The neurotrophic factor family

Neurotrophins

NGF was the first member of the neurotrophin family to be characterized, by Levi-Montalcini and Hammburger (Levi-Montalcini, 1987). Three other members have been isolated in mammals: BDNF, NT-3 and NT-4/5.

NGF was purified as a survival factor for sympathetic and sensory neurons in culture (Levi-Montalcini, 1987). Besides its trophic effects on the peripheral nervous system (PNS), NGF is also a survival and differentiating factor for cholinergic neurons in the central nervous system (CNS). Barde and colleagues (1982) isolated a second protein, BDNF, with neurotrophic activities in sensory neurons (Barde et al., 1982). Nowadays, it is well documented that this neurotrophin exerts effects on the sprouting and survival of other PNS and CNS neurons (Huang and Reichardt, 2001). This neurotrophic factor was initially cloned from pig, rat and human brain and its sequence led to the finding that the mature BDNF and NGF proteins show striking amino acid similarities.

Based on cloning strategies, the use of conserved regions of NGF and BDNF allowed the characterization of other neurotrophins. Thus, in 1990, the third member of the neurotrophin family, NT-3, was characterized by molecular techniques and its DNA sequence confirmed the homology with NGF and BDNF (Ernfors et al., 1990; Maisonpierre et al., 1990). NT-3 induces profuse neuritic outgrowth from dorsal root, nodose and sympathetic ganglion (Maisonpierre et al., 1990).

A year later two new neurotrophins were described in *Xenopus* and human simultaneously, which were called NT-4 and NT-5, respectively. However, the homologous sequence of NT-4 in mammals and the biological activity of both neurotrophins showed that they were homologues of a single factor expressed in different animal species, and so they were called NT-4/5. This neurotrophin has biological effects on sensory and sympathetic neurons (Berkemeier et al., 1991).

The neurotrophin receptors, a dual system

Neurotrophins exert their trophic and topic actions through the interaction with two receptor types. All four neurotrophins bind to the p75 neurotrophin receptor (p75^{NTR}) and a specific member of the tyrosine kinase (Trk) receptor family. Binding studies have indicated that neurotrophins interact with neurons with high-affinity (Kd $\sim 10^{-11}$ M), however, independently p75^{NTR} and Trk receptors exhibit low-affinity binding sites with similar Kd of about 10^{-9} M (Chao, 1994). Therefore, p75^{NTR} and Trk receptors can be considered as subunits in neurotrophin receptor signaling (Barbacid, 1994). Besides the actions of p75^{NTR} as coreceptor for Trk receptors, p75^{NTR} can also signal independently to stimulate ceramine production, NF-kβ and c-Jun kinase activity (Casaccia-Bonnefil et al., 1999) and, to induce neuronal death via apoptosis. In fact, p75^{NTR} was the first described member of the family of the tumor necrosis factor receptors (Chao, 1994), which now contains about 25 receptors (Roux and Barker, 2002).

TrkA was the first described member of the Trk family (Martin-Zanca et al., 1986; Kaplan et al., 1991) that also includes TrkB (Klein et al., 1991b; Middlemas et al., 1991) and TrkC (Lamballe et al., 1991). Each neurotrophin binds specifically to one Trk receptor, thus, NGF binds to TrkA (Hempstead et al., 1991; Klein et al., 1991a), BDNF and NT-4/5 act through TrkB (Berkemeier et al., 1991; Klein et al., 1991b, 1992; Soppet et al., 1991; Squinto et al., 1991), and TrkC is the preferred receptor for NT-3 (Lamballe et al., 1991). However, this description is oversimplified because in some experimental conditions NT-3 may also activate TrkA and TrkB, and NT-4/5 can induce autophosphorylation of TrkA (Cordon-Cardo et al., 1991; Ip et al., 1993b; Chao, 1994). Trk alone confers responsiveness to the neurotrophin; however, nowadays it is well accepted that the relative amount of p75^{NTR} with respect to Trk is crucial for neurotrophin-receptor specificity. Thus, in the presence of p75^{NTR}, only BDNF provides a functional activation of TrkB (Bibel et al., 1999, Lee et al., 2001). Similarly, p75^{NTR} restricts signaling of TrkA to NGF, but not to NT-3 (Benedetti et al., 1993).

Differential splicing of the mRNA^{Trk} results in the expression of proteins with differences in their extracellular domains that affect ligand interactions (Barbacid, 1994; Shelton et al., 1995). The TrkA isoform that lacks these amino acids is much more specifically activated by NGF (Clary and Reichardt, 1994), and mainly localized in nonneuronal cells whereas the longest isoform is expressed by neuronal populations (Barker et al., 1993). In addition, the TrkB receptor isoform that lacks a short amino acid sequence can only be activated by BDNF and not by NT-4/5 or NT-3 (Strohmaier et al., 1996). Furthermore, the differential splicing can also affect exons encoding for tyrosine kinase domains that are determinant for neurotrophin-responsiveness, generating truncated forms of Trk receptors. TrkB and TrkC receptors present both full-length and truncated isoforms, whereas no isoform lacking the tyrosine kinase domain has been described for TrkA (Barbacid, 1994). Different functions have been attributed to these short isoforms of TrkB and TrkC.

It has been hypothesized that the expression of these receptors by nonneuronal cells may be involved in ligand recruitment and presentation to the catalytic isoforms during regeneration following damage (Beck et al., 1993). Within neurons, these truncated isoforms might act as dominant negative inhibitors of their tyrosine kinase isoforms, thereby attenuating neurotrophin responses (Barbacid, 1994; Eide et al., 1996; Yacoubian and Lo, 2000). In addition, it has been demonstrated that truncated isoforms of TrkB and TrkC receptors can directly activate intracellular signaling pathways (Baxter et al., 1997; Hapner et al., 1998). Another isoform has been described in TrkA and TrkC that contains an amino acid insert in the tyrosine kinase domain which appears to modify the substrate specificity without altering the catalytic activity of TrkC (Guiton et al., 1995; Meakin et al., 1997).

Binding of neurotrophin homodimers causes the receptor dimerization, followed by phosphorylation of different tyrosine residues (Barbacid, 1994). Once the intracellular Trk domain is autophosphorylated, different signaling cascades are activated, including the PI3K/Akt kinase pathway, mitogen-activated protein kinase MAPK/ERK pathway, and the PLC-γ1 pathway. The activation of these distinct pathways has different functions (Segal and Greenberg, 1996). PI3K stimulates the activation of many signaling proteins such as the serine/threonine Akt, which in turn regulates several proteins involved in neuronal survival. The MAPK/ERK pathway has been related to the survival and differentiation effects of neurotrophins, but it plays many other roles in neurons including plasticity and long term potentiation. PLC-γ1 is another substrate phosphorylated by Trk kinase. Activated PLC-γ1 hydrolyzes phosphatidylinositides to generate diacylglycerol and inositol 1,4,5 triphosphate, which induces an increase in the cytoplasmic Ca^{2+} that regulates many intracellular pathways (Kaplan and Miller, 2000; Brunet et al., 2001; Patapoutian and Reichardt, 2001).

Distribution of neurotrophins and their receptors in the basal ganglia

The expression of neurotrophins has been demonstrated in the basal ganglia. NGF is localized in the striatum, and high mRNA levels are detected between postnatal day (P) 2 through P30 (Mobley et al., 1989). The increase in NGF protein and high-affinity NGF binding is coincident with cholinergic neurochemical differentiation (Mobley et al., 1989; Pérez-Navarro and Alberch, 1995). NGF is also localized in the cerebral cortex, where mRNA^{NGF} expression peaks between P10 and P30 (Large et al., 1986). The cellular localization of NGF protein in the cortex shows the highest levels in deep layer V and moderate levels in layers II/III, upper V and VI, whereas low expression is detected in layer IV (Pitts and Miller, 2000).

The presence of BDNF expression in the striatum has been described using Northern blot (Hofer et al., 1990) or RNase protection assay (Canals et al., 1998; Checa et al., 2000), but not by in situ hybridization (Hofer et al., 1990). The levels of mRNA^{BDNF} in the striatum increase in the third postnatal week when they reach their maximum (Checa et al., 2000). The presence of BDNF protein in this nucleus is controversial because of the low specificity of the available antibodies. Using an antibody against the recombinant human BDNF, which does not show immunoreactivity in BDNF KO mice, Yan et al. (1997b) did not find signal in the adult striatum. In contrast, medium-sized positive neurons are labeled in the striatum of rat (Kawamoto et al., 1996) or monkey (Kawamoto et al., 1999) with an antibody directed against a synthetic peptide, which only shows a specific band by Western blot. However, BDNF protein is detected in the striatum by ELISA at a concentration of 7 ng per gram of tissue (Yurek and Fletcher-Turner, 2001). Furthermore, it has also been shown that BDNF levels in the striatum decrease by half in aged rats (Katoh-Semba et al., 1998).

BDNF is also expressed in the main connecting areas of the striatum: cerebral cortex and SN. The levels of mRNA^{BDNF} in the cerebral cortex increase in postnatal development, reaching the maximal expression at P21 (Checa et al., 2000). In the adult neocortex, mRNA^{BDNF} is mainly restricted to excitatory pyramidal neurons, but not in GABAergic inhibitory interneurons (Cellerino et al., 1996; Marty et al., 1997; Canals et al., 2001). BDNF protein is localized in layers II–III, V and VI, and low or no signal is detected in layer IV (Kawamoto et al., 1996,

1999; Yan et al., 1997b; Friedman et al., 1998; Pitts and Miller, 2000). BDNF immunoreactivity in layers V and VI appears in medium-sized stellate neurons (Yan et al., 1997b), indicating that pyramidal excitatory neurons express and release BDNF, which is taken up by stellate GABAergic neurons (Marty et al., 1997). Immunoelectron microscopy of the cerebral cortex revealed that BDNF is localized in postsynaptic densities of axo-spinous asymmetric synaptic junctions, which are the morphological correlates of excitatory glutamatergic synapses (Aoki et al., 2000). BDNF has also been detected in cortical glial cells (Pitts and Miller, 2000) as well as in the pallidum (Kawamoto et al., 1996) and substantia nigra (Kawamoto et al., 1996, 1999; Yan et al., 1997b). A transient increase of mRNA^{BDNF} levels is detected in the SN at P7 and then decrease progressively to adult levels after the first month (Friedman et al., 1991; Checa et al., 2000). Within the SN pars compacta, BDNF staining is detected in the neuronal cell body and the proximal processes (Kawamoto et al., 1999). In the pars reticulata, immunopositive neurons are scattered (Kawamoto et al., 1999) and numerous immunolabeled nerve fibers form dense networks (Kawamoto et al., 1996, 1999). The levels of BDNF protein have been quantified in the ventral midbrain showing a concentration of 15 ng per gram of tissue (Yurek and Fletcher-Turner, 2001).

The expression of NT-3 is differentially regulated in the striatum and its connecting areas during development. mRNA^{NT-3} levels remain constant during striatal postnatal development, meanwhile a downregulation or an upregulation is observed in cortex and SN, respectively (Checa et al., 2000). In the adult brain, NT-3 expression can be detected in all three cerebral structures (Hohn et al., 1990; Canals et al., 1998; Checa et al., 2000). The levels of NT-3 protein in the striatum and ventral midbrain are similar, showing around 3.5 and 4 ng per gram of tissue respectively (Yurek and Fletcher-Turner, 2001). In the cerebral cortex, levels of NT-3 protein are low and mainly located in pyramidal neurons. The highest expression of NT-3 in this brain area is detected in the cingulate cortex (Friedman et al., 1998; Pitts and Miller, 2000). However, NT-3 protein increases in aged rats (Katoh-Semba et al., 1998).

The localization of NT-4/5 in the brain is difficult to study because its mRNA is expressed at much lower levels than any other neurotrophin (Ibañez, 1996). Northern blot analysis shows that NT-4/5 is not expressed in fetal or adult human brain (Ip et al., 1992), but it is detected in the adult rat brain (Berkemeier et al., 1991). Using the RNase protection assay, it has been shown that NT-4/5 is expressed in the rat brain during embryonic development and more so during postnatal stages (Timmusk et al., 1993). In the adult rat, similar levels of mRNA$^{NT-4/5}$ have been detected in the striatum, cortex and midbrain (Timmusk et al., 1993). Antibodies against a specific peptide of the NT-4/5 sequence located immunoreactivity in the cerebral cortex (Friedman et al., 1998). The strongest NT-4/5 signal is observed in the outer pyramidal layer in the neonate and in the inner pyramidal layer in the adult cortex (Friedman et al., 1998).

p75^{NTR} gene is expressed in a wide variety of cell populations during early stages of CNS development (Ernfors et al., 1988). In the basal ganglia, mRNA^{p75NTR} is detected in the developing striatum (Mobley et al., 1989; Koh and Higgins, 1991), but little if any expression of this receptor has been detected by in situ hybridization or by immunohistochemistry in the adult (Holtzman et al., 1992).

TrkA gene is expressed in defined structures of the CNS, whereas TrkB (Klein et al., 1990a, 1990b) and TrkC (Tessarollo et al., 1993; Lamballe et al., 1994) genes are widely expressed. The onset of Trk gene expression takes place during early stages of neurogenesis (between E7.5 and E9.5) (Klein et al., 1990a; Martin-Zanca et al., 1990; Tessarollo et al., 1993; Barbacid, 1994). Within the basal ganglia, mRNA^{TrkA} is confined to the cholinergic population of the striatum (Holtzman et al., 1992; Merlio et al., 1992; Steininger et al., 1993; Fagan et al., 1997). In this nucleus, the levels of TrkA increase during postnatal development, and are high from the second postnatal week until adulthood (Ringstedt et al., 1993). TrkA expression is not detected in other neuronal populations of the cortico-basal ganglia-SN system (Barbacid, 1994).

mRNA^{TrkB} levels increase during striatal embryonic stages until E20, and then gradually decline in the postnatal development (Jung and Bennett, 1996). TrkB immunohistochemistry reveals that positive cells and fibers begin to appear on patches at E19 where TrkB receptor is localized on the surface of striatal projection neurons (Costantini et al., 1999). This distribution persists until P14, when it progressively becomes more homogenous (Costantini et al., 1999). Both full-length and truncate isoforms of TrkB receptor are uniformly distributed in striatal medium-sized neurons, whereas large interneurons only express the full-length TrkB isoform (Merlio et al., 1992; Yan et al., 1997a; Costantini et al., 1999). During cortical development, mRNA^{TrkB} for the catalytic form exhibits the highest levels at E20 in the cingulate cortex and at P1 in the parietal cortex (Jung and Bennett, 1996), while the truncated TrkB form displays the highest levels in the adult (Checa et al., 2001). All cortical neurons express TrkB receptor although its expression is higher in inhibitory GABAergic interneurons (Huang and Reichardt, 2001). Like BDNF, electron microscopy revealed that TrkB is localized in postsynaptic densities, suggesting that this neurotrophin participates in synaptic plasticity (Aoki et al., 2000). In the ventral pallidum, small cells show strong expression of TrkB (Merlio et al., 1992). Within the midbrain, neurons in the SN pars compacta are intensely stained whereas in the pars reticulata the signal is low for TrkB protein and mRNA^{TrkB} (Merlio et al., 1992; Yan et al., 1997a; Checa et al., 2001). In situ hybridization studies also show TrkB expression in glial cells, including microglial, oligodendrocytes and astrocytes (Merlio et al., 1992).

TrkC expression is first detected in the striatal primordia at E16, reaching the maximal levels two days later (Jung and Bennett, 1996). The high levels of TrkC mRNA observed at E18 remain constant until the first postnatal week, when they decrease to adult levels (Jung and Bennett, 1996; Checa et al., 2001). In the adult striatum, the cellular distribution of mRNA^{TrkC} is similar to that of mRNA^{TrkB} (Merlio et al., 1992). In the cerebral cortex, mRNA^{TrkC} peaks at E20 and in the early postnatal stages it decreases more than three times to adult levels (Jung and Bennett, 1996; Checa et al., 2001). The distribution of TrkC overlaps with TrkB in some areas of the adult cerebral cortex, showing intense expression in layers II/III and VI (Barbacid, 1994). In the SN, a very low decrease is observed in the first postnatal week (Checa et al., 2001). In the

adult, mRNA^{TrkC} is only detected in half of the large and medium-sized neurons in the SN pars compacta (Merlio et al., 1992), and also in GFAP-positive astrocytes (Checa et al., 2001). The expression of the truncate TrkC NC2 isoform has been detected since E15.5 in the cortex and striatum, and since E17.5 in the mesencephalon (Menn et al., 2000). In the adult mouse, this TrkC isoform is expressed at high levels in layers II/III of the cerebral cortex and at low levels in the striatum. However, TrkC NC2 protein was demonstrated in apical and proximal basal dendrites of the pyramidal neurons of layer V (Menn et al., 2000).

Functional implications of neurotrophins in the basal ganglia

The patterns of developmental regulation of mRNA^{NGF} and NGF protein suggest that NGF does mediate the development of cholinergic neurons in the striatum (Mobley et al., 1989). In fact, intracerebroventricular NGF administration has been shown to increase cell body size and to enhance the neurochemical differentiation of cholinergic neurons of the striatum (Hagg et al., 1989; Vahlsing et al., 1991). In addition, injection of NGF during prenatal development specifically increases the number of cholinergic neurons in the matrix compartment of the striatum, but not in the patch compartment, suggesting that NGF is involved in the differential maturation of these compartments (Van Vulpen and Van der Kooy, 1999). In vitro experiments show that NGF specifically induces ChAT activity and promotes the survival of cholinergic striatal neurons (Martinez et al., 1985; Mobley et al., 1985; Abiru et al., 1996).

Controversially, it was initially reported that striatal cholinergic neurons were present in NGF and TrkA KO mice (Crowley et al., 1994; Smeyne et al., 1994). However, quantitative analyzes of striatal cholinergic neurons in postnatal TrkA KO mice reveal atrophy of these neurons (Fagan et al., 1997). NGF KO animals also displayed reduced synthesis of ChAT (Crowley et al., 1994). It has been proposed that cholinergic neurons that express p75^{NTR} but not TrkA die during postnatal development. The observations of increased numbers of cholinergic neurons in p75[NTR] KO mice are consistent with the hypothesis that p75[NTR] induces apoptosis (Van der Zee et al., 1996).

A large population of CNS neurons respond to TrkB ligands, BDNF and NT-4/5, including cortical neurons, striatal projection neurons and nigral dopaminergic neurons. In the neocortex, BDNF signaling through TrkB has been implicated in both development and maintenance of cortical circuitry (Ghosh et al., 1994; Huang and Reichardt, 2001). The expression of mRNA[BDNF] in pyramidal cells and the localization of BDNF protein on stellate cells suggest that this neurotrophin maintains cortical GABAergic interneurons (Marty et al., 1997). Indeed, the expression of calcium-binding proteins and neuropeptide Y (NPY) is reduced in BDNF KO animals, suggesting that BDNF regulates the post-natal differentiation of cortical interneurons (Jones et al., 1994). The results of in vitro studies are consistent with these findings since it has been demonstrated that the application of BDNF to cortical cultures induces the expression of calbindin (Ip et al., 1993a; Widmer and Hefti, 1994; Fiumelli et al., 2000). However, the levels of calretinin are decreased in cortical cultures exposed to BDNF (Fiumelli et al., 2000). Thus, BDNF exerts trophic actions on specific subpopulations of GABAergic interneurons of the cerebral cortex that express different calcium binding proteins. However, the analysis of conditional null mutant TrkB mice shows a selective dendrite degeneration in pyramidal neurons of cortical layers II/III and V, without affecting layers I, IV and VI (Xu et al., 2000). These results are in concordance with those showing that BDNF, interacting with extracellular TrkB receptors, induces dramatic sprouting of basal dendrites of cortical pyramidal neurons in organo-typic cultures (Horch, et al., 1999). However, by comparing cortical layers McAllister et al. (1995) demonstrated that neurons in layer IV respond strongly to BDNF, whereas neurons in layers V and VI respond to NT-4/5. Besides these local actions, cortical BDNF can also act as a target-derived neurotrophic factor. Thus, BDNF regulates the maturation of glutamatergic thalamo-cortical synapses in the somatosensory cortex (Itami et al., 2000).

BDNF is present in the striatum, but the low levels of mRNA[BDNF] have led researchers to look for other sources of this neurotrophin to the striatum (Altar et al., 1997). The anterograde transport hypothesis in the corticostriatal pathway is based on the absence of BDNF expression in the striatum, and on other indirect evidence. Studies using intracerebral injec-tion of recombinant human BDNF have shown anterograde transport in the hippocampus but not in the corticostriatal pathway (Sobreviela et al., 1996). It has been speculated that increased levels of cortical BDNF protein after colchicine injection are due to the lack of anterograde transport to the striatum (Altar et al., 1997). However, it has recently been shown that the injection of colchicine in the striatum increases BDNF expression in cortical neurons, which leads to the enhancement of cortical BDNF protein after striatal lesion (Canals et al., 2001). Although anterograde transport must be considered, the presence and differential regulation of BDNF expression in the striatum after different types of injury (Salin et al., 1995; Wong et al., 1997; Canals et al., 1998) suggest a local action of this neuro-trophin in striatal neurons. In fact, BDNF is a potent survival factor for striatal neurons (Ventimiglia et al., 1995). Furthermore, it induces the differentiation of striatal GABA- (Mizuno et al., 1994; Ventimiglia et al., 1995), calbindin- (Ventimiglia et al., 1995; Ivkovic and Ehrlich, 1999) and DARPP-32-positive cells (Nakao et al., 1995; Ivkovic and Ehrlich, 1999). In addition, BDNF null mutant animals present a reduction of striatal projection neurons, as assessed by calbindin (Jones et al., 1994) or DARPP-32 staining (Ivkovic et al., 1997), and of cholinergic interneurons (Ward and Hagg, 2000). It has also been shown that the effect of other trophic factors, such as bone morphogenetic protein-2, on the differentiation of calbindin-positive cells is mediated by endogenous BDNF in striatal cultures (Gratacos et al., 2001a).

Endogenous striatal BDNF may also be a target-derived source for both cortical and nigral neurons that project to the striatum, since intrastriatal injection of BDNF shows that it is retrogradely transported to these brain areas in rat and monkey (Mufson et al., 1994, 1996). The retrograde signal was specifically localized to pyramidal neurons of cortical layer V (Mufson et al., 1994) and in a subpopulation of dopaminergic neurons of the SN

(Mufson et al., 1996). Moreover, it is well documented that BDNF promotes the survival and differentiation of nigral dopaminergic neurons (Hyman et al., 1991). Although the number of dopaminergic neurons is not reduced in the null BDNF mutant animal (Jones et al., 1994), striatal dopamine functions are altered in the heterozygous BDNF animal (Dluzen et al., 2001, 2002). These results are in concordance with in vitro studies showing that BNDF increases levels (Hyman et al., 1994; Spenger et al., 1995; Hoglinger et al., 1998), release and uptake of dopamine (Hyman et al., 1994; Blochl and Sirrenberg, 1996).

NT-3 has also neurotrophic effects in striatal, cortical and nigral neurons. NT-3 increases the number of GABA- and calbindin-positive neurons in striatal cultures (Ventimiglia et al., 1995). This neurotrophin also promotes the survival and morphological differentiation of DARPP-32-positive neurons, increasing the length of neurites and the soma area (Nakao et al., 1996). In the cerebral cortex, the formation and maintenance of neuronal circuits are mediated by equilibrium between NT-3 and BDNF effects. NT-3 and BDNF oppose one another in regulating axonal and dendritic growth of pyramidal neurons (McAllister et al., 1997). Thus, BDNF stimulates dendritic growth in layer IV and inhibits it in layer VI, whereas NT-3 does the opposite, inhibiting sprouting in layers II/III and IV, and stimulating branching in layer VI (McAllister et al., 1997; Castellani and Bolz, 1999). In the SN, NT-3 has similar effects to BDNF increasing dopaminergic activity (Hyman et al., 1994). Furthermore, both NT-3 and TrkC KO animals have an extraordinary phenotype of abnormal movements and postures. However, the basal ganglia that might contribute to this abnormal phenotype have not yet been fully analyzed (Snider, 1994).

NT-4/5 shares with BDNF many trophic effects since it also acts through TrkB. Thus, NT-4/5 increases the survival and differentiation of GABA-, calbindin- and calretinin-positive neurons in striatal cultures (Widmer and Hefti, 1994; Ventimiglia et al., 1995). Furthermore, NT-4/5 induces the same regulation of calcium-binding proteins as BDNF does on cortical cultures (Fimmelli et al., 2000). However, in contrast to BDNF, NT-4/5 elicits a high increase in the number of dopaminergic neurons in culture, but it does not promote dopamine uptake (Hyman et al., 1994) (Table 1).

The GDNF family

GDNF family members

The GDNF family is a group of four proteins distantly related to the transforming growth factor-β superfamily. GDNF, the first trophic factor of this family to be discovered, was purified from a glioma cell line for its ability to promote the survival of midbrain dopaminergic neurons in vitro (Lin et al., 1993). Neurturin (NRTN), the second member of the family, was biochemically purified three years later as a trophic factor capable of regulating sympathetic neurons in vitro (Kotzbauer et al., 1996). The sequence similarities between NRTN and GDNF (about 42%) led to the identification of the GDNF family ligands, which allowed the cloning of the other two members of this family: persephin (PSPN) (Milbrandt et al., 1998) and artemin (ARTN) (Baloh et al., 1998b).

GDNF, NRTN, PSPN and ARTN exert neurotrophic effects on several neuronal populations of the CNS. All of them promote the survival of cultured ventral midbrain dopaminergic neurons (Lin et al., 1993; Baloh et al., 1998b; Horger et al., 1998; Milbrandt et al., 1998; Akerud et al., 1999, 2002; Rosenblad et al., 2000). The GDNF family proteins also regulate several other aspects of midbrain dopaminergic neurons. GDNF and PSPN, but not NRTN, induce neurite outgrowth of dopaminergic neurons in vitro (Akerud et al., 1999, 2002). In cultured neurons from the ventral tegmental area, GDNF treatment regulates the establishment of functional synaptic terminals that may enhance dopamine release (Bourque and Trudeau, 2000). Furthermore, in vivo studies have shown that intranigral GDNF or intrastriatal NRTN, GDNF or PSPN delivery induces behavioral and biochemical changes that can be associated with functional upregulation of nigral dopaminergic neurons (Hebert et al., 1996; Horger et al., 1998; Akerud et al., 2002).

Another neurotrophic action shared by GDNF, NRTN and PSPN is the survival-promoting effect on motor neurons (Henderson et al., 1994; Oppenheim

TABLE 1

Expression of mRNA for neurotrophins, GNDF family members, CNTF and their receptors in the striatum, ventral midbrain and cerebral cortex

	Striatum		Ventral midbrain		Cerebral cortex	
	Development	Adult	Development	Adult	Development	Adult
NGF	+ +	+	−	−	+ +	+
BDNF	+	+	+ +	+	+	+ +
NT-3	+	+	+ +	+	+ + +	+
NT-4/5	n.d.	+	n.d.	+	n.d.	+
P75^{NTR}	+	−	−	−	−	−
TrkA	+	+ +	−	−	−	−
TrkB	+ +	+	+ +	+	+ + +	+
TrkC	+ + +	+	+	+	+	+
GDNF	+ +	+	−	+	+	+
NRTN	+ +	+	+	+	+ +	+
ART	+	+	+	+	n.d.	n.d.
PSPN	+ +	+	+ +	+	+	n.d.
Ret	−	+	+ + +	+ + +	−	−
GFRα1	+	+	+ + +	+ + +	+	+
GFRα2	−	+	+	+	+	+
GFRα3	−	+	−	n.d.	−	+
GFRα4	n.d.	n.d.	n.d.	+	n.d.	+

n.d.: non determined, −: non detected, +: low levels, + +: moderate levels, + + +: high levels.

et al., 1995; Milbrandt et al., 1998; Bilak et al., 1999). GDNF and NRTN also induce neurite outgrowth of noradrenergic neurons from the locus coeruleus in vitro (Holm et al., 2002) while GDNF promotes the survival of GABAergic septal neurons (Price et al., 1996), basal forebrain cholinergic neurons (Ha et al., 1996) and Purkinje cerebellar neurons (Mount et al., 1995) during development. Furthermore, GDNF enhances the synthesis of SP (Humpel et al., 1996) and induces calretinin phenotype in cultured striatal neurons (Farkas et al., 1997). Although the wide array of effects induced by the GDNF family members on developing neurons from the CNS, mice lacking different GDNF family proteins or receptors showed only minor or no deficits in the brain (Baloh et al., 2000; Airaksinen and Saarma, 2002).

The receptors for the GDNF family members

All GDNF family members signal by binding to a common receptor Ret (Jing et al., 1996; Treanor et al.,

1996; Trupp et al., 1996). Ret can only be activated if the GDNF family ligands have previously bound to an accessory protein linked to the plasma membrane by a glycosyl phosphatidylinositol anchor. These classes of proteins were named GDNF family receptor α (GFRα) and consist of four coreceptors (GFRα1-4). GFRαs are believed to provide certain ligand specificity, GFRα1 being the preferred receptor for GDNF (Jing et al., 1996; Treanor et al., 1996), GFRα2 to NRTN (Baloh et al., 1997; Buj-Bello et al., 1997; Jing et al., 1997; Klein et al., 1997), GFRα3 to ARTN (Baloh et al., 1998a; Masure et al., 1998; Naveilhan et al., 1998; Worby et al., 1998) and GFRα4 for PSPN (Enokido et al., 1998; Thompson et al., 1998; Masure et al., 2000). However, there is some cross-interaction between the GDNF family proteins and these receptors (Baloh et al., 1997, 1998b; Cacalano et al., 1998; Scott and Ibañez, 2001). Furthermore, it has been shown that GFRα1 released, as well as anchored to the membrane of neighboring cells, can mediate activation of Ret in *trans* configuration in the presence of GDNF (Trupp et al., 1997; Yu et al., 1998; Paratcha et al., 2001).

Ret is a tyrosine kinase receptor that presents intracytoplasmic tyrosine residues susceptible of being phosphorylated. When Ret binds to a GFRα-GDNF family ligand complex, it activates several intracellular signaling pathways, such as the one mediated by MAPK, PI3K, JNK and PLC-γ. These pathways regulate different cellular functions such as survival, differentiation, proliferation and migration (Airaksinen and Saarma, 2002). Signaling through PI3K or MAPK pathways has been suggested to promote both neuronal survival and differentiation (Pong et al., 1998; Van Weering et al., 1998; Soler et al., 1999; Coulpier et al., 2002; Pelicci et al., 2002). Ret signaling has also been shown to play a role in the migration, axonal growth and guidance of developing sympathetic neurons (Enomoto et al., 2001). Recent data show that Ret can also be phosphorylated on a Ser residue by PKA, following cAMP elevation, and this phosphorylation modulates actin dynamics in neurons (Fukuda et al., 2002). Furthermore, it has also been suggested that GDNF, by binding GFRα1, can activate intracellular signaling independently of Ret (Poteryaev et al., 1999; Trupp et al., 1999).

Expression of GDNF family ligands and receptors in the basal ganglia

GDNF family ligands and receptors are expressed in the nervous system as well as in many peripheral tissues. Among several structures in the brain, GDNF is expressed in the striatum and also in the cerebral cortex and substantia nigra, but it shows different patterns of expression during development. GDNF expression is first detected in the cortex at E14 and in the striatum at E16, while it is not expressed in the midbrain during embryonic development (Nosrat et al., 1996; Golden et al., 1999). In the adult rat and mouse brain, GDNF is detected in the striatum, mesencephalon and cortex (Pochon et al., 1997; Trupp et al., 1997; Golden et al., 1998; Naveilhan et al., 1998; Marco et al., 2002a). In the striatum, during postnatal development (P7) GDNF immunoreactivity shows a patchy distribution (Lopez-Martin et al., 1999) whereas in the adult, mRNA^{GDNF} and GDNF protein are mainly detected in cholinergic interneurons and, to a lesser extent, in

medium-sized spiny projection neurons (Pochon et al., 1997; Trupp et al., 1997; Bizon et al., 1999; Bresjanac and Antauer, 2000). The adult SN is divided in two distinct populations depending on the GDNF expression level: pars compacta dopaminergic neurons show intense GDNF signal while in pars reticulata only a few scattered neurons express GDNF (Pochon et al., 1997). In adult cerebral cortex, GDNF expression is detected in all the cortical layers except layer I (Pochon et al., 1997; Golden et al., 1999). Immunohistochemical analysis has shown a similar distribution of GDNF protein in human brain (Kawamoto et al., 2000).

NRTN is expressed in the same areas as GDNF. During development NRTN expression appears sequentially in the midbrain and striatum. NRTN expression first appears in the midbrain during embryonic development (Horger et al., 1998; Golden et al., 1999) and maintains its expression at moderate levels during early postnatal life and in the adult (Horger et al., 1998; Naveilhan et al., 1998; Akerud et al., 1999). In the striatum, NRTN signal is detected at P0 and is maintained until adulthood (Horger et al., 1998; Naveilhan et al., 1998; Akerud et al., 1999; Marco et al., 2002a). In this nucleus, NRTN expression seems to be neuronal but the signal is not distributed in a patch or matrix pattern (Widenfalk et al., 1997). NRTN has also been detected in the cerebral cortex by E14 at low levels (Golden et al., 1999); it reached high levels at P15 and then decreased to adult levels (Holm et al., 2002).

The mRNA of the third member of the GDNF family, ARTN, is not detected in rat CNS during embryonic development, but it has been localized to the striatum and SN, in fetal and adult human brain (Baloh et al., 1998a).

Finally, using RT-PCR, PSPN has been demonstrated at moderate levels in the striatum, mesencephalon and cerebral cortex of P0 rats (Jaszai et al., 1998). With the same technique, PSPN expression has been examined in the ventral mesencephalon and striatum during development. In the ventral mesencephalon, PSPN is detected during embryonic development, beginning at E15, until adulthood with a peak of expression at E19 (Akerud et al., 2002). In the striatum, PSPN is found in the same developmental stages but it peaks at P0 (Akerud et al., 2002).

The receptors for the GDNF family are also expressed in the CNS showing complementary as well as overlapping patterns. The common receptor Ret is expressed in the SN and striatum, but not in the cortex. In the adult striatum, mRNA^{Ret} levels are very low and its expression has only been detected using a very sensitive technique (Pérez-Navarro et al., 1999b; Marco et al., 2002a). Immunohistochemical studies disclosed that striatal neurons expressed Ret both in vivo (Marco et al., 2002a) and in vitro (Gratacos et al., 2001b). Ret has been detected in the ventral midbrain at E10 and its expression continued at very high levels throughout embryonic and post-natal development until adulthood (Nosrat et al., 1997; Widenfalk et al., 1997; Yu et al., 1998; Burazin and Gundlach, 1999; Golden et al., 1999; Marco et al., 2002c). mRNA^{Ret} and Ret protein have been found in dopaminergic neurons of the SN pars compacta and in neurons of the ventral tegmental area (Nosrat et al., 1997; Widenfalk et al., 1997, Glazner et al., 1998). In the SN pars reticulata, mRNA^{Ret} has only been detected at P14, with high levels of expression (Burazin and Gundlach, 1999).

As observed for Ret, mRNA$^{GFR\alpha1}$ is expressed in the striatum and in the ventral mesencephalon. Its levels are very low in the striatum but can be detected by in situ hybridization during development (Nosrat et al., 1997; Widenfalk et al., 1997; Golden et al., 1999) and by the RNase protection assay in adulthood (Naveilhan et al., 1998; Pérez-Navarro et al., 1999b; Marco et al., 2002a). GFRα1 protein can be easily detected by immunohistochemistry in the adult rat striatum, showing the same pattern of localization as GDNF and Ret, with high signal in large striatal interneurons and weak signal in medium-sized striatal projection neurons (Bresjanac and Antauer, 2000; Marco et al., 2002a). Furthermore, striatal neurons in vitro also express GFRα1 (Gratacos et al., 2001b). In the ventral midbrain, GFRα1 is shown at high levels in the SN pars compacta and in the ventral tegmental area during embryonic and postnatal development until adulthood (Nosrat et al., 1997; Glazner et al., 1998; Horger et al., 1998; Yu et al., 1998; Burazin and Gundlach, 1999; Golden et al., 1999; Marco et al., 2002c) similar to what is observed for Ret expression. In the SN pars reticulata, this receptor expression has also been detected during postnatal development and in the

adult (Burazin and Gundlach, 1999; Sarabi et al., 2001). Double labeling studies disclosed that most of the dopaminergic neurons in the SN pars compacta expressed GFRα1 (Horger et al., 1998; Sarabi et al., 2001), while in the SN pars reticulata neurons expressing these receptors were GABAergic (Sarabi et al., 2001). In contrast to that observed for Ret expression, mRNA$^{GFR\alpha1}$ has been demonstrated in neurons from layers IV, V and VI of the cerebral cortex during embryonic development until adulthood (Glazner et al., 1998; Burazin and Gundlach, 1999; Golden et al., 1999). Double labeling analysis has recently shown that only a very small percentage of cortical GABAergic neurons express GFRα1 (Sarabi et al., 2003).

GFRα2 is expressed at low levels in the striatum and its mRNA has been detected by the RNase protection assay in the adult mouse and rat striatum (Naveilhan et al., 1998; Marco et al., 2002a). Furthermore, immunohistochemical analysis has shown that this receptor is located in the cell body and neurites of striatal neurons in vitro (Gratacos et al., 2001b). The mRNA for this receptor has also been detected in the ventral midbrain during embryonic development until adulthood (Widenfalk et al., 1997; Horger et al., 1998; Burazin and Gundlach, 1999; Golden et al., 1999; Marco et al., 2002a). In adult rat brain, little if any, mRNA$^{GFR\alpha2}$ has been found in nigral dopaminergic neurons (Horger et al., 1998; Marco et al., 2002c), which is also consistent with results obtained during embryonic development (Wang et al., 2000). In contrast, regions immediately adjacent to nigral dopaminergic neurons show a strong signal for this receptor mRNA (Horger et al., 1998). In the cerebral cortex, mRNA$^{GFR\alpha2}$ is detected in cortical layers III, V and VI during embryonic and postnatal development as well as in the adult (Widenfalk et al., 1997; Wang et al., 1998; Burazin and Gundlach, 1999; Golden et al., 1999).

mRNA$^{GFR\alpha3}$ is not expressed by in situ hybridization in the developing CNS (Baloh et al., 1998a; Trupp et al., 1998; Worby et al., 1998; Yu et al., 1998) but is detected in the mouse brain during embryonic development by RNase protection assay (Naveilhan et al., 1998). In human brain, mRNA$^{GFR\alpha3}$ is found at low levels during development, and in the adult it is detected in the cerebral cortex with low levels of expression (Masure et al., 1998).

GFRα4 has been shown by northern blot in mouse brain during embryonic development and in adulthood, and also in adult rat brain (Masure et al., 2000). Using in situ hybridization, this receptor has been localized in the cerebral cortex and in the ventral midbrain of the adult mouse. In the ventral midbrain, mRNA^{GFRα4} labeling is localized to the ventral tegmental area, in SN pars compacta neurons, and, to a lesser extent, in neurons of the SN pars reticulata (Masure et al., 2000). These findings have recently been extended to the adult rat brain, where mRNA^{GFRα4} has been detected in the ventral tegmental area and in nigrostriatal dopaminergic neurons (Akerud et al., 2002).

Ciliary neurotrophic factor

Ciliary neurotrophic factor

CNTF was discovered as a component of embryonic chick eye that supported the survival of isolated ciliary neurons in culture (Adler et al., 1979; Nishi and Berg, 1981; Barbin et al., 1984; Lin et al., 1989; Stockli et al., 1989). Subsequent cloning and sequencing revealed that CNTF is a member of an α-helical cytokine superfamily that includes cytokines, such as interleukin-6 (IL-6), leukemia inhibitory factor (LIF) and leptin (Sendtner et al., 1994; Ip and Yancopoulos, 1996). CNTF is expressed in glial cells within the CNS and PNS (Lisovoski et al., 1997). CNTF stimulates gene expression, cell survival or differentiation in a variety of neuronal cell types (Manthorpe et al., 1992; Ip and Yancopoulos, 1996). In addition to ciliary neurons, CNTF inhibits the proliferation of sympathetic precursors (Ernsberger et al., 1989), affects the differentiation of developing sympathetic neurons and induces the expression of neuropeptide genes in neuronal cell lines (Saadat et al., 1989; Rao et al., 1992; Symes et al., 1993). It also enhances survival in preganglionic sympathetic neurons (Blottner et al., 1989), and sensory neurons (Barbin et al., 1984). Other studies have reported that cultured motor neurons are very sensitive to the trophic influences of CNTF (Arakawa et al., 1990; Oppenheim et al., 1991).

CNTF can also participate in other functions such as oligodendrocyte maturation inducing a strong promyelinating effect (Linker et al., 2002; Stankoff et al., 2002). Moreover, CNTF regulates voltage-gated ion channels in neuroblastoma cells and potentiates the release of neurotransmitters (Lesser and Lo, 1995; Stoop and Poo, 1995). Treatment of humans and animals with CNTF is also known to induce weight loss characterized by a preferential loss of body fat. When administered systemically, CNTF activates downstream signaling molecules in areas of the hypothalamus, which regulates food intake (Gloaguen et al., 1997). In addition to its neural actions, CNTF and analogs act on nonneuronal cells such as hepatocytes, skeletal muscle cells, embryonic stem cells and bone marrow stromal cells (Sleeman et al., 2000).

Several lines of evidence indicate that CNTF may be a key player in the response to injury by the nervous system. CNTF itself lacks a classical signal peptide sequence of a secreted protein (Lin et al., 1989; Stockli et al., 1989), but it is thought to convey its cytoprotective effects after release from adult glial cells by some mechanisms induced by injury. Dramatic changes in CNTF expression levels occur following neural trauma (Ip et al., 1993c; Asada et al., 1995) or ischemia (Park et al., 2000). The increased expression of CNTF is restricted to astrocytes adjacent to the site of damage, supporting the idea that astrocytes play an important role in maintaining or restoring neuronal function following brain injury (Rudge et al., 1995). Interestingly, mice that are homozygous for an inactivated CNTF gene develop normally and initially thrive. Only later in adulthood do they exhibit a mild loss of motor neurons with resulting muscle weakness, leading to the suggestion that CNTF is not essential for neural development, but instead acts in response to injury or other stresses (Masu et al., 1993; Sendtner et al., 1996). Thus, several studies led CNTF to be considered as an 'injury factor' in the CNS and PNS. It has been reported that endogenous CNTF located in the cytosol of myelinating Schwann cells can act as a lesion factor once it is released after motoneuron axotomy in adult mice (Sendtner et al., 1997). Furthermore, CNTF administration prevents the loss of cholinergic (Hagg et al., 1992), dopaminergic (Hagg and Varon, 1993), GABAergic

(Anderson et al., 1996) and thalamocortical (Clatterbuck et al., 1993) neurons in various models of neurodegenerative diseases, suggesting that delivery of CNTF to the damaged nervous system may be one means of treating the behavioral and neuroanatomical consequences of neurological diseases.

CNTF receptor

CNTF acts through a receptor complex (CNTFRα, gp130 and LIFRβ subunits) that is closely related to and shares subunits with the receptor complexes for IL-6 and LIF (Davis et al., 1991; Davis and Yancopoulos, 1993; Stahl and Yancopoulos, 1994). In contrast to other known cytokine receptors, CNTFRα has neither a transmembrane nor a cytoplasmatic region, but is anchored to the cell membrane by a glycosyl-phosphatidylinositol linkage. The CNTFRα can function in either the membrane-bound form or soluble form, the latter being produced by phospholipase C-mediated cleavage of the membrane-bound receptor (Davis et al., 1993). Mice lacking CNTFRα, gp130 or LIFRβ die perinatally (DeChiara et al., 1995; Li et al., 1995; Ware et al., 1995), indicating there may be a second developmentally important CNTF-like ligand. CNTFRα does not play a direct role in signaling, but instead forms a complex with CNTF that promotes its binding to the signal transduction 'β' receptor components, gp130 and LIFRβ, forming a tripartite receptor complex (Stahl and Yancopoulos, 1994). CNTF-induced heterodimerization of the β receptor subunits leads to tyrosine phosphorylation (through constitutively associated JAKs), and the activated receptor provides docking sites for SH2-containing signaling molecules, such as STAT proteins (Stahl et al., 1995). Activated STATs dimerize and translocate to the nucleus to bind specific DNA sequences, resulting in enhanced transcription of responsive genes. In addition to the JAK/STAT pathway, CNTF and other cytokines can also induce tyrosine phosphorylation of PLCγ, PTP1D, pp120, Shc, Grb2, Raf-1 and ERK1 and ERK2 (Boulton et al., 1994; Hirano et al., 1997).

Neuroprotection by neurotrophins in HD

Endogenous regulation of neurotrophins in HD

Excitotoxicity has been proposed as the mechanism leading to the differential vulnerability between projection neurons and interneurons in HD (Beal, 1994). The distribution of glutamate receptor subtypes and subunits, together with the activity of the corticostriatal pathway, could be important to define the threshold of vulnerability of each neuronal population to excitotoxic lesion (Landwehrmeyer et al., 1995; Chen et al., 1996; Gotz et al., 1997). Several studies have reported the differential trophic response of striatal cells to the excitotoxic damage induced by distinct glutamate receptor agonists (Alberch et al., 2002). The expression of neurotrophins and their receptors are differentially regulated depending on the glutamate receptor stimulated (Canals et al., 1998, 1999). mRNA^{NGF} levels are regulated in the striatum after intraparenchymal injection of various glutamate receptor agonists, showing a specific pattern of expression (Canals et al., 1998). Intrastriatal injection of quinolinic acid produces a large increase in mRNA^{NGF} and NGF protein during development (Pérez-Navarro and Alberch, 1995) and in adulthood (Pérez-Navarro et al., 1994; Canals et al., 1998). The injection of non-NMDA agonists shows that AMPA, but not kainic acid, induces an early and transient increase of mRNA^{NGF} (Canals et al., 1998). However, a late effect of kainic acid increasing NGF protein levels is observed two days postlesion, which is coincident with excitotoxin-induced reactive astrogliosis (Strauss et al., 1994; Pérez-Navarro and Alberch, 1995). Furthermore, intrastriatal injection of basic fibroblast growth factor (bFGF), which is a mitogenic agent for astrocytes (Morrison and de Vellis, 1981), simultaneously with quinolinic acid potentiates the increase in NGF levels (Pérez-Navarro et al., 1994). These results suggest that local reactive astrocytes may be responsible for the changes in NGF levels. The biological effects of this regulation could be related to the effect of NGF on the stimulation of ChAT activity in the striatum (Martinez et al., 1985; Mobley et al., 1985). Therefore, the enhancement of mRNA^{NGF} and NGF protein levels after excitotoxicity can be considered as an endogenous neuroprotective mechanism for striatal cholinergic interneurons (Pérez-Navarro et al., 1994; Strauss et al., 1994; Canals et al., 1998).

The high-affinity receptor for NGF, TrkA, is also specifically regulated by different glutamate receptor agonists (Canals et al., 1999). Intrastriatal injection of quinolinic acid produces a downregulation of mRNA^{TrkA} (Venero et al., 1994; Canals et al., 1999), which correlates with the decrease of cholinergic neurons (Venero et al., 1994). On the other hand, TrkA expression is increased after the treatment with kainic acid or a general agonist of the metabotropic glutamate receptors, 1S,3R-1-aminocyclopentane-1,3-dicarboxylic acid (ACPD) (Canals et al., 1999). The lack of correlation between the regulation of the neurotrophin and its receptor suggests the presence of different mechanisms that could regulate the expression of ligand or receptor independently, such as neuronal activity or the extent of the lesion (Canals et al., 1998, 1999; Checa et al., 2000, 2001).

The expression of BDNF is also regulated after excitotoxic stimuli. Intrastriatal injection of quinolinic acid induces a local increase in mRNA^{BDNF} levels during the first postnatal week and at P21 (Checa et al., 2000), but not in the adult (Canals et al., 1998). In contrast, the stimulation of both non-NMDA receptor types, AMPA and kainate, increases in mRNA^{BDNF} levels in the adult striatum (Canals et al., 1998). However, TrkB expression is regulated independently from the ligand since the activation of both NMDA and nonNMDA glutamate receptors increases striatal mRNA^{TrkB} (Canals et al., 1999).

Besides these local actions, transneuronal regulation of the expression of BDNF and TrkB in the cerebral cortex is observed in the excitotoxic model of HD (Checa et al., 2000; Canals et al., 2001). Thus, the regulation by excitotoxicity of endogenous corticostriatal trophic relations has been described during postnatal development (Checa et al., 2000) and in adulthood (Canals et al., 2001). Intrastriatal injection of quinolinic acid produces a biphasic enhancement in cortical mRNA^{BDNF} levels in the early neonatal stage (P5) and from P21 to the adult (Checa et al., 2000). In situ hybridization studies show that, in the adult, intrastriatal injection of quinolinic or kainic acid upregulates BDNF in neurons of cortical layers II/III, V and VI (Canals et al., 2001). Canals and colleagues (2001) followed different technical approaches to show that mRNA^{BDNF} is regulated in response to the damage of the striatal target area. Hence, cortical BDNF upregulation is prevented when the striatum is protected by grafting a BDNF-secreting cell line before the striatal excitotoxic injury. In addition, the expression of this neurotrophin in the cortex is also upregulated in rats lesioned with 3-nitropropionic acid or colchicine (Canals et al., 2001). Furthermore, the levels of full-length and truncated mRNA^{TrkB} are increased in cortical neurons of the ipsilateral cortex of quinolinic acid injected striata (Checa et al., 2001). These results suggest that the upregulation of cortical BDNF and TrkB after the lesion of the target area, the striatum, may be an endogenous trophic mechanism to protect cortical neurons.

Intrastriatal injection of quinolinic or kainic acid does not affect endogenous NT-3 in the striatum during postnatal development and in adulthood (Canals et al., 1998; Checa et al., 2000). However, AMPA and ACPD downregulate and increase striatal mRNA^{NT-3} levels, respectively (Canals et al., 1998). In contrast to BDNF, the cerebral cortex ipsilateral to quinolinic acid-injected striatum shows a decrease in the levels of NT-3 in early postnatal development and in the adult (Checa et al., 2000). The levels of striatal mRNA^{TrkC} are decreased when P5 rats are intrastriatally lesioned with quinolinic acid and no changes are observed in other postnatal ages (Checa et al., 2001) and after treatment with other glutamate receptor agonists (Canals et al., 1999).

The involvement of BDNF in the maintenace of the corticostriatal pathway in HD has recently been confirmed by studies performed in HD transgenic mice or cellular models of HD. Zuccato and colleagues (2001) described that wild-type huntingtin transfected to striatal inmortalized neural stem cells upregulates the transcription of BDNF, whereas transfection with the mutant huntingtin decreases the levels of this neurotrophin. Mutant huntingtin also induces a reduction of BDNF in cerebral cortex and striatum in the transgenic mice overexpressing mutant full-length huntingtin (YAC72) (Zuccato et al., 2001). However, changes in NGF or NT-3 are not observed in any of these HD models (Zuccato et al., 2001).

Crucial data on the study of trophic factors in HD should be provided by the analysis of brain samples from different human diseased states. Recently, studies have shown that some neurotrophins are modified in autopsy brain tissue from HD patients (Ferrer et al., 2000; Zuccato et al., 2001). The decrease in BDNF protein in the caudate-putamen nucleus of HD patients ranges from 53 to 82% (Ferrer et al., 2000; Zuccato et al., 2001). However, controversial results have been published on the levels of BDNF in the cerebral cortex of HD brains. Zuccato and colleagues (2001) showed that mRNA^{BDNF} is decreased in samples obtained from the frontoparietal cortex of two patients (grade II and III), whereas Ferrer and colleagues (2000) demonstrated that samples from four HD patients (grade III) have preserved levels of BDNF protein in the parietal and temporal cortex. Other neurotrophic factors have also been examined in the cerebral cortex of HD human brains, showing a decrease in mRNA^{NT-3} levels, whereas NGF expression is unchanged (Zuccato et al., 2001).

Exogenous administration of neurotrophins in HD

Several groups have characterized the action of exogenous administration of NGF in vivo in the protection of striatal cholinergic neurons in an excitotoxic model of HD. A single coinjection of NGF together with quinolinic or kainic acid selectively prevents cell death of cholinergic interneurons, and the reduction in ChAT activity in the striatum (Altar et al., 1992; Davies and Beardsall, 1992; Pérez-Navarro et al., 1994, Pérez-Navarro and Alberch, 1995). Pretreatment with NGF before quinolinic acid injection prevents some of the disturbances induced by excitotoxicity, including increase of nitric oxide (Maksimovic et al., 2002) and decrease of glutathione metabolism (Cruz-Aguado et al., 2000). Another group has administrated NGF in the striatum daily for one week (Venero et al., 1994), and this continuous NGF infusion prevents the reduction in ChAT activity in cells expressing TrkA, but not the decrease in glutamate decarboxylase mRNA levels (Venero et al., 1994), suggesting a specific action of NGF on striatal cholinergic neurons.

Other systems to deliver NGF have been developed. Hence, NGF-producing cells have been implanted to protect the striatum against excitotoxicity

(Schumacher et al., 1991; Frim et al., 1993; Martínez-Serrano and Björklund, 1996). Grafting of fibroblasts genetically modified to express high levels of NGF provides marked neuroprotection against striatal damage, and in cholinergic and NADPH-diaphorase neurons (Schumacher et al., 1991). Furthermore, when NGF-secreting stem cells are implanted into the striatum one week before the injection of quinolinic acid, striatal projection neurons are also protected (Martínez-Serrano and Björklund, 1996). Intrastriatal preimplantation of NGF-secreting fibroblasts also prevents 3-nitrotyrosine formation in a 3-nitropropionic acid model of HD (Galpern et al., 1996).

The use of microspheres for protein delivery in vivo has also been tested for NGF administration. Morphological analysis after the implant of NGF-loaded microspheres shows that animals tolerate the surgical procedure and that microspheres still contain NGF, 2.5 months later (Gouhier et al., 2000; Menei et al., 2000). After quinolinic acid injection, the lesion size in the group treated with NGF-releasing microspheres is reduced by 40% compared to control groups (Menei et al., 2000). The use of poly-D,L-lactide-co-glycolide microspheres loaded with NGF also shows a slight neuroprotection, evaluated by ex vivo autoradiography of D2 dopaminergic receptors (Gouhier et al., 2000).

Intraparenchymal administration of NGF following the above mentioned techniques require invasive neurosurgical procedures. Some groups have developed noninvasive systems to deliver NGF. Friden et al. (1993) showed that when this neurotrophin is conjugated to an antibody directed against transferrin receptor, it crosses the blood-brain barrier and prevents the degeneration of cholinergic neurons in a model of HD (Kordower et al., 1994). However, following this protocol, intravenous administration of NGF must be performed two days prior to striatal lesion and every two days for 14 days (Kordower et al., 1994).

BDNF is one of the main candidates for gene therapy strategies in HD because of its protective effects on striatal projection neurons. Intrastriatal administration of BDNF by various routes, including cell and viral delivery, protects striatal cells from excitotoxicity in vivo (Martínez-Serrano and Björklund, 1996; Bemelmans et al., 1999;

Pérez-Navarro et al., 2000b; Gratacos et al., 2001b). Viral administration of exogenous BDNF, two weeks before intrastriatal injection of quinolinic acid, protects DARPP-32-positive neurons with a 55% reduction in lesion size (Bemelmans et al., 1999). However, this effect is only slightly observed when neural stem cells genetically modified to release BDNF are used (Martínez-Serrano and Björklund, 1996). However, intrastriatal grafting of BDNF-secreting fibroblasts induces an important reduction of the lesion area and the number of TUNEL-positive nuclei in the quinolinic acid-injected striatum (Pérez-Navarro et al., 2000b). In addition, in situ hybridization analysis demonstrates that GAD-, enkephalin-, SP- and dynorphin-positive neurons are protected in the excitotoxic model of HD when the striatum is previously transplanted with BDNF-secreting cells (Pérez-Navarro et al., 2000b). Pérez-Navarro and colleagues (1999a) demonstrated that BDNF also regulates the soma area of SP-positive neurons in nonlesioned striatum and prevents the cell atrophy of striatopallidal and striatonigral neurons. Furthermore, BDNF-secreting fibroblasts also prevent the reduction in the number of striatal GABAergic neurons and cholinergic interneurons induced by kainic acid injection (Gratacos et al., 2001b).

Similarly, the effects of NT-3-secreting fibroblasts have also been analyzed in quinolinic or kainic acid injected striatum (Pérez-Navarro et al., 1999a, 2000b; Gratacos et al., 2001b). The transplant of NT-3-secreting fibroblasts completely prevents striatal projection neuron atrophy induced by quinolinic acid treatment, and this neurotrophin is more efficient than BDNF or NT-4/5 (Pérez-Navarro et al., 1999a). NT-3 also prevents the striatal cell death induced by quinolinic acid in vivo, but its protective effect is lower than that of BDNF (Pérez-Navarro et al., 2000b). In the kainic acid injected striatum, NT-3 protects projection neurons and interneurons (Gratacos et al., 2001b). The grafting of fibroblasts secreting NT-4/5 has similar efficacy to NT-3 preventing apoptotic cells after quinolinic acid lesion (Pérez-Navarro et al., 2000b). However, the analysis of neuronal populations in the striatum revealed that NT-4/5 has low protective effects on GAD- or enkephalin-positive neurons, whereas it has the same efficiency as BDNF or NT-3 on

SP-positive neurons (Pérez-Navarro et al., 2000b). Similar effects were observed by Alexi and colleagues (1997) by continuous intrastriatal injection of NT-4/5 for seven days postlesion (Table 2).

In contrast to the excitotoxic models, the neuroprotective effects of neurotrophins have not been well characterized in transgenic HD models. The relationship between BDNF and huntingtin has been studied in cellular models of HD, showing that BDNF can rescue striatal neurons from the apoptosis induced by the transfection of mutant huntingtin (Saudou et al., 1998).

Neuroprotection by GDNF family members in HD

Endogenous regulation of GDNF family members in HD models

The endogenous regulation of the expression of GDNF, NRTN and their receptors has been examined in excitotoxic models of HD (Alberch et al., 2002). Intrastriatal injection of quinolinic or kainic acid induces differential regulation of the mRNA levels of these factors and their receptors. GDNF, Ret and mRNA$^{GFR\alpha1}$ levels are increased by intrastriatal injection of quinolinic or kainic acid. NRTN expression is only increased by quinolinic acid, whereas mRNA$^{GFR\alpha2}$ levels are not modified by any of these glutamate receptor agonists (Marco et al., 2002a). Furthermore, immunohistochemical analysis disclosed that GDNF and GFRα1, but not Ret protein, change their cellular localization in response to excitotoxicity. In control animals, GDNF, GFRα1 and Ret are mainly located in large striatal neurons while weak staining is observed in medium-sized neurons (Pochon et al., 1997; Bizon et al., 1999; Marco et al., 2002a). One day after lesion, only reactive astrocytes are stained for GDNF, and GFRα1 is still localized in neurons (Marco et al., 2002a). However, at later time points after quinolinic acid lesion, GFRα1 can also be detected in astroglial cells (Bresjanac and Antauer, 2000; Marco et al., 2002a). It has been suggested that GFRα1 could capture and concentrate diffusible GDNF, and presented it to Ret-expressing cells (Trupp et al., 1997; Yu et al., 1998). Thus, GFRα1

TABLE 2
Protective effects of neurotrophins, GDNF family members and CNTF in animal models of HD

	Neural population	Type of lesion	Delivery system
NGF	Cholinergic intereurons	Quinolinic acid	Single intrastriatal injection (1–4)
			Continous intrastriatal injection (5)
			Grafting of fibroblasts-secreting NGF (6)
			Microspheres (7,8)
			Intravenous NGF-conjugated to transferrin antibody (9)
		Kainic acid	Single intrastriatal injection (4)
	GABAergic projection neurons	Quinolinic acid	Grafting of stem cells-secreting NGF (10)
	NADPHd-positive interneurons	Quinolinic acid	Microspheres (7,8)
BDNF	GABAergic projection neurons	Quinolinic acid	Grafting of stem cells-secreting BDNF (10)
			Grafting of fibroblasts-secreting BDNF (11)
			Viral administration
		Kainic acid	Grafting of fibroblasts-secreting BDNF (13)
	GABAergic interneurons	Kainic acid	Grafting of fibroblasts-secreting BDNF (13)
	Cholinergic interneurons	Kainic acid	Grafting of fibroblasts-secreting BDNF (13)
NT-3	GABAergic projection neurons	Quinolinic acid	Grafting of fibroblasts-secreting NT-3 (11)
			Continous intrastriatal injection (14)
		Kainic acid	Grafting of fibroblasts-secreting NT-3 (13)
	GABAergic interneurons	Kainic acid	Grafting of fibroblasts-secreting NT-3 (13)
	Cholinergic interneurons	Quinolinic acid	Continous intrastriatal injection (14)
		Kainic acid	Grafting of fibroblasts-secreting NT-3 (13)
NT-4/5	GABAergic projection neurons	Quinolinic acid	Continous intrastriatal injection (15)
			Grafting of fibroblasts-secreting NT-4/5 (11)
	GABAergic interneurons	Quinolinic acid	Continuous intrastriatal injection (15)
GDNF	GABAergic projection neurons	Quinolinic acid	Single i.c.v injection (14)
			Continuous i.c.v injection (14)
			Grafting of fibroblasts-secreting GDNF (16)
		Kainic acid	Grafting of fibroblasts-secreting GDNF (13)
		3-nitropropionic acid	Single i.c.v injection (17)
	Stratonigral neurons	Quinolinic acid	Grafting of fibroblasts-secreting GDNF (18)
	Cholinergic interneurons	Quinolinic acid	Grafting of fibroblasts-secreting GDNF (19)
		Kainic acid	Grafting of fibroblasts-secreting GDNF (13)
		3-nitropropionic acid	Single i.c.v injection (17)
	GABAergic interneurons	Kainic acid	Grafting of fibroblasts-secreting GDNF (13)
NRTN	GABAergic projection neurons	Quinolinic acid	Grafting of fibroblasts-secreting NRTN (19)
		Kainic acid	Grafting of fibroblasts-secreting NRTN (13)
	Striatopallidal neurons	Quinolinic acid	Grafting of fibroblasts-secreting NRTN (20)
	GABAergic interneurons	Kainic acid	Grafting of fibroblasts-secreting GDNF (13)
	Cholinergic interneurons	Kainic acid	Grafting of fibroblasts-secreting GDNF (13)
CNTF	GABAergic projection neurons	Quinolinic acid	Continous intrastriatal injection (21)
			Daily i.c.v injection (21)
			Encapsulated fibroblasts secreting CNTF (22–24)
			Intrastriatal viral vectors (25)
		3-nitropropionic acid	Encapsulated fibroblasts secreting CNTF (26)
			Intrastriatal viral vectors (27)
	Cholinergic interneurons	Quinolinic acid	Encapsulated fibroblasts secreting CNTF (23)
			Intrastriatal viral vectors (25)
	NADPHd-positive interneurons	Quinolinic acid	Encapsulated fibroblasts secreting CNTF (22)
			Intrastriatal viral vectors (25)
	Cortical neurons	Quinolinic acid	Encapsulated fibroblasts secreting CNTF (24)

(1) Altar et al., 1992; (2) Davies and Beardsall, 1992; (3) Pérez-Navarro et al., 1994; (4) Pérez-Navarro and Alberch, 1995; (5) Venero et al., 1994; (6) Schumacher et al., 1991; (7) Gouhier et al., 2000; (8) Menei et al., 2000; (9) Kordower et al., 1994; (10) Martínez-Serrano and Björklund, 1996; (11) Pérez-Navarro et al., 2000b; (12) Bemelmans et al., 1999; (13) Gratacos et al., 2001b; (14) Araujo and Hilt, 1997; (15) Alexi et al., 1997; (16) Pérez-Navarro et al., 1996; (17) Araujo and Hilt, 1998; (18) Pérez-Navarro et al., 1999b; (19) Pérez-Navarro et al., 2000b; (20) Marco et al., 2002c; (21) Anderson et al., 1996; (22) Emerich et al., 1996; (23) Emerich et al., 1997b; (24) Kordower et al., 1999; (25) Pereira de Almeida, 2001; (26) Mittoux et al., 2000; (27) Mittoux et al., 2002.

expression on reactive astrocytes may account for the enhancement of the biological effects of GDNF on surviving striatal neurons.

Neuroprotection by exogenous factors

Neuroprotective effects of GDNF and NRTN have been examined in various models of HD. Intrastriatal grafting of fibroblast cell lines secreting high levels of GDNF (Pérez-Navarro et al., 1996) or NRTN (Pérez-Navarro et al., 2000a) prevents the decrease in the number of striatal GABAergic projection neurons (assessed by immunohistochemistry against calbindin) but not the decrease in the number of striatal GABAergic interneurons (parvalbumin-positive) induced by quinolinic acid. However, the effects of these factors are different, NRTN being more efficient than GDNF in protecting calbindin-positive neurons. Analysis of striatal neuronal populations by in situ hybridization shows that GDNF and NRTN protective effects are selective, since GDNF only prevents the degeneration of striatonigral projection neurons (Pérez-Navarro et al., 1999b), while NRTN acts on striatopallidal projection neurons (Marco et al., 2002b). GDNF (Pérez-Navarro et al., 1999b) and NRTN (Marco et al., 2002b) prevent not only the decrease in the number of these neuronal populations but also the phenotypic changes induced by intrastriatal quinolinic acid injection. Furthermore, intrastriatal grafting of the GDNF-secreting cell line in nonlesioned striata induces an increase in the soma area of striatonigral neurons and also in preprotachykinin-A mRNA levels (Pérez-Navarro et al., 1999b). These neurotrophic factors also show differential effects on striatal cholinergic interneurons, since GDNF but not NRTN protects them from the decrease in ChAT activity induced by intrastriatal injection of quinolinic acid (Pérez-Navarro et al., 2000a). All together, these results suggest that GDNF and NRTN could bind to different receptors localized in specific striatal neuronal populations, resulting in selective neuroprotection.

GDNF and NRTN are also neuroprotective against kainic acid-induced excitotoxic damage in primary striatal cultures and in vivo (Gratacos et al., 2001b). However, the neuroprotective effects of the two factors after intrastriatal injection of kainic or quinolinic acid are quite different. GABAergic striatal interneurons are protected by both factors against kainic acid (Gratacos et al., 2001b), but not against quinolinic acid-induced excitotoxicity (Pérez-Navarro et al., 1996, 2000a). Furthermore, GDNF, but not NRTN, prevents cholinergic interneurons degeneration induced by quinolinic acid (Pérez-Navarro et al., 2000a), whereas both factors protect these neurons against kainic acid lesion (Gratacos et al., 2001b). The differential protective effects of GDNF and NRTN against quinolinic- or kainic acid-induced damage could be explained by the distinct intracellular signaling cascades activated by these glutamate receptor agonists.

Other GDNF delivery systems have also been used in the excitotoxic model of HD: a single cerebroventricular injection of GDNF (10 µg) 30 min before intrastriatal quinolinic acid injection, or a continuous infusion of GDNF for two weeks (10 µg/day). Both types of treatment prevent the decrease in D_1 and D_2 dopamine binding sites in the striatum, [^{3}H]GABA uptake, SP, enkephalin and dynorphin levels in striatal target tissues (SN and globus pallidus), and the reduction in ChAT activity induced by quinolinic acid injection (Araujo and Hilt, 1997). In the 3-nitropropionic acid model, intracerebroventricular injection of GDNF also prevents changes in GABAergic and cholinergic neuronal markers (Araujo and Hilt, 1998). Furthermore, GDNF treatment attenuates changes in motor activity induced by quinolinic acid (Araujo and Hilt, 1997) or 3-nitropropionic acid treatment (Araujo and Hilt, 1998).

The other members of the GDNF family, ARTN and PSPN, have not been tested in animal models of HD. However, it has recently been shown that mice lacking PSPN are hypersensitive to cerebral ischemia and that exogenous administration of PSPN can reduce the infarct volume in KO mice and in rats (Tomac et al., 2002). Furthermore, this study shows that PSPN can prevent glutamate-induced damage in cortical cultures, suggesting PSPN as a potent modulator of excitotoxicity in the CNS. Since PSPN is expressed in rat striatum, it is tempting to speculate that this trophic factor could also protect striatal neurons against quinolinic- or kainic acid-induced excitotoxicity.

Neuroprotection by CNTF in HD

CNTF in the striatum is undetectable by ELISA (Pereira de Almeida et al., 2001), and it has only been detected by immunohistochemistry in glial cells after quinolinic acid treatment (Lisovoski et al., 1997). However, a number of studies have found that CNTF is neuroprotective in the quinolinic acid model of HD. Direct infusion of CNTF (9.4 µg/day) using osmotic pumps or daily injections beginning 3–4 days prior to quinolinic acid lesion partially improves medium-sized striatal neuron survival when assessed by Nissl staining 8–9 days after excitotoxic damage (Anderson et al., 1996). Similar results were obtained using a polypeptide CNTFR agonist, Axokine 1 (4.8 µg/day) (Anderson et al., 1996). Other studies used a different drug delivery route to confirm the neuroprotective effects of CNTF in the striatum, for example, via implanted encapsulated cells engineered to produce CNTF (Emerich et al., 1996, 1997a, 1997b). Baby hamster kidney (BHK) fibroblasts were genetically modified to synthesize and secrete human CNTF and then encapsulated in polymer membranes. This approach allows continuous diffusion of CNTF into the brain parenchyma, but prevents the cells from contacting host tissue in order to both protect cells from host immune destruction and the host from potential tumor formation by grafted cells. Unilateral grafting of these cells in the ventricle avoids the loss of Nissl-, GAD-, NADPH-diaphorase and ChAT-stained neurons (Emerich et al., 1996). The neuronal protection afforded by the CNTF-secreting cells is sufficient to prevent aberrant motor behavior induced by the lesion. Rats receiving BHK-CNTF implants do not display the apomorphine rotational behavior induced by unilateral quinolinic acid injection (Emerich et al., 1996). However, lesion-induced deficits on a fine motor task are unaffected by cellularly delivered CNTF. Another study performed a battery of behavioral tests to characterize the behavioral protection after grafting of CNTF-secreting cells. It demonstrated a prevention of motor deficits, such as hyperactivity and the inability to use the forelimbs to retrieve food pellets in a staircase test, induced by bilateral quinolinic acid lesions (Emerich et al., 1997a); an improvement of cognitive function was also observed in this study.

The neuroprotective effects of CNTF on GABAergic striatal neurons in the rodent model of HD led to research in nonhuman primates, which is the previous step to clinical trails. Polymer capsules containing BHK-CNTF cells were grafted unilaterally into the putamen and the caudate nucleus of rhesus monkeys under MRI guidance, one week before quinolinic acid was injected (Emerich et al., 1997b; Kordower et al., 1999). All animals showed significant lesions that are prevented by grafting of CNTF-secreting cells. Three- and seven-fold increases in GAD-positive neurons in the caudate nucleus and putamen, respectively, were observed in CNTF-grafted animals relative to controls. Similarly, there was a 2.5- and a 4-fold increase in ChAT-positive neurons in caudate and putamen, respectively (Emerich et al., 1997b). Besides the preservation of GABAergic somata, the innervation is also spared (Kordower et al., 1999). DARPP-32 was used to analyze the integrity of striatal connectivity with the SN and the globus pallidus. Striatal excitotoxic lesion induces a decrease in DARPP-32 immunoreactivity within the globus pallidus and SN in BHK-control animals. In contrast, the lesion-induced decrease in GABAergic innervation of both these areas is prevented in BHK-CNTF-grafted monkeys (Kordower et al., 1999). The integrity of the corticostriatal pathway has also been evaluated in this HD model. Although the quinolinic acid lesion does not affect the number of neurons in cortical layer V, the cross-sectional area is reduced by 23%. This atrophy of cortical neurons is almost completely prevented by CNTF grafts. Thus, the rescue of striatal cells may help to avoid the cortical degeneration observed in HD. Therefore, these studies show not only that cell survival is improved, but also that basal ganglia circuitry is spared after grafting of CNTF-secreting cells in the excitotoxic model of HD. This protective effect of CNTF was further confirmed in the 3-nitropropionic primate model of HD. Intracerebral grafting of BHK-CNTF cells at the time of appearance of striatal dysfunction in six monkeys chronically treated with 3-nitropropionic acid not only protects neurons from degeneration but also restores motor and cognitive symptoms (Mittoux et al., 2000).

Over the past few years, neuroprotective strategies rely on the development of effective delivery systems leading to robust and long-term expression of the transgene, and the presence of the protein in a large area of the striatum. Thus, recently, in vivo gene therapy approaches with viral vector have been assessed (Pereira de Almeida et al., 2001; Mittoux et al., 2002). The sterotaxical injection of lentiviruses encoding for CNTF into the striatum reduces apomorphine-induced rotations ipsilateral to the lesion and protects DARPP-32, ChAT and NADPH-diaphorase neuronal populations against quinolinic acid excitotoxicity (Pereira de Almeida et al., 2001). In the 3-nitropropionic model, the single injection of CNTF adenovirus protects striatal neurons. Moreover, retrograde and anterograde transport of the transgene is also observed with important neuroprotective effects on the cortico-striato-pallidal pathway, which is severely affected in HD (Mittoux et al., 2002).

In a cellular model of HD, a protective effect of CNTF has also been observed. The transfection of a fragment of huntingtin that includes the polyglutamine sequence induces degeneration of striatal neurons. The treatment with CNTF completely protects striatal neurons against mutant huntingtin-induced apoptosis (Saudou et al., 1998).

Clinical trials have been initiated to determine the protective effects of CNTF-producing cells implanted into the lateral ventricle of HD patients (Bachoud-Levi et al., 2000). An ex vivo gene therapy approach based on encapsulated genetically modified BHK cells has been used for the continuous and long-term intracerebral delivery of CNTF. A device, containing up to 106 human CNTF-producing (0.15–0.5 μg CNTF/day) BHK cells surrounded by a semipermeable membrane, was implanted into the right lateral ventricle of six patients. In this phase-I study, the principal goal was to evaluate the safety and tolerability of the procedure (Bachoud-Levi et al., 2000).

Conclusion

The interest in HD is to find protective agents for striatal GABAergic projection neurons. However, the preservation of the corticostriatal pathway is also critical, since cortical atrophy is observed in HD. Excitotoxicity and metabolic impairment have been used extensively as models of HD, and they have helped to characterize the trophic requirements of the different neuronal populations in the striatum. Mutant huntingtin is the starting point of an unknown cascade of intracellular events that lead to selective death of striatal neurons. Neurotrophic factors may be involved in these mechanisms. Recently, several lines of evidence have linked huntingtin with neurotrophic factors and excitotoxicity. Huntingtin regulates transcription of BDNF and could be relevant for maintaining corticostriatal connectivity (Zuccato et al., 2001). In transgenic HD mice, a decrease in BDNF is observed (Zuccato et al., 2001) and also an increase in the sensitivity of NMDA-mediated excitotoxicity (Zeron et al., 2002). Furthermore, excitotoxicity regulates BDNF expression in cortical and striatal neurons (Canals et al., 1998, 2001). Therefore, BDNF can play a pivotal role, probably also interacting with other trophic factors, in the development of therapeutic strategies to prevent striatal degeneration induced by mutant huntingtin.

Neurotrophins, GDNF family members and CNTF are the neurotrophic factors with the strongest neuroprotective effects in these animal models. Neurotrophins, especially BDNF and NT-3, have neuroprotective effects in GABAergic projection neurons, but with different properties. Hence, BDNF has a strong effect on survival, whereas NT-3 is a potent inducer of neuronal phenotype. The neurotrophin protecting effects do not differentiate between striatopallidal or striatonigral neurons. However, GDNF family members can discriminate between these two populations of projection neurons. GDNF only protects striatonigral neurons, while NTRN is specific for striatopallidal neurons. Therefore, the use of one of these neurotrophic factors may help to recover the equilibrium between the direct and indirect pathways that is affected in some neurological disorders of the basal ganglia, such as HD or multiple system atrophy. The success of CNTF protection striatal neurons has moved the focus of research to primate models and then to humans, although the results of this clinical trial have not yet been presented.

The initial failure of clinical trials with neurotrophic factors in different neurodegenerative disorders has revealed that the manner and site of administration are of critical importance (Thoenen and Sendtner, 2002). In the last few years, various systems have been developed to improve the delivery and specificity of neurotrophic factors. Injections of viral vectors or transplantation of engineered cells, including neuronal progenitors, represent promising perspectives.

The recent knowledge of the trophic map of striatal and cortical neurons and further studies using the transgenic HD mouse model and new delivery systems can provide new information to develop a neurotrophic factor therapy for HD.

Acknowledgments

Financial support was provided by CICYT (SAF99-0019; SAF2002-00314; SAF2002-00311 Ministerio de Ciencia y Tecnología, Spain), Generalitat de Catalunya, Fundació La Marató de TV3 (990310), Fundació La Caixa and Fundación Ramón Areces.

Abbreviations

ACPD	1S, R-1-aminocyclopentane-1, 3-dicarboxylic acid
ARTN	artemin
BDNF	brain-derived neurotrophic factor
BHK	baby hamster kidney
ChAT	choline acetyltransferase
CNTF	ciliary neurotrophic factor
CNTFR	ciliary neurotrophic factor receptor
CNS	central nervous system
E	embryonic day
GABA	γ-aminobutyric acid
GAD	glutamic acid decarboxylase
GDNF	glial cell line-derived neurotrophic factor
GFRα	GDNF family receptor α
HD	Huntington's disease
IL-6	interleukin-6
KO	knockout
LIF	leukemia inhibitory factor
NGF	nerve growth factor
NMDA	N-methyl-D-aspartate
NT-3	neurotrophin-3
NT-4/5	neurotrophin-4/5
NTRN	neurturin
P	postnatal day
PNS	peripheral nervous system
p75^{NTR}	p75 neurotrophin receptor
PSPN	persephin
SN	substantia nigra
SP	substance P

References

Abiru, Y., Nishio, C. and Hatanaka, H. (1996) The survival of striatal cholinergic neurons cultured from postnatal 2-week-old rats is promoted by neurotrophins. Dev. Brain Res., 91: 260–267.

Adler, R., Landa, K.B., Manthorpe, M. and Varon, S. (1979) Cholinergic neuronotrophic factors: intraocular distribution of trophic activity for ciliary neurons. Science, 204: 1434–1436.

Airaksinen, M.S. and Saarma, M. (2002) The GDNF family: signalling, biological functions and therapeutic value. Nat. Rev. Neurosci., 3: 383–394.

Akerud, P., Alberch, J., Eketjall, S., Wagner, J. and Arenas, E. (1999) Differential effects of glial cell line-derived neurotrophic factor and neurturin on developing and adult substantia nigra dopaminergic neurons. J. Neurochem., 73: 70–78.

Akerud, P., Canals, J.M., Snyder, E.Y. and Arenas, E. (2001) Neuroprotection through delivery of glial cell line-derived neurotrophic factor by neural stem cells in a mouse model of Parkinson's disease. J. Neurosci., 21: 8108–8118.

Akerud, P., Holm, P.C., Castelo-Branco, G., Sousa, K., Rodriguez, F.J. and Arenas, E. (2002) Persephin-overexpressing neural stem cells regulate the function of nigral dopaminergic neurons and prevent their degeneration in a model of Parkinson's disease. Mol. Cell. Neurosci., 21: 205–222.

Akkina, S.K., Patterson, C.L. and Wright, D.E. (2001) GDNF rescues nonpeptidergic unmyelinated primary afferents in streptozotocin-treated diabetic mice. Exp. Neurol., 167: 173–182.

Alberch, J., Perez-Navarro, E. and Canals, J.M. (2002) Neuroprotection by neurotrophins and GDNF family members in the excitotoxic model of Huntington's disease. Brain Res. Bull., 57: 817–822.

Alexi, T., Venero, J.L. and Hefti, F. (1997) Protective effects of neurotrophin-4/5 and transforming growth factor-alpha on striatal neuronal phenotypic degeneration after excitotoxic lesioning with quinolinic acid. Neuroscience, 78: 73–86.

Alexi, T., Borlongan, C.V., Faull, R.L., Williams, C.E., Clark, R.G., Gluckman, P.D. and Hughes, P.E. (2000) Neuroprotective strategies for basal ganglia degeneration: Parkinson's and Huntington's diseases. Prog. Neurobiol., 60: 409–470.

Altar, C.A., Armanini, M., Dugich-Djordjevic, M., Bennett, G.L., Williams, R., Feinglass, S., Anicetti, V.,

Sinicropi, D. and Bakhit, C. (1992) Recovery of cholinergic phenotype in the injured rat neostriatum: roles for endogenous and exogenous nerve growth factor. J. Neurochem., 59: 2167–2177.

Altar, C.A., Cai, N., Bliven, T., Juhasz, M., Conner, J.M., Acheson, A.L., Lindsay, R.M. and Wiegand, S.J. (1997) Anterograde transport of brain-derived neurotrophic factor and its role in the brain. Nature, 389: 856–860.

Anderson, K.D., Panayotatos, N., Corcoran, T.L., Lindsay, R.M. and Wiegand, S.J. (1996) Ciliary neurotrophic factor protects striatal output neurons in an animal model of Huntington disease. Proc. Natl. Acad. Sci. USA, 93: 7346–7351.

Aoki, C., Wu, K., Elste, A., Len, G., Lin, S., McAuliffe, G. and Black, I.B. (2000) Localization of brain-derived neurotrophic factor and TrkB receptors to postsynaptic densities of adult rat cerebral cortex. J. Neurosci. Res., 59: 454–463.

Apfel, S.C. (1996) Neurotrophic factors in the treatment of neurotoxicity: an overview. Neurotoxicology, 17: 839–844.

Apfel, S.C. (1999) Neurotrophic factors and diabetic peripheral neuropathy. Eur. Neurol., 41 (Suppl. 1): 27–34.

Apfel, S.C. (2002) Is the therapeutic application of neurotrophic factors dead?. Ann. Neurol., 51: 8–11.

Arakawa, Y., Sendtner, M. and Thoenen, H. (1990) Survival effect of ciliary neurotrophic factor (CNTF) on chick embryonic motoneurons in culture: comparison with other neurotrophic factors and cytokines. J. Neurosci., 10: 3507 3515.

Araujo, D.M. and Hilt, D.C. (1997) Glial cell line-derived neurotrophic factor attenuates the excitotoxin- induced behavioral and neurochemical deficits in a rodent model of Huntington's disease. Neuroscience, 81: 1099–1110.

Araujo, D.M. and Hilt, D.C. (1998) Glial cell line-derived neurotrophic factor attenuates the locomotor hypofunction and striatonigral neurochemical deficits induced by chronic systemic administration of the mitochondrial toxin 3-nitropropionic acid. Neuroscience, 82: 117–127.

Asada, H., Ip, N.Y., Pan, L., Razack, N., Parfitt, M.M. and Plunkett, R.J. (1995) Time course of ciliary neurotrophic factor mRNA expression is coincident with the presence of protoplasmic astrocytes in traumatized rat striatum. J. Neurosci. Res., 40: 22–30.

Bachoud-Levi, A.C., Deglon, N., Nguyen, J.P., Bloch, J., Bourdet, C., Winkel, L., Remy, P., Goddard, M., Lefaucheur, J.P., Brugieres, P., Baudic, S., Cesaro, P., Peschanski, M. and Aebischer, P. (2000) Neuroprotective gene therapy for Huntington's disease using a polymer encapsulated BHK cell line engineered to secrete human CNTF. Hum. Gene Ther., 11: 1723–1729.

Baloh, R.H., Tansey, M.G., Golden, J.P., Creedon, D.J., Heuckeroth, R.O., Keck, C.L., Zimonjic, D.B., Popescu, N.C., Johnson, E.M.J. and Milbrandt, J. (1997) TrnR2, a novel receptor that mediates neurturin and GDNF signaling through Ret. Neuron, 18: 793–802.

Baloh, R.H., Gorodinsky, A., Golden, J.P., Tansey, M.G., Keck, C.L., Popescu, N.C., Johnson, E.M.J. and Milbrandt, J. (1998a) GFRα3 is an orphan member of the GDNF/neurturin/persephin receptor family. Proc. Natl. Acad. Sci. USA, 95: 5801–5806.

Baloh, R.H., Tansey, M.G., Lampe, P.A., Fahrner, T.J., Enomoto, H., Simburger, K.S., Leitner, M.L., Araki, T., Johnson, E.M.J. and Milbrandt, J. (1998b) Artemin, a novel member of the GDNF ligand family, supports peripheral and central neurons and signals through the GFRα3-RET receptor complex. Neuron, 21: 1291–1302.

Baloh, R.H., Enomoto, H., Johnson, E.M.J. and Milbrandt, J. (2000) The GDNF family ligands and receptors-implications for neural development. Curr. Opin. Neurobiol., 10: 103–110.

Barbacid, M. (1994) The Trk family of neurotrophin receptors. J. Neurobiol., 25: 1386–1403.

Barbin, G., Manthorpe, M. and Varon, S. (1984) Purification of the chick eye ciliary neuronotrophic factor. J. Neurochem., 43: 1468–1478.

Barde, Y.A., Edgar, D. and Thoenen, H. (1982) Purification of a new neurotrophic factor from mammalian brain. EMBO J., 1: 549–553.

Barker, P.A., Lomen-Hoerth, C., Gensch, E.M., Meakin, S.O., Glass, D.J. and Shooter, E.M. (1993) Tissue-specific alternative splicing generates two isoforms of the trkA receptor. J. Biol. Chem., 268: 15150–15157.

Baxter, G.T., Radeke, M.J., Kuo, R.C., Makrides, V., Hinkle, B., Hoang, R., Medina-Selby, A., Coit, D. and Valenzuela, P. (1997) Signal transduction mediated by the truncated trkB receptor isoforms, trkB.T1 and trkB.T2. J. Neurosci., 17: 2683–2690.

Beal, M.F., Kowall, N.W., Ellison, D.W., Mazurek, M.F., Swartz, K.J. and Martin, J.B. (1986) Replication of the neurochemical characteristics of Huntington's disease by quinolinic acid. Nature, 321: 168–171.

Beal, M.F., Brouillet, E., Jenkins, B.G., Ferrante, R.J., Kowall, N.W., Miller, J.M., Storey, E., Srivastava, R., Rosen, B.R. and Hyman, B.T. (1993) Neurochemical and histologic characterization of striatal excitotoxic lesions produced by the mitocondrial toxin 3-nitropropionic acid. J. Neurosci., 13: 4181–4192.

Beal, M.F. (1994) Neurochemistry and toxin models in Huntington's disease. Curr. Opin. Neurol., 7: 542–547.

Beck, K.D., Lamballe, F., Klein, R., Barbacid, M., Schauwecker, P.E., McNeill, T.H., Finch, C.E., Hefti, F. and Day, J.R. (1993) Induction of noncatalytic TrkB neurotrophin receptors during axonal sprouting in the adult hippocampus. J. Neurosci., 13: 4001–4014.

Bemelmans, A.P., Horellou, P., Pradier, L., Brunet, I., Colin, P. and Mallet, J. (1999) Brain-derived neurotrophic factor-mediated protection of striatal neurons in an excitotoxic rat model of Huntington's disease, as demonstrated by adenoviral gene transfer. Hum. Gene Ther., 10: 2987–2997.

218

Benedetti, M., Levi, A. and Chao, M.V. (1993) Differential expression of nerve growth factor receptors leads to altered binding affinity and neurotrophin responsiveness. Proc. Natl. Acad. Sci. USA, 90: 7859–7863.

Berkemeier, L.R., Winslow, J.W., Kaplan, D.R., Nikolics, K., Goeddel, D.V. and Rosenthal, A. (1991) Neurotrophin-5: a novel neurotrophic factor that activates trk and trkB. Neuron, 7: 857–866.

Bibel, M., Hoppe, E. and Barde, Y.A. (1999) Biochemical and functional interactions between the neurotrophin receptors trk and p75NTR. EMBO J., 18: 616–622.

Bilak, M.M., Shifrin, D.A., Corse, A.M., Bilak, S.R. and Kuncl, R.W. (1999) Neuroprotective utility and neurotrophic action of neurturin in postnatal motor neurons: comparison with GDNF and persephin. Mol. Cell. Neurosci., 13: 326–336.

Bizon, J.L., Lauterborn, J.C. and Gall, C.M. (1999) Subpopulations of striatal interneurons can be distinguished on the basis of neurotrophic factor expression. J. Comp. Neurol., 408: 283–298.

Blesch, A. and Tuszynski, M.H. (2002) Spontaneous and neurotrophin-induced axonal plasticity after spinal cord injury. Prog. Brain Res., 137: 415–423.

Blochl, A. and Sirrenberg, C. (1996) Neurotrophins stimulate the release of dopamine from rat mesencephalic neurons via Trk and p75LNTR receptors. J. Biol. Chem., 271: 21100–21107.

Blottner, D., Bruggemann, W. and Unsicker, K. (1989) Ciliary neurotrophic factor supports target-deprived preganglionic sympathetic spinal cord neurons. Neurosci. Lett., 105: 316–320.

Boulton, T.G., Stahl, N. and Yancopoulos, G.D. (1994) Ciliary neurotrophic factor/leukemia inhibitory factor/interleukin 6/ oncostatin M family of cytokines induces tyrosine phosphorylation of a common set of proteins overlapping those induced by other cytokines and growth factors. J. Biol. Chem., 269: 11648–11655.

Bourque, M.J. and Trudeau, L.E. (2000) GDNF enhances the synaptic efficacy of dopaminergic neurons in culture. Eur. J. Neurosci., 12: 3172–3180.

Bresjanac, M. and Antauer, G. (2000) Reactive astrocytes of the quinolinic acid-lesioned rat striatum express GFRα1 as well as GDNF in vivo. Exp. Neurol., 164: 53–59.

Brouillet, E., Conde, F., Beal, M.F. and Hantraye, P. (1999) Replicating Huntington's disease phenotype in experimental animals. Prog. Neurobiol., 59: 427–468.

Brunet, A., Datta, S.R. and Greenberg, M.E. (2001) Transcription-dependent and -independent control of neuronal survival by the PI3K-Akt signaling pathway. Curr. Opin. Neurobiol., 11: 297–305.

Buj-Bello, A., Adu, J., Pinon, L.G., Horton, A., Thompson, J., Rosenthal, A., Chinchetru, M., Buchman, V.L. and Davies, A.M. (1997) Neurturin responsiveness requires a GPI-linked receptor and the Ret receptor tyrosine kinase. Nature, 387: 721–724.

Burazin, T.C. and Gundlach, A.L. (1999) Localization of GDNF/neurturin receptor (c-ret, GFRα-1 and α-2) mRNAs in postnatal rat brain: differential regional and temporal expression in hippocampus, cortex and cerebellum. Mol. Brain Res., 73: 151–171.

Cacalano, G., Fariñas, I., Wang, L.C., Hagler, K., Forgie, A., Moore, M., Armanini, M., Phillips, H., Ryan, A.M., Reichardt, L.F., Hynes, M., Davies, A., Rosenthal, A. (1998) GFRα1 is an essential receptor component for GDNF in the developing nervous system and kidney. Neuron, 21: 53–62.

Canals, J.M., Marco, S., Checa, N., Michels, A., Pérez-Navarro, E., Arenas, E. and Alberch, J. (1998) Differential regulation of the expression of NGF, BDNF and NT-3 after excitotoxicity in a rat model of Huntington's disease. Neurobiol. Dis., 5: 357–364.

Canals, J.M., Checa, N., Marco, S., Michels, A., Pérez-Navarro, E. and Alberch, J. (1999) The neurotrophin receptors trkA, trkB and trkC are differentially regulated after excitotoxic lesion in rat striatum. Mol. Brain Res., 69: 242–248.

Canals, J.M., Checa, N., Marco, S., Åkerud, P., Michels, A., Pérez-Navarro, E., Tolosa, E., Arenas, E. and Alberch, J. (2001) Expression of brain-derived neurotrophic factor in cortical neurons is regulated by striatal target area. J. Neurosci., 21: 117–124.

Casaccia-Bonnefil, P., Gu, C., Khursigara, G. and Chao, M.V. (1999) p75 neurotrophin receptor as a modulator of survival and death decisions. Microsc. Res. Tech., 45: 217–224.

Castellani, V. and Bolz, J. (1999) Opposite roles for neurotrophin-3 in targeting and collateral formation of distinct sets of developing cortical neurons. Development, 126: 3335–3345.

Cattaneo, E., Rigamonti, D., Goffredo, D., Zuccato, C., Squitieri, F. and Sipione, S. (2001) Loss of normal huntingtin function: new developments in Huntington's disease research. Trends Neurosci., 24: 182–188.

Cellerino, A., Maffei, L. and Domenici, L. (1996) The distribution of brain-derived neurotrophic factor and its receptor trkB in parvalbumin-containing neurons of the rat visual cortex. Eur. J. Neurosci., 8: 1190–1197.

Chao, M.V. (1994) The p75 neurotrophin receptor. J. Neurobiol., 25: 1373–1385.

Checa, N., Canals, J.M. and Alberch, J. (2000) Developmental regulation of BDNF and NT-3 expression by quinolinic acid in the striatum and its main connections. Exp. Neurol., 165: 118–124.

Checa, N., Canals, J.M., Gratacos, E. and Alberch, J. (2001) TrkB and TrkC are differentially regulated by excitotoxicity during development of the basal ganglia. Exp. Neurol., 172: 282–292.

Chen, Q., Veenman, C.L. and Reiner, A. (1996) Cellular expression of ionotropic glutamate receptors subunits on specific striatal neuron types and its implications for striatal vulnerability in glutamate receptor-mediated excitotoxicity. Neuroscience, 73: 715–731.

Clary, D.O. and Reichardt, L.F. (1994) An alternatively spliced form of the nerve growth factor receptor TrkA confers an enhanced response to neurotrophin 3. Proc. Natl. Acad. Sci. USA, 91: 11133–11137.

Clatterbuck, R.E., Price, D.L. and Koliatsos, V.E. (1993) Ciliary neurotrophic factor prevents retrograde neuronal death in the adult central nervous system. Proc. Natl. Acad. Sci. USA, 90: 2222–2226.

Conneally, P.M. (1984) Huntington disease: genetics and epidemiology. Am. J. Hum. Genet., 36: 506–526.

Conner, J.M. and Varon, S. (1996) Maintenance of sympathetic innervation into the hippocampal formation requires a continuous local availability of nerve growth factor. Neuroscience, 72: 933–945.

Cooper, J.D., Skepper, J.N., Berzaghi, M.D., Lindholm, D. and Sofroniew, M.V. (1996) Delayed death of septal cholinergic neurons after excitotoxic ablation of hippocampal neurons during early postnatal development in the rat. Exp. Neurol., 139: 143–155.

Cordon-Cardo, C., Tapley, P., Jing, S.Q., Nanduri, V., O'Rourke, E., Lamballe, F., Kovary, K., Klein, R., Jones, K.R. and Reichardt, L.F. (1991) The trk tyrosine protein kinase mediates the mitogenic properties of nerve growth factor and neurotrophin-3. Cell, 66: 173–183.

Corse, A.M., Bilak, M.M., Bilak, S.R., Lehar, M., Rothstein, J.D. and Kuncl, R.W. (1999) Preclinical testing of neuroprotective neurotrophic factors in a model of chronic motor neuron degeneration. Neurobiol. Dis., 6: 335–346.

Costantini, L.C., Feinstein, S.C., Radeke, M.J. and Snyder-Keller, A. (1999) Compartmental expression of trkB receptor protein in the developing striatum. Neuroscience, 89: 505–513.

Coulpier, M., Anders, J. and Ibanez, C.F. (2002) Coordinated activation of autophosphorylation sites in the RET receptor tyrosine kinase: importance of tyrosine 1062 for GDNF mediated neuronal differentiation and survival. J. Biol. Chem., 277: 1991–1999.

Coyle, J.T. and Puttfarcken, P. (1993) Oxidative stress, glutamate, and neurodegenerative disorders. Science, 262: 689–695.

Crowley, C., Spencer, S.D., Nishimura, M.C., Chen, K.S., Pitts-Meek, S., Armanini, M.P., Ling, L.H., MacMahon, S.B., Shelton, D.L. and Levinson, A.D. (1994) Mice lacking nerve growth factor display perinatal loss of sensory and sympathetic neurons yet develop basal forebrain cholinergic neurons. Cell, 76: 1001–1011.

Cruz-Aguado, R., Turner, L.F., Diaz, C.M. and Pinero, J. (2000) Nerve growth factor and striatal glutathione metabolism in a rat model of Huntington's disease. Restor. Neurol. Neurosci., 17: 217–221.

Cudkowicz, M. and Kowall, N.W. (1990) Degeneration of pyramidal projection neurons in Huntington's disease cortex. Ann. Neurol., 27: 200–204.

Davies, S.W. and Beardsall, K. (1992) Nerve growth factor selectively prevents excitotoxin induced degeneration of striatal cholinergic neurons. Neurosci. Lett., 140: 161–164.

Davis, S., Aldrich, T.H., Valenzuela, D.M., Wong, V.V., Furth, M.E., Squinto, S.P. and Yancopoulos, G.D. (1991) The receptor for ciliary neurotrophic factor. Science, 253: 59–63.

Davis, S., Aldrich, T.H., Ip, N.Y., Stahl, N., Scherer, S., Farruggella, T., DiStefano, P.S., Curtis, R., Panayotatos, N., Gascan, H. (1993) Released form of CNTF receptor α component as a soluble mediator of CNTF responses. Science, 259: 1736–1739.

Davis, S. and Yancopoulos, G.D. (1993) The molecular biology of the CNTF receptor. Curr. Opin. Cell. Biol., 5: 281–285.

Dawbarn, D., D. Quidt, M.E. and Emson, P.C. (1985) Survival of basal ganglia neuropeptide Y-somatostatin neurones in Huntington's disease, Brain Res., 340: 251–260.

DeChiara, T.M., Vejsada, R., Poueymirou, W.T., Acheson, A., Suri, C., Conover, J.C., Friedman, B., McClain, J., Pan, L. and Stahl, N. (1995) Mice lacking the CNTF receptor, unlike mice lacking CNTF, exhibit profound motor neuron deficits at birth. Cell, 83: 313–322.

DiFiglia, M., Sapp, E., Chase, K.O., Davies, S.W., Bates, G.P., Vonsattel, J.P. and Aronin, N. (1997) Aggregation of huntingtin in neuronal intranuclear inclusions and dystrophic neurites in brain. Science, 277: 1990–1993.

Dluzen, D.E., Gao, X., Story, G.M., Anderson, L.I., Kucera, J. and Walro, J.M. (2001) Evaluation of nigrostriatal dopaminergic function in adult +/+ and +/− BDNF mutant mice. Exp. Neurol., 170: 121–128.

Dluzen, D.E., Anderson, L.I., McDermott, J.L., Kucera, J. and Walro, J.M. (2002) Striatal dopamine output is compromised within +/− BDNF mice. Synapse, 43: 112–117.

Duyao, M.P., Auerbach, A.B., Ryan, A., Persichetti, F., Barnes, G.T., McNeil, S.M., Ge, P., Vonsattel, J.P., Gusella, J.F. and Joyner, A.L. (1995) Inactivation of the mouse Huntington's disease gene homolog Hdh. Science, 269: 407–410.

Eide, F.F., Vining, E.R., Eide, B.L., Zang, K. and Reichardt, L.F. (1996) Naturally occurring truncated trkB receptors have dominant inhibitory effects on brain-derived neurotrophic factor signaling. J. Neurosci., 16: 3123–3129.

Emerich, D.F., Lindner, M.D., Winn, S.R., Chen, E.Y., Frydel, B.R. and Kordower, J.H. (1996) Implants of encapsulated human CNTF-producing fibroblasts prevent behavioral deficits and striatal degeneration in a rodent model of Huntington's disease. J. Neurosci., 16: 5168–5181.

Emerich, D.F., Cain, C.K., Greco, C., Saydoff, J.A., Hu, Z.Y., Liu, H. and Lindner, M.D. (1997a) Cellular delivery of human CNTF prevents motor and cognitive dysfunction in a rodent model of Huntington's disease. Cell Transplant., 6: 249–266.

Emerich, D.F., Winn, S.R., Hantraye, P.M., Peschanskl, M., Chen, E.-Y., Chu, Y., McDermott, P., Baetge, E.E. and Kordower, J.H. (1997b) Protective effect of encapsulated cells producing neurotrophic factor CNTF in a monkey model of Huntington's disease. Nature, 386: 395–399.

Enokido, Y., de Sauvage, F., Hongo, J.A., Ninkina, N., Rosenthal, A., Buchman, V.L. and Davies, A.M. (1998) GFR α-4 and the tyrosine kinase Ret form a functional receptor complex for persephin. Curr. Biol., 8: 1019–1022.

Enomoto, H., Crawford, P.A., Gorodinsky, A., Heuckeroth, R.O., Johnson, E.M.J. and Milbrandt, J. (2001) RET signaling is essential for migration, axonal growth and axon guidance of developing sympathetic neurons. Development, 128: 3963–3974.

Ernfors, P., Hallbook, F., Ebendal, T., Shooter, E.M., Radeke, M.J., Misko, T.P. and Persson, H. (1988) Developmental and regional expression of beta-nerve growth factor receptor mRNA in the chick and rat. Neuron, 1: 983–996.

Ernfors, P., Wetmore, C., Olson, L. and Persson, H. (1990) Identification of cells in rat brain and peripheral tissues expressing mRNA for members of the nerve growth factor family. Neuron, 5: 511–526.

Ernsberger, U., Sendtner, M. and Rohrer, H. (1989) Proliferation and differentiation of embryonic chick sympathetic neurons: effects of ciliary neurotrophic factor. Neuron, 2: 1275–1284.

Fagan, A.M., Garber, M., Barbacid, M., Silos-Santiago, I. and Holtzman, D.M. (1997) A role for TrkA during maturation of striatal and basal forebrain cholinergic neurons in vivo. J. Neurosci., 17: 7644–7654.

Farkas, L.M., Suter-Crazzolara, C. and Unsicker, K. (1997) GDNF induces the calretinin phenotype in cultures of embryonic striatal neurons. J. Neurosci. Res., 50: 361–372.

Ferrante, R.J., Kowall, N.W., Beal, M.F., Richardson, E.P.J., Bird, E.D. and Martin, J.B. (1985) Selective sparing of a class of striatal neurons in Huntington's disease. Science, 230: 561–563.

Ferrer, I., Goutan, E., Marin, C., Rey, M.J. and Ribalta, T. (2000) Brain-derived neurotrophic factor in Huntington's disease. Brain Res., 866: 257–261.

Fischer, W., Wictorin, K., Bjorklund, A., Williams, L.R., Varon, S. and Gage, F.H. (1987) Amelioration of cholinergic neuron atrophy and spatial memory impairment in aged rats by nerve growth factor. Nature, 329: 65–68.

Fiumelli, H., Kiraly, M., Ambrus, A., Magistretti, P.J. and Martin, J.L. (2000) Opposite regulation of calbindin and calretinin expression by brain-derived neurotrophic factor in cortical neurons. J. Neurochem., 74: 1870–1877.

Friden, P.M., Walus, L.R., Watson, P., Doctrow, S.R., Kozarich, J.W., Backman, C., Bergman, H., Hoffer, B., Bloom, F. and Granholm, A.C. (1993) Blood-brain barrier penetration and in vivo activity of an NGF conjugate. Science, 259: 373–377.

Friedman, W.J., Olson, L. and Persson, H. (1991) Cells that express brain-derived neurotrophic factor mRNA in the developing postnatal rat brain. Eur. J. Neurosci., 3: 688–697.

Friedman, W.J., Black, I.B. and Kaplan, D.R. (1998) Distribution of the neurotrophins brain-derived neurotrophic factor, neurotrophin-3, and neurotrophin-4/5 in the postnatal rat brain: an immunocytochemical study. Neuroscience, 84: 101–114.

Frim, D.M., Uhler, T.A., Short, M.P., Ezzedine, Z.D., Klagsbrum, M., Breackfield, X.O. and Isacson, O. (1993) Effects of biologically delivered NGF, BDNF and bFGF on striatal excitotoxic lesions. NeuroReport, 4: 367–370.

Fukuda, T., Kiuchi, K. and Takahashi, M. (2002) Novel mechanism of regulation of Rac activity and lamellipodia formation by RET tyrosine kinase. J. Biol. Chem., 277: 19114–19121.

Galpern, W.R., Matthews, R.T., Beal, M.F. and Isacson, O. (1996) NGF attenuates 3-nitrotyrosine formation in a 3-NP model of Huntington's disease. NeuroReport, 7: 2639–2642.

Gerfen, C.R. (1992) The neostriatal mosaic: multiple levels of compartmental organization in the basal ganglia. Annu. Rev. Neurosci., 15: 285–320.

Gervais, F.G., Singaraja, R., Xanthoudakis, S., Gutekunst, C.A., Leavitt, B.R., Metzler, M., Hackam, A.S., Tam, J., Vaillancourt, J.P., Houtzager, V., Rasper, D.M., Roy, S., Hayden, M.R. and Nicholson, D.W. (2002) Recruitment and activation of caspase-8 by the Huntingtin-interacting protein Hip-1 and a novel partner Hippi. Nat. Cell Biol., 4: 95–105.

Ghosh, A., Carnahan, J. and Greenberg, M.E. (1994) Requirement for BDNF in activity-dependent survival of cortical neurons. Science, 263: 1618–1623.

Glazner, G.W., Mu, X. and Springer, J.E. (1998) Localization of glial cell line-derived neurotrophic factor receptor alpha and c-ret mRNA in rat central nervous system. J. Comp. Neurol., 391: 42–49.

Gloaguen, I., Costa, P., Demartis, A., Lazzaro, D., Di Marco, A., Graziani, R., Paonessa, G., Chen, F., Rosenblum, C.I., Van der Ploeg, L.H., Cortese, R., Ciliberto, G. and Laufer, R. (1997) Ciliary neurotrophic factor corrects obesity and diabetes associated with leptin deficiency and resistance. Proc. Natl. Acad. Sci. USA, 94: 6456–6461.

Golden, J.P., Baloh, R.H., Kotzbauer, P.T., Lampe, P.A., Osborne, P.A., Milbrandt, J. and Johnson, E.M., Jr. (1998) Expression of neurturin, GDNF, and their receptors in the adult mouse CNS. J. Comp. Neurol., 398: 139–150.

Golden, J.P., DeMaro, J.A., Osborne, P.A., Milbrandt, J. and Johnson, E.M.J. (1999) Expression of neurturin, GDNF, and

GDNF family-receptor mRNA in the developing and mature mouse. Exp. Neurol., 158: 504–528.

Gotz, T., Kraushaar, U., Geiger, J., Lubke, J., Berger, T. and Jonas, P. (1997) Functional properties of AMPA and NMDA receptors expressed in identified types of basal ganglia neurons. J. Neurosci., 17: 204–215.

Gouhier, C., Chalon, S., Venier-Julienne, M.C., Bodard, S., Benoit, J., Besnard, J. and Guilloteau, D. (2000) Neuroprotection of nerve growth factor-loaded microspheres on the D2 dopaminergic receptor positive-striatal neurones in quinolinic acid-lesioned rats: a quantitative autoradiographic assessment with iodobenzamide. Neurosci. Lett., 288: 71–75.

Gratacos, E., Checa, N., Pérez-Navarro, E. and Alberch, J. (2001a) Brain-derived neurotrophic factor (BDNF) mediates bone morphogenetic protein-2 (BMP-2) effects on cultured striatal neurones. J. Neurochem., 79: 747–755.

Gratacos, E., Pérez-Navarro, E., Tolosa, E., Arenas, E. and Alberch, J. (2001b) Neuroprotection of striatal neurons against kainate excitotoxicity by neurotrophins and GDNF family members. J. Neurochem., 78: 1287–1296.

Graybiel, A.M. (2000) The basal ganglia. Curr. Biol., 10: R509–R511

Greenamyre, J.T. and Shoulson, I. (1994) Huntington's Disease. WB Saunders, Philadelphia, pp. 685–704.

Grill, R., Murai, K., Blesch, A., Gage, F.H. and Tuszynski, M.H. (1997) Cellular delivery of neurotrophin-3 promotes corticospinal axonal growth and partial functional recovery after spinal cord injury. J. Neurosci., 17: 5560–5572.

Guiton, M., Gunn-Moore, F.J., Glass, D.J., Geis, D.R., Yancopoulos, G.D. and Tavare, J.M. (1995) Naturally occurring tyrosine kinase inserts block high affinity binding of phospholipase C gamma and Shc to TrkC and neurotrophin-3 signaling. J. Biol. Chem., 270: 20384–20390.

Gusella, J.F. and MacDonald, M.E. (1998) Huntingtin: a single bait hooks many species. Curr. Opin. Neurobiol., 8: 425–430.

Ha, D.H., Robertson, R.T., Ribak, C.E. and Weiss, J.H. (1996) Cultured basal forebrain cholinergic neurons in contact with cortical cells display synapses, enhanced morphological features, and decreased dependence on nerve growth factor. J. Comp. Neurol., 373: 451–465.

Hagg, T., Hagg, F., Vahlsing, H.L., Manthorpe, M. and Varon, S. (1989) Nerve growth factor effects on cholinergic neurons of neostriatum and nucleus accumbens in the adult rat. Neuroscience, 30: 95–103.

Hagg, T., Quon, D., Higaki, J. and Varon, S. (1992) Ciliary neurotrophic factor prevents neuronal degeneration and promotes low affinity NGF receptor expression in the adult rat CNS. Neuron, 8: 145–158.

Hagg, T. and Varon, S. (1993) Ciliary neurotrophic factor prevents degeneration of adult rat substantia nigra dopaminergic neurons in vivo. Proc. Natl. Acad. Sci. USA, 90: 6315–6319.

Hapner, S.J., Boeshore, K.L., Large, T.H. and Lefcort, F. (1998) Neural differentiation promoted by truncated trkC receptors in collaboration with p75(NTR). Dev. Biol., 201: 90–100.

Harper, P.S. (1992) The epidemiology of Huntington's disease. Hum. Genet., 89: 365–376.

Hebert, M.A., Van Horne, C.G., Hoffer, B.J. and Gerhardt, G.A. (1996) Functional effects of GDNF in normal rat striatum: presynaptic studies using in vivo electrochemistry and microdialysis. J. Pharmacol. Exp. Ther., 279: 1181–1190.

Hefti, F. (1994) Development of effective therapy for Alzheimer's disease based on neurotrophic factors. Neurobiol. Aging, 15 (Suppl 2): S193–S194.

Hempstead, B.L., Martin-Zanca, D., Kaplan, D.R., Parada, L.F. and Chao, M.V. (1991) High-affinity NGF binding requires coexpression of the trk proto-oncogene and the low-affinity NGF receptor. Nature, 350: 678–683.

Henderson, C.E., Phillips, H.S., Pollock, R.A., Davies, A.M., Lemeulle, C., Armanini, M., Simpson, L.C., Moffet, B., Vandlen, R.A., Koliatsos, V.E. and Rosenthal, A. (1994) GDNF: A potent survival factor motoneurons present in peripheral nerve and muscle. Science, 266: 1062–1064.

Hirano, T., Nakajima, K. and Hibi, M. (1997) Signaling mechanisms through gp130: a model of the cytokine system. Cytokine Growth Factor Rev., 8: 241–252.

Hodgson, J.G., Agopyan, N., Gutekunst, C.A., Leavitt, B.R., LePiane, F., Singaraja, R., Smith, D.J., Bissada, N., McCutcheon, K., Nasir, J., Jamot, L., Li, X.J., Stevens, M.E., Rosemond, E., Roder, J.C., Phillips, A.G., Rubin, E.M., Hersch, S.M. and Hayden, M.R. (1999) A YAC mouse model for Huntington's disease with full-length mutant huntingtin, cytoplasmic toxicity, and selective striatal neurodegeneration. Neuron, 23: 181–192.

Hofer, M., Pagliusi, S.R., Hohn, A., Leibrock, J. and Barde, Y.A. (1990) Regional distribution of brain-derived neurotrophic factor mRNA in the adult mouse brain. EMBO J., 9: 2459–2464.

Hoglinger, G.U., Sautter, J., Meyer, M., Spenger, C., Seiler, R.W., Oertel, W.H. and Widmer, H.R. (1998) Rat fetal ventral mesencephalon grown as solid tissue cultures: influence of culture time and BDNF treatment on dopamine neuron survival and function. Brain Res., 813: 313–322.

Hohn, A., Leibrock, J., Bailey, K. and Barde, Y.-A. (1990) Identification and characterization of a novel member of the nerve growth factor/brain-derived neurotrophic factor family. Nature, 344: 339–341.

Holm, P.C., Akerud, P., Wagner, J. and Arenas, E. (2002) Neurturin is a neuritogenic but not a survival factor for developing and adult central noradrenergic neurons. J. Neurochem., 81: 1318–1327.

Holtzman, D.M., Li, Y., Parada, L.F., Kinsman, S., Chen, C.-K., Valletta, J.S., Zhou, J., Long, J.B. and Mobley, W.C. (1992) p140trk mRNA marks NGF responsive forebrain neurons: Evidence that trk gene expression is induced by NGF. Neuron, 9: 465–478.

Holtzman, D.M., Sheldon, R.A., Jaffe, W., Cheng, Y. and Ferriero, D.M. (1996) Nerve growth factor protects the neonatal brain against hypoxic-ischemic injury. Ann. Neurol., 39: 114–122.

Horch, H.W., Krüttgen, A., Portbury, S.D. and Katz, L.C. (1999) Destabilization of cortical dendrites and spines by BDNF.. Neuron, 23: 353–364.

Horger, B.A., Nishimura, M.C., Armanini, M.P., Wang, L.C., Poulsen, K.T., Rosenblad, C., Kirik, D., Moffat, B., Simmons, L., Johnson, E., Jr., et al. (1998) Neurturin exerts potent actions on survival and function of midbrain dopaminergic neurons. J. Neurosci., 18: 4929–4937.

Huang, E.J. and Reichardt, L.F. (2001) Neurotrophins: roles in neuronal development and function. Annu. Rev. Neurosci., 24: 677–736.

Humpel, C., Marksteiner, J. and Saria, A. (1996) Glial-cell-line-derived neurotrophic factor enhances biosynthesis of substance P in striatal neurons in vitro. Cell Tissue Res., 286: 249–255.

Hyman, C., Hofer, M., Barde, Y.A., Juhasz, M., Yancopoulos, G.D., Squinto, S.P. and Lindsay, R.M. (1991) BDNF is a neurotrophic factor for dopaminergic neurons of the substantia nigra. Nature, 350: 230–232.

Hyman, C., Juhasz, M., Jackson, C., Wright, P., Ip, N.Y. and Lindsay, R.M. (1994) Overlapping and distinct actions of the neurotrophins BDNF, NT-3, and NT-4/5 on cultured dopaminergic and GABAergic neurons of the ventral mesencephalon. J. Neurosci., 14: 335–347.

Ibañez, C.F. (1996) Neurotrophin-4: the odd one out in the neurotrophin family. Neurochem. Res., 21: 787–793.

Ip, N.Y., Ibañez, C.F., Nye, S.H., McClain, J., Jones, P.F., Gies, D.R., Belluscio, L., Le Beau, M.M., Espinosa, R. and Squinto, S.P. (1992) Mammalian neurotrophin-4: structure, chromosomal localization, tissue distribution, and receptor specificity. Proc. Natl. Acad. Sci. USA, 89: 3060–3064.

Ip, N.Y., Li, Y., Yancopoulos, G.D. and Lindsay, R.M. (1993a) Cultured hippocampal neurons show responses to BDNF, NT-3, and NT-4, but not NGF. J. Neurosci., 13: 3394–3405.

Ip, N.Y., Stitt, T.N., Tapley, P., Klein, R., Glass, D.J., Fandl, J., Greene, L.A., Barbacid, M. and Yancopoulos, G.D. (1993b) Similarities and differences in the way neurotrophins interact with the Trk receptors in neuronal and nonneuronal cells. Neuron, 10: 137–149.

Ip, N.Y., Wiegand, S.J., Morse, J. and Rudge, J.S. (1993c) Injury-induced regulation of ciliary neurotrophic factor mRNA in the adult rat brain. Eur. J. Neurosci., 5: 25–33.

Ip, N.Y. and Yancopoulos, G.D. (1996) The neurotrophins and CNTF: two families of collaborative neurotrophic factors. Annu. Rev. Neurosci., 19: 491–515.

Itami, C., Mizuno, K., Kohno, T. and Nakamura, S. (2000) Brain-derived neurotrophic factor requirement for activity-dependent maturation of glutamatergic synapse in developing mouse somatosensory cortex. Brain Res., 857: 141–150.

Ivkovic, S., Polonskaia, O., Fariñas, I. and Ehrlich, M.E. (1997) Brain-derived neurotrophic factor regulates maturation of the DARPP-32 phenotype in striatal medium spiny neurons: studies in vivo and in vitro. Neuroscience, 79: 509–516.

Ivkovic, S. and Ehrlich, M.E. (1999) Expression of the striatal DARPP-32/ARPP-21 phenotype in GABAergic neurons requires neurotrophins in vivo and in vitro. J. Neurosci., 19: 5409–5419.

Jaszai, J., Farkas, L., Galter, D., Reuss, B., Strelau, J., Unsicker, K. and Krieglstein, K. (1998) GDNF-related factor persephin is widely distributed throughout the nervous system. J. Neurosci. Res., 53: 494–501.

Jing, S., Wen, D., Yu, Y., Holst, P.L., Luo, Y., Fang, M., Tamir, R., Antonio, L., Hu, Z., Cupples, R., Louis, J.C., Hu, S., Altrock, B.W. and Fox, G.M. (1996) GDNF-induced activation of the ret protein tyrosine kinase is mediated by GDNFR-α, a novel receptor for GDNF. Cell, 85: 1113–1124.

Jones, K.R., Fariñas, I., Backus, C. and Reichardt, L.F. (1994) Target disruption of the BDNF gene perturbs brain and sensory neuron development but not motor neuron development. Cell, 76: 989–999.

Jung, A.B. and Bennett, J.P.J. (1996) Development of striatal dopaminergic function. III: Pre- and postnatal development of striatal and cortical mRNAs for the neurotrophin receptors trkBTK+ and trkC and their regulation by synaptic dopamine. Dev. Brain Res., 94: 133–143.

Kaplan, D.R., Hempstead, B.L., Martin-Zanca, D., Chao, M.V. and Parada, L.F. (1991) The trk proto-oncogene product: a signal transducing receptor for nerve growth factor. Science, 252: 554–558.

Kaplan, D.R. and Miller, F.D. (2000) Neurotrophin signal transduction in the nervous system. Curr. Opin. Neurobiol., 10: 381–391.

Katoh-Semba, R., Semba, R., Takeuchi, I.K. and Kato, K. (1998) Age-related changes in levels of brain-derived neurotrophic factor in selected brain regions of rats, normal mice and senescence-accelerated mice: a comparison to those of nerve growth factor and neurotrophin-3. Neurosci. Res., 31: 227–234.

Kawamoto, Y., Nakamura, S., Nakano, S., Oka, N., Akiguchi, I. and Kimura, J. (1996) Immunohistochemical localization of brain-derived neurotrophic factor in adult rat brain. Neuroscience, 74: 1209–1226.

Kawamoto, Y., Nakamura, S., Kawamata, T., Akiguchi, I. and Kimura, J. (1999) Cellular localization of brain-derived neurotrophic factor-like immunoreactivity in adult monkey brain. Brain Res., 821: 341–349.

Kawamoto, Y., Nakamura, S., Matsuo, A., Akiguchi, I. and Shibasaki, H. (2000) Immunohistochemical localization of glial cell line-derived neurotrophic factor in the human central nervous system. Neuroscience, 100: 701–712.

Klein, R., Conway, D., Parada, L.F. and Barbacid, M. (1990a) The trkB tyrosine protein kinase gene codes for a second neurogenic receptor that lacks the catalytic kinase domain. Cell, 61: 647–656.

Klein, R., Martin-Zanca, D., Barbacid, M. and Parada, L.F. (1990b) Expression of the tyrosine kinase receptor gene trkB is confined to the murine embryonic and adult nervous system. Development, 109: 845–850.

Klein, R., Jing, S.Q., Nanduri, V., O'Rourke, E. and Barbacid, M. (1991a) The trk proto-oncogene encodes a receptor for nerve growth factor. Cell, 65: 189–197.

Klein, R., Nanduri, V., Jing, S.A., Lamballe, F., Tapley, P., Bryant, S., Cordon-Cardo, C., Jones, K.R., Reichardt, L.F. and Barbacid, M. (1991b) The trkB tyrosine protein kinase is a receptor for brain-derived neurotrophic factor and neurotrophin-3. Cell, 66: 395–403.

Klein, R., Lamballe, F., Bryant, S. and Barbacid, M. (1992) The trkB tyrosine protein kinase is a receptor for neurotrophin-4. Neuron, 8: 947–956.

Klein, R.D., Sherman, D., Ho, W.H., Stone, D., Bennett, G.L., Moffat, B., Vandlen, R., Simmons, L., Gu, Q., Hongo, J.A., Devaux, B., Poulsen, K., Armanini, M., Nozaki, C., Asai, N., Goddard, A., Phillips, H., Henderson, C.E., Takahashi, M. and Rosenthal, A. (1997) A GPI-linked protein that interacts with Ret to form a candidate neurturin receptor. Nature, 387: 717–721.

Koh, S. and Higgins, G.A. (1991) Differential regulation of the low-affinity nerve growth factor receptor during postnatal development of the rat brain. J. Comp. Neurol., 313: 494–508.

Kordower, J.H., Charles, V., Bayer, R., Bartus, R.T., Putney, S., Walus, L.R. and Friden, P.M. (1994) Intravenous administration of a transferrin receptor antibody-nerve growth factor conjugate prevents the degeneration of cholinergic striatal neurons in a model of Huntington disease. Proc. Natl. Acad. Sci. USA, 91: 9077–9080.

Kordower, J.H., Isacson, O. and Emerich, D.F. (1999) Cellular delivery of trophic factors for the treatment of Huntington's disease: is neuroprotection possible?. Exp. Neurol., 159: 4–20.

Kordower, J.H., Emborg, M.E., Bloch, J., Ma, S.Y., Chu, Y., Leventhal, L., McBride, J., Chen, E.Y., Palfi, S., Roitberg, B.Z., Brown, W.D., Holden, J.E., Pyzalski, R., Taylor, M.D., Carvey, P., Ling, Z., Trono, D., Hantraye, P., Deglon, N. and Aebischer, P. (2000) Neurodegeneration prevented by lentiviral vector delivery of GDNF in primate models of Parkinson's disease. Science, 290: 767–773.

Kotzbauer, P.T., Lampe, P.A., Heuckeroth, R.O., Golden, J.P., Creedon, D.J., Johnson, E.M.J. and Milbrandt, J. (1996) Neurturin, a relative of glial-cell-line-derived neurotrophic factor. Nature, 384: 467–470.

Laforet, G.A., Sapp, E., Chase, K., McIntyre, C., Boyce, F.M., Campbell, M., Cadigan, B.A., Warzecki, L., Tagle, D.A., Reddy, P.H., Cepeda, C., Calvert, C.R., Jokel, E.S., Klapstein, G.J., Ariano, M.A., Levine, M.S., DiFiglia, M. and Aronin, N. (2001) Changes in cortical and striatal neurons predict behavioral and electrophysiological

abnormalities in a transgenic murine model of Huntington's disease. J. Neurosci., 21: 9112–9123.

Lamballe, F., Klein, R. and Barbacid, M. (1991) trkC, a new member of the trk family of tyrosine protein kinases, is a receptor for neurotrophin-3. Cell, 66: 967–979.

Lamballe, F., Smeyne, R.J. and Barbacid, M. (1994) Developmental expression of trkC, the neurotrophin-3 receptor, in the mammalian nervous system. J. Neurosci., 14: 14–28.

Landwehrmeyer, G.B., Standaert, D.G., Testa, C.M., Penney, J.B.J. and Young, A.B. (1995) NMDA receptor subunit mRNA expression by projection neurons and interneurons in rat striatum. J. Neurosci., 15: 5297–5307.

Large, T.H., Bodary, S.C., Clegg, D.O., Weskamp, G., Otten, U. and Reichardt, L.F. (1986) Nerve growth factor gene expression in the developing rat brain. Science, 234: 352–355.

Lee, F.S., Kim, A.H., Khursigara, G. and Chao, M.V. (2001) The uniqueness of being a neurotrophin receptor. Curr. Opin. Neurobiol., 11: 281–286.

Lesser, S.S. and Lo, D.C. (1995) Regulation of voltage-gated ion channels by NGF and ciliary neurotrophic factor in SK-N-SH neuroblastoma cells. J. Neurosci., 15: 253–261.

Levi-Montalcini, R. (1987) The nerve growth factor 35 years later. Science, 237: 1154–1162.

Li, M., Sendtner, M. and Smith, A. (1995) Essential function of LIF receptor in motor neurons. Nature, 378: 724–727.

Lin, L.F., Mismer, D., Lile, J.D., Armes, L.G., Butler, E.T., Vannice, J.L. and Collins, F. (1989) Purification, cloning, and expression of ciliary neurotrophic factor (CNTF). Science, 246: 1023–1025.

Lin, L.-F.H., Doherty, D., Lile, J., Bektesh, S. and Collins, F. (1993) GDNF: a glial cell line-derived neurotrophic factor for midbrain dopaminergic neurons. Science, 260: 1130–1132.

Linker, R.A., Maurer, M., Gaupp, S., Martini, R., Holtmann, B., Giess, R., Rieckmann, P., Lassmann, H., Toyka, K.V., Sendtner, M. and Gold, R. (2002) CNTF is a major protective factor in demyelinating CNS disease: a neurotrophic cytokine as modulator in neuroinflammation. Nat. Med., 8: 620–624.

Lisovoski, F., Akli, S., Peltekian, E., Vigne, E., Haase, G., Perricaudet, M., Dreyfus, P.A., Kahn, A. and Peschanski, M. (1997) Phenotypic alteration of astrocytes induced by ciliary neurotrophic factor in the intact adult brain, as revealed by adenovirus-mediated gene transfer. J. Neurosci., 17: 7228–7236.

Lopez-Martin, E., Caruncho, H.J., Rodriguez-Pallares, J., Guerra, M.J. and Labandeira-Garcia, J.L. (1999) Striatal dopaminergic afferents concentrate in GDNF-positive patches during development and in developing intrastriatal striatal grafts. J. Comp. Neurol., 406: 199–206.

MacDonald, V. and Halliday, G. (2002) Pyramidal cell loss in motor cortices in Huntington's disease. Neurobiol. Dis., 10: 378–386.

Maisonpierre, P.C., Belluscio, L., Squinto, S., Ip, N.Y., Furth, M.E., Lindsay, R.M. and Yancopoulos, G.D. (1990) Neurotrophin-3: a neurotrophic factor related to NGF and BDNF. Science, 247: 1446–1451.

Maksimovic, I.D., Jovanovic, M.D., Malicevic, Z., Colic, M. and Ninkovic, M. (2002) Effects of nerve and fibroblast growth factors on the production of nitric oxide in experimental model of Huntington's disease. Vojnosanit. Pregl., 59: 119–123.

Mangiarini, L., Sathasivam, K., Seller, M., Cozens, B., Harper, A., Hetherington, C., Lawton, M., Trottier, Y., Lehrach, H., Davies, S.W., Bates, G.P. (1996) Exon 1 of the HD gene with an expanded CAG repeat is sufficient to cause a progressive neurological phenotype in transgenic mice. Cell, 87: 493–506.

Manthorpe, M., Hagg, T. and Varon, S. (1992) Ciliary neurotrophic factor. In: Fallon J.A. and Loughlin S.E. (Eds.), Neurotrophic Factors. Academic Press, New York, pp. 443–473.

Marco, S., Canudas, A.M., Canals, J.M., Gavalda, N., Perez-Navarro, E. and Alberch, J. (2002a) Excitatory amino acids differentially regulate the expression of GDNF, neurturin, and their receptors in the adult rat striatum. Exp. Neurol., 174: 243–252.

Marco, S., Perez-Navarro, E., Tolosa, E., Arenas, E. and Alberch, J. (2002b) Striatopallidal neurons are selectively protected by neurturin in an excitotoxic model of Huntington's disease. J. Neurobiol., 50: 323–332.

Marco, S., Saura, J., Perez-Navarro, E., Jose, M.M., Tolosa, E. and Alberch, J. (2002c) Regulation of c-Ret, GFRα1, and GFRα2 in the substantia nigra pars compacta in a rat model of Parkinson's disease. J. Neurobiol., 52: 343–351.

Martin-Zanca, D., Hughes, S.H. and Barbacid, M. (1986) A human oncogene formed by the fusion of truncated tropomyosin and protein tyrosine kinase sequences. Nature, 319: 743–748.

Martin-Zanca, D., Barbacid, M. and Parada, L.F. (1990) Expression of the trk proto-oncogene is restricted to the sensory cranial and spinal ganglia of neural crest origin in mouse development. Genes Dev., 4: 683–694.

Martin, J.B. and Gusella, J.F. (1986) Huntington's disease. Pathogenesis and management. N. Engl. J. Med., 315: 1267–1276.

Martinez, H.J., Dreyfus, C.F., Jonakait, G.M. and Black, I.B. (1985) Nerve growth factor promotes cholinergic development in brain striatal cultures. Proc. Natl. Acad. Sci. USA, 82: 7777–7781.

Martínez-Serrano, A. and Björklund, A. (1996) Protection of the neostriatum against excitotoxic damage by neurotrophin-producing, genetically modified neural stem cells. J. Neurosci., 16: 4604–4616.

Marty, S., Berzaghi, M.P. and Berninger, B. (1997) Neurotrophins and activity-dependent plasticity of cortical interneurons. Trends Neurosci., 20: 198–202.

Masu, Y., Wolf, E., Holtmann, B., Sendtner, M., Brem, G. and Thoenen, H. (1993) Disruption of the CNTF gene results in motor neuron degeneration. Nature, 365: 27–32.

Masure, S., Cik, M., Pangalos, M.N., Bonaventure, P., Verhasselt, P., Lesage, A.S., Leysen, J.E. and Gordon, R.D. (1998) Molecular cloning, expression and tissue distribution of glial-cell-line-derived neurotrophic factor family receptor α-3 (GFRα-3). Eur. J. Biochem., 251: 622–630.

Masure, S., Cik, M., Hoefnagel, E., Nosrat, C.A., Van, D.L.I, Scott, R., Van Gompel, P., Lesage, A.S., Verhasselt, P., Ibanez, C.F. and Gordon, R.D. (2000) Mammalian GFR?-4, a divergent member of the GFRα family of coreceptors for glial cell line-derived neurotrophic factor family ligands, is a receptor for the neurotrophic factor persephin. J. Biol. Chem., 275: 39427–39434.

McAllister, A.K., Lo, D.C. and Katz, L.C. (1995) Neurotrophins regulate dentritic growth in developing visual cortex. Neuron, 15: 791–803.

McAllister, A.K., Katz, L.C. and Lo, D.C. (1997) Opposing roles for endogenous BDNF and NT-3 in regulating cortical dentritic growth. Neuron, 18: 767–778.

McMurray, C.T. (2001) Huntington's disease: new hope for therapeutics. Trends Neurosci., 24: S32–S38.

Meakin, S.O., Gryz, E.A. and MacDonald, J.I. (1997) A kinase insert isoform of rat TrkA supports nerve growth factor-dependent cell survival but not neurite outgrowth. J. Neurochem., 69: 954–967.

Menei, P., Pean, J.M., Nerriere-Daguin, V., Jollivet, C., Brachet, P. and Benoit, J.P. (2000) Intracerebral implantation of NGF-releasing biodegradable microspheres protects striatum against excitotoxic damage. Exp. Neurol., 161: 259–272.

Menn, B., Timsit, S., Represa, A., Mateos, S., Calothy, G. and Lamballe, F. (2000) Spatiotemporal expression of noncatalytic TrkC NC2 isoform during early and late CNS neurogenesis: a comparative study with TrkC catalytic and p75NTR receptors. Eur. J. Neurosci., 12: 3211–3223.

Merlio, J.P., Ernfors, P., Jaber, M. and Persson, H. (1992) Molecular cloning of rat trkC and distribution of cells expressing messenger RNA for members of the trk family in the rat central nervous system. Neuroscience, 51: 513–532.

Middlemas, D.S., Lindberg, R.A. and Hunter, T. (1991) trkB, a neural receptor protein-Tyrosine Kinase: evidence for a full-length and two truncated receptors. Mol. Cell. Biol., 11: 143–153.

Milbrandt, J., de Sauvage, F.J., Fahrner, T.J., Baloh, R.H., Leitner, M.L., Tansey, M.G., Lampe, P.A., Heuckeroth, R.O., Kotzbauer, P.T., Simburger, K.S., Golden, J.P., Davies, J.A., Vejsada, R., Kato, A.C., Hynes, M., Sherman, D., Nishimura, M., Wang, L.C., Vandlen, R., Moffat, B., Klein, R.D., Poulsen, K., Gray, C., Garces, A. and

Johnson, E.M.J. (1998) Persephin, a novel neurotrophic factor related to GDNF and neurturin. Neuron, 20: 245–253.

Mitsumoto, H., Ikeda, K., Klinkosz, B., Cedarbaum, J.M., Wong, V. and Lindsay, R.M. (1994) Arrest of motor neuron disease in wobbler mice cotreated with CNTF and BDNF. Science, 265: 1107–1110.

Mittoux, V., Joseph, J.M., Conde, F., Palfi, S., Dautry, C., Poyot, T., Bloch, J., Deglon, N., Ouary, S., Nimchinsky, E.A., Brouillet, E., Hof, P.R., Peschanski, M., Aebischer, P. and Hantraye, P. (2000) Restoration of cognitive and motor functions by ciliary neurotrophic factor in a primate model of Huntington's disease. Hum. Gene Ther., 11: 1177–1187.

Mittoux, V., Ouary, S., Monville, C., Lisovoski, F., Poyot, T., Conde, F., Escartin, C., Robichon, R., Brouillet, E., Peschanski, M. and Hantraye, P. (2002) Corticostriatopallidal neuroprotection by adenovirus-mediated ciliary neurotrophic factor gene transfer in a rat model of progressive striatal degeneration. J. Neurosci., 22: 4478–4486.

Mizuno, K., Carnahan, J. and Nawa, H. (1994) Brain-derived neurotrophic factor promotes differentiation of striatal GABA-ergic neurons. Dev. Biol., 165: 243–256.

Mobley, W.C., Rutkowski, J.L., Tennekoon, G.I., Buchanan, K. and Johnston, M.V. (1985) Choline acetyltransferase activity in striatum of neonatal rats increased by nerve growth factor. Science, 229: 284–287.

Mobley, W.C., Woo, J.E., Edwwards, R.H., Riopelle, R.J., Longo, F.M., Weskamp, G., Otten, U., Valletta, J.S. and Johnston, M.V. (1989) Development regulation of nerve growth factor and its receptor in the rat caudate-putamen. Neuron, 3: 655–664.

Morrison, R.S. and de Vellis, J. (1981) Growth of purified astrocytes in a chemically defined medium. Proc. Natl. Acad. Sci. USA, 78: 7205–7209.

Mount, H.T.J., Dean, D.O., Alberch, J., Dreyfus, C.F. and Black, I.B. (1995) Glial cell line-derived neurotrophic factor promotes the survival and morphological differentiation of Purkinge cells. Proc. Natl. Acad. Sci. USA, 92: 9092–9096.

Mufson, E.J., Kroin, J.S., Sobreviela, T., Burke, M.A., Kordower, J.H., Penn, R.D. and Miller, J.A. (1994) Intrastriatal infusions of brain-derived neurotrophic factor: retrograde transport and colocalization with dopamine containing substantia nigra neurons in rat. Exp. Neurol., 129: 15–26.

Mufson, E.J., Kroin, J.S., Liu, Y.T., Sobreviela, T., Penn, R.D., Miller, J.A. and Kordower, J.H. (1996) Intrastriatal and intraventricular infusion of brain-derived neurotrophic factor in the cynomologous monkey: distribution, retrograde transport and co-localization with substantia nigra dopamine-containing neurons. Neuroscience, 71: 179–191.

Nakao, N., Kokaia, Z., Odin, P. and Lindvall, O. (1995) Protective effects of BDNF and NT-3 but not PDGF against hypoglycemic injury to cultured striatal neurons. Exp. Neurol., 131: 1–10.

Nakao, N., Odin, P., Lindvall, O. and Brundin, P. (1996) Differential trophic effects of basic fibroblast growth factor, insulin-like growth factor-1, and neurotrophin-3 on striatal neurons in culture. Exp. Neurol., 138: 144–157.

Nasir, J., Floresco, S.B., O'Kusky, J.R., Diewert, V.M., Richman, J.M., Zeisler, J., Borowski, A., Marth, J.D., Phillips, A.G. and Hayden, M.R. (1995) Targeted disruption of the Huntington's disease gene results in embryonic lethality and behavioral and morphological changes in heterozygotes. Cell, 81: 811–823.

Naveilhan, P., Baudet, C., Mikaels, Å., Shen, L., Westphal, H. and Ernfors, P. (1998) Expression and regulation of GFRα3, a glial cell line-derived neurotrophic factor family receptor. Proc. Natl. Acad. Sci. USA, 95: 1295–1300.

Nishi, R. and Berg, D.K. (1981) Two components from eye tissue that differentially stimulate the growth and development of ciliary ganglion neurons in cell culture. J. Neurosci., 1: 505–513.

Nosrat, C.A., Tomac, A., Lindqvist, E., Lindskog, S., Humpel, C., Stromberg, I., Ebendal, T., Hoffer, B.J. and Olson, L. (1996) Cellular expression of GDNF mRNA suggests multiple functions inside and outside the nervous system. Cell Tissue Res., 286: 191–207.

Nosrat, C.A., Tomac, A., Hoffer, B.J. and Olson, L. (1997) Cellular and developmental patterns of expression of Ret and glial cell line-derived neurotrophic factor receptor alpha mRNAs. Exp. Brain Res., 115: 410–422.

Oppenheim, R.W., Prevette, D., Yin, Q.W., Collins, F. and MacDonald, J. (1991) Control of embryonic motoneuron survival in vivo by ciliary neurotrophic factor. Science, 251: 1616–1618.

Oppenheim, R.W., Houenou, L.J., Johnson, J.E., Lin, L.F., Li, L., Lo, A.C., Newsome, A.L., Prevette, D.M. and Wang, S. (1995) Developing motor neurons rescued from programmed and axotomy-induced cell death by GDNF. Nature, 373: 344–346.

Paratcha, G., Ledda, F., Baars, L., Coulpier, M., Besset, V., Anders, J., Scott, R. and Ibanez, C.F. (2001) Released GFRα potentiates downstream signaling, neuronal survival, and differentiation via a novel mechanism of recruitment of c-Ret to lipid rafts. Neuron, 29: 171–184.

Park, C.K., Ju, W.K., Hofmann, H.D., Kirsch, M., Ki, K.J., Chun, M.H. and Lee, M.Y. (2000) Differential regulation of ciliary neurotrophic factor and its receptor in the rat hippocampus following transient global ischemia. Brain. Res., 861: 345–353.

Patapoutian, A. and Reichardt, L.F. (2001) Trk receptors: mediators of neurotrophin action. Curr. Opin. Neurobiol., 11: 272–280.

Pelicci, G., Troglio, F., Bodini, A., Melillo, R.M., Pettirossi, V., Coda, L., De Giuseppe, A., Santoro, M. and Pelicci, P.G. (2002) The neuron-specific Rai (ShcC) adaptor protein inhibits apoptosis by coupling Ret to the phosphatidylinositol 3-kinase/Akt signaling pathway. Mol. Cell. Biol., 22: 7351–7363.

Pereira de Almeida, L., Zala, D., Aebischer, P. and Deglon, N. (2001) Neuroprotective effect of a CNTF-expressing lentiviral vector in the quinolinic acid rat model of Huntington's disease. Neurobiol. Dis., 8: 433–446.

Pérez-Navarro, E., Alberch, J., Arenas, E., Calvo, N. and Marsal, J. (1994) Nerve growth factor and basic fibroblast growth factor protect cholinergic neurons against quinolinic acid excitotoxicity in rat neostriatum. Eur. J. Neurosci., 6: 706–711.

Pérez-Navarro, E. and Alberch, J. (1995) Protective role of Nerve growth factor against excitatory amino acids injury during neostriatal cholinergic neurons postnatal development. Exp. Neurol., 135: 146–152.

Pérez-Navarro, E., Arenas, E., Reiriz, J., Calvo, N. and Alberch, J. (1996) Glial cell line-derived neurotrophic factor protects striatal calbindin-immunoreactive neurons from excitotoxic damage. Neuroscience, 75: 345–352.

Pérez-Navarro, E., Alberch, J., Neveu, I. and Arenas, E. (1999a) BDNF, NT-3 and NT-4/5 differentially regulate the phenotype and prevent degenerative changes of striatal projection neurons after excitotoxicity in vivo. Neuroscience, 91: 1257–1264.

Pérez-Navarro, E., Arenas, E., Marco, S. and Alberch, J. (1999b) Intrastriatal grafting of a GDNF-producing cell line protects striatonigralneurons from quinolinic acid excitotoxicity in vivo. Eur. J. Neurosci., 11: 241–249.

Pérez-Navarro, E., Akerud, P., Marco, S., Canals, J.M., Tolosa, E., Arenas, E. and Alberch, J. (2000a) Neurturin protects striatal projection neurons but not interneurons in a rat model of Huntington's disease. Neuroscience, 98: 89–96.

Pérez-Navarro, E., Canudas, A.M., Akerund, P., Alberch, J. and Arenas, E. (2000b) Brain-derived neurotrophic factor, neurotrophin-3, and neurotrophin-4/5 prevent the death of striatal projection neurons in a rodent model of Huntington's disease. J. Neurochem., 75: 2190–2199.

Petersen, A., Mani, K. and Brundin, P. (1999) Recent advances on the pathogenesis of Huntington's disease. Exp. Neurol., 157: 1–18.

Pitts, A.F. and Miller, M.W. (2000) Expression of nerve growth factor, brain-derived neurotrophic factor, and neurotrophin-3 in the somatosensory cortex of the mature rat: coexpression with high affinity neurotrophin receptor. J. Comp. Neurol., 418: 241–254.

Pochon, N.A., Menoud, A., Tseng, J.L., Zurn, A.D. and Aebischer, P. (1997) Neuronal GDNF expression in the adult rat nervous system identified by in situ hybridization. Eur. J. Neurosci., 9: 463–471.

Pong, K., Xu, R.Y., Baron, W.F., Louis, J.C. and Beck, K.D. (1998) Inhibition of phosphatidylinositol 3-kinase activity blocks cellular differentiation mediated by glial cell line-derived neurotrophic factor in dopaminergic neurons. J. Neurochem., 71: 1912–1919.

Poteryaev, D., Titievsky, A., Sun, Y.F., Thomas-Crusells, J., Lindahl, M., Billaud, M., Arumae, U. and Saarma, M. (1999) GDNF triggers a novel ret-independent Src kinase family-coupled signaling via a GPI-linked GDNF receptor alpha1. FEBS Lett., 463: 63–66.

Price, M.L., Hoffer, B.J. and Granholm, A.C. (1996) Effects of GDNF on fetal septal forebrain transplants in oculo. Exp. Neurol., 141: 181–189.

Rao, M.S., Tyrrell, S., Landis, S.C. and Patterson, P.H. (1992) Effects of ciliary neurotrophic factor (CNTF) and depolarization on neuropeptide expression in cultured sympathetic neurons. Dev. Biol., 150: 281–293.

Reddy, P.H., Williams, M. and Tagle, D.A. (1999) Recent advances in understanding the pathogenesis of Huntington's disease. Trends Neurosci., 22: 248–255.

Reiner, A., Albin, R.L., Anderson, K.D., D'Amato, C.J., Penney, J.B. and Young, A.B. (1988) Differential loss of striatal projection neurons in Huntington disease. Proc. Natl. Acad. Sci. USA, 85: 5733–5737.

Richfield, E.K., Maguire-Zeiss, K.A., Vonkeman, H.E. and Voorn, P. (1995) Preferential loss of preproenkephalin versus preprotachykinin neurons from the striatum of Huntington's disease patients. Ann. Neurol., 38: 852–861.

Rigamonti, D., Bauer, J.H., De-Fraja, C., Conti, L., Sipione, S., Sciorati, C., Clementi, E., Hackam, A., Hayden, M.R., Li, Y., Cooper, J.K., Ross, C.A., Govoni, S., Vincenz, C. and Cattaneo, E. (2000) Wild-type huntingtin protects from apoptosis upstream of caspase-3. J. Neurosci., 20: 3705–3713.

Ringstedt, T., Lagercrantz, H. and Persson, H. (1993) Expression of members of the trk family in the developing postnatal rat brain. Dev. Brain Res., 72: 119–131.

Rosenblad, C., Gronborg, M., Hansen, C., Blom, N., Meyer, M., Johansen, J., Dago, L., Kirik, D., Patel, U.A., Lundberg, C., Trono, D., Bjorklund, A. and Johansen, T.E. (2000) In vivo protection of nigral dopamine neurons by lentiviral gene transfer of the novel GDNF-family member neublastin/artemin. Mol. Cell. Neurosci., 15: 199–214.

Ross, C.A. (2002) Polyglutamine pathogenesis: emergence of unifying mechanisms for Huntington's disease and related disorders. Neuron, 35: 819–822.

Roux, P.P. and Barker, P.A. (2002) Neurotrophin signaling through the p75 neurotrophin receptor. Prog. Neurobiol., 67: 203–233.

Rudge, J.S., Pasnikowski, E.M., Holst, P. and Lindsay, R.M. (1995) Changes in neurotrophic factor expression and receptor activation following exposure of hippocampal neuron/astrocyte cocultures to kainic acid. J. Neurosci., 15: 6856–6867.

Saadat, S., Sendtner, M. and Rohrer, H. (1989) Ciliary neurotrophic factor induces cholinergic differentiation of rat sympathetic neurons in culture. J. Cell Biol., 108: 1807–1816.

Salin, T., Mudò, G., Jiang, X.H., Timmusk, T., Metsis, M. and Belluardo, N. (1995) Up-regulation of trkB mRNA expression in the rat striatum after seizures. Neurosci. Lett., 194: 181–184.

Sarabi, A., Hoffer, B.J., Olson, L. and Morales, M. (2001) GFRα-1 mRNA in dopaminergic and nondopaminergic neurons in the substantia nigra and ventral tegmental area. J. Comp. Neurol., 441: 106–117.

Sarabi, A., Hoffer, B.J., Olson, L. and Morales, M. (2003) Glial cell line neurotrophic factor-family receptor a-1 is present in central neurons with distinct phenotypes, *Neuroscience*, 116: 261–271.

Saudou, F., Finkbeiner, S., Devys, D. and Greenberg, M.E. (1998) Huntingtin acts in the nucleus to induce apoptosis but death does not correlate with the formation of intranuclear inclusions. Cell, 95: 55–66.

Schumacher, J.M., Short, M.P., Hyman, B.T., Breakefield, X.O. and Isacson, O. (1991) Intracerebral implantation of nerve growth factor-producing fibroblasts protects striatum against neurotoxic levels of excitatory amino acids. Neuroscience, 45: 561–570.

Scott, R.P. and Ibañez, C.F. (2001) Determinants of ligand binding specificity in the glial cell line-derived neurotrophic factor family receptor αs. J. Biol. Chem., 276: 1450–1458.

Segal, R.A. and Greenberg, M.E. (1996) Intracellular signaling pathways activated by neurotrophic factors. Annu. Rev. Neurosci., 19: 463–489.

Sendtner, M., Carroll, P., Holtmann, B., Hughes, R.A. and Thoenen, H. (1994) Ciliary neurotrophic factor. J. Neurobiol., 25: 1436–1453.

Sendtner, M., Gotz, R., Holtmann, B., Escary, J.L., Masu, Y., Carroll, P., Wolf, E., Brem, G., Brulet, P. and Thoenen, H. (1996) Cryptic physiological trophic support of motoneurons by LIF revealed by double gene targeting of CNTF and LIF. Curr. Biol., 6: 686–694.

Sendtner, M., Gotz, R., Holtmann, B. and Thoenen, H. (1997) Endogenous ciliary neurotrophic factor is a lesion factor for axotomized motoneurons in adult mice. J. Neurosci., 17: 6999–7006.

Shelton, D.L., Sutherland, J., Gripp, J., Camerato, T., Armanini, M.P., Phillips, H.S., Carroll, K., Spencer, S.D. and Levinson, A.D. (1995) Human trks: molecular cloning, tissue distribution, and expression of extracellular domain immunoadhesins. J. Neurosci., 15: 477–491.

Shoulson, I. (1998) Experimental therapeutics of neurodegenerative disorders: unmet needs. Science, 282: 1072–1074.

Sleeman, M.W., Anderson, K.D., Lambert, P.D., Yancopoulos, G.D. and Wiegand, S.J. (2000) The ciliary neurotrophic factor and its receptor, CNTFR α. . Pharm. Acta Helv., 74: 265–272.

Smeyne, R.J., Klein, R., Schnapp, A., Long, L.K., Bryant, S., Lewin, A., Lira, S.A. and Barbacid, M. (1994) Severe sensory and sympathetic neuropathies in mice carrying a disrupted Trk/NGF receptor gene. Nature, 368: 246–249.

Snider, W.D. (1994) Functions of the neurotrophins during nervous system development: what the knockouts are teaching us. Cell, 77: 627–638.

Sobreviela, T., Pagcatipunan, M., Kroin, J.S. and Mufson, E.J. (1996) Retrograde transport of BDNF following infusion in neo- and limbic cortex in rat: relationship to BDNF mRNA expressing neurons. J. Comp. Neurol., 375: 417–444.

Sofroniew, M.V., Galletly, N.P., Isacson, O. and Svendsen, C.N. (1990) Survival of adult basal forebrain cholinergic neurons after loss of target neurons. Science, 247: 338–342.

Soler, R.M., Dolcet, X., Encinas, M., Egea, J., Bayascas, J.R. and Comella, J.X. (1999) Receptors of the glial cell line-derived neurotrophic factor family of neurotrophic factors signal cell survival through the phosphatidylinositol 3-kinase pathway in spinal cord motoneurons. J. Neurosci., 19: 9160–9169.

Soppet, D., Escandon, E., Maragos, J., Middlemas, D.S., Reid, S.W., Blair, J., Burton, L.E., Stanton, B.R., Kaplan, D.R. and Hunter, T. (1991) The neurotrophic factors brain-derived neurotrophic factor and neurotrophin-3 are ligands for the trkB tyrosine kinase receptor. Cell, 65: 895–903.

Spenger, C., Hyman, C., Studer, L., Egli, M., Evtouchenko, L., Jackson, C., Dahl-Jorgensen, A., Lindsay, R.M. and Seiler, R.W. (1995) Effects of BDNF on dopaminergic, serotonergic, and GABAergic neurons in cultures of human fetal ventral mesencephalon. Exp. Neurol., 133: 50–63.

Squinto, S.P., Stitt, T.N., Aldrich, T.H., Davis, S., Bianco, S.M., Radziejewski, C., Glass, D.J., Masiakowski, P., Furth, M.E. and Valenzuela, D.M. (1991) trkB encodes a functional receptor for brain-derived neurotrophic factor and neurotrophin-3 but not nerve growth factor. Cell, 65: 885–893.

Stahl, N. and Yancopoulos, G.D. (1994) The tripartite CNTF receptor complex: activation and signaling involves components shared with other cytokines. J. Neurobiol., 25: 1454–1466.

Stahl, N., Farruggella, T.J., Boulton, T.G., Zhong, Z., Darnell, J.E.J. and Yancopoulos, G.D. (1995) Choice of STATs and other substrates specified by modular tyrosine-based motifs in cytokine receptors. Science, 267: 1349–1353.

Stankoff, B., Aigrot, M.S., Noel, F., Wattilliaux, A., Zalc, B. and Lubetzki, C. (2002) Ciliary neurotrophic factor (CNTF) enhances myelin formation: a novel role for CNTF and CNTF-related molecules. J. Neurosci., 22: 9221–9227.

Steininger, T.L., Wainer, B.H., Klein, R., Barbacid, M. and Palfrey, H.C. (1993) High-affinity nerve growth factor receptor (Trk) immunoreactivity is localized in cholinergic neurons of the basal forebrain and striatum in the adult rat brain. Brain Res., 612: 330–335.

Stockli, K.A., Lottspeich, F., Sendtner, M., Masiakowski, P., Carroll, P., Gotz, R., Lindholm, D. and Thoenen, H. (1989)

Molecular cloning, expression and regional distribution of rat ciliary neurotrophic factor. Nature, 342: 920–923.

Stoop, R. and Poo, M.M. (1995) Potentiation of transmitter release by ciliary neurotrophic factor requires somatic signaling. Science, 267: 695–699.

Strauss, S., Otten, U., Joggerst, B., Plüss, K. and Volk, B. (1994) Increased levels of nerve growth factor (NGF) protein and mRNA and reactive gliosis following kainic acid injection ibto the striatum. Neurosci. Lett., 168: 193–196.

Strohmaier, C., Carter, B.D., Urfer, R., Barde, Y.A. and Dechant, G. (1996) A splice variant of the neurotrophin receptor trkB with increased specificity for brain-derived neurotrophic factor. EMBO J., 15: 3332–3337.

Symes, A.J., Rao, M.S., Lewis, S.E., Landis, S.C., Hyman, S.E. and Fink, J.S. (1993) Ciliary neurotrophic factor coordinately activates transcription of neuropeptide genes in a neuroblastoma cell line. Proc. Natl. Acad. Sci. USA, 90: 572–576.

Tessarollo, L., Tsoulfas, P., Martin-Zanca, D., Gilbert, D.J., Jenkins, N.A., Copeland, N.G. and Parada, L.F. (1993) TrkC, a receptor for neurotrophin-3, is widely expressed in the developing nervous system and in non-neuronal tissues. Development, 118: 463–475.

The Huntington's Disease Collaborative Research Group (1993) A novel gene containing a trinucleotide repeat that is expanded and unstable on Huntington's disease chromosomes. Cell, 72: 971–983.

Thoenen, H. and Sendtner, M. (2002) Neurotrophins: from enthusiastic expectations through sobering experiences to rational therapeutic approaches, Nat. Neurosci., 5 Suppl: 1046–1050.

Thompson, J., Doxakis, E., Pinon, L.G., Strachan, P., Buj-Bello, A., Wyatt, S., Buchman, V.L. and Davies, A.M. (1998) GFRα-4, a new GDNF family receptor. Mol. Cell. Neurosci., 11: 117–126.

Timmusk, T., Belluardo, N., Metsis, M. and Persson, H. (1993) Widespread and developmentally regulated expression of neurotrophin-4 mRNA in rat brain and peripheral tissues. Eur. J. Neurosci., 5: 605–613.

Tomac, A.C., Agulnick, A.D., Haughey, N., Chang, C.F., Zhang, Y., Backman, C., Morales, M., Mattson, M.P., Wang, Y., Westphal, H. and Hoffer, B.J. (2002) Effects of cerebral ischemia in mice deficient in Persephin. Proc. Natl. Acad. Sci. USA, 99: 9521–9526.

Treanor, J.J., Goodman, L., de Sauvage, F., Stone, D.M., Poulsen, K.T., Beck, C.D., Gray, C., Armanini, M.P., Pollock, R.A., Hefti, F., Phillips, H.S., Goddard, A., Moore, M.W., Buj-Bello, A., Davies, A.M., Asai, N., Takahashi, M., Vandlen, R., Henderson, C.E. and Rosenthal, A. (1996) Characterization of a multicomponent receptor for GDNF. Nature, 382: 80–83.

Trupp, M., Arenas, E., Fainzilber, M., Nilsson, A.S., Sieber, B.A., Grigoriou, M., Kilkenny, C., Salazar-Grueso, E., Pachnis, V.

and Arumae, U. (1996) Functional receptor for GDNF encoded by the c-ret proto-oncogene. Nature, 381: 785–789.

Trupp, M., Belluardo, N., Funakoshi, H. and Ibañez, C.F. (1997) Complementary and overlapping expression of glial cell line-derived neurotrophic factor (GDNF), c-ret proto-oncogene, and GDNF receptor-α indicates multiple mechanisms of trophic actions in the adult rat CNS. J. Neurosci., 1715: 3527–3554.

Trupp, M., Scott, R., Whittemore, S.R. and Ibañez, C.F. (1999) Ret-dependent and -independent mechanisms of glial cell line-derived neurotrophic factor signaling in neuronal cells. J. Biol. Chem., 274: 20885–20894.

Vahlsing, H.L., Hagg, T., Spencer, M., Conner, J.M., Manthorpe, M. and Varon, S. (1991) Dose-dependent responses to nerve growth factor by adult rat cholinergic medial septum and neostriatum neurons. Brain Res., 552: 320–329.

Van der Zee, C.E., Ross, G.M., Riopelle, R.J. and Hagg, T. (1996) Survival of cholinergic forebrain neurons in developing p75NGFR-deficient mice. Science, 274: 1729–1732.

Van Vulpen, E.H. and Van D. Kooy, D. (1999) NGF facilitates the developmental maturation of the previously committed cholinergic interneurons in the striatal matrix. J. Comp. Neurol., 411: 87–96.

Van Weering, D.H., de Rooij, J., Marte, B., Downward, J., Bos, J.L. and Burgering, B.M. (1998) Protein kinase B activation and lamellipodium formation are independent phosphoinositide 3-kinase-mediated events differentially regulated by endogenous Ras. Mol. Cell. Biol., 18: 1802–1811.

Venero, J.L., Knüsel, B., Beck, K.D. and Hefti, F. (1994) Expression of neurotrophin and trk receptor genes in adult rats with fimbria transections: effect of intraventricular nerve growth factor and brain-derived neurotrophic factor administration. Neuroscience, 59: 797–815.

Ventimiglia, R., Mather, P.E., Jones, B.E. and Lindsay, R.M. (1995) The neurotrophins BDNF, NT-3 and NT-4/5 promote survival and morphological and biochemical differentiation of striatal neurons. Eur. J. Neurosci., 7: 213–222.

Wang, C.Y., Ni, J., Jiang, H., Hsu, T.A., Dugich-Djordjevic, M., Feng, L., Zhang, M., Mei, L., Gentz, R. and Lu, B. (1998) Cloning and characterization of glial cell line-derived neurotrophic factor receptor-B: a novel receptor for members of glial cell line-derived neurotrophic factor family of neurotrophic factors. Neuroscience, 83: 7–14.

Wang, L.C., Shih, A., Hongo, J., Devaux, B. and Hynes, M. (2000) Broad specificity of GDNF family receptors GFRα1 and GFRα2 for GDNF and NTN in neurons and transfected cells. J. Neurosci. Res., 61: 1–9.

Ward, N.L. and Hagg, T. (2000) BDNF is needed for postnatal maturation of basal forebrain and neostriatum cholinergic neurons in vivo. Exp. Neurol., 162: 297–310.

Ware, C.B., Horowitz, M.C., Renshaw, B.R., Hunt, J.S., Liggitt, D., Koblar, S.A., Gliniak, B.C., McKenna, H.J.,

Papayannopoulou, T. and Thoma, B. (1995) Targeted disruption of the low-affinity leukemia inhibitory factor receptor gene causes placental, skeletal, neural and metabolic defects and results in perinatal death. Development, 121: 1283–1299.

Widenfalk, J., Nosrat, C., Tomac, A., Westphal, H., Hoffer, B. and Olson, L. (1997) Neurturin and glial cell line-derived neurotrophic factor receptor-β (GDNFR-β), novel proteins related to GDNF and GDNFR-α with specific cellular patterns of expression suggesting roles in the developing and adult nervous system and in the peripheral organs. J. Neurosci., 17: 8506–8519.

Widmer, H.R. and Hefti, F. (1994) Stimulation of GABAergic neuron differentiation by NT-4/5 in cultures of rat cerebral cortex. Dev. Brain Res., 80: 279–284.

Wong, J.Y., Liberatore, G.T., Donnan, G.A. and Howells, D.W. (1997) Expression of brain-derived neurotrophic factor and TrkB neurotrophin receptors after striatal injury in the mouse. Exp. Neurol., 148: 83–91.

Worby, C.A., Vega, Q.C., Chao, H.H., Seasholtz, A.F., Thompson, R.C. and Dixon, J.E. (1998) Identification and characterization of GFRα-3, a novel Co-receptor belonging to the glial cell line-derived neurotrophic receptor family. J. Biol. Chem., 273: 3502–3508.

Xu, B., Zang, H., Ruff, N.L., Zhang, Y.A., McConnell, S.K., Stryker, M.P. and Reichardt, L.F. (2000) Cortical degeneration in the absence of neurotrophin signaling: dendritic retraction and neuronal loss after removal of the receptor TrkB. Neuron, 26: 233–245.

Yacoubian, T.A. and Lo, D.C. (2000) Truncated and full-length TrkB receptors regulate distinct modes of dendritic growth. Nat. Neurosci., 3: 342–349.

Yamamoto, A., Lucas, J.J. and Hen, R. (2000) Reversal of neuropathology and motor dysfunction in a conditional model of Huntington's disease. Cell, 101: 57 66.

Yan, Q., Radeka, M.J., Matheson, C.R., Talvenheimo, J., Welcher, A.A. and Feinstein, S.C. (1997a) Immunocytochemical localization of TrkB in the central nervous system of the adult rat. J. Comp. Neurol., 378: 135–157.

Yan, Q., Rosenfeld, R.D., Matheson, C.R., Hawkins, N., Lopez, O.T., Bennet, L. and Welcher, A.A. (1997b) Expression of brain-derived neurotrophic factor protein in the adult rat central nervous system. Neuroscience, 78: 431–448.

Yu, T., Scully, S., Yu, Y., Fox, G.M., Jing, S. and Zhou, R. (1998) Expression of GDNF family receptor components during development: implications in the mechanisms of interaction. J. Neurosci., 18: 4684–4696.

Yurek, D.M. and Fletcher-Turner, A. (2001) Differential expression of GDNF, BDNF, and NT-3 in the aging nigrostriatal system following a neurotoxic lesion. Brain Res., 891: 228–235.

Zeron, M.M., Hansson, O., Chen, N., Wellington, C.L., Leavitt, B.R., Brundin, P., Hayden, M.R. and Raymond, L.A. (2002) Increased sensitivity to N-methyl-d-aspartate receptor-mediated excitotoxicity in a mouse model of Huntington's disease. Neuron, 33: 849–860.

Zuccato, C., Ciammola, A., Rigamonti, D., Leavitt, B.R., Goffredo, D., Conti, L., MacDonald, M.E., Friedlander, R.M., Silani, V., Hayden, M.R., Timmusk, T., Sipione, S. and Cattaneo, E. (2001) Loss of huntingtin-mediated BDNF gene transcription in Huntington's disease. Science, 293: 493–498.

Neurotrophins, PNS and peripheral tissues

Progress in Brain Research, Vol. 146
ISSN 0079-6123

CHAPTER 15

Neural crest development and neuroblastoma: the genetic and biological link

Akira Nakagawara*

Division of Biochemistry, Chiba Cancer Center Research Institute, Chiba 260-8717, Japan

Abstract: Neuroblastoma is one of the most common pediatric solid tumors originating from the sympathoadrenal lineage of neural crest. The tumor shows extremely different clinical phenotypes such as spontaneous regression on one hand and aggressive growth on the other hand. The different biological behavior of neuroblastoma appears to be determined by the genetic abnormalities including amplification of *MYCN* oncogene, DNA ploidy and some allelic imbalances. However, the spontaneous regression of neuroblastoma mimics the programmed cell death normally occurring in developing sympathetic cells expressing both TrkA tyrosine kinase A and p75^{NTR} neurotrophin receptor. Indeed, TrkA expression is the most important factor related to the induction of tumor cell differentiation and/or programmed cell death because without its expression spontaneous regression of neuroblastoma never occurs. Thus, the enigmatic clinical behaviors of neuroblastoma are strictly linked to the molecular mechanism of neural crest development.

Keywords: neuroblastoma; NGF; TrkA; p75^{NTR}; *MYCN* oncogene; MYCN oncoprotein; stem cells

Neuroblastoma, a neural crest tumor in childhood

Neuroblastoma is an embyonic tumor originating from the sympathoadrenal lineage of neural crest and one of the most common solid tumors found in children (Bolande, 1974). Its incidence is about 1/8000 births and there is no significant difference among U.S., Europe and Japan. However, after beginning the mass screening to test the urine for the levels of catecholamine metabolites (VMA and HVA) in Japan in 1985 (Sawada et al., 1984), the incidence of neuroblastoma has almost doubled without decreasing the number of the sporadic tumors (Bessho, 1996). This strongly suggested the actual presence of '*in situ* neuroblastoma', which was first

proposed by Beckwith and Perrin (1963), during the development of sympathetic neurons in human fetuses. They described the detection of '*in situ* neuroblastoma' in developing human embryos at the incidence of more than 40 times that of sporadic neuroblastomas, but most of them regressed spontaneously. Therefore, it is highly possible that we detect a part of the '*in situ* neuroblastomas' by mass screening, most of which otherwise regress without giving any therapy. However, at this moment it is unclear whether the regression of *in situ* neuroblastoma is due to the developmentally regulated programmed cell death of neuronal cells.

The sporadic neuroblastomas clinically found are divided into several subsets according to the clinical behavior, biological markers and genetic abnormalities (Brodeur, 2003). One of the most important clinical factors is the patient's age. The tumors found in the patients under one year of age are usually favorable and take a good clinical course to cure. On the other hand, many of the tumors symptomatically

*Correspondence to: A. Nakagawara, Division of Biochemistry, Chiba Cancer Center Research Institute, 666-2 Nitona, Chuoh-ku, Chiba 260-8717, Japan. Tel.: +81-43-264-5431; Fax: +81-43-265-4459; E-mail: akiranak@chiba-ccri.chuo.chiba.jp

DOI: 10.1016/S0079-6123(03)46015-9

found in the patients over one year of age are poor prognostic and eventually kill the patients. Among the biological markers so far found, expression of the TrkA tyrosine kinase A receptor, as well as p75 neurotrophin receptor (p75^{NTR}) expression, is the most important indicator of prognosis (Nakagawara et al., 1992 , 1993). TrkA is a high-affinity receptor for nerve growth factor (NGF), and p75^{NTR} is its low-affinity receptor. The high levels of TrkA expression are strongly associated with favorable prognosis, whereas its decreased levels are significantly correlated with poor prognosis (Nakagawara et al., 1993). The important genetic markers include DNA ploidy, amplification of the *MYCN* oncogene and an allelic loss of the distal region of chromosome 1p (1p36) (Westermann and Schwab, 2002; Brodeur, 2003). Contrary to the other cancers, neuroblastomas

with hyperdiploid karyotype show a good prognosis, while those with *MYCN* amplification and/or deletion of chromosome 1p36 are strongly associated with poor prognosis. The combination of these strong prognostic indicators segregates the subsets of neuroblastoma with different clinical behavior.

Figure 1 shows three types of neuroblastoma subset. Fig. 1 (left) demonstrates a stage 1 tumor originated from the adrenal gland in a patient under one year of age. The tumor is well encapsulated without metastasis. This type of neuroblastoma usually expresses high levels of TrkA and shows triploid DNA pattern with a single copy of *MYCN*. It clinically regresses spontaneously but very slowly. The baby in Fig. 1 (middle) is the patient with stage 4s neuroblastoma. The immature tumor cells occupy

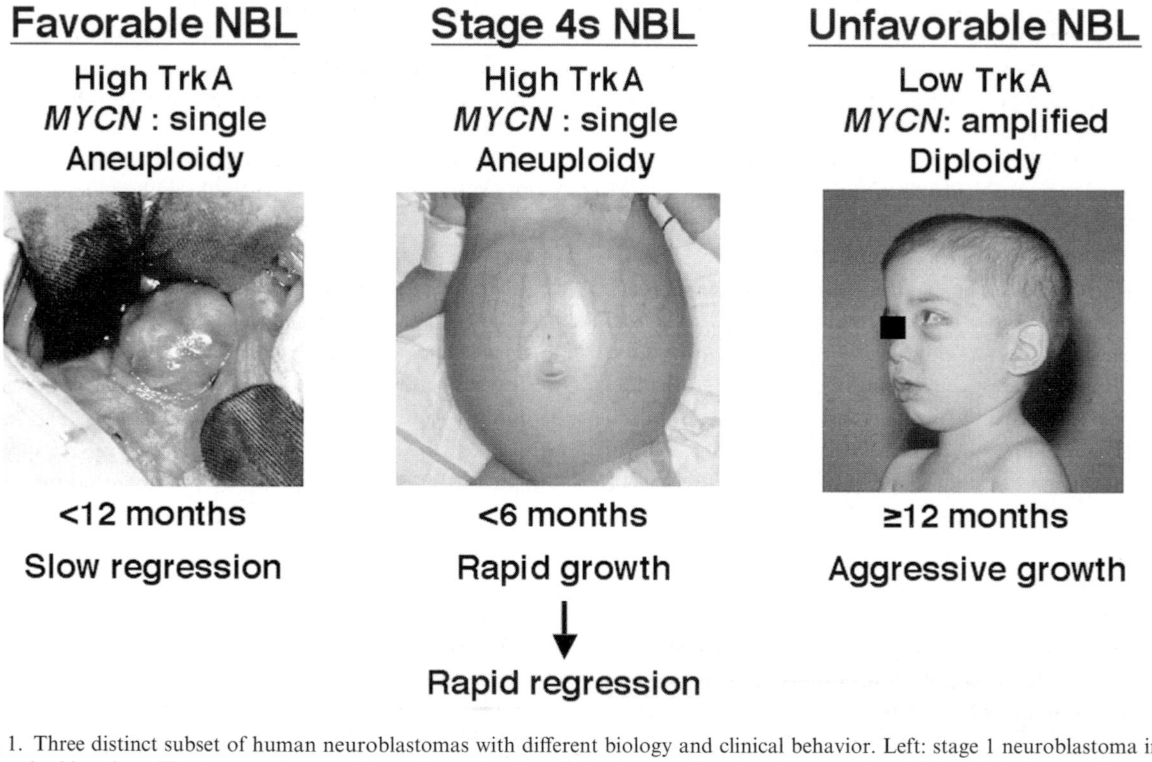

Favorable NBL	Stage 4s NBL	Unfavorable NBL
High TrkA	High TrkA	Low TrkA
MYCN : single	*MYCN* : single	*MYCN*: amplified
Aneuploidy	Aneuploidy	Diploidy

<12 months	<6 months	≥12 months
Slow regression	Rapid growth	Aggressive growth

↓

Rapid regression

Fig. 1. Three distinct subset of human neuroblastomas with different biology and clinical behavior. Left: stage 1 neuroblastoma in a 7-month-old patient. The tumor originated from the right adrenal gland is small and well encapsulated. This kind of neuroblastoma usually regresses spontaneously. Middle: stage 4s neuroblastoma in a one-month-old baby. The primary tumor is located at the left adrenal gland. The liver is extremely enlarged and occupied by the tumor cells. The neuroblastoma cells are also positive in the bone marrow. The abdominal distension often oppresses the diaphragm to induce dyspnea. In a typical stage 4s neuroblastoma, the rapid tumor growth suddenly stops and starts to regress spontaneously. Right: stage 4 neuroblastoma in a 3-year-old boy. The tumor cells originated from the adrenal gland metastasize to long bones, skull and orbita with protrusion of the eye. The tumor cells show low TrkA expression, amplification of *MYCN*, diploid karyotype and deletion of the distal region of chromosome 1p.

the adrenal gland, liver and bone marrow (sometimes even skin), and rapidly grow at an early clinical stage. However, one day the tumor cells suddenly stop growing and start to regress spontaneously. This seems like just a miracle. The stage 4s tumor also shows high TrkA expression, triploidy and no amplification of *MYCN*. In contrast, the advanced stage of neuroblastoma shown in Fig. 1 (right) usually occurs in the patient over one year of age and metastasizes to bones and distant lymph nodes and eventually kill the patient. In this type of neuroblastoma, TrkA expression is strongly downregulated, the DNA ploidy pattern is diploid, and *MYCN* is amplified.

Genetic abnormalities of neuroblastoma

Neuroblastoma has many types of genetic abnormalities including chromosomal aneuploidy, gene amplifications, deletions, mutations, and deregulated DNA methylations. However, the pattern of the genetic aberration is different among the subsets, especially between those with favorable and unfavorable prognosis (Westermann and Schwab, 2002; Brodeur, 2003). The tumors with a tendency to regress spontaneously usually have triploidy but few abnormalities in the genome. On the other hand, the tumors with aggressive growth show a diploid or tetraploid karyotype, frequent amplification of *MYCN* oncogene, and chromosomal deletion of 1p36. The frequent gain of the chromosome 17q is reported to be associated with poor prognosis, however, it is also

commonly observed in the tumors with favorable prognosis (Tomioka et al., 2003). The loss of heterozygosity at the chromosome 11q23 is reported to be frequent in the intermediate type of neuroblastoma in advanced stages with a single copy of *MYCN* and variable levels of TrkA expression (Guo et al., 1999). Thus, the subsets with different clinical behavior may be defined by the combination of the genomic aberrations.

Molecular and biological bases of neuroblastoma

Figure 2 shows a scheme of migration of the developing neural crest-derived cells, which segregate into several lineages such as melanocytes, sensory neurons, enteric neurons, and sympathetic ganglion cells. However, neuroblastoma never occurs in the other tissues than sympathetic ganglion or adrenal medulla. This suggests that the genetic events to cause neuroblastoma occur after the cell fate determination directing to sympathetic differentiation. The most likely candidate molecule to decide the direction of sympathetic differentiation at this moment is a basic helix-loop-helix transcription factor MASH1 which is transiently expressed during the neural development (Guillemot et al., 1993). In human neuroblastomas, MASH1/hASH1 is kept overexpressed (Soderholm et al., 1999; Ichimiya, et al., 2001). Interestingly, induction of neuroblastoma cell differentiation in culture by treating with retinoic acid decreases the level of MASH1 mRNA. These suggest that the

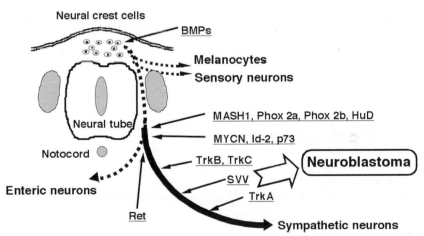

Fig. 2. Neuroblastoma occurs only from the sympathoadrenal lineage of neural crest. The important molecules regulating the sympathetic development are shown.

oncogenic events occurred in the early stage maintain the cells under arrest of differentiation by keeping the MASH1 expression at high levels, although this hypothesis must be proved. Even though the precise mechanism to regulate the oncogenic events of neuroblastoma and normal sympathetic differentiation is still elusive, it may be true that the targeting of the specific genes in such events is strictly controlled by the developmental program.

According to the accumulating evidence, it is clear that many important genes regulating normal development of sympathetic neurons are targeted to cause neuroblastoma or to modulate its biology. They include the *MYCN* gene encoding a basic helix-loop-helix transcription factor (Brodeur et al., 1984; Schwab et al., 1984), Id-2, a target of MYCN and a negative regulator of basic helix-loop-helix transcription factors (Iasorella et al., 2000), MASH1 (Soderholm et al., 1999; Ichimiya et al., 2001), Phox2a and Phox2b, the homeotic proteins functioning with MASH1 (Kuno et al., unpublished data), and the downstream receptors such as Trk family members (Nakagawara et al., 1993) and Ret (Hishiki et al., 1998). These suggest that the important regulators of sympathetic differentiation are targeted to cause or maintain the cancerous status of neuroblastoma. This idea can be extended to the possible link between developmentally regulated programmed cell death of sympathetic neurons and spontaneous regression of neuroblastoma, because in both phenomena, expression of TrkA receptor is necessary (Nakagawara, 1998a, 2001). In other words, TrkA expression is almost exclusively required to induce spontaneous regression of neuroblastoma.

NGF family signaling in neuroblastoma

The NGF signals and their depletion strongly regulate survival and death of the normal sympathetic neurons, respectively. Similarly, it has recently become obvious that the NGF family signals strongly regulate the biology of neuroblastoma. Most neuroblastomas with favorable prognosis express high levels of both TrkA and p75[NTR] and functionally respond to exogenous NGF by extending neurites and promoting survival in primary culture (Nakagawara et al., 1993). The association between high levels of expression of TrkA and/or p75[NTR] and

favorable outcome is statistically significant in primary human neuroblastomas. On the contrary, in aggressive neuroblastomas with *MYCN* amplification in advanced stages, expression of TrkA is extremely downregulated. The many studies about the role of Trk signaling in neuroblastoma cell lines also suggest that the intracellular TrkA signal is disturbed even though autophosphorylation of TrkA is induced by addition of NGF (Nakagawara et al., 1994). Thus, for the gain of growth advantage, the aggressive neuroblastoma cells appear to shut off the TrkA signal. Instead, they utilize a functional brain-derived neurotrophic factor (BDNF) and/or neurotrophin-4 (NT-4)/TrkB signaling system in an autocrine manner (Nakagawara et al., 1994). This BDNF/TrkB autocrine system also promotes invasion and metastasis in advanced tumors (Matsumoto et al., 1995). These suggest that spontaneous regression occurs only in neuroblastoma with high levels of TrkA expression and is induced by depletion of NGF within the tumor. The aggressive neuroblastoma cells seem to escape from the control by NGF, but to take advantage of the BDNF/TrkB autocrine loop for promotion of survival.

The family of glial cell line-derived neurotrophic factor (GDNF) mediates another important extracellular signal to regulate the survival of sympathetic neurons. Many neuroblastoma cells express the GDNF family receptors (Ret, GFRα-1, -2 and -3) and functionally respond to their ligands (GDNF, neurturin and artemin) in the primary culture (Hishiki et al., 1998). However, their expression and the responsiveness to the ligands are not associated with the disease stages or prognosis.

The other neurotrophic factors, pleiotrophin (PTN) and midkine (MK), may also be important in regulating neuroblastoma biology (Nakagawara et al., 1995). The expression of PTN is high in favorable neuroblastomas, whereas that of MK is high in all primary neuroblastomas. However, their functional roles in neuroblastoma are currently unknown.

Role of p53 and p73 in life and death of neuroblastoma

Pozniak et al. (2000) have recently reported about the crucial role of the tumor suppressor p53 and its family member p73 in regulating survival and apoptosis during the induction of programmed cell death in mouse sympathetic neurons. Life and death of the sympathetic cervical ganglion (SCG) neurons are regulated by the balance between the levels of p53 and ΔNp73, an NH$_2$-terminally truncated dominant-negative form of p73. In neuroblastoma, p53 is not mutated but localized in the cellular cytoplasm especially in advanced stage tumors (Moll et al., 1995). Just recently, the anchoring molecule of p53 in the cytoplasm has been identified as Parc which is a structurally E3 ubiquitin ligase but binds to and stabilizes p53 (Nikolaev et al., 2003). It is interesting that the apoptosis-inducing stresses often trigger nuclear translocation of cytoplasmic p53 in neuro-blasatoma cell lines (Ostermeyer et al., 1996).

p73 is the first family member of p53 and has occasionally been discovered as a candidate tumor suppressor of neuroblastoma (Kaghad et al., 1997). It is mapped to chromosome 1p36.2-3 which is commonly deleted in many aggressive neuroblas-tomas with *MYCN* amplification. The extensive mutation search has revealed that p73 is not mutated in many cancers including neuroblastoma (Ikawa et al., 1999). However, we found two mutations of the COOH-terminally located proline residues, one was somatic and the other germline. Nevertheless, most primary neuroblastomas have no mutation of p73 (Ichimiya et al., 1999).

Interestingly, in many malignant solid tumors, p73 has satisfactorily shown to be upregulated, though it has functionally the apoptosis-inducing ability like p53. We and the other investigators have recently found that p73 can bind to the Δ*Np73* proper pro-moter and induce transcription of which possesses the oncogenic function (Nakagawa et al., 2002). In addition, Δ*Np73* binds to both wild type p53 and p73 to suppress their apoptosis-inducing function (Nakagawa et al., 2002). These observations are very important because they might at least in part explain how the cancers without p53 mutation do develop the tumors with poor prognosis. In neuroblastoma, Casciano et al. (2002) have reported that both *p73* and Δ*Np73* are highly expressed in aggressive rather than favorable tumors.

Figure 3 shows the current summary of the signals for induction of neuronal apoptosis. Both p53 and p73 as well as ΔNp73 might be cooperatively func-tioning to regulate the programmed cell death of sympathetic as well as neuroblastoma cells.

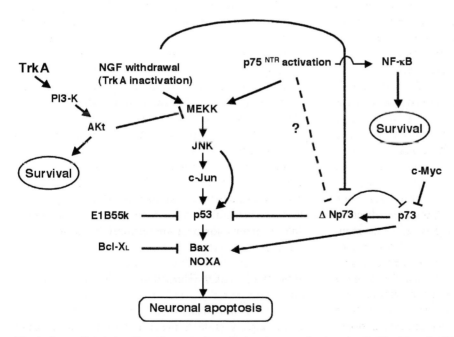

Fig. 3. Intracellular signaling of neuronal survival and apoptosis: the role of p53, p73 and ΔNp73.

Comprehensive genomics to identify the novel genes

To date, several genes functioning as landmark regulators in different subsets of neuroblastoma, such as *MYCN*, *MASH1*, *Trk*, *p53* and *p73*, have been identified. However, in the present postgenome era, we can try comprehensive approach to identify the important genes in a mass scale. For that purpose, we generated oligo-capping cDNA libraries from the primary neuroblastoma tissues of three different subsets as shown in Fig. 1. In total, we obtained 6252 gene clusters from 9729 clones randomly picked up from the cDNA libraries, among which 34% were novel genes with unknown function. The expression profiles of each subset of neuroblastoma were extremely different. By using the semi-quantitative reverse transcriptase (RT)-PCR, we have identified 757 genes differentially expressed between favorable (stage 1, high expression of *TrkA* and a single copy of *MYCN*) and unfavorable (stage 3 or 4, decreased levels of *TrkA* expression and amplification of *MYCN*) neuroblastomas. Among them, 502 are novel genes. [The results of our neuroblastoma cDNA project excluding those obtained from the stage 4s cDNA libraries were published elsewhere (Ohira et al., 2000; Ohira et al., in press).]

The expression profile of known genes was very different among the three subsets of neuroblastoma. The favorable subset frequently expressed neuronal specific genes including those related to neural differentiation, synapse, catecholamine metabolism and protein degradation. On the other hand, the unfavorable subset expressed many genes related to cell cycle control, protein synthesis and transcriptional regulation. The 4s tumor contained apoptosis-related genes, oncogenes and HLA family members which might be derived from the infiltrated lymphocytes into the tumor.

The 757 differentially expressed genes were strongly implicated in understanding of neuroblastoma biology. Of interest, vast majority of those genes was expressed at higher levels in the favorable subset as compared to the unfavorable one. The genes highly expressed in the favorable subset contained those related to neuronal differentiation, migration, cell–cell interaction, protein degradation, synaptic vesicles, catecholamine metabolism and intracellular signaling (Ohira et al., 2000; Ohira et al., in press).

Most of them define the neuronal-specific phenotype and maintain the neuronal function. They also included heat shock proteins and ubiquitin/proteasome-related molecules that might sense the stress. On the other hand, only about 10% of the differential genes were expressed at high levels in the unfavorable subset. The protein products of such known genes contained many transcriptional and translational regulators including oncoproteins.

We also applied the primary culture of newborn mouse SCG neurons for screening those genes which change during the NGF-induced differentiation and/or the NGF depletion-induced apoptosis. This approach has identified 33 genes related to the former and 56 genes changeable during the latter (Isogai et al., unpublished data).

Our unique approach has identified more than several interesting genes as well as their products which include Nbla0219/BMCC1, a novel proapoptotic molecule with BCH domain, P-loop and coiled-coil domain, and Nbla0078/NEDL1, a novel E3 ubiquitin ligase with the HECT domain. The other interesting genes whose analyses have been published during our studies also include human *RIM/Nbla0761*, a Rab3-interacting molecule in the synaptic vesicles, *XCE/Nbla3145*, a new endothelin-converting enzyme and *FOG2/Nbla3139*, a coactivator of GATA transcription factor. Currently, a total of 7000 genes we cloned from the primary neuroblastomas are being fixed on the slide glass for cDNA microarray analysis.

Thus, our neuroblastoma cDNA project has provided enormous information and the gene materials for understanding of neuroblastoma biology as well as the molecular mechanism of neural crest development.

Developmental time axis and oncogenic events

Our neuroblastoma cDNA project has provided us with tremendous information about the genes expressed in different subsets with characteristic biology (Ohira et al., 2000; Ohira et al., in press). It suggested the presence of a kind of rule in the expression patterns of the subset-specific genes. Figure 4 shows the groups of genes expressed along the time axis of sympathetic neuron development. During the early

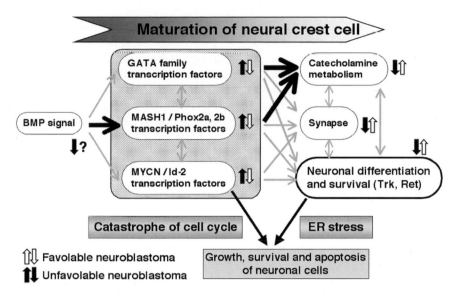

Fig. 4. The gene expression cascade along the time axis during the neural crest development, which is expected from the results obtained from the neuroblastoma cDNA project. Many transcription factors are upregulated in the unfavorable neuroblastoma, whereas the genes related to the terminal differentiation of neuron are upregulated in the favorable neuroblastoma. ER, endoplasmic reticulum.

stages of development, many transcription factors seem to function in deciding the direction of differentiation as well as regulating cell growth and survival of neural crest-derived cells. Interestingly, many genes highly expressed in unfavorable NBLs contain transcription factors and the components of their complexes. They include MYCN and Id family transcription factors that link to the regulation of Rb and p53 and regulate cell growth and apoptosis (Lasorella et al., 2000). The another basic helix-loop-helix transcription factor, MASH1, is constitutively activated in neuroblastoma, and by collaborating with Phox2a and Phox2b, it may regulate the arrest of differentiation in an unfavorable neuroblastomas (Kuno et al., unpublished data). Our neuroblastoma cDNA project has also revealed that there may be a neuronal cassette of GATA transcription factor complex that controls growth and differentiation of sympathetic progenitor cells (Ohira et al., 2003). Some molecules in this complex are upregulated in unfavorable neuroblastomas (Aoyama et al., manuscript in preparation). Thus, many important components in the transcriptional regulators appear to be highly expressed in unfavorable neuroblastomas and function to regulate the tumor cell growth or the status of de-differentiation.

On the other hand, most of a remarkable number of the genes expressed at high levels in favorable neuroblastomas encode the molecules that are necessary to maintain the neuronal function. They may be necessary for keeping catecholamine metabolism, synapse formation, neuronal cell survival, etc. We have also found many genes related to the ubiquitin-proteasome pathway and heat shock proteins in favorable neuroblastomas. They might be involved in induction of apoptosis triggered by endoplasmic reticulum stress.

Thus, the pattern of the differentially expressed genes in neuroblastoma subsets suggests the changes in the developmentally regulated gene expression along the time axis.

The hypothesis of neuroblastoma stem cells

According to the result of neuroblastoma mass screening, it may be true that most of the early stage neuroblastomas do not progress to the advanced tumors. In addition, the study of molecular mechanism linking neural development and neuroblastoma has revealed that the aggressive neuroblastoma occuring in an older patient seems to be arrested at the earlier stage of differentiation than the favorable tumor found in younger patient (Nakagawara, 1998b). These suggest a difference between the progenitor cells, or the cancer stem cells, of neuroblastomas with dissimilar genetic and biological characteristics. The classic observation by Beckwith and Perrin (1963) suggests that the migrating neural crest cells destined to differentiate to sympathetic neurons first enter the sympathetic ganglion, and a part of them further migrate to reach the adrenal medulla where the concentration of glucocorticoid is kept high. The detailed investigations done by neurobiologists have shown high expression levels of both TrkA and p75^{NTR} in almost all of the neurons in the sympathetic ganglion as well as their dependence on the target-derived NGF for survival (trophic theory) (McMahon et al., 1994). This may be reflected on that neuroblastomas originated from the sympathetic ganglia show better prognosis than those from the adrenal medulla and that the former tumors usually express high levels of TrkA and p75^{NTR}. On the other hand, aggressive neuroblastomas with *MYCN* amplification almost exclusively originate from the adrenal medulla in patients older than one year (Nakagawara et al., 1990). The most imporant fact is the hyperdiploid karyotype of more

than half of the favorable neuroblastomas (Brodeur, 2003), suggesting that the mitotic instability or dysfunction is the main event during the tumorigenesis. On the other hand, in aggressive tumors occurring in the adrenal medulla, the DNA ploidy is diploid and the regional abnormalities of the chromosome induced by the genetic instability are the main phenotypes of the genome. Those tumor cells usually have an autocrine signaling loop of BDNF and TrkB receptor, the dependency on which has been reported to be the rather immature phenotype of developing sympathetic neurons (Pinon et al., 1996).

Figure 5 shows the hypothetical scheme of the cancer stem cells A and B of neuroblastoma. The cancer stem cells A, which express high levels of TrkA and p75^{NTR} and are dependent on NGF for survival, are present in both sympathetic ganglia and adrenal medulla. They are prone to have a mitotic dysfunction to get a proliferating advantage. However, those cancer stem cells are still dependent on NGF to survive and to differentiate. The hypothesis recently proposed by Kaneko and Knudson (2000) in terms of the DNA ploidy pattern in neuroblastoma may be attractive to explain the mitotic dysfunction found in favorable neuroblastomas. In addition, those stem cells may sustain the function at the late stage

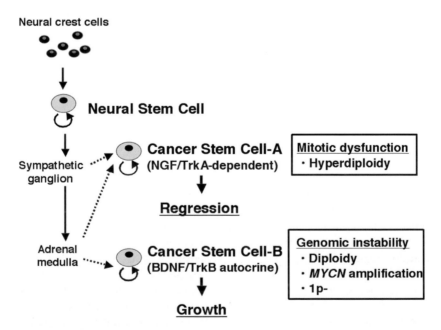

Fig. 5. The hypothesis of cancer stem cells in neuroblastoma.

of sympathetic development by maintaining the NGF-dependency. On the other hand, the cancer stem cells B may be localized in the adrenal medulla and are prone to have a genetic instability to cause *MYCN* amplification and allelic loss of chromosome 1p36. They may retain the phenotype of rather immature stage of sympathetic development by expressing TrkB but not TrkA. The comprehensive approach such as our neuroblastoma cDNA project and an array CGH methodology may help to prove the above hypothesis in the near future.

Conclusion

Neuroblastoma is an enigmatic tumor showing multiple clinical behaviors. However, the recent advances in neuroblastoma research have revealed that the molecular mechanism of neural crest development may strictly regulate the tumorigenesis as well as biology of different subsets of neuroblastoma. Indeed, the Trk family receptors and MYCN oncoprotein appear to link between neural development and cancer. Thus, unveiling the molecular mechanism of neuroblastoma should extremely help to understand how the neurons develop, differentiate, survive and die.

Abbreviations

ARTN	artemin
BDNF	brain-derived neurotrophic factor
GDNF	glial cell line-derived neurotrophic factor
MK	midkine
NGF	nerve growth factor
NT-4	neurotrophin-4
NTRN	neurturin
p75NTR	p75 neurotrophin receptor
PTN	pleiotrophin
SCG	superior cervical ganglion
Trk	tyrosine kinase

References

Beckwith, J.B. and Perrin, E.V. (1963) In situ neuroblastomas: A contribution to the natural history of neural crest tumors. Am. J. Pathol., 43: 1089–1101.

Bessho, F. (1996) Where should neuroblastoma mass screening go? Lancet, 348: 1682–1687.

Bolande, R.P. (1974) The neurocristopathies: A unifying concept of disease arising in neural crest maldevelopment. Human Pathol., 5: 409–429.

Brodeur, G.M., Seeger, R.C., Schwab, M., Varmus, H.E. and Bishop, J.M. (1984) Amplification of N-myc in untreated human neuroblastomas correlates with advanced disease stage. Science, 224: 1121–1124.

Brodeur, G.M. (2003) Neuroblastoma: biological insights into a clinical enigma. Nat. Rev. Cancer, 3: 203–216.

Casciano, I., Mazzocco, K., Boni, L., Pagnan, G., Banelli, B., Allemanni, G., Ponzoni, M., Tonini, G.P. and Romani, M. (2002) Expression of DeltaNp73 is a molecular marker for adverse outcome in neuroblastoma patients. Cell Death Differ., 9: 246–251.

Guillemot, F., Lo, L.C., Johnson, J.E., Auerbach, A., Anderson, D.J. and Joyner, A.L. (1993) Mammalian achaete-scute homolog 1 is required for the early development of olfactory and autonomic neurons. Cell, 75: 463–476.

Guo, C., White, P.S., Weiss, M.J., Hogarty, M.D., Thompson, P.M., Stram, D.O., Gerbing, R., Matthay, K.K., Seeger, R.C., Brodeur, G.M. and Maris, J.M. (1999) Allelic deletion at 11q23 is common in MYCN single copy neuroblastomas. Oncogene, 18: 4948–4957.

Hishiki, T., Nimura, Y., Isogai, E., Kondo, K., Ichimiya, S, Nakamura, Y., Ozaki, T., Sakiyama, S., Hirose, M., Seki, N., Takahashi, H., Ohnuma, N., Tanabe, M. and Nakagawara, A. (1998) Glial cell line-derived neurotrophic factor/neurturin-induced differentiation and its enhancement by retinoic acid in primary human neuroblastomas expressing c-Ret, GFR alpha-1, and GFR alpha-2. Cancer Res., 58. 2158–2165.

Ichimiya, S., Nimura, Y., Kageyama, H., Takada, N., Sunahara, M., Shishikura, T., Nakamura, Y., Sakiyama, S., Seki, N., Ohira, M., Kaneko, Y., McKeon, F., Caput, D. and Nakagawara, A. (1999) p73 at chromosome 1p36.3 is lost in advanced stage neuroblastoma but its mutation is infrequent. Oncogene, 18: 1061–1066.

Ichimiya, S., Nimura, Y., Seki, N., Ozaki, T., Nagase, T. and Nakagawara, A. (2001) Downregulation of hASH1 is associated with the retinoic acid-induced differentiation of human neuroblastoma cell lines. Med. Pediatr. Oncol., 36: 132–134.

Ikawa, S., Nakagawara, A. and Ikawa, Y. (1999) p53 family genes: structural comparison, expression and mutation. Cell Death Differ., 6: 1154–1161.

Kaghad, M., Bonnet, H., Yang, A., Creancier, L., Biscan, J.C., Valent, A., Minty, A., Chalon, P., Lelias, J.M., Dumont, X., Ferrara, P., McKeon, F. and Caput, D. (1997) Monoallelically expressed gene related to p53 at 1p36, a region frequently deleted in neuroblastoma and other human cancers. Cell, 90: 809–819.

Kaneko, Y. and Knudson, A.G. (2000) Mechanism and relevance of ploidy in neuroblastoma. Genes Chrom. Cancer, 29: 89–95.

Lasorella, A., Noseda, M., Beyna, M., Yokota, Y. and Iavarone, A. (2000) Id2 is a retinoblastoma protein target and mediates signaling by Myc oncoproteins. Nature, 407: 592–598.

Matsumoto, K., Wada, R.K., Yamashiro, J.M., Kaplan, D.R. and Thiele, C.J. (1995) Expression of brain-derived neurotrophic factor and p145TrkB affects survival, differentiation, and invasiveness of human neuroblastoma cells. Cancer Res., 55: 1798–1806.

McMahon, S.B., Armanini, M.P., Ling, L.H. and Phillips, H.S. (1994) Expression and coexpression of Trk receptors in subpopulations of adult primary sensory neurons projecting to identified peripheral targets. Neuron, 12: 1161–1171.

Moll, U.M., LaQuaglia, M., Benard, J. and Riou, G. (1995) Wild-type p53 protein undergoes cytoplasmic sequestration in undifferentiated neuroblastomas but not in differentiated tumors. Proc. Natl. Acad. Sci. USA, 92: 4407–4411.

Nakagawa, T., Takahashi, M., Ozaki, T., Watanabe, K.K., Todo, S., Mizuguchi, H., Hayakawa, T. and Nakagawara, A. (2002) Autoinhibitory regulation of p73 by Delta Np73 to modulate cell survival and death through a p73-specific target element within the Delta Np73 promoter. Mol. Cell. Biol., 22: 2575–2585.

Nakagawara, A., Ikeda, K., Higashi, K. and Sasazuki, T. (1990) Inverse correlation between N-myc amplification and catecholamine metabolism in children with advanced neuroblastoma. Surgery, 107: 43–49.

Nakagawara, A., Arima, M., Azar, C.G., Scavarda, N.J. and Brodeur, G.M. (1992) Inverse relationship between trk expression and N-myc amplification in human neuroblastomas. Cancer Res., 52: 1364–1368.

Nakagawara, A., Arima-Nakagawara, M., Scavarda, N.J., Azar, C.G., Cantor, A.B. and Brodeur, G.M. (1993) Association between high levels of expression of the TRK gene and favorable outcome in human neuroblastoma. N. Engl. J. Med., 328: 847–854.

Nakagawara, A., Azar, C.G., Scavarda, N.J. and Brodeur, G.M. (1994) Expression and function of TRK-B and BDNF in human neuroblastomas. Mol. Cell. Biol., 14: 759–767.

Nakagawara, A., Milbrandt, J., Muramatsu, T., Deuel, T.F., Zhao, H., Cnaan, A. and Brodeur, G.M. (1995) Differential expression of pleiotrophin and midkine in advanced neuroblastomas. Cancer Res., 55: 1792–1797.

Nakagawara, A. (1998a) Molecular basis of spontaneous regression of neuroblastoma: role of neurotrophic signals and genetic abnormalities. Hum. Cell, 11: 115–124.

Nakagawara, A. (1998b) The NGF story and neuroblastoma. Med. Pediatr. Oncol., 31: 113–115.

Nakagawara, A. (2001) Trk receptor tyrosine kinases: a bridge between cancer and neural development. Cancer Lett., 169: 107–114.

Nikolaev, A.Y., Li, M., Puskas, N., Qin, J. and Gu, W. (2003) Parc: a cytoplasmic anchor for p53. Cell, 112: 29–40.

Ohira, M., Shishikura, T., Kawamoto, T., Inuzuka, H., Morohashi, A., Takayasu, H., Kageyama, H., Takada, N., Takahashi, M., Sakiyama, S., Suzuki, Y., Sugano, S., Kuma, H., Nozawa, I. and Nakagawara, A. (2000) Hunting the subset-specific genes of neuroblastoma: expression profiling and differential screening of the full-length-enriched oligo-capping cDNA libraries. Med. Pediatr. Oncol., 35: 547–549.

Ohira, M., Morohashi, A., Inuzuka, H., Shishikura, T., Kawamoto, T., Kageyama, H., Nakamura, Y., Isogai, E., Takayasu, H., Sakiyama, S., Suzuki, Y., Sugano, S., Goto, T., Sato, S. and Nakagawara, A. (2003) Expression profiling and characterization of 4200 genes cloned from primary neuroblastomas: Identification of 305 genes differentially expressed between favorable and unfavorable subsets. Oncogene, in press.

Ostermeyer, A.G., Runko, E., Winkfield, B., Ahn, B. and Moll, U.M. (1996) Cytoplasmically sequestered wild-type p53 protein in neuroblastoma is relocated to the nucleus by a C-terminal peptide. Proc. Natl. Acad. Sci. USA, 93: 15190–15194.

Pinon, L.G., Minichiello, L., Klein, R. and Davies, A.M. (1996) Timing of neuronal death in trkA, trkB and trkC mutant embryos reveals developmental changes in sensory neuron dependence on Trk signaling. Development, 122: 3255–3261.

Pozniak, C.D., Radinovic, S., Yang, A., McKeon, F., Kaplan, D.R. and Miller, F.D. (2000) An anti-apoptotic role for the p53 family member, p73, during developmental neuron death. Science, 289: 304–306.

Sawada, T., Hirayama, M., Nakata, T., Takeda, T., Takasugi, N., Mori, T., Maeda, K., Koide, R., Hanawa, Y., Tsunoda, A., et al. (1984) Mass screening for neuroblastoma in infants in Japan. Interim report of a mass screening study group. Lancet, 2: 271–273.

Schwab, M., Varmus, H.E., Bishop, J.M., Grzeschik, K.H., Naylor, S.L., Sakaguchi, A.Y., Brodeur, G. and Trent, J. (1984) Chromosome localization in normal human cells and neuroblastomas of a gene related to c-myc. Nature, 308: 288–291.

Soderholm, H., Ortoft, E., Johansson, I., Ljungberg, J., Larsson, C., Axelson, H. and Pahlman, S. (1999) Human achaete-scute homologue 1 (HASH-1) is downregulated in differentiating neuroblastoma cells. Biochem. Biophys. Res. Commun., 256: 557–563.

Tomioka, N., Kobayashi, H., Kageyama, H., Ohira, M., Nakamura, Y., Sasaki, F., Todo, S., Nakagawara, A. and Kaneko, Y. (2003) Chromosomes that show partial loss or gain in near-diploid tumors coincide with chromosomes that show whole loss or gain in near-triploid tumors: evidence suggesting the involvement of the same genes in the tumorigenesis of high- and low-risk neuroblastomas. Genes Chrom. Cancer, 36: 139–150.

Westermann, F. and Schwab, M. (2002) Genetic parameters of neuroblastomas. Cancer Lett., 184: 127–147.

Progress in Brain Research, Vol. 146
ISSN 0079-6123

CHAPTER 16

Neurotrophin-3 in the development of the enteric nervous system

Alcmène Chalazonitis*

*Department of Anatomy and Cell Biology, Columbia University, College of Physicians and Surgeons,
630W, 168th Street, New York, NY 10032, USA*

Abstract: To date, the only neurotrophin that has been shown to influence the development of the enteric nervous system (ENS) is neurotrophin-3 (NT-3). NT-3 plays an essential role in the development of both the neural-crest-derived peripheral nervous system and the central nervous system (i.e., Chalazonitis, 1996, Mol. Neurobiol., 12: 39–53; Sieber-Blum, 1999, Neurotrophins and the Neural Crest, CRC Press, Boca Raton). This review integrates data obtained from our laboratory and from our collaboration with other investigators that demonstrate a late-acting role for NT-3 in the development of enteric neurons in vitro and in vivo. Studies of the biological actions of NT-3 on enteric neuronal precursors in vitro demonstrate that NT-3 acts directly on the precursor cells and that it also acts in combination with other neurotrophic factors such as glial cell line-derived neurotrophic factor and a ciliary neurotrophic factor-like molecule, to promote the survival and differentiation of enteric neurons and glia. Importantly, bone morphogenetic protein-2 (BMP-2) and BMP-4, members of the transforming growth factor-beta (TGF-β) superfamily, regulate the onset of action of NT-3 during fetal gut development. Analyzes performed on mice deficient in the genes encoding NT-3 or its transducing tyrosine kinase receptor, TrkC, and conversely on transgenic mice that overexpress NT-3 substantiate a physiological role for NT-3 in the development and maintenance of a subset of enteric neurons. There is loss of neurons in both the myenteric and submucosal plexuses of mice lacking NT-3/TrkC signaling and selective hyperplasia in the myenteric plexus of mice overexpressing NT-3. Analyzes performed on transgenic mice that overexpress noggin, a specific BMP-4 antagonist, show significant decreases in the density of TrkC-expressing neurons but significant increase in overall neuronal density of both plexuses. Conversely, overexpression of BMP-4 is sufficient to produce, an increase in the proportion of TrkC-expressing neurons in both plexuses. Overall, our data point to a regulatory role of BMP-4 in the responses of subsets of myenteric and submucosal neurons to NT-3. NT-3 is required for the differentiation, maintenance and proper physiological function of late-developing enteric neurons that are important for the control of gut peristalsis.

Keywords: neural crest; immunoselection; NT-3; TrkC; GDNF; neuropoietic cytokines; bone morphogenetic proteins; noggin; transgenic mice; gut

The enteric nervous system, its organization and the origin of its diversity

Organization and function

The enteric nervous system (ENS) is an independent division of the peripheral nervous system (PNS) and in contrast to the PNS, can function autonomously. The ENS controls the synchronized activity of the longitudinal and circular smooth muscles that regulate peristalsis and controls the secretomotor reflexes of the gut. The pacemaker activity of the interstitial cells of Cajál (ICC) also contributes to intestinal motility (Huizinga et al., 1995; Sanders et al., 1999). Furthermore, an extrinsic innervation of the gut including sensory efferent fibers, and sympathetic and parasympathetic afferent fibers develops later than the ENS. The number of intrinsic

*Correspondence to: A. Chalazonitis, 630W, 168th Street, New York, NY 10032, USA. Tel.: + 1-212-305-8412; Fax: + 1-212-305-3970; E-mail: ac83@columbia.edu

DOI: 10.1016/S0079-6123(03)46016-0

enteric neurons is quite large compared to that of the extrinsic innervation. For example, the human ENS contains more than a 100 millions neurons in the intestine alone (Gershon, 1998), and its extrinsic innervation of ~3000 nerve fibers is tenuous in comparison. In the absence of extrinsic innervation, peristalsis persists along the gut tube, due to the actions of the ENS and ICC cells. Thus, the extrinsic innervation modulates rather than controls peristalsis. The mature ENS is organized into two interconnected plexuses of ganglia: (i) the myenteric/Auerbach's plexus that develops first, and (ii) the submucosal/Meissner's plexus that develops later (Pham et al., 1991). The vast number of neurons that form the ganglia of each of the plexuses, the diversity of their neuronal phenotypes (up to 13 distinct types; Lomax and Furness, 2000) including the enteric astrocytes (Gabela and Trigg, 1984) and the networks of ganglia laid out in a 'chicken wire' pattern repetitively along the entire length of gut, suggest unique developmental regulatory mechanisms. Understanding the signaling mechanisms that govern the development of such a complex nervous system represents a challenge and a fascination. Multiple reviews have previously described the organization of the mature ENS with respect to its functional anatomy (Furness and Costa, 1987; Gershon et al., 1994), its physiology and its neurotransmitter phenotypes (Furness et al., 1988; Sang and Young, 1998; Lomax and Furness, 2000).

Neural crest origin and migration of the enteric precursor cells

It has been known for 30 years that the ENS is entirely derived from the neural crest (Le Douarin, 1982; Le Douarin and Kalcheim, 1999). Neural crest cells follow two major pathways of migration, the vagal and sacral pathways. Crest cells that originate from somites levels C3–C6 follow the vagal pathway and are destined to colonize the entire gut. A second contingent of crest-derived cells that originates from somite levels S28 caudad, follows the sacral pathway, and joins the vagal crest-derived cells to colonize the postumbilical gut (Pomeranz et al., 1991; Burns and Le Douarin, 1998). A third contingent of crest-cells originates at the truncal level and follows a different

migratory route to colonize the esophagus and the cardiac stomach (Durbec et al., 1996).

Enteric crest-derived cells of vagal and sacral, but not truncal origin require GDNF, Ret and GFRα-1 to form the ENS

Vagal and sacral cells both express the tyrosine kinase receptor Ret, which is the transducing receptor for the glial cell line-derived neurotrophic factor (GDNF) family of growth factors. GDNF does not activate Ret directly, but acts by binding to the GPI-anchored receptor GFRα-1 associated with Ret (Jing et al., 1996). In familial forms of Hirschprung's disease (HD), characterized by a terminal colon that is aganglionic, the gene for Ret is mutated (Edery et al., 1994; Romeo et al., 1994; Pasini et al., 1995). Mutation in the genes for GDNF and GFRα-1 can also cause HD (Ivanchuk et al., 1996; Angrist et al., 1998). In addition, mutations in genes encoding other proteins such as endothelin-3 (ET-3) or the endothelin receptor B (ET$_B$) can lead to HD (Puffenberger et al., 1994; Edery et al., 1996). Mice lacking either functional c-ret (Schuchardt et al., 1994), GDNF (Moore et al., 1996; Pichel et al., 1996; Sánchez et al., 1996) or GFRα-1 (Enomoto et al., 1998) lack an ENS caudad to the cardiac stomach. In contrast to the enteric precursors of vagal and sacral origins, the precursors of truncal origin are Ret-independent, since esophagus and cardiac stomach remain innervated in Ret-deficient mice (Durbec et al., 1996).

Diversification of the common, Ret-expressing enteric precursors into neuronal and glial lineages

Although the enteric crest-derived cells that colonize the fetal gut are pluripotent, they are already restricted when compared with the initial population of primary crest cells that migrate out of the neural tube (Rothman et al., 1990, 1993). As the enteric crest-derived cells invade the immature gut (E8 in mice and E9 in rat), all require the master transcription factor Phox2b (Pattyn et al., 1999). Those precursors in gut that originate from vagal and sacral somite levels express nestin, an intermediate filament marker of undifferentiated precursors (Lendhal et al., 1990), and respond to GDNF by proliferating and then differentiating into neurons and glia. As the fetal gut matures, the rate of proliferation of the crest-derived precursors declines, and eventually stops. The arrest of proliferation is staggered so that neuronal precursors stop divide and differentiate by approximately one month after birth (Pham et al., 1991), whereas glial precursors continue to divide even in the adult and enter a quiescent state of differentiation.

Recent experiments have shown that injection of a single Ret-expressing enteric precursor into an aganglionic gut isolated from a *ret* knockout mouse, can generate neurons as well as glia (Natarajan et al., 1999). These observations suggest that bipotent neuron/glia precursors, which express Ret, split as the gut matures, into a pan-neuronal lineage and a glial lineage. The precursors may downregulate expression of nestin, and can then be recognized by either neuronal or glial markers. The neuronal lineage continues to respond to GDNF and switches from nestin-expressing to neuronal filament-expressing (i.e., peripherin) cells. The 'pan-neuronal' lineage, which is Phox-2b- and GDNF-dependent, diverges subsequently into various sublineages that generate distinct neuronal subsets with neurons being born from proliferating precursor cells in staggered rather than synchronous fashion. For instance, serotonergic neurons are the earliest born enteric neurons while CGRP-containing neurons are the latest born (Pham et al., 1991). Moreover, some neurotransmitter phenotypes may be expressed only transiently by neuronal precursors. For instance, neuronal precursors of the 'sympatho-enteric' lineage, controlled by the transcription factor Mash-1 (*M*ammalian *a*chete *s*cute *h*omolog), transiently express the neurotransmitter enzyme tyrosine hydroxylase (TH) (Blaugrund et al., 1996). These 'transient catecholaminergic' (TC) cells, subsequently acquire other stable characteristics such as synthesis of 5-HT, NOS or NPY, or SP or CGRP (Fig. 1).

When enteric crest-derived cells commit to the glial lineage they cease to respond to GDNF, as shown in the more mature crest-derived cells isolated at E14 (Chalazonitis et al., 1998b), and express the glial markers S-100 or GFAP. It is unclear whether all enteric glia loose expression of nestin (Vanderwinden et al., 2002). The loss of responsiveness to GDNF by enteric glia is likely due to a downregulation or loss of Ret expression (Bär et al., 1997) and a switch to

dependency on other factors that promote glial cell development such as the neuregulins (NRG) which bind to the ErbB3 receptor (Meyer and Birchmeier, 1995).

Regulation of responsiveness to neurotrophic factors is necessary, if not sufficient, to support the diversification of enteric lineages

In contrast to other ganglia of the PNS, the enteric ganglia do not form outside of, but rather *within* their targets of innervation. Recently, several studies have focused on the nature and biological actions of the growth and differentiation signals localized within the microenvironment of the bowel that influence and drive the large diversity of enteric neuronal and astrocytic phenotypes (Gershon et al., 1992). The developmental period chosen for these studies follows the start of colonization of the gut by undifferentiated neuronal and glia enteric precursors and spans roughly from mid-gestation (E11–E12 in rodent) until after birth. The pluripotent proliferating neural crest-derived cells, respond to multiple signals known to be present within the bowel. Such signals include extracellular matrix proteins such as laminin (Gershon et al., 1992; Pomeranz et al., 1993; Chalazonitis et al., 1997), diffusible factors, such as NT-3 (Gershon et al., 1993; Chalazonitis et al., 1994) and GDNF (Chalazonitis et al., 1998b), and neuropoietic cytokines such as ciliary neurotrophic factor (CNTF), or leukemia inhibitory factor (LIF) (Chalazonitis et al., 1998a). Another CNTF-like factor that may act at specific stages in combination with NT-3 has also been identified (Elson et al., 2000). Other factors such as ET-3 modulate development of enteric neurons (Baynash et al., 1994; Hearn et al., 1998; Wu et al., 1999) and the neurotransmitter 5-HT has also been reported to exert trophic effects on enteric neurons (Fiorica-Howells et al., 2000).

Overall these studies reveal that the enteric neural crest-derived precursors are subjected to a program of differentiation that appears to follow sequential rather than simultaneous steps. The responsiveness of the enteric neuronal precursors to one growth factor is neither uniform in time, nor do all precursors respond simultaneously to all of the factors present in the gut microenvironment. Instead subsets of

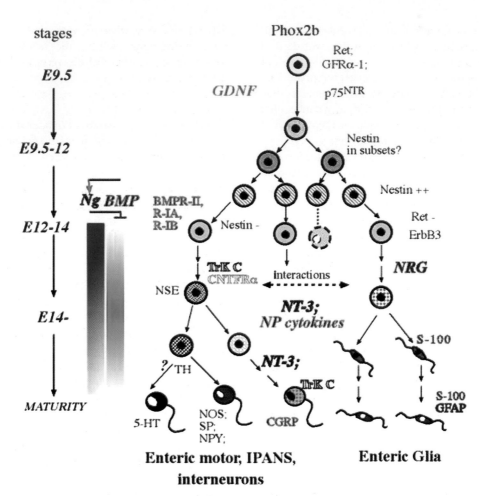

Fig. 1. Temporal model describing the role of NT-3 during development of the ENS (see text). The embryonic and fetal stages examined are defined along the vertical time line on the left. The factors (ligands), distinguished by color code, are indicated in bold capital italics. Their receptors are indicated in bold or plain capitals with the same color code as the corresponding ligands. Phenotypic markers for precursor cells and for the differentiating neurons or glia are indicated in plain capitals. A 'minus' sign depicts markers no longer expressed. A representation of the inhibition of BMP-2 and -4 signaling (faded green) by noggin (Ng), and of the activation of Ng signaling (intense red) by the BMPs in feedback loop, is depicted by two rectangles colored in reverse gradients. Both gradients fade at maturity when noggin and BMP-2 and BMP-4 expression are downregulated The stages from E9.5 to E12 are characterized by the proliferative effects of GDNF on the pluripotent nestin-expressing enteric precursors as they enter and colonize the fetal bowel. At the stage E12–14, when BMP signaling is known to be active, appropriate concentrations of BMP-2 or BMP-4 (intense green) promote crest-derived cells to exit from the cell cycle, differentiate into neurons and increase expression of TrkC. Some of these cells undergo apoptosis (dotted cell body). By stage E14, the common neuron-glia crest-derived cells are divided into the committed pan-neuronal (marked as an example by NSE), and glial lineages. From E14, through late gestation, postnatal and maturity stages, the pan-neuronal lineage further diverges into many sub-lineages that become dependent on late-acting factors such as NT-3 or the neuropoietic (NP) cytokines; in contrast the glial lineage that also responds to the NP cytokines, becomes dependent on neuregulins (NRG). Interactions between neurons and glia increase in importance and are represented by a bidirectional dotted arrow.

neuronal or glial precursors respond to the growth factors at distinct stages of development so that their responsiveness is regulated. For instance, enteric precursors committed to the neuronal lineage when isolated at E14, no longer proliferate in response to GDNF but differentiate instead. In contrast, enteric precursors committed to the glial lineage when isolated at E12 are no longer supported by GDNF and may require other factors such as the neuregulin, glial growth factor 2 (GGF2), which is expressed in developing gut (Chalazonitis, D'Autréaux and Gershon, unpublished observation). ErbB-3, the binding receptor subunit for GGF2 forms an heterodimer with ErbB2, so that ErbB2 activates signal transduction. ErbB3 is expressed in enteric precursors (Pattyn et al., 1999) and mice lacking the gene encoding ErbB3, display a selective loss of enteric astrocytes while neurons numbers are apparently unaffected (Riethmacher et al., 1997).

Determination of the molecular mechanisms that regulate responsiveness of enteric precursors to the multitude of growth factors that they encounter is therefore paramount in understanding phenotypic diversification within the ENS. It is already known that commitment to neuronal 'lineages' is regulated by transcription factors such as Pax 3 and Sox 10 that modulate expression of c-Ret in enteric ganglia (Lang et al., 2000). Other examples of important transcription factors include *H*eart *a*utonomic nervous system and *n*eural crest *d*erivatives-2 (HAND-2), a neurogenic transcription factor of the bHLH family (Howard et al., 2000), and Mash-1, required for enteric and sympathetic neurons and specifically for the TC cells and serotonergic neurons in the ENS (Lo et al., 1991; Blaugrund et al., 1996). Regulatory molecules such as the bone morphogenetic proteins (BMPs) also have been shown to profoundly affect the development of many tissues including the bowel. The BMPs exert different biological effects at different concentrations; such effects include proliferation, promotion of neural crest-cell fates, neuronal differentiation and apoptosis (Hogan, 1996; Mehler et al., 1997; LaBonne and Bronner-Fraser, 1998; Von Bubnoff and Cho, 2001; Ten Dijke et al., 2002). In gut BMP-4 specify the anterior–posterior and radial axes of the bowel and the development of its mesenchyme (Roberts, 2000). Furthermore, BMP-4 may also modulate enteric neuron development (Sukegawa et al., 2000). We have hypothesized that members of the BMP family may also regulate the responsiveness of enteric precursors to specific growth factors and thus influence lineage diversification.

Although the majority of enteric neurons belonging to various distinct sublineages respond to the same factor, e.g., GDNF, the data obtained thus far show that a lineage distinguished by its specific responsiveness to NT-3, does not correspond to a single functional mature phenotype of enteric neuron. Moreover, several neurotrophic factors such as NT-3 and the neuropoietic cytokines, may act on glial as well as neuronal lineages.

In vitro strategies

Isolation of enteric crest-derived precursors by immunoselection to analyze the biological actions of growth factors

When primary crest cells that migrate to form the sensory, sympathetic or parasympathetic neurons of the PNS reach their final destinations they aggregate and form pairs of bilateral ganglia. However, these ganglia are remote and *outside their targets* of innervation. In contrast, the crest cells that migrate to the gut *invade their target*, the bowel, and are contained within it. Therefore, a strategy was developed to isolate these crest-derived cells from their target tissues to facilitate analysis of the biological effects of growth factors endogenous to the bowel that are being exerted on these precursors. The target tissues include circular and longitudinal smooth muscles or the mucosal epithelium, which are rich sources of diffusible factors and extracellular matrix components such as laminin. The strategy for isolation was to immunoselect the crest-derived cells using antibodies against surface markers that are expressed selectively, and in a stable fashion, throughout the developing fetal gut. The developmental times E12, E14 and E16 were selected because enteric crest-derived precursors are changing in their properties and state of differentiation throughout these stages. The yield of crest-derived cells that can be isolated at these ages is sufficient to provide enough cells for experimentation in culture while the yield of noncrest-derived cells is not so large as to compromise the efficiency of separation. It should be noted that as the fetal gut matures, the noncrest-derived cells continue to proliferate whereas the rate of proliferation of the crest-derived cells slows down and eventually stops.

One neural crest cell marker is the *h*uman-*n*atural *k*iller cell antigen (HNK-1), which was first identified in avian as the '*n*eural *c*rest' antigen (NC-1) and subsequently identified in rodent embryos as HNK-1. Immunoselection using antibodies to HNK-1, was thus used for our early studies (Chalazonitis et al., 1994). However, this marker tends to be expressed only in subpopulations of neural crest cells in rodents. Another marker selectively and stably expressed by the full set of enteric-crest-derived cells throughout the developmental stages we have studied, is the pan-neurotrophin receptor p75 (p75^{NTR}) (Baetge et al., 1990; Young et al., 1999). Thus more recently, we have used p75^{NTR} antibodies routinely to isolate the ENS precursors. The method of immunoselection has been described in detail (Chalazonitis et al., 1998a,b, 2001). Briefly, the fetal gut is dissociated into a single cell suspension, exposed first to p75^{NTR} (or to HNK-1/NC-1) antibodies and then to species-specific secondary antibodies bound to magnetic beads held in a magnetized column. Following incubation with the secondary antibodies, the 'noncrest-derived' or 'residual' cells, which are not bound to the magnetic beads and are therefore not retained on the column, are eluted out as the 'p75^{NTR}-negative' fraction. Once the column is removed from the magnetic field, the crest-derived cells retained on the column are flushed out and recovered as the 'p75^{NTR}-positive' fraction. The average proportions of positive neural crest-cells obtained immediately after isolation from the cell suspension of dissociated fetal gut has been determined to be 15.7% (at E12), 17.8% (at E14) and 20.1% (at E16) of total gut cells. These proportions do not differ statistically at the three developmental ages. RT-PCR analyzes of the separated populations indicate that the neural crest-derived population expresses mRNA^{p75NTR} while the nonneural crest-derived population does not; conversely the nonneural crest-derived population expresses mRNA for SMα-actin, while the neural crest-derived population does not (see Table 1). Thus, the immunoselection technique using the p75^{NTR} antibodies fulfills criteria required for a clean separation. The p75^{NTR} antigen tends to be lost within a few days in culture. Eighteen hours after plating, the proportion of p75^{NTR}-immunoreactive cells was >51% of the crest-derived population and was <6% of the noncrest-derived 'residual' population.

In vivo strategies

Null mutations and overexpression of genes in the ENS

Mice carrying targeted deletions of genes encoding NT-3 (Tessarollo et al., 1994) or TrkC (Tessarollo et al., 1997) were generated and subsequently backcrossed in the same genetic background. TrkC knockout (KO) mice often survive for 3 weeks postnatally, whereas NT-3 KO mice die within the first week. Other mice were generated that overexpress NT-3, under control of the human dopamine beta hydroxylase (DBH) promoter. In these animals, NT-3 is expressed in the colonizing crest-derived cells and the expression is sustained throughout gut development (Kapur et al., 1991; Rice et al., 2000; Chalazonitis et al., 2001). These mice survive and grow normally. Transgenic mice that overexpress BMP-4 were generated under the control of the neuron-specific enolase (NSE) promoter, thereby directing expression at the time neurons become postmitotic. Overexpression of BMP-4 was significant by E16, and was maintained in the adult in all nervous tissues tested (Gomes et al., 2003). These mice grow without apparent gross abnormalities, although they are smaller than WT animals. Transgenic mice overexpressing the BMP-4 antagonist noggin were also generated under control of the NSE promoter. These mice develop normally, although they display an uneven coat of hair with a

TABLE 1

Markers identifying enteric crest-derived cells differentiating into neurons or glia, and enteric noncrest-derived cells differentiating into smooth muscle

Marker	Enteric cells identified
p75^{NTR}; NC-1/HNK-1	Neural crest-derived cells
Nestin	Neuron/glia bipotent precursors
PGP9.5	All neuronal precursors; early
MAP1b; NFM; NFL	Subsets of enteric neurons
NSE	Postmitotic neurons
Peripherin	80% of enteric neurons
Cuprolinic Blue	All mature enteric neurons
S100	Developing and mature glia
GFAP	Mature glia
Absence of p75^{NTR}	Nonneural crest-derived cells
SMα actin	Smooth muscle
SMγ actin	Enteric smooth muscle

'wet look' (Guha et al., 2001 and see Chalazonitis et al., 2002, 2003)

Neuronal densities in myenteric and submucosal ganglia were determined using cuprolinic blue, a reagent that stains single stranded RNA, allowing the selective visualization of neuronal cell bodies rich in ribosomes (Holst and Powley, 1995; Karaosmanoglu et al., 1996). The densities of specific neuronal phenotypes (serotonergic, CGRP-immunoreactive or TrkC-immunoreactive) in myenteric and submucosal plexuses in intestines of KO and transgenic animals were determined/mm^2 and compared with their respective wild type littermates.

Immunocytochemical identification of enteric crest-derived and noncrest-derived cells phenotypes, their growth factor receptors and some of their signal transduction effectors

A battery of recognition markers utilized to identify enteric crest-derived undifferentiated precursors, differentiated precursors (early and mature neurons, early and mature glia) as well as markers that identify noncrest-derived cells are listed in Table 1.

Antibodies that recognize specific transducing receptors and/or their binding subunits that are expressed by specific crest-derived cell populations, as well as those that recognize downstream signaling molecules that become activated after growth factor binding, including some, not utilized in our studies, are listed in Table 2.

NT-3 is the only neurotrophin that promotes neuronal differentiation of a proportion of enteric precursors, isolated from E14 gut

Earlier studies with fetal ENS explants showed that neither exogenous nerve growth factor (NGF) nor antibodies that block NGF action affect development of enteric neurons (Dreyfus et al., 1977a,b). Since the growth medium used in these analyzes included chick embryo extract and fetal calf serum, a critical test of effects of neurotrophins in gut had to be carried out in cultures maintained in a defined medium. The chemically defined medium of choice is the basic Brazeau medium (BBM) (Brazeau et al., 1981) because it has previously been used to successfully grow neural crest precursors (Ziller et al., 1983). The medium contains a mixture of Ham's F12, DMEM and BGJb media. It is supplemented with mitogenic growth factors (bFGF, EGF), and hormones (insulin, tri-iodothyronine, cortisol, parathyroid hormone,

TABLE 2

Growth and secreted factors identified in developing gut and their receptors and downstream effectors identified in enteric crest-derived cells

Factors expressed in gut	Receptors expressed by crest-derived cells		Intracellular effectors molecules
	Binding receptors	Transducing receptors	
NT-3		TrkC: canonical isoform K1	c-Fos
		Isoform K3	
CNTF	CNTFRα	gp130	STAT-3
CLF/CLC?	CNTFRα	LIFRβ	
LIF		gp130	STAT-3
		LIFRβ	
GDNF	GFRα1	Ret[a]	PI3-K[b]
	GFRα2		Shc[b]
NRG (GGF2)	ErbB3[c]	ErbB2	PI3-K[d]
BMP-2 or-4	BMPR-II	BMPR-IA	Smad-1P
		BMPR-IB	
α1 subunit of laminin-1 (soluble form)		LBP110 (binds to the IKVAV domain)	c-Fos

[a]Pachnis et al. (1993), Lo and Anderson (1995).
[b]Mograbi et al. (2001), Focke et al. (2001).
[c]Pattyn et al. (1999).
[d]Buonanno and Fischbach (2001).

glucagon and transferrin) that are essential for precursor cells survival. It is a relatively simpler medium than media used for the growth of trunk primary crest-cells (Stemple and Anderson, 1992) and does not contain retinoic acid, vitamin E or other agents that might influence *enteric* crest-derived cells to differentiate into particular neuronal lineages such as the sympatho-adrenal lineage.

Enteric crest-derived cells grown in the defined BBM medium were incubated with each of the members of the neurotrophin family: NGF, brain-derived neurotrophic factor (BDNF), neurotrophin-3 (NT-3) and NT-4/5. While all of the neurotrophins bind to $p75^{NTR}$ with low affinity, they each bind with high affinity to a specific tyrosine kinase (Trk) that transduces their biological activities: NGF binds to TrkA, NT-3 to TrkC and BDNF or NT-4/-5 binds to TrkB. Each of the neurotrophins was applied at physiological concentrations for a 6-day period to enteric-crest-derived cultures isolated from E14 rat gut.

NT-3 promoted a 2.5-fold increase in the number of neurons (NSE-immunoreactive cells) compared to treatment with vehicle. This effect, although modest is significant and has been consistently reproduced using other neuronal markers. NT-3 has no proliferative effect at this stage, but does promote a striking increase of neuritic outgrowth by postmitotic neurons; thus NT-3 stimulates differentiation. Using the RT-PCR method of detection, the NT-3 effect is correlated with the clear expression of TrkC mRNA in intact E14 and E16 fetal rat gut and in the freshly immunoselected E14 enteric crest-derived populations (Chalazonitis et al., 1994). TrkC-immunoreactive neurons have been identified in cultures of enteric crest-derived cells isolated from E14 fetal gut and NT-3 treatment increases the proportion of such neurons (Chalazonitis et al., 1998b, 2001).

In contrast, NGF treatment does not promote differentiation of either enteric neurons or glia, confirming the earlier negative observations in NGF-treated organotypic explants of fetal gut. The lack of effect of NGF is corroborated by the lack of TrkA expression in fetal gut. BDNF and NT-4/5 also exert no significant effects on the enteric crest-derived cells isolated from E14 gut probably reflecting the fact that their specific transducing receptor TrkB develops late in ontogeny. TrkB-immunoreactivity is restricted in vivo, to subsets of myenteric neurons at E16, and peak expression is not reached until before birth (Levanon et al., 2003).

NT-3 in contrast to GDNF, promotes glial differentiation of enteric precursors isolated from E14 and E16 fetal rat gut

Treatment with NT-3 (40 ng/ml) for one week promotes a 235% increase in the proportion of S-100-immunoreractive cells in cultured enteric precursors isolated from E14 to E16 fetal gut. The trophic effect of NT-3 on glia contrasts with a lack of such an effect by GDNF in comparable cultures. GDNF actually decreases the proportion of cultured glial cells by 48%. These data suggest that GDNF may, at these later stages, selectively act on the neuronal lineage at the expense of the glial lineage, while NT-3 has an effect on both lineages. NT-3 could act either directly on the glial precursors or indirectly via modulation of glial growth factors. Interestingly, NT-3 rapidly upregulates expression of neuregulin (GGF2) at both the mRNA and protein levels in cultured motoneurons, isolated during late gestation, when synaptogenesis is occurring (Loeb and Fischbach, 1997).

Enteric neuronal precursors isolated from E14 gut depend on NT-3 for survival

Enteric crest-derived cells immunoselected with anti-$p75^{NTR}$ have been shown to develop into neurons in vitro, in the presence of NT-3. Do these neurons become dependent on this neurotrophin for their continued survival in the adult ENS? NT-3 was withdrawn for 18 h (this period of time is short enough for preservation of neuronal morphology but long enough for Tunel labeling of fragmented DNA to become detectable) and the proportion of live TrkC neurons to TrkC neurons undergoing apoptosis was compared in cultures in which NT-3 had not been withdrawn. Dependency on NT-3 was demonstrated by a significant loss of live TrkC-immunoreactive neurons concomitant with an increase in the proportion of TrkC neurons undergoing apoptosis in the cultures where NT-3 was withdrawn (Chalazonitis et al., 2001).

The enteric mesenchyme constitutes a source of NT-3 that promotes the development of enteric neurons

Evidence in culture

Treatment with a blocking antibody to NT-3

Shortly after NT-3 was cloned, its expression was reported in fetal rat stomach (Scarisbrick et al., 1993) and in avian fetal gut mesenchyme (Chalazonitis, 1996; Le Douarin and Kalcheim, 1999). To determine which population, crest-derived cells or the enteric mesenchyme, is the source of NT-3, experiments were performed using cultures of E14-derived fetal gut. Crest-derived and noncrest-derived cells were grown either as mixed or separated populations and exposed to an antibody that blocks NT-3 function. A significant decrease in the proportion of neurons (peripherin-immunorective cells) and a paucity of neuritic arborization occurred in mixed cultures treated with anti-NT-3 compared to replicate cultures treated with vehicle. If, on the other hand, cultures of immunoselected enteric crest-derived cells were treated with anti-NT-3, no significant decrease in survival of the neurons was detected compared to cultures treated with vehicle. Similarly, isolated primary neural crest from quail, treated with the same anti-NT-3 antibodies did not show a decrease in neuronal survival (Ren et al., 2001). These results suggest that endogenous NT-3 is secreted by the noncrest-derived enteric mesenchyme rather than by the enteric crest-derived cells themselves. Thus, in the mixed cultures it is the NT-3 made by the enteric mesenchyme rather than by the crest-derived cells that is neutralized by anti-NT-3 treatment, resulting in a diminished differentiation/survival of enteric neurons derived from the p75^{NTR}-immunoselected population. These data are consistent with the earlier reports of NT-3 expression in the mesenchyme of fetal rat and avian gut.

Evidence in vivo

Retrograde transport of NT-3 by subpopulations of myenteric and submucosal neurons

Retrograde transport of a specific neurotrophin is the hallmark of neurons that have become dependent on it when it is synthesized by the target tissues these neurons innervate. Thus retrograde transport of NT-3 was utilized to identify NT-3-dependent enteric neurons in vivo. NT-3 immunoreactivity has been localized in adult rat gut, to the intestinal villi of the epithelium that are innervated by neurons of the submucosal plexus, and to the longitudinal muscle layers of the gut that are innervated by neurons of the myenteric plexus (Zhou and Rush, 1993). In adult rat gut, retrograde transport of ^{125}I-NT-3 injected into the mucosa labels neurons in ganglia of the submucosal, but not the myenteric plexus. In contrast, injections of ^{125}I-NT-3 into myenteric ganglia, the tertiary plexus, and muscle, label neurons in underlying submucosal and distant myenteric ganglia. The labeling patterns suggest that NT-3-dependent neurons in the submucosal ganglia correspond to the intrinsic primary afferent neurons (IPANS) and/or secretomotor neurons, whereas the NT-3-dependent myenteric neurons innervate other myenteric ganglia and/or the longitudinal muscle (Chalazonitis et al., 2001). These data are consistent with the report that NT-3-immunoreactivity is present in vasoactive intestinal peptide (VIP)-containing enteric neurons, and is localized to ganglia and processes innervating the mucosa, and occasionally, the muscle layer and vasculature (De Giorgio et al., 2000).

Overexpression of NT-3 directed to the ENS by the DBH promoter results in hyperplasia of myenteric ganglia in transgenic mice

DBH-NT-3 transgenic mice survive and gain weight normally. Enteric neurons identified using cuprolinic blue were quantified in laminar preparations of the myenteric and submucosal plexuses. Neurons in the myenteric ganglia were on average larger in the transgenic mice compared to those in WT animals. Also the number and packing density of the neurons were significantly greater per ganglia. In contrast, the number of neurons in the submucosal plexus was not significantly different in DBH-NT-3 and wild type mice. The increased number of myenteric neurons is likely to develop because of the excessive NT-3 being produced by the transient catecholaminergic (TC), Mash-1-dependent, cells as well as by the noncatecholaminergic derived neurons that continue to express DBH. The lack of change in the final number of submucosal neurons in the transgenic mice may result from a lower level of transgene expression in the submucosal plexus, particularly since the latter is derived from non-TC progenitors that mature into neurons that are born later than the myenteric TC-derived neurons. For example, the transgene was not localized to CGRP-immunoreactive neurons.

The absence of NT-3/TrkC signaling in mice lacking functional NT-3 or TrkC genes results in losses of enteric neurons in both plexuses; the losses affect subsets of neurons

The first study that described the phenotype of the ENS in NT-3 knockout mice reported an 'apparently' normal ENS (Fariñas et al., 1994). This study had followed the report of an in vitro effect exerted by NT-3 on enteric crest-derived precursors (Chalazonitis et al., 1994). Since in the latter study NT-3 only affected a minor proportion of the precursors, a more detailed analysis of the ENS in mice lacking the genes encoding either TrkC or NT-3 was carried out. No TrkC-immunoreactive neurons were observed in either plexus in TrkC−/− mice, confirming that expression of TrkC is lacking in these KO animals (Klein et al., 1994; Tessarollo et al., 1997). Indeed, the ENS in these mice have partial but significant defects (about 30% less neurons compared to WT) in myenteric ganglia. Generally, the neuronal deficit is more pronounced in the submucosal (40–50%) than in the myenteric plexus. The most severe deficit occurs in the submucosal plexus of the NT-3 KO mice (75%) as compared to the decrease observed in the TrkC KO animals (50%). It is thus possible that NT-3 may interact with another receptor, such as TrkB, that may partially compensate for the absence of TrkC (Tessarollo et al., 1997; Fariñas et al., 1998). Alternatively, in the absence of NT-3/TrkC signaling, intrinsic apoptotic effects of p75^{NTR} may become unmasked (Bredesen and Rabizadeh, 1997). These limited defects, that particularly affect neurons of the submucosal plexus, imply that NT-3 exerts a physiological role in development and survival of enteric neurons at a much later time than GDNF (Durbec et al., 1996)

and that in contrast to GDNF, NT-3 affects only subpopulations of ENS neurons.

Neuropoietic cytokines such as CNTF or LIF may in part substitute for NT-3 in the development of enteric neurons and glia

The limited deficits that occur in the ENS of TrkC or NT-3 KO mice could also imply that neurotrophic factors other than NT-3 are expressed within the fetal gut, may also promote neuronal and glial development. Such factors include the neuropoietic cytokines CNTF and LIF or a CNTF-like factor that may be the *c*ytokine-*l*ike *f*actor-1/*c*ardiotrophin-*l*ike *c*ytokine complex. CLF/CLC interacts with the tri-partite CNTF receptor complex composed of the GPI-anchored binding subunit CNTFRα and the transducing subunits gp130 and LIFRβ (Elson et al., 2000). Mice lacking CNTF or LIF do not show abnormalities in the ENS, in contrast to mice lacking the genes for CNTFRα or LIFRβ which show deficits in enteric motor neurons (Gershon, 1997). The CLF/CLC complex rather than LIF or CNTF, could be the physiological ligand implicated in ENS development.

CNTF and LIF promote both neuronal and glial development in vitro

The tripartite CNTF receptor complex is expressed in fetal bowel (Ip et al., 1993) and the transducing subunits gp130 and LIFRβ are expressed both by noncrest and by crest-derived cells isolated from E14 gut. However, the binding subunit CNTFRα is expressed by the crest-derived cells only. Moreover, CNTFRα immunoreactivity is localized to myenteric ganglia in E16 and E18 fetal gut (Chalazonitis et al., 1998a). Thus crest-derived cells express all of the receptor components to enable them to respond to CNTF or LIF. Accordingly, CNTF and LIF activate the JAK/STAT signal transduction pathway with nuclear translocation of STAT-3 in p75^{NTR}-expressing cells isolated from E14 gut (Chalazonitis et al., 1998a). CNTF as well as LIF promotes, in a dose-dependent manner, an increase of enteric neuron differentiation. The proportion of peripherin-immunoreactive cells increased by about 2-fold and 3-fold in the presence of CNTF and LIF, respectively. The two cytokines did not exert additive effects, indicating that the population of responding crest-derived cells overlaps. Furthermore, neither CNTF nor LIF affected proliferation since the total number of cultured crest-derived cells remained unchanged compared with no cytokine treatment. Development of cells expressing glial markers such as S-100 or GFAP increased significantly, by $\sim$3-fold in the presence of 10 ng/ml of CNTF. Noticeably, as CNTF and LIF promoted neuronal and glial development, a concomitant decrease in the proportion of nestin-expressing cells occurred. This decrease was maximized in cultures treated with 1 ng/ml of CNTF but about 10% of the total cells still expressed nestin (Chalazonitis et al., 1998a). These data imply that the neuropoietic cytokines affect the differentiation of cells of both the neuronal and glial-committed enteric lineages.

CNTF and LIF can enhance or diminish the neurotrophic effects of NT-3 on development of a subset of enteric neurons in vitro

Simultaneous treatment with NT-3 and CNTF have additive effects

When enteric-crest-derived precursors, immunoselected with anti-p75[NTR] from E14 gut, are treated simultaneously with saturating concentrations of either (CNTF + NT-3) or (LIF + NT-3), the number of enteric neurons that developed was much larger than treatment with NT-3 alone (Chalazonitis et al., 1998a). These data suggest that the subpopulations of neurons responding to NT-3 do not overlap with those responding to the neuropoietic cytokines. Because there are fewer enteric motor NOS-expressing neurons in mice deficient for the gene encoding CNTFRα (Gershon, 1997), NOS-immunoreactivity was determined in cultures of enteric precursors immunoselected from E14 gut and treated with NT-3, CNTF or with both. NT-3 was more efficacious in promoting development of NOS-expressing neurons (31% of the cells) than CNTF (13% of the cells) and the combined treatment of CNTF and NT-3 was not additive. Thus CNTF promoted development of a

subset of the NOS neurons, responsive to NT-3 (Chalazonitis et al., 1998a).

Sequential treatment with NT-3 followed by CNTF reveals that NT-3 enhances both differentiation and survival and CNTF acts only as a differentiating factor

Sequential treatment of cultures with NT-3 for the first 5 days of an 8-day period, followed by CNTF for the remaining 3 days results in a significantly higher number of developing neurons than when the cultures are treated with a single treatment of either NT-3 or CNTF for 8 days. In contrast, if the sequential treatment is reversed so that CNTF is present for the first 5 days of the 8-day period, followed by NT-3 for the remaining 3 days, the number of developing neurons is similar to that developing with an 8-day treatment with either factor alone. Thus, the development of enteric neuronal precursors isolated from E14 gut is enhanced when the cells are initially exposed to NT-3 and secondarily to CNTF, while no enhancement occurs with initial exposures to CNTF and secondarily to NT-3. In addition, withdrawal of NT-3 for the last 3 days of an 8-day treatment induces loss of neurons (Chalazonitis et al., 2001), while withdrawal of CNTF for the last 3 days of an 8-day treatment, does not (Chalazonitis et al., 1998a). These observations suggest that while the E14 gut neuronal precursors require NT-3 to differentiate *and* survive, CNTF does not seem to be required for their survival, but acts only as a differentiation factor.

Accentuation of enteric neuron dependency on NT-3 by CNTF

To determine whether enteric neuronal precursors, isolated from E14 gut exhibit increased dependency on NT-3 because of exposure to CNTF, precursors were treated with combined NT-3 + CNTF for the first 5 days of an 8-day culture period and then both factors were withdrawn for the last 3 days. Neuronal development was expressed as percentage of that promoted with NT-3 for 8 days. Fewer neurons developed (160%) compared with 250% when both factors were maintained for 8 days (Chalazonitis et al., 2001). The double withdrawal paradigm does not distinguish between the differentiation and/or survival effects of each of the factors. To do so, enteric neuronal development was assayed following cotreatment for the first 5 days followed by selective withdrawal of either NT-3 or CNTF for the remaining 3 days. When this withdrawal paradigm was applied a striking difference between the effect of the factors in maintaining enteric neuronal development became apparent. When CNTF was withdrawn for the last 3 days of the 8-day period while *NT-3 was maintained*, the number of differentiating neurons was *unaffected* compared to continuous treatment with both factors. In contrast, when NT-3 was withdrawn for the last 3 days, while *CNTF was maintained*, the number of neurons surviving was significantly *less* (100%) than when both factors were withdrawn (160%) (Chalazonitis et al., 2001).

Taken together these observations suggest that: (i) NT-3 is sufficient to maintain *all* of the neurons that were exposed to both factors from the beginning of the culture period, (ii) CNTF promotes neuronal differentiation but is not required for survival of enteric neurons that have initially differentiated with both factors, and (iii) under conditions of selective NT-3 withdrawal, the continuous presence of CNTF is actually detrimental to neuronal development (Chalazonitis et al., 2001). In conclusion, although enteric neuronal precursors differentiate in the presence of CNTF, this cytokine accentuates their dependence on NT-3 for their subsequent maintenance. Neuropoietic cytokines have been previously reported not to support the survival of embryonic sympathetic neurons (Kotzbauer et al., 1994) and even to induce in these neurons programmed cell death in vitro (Kessler et al., 1993).

NT-3 exerts late and restricted developmental effects in the developing gut

NT-3 does not have a neurotrophic effect on enteric crest-derived cells isolated from E12 gut

When crest-derived precursor cells are immunoselected from E12 gut, a striking lack of responsiveness to NT-3 occurs in vitro. NT-3 exerts neither a proliferative effect on the nestin-expressing precursor population, nor is neuronal differentiation significantly increased. Concomitantly, TrkC-immunoreactivity expressed by the neurons in these cultures is rare (< 2% of the total cells per culture) (Chalazonitis et al., 1998b). As described above, the developmental effects of NT-3 become significant in cultured crest-derived precursors isolated from more mature E14 gut. The report that postnatal rat myenteric neurons continue to respond to NT-3 in vitro (Saffrey et al., 2000) is consistent with these data.

NT-3 affects a subset of enteric neurons that develop late

The deficit in neuronal numbers in mice with targeted deletions of genes encoding NT-3 or TrkC was more severe in the submucosal rather than in the myenteric plexus, indicating that the submucosal neurons that develop later than the myenteric neurons are those more affected by the absence of NT-3 signaling. Furthermore, these deficits decreased in severity from proximal to distal intestine. The numbers of neurons in both wild type and KO mice are not significantly different in the myenteric plexus of the distal intestine. Similarly, deficits in the submucosal plexus are progressively more severe in homozygote > heterozygote > wild type in the proximal and mid-intestine. Noticeably, the largest neuronal deficit (2/3 of neurons missing) occurred in the submucosal plexus of the proximal intestine in NT-3 KO mice. Since there is a rostro-caudal delay during ENS development, the enteric neurons located in the proximal region of the intestine, being the most mature, would be expected to include those expressing TrkC, and thus be the most affected by the absence of NT-3 signaling.

Neurotransmitter-defined phenotypes of enteric neurons were analyzed to determine whether the neuronal deficits could be accounted for by the selective absence of a particular subset of neurons. Studies were performed with antibodies to 5-HT and CGRP which are markers for the early born, Mash-1-dependent, and the late-born, Mash-1-independent lineages of neurons respectively (Pham et al., 1991; Blaugrund et al., 1996). In the gut of mice lacking

TrkC, 5-HT and CGRP immunoreactive neurons are present at P11 and P22. The density of 5HT-immunoreactive neurons located in the myenteric plexus was not significantly affected compared to WT animals at either P11 or P22, indicating that this subset of the earliest born neurons is not NT-3-dependent. In contrast, the density of CGRP-immunoreactive neurons of P22 animals was significantly decreased compared to WT (Chalazonitis et al., 2001). The late onset of NT-3-responsiveness by enteric neurons is consistent with the observation that their TrkC expression does not become significant until about mid-gestation. Submucosal intrinsic primary afferent neurons (IPANS) express CGRP and mediate the peristaltic reflex of the gut (Grider, 1994; Grider et al., 1996; Pan and Gershon, 2000). The CGRP-immunoreactive submucosal neurons that are selectively decreased in the TrkC KO mice may well correspond to the subset of IPANS documented to transport NT-3 in a retrograde manner after injection of ^{125}I-NT-3 into the mucosa or the myenteric plexus, or the smooth muscle and tertiary plexus (Chalazonitis et al., 2001). Consistent with these data is the report that Trk receptors were identified using a pan-Trk antibody, in numerous submucosal and myenteric ganglion cells and their processes in the adult rat intestinal wall (De Giorgio et al., 2000). Moreover, TrkC-immunoreactivity is not expressed until the end of the fourth month of gestation in the developing ENS in humans (Hoehner et al., 1996a). Taken together these data indicate that developmental effects of NT-3 are delayed compared to those of GDNF and affect restricted subsets of neurons that tend to develop late.

Developmental regulation of the timing of NT-3-responsiveness in the ENS

Since NT-3 is a neurotrophic factor acting relatively late in development of the ENS, molecular signals must be implicated in regulating the timing of responsiveness of the enteric precursors to NT-3. Therefore, we examined whether factors expressed at earlier stages than E14 in gut may increase the expression of TrkC in the enteric precursors isolated at E12.

Evidence in vitro

GDNF increases the proportion of TrkC-expressing neurons in cultures of enteric precursors isolated from E12 gut

Enteric precursors rarely express TrkC when they are isolated from E12 gut (see above). However, at this stage, the nestin-expressing precursors cells proliferate in the presence of GDNF. A >3-fold increase in the proportion of cells expressing the proliferating cell nuclear antigen (PCNA) occurs in response to GDNF along with a 15-fold increase in the proportion of nestin-expressing enteric precursors (Chalazonitis et al., 1998b). Since GDNF treatment expands the pool of enteric precursors, an increase in the subset expressing TrkC also might be expected to occur. Expression of TrkC-immunoreactive cells was determined, following a 7-day exposure to GDNF by crest-derived cells isolated from gut at E12. A large, 12-fold increase of TrkC-expressing cells exhibiting clear neuronal morphology occurred compared to vehicle treated cultures (Chalazonitis et al., 1998b). These data indicate that GDNF action is a prerequisite for allowing development of precursor subtypes that are already committed to be NT-3-responsive. Cotreatment with NT-3 and GDNF did not promote a greater developmental effect than with GDNF alone, indicating that the NT-3-responsive neurons constitute a subset of the larger pool of GDNF-responsive precursors that differentiate into neurons.

BMP-4 and BMP-2 induce precocious TrkC-expression and enhance NT-3 dependency of enteric neurons

The proliferative action of GDNF has been shown to amplify the subset of enteric precursors that express TrkC. In addition to GDNF, other secreted factors may also contribute to the differentiation of E12 enteric precursors into TrkC-expressing subsets. The bone morphogenetic proteins (BMP), members of the transforming growth factor-beta (TGF-β) superfamily (Hogan 1996; Von Bubnoff and Cho, 2001), are candidate molecules that could affect TrkC expression since they influence the fate of many cell types including neural crest cells (Liem et al., 1995; Shah et al., 1996; Varley and Maxwell, 1996, LaBonne and Bronner-Fraser, 1998; Marchant et al., 1998; Sela-Donenfeld and Kalcheim, 1999) and CNS progenitors of the subventricular zone (Mehler et al., 1997, 2000). BMP-2 and BMP-4 are expressed in E12 and E14 gut and their expression is developmentally regulated. Furthermore isolated enteric crest-derived cells also express the three receptor subunits BMPR-II, -IA and -IB (Chalazonitis et al., 2002, 2003) that specifically bind and transduce the actions of BMP-2 and BMP-4 (Ten Dijke et al., 1994). Treatment of isolated E12 enteric p75^{NTR+} crest-derived cells with BMP-2 or BMP-4 at high concentration (20 ng/ml) promotes substantial increases in the proportion of TrkC-expressing cells. This effect diminishes significantly when enteric crest-derived cells are isolated at the later developmental stage of E14. Thus, the least mature enteric precursors are, the most sensitive to BMP-2 and BMP-4. NT-3 does not enhance BMPs-mediated increases in TrkC expression by precursors isolated from E12 gut but becomes additive with the BMPs by promoting the survival of the more mature, TrkC-expressing neurons cultured from E14 gut (Chalazonitis, D'Autréaux, Guha, Pham, Faure, Chen, Roman, Rothman, Kessler and Gershon, unpublished data). BMP-2 and -4 promote an increase in dependency of these neurons on NT-3, and the number of TrkC-neurons undergoing apoptosis increases when NT-3 is withdrawn in the continued presence of the BMPs. Conversely, addition of NT-3 decreases the proportion of apoptotic TrkC-expressing neurons following pre-treatment with the BMPs (Chalazonitis et al., unpublished observation). Thus, by inducing precocious TrkC-expression in early enteric, non-NT-3 responsive precursors isolated from E12 gut, BMP-2 and BMP-4 regulate the specification of subsets of NT-3-responsive neurons. Similarly, BMP-2 has also been shown to regulate TrkC expression in sympathetic ganglion neurons (Zhang et al., 1998).

Evidence in vivo

To determine whether BMP-2 and BMP-4 signaling exerts physiological effects during ENS development in vivo, the overall neuronal densities and the densities and proportions of neurons expressing TrkC were quantified in ganglia of the myenteric and submucosal plexuses of transgenic mice that overexpress either BMP-4 or the BMP-4 antagonist noggin, under the control of the NSE promoter.

Overexpression of BMP-4 increases the proportion of TrkC-expressing neurons in myenteric plexus of the duodenum

The ENS phenotype of mice that overexpress BMP-4 was analyzed in 4 weeks-old heterozygous animals (homozygotes often do not survive after birth, W. Gomes, personal communication) and the cuprolinic blue staining method used to analyze neuronal density in the plexuses, requires enteric neurons to be mature. The average numbers of neurons did not change significantly in either the myenteric or submucosal plexuses of duodenum but decreased in the ileum of transgenic animals. The decrease of neuronal density observed in the BMP-4 overexpressing animals was small compared to the large neuronal loss that occurred with high concentrations of BMP-4 in cultured enteric crest-derived cells (Chalazonitis et al., unpublished data). This may reflect physiologic limitation of BMP-4 signaling through mechanisms such as a feedback loop in which BMP signaling increases noggin transcription in a dose-dependent fashion which in turn, limits BMP-4 effects (Gazzerro et al., 1998). However, significant increases in the density of TrkC-expressing neurons occurred in myenteric and submucosal plexuses in the duodenum of NSE-BMP-4 animals (Chalazonitis, Gomes, Pham, Kessler and Gershon, unpublished data). These observations are consistent with the stimulatory effects of BMP4 on the expression of TrkC in cultured enteric crest-derived cells (Chalazonitis et al., 2002).

Overexpression of the BMP-4 antagonist, noggin, decreases the density of TrkC-expressing neurons in both myenteric and submucosal plexuses and increases neuronal densities in duodenum, jejunum and ileum

Noggin antagonizes BMP signaling by sequestering BMP-2 and BMP-4 (Lim et al., 2000). Noggin is expressed in fetal gut at E12 and E14 but its expression is downregulated in adult gut (Chalazonitis et al., 2002). By contrast, when noggin is over-expressed under the control of the NSE promoter levels of noggin increase from E14 to late gestational stages as the enteric neurons become postmitotic and express NSE, and noggin levels continue to be elevated at postnatal ages. Significant decreases in the densities of TrkC-expressing neurons occurred in both myenteric and submucosal plexuses of 1 month-old animals in all three regions of the gut and were similar in homozygous and heterozygous NSE-noggin animals (Chalazonitis et al., 2002, 2003). Neuronal densities in myenteric plexuses were higher in homozygotes compared to heterozygotes whereas in submucosal plexus the increases in neuronal densities were equivalent in homozygotes and heterozygotes (Chalazonitis et al., 2002, 2003). Thus the magnitude of changes in phenotypes (increase in neuronal densities and decrease in TrkC-expressing neurons) were different in myenteric and in sub-mucosal plexuses. This could possibly reflect differences in the regulation of BMP signaling by feedback loops as well as by other BMP antagonists known to be expressed in fetal gut (Chalazonitis et al., 2002). In summary, in the developing gut of transgenic animals, alterations in BMP signaling, consequent to either excess BMP-4 or excess noggin, are regulated locally and in a time-dependent manner.

Overexpression of noggin results in a substantial decrease in the proportion of TrkC-expressing neurons, particularly in the subset that expresses CGRP

The increase in overall numbers of enteric neurons coupled with the decrease in the density of neurons expressing TrkC in noggin-over expressing animals results in an overall 50% reduction, in the proportion of TrkC-expressing neurons. This decrease occurs in both plexuses and in all three regions of the gut examined (Chalazonitis et al., 2002, 2003). Analysis of one particular subset of neurons that requires NT 3 for their development, the late-born, CGRP-expressing, IPANS located in submucosal and myenteric plexuses (Chalazonitis et al., 2001) demonstrates decreases in the density and proportion

of CGRP-immunoreactive neurons in both myenteric and submucosal plexuses in jejunum and ileum of the NSE-noggin animals (Chalazonitis et al., 2002).

The overall data obtained in the transgenic animals show that increasing or lowering of BMP-4 signaling in vivo, at a time when neuronal precursors begin to express NSE, leads to either an excess or a loss of NT-3-dependent neurons, respectively. In either case, disruption of the critical level of BMP signaling needed to regulate the final number of NT-3-dependent neurons, including the CGRP-expressing subset of IPANS, can lead to abnormalities in control of the peristaltic reflex and proper functioning of the gut. Exogenous recombinant NT-3, when injected subcutaneously in humans, accelerates colonic transit (Coulie et al., 2000). NT-3 levels and TrkC-immunoreactivity are no longer detected in ganglion cells at the 'transitional zone' of intestine, in HD patients (Hoehner et al., 1996b) and the total levels of NT-3 are significantly lower in infantile hypertrophic pyloric stenosis (Guarino et al., 2001).

A temporal model for the role of NT-3 in development of the ENS

The model shown in Fig. 1 describes the actions of NT-3 and of the various factors that interact with NT-3 during development of the ENS. These factors include GDNF, members of the TGF β superfamily of molecules (BMP-2, -4) that regulate NT-3-responsiveness, and neuropoietic cytokines such as CNTF and LIF. The model commences as the enteric crest-derived precursors start colonizing the embryonic gut and continues through fetal, prenatal and postnatal stages until formation of the mature ENS is complete. The model describes developmental events that take place in the anterior gut since this region is colonized first, and since the crest-derived cells that have been studied in vitro are primarily derived from this region. In addition, all of the assessments of ENS neuronal phenotypes in mice deficient for NT-3 and TrkC as well as in transgenic animals with altered BMP signaling, were in duodenum, jejunum and ileum. The model emphasizes the development and diversification of the enteric precursors that in the early stages start expressing TrkC, then in the later stages respond to NT-3 by differentiating, and then require NT-3 for their ongoing maintenance in adulthood. The specific temporal aspects of the model apply to both rats and mice. The time line on the left is for rat, which follows that of the developing mouse by an approximate delay of 1 to 1.5 days. At E9.5 pluripotent crest-derived precursors enter the gut (Baetge and Gershon, 1989; Baetge et al., 1990) and express the master transcription factor Phox 2b and the pan-neurotrophin receptor p75^{NTR}. From E9.5 through E12, they also coexpress Ret and GFRα-1, and respond to GDNF by proliferating massively. The proliferating precursors cells, which migrate in a rostro-caudal fashion, are sufficient in number to completely colonize the bowel. It is assumed that by E12 most of these precursors express the marker nestin. From E12 through E14, the rate of proliferation of enteric precursors slows down as separate neuronal and glial lineages develop. GDNF continues to promote differentiation/survival of the pan-neuronal lineage, causing a switch from expression of nestin to expression of neuronal markers such as NSE or peripherin and/or the neurofilament triplet. GDNF also promotes the survival of a sparse, TrkC-expressing subset of neurons. At this stage the precursors express the three BMP receptor subunits, -II required for binding and -IA, and -IB required for activation and transduction after BMP-2 and BMP-4 binding. As levels of BMP signaling increase (represented by the graded intensities of green in the BMP-2, -4 rectangle), the proliferative action of GDNF abates, and precursors exit the cell cycle. A proportion of these newly born precursors may also become apoptotic. Critical levels of BMPs greatly increase TrkC expression (Chalazonitis et al., unpublished data). When at a sufficiently high level, BMP signaling can activate noggin expression. As represented by the graded intensities of red in the noggin (Ng) rectangle, BMP signaling becomes antagonized by noggin (and/or other antagonists such as gremlin, follistatin or chordin) in a feedback loop. In NSE-noggin transgenic mice, the level of BMP signaling is decreased compared to wild type, and thus more neuronal precursors continue to proliferate and develop, while less of them express TrkC. By E14, BMP-2 and -4 heighten dependency of neuronal precursors on NT-3 (Chalazonitis et al., unpublished

observations). The glial lineage is no longer supported by GDNF and hypothetically, it may be induced by the BMP to depend on other factors such as the neuregulins. From E14 until prenatal stages, further diversification of the pan-neuronal lineage takes place (i.e., Mash-1-dependent vs Mash-1-independent). Among the Mash-1 dependent lineage, TH expression is lost and replaced by other neurotransmitter phenotypes such as 5-HT. The final neuronal phenotypes are established in a temporal hierarchy. First the early (i.e., serotonergic), then the intermediate and finally the late enteric neurons are born. Subsets of the late-born neurons may also respond to neuropoietic cytokines since the tripartite component of the CNTF receptor is expressed by enteric neurons at this stage. Interactions with NT-3, the neuropoietic cytokines and the neuregulins may determine the final numbers of neuronal subsets such as those expressing NOS or CGRP, as well as of enteric glia. From postnatal to adult stages, NT-3 continues to support the survival of subpopulations of myenteric and submucosal neurons. As more and more neuron–glia interactions (bidirectional stippled arrow in Fig. 1) take place, the final numbers of differentiated astrocytes become established.

Acknowledgments

The author wishes to thank Dr. Michael Gershon for his continuing support and interaction through the years without which this work would not have been possible. Thanks are due to Dr. John A. Kessler, Dr. Taube Rothman and Dr. Lloyd Greene for helpful comments on the manuscript. This work is currently supported by NIH grants NIDDK 58056 (AC), NS15547 (MDG) NS20013, 20778 (JAK).

Abbreviations

BDNF	brain-derived neurotrophic factor
BMP	bone morphogenetic protein
CNTF	ciliary neurotrophic factor
CLF/CC	cytokine-like factor/cardiotrophin-like cytokine
DBH	dopamine beta hydroxylase
ENS	enteric nervous system

GDNF	glial cell line-derived neurotrophic factor
GGF2	glial growth factor 2
HD	Hirschprung's disease
ICC	interstitial cells of Cajál
KO	knockout
LIF	leukemia inhibitory factor
Mash-1	*M*ammalian *a*chete *s*cute *h*omolog, a transcription factor
NGF	nerve growth factor
NSE	neuron-specific enolase
NT-3	neurotrophin-3
p75NTR	p75 neurotrophin receptor
TC	'transient catecholaminergic' cells
TGF-beta	transforming growth factor-β
TH	tyrosine hydroxylase
Trk	tyrosine kinase

References

Angrist, M., Jing, S., Bolk, S., Bentley, K., Nallasamy, S., Haluska, M., Fox, G.M. and Chakravarti, A. (1998) Human GFRA1: cloning, mapping, genomic structure, and evaluation as a candidate gene for Hirschsprung's disease susceptibility. Genomics, 15: 354–362.

Baetge, G. and Gershon, M.D. (1989) Transient catecholaminergic (TC) cells in the vagus nerves and bowel of fetal mice: relationship to the development of enteric neurons. Dev. Biol., 132: 189–211.

Baetge, G., Pintar, J.E. and Gershon, M.D. (1990) Transiently catecholaminergic (TC) cells in the bowel of fetal rats and mice: Precursors of non-catecholaminergic enteric neurons. Dev. Biol., 141: 353–380.

Bär, K.J., Facer, P., Williams, N.S., Tam, P.K.H. and Anand, P. (1997) Glial-derived neurotrophic factor in human adult and fetal intestine and Hirschsprung's disease. Gastroenterology, 112: 1381–1385.

Baynash, A.G., Hosoda, K., Giaid, A., Richardson, J.A., Emoto, N., Hammer, R.E. and Yanagisawa, M. (1994) Interaction of endothelin-3 with endothelin-B receptor is essential for development of epidermal melanocytes and enteric neurons. Cell, 79: 1277–1285.

Blaugrund, E., Pham, T.D., Tennyson, V.M., Lo, L., Sommer, L., Anderson, D.J. and Gershon, M.D. (1996) Distinct subpopulations of enteric neuronal progenitors defined by time of development, sympathoadrenal lineage markers and Mash-1-dependence. Development, 122: 309–320.

Brazeau, P., Ling, N., Esch, F., Böhlen, P., Benoit, R. and Guillemin, R. (1981) High biological activity of the synthetic replicates of somatostatin-28 and somatostatin-25. Reg. Peptides., 1: 225–264.

Bredesen, D.E. and Rabizadeh, S. (1997) p75[NTR] and apoptosis: Trk-dependent and Trk-independent effects. Trends. Neurosci., 20: 287–290.

Buonanno, A. and Fischbach, G.D. (2001) Neuregulin and ErbB receptor signaling pathways in the nervous system. Curr. Opin. Neurobiol., 11: 287–296.

Burns, A.J. and Le Douarin, N.M. (1998) The sacral neural crest contributes neurons and glia to the post-umbilical gut: spatiotemporal analysis of the development of the enteric nervous system. Development, 125: 4335–4347.

Chalazonitis, A., Rothman, T.P., Chen, J., Lamballe, F., Barbacid, M. and Gershon, M.D. (1994) Neurotrophin-3 induces neural crest-derived cells from fetal rat gut to develop in vitro as neurons or glia. J. Neurosci., 14: 6571–6584.

Chalazonitis, A. (1996) Neurotrophin-3 as an essential signal for the developing nervous system. Mol. Neurobiol., 12: 39–53.

Chalazonitis, A., Tennyson, V.M., Kibbey, M.C., Rothman, T.P. and Gershon, M.D. (1997) The alpha1 subunit of laminin-1 promotes the development of neurons by interacting with LBP110 expressed by neural crest-derived cells immunoselected from the fetal mouse gut. J. Neurobiol., 33: 118–138.

Chalazonitis, A., Rothman, T.P., Chen, J., Vinson, E.N., MacLennan, A.J. and Gershon, M.D. (1998a) Promotion of the development of enteric neurons and glia by neuropoietic cytokines: interactions with neurotrophin-3. Dev. Biol., 198: 343–365.

Chalazonitis, A., Rothman, T.P., Chen, J. and Gershon, M.D. (1998b) Age-dependent differences in the effects of GDNF and NT-3 on the development of neurons and glia from neural crest-derived precursors immunoselected from the fetal rat gut: expression of GFRalpha-1 in vitro and in vivo. Dev. Biol., 204: 385–406.

Chalazonitis, A., Pham, T.D., Rothman, T.P., DiStefano, P.S., Bothwell, M., Blair-Flynn, J., Tessarollo, L. and Gershon, M.D. (2001) Neurotrophin-3 is required for the survival-differentiation of subsets of developing enteric neurons. J. Neurosci., 21: 5620–5636.

Chalazonitis, A., Guha, U., Pham, T.D., D'Autréaux, F., Chen, J., Rothman, T.P., Kessler, J.A. and Gershon, M.D. (2002) Expression of TrkC by enteric neuronal progenitors: specification in vitro by BMPs 2 and 4 and inhibition by noggin overexpression in transgenic mice. Seventh International Conference on NGF and Related Molecules, 15–19 May, 2002, Modena, Italy. Abstract (B), p. 24.

Chalazonitis, A., D'Autréaux, F., Guha, U., Pham, T.D., Faure, C., Chen, J., Rothman, T.P., Kessler, J.A. and Gershon, M.D. (2003) regulation of enteric neuronal development and specification of NT-3 dependence by bone morphogenetic proteins (BMPs)-2 and -4. Neurogastroenterol. Motil., 15: 196–197.

Coulie, B., Szarka, L.A., Camilleri, M., Burton, D.B., McKinzie, S., Stambler, N. and Cedarbaum, J.M. (2000)

Recombinant human neurotrophic factors accelerate colonic transit and relieve constipation in humans. Gastroenterology, 119: 41–50.

De Giorgio, R., Arakawa, J., Wetmore, C.J. and Sternini, C. (2000) Neurotrophin-3 and neurotrophin receptor immunoreactivity in peptidergic enteric neurons. Peptides, 21: 1421–1426.

Dreyfus, C.F., Bornstein, M.B. and Gershon, M.D. (1977a) Synthesis of serotonin by neurons of the myenteric plexus in situ and in organotypic tissue culture. Brain Res., 128: 125–139.

Dreyfus, C.F., Sherman, D. and Gershon, M.D. (1977b) Uptake of serotonin by intrinsic neurons of the myenteric plexus grown in organotypic tissue culture. Brain Res., 128: 109–123.

Durbec, P.L., Larsson-Blomberg, L.A., Schuchardt, A., Costantini, F. and Pachnis, V. (1996) Common origin and developmental dependence on *c-ret* of subsets of enteric and sympathetic neuroblasts. Development, 122: 349–358.

Edery, P., Attie, T., Amiel, J., Pelet, A., Eng, C., Hofstra, R.M., Martelli, H., Bidaud, C., Munnich, A. and Lyonnet, S. (1996) Mutation of the endothelin-3 gene in the Waardenburg-Hirschprung disease (Shah-Waardenburg syndrome). Nat. Genet., 12: 442–444.

Edery, P., Lyonnet, S., Mulligan, L.M., Pelet, A., Dow, E., Abel, L., Holder, S., Nihoul-Fékété, C., Ponder, B.A.J. and Munnich, A. (1994) Mutations of the RET proto-oncogene in Hirschsprung's disease. Nature, 367: 378–380.

Elson, G.C., Lelievre, E., Guillet, C., Chevalier, S., Plun-Favreau, H., Froger, J., Suard, I., de Coignac, A.B., Delneste, Y., Bonnefoy, J.Y., Gauchat, J.F. and Gascan, H. (2000) CLF associates with CLC to form a functional heteromeric ligand for the CNTF receptor. Nat. Neurosci., 3: 867–872.

Enomoto, H., Araki, T., Jackman, A., Heuckeroth, R., Snider, W.D., Johnson, E.M. and Milbrandt, J. (1998) GFRa1-deficient mice have deficits in the enteric nervous system and kidneys. Neuron, 21: 317–324.

Fariñas, I., Jones, K.R., Backus, C., Wang, X.-Y. and Reichardt, L.F. (1994) Severe sensory and sympathetic deficits in mice lacking neurotrophin-3. Nature, 369: 658–661.

Fariñas, I., Wilkinson, G.A., Backus, C., Reichardt, L.F. and Patapoutian, A. (1998) Characterization of neurotrophin and Trk receptor functions in developing sensory ganglia: direct NT-3 activation of TrkB neurons in vivo. Neuron, 21: 325–334.

Fiorica-Howells, E., Maroteaux, L. and Gershon, M.D. (2000) Serotonin and the 5-HT$_{2B}$ receptor in the development of enteric neurons. J. Neurosci., 20: 294–305.

Focke, P.J., Schiltz, C.A., Jones, S.J., Watteers, J.J. and Epstein, M.L. (2001) Enteric neuroblasts require the phosphatidylinositol 3-kinase pathway for GDNF-stimulated proliferation. J. Neurobiol., 47: 306–317.

Furness, J.B. and Costa, M. (1987). The Enteric Nervous System. Churchill Livingstone, New York, NY.

Furness, J.B., Llewellyn-Smith, J.C., Bornsteijn, J.C. and Costa, M. (1988). Chemical neuronanatomy and the analysis of neuronal circuitry in the enteric nervous system. In: A. Björklund, T. Hökfelt and C. Owman (Eds.), Handbook of Chemical Neuroanatomy. The Peripheral Nervous System, Vol. 6, Elsevier Science Publishers, B.V., pp. 161–218.

Gabela, G. and Trigg, P. (1984) Size of neurons and glial cells in the enteric ganglia of mice, guinea pigs, rabbits and sheep. J. Neurocytol., 13: 49–71.

Guarino, N., Yoneda, A., Shima, H. and Puri, P. (2001) Selective neurotrophin deficiency in infantile hypertrophic pyloric stenosis. J. Pediatr. Surg., 36: 1280–1284.

Gazzerro, E., Gangji, V. and Canalis, E. (1998) Bone morphogenetic proteins induce the expression of noggin, which limits their activity in cultured rat osteoblasts. J. Clin. Invest., 102: 2106–2114.

Gershon, M.D., Chalazonitis, A. and Rothman, T.P. (1992). Ontogeny of neurons and glia of the enteric nervous system: sources and signals that shape development. In: G.E. Holle and J.D. Wood (Eds.), Advances in the Innervation of the Gastrointestinal Tract. Elsevier Science Publishers, B.V., pp. 19–34.

Gershon, M.D., Chalazonitis, A. and Rothman, T.P. (1993) From Neural Crest to Bowel: Development of the Enteric Nervous System. J. Neurobiol., 24: 199–214.

Gershon, M.D., Kirchgessner, A.L. and Wade, P.R. (1994). Functional anatomy of the enteric nervous system. In: D.H. Johnson, L.R. Alpers, E.D. Jacobson and J.H. Walsh (Eds.), Physiology of the Gastrointestinal Tract, Third Edition, Vol. 1, Raven Press, New York, pp. 381–422.

Gershon, M.D. (1997) Genes and lineages in the formation of the enteric nervous system. Curr. Opin. Neurobiol., 7: 101–109.

Gershon, M.D. (1998). The Second Brain. Harper Collins, New York, NY.

Gomes, W.A., Mehler, M.F. and Kessler, J.A. (2003) Transgenic overexpression of BMP4 increases astroglial and decreases oligodendroglial lineage commitment. Dev. Biol., 255: 164–177.

Grider, J.R. (1994) CGRP as a transmitter in the sensory pathway mediating peristaltic reflex. Am. J. Physiol., 266: G1139–1145.

Grider, J.R., Kuemmerle, J.F. and Jin, J.G. (1996) 5-HT released by mucosal stimuli initiates peristalsis by activating 5-HT4/5-HT1p receptors on sensory CGRP neurons. Am. J. Physiol., 270: G778–782.

Guha, U., Gomes, W., Gupta, M., Rice, F.L. and Kessler, J.A. (2001) Effects of transgenic overexpression of noggin and BMP4 on the development and maintenance of cutaneous innervation. Soc. Neurosci. Abstr., 27: 939.

Hearn, C.J., Murphy, M. and Newgreen, D. (1998) GDNF and ET-3 differentially modulate the numbers of avian neural crest cells and enteric neurons *in vitro*. Dev. Biol., 197: 93–105.

Hoehner, J.C., Wester, T., Påhlman, S. and Olsen, L. (1996a) Localization of neurotrophins and their high-affinity receptors during human enteric nervous system development. Gastroenterology, 110: 765–767.

Hoehner, J.C., Wester, T., Påhlman, S. and Olsen, L. (1996b) Alterations in neurotrophin and neurotrophin-receptor localization in Hirschsprung's disease. J. Pediatr. Surg., 31: 1524–1529.

Hogan, B.L.M. (1996) Bone morphogenetic proteins: multi-functional regulators of vertebrate development. Gen. Dev., 10: 1580–1594.

Holst, M.C. and Powley, T.L. (1995) Cuprolinic blue (quino-linic phthalocyanine) counterstaining of enteric neurons for peroxidase immunocytochemistry. J. Neurosci. Methods, 62: 121–127.

Howard, M.J., Stanke, M., Schneider, C., Wu, X. and Rohrer, H. (2000) The transcription factor dHAND is a downstream effector of BMPs in sympathetic neuron specification. Development, 127: 4073–4081.

Huizinga, J.D., Thuneberg, L., Klüppel, M., Malysz, J., Mikkelsen, H.B. and Bernstein, A. (1995) W/kit gene required for interstitial cells of Cajal and for intestinal pacemaker activity. Nature, 373: 347–349.

Ip, N.Y., McClain, J., Barrezueta, N.X., Aldrich, T.H., Pan, L., Li, Y., Wiegand, S.J., Friedman, B., Davis, S., Yancopoulos, G.D. (1993) The α component of the CNTF receptor is required for signaling and defines potential CNTF targets in the adult and during development. Neuron, 10: 89–102.

Ivanchuk, S.M., Myers, S.M., Eng, C. and Mulligan, L.M. (1996) De novo mutation of GDNF, ligand for the RET/GDNF-alpha receptor complex in Hirschsprung's disease. Hum. Mol. Genet., 5: 2020–2026.

Jing, S., Wen, D., Yu, Y., Holst, P.L., Luo, Y., Fang, M., Tamir, R., Antonio, L., Hu, Z., Cupples, R., Louis, J.-C., Hu, S., Altrock, B.W. and Fox, G.M. (1996) GDNF-induced.

Kapur, R.P., Hoyle, G.W., Mercer, E.H., Brinster, R.I. and Palmiter, R.D. (1991) Some neuronal cell populations express human dopamine b-hydroxylase-*lacZ* transgenes transiently during embryonic development. Neuron, 7: 717–727.

Karaosmanoglu, T., Aygun, B., Wade, P.R. and Gershon, M.D. (1996) Regional differences in the number of neurons in the myenteric plexus of the guinea pig small intestine and colon: an evaluation of markers used to count neurons. Anat. Rec., 244: 470–480.

Kessler, J.A., Ludlam, W.H., Freidlin, M.M., Hall, D.H., Michaelson, M.D., Spray, D.C., Dougherty, M. and Batter, D.K. (1993) Cytokine-induced programmed death of cultured sympathetic neurons. Neuron, 11: 1123–1132.

Klein, R., Silos-Santiago, I., Smeyne, R.J., Lira, S.A., Brambilla, R., Bryant, S., Zhang, L., Snider, W.D. and

Barbacid, M. (1994) Disruption of the neurotrophin-3 receptor trkC eliminates Ia muscle afferents and results in abnormal movements. Nature, 368: 249–251.

Kotzbauer, P.T., Lampe, P.A., Astus, S., Milbrandt, J. and Johnson, E.M. (1994) Postnatal development of survival responsiveness in rat sympathetic neurons to leukemia inhibitory factor and ciliary neurotrophic factor. Neuron, 12: 763–773.

LaBonne, C. and Bronner-Fraser, M. (1998) Neural crest induction in *Xenopus*: evidence for a two-signal model. Development, 125: 2403–2414.

Lang, D.B., Chen, F., Milewski, R., Li, J., Lu, M.M. and Epstein, J.A. (2000) Pax 3 is required for enteric ganglia formation and functions with Sox10 to modulate expression of *c-ret*. J. Clin. Invest., 106: 963–971.

Le Douarin, N.M. (1982). The Neural Crest. Cambridge University Press, Cambridge, UK.

Le Douarin, N.M. and Kalcheim, C. (1999). The Neural Crest. Cambridge University Press, Cambridge, UK.

Lendhal, M.P., Zimmermann, L.B. and McKay, R.D.G. (1990) CNS stem cells express a new class of intermediate filament protein. Cell, 60: 585–595.

Levanon, D., D'Autreaux, F., Li, Z. and Gershon, M.D. TrkB expression in the developing enteric nervous system (ENS) of rats and mice. Neurogastroenterol. Motil., 15: 199.

Lo, L.-C., Johnson, J.E., Wuenschell, C.W., Saito, T. and Anderson, D.J. (1991) Mammalian *achaete-scute* homolog 1 is transiently expressed by spatially restricted subsets of early neuroepithelial and neural crest cells. Genes Dev., 5: 1524–1537.

Lo, L. and Anderson, D.J. (1995) Postmigratory neural crest cells expressing c-RET display restricted developmental and proliferative capacities. Neuron, 15: 527–539.

Liem, K.F., Tremmi, G., Roelink, H. and Jessell, T.M. (1995) Dorsal differentiation of neural plate cells induced by BMP-mediated signals from epidermal ectoderm. Cell, 82: 969–979.

Loeb, J.A. and Fischbach, G.D. (1997) Neurotrophic factors increase neuregulin expression in embryonic ventral spinal cord neurons. J. Neurosci., 17: 1416–1424.

Lomax, A.E. and Furness, J.B. (2000) Neurochemical classification of enteric neurons in the guinea-pig distal colon. Cell. Tissue. Res., 305: 59–72.

Marchant, L., Linker, C., Ruiz, P., Guerrero, N. and Mayor, R. (1998) The inductive properties of mesoderm suggest that the neural crest cells are specified by a BMP gradient. Dev. Biol., 198: 319–329.

Mehler, M.F., Mabie, P.C., Zhang, D. and Kessler, J.A. (1997) Bone morphogenetic proteins in the nervous system. Trends Neurosci., 20: 309–317.

Mehler, M.F., Mabie, P.C., Zhu, G., Gokhan, S. and Kessler, J.A. (2000) Developmental changes in progenitor cell responsiveness to bone morphogenetic proteins differentially modulate progressive CNS lineage fate. Dev. Neurosci., 22: 74–85.

Meyer, D. and Birchmeier, C. (1995) Multiple essential functions of neuregulin in development. Nature, 378: 386–390.

Mograbi, B., Bocciardi, R., Bourget, I., Busca, R., Rochet, N., Farahi-Far, D., Juhel, R.T. and Rossi, B. (2001) Glial cell line-derived neurotrophic factor-stimulated phosphatidylinositol 3-Kinase and Akt activities exert opposing effects on the ERK pathway. J. Biol. Chem., 276: 45307–45319.

Moore, M.W., Klein, R.D., Fariñas, I., Sauer, H., Armanini, M., Phillips, H., Reichardt, L.F., Ryan, A.M., Carver-Moore, K., Rosenthal, A. (1996) Renal and neuronal abnormalities in mice lacking GDNF. Nature, 382: 76–79.

Natarajan, D., Grigoriou, M., Marcos-Gutierrez, C.V., Atkins, C. and Pachnis, V. (1999) Multipotential progenitors of the mammalian enteric nervous system capable of colonizing aganglionic bowel in organ culture. Development, 126: 157–168.

Pachnis, V., Mankoo, B. and Costantini, F. (1993) Expression of the c-ret proto-oncogene during mouse embryogenesis. Development, 119: 1005–1017.

Pan, H. and Gershon, M.D. (2000) Activation of intrinsic afferent pathways in submucosal ganglia of the guinea pig small intestine. J. Neurosci., 20: 3295–3309.

Pasini, B., Borrello, M.G., Greco, A., Bongarzone, I., Luo, Y., Mondellini, P., Alberti, L., Miranda, C., Arighi, E., Bocciardi, R., Seri, M., Barone, V., Radice, M.T., Romeo, G. and Pierotti, M.A. (1995) Loss of function effect of RET mutations causing Hirschsprung disease. Nat. Genet., 10: 35–40.

Pattyn, A., Morin, X., Cremer, H., Goridis, C. and Brunet, J.-F. (1999) The homeobox gene Phox2b is essential for the development of autonomic neural crest derivatives. Nature, 399: 366–370.

Pham, T.D., Gershon, M.D. and Rothman, T.P. (1991) Time of origin of neurons in the murine enteric nervous system. J. Comp. Neurol., 314: 789–798.

Pomeranz, H.D., Rothman, T.P. and Gershon, M.D. (1991) Colonization of the post-umbilical bowel by cells derived from the sacral neural crest: direct tracing of cell migration using an intercalating probe and a replication-deficient retrovirus. Development, 111: 647–655.

Pomeranz, H.D., Rothman, T.P., Chalazonitis, A., Tennyson, V.M. and Gershon, M.D. (1993) Neural crest-derived cells isolated from the gut by immunoselection develop neuronal and glial phenotypes when cultured on laminin. Dev. Biol., 156: 341–361.

Pichel, J.G., Shen, L., Sheng, H.Z., Granholm, A.-C., Drago, J., Grinberg, A., Lee, E.J., Huang, S.B., Saarma, M., Hoffer, B.J., Sariola, H. and Westphal, H. (1996) Defects in enteric innervation and kidney development in mice lacking GDNF. Nature, 382: 73–76.

Puffenberger, E.G., Hosoda, K., Washington, S.S., Nakao, K., De Wit, D., Yanagisawa, M. and Chakravarti, A. (1994) A missense mutation of the endothelin-receptor gene in mutagenic Hirschprung's disease. Cell, 79: 1257–1266.

Ren, Z.G., Porzgen, P., Zhang, J.M., Chen, X.R., Amara, S.G., Blakely, R.D. and Sieber-Blum, M. (2001) Autocrine regulation of norepinephrine transporter expression. Mol. Cell. Neurosci., 17: 539–550.

Rice, J., Doggett, B., Sweetser, D.A., Yanagisawa, H., Yanagisawa, M. and Kapur, R.P. (2000) Transgenic rescue of aganglionosis and piebaldism in lethal spotted mice. Dev. Dyn., 1: 120–132.

Riethmacher, D., Sonnenberg-Riethmacher, E., Brinkmann, V., Yamaai, T., Lewin, G.R. and Birchmeier, C. (1997) Severe neuropathies in mice with targeted mutations in the ErbB3 receptor. Nature, 389: 725–730.

Roberts, D.J. (2000) Molecular mechanisms of development of the gastrointestinal tract. Dev. Dyn., 219: 109–120.

Romeo, G., Ronchetto, P., Luo, Y., Barone, V., Seri, M., Ceccherini, I., Pasini, B., Bocciardi, R., Lerone, M., Kääriäinen, H., Martucciello, G. (1994) Point mutations affect the tyrosine kinase domain of the RET proto-oncogene in Hirschsprung's disease. Nature, 367: 377–378.

Rothman, T.P., Le Douarin, N.M., Fontaine-Pérus, J.C. and Gershon, M.D. (1990) Developmental potential of neural crest-derived cells migrating from segments of developing quail bowel back-grafted into younger chick host embryos. Development, 109: 411–423.

Rothman, T.P., Le Douarin, N.M., Fontaine-Pérus, J.C. and Gershon, M.D. (1993) Colonization of the bowel by neural crest-derived cells re-migrating from foregut backtransplanted to vagal or sacral regions of host embryos. Dev. Dyn., 196: 217–233.

Saffrey, M.J., Wardhaugh, T., Walker, T., Daisley, J. and Silva, A.T. (2000) Trophic actions of neurotrophin-3 on postnatal rat myenteric neurons in vitro. Neurosci. Lett., 278: 133–136.

Sánchez, M., Silos-Santiago, I., Frisén, J., He, B., Lira, S. and Barbacid, M. (1996) Renal agenesis and the absence of enteric neurons in mice lacking GDNF. Nature, 382: 70–73.

Sanders, K.M., Ördög, T., Koh, S.D., Torishashi, S. and Ward, S.M. (1999) Development and plasticity of interstitial cells of Cajal. Neurogastroenterol. Mot., 11: 311–338.

Sang, Q. and Young, H.M. (1998) The identification and chemical coding of cholinergic neurons in the small and large intestine of the mouse. Anat. Rec., 251: 185–199.

Scarisbrick, I.A., Isackson, P.J. and Jones, E.G. (1993) Expression of NGF, BDNF and NT-3 at sites of major tissue interactions during development. Soc. Neurosci. Abstr., 19: 251.

Schuchardt, A., D'Agati, V., Larsson-Blomberg, L., Costantini, F. and Pachnis, V. (1994) Defects in the kidney and enteric nervous system of mice lacking the tyrosine kinase receptor Ret. Nature, 367: 380–383.

Shah, N.M., Groves, A.K. and Anderson, D.J. (1996) Alternative neural crest cell fates are instructively promoted by TGFβ superfamily members. Cell, 85: 331–343.

Sela-Donenfeld, D. and Kalcheim, C. (1999) Regulation of the onset of neural crest migration by coordinated activity of BMP4 and Noggin in the dorsal neural tube. Development, 126: 4749–4762.

Sieber-Blum, M. (1999). In: M. Sieber-Blum (Ed.), Neurotrophins and the Neural Crest. CRC Press LLC, Boca Raton, Boston, London, New York, Washington, D.C.

Stemple, D. and Anderson, D.J. (1992) Isolation of a stem cell for neurons and glia from the mammalian neural crest. Cell, 71: 973–985.

Sukegawa, A., Narita, T., Kameda, T., Saitoh, K., Nohno, T., Iba, H., Yasugi, S. and Fukuda, K. (2000) The concentric structure of the developing gut is regulated by Sonic hedgehog derived from endodermal epithelium. Development, 127: 1971–1980.

Ten Dijke, P., Yamashita, H., Sampath, T.K., Reddi, A.H., Estevez, M., Riddle, D.L., Ichijo, H., Heldin, C.-H. and Miyazono, K. (1994) Identification of type I receptors for osteogenic protein-1 and bone morphogenetic protein-4. J. Biol. Chem., 269: 16985–16988.

Ten Dijke, P., Goumans, M.J., Itoh, F. and Itoh, S. (2002) Regulation of cell proliferation by Smad proteins. J. Cell. Physiol., 191: 1–16.

Tessarollo, L., Vogel, K.S., Palko, M.E., Reid, S.W. and Parada, L.F. (1994) Targeted mutation in the neurotrophin-3 gene results in loss of muscle sensory neurons. Proc. Natl. Acad. Sci. USA., 91: 11844–11848.

Tessarollo, L., Tsoulfas, P., Donovan, M.J., Palko, M.E., Blair-Flynn, J., Hempstead, B.L. and Parada, L.F. (1997) Targeted deletion of all isoforms of the trkC gene suggests the use of alternate receptors by its ligand neurotrophin-3 in neuronal development and implicates trkC in normal cardiogenesis. Proc. Natl. Acad. Sci. USA., 94: 14776–14781.

Vanderwinden, J. M., Gillard, K., De laet, M.-H., Messam, C.A. and Schiffmann, S.N. (2002) Distribution of the intermediate filament nestin in the muscularis propria of the human gastrointestinal tract. Cell Tissue Res., 309: 261–268.

Varley, J.E. and Maxwell, G.D. (1996) BMP-2 and BMP-4 but not BMP-6, increase the number of adrenergic cells which develop in quail trunk neural crest cultures. Exp. Neurol., 140: 84–94.

Von Bubnoff, A. and Cho, K.W. (2001) Intracellular BMP signaling regulation in vertebrates: pathway or network? Dev. Biol., 239: 1–14.

Wu, J.J., Chen, J.-X., Rothman, T.P. and Gershon, M.D. (1999) Inhibition of in vitro enteric neuronal development by endothelin-3: mediation by endothelin B receptors. Development, 126: 1161–1173.

Young, H.M., Ciampoli, D., Hsuan, J. and Canty, A.J. (1999) Expression of Ret-, p75(NTR)-, Phox2a-, Phox2b-, and tyrosine hydroxylase-immunoreactivity by undifferentiated neural crest-derived cells and different classes of enteric neurons in the embryonic mouse gut. Dev. Dyn., 216: 137 152.

Zhang, D., Mehler, M., Song, Q. and Kessler, J.A. (1998) Bone morphogenetic protein receptors in the nervous system and possible roles in regulating trkC expression. J. Neurosci., 18: 3314–3326.

Zhou, X.-F. and Rush, R.A. (1993) Localization of neurotrophin-3-like immunoreactivity in peripheral tissues of the rat. Brain. Res., 621: 189–199.

Ziller, C., Dupin, E., Brazeau, P., Paulin, D. and Le Douarin, N.M. (1983) Early segregation of a neuronal precursor cell line in the neural crest as revealed by culture in a chemically defined medium. Cell, 32: 627–638.

Progress in Brain Research, Vol. 146
ISSN 0079-6123

CHAPTER 17

Neurotrophins in the ear: their roles in sensory neuron survival and fiber guidance

Bernd Fritzsch[1],*, Lino Tessarollo[2], Enzo Coppola[2] and Louis F. Reichardt[3]

[1]*Creighton University, Department of Biomedical Sciences, Omaha, NE 68178, USA*
[2]*NCI-FCDRC, Frederick, MD, USA*
[3]*Howard Hughes Medical Institute, University of California, San Francisco, CA, USA*

Abstract: We review the history of neurotrophins in the ear and the current understanding of the function of neurotrophins in ear innervation, development and maintenance. Only two neurotrophins, brain-derived neurotrophic factor (BDNF) and neurotrophin-3 (NT-3), and their receptors, tyrosine kinase B (TrkB) and TrkC, appear to provide trophic support for inner ear sensory neuron afferents. Mice lacking either both receptors or both ligands lose essentially all sensory innervation of targets in the vestibular and auditory systems of the ear. Analyzes of single mutants show less complete and differential effects on innervation of the different sensory organs within the ear. BDNF and TrkB are most important for survival of vestibular sensory neurons whereas NT-3 and TrkC are most important for survival of cochlear sensory neurons. The largely complementary roles of BDNF to TrkB and NT-3 to TrkC signaling do not reflect specific requirements for innervation of different classes of hair cells. Most neurons express both receptors. Instead, the losses observed in single mutants are related to the spatio-temporal expression pattern of the two neurotrophins. In an area where only one neurotrophin is expressed at a particular time in development, the other neurotrophin is not present to compensate for this absence, resulting in death of neurons innervating that region. Decisive evidence for this suggestion is provided by transgenic mice in which the BDNF coding region has been inserted into the NT-3 gene, resulting in expression of BDNF instead of NT-3. The expression of BDNF in the spatio-temporal pattern of NT-3 results in survival of almost all neurons that are normally lost in the NT-3 mutant. Thus, BDNF and NT-3 have a high level of functional equivalence for inner ear sensory neuron survival. Further analysis of the patterns of afferent fiber losses in mutations that do not develop differentiated hair cells shows that the expression of neurotrophins is remarkably strong and can support afferent innervation. Indeed, BDNF may be one of the earliest genes expressed selectively in hair cells and it appears to be regulated somewhat independently of the genes needed for hair cell differentiation.

Keywords: BDNF; NT-3; TrkB; TrkC; ear innervation; mutant mice

Neurotrophins and the ear: an historical perspective

The existence of neurotrophins to guide fibers to the sensory epithelia goes back to the work of Santiago Ramon y Cajal (Cajal, 1919). Based on his Golgi impregnations that showed what appeared to be directed growth of sensory fibers to hair cells, he

*Correspondence to: B. Fritzsch, Creighton University, Department of Biomedical Sciences, Omaha, NE 68178, USA. Tel.: +1-402-280-2915; Fax: +1-402-280-5556; E-mail: Fritzsch@Creighton.edu

postulated that a neurotrophic substance might be released from the differentiating hair cells and that this factor might guide fibers to grow toward these cells, causing them to engage specifically in contacts only with those cells that release these substance(s). This proposal for a function of neurotrophic factor(s) in ear development was revived in 1949, when Rita Levi-Montalcini published a paper in which she studied the effect of extirpation of the otocyst on the maturation and the survival of central vestibular and cochlear second order neurons in the chick (Levi-Montalcini, 1949). In part building on the earlier

DOI: 10.1016/S0079-6123(03)46017-2

suggestions of Ramon y Cajal, Levi-Montalcini postulated over 50 years ago that the peripheral nervous system trophically supports and maintains second order vestibular and, even more clearly, second order auditory neurons.

Despite these early insights, the use of the ear as a model system of development almost disappeared from 1950 until 1990, so studies on this organ were not important in the development of the modern version of the neurotrophic theory. In fact, it is at present unclear which neurotrophic substance(s) is released by the afferent fibers to support second order vestibular and cochlear neurons in the brainstem. Given the transcellular transport of neurotrophins across cells, it is possible that neurotrophins released from hair cells in the ear ultimately contribute directly to the survival of second order neurons in the brainstem. But this will be hard to test as lack of neurotrophins in hair cells is always correlated with sensory neuron loss or reduction.

As it turned out, NGF, the famous neurotrophic factor discovered by Rita Levi-Montalcini after her initial work in the ear appears to play no role in the development on inner ear sensory neurons. Neither in situ hybridization studies (Pirvola et al., 1992) nor genetics (Fritzsch et al., 1999) has provided any data in support of the hypothesis that NGF has an essential function in the ear. Although a transient expression of TrkA has been detected in some developing sensory neurons early in development (Fritzsch et al., 1999), the absence of TrkA has no discernable phenotype on the development of inner ear afferent innervation. However, the autonomic innervation of the ear is likely lost in TrkA null mutants (Fritzsch et al., 1997c).

Initial studies made it seem very likely that NGF would be important in ear development. The earliest investigations into the molecular basis of neuro-trophic support of inner ear sensory neurons seemed to indicate that NGF was a major player. Data supporting this idea were obtained from tissue culture experiments and from analyzes of neurotrophin gene expression patterns (Bernd and Represa, 1989). Furthermore, data obtained again from tissue culture experiments clearly demonstrated the existence of diffusible substances that seemed to help directing fiber growth to the ear and to support the survival of primary sensory neurons (Bianchi and Cohan, 1991).

Yet other experiments indicated that another trophic factor, brain-derived neurotrophic factor (BDNF), was a candidate to promote the survival of inner ear sensory neurons (Lindsay et al., 1985). Indeed, as early as 1985 there were suggestions that BDNF was present in the ear, but these were not followed up until 1990 when the isolation of BDNF cDNAs made it possible to detect it reliably.

Only two neurotrophins and their high-affinity receptors mediate survival of all inner ear sensory neurons to all six sensory epithelia

Within the past ten years, two neurotrophic factors, BDNF and NT-3, and their high-affinity receptors, TrkB and TrkC, have been identified by in situ hybridization and other molecular techniques in the ear (Pirvola et al., 1992; Wheeler et al., 1994; Fritzsch et al., 1999; Farinas et al., 2001). Targeted mutations of each of these neurotrophins and receptors have clarified their relative contributions to the survival of different sensory neurons in the ear (Fritzsch et al., 1999). These data have shown that in newborn (P0) NT-3 null mutants, there is a dramatic loss of 85% of cochlear (spiral) sensory neurons (Fig. 1) and 23% of vestibular sensory neurons (Farinas et al., 1994; Ernfors et al., 1995). This extensive loss of sensory neurons in the cochlea happens within 2–3 days after the fibers have extended to the sensory epithelia (Farinas et al., 2001). In contrast, TrkC kinase mutants lose only 16–29% loss of vestibular and 51–66% of spiral sensory (Schimmang et al., 1995; Fritzsch et al., 2001). This difference has been attributed to the limited capacity of NT-3 to signal through the TrkB receptor (Fritzsch et al., 1997b). However, the initial TrkC mutation did not prevent expression of truncated receptors lacking a kinase domain (*trkC kin* −/−) (Klein et al., 1994). Indeed, differences have been reported in other systems between the phenotypes of mice lacking kinase-containing isoforms of TrkC and mice lacking all isoforms of TrkC (*trkC-FL*−/−) (Tessarollo et al., 1997). Our data on the selective loss of specific patterns of innervation in the cochlea suggest that in the *trkC-FL* null mutants the loss of afferent innervation is as extensive as in the *NT-3* null mutants (see below).

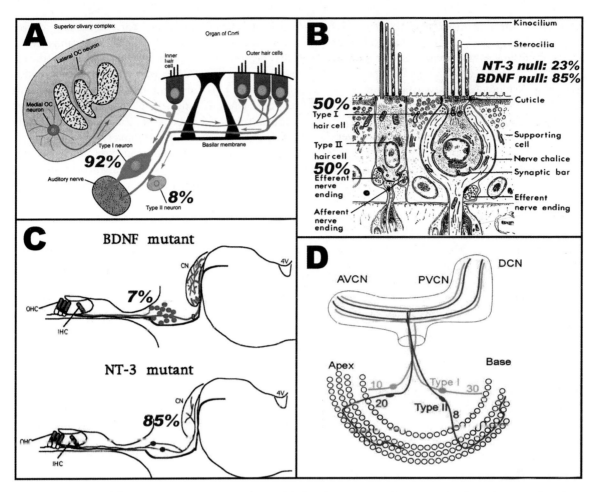

Fig. 1. Innervation of the cochlea and vestibular sensory epithelia. The two types of cochlea hair cells are innervated by three types of fibers (**A**): afferent fibers originating from two types of spiral ganglia (92% type I neurons to inner hair cells, 8% type II neurons to outer hair cells), efferent fibers originating from the superior olivary complex in the brainstem and autonomic fibers originating from the superior cervical ganglion (not shown). The two types of vestibular hair cells (**B**) are innervated by three types of afferent fibers distributed rather equally to calyceal endings around type I hair cells and bouton endings at type II hair cells. Losses of sensory neurons in single neurotrophin mutants (**C**) appear to correlate reasonable well with sensory neuron types (**A**) in the cochlea. There is, however, no apparent correlation with types of afferent endings in the vestibular epithelia (**B**). Closer examination shows that the distribution of afferents to the cochlea is not only segregated into inner and outer hair cell patterns but shows an additional longitudinal gradient such that density of innervation of inner hair cells is highest in the base whereas frequency of synapses on outer hair cells is highest in the apex. Modified after Wersaell and Bagger-Sjoebaeck (1974), Ernfors et al. (1995), Brown (1997), and Rubel and Fritzsch (2002).

Similar to the apparently specific requirement for NT-3 to support survival of spiral sensory neurons, vestibular neurons have an equally strong requirement for BDNF with deficiencies in the *BDNF* null mutant of 80–85% of vestibular neurons and 7% of spiral neurons (Jones et al., 1994; Ernfors et al., 1995; Schimmang et al., 1995; Bianchi et al., 1996). In contrast, *trkB* null mutant mice are reported to lose 56–85% of vestibular neurons and 15–20% spiral neurons (Fritzsch et al., 2001). Detailed counting has shown that neuronal loss happens within 2–3 days after the fibers have first extended toward the sensory epithelia (Bianchi et al., 1996). Together these data suggest that vestibular and cochlear (spiral) sensory neurons have distinct, but complementary neurotrophin requirements (Ernfors et al., 1995).

In the cochlea there are two types of hair cells (named inner and outer hair cells) that have different functions in audition. Each type of hair cell is innervated by a specific type of afferent fiber in a nonoverlapping fashion. Thus, approximately 92% of cochlear sensory neurons that form type I afferents converge on the inner hair cells whereas the 8% of these neurons that form type II afferents innervate instead the more numerous outer hair cells (Rubel and Fritzsch, 2002). The comparative numbers of type I and II sensory neurons are similar to the magnitudes of the spiral sensory neuron deficiencies reported for NT-3 and BDNF null mutations (85% and 7%, respectively), raising the possibility that the two classes of spiral sensory neurons require different neurotrophins for survival. This proposal, however, is not well supported by analysis of other sensory organs in the inner ear. Each vestibular sensory epithelium in mammals also consists of two types of hair cells, 50% type I and 50% type II, with distinct patterns of distribution, polarity and physiology (Lysakowski and Goldberg, 2003). These hair cells are innervated by three different types of afferent fibers; thick fibers forming large, calyceal endings around type I hair cells, thin fibers which end exclusively as boutons on type II hair cells and medium sized fibers that innervate type I hair cells with a calyx but also innervate type II hair cells with branches that form terminal boutons. The third type of afferent is referred to as dimorphic. Neither the sensory neuron loss reported in the *BDNF* mutant (80–85%) nor that observed in the *NT-3* mutant (23%) match these cellular distributions. In addition, BDNF appears to be expressed in all classes of differentiated hair cells in each of the sensory organs in the perinatal inner ear (Pirvola et al., 1992).

The relative distribution of neurotrophins with more prominent expression of BDNF in the vestibular system and of NT-3 in the cochlea does support the evidence for complementary roles of these neurotrophins in the vestibular and cochlear sensory epithelia, respectively (Pirvola et al., 1992; Farinas et al., 2001). As will be described below, the initial investigations did not examine the dynamics of neurotrophin expression throughout embryonic development and based conclusions on the assumption that the pattern of expression postnatally

provided strong evidence for functions during embryogenesis (Pirvola et al., 1992; Farinas et al., 2001). Moreover, the pattern of innervation in the ear raises several issues concerning the possible function of neurotrophins for regulating numerical matching between neurons and their targets. There is no uniform relationship between afferents and hair cells which can vary from 30–1 (convergence on a single inner hair cell) to 1–30 (divergence on outer hair cells and some vestibular fibers). It remains questionable how a single neurotrophin, such as BDNF, distributed fairly uniform in all hair cells, should be able to mediate these differences. Clearly, quantitative data on specific amounts of BDNF expressed in different types of hair cells are needed to evaluate this aspect of ear innervation.

While specific null mutations showed significant effects on sensory neurons survival to distinct endorgans of the ear, it remained unclear whether more neurotrophins or other neurotrophic factors might add to the survival of inner ear sensory neurons. However, double mutant mice which lack both the neurotrophin receptors TrkB and TrkC or both neurotrophins BDNF and NT-3 have no surviving sensory neurons in the inner ear at birth (Ernfors et al., 1995; Liebl et al., 1997; Silos-Santiago et al., 1997). This dramatic effect of double mutations on ear innervation lays therefore to rest any speculations about additional neurotrophins and neurotrophin receptors requirements. These data on double mutants also show that even if other ligands and receptors are present, like the ubiquitous p75 neurotrophin receptor, their function for the development of the inner ear sensory neurons is not critical as compared to BDNF/NT-3 and their receptors TrkB/TrkC. We recently followed up on earlier suggestions that at least some afferent fibers might survive in NT-3/BDNF double null mutations until birth (Ernfors et al., 1995). Indeed, in a small percentage of newborn NT-3/BDNF double null mutations some fibers to the cochlear apex survive (Coppola et al., unpublished observations). However, it remains unclear how long these cells and fibers would survive in the complete absence of BDNF and NT-3 since BDNF/NT-3 double mutants die soon after birth. It also remains unclear what additional neurotrophin or other neurotrophic substance is able to rescue at least in some NT-3/BDNF double null mutations those neurons. A likely candidate is glial cell line-derived neurotrophic factor (Hashino et al., 1999) which is expressed in the late embryonic mammalian cochlea (Qun et al., 1999).

Interestingly, neurotrophins are apparently downregulated in neonates (Wheeler et al., 1994; Fritzsch et al., 1999) and they appear to be largely lost in adults, despite the fact that their Trk receptors are still expressed in the sensory neurons (Fritzsch et al., 1999). This has led to the suggestion that other neurotrophic factors may play a role in the neonatal death of sensory neurons (Hashino et al., 1999; Echteler and Nofsinger, 2000), a suggestion that requires further experimental verification in mutants with conditional targeting of neurotrophin genes.

Specific losses of distinct cochlear and vestibular afferents occur in single neurotrophin and neurotrophin receptor null mutant mice

Initial data on NT-3 null mutant mice suggested an apparently uniform loss of afferent innervation only to the inner hair cells in the cochlea (Ernfors et al., 1995). In contrast, data generated with specific tracing techniques that could distinguish afferent from efferent fibers and autonomic fibers suggested a very different effect (Fritzsch et al., 1997a). The latter authors reported loss of afferent growth to outer hair cells and a selective loss of afferents in the basal turn of the cochlea. These suggestions were confirmed by follow up studies that included all three independently generated NT-3 mutant lines, including the original NT-3 null mutation on which this suggestion was based (Coppola et al., 2001; Farinas et al., 2001). While minor differences might exist between these mouse lines, it is now clear that all NT-3 null mutants lose all of the sensory neurons that normally innervate the basal turn of the cochlea, which is innervated instead by tangential extension along the inner hair cell row of sensory fibers derived from neurons that normally innervate only the middle turn (Fig. 2). Interestingly, when BDNF expression is reduced by introduction of one allele of the BDNF mutant into NT-3−/− embryos, the loss of cochlear afferent neurons expands further into the apical region, resulting in an even more exaggerated

elongation of apical afferent fibers along the inner hair cells of the middle and basal turn (Fig. 2).

Similar suggestions of inner hair cell specific effects as for NT-3 were raised by others in their studies of TrkC kinase-deficient mice (Schimmang et al., 1995; Schimmang et al., 1997). However, data generated with specific tracing techniques in the same mutant line suggested also loss of basal turn sensory neurons (Fritzsch et al., 1995). Interestingly, this effect was much less pronounced than in NT-3 null mutants and suggested, together with the less severe loss of sensory neurons, that the *trkC kin* −/− mutation effects might be mitigated by NT-3 signaling through TrkB. We recently investigated the effects of a TrkC-full length null mutation (Tessarollo et al., 1997) in the ear and found that in these mutants there is a loss of basal turn afferent innervation that is very similar to that observed in NT-3 null mutants (Fig. 2). This result suggests that truncated TrkC receptors, which continued to be expressed in the original TrkC kinase mutation, can support the survival of a limited subset of afferent fibers to the cochlea, probably in cooperation with TrkB (Postigo et al., 2002). Essentially, these data in the ear do not support the proposal that NT-3 signaling through TrkB plays a significant role in vestibular or spiral ganglion neurons development in vivo although there is convincing evidence that NT-3 signaling through TrkB is significant in other regions of the sensory nervous system.

A partially independent confirmation of the conclusions reached through analysis of the neurotrophin receptor null mutants has recently been obtained in other mutant mice. Mutation of the bHLH transcription factor gene NeuroD reduces expression of TrkC in spiral sensory neurons (Kim et al., 2001) and also results in a specific loss of basal turn neurons. Another mutation that is known to reduce the expression of TrkC, Brn3a, also resulted in a reduction in basal turn innervation (Huang et al., 2001). Together these data strongly support the hypothesis of a regional specific dependence of spiral sensory neurons on NT-3-promoted signaling through its receptor TrkC.

In apparent agreement with their suggestion of a selective loss of inner hair cell innervation in NT-3 null mutant mice, Ernfors et al. (1995) reported that all afferent innervation to outer hair cells is lost in

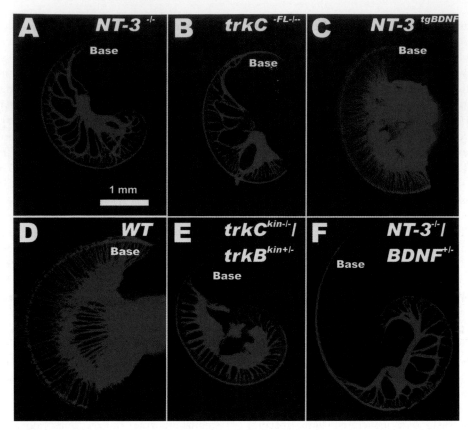

Fig. 2. The patterns of cochlear innervation in control and in various mutant mice. Innervation of the cochlea is fairly uniform along the entire length in P0 control mice (**D**). In *NT-3* null mutant mice there is loss of basal turn spiral sensory neurons (**A**). However, middle turn neurons extend process along the inner hair cells to reach the basal hair cells (**A**). Mice homozygous null for a mutant lacking all isoforms of TrkC (*trkC-FL−/−*) show an almost identical loss of basal turn sensory neurons and extension of afferents along the inner hair cells toward the base (**B**). In contrast, *trkC kin −/−* mice show a less severe phenotype (data not shown). However, introducing a single allele of *trkB kin-* into the *trkC kin −/−* background (**E**) mimics in great detail loss of the basal turn spiral sensory neurons found in *trkC-FL−/−* (**B**) and *NT-3 −/−* mutants (**A**). Introducing a single allele of a BDNF mutant (BDNF +/−) into the homozygous *NT-3* null (*NT3−/−*) background enhances the phenotype of the *NT3−/−*, resulting in loss of spiral neurons throughout the basal and middle turn. The remaining neurons in the apex extend along the middle and basal turn to innervate the inner hair cells. Inserting the BDNF coding region into the NT3 gene prevents the loss of spiral sensory neurons in the basal turn that is observed in the *NT-3* homozygotes (**C**). Together these data show that NT-3 and TrkC loss result in a regional specific loss of spiral sensory neurons that can be extended toward the apex by combining homozygous alleles of *NT-3* or *trkC* with heterozygous mutations of *BDNF* or *trkB*, respectively. Driving expression of BDNF under NT-3 promoter control results in rescue of the NT-3 phenotype, indicating that BDNF and NT-3 are functionally equivalent in their abilities to promote the survival of spiral sensory neurons. Modified after Fritzsch et al. (1998), Coppola et al. (2001) and Farinas et al. (2001).

BDNF null mutant mice. However, others found that BDNF null mutant mice retain outer hair cell innervation in the basal turn but have a reduced density of innervation in the apical turn of the cochlea including an apparent loss of afferents to the outer hair cells (Bianchi et al., 1996). These data were later confirmed by Farinas and colleagues (2001).

Data on TrkB kinase mutant mice were also in disagreement. In line with the suggestion of preferential effects of TrkC on inner hair cell innervation, one group reported loss of innervation of outer hair cells (Schimmang et al., 1995; 1997). In contrast, others working on the same mutation found a reduction of afferent innervation to the apex (Fritzsch et al., 1998). Moreover, basal turn spiral neurons may be the only neurons present at birth in the cochlea of a *trkB kin*$^{-/-}$ mutant mouse combined with *trkC kin*$^{+/-}$ heterozygosity. In addition, the *trkB kin*$^{-/-}$ mutation combined with *trkC kin*$^{+/-}$ heterozygosity shows a patchy and variable loss of middle turn spiral neurons in mice of different litters. It is possible that this heterogeneity of the TrkC kinase heterozygosity might reflect again a limited signaling capacity of the truncated TrkC receptor present in these mouse lines or it may represent an effect of the genetic variation of mice derived from different strains.

In the vestibular system, BDNF and its receptor TrkB are more important for neuronal survival and neurite growth than NT-3 and its receptor TrkC. While all reports agreed on this, slight differences in the pattern of losses were reported by different groups. Clearly, the three canal sensory epithelia show a complete or almost complete loss of innervation in either BDNF or TrkB null mutant mice (Ernfors et al., 1995; Fritzsch et al., 1995; Bianchi et al., 1996; Fritzsch et al., 1998). Reports on incomplete loss of afferent innervation to the canal sensory epithelia are based on nonspecific nerve fiber staining and might have erroneously identified efferent or autonomic nerve fibers known to exist in the ear (Fritzsch et al., 1998; Karis et al., 2001). While the organization of hair cells and the basic pattern of innervation is identical among all mammalian vestibular sensory epithelia (Lysakowski and Goldberg, 2003), there is a clear difference between the gravistatic receptors (saccule and utricle) and angular acceleration epithelia (the three canal cristae): gravistatic receptors show a limited afferent innervation whereas no afferent fibers reach the canal cristae at birth in BDNF or TrkB mutant mice (Fritzsch et al., 1998). As with the cochlea, losses to the other vestibular epithelia, already reduced in single BDNF or TrkB mutations, are further increased by the loss of one NT-3 or TrkC allele (Fritzsch et al., 1998).

Together these data on individual NT-3, BDNF, TrkB kinase and TrkC kinase deficient mice suggest that NT-3 regulates exclusively the survival of basal turn cochlear sensory neurons while BDNF is a major, but not an exclusive, player for the apical turn sensory neuron survival. In contrast, the canal cristae innervation depends exclusively on BDNF to TrkB signaling whereas the saccular and utricular afferents depend in part on NT-3/TrkC as some innervation is retained in BDNF or TrkB mutants.

Neurotrophins in the ear show a highly dynamic pattern of expression

Before we can correlate the specific losses in afferent innervation to further evaluate the suggestions of specific hair cell attraction made by Cajal, we need to investigate in more detail the spatio-temporal pattern of expression of the two neurotrophins essential for the ear sensory neuron survival, BDNF and NT-3, and their receptors, TrkB and TrkC.

Both during development and in the adult sensory neurons there appears to be a uniform expression of both TrkB and TrkC on all of the sensory neurons in the ear (Fritzsch et al., 1999; Farinas et al., 2001). While no quantitative data exist on the expression levels of those two receptors in various vestibular and auditory sensory neurons, it appears likely that neither spatial nor temporal differences in TrkB and TrkC expression contribute significantly to the complex pattern of spatio-temporal losses reported in neurotrophin and neurotrophin receptor mutant mice (Fritzsch et al., 1995; Bianchi et al., 1996; Farinas et al., 2001). Therefore, the loss of afferents should predominantly relate to differential distribution of ligands. In our attempt to correlate the patterns of afferent fiber loss with neurotrophin expression, we employed mice with a lacZ reporter inserted in the *BDNF* and *NT-3* loci (Farinas et al., 2001; Fritzsch et al., 2002).

Primary sensory neurons express neurotrophins soon after delamination is initiated from the placodal epithelium (Fig. 3). Furthermore, a given primary sensory neuron expresses the same neurotrophin which is present in the area of the otocyst from which it delaminated (Fritzsch et al., 2002). Recent experiments show that such delamination of putative neural precursors occurs in *math1* homozygotes, even though no hair cells are ever formed in this mutant (Fritzsch et al., submitted). It is likely that all neurotrophin positive delaminating cells are sensory neuron precursors as no delaminating *BDNF*-LacZ positive cells are observed in *ngn1* mutants, animals that never form any inner ear sensory

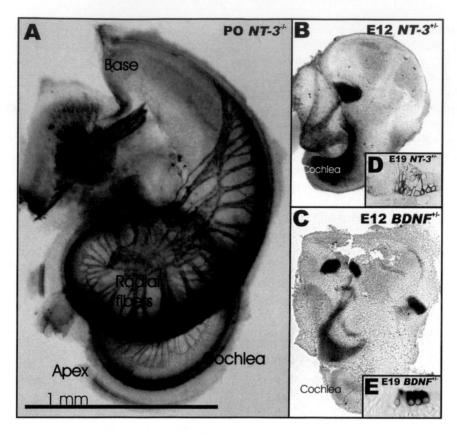

Fig. 3. Correlation of afferent loss with neurotrophin expression. Combining *NT-3*lacZ expression (green in **A**) with anti-tubulin stain for nerve fibers (brown) shows no correlation of nerve fiber loss in *NT-3* mutant homozygotes with the expression pattern of *NT-3* lacZ at birth. Similarly, there is no correlation between nerve fiber loss in *BDNF* mutant homozygotes and the pattern of BDNF-lacZ expression at birth (Farinas et al., 2001). However, examination of younger embryos shows distinct differential distribution of BDNF and NT-3 as revealed by the LacZ reporter reaction product (**B**, **C**). Closer examination shows that the differential distribution of neurotrophins extends to the cellular level and shows expression of BDNF almost exclusively in hair cells (**E**) whereas NT-3 is almost exclusively expressed in supporting cells (**D**). Data modified after Farinas et al. (2001).

neurons (Ma et al., 1998, 2000). Comparison of *BDNF-* and *NT-3*-LacZ positive cells with delaminating neurons marked by NeuroD-LacZ also suggests that the delaminating BDNF or NT-3 positive precursors are in fact NeuroD-expressing neuronal precursors (Liu et al., 2000; Kim et al., 2001; Fritzsch et al., 2002) but proof will require colabeling for NeuroD and each of the neurotrophins. Alternatively, one should predict that the delaminating cells that are nonneuronal will not be affected by the absence of NeuroD or Ngn1 during later development, a prediction that could be examined by tracing the fates of these neurotrophin-positive cells.

Detailed analysis of NT-3- and BDNF-lacZ-positive cells in combination with immunocytochemistry shows that as soon as these cells intiate expression of neuronal markers such as tubulin or the Trk receptors, they cease expression of the neurotrophins (Fritzsch et al., 2002). These data also suggest that fibers that develop from these sensory neurons project toward sensory epithelia along the delaminating neurotrophin positive precursors. Whether the expression of neurotrophins in these precursors is biologically meaningful will require use of floxed mice where deletion of each neurotrophin can be restricted to the otocyst. Below we briefly report on an approach that uses mutants with severe impairments in hair cell differentiation in an effort to achieve this otocyst and hair cell specific loss of BDNF.

Beyond the possible role this expression might play in delaminating neuron precursors, detailed analysis of peripheral projections suggests that the initial projections of sensory neurons is rather normal in mice lacking only one neurotrophin (Fritzsch et al., 1995; Bianchi et al., 1996; Farinas et al., 2001). However, within two to three days, fibers and shortly afterwards primary sensory neurons disappear, generating the specific patterns of loss observed in these mutants (Fritzsch et al., 1995, 1997b, 1999) with all afferents and neurons lost in the double null mutations (Silos-Santiago et al., 1997).

These data indicate that initial fiber growth occurs normally in the absence of neurotrophins. Indeed, initial fiber growth may use the same molecular cues recognized by the delaminating primary sensory neuron precursors. However, there is subsequently a critical period of neurotrophin dependency that will result in elimination of all connections that reach areas normally or experimentally deprived of specific neurotrophins. Such aberrantly growing fibers have been reported in the developing, but not in the adult ear (Lorente de No, 1926). The partially overlapping expression of neurotrophins reported in the developing mammalian ear (Pirvola et al., 1992; Farinas et al., 2001) seems to translate into a spatio-temporal loss of primary sensory neurons in specific mutants (Farinas et al., 2001). Particularly, in the cochlea there is a delayed upregulation of BDNF expression in the basal turn, leaving all basal turn neurons solely dependent on NT-3 during a brief, but critical period of embryogenesis. Thus, if NT-3 is absent, there is a progressive loss of spiral neurons, especially in the basal turn, where BDNF is not present to compensate for the absence of NT-3. Overall, sensory neuron loss will occur in an embryo with a targeted mutation in BDNF or NT-3 only where the other is not present to compensate for its absence.

This suggestion has led to the prediction that in the ear, NT-3 and BDNF are functionally equivalent and can be substituted for each other without compromising the survival and development of sensory neurons. Consistent with this, data show that the topological loss of sensory neurons in the basal turn in the NT-3 mutant can be rescued by transgenic expression of BDNF under the control of NT-3 gene regulatory elements (Coppola et al., 2001). Whether the corresponding transgenic animal in which the NT-3 coding region is inserted into the BDNF gene is equally effective in rescuing the BDNF phenotype remains to be seen.

Overall, these data support a role for a very early onset of elimination of exuberant or unconnected afferents and the primary sensory neurons that generate these axons. This verification of proper connections occurs immediately after the fibers have reached and start to invade their target organs (as early as E11 in the canal epithelia and E13 in the basal turn of the cochlea). These early survival effects do not preclude a subsequent or even concomitant role in fiber guidance, by neurotrophic factors.

Indeed, in transgenic animals rescues in which BDNF is expressed under NT-3 promoter control, more afferents are observed extending to the basal turn of the cochlea where they innervate outer hair cells at abnormally high density (Figs. 2 and 4). Some of these afferents appear to be vestibular fibers rerouted from the nerve branch to the posterior crista (Fig. 4). In addition, some afferents seem to get misrouted underneath the basilar membrane without ever reaching the hair cells of the organ of Corti (Coppola et al., 2001). It is conceivable that such 'disoriented fiber growth' is the consequence of conflicts between pathfinding and neurotrophin signals. Clearly, such transgenic animals have to be studied in more detail to reveal the nature of these unusual pathfinding properties.

The most interesting and striking effect of the single neurotrophin null mutation is the striking dependence of the basal turn cochlear neurons on NT-3. This is due to the fact that BDNF shows a delayed expression in the basal turn. Clearly, BDNF can not only compensate for NT-3 and rescue the basal turn neurons but can also attract vestibular fibers from the nearby nerve to the posterior crista to innervate the basal turn instead (Fig. 4). It is conceivable that evolutionary pressures have resulted in the delayed expression of BDNF in the basal turn because of the need to avoid misrouting of vestibular afferents. Misrouting of vestibular fibers to cochlear hair cells, assuming they maintain their normal connections in the CNS, will result in auditory

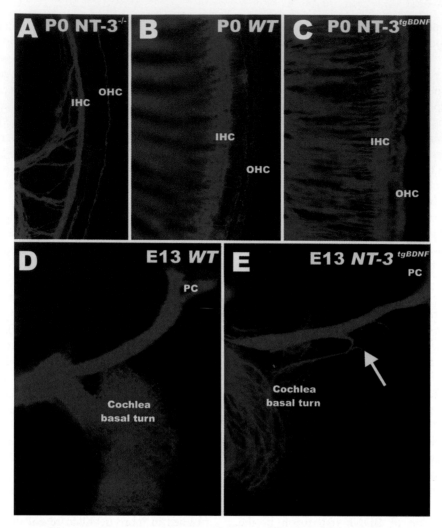

Fig. 4. *NT-3tgBDNF* rescues the *NT-3* phenotype and causes afferent rerouting. This micrographs show the almost complete loss of afferents in the basal turn in NT-3 null mutants, except for inner hair cells (**A**). Control littermates show the normal pattern of dense innervation of inner hair cells and less dense, but continuous innervation of three rows of afferent fibers to outer hair cells (**B**), *NT-3 tgBDNF* mice show rescue of innervation to the basal turn, but also a much more exuberant growth of fibers to the outer hair cell region (**C**) than is observed in the control (**B**). Closer examination of 13-day-old embryos suggest that at least some vestibular fibers to the posterior canal are rerouted toward the basal turn (**D**, **E**) and may contribute to the denser innervation of this part of the cochlea in *NT-3 tgBDNF* homozygous mice. Modified after Coppola et al. (2001) and Farinas et al. (2001).

information interfering with perceptions of position and motion.

BDNF expression is present in mutants defective in hair cell differentiation

As indicated above, one way of eliminating neurotrophins in the target of the inner ear afferents might be to eliminate hair cells or prevent their differentiation. This is expected to eliminate expression of BDNF because in older embryos BDNF is expressed exclusively in hair cells within the ear (Fig. 3). In addition, if no hair cells form it is likely that supporting cells will not form normally as both are apparently linked in their differentiation via reciprocal interactions mediated through the delta-notch regulatory system (Zine et al., 2001). Recent investigations into two mutations that result in undifferentiated hair cells provide, however, no indication for losses of sensory neurons in these mutants resulting from neurotrophin deficiency although there are examples of sensory neuron loss that appear to be caused by abnormally low expression of the Trk receptors in other mutants (Liu et al., 2000; Huang et al., 2001; Kim et al., 2001).

Mice lacking the Pou-domain-containing transcription factor, Brn3c, develop only a limited complement of morphologically undifferentiated hair cells which can be identified as hair cells only because they express certain hair cell-specific molecular markers (Xiang et al., 1998). Closer examination of afferent innervation showed no correlation of fiber loss with the formation of these morphologically undifferentiated hair cells (Xiang et al., 2003). Specifically, a robust sensory innervation persists through embryogenesis into early neonatal life. The survival throughout embryogenesis is apparently mediated by the limited expression of both neurotrophins in undifferentiated sensory epithelia as revealed with in situ hybridization. Even several months old animals had a considerable innervation of the apical turn of the cochlea. However, this long term retention of cochlear innervation is likely not mediated by neurotrophins as these are down-regulated in neonatal animals (Wheeler et al., 1994).

It also seemed possible that examination of mutants lacking the bHLH transcription factor Math1 would make it possible to test the possible functions of guidance and survival factors released from hair cells on the guidance and survival of sensory afferent neurons. Analysis shows that Math1 is required for hair cell differentiation and probably acts upstream of Brn3c (Bermingham et al., 1999; Fritzsch et al., 2000). Surprisingly, there is very little effect of this mutation on the initial fiber growth of sensory neurons (Fritzsch et al., submitted). However, older embryos show a severe reduction of afferents that does not fit to the pattern of loss observed in neurotrophin null mutations. Closer examination of the expression of BDNF using the BDNF-LacZ reporter showed that even in *math1* null mutants some undifferentiated hair cell precursors form and express BDNF. Thus, at least in some hair cell precursors, BDNF expression does nor require Math1-mediated hair cell differentiation.

Together the data summarized above show that at present none of the attempts to eliminate neurotrophin expression in the ear through mutation of essential transcription factors has made it possible to test the proposal of Ramon y Cajal that hair cells secrete neurotrophic substance(s) that attract sensory afferents. Unfortunately, the recent finding of neurotrophin expression in delaminating sensory neurons has made the interpretation of neurotrophin effects even more complex. Clearly, the limited expression of BDNF in the undifferentiated hair cell precursors of Math1-null mutant mice is apparently enough to support many afferents throughout embryonic life. Consequently, we cannot exclude a biologically significant effect of the limited expression of neurotrophins within delaminating sensory neurons. Thus, none of the mutations described above has critically tested the role exclusively played by hair cells in attraction and maintenance of inner ear sensory neurons. Further work on more targeted mutations eliminating neurotrophins only in the sensory epithelia, but not in the delaminating sensory neurons, will be needed to test the proposal made by Ramon y Cajal almost 100 years ago.

Conclusions

The data at hand allow the following conclusions to be drawn. Neurotrophins are not associated with specific hair cell types. All hair cells of all sensory epithelia express BDNF in late embryonic life, even when hair cells differentiation is severely impaired. In contrast, expression of NT-3 is largely restricted to supporting cells of the cochlea, saccule and utricle. The topologically restricted spatio-temporal expression pattern of neurotrophins corresponds well to the different patterns of neuron and fiber loss reported for different mutants. In contrast, there is no evidence that Trk receptors are differentially expressed in spiral or vestibular sensory neurons. Given that the entire pattern of sensory neuron loss appears to reflect the patterns of neurotrophin expression, it is not surprising that expression of BDNF can compensate for the absence of NT-3 when BDNF is expressed under NT-3 promoter control. Thus, in the cochlea, the two neurotrophins have equal potential for supporting neuronal survival and can be substituted for each other. Their differential effect in single null mutations reflects the spatio-temporal expression differences and not any specific function of each neurotrophin

Acknowledgments

Supported by NIH (PPG 2P01 DC00215; RO1 DC005950; BF) and the Howard Hughes Medical Institute (LFR). LFR is an investigator of the Howard Hughes Medical Institute.

Abbreviations

BDNF   brain-derived neurotrophic factor
NT-3   neurotrophin-3
TrkB   tyrosine kinase receptor B
TrkC   tyrosine kinase receptor C

References

Bermingham, N.A., Hassan, B.A., Price, S.D., Vollrath, M.A., Ben-Arie, N., Eatock, R.A., Bellen, H.J., Lysakowski, A. and Zoghbi, H.Y. (1999). Math1: an essential gene for the generation of inner ear hair cells. Science, 284: 1837–1841.

Bernd, P. and Represa, J. (1989). Characterization and localization of nerve growth factor receptors in the embryonic otic vesicle and cochleo-vestibular ganglion. Dev. Biol., 134: 11–20.

Bianchi, L.M. and Cohan, C.S. (1991). Developmental regulation of a neurite-promoting factor influencing statoacoustic neurons. Brain Res. Dev. Brain Res., 64: 167–174.

Bianchi, L.M., Conover, J.C., Fritzsch, B., DeChiara, T., Lindsay, R.M. and Yancopoulos, G.D. (1996). Degeneration of vestibular neurons in late embryogenesis of both heterozygous and homozygous BDNF null mutant mice. Development, 122: 1965–1973.

Brown, M.C. (1997). Auditory physiology. In: G.M. English (Ed.), Otolaryngology. Lippincott-Raven, Philadelphia, pp. 1–54.

Cajal, S.R. (1919). Accion neurotropica de los epitelios. Trab. del Lab. de Invest. biol., 17: 1–153.

Coppola, V., Kucera, J., Palko, M.E., Martinez-De Velasco, J., Lyons, W.E., Fritzsch, B. and Tessarollo, L. (2001). Dissection of NT3 functions in vivo by gene replacement strategy. Development, 128: 4315–4327.

Echteler, S.M. and Nofsinger, Y.C. (2000). Development of ganglion cell topography in the postnatal cochlea. J. Comp. Neurol., 425: 436–446.

Ernfors, P., Van De Water, T., Loring, J. and Jaenisch, R. (1995). Complementary roles of BDNF and NT-3 in vestibular and auditory development. Neuron, 14: 1153–1164.

Farinas, I., Jones, K.R., Backus, C., Wang, X.Y. and Reichardt, L.F. (1994). Severe sensory and sympathetic deficits in mice lacking neurotrophin-3. Nature, 369: 658–661.

Farinas, I., Jones, K.R., Tessarollo, L., Vigers, A.J., Huang, E., Kirstein, M., de Caprona, D.C., Coppola, V., Backus, C., Reichardt, L.F., Fritzsch, B. (2001). Spatial shaping of cochlear innervation by temporally regulated neurotrophin expression. J. Neurosci., 21: 6170–6180.

Fritzsch, B., Barbacid, M. and Silos-Santiago, I. (1998). The combined effects of trkB and trkC mutations on the innervation of the inner ear. Int. J. Dev. Neurosci., 16: 493–505.

Fritzsch, B., Beisel, K.W. and Bermingham, N.A. (2000). Developmental evolutionary biology of the vertebrate ear: conserving mechanoelectric transduction and developmental pathways in diverging morphologies. Neuroreport, 11: R35–R44.

Fritzsch, B., Beisel, K.W., Jones, K., Farinas, I., Maklad, A., Lee, J. and Reichardt, L.F. (2002). Development and evolution of inner ear sensory epithelia and their innervation. J. Neurobiol., 53: 143–156.

Fritzsch, B., Farinas, I. and Reichardt, L.F. (1997a). Lack of neurotrophin 3 causes losses of both classes of spiral ganglion neurons in the cochlea in a region-specific fashion. J. Neurosci., 17: 6213–6225.

Fritzsch, B., Pirvola, U. and Ylikoski, J. (1999). Making and breaking the innervation of the ear: neurotrophic support during ear development and its clinical implications. Cell Tissue Res., 295: 369–382.

Fritzsch, B, Silos-Santiago, I., Bianchi, L.M. and Farinas, I. (1997b). The role of neurotrophic factors in regulating the development of inner ear innervation. Trends Neurosci., 20: 159–164.

Fritzsch, B., Silos-Santiago, I., Bianchi, L.M. and Farinas, I. (1997c). The role of neurotrophic factors in regulating the development of inner ear innervation. Trends Neurosci., 20: 159–164.

Fritzsch, B., Silos-Santiago, I., Farinas, I. and Jones, K.R. (2001). Neurotrophins and neurotrophin receptors involved in supporting afferent inner ear innervation. In: I. Mocchetti (Ed.), Neurobiology of the Neurotrophins. FP Graham Publishing Co., Johnson City, TN, pp. 149–163.

Fritzsch, B., Silos-Santiago, I., Smeyne, R., Fagan, A.M. and Barbacid, M. (1995). Reduction and loss of inner ear innervation in trkB and trkC receptor knockout mice: A whole mount DiI and scanning electron microscopic analysis. Audit. Neurosci., 1: 401–417.

Hashino, E., Dolnick, R.Y. and Cohan, C.S. (1999). Developing vestibular ganglion neurons switch trophic sensitivity from BDNF to GDNF after target innervation. J. Neurobiol., 38: 414–427.

Huang, E.J., Liu, W., Fritzsch, B., Bianchi, L.M., Reichardt, L.F. and Xiang, M. (2001). Brn3a is a transcriptional regulator of soma size, target field innervation and axon pathfinding of inner ear sensory neurons. Development, 128: 2421–2432.

Jones, K.R., Farinas, I., Backus, C. and Reichardt, L.F. (1994). Targeted disruption of the BDNF gene perturbs brain and sensory neuron development but not motor neuron development. Cell, 76: 989–999.

Karis, A., Pata, I., van Doorninck, J.H., Grosveld, F., de Zeeuw, C.I., de Caprona, D. and Fritzsch, B. (2001). Transcription factor GATA-3 alters pathway selection of olivocochlear neurons and affects morphogenesis of the ear. J. Comp. Neurol., 429: 615–630.

Kim, W.Y., Fritzsch, B., Serls, A., Bakel, L.A., Huang, E.J., Reichardt, L.F., Barth, D.S. and Lee, J.E. (2001). NeuroD-null mice are deaf due to a severe loss of the inner ear sensory neurons during development. Development, 128: 417–426.

Klein, R., Silos-Santiago, I., Smeyne, R.J., Lira, S.A., Brambilla, R., Bryant, S., Zhang, L., Snider, W.D. and Barbacid, M. (1994). Disruption of the neurotrophin-3 receptor gene trkC eliminates Ia muscle afferents and results in abnormal movements. Nature, 368: 249–251.

Levi-Montalcini, R. (1949). The development of the acousticovestibular centers in the chick embryo in the absence of the afferent root fibers and of descending fiber tracts. J. Comp. Neurol., 91: 209–241.

Liebl, D.J., Tessarollo, L., Palko, M.E. and Parada, L.F. (1997). Absence of sensory neurons before target innervation in brain-derived neurotrophic factor-, neurotrophin3-, and trkC-deficient embryonic mice. J. Neurosci., 17: 9113–9127.

Lindsay, R.M., Barde, Y.A., Davies, A.M. and Rohrer, H. (1985). Differences and similarities in the neurotrophic growth factor requirements of sensory neurons derived from neural crest and neural placode. J. Cell Sci., 3: 115–129.

Liu, M., Pereira, F.A., Price, S.D., Chu, M.J., Shope, C., Himes, D., Eatock, R.A., Brownell, W.E., Lysakowski, A. and Tsai, M.J. (2000). Essential role of BETA2/NeuroD1 in development of the vestibular and auditory systems. Genes Dev., 14: 2839–2854.

Lorente, de No R. (1926) Etudes sur l'anatomie et la physiologie du labyrinthe de l'oreille et du VIII nerf. II. Quelques donnees au sujet de l'anatomie des organes sensoriels du labyrinthe. Trav. Lab. Rech. Biol. Univ. Madrid, 24: 53–153.

Lysakowski, A.L. and Goldberg, J.M. (2003) Morphophysiology of the vestibular periphery. In: S. Highstein, A.N. Popper and R.R. Fay (Eds.), The Anatomy and Physiology of the Central and Peripheral Vestibular System. Springer, New York pp. 57–152.

Ma, Q., Anderson, D.J. and Fritzsch, B. (2000). Neurogenin 1 null mutant ears develop fewer, morphologically normal hair cells in smaller sensory epithelia devoid of innervation. J. Assoc. Res. Otolaryngol., 1: 129–143.

Ma, Q., Chen, Z., del Barco Barrantes, I., de la Pompa, J.L. and Anderson, D.J. (1998). Neurogenin 1 is essential for the determination of neuronal precursors for proximal cranial sensory ganglia. Neuron, 20: 469–482.

Pirvola, U., Ylikoski, J., Palgi, J., Lehtonen, E., Arumae, U. and Saarma, M. (1992). Brain-derived neurotrophic factor and neurotrophin 3 mRNAs in the peripheral target fields of developing inner ear ganglia. Proc. Natl. Acad. Sci. USA, 89: 9915–9919.

Postigo, A., Calella, A.M., Fritzsch, B., Knipper, M., Katz, D., Eilers, A., Schimmang, T., Lewin, G.R., Klein, R. and Minichiello, L. (2002). Distinct requirements for TrkB and TrkC signaling in target innervation by sensory neurons. Genes Dev., 16: 633–645.

Qun, L.X., Pirvola, U., Saarma, M. and Ylikoski, J. (1999). Neurotrophic factors in the auditory periphery. Ann. N.Y. Acad. Sci., 884: 292–304.

Rubel, E.W. and Fritzsch, B. (2002). Auditory system development: primary auditory neurons and their targets. Annu. Rev. Neurosci., 25: 51–101.

Schimmang, T., Alvarez-Bolado, G., Minichiello, L., Vazquez, E., Giraldez, F., Klein, R. and Represa, J. (1997). Survival of inner ear sensory neurons in trk mutants. Mech. Dev., 64: 77–85.

Schimmang, T., Minichiello, L., Vazquez, E., San Jose, I., Giraldez, F., Klein, R. and Represa, J. (1995). Developing inner ear sensory neurons require TrkB and TrkC receptors for innervation of their peripheral targets. Development, 121: 3381–3389.

Silos-Santiago, I., Fagan, A.M., Garber, M., Fritzsch, B. and Barbacid, M. (1997). Severe sensory deficits but normal CNS development in newborn mice lacking TrkB and TrkC tyrosine protein kinase receptors. Eur. J. Neurosci., 9: 2045–2056.

Tessarollo, L., Tsoulfas, P., Donovan, M.J., Palko, M.E., Blair-Flynn, J., Hempstead, B.L. and Parada, L.F. (1997). Targeted deletion of all isoforms of the trkC gene suggests the use of alternate receptors by its ligand neurotrophin-3 in neuronal development and implicates trkC in normal cardiogenesis. Proc. Natl. Acad. Sci. USA, 94: 14776–14781.

Wersaell, J. and Bagger-Sjoebaeck, D. (1974). Morphology of the vestibular sense organ. In: H.H. Kornhuber (Ed.), Handbook of Sensory Physiology. Springer Verlag, Berlin, pp. 123–170.

Wheeler, E.F., Bothwell, M., Schecterson, L.C. and von Bartheld, C.S. (1994). Expression of BDNF and NT-3 mRNA in hair cells of the organ of Corti: quantitative analysis in developing rats. Hear Res., 73: 46–56.

Xiang, M., Gao, W.Q., Hasson, T. and Shin, J.J. (1998). Requirement for Brn-3c in maturation and survival, but not in fate determination of inner ear hair cells. Development, 125: 3935–3946.

278

Xiang, M., Maklad, A., Pirvola, U. and Fritzsch, B. (2003) Brn3c null mutant mice show long-term, incomplete retention of some afferent inner ear innervation. Dev. BMC Neurosci, 4: 1–16.

Zine, A., Aubert, A., Qiu, J., Therianos, S., Guillemot, F., Kageyama, R. and de Ribaupierre, F. (2001). Hes1 and Hes5 activities are required for the normal development of the hair cells in the mammalian inner ear. J. Neurosci., 21: 4712–4720.

Progress in Brain Research, Vol. 146
ISSN 0079-6123

CHAPTER 18

Neurotrophin presence in human coronary atherosclerosis and metabolic syndrome: a role for NGF and BDNF in cardiovascular disease?

George N. Chaldakov[1], Marco Fiore[2], Ivan S. Stankulov[1], Luigi Manni[2],
Mariyana G. Hristova[1], Alessia Antonelli[2], Peter I. Ghenev[3] and Luigi Aloe[2,*]

[1]*Division of Cell Biology, Department of Forensic Medicine, Medical University, Varna, Bulgaria*
[2]*Institute of Neurobiology and Molecular Medicine, Consiglio Nazionale delle Ricerche (CNR), Viale Marx 15/43, I-00137 Rome, Italy*
[3]*Department of General and Clinical Pathology, Medical University, Varna, Bulgaria*

Abstract: The development of atherosclerotic cardiovascular disease is a common comorbidity in patients with the metabolic syndrome, a concurrence of cardiovascular risk factors in one individual. While multiple growth factors and adipokines are identified in atherosclerotic lesions, as well as neurotrophins implicated in both cardiac ischemia and lipid and glucose metabolism, the potential role of neurotrophins in human coronary atherosclerosis and in the metabolic syndrome still remains to be elucidated. Here we describe and discuss our results that represent a novel attempt to study the cardiovascular and metabolic biology of nerve growth factor (NGF), brain-derived neurotrophic factor (BDNF) and mast cells (MC). The local amount of NGF, the immunolocalization of p75 neurotrophin receptor (p75^{NTR}) and the number of MC were correlatively examined in coronary vascular wall and in the surrounding subepicardial adipose tissue, obtained from autopsy cases in humans with advanced coronary atherosclerosis. We also analyzed the plasma levels of NGF, BDNF and leptin and the number of MC in biopsies from abdominal subcutaneous adipose tissue in patients with a severe form of the metabolic syndrome. The results demonstrate that NGF levels are decreased in atherosclerotic coronary vascular tissue but increased in the subepicardial adipose tissue, whereas both tissues express a greater number of MC and a stronger p75^{NTR} immunoreactivity, compared to controls. Metabolic syndrome patients display a significant hyponeurotrophinemia and an increased number of adipose MC; the later correlates with elevated plasma leptin levels. In effect, we provide the first evidence for (i) an altered presence of NGF, p75^{NTR} and MC in both coronary vascular and subepicardial adipose tissue in human coronary atherosclerosis, and (ii) a significant decrease in plasma NGF and BDNF levels and an elevated amount of plasma leptin and adipose MC in metabolic syndrome patients. Together our findings suggest that neuroimmune mediators such as NGF, BDNF, leptin and MC may be involved in the development of cardiovascular disease and related disorders.

Keywords: p75^{NTR}; adipose tissue; mast cells; leptin

Introduction

Atherosclerosis is an inflammatory vascular disease affecting large and medium-sized arteries (reviewed

*Correspondence to: L. Aloe, Institute of Neurobiology and Molecular Medicine, Consiglio Nazionale delle Ricerche (CNR), Viale Marx 15/43, I-00137 Rome, Italy.
Tel.: +39-06-8292-592; Fax: +39-06-8609-0269;
E-mail: aloe@in.rm.cnr.it

by Ross, 1999; Scott, 2002). Coronary atherosclerosis causes ischemic heart disease and acute coronary syndromes, including myocardial infarction, whereas cerebral atherosclerosis causes stroke, and peripheral atherosclerosis contributes to gangrene. According to the World Health Organization's prognosis for the year 2020, atherosclerotic vascular disease will be one of the most prevalent diseases that has reached epidemic proportions worldwide (see Scott, 2002).

DOI: 10.1016/S0079-6123(03)46018-4

The metabolic syndrome is a concurrence of cardiovascular risk factors (visceral obesity, disturbed glucose and insulin metabolism, dyslipidemia, and hypertension) in one individual. The development of atherosclerotic cardiovascular disease is a common comorbidity in metabolic syndrome patients (Lakka et al., 2002; Meigs, 2002). Twenty to 25% (about 47 million) of the adult population in the United States have metabolic syndrome, and in some older groups this prevalence approaches 50% (Ford et al., 2002; Keller and Lemberg, 2003). Overall, the human and economic burden of both atherosclerosis and metabolic syndrome is obvious (Turtle, 2000; Keller and Lemberg, 2003).

Neurotrophins and immune cells: a rational for correlative implication in cardiovascular disease

Life at the cellular level requires growth-promoting trophic support. Work initiated by the discovery of nerve growth factor (NGF) in the early 1950's and later embodied in the neurotrophic theory has brought an increasing impact on the progress of neuroscience (Levi-Montalcini, 1997). NGF is a member of the neurotrophin family of proteins, including brain-derived neurotrophic factor (BDNF), neurotrophin-3 (NT-3), NT-4/5, NT-6 and NT-7 (reviewed by Sofroniew et al., 2001). Biological actions of NGF are mediated by the initial ligation of two different cell surface receptors: (i) the low-affinity p75 neurotrophin receptor (p75^{NTR}), and (ii) the high-affinity receptor tyrosine kinase A (TrkA) (Sofroniew et al., 2001).

The NGF field has witnessed a number of breakthroughs in recent years. There is compelling evidence now that NGF and other neurotrophins, in addition to their neurotrophic function, enhance survival and activity of a large number of nonneuronal cells. Moreover, neurotrophins, particularly NGF and BDNF, are synthesized, stored, and released not only by directly innervated cells but also by immune cells (reviewed by Aloe et al., 2001), vascular endothelial cells (Nakahashi et al., 2000), platelets (Fujimura et al., 2002), and fibroblasts/myofibroblasts (Micera et al., 2001). All these cell types, together with perivascular nerves and the

TABLE 1

Cellular targets and sources for neurotrophins as potentially related to atherogenesis

Immune cells	Other cells
Mast cells	Endothelial cells
Lymphocytes	Vascular smooth muscle cells
Macrophages	Fibroblasts/myofibroblasts
Dendritic cells	Platelets
Neutrophils	Adipocytes
	Perivascular nerves

classical NGF-secreting target cells vascular smooth muscle cells (Creedon and Tuttle, 1997; Cowen and Gavazzi, 1998) and adipocytes (Nisoli et al., 1996, 1998; Chaldakov et al., 2001a; Paul Trayhurn, personal communication), are increasingly implicated in the process of atherogenesis (Chaldakov and Vankov, 1986; Scott et al., 1992; Kacem et al., 1997; Ross, 1999; Chaldakov et al., 2000) (Table 1).

Multiple growth factors and adipokines are identified in atherosclerotic lesions (Ross, 1999; Scott, 2002; Young et al., 2002) and in the metabolic syndrome (Matsuzawa et al., 1999; Loskutoff et al., 2000; Trayhurn and Beattie, 2001; Grimble, 2002). Neurotrophins and immune cells are viewed at present as a complex network of pleiotropic mediators reportedly implicated in both neurological and nonneurological diseases (Villoslada et al., 2000; Aloe et al., 2001; Sofroniew et al., 2001). In this context, NGF and BDNF are known as crucially involved in (i) neuroimmune and inflammatory responses (Aloe et al., 2001; also see this issue), (ii) vascular responses (Donovan et al., 1995; Abe et al., 1997; Hiltunen et al., 2001; Qin et al., 2002), and (iii) metabolic responses (Rios et al., 2001; Nakagawa et al., 2002; Tsuchida et al., 2002; Hanyu et al., 2003; Kuroda et al., 2003). Although such events might also be associated with the initiation and development of both atherosclerosis and metabolic syndrome, the potential importance of neurotrophins in these diseases has only recently emerged (Donovan et al., 1995; Creedon and Tuttle, 1997; Nemono et al., 1998; McCaffrey et al., 2000; Wang et al., 2000; Chaldakov et al., 2001a,b).

Connecting some of the dots (described above) may provide a strong rational for correlative study of the potential roles of neurotrophins and immune cells in coronary atherosclerosis and in the metabolic syndrome. Here we describe and discuss the results of our first steps in addressing this issue.

NGF and mast cells in coronary atherosclerosis

The specimens of human coronary arteries were obtained from 21 autopsy cases performed at the Department of Forensic Medicine and Cell Biology, Medical University, Varna, Bulgaria, in accordance with the Helsinki Convention, and carried out following the Bulgarian and Italian legislations on biomedical research. The proximal 15–20 mm of the left anterior descending coronary artery (LAD), together with the surrounding subepicardial adipose tissue (SEAT) and the outermost layers of epimyocardium, were removed from the heart and cut into three consecutive portions, each measuring 5–7 mm in length, used for histological, NGF amount and immunohistochemical analyses, respectively. We examined LAD coronary vascular wall proper and the adjacent SEAT and myocardium of age-matched men without atherosclerotic lesions ($n = 9$; herein referred to as controls) and with atherosclerotic lesions ($n = 12$), including type II lesions ($n = 1$), type V lesions ($n = 4$), and type VI lesions ($n = 7$), according to Stary et al. (1995), thus most lesions were in advanced stages of atherosclerosis.

For a histological evaluation of MC, specimens from coronary arteries were fixed in 4% paraformaldehyde, cut at 15 μm thickness using a cryomicrotome, and stained with 0.5% toluidine blue (pH 2.5).

The content (pg/g) of NGF in coronary vascular wall and the adjacent SEAT and myocardium was measured by a highly sensitive two-site immunoenzymatic assay (ELISA) which recognizes human and murine NGF and does not cross react with BDNF, as previously described (Bracci-Laudiero et al., 1992). For immunohistochemistry, we used antibodies against human p75[NTR].

NGF amount in coronary arteries

We found that the level of NGF in atherosclerotic coronary wall was significantly lower than in controls ($p < 0.01$). In the SEAT, NGF level showed a tendency for an increase in atherosclerotic specimens compared to controls, whereas in the adjacent myocardium, the NGF levels in atherosclerotic specimens were comparable with that of the controls (Fig. 1). Since these findings represent the first quantitative evidence for NGF in both human coronary vascular tissue and SEAT, one may ask where the NGF is produced. The fact that these NGF amounts in the adjacent SEAT and myocardium are significantly different from those found in the coronary vascular tissue, suggests that NGF is produced locally. This may support the hypothesis that NGF synthesis is processed in a tissue specific manner in different compartments of the human

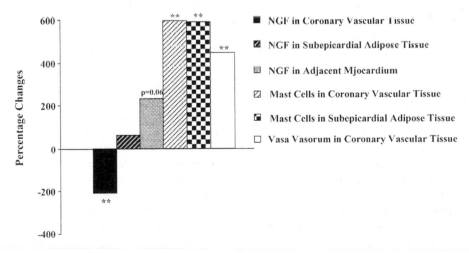

Fig. 1. NGF and mast cell changes in selected human cardiac tissues expressed as percentage of controls.

cardiovascular tissue (Abe et al., 1997; Kaye et al., 2000; Hiltunen et al., 2001; Qin et al., 2002).

$p75^{NTR}$ immunoreactivity in coronary arteries

To further assess whether NGF is locally and differentially produced and to identify possible NGF-receptive fields, parallel immunohistochemical studies were undertaken with the coronary arteries. The results showed that coronary vascular tissue, particularly the adventitia (Chaldakov et al., 2001a), SEAT and myocardium (not shown) expressed a stronger $p75^{NTR}$ immunoreactivity in atherosclerotic compared to control arteries. Why $p75^{NTR}$ expression is preferentially increased in the adventitial compartment of atherosclerotic coronary wall is at present unclear. Since $p75^{NTR}$ is a multifunctional receptor that plays a crucial part in cell responsiveness to NGF and other neurotrophins, it is possible that its overexpression may influence the signaling by BDNF and NT-3, found in human coronary atherosclerosis (see Donovan et al., 1995).

Mast cells in coronary arteries

Since MC are known to be a cellular component of the coronary artery, and since these cells not only respond to the action of NGF, but also secrete this neurotrophin (Aloe and Micera, 1999), we investigated the distribution of MC in atherosclerotic and control coronary arteries. In atherosclerotic coronary vascular tissue, MC (number per cross section) were much numerous and were found mainly in the adventitia, in association with vasa vasorum which presence was also increased (Fig. 1), as compared to controls ($p < 0.01$). Likewise, adventitial MC and lymphocytes were found to be more frequently positioned close to or within the perivascular nerves, as compared to controls (Ghenev and Chaldakov, 1997); such a link was also reported by Laine et al. (2000). The 'atherosclerotic' SEAT also displayed a greater number/mm² of MC than its counterpart in the controls ($p < 0.01$) (Fig. 1).

The fact that the number of MC increases whereas the availability of NGF decreases suggests that MC do not play a decisive role in NGF synthesis in atherosclerotic vascular tissue. There is also a possibility that the content of NGF in atherosclerotic coronary wall is reduced because medial smooth muscle cells undergo significant atrophy (present and other studies) and/or because of endothelial dysfunction (see Ross, 1999; cf. Nakahashi et al., 2000). Note that denervation leads to reduced NGF levels in rat vascular tissue (Liu et al., 1996), whereas decreased perivascular innervation correlates with the development of atherosclerosis (Fronek and Turner, 1980; Scott et al., 1992; Kacem et al., 1997). Yet, the main questions remain: (i) whether the reduced NGF levels in coronary vascular wall are markers or makers/mediators of advanced developmental stages in human atherosclerosis, and (ii) whether the enhanced amount of both NGF and MC found in the 'atherosclerotic' SEAT is a NGF-mediated, vasculoprotective (atheroprotective) effort of this tissue (see Abe et al., 1997; also in this issue Aloe, and Kawamoto and Madsuda for beneficial effects of NGF treatment in wound healing in various tissues; cf. Libby and Aikawa, 2002 for atherosclerotic lesions that are largely viewed now as vascular wound healing process). Clearly, studies aimed at NGF evaluation in the earlier atherosclerotic lesions become mandatory, since they may provide information about possible dynamic changes in the NGF expression during the development of atherosclerosis (see Chaldakov et al., 2001a). Nevertheless, neurotrophin/receptor-based, genetically-modified mice, and possibly their double mutant, for e.g., with the atherosclerosis-prone apo-E-deficient mice, or with MC-deficient, W/WV mice (see Hatanaka et al., 1986) fed with high-cholesterol diet may help to answer this and related questions.

NGF, BDNF and mast cells in metabolic syndrome

We studied 23 patients with metabolic syndrome, 20 females and 3 males. The control group comprised 10 age-matched healthy subjects (7 females and 3 males). The diagnosis of all patients examined was concerned with the definition of metabolic syndrome proposed by United States National Cholesterol Education Program's Adult Treatment Panel III (ATP III) (Lakka et al., 2002; Meigs, 2002). The visceral obesity of these patients was documented by measurements

TABLE 2
Clinical characteristics of the study groups

	Controls	Metabolic Syndrome	
Age	42.50 ± 2.75	45.69 ± 2.18	$p = 0.40$
Body weight (kg)	64.80 ± 1.98	100.47 ± 3.43	$p < 0.01$
Body mass index (kg $\times$ m^{-2})	22.00 ± 0.57	39.56 ± 1.00	$p < 0.01$
Waist circumference (cm)	70.90 ± 2.34	107.10 ± 3.45	$p < 0.01$
Hip circumference (cm)	91.81 ± 1.92	125.72 ± 1.49	$p < 0.01$
Waist-to-hip ration	0.72 ± 0.07	0.85 ± 0.55	$p < 0.05$
Visceral adipose tissue (CAT/L3)			
Density (Hunsfield units)	$13,649.18 \pm 4099.60$	$19,109.68 \pm 3066.50$	$p < 0.05$
Surface (cm^2)	100.65 ± 34.29	157.24 ± 30.14	$p < 0.05$
Systolic blood pressure (mm/Hg)	118.50 ± 4.01	146.52 ± 0.47	$p < 0.01$
Diastolic blood pressure (mm/Hg)	81.00 ± 1.45	96.52 ± 1.52	$p < 0.01$
Triglycerides (mmol $\times$ l^{-1})	1.26 ± 0.09	2.96 ± 0.74	$p = 0.14$
Total cholesterol (mmol $\times$ l^{-1})	4.73 ± 0.18	6.34 ± 0.55	$p = 0.06$
LDL cholesterol (mmol $\times$ l^{-1})	2.77 ± 0.18	3.98 ± 0.34	$p < 0.05$
HDL cholesterol (mmol $\times$ l^{-1})	1.07 ± 0.08	1.16 ± 0.07	$p = 0.47$
Glucose (mmol $\times$ l^{-1})	4.67 ± 0.13	8.02 ± 0.65	$p < 0.01$
Cortisol (nmol $\times$ l^{-1})	368.00 ± 14.89	525.08 ± 33.83	$p < 0.01$
Insulin (μU/ml)	25.42 ± 1.76	35.42 ± 0.95	$p < 0.01$

Means $\pm$ SE. p indicates differences between groups in the statistical analysis (ANOVA).

of waist and hip circumferences and by computer axial tomography (CAT) measurement at the level of lumbar vertebrae 3 (L3). The metabolic syndrome patients were with type 2 diabetes mellitus, thus viewed as affected by a severe form of the metabolic syndrome, according to ATP III's criteria. This study was carried out in accordance with the Helsinki Convention, and following the Bulgarian and Italian legislation on biomedical research. Informed consent was obtained from all the subjects studied. Clinical characteristics of the patients are presented in Table 2.

Plasma levels (pg/ml) of NGF were measured by ELISA, as indicated above. The levels (pg/ml) of BDNF were measured using an ELISA kit 'BDNF Emaxtm ImmunoAssay System, Cat # G6891' by Promega, Madison, WI, USA following the instructions suggested by the manufacturer. Plasma concentrations of glucose, total cholesterol, triglycerides, LDL and high-density lipoprotein (HDL) were measured with routine enzymatic methods, and insulin and cortisol by radioimmunologic method. The concentration (ng/ml) of leptin present in the plasma was measured by radioimmunoassay using the kit by Alpha Diagnostics, San Antonio, TX, USA, following the instructions described by the manufacturer.

For histological analysis, abdominal subcutaneous adipose tissue was collected by biopsies from the same patients and the same control subjects. Specimens were fixed in 4% paraformaldehyde, cut at 15 μm thickness using a cryomicrotome, and stained with 0.5% toluidine blue (pH 2.5) for MC (number/mm^2).

We found that the circulating levels of NGF and BDNF were lower in patients affected by metabolic syndrome as compared to the levels found in the blood plasma of controls ($p < 0.05$) (Fig. 2). Figure 2 also illustrates that the amount of plasma leptin was markedly higher and the number of adipose MC greater in metabolic syndrome patients compared to controls ($p < 0.01$). A significant correlation was observed between plasma leptin levels and adipose MC number in metabolic syndrome patients ($r = 0.71$, $p < 0.01$) but not in controls ($r = 0.44$, $p < 0.23$).

Neurotrophins in the metabolic syndrome

Our study demonstrates that the plasma levels of NGF and BDNF are both significantly reduced in patients with the metabolic syndrome, as compared to controls. This finding seems to support our

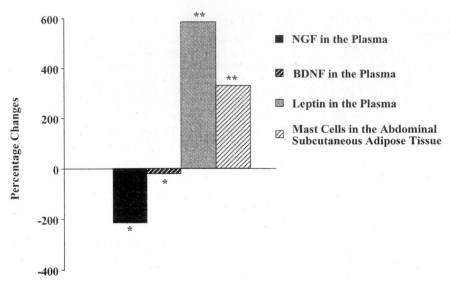

Fig. 2. Changes in plasma NGF, BDNF levels and mast cell distribution in the abdominal subcutaneous adipose tissues, expressed as percentage of controls.

working hypothesis that certain metabotrophic deficit due to hyponeurotrophinemia may operate in this metabopathy. Examples of metabotrophic potentials of NGF include: (i) NGF shares a striking structural homology with proinsulin (Mukherjee and Mukherjee, 1982) and exerts certain insulin-like effects on lipid metabolism (Mukherjee and Mukherjee, 1982; Ng and Wong, 1985) and on energy homeostasis (Salton, 2003), (ii) NGF, via autocrine/paracrine pathway, enhances glucose-induced insulin secretion (Hiriat et al., 2001; for BDNF in the pancreas, see Hanyu et al., 2003; Lucini et al., 2003), and (iii) NGF upregulates the expression of LDL receptor-related protein (Bu et al., 1998), a member of the LDL receptor gene family whose malfunction is causally related to atherosclerosis (Ross, 1999). BDNF also exerts various metabotrophic effects. For example, this neurotrophin improves glucose and lipid metabolism in obese diabetic mice and controls energy homeostasis and feeding behavior (Ono et al., 2000; Rios et al., 2001; Nakagawa et al., 2002; Tsuchida et al., 2002; Kuroda et al., 2003). Moreover, a popularly consumed high-fat/high-carbohydrate diet, which is considered a major risk factor for atherosclerosis, reduces hippocampal BDNF content (Molteni et al., 2002), and so does (for NGF) dietary omega-3 fatty acid deficiency (Ikemoto et al., 2000), which is also an atherogenic

risk factor. Such a lifestyle-neurotrophin link deserves further investigation.

A question raised by a downregulation of circulating NGF and BDNF levels in the metabolic syndrome also regards the cellular source(s) of these endogenous neurotrophins (see Kasayama and Oka, 1989). It is possible that this hyponeurotrophinemia might be due to a reduced synthesis/accelerated degradation of NGF and BDNF in (atherosclerotic) vascular tissue, as discussed above.

Our study also shows a significant increase in plasma leptin levels which resulted to be associated with an elevated presence of MC in the abdominal subcutaneous adipose tissue of metabolic syndrome patients; noteworthy, though the decreased circulating levels of NGF, a potential MC growth factor (Aloe et al., 2001). This finding raises the question as to whether leptin plays a role in MC growth (Hristova et al., 2001), as it is reviewed for other immune cells and for other adipose tissue-secreted molecules (adipokines) (Chaldakov et al., 2003).

Conclusion and perspective

To our knowledge the study described herein represents a novel attempt to correlatively investigating the cardiovascular and metabolic biology of neurotrophins and immune cells. We demonstrated that advanced coronary atherosclerosis and a severe form of the metabolic syndrome display reduced vascular tissue NGF and circulating NGF and BDNF levels, respectively. Also shown was an increased presence of MC in vascular and adipose tissue, possibly associated with elevated plasma levels of leptin, a pleiotropic adipokine sharing some inflammation-, immune- and sympathetic nerve-related actions with NGF (for leptin see Nisoli et al., 1996, 1998; Fantuzzi and Faggioni, 2000; Mark et al., 2003; and the present study that suggests a novel, stimulatory role for leptin in MC growth). Further, we provide the first evidence for an altered presence of NGF/p75^{NTR} and MC in SEAT surrounding the atherosclerotic coronary arteries. This latter finding supports the hypothesis that SEAT, via paracrine way, may appear to be an important target for studying adipobiology of coronary atherosclerosis (Arbustini and Roberts, 1996; Chaldakov et al., 2000, 2001c, 2003).

In vascular biology, considerable evidence exists at present for the involvement of neurotrophins in the pathogenesis of hypertension (Kageyama et al., 1996; Bell, 1996). In addition, several recent publications deal with giant cell arteritis (Saldanha et al., 1999), heart failure (Kaye et al., 2000), and ischemic disorders in heart (Abe et al., 1997; Schmid et al., 1999; Chen et al., 2001; Hiltunen et al., 2001), in striated muscle (Chiaretti et al., 2002; Turrini et al., 2002), and in brain (Schabitz et al., 2000). Atherosclerosis and metabolic syndrome may also be included in this emerging list of neurotrophin-associated vascular diseases.

From the evidence presented here the following main conclusion may be drawn: as in other inflammatory diseases, NGF, BDNF and MC, could, via their pleiotropic potentials, be implicated in the development of atherosclerotic cardiovascular disease and related disorders. Indeed, a growing body of evidence supports the hypothesis that the cell biology of atherosclerosis, including acute coronary syndromes, shares many similarities with various inflammatory-fibroproliferative diseases, such as liver cirrhosis, systemic sclerosis, pulmonary fibrosis, rheumatoid arthritis alike (Ross, 1999), and with bronchial asthma, multiple sclerosis, psoriasis, and wound healing (this issue). Likewise, inflammatory events strongly, and possibly causally, are associated with the development of metabolic syndrome (Grimble, 2002; Ridker et al., 2003). It is noteworthy therefore that, emphasized by the current studies of neurotrophins and immune cells, NGF-based neuroimmune mechanisms are implicated in a large number of inflammatory diseases (Aloe and Micera, 1999; Villoslada et al., 2000; Aloe et al., 2001). Perhaps, the field of cardiovascular and metabolic biology of neurotrophins appears to be shaped. Supportively, neurotrophins and their receptors are implicated in cardiac morphogenesis (reviewed by Hempstead, 2001). It has also been demonstrated that brief myocardial ischemia induces a dysfunction of sympathetic cardiac innervation accompanied by an increase in NGF cardiac levels whereas exogenous administration of NGF protects against such a postischemic dysfunction (Abe et al., 1997). In this context, it is worth noting that (i) NGF-induced increase in brain T lymphocyte interleukin-10 (IL-10) production provides therapeutic benefit in a model of multiple sclerosis (Villoslada et al., 2000), and (ii) T lymphocyte IL-10 overexpression inhibits atherogenesis in LDL receptor-deficient mice (Pinderski et al., 2002). This pressingly invites studies on NGF-IL-10-T cells in cardiovascular disease.

Of course, many questions remain to be answered, and future studies scheduled, in order to evaluate, for example: (i) plasma levels of NGF and BDNF in patients with advanced atherosclerosis, including acute coronary syndromes, (ii) circulating immune cells, also platelets, as potential peripheral markers for alterations in neurotrophins/receptors in patients with atherosclerosis and with metabolic syndrome, and (iii) the importance of metabotrophic activity exerted by other neurotrophic factors, particularly ciliary neurotrophic factor (Lambert et al., 2001; O'Dell et al., 2002; Ott et al., 2002; Zvonic et al., 2003), leukemia inhibitory factor (Moran et al., 1997; Beretta et al., 2002), bone morphogenetic proteins (Chen et al., 2003), and metallothioneins (Beattie et al., 1998; Trayhurn and Beattie, 2001) in the pathogenesis of metabolic syndrome.

Indeed, 'the NGF found its rightful place on the neuroscience chess board', Rita Levi-Montalcini (2003) stated in her opening lecture at the Modena NGF2002 Conference. Hopefully, there will be also a place for NGF and related molecules on the cardiovascular science chess board. Ongoing and future studies will reduce or reinforce this hypothesis.

Acknowledgments

This study was supported in part by grant nr. 1-2001-87 from JDRFI, by PF 2002 ISS, "Invecchiamento cerebrale e ruolo del NGF (ICG 120/4RA00-90), Progetto Biotecnologie MURST/CNR, and PF, ISS, ICS 120.5/PF 00.123 to L. Aloe.

Abbreviations

BDNF	brain-derived neutrophic factor
CAT	computer axial tomography
HDL	high-density lipoprotein
IL-10	interleukin-10
LAD	left anterior descending coronary artery
LDL	low-density lipoprotein
MC	mast cells
NGF	nerve growth factor
NT-3	neurotrophin-3
p75NTR	p75 neurotrophin receptor
SEAT	subepicardial adipose tissue

References

Abe, T., Morgan, D.A. and Gutterman, D.D. (1997) Protective role of nerve growth factor against postischemic dysfunction of sympathetic coronary innervation. Circulation, 95: 213–220.

Aloe, L. and Micera, A. (1999) Nerve growth factor: basic findings and clinical trials. In: L. Aloe and G.N. Chaldakov (Eds.), Nerve Growth Factor in Health and Disease. Biomedical Reviews, Vol. 10, The Bulgarian-American Center, Varna, Bulgaria, pp. 3–14.

Aloe, L., Tirassa, P. and Bracci-Laudiero, L. (2001) NGF in neurological and non-neurological diseases: basic findings and emerging pharmaceutical perspectives. Curr. Pharm. Des., 7: 113–123.

Arbustini, E. and Roberts, W.C. (1996) Morphological observations in the epicardial coronary arteries and their surroundings late after cardiac transplantation. Am. J. Cardiol., 78: 814–820.

Beattie, J.H., Wood, A.M., Newman, A.M., Bremner, I., Choo, K.H., Michalska, A.E., Duncan, J.S. and Trayhurn, P. (1998) Obesity and hyperleptinemia in metal-lothionein (-I and -II) null mice. Proc. Natl. Acad. Sci. USA, 95: 358–363.

Bell, C. (1996) Neurotrophic abnormalities and development of high blood pressure in genetically hypertensive rats. Biomed. Rev., 6: 43–55.

Beretta, E., Dhillon, H., Kalra, P.S. and Karla, S.P. (2002) Central LIF gene therapy suppresses food intake, body weight, serum leptin and insulin for extended periods. Peptidies, 23: 975–984.

Bracci-Laudiero, L., Aloe, L. and Levi-Montalcini, R. (1992) Multiple sclerosis patients express increased levels of beta-nerve growth factor in cerebrospinal fluid. Neurosci. Lett., 147: 9–12.

Bu, G., Sun, Y., Schwartz, A.L. and Holtzman, D.M. (1998) Nerve growth factor induces rapid increases in functional cell surface low density lipoprotein-related protein. J. Biol. Chem., 273: 13359–13365.

Chaldakov, G.N. and Vankov, V.N. (1986) Morphological aspects of secretion in the arterial smooth muscle cell, with special reference to the Golgi complex and microtubular cytoskeleton. Atherosclerosis, 61: 175–192.

Chaldakov, G.N., Fiore, M., Ghenev, P.I., Stankulov, I.S. and Aloe, L. (2000) Atherosclerotic lesions: possible interactive involvement of intima, adventitia and associated adipose tissue. Int. Med. J., 7: 43–49.

Chaldakov, G.N., Stankulov, I.S., Fiore, M., Ghenev, P.I. and Aloe, L. (2001a) Nerve growth factor levels and mast cell distribution in human coronary atherosclerosis. Atherosclerosis, 159: 57–66.

Chaldakov, G.N., Fiore, M., Stankulov, I.S., Hristova, M., Antonelli, A., Manni, L., Ghenev, P.I., Angelucci, F. and Aloe, L. (2001b) NGF, BDNF, leptin, and mast cells in human coronary atherosclerosis and metabolic syndrome. Arch. Physiol. Biochem., 109: 357–360.

Chaldakov, G.N., Stankolov, I.S. and Aloe, L. (2001c) Subepicardial adipose tissue in human coronary atherosclerosis: another neglected phenomenon. Atherosclerosis, 154: 237–238.

Chaldakov, G.N., Stankulov, I.S., Hristova, M. and Ghenev, P.I. (2003) Adipobiology of disease: adipokines and adipokine-targeted pharmacology. Curr. Pharm. Des., 9: 789–794.

Chen, C., Grzegorzewski, K.J., Barash, S., Zhao, Q., Schneider, H., Wang, Q., Singh, M., Pukac, L., Bell, A.C., Duan, R., Coleman, T., Duttaroy, A., Cheng, S., Hirsch, J., Zhang, L., Lazard, Y., Fischer, C., Barber, M.C., Ma, Z.D., Zhang, Y.Q., Reavey, P., Zhong, L., Teng, B., Sanyal, I., Ruben, O. and Birse, C.E. (2003) An intergrated functional genomics screening program reveals a role for BMP-9 in glucose homeostasis. Nat. Biotechnol., 21: 294–301.

Chen, P.S., Chen, L.S., Cao, J.M., Shariffi, B., Karagueusian, H.S. and Fishbein, M.C. (2001) Sympathetic nerve sprouting, electrical remodeling and the mechansims of sudden cardiac death. Cardiovas. Res., 50: 409–416.

Chiaretti, A., Piastra, M., Caresta, E., Nanni, L. and Aloe, L. (2002) Improving ischaemic skin revascularisation by nerve growth factor in a child with crush syndrome. Arch. Dis. Child., 87: 446–448.

Cowen, T. and Gavazzi, I. (1998) Plasticity in adult and ageing sympathetic neurons. Prog. Neurobiol., 54: 249–288.

Creedon, D.J. and Tuttle, J.B. (1997) Synergistic increase in nerve growth factor secretion by cultured vascular smooth muscle cells treated with injury-related growth factors. J. Neurosci. Res., 47: 277–286.

Donovan, M.J., Miranda, R.C., Kraemer, R., McCaffrey, T.A., Tessarollo, L., Mahadeo, D., Sharif, S., Kaplan, D.R., Tsoulfas, P., Parada, L. and Hempstead, B.L. (1995) Neurotrophin and neurotropin receptors in vascular smooth muscle cells. Regulation of expression in response to injury. Am. J. Pathol., 147: 309–324.

Fantuzzi, G. and Faggioni, R. (2000) Leptin in the regulation of immunity, inflammation, and hematopoiesis. J. Leuk. Biol., 68: 437–446.

Ford, E.S., Giles, W.H. and Dietz, W.H. (2002) Prevelence of the metabolic syndrome among US adults: findings from the third National Health and Nutritional Examination Survey. JAMA, 287: 356–359.

Fronek, K. and Turner, J.D. (1980) Combined effect of cholesterol feeding and sympathectomy on the lipid content in rabbit aorta. Atherosclerosis, 37: 521–528.

Fujimura, H., Altar, C.A., Chen, R., Nakamura, T., Nakahashi, T., Kambayashi, J., Sun, B. and Tandon, N.N. (2002) Brain-dervied neurotrophic factor is stored in human platelets and released by agonist stimulation. Thromb. Haemost., 87: 728–734.

Ghenev, P.I. and Chaldakov, G.N. (1997) Neural-immune links in adventitial remodeling in human coronary atherosclerosis. Circulation, 96: 2083–2084.

Grimble, R.F. (2002) Inflammatory status and insulin resistance. Curr. Opin. Clin. Nutr. Metab. Care, 5: 551–559.

Hanyu, O., Yamatani, K., Ikarashi, T., Soda, S., Maruyama, S., Kamimura, T., Kaneko, S., Hirayama, S., Suzuki, K., Nakagawa, O., Nawa, H. and Aizawa, Y. (2003) Brain-derived neurotrophic factor modulates glucagon secretion from pancreatic alpha cells: its contribution to glucose metabolism. Diabetes Obes. Metab., 5: 27–37.

Hatanaka, K., Tanishita, H., Ishibashi-Ueda, H. and Yamamoto, A. (1986) Hyperlipidemia in mast cell-deficient W/WV mice. Biochem. Biophys. Acta, 878: 440–445.

Hempstead, B. (2001) Role of neurotrophins in cardiovascular development. In: I. Mocchetti (Ed.), Neurobiology of the Neurotrophins. FP Graham Publishing Co., Johnson City, TN, pp. 141–148.

Hiltunen, J.O., Laurikainen, A., Vakeva, A., Meri, S. and Saarma, M. (2001) Nerve growth factor and brain-derived neurotrophic factor mRNAs are regulated in dinstic cell populations of art heart after ischaemia and reperfusion. J. Pathol., 194: 247–253.

Hiriat, M., Vidaltamayo, R. and Sanchez-Soto, M.C. (2001) Nerve and fibroblast growth factors as modulators of pancreatic beta cell plasticity and insulin secretion. Isr. Med. Assoc. J., 3: 114–116.

Hristova, M., Aloe, L., Ghenev, P.I., Fiore, M. and Chaldakov, G.N. (2001) Leptin and mast cells: A novel adipoimmune link. Turk. J. Med. Sci., 31: 581–583.

Ikemoto, A., Nitta, A., Furukawa, S., Ohishi, M., Nakamura, A., Fujii, Y. and Okuyama, H. (2000) Dietary n-3 fatty acid deficiency decreases nerve growth factor content in rat hippocampus. Neurosci. Lett., 285: 99–102.

Kacem, K., Bonvento, G. and Seylaz, J. (1997) Effect of sympathectomy on the phenotype of smooth muscle cells of middle cerebral and ear arteries of hyperlipidemic rabbits. Histochem. J., 29: 279–286.

Kageyama, H., Nemoto, K. and Nemoto, F. (1996) Mutation of the trkB gene encoding the high-affinity receptor for brain-derived neurotrophic factor in stroke-prone spontaneously hypertensive rats. Biochem. Biophys. Res. Commun., 229: 713–718.

Kasayama, S. and Oka, T. (1989) Impaired production of nerve growth factor in the submandibular gland of diabetic mice. Am. J. Physiol., 257 (3 Pt 1): E400–E404.

Kaye, D.M., Vaddadi, G., Gruski, S.L., Du, X.-J. and Esler, M.D. (2000) Reduced myocardial nerve growth factor expression in human and experimental heart failure. Circ. Res., 86: e80–e84.

Keller, K.B. and Lemberg, L. (2003) Obesity and the metabolic syndrome. Am. J. Crit. Care, 12: 167–170.

Kuroda, A., Yamasaki, Y., Matsuhisa, M., Kubota, M., Nakahara, I., Nakatani, Y., Hoshi, A., Gorogawa, S., Umayahara, Y., Itakura, Y., Nakagawa, T., Taijii, M., Kajimoto, Y. and Hori, M. (2003) Brain-derived neurotrophic fator ameliorates hepatic insulin resistance in Zucker fatty rats. Metabolism, 52: 203–208.

Laine, P., Naukkarinen, A., Heikkila, L., Penttila, A. and Kovanen, P.T. (2000) Adventitial mast cells connect with sensory nerve fibers in atherosclerotic coronary arteries. Circulation, 101: 1665–1669.

Lakka, H.M., Laaksonen, D.E., Lakka, T.A., Niskanen, L.K., Kumpusalo, E., Tuomiletho, J. and Salonen, J.T. (2002) The metabolic syndrome and total and cardiovascular disease mortality in middle-aged men. JAMA, 288: 2709–2716.

Lambert, P.D., Anderson, K.D., Sleeman, M.W., Yong, V., tan, J., Hijarunguru, A., Corcoran, T., Murray, J.D., Thabet, K.E., Yancopoulos, G.D. and Wiegand, S.J. (2001) Ciliary neurotrophic factor activates leptin-like pathways and reduces body fat, without cachexia or rebound wight gain, even in leptin-resistant obesity. Proc. Natl. Acad. Sci. USA, 98: 4652–4657.

Levi-Montalcini R. (1997) The Saga of the Nerve Growth Factor. Preliminary Studies, Discovery, Further Development. World Scientific, Singapore.

Levi-Montalcini R. (2003) The nerve growth factor and the neuroscience chess board. In: L. Aloe and L. Calza (Eds.),

288

NGF and Related Molecules in Health and Disease. Progress in Brain Research, Elsevier, Amsterdam.

Libby, P. and Aikawa, M. (2002) Stabilization of atherosclerotic plaques: New mechanisms and clinical targets. Nat. Med., 8: 1257–1262.

Liu, D.T., Reid, M.T., Bridges, D.C.M.cL. and Rush, R.A. (1996) Denervation, but not decentralization, reduces nerve growth factor content of the mesenteric artery. J. Neurochem., 66: 2295–2299.

Loskutoff, D.J., Fujisawa, K. and Samad, F. (2000) The fat mouse. A powerful genetic model to study hemostatic gene expression in obesity/NIDDM. Ann. N.Y. Acad. Sci., 902: 272–281.

Lucini, C., Maruccio, L., De Girolamo, P. and Castaldo, L. (2003) Brain-dervied neurotrophic factor in higher vertebrate pancreas: immunolocalization in glucagon cells. Anat. Embryol. (Berl.), 206: 311–318.

Mark, A.L., Rahmouni, K., Correia, M. and Haynes, W.G. (2003) A leptin-sympathetic-leptin feedback loop: potential implications for regulation of arterial pressure and body fat. Acta Physiol. Scand., 177: 345–349.

Matsuzawa, Y., Funahashi, T. and Nakamura, T. (1999) Molecular mechanism of metabolic syndrome X: contribution of adipocytokines, adipocyte-derived bioactive substances. Ann. N.Y. Acad. Sci., 892: 146–154.

McCaffrey, T.A., Fu, C., Du, B., Eksinar, S., Kent, K.C. and Bush, H. Jr. et al. (2000) High-level expression of Egr-1 and Egr-1-inducible genes in mous and human atherosclerosis. J. Clin. Invest., 105: 653–662.

Meigs, J.B. (2002) Epidemiology of the metabolic syndrome. Am. J. Manag. Care, 8(11 Suppl): S283–S292.

Micera, A., Vigneti, E., Pickholtz, D., Reich, R., Pappo, O., Bonini, S., Maquart, F.X., Aloe, L. and Levi-Schaffer, F. (2001) Nerve growth factor displays stimulatory effects on human skin and lung fibroblasts, demonstrating a direct role for this factor in tissue repair. Proc. Natl. Acad. Sci. USA, 98: 6162–6167.

Molteni, R., Barnard, R.J., Ying, Z., Roberts, C.K. and Gomez-Pinilla, F. (2002) A high-fat, refined sugar diet reduces hippocampal BDNF, neuronal plasticity, and learning. Neuroscience, 112: 803–814.

Moran, C.S., Campbell, J.H. and Campbell, G.R. (1997) Human leukemia inhibitory factor upregulates LDL receptors on liver cells and decreases serum cholesterol in the cholesterol-fed rabbit. Arterioscler. Thromb. Vasc. Biol., 17: 1267–1273.

Mukherjee, S.P. and Mukherjee, C. (1982) Similar activities of nerve growth factor and its homologue proinsulin in intracellular hydrogen peroxide production and metabolism in adipocytes. Transmembrane signalling relative to insulin-mimicking effects. Biochem. Pharmacol., 31: 3163–3172.

Nakagawa, T., Ono-Kishino, M., Sugaru, E., Yamanaka, M., Taiji, M. and Noguchi, H. (2002) Brain-derived neurotrophic factor (BDNF) regulates glucose and energy metabolism in diabetic mice. Diabetes Metab. Res. Rev., 18: 185–191.

Nakahashi, T., Fujimura, H. and Altar, C.A. (2000) Vascular endothelial cells synthesize and secrete brain-derived neurotrophic factor. FEBS Lett., 470: 113–117.

Nemono, K., Fukamachi, K., Nemoto, F., Miyata, S., Hamada, M. and Nakamura, Y., et al. (1998) Gene expression of neurotrophins and their receptors in cultured rat vascular smooth muscle cells. Biochem. Biophys. Res. Commun., 245: 284–288.

Ng, T.B. and Wong, C.M. (1985) Epidermal and nerve growth factors manifest antilipolytic and lipogenic activities in isolated rat adipocytes. Comp. Biochem. Physiol. (B), 81: 687–689.

Nisoli, E., Tonello, C., Briscini, L., Flaim, R. and Carruba, M.O. (1996) Leptin and nerve growth factor regulate adipose tissue. Nat. Med., 2: 130.

Nisoli, E., Tonello, C. and Carruba, M.O. (1998) Nerve growth factor, beta3-adrenoceptor and uncoupling protein 1 expression in rat brown fat during postnatal development. Neurosci. Lett., 246: 5–8.

O'Dell, S.D., Syddall, H.E., Sayer, A.A., Cooper, C., Fall, C.H., Dennison, E.M., Phillips, D.I., Gaunt, T.R., Briggs, P.J., Day, I.N. (2002) Null mutation in human ciliary neurotrophic factor confers highy body mass index in males. Eur. J. Hum. Genet., 10: 749–752.

Ono, M., Itakura, Y. and Nonomura, T. (2000) Intermittent administration of brain-derived neurotrophic factor ameliorates glucose metabolism in obese diabetic mice. Metabolism, 49: 129–133.

Ott, V., Fasshauer, M., Dalski, A., Klein, H.H. and Klein, J. (2002) Direct effects of ciliary neurotrophic factor on brown adipocytes: evidence for a role in peripheral regulation of energy homeostasis. J. Endocrinol., 173: R1–R8.

Pinderski, L.J., Fischbein, M.P., Subbanagounder, G., Fishbein, M.C., Kubo, N., Cheroutre, H., Curtiss, L.K., Berliner, J.A. and Boisvert, W.A. (2002) Overexpression of interleukin-10 by activated T lymphocytes inhibits atherosclerosis in LDL receptor-deficient mice by altering lymphocyte and macrophage phenotypes. Cir. Res., 90: 1064–1071.

Qin, F., Vulapalli, R.S., Stevens, S.Y. and Liang, C.-S. (2002) Loss of cardiac sympathetic neurotransmitters in heart failure and NE infusion is associated with reduced NGF. Am. J. Physiol. Heart Circ. Physiol., 282: H363–H371.

Ridker, P.M., Buring, J.E., Cook, N.R. and Rifai, N. (2003) C-reactive protein, the metabolic syndrome, and risk of incident cardiovascular events: an 8-year follow-up of 14 719 initially healthy American women. Circulation, 107: 391–397.

Rios, M., Fan, G., Fekete, C., Kelly, J., Bates, B., Kuehn, R., Lechan, R.M. and Jaenisch, R. (2001) Conditional deletion of brain-dervied neurotrophic factor in the postnatal brain leads to obesity and hyperactivity. Mol. Endocrinol., 15: 1748–1757.

Ross, R. (1999) Mechanisms of disease: Atherosclerosis—an inflammatory disease. N. Engl. J. Med., 340: 115–126.

Saldanha, G., Hongo, J., Plant, G., Acheson, J., Levy, I. and Anand, P. (1999) Decreased CGRP, but preserved Trk A immunoreactivity in nerve fibers in inflamed human superficial temporal arteritis. J. Neurol. Neurosurg. Psychiatry, 66: 390–392.

Salton, S.R. (2003) Neurotrophins, growth-factor-regulated genes and the control of energy balance. Mt. Sinai J. Med., 70: 93–100.

Schabitz, W.R., Sommer, C., Zoder, W., Kiessling, M., Schwaninger, M. and Schwab, S. (2000) Intravenous brain-derived neurotrophic factor reduces infarct size and counter-regulates Bax and Bcl-2 expression after temporal focal cerebral ischemia. Stroke, 31: 2212–2217.

Schmid, H., Forman, L.A., Cao, X., Sherman, P.S. and Stevens, M.J. (1999) Heterogeneous cardiac sympathetic denervation and decreased myocardial nerve growth factor in streptozotocin-induced diabetic rats: implications for cardiac sympathetic dysinnervation complicating diabetes. Diabetes, 48: 603–608.

Scott, T.M., Honey, A.C., Martin, J.F. and Booth, R.F. (1992) Perivascular innervation is lost in experimental atherosclerosis. Cardioscience, 3: 145–153.

Scott, J. (2002) The pathogenesis of atherosclerosis and new opportunities for treatment and prevention. J. Neural Transm., 63 (Suppl.): 1–17.

Sofroniew, M.V., Howe, C.L. and Mobley, W.C. (2001) Nerve growth factor signaling, neuroprotection, and neural repair. Annu. Rev. Neurosci., 24: 1217–1281.

Stary, H.C., Chandler, A.B., Dinsmore, R.E., Fuster, V., Glagov, S., Insull, W. Jr., Rosenfeld, M.E., Schwartz, C.J., Wagner, W.D. and Wissler, R.W. (1995). A definition of advanced types of atherosclerotic lesions and a histological classification of atherosclerosis. A report from the Committee on Vascular Lesions of the Council on Arteriosclerosis, American Heart Association. Circulation, 92: 1355–1374.

Trayhurn, P. and Beattie, J.H. (2001) Physiological role of adipose tissue: white adipose tissue as an endocrine and secretory organ. Proc. Nutr. Soc., 60: 329–339.

Tsuchida, A., Nonomura, T., Nakagawa, T., Itakura, Y., Ono-Kishino, M., Yamanaka, M., Sugaru, E., Taiji, M. and Noguchi, H. (2002) Brain-derived neurotrophic factor ameliorates lipid metabolism in diabetic mice. Diabetes Obes. Metab., 4: 262–269.

Turrini, P., Gaetana, C., Antonelli, A., capogrosso, M.C. and Aloe, L. (2002) nerve growth factor induces angiogenic activity in a mouse model of hindlimb ishemia. J. Neurosci. Lett., 323: 109–112.

Turtle, J.R. (2000) The economic burden of insulin resistance. Int. J. Clin. Pract., 113 Suppl.: 23–28.

Villoslada, P., Hauser, S.L., Bartke, I., Unger, J., Heald, N., Rosenberg, D., Cheung, S.W., Mobley, W.C., Fisher, S., Genain, C. (2000) Human nerve growth factor protects common marmosets against autoimmune encephalomyelitis by switching the balance of T helper cell type 1 and 2 cytokines within the central nervous system. J. Exp. Med., 191: 1799–1806.

Wang, S., Bray, P., McCaffrey, T., March, K., Hempstead, B.L. and Kraemer, R. (2000) p75(NTR) mediates neurotrophin apoptosis of vascular smooth muscle cells. Am. J. Pathol., 157: 1247–1258.

Young, J.L., Libby, P. and Schonbeck, U. (2002) Cytokines in the pathogenesis of atherosclerosis. Thromb. Haemost., 88: 554–567.

Zvonic, S., Cornelius, P., Stewart, W.C., Mynatt, R.L. and Stephens, J.M. (2003) The regulation and activation of ciliary neurotrophic factor signaling proteins in adipocytes. J. Biol. Chem., 278: 2228–2235.

Progress in Brain Research, Vol. 146
ISSN 0079-6123

CHAPTER 19

Neurotrophins in spinal cord nociceptive pathways

Adalberto Merighi[1,*], Giorgio Carmignoto[2], Sara Gobbo[2], Laura Lossi[1], Chiara Salio[1], Angela Maria Vergnano[1] and Michaela Zonta[2]

[1]*Department of Veterinary Morphophysiology and Rita Levi-Montalcini Center for Brain Repair, Via Leonardo da Vinci 44, 10095 Grugliasco, Turin, Italy*
[2]*CNR Institute of Neuroscience and Department of Experimental Biomedical Experimental Sciences, Viale G. Colombo 3, 35121 Padova, Italy*

Abstract: Neurotrophins are a well-known family of growth factors for the central and peripheral nervous systems. In the course of the last years, several lines of evidence converged to indicate that some members of the family, particularly NGF and BDNF, also participate in structural and functional plasticity of nociceptive pathways within the dorsal root ganglia and spinal cord. A subpopulation of small-sized dorsal root ganglion neurons is sensitive to NGF and responds to peripheral NGF stimulation with upregulation of BDNF synthesis and increased anterograde transport to the dorsal horn. In the latter, release of BDNF appears to modulate or even mediate nociceptive sensory inputs and pain hypersensitivity. We summarize here the status of the art on the role of neurotrophins in nociceptive pathways, with special emphasis on short-term synaptic and intracellular events that are mediated by this novel class of neuromessengers in the dorsal horn. Under this perspective we review the findings obtained through an array of techniques in naïve and transgenic animals that provide insight into the modulatory mechanisms of BDNF at central synapses. We also report on the results obtained after immunocytochemistry, in situ hybridization, and monitoring intracellular calcium levels by confocal microscopy, that led to hypothesize that also NGF might have a direct central effect in pain modulation. Although it is unclear whether or not NGF may be released at dorsal horn endings of certain nociceptors in vivo, we believe that these findings offer a clue for further studies aiming to elucidate the putative central effects of NGF and other neurotrophins in nociceptive pathways.

Keywords: spinal cord; dorsal horn; pain; neurotrophins; NGF; BDNF; synapse; receptor

Introduction

Pain is the perception of an aversive or unpleasant sensation originating from a given region of the body and/or viscera. The International Association for the Study of Pain has given the following definition: *Pain is an unpleasant sensory and emotional experience associated with actual or potential tissue damage or described in terms of such damage*. Therefore, pain is highly subjective and comprises affective (emotional) and sensory discriminative components, involving integration and elaboration of peripheral inputs conveyed to different areas of the central nervous system (CNS) along to the so called nociceptive pathways. These latter are initially activated by a specific subset of peripheral sensory receptors, the nociceptors, that provide information about tissue damage from somatic or visceral structures. Noxious stimuli encoded by nociceptors are first conveyed to the dorsal horn (DH) of the spinal cord or the main spinal nucleus of the trigeminal nerve, then they travel to supraspinal relay centers, such as the thalamus, and finally reach the corticolimbic structures where they are perceived as pain.

*Correspondence to: Adalberto Merighi, Department of Veterinary Morphophysiology and Rita Levi-Montalcini Center for Brain Repair, Via Leonardo da Vinci 44, 10095 Grugliasco, Turin, Italy. Tel.: +39-01-1670-9118; Fax: +39-01-1236-9118; E-mail: adalberto.merighi@unito.it

DOI: 10.1016/S0079-6123(03)46019-6

We will give here an account of the current knowledge about the role of neurotrophins (NTs) in the primary processing of noxious stimuli in the spinal cord DH. Before doing so we will briefly mention some general features of the anatomical, physiological, and neurochemical organization of primary afferent pathways, to put things under the right perspective.

Organization of the primary afferent pathways in the spinal cord

The organization and physiological properties of primary afferent pathways in the spinal cord have been extensively and authoritatively reviewed elsewhere (Ruda et al., 1986; Willis and Coggeshall, 1991; Todd and Spike, 1993; Ribeiro-Da-Silva, 1994; Schadrack and Zieglgansberger, 1998; Alvares and Fitzgerald, 1999; Dubner and Ren, 1999; Millan, 1999; Petersen-Zeitz and Basbaum, 1999; Burnstock, 2000; Gerber et al., 2000a; Herrero et al., 2000; McHugh and McHugh, 2000; Hunt and Mantyh, 2001; Ji and Woolf, 2001; Morisset et al., 2001; Willis, 2001; Mendell and Arvanian, 2002; Willis, 2002), and cutaneous primary afferent fibers (PAFs) have been studied in deeper detail. Therefore, the following short description will be mainly based upon the data available for this type of fibers.

General properties of PAFs

According to their diameter, structure, and functional properties, cutaneous PAFs can be grouped into three categories: (i) C fibers, that are thin, unmyelinated and display slow-conducting velocity, (ii) Aδ fibers, that are of intermediate diameter, scarcely myelinated and have intermediate conduction velocity, and (iii) Aβ fibers, that are of large size, myelinated and display fast conduction velocity. In the skin these three categories of fibers are typically present in proportions of about 70, 10 and 20% respectively, although some variations may occur. All types of cutaneous afferents can transmit nonnociceptive information, but, under normal circumstances, only C and Aδ fibers transmit nociceptive information. C and Aδ fibers originate from small size type B neurons in the dorsal root ganglia (DRGs) and reach the spinal cord via the dorsal roots, although unmyelinated fibers originating from the DRGs are present in the ventral root as well.

In general terms, upon exposure of the skin to a noxious stimulus, Aδ fibers elicit a first rapid phase of *sharp* pain, whereas C fibers evoke a second wave of *dull* pain (Treede et al., 1992; Millan, 1999). There are multiple classes of both C and Aδ fibers, and their characterization is somewhat complicated by the existence of interspecies differences, heterogeneity in the properties of these fibers in different tissues, and terminological ambiguities. We will try to simplify this issue as follows. C fibers comprise: (i) nonnociceptive thermoreceptors, (ii) low threshold mechanoceptors responsive to pressure, and (iii) nociceptors (Willis, 1991; Treede et al., 1992). The C fiber nociceptors represent a heterogeneous group of receptors among which are comprised chemoreceptors, high threshold thermoreceptors and mechanoreceptors. Some of these may be specifically activated by chemical irritants, heath or mechanical stimuli, but others may respond to different types of stimuli and are thus referred to as polymodal or CMH (chemical, mechanical, heath) nociceptors. Polymodal nocire-sponsive C fibers have a crucial role in the sensitization of spinal neurons, a process underlying chronic painful states.

Aδ fibers comprise two classes of mechanoreceptors: one is normally activated by high intensity mechanical stimuli above the noxious threshold, while the other displays a lower threshold (Treede et al., 1992; Beydoun et al., 1996, 1997). Members of the first class are usually referred to as Type I mechanoreceptors. These high threshold rapidly conducting receptors are, under normal conditions, weakly responsive to high intensity heat, cold and chemical stimuli. However, upon repetitive thermal stimulation these fibers may become sensitized, and in the presence of tissue damage, they respond to heat with sustained responses of long duration and slow latency.

The Type II Aδ mechanoreceptors have been primarily described in the primate hairy skin. They display a lower threshold to noxious heat stimuli than the corresponding Type I fibers (Treede et al., 1992; Beydoun et al., 1996, 1997). Finally there are some Aδ fibers more responsive to cooling than mechanical stimuli (Simone and Kajander, 1996, 1997).

Aβ cutaneous fibers originate from large-sized type A DRG neurons and they also reach the spinal cord via the dorsal roots. In the absence of tissue or nerve injury Aβ fibers respond only to touch, vibration, pressure and other modes on nonnoxious, low intensity mechanical stimulation of the skin (Willis and Coggeshall, 1991; Mense, 1993; Woolf et al., 1994b). Cutaneous nociceptors can encode the intensity of noxious thermal and mechanical stimuli and the spatial localization of noxious stimuli (Treede et al., 1992; Dubner and Ren, 1999). This indicates the existence of central mechanisms of spatial and temporal summation of pain signaling (Vierck et al., 1997). However, the prompt response of nociceptors to noxious stimuli underlies the so called *alerting role of acute pain*, and very likely occurs with little modulation at peripheral or central levels (Handwerker and Kobal, 1993). Thus, in the absence of tissue damage, repetitive noxious stimulation may be associated with a decreased response of some polymodal C fibers, a form of receptor adaptation observed in other sensory modalities. However, when tissue damage occurs nociceptor adaptation is largely outweighed by central and peripheral mechanisms of sensitization.

Pain originating in organs other than the skin displays some distinguishing features with respect to those outlined above: first, the distinction of a sharp and dull phase of pain perception is not obvious; second, deep somatic and visceral pain is often associated with a pronounced autonomic component in which hypotension, nausea, sweating and other clinical signs related to stimulation of sympathetic and parasympathetic pathways may become apparent; and third, deep and visceral pain is often diffuse, i.e., it may be difficult to exactly localize its source, and it is frequently subjectively referred to other intact tissues.

Nociceptive fibers in muscles are free nerve endings distributed within the connective tissue between individual muscle fibers and investing fascial envelopes. They are small-sized group III myelinated fibers (equivalent to cutaneous Aδ fibers) and group IV unmyelinated fibers (equivalent to C fibers). The first are activated by mechanical stimuli and respond to muscle stretch and contraction, besides nonnoxious pressure. However, about one-third of them are nociceptors, typically being activated by hypoxia and/or ischemia, and localized noxious rise in pressure (Hoheisel et al., 1993; Mense, 1993;

Marchettini et al., 1996, 2000). Group III fibers can be sensitized by thermal and chemical stimuli. Group IV fibers share several characteristics of group III and about 50% of them are nociceptive.

Nociceptive fibers in joints mainly consist of group III and IV free nerve endings and a small subset of large corpuscolar myelinated Aδ-like fibers. These nociceptors respond to anomalous mechanical solicitation of joint beyond the normal range of movement.

Nociceptive information from the thoracic and upper abdominal viscera reaches the CNS via the sympathetic chains, whereas that from the lower gut and bladder follows the route of pelvic parasympathetic nerves. There is evidence for the existence of polymodal C and Aδ fibers in the heart and gut (Cervero and Connell, 1984; Meller and Gebhart, 1994; Cervero, 1995; Gebhart, 1995; Blackshaw and Gebhart, 2002). Genuine Aβ fibers are absent in viscera, suggesting that nonnoxious information in viscera must be transmitted by Aδ and C fibers. Therefore, according to the pattern theory, visceral pain is intensity-encoded by nonspecific fibers likewise responsive to innocuous stimulation (Handwerker and Kobal, 1993; Cervero, 1995).

Termination of PAFs in the spinal gray matter

In thick sections, the gray matter of the spinal cord has a cytoarchitectonic laminar organization in which 10 different laminae have been recognized (Rexed, 1954). Lamina II, the substantia gelatinosa Rolandi, is further subdivided into an outer (II$_o$) and an inner (II$_i$) portion.

The pattern of termination of PAF in the cord is mainly ipsilateral, although the presence of contralateral projections has been reported by several investigators, and tract-tracing studies have demonstrated the existence of preferential laminar terminations for different types of PAFs. Moreover, in the dorsal root entry zone close to the surface of the spinal cord, PAFs originating from different peripheral territories become segregated into medial and lateral streams. The medial stream comprises the large-sized Aδ fibers which divide within the dorsal funiculus into ascending and descending branches. These latter swing into the gray matter and make synapses in laminae II–VI of the DH and lamina X, the region surrounding the central canal. The largest ascending branches run all the way to the dorsal column nuclei in the medulla oblongata. The lateral stream comprises Aδ and C fibers, which, upon entry, divide into short ascending and descending branches within the dorsolateral tract of Lissauer. The longitudinal projections of Aδ and C fibers extend along several consecutive segments of the cord (Willis and Coggeshall, 1991). Nonnociceptive Aδ fibers terminate primarily in laminae II$_i$ and III, whereas nociceptive Aδ fibers terminate in laminae I, II$_o$, V and X. C fibers terminate in laminae I–II and X.

In the superficial DH, three main types of PAF terminal configurations have been defined. In two of these, a PAF terminal forms the core of a multi-synaptic complex commonly referred to as a synaptic glomerulus. The third type of synaptic configuration made by PAF is a conventional axo-dendritic synapse, which is however particularly enriched in dense-cored, large granular vesicles (LGVs) which are known to contain one or more neuropeptides, and are thus also referred to as P-type vesicles (Merighi, 2002). Synaptic glomeruli are mostly peculiar to lamina II. There are two different types of glomeruli (type I and type II) with different and characteristic ultrastructural and neurochemical features. It is now clearly demonstrated that the central terminal of type I glomeruli originates from a C fiber, whereas the terminals of Aδ fibers are engaged in type II glomeruli (Ribeiro-Da-Silva, 1994; Valtschanoff et al., 1994).

Neurochemistry of nociceptive PAFs

The issue of the neurochemical characterization of nociceptive PAFs is a complex one since differences are expected to derive from: (i) the existence of heterogeneous functional types of PAF (as outlined above), i.e., C fibers compared to Aδ fibers; (ii) the different tissue targets of innervation, although this point needs further clarification; and (iii) the phenotypic switch that occurs between the normal intact state vs inflammation and/or tissue/nerve injury.

In general terms, PAF contain and release a cocktail of biologically active molecules rather than a single neuromessenger. This seems to be a general rule rather than an exception since the costorage and coexistence of multiple messengers in neurons is a widespread phenomenon throughout the CNS and PNS (Lundberg, 1996; Hökfelt et al., 2000; Merighi, 2002). DRG cells are unipolar neurons giving rise to a common stem axon that immediately bifurcates into a centrifugal process reaching the peripheral tissues, and a centripetal process that enters the spinal cord. Although some caution should be exercised (Merighi, 2002), one expects that the same combination of messengers is found at central and peripheral endings of these neurons. A combination of excitatory amino acids (EAAs) and peptides is usually found in small- to medium-sized DRG neurons that give rise to nociceptive C and Aδ fibers (Hökfelt, 1991; Lundberg, 1996; Hökfelt et al., 2000; Merighi, 2002).

Glutamate was one of the first neurotransmitter proposed for PAFs (Johnson, 1972; Hutchinson et al., 1978), and nowadays the role of this amino acid as a major excitatory neurotransmitter in the spinal cord (and throughout the CNS) is fully established. The availability of antisera directed against amino acids fixed in tissue sections has made possible the visualization of glutamate (and other amino acids) in specific neurons and pathways that use them as neurotransmitters (Ottersen and Störm-Mathisen, 1984, 1987). Light immunocytochemical studies revealed that glutamate is detected in both type A (large) and B (medium and small) DRG neurons and in a rich plexus of fibers in the superficial laminae of the DH (Battaglia and Rustioni, 1988; Rustioni and Weinberg, 1989). At the ultrastructural level, glutamate immunoreactivity was localized in terminals which formed the core of type I and II glomeruli or were engaged in simple axo-dendritic synapses (De Biasi and Rustioni, 1988; Maxwell et al., 1990; Merighi et al., 1991; Valtschanoff et al., 1994). At the subcellular level staining was selectively localized to small, clear, round synaptic vesicles (Merighi et al., 1991; Valtschanoff et al., 1994). Aspartate has also been implicated as a putative neurotransmitter of PAFs (Rustioni and Weinberg, 1989). This is of relevance since aspartate has been demonstrated to be a selective agonist of the N-methyl-D-aspartate

(NMDA) glutamate receptor (Curras and Dingledine, 1992). Glutamate and aspartate activate both metabotropic (mGlu) receptors coupled via G-proteins to soluble second messengers, and ionotropic receptors directly coupled to cation-permeable ion channels. All these receptors show a complex pattern of localization in different neuronal types of the DH. Three major groups of mGlu receptors have been recognized (Conn and Pin, 1997; Valerio et al., 1997): group I receptors (mGlu$_1$ and mGlu$_5$) are localized to the superficial laminae of the DH and positively coupled to phospholipase C (PLC), and possibly nitric oxide synthase (NOS), whereas group II (mGlu$_2$ and mGlu$_3$) and III (mGlu$_4$, mGlu$_6$ and mGlu$_8$) are negatively coupled to adenyl cyclase (AC). Specific receptor types differentially alter intracellular calcium concentration ($[Ca^{2+}]_i$) and transduction mechanisms, thereby modulating neuronal excitability by, for example, ion channel/receptor phosphorylation and modulation of gene transcription.

DL-α-NH$_2$-2,3,-dihydro-5-methyl-3-oxo-4-isoxazolepropanoic acid (AMPA) receptor isoforms are made of several subunits, among which the Glu R2 plays a major role in controlling Ca^{2+} permeability and conductance. The different isoforms have specific patterns of distribution in the spinal cord (Popratiloff et al., 1998; Albuquerque et al., 1999; Engelman et al., 1999; Akesson et al., 2000; Lu et al., 2002). The conductance of AMPA receptors localized on DH nociresponsive neurons is predominantly to Na$^+$, and these receptors show low affinity for glutamate (compared to NMDA), low voltage-dependence, rapid kinetics and desensitization upon selective stimulation (Stanfa and Dickenson, 1999; Li et al., 1999a; Szekely et al., 2002). NMDA receptors are composed of heteromultimeric subunits and can be classified into several different types differentially distributed throughout the CNS. In the DH the NMDA R$_1$ is the predominant form (Tölle et al., 1993; Bardoni et al., 2000; Bardoni, 2001). All classes of NMDA receptors display slow channel kinetics, a variable degree of voltage-dependent Mg^{2+} block, and marked permeability to Ca^{2+}.

Neuropeptides are highly concentrated in the superficial DH and lamina X. An impressive number of immunocytochemical studies in the last decades provided a detailed mapping of their distribution.

For the purpose of the present work we will restrict our discussion to the tackykinins and the calcitonin gene-related peptide (CGRP) that coexist with EAAs in at least a subset of PAF terminals (Merighi, 2002).

The tackykinins are a family of structurally related peptides derived from two precursor proteins (Helke et al., 1990). The most widely known member of the family is substance P (SP) which has been implicated in nociceptive neurotransmission since the early 1950s (Lembeck, 1953). Other members of the family include neurokinin A (NKA), neurokinin B (NKB) and neuropeptide K (Helke et al., 1990). Light and electron microscopy distribution studies have demonstrated that: (i) SP and NKA are widely distributed in PAF terminals in laminae I–III, and X of the spinal cord (Ruda et al., 1986), although the existence of intrinsic spinal neurons containing SP is also widely recognized (Rustioni and Weinberg, 1989; Todd and Spike, 1993; Ribeiro-Da-Silva, 1994); (ii) they coexist with EAAs in single PAF terminals where they are segregated in LGVs; and (iii) they may be costored together and/or with other peptides, namely CGRP, in individual LGVs (Merighi, 2002).

Three different G protein-coupled receptors named NK$_1$, NK$_2$ and NK$_3$ have been so far described in mammals with maximal affinity for SP, NKA and NKB respectively (Regoli et al., 1994). Nonetheless, NKA may also activate NK$_1$ receptors, and SP NK$_2$ sites. Both NK$_1$ and NK$_2$ receptors are positively coupled with PLC (Bentley and Gent, 1995; Mantyh et al., 1997). NK$_1$ receptor and its mRNA are widely distributed in the DH (Bleazard et al., 1994; Nakaya et al., 1994; Liu et al., 1994a; Brown et al., 1995) where they are present in numerous cell bodies mainly localized in laminae I and III, but only a few in lamina II.

The pronociceptive peptide CGRP is also particularly abundant in sensory pathways, where the predominant form is (α)-CGRP (Amara et al., 1985). As mentioned above, in PAF terminals CGRP is often costored with SP (or NKA) within individual LGVs. CGRP acts on at least two different types of receptors named CGRP$_1$ and CGRP$_2$ which are both coupled to AC (Wimalawansa et al., 1993; Wimalawansa, 1996). The mechanism of action of CGRP at central PAF endings in the DH remains to be elucidated.

Functional interactions of EAAs, SP and CGRP

The cooperative effects of glutamate, SP and CGRP at central and peripheral endings of nociceptors have been the object of extensive studies. At central endings, glutamate regulates nociceptive neurotransmission by both pre- and postsynaptic mechanisms. NMDA and different types of mGlu receptors have been localized to DH neurons (i.e., postsynaptically to PAFs) as well as to PAFs (i.e., presynaptically). Activation of presynaptic NMDA receptors exerts a feedback positive action on the release of EAAs and SP from PAF terminals, thereby enhancing synaptic transmission (Liu et al., 1994b; Liu and Sandkuhler, 1998; Boyce et al., 1999; Lu et al., 2003). On the other hand the role of individual mGlu receptor subtypes presynaptically localized to PAF terminals remains to be elucidated, since they might either increase or decrease synaptic release in relation to their intracellular signal transduction mechanisms. At present, evidence for an inhibitory role of group II and III mGlu receptors has been provided (Zhong et al., 2000; Gerber et al., 2000b; Zhou et al., 2001).

High-intensity stimulation of PAFs produces a fast, AMPA/kainate-receptor-mediated excitatory postsynaptic spontaneous potential (EPSC) in the superficial laminae of the DH. Activation of low-threshold afferent fibers generates typical AMPA-receptor-mediated EPSCs only, indicating that kainate receptors may be restricted to synapses formed by high-threshold nociceptive and thermo-receptive PAFs (Li et al., 1999b). Although the histological localization of kainate receptors remains uncertain, it seems likely that at least part of them is located on PAF terminals (Kerchner et al., 2002), and that their desensitization reduces the mechanical allodynia and thermal hyperalgesia that follows PAF injury (Zhou et al., 1997; Kerchner et al., 2001b). Presynaptic kainate receptors are also linked to changes in GABA/glycine release from spinal cord interneurons (Kerchner et al., 2001a).

Similarly to glutamate, SP is likely to act both pre- and postsynaptically onto NK_1 receptors at synapses between PAF terminals and DH neurons (Routh and Helke, 1995; Heppenstall and Fleetwood-Walker 1997; Todd, 2002). The cooperative action of glutamate, tackykinins and CGRP on DH neurons follows a well established temporal pattern, and their respective roles depend upon the nature and duration of noxious stimulation. Initial triggering is provided by AMPA receptors that display extremely rapid kinetics, and their activation mediates a rapid depolarization of DH neurons over a few ms. Slower and more sustained EPSPs lasting up to 10 s are consequent to activation of NMDA, group I mGlu and NK_1 (NK_2 and CGRP) receptors.

At peripheral endings, antidromic action potentials (related to the efferent function of nociceptive PAFs) may be triggered in collateral fibers giving rise to the so-called *axon reflex*. This leads to release of EAAs, SP, CGRP that, together with other mediators, enhance nociceptive transmission by a feedback action on PAF terminals, and by acting onto the surrounding tissues thereby increasing inflammation and pain (Wood and Docherty, 1997). The peripheral terminals of small caliber PAFs express AMPA, NMDA and kainate receptors. Local administration of ketamine, a NMDA channel blocker, inhibits primary and secondary hyperalgesia in humans. The algogenic action of glutamate at periphery is also likely to be linked to its capability to stimulate release of SP from PAFs or sympathetic terminals. In keeping with these findings, activation of peripheral NMDA receptors on PAFs triggers the release of SP at their central endings (Liu et al., 1994a). This process may be mediated by NO (Sorkin, 1993; Aimar et al., 1998).

Finally, it is of interest to note that large caliber $A\beta$ fibers also display AMPA and NMDA receptors at both their central and peripheral endings (Wood and Docherty, 1997). This raises the possibility that this category of fibers is also modulated by EAAs, that can thus play a role in the induction of mechanical allodynia.

Functional and neurochemical characterization of DH neurons involved in nociception

Although there is a variety of morphological neuronal types in the spinal cord, and particularly in the DH, three main classes of neurons can be defined in relation to their response to primary afferent sensory input (Millan, 1999).

The first class is represented by neurons that specifically respond to noxious stimuli. These nociceptive-specific neurons are silent under normal circumstances, become activated only by high intensity noxious stimuli conveyed to the cord by C and Aδ fibers, and display a limited capability to encode stimulus intensity. They are primarily located in laminae I and II$_o$, but are also found in laminae V–VI. The second class is represented by the multireceptorial or wide-dynamic range (WDR) neurons. The term WDR refers to their property of responding to a wide range of stimulus intensities, from innocuous to noxious, with a direct stimulus-response relationship. These cells represent the principal neuronal type capable to encode stimulus intensity. WDR neurons are activated by thermal, mechanical and chemical stimuli conveyed to the cord by either C, Aδ and Aβ fibers. They are mainly located in laminac IV–VI, but are also detected in laminae I, II$_o$ and X (and in the ventral horn). The third class is represented by nonnociceptive neurons, that are found primarily in laminae II$_i$, III and IV, although a few also occur in lamina I.

On the basis of their output destination, DH neurons may be instead classified as projection neurons, that send their axons outside the gray matter to supraspinal centers, and interneurons, whose axonal arborizations are confined within the spinal cord. Both can be activated by nociceptive PAFs, and comprise any of the functional types (nociceptive-specific, WDR and nonnociceptive) described above. Spinal cord interneurons play a crucial role in the modulation and integration of nociceptive stimuli to be relayed to higher centers, and thus are likely implicated in the process of neuronal sensitization and referred pain. Interneurons of lamina II are of particular relevance under this aspect. For the sake of simplicity, interneurons can be further subdivided into excitatory and inhibitory, according to their modulatory role on nociceptive inputs. However it must be kept in mind that some of them may be able to exert both effects through activation of different receptor types. Excitatory interneurons have a role in the indirect polysynaptic activation of projection neurons, and also form positive feedback circuits onto PAF terminals. In a similar fashion, inhibitory interneurons exert their roles through pre- and postsynaptic mechanisms. It seems likely that the majority of inhibitory interneurons directly target projection neurons of different functional types (nociceptive-specific and WDR), as well as PAF terminals. An additional possibility is the existence of a direct circuitry connecting inhibitory and excitatory interneurons (Fields et al., 1991; McHugh and McHugh, 2000; Le Bars, 2002). The neurotransmitters utilized by excitatory interneurons are not fully characterized, but several lines of evidence indicate EAAs and neuropeptides as likely candidates. On the other hand, inhibitory interneurons have been extensively characterized in terms of their neurotransmitter content. These neurons utilize three main families of neuromessengers (acetylcholine, opioid peptides and inhibitory amino acids) that may show more or less complex patterns of coexistence/costorage. Most pertinent to the present discussion, the inhibitory interneurons that use γ-amino-butyric acid (GABA) and/or glycine as their (principal) neurotransmitter(s) are widely distributed in the superficial laminae of the dorsal horn (Kerkut and Bagust, 1995; Todd et al., 1995, 1996).

GABA acts on two types of receptors named GABA$_A$ and GABA$_B$. GABA$_A$ receptors control a chloride-permeable ion channel, that upon opening usually leads to cell hyperpolarization. GABA$_B$ receptors are G protein coupled. Upon activation they inhibit AC, decrease Ca^{2+} currents, increase K^+ currents and thus hyperpolarize neurons and reduce transmitter release (Malcangio and Bowery, 1996). Glycine is often coexisting with GABA, and acts on strychnine-sensitive glycine$_A$ receptors (Laing et al., 1994; Todd et al., 1996). Neurons expressing GABA and glycine receptors are often postsynaptic to PAF terminals, but presynaptic GABA receptors have also been localized on these terminals (Xi and Akasu, 1996; Rudomin, 1999).

NTs in nociceptive pathways

During development, sensory modality in DRG neurons and their fibers is linked to specific growth factor requirements. A large percentage of type B DRG neurons and their C fibers is sensitive to capsaicin, the pungent alkaloid of hot peppers and vanilloid. These capsaicin-sensitive fibers can be further subdivided into two different populations (Bennett et al., 1998). The first contains CGRP and

SP and depends upon nerve growth factor (NGF) for its initial survival and development (see below). The second, that can be labeled by IB-4 lectin, expresses a subclass of ATP receptors and is dependent upon glial cell-derived neurotrophic factor (GDNF). Cutaneous mechanoceptors and muscle proprioceptive fibers depend upon brain-derived neurotrophic factor (BDNF) and neurotrophin-3 (NT-3) respectively, but in the absence of neurotrophin-4 (NT-4) function, precursor cells intended to become BDNF-dependent mechanoreceptors instead differentiate into NT-3-dependent proprioceptive neurons (Liebl et al., 2000). It is now clearly established that the role of NTs in nociceptive pathways is not limited to development. For example, NGF not only functions as a trophic factor for a large population of DRG neurons, but also plays a key role in mediating alterations of the C fiber phenotype in certain conditions of enhanced pain stimulation (such as those following inflammation) leading to hyperalgesia.

In a more general context, NTs also appear to be able to interfere with short-term synaptic events in the mature CNS. Indeed, a growing body of data indicates that NTs are directly involved in the exchange of information between central neurons in several areas of the brain and spinal cord. NGF, BDNF, NT-3 or NT-4 have all been implicated in synaptic potentiation and plasticity (Gottschalk et al., 1998; Pozzo-Miller et al., 1999; Xu et al., 2000; Kaplan and Cooper, 2001; Yang et al., 2001), and long-term potentiation (LTP) and/or long-term depression (LTD) in the hippocampus (Saarelainen et al., 2000; Xie et al., 2000), visual system (Huang et al., 1999; Pesavento et al., 2000; Sermasi et al., 2000), and, most related to the present discussion, spinal cord (Arvanov et al., 2000).

The effects of NTs at peripheral and central endings of PAFs should be considered in an overall organic frame (Bennett, 2001). In general, locally produced NTs are taken up by peripheral endings of nociceptors, exert a series of biological effects at these sites, and are then retrogradely transported to DRGs. An anterograde transport along the dorsal roots has been documented for at least some members of the family, that may be therefore released at central endings of PAFs within the spinal cord.

We will discuss below the evidence available so far for each individual NT for a role in the modulation of nociceptive inputs that are relayed to the spinal cord. Nonetheless, as it will become apparent in the following, there are several types of interactions among members of this family of neurotrophic factors in nociceptive pathways.

Before discussing these issues, we will briefly mention some general concepts on the induction and regulation of NT synthesis and release from neurons, to put things in the right perspective. However, one should keep in mind that most information regarding this issue has been obtained in vitro or in brain areas other than the spinal cord.

Neuronal synthesis and release of NTs

NTs can be released from neurons by an unconventional mechanism that involves an increase in $[Ca^{2+}]_i$ from intracellular stores and is independent from extracellular Ca^{2+}, but rather depends upon extracellular Na^+ influx via voltage gated sodium channels (Blochl and Thoenen, 1995, 1996). It seems possible that a rise in cAMP also triggers NT release (Thoenen, 1995). Activation of different types of glutamate/GABA receptors (that, as already mentioned, are widely expressed in nociceptive pathways) may stimulate or inhibit synthesis and/or release of NTs, particularly NGF and BDNF. Although production of these NTs may be regulated by neuronal activity, this does not appear to be the case for NT-3 (Lindholm et al., 1994). Glutamate receptors display different but still not completely understood interactions with NTs in vitro. Some authors have reported that NMDA receptors may be involved in BDNF upregulation (Favaron et al., 1993; Kokaia et al., 1993), but this has been denied by others (Zafra et al., 1990; Wetmore et al., 1994; Lauterborn et al., 2000). Moreover, in these studies it was observed that positive modulation of AMPA or kainate receptors increases NT expression in hippocampal and cortical neurons (Zafra et al., 1990; Lauterborn et al., 2000).

The effect of GABA receptor activation on NT expression is also far from being clear. Under certain conditions activation of GABA receptors seems to downregulate NT production (Zafra et al., 1991;

Thoenen, 1995), whereas an increase in NTs and their mRNAs was observed in other experimental contexts (Heese et al., 2000; Obrietan et al., 2002). This issue is further complicated by the depolarizing effect of GABA during very early stage of development (Berninger et al., 1995; Marty et al., 1996; Obrietan et al., 2002).

Nonetheless, given that transmitter amino acids are likely to influence NT levels in the CNS, in a reciprocal fashion NTs exert a marked influence upon gene expression and synaptic transmission. These actions are not limited to long term plastic changes in synapses, which in the case of the nociceptive pathways are much better characterized and understood, but also comprise a number of trk-mediated rapid effects via multiple intracellular signal transduction pathways. NT signaling through trks not only occurs postsynaptically, but also in a retrograde fashion upon activation of presynaptic receptors at the same synapses from which NTs have been released. The dissection of these trk-mediated mechanisms in nociceptive pathways is still far from being complete. Postsynaptic actions are linked to phosphorylation of ion channels and transmitter receptors which potentiates EPSPs, in particular NMDA receptor-mediated currents (Kaplan and Stephens, 1994; Levine et al., 1998). The intracellular mechanism involved likely consists in a rise of $[Ca^{2+}]_i$ with activation of PLC or other protein kinases (Sakai et al., 1997; Suen et al., 1997; Yu et al., 1997).

An alternative or additional mechanism has been proposed for BDNF that can enhance NMDA receptor-mediated currents by acting at the allosteric glicine$_B$ site via a retrograde presynaptic mechanism (Jarvis et al., 1997).

Presynaptic actions of NTs have been shown to consist in an increase of synaptic currents, again linked to a rise of $[Ca^{2+}]_i$ that is responsible for an augmented release of transmitters, among which glutamate (Berninger and Poo, 1996). Rise in $[Ca^{2+}]_i$ is dependent upon extracellular Ca^{2+} and likely involves L-type voltage-dependent Ca^{2+} channels, which can be phosphorylated by different protein kinases that may be induced via trk receptors (Berninger and Poo, 1996; Lesser et al., 1997; Sakai et al., 1997; Sherwood et al., 1997; Baldelli et al., 1999, 2002).

NGF

NGF was initially characterized for its growing effects on sensory neurons (Levi-Montalcini and Angeletti, 1968). These effects are exerted upon binding to high-affinity TrkA receptors which are expressed in approximately 40% of adult rat DRG neurons (Richardson et al., 1986; Verge et al., 1989; Smeyne et al., 1994). During development NGF appears to be necessary for survival of small diameter DRG neurons giving rise to nociceptive C fibers projecting to laminae I–II of the DH (Otten et al., 1980, 1982; Ruit et al., 1992). NGF also influences the phenotypic development of cutaneous high threshold mechanoreceptors of the Aδ type (Ritter et al., 1991; Lewin et al., 1992; Ritter and Mendell, 1992). From development to adulthood, there is a shift in the role of NGF that initially acts as a survival factor (in particular for outgrowth and proliferation in sensory axons), and becomes then indispensable for the maintenance of a differentiated phenotype of DRG neurons (Carroll et al., 1992; Lewin et al., 1992; Ruit et al., 1992).

Functional role of NGF in adult DRG neurons

In general terms, reduction of the NGF supply to DRGs that follows a peripheral nerve damage in adult animals leads to a diminution of SP (and other peptides) production in *neuropathic pain*, whereas an accumulation of NGF results in increased SP (and other peptides) synthesis in *inflammatory pain*. A remarkable indication of this role of NGF on plasticity of primary sensory neurons expressing SP comes from the results of the surgical rerouting of muscle sensory fibers to skin (Lewin and McMahon, 1991). It is well known that the levels of endogenous NGF are by far higher in skin than in muscle, and, accordingly, reoriented PAFs display an increased content of SP.

NGF counteracts the loss of CGRP and SP expressing DRG neurons following injury (Fitzgerald et al., 1985; Inaishi et al., 1992), and regulates the expression of these two neuropeptides in vitro (Lindsay et al., 1989; Lindsay and Harmar, 1989). Other studies in which a synthetic protein, trkA-IgG, was used to sequester endogenous NGF and block the survival effects of NGF on cultured sensory neurons, showed that administration of trkA-IgG produces a sustained thermal and chemical hypoalgesia and leads to a downregulation of the sensory neuropeptide CGRP (McMahon et al., 1995).

Experiments in vivo confirmed and extended these observations. Upon basal conditions lumbar DRG neurons express high levels of CGRP, SP and somatostatin, while there are only limited amounts of vasoactive intestinal peptide (VIP), cholecystokinin (CCK), neuropeptide tyrosine (NPY) and galanin. Following sciatic nerve transection there is a *phenotypic shift* in peptide expression with substantial decrease in expression of CGRP, SP, somatostatin and their mRNAs in parallel with increased VIP, CCK and galanin (Munglani et al., 1996; Ramer et al., 1998; Hökfelt et al., 2000). In parallel, peripheral axotomy causes Aα fibers to sprout into lamina II, a region from which they are normally excluded (Bennett et al., 1996). Intrathecal NGF administration counteracts the decrease in expression of CGRP and SP, but not somatostatin, and reduces the numbers of VIP-, CCK-, NPY- and galanin-expressing neurons after injury (Verge et al., 1995). The action of NGF on SP may be explained by the presence of a NGF-responsive element controlling mRNA^{SP} transcription in the promoter region of the preprotachykinin gene (Heumann, 1994). In addition, the lack of NGF effects on somatostatin producing DRG cells is consistent with the absence of TrkA receptor protein and mRNA in these neurons (Verge et al., 1995). Moreover, NGF (but not NT-3 or BDNF) prevents both the axotomy-induced reduction in CGRP staining within lamina II and the sprouting of Aα fibers into this region. It is likely that the prevention of Aβ fiber sprouting is a secondary consequence of NGF rescuing small fibers originating from TrkA-expressing DRG neurons (Bennett et al., 1996). Similarly, expression of NGF by recombinant adenovirus is capable to induce a robust axonal regeneration into normal as well as ectopic locations within the DH after dorsal root injury (Romero et al., 2001).

Behavioral studies, in which rats treated with NGF (and GDNF but not BDNF) recover sensitivity to noxious heat and pressure following axotomy, confirm the importance of NGF in maintenance of adult sensory neurons. Moreover, they confirmed that NGF and BDNF have different effects in vivo, on the ground of the their well known actions on distinct subpopulations of DRG neurons (Ramer et al., 2000).

NGF, hyperalgesia and inflammation

Administration of excess NGF to neonatal or mature animals can lead to a profound behavioral hyperalgesia (Lewin et al., 1993). Neonatal NGF treatment results in a marked mechanical hyperalgesia that persists until maturity. This hyperalgesia is explained by an NGF-mediated sensitization of Aδ nociceptive PAFs to mechanical stimuli. Treatment of juvenile animals leads to a very similar behavioral hyperalgesia, but there is not sensitization of Aδ nociceptors to mechanical stimuli. Adult animals also develop mechanical hyperalgesia upon NGF treatment, but without sensitization of Aδ nociceptive afferents. In addition, adult animals develop heat hyperalgesia. Therefore, it appears that the NGF-induced mechanical hyperalgesia is brought about by different mechanisms in neonatal and adult rats. Furthermore, NGF-induced mechanical hyperalgesia in adult animals may be due to central changes, whereas heat hyperalgesia is likely to result from sensitization of peripheral receptors to heat. In keeping, adult rats that received intraplantar injections of NGF develop thermal hyperalgesia linked to reduction of thermal nociceptive threshold in capsaicin sensitive PAFs (Amann et al., 1995). Similar effects are observed in humans after intradermal injection of recombinant NGF (Dyck et al., 1997).

Numerous experiments with transgenic mice lacking or overexpressing NGF clearly confirmed the occurrence of perturbations in nociceptive transmission in the presence of anomalous levels of this NT. When the basal levels of skin NGF are reduced by producing transgenic animals that expressed a fusion gene construct containing an antisense NGF cDNA linked to the K14 keratin promoter, mice display a profound hypoalgesia to noxious and mechanical stimuli (Davis et al., 1993). Similar results are obtained after mis-expression and deletion studies of the *ngf* gene (Akopian et al., 1996). In keeping, mice with null mutation of the *TrkA* gene exhibit a severe deficit in nociception following thermal and mechanical stimuli (Smeyne et al., 1994). On the other hand, increase of NGF basal levels in the skin (Davis et al., 1993) or the spinal white matter (Ribeiro-Da-Silva et al., 2000) in transgenic animals leads to hyperalgesia and/or allodynia. Excess target-derived NGF does not alter physiological response properties or the types of neurons containing CGRP, but rather induces an increase in the relative proportions of myelinated nociceptors (Ritter et al., 2000). Therefore, all these data clearly demonstrate the capability of experimentally administered NGF to modulate the phenotype of DRG neurons, and that endogenous production of NGF regulates the sensitivity of nociceptive systems (McMahon, 1996). Although the aforementioned data are mainly referred to cutaneous afferents, visceral nociceptive afferents also express trkA receptors and respond to NGF in a similar fashion (Dmitrieva and McMahon, 1996; Dmitrieva et al., 1997).

Peripheral inflammation leads to profoundly increased pain sensitivity: noxious stimuli generate a greater response, and stimuli that are normally innocuous elicit pain, as a consequence of an increased local sensitivity of peripheral terminals of Aδ and C fibers. Since the endogenous levels of NGF raise substantially in inflamed tissues (Weskamp and Otten, 1987; Donnerer et al., 1992; Woolf et al., 1994a; McMahon, 1996), it was not surprising that NGF turned out to be a cardinal link between inflammation and hyperalgesia. NGF produced in peripheral tissues (or administered locally) acts on cells expressing trkA receptors such as inflammatory cells, sympathetic neurons, and PAF terminals originating from NGF-sensitive DRG neurons.

In general, hyperalgesia associated with experimental inflammation is blocked by the pharmacological antagonism of NGF in several animal models (McMahon, 1996). Experimental inflammation results in local sensory hypersensitivity and upregulates SP and CGRP in DRG neurons innervating the inflamed tissue (Woolf et al., 1994a; Amann et al., 1995). Systemic administration of anti-NGF neutralizing antibodies prevents behavioral changes (McMahon et al., 1995), neuropeptide upregulation, and the inflammation-induced expression of the immediate early gene c-fos in DH neurons (Woolf et al., 1994a). Following carrageenan inflammation, which raises the endogenous levels of NGF in inflamed tissues,

a marked increase in the proportion of active Aδ and C nociceptors is observed (Koltzenburg et al., 1999). Spontaneously active fibers are sensitized to heat, but the mechanical threshold of nociceptive PAFs remains unchanged. When the NGF-neutralizing molecule trkA-IgG is administered together with carrageenan, primary afferent nociceptors do not sensitize and display essentially normal response properties.

Under physiological conditions, NGF in inflamed tissue derives from different sources including PAF terminals, Schwann cells, sympathetic neurons, mast cells, macrophages, fibroblast and skin keratinocytes (Brown et al., 1991). When the relative contribution of these different cell types to the hyperalgesic action of NGF in inflammation was examined, it was observed that sympathetic neurons transiently contribute to inflammatory hyperalgesia. However, mast cells (which are a potential source of NGF) and sensory neurons are important for the sustained action of NGF in increased nociceptive transmission (Woolf et al., 1996). The hyperalgesic actions of NGF are in part a consequence of the sensitization of the peripheral terminals of high threshold nociceptors, either as a result of a direct action of NGF on trkA expressing PAFs, or indirectly, via the release of sensitizing mediators from trkA expressing inflammatory cells and postganglionic sympathetic neurons (Amann et al., 1995). Indeed chronic degranulation of mast cells significantly impairs the rise of NGF levels in inflamed tissues (Woolf et al., 1996), possibly as a consequence of an impairment in synthesis, storage or release of NGF from these inflammatory cells, or a reduction in the release of tumor necrosis factor α or other cytokines from mast cells, which, in turn, stimulate other cell types to produce NGF.

In animals subjected to Freund's adjuvant-induced arthritis (AIA), a model of long-lasting inflammatory pain, there is an increased immunoreactivity for the NGF TrkA receptor and the low-affinity p75 neurotrophin receptor (p75^{NTR}) in NGF-sensitive, but not GDNF-sensitive, DRG neurons. The rise of p75^{NTR} occurs immediately upon Freund's adjuvant administration, and is followed by increase of TrkA immunoreactivity linked to the development of a long-lasting inflammatory response. In parallel, the two receptors are upregulated at central endings of NGF-sensitive nociceptors (Pezet et al., 2001). Moreover, blockade of NGF bioactivity using an antagonist of TrkA and p75^{NTR} results in suppression of inflammatory pain (Owolabi et al., 1999). These results implicate NGF in long-term mechanisms accompanying chronic inflammatory pain, via the regulation of its high-affinity receptor TrkA.

As previously mentioned, a growing body of evidence suggests that chronic pathologic pain results from long-term plasticity of central nociceptive pathways (Woolf and Costigan, 1999). Diverse peripheral neuropathies are characterized by a loss or reduction of NGF supply to PAFs and DRGs (Woolf, 1996). Several types of PAF lesions and diabetic neuropathy are characterized by decrease of the mRNA levels for trkA and p75^{NTR} in DRG neurons with a further reduction of NGF availability to these cells (Verge et al., 1995; Delcroix et al., 1997, 1998). Reduction of NGF availability could be a cause of cisplatin-induced peripheral neuropathies, and exogenous administration of NGF may be helpful to prevent or reduce cisplatin neurotoxicity in cancer patients (Aloe et al., 2000). Also, NGF levels are reduced in the spinal cord during experimental allergic encephalomyelitis (Calzà et al., 1997, 2002), and the clinical course of the disease ameliorates following NGF administration (Villoslada et al., 2000). Very recently, precursors of NGF have been shown to be the predominant forms of this NT in brain and peripheral tissues, and evidence has been provided that proNGF preferentially binds p75^{NTR} with high affinity. In human and rat skin and nerve extracts, a 53 kD band was detected by Western blot using antibodies against rhNGF or preproNGF, that could correspond to a previously described modified preproNGF-like molecule (Yiangou et al., 2002). Expression of these molecules is markedly reduced in skin extracts from patients with subclinical diabetic neuropathy, but is increased in extracts of inflamed colon from patients with Crohn's disease. Antibodies to both rhNGF and preproNGF immunostain basal keratinocytes in tissue sections of normal human and rat skin, and show accumulation of immunoreactive material in nerve fibers distally to sciatic nerve ligation in rats. PreproNGF antibodies immunostain rat large/medium sized DRG neurons, whereas only small neurons are stained with antibodies to mature

rhNGF. These observation suggest that preproNGF may be preferentially taken up and transported by p75^{NTR}. On these bases, the different molecular forms derived from preproNGF are likely to be of importance in sensory mechanisms, and deserve further investigation (Yiangou et al., 2002).

As repeatedly mentioned before, PAFs not only express the high affinity trkA receptor for NGF, but also the low affinity p75^{NTR} (Henry et al., 1994; Kitzman et al., 1998). The level of expression of p75^{NTR} in DRG neurons appears to be modulated by NGF level in target tissues (Kitzman et al., 1998). The interrelationships of p75^{NTR} and NGF (and other NTs) in determining the biological functions of the molecule(s) are still poorly understood. Directly related to the present discussion, p75^{NTR} enhances the retrograde flow of NGF from PAF terminals to DRGs. In fact, NGF is retrogradely transported in sensory neurons within DRGs, where it alters transcription of a number of proteins and peptides. NGF-mediated modification of gene expression in DRG neurons during inflammation is central to the pathophysiology of persistent pain. The phenotype changes produced by NGF during inflammation include the aforementioned downregulation/upregulation of several neuropeptides, which may amplify sensory input signals in the spinal cord and augment neurogenic inflammation in the periphery, but also upregulation of receptors, ion channels and growth related molecules which may lead to a hyperinnervation of injured tissue by promoting terminal sprouting.

Both the endosome-assisted dispatch of activated TrkA receptors to DRGs as well as transport of a NGF-TrkA complex may be responsible for these NGF-mediated actions at the DRG level (Ehlers et al., 1995; Grimes et al., 1996; Riccio et al., 1997). By these mechanisms, NGF can enhance the production of several nociception-related molecules in C fiber terminals besides SP and CGRP. Among these, certain types of Na$^+$ channels, proton-activated ion channels and vanilloid receptors (Guo et al., 2001) are of particular interest. Also, it was recently shown that intrathecal NGF administration induces novel purinergic receptor P2X(3) expression in DRGs and spinal gray matter, with intense immunoreactivity in axons projecting to laminae I, II$_o$ and X (Ramer et al., 2001). In DRG neurons the purinergic receptor P2X(3) is found predominantly in GDNF-sensitive nociceptors. Therefore, de novo expression of P2X(3) in NGF-sensitive nociceptors may contribute to chronic inflammatory pain (Ramer et al., 2001). Moreover, peripheral production of NGF during inflammation and its retrograde transport to DRGs induces p38 MAPK activation in the soma of C fiber nociceptors. Inflammation also increases protein (but not mRNA) levels of the heat-gated ion channel TRPV1 (VR1) in these neurons, which is then transported to peripheral but not central C fiber terminals. Inhibiting p38 activation in DRGs reduces the increase in TRPV1 in the DRG and inflamed skin, and diminishes inflammation-induced heat hypersensitivity. Likely, activation of p38 in the DRG following retrograde NGF transport, by increasing TRPV1 levels in nociceptor peripheral terminals in a transcription-independent fashion, contributes to the maintenance of inflammatory heat hypersensitivity (Ji et al., 2002). Therefore, following peripheral inflammation a complex array of long-term phenotypic modifications of NGF-sensitive nociceptors is likely to be responsible of a series of cellular and molecular events eventually leading to the onset of pathologic pain. Under this perspective and of particular relevance to the present discussion, NGF has also been demonstrated to be capable to induce expression of BDNF and its mRNA in trkA bearing DRG neurons after intrathecal treatment (Apfel et al., 1996; Cho et al., 1997; Michael et al., 1997). This issue will be discussed below in the section of this chapter dedicated to BDNF.

As mentioned above, there is evidence that the delay of several hours in the onset of mechanical hyperalgesia evoked by systemic or spinal administration of NGF (Lewin et al., 1994), is related to upregulation of SP, CGRP and NKA synthesis in DRG, and transport to central terminals in the DH. In keeping, antibodies against NGF or trkA fusion proteins, not only block inflammatory hyperalgesia at periphery, but also the increase of sensory neuropeptide levels in DRG neurons and PAF central terminals (Donnerer et al., 1993; Lewin and Mendell, 1993; Lewin et al., 1993; Croll et al., 1994; Woolf et al., 1994a; McMahon et al., 1995; Malcangio et al., 1997). Moreover, NGF-induced hyperalgesia and wind-up of DH neurons are abolished by NK_1

receptor antagonists, in accord with the idea that the longer-term pronociceptive actions of NGF are of an *indirect* type, and mediated through the sensitization of SP-responsive DH neurons (Thompson et al., 1995; McMahon, 1996; Ma and Bisby, 1998). A further confirmation of this indirect role of NGF as a central mediator of hyperalgesia comes from the observation that its pronociceptive capabilities are limited by the accumulation of N-terminal SP metabolites (Larson and Kitto, 1997).

Nonetheless, TrkA and $p75^{NTR}$ are expressed in the spinal gray matter with a laminar pattern corresponding to that of nociceptive C and Aδ fibers (Fig. 1A, B, E). By using real time confocal imaging of calcium signal in acute spinal cord slice preparations from young rats, we have obtained preliminary evidence that NGF acutely increases $[Ca^{2+}]_i$ in a subpopulation of lamina II neurons (Merighi et al., 1999; Fig. 2). The possibilty that NGF has central effects in the DH is confirmed by observation that systemic treatment with NGF induces a significant increase in evoked release of substance P in an isolated spinal cord preparation (Malcangio et al., 1996). However, acute superfusion of NGF through a naïve rat spinal cord preparation does not alter basal or electrically evoked release of SP-like immunoreactivity (Malcangio et al., 1997). Anterograde transport of NGF to the DH has not been demonstrated (Anand et al., 1995), and thus the in vivo relevance of these findings remains to be established.

BDNF

It is well known that phenotype alterations in C fibers may be paralleled by the so-called *phenotypic switch* in Aβ fibers, which, under inflammatory conditions, begin to synthesize and release SP (Woolf et al., 1992; Neumann et al., 1996). In initial experiments, NGF was hypothesized to have a role in this phenotypic switch, since it was shown to affect the central terminals of PAFs by switching the sensory modalities of Aβ fibers and modifying the pattern of their responses to electrophysiological stimulation (Lewin and Mendell, 1994; Thompson et al., 1995). Also, these NGF-induced modifications were reduced following treatment with the NK_1 receptor

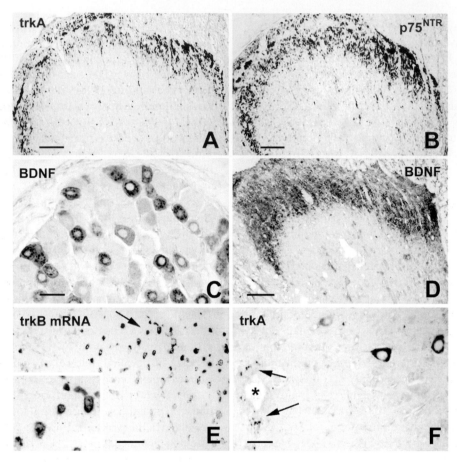

Fig. 1. Distribution of NTs and their receptors in nociceptive pathways. **A–B**: Immunocytochemical localization of the high-affinity NGF receptor trkA (**A**) and the low-affinity p75^{NTR} (**B**) in fibers of the superficial laminae of the DH. **C–D**: localization of BDNF immunoreactivity in DRG neurons (**C**) and the superficial dorsal horn (**D**). **E**: In situ hybridization of mRNA^{TrkB} in the dorsal horn. Note the presence of numerous positive neurons in lamina II. The arrow indicates the area shown at higher magnification in the insert. **F**: TrkA-immunoreactive cell bodies and fibers (arrows) in the gray matter (lamina X) surrounding the central canal (asterisk). Bars: A, B, D, E = 200_m; C–F = 50_m.

antagonist RP675880 thus reinforcing the idea of a close link between NGF and SP in mediating changes in nociceptive transmission (Thompson et al., 1995). However, large diameter Aβ fibers do not express trkA receptors (McMahon et al., 1994) and thus NGF could not be *directly* involved in the phenotypic shift. On the other hand, trkB receptors for BDNF are found in DRG neurons of different diameters, including a population of mechanical nociceptors (Wright and Snider, 1995; Lewin and Barde, 1996; Koltzenburg, 1999). In contrast to cutaneous afferents, about 90% of visceral afferents coexpress trkB and trkA receptors (McMahon et al., 1994), and BDNF has been demonstrated to enhance the

sensitivity to capsaicin of vagal PAFs containing SP (Winter, 1998). Therefore it seems possible that BDNF is peripherally upregulated in visceral PAFs upon inflammation, and might be able to modulate the phenotype of these fibers in visceral pain (McMahon et al., 1994; Lewin and Barde, 1996).

Modulation of nociceptor central synapses by BDNF

Unlike NGF, that appears to be primarily involved in the modulation of nociceptive PAFs at periphery, BDNF has a direct role in the modulation of

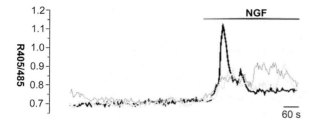

Fig. 2. Kinetics of the $[Ca^{2+}]_i$ change in three neurons from the P8 rat substantia gelatinosa following acute NGF (500 ng/ml) administration, as expressed by the ratio between Indo-1 emission wavelength at 405 and 485 nm. A total of nine cells were recorded during this experiment. We have recorded a total of 36 cells in three different experiments, and 14 of them were NGF-responsive (mean ± SEM: 42.7 ± 11.3).

central synapses of nociceptors (Pezet et al., 2002c; Malcangio and Lessmann, 2003). Indeed, there is now compelling evidence that BDNF is capable of affecting central neuronal activity at least in the hippocampus, visual system and, related to the present discussion, the spinal cord. As we will discuss below, BDNF intervenes in central sensitization, that, as repeatedly mentioned, is responsible of several types of hyperalgesia and can be considered a particular form of synaptic plasticity similar to LTP or LTD. In sensory pathways, BDNF and mRNA^{BDNF} are, under normal conditions, localized in several different DRG neuron subtypes (Fig. 1C), the majority of which corresponds to small- to medium-sized cells expressing trkA and CGRP (Apfel et al., 1996; Cho et al., 1996; Michael et al., 1997). This close relationship between BDNF and TrkA/CGRP expressing DRG neurons is mirrored at their central endings in the spinal cord (Fig. 1D), where BDNF-immunoreactivity is restricted to CGRP-containing PAFs. BDNF immunoreactivity is not detected in IB4- or trkC-labeled DRG neurons, although a few of these cells may express the BDNF mRNA (Michael et al., 1997). Consistent with these observations, BDNF immunoreactivity is absent from the termination zones of large caliber and IB4-labeled PAFs in the spinal cord (Michael et al., 1997). The fact that about one-third of DRG neurons constitutively express BDNF in normal adult animals raises the question of which is the biological role of this NT in primary afferent pathways. In vitro, a BDNF autocrine loop has been shown to act on survival of a subpopulation of adult DRG neurons

(Acheson et al., 1995). However this seems unlikely to take place under normal circumstances in vivo, since only 1% of cells expressing mRNA^{BDNF} also express TrkB, and only 2% of TrkB cells contained mRNA^{BDNF} (Michael et al., 1997). The low degree of coexistence between BDNF immunoreactivity and trkB labeling in DRG neurons is somewhat surprising, considering that target-derived BDNF is believed to be retrogradely transported to the soma of TrkA-expressing neurons (Distefano et al., 1992). On the other hand, it appears that there is little retrograde BDNF transport from peripheral terminals of TrkB-expressing PAFs, and target-derived BDNF might only represent a small fraction of the total BDNF protein within DRGs (Michael et al., 1997). As previously mentioned in the introduction to this review, there is a complex pattern of messenger coexistence/costorage in PAF central terminals (Merighi, 2002). Therefore, it seems obvious that BDNF may modulate EAA and/or SP actions in the DH.

In PAF terminals of the naïve spinal cord, BDNF immunoreactivity appears to be localized to LGVs, where it might be costored with SP (Michael et al., 1997), although no direct morphological evidence for it has been provided yet, and BDNF and SP can be released independently after capsaicin administration to the isolated DH (Lever et al., 2001).

Since SP release from PAF terminals in the DH is under the inhibitory control of endogenous GABAergic and opioid interneurons that antagonize NMDA-evoked facilitation, the effects of bicucullin and naloxone on BDNF release have been evaluated, coming to the conclusion that release of endogenous BDNF is not modulated by the GABAergic inhibitory system (Lever et al., 2001). Conversely, release of endogenous BDNF in the isolated rat DH appears to be mediated by NMDA receptor activation after chemical stimulation by capsaicin or electrical stimulation of dorsal roots at C fiber strength. These experiments clearly demonstrate that endogenous BDNF can be released at central synapses of nociceptors. However the cellular mechanisms that intervene in the regulation of such a release remain, for the most, obscure. Lever and colleagues (2001) have shown that after short bursts of high-frequency stimulation of dorsal roots (300 pulses in 75 trains, 100 Hz) BDNF is released together with glutamate and SP. Interestingly, BDNF release is inhibited by the NMDA receptor antagonist D(-)2-amino-5-phosphonopentanoate (D-AP-5) but not by the AMPA/kainate (non-NMDA) antagonist 6-cyano-7-nitro-quinoxaline-2,3 dione (CNQX), indicating that glutamate stimulates BDNF release by activation of NMDA receptors. Nonetheless tetanic high-frequency stimulation (300 pulses in 3 trains, 100 Hz) does not evoke significant BDNF release, suggesting that glutamate alone is not sufficient to induce release of the NT. Release of endogenous BDNF in the DH is likely to occur mainly, if not exclusively, from the central terminals of nociceptive DRG neurons on at least three bases: (i) BDNF immunoreactivity in the DH is substantially reduced after capsaicin application (Lever et al., 2001), (ii) there is no histological evidence for intrinsic or descending BDNF neurons/pathways in the DH (Apfel et al., 1996; Michael et al., 1997; Heppenstall and Lewin, 2001), and (iii) results of the aforementioned electrophysiological experiments are consistent with the pattern of NMDA glutamate receptor distribution in the DH. Therefore, it seems possible that release of BDNF from PAF central terminals is modulated, among others, by presynaptic NMDA receptors on these fibers (Liu et al., 1994b; Marvizon et al., 1997; Malcangio et al., 1998) and/or postsynaptic NMDA receptors via a retrograde signaling mechanism involving diffusible messengers such as NO (Merighi et al., 2000).

An obvious question to be raised regards the physiological significance of BDNF release in the naïve DH. In an attempt to answer this question and to better characterize the mechanism(s) of action of this NT in the spinal cord, sequestration of endogenous BDNF (and NT-4) by a trkB fusion protein or analysis of BDNF-deficient mice has been employed. However, after BDNF sequestration no changes in either basal pain sensitivity or mechanical hypersensitivity induced by peripheral capsaicin administration (a measure of C fiber-mediated central sensitization) are observed (Mannion et al., 1999). On the other hand, in BDNF-deficient mice selective deficits in the ventral root potential (VPR) evoked by stimulating nociceptive PAFs are observed, whereas the nonnociceptive portion of the VPR remains unaltered (Heppenstall and Lewin, 2001). Although there may be some concern in the interpretation of these results, since a decrease of about 30% of the total number of DRG neurons was previously observed in BDNF null mutants (Ernfors et al., 1994a), Heppenstall and Lewin have shown no obvious alterations in the number and connectivity of nociceptive neurons in their mutants, and come to conclude that BDNF released from nociceptive PAFs has a role in modulating or even mediating reflex excitability, and that BDNF is necessary for normal baseline nociceptive responses in the spinal cord.

Whereas all these observations converge to support a pronociceptive effect of endogenous BDNF in the spinal cord, the results obtained after application of pharmacological concentrations of exogenous BDNF in different experimental paradigms, are subject to a more complex array of interpretations. Initial studies in which the NT was infused into the midbrain came to the conclusion that BDNF has an antinociceptive effect (Siuciak et al., 1994; 1995). Similarly, after engrafting of genetically engineered BDNF-secreting neurons or recombinant adeno-associated viral-mediated over-expression of BDNF in the spinal cord, attenuation of allodynia and hyperalgesia is observed after sciatic nerve constriction (Cejas et al., 2000; Hains et al., 2001a,b; Eaton et al., 2002). Also, topical application of BDNF to the adult rat isolated DH inhibits the electrically evoked release of SP from sensory neurons, and increases K^+-stimulated release of GABA (Pezet et al., 2002a).

On the other hand, others have provided evidence that, in the isolated spinal cord, exogenous BDNF selectively enhances the C fiber-mediated component of the flexor withdrawal reflex and NMDA-evoked responses recorded from ventral roots (Kerr et al., 1999). In keeping with the idea that BDNF has a pronociceptive effect in the DH, we have recently observed that BDNF alters $[Ca^{2+}]_i$ in lamina II neurons after confocal imaging of acute spinal cord slices (Merighi et al., 1999). After capsaicin administration, we came to hypothesize that rise of $[Ca^{2+}]_i$ in these neurons is not directly due to a postsynaptic effect of BDNF, but rather mediated by a presynaptic mechanism involving release of SP from PAF terminals (Carmignoto et al., 2003).

Upregulation of BDNF following inflammation or injury

A correct understanding of the role of BDNF in nociception is further complicated by plasticity of PAFs that occurs under pathophysiological and/or experimental conditions. Indeed, the DRG content of BDNF is markedly upregulated in an NGF-dependent fashion upon different experimental manipulations leading to inflammation and/or injury (Thompson et al., 1999).

After intrathecal NGF treatment there is up to 80–90% increase of BDNF and its mRNA in TrkA/CGRP expressing DRG neurons, whereas NGF does not increase BDNF expression in those neurons lacking TrkA (Apfel et al., 1996; Michael et al., 1997). Similar results are obtained after systemic NGF treatment, a procedure that mimics peripheral inflammatory states (Kerr et al., 1999). Moreover in a number of experimental models that are known to upregulate peripheral NGF, such as experimental inflammation after intraplantar formalin/carageenan (Kerr et al., 1999; Mannion et al., 1999) or neuritis of the nerve roots after lumbar disc herniation (Obata et al., 2002), the pattern of BDNF expression is modified in a similar fashion. C-fiber electrical activity also increases BDNF expression in DRGs, and both inflammation and activity increase full-length TrkB receptor levels in the DH (Mannion et al., 1999). In addition, peripheral nerve injury leads to the same kinds of plastic changes in DRGs and spinal cord. Following a chronic sciatic nerve lesion or axotomy there is a long-lasting ipsilateral increase (2–4 weeks) in mRNA^{BDNF} and BDNF protein in large diameter TrkB- and TrkC-expressing DRGs, whereas BDNF level is unchanged or reduced in most TrkA expressing neurons (Michael et al., 1999; Zhou et al., 1999; Fukuoka et al., 2001; Ha et al., 2001). Under these conditions, BDNF-immunoreactive nerve terminals in the ipsilateral spinal cord increase in laminae II–IV and nucleus gracile. These studies show that in the DRGs, small TrkA neurons switch off their normal synthesis of BDNF, whereas large trkB/C neurons switch to a BDNF phenotype. Plastic changes affecting sensory pathways are associated to thermal hyperalgesia, a sign of neuropathic pain, and appear to be reversed upon relief of behavioral changes. Namely, disappearance of thermal hyperalgesia is associated with downregulation to baseline levels of BDNF in DRGs and spinal cord (Miletic and Miletic, 2002).

Thus, a growing body of evidence indicates that chronic/neuropathic pain due to inflammation and/or peripheral nerve injury is mediated by BDNF at central synapses in the DH, upon a NGF-dependent phenotypic shift in the DRGs that leads to BDNF upregulation. In keeping, BDNF-induced behavioral changes indicative of neuropathic pain are abolished upon administration of a BDNF sequestering TrkB-IgG fusion protein (Kerr et al., 1999; Mannion et al., 1999; Thompson et al., 1999), anti-BDNF antibodies to the spinal cord (Fukuoka et al., 2001), or peripherally applied anti-NGF antibodies (Fukuoka et al., 2001).

Despite the extensive work done on this issue, the mechanism of action of BDNF within the spinal cord remains elusive. In a recent review, four different states of sensory processing have been described in relation to plasticity of the sensory system (Woolf and Costigan, 1999). The first state corresponds to normal nociceptive transmission. Under these conditions the system operates at baseline sensitivity, and only an intense stimulation of C and Aδ fibers produces short-lasting pain. The second state occurs in a time scale of minutes and is linked to the onset of posttranslational changes within the DH. These changes lead to central sensitization and modify the basal threshold of sensitivity resulting in hyperalgesia and allodynia. The third state is a consequence of activity-dependent transcriptional changes within DRGs and DH, occurring in a time scale of hours. In this case the system is further potentiated with increased responsiveness to C fiber inputs. The fourth state is a consequence of peripheral inflammation and occurs in a time scale of hours/days. In this case potentiation of the system is paralleled by DRG phenotypic switch, leading to highest central sensitization.

All lines of evidence available so far converge to the conclusion that BDNF physiologically acts starting from stage two onward. At this stage (which is *not* associated to inflammation) relatively brief C fiber inputs can sequentially evoke augmented membrane excitability, windup and central sensitization (Woolf and Wall, 1986). Windup is a consequence of the removal of Mg^{2+} block of the NMDA receptor with amplification of each subsequent input (Woolf and Costigan, 1999). BDNF likely plays a role in central sensitization upon release from PAF terminals and subsequent binding to TrkB-expressing DH interneurons. The activation of trkB (and NMDA, mGluR, NK$_1$) receptors gives rise to an increase in $[Ca^{2+}]_i$ (Heath et al., 1994; Carmignoto et al., 2003) following extracellular influx and/or release from internal stores. Indeed, the action of BDNF on $[Ca^{2+}]_i$ has been proposed to be mediated by the trkB-induced activation of PLCγ with the consequent production of inositol trisphosphate (IP3) and release of Ca^{2+} from internal Ca^{2+} stores (Zirrgiebel et al., 1995; Li et al., 1998). Increase in $[Ca^{2+}]_i$ leads to activation of other tyrosine kinases such as src and PKC (Zirrgiebel et al., 1995). The major targets of these kinases are the NR1 and NR2 subunits of the NMDA receptor (Yu et al., 1997; Suen et al., 1997b). These posttranslational modifications eventually lead to reduction of the NMDA receptor Mg^{2+} block.

Stages three and four are respectively associated with early and late changes in transcription within DRG and DH neurons. Two hours after C fiber stimulation, mRNA^{BDNF} levels increase in DRGs, and the mRNA^{TrkB} (Fig. 1F) augments in the DH (Woolf and Costigan, 1999). Namely, although in situ hybridization reveals the existence of a large population of DH neurons with mRNA^{TrkB}, immunohistochemistry demonstrates a low number of lamina II neurons expressing the full-length trkB protein in the naïve spinal cord (Mannion et al., 1999; Michael et al., 1999), and higher levels were only observed following episodes of high neuronal activity, inflammation and/or axotomy (Mannion et al., 1999; Michael et al., 1999; Thompson et al., 1999). Increase in the amount of BDNF in DRGs and central PAF terminals, together with augmented expression of its high affinity receptor in the DH leads to potentiation of the system and further transcriptional changes. Acute BDNF stimulation induces a robust phosphorylation of the transcription factor cAMP response element-binding protein (CREB) in acute hippocampus and visual cortex slices (Pizzorusso et al., 2000). It is therefore conceivable that at least part of the activity-dependent upregulation of BDNF within the DRGs is also CREB-dependent. Moreover in hippocampus and visual

cortex (Pizzorusso et al., 2000), spinal cord (Thompson et al., 1995) and DRGs (Kim et al., 2000) increased expression of c-fos is observed after BDNF in vitro or in vivo. Also noxious stimulation induces trkB receptor and downstream ERK phosphorylation in DH (Pezet et al., 2002b).

Whereas all the above observation give a rather coherent explanation of the postsynaptic effects of BDNF on DH neurons, a presynaptic action onto trkB-expressing PAF terminals cannot be completely ruled out from observation of the time course of the BDNF acute effect on substantia gelatinosa neurons after real time confocal imaging of $[Ca^{2+}]_i$. Indeed, in slices preincubated with BDNF, $[Ca^{2+}]_i$ oscillations progressively increase with time both in frequency and amplitude (with a peak of activity 2–3 h after incubation), and eventually affect an augmented number of cells (Merighi et al., 1999). This observation, suggests that the effect of BDNF is not linked to trkB-mediated formation of IP3. Indeed, activation of PLCγ, that may result from activation of TrkB receptors by BDNF, rapidly triggers the formation of IP3, and then the release of Ca^{2+} from intracellular compartments. The ensuing $[Ca^{2+}]_i$ elevations strictly depend on the continuous formation of IP3, and they rapidly fade away in the absence of the stimulus that activates the IP3-coupled receptor (Pozzan et al., 1994). The direct activation of trkB receptors on DH neurons can, therefore, hardly account for the long-lasting increase in $[Ca^{2+}]_i$ oscillation frequency and amplitude that are observed in slices subjected to a relatively short preincubation with BDNF before recording, rather than continuous superfusion for several hours. Accordingly, this latter type of BDNF action is more compatible with an indirect presynaptic mechanism that requires the recruitment of additional (retrograde?) messenger(s) that in turn trigger $[Ca^{2+}]_i$ oscillations in lamina II. The existence of such a presynaptic mechanism needs to be substantiated by an unequivocal localization of trkB receptors upon PAF terminals, localization that is at present still under debate.

NT-3

Neurotrophin-3 (NT-3) has a well documented role as a survival factor during development of muscle spindle afferent (group Ia) fibers. Selective survival of muscle sensory neurons occurs in vitro in the presence of NT-3 (Hory-Lee et al., 1993). Moreover, there is a selective loss of muscle afferents after treatment with anti-NT-3 antibodies before birth (Oakley et al., 1995), and muscle PAFs do not survive in mice with null mutation of NT-3 (Snider, 1994). In keeping, these fibers are rescued after reintroducing a NT-3 gene (Wright et al., 1997), and animals lacking the NT-3/trkC signaling pathway do not develop muscle PAFs (Ernfors et al., 1994b). In addition, several lines of evidence demonstrate a role for NT-3 in the rescue of spindle afferent function after axotomy or chemically induced neuropathies (Mendell et al., 2001).

NT-3 exerts its biological effects via trkC receptors that are primarily localized to large caliber myelinated $A\beta$ fibers, although trkC has been localized throughout the spinal cord, including the superficial laminae of the DH (Zhou and Rush, 1994), and descending serotoninergic fibers. Large myelinated muscle mechanosensitive PAFs retrogradely transport NT-3 from periphery to DRG neuronal cell bodies, but a centripetal transport to the spinal cord remains to be demonstrated (Distefano et al., 1992). If indeed endogenous NT-3 reaches the naïve spinal cord, as it could be inferred from the observation that NT-3 is necessary for development of the mono-synaptic stretch reflex after birth (Seebach et al., 1999), the mechanism(s) through which it acts at central synapses has been in part elucidated using the monosynaptic stretch reflex as a model system (Mendell et al., 2001). Application of exogenous NT-3 increases the amplitude of dorsal root-evoked monosynaptic AMPA/kainate receptor-mediated EPSP within minutes, and requires the presence of active NMDA receptors on the motorneuron membrane (Arvanov et al., 2000). The action of NT-3 is specifically exerted on this type of synapses and not on other inputs that reach the motorneurons, and disappears in animals older that one week (Mendell et al., 2001). Although the physiological role of NT-3 does not seem to be restricted to muscle afferents, since excess NT-3 enhances the separation of dermatomes that occurs postnatally (Ritter et al., 2001), the existence and origin(s) of a putative pool of NT-3 in the DH of mature animals remain unclear, and

a precise role of NT-3 in nociception still has to be established.

Initial studies in which NT-3 was infused into the midbrain indicate, on a behavioral basis, that it has antinociceptive effects (Siuciak et al., 1994). This is further confirmed in an isolated spinal cord preparation in which a single systemic injection of NT-3 induces mechanical (but not thermal) hypoalgesia (Malcangio et al., 1997). Moreover, NT-3 inhibits the release of SP, suggesting that this mechanism may be responsible for NT-3-induced antinociception. In this experimental context NT-3, although not modifying SP basal outflow, dose-dependently inhibits the electrically evoked, but not capsaicin-induced, release of the peptide. NT-3-induced inhibition persists even in the presence of pharmacological concentrations of NGF in the perfusion fluid, and is still significant when the evoked release of SP is enhanced by a prolonged in vivo treatment with NGF. It seems likely that inhibition of SP release by NT-3 is mediated by activation of $GABA_A$ receptors onto DH interneurons since naloxone, but not CGP 36742 (a $GABA_B$ antagonist), abolishes the NT-3 effect on SP release (Malcangio et al., 1997). In keeping, locally administered NT-3 appears to reduce the peripheral sensitization that follows a chemically-induced neuropathy (Helgren et al., 1997). More recently, however, it was reported that NT-3 increases C-fibre stimulation-evoked SP release, and capsaicin superfusion-induced SP release in the isolated spinal cord. From these observations it was concluded that increased SP-release from the spinal cord is not necessarily associated with behavioral hyperalgesia in NT-3 treated rats (Malcangio et al., 2000). Similarly, a number of other experiments lead to support a pronociceptive role of NT-3. Transganglionic labeling of $A\beta$ fibers with choler-agenoid-horseradish peroxidase (C-HRP) in animals treated intrathecally with NT-3 for 14 days via an osmotic pump shows that the area of C-HRP label expands into lamina II. These NT-3-treated animals have a significant decrease in mechanical nociceptive threshold (White, 1998). Consistent with this finding, intrathecal administration of NT-3 antisense nucleo-tides attenuates the density of C-HRP labeling in lamina II in nerve injured animals, and significantly attenuates nerve injury-induced allodynia (White, 2000).

NT-4

In situ hybridization studies demonstrate that NT-4 is synthesized by DRG and DH neurons (Heppenstall and Lewin, 2001). Very little work has been done to characterize the effects, if any, of NT-4 in central nociceptive pathways. In NT-4 null mice there are not obvious deficits in flexion reflex (Heppenstall and Lewin, 2001). However in these mutants the anti-nociceptive effect of morphine appears to be partially mediated by NT-4-induced TrkB receptor activation (Lucas et al., 2003).

Conclusion

A significant advancement of our understanding of the physiological role of NTs in nociceptive pathways has been achieved in the last years. Nonetheless a number of questions are still unresolved mainly linked to the mechanism(s) through which NTs acutely modulate synaptic transmission at central synapses in the DH. Our possibilities of comprehension of this aspect of NT neurobiology are further hampered by the fact that this family of growth factors exert a wide array of functions during development in the survival and maintenance of nerve cells, and that substantial maturation of the nociceptive pathways still occurs in postnatal life. The recent observation that NTs can be sorted into either the constitutive or regulated secretory pathways (Mowla et al., 1999), explains how NTs can act both as survival factors (when entering the constitutive pathway) and neuromessenger (when entering the regulated pathway). Similarly to what happened with NGF, that was initially considered as the prototype of neurotrophic factors, BDNF is now emerging as the prototype of a novel class of neuromessengers in several areas of the CNS. In keeping, BDNF is processed in the regulated secretory pathway within brain neurons and secreted in an activity-dependent manner (Mowla et al., 1999), and in BDNF knockout mice there is an impairment of synaptic vesicle docking (Pozzo-Miller et al., 1999), confirming that it is released at synapses and influences neuronal activity. In the nociceptive pathways plastic changes mediated by NTs, particularly NGF, have been extensively documented, and the cellular and molecular mechanisms that underlie these changes have been basically unraveled. Nonetheless the acute modulatory effects of NTs at central synapses in the DH are still far to be understood, since some effects of BDNF (and perhaps other members of the family) are hardly explained only on the basis of posttranslational and/or activity-dependent transcriptional modifications of the nociceptors and/or their postsynaptic target neurons in the DH.

Clearly much work still needs to be done to clarify the subcellular site of storage of NTs at central synapses, their pattern of coexistence with EAAs and/or neuropeptides, and functional interactions these nociceptive messengers. This information, together with an in-depth ultrastructural and functional analysis of receptor distribution in the DH, will be fundamental to fully understand fast synaptic events that are mediated by NTs in nociceptive pathways.

Acknowledgments

The original work described in this paper has been funded by grants of the Italian MIUR (Fondi FIRB 2001 and COFIN 2002 to AM and GC) and by local grants from the University of Torino.

Abbreviations

AIA	adjuvant-induced arthritis
AMPA	DL-α-NH$_2$-2,3,-dihydro-5-methyl-3-oxo-4-isoxazolepropanoic acid
AC	adenyl cyclase
$[Ca^{2+}]_i$	intracellular calcium concentration
BDNF	brain-derived neurotrophic factor
CCK	cholecystokinin
CGRP	calcitonin gene-related peptide
C-HRP	choleragenoid-horseradish peroxidase
CNQX	6-cyano-7-nitro-quinoxaline-,2,3 dione
CNS	central nervous system
CREB	cAMP response element-binding protein
D-AP-5	D(-)2-amino-5-phosphonopentanoate
DH	dorsal horn
DRG	dorsal root ganglion
EAA	excitatory amino acid
EPSP	excitatory postsynaptic spontaneous potential

GABA	γ-amino-butyric acid
GDNF	glial cell line-derived neurotrophic factor
IP3	inositol trisphosphate
LTD	long term depression
LTP	long term potentiation
NGF	nerve growth factor
NK	neurokinin
NKA	neurokinin A
NKB	neurokinin B
NMDA	N-methyl-D-aspartate
NO	nitric oxide
NOS	nitric oxide synthase
NPY	neuropeptide tyrosine
NT	neurotrophin
NT-3	neurotrophin-3
NT-4	neurotrophin-4
PAF	primary afferent fiber
PLC	phospholipase C
$p75^{NTR}$	p75 neurotrophin receptor
SP	substance P
VIP	vasoactive intestinal polypeptide
VPR	ventral root potential
WDR	wide dynamic range

References

Acheson, A., Conover, J.C., Fandl, J.P., DeChiara, T.M., Russell, M., Thadani, A., Squinto, S.P., Yancopoulos, G.D. and Lindsay, R.M. (1995) A BDNF autocrine loop in adult sensory neurons prevents cell death. Nature, 374: 450–453.

Aimar, P., Pasti, L., Carmignoto, G. and Merighi, A. (1998) Nitric oxide-producing islet cells modulate the release of sensory neuropeptides in the rat substantia gelatinosa. J. Neurosci., 18: 10375 10388.

Akesson, E., Kjaeldgaard, A., Samuelsson, E.B., Seiger, A. and Sundstrom, E. (2000) Ionotropic glutamate receptor expression in human spinal cord during first trimester development. Brain Res. Dev. Brain Res., 119: 55–63.

Akopian, A.N., Abson, N.C. and Wood, J.N. (1996) Molecular genetic approaches to nociceptor development and function. Trends Neurosci., 19: 240–246.

Albuquerque, C., Lee, C.J., Jackson, A.C. and MacDermott, A.B. (1999) Subpopulations of GABAergic and non-GABAergic rat dorsal horn neurons express Ca2+-permeable AMPA receptors. Eur. J. Neurosci., 11: 2758–2766.

Aloe, L., Manni, L., Properzi, F., De Santis, S. and Fiore, M. (2000) Evidence that nerve growth factor promotes the recovery of peripheral neuropathy induced in mice by cisplatin: behavioral, structural and biochemical analysis. Auton. Neurosci., 86: 84–93.

Alvares, D. and Fitzgerald, M. (1999) Building blocks of pain: the regulation of key molecules in spinal sensory neurones during development and following peripheral axotomy. Pain, Suppl 6: S71–S85.

Amann, R., Schuligoi, R., Herzeg, G. and Donnerer, J. (1995) Intraplantar injection of nerve growth factor into the rat hind paw: local edema and effects on thermal nociceptive threshold. Pain, 64: 323–329.

Amara, S.G., Arriza, J.L., Leff, S.E., Swanson, L.W., Evans, R.M. and Rosenfeld, M.G. (1985) Expression in brain of a messenger RNA encoding a novel neuropeptide homologous to calcitonin gene-related peptide. Science, 229: 1094–1097.

Anand, P., Parrett, A., Martin, J., Zeman, S., Foley, P., Swash, M., Leigh, P.N., Cedarbaum, J.M., Lindsay, R.M., Williams-Chestnut, R.E., Sinicropi, D.V. (1995) Regional changes of ciliary neurotrophic factor and nerve growth factor levels in post mortem spinal cord and cerebral cortex from patients with motor disease. Nature Med., 1: 168–172.

Apfel, S.C., Wright, D.E., Wiideman, A.M., Dormia, C., Snider, W.D. and Kessler, J.A. (1996) Nerve growth factor regulates the expression of brain-derived neurotrophic factor mRNA in the peripheral nervous system. Mol. Cell Neurosci., 7: 134–142.

Arvanov, V.L., Seebach, B.S. and Mendell, L.M. (2000) NT-3 evokes an LTP-like facilitation of AMPA/kainate receptor-mediated synaptic transmission in the neonatal rat spinal cord. J. Neurophysiol., 84: 752–758.

Baldelli, P., Magnelli, V. and Carbone, E. (1999) Selective up-regulation of P- and R-type Ca2+ channels in rat embryo motoneurons by BDNF. Eur. J. Neurosci., 11: 1127–1133.

Baldelli, P., Novara, M., Carabeli, V., Hernandez-Guijo, J.M. and Carbone, E. (2002) BDNF up-regulates evoked GABAergic transmission in developing hippocampus by potentiating presynaptic N- and P/Q-type Ca2+ channels signaling. Eur. J. Neurosci., 16: 2297–2310.

Bardoni, R. (2001) Excitatory synaptic transmission in neonatal dorsal horn: NMDA and ATP receptors. News Physiol. Sci., 16: 95–100.

Bardoni, R., Magherini, P.C. and MacDermott, A.B. (2000) Activation of NMDA receptors drives action potentials in superficial dorsal horn from neonatal rats. Neuroreport, 11: 1721–1727.

Battaglia, G. and Rustioni, A. (1988) Coexistence of glutamate and substance P in dorsal root ganglion neurons of the rat and monkey. J. Comp. Neurol., 277: 302–312.

Bennett, D.L. (2001) Neurotrophic factors: important regulators of nociceptive function. Neuroscientist, 7: 13–17.

Bennett, D.L., French, J., Priestley, J.V. and McMahon, S.B. (1996) NGF but not NT-3 or BDNF prevents the A fiber sprouting into lamina II of the spinal cord that occurs following axotomy. Mol. Cell Neurosci., 8: 211–220.

Bennett, D.L., Michael, G.J., Ramachandran, N., Munson, J.B., Averill, S., Yan, Q., McMahon, S.B. and Priestley, J.V. (1998) A distinct subgroup of small DRG cells express GDNF receptor components and GDNF is protective for these neurons after nerve injury. J. Neurosci., 18: 3059–3072.

Bentley, G.N. and Gent, J.P. (1995) Neurokinin actions on substantia gelatinosa neurones in an adult longitudinal spinal cord preparation. Brain Res., 673: 101–111.

Berninger, B. and Poo, M. (1996) Fast actions of neurotrophic factors. Curr. Opin. Neurobiol., 6: 324–330.

Berninger, B., Marty, S., Zafra, F., da Penha, B.M., Thoenen, H. and Lindholm, D. (1995) GABAergic stimulation switches from enhancing to repressing BDNF expression in rat hippocampal neurons during maturation in vitro. Development, 121: 2327–2335.

Beydoun, A., Dyke, D.B., Morrow, T.J. and Casey, K.L. (1996) Topical capsaicin selectively attenuates heat pain and A delta fiber-mediated laser-evoked potentials. Pain, 65: 189–196.

Beydoun, A., Morrow, T.J. and Casey, K.L. (1997) Pain-related laser-evoked potentials in awake monkeys: identification of components, behavioral correlates and drug effects. Pain, 72: 319–324.

Blackshaw, L.A. and Gebhart, G.F. (2002) The pharmacology of gastrointestinal nociceptive pathways. Curr. Opin. Pharmacol., 2: 642–649.

Bleazard, L., Hill, R.G. and Morris, R. (1994) The correlation between the distribution of the NK_1 receptor and the actions of tachykinin agonists in the dorsal horn of the rat indicates that substance P does not have a functional role on substantia gelatinosa (lamina II) neurons. J. Neurosci., 14: 7655–7664.

Blochl, A. and Thoenen, H. (1995) Characterization of nerve growth factor (NGF) release from hippocampal neurons: evidence for a constitutive and an unconventional sodium-dependent regulated pathway. Eur. J. Neurosci., 7: 1220–1228.

Blochl, A. and Thoenen, H. (1996) Localization of cellular storage compartments and sites of constitutive and activity-dependent release of nerve growth factor (NGF) in primary cultures of hippocampal neurons. Mol. Cell Neurosci., 7: 173–190.

Boyce, S., Wyatt, A., Webb, J.K., O'Donnell, R., Mason, G., Rigby, M., Sirinathsinghji, D., Hill, R.G. and Rupniak, N.M. (1999) Selective NMDA NR2B antagonists induce antinociception without motor dysfunction: correlation with restricted localisation of NR2B subunit in dorsal horn. Neuropharmacology, 38: 611–623.

Brown, J.L., Liu, H., Maggio, J.E., Vigna, S.R., Mantyh, P.W. and Basbaum, A.I. (1995) Morphological characterization of substance P receptor-immunoreactive neurons in the rat spinal cord and trigeminal nucleus caudalis. J. Comp. Neurol., 356: 327–344.

Brown, M.C., Perry, V.H., Lunn, E.R., Gordon, S. and Heumann, R. (1991) Macrophage dependence of peripheral sensory nerve regeneration: possible involvement of nerve growth factor. Neuron, 6: 359–370.

Burnstock, G. (2000) P2X receptors in sensory neurones. Br. J. Anaesth., 84: 476–488.

Calzà, L., Giardino, L., Pozza, M., Micera, A. and Aloe, L. (1997) Time-course changes of nerve-growth factor, corticotropin-releasing hormone, and nitric oxide synthase isoforms and their possible role in the development of inflammatory response in experimental allergic encephalomyelitis. Proc. Natl. Acad. Sci. USA, 94: 3368–3373.

Calzà, L., Fernandez, M., Giuliani, A., Aloe, L. and Giardino, L. (2002) Thyroid hormone activates oligodendrocyte precursors and increases a myelin-forming protein and NGF content in the spinal cord during experimental allergic encephalomyelitis. Proc. Natl. Acad. Sci. USA, 99: 3258–3263.

Carmignoto, G., Zonta, M., Gobbo, S., Lossi, L., Gustincich, S., Salio, C. and Merighi, A. (2003) Pre-synaptic modulatory effect of brain-derived neurotrophic factor (BDNF) on the release of sensory neuromessengers in rat spinal cord slices. Pain, submitted.

Carroll, S.L., Silos-Santiago, I., Frese, S.E., Ruit, K.G., Milbrandt, J. and Snider, W.D. (1992) Dorsal root ganglion neurons expressing trk are selectively sensitive to NGF deprivation in utero. Neuron, 9: 779–788.

Cejas, P.J., Martinez, M., Karmally, S., McKillop, M., McKillop, J., Plunkett, J.A., Oudega, M. and Eaton, M.J. (2000) Lumbar transplant of neurons genetically modified to secrete brain-derived neurotrophic factor attenuates allodynia and hyperalgesia after sciatic nerve constriction. Pain, 86: 195–210.

Cervero, F. (1995) Visceral pain: Mechanisms of peripheral and central sensitization. Ann. Med., 27: 235–239.

Cervero, F. and Connell, L.A. (1984) Distribution of somatic and visceral primary afferent fibres within the thoracic spinal cord of the cat. J. Comp. Neurol., 230: 88–98.

Cho, H.J., Park, E.H., Bae, M.A. and Kim, J.K. (1996) Expression of mRNAs for preprotachykinin and nerve growth factor receptors in the dorsal root ganglion following peripheral inflammation. Brain Res., 716: 197–201.

Cho, H.J., Kim, J.K., Zhou, X.F. and Rush, R.A. (1997) Increased brain-derived neurotrophic factor immunoreactivity in rat dorsal root ganglia and spinal cord following peripheral inflammation. Brain Res., 764: 269–272.

Conn, P.J. and Pin, J.P. (1997) Pharmacology and functions of metabotropic glutamate receptors. Ann. Rev. Pharmacol., 37: 205–237.

Croll, S.D., Wiegand, S.J., Anderson, K.D., Lindsay, R.M. and Nawa, H. (1994) Regulation of neuropeptides in adult rat forebrain by the neurotrophins BDNF and NGF. Eur. J. Neurosci., 6: 1343–1353.

Curras, M.C. and Dingledine, R.J. (1992) Selectivity of amino acid transmitters acting at N-methyl D-aspartate and amino-3-hydroxy-5-methyl-4-isoxazolypropionate receptors. Mol. Pharmacol., 41: 520–526.

Davis, B.M., Lewin, G.R., Mendell, L.M., Jones, M.E. and Albers, K.M. (1993) Altered expression of nerve growth factor in the skin of transgenic mice leads to changes in response to mechanical stimuli. Neuroscience, 56: 789–792.

De Biasi, S. and Rustioni, A. (1988) Glutamate and substance P coexist in primary afferent terminals in the superficial laminae of the spinal cord. Proc. Natl. Acad. Sci. USA, 85: 7820–7824.

Delcroix, J.D., Tomlinson, D.R. and Fernyhough, P. (1997) Diabetes and axotomy-induced deficits in retrograde axonal transport of nerve growth factor correlate with decreased levels of p75LNTR protein in lumbar dorsal root ganglia. Brain Res. Mol. Brain Res., 51: 82–90.

Delcroix, J.D., Michael, G.J., Priestley, J.V., Tomlinson, D.R. and Fernyhough, P. (1998) Effect of nerve growth factor treatment on p75NTR gene expression in lumbar dorsal root ganglia of streptozocin-induced diabetic rats. Diabetes, 47: 1779–1785.

Distefano, P.S., Friedman, B., Radziejewski, C., Alexander, C., Boland, P., Schick, C.M., Lindsay, R.M. and Wiegand, S.J. (1992) The neurotrophins BDNF, NT-3, and NGF display distinct patterns of retrograde axonal transport in peripheral and central neurons. Neuron, 8: 983–993.

Dmitrieva, N. and McMahon, S.B. (1996) Sensitisation of visceral afferents by nerve growth factor in the adult rat. Pain, 66: 87–97.

Dmitrieva, N., Shelton, D., Rice, A.S. and McMahon, S.B. (1997) The role of nerve growth factor in a model of visceral inflammation. Neuroscience, 78: 449–459.

Donnerer, J., Schuligoi, R. and Stein, C. (1992) Increased content and transport of substance P and calcitonin gene-related peptide in sensory nerves innervating inflamed tissue: evidence for a regulatory function of nerve growth factor in vivo. Neuroscience, 49: 693–698.

Donnerer, J., Schuligoi, R., Stein, C. and Amann, R. (1993) Upregulation, release and axonal transport of substance P and calcitonin gene-related peptide in adjuvant inflammation and regulatory function of nerve growth factor. Regul. Pept., 46: 150–154.

Dubner, R. and Ren, K. (1999) Endogenous mechanisms of sensory modulation. Pain, Suppl. 6: S45–S53.

Dyck, P.J., Peroutka, S., Rask, C., Burton, E., Baker, M.K., Lehman, K.A., Gillen, D.A., Hokanson, J.L. and O'Brien, P.C. (1997) Intradermal recombinant human nerve growth factor induces pressure allodynia and lowered heat-pain threshold in humans. Neurology, 501–505.

Eaton, M.J., Blits, B., Ruitenberg, M.J., Verhaagen, J. and Oudega, M. (2002) Amelioration of chronic neuropathic pain after partial nerve injury by adeno-associated viral (AAV) vector-mediated over-expression of BDNF in the rat spinal cord. Gene Ther., 9: 1387–1395.

Ehlers, M.D., Kaplan, D.R., Price, D.L. and Koliatsos, V.E. (1995) NGF-stimulated retrograde transport of trkA in the mammalian nervous system. J. Cell Biol., 130: 149–156.

Engelman, H.S., Allen, T.B. and MacDermott, A.B. (1999) The distribution of neurons expressing calcium-permeable AMPA receptors in the superficial laminae of the spinal cord dorsal horn. J. Neurosci., 19: 2081–2089.

Ernfors, P., Lee, K.F. and Jaenisch, R. (1994a) Mice lacking brain-derived neurotrophic factor develop with sensory deficits. Nature, 368: 147–150.

Ernfors, P., Lee, K.-F., Kucera, J. and Jaenisch, R. (1994b) Lack of neurotrophin-3 leads to deficiencies in the peripheral nervous system and loss of limb proprioceptive afferents. Cell, 77: 503–512.

Favaron, M., Manev, R.M., Rimland, J.M., Candeo, P., Beccaro, M. and Manev, H. (1993) NMDA-stimulated expression of BDNF mRNA in cultured cerebellar granule neurones. Neuroreport, 4: 1171–1174.

Fields, H.L., Heinricher, M.M. and Mason, P. (1991) Neurotransmitters in nociceptive modulatory circuits. Annu. Rev. Neurosci., 14. 219–245.

Fitzgerald, M., Wall, P.D., Goedert, M. and Emson, P.C. (1985) Nerve growth factor counteracts the neurophysiological and neurochemical effects of chronic sciatic nerve section. Brain Res., 332: 131–141.

Fukuoka, T., Kondo, E., Dai, Y., Hashimoto, N. and Noguchi, K. (2001) Brain-derived neurotrophic factor increases in the uninjured dorsal root ganglion neurons in selective spinal nerve ligation model. J. Neurosci., 21: 4891–4900.

Gebhart, G.F. (1995) Visceral nociception: consequences, modulation and the future. Eur. J. Anaesthesiol, Suppl. 10: 24–27.

Gerber, G., Youn, D.H., Hsu, C.H., Isaev, D. and Randic, M. (2000a) Spinal dorsal horn synaptic plasticity: involvement of group I metabotropic glutamate receptors. Prog. Brain Res., 129: 115–134.

Gerber, G., Zhong, J., Youn, D. and Randic, M. (2000b) Group II and group III metabotropic glutamate receptor agonists depress synaptic transmission in the rat spinal cord dorsal horn. Neuroscience, 100: 393–406.

Gottschalk, W., Pozzo-Miller, L.D., Figurov, A. and Lu, B. (1998) Presynaptic modulation of synaptic transmission and plasticity by brain-derived neurotrophic factor in the developing hippocampus. J. Neurosci., 18: 6830–6839.

Grimes, M.L., Zhou, J., Beattie, E.C., Yuen, E.C., Hall, D.E., Valletta, J.S., Topp, K.S., LaVail, J.H., Bunnett, N.W., Mobley, W.C. (1996) Endocytosis of activated TrkA: evidence that nerve growth factor induces formation of signaling endosomes. J. Neurosci., 16: 7950–7964.

Guo, A., Simone, D.A., Stone, L.S., Fairbanks, C.A., Wang, J. and Elde, R. (2001) Developmental shift of vanilloid receptor 1 (VR1) terminals into deeper regions of the superficial dorsal horn: correlation with a shift from TrkA to Ret expression by dorsal root ganglion neurons. Eur. J. Neurosci., 14: 293–304.

Ha, S.O., Kim, J.K., Hong, H.S., Kim, D.S. and Cho, H.J. (2001) Expression of brain-derived neurotrophic factor in rat dorsal root ganglia, spinal cord and gracile nuclei in experimental models of neuropathic pain. Neuroscience, 107: 301–309.

Hains, B.C., Fullwood, S.D., Eaton, M.J. and Hulsebosch, C.E. (2001a) Subdural engraftment of serotonergic neurons following spinal hemisection restores spinal serotonin, down-regulates serotonin transporter, and increases BDNF tissue content in rat. Brain Res., 913: 35–46.

Hains, B.C., Johnson, K.M., McAdoo, D.J., Eaton, M.J. and Hulsebosch, C.E. (2001b) Engraftment of serotonergic precursors enhances locomotor function and attenuates chronic central pain behavior following spinal hemisection injury in the rat. Exp. Neurol., 171: 361–378.

Handwerker, H.O. and Kobal, G. (1993) Psychophysiology of experimentally induced pain. Physiol. Rev., 73: 639–671.

Heath, M.J.S., Womack, M.D. and MacDermott, A.B. (1994) Substance P elevates intracellular calcium in both neurons and glial cells from the dorsal horn of the spinal cord. J. Neurophysiol., 72: 1192–1198.

Heese, K., Otten, U., Mathivet, P., Raiteri, M., Marescaux, C. and Bernasconi, R. (2000) GABA(B) receptor antagonists elevate both mRNA and protein levels of the neurotrophins nerve growth factor (NGF) and brain-derived neurotrophic factor (BDNF) but not neurotrophin-3 (NT-3) in brain and spinal cord of rats. Neuropharmacology, 39: 449–462.

Helgren, M.E., Cliffer, K.D., Torrento, K., Cavnor, C., Curtis, R., Distefano, P.S., Wiegand, S.J. and Lindsay, R.M. (1997) Neurotrophin-3 administration attenuates deficits of pyridoxine-induced large-fiber sensory neuropathy. J. Neurosci., 17: 372–382.

Helke, C.J., Krause, J.E., Mantyh, P.W. and Couture, R. (1990) Diversity of tachykinin peptidergic neurons: multiple peptides, receptors and regulatory mechanisms. FASEB J., 4: 1606–1615.

Henry, M.A., Westrum, L.E., Bothwell, M. and Press, S. (1994) Electron microscopic localization of nerve growth factor receptor (p75)-immunoreactivity in pars caudalis/medullary dorsal horn of the cat. Brain Res., 642: 137–145.

Heppenstall, P.A. and Fleetwood-Walker, S.M. (1997) The glycine site of the NMDA receptor contributes to neuroki-nin1 receptor agonist facilitation of NMDA receptor agonist-evoked activity in rat dorsal horn neurons. Brain Res., 744: 235–245.

Heppenstall, P.A. and Lewin, G.R. (2001) BDNF but not NT-4 is required for normal flexion reflex plasticity and function. Proc. Natl. Acad. Sci. USA, 98: 8107–8112.

Herrero, J.F., Laird, J.M. and Lopez-Garcia, J.A. (2000) Wind-up of spinal cord neurones and pain sensation: much ado about something? Prog. Neurobiol., 61: 169–203.

Heumann, R. (1994) Neurotrophin signaling. Curr. Opin. Neurobiol., 4: 668–679.

Hoheisel, U., Mense, S., Simons, D.G. and Yu, X.-M. (1993) Appearance of new receptive fields in rat dorsal horn neurons following noxious stimulation of skeletal muscle: A model for referral of muscle pain. Neurosci. Lett., 153: 9–12.

Hökfelt, T. (1991) Neuropeptide in perspective: the last ten years. Neuron, 7: 867–879.

Hökfelt, T., Broberger, C., Xu, Z.-Q.D., Sergeyev, V., Ubink, R. and Diez, M. (2000) Neuropeptides—an overview. Neuropharmacology, 39: 1337–1356.

Hory-Lee, F., Russell, M., Lindsay, R.M. and Frank, E. (1993) Neurotrophin 3 supports the survival of developing muscle sensory neurons in culture. Proc. Natl. Acad. Sci. USA, 90: 2613–2617.

Huang, Z.J., Kirkwood, A., Pizzorusso, T., Porciatti, V., Morales, B., Bear, M.F., Maffei, L. and Tonegawa, S. (1999) BDNF regulates the maturation of inhibition and the critical period of plasticity in mouse visual cortex. Cell, 98: 739–755.

Hunt, S.P. and Mantyh, P.W. (2001) The molecular dynamics of pain control. Nat. Rev. Neurosci., 2: 83–91.

Hutchinson, G.B., McLennan, H. and Wheal, H.V. (1978) The responses of Renshaw cells and spinal interneurons of the rat to L-glutamate and L-aspartate. Brain Res., 141: 129–136.

Inaishi, Y., Kashihara, Y., Sakaguchi, M., Nawa, H. and Kuno, M. (1992) Cooperative regulation of calcitonin gene-related peptide levels in rat sensory neurons via their central and peripheral processes. J. Neurosci., 12: 518–524.

Jarvis, C.R., Xiong, Z.G., Plant, J.R., Churchill, D., Lu, W.Y., MacVicar, B.A. and MacDonald, J.F. (1997) Neurotrophin modulation of NMDA receptors in cultured murine and isolated rat neurons. J. Neurophysiol., 78: 2363–2371.

Ji, R.R. and Woolf, C.J. (2001) Neuronal plasticity and signal transduction in nociceptive neurons: implications for the initiation and maintenance of pathological pain. Neurobiol. Dis., 8: 1–10.

Ji, R.R., Samad, T.A., Jin, S.X., Schmoll, R. and Woolf, C.J. (2002) p38 MAPK activation by NGF in primary sensory neurons after inflammation increases TRPV1 levels and maintains heat hyperalgesia. Neuron, 36: 57–68.

Johnson, J.L. (1972) Glutamic acid as a synaptic transmitter candidate in dorsal sensory neuron—reconsideration. Life Sci., 20: 1637–1644.

Kaplan, D.R. and Cooper, E. (2001) PI-3 kinase and IP3: partners in NT3-induced synaptic transmission. Nat. Neurosci., 4: 5–7.

Kaplan, D.R. and Stephens, R.M. (1994) Neurotrophin signal transduction by the Trk receptor. J. Neurobiol., 25: 1404–1417.

Kerchner, G.A., Wang, G.D., Qiu, C.S., Huettner, J.E. and Zhuo, M. (2001a) Direct presynaptic regulation of GABA/glycine release by kainate receptors in the dorsal horn: an ionotropic mechanism. Neuron, 32: 477–488.

Kerchner, G.A., Wilding, T.J., Li, P., Zhuo, M. and Huettner, J.E. (2001b) Presynaptic kainate receptors regulate spinal sensory transmission. J. Neurosci., 21: 59–66.

Kerchner, G.A., Wilding, T.J., Huettner, J.E. and Zhuo, M. (2002) Kainate receptor subunits underlying presynaptic regulation of transmitter release in the dorsal horn. J. Neurosci., 22: 8010–8017.

Kerkut, G.A. and Bagust, J. (1995) The isolated mammalian spinal cord. Prog. Neurobiol., 46: 1–48.

Kerr, B.J., Bradbury, E.J., Bennett, D.L., Trivedi, P.M., Dassan, P., French, J., Shelton, D.B., McMahon, S.B. and Thompson, S.W. (1999) Brain-derived neurotrophic factor modulates nociceptive sensory inputs and NMDA-evoked responses in the rat spinal cord. J. Neurosci., 19: 5138–5148.

Kim, H.S., Lee, S.J., Kim, D.S. and Cho, H.J. (2000) Effects of brain-derived neurotrophic factor and neurotrophin-3 on expression of mRNAs encoding c-Fos, neuropeptides and glutamic acid decarboxylase in cultured spinal neurons. Neuroreport, 11: 3873–3876.

Kitzman, P.H., Perrone, T.N., LeMaster, A.M., Davis, B.M. and Albers, K.M. (1998) Level of p75 receptor expression in sensory ganglia is modulated by NGF level in the target tissue. J. Neurobiol., 35: 258–270.

Kokaia, Z., Gido, G., Ringstedt, T., Bengzon, J., Kokaia, M., Siesjo, B.K., Persson, H. and Lindvall, O. (1993) Rapid increase of BDNF mRNA levels in cortical neurons following spreading depression: regulation by glutamatergic mechanisms independent of seizure activity. Brain. Res. Mol. Brain Res., 19. 277–286.

Koltzenburg, M. (1999) The changing sensitivity in the life of the nociceptor. Pain, Suppl. 6: S93–S102.

Koltzenburg, M., Bennett, D.L., Shelton, D.L. and McMahon, S.B. (1999) Neutralization of endogenous NGF prevents the sensitization of nociceptors supplying inflamed skin. Eur. J. Neurosci., 11: 1698–1704.

Laing, I., Todd, A.J., Heizmann, C.W. and Schmidt, H.H.H.W. (1994) Subpopulations of GABAergic neurons in laminae I–III of rat spinal dorsal horn defined by coexistence with classical transmitters, peptides, nitric oxide synthase or parvalbumin. Neuroscience, 61: 123–132.

Larson, A.A. and Kitto, K.F. (1997) Mutual antagonism between nerve growth factor and substance P N-terminal activity on nociceptive activity in mice. J. Pharmacol. Exp. Ther., 282: 1345–1350.

Lauterborn, J.C., Lynch, G., Vanderklish, P., Arai, A. and Gall, C.M. (2000) Positive modulation of AMPA receptors increases neurotrophin expression by hippocampal and cortical neurons. J. Neurosci., 20: 8–21.

Le Bars, D. (2002) The whole body receptive field of dorsal horn multireceptive neurones. Brain Res. Brain Res. Rev., 40: 29 44.

Lembeck, F. (1953) Zur Frage den zentralen bertragung afferenter Impulse. Das Vorkommen und die Bedeutung der Substanz P in den dorsalen Wurzeln des Rüchkenmarks.

Naunyn-Schmiedeberg's. Arch. Exp. Pathol. Pharm., 219: 197–213.

Lesser, S.S., Sherwood, N.T. and Lo, D.C. (1997) Neurotrophins differentially regulate voltage-gated ion channels. Mol. Cell Neurosci., 10: 173–183.

Lever, I.J., Bradbury, E.J., Cunningham, J.R., Adelson, D.W., Jones, M.G., McMahon, S.B., Marvizon, J.C. and Malcangio, M. (2001) Brain-derived neurotrophic factor is released in the dorsal horn by distinctive patterns of afferent fiber stimulation. J. Neurosci., 21: 4469–4477.

Levi-Montalcini, R. and Angeletti, P.U. (1968) Nerve growth factor. Physiol. Rev., 48: 534–569.

Levine, E.S., Crozier, R.A., Black, I.B. and Plummer, M.R. (1998) Brain-derived neurotrophic factor modulates hippocampal synaptic transmission by increasing N-methyl-D-aspartic acid receptor activity. Proc. Natl. Acad. Sci. USA, 95: 10235–10239.

Lewin, G.R. and Barde, Y.-A. (1996) Physiology of neurotrophins. Ann. Rev. Neurosci., 19: 289–317.

Lewin, G.R. and McMahon, S.B. (1991) Dorsal horn plasticity following re-routing of peripheral nerves: evidence for tissue-specific neurotrophic influences from the periphery. Eur. J. Neurosci., 3: 1112–1122.

Lewin, G.R. and Mendell, L.M. (1993) Nerve growth factor and nociception. Trends Neurosci., 16: 353–359.

Lewin, G.R. and Mendell, L.M. (1994) Regulation of cutaneous C-fiber heat nociceptors by nerve growth factor in the developing rat. J. Neurophysiol., 71: 941–949.

Lewin, G.R., Ritter, A.M. and Mendell, L.M. (1992) On the role of nerve growth factor in the development of myelinated nociceptors. J. Neurosci., 12: 1896–1905.

Lewin, G.R., Ritter, A.M. and Mendell, L.M. (1993) Nerve growth factor-induced hyperalgesia in the neonatal and adult rat. J. Neurosci., 13: 2136–2148.

Lewin, G.R., Rueff, A. and Mendell, L.M. (1994) Peripheral and central mechanisms of NGF-induced hyperalgesia. Eur. J. Neurosci., 6: 1903–1912.

Li, P., Kerchner, G.A., Sala, C., Wei, F., Huettner, J.E., Sheng, M. and Zhuo, M. (1999a) AMPA receptor-PDZ interactions in facilitation of spinal sensory synapses. Nat. Neurosci., 2: 972–977.

Li, P., Wilding, T.J., Kim, S.J., Calejesan, A.A., Huettner, J.E. and Zhuo, M. (1999b) Kainate-receptor-mediated sensory synaptic transmission in mammalian spinal cord. Nature, 397: 161–164.

Li, Y.X., Zhang, Y., Ju, D., Lester, H.A., Shuman, E.M. and Davidson, N. (1998) Expression of a dominant negative TrkB receptor, T1, reveals a requirement for presynaptic signaling in BDNF-induced synaptic potentiation in cultured hippocampal neurons. Proc. Natl. Acad. Sci. USA, 95: 10884 10889.

Liebl, D.J., Klesse, L.J., Tessarollo, L., Wohlman, T. and Parada, L.F. (2000) Loss of brain-derived neurotrophic factor-dependent neural crest-derived sensory neurons in

316

neurotrophin-4 mutant mice. Proc. Natl. Acad. Sci. USA, 97: 2297–2302.

Lindholm, D., Castren, E., Berzaghi, M., Blochl, A. and Thoenen, H. (1994) Activity-dependent and hormonal regulation of neurotrophin mRNA levels in the brain—implications for neuronal plasticity. J. Neurobiol., 25: 1362–1372.

Lindsay, R.M. and Harmar, A.J. (1989) Nerve growth factor regulates expression of neuropeptide genes in adult sensory neurons. Letters to Nature, 337: 362–364.

Lindsay, R.M., Lockett, C., Sternberg, J. and Winter, J. (1989) Neuropeptide expression in cultures of adult sensory neurons: modulation of substance P and calcitonin gene-related peptide levels by nerve growth factor. Neuroscience, 33: 53–53.

Liu, H., Brown, J.L., Jasmin, L., Maggio, J.E., Vigna, S.R., Mantyh, P.W. and Basbaum, A.I. (1994a) Synaptic relationship between substance P and the substance P receptor: light and electron microscopic characterization of the mismatch between neuropeptides and their receptors. Proc. Natl. Acad. Sci. USA, 91: 1009–1013.

Liu, H., Wang, H., Sheng, M., Jan, L.Y., Jan, Y.N. and Basbaum, A.I. (1994b) Evidence for presynaptic N-methyl-D-aspartate autoreceptors in the spinal cord dorsal horn. Proc. Natl. Acad. Sci. USA, 91: 8383–8387.

Liu, X.G. and Sandkuhler, J. (1998) Activation of spinal N-methyl-D-aspartate or neurokinin receptors induces long-term potentiation of spinal C-fibre-evoked potentials. Neuroscience, 86: 1209–1216.

Lu, C.R., Hwang, S.J., Phend, K.D., Rustioni, A. and Valtschanoff, J.G. (2002) Primary afferent terminals in spinal cord express presynaptic AMPA receptors. J. Neurosci., 22: 9522–9529.

Lu, C.R., Hwang, S.J., Phend, K.D., Rustioni, A. and Valtschanoff, J.G. (2003) Primary afferent terminals that express presynaptic NR1 in rats are mainly from myelinated, mechanosensitive fibers. J. Comp. Neurol., 460: 191–202.

Lucas, G., Hendolin, P., Harkany, T., Agerman, K., Paratcha, G., Holmgren, C., Zilberter, Y., Sairanen, M., Minichiello, L., Castren, E., Ernfors, P. (2003) Neurotrophin-4 mediated TrkB activation reinforces morphine-induced analgesia. Nat. Neurosci., 6: 221–222.

Lundberg, J.M. (1996) Pharmacology of cotransmission in the autonomic nervous system: integrative aspects on amines, neuropeptides, adenosine triphosphate, amino acids and nitric oxide. Pharmacol. Rev., 48: 113–178.

Ma, W. and Bisby, M.A. (1998) Increase of preprotachykinin mRNA and substance P immunoreactivity in spared dorsal root ganglion neurons following partial sciatic nerve injury. Eur. J. Neurosci., 10: 2388–2399.

Malcangio, M. and Bowery, N.G. (1996) GABA and its receptors in the spinal cord. Trends Pharmacol. Sci., 17: 457–462.

Malcangio, M., Bowery, N.G., Flower, R.J. and Perretti, M. (1996) Effect of interleukin-1β on the release of substance P from rat isolated spinal cord. Eur. J. Pharmacol., 299: 113–118.

Malcangio, M., Fernandes, K. and Tomlinson, D.R. (1998) NMDA receptor activation modulates evoked release of substance P from rat spinal cord. Br. J. Pharmacol., 125: 1625–1626.

Malcangio, M., Garrett, N.E., Cruwys, S. and Tomlinson, D.R. (1997) Nerve growth factor- and neurotrophin-3-induced changes in nociceptive threshold and the release of substance P from the rat isolated spinal cord. J. Neurosci., 17: 8459–8467.

Malcangio, M. and Lessmann, V. (2003) A common thread for pain and memory synapses? Brain-derived neurotrophic factor and trkB receptors. Trends Pharmacol. Sci., 24: 116–121.

Malcangio, M., Ramer, M.S., Boucher, T.J. and McMahon, S.B. (2000) Intrathecally injected neurotrophins and the release of substance P from the rat isolated spinal cord. Eur. J. Neurosci., 12: 139–144.

Mannion, R.J., Costigan, M., Decosterd, I., Amaya, F., Ma, Q.P., Holstege, J.C., Ji, R.R., Acheson, A., Lindsay, R.M., Wilkinson, G.A., Woolf, C.J. (1999) Neurotrophins: peripherally and centrally acting modulators of tactile stimulus-induced inflammatory pain hypersensitivity. Proc. Natl. Acad. Sci. USA, 96: 9385–9390.

Mantyh, P.W., Rogers, S.D., Honoré, P., Allen, B.J., Ghilardi, J.R., Li, J., Daughters, R.S., Lappi, D.L., Wiley, R.G., Simone, D.A. (1997) Inhibition of hyperalgesia by ablation of lamina I spinal neurons expressing substance P receptor. Science, 278: 275–279.

Marchettini, P., Simone, D.A., Caputi, G. and Ochoa, J.L. (1996) Pain from excitation of identified muscle nociceptors in humans. Brain Res., 740: 109–116.

Marchettini, P., Lacerenza, M. and Formaglio, F. (2000) Sympathetically maintained pain. Curr. Rev. Pain, 4: 99–104.

Marty, S., Berninger, B., Carroll, P. and Thoenen, H. (1996) GABAergic stimulation regulates the phenotype of hippocampal interneurons through the regulation of brain-derived neurotrophic factor. Neuron, 16: 565–570.

Marvizon, J.C., Martinez, V., Grady, E.F., Bunnett, N.W. and Mayer, E.A. (1997) Neurokinin 1 receptor internalization in spinal cord slices induced by dorsal root stimulation is mediated by NMDA receptors. J. Neurosci., 17: 8129–8136.

Maxwell, D.J., Christie, W.H., Short, A.D., Störm-Mathisen, J. and Ottersen, O.P. (1990) Central boutons of glomeruli in the spinal cord of the cat are enriched with L-glutamate-like immunoreactivity. Neuroscience, 36: 83–104.

McHugh, J.M. and McHugh, W.B. (2000) Pain: neuroanatomy, chemical mediators, and clinical implications. AACN. Clin. Issues, 11: 168–178.

McMahon, S.B. (1996) NGF as a mediator of inflammatory pain. Philos. Trans. R. Soc. Lond. B. Biol. Sci., 351: 431–440.

McMahon, S.B., Armanini, M.P., Ling, L.H. and Phillips, H.S. (1994) Expression and coexpression of Trk receptors in subpopulations of adult primary sensory neurons projecting to identified peripheral targets. Neuron, 12: 1161–1171.

McMahon, S.B., Bennett, D.L., Priestley, J.V. and Shelton, D.L. (1995) The biological effects of endogenous nerve growth factor on adult sensory neurons revealed by a trkA-IgG fusion molecule. Nat. Med., 1: 774–780.

Meller, S.T. and Gebhart, G.F. (1994) Spinal mediators of hyperalgesia. Drugs, 47: 10–20.

Mendell, L.M. and Arvanian, V.L. (2002) Diversity of neurotrophin action in the postnatal spinal cord. Brain Res. Brain Res. Rev., 40: 230–239.

Mendell, L.M., Munson, J.B. and Arvanian, V.L. (2001) Neurotrophins and synaptic plasticity in the mammalian spinal cord. J. Physiol., 533: 91–97.

Mense, S. (1993) Nociception from skeletal muscle in relation to clinical muscle pain. Pain, 54: 241–289.

Merighi, A. (2002) Costorage and coexistence of neuropeptides in the mammalian CNS. Progr. Neurobiol., 66: 161–190.

Merighi, A., Polak, J.M. and Theodosis, D.T. (1991) Ultrastructural visualization of glutamate and aspartate immunoreactivities in the rat dorsal horn with special reference to the co-localization of glutamate, substance P and calcitonin gene-related peptide. Neuroscience, 40: 67–80.

Merighi, A., Zonta, M. and Carmignoto, G. (1999) Neurotrophins acutely alter intracellular calcium levels in dorsal horn neurons of the rat spinal cord. Abstracts of the Society of Neuroscience USA, 25: 1043–1043.

Merighi, A., Aimar, P., Pasti, L., Lossi, L. and Carmignoto, G. (2000). Neuromodulatory effects of nitric oxide in pain perception. In: G. Poli, E. Cadenas and L. Packer (Eds.), Free Radicals in Brain Pathophysiology. Marcel Dekker, New York, pp. 17–53.

Michael, G.J., Averill, S., Nitkunan, A., Rattray, M., Bennett, D.L.H., Yan, Q. and Priestley, J.V. (1997) Nerve growth factor treatment increases brain-derived neurotrophic factor selectively in trk-A expressing dorsal root ganglion cells and in their central terminations within the spinal cord. J. Neurosci., 17: 8476–8490.

Michael, G.J., Averill, S., Shortland, P.J., Yan, Q. and Priestley, J.V. (1999) Axotomy results in major changes in BDNF expression by dorsal root ganglion cells: BDNF expression in large trkB and trkC cells, in pericellular baskets, and in projections to deep dorsal horn and dorsal column nuclei. Eur. J. Neurosci., 11: 3539–3551.

Miletic, G. and Miletic, V. (2002) Increases in the concentration of brain derived neurotrophic factor in the lumbar spinal dorsal horn are associated with pain behavior following chronic constriction injury in rats. Neurosci. Lett., 319: 137–140.

Millan, M.J. (1999) The induction of pain: an integrative review. Prog. Neurobiol., 57: 1–164.

Morisset, V., Ahluwalia, J., Nagy, I. and Urban, L. (2001) Possible mechanisms of cannabinoid-induced antinociception in the spinal cord. Eur. J. Pharmacol., 429: 93–100.

Mowla, S.J., Pareek, S., Farhadi, H.F., Petrecca, K., Fawcett, J.P., Seidah, N.G., Morris, S.J., Sossin, W.S. and Murphy, R.A. (1999) Differential sorting of nerve growth factor and brain-derived neurotrophic factor in hippocampal neurons. J. Neurosci., 19: 2069–2080.

Munglani, R., Harrison, S.M., Smith, G.D., Bountra, C., Birch, P.J., Elliot, P.J. and Hunt, S.P. (1996) Neuropeptide changes persist in spinal cord despite resolving hyperalgesia in a rat model of mononeuropathy. Brain Res., 743: 102–108.

Nakaya, Y., Kaneko, T., Shigemoto, R., Nakanishi, S. and Mizuno, N. (1994) Immunohistochemical localization of substance P receptor in the central nervous system of the adult rat. J. Comp. Neurol., 347: 249–274.

Neumann, S., Doubell, T.P., Leslie, T. and Woolf, C.J. (1996) Inflammatory pain hypersensitivity mediated by phenotypic switch in myelinated primary sensory neurons. Nature, 384: 360–364.

Oakley, R.A., Garner, A.S., Large, T.H. and Frank, E. (1995) Muscle sensory neurons require neurotrophin-3 from peripheral tissues during the period of normal cell death. Development, 121: 1341–1350.

Obata, K., Tsujino, H., Yamanaka, H., Yi, D., Fukuoka, T., Hashimoto, N., Yonenobu, K., Yoshikawa, H. and Noguchi, K. (2002) Expression of neurotrophic factors in the dorsal root ganglion in a rat model of lumbar disc herniation. Pain, 99: 121–132.

Obrietan, K., Gao, X.B. and van den Pol, A.N. (2002) Excitatory actions of GABA increase BDNF expression via a MAPK-CREB-dependent mechanism—a positive feedback circuit in developing neurons. J. Neurophysiol., 88: 1005–1015.

Otten, U., Goedert, M., Mayer, N. and Lembeck, F. (1980) Requirement of nerve growth factor for development of substance P-containing sensory neurones. Nature, 287: 158–159.

Otten, U., Ruegg, U.T., Hill, R.C., Businger, F. and Peck, M. (1982) Correlation between substance P content of primary sensory neurones and pain sensitivity in rats exposed to antibodies to nerve growth factor. Eur. J. Pharmacol., 85: 351–353.

Ottersen, O.P. and Störm-Mathisen, J. (1984) Glutamate- and GABA-containing neurons in the mouse and rat brain as demonstrated with a new immunocytochemical technique. J. Comp. Neurol., 229: 374–392.

Ottersen, O.P. and Störm-Mathisen, J. (1987) Localization of amino acid neurotransmitters by immunocytochemistry. Trends Neurosci., 10: 250–255.

Owolabi, J.B., Rizkalla, G., Tehim, A., Ross, G.M., Riopelle, R.J., Kamboj, R., Ossipov, M., Bian, D., Wegert, S., Porreca, F. and Lee, D.K. (1999) Characterization of antiallodynic actions of ALE-0540, a

318

novel nerve growth factor receptor antagonist, in the rat. J. Pharmacol. Exp. Ther., 289: 1271–1276.

Pesavento, E., Margotti, E., Righi, M., Cattaneo, A. and Domenici, L. (2000) Blocking the NGF-TrkA interaction rescues the developmental loss of LTP in the rat visual cortex: role of the cholinergic system. Neuron, 25: 165–175.

Petersen-Zeitz, K.R. and Basbaum, A.I. (1999) Second messengers, the substantia gelatinosa and injury-induced persistent pain. Pain, Suppl. 6: S5–S12.

Pezet, S., Onteniente, B., Jullien, J., Junier, M.P., Grannec, G., Rudkin, B.B. and Calvino, B. (2001) Differential regulation of NGF receptors in primary sensory neurons by adjuvant-induced arthritis in the rat. Pain, 90: 113–125.

Pezet, S., Cunningham, J., Patel, J., Grist, J., Gavazzi, I., Lever, I.J. and Malcangio, M. (2002a) BDNF modulates sensory neuron synaptic activity by a facilitation of GABA transmission in the dorsal horn. Mol. Cell Neurosci., 21: 51–62.

Pezet, S., Malcangio, M., Lever, I.J., Perkinton, M.S., Thompson, S.W., Williams, R.J. and McMahon, S.B. (2002b) Noxious stimulation induces Trk receptor and downstream ERK phosphorylation in spinal dorsal horn. Mol. Cell Neurosci., 21: 684–695.

Pezet, S., Malcangio, M. and McMahon, S.B. (2002c) BDNF: a neuromodulator in nociceptive pathways? Brain Res. Brain Res. Rev., 40: 240–249.

Pizzorusso, T., Ratto, G.M., Putignano, E. and Maffei, L. (2000) Brain-derived neurotrophic factor causes cAMP response element-binding protein phosphorylation in absence of calcium increases in slices and cultured neurons from rat visual cortex. J. Neurosci., 20: 2809–2816.

Popratiloff, A., Weinberg, R.J. and Rustioni, A. (1998) AMPA receptors at primary afferent synapses in substantia gelatinosa after sciatic nerve section. Eur. J. Neurosci., 10: 3220–3230.

Pozzan, T., Rizzuto, R., Volpe, P. and Meldolesi, J. (1994) Molecular and cellular physiology of intracellular calcium stores. Physiol. Rev., 74: 595–636.

Pozzo-Miller, L.D., Gottschalk, W., Zhang, L., McDermott, K., Du, J., Gopalakrishnan, R., Oho, C., Sheng, Z.H. and Lu, B. (1999) Impairments in high-frequency transmission, synaptic vesicle docking, and synaptic protein distribution in the hippocampus of BDNF knockout mice. J. Neurosci., 19: 4972–4983.

Ramer, M.S., Ma, W., Murphy, P.G., Richardson, P.M. and Bisby, M.A. (1998) Galanin expression in neuropathic pain: friend or foe?. Ann. N. Y. Acad. Sci., 863: 390–401.

Ramer, M.S., Priestley, J.V. and McMahon, S.B. (2000) Functional regeneration of sensory axons into the adult spinal cord. Nature, 403: 312–316.

Ramer, M.S., Bradbury, E.J. and McMahon, S.B. (2001) Nerve growth factor induces P2X(3) expression in sensory neurons. J. Neurochem., 77: 864–875.

Regoli, D., Boudon, A. and Fauchére, J.-L. (1994) Receptors and antagonists for substance P and related peptides. Pharmacol. Rev., 46: 551–599.

Rexed, B. (1954) A cytoarchitectonic atlas of the spinal cord in the cat. J. Comp. Neurol., 100: 297–379.

Ribeiro-Da-Silva, A. (1994). Substantia gelatinosa of spinal cord. In: G. Paxinos (Ed.), The Rat Nervous System. Academic Press, New York.

Ribeiro-Da-Silva, A., Cuello, A.C. and Henry, J.L. (2000) NGF over-expression during development leads to permanent alterations in innervation in the spinal cord and in behavioral responses to sensory stimuli. Neuropeptides, 34: 281–291.

Riccio, A., Pierchala, B.A., Ciarallo, C.L. and Ginty, D.D. (1997) An NGF-TrkA-mediated retrograde signal to transcription factor CREB in sympathetic neurons. Science, 277: 1097–1100.

Richardson, P.M., Issa, V.M. and Riopelle, R.J. (1986) Distribution of neuronal receptors for nerve growth factor in the rat. J. Neurosci., 6: 2312–2321.

Ritter, A.M. and Mendell, L.M. (1992) Somal membrane properties of physiologically identified sensory neurons in the rat: effects of nerve growth factor. J. Neurophysiol., 68: 2033–2041.

Ritter, A.M., Lewin, G.R., Kremer, N.E. and Mendell, L.M. (1991) Requirement for nerve growth factor in the development of myelinated nociceptors in vivo. Nature, 350: 500–502.

Ritter, A.M., Woodbury, C.J., Albers, K., Davis, B.M. and Koerber, H.R. (2000) Maturation of cutaneous sensory neurons from normal and NGF-overexpressing mice. J. Neurophysiol., 83: 1722–1732.

Ritter, A.M., Woodbury, C.J., Davis, B.M., Albers, K. and Koerber, H.R. (2001) Excess target-derived neurotrophin-3 alters the segmental innervation of the skin. Eur. J. Neurosci., 14: 411–418.

Romero, M.I., Rangappa, N., Garry, M.G. and Smith, G.M. (2001) Functional regeneration of chronically injured sensory afferents into adult spinal cord after neurotrophin gene therapy. J. Neurosci., 21: 8408–8416.

Routh, V.H. and Helke, C.J. (1995) Tachykinin receptors in the spinal cord. Prog. Brain Res., 104: 93–108.

Ruda, M.A., Bennett, G.J. and Dubner, R. (1986) Neurochemistry and neural circuitry in the dorsal horn. Progr. Brain Res., 66: 219–268.

Rudomin, P. (1999) Presynaptic selection of afferent inflow in the spinal cord. J. Physiol. Paris, 93: 329–347.

Ruit, K.G., Elliott, J.L., Osborne, P.A., Yan, Q. and Snider, W.D. (1992) Selective dependence of mammalian dorsal root ganglion neurons on nerve growth factor during embryonic development. Neuron, 8: 573–587.

Rustioni, A. and Weinberg, R.J. (1989). The somatosensory system. In: A. Björklund, T. Hökfelt and L.W. Swanson (Eds.), Handbook of Chemical Neuroanatomy. Integrated

Systems of the CNS, Part II, Vol. 7, Elsevier, Amsterdam, New York, pp. 219–321.

Saarelainen, T., Pussinen, R., Koponen, E., Alhonen, L., Wong, G., Sirvio, J. and Castren, E. (2000) Transgenic mice overexpressing truncated trkB neurotrophin receptors in neurons have impaired long-term spatial memory but normal hippocampal LTP. Synapse, 38: 102–104.

Sakai, N., Yamada, M., Numakawa, T., Ogura, A. and Hatanaka, H. (1997) BDNF potentiates spontaneous Ca^{2+} oscillations in cultured hippocampal neurons. Brain Res., 778: 318–328.

Schadrack, J. and Zieglgansberger, W. (1998) Pharmacology of pain processing systems. Z. Rheumatol., 57(Suppl. 2): 1–4.

Seebach, B.S., Arvanov, V. and Mendell, L.M. (1999) Effects of BDNF and NT-3 on development of Ia/motoneuron functional connectivity in neonatal rats. J. Neurophysiol., 81: 2398–2405.

Sermasi, E., Margotti, E., Cattaneo, A. and Domenici, L. (2000) Trk B signaling controls LTP but not LTD expression in the developing rat visual cortex. Eur. J. Neurosci., 12: 1411–1419.

Sherwood, N.T., Lesser, S.S. and Lo, D.C. (1997) Neurotrophin regulation of ionic currents and cell size depends on cell context. Proc. Natl. Acad. Sci. USA, 94: 5917–5922.

Simone, D.A. and Kajander, K.C. (1996) Excitation of rat cutaneous nociceptors by noxious cold. Neurosci. Lett., 213: 53–56.

Simone, D.A. and Kajander, K.C. (1997) Responses of cutaneous A-fiber nociceptors to noxious cold. J. Neurophysiol., 77: 2049–2060.

Siuciak, J.A., Altar, C.A., Wiegand, S.J. and Lindsay, R.M. (1994) Antinociceptive effect of brain-derived neurotrophic factor and neurotrophin-3. Brain Res., 633: 326–330.

Siuciak, J.A., Wong, V., Pearsall, D., Wiegand, S.J. and Lindsay, R.M. (1995) BDNF produces analgesia in the formalin test and modifies neuropeptide levels in rat brain and spinal cord areas associated with nociception. Eur. J. Neurosci., 7: 663–670.

Smeyne, R.J., Klein, R., Schnapp, A., Long, L.K., Bryant, S., Lewin, A., Lira, S.A. and Barbacid, M. (1994) Severe sensory and sympathetic neuropathies in mice carrying a disrupted Trk/NGF receptor gene. Nature, 368: 246–249.

Snider, W.D. (1994) Functions of the neurotrophins during nervous system development: what the knockouts are teaching us. Cell, 77: 627–638.

Sorkin, L.S. (1993) NMDA evokes an L-NAME sensitive spinal release of glutamate and citrulline. Neuroreport, 4: 479–482.

Stanfa, L.C. and Dickenson, A.H. (1999) The role of non-N-methyl-D-aspartate ionotropic glutamate receptors in the spinal transmission of nociception in normal animals and animals with carrageenan inflammation. Neuroscience, 93: 1391–1398.

Suen, P.C., Wu, K., Levine, E.S., Mount, H.T., Xu, J.L., Lin, S.Y. and Black, I.B. (1997) Brain-derived neurotrophic factor rapidly enhances phosphorylation of the postsynaptic N-methyl-D-aspartate receptor subunit 1. Proc. Natl. Acad. Sci. USA, 94: 8191–8195.

Szekely, J.I., Torok, K. and Mate, G. (2002) The role of ionotropic glutamate receptors in nociception with special regard to the AMPA binding sites. Curr. Pharm. Des., 8: 887–912.

Thoenen, H. (1995) Neurotrophins and neuronal plasticity. Science, 270: 593–598.

Thompson, S.W., Dray, A., McCarson, K.E., Krause, J.E. and Urban, L. (1995) Nerve growth factor induces mechanical allodynia associated with novel A fibre-evoked spinal reflex activity and enhanced neurokinin-1 receptor activation in the rat. Pain, 62: 219–231.

Thompson, S.W., Bennett, D.L., Kerr, B.J., Bradbury, E.J. and McMahon, S.B. (1999) Brain-derived neurotrophic factor is an endogenous modulator of nociceptive responses in the spinal cord. Proc. Natl. Acad. Sci. USA, 96: 7714–7718.

Todd, A.J. (2002) Anatomy of primary afferents and projection neurones in the rat spinal dorsal horn with particular emphasis on substance P and the neurokinin 1 receptor. Exp. Physiol., 87: 245–249.

Todd, A.J. and Spike, R.C. (1993) The localization of classical transmitters and neuropeptides within neurons in laminae I–III of the mammalian spinal dorsal horn. Prog. Neurobiol., 41: 609–638.

Todd, A.J., Spike, R.C., Chong, D. and Neilson, M. (1995) The relationship between glycine and gephyrin in synapses of the rat spinal cord. Eur. J. Neurosci., 7: 1–11.

Todd, A.J., Watt, C., Spike, R.C. and Sieghart, W. (1996) Colocalization of GABA, glycine, and their receptors at synapses in the rat spinal cord. J. Neurosci., 16: 974–982.

Tölle, T.R., Berthele, A., Zieglgänsberger, W., Seeburg, P.H. and Wisden, W. (1993) The differential expression of 16NMDA and non-NMDA receptor subunits in the rat spinal cord and in periaqueductal gray. J. Neurosci., 13: 5009–5028.

Treede, R.-D., Meyer, R.A., Raja, S.N. and Campbell, J.N. (1992) Peripheral and central mechanisms of cutaneous hyperalgesia. Progr. Neurobiol., 58: 397–421.

Valerio, A., Paterlini, M., Boifava, M., Memo, M. and Spano, P.F. (1997) Metabotropic glutamate receptors mRNA expression in rat spinal cord. Neuroreport, 8: 2695–2699.

Valtschanoff, J.G., Phend, K.D., Bernardi, P.S., Weinberg, R.J. and Rustioni, A. (1994) Amino acid immunocytochemistry of primary afferent terminals in the rat dorsal horn. J. Comp. Neurol., 346: 237–252.

Verge, V.M.K., Richardson, P.M., Benoit, R. and Riopelle, R.J. (1989) Histochemical characterization of sensory neurons with high-affinity receptors for nerve growth factor. J. Neurocytol., 18: 583–591.

Verge, V.M., Richardson, P.M., Wiesenfeld-Hallin, Z. and Hökfelt, T. (1995) Differential influence of nerve growth factor on neuropeptide expression in vivo: a novel role in peptide suppression in adult sensory neurons. J. Neurosci., 15: 2081–2096.

Vierck, C.J., Jr., Cannon, R.L., Fry, G., Maixner, W. and Whitsel, B.L. (1997) Characteristics of temporal summation of second pain sensations elicited by brief contact of glabrous skin by a preheated thermode. J. Neurophysiol., 78: 992–1002.

Villoslada, P., Hauser, S.L., Bartke, I., Unger, J., Heald, N., Rosenberg, D., Cheung, S.W., Mobley, W.C., Fisher, S., Genain, C.P. (2000) Human nerve growth factor protects common marmosets against autoimmune encephalomyelitis by switching the balance of T helper cell type 1 and 2 cytokines within the central nervous system. J. Exp. Med., 191: 1799–1806.

Weskamp, G. and Otten, U. (1987) An enzyme-linked immunoassay for nerve growth factor (NGF): a tool for studying regulatory mechanisms involved in NGF production in brain and in peripheral tissues. J. Neurochem., 48: 1779–1786.

Wetmore, C., Olson, L. and Bean, A.J. (1994) Regulation of brain-derived neurotrophic factor (BDNF) expression and release from hippocampal neurons is mediated by nonNMDA type glutamate receptors. J. Neurosci., 14: 1688–1700.

White, D.M. (1998) Contribution of neurotrophin-3 to the neuropeptide Y-induced increase in neurite outgrowth of rat dorsal root ganglion cells. Neuroscience, 86: 257–263.

White, D.M. (2000) Neurotrophin-3 antisense oligonucleotide attenuates nerve injury-induced Abeta-fibre sprouting. Brain Res., 885: 79–86.

Willis, W.D. (1991). Sensory Mechanisms of the Spinal Cord. Oxford University Press, New York.

Willis, W.D. (2001) Role of neurotransmitters in sensitization of pain responses. Ann. N. Y. Acad. Sci., 933: 142–156.

Willis, W.D. (2002) Long-term potentiation in spinothalamic neurons. Brain Res. Brain Res. Rev., 40: 202–214.

Willis, W.D., Jr. and Coggeshall, R.E. (1991). Sensory Mechanisms of the Spinal Cord. Plenum Press, New York and London.

Wimalawansa, S.J. (1996) Calcitonin gene-related peptide and its receptors: molecular genetics, physiology, pathophysiology, and therapeutic potentials. Endocrine Rev., 17: 533–585.

Wimalawansa, S.J., Gunasekera, R.D. and Zhang, F. (1993) Isolation, purification, and characterization of calcitonin gene-related peptide receptor. Peptides, 14: 691–699.

Winter, J. (1998) Brain derived neurotrophic factor, but not nerve growth factor, regulates capsaicin sensitivity of rat vagal ganglion neurones. Neurosci. Lett., 241: 21–24.

Wood, J.N. and Docherty, R. (1997) Chemical activators of sensory neurons. Annu. Rev. Physiol., 59: 457–482.

Woolf, C.J. (1996) Phenotypic modification of primary sensory neurons: the role of nerve growth factor in the production of persistent pain. Philos. Trans. R. Soc. Lond B. Biol. Sci., 351: 441–448.

Woolf, C.J. and Costigan, M. (1999) Transcriptional and posttranslational plasticity and the generation of inflammatory pain. Proc. Natl. Acad. Sci. USA, 96: 7723–7730.

Woolf, C.J. and Wall, P.D. (1986) The effectiveness of C-primary afferent fibres of different origins in evoking a prolonged facilitation of the flexor reflex in the rat. J. Neurosci., 6: 221–232.

Woolf, C.J., Shortland, P. and Coggeshall, R.E. (1992) Peripheral nerve injury triggers central sprouting of myelinated afferents. Nature, 355: 75–78.

Woolf, C.J., Safieh-Garabedian, B., Ma, Q.P., Crilly, P. and Winter, J. (1994a) Nerve growth factor contributes to the generation of inflammatory sensory hypersensitivity. Neuroscience, 62: 327–331.

Woolf, C.J., Shortland, P. and Sivilotti, L.G. (1994b) Sensitization of high mechanothreshold superficial dorsal horn and flexor motor neurones following chemosensitive primary afferent activation. Pain, 58: 141–155.

Woolf, C.J., Ma, Q.-P., Allchorne, A. and Poole, S. (1996) Periphreal cell types contributing to the hyperalgesic action of nerve growth factor in inflammation. J. Neurosci., 16: 2716–2723.

Wright, D.E. and Snider, W.D. (1995) Neurotrophin receptor mRNA expression defines distinct populations of neurons in rat dorsal root ganglia. J. Comp. Neurol., 351: 329–338.

Wright, D.E., Zhou, L., Kucera, J. and Snider, W.D. (1997) Introduction of a neurotrophin-3 transgene into muscle selectively rescues proprioceptive neurons in mice lacking endogenous neurotrophin-3. Neuron, 19: 503–517.

Xi, Z.X. and Akasu, T. (1996) Presynaptic GABAA receptors in vertebrate synapses. Kurume Med. J., 43: 115–122.

Xie, C.W., Sayah, D., Chen, Q.S., Wei, W.Z., Smith, D. and Liu, X. (2000) Deficient long-term memory and long-lasting long-term potentiation in mice with a targeted deletion of neurotrophin-4 gene. Proc. Natl. Acad. Sci. USA, 97: 8116–8121.

Xu, B., Gottschalk, W., Chow, A., Wilson, R.I., Schnell, E., Zang, K., Wang, D., Nicoll, R.A., Lu, B., Reichardt, L.F. (2000) The role of brain-derived neurotrophic factor receptors in the mature hippocampus: modulation of long-term potentiation through a presynaptic mechanism involving TrkB. J. Neurosci., 20: 6888–6897.

Yang, F., He, X., Feng, L., Mizuno, K., Liu, X.W., Russell, J., Xiong, W.C. and Lu, B. (2001) PI-3 kinase and IP3 are both necessary and sufficient to mediate NT3-induced synaptic potentiation. Nat. Neurosci., 4: 19–28.

Yiangou, Y., Facer, P., Sinicropi, D.V., Boucher, T.J., Bennett, D.L., McMahon, S.B. and Anand, P. (2002) Molecular forms of NGF in human and rat neuropathic tissues: decreased NGF precursor-like immunoreactivity in human diabetic skin. J. Peripher. Nerv. Syst., 7: 190–197.

Yu, X.M., Askalan, R., Keil, G.J. and Salter, M.W. (1997) NMDA channel regulation by channel-associated protein tyrosine kinase Srccc. Science, 275: 674–678.

Zafra, F., Hengerer, B., Leibrock, J., Thoenen, H. and Lindholm, D. (1990) Activity dependent regulation of BDNF and NGF mRNAs in the rat hippocampus is mediated by non-NMDA glutamate receptors. EMBO J., 9: 3545–3550.

Zafra, F., Castren, E., Thoenen, H. and Lindholm, D. (1991) Interplay between glutamate and gamma-aminobutyric acid transmitter systems in the physiological regulation of brain-derived neurotrophic factor and nerve growth factor synthesis in hippocampal neurons. Proc. Natl. Acad. Sci. USA, 88: 10037–10041.

Zhong, J., Gerber, G., Kojic, L. and Randic, M. (2000) Dual modulation of excitatory synaptic transmission by agonists at group I metabotropic glutamate receptors in the rat spinal dorsal horn. Brain Res., 887: 359–377.

Zhou, X.F. and Rush, R.A. (1994) Localization of neurotrophin-3-like immunoreactivity in the rat central nervous system. Brain Res., 643: 162–172.

Zhou, L.M., Gu, Z.Q., Costa, A.M., Yamada, K.A., Mansson, P.E., Giordano, T., Skolnick, P. and Jones, K.A. (1997) (2S,4R)-4-methylglutamic acid (SYM 2081): a selective, high-affinity ligand for kainate receptors. J. Pharmacol. Exp. Ther., 280: 422–427.

Zhou, S., Komak, S., Du, J. and Carlton, S.M. (2001) Metabotropic glutamate 1alpha receptors on peripheral primary afferent fibers: their role in nociception. Brain Res., 913: 18–26.

Zhou, X.F., Chie, E.T., Deng, Y.S., Zhong, J.H., Xue, Q., Rush, R.A. and Xian, C.J. (1999) Injured primary sensory neurons switch phenotype for brain-derived neurotrophic factor in the rat. Neuroscience, 92: 841–853.

Zirrgiebel, U., Ohga, Y., Carter, B., Berninger, B., Inagaki, N., Thoenen, H. and Lindholm, D. (1995) Characterization of TrkB receptor-mediated signaling pathways in rat cerebellar granule neurons: involvement of protein kinase C in neuronal survival. J. Neurochem., 65: 2241–2250.

Neurotrophins and the immune system

Neuropeptides and the immune system

Progress in Brain Research, Vol. 146
ISSN 0079-6123

CHAPTER 20

The role of neurotrophins in bronchial asthma: contribution of the pan-neurotrophin receptor p75[1]

Harald Renz*, Sebastian Kerzel and Wolfgang Andreas Nockher

Department of Clinical Chemistry and Molecular Diagnostics, Central Laboratory, Hospital of the Philipps-University, D-35033 Marburg, Germany

Abstract: Allergic bronchial asthma is characterized by chronic inflammation of the airways, development of airway hyperreactivity and recurrent reversible airway obstruction. Target and effector cells responsible for airway hyperresponsiveness and airway obstruction include sensory and motor neurons as well as epithelial and smooth muscle cells. Although it is well established that the inflammatory process is controlled by T-helper-2 (Th2) cells, the mechanisms by which immune cells interact with neurons, epithelial cells or smooth muscle cells still remain uncertain. Due to growing evidence for extensive communication between neurons and immune cells, the mechanisms of this neuroimmune crosstalk in lung and airways of asthmatic patients are becoming the focus of asthma research. Neurotrophins represent molecules potentially responsible for regulating and controlling the crosstalk between the immune and peripheral nervous system. They are constitutively expressed by resident lung cells and produced in increasing concentrations by immune cells invading the airways under pathological conditions. Neurotrophins modify the functional activity of sensory and motor neurons, leading to enhanced and altered neuropeptide and tachykinin production. These effects are defined as neuronal plasticity. The consequences are the development of neurogenic inflammation.

Keywords: neurotrophins; bronchial asthma; airway; inflammation; airway hyperresponsiveness; NGF; BDNF; p75^{NTR}

Airway inflammation in bronchial asthma

Bronchial asthma is characterized by chronic airway inflammation, development of airway hyperreactivity and subsequently of airway remodeling. In allergic bronchial asthma, recurrent and reversible airway obstruction is induced by allergen inhalation (MacLean et al., 1988). In the late 1980s and early 1990s, the concept emerged that Th-2 T-cells orchestrate many aspects of pathologic immune responses including effector functions of B-cells, mast cells and eosinophils. Th-2 cells produce an array of cytokines such as interleukin-4 (IL-4), IL-5, IL-9 and IL-13. In B-cells, interleukin-4 and interleukin-13 are involved in isotype switching towards immunoglobulin-E (IgE), while IL-5 processes pro-inflammatory properties, for example the development, differentiation, recruitment and survival of eosinophils. The importance of T-cells and T-cell-driven processes in these diseases is further underlined by the effectiveness of anti-inflammatory therapies involving corticosteroids and others (Joos et al., 1997).

Airway hyperreactivity

Although the currently available anti-inflammatory drugs effecting reduce symptoms and improve patients' quality of life, they do not completely

[1]This chapter represents a short version of Am. J. Respir. Cell. Mol. Biol., 2003; 28: 170–178, and Eur. J. Pharmacol., 2001; 429: 231–237

*Correspondence to: H. Renz, Department of Clinical Chemistry and Molecular Diagnostics, Central Laboratory, Hospital of the Philipps University, Baldingerstr, D-35033 Marburg, Germany. Tel.: +49-6421-286-6234/5; Fax: +49-6421-286-5594; E-mail: renzh@med.uni-marburg.de

DOI: 10.1016/S0079-6123(03)46020-2

eliminate all symptoms (Ichinose et al., 1990). This observation suggests that in addition to the immunological dysregulation in allergy and asthma, other aspects need to be considered in the pathology of this complex disease. Airway hyperreactivity, for example, represents an important hallmark in the pathogenesis of allergy and asthma. Nonspecific bronchial hyperresponsiveness may be defined as an increase in the ease and degree of airway narrowing in response to a wide range of bronchoconstrictor stimuli (Larsen et al., 1994). The development of airway hyperresponsiveness is mediated by multiple independent and additive pathways working in concert. The airway changes leading to airway narrowing mainly include: (i) altered neuronal regulation of airway tone, and (ii) increase in muscle content or function (Iwamoto et al., 1993; Larsen et al., 1994). Therefore, the mode of measuring airway hyperresponsiveness is critical for identifying the underlying mechanisms. Depending on the measurement method chosen, different pathways can be distinguished. It has been shown previously that in vitro electrical field stimulation of tracheal smooth muscle segments reflects specific neuronal airway dysfunction since the addition of both atropin (dysruption of cholinergic pathways) and capsaicin (neuropeptide depletion of sensoring neurons) completely blocks any reaction of the airway (Virchow et al., 1995; Wong and Koh, 2000). Metacholine acts directly via muscarinic-M_3-receptor on smooth muscle cells. Histamine stimulates histamine-H-receptors expressed on smooth muscle cells and sensory and motor neurons. Serotonin (5-hydroxytryptamine-5-HT) is an agonist of 5-HT receptors expressed on sensory and motor neurons (Undem et al., 1993; Weinreich et al., 1997). Many of these pharmacological stimuli have been used to assess the presence of airway hyperresponsiveness in patients with bronchial asthma. The complexity of the pathogenic mechanisms underlying airway hyperresponsiveness is further emphasized by the fact that these and other pharmacological stimuli act on various neuronal and nonneuronal cell populations present in the airways (Fig. 1).

The concept of neurogenic inflammation

Growing evidence indicates neuronal dysregulation on several levels in bronchial asthma (Sanico et al.,

2000). First of all, cholinergic nerves represent the dominant bronchoconstrictory pathway and they exhibit increased activity. Possible underlying mechanisms include enhanced cholinergic reflex activity, increased acetylcholine release, enhanced sensitivity of smooth muscle to acetylcholine or increased sensitivity of muscarinic receptors on airway smooth muscle cells. It has also been shown that sensory nerves are able to modulate cholinergic functions. Cholinergic activities were shown to be increased by tachykinins (Fischer et al., 1996; Laurenzi et al., 1998). In animal models of hyperreactivity and asthma, increased release of acetylcholine has been demonstrated (Kannan et al., 1993) (Fig. 2).

The excitatory nonadrenergic, noncholinergic (e-NANC) system exhibits a high degree of plasticity in inflammatory conditions. Substance P (SP) and neurokinin-A (NK-A) are closely related members of the neuropeptide family termed tachykinins. They are preferentially released by sensory C-fibers and are synthesized preferentially in cell bodies of the sensory ganglia by a complex biosynthetic pathway. These neuropeptides are then transferred via axonal transport not only to presynaptic axon endings in the spinal cord and the nucleus of the solitary tract, but also to peripheral sensory nerve endings (Weinreich et al., 1995). Upon stimulation by mechanical, thermal, chemical or inflammatory conditions, tachykinins are released from nerve cells

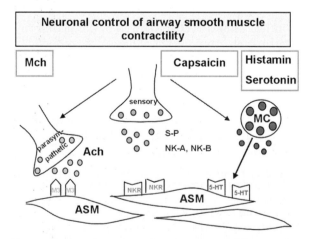

Fig. 1. Neuronal control of airway smooth muscle contractility.

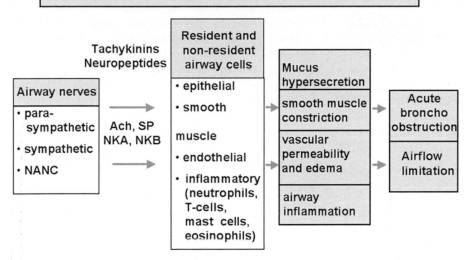

Fig. 2. The concept of neurogenic inflammation.

through a local (axon) reflex mechanism (Kay, 1996). Tachykinins act in a dual fashion as afferent neurotransmitters to the central nervous system as well as efferent neurosecretory mediators. Increased levels of SP have been detected in the airways of asthmatic patients (Herz et al., 1998) and allergen challenge increased NK-A levels in bronchoalveolar lavage fluids of asthmatic patients (Dmitrieva et al., 1997). There is evidence for an increase in both the number and length of SP-immunoreactive nerve fibres in airways of bronchial asthma patients as compared to airways of healthy subjects (Perretti and Manzini, 1993). Reasons for increased levels of SP and NK-A in lung and airways of asthmatic patients may be related either to enhanced production and release of these neuropeptides or to impaired degradation of tachykinins. Tachykinis are degraded and inactivated by neutral endopeptidase, a membrane-bound metallopeptidase, located mainly on the surface of airway epithelial cells and also present in airway smooth muscle cells, submucosal glands and fibroblasts. Allergen exposure, inhalation of cigarette smoke and other respiratory irritants have been associated with a reduced neutral endopeptidase activity thus enhancing the effects of tachykinins within the airways (van Hagen et al., 1994; Fujino

et al., 1997; Kaltreider et al., 1997; van der Velden and Hulsmann, 1999;).

Functional consequences of neuropeptide and neurotransmitter activities in bronchial asthma include mucus hypersecretion, recruitment of inflammatory cells via upregulation of endothelial adhesion molecules, plasma extravasation and formation of edema via enhanced vascular permeability. Activation of endothelial cells and smooth muscle bronchoconstriction are also immediate consequences of neurotransmitter and neuropeptide activity. The enhanced activity of neuropeptides and neurotransmitters followed by their functional effects on epithelial cells, airway smooth muscle cells and inflammatory cells resulted in the concept of neurogenic inflammation (Fig. 2 and Fig. 4). One important question which has been left unanswered so far is related to the mechanism of increased neuronal activity in bronchial asthma.

Neuronal plasticity and airway hyperresponsiveness in response to neurotrophins

Neuronal plasticity in the peripheral nervous system is as of yet not well characterized. To some extent,

328

Neuronal plasticity

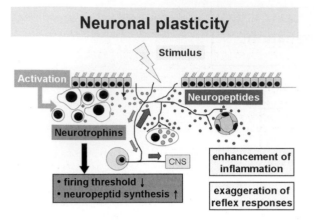

Fig. 3. The axon reflex as a central component of neurogenic inflammation in bronchial asthma.

Neurogenic inflammation

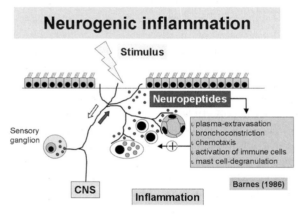

Fig. 4. Neuronal plasticity in the periphery mediated by neurotrophins.

inflammation induced hyperalgesia shows remarkable similarities to airway hyperresponsiveness, particularly with respect to the effects of neurotrophins. Hyperalgesia can be defined as a decrease in the threshold for painful stimuli and heightened reflex pathways in sensory neurons (Carr et al., 2001). It is well established that neurotrophins play a central role in inflammation induced hyperalgesia, Fig. 3 (Safieh-Garabedian et al., 1995; Woolf et al., 1997).

Nerve growth factor (NGF) is a member of the neurotrophin family of proteins. These structurally related proteins exert their effects primarily as target-derived paracrine and autocrine factors and were originally noted for their ability to promote survival, growth and differentiation of neurons. In this

context, the role of neurotrophins is well defined (Lewin and Barde, 1996). Initial evidence for an important role of neurotrophins in allergic bronchial asthma came from the finding of increased NGF levels in serum and bronchoalveolar lavage fluids (BALF) from asthmatics (Bonini et al., 1996). Further experiments demonstrated that allergen challenge induced neurotrophin production by a variety of nonneuronal cells including immune cells (Virchow et al., 1998; Braun et al., 2000).

The biological effects of neurotrophins are mediated by binding either to the specific high-affinity (Kd 10^{-11}) tyrosine kinase receptors TrkA (for NGF), TrkB (for BDNF), TrkC (for NT-3) or the low-affinity (Kd 10^{-9}) pan-neurotrophin receptor p75 ($p75^{NTR}$). It was the aim of this study to examine the expression patterns of $p75^{NTR}$ in the normal and inflamed lung and to assess the role of $p75^{NTR}$ in a murine model of allergic airway inflammation.

There is some evidence suggesting that sensory neurons innervating the lung are also responsive to neurotrophins since local increase of neurotrophins in the lung could mediate similar neuronal changes in animal models as seen during allergic inflammation (Undem et al., 1999; Hunter et al., 2000) (Fig. 5). It has been well established that visceral sensory neurons localized in the nodose and dorsal root ganglia require neurotrophins for survival during development (Snider, 1994). In adults, functional properties of neurons are also affected by neurotrophins (Chalazonitis et al., 1987). NGF was shown to upregulate neuropeptide production in sensory neurons and to contribute to inflammatory hypersensitivity (Donnerer et al., 1992). Though cultured nodose ganglion neurons do not require NGF for survival, their SP production is regulated by NGF (MacLean et al., 1988). In transgenic mice overexpressing NGF in airway restricted Clara-cells, a marked sensory and sympathetic hyperinnervation and increased neuropeptide content was observed in projecting sensory neurons (Hoyle et al., 1998). In addition, these mice demonstrated airway hyperresponsiveness in response to capsaicin. In a guinea pig model, tracheal injection of NGF induced SP production in mechanically sensitive 'Aδ' fibres that do not produce SP under physiological conditions (Hunter et al., 2000). These NGF mediated effects are comparable to neuronal changes observed during

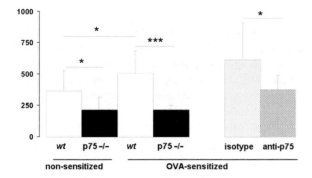

Fig. 5. To assess the neuronal hyperreactivity, the response to inhaled capsaicin was determined. Wild-type controls showed a response which was enhanced in OVA-sensitized wild-type mice. This sensory hyperresponsiveness was abolished in $p75^{NTR}$ −/− mice. To confirm these observations, wild-type animals were either treated with anti-p75^{NTR} antibodies or isotype controls (both intranasally). The controls showed a strong response to inhaled capsaicin which was markedly diminished in anti-75^{NTR} treated animals.

TABLE 1
Airway inflammation in p75^{NTR} −/− mice and in anti-p75^{NTR} antibody treated animals

	wt/isotope	p75^{NTR}−/−	anti-p75^{NTR}
Eosinophils	+ + +	+	+
Neutrophils	+	+	+
Lymphocytes	+ +	+	+
Macrophages	=	=	=
IL-5	+	+	+

(1) To assess the airway inflammation, the influx of immune cells and the concentration of interleukin-5 (IL-5) were determined in bronchoalveolar lavage fluids (BALF). (2) All mice were sensitized intraperitoneally to ovalbumin (OVA) and challenged with OVA-aerosol. Analysis was performed 24 h after the last allergen challenge.

allergic inflammation (Undem et al., 1999). The induction of neuropeptides in mechanically sensitive neurons may lead to exaggerated reflex responses to innocuous stimuli (Hunter et al., 2000). In a murine model of allergic airway inflammation, we were able to demonstrate that blocking of NGF by local treatment with anti-NGF antibodies prevented the development of airway hyperresponsiveness (Braun et al., 1998). Therefore, it is not surprising that NGF treatment induces airway hyperresponsiveness in the guinea pig. Along this line, de Vries et al. (1999) could block NGF induced airway hyperresponsiveness to histamine with the specific tachykinin NK-1 receptor antagonist SR 140333, thus pointing again to the pivotal role played by tachykinins in this condition. Taken together, these data provide further evidence that neurotrophins are central signaling molecules in immune-nerve cell communication as it occurs in pathophysiological conditions including bronchial asthma (Braun et al., 2000).

The contribution of the pan-neurotrophin receptor p75

OVA-sensitized wild-type animals developed the typical allergic inflammation, characterized by the influx of eosinophils and lymphocytes into the airways. In $p75^{NTR}$ −/− mice, the number of eosinophils and lymphocytes recruited into the airways were significantly lower ($p < 0.0001$ and $p < 0.05$, respectively) (Table 1). In contrast, a slightly higher number of macrophages was detected in $p75^{NTR}$ −/− mice as compared to wild-type animals ($p < 0.05$). In nonsensitized but challenged mice, neither IL-4, IL-5 nor IFN-γ were detectable in any study group (data not shown). In OVA-sensitized and challenged animals, IL-5 concentrations were elevated. However, sensitized $p75^{NTR}$ −/− mice showed a tendency to lower levels ($p = 0.067$) compared to sensitized wild-type controls.

To further substantiate these results, mice were treated with anti-p75^{NTR} blocking antibodies. Blocking p75^{NTR} in sensitized and challenged wild-type mice reduced the influx of eosinophils by about 50% as compared to controls ($p < 0.05$). A similar effect was seen on the number of lymphocytes in BAL fluids. As in $p75^{NTR}$ −/− mice, the number of macrophages increased, whereas the number of neutrophils remained unaffected. As observed in $p75^{NTR}$ −/− mice there was a tendency towards lower levels of IL-5 in BAL fluids from anti-p75^{NTR} treated mice ($p = 0.08$).

In order to assess the contribution of p75^{NTR} on the development of airway reactivity (AR), we used head-out body plethysmography, allowing evaluation of the breathing pattern of nonanesthetized, spontaneously breathing mice in response to different stimuli. Differences in sensory nerve activities were measured using inhalation of capsaicin. A characteristic feature of the so-caused sensory irritation is lengthening of the pause prior to expiration (time of break, TB) in the airflow curve (Remmers et al., 1986). All mice demonstrated a slight increase in TB following administration of vehicle control (VC) alone (Fig. 5), reflecting some unspecific stimulation of sensory nerves by the diluent itself. Following capsaicin challenges, wild-type mice showed a considerable dose-dependent rise in TB. At concentrations higher than 100 μg/ml, OVA-sensitized and challenged mice reacted significantly stronger than nonsensitized animals, indicating hyperreactivity to capsaicin. In contrast, the capsaicin response was almost completely abolished in $p75^{NTR}$ −/− mice. The difference to the wild-types was significant in the nonsensitized groups ($p < 0.05$) and highly significant in the OVA-sensitized groups ($p < 0.001$).

To further explore the potential role of p75^{NTR} in hyperreactivity, wild-type mice treated with anti-p75^{NTR} antibodies were examined. Sensitized and challenged animals treated with isotype control antibody showed a similar dose-dependent rise in TB as observed in OVA-sensitized and challenged nontreated mice. However, this increase was significantly reduced in mice treated with the anti-p75^{NTR} antibody, again indicating that airway neuronal hyperreactivity depends on expression of functional p75^{NTR}.

Conclusion and summary

The contribution of neurotrophins on allergic airway inflammation and airway hyperresponsiveness has been established. Recently, however, their mechanism of action is not completely understood. Here, we were able to demonstrate that p75^{NTR} is required for the development of several characteristic features of allergic bronchial asthma. Deactivation of p75^{NTR}, either by genetic targeting or by treatment of wild-type animals with a blocking antibody, revealed several major effects. First, there was a marked reduction of allergic airway inflammation, characterized by the lowered influx of eosinophils and lymphocytes. These findings receive strong support from the observations by Tokuoka et al. (2001), who also

found that the number of eosinophils and level of IL-5 in BAL fluids diminished in $p75^{NTR}$ −/− mice. Second, the neuronal hyperreactivity to capsaicin, following allergic airway inflammation, was almost completely abolished. In contrast, airway hyperresponsiveness to methacholine remained unaffected (data not shown).

A key feature in the pathophysiology of allergic bronchial asthma is a qualitative and quantitative change in the functional activity of peripheral sensory neurons, summarized by the term 'neuronal plasticity' (Lundberg, 1995). This comprises an increase in mechano-sensitivity of sensory nerve endings, resulting in exaggerated neuronal excitability. Neurotrophins have the capacity to induce this modulation of the nociceptive system. Polymodal nociceptor neurons (unmyelinated C-fibers and A-δ-fibers) are very sensitive to capsaicin, an ingredient of hot pepper, which acts specifically via the vanilloid receptor 1 (VR1).

A characteristic feature of activation of sensory neurons is reflex prolongation of the breathing pause prior to expiration (time of break, TB). Remmers and colleagues were able to show that this postinspiratory apnoeic state, termed 'stage I of expiration', is caused by stimulation of sensory airway nerves (Remmers et al., 1986). It is due to a central chemoreflex mediated via the nucleus of the solitary tract. Electrical or chemical stimulation of laryngeal, vagal or carotid sinus nerves were shown to prolong the period of depolarization in postinspiratory neurons without changing the duration of stage II expiratory or inspiratory inhibition, indicating a fairly selective prolongation of the first stage of expiration (Remmers et al., 1986). Validation of body plethysmographic measurement was performed in a series of studies (Alarie, 1998) and recently adapted for asthma research in the mouse (Braun et al., 1999), (Finotto et al., 2001). Our findings suggest that blocking neurotrophin signaling to the sensory neurons via p75^{NTR} prevents the subsequent neuronal plasticity. This may be related to inhibition of tachykinin synthesis and release. This hypothesis is strongly supported by the very recent observation of Hoyle and colleagues, who found that the amount of SP in the lungs of $p75^{NTR}$ −/− mice was reduced by approximately 50% compared to wild-type mice (Graham et al., 2001). SP and other tachykinins elicit a broad range of proinflammatory actions on immune cells (reviewed by van der Velden and Hulsmann, 1999). For example, SP has a degranulating effect on eosinophils and induces human eosinophil migration in vitro. In an in vivo study with allergic rhinitis patients, it was shown that SP administered after repeated allergen challenge enhanced the recruitment of eosinophils (van der Velden and Hulsmann, 1999).

A further key finding of this study is that the inflammation-induced neuronal hyperreactivity to capsaicin was almost completely prevented in $p75^{NTR} -/-$ mice. This observation was confirmed by experiments with wild-type mice treated with a blocking anti-p75^{NTR} antibody. These results suggest for the first time that the pan-neurotrophin-receptor p75^{NTR} plays an important role in the development of the hyperalgesia-like state in neuronal hyperreactivity. Hoyle and colleagues (1998) showed that transgenic mice with Clara cell specific (CCSP) overexpression of NGF (NGF-tg) are more sensitive to inhaled capsaicin compared to wild-type mice. Consistently, we recently demonstrated that NGF-tg mice show a remarkable higher rise in TB than wild type controls following aerosolic capsaicin challenge (Path et al., 2002).

Abbreviations

BAL (F)	broncho-alveolar lavage (fluids)
CCSP	clara-cell specific
IgE	immunoglobulin E
NANC	nonadrenergic, noncholinergic
OVA	ovalbumin
SP	substance P
TB	time of break
Th	T helper
Trk	tyrosine kinase receptor
VR1	vanilloid-receptor 1

References

Alarie, Y. (1998) Computer-based bioassay for evaluation of sensory irritation of airborne chemicals and its limit of detection. Arch. Toxicol., 72: 277–282.

Bonini, S., Lambiase, A., Angelucci, F., Magrini, L., Manni, L. and Aloc, L. (1996) Circulating nerve growth factor levels are increased in humans with allergic diseases and asthma. Proc. Natl. Acad. Sci. USA, 93: 10955–10960.

Braun, A., Appel, E., Baruch, R., Herz, U., Botchkarev, V., Paus, R., Brodie, C. and Renz, H. (1998) Role of nerve growth factor in a mouse model of allergic airway inflammation and asthma. Eur. J. Immunol., 28: 3240–3251.

Braun, A., Lommatzsch, M., Mannsfeldt, A., Neuhaus-Steinmetz, U., Fischer, A., Schnoy, N., Lewin, G.R. and Renz, H. (1999) Cellular sources of enhanced brain-derived neurotrophic factor production in a mouse model of allergic inflammation. Am. J. Respir. Cell. Mol. Biol., 21: 537–546.

Braun, A., Lommatzsch, M. and Renz, H. (2000) The role of neurotrophins in allergic bronchial asthma. Clin. Exp. Allergy, 30: 178–186.

Carr, M.J., Hunter, D.D. and Undem, B.J. (2001) Neurotrophins and asthma. Curr. Opin. Pulm. Med., 7: 1–7.

Chalazonitis, A., Peterson, E.R. and Crain, S.M. (1987) Nerve growth factor regulates the action potential duration of mature sensory neurons. Proc. Natl. Acad. Sci. USA, 84: 289–293.

de Vries, A., Dessing, M.C., Engels, F., Henricks, P.A. and Nijkamp, F.P. (1999) Nerve growth factor induces a neurokinin-1 receptor- mediated airway hyperresponsiveness in guinea pigs. Am. J. Respir. Crit. Care Med., 159: 1541–1544.

Dmitrieva, N., Shelton, D., Rice, A.S. and McMahon, S.B. (1997) The role of nerve growth factor in a model of visceral inflammation. Neuroscience, 78: 449–459.

Donnerer, J., Schuligoi, R. and Stein, C. (1992) Increased content and transport of substance P and calcitonin gene-related peptide in sensory nerves innervating inflamed tissue: evidence for a regulatory function of nerve growth factor in vivo. Neuroscience, 49: 693–698.

Finotto, S., De Sanctis, G.T., Lehr, H.A., Herz, U., Buerke, M., Schipp, M., Bartsch, B., Atreya, R., Schmitt, E., Galle, P.R., Renz, H., Neurath, M.F. (2001) Treatment of allergic airway inflammation and hyperresponsiveness by antisense-induced local blockade of GATA-3 expression. J. Exp. Med., 193: 1247–1260.

Fischer, A., McGregor, G.P., Saria, A., Philippin, B. and Kummer, W. (1996) Induction of tachykinin gene and peptide expression in guinea pig nodose primary afferent neurons by allergic airway inflammation. J. Clin. Invest., 98: 2284–2291.

Fujino, H., Kitamura, Y., Yada, T., Uehara, T. and Nomura, Y. (1997) Stimulatory roles of muscarinic acetylcholine receptors on T cell antigen receptor/CD3 complex-mediated interleukin-2 production in human peripheral blood lymphocytes. Mol. Pharmacol., 51: 1007–1014.

Graham, R.M., Friedman, M. and Hoyle, G.W. (2001) Sensory nerves promote ozone-induced lung inflammation in mice. Am. J. Respir. Crit. Care Med., 164: 307–313.

332

Herz, U., Braun, A., Ruckert, R. and Renz, H. (1998) Various immunological phenotypes are associated with increased airway responsiveness. Clin. Exp. Allergy, 28: 625–634.

Hoyle, G.W., Graham, R.M., Finkelstein, J.B., Nguyen, K.P., Gozal, D. and Friedman, M. (1998) Hyperinnervation of the airways in transgenic mice overexpressing nerve growth factor. Am. J. Respir. Cell. Mol. Biol., 18: 149–157.

Hunter, D.D., Myers, A.C. and Undem, B.J. (2000) Nerve growth factor-induced phenotypic switch in guinea pig airway sensory neurons. Am. J. Respir. Crit. Care Med., 161: 1985–1990.

Ichinose, M., Belvisi, M.G. and Barnes, P.J. (1990) Histamine H3-receptors inhibit neurogenic microvascular leakage in airways. J. Appl. Physiol., 68: 21–25.

Iwamoto, I., Nakagawa, N., Yamazaki, H., Kimura, A., Tomioka, H. and Yoshida, S. (1993) Mechanism for substance P-induced activation of human neutrophils and eosinophils. Regul. Pept., 46: 228–230.

Joos, G.F., Lefebvre, R.A., Bullock, G.R. and Pauwels, R.A. (1997) Role of 5-hydroxytryptamine and mast cells in the tachykinin-induced contraction of rat trachea in vitro. Eur. J. Pharmacol., 338: 259–268.

Kaltreider, H.B., Ichikawa, S., Byrd, P.K., Ingram, D.A., Kishiyama, J.L., Sreedharan, S.P., Warnock, M.L., Beck, J.M. and Goetzl, E.J. (1997) Upregulation of neuropeptides and neuropeptide receptors in a murine model of immune inflammation in lung parenchyma. Am. J. Respir. Cell. Mol. Biol., 16: 133–144.

Kannan, Y., Matsuda, H., Ushio, H., Kawamoto, K. and Shimada, Y. (1993) Murine granulocyte-macrophage and mast cell colony formation promoted by nerve growth factor. Int. Arch. Allergy Immunol., 102: 362–367.

Kay, A.B. (1996) Pathology of mild, severe, and fatal asthma. Am. J. Respir. Crit. Care Med., 154: S66–S69.

Larsen, G.L., Fame, T.M., Renz, H., Loader, J.E., Graves, J., Hill, M. and Gelfand, E.W. (1994) Increased acetylcholine release in tracheas from allergen-exposed IgE-immune mice. Am. J. Physiol., 266: L263–L270.

Laurenzi, M.A., Beccari, T., Stenke, L., Sjolinder, M., Stinchi, S. and Lindgren, J.A. (1998) Expression of mRNA encoding neurotrophins and neurotrophin receptors in human granulocytes and bone marrow cells-enhanced neurotrophin-4 expression induced by LTB4. J. Leukoc. Biol., 64: 228–234.

Lewin, G.R. and Barde, Y.A. (1996) Physiology of the neurotrophins. Annu. Rev. Neurosci., 19: 289–317.

Lundberg, J.M. (1995) Tachykinins, sensory nerves, and asthma-an overview. Can. J. Physiol. Pharmacol., 73: 908–914.

MacLean, D.B., Lewis, S.F. and Wheeler, F.B. (1988) Substance P content in cultured neonatal rat vagal sensory neurons: the effect of nerve growth factor. Brain Res., 457: 53–62.

Path, G., Braun, A., Meents, N., Kerzel, S., Quarcoo, D., Raap, U., Hoyle, G.W., Nockher, W.A. and Renz, H. (2002)

Augmentation of allergic early-phase reaction by nerve growth factor. Am. J. Respir. Crit. Care Med., 166: 818–826.

Perretti, F. and Manzini, S. (1993) Activation of capsaicin-sensitive sensory fibers modulates PAF-induced bronchial hyperresponsiveness in anesthetized guinea pigs. Am. Rev Respir. Dis., 148: 927–931.

Remmers, J.E., Richter, D.W., Ballantyne, D., Bainton, C.R. and Klein, J.P. (1986) Reflex prolongation of stage I of expiration. Pflugers Arch., 407: 190–198.

Safieh-Garabedian, B., Poole, S., Allchorne, A., Winter, J. and Woolf, C.J. (1995) Contribution of interleukin-1 beta to the inflammation-induced increase in nerve growth factor levels and inflammatory hyperalgesia. Br. J. Pharmacol., 115: 1265–1275.

Sanico, A.M., Stanisz, A.M., Gleeson, T.D., Bora, S., Proud, D., Bienenstock, J., Koliatsos, V.E. and Togias, A. (2000) Nerve growth factor expression and release in allergic inflammatory disease of the upper airways. Am. J. Respir. Crit. Care Med., 161: 1631–1635.

Snider, W.D. (1994) Functions of the neurotrophins during nervous system development: what the knockouts are teaching us. Cell, 77: 627–638.

Tokuoka, S., Takahashi, Y., Masuda, T., Tanaka, H., Furukawa, S. and Nagai, H. (2001) Disruption of antigen-induced airway inflammation and airway hyper-responsiveness in low affinity neurotrophin receptor p75 gene deficient mice. Br. J. Pharmacol., 134: 1580–1586.

Undem, B.J., Hubbard, W. and Weinreich, D. (1993) Immunologically induced neuromodulation of guinea pig nodose ganglion neurons. J. Auton. Nerv. Syst., 44: 35–44.

Undem, B.J., Hunter, D.D., Liu, M., Haak-Frendscho, M., Oakragly, A. and Fischer, A. (1999) Allergen-induced sensory neuroplasticity in airways. Int. Arch. Allergy Immunol., 118: 150–153.

van der Velden, V.H. and Hulsmann, A.R. (1999) Peptidases: structure, function and modulation of peptide-mediated effects in the human lung. Clin. Exp. Allergy, 29: 445–456.

van Hagen, P.M., Krenning, E.P., Kwekkeboom, D.J., Reubi, J.C., Anker-Lugtenburg, P.J., Lowenberg, B. and Lamberts, S.W. (1994) Somatostatin and the immune and haematopoetic system; a review. Eur. J. Clin. Invest., 24: 91–99.

Virchow, J.C., Jr., Walker, C., Hafner, D., Kortsik, C., Werner, P., Matthys, H. and Kroegel, C. (1995) T cells and cytokines in bronchoalveolar lavage fluid after segmental allergen provocation in atopic asthma. Am. J. Respir. Crit. Care Med., 151: 960–968.

Virchow, J.C., Jr., Julius, P., Lommatzsch, M., Luttmann, W., Renz, H. and Braun, A. (1998) Neurotrophins are increased in bronchoalveolar lavage fluid after segmental allergen provocation. Am. J. Respir. Crit. Care Med., 158: 2002–2005.

Weinreich, D., Moore, K.A. and Taylor, G.E. (1997) Allergic inflammation in isolated vagal sensory ganglia unmasks silent NK-2 tachykinin receptors. J. Neurosci., 17: 7683–7693.

Weinreich, D., Undem, B.J., Taylor, G. and Barry, M.F. (1995) Antigen-induced long-term potentiation of nicotinic synaptic transmission in the superior cervical ganglion of the guinea pig. J. Neurophysiol., 73: 2004–2016.

Wong, W.S. and Koh, D.S. (2000) Advances in immunopharmacology of asthma. Biochem. Pharmacol., 59: 1323–1335.

Woolf, C.J., Allchorne, A., Safieh-Garabedian, B. and Poole, S. (1997) Cytokines, nerve growth factor and inflammatory hyperalgesia: the contribution of tumour necrosis factor alpha. Br. J. Pharmacol., 121: 417–424.

Progress in Brain Research, Vol. 146
ISSN 0079-6123

CHAPTER 21

Expression of nerve growth factor in the airways and its possible role in asthma

Véronique Freund* and Nelly Frossard

Institut National de la Santé et de la Recherche Médicale (INSERM), Unité 425, Neuroimmunopharmacologie Pulmonaire, Université Louis Pasteur, Strasburg I, Illkirch, France

Abstract: Nerve growth factor (NGF), in addition to its essential role in neuronal growth and survival, may also act as an inflammatory mediator. As several animal studies have shown, NGF appears to play a part in the development of airway hyperresponsiveness and in the increased sympathetic and sensory innervation of the lung. It also has a profound effect on airway inflammation and asthma-related symptoms. Sources of NGF in the airways are numerous: inflammatory cells infiltrated into the bronchial mucosa, and structural cells including lung fibroblasts, airway epithelial and smooth muscle cells. These cells, by releasing more NGF in inflammatory conditions, may contribute to the increased NGF levels observed in bronchoalveolar lavage fluid and serum from patients with asthma. Taken together, these results suggest that NGF is an important mediator in both inflammation and asthma.

Keywords: airway; allergy; asthma; inflammation; mast cell; nerve growth factor; neurotrophin

Introduction

Nerve growth factor (NGF) was discovered by Rita Levi-Montalcini and co-workers more than 50 years ago (reviewed by Levi-Montalcini et al., 1987, 1995, 1996). It belongs to the neurotrophin family of proteins, which also includes brain-derived neurotrophic factor (BDNF) and neurotrophin-3 (NT-3) and NT-4/5 (Barde, 1990). NGF is essential for the survival and development of peripheral and brain neurons (Levi-Montalcini, 1966; Levi-Montalcini and Aloe, 1985; Levi-Montalcini, 1987; Thoenen et al., 1987). It may also be involved in inflammation. This review examines the current literature about NGF expression and regulation in the airways and summarizes the findings that suggest that it may have a role in asthma.

NGF expression and regulation in the airways

Sources of NGF in the airways

Mouse salivary glands, mouse sarcoma and snake venom were the first sources of NGF to be discovered (Aloe and Levi-Montalcini, 1980; reviewed by Levi-Montalcini, 1987), but studies since then have found a broad range of cells able to synthesize and secrete NGF. Its sources in the airways include inflammatory cell infiltrates of the bronchial mucosa as well as structural cells (Fig. 1).

In vitro studies

Inflammatory cells

NGF production was first discovered in cells from the peripheral and central nervous system, but studies in

*Correspondence to: V. Freund, Institut National de la Santé et de la Recherche Médicale (INSERM), Unité 425, Neuroimmunopharmacologie pulmonaire, Université Louis Pasteur, Strasburg I, Faculté de Pharmacie, 74 Route du Rhin, B.P. 24, F-67401 Illkirch Cedex, France.
Tel.: +33-3-9024-4200; Fax: +33-3-9024-4309;
E-mail: vfreund@pharma.u-strasbg.fr

DOI: 10.1016/S0079-6123(03)46021-4

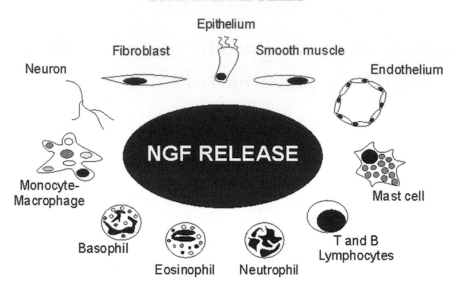

STRUCTURAL CELLS

Neuron
Fibroblast
Epithelium
Smooth muscle
Endothelium

NGF RELEASE

Monocyte-
Macrophage

Mast cell

Basophil
Eosinophil
Neutrophil
T and B
Lymphocytes

INFLAMMATORY CELLS

Fig. 1. Cellular sources of NGF in the airways. NGF may be expressed and released by airway cells, structural and inflammatory cells infiltrated within the bronchial mucosa.

the 1990s also reported that it is secreted by cells of the immune system (Aloe et al., 1994; Levi-Montalcini et al., 1995, 1996). Indeed, T and B lymphocytes (Ehrhard et al., 1993; Santambrogio et al., 1994; Torcia et al., 1996; Lambiase et al., 1997), mast cells (Leon et al., 1994; Nilsson et al., 1997), eosinophils (Solomon et al., 1998; Kobayashi et al., 2002) and macrophages (Ricci et al., 2000; Barouch et al., 2001) synthesize, store, and release NGF. The increased number of inflammatory cells infiltrating the bronchial mucosa in asthma are therefore probably an important source of the increased NGF levels in the airways.

Structural cells

Recent in vitro experiments report NGF expression in cultured lung and airway structural cells: human lung fibroblasts (Olgart and Frossard, 2001), an alveolar epithelial cell line A549 (Fox et al., 2001; Pons et al., 2001) and airway smooth muscle cells (Freund et al., 2002). NGF mRNA and protein were identified and quantified with real-time quantitative PCR (15 fg NGF cDNA/pg GAPDH cDNA in airway smooth muscle cells) (Freund et al., 2002) and with enzyme-linked immunosorbent assay (between 10 and 25 pg NGF/ml culture supernatant) (Fig. 2). These findings are consistent with results showing NGF expression in structural cells of other origins: vascular (Ueyama et al., 1993; Donovan et al., 1995; Nemoto et al., 1998) and bladder (Tanner et al., 2000) smooth muscle cells, human retinal (Ishida et al., 1997) and intestinal epithelial cells (Varilek et al., 1995), and human fibroblast cell lines (Murase et al., 1992). The NGF expression that our team detected in lung structural cells is consistent with the release of other neurotrophins by structural cells reported by other studies: human keratinocytes release NT-4/5 (Grewe et al., 2000), and vascular (Nemoto et al., 1998) and airway (Kemi et al., 2002) smooth muscle cells secrete NT-3 and BDNF.

In vivo studies

The in vitro secretion of NGF in cultured cells accords fully with reports of NGF-positive immunolabeling

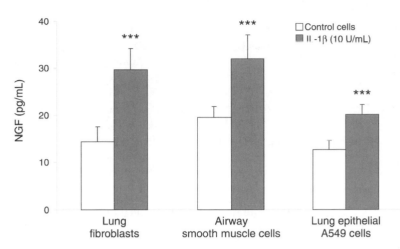

Fig. 2. Constitutive secretion of NGF by human airway structural cells and its upregulation in inflammatory conditions. Lung fibroblasts (left), lung epithelial A549 cells (middle) and airway smooth muscle cells (right) in culture constitutively secrete NGF. This secretion increases significantly in the presence of the proinflammatory cytokine IL-1β (10 U/ml), which mimics inflammatory conditions in vitro. NGF levels were assessed by ELISA and expressed in pg/ml culture supernatant. For lung fibroblasts, airway smooth muscle cells and lung epithelial A549 cells, values are means $\pm$ SEM of $n = 8$, 6 and 12 experiments, respectively, performed in duplicate on cells from 4 to 6 different donors. ***: $p < 0.001$.

in bronchial biopsy samples from patients with or without asthma (Kassel et al., 2001; Olgart-Höglund et al., 2002). NGF immunolabeling is intense in bronchial epithelium and the bronchial smooth muscle layer, but less intense in airway fibroblasts (Fig. 3a,b). In bronchial biopsy samples from asthmatic patients, the inflammatory cell infiltrate also shows NGF immunolabeling (Kassel et al., 2001; Olgart-Höglund et al., 2002) (Fig. 3b) and inflammatory cells express NGF in the bronchoalveolar lavage fluid of a mouse model of asthma (Braun et al., 1998). These in vivo results thus support the hypothesis that structural cells may be another source of increased NGF expression in the airways.

Regulation of NGF expression in the airways

The mechanisms regulating the production and release of NGF in the airways in vivo require further study, but for the time being, in vitro experiments provide some information. Examination of the regulation of NGF expression during cell proliferation and quiescence shows that NGF secretion in lung fibroblasts (Olgart and Frossard, 2001) and in airway smooth muscle cells (Freund et al., unpublished data) depends on cell density. That is, NGF

secretion is high when the cell culture begins and cell contacts are sparse; it decreases progressively as the number of cells increases (Olgart and Frossard, 2001). This result accords with findings for other types of structural cells, including epithelial cells (Di Marco et al., 1993) and cutaneous fibroblasts (Murase et al., 1992), and suggests that NGF is produced in larger quantities by cells with few cell-to-cell contacts than by cells approaching confluence. Moreover, our finding that airway structural cells express NGF receptors (Freund et al., unpublished data) suggests that NGF has an autocrine effect.

Upregulation of NGF in inflammatory conditions

NGF expression also appears to be modulated by inflammatory mediators in the airways. Inflamed airways contain increased levels of NGF (Sanico et al., 2000; Olgart-Höglund et al., 2002), perhaps due in part to an increase in the number of inflammatory cells that infiltrate the bronchial mucosa and release NGF (reviewed by Olgart and Frossard, 2002). Autocrine or paracrine NGF may activate many or most of these inflammatory cells, which then release proinflammatory cytokines, which may in turn stimulate NGF production in other cells (Lindholm

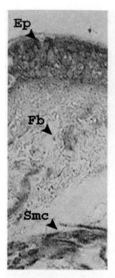

A) Control B) Asthmatic patient

Fig. 3. NGF immunolocalization on bronchial biopsy sections from a control subject (A) and an asthmatic patient (B). Intense labeling was found in the epithelium (Ep) and bronchial smooth muscle cells (Smc) of the control and the asthmatic patient. NGF labeling was also observed in fibroblasts (Fb), but less intensely. An intense NGF signal was also found in the inflammatory cell infiltrate within the bronchial mucosa of asthmatic airways (Ic).

et al., 1987, 1988; Hattori et al., 1993, 1994; Levi-Montalcini et al., 1996). For example, Th2-derived cytokines, such as IL-4 or IL-5, enhance NGF secretion in other cell types, such as astrocytes (Awatsuji et al., 1993; Brodie, 1996; Brodie et al., 1998). In addition, the pro-inflammatory cytokines interleukin-1beta (IL-1β) and tumor necrosis factor-alpha (TNF-α), which are present in large quantities in asthmatic airways (Barnes et al., 1998; Tillie-Leblond et al., 1999), can upregulate NGF secretion in airway structural cells; this has been shown in cultures of human lung fibroblasts (Olgart and Frossard, 2001), lung epithelial A549 cells (Pons et al., 2001) and airway smooth muscle cells (Freund et al., 2002; Fig. 2) as well as in human bronchial ring in vitro (Frossard et al., 2001). Some studies also show that pro-inflammatory cytokines can act synergistically to enhance NGF secretion; examples include TNFα with IL-1β and IFNγ in fibroblasts

(Hattori et al., 1994), or TNFα with IL-4 in astrocytes (Brodie et al., 1998).

Upregulation under inflammatory conditions also occurs in vivo: NGF levels are upregulated in cutaneous inflammation (Safieh-Garabedian et al., 1995) and in the airways in asthma patients (Bonini et al., 1996; Virchow et al., 1998; Olgart-Höglund et al., 2002). Taken together, these results suggest that NGF production may be enhanced in inflamed airways by cytokine action in many types of cells.

Downregulation of NGF by anti-inflammatory drugs

Because corticosteroids are known to have anti-inflammatory effects and to be clinically effective in suppressing inflammation in the lungs and airways, researchers have examined the effect of two of them, budesonide and dexamethasone, on NGF secretion. Lindholm et al. (1990) reported that glucocorticoids downregulate cytokine-stimulated NGF expression by repressing NGF gene transcription in cultured sciatic rat fibroblasts. Our team found that glucocorticoids downregulate cytokine-induced NGF secretion in vitro in human lung fibroblasts (Olgart and Frossard, 2001) and in lung epithelial A549 cells (Pons et al., 2001) (Fig. 4). These data suggest that glucocorticoids inhibit NGF secretion in inflammatory conditions in the airways. Accordingly, airway structural cells, such as fibroblasts and/or epithelial cells, may be important therapeutic targets for this anti-inflammatory action.

Effect of NGF in the airways

Action on neuronal cells

Transgenic mice that overexpress NGF in the lungs show marked airway hyperinnervation of sensory and sympathetic fibers; under electric field stimulation, they release increased quantities of substance P (SP) (Hoyle et al., 1998). This result is consistent with findings that NGF can stimulate neuron development, growth and survival, particularly of the sensory and sympathetic innervation (Levi-Montalcini, 1987). Indeed, these neurons express NGF receptors (Verge et al., 1989; Lee et al., 1992; Schatteman et al., 1993;

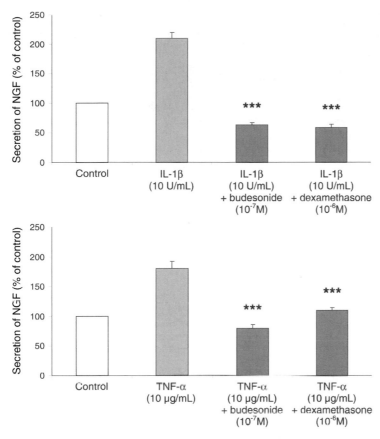

Fig. 4. Reduction of the inflammatory cytokines-induced increase in NGF secretion in human lung fibroblasts by glucocorticoids. Budesonide (0.1 µM) and dexamethasone (1 µM) reduced the IL-1β (10 U/ml)- and TNF-α (10 ng/ml)-induced increase in NGF secretion by lung fibroblasts in culture. NGF levels were assessed by ELISA and expressed as a percentage of the respective controls. Values are means ± SEM of $n = 8$ experiments performed in duplicate on cells from four different donors. ***: $p < 0.001$.

Smeyne et al., 1994; Muragaki et al., 1995). NGF also causes sensory neurons to increase their secretion of tachykinins, such as SP or calcitonin gene-related peptide (CGRP) (Lindsay and Harmar, 1989; Donnerer et al., 1992; Vedder et al., 1993). These findings denote the importance of NGF in airway neuron activity.

Action on inflammatory cells

NGF activates cells through two types of cell surface receptors: the receptor tyrosine kinase A (TrkA), which has a high affinity for NGF, and the p75 neurotrophin receptor (p75^{NTR}), with low affinity for NGF (Barbacid, 1995; Greene and Kaplan, 1995; Segal and Greenberg, 1996; Kaplan and Miller, 1997). Most inflammatory cells can produce NGF and also express both of these NGF receptors. For example, TrkA receptor has been found on human mast cells (Tam et al., 1997; Nilsson et al., 1997), in particular, human bronchial mast cells (Kassel et al., 2001), human T-lymphocytes (Ehrhard et al., 1993; Lambiase et al., 1997) and B-lymphocytes (Torcia et al., 1996), human alveolar macrophages (Ricci et al., 2000), and human basophils (Burgi et al., 1996). These TrkA receptors and their signaling pathways have a central role in inflammatory cells (Ehrhard et al., 1993, 1994; Kawamoto et al., 1995; Franklin et al., 1995; Burgi et al., 1996; Melamed et al., 1996; Lambiase et al., 1997; Nilsson et al., 1997; Tam et al., 1997; Welker et al., 1998; Melamed et al., 1999).

Mast cells were the first cells found to respond to NGF in vivo (Levi-Montalcini et al., 1995; Aloe and Levi-Montalcini, 1977) and in vitro (Böhm et al., 1986). Since then, many studies have shown that NGF modulates most inflammatory cell responses (Table 1): it promotes differentiation and survival of mast cells (Aloe and De Simone, 1989; Matsuda et al., 1991; Horigome et al., 1994; Kawamoto et al., 1995; Bullock and Johnson, 1996; Kanbe et al., 2000) and induces differentiation and proliferation of T- and B-cells (Thorpe and Perez-Polo, 1987; Otten et al., 1989; Brodie and Gelfand, 1992; Kronfeld et al., 2002) as well as proliferation of eosinophils (Hamad et al., 1996) and monocytes (La Sala et al., 2000). Moreover, NGF activates inflammatory cells; specifically, it induces the release of mediator from mast cells (Bruni et al., 1982; Sugiyama et al., 1985; Pearce and Thompson, 1986; Bullock and Johnson, 1996; Murakami et al., 1997; Tal and Liberman, 1997; Kawamoto et al., 2002), neutrophils (Kannan et al., 1991), eosinophils (Takafuji et al., 1992; Solomon et al., 1998), basophils (Matsuda et al., 1988; Bischoff and Dahinden, 1992; Kawamoto et al., 1995; Heinemann et al., 2003), and macrophages (Susaki et al., 1996; Barouch et al., 2001). It also induces B-lymphocytes to produce immunoglobulin (Otten et al., 1989; Kimata et al., 1991). Finally, NGF is involved in chemotaxis, viability, and other functional properties of human neutrophils in vitro (Gee et al., 1983; Kannan et al., 1991) and in vivo (Boyle et al., 1985).

NGF action on structural cells

Although the action of NGF on airway structural cells remains to be further elucidated, recent studies certainly suggest that there are effects (Table 1). NGF stimulates the contraction and migration of lung fibroblasts (Micera et al., 2001, Kohyama et al., 2002) and vascular smooth muscle cells (Kraemer et al., 1998); it also induces the proliferation of airway smooth muscle cells (Freund et al., 2003). These results suggest that NGF plays a role in the airway remodeling that occurs in asthma. Observations of

TABLE 1
Effect of NGF on airway structural and inflammatory cells

NGF effect	Cell type
Differentiation	Neuron
	Mast cell
	T- and B-lymphocytes
Survival	Mast cell
	Eosinophil
	Neuron
	Monocyte
	Neutrophil
Proliferation	Neuron
	Airway smooth muscle cell
	T-Lymphocyte
	CD34$^+$ cell
Migration	Vascular smooth muscle cell
	Lung fibroblast
Activation	Mast cell (degranulation)
	Basophil (mediator release)
	T-Lymphocyte (mediator release)
	B-Lymphocyte (immunoglobulin production)
	Macrophage (mediator production)
	Neutrophil (superoxide anion production)

the expression of both NGF receptors, TrkA and p75^{NTR}, in these structural cells (Donovan et al., 1995; Micera et al., 2001; Freund et al., 2003) indicates the probability of an autocrine action of NGF.

Possible role of NGF in asthma

Accumulating studies suggest that NGF is involved in the development of inflammation in airway diseases, asthma in particular.

NGF expression and release in inflammation and asthma

Increased levels of NGF have been observed in allergic and inflammatory diseases and related to the severity of asthma (Bonini et al., 1996). Samples of both serum and bronchoalveolar lavage fluid from patients with asthma contain higher NGF levels than samples from nonasthmatic controls (Bonini et al., 1996; Undem et al., 1999; Olgart-Höglund et al., 2002); these findings suggest local upregulation of NGF secretion in asthma. This increase is even greater after segmental allergenic challenge (Virchow et al., 1998) and after low-dose allergen administration (Kassel et al., 2001). These results are similar to those involving the upper airways, where nasal lavage fluid from patients with allergic rhinitis has been found to have higher NGF levels than fluid from controls; these differences increased further after nasal allergen challenge (Sanico et al., 2000). Immunohistochemical studies show that cell sources of NGF in human asthmatic airways include the bronchial epithelium, fibroblasts and airway smooth muscle cells, as well as infiltrated inflammatory cells (Kassel et al., 2001; Olgart-Höglund et al., 2002).

NGF contribution to bronchial hyperresponsiveness

Recent studies suggest that NGF is involved in the bronchial hyperresponsiveness observed in asthmatic airways. The first evidence came from the findings that anti-NGF blocking antibodies can abolish hyperresponsiveness to the endogenous neuropeptides released by electrical field stimulation in sensitized and challenged mice (Braun et al., 1998). In addition, NGF pretreatment induces hyperresponsiveness to histamine in vivo in guinea pigs (De Vries et al., 1999; Friberg et al., 2001) and in vitro in isolated guinea pig tracheal rings (De Vries et al., 2001). Pretreatment with anti-NGF blocking antibodies reduces the severe acute bronchoconstriction induced in vivo in guinea pigs after allergenic challenge (De Vries et al., 2002). Moreover, a neurokinin 1 (NK1) receptor antagonist blocks the bronchial hyperresponsiveness in guinea pigs that follows NGF pretreatment; this finding suggests that tachykinins play a role in this phenomenon (De Vries et al., 1999). NGF increases the sensitivity and excitability of peripheral neurons (Lewin et al., 1993, 1994; Levi-Montalcini et al., 1996; McMahon, 1996), and anti-NGF neutralizing antibodies (Woolf et al., 1994) or TrkA-IgG fusion proteins (McMahon, 1996) reverse the experimental inflammation that induces local sensory hypersensitivity and upregulation of the expression of such neuropeptides as SP or CGRP. A neurogenic inflammation induced by NGF may thus be a part of the mechanism by which NGF induces bronchial hyperresponsiveness (Kerzel et al., 2003; reviewed by Renz, 2001a,b; Jacoby, 2003). This

mechanism may be direct or may involve an indirect component, such as mast cells, which are thought to contribute to the hyperalgesic actions of NGF in inflammation (Woolf et al., 1996).

NGF contribution to airway inflammation

In a mouse model of asthma, anti-NGF blocking antibodies reduce the levels of IL-4 and IL-5, both Th2 cell-derived cytokines, but do not affect the number of inflammatory cells infiltrating the airways (Braun et al., 1998). A Th2-related response takes part in the development of asthma symptoms, and IL-4 and IL-5 are essential in increasing immunoglobulin E secretion (Braun et al., 1998). This result suggests that NGF is only indirectly involved in recruiting inflammatory cells in the bronchial mucosa; rather it primes inflammatory cells once they have infiltrated the airways (Braun et al., 1998). A recent study also reports the potential role of NGF in the acute early phase of allergy. The use of transgenic mice with tissue-specific NGF overexpression in Clara cells uncovered enhanced allergic early-phase reaction after sensitization and allergenic challenge (Path et al., 2002). Additionally, since the $p75^{NTR}$ is expressed both in normal and asthmatic lungs and airways, analysis of $p75^{NTR}-/-$ mice as well as in vivo blocking of $p75^{NTR}$ recently revealed that airway inflammation is, to a large extent, dependent upon functional receptor expression, and that neuronal hyperreactivity depends on this receptor (Kerzel et al., 2003). These results suggest NGF has a functional role in the development of allergic early reactions in the airways.

Conclusion

In summary, NGF levels in the airways increase in asthma and are further upregulated after allergenic challenge. NGF also appears to contribute to inflammation and bronchial hyperresponsiveness in asthmatic airways. Because in vitro studies show that NGF is upregulated by inflammatory cytokines and decreased by anti-inflammatory glucocorticoids, and because NGF effects are demonstrated in the airway remodeling, this neurotrophin should be considered a new mediator in airway inflammation. In particular, a better understanding of the role NGF plays in the pathogenesis of asthma might improve its treatment.

Abbreviations

BAL	bronchoalveolar lavage
BDNF	brain-derived neurotrophic factor
CGRP	calcitonin gene-related protein
EFS	electrical field stimulation
ELISA	enzyme-linked immunosorbent assay
IL-1β	interleukin-1β
IL-4/5/10	interleukin-4/5/10
NGF	nerve growth factor
NK1	neurokinin 1
NT-3	neurotrophin-3
NT-4/5	neurotrophin-4/5
NTR	neurotrophin
PCR	polymerase chain reaction
p75^{NTR}	p75 neurotrophin receptor
SP	substance P
TNF-α	tumor necrosis factor-alpha
TrkA	receptor tyrosine kinase A

References

Aloe, L. and Levi-Montalcini, R. (1977) Mast cell increase in tissues in neo-natal rats injected with the nerve growth factor. Brain Res., 133: 358–363.

Aloe, L. and Levi-Montalcini, R. (1980) Comparative studies on testosterone and L-thyroxine. Effects on the synthesis of nerve growth factor in mouse submaxillary salivary glands. Exp. Cell Res., 125: 15–22.

Aloe, L. and De Simone, R. (1989) NGF primed spleen cells injected in brain of developing rats differentiate into mast cells. Int. J. Dev. Neurosci., 7: 565–573.

Aloe, L., Skaper, S.D., Leon, A. and Levi-Montalcini, R. (1994) Nerve growth factor and autoimmune diseases. Autoimmunity, 19: 141–150.

Awatsuji, H., Furukawa, Y., Hiroto, M., Murakami, Y., Nii, S., Furukawa, S. and Hayashi, K. (1993) Interleukin-4 and -5 as modulators of nerve growth factor synthesis/secretion in astrocytes. J. Neurosci. Res., 34: 539–545.

Barbacid, M. (1995) Neurotrophic factors and their receptors. Curr. Opin. Cell Biol., 7: 148–155.

Barde, Y.A. (1990) The nerve growth factor family. Prog. Growth Factor Res., 2: 237–248.

Barnes, P.J., Chung, K.F. and Page, C.P. (1998) Inflammatory mediators of asthma: an update. Pharmacol. Rev., 50: 515–596.

Barouch, R., Appel, E., Kazimirsky, G. and Brodie, C. (2001) Macrophages express neurotrophins and neurotrophin receptors. Regulation of nitirc oxide production by NT-3. J. Neuroimmunol., 112: 72–77.

Bischoff, S. and Dahinden, C.A. (1992). Effect of nerve growth factor on the release of inflammatory mediators by mature human basophils, 79: 2662–2669.

Böhm, A., Aloe, L. and Levi-Montalcini, R. (1986) Nerve growth factor enhances precocious differentiation an numerical increase in mast cells in culture of rat splenocytes. Accad. Naz. Lincei, 80: 1–6.

Bonini, S., Lambiase, A., Bonini, S., Angelucci, F., Magrini, L., Manni, L. and Aloe, L. (1996) Circulating nerve growth factor levels are increased in humans with allergic diseases and asthma. Proc. Natl. Acad. Sci. USA, 93: 10955–10960.

Boyle, M.D.P., Lawman, M.J.P., Gee, A.P. and Young, M. (1985) Nerve growth factor: a chemotactic factor for polymorphonuclear leukocytes in vivo. J. Immunol., 134: 564–568.

Braun, A., Appel, E., Baruch, R., Herz, U., Botchkarev, V., Paus, R., Brodie, C. and Renz, H. (1998) Role of nerve growth factor in a mouse model of allergic airway inflammation and asthma. Eur. J. Immunol., 28: 3240–3251.

Brodie, C. and Gelfand, E.W. (1992) Functional nerve growth factor receptors on human B lymphocytes. Interaction with IL-2. J. Immunol., 148: 3492–3497.

Brodie, C. (1996) Differential effects of Th1 and Th2 derived cytokines on NGF synthesis by mouse astrocytes. FEBS Lett., 394: 117–120.

Brodie, C., Goldreich, N., Haiman, T. and Kazimirsky, G. (1998) Functional IL-4 receptors on mouse astrocytes: IL-4 inhibits astrocyte activation and induces NGF secretion. J. Neuroimmunol., 81: 20–30.

Bruni, A., Bigon, E., Boarato, E., Mietto, L., Leon, A. and Toffano, G. (1982) Interaction between nerve growth factor and lysophosphatidylserine on rat peritoneal mast cells. FEBS Lett., 138: 190–192.

Bullock, E.D. and Johnson, E.M. (1996) Nerve growth factor induces the expression of certain cytokine genes and bcl-2 in mast cells. Potential role in survival promotion. J. Biol. Chem., 271: 27500–27508.

Burgi, B., Otten, U.H., Ochensberger, B., Rihs, S., Heese, K., Ehrhard, P.B., Ibanez, C.F. and Dahinden, C.A. (1996) Basophil priming by neurotrophic factors: activation through the trk receptor. J. Immunol., 157: 5582–5588.

De Vries, A., Dessing, M.C., Engels, F., Henricks, P.A.J. and Nijkamp, F.P. (1999) Nerve growth factor induces a neurokinin-1 receptor-mediated airway hyperresponsiveness in guinea pigs. Am. J. Respir. Crit. Care Med., 159: 1541–1544.

De Vries, A., Van Rijnsoever, C., Engels, F., Henricks, P.A. and Nijkamp, F.P. (2001) The role of sensory nerve endings in nerve growth factor-induced airway hyperresponsiveness to histamine in guinea pigs. Br. J. Pharmacol., 134: 771–776.

De Vries, A., Engels, F., Henricks, P.A., Leusink-Muis, T., Fischer, A. and Nijkamp, F.P. (2002) Antibodies directed against nerve growth factor inhibit the acute bronchiconstriction due to allergen challenge in guinea-pigs. Clin. Exp. Allergy, 32: 325–328.

Di Marco, E., Mathor, M., Bondanza, S., Cutuli, N., Marchisio, P.C., Cancedda, R. and De Luca, M. (1993) Nerve growth factor binds to normal human keratinocytes through high and low affinity receptors and stimulate their growth by a novel autocrine loop. J. Biol. Chem., 268: 22838–22846.

Donnerer, J., Schuligoi, R. and Stein, C. (1992) Increased content and transport of substance P and calcitonin gene-related peptide in sensory nerves innervating inflamed tissue: evidence for a regulatory function of nerve growth factor in vivo. Neuroscience, 49: 693–698.

Donovan, M.J., Miranda, R.C., Kraemer, R., McCaffrey, T.A., Tessarollo, L., Mahadeo, D., Sharif, S., Kaplan, D.R., Tsoulfas, P., Parada, L. and Hempstead, B.L. (1995) Neurotrophin and neurotropin receptors in vascular smooth muscle cells. Regulation of expression in response to injury. Am. J. Pathol., 147: 309–324.

Ehrhard, P.B., Ganter, U., Stalder, A., Bauer, J. and Otten, U. (1993) Expression of functional trk protooncogen in human monocytes. Proc. Natl. Acad. Sci. USA, 90: 5423–5427.

Ehrhard, P.B., Erb, P., Graumann, U., Schmutz, B. and Otten, U. (1994) Expression of functional trk tyrosine kinase receptors after T cell activation. J. Immunol., 152(6): 2705–2709.

Fox, A.J., Patel, H.J., Barnes, P.J. and Belvisi, M.G. (2001) Release of nerve growth factor by human pulmonary epithelial cells: role in airway inflammatory diseases. Eur. J. Pharmacol., 424: 159–162.

Franklin, R.A., Brodie, C., Melamed, I., Terada, N., Lucas, J.J. and Gelfand, E.W. (1995) Nerve growth factor induces activation of MAP-kinase and p90rsk in human B lymphocytes. J. Immunol., 154: 4965–4972.

Freund, V., Pons, F., Joly, V., Mathieu, E., Martinet, N. and Frossard, N. (2002) Upregulation of nerve growth factor expression by hyman airway smooth muscle cells in inflammatory conditions. Eur. Respir. J., 20: 458–463.

Freund., V., Diop, L., Bertrand, C. and Frossard, N. (2003) Neurotrophin receptors involved in the NGF-induced proliferation of human airway smooth muscle cells. Am. J. Respir. Crit. Care Med., 67: A30.

Friberg, S.G., Olgart, C. and Gustafsson, L.E. (2001) Nerve growth factor increases airway responsiveness and decreases levels of exhaled nitric oxide during histamine challenge in an in vivo guinea pig model. Acta Physiol. Scand., 173: 239–245.

Frossard, N., Naline, E., Olgart, C., Mathieu, E., Emonds-Alt, X. and Advenier, C. (2001) Increased production of NGF by IL-1β in the human bronchus in vitro is related to bronchial hyperresponsiveness. Effects of SR142801. Am. J. Respir. Crit. Care Med., 163: A55.

Gee, A.P., Boyle, M.D.P., Munger, K.L., Lawman, M.J.P. and Young, M. (1983) Nerve growth factor: stimulation of polymorphonuclear leukocyte chemotaxis in vitro. Proc. Natl. Acad. Sci. USA, 80: 7215–7218.

Greene, L.A. and Kaplan, D.R. (1995) Early events in neurotrophin signaling via Trk and p75 receptors. Curr. Opin. Neurobiol., 5: 579–587.

Grewe, M., Vogelsgang, K., Ruzicka, T., Stege, H. and Krutmann, J. (2000) Neurotrophin-4 production by human epidermal keratonicytes increased expression in atopic dermatitis. J. Invest. Dermatol., 114: 1108–1112.

Hamad, A., Watanabe, N., Ohtomo, H. and Matsuda, H. (1996) Nerve growth factor enhances survival and cytotoxic activity of human eosinophils. Br. J. Haematol., 93: 299–302.

Hattori, A., Tanaka, E., Murase, K., Ishida, N., Chatani, Y., Tsujimoto, M., Hayashi, K. and Kohno, M. (1993) Tumor necrosis factor stimulates the synthesis and secretion of biologically active nerve growth factor in non-neuronal cells. J. Biol. Chem., 268: 2577–2582.

Hattori, A., Iwazaki, S., Murase, K., Tsujimoto, M., Sato, M., Hayashi, K. and Kihono, M. (1994) Tumor necrosis factor is markedly synergistic with interleukin-1 and interferon gamma in stimulating the production of nerve growth facotr in fibroblasts. FEBS Lett., 340: 177–180.

Heinemann, A., Ofner, M., Amann, R. and Paskar, B.A. (2003) A novel assay to measure the calcium flux in human basophils: effects of chemokines and nerve growth factor. Pharmacology, 67: 49–54.

Horigome, K., Bullock, E.D. and Johnson, E.M. (1994) Effects of nerve growth factor on rat peritoneal mast cells: survival promotion and immediate-early gene induction. J. Biol. Chem., 269: 2695–2702.

Hoyle, G.W., Graham, R.M., Finkelstein, J.B., Nguyen, K.P.T., Gozal, D. and Friedman, M. (1998) Hyperinnervation of the airways in transgenic mice overexpressing nerve growth factor. Am. J. Respir. Cell. Mol. Biol., 18: 149–157.

Ishida, K., Yoshimura, N., Yoshida, M., Honda, Y., Murase, K. and Hayashi, K. (1997) Expression of neurotrophic factors in cultures human retinal pigment epithelial cells. J. Biochem., 16: 96–101.

Jacoby, D.B. (2003) Airway neural plasticity—The nerves they are a-changin'. Am. J. Respir. Cell Mol. Biol., 28: 138–141.

Kanbe, N., Kurosawa, M., Miyachi, Y., Kanbe, M., Saitoh, H. and Matsuda, H. (2000) Nerve growth factor prevents apoptosis of cord blood-derived human cultured mast cells synergistically with stem cell factor. Clin. Exp. Allergy, 30: 1113–1120.

Kannan, Y., Ushio, H., Koyama, H., Okada, M., Oikawa, M., Yoshihara, T., Kaneko, M. and andMatsuda, H. (1991) 2.5S nerve growth factor enhances survival, phagocytosis, and superoxide production of murine neutrophils. Blood, 77: 1320–1325.

344

Kaplan, D.R. and Miller, F.D. (1997) Signal transduction by the neurotrophin receptors. Curr. Opin. Cell Biol., 9: 213–221.

Kassel, O., De Blay, F., Duvernelle, C., Olgart, C., Israel-Biet, D., Krieger, P., Moreau, L., Muller, C., Pauli, G. and Frossard, N. (2001) Local increase in the number of mast cells and expression of nerve growth factor in the bronchus of asthmatic patients after repeated inhalation of allergen at low-dose. Clin. Exp. Allergy, 31: 1432–1440.

Kawamoto, K., Okada, T., Kannan, Y., Ushio, H., Matsumoto, M. and Matsuda, H. (1995) Nerve growth factor prevents apoptosis of rat peritoneal mast cells through the trk proto-oncogene receptor. Blood, 86: 4638–4644.

Kawamoto, K., Aoki, J., Tanaka, A., Itakura, A., Hosono, H., Arai, H., Kiso, Y. and Matsuda, H. (2002) Nerve growth factor activates mast cells through the collaborative interaction with lysophosphatidylserine expressed on the membrane surface of activated platelets. J. Immunol., 168: 6412–6419.

Kemi, C., Grunewald, J., Eklund, A. and Olgart-Höglund, C. (2002) Differential regulation of neurotrophin expression in human airway smooth muscle cells by pro-inflammatory cytokines, Eur. J. Respir., 20: 383S. Abstract.

Kerzel, S., Päth, G., Nockher, W.A., Quarcoo, D., Raap, U., Groneberg, D.A., Dinh, Q.T., Fischer, A., Braun, A. and Renz, H. (2003) Pan-neurotrophin receptor p75 contributes to neuronal hyperreactivity and airway inflammation in a murine model of experimental asthma. Am. J. Respir. Cell Mol. Biol., 28: 170–178.

Kimata, H., Yoshida, A., Ishioka, C., Kusunoki, T., Hosoi, S. and Mikawa, H. (1991) Nerve growth factor specifically induces human IgG4 production. Eur. J. Immunol., 21: 137–141.

Kobayashi, H., Gleich, G.J., Butterfield, J.H. and Kita, H. (2002) Human eosinophils produce neurotrophins and secrete nerve growth factor on immunologic stimuli. Blood, 99: 2214–2220.

Kohyama, T., Liu, X., Wen, F.Q., Kobayashi, T., Abe, S., Ertl, R. and Rennard, S.I. (2002) Nerve growth factor stimulates fibronectin-induced fibroblast migration. J. Lab. Clin. Med., 140: 329–335.

Kraemer, R., Nguyen, H., March, K.L. and Hempstead, B. (1998) NGF activates similar intracellular signaling pathways in vascular smooth muscle cells as PDGF-BB but elicits different biological responses. Arterioscler. Thromb. Vasc. Biol., 19: 1041–1050.

Kronfeld, I., Kazimirsky, G., Gelfand, E.W. and Brodie, C. (2002) NGF rescues human B lymphocytes from anti-IgM induced apoptosis by activation of PKCzeta. Eur. J. Immunol., 32: 136–143.

La Sala, A., Corinti, S., Federici, M., Saragovi, H.U. and Girolomoni, G. (2000) Ligand activation of nerve growth factor receptor TrkA protects monocytes from apoptosis. J. Leukoc. Biol., 68: 104–110.

Lambiase, A., Bracci-Laudiero, L., Bonini, S., Bonini, S., Starace, G., D'Elios, M.M., De Carli, M. and Aloe, L. (1997) Human CD4+ T cell clones produce and release nerve growth factor and express high affinity nerve growth factor receptors. J. Allergy Clin. Immunol., 100: 408–414.

Lee, K.F., Li, E., Huber, J., Landis, S.C., Sharpe, A.H., Chao, M.V. and Jaenisch, R. (1992) Targeted mutation of the gene encoding the low affinity NGF receptor p75 leads to deficits in the peripheral sensory nervous system. Cell, 69: 737–749.

Leon, A., Buriani, A., Dal Toso, R., Fabris, M., Romanello, S., Aloe, L. and Levi-Montalcini, R. (1994) Mast cells synthesize, store, and release nerve growth factor. Proc. Natl. Acad. Sci. USA, 91: 3739–3743.

Levi-Montalcini, R. (1966) The nerve growth factor: its mode of action on sensory and sympathetic nerve cells. Harvey Lect., 60: 217–259.

Levi-Montalcini, R. and Aloe, L. (1985) Differentiating effects of murine nerve growth factor in the peripheral and central nervous systems of Xenopus laevis tadpoles. Proc. Natl. Acad. Sci. USA, 82: 7111–7115.

Levi-Montalcini, R. (1987) The nerve growth factor 35 years later. Science, 237: 1154–1162.

Levi-Montalcini, R., Dal Toso, R., Della Valle, F., Skaper, S.D. and Leon, A. (1995) Update of the NGF saga. J. Neurol. Sci., 130: 119–127.

Levi-Montalcini, R., Skaper, S.D., Dal Toso, R., Petrelli, L. and Leon, A. (1996) Nerve growth factor: from neurotrophin to neurokine. Trends Neurosci., 19: 514–520.

Lewin, G.R., Ritter, A.M. and Mendell, L.M. (1993) NGF induced hyperalgesia in the neonatal and adult. J. Neurosci., 13: 2136–2148.

Lewin, G.R., Rueff, A. and Mendell, L.M. (1994) Peripheral and central mechanisms of NGF-induced hyperalgesia. Eur. J. Neurosci., 6: 1903–1912.

Lindholm, D., Heumann, R., Meyer, M. and Thoenen, H. (1987) Interleukin-1 regulates synthesis of nerve growth factor in non-neuronal cells of rat sciatic nerve. Nature, 330: 658–659.

Lindholm, D., Heumann, R., Hengerer, B. and Thoenen, H. (1988) Interleukin-1 increases stability and transcription of mRNA encoding nerve growth factor in cultured rat fibroblasts. J. Biol. Chem., 263: 16348–16351.

Lindholm, D., Hengerer, B., Heumann, R., Carroll, P. and Thoenen, H. (1990) Glucocorticoid hormones negatively regulate nerve growth factor expression in vivo and in cultured rat fibroblasts. Eur. J. Neurosci., 2: 795–801.

Lindsay, R.M. and Harmar, A.J. (1989) Nerve growth factor regulates expression of neuropeptide genes in adult sensory neurones. Nature, 337: 362–364.

Matsuda, H., Switzer, J., Coughlin, M.D., Bienenstock, J. and Denburg, J.A. (1988) Human basophilic cell differentiation promoted by 2.5S nerve growth factor. Int. Arch. Allergy Appl. Immunol., 86: 453–457.

Matsuda, H., Kannan, Y., Ushio, H., Kiso, Y., Kanemoto, T., Suzuki, H. and Kitamura, Y. (1991) Nerve growth factor induces development of connective tissue-type mast cells in vitro from murine bone marrow cells. J. Exp. Med., 174: 7–14.

McMahon, S.B. (1996) NGF as a mediator of pain. Philos. Trans. R. Soc. Lond. B. Biol. Sci., 351: 431–440.

Melamed, I., Kelleher, C.A., Franklin, R.A., Hempstead, B., Kaplan, D. and Gelfand, E.W. (1996) Nerve growth factor signal transduction in B lymphocytes is mediated by gp140trk. Eur. J. Immunol., 26: 1985–1992.

Melamed, I., Patel, H., Brodie, C. and Gelfand, E.W. (1999) Activation of Vav and Ras through the nerve growth factor and B cell receptors by different kinases. Cell Immunol., 191: 83–89.

Micera, A., Vigneti, E., Pickholtz, D., Reich, R., Paoo, O., Bonini, S., Maquart, F.X., Aloe, L. and Levi-Schaffer, F. (2001) Nerve growth factor displays stimulatory effects on human skin and lung fibroblasts, demonstrating a direct role for this factor in tissue repair. Proc. Natl. Acad. Sci. USA, 98: 6162–6167.

Muragaki, Y., Timothy, N., Leight, S., Hempstead, B.L., Chao, M.V. and Trojanowski, J.Q. and Lee, V.M. (1995) Expression of trk receptors in the developing and adult human central and peripheral nervous system. J. Comp. Neurol., 356: 387–397.

Murakami, M., Tada, K., Nakajima, K. and Kudo, I. (1997) Cyclooxygenase-2-dependent prostaglandin D2 generation is initiated by nerve growth factor in rat peritoneal mast cells: its augmentation by extracellular type II secretory phospholipase A2. J. Immunol., 159: 439–446.

Murase, K., Murakami, Y., Takayanagi, K., Furukawa, Y. and Hayashi, K. (1992) Human fibroblast cells synthezise and secrete nerve growth factor in culture. Biochem. Biophys. Res. Commun., 184: 373–379.

Nemoto, K., Fukamachi, K., Nemoto, F., Miyata, S., Hamada, M., Nakamura, Y., Senba, E. and Ueyama, T. (1998) Gene expression of neurotrophins and their receptors in cultured rat vascular smooth muscle cells. Biochem. Biophys. Res. Commun., 245: 284–288.

Nilsson, G., Forsberg-Nilsson, K., Xiang, Z., Hallbààk, F., Nilsson, K. and Metcalfe, D.D. (1997) Human mast cells express functional TrkA and are a source of nerve growth factor. Eur. J. Immunol., 27: 2295–2301.

Olgart, C. and Frossard, N. (2001) Human lung fibroblasts secrete nerve growth factor. Effect of inflammatory cytokines and glucocorticoids. Eur. Respir. J., 18: 115–121.

Olgart, C. and Frossard, N. (2002) Nerve growth factor and asthma. Pulm. Pharm. Ther., 15: 51–60.

Olgart-Höglund, C., De Blay, F., Oster, J.P., Duvernelle, C., Kassel, O., Pauli, G. and Frossard, N. (2002) Nerve growth factor levels and localisation in human asthmatic bronchi. Eur. Respir. J., 20: 1110–1116.

Otten, U., Ehrhard, P. and Peck, R. (1989) Nerve growth factor induces growth and differentiation of human B lymphocytes. Proc. Natl. Acad. Sci. USA, 86: 10059–10063.

Path, G., Braun, A., Meents, N., Kerzel, S., Quarcoo, D., Raap, U., Hoyle, G.W., Nockher, W.A. and Renz, H. (2002) Augmentation of allergic early-phase reaction by nerve growth factor. Am. J. Respir. Crit. Care Med., 166: 818–826.

Pearce, F.L. and Thompson, H.L. (1986) Some characteristics of histamine secretion from rat peritoneal mast cells stimulated with nerve growth factor. J. Physiol., 372: 379–393.

Pons, F., Freund, V., Kuissu, H., Mathieu, E., Olgart, C. and Frossard, N. (2001) Nerve growth factor secretion by human lung epithelial A549 cells in pro- and anti-inflammatory conditions. Eur. J. Pharmacol., 428: 365–369.

Renz, H. (2001a) Neurotrophins in bronchial asthma. Respir. Res., 2: 265–268.

Renz, H. (2001b) The role of neurotrophins in bronchial asthma. Eur. J. Pharmacol., 429: 231–237.

Ricci, A., Greco, S., Mariotta, S., Felici, L., Amenta, F. and Bronzetti, E. (2000) Neurotrophin and neurotrophin receptor expression in alveolar macrophages : an immunocytochemical study. Growth Factors, 18: 193–202.

Safieh-Garabedian, B., Poole, S., Allchorne, A., Winter, J. and Woolf, C.J. (1995) Contribution of interleukin-1β to the inflammation-induced increase in nerve growth factor levels and inflammatory hyperalgesia. Br. J. Pharmacol., 115: 1265–1275.

Sanico, A.M., Stanisz, A.M., Gleeson, T.D., Bora, S., Proud, D., Bienenstock, J., Koliatsos, V.E. and Togias, A. (2000) Nerve growth factor expression and release in allergic inflammatory disease of the upper airways. Am. J. Respir. Crit. Care Med., 161: 1631–1635.

Santambrogio, L., Benedetti, M., Chao, M.V., Muzaffar, R., Kulig, K., Gabelini, N. and Hochwald, G. (1994) Nerve growth factor production by lymphocytes. J. Immunol., 153: 4488–4495.

Schatteman, G.C., Langer, T., Lanahan, A.A. and Bothwell, M.A. (1993) Distribution of the 75-kD low-affinity nerve growth factor receptor in the primate peripheral nervous system. Somatosens. Mot. Res., 10: 415–432.

Segal, R.A. and Greenberg, M.E. (1996) Intracellular signaling pathways activated by neurotrophic factors. Annu. Rev. Neurosci., 19: 463–489.

Smeyne, R.J., Klein, R., Schnapp, A., Long, L.K., Bryant, S., Lewin, A., Lira, S.A. and Barbacid, M. (1994) Severe sensory and sympathetic neuropathies in mice carrying a disrupted Trk/NGF receptor gene. Nature, 368: 246–249.

Solomon, A., Aloe, L., Pe'er, J., Frucht-Pery, J., Bonini, S., Bonini, S. and Levi-Schaffer, F. (1998) Nerve growth factor is preformed in and activates human peripheral blood eosinophils. J. Allergy Clin. Immunol., 102: 454–460.

Sugiyama, K., Suzuki, Y. and Furuta, H. (1985) Histamine-release induced by 7S nerve growth factor of mouse submandibular salivary glands. Arch. Oral Biol., 30: 93–95.

Susaki, Y., Shimizu, S., Katakura, K., Watanabe, N., Matsumoto, M., Tsudzuki, M., Furusaka, T., Kitamura, Y. and Matsuda, H. (1996) Functional properties of murine

346

macrophages promoted by nerve growth factor. Blood, 88: 4630–4637.

Takafuji, S., Bischoff, S.C., De Weck, A.L. and Dahinden, C.A. (1992) Opposing effects of tumor necrosis factor alpha and nerve growth factor upon leukotriene C4 production by human eosinophils triggered with N-formyl-methionyl-leuyl-phenylalanine. Eur. J. Immunol., 22: 969–974.

Tal, M. and Liberman, R. (1997) Local injection of nerve growth factor (NGF) triggers degranulation of mast cells in rat paw. Neurosci. Lett., 22: 129–132.

Tam, S.Y., Tsai, M., Yamaguchi, M., Yano, K., Butterfield, J.H. and Galli, S.J. (1997) Expression of functional TrkA receptor tyrosine kinase in the HMC-1 human mast cells. Blood, 90: 1807–1820.

Tanner, R., Chambers, P., Khadra, M.H. and Gillepsie, J.L. (2000) The production of nerve growth factor by human bladder smooth muscle cells in vivo and in vitro. B.J.U., 85: 1115–1119.

Thoenen, H., Brandtlow, C. and Heumann, R. (1987) The physiological function of nerve growth factor in the central nervous system: comparison with the periphery. Rev. Physiol. Biochem., 109: 145–178.

Thorpe, L.W. and Perez-Polo, J.R. (1987) The influence of nerve growth factor on the in vitro proliferative response of rat spleen lymphocytes. J. Neurosci. Res., 18: 134–139.

Tillie-Leblond, I., Pugin, J., Marquette, C.H., Lamblin, C., Saulnier, F., Brichet, A., Wallaert, B., Tonnel, A.B. and Gosset, P. (1999) Balance between proinflammatory cytokines and their inhibitors in bronchial lavage from patients with status asthmaticus. Am. J. Respir. Crit. Care Med., 159: 487–494.

Torcia, M., Bracci-Laudiero, L., Lucibello, M., Nencioni, L., Labardi, D., Rubartelli, A., Cozzolino, F., Aloe, L. and Garaci, E. (1996) Nerve growth factor is an autocrine survival factor for memory B lymphocytes. Cell, 85: 345–356.

Ueyama, T., Hamada, M., Hano, T., Nishio, I., Masuyama, Y. and Furukawa, S. (1993) Production of nerve growth factor by cultured vascular smooth muscle cells from spontanously hypertensive and Wistar-Kyoto rats. J. Hypertens, 11: 1061–1065.

Undem, B.J., Hunter, D.D., Liu, M., Haak-Fredscho, M., Oakragly, A. and Fischer, A. (1999) Allergen-induced sensory neuroplasticity in airways. Int. Arch. Allergy. Immunol., 118: 150–153.

Varilek, G.W., Neil, G.A., Bishop, W.P., Lin, J. and Panttazis, N.J. (1995) Nerve growth factor synthesis by intestinal epithelial cells. Am. J. Physiol., 269: G445–G452.

Vedder, H., Affolter, H.U. and Otten, U. (1993) Nerve growth factor (NGF) regulates tachykinin gene expression and biosynthesis in rat sensory neurones during early postnatal development. Neuropeptides, 24: 351–357.

Verge, V.M., Richardson, P.M., Benoit, R. and Riopelle, R.J. (1989) Histochemical characterization of sensory neurones with high-affinity receptors for nerve growth factor. J. Neurocytol., 18: 583–591.

Virchow, J.C., Julius, P., Lommatzsch, M., Luttmann, W., Renz, H. and Braun, A. (1998) Neurotrophins are increased in bronchoalveolar lavage fluid after segmental allergen provocation. Am. J. Respir. Crit. Care Med., 158: 2002–2005.

Welker, P., Grabbe, J., Grutzkau, A. and Henz, B.M. (1998) Effects of nerve growth factor (NGF) and other fibroblast-derived growth factors on immature mast cells (HMC-1). Immunol., 94: 310–317.

Woolf, C.J., Safieh-Garabedian, B., Ma, Q.P., Crilly, P. and Winter, J. (1994) Nerve growth factor contributes to the generation of inflammatory sensory hypersensitivity. Neuroscience, 62: 327–331.

Woolf, C.J., Ma, Q.P., Allchorne, A. and Poole, S. (1996) Peripheral cell types contributing to the hyperalgesic action of nerve growth factor in inflammation. J. Neurosci., 16: 2716–2723.

Progress in Brain Research, Vol. 146
ISSN 0079-6123

CHAPTER 22

Neurotrophins and neurotrophin receptors in allergic asthma

Christina Nassenstein[1], Sebastian Kerzel[2] and Armin Braun[1,*]

[1]*Immunology and Allergology, Fraunhofer Institute of Toxicology and Experimental Medicine, D-30625 Hannover, Germany*
[2]*Department of Clinical Chemistry and Molecular Diagnostics, Central Laboratory, Hospital of the Philipps-University, Marburg University, Germany*

Abstract: The neurotrophins nerve growth factor, brain-derived neurotrophic factor, neurotrophin-3 (NT-3) and NT-4 play a pivotal role in the development of the nervous system. Despite their well-known effects on neurons, elevated neurotrophin concentrations have been observed under pathological conditions in sera of patients with inflammatory disorders. Patients with asthma feature both airway inflammation and an abnormal airway reactivity to many unspecific stimuli, referred to as airway hyperresponsiveness, which is, at least partly, neuronally controlled. Interestingly, these patients show increased levels of neurotrophins in the blood as well as locally in the lung. It has been demonstrated that neurotrophin release from immune cells is triggered by allergen contact. The presence of neurotrophins and the neurotrophin receptors p75 (p75^{NTR}), tyrosine kinase A (TrkA), TrkB and TrkC have been described in several immune cells. There is strong evidence for an involvement of neurotrophins in regulation of hematopoiesis and, in addition, in modulation of immune cell function in mature cells circulating in blood or resting in lymphatic organs and peripheral tissues. The aim of this review is to demonstrate possible roles of neurotrophins during an allergic reaction in consideration of the temporospatial compartmentalization.

Keywords: airway hyperresponsiveness; neurogenic inflammation; allergy; sensory neurons; NGF; BDNF

Introduction

There is growing evidence for a strong interaction between the nervous and the immune systems. Under physiological conditions, both systems are responsible for the induction and control of an adequate inflammatory response to pathologic stimuli such as viruses and bacteria. However, there is also evidence for a pathologic dysfunction of this neuro-immune system resulting in acute or chronic inflammation in diseases such as allergic bronchial asthma (Tracey, 2002).

*Correspondence to: A. Braun, Immunology and Allergology, Fraunhofer Institute of Toxicology and Experimental Medicine, Nikolai-Fuchs-Str. 1, D-30625 Hannover, Germany. Tel.: +49 (0) 511-5350-263; Fax: +49 (0) 511-5350-155; E-mail: braun@item.fraunhofer.de

More than 5% of young adults and more than 10% of children suffer from allergic asthma (Janson et al., 1997; ISAAC Study Report, 1998). For unknown reasons, there has been an obvious increase in morbidity, mortality and enhanced cost of illness over the past decades (Weiss et al., 2000; Gergen, 2001). Allergic asthma is clinically characterized by recurrent and reversible bronchial obstruction, chronic inflammation of the lung as well as airway hyperreactivity (AHR). The allergic inflammation is due to several mediators released by activated immune cells infiltrating the lung in response to normally harmless environmental antigens. AHR describes an abnormal bronchoconstriction in response to a broad range of unspecific stimuli, including cold air, cigarette smoke, methacholine, histamine and hypertonic saline, which are tolerated

DOI: 10.1016/S0079-6123(03)46022-6

without symptoms by nonasthmatic individuals (Sterk, 1995; van Schoor et al., 2000). First evidence that neurotrophins are involved in the neuro-immune crosstalk in asthma was published by Aloe and colleagues who found that patients with asthma show increased levels of neurotrophins in the blood (Bonini et al., 1996). In addition, it has been demonstrated that neurotrophins are enhanced in bronchoalveolar lavage fluid as well (Undem et al., 1999), and local allergen challenge elevates pulmonary neurotrophin concentrations in sensitized patients with asthma (Virchow et al., 1998).

Nerve growth factor (NGF), brain-derived neurotrophic factor (BDNF), neurotrophin-3 (NT-3) and NT-4 belong to the family of proteins named neurotrophins. The neurotrophin signals are mediated either by the low-affinity pan-neurotrophin receptor p75 ($p75^{NTR}$) to which all members of the neurotrophin family bind with similar affinity (K_d 10^{-9}) (Johnson et al., 1986; Radeke et al., 1987) and/or the tyrosine protein kinase (Trk) receptors of the Trk family to which neurotrophins bind with high affinity (K_d 10^{-11}). TrkA has been identified as the preferred receptor for NGF (Kaplan et al., 1991; Klein et al., 1991a) and TrkB has been reported to exert the effects of both BDNF and NT-4 (Klein et al., 1991b; Ip et al., 1992). For NT-3, TrkC plays a central role in cellular signaling (Lamballe et al., 1993).

In several in vivo animal studies, a functional association between inflammatory changes and increased levels of neurotrophins has been observed. Treatment of OVA-sensitized and aerosol-challenged BALB/c mice with blocking antibodies against NGF suppresses airway inflammation. In addition, sensitized and challenged Clara cell secretory protein promoter-NGF-tg (CCSP-NGF-tg) mice constitutively overexpressing NGF in the airways under the Clara cell-specific promotor (CCSP) show an augmented airway inflammation as compared to wild type mice (Path et al., 2002). Beside these effects, NGF is involved in the development of bronchial hyperresponsiveness, as has been demonstrated in several animal models of asthma. Intranasal instillation of NGF in mice induces AHR measured by electrical field stimulation without previous sensitization or allergen challenge. Furthermore, an increased AHR has been described in CCSP-NGF-tg mice

following treatment with allergens, as compared to wild type mice (Braun et al., 2001). Even in guinea pigs, peripherally administered NGF potentiates the histamine-induced bronchoconstriction (de Vries et al., 1999). To understand the role of neurotrophins on an immunologic basis in allergic inflammation and AHR, the underlying cellular mechanisms of pathophysiological changes in the airways are described herein first.

Pathophysiology of asthma

Mechanisms of allergen sensitization in allergic asthma

After repeated exposure to a commonly inhaled allergen, individuals with an atopic predisposition develop specific IgE antibodies to various allergens. This process is also termed sensitization, and includes a sequence of reactions in which different cell types are involved. First of all, allergens are taken up, processed and presented by MHC molecules of antigen-presenting cells, mainly from myeloid dendritic cells (DC) in the respiratory mucosa. Although recent studies suggest that the majority of antigen-loaded APC remain within the lung tissues (Constant et al., 2002), it is commonly accepted that a fraction of airway DC differentiate into professional antigen-presenting cells and migrate into the paracortical T-cell zone of the draining regional lymph nodes. Dependent on the costimulatory activity, these DCs favor the differentiation of naive Th0 cells into Th2 cells which secrete interleukin-4 (IL-4), IL-5, IL-10, and IL-13. These cytokines induce B cells to undergo class-switching to IgE production. Two signals have been described to be essential. The first one is provided by the Th2 cytokines IL-4 or IL-13, which interact with receptors on the surface of B cells and cause a phosphorylation of the transcriptional regulator STAT6 via activation of the Janus family tyrosine kinases JAK1 and JAK3 (Hoey and Grusby, 1999). The second signal for IgE class switching is a costimulatory interaction between CD40 ligand on the T-cell surface with CD40 on the B cell surface (Bacharier and Geham, 2000). Overall, Th2 cells, present in respiratory mucosa and regional lymphoid tissues of individuals with allergic asthma, can switch the antibody isotype from IgM to IgE, or they can cause switching to IgG2 and IgG4 (human) or IgG1 and IgG3 (mouse).

Early phase response

Subsequent reexposure to the same agent causes a biphasic, reversible airflow obstruction referred to as the early and late phase response. The early phase response (within minutes) is characterized by a rapid onset of mucosal edema, and an increase in airway smooth muscle tone which is associated with mast cell activation and degranulation following allergen cross-linking of the IgE molecules bound to the constitutively expressed high-affinity IgE receptor FcεRI on the surface of mast cells (Pauwels, 1989; Turner and Kinet, 1999). Mast cells are highly specialized cells which are located in subendothelial and mucosal tissues, as well as in epithelial tissues in the vicinity of small blood vessels and postcapillary venules. The secretion of preformed factors (histamine, enzymes, growth factors and toxic proteins) and newly generated mediators (chemokines, lipid mediators, and cytokines) from mast cells is thought to be the initial step for the allergic late phase reaction (Wasserman, 1984).

Late phase response

The allergic late phase response occurs hours after exposure and may persist for several days without therapy. It is characterized by airway narrowing and an influx of neutrophils, eosinophils and lymphocytes from the blood into the lung parenchyma and airway epithelium (de Monchy et al., 1985; Cockcroft, 1988; Pauwels, 1989; Bousquet et al., 1990). The mast cell-derived mediators have multiple functions in this respect: chymase and tryptase and other serine proteases can activate matrix metalloproteinases causing tissue destruction. The synthesis and release of cytokines, particularly tumor necrosis factor-alpha (TNF-α), activate endothelial cells causing increased expression of adhesion molecules, which promotes the influx of inflammatory cells into tissues and thereby further enhances the chemoattractive feature of chemokines, cytokines and lipid mediators over several hours (Metcalfe et al., 1997; Mekori and

Metcalfe, 1999; Bingham and Austen, 2000; Williams and Galli, 2000). Eosinophils, as well as neutrophils, basophils, and lymphocytes, are increased in airway tissue, including the submucosa, epithelium and airway lumen (Bascom et al., 1988; Bachert et al., 1990; Calderón et al., 1994; Horwitz and Busse, 1995). Eosinophils, mast cells, T cells and basophils can interact with each other. Eosinophils appear to generate a variety of proinflammatory mediators and cytolytic enzymes that disrupt epithelial integrity, inflict substantial damage to the extracellular matrix, neurons and airway inflammatoy cells, and are also known to stimulate degranulation of mast cells and basophils (Pearlman, 1999). Once activated and recruited to the airways, the inflammatory cells can easily induce a chronic inflammatory response followed by structural changes of the lung (airway remodeling).

Neurotrophins and asthma

Cellular sources of neurotrophins in the immune system

The recruitment of inflammatory cells from the bone marrow may play an important role in the induction of allergic inflammation and asthma (Denburg et al., 2000). Before these cells infiltrate the lung and lymphatic tissues, they circulate in the peripheral blood.

In the past years, several cell types have been identified as cellular sources of neurotrophins. There is evidence for local neurotrophin synthesis in the bone marrow. Laurenzi et al. (1998) showed mRNA[BDNF] and mRNA[NT-4] in human bone marrow preparations, but the detection of NGF- and NT-3-specific mRNA transcripts failed. In contrast, Labouyrie et al. (1999) have demonstrated transcripts for NGF, BDNF, NT-3 and NT-4 in comparable human samples. On a cellular basis, murine stromal cells in long-term bone marrow cultures have been identified for NGF and BDNF synthesis (Dormady et al., 2001). NGF is produced by murine bone marrow-derived cultured mast cells and released upon activation through cross-linkage of the high affinity IgE receptors (Xiang and Nilsson, 2000).

Human megakaryocytes are also cellular source for BDNF (Fujimura et al., 2002).

NGF is elevated in the sera of patients with allergic diseases, and highest levels were found in patients suffering from asthma (Bonini et al., 1996), pointing to a neurotrophin synthesis in peripheral blood cells. In fact, neurotrophins can be synthesized by a wide range of white blood cells and the production is often regulated by inflammatory signals.

Human purified monocytes/macrophages derived from peripheral blood (PB) express NGF constitutively. Stimulation with lipopolysaccharide (LPS) (Caroleo et al., 2001) and infection with HIV (Garaci et al., 1999) caused a strong upregulation of NGF synthesis in these cells. In contrast, Ehrhard et al. (1993) have reported that NGF is not detectable constitutively nor in *Staphylococcus aureus* Cowain strain I (SAC)-stimulated cells. Furthermore, human monocytes, which became activated through an isolation process, express BDNF (Kerschensteiner et al., 1999).

For T cells, conflicting data are available. In human purified T cells, NGF synthesis was not detected (Otten et al., 1989). In contrast, Lambiase et al. (1997) have demonstrated a constitutive mRNA^{NGF} expression in human CD4/Th1 T cell clones and Th2 clones, whereas PHA stimulation is required in Th0 cell clones to produce NGF; NGF synthesis in Th2 clones was further increased by PHA stimulation. BDNF is expressed by both human Th1 and Th2 polarized T cells as well as in CD8 T-lymphocytes and can be stimulated by coincubation with allogenic PBMCs and PHA (Kerschensteiner et al., 1999). Human B cells are also a source of NGF. SAC-stimulation increases NGF production (Torcia et al., 1996) and also induces BDNF synthesis (Kerschensteiner et al., 1999). Furthermore, Besser and Wank (1999) demonstrated mRNA^{NT-3} expression after IL-4 or PHA stimulation in a human B cell line.

Moreover, human granulocytes contribute to neurotrophin synthesis in peripheral blood. Laurenzi et al. (1998) observed mRNA coding for NGF, BDNF, and NT-4, but not NT-3, in granulocyte preparations. NT-4 production was elevated after stimulation with leukotriene B4. Constitutive expression and intracellular storage of NGF and NT-3 was seen in human eosinophils (Solomon et al., 1998; Kobayashi et al., 2002). New synthesis of NGF is enhanced by Fc-receptor-mediated stimuli, such as immunoglobulin (Ig)A and IgG immune complexes and IL-5 (Kobayashi et al., 2002). Another relevant source of neurotrophins for allergic disease are mast cells. Human umbilical cord mast cells express NGF and BDNF, but no NT-3 (Tam et al., 1997).

In the lung, macrophages, alveolar cells type I, T-cells, mast cells, fibroblasts and epithelial cells have been identified to synthesize neurotrophins. Lung macrophages can be divided into two major subclasses, located in various anatomical compartments in the lung: the interstitial macrophages and the alveolar macrophages. Some authors revealed that interstitial macrophages may be the immediate precursors of some alveolar macrophages (Bowden and Adamson, 1972). Although differences in morphology, the pattern of surface receptors and function have been shown, recent studies indicate that also interstitial macrophages are highly immunocompetent cells (Sebring and Lehnert, 1992; Prokhorova et al., 1994; Johansson et al., 1997).

Comparing neurotrophin expression in macrophages, it becomes clear that there are differences in the expression pattern. Both alveolar and interstitial macrophages express NGF protein after allergen provocation (Braun et al., 1998). In this study, NGF was determined in bronchoalveolar lavage fluid (BALF) cells as well as in bronchial biopsies after allergen challenge in OVA-sensitized and challenged mice in comparison to control animals. The observed increased NGF levels in BALF from OVA-treated mice revealed, besides an allergen-dependent augmentation in NGF production, an allergen-dependent release of NGF. In contrast, BDNF is constitutively expressed in interstitial murine macrophages (Hikawa et al., 2002), but synthesis of BDNF in alveolar macrophages requires activation as induced by allergen challenge in a mouse model of asthma (Braun et al., 1999a). Investigations for NT-3 production demonstrated a restricted, constitutive synthesis in murine alveolar macrophages, whereas NT-4 was expressed constitutively both from alveolar macrophages as well as from interstitial macrophages (Hikawa et al., 2002). In addition to alveolar cells type I (Hikawa et al., 2002), also T-cells as well as epithelial cells in the airways serve as a source for BDNF in the mouse (Braun et al., 1999a; Lommatzsch et al., 1999). Epithelial cells constitutively express BDNF, whereas this neurotrophin was observed in tissue T cells exclusively after allergen challenge in sensitized mice (Braun et al., 1998). In human biopsies, NGF immunoreactivity has been observed in the epithelium also (Kassel et al., 2001), but has been detected neither in constitutive nor in inducible form in mice (Braun et al., 1998). In human biopsies of patients with asthma, fibroblasts, blood vessels and 'a few infiltrating cells' showed a positive NGF immunoreactivity. Low dose allergen challenge increased the $mRNA^{NGF}$ content (Kassel et al., 2001).

The data about neurotrophin production in lymph nodes are rare. It could be demonstrated by Torcia et al. (1996) that human monocytes and T cells, in contrast to B cells, could not express NGF. Even stimulation with SAC has no effect on T lymphocytes and monocytes, whereas NGF was increased in B-cells. In rat inguinal lymph nodes, T cells as well as infiltrating NK-cells were identified for BDNF and NT-3 production (Hammarberg et al., 2000).

Overall, it has been shown that neurotrophin synthesis occurs in immune cells and is partly dependent on inflammation. However, the available information about regulation and in particular allergen-dependent regulation of these factors in vivo is not satisfactory.

The role of neurotrophins in allergic inflammation

The role of neurotrophins in asthma-associated inflammation has not yet been well characterized. However, the inflammatory response to allergens is a complex spatiotemporal behavior, which involves several subtypes of immunocompetent cells and a wide variety of mediators. No studies have been published describing the impact of inhibiting neurotrophin actions on airway inflammation in patients with asthma or the effects of exogenously applied neurotrophins in the human lung (see Carr et al., 2001).

Neurotrophins may influence inflammatory responses either directly by regulating immune cell functions (Aloe et al., 1997), or indirectly through modulation of the synthesis of sensory neuropeptides

(Goedert et al., 1981; Lindsay and Harmar, 1989; Verge et al., 1995), including substance P (SP) and calcitonin gene-related peptide (CGRP). The enhanced release of these neuropeptides is known to modulate immune cell functions (Schaffer et al., 1998). This is summarized by the concept of 'neurogenic inflammation'. A variety of studies have revealed the involvement of neurogenic inflammation in the pathophysiology of inflammatory diseases (Levine et al., 1984; Payan, 1992).

In the following, we will first focus on the direct effects of neurotrophins on immune cells and then, after discussing the role of neurotrophins in neuronal AHR, describe their involvement in the crosstalk between neurons and immune cells in asthma.

Most of the involved immune cells mature in the bone marrow before circulating in the blood and then being recruited to the lung and lymph nodes during immune reactions and inflammation. Therefore, the action of neurotrophins must be described on the levels of sensitization, early phase response and late phase response to completely characterize their functions in asthma.

Neurotrophins and allergic sensitization

As demonstrated above, antigen-presenting cells, including dendritic cells and macrophages, as well as T-cells and B-cells are the most relevant cells during sensitization. The cells are bone marrow derived and predominantly localized in the lung and the local draining lymph nodes. Both organs are innervated by different types of nerve fibers including sensory nerves (Popper et al., 1988; Hukkanen et al., 1992; Lamb and Sparrow, 2002) pointing out the potential presence of an intensive crosstalk between inflammatory cells and the peripheral nervous system.

The expression of neurotrophin receptors in antigen-presenting cells is a mandatory prerequisite for a biologic effect of neurotrophins, controversially discussed and probably dependent on localization as well as on the maturation status of the cells. The myeloid progenitor is the precursor of granulocytes, macrophages, dendritic cells and mast cells. It has been shown that promyelocytes, myelocytes and metamyelocytes in the bone marrow of human individuals express TrkC truncated and TrkC kinase, but have no immunoreactivity for p75[NTR], TrkA or TrkB (Labouyrie et al., 1999). Macrophages in the stroma of human bone marrow samples show a positive immunoreactivity for TrkA, TrkB kinase, TrkB truncated, TrkC kinase and TrkC truncated, but no staining for p75[NTR] (Labouyrie et al., 1999). Receptor expression and the first functional observations of neurotrophin effects on hematopoiesis (stimulation with NGF enhances granulocyte-macrophage colony formation from murine bone marrow cells in cooperation with hemopoietic factor/s) (Kannan et al., 1993) suggest that in addition to NGF also BDNF, NT-3 and NT-4 may play a central role in hematopoiesis.

In human peripheral blood, monocytes/macrophages constitutively express TrkA and p75[NTR]. Stimulation with LPS caused a strong upregulation of TrkA but had no effect on p75[NTR] expression. Treatment with antibodies against NGF to interrupt the autocrine effect of this neurotrophin, markedly reduced TrkA expression and caused an increase in p75[NTR] expression, accompanied by enhanced apoptosis. These results indicate that NGF is an autocrine survival factor which exerts its effects via TrkA (Caroleo et al., 2001). The results from la Sala et al. (2000) are in line with this suggestion. In this study, application of TrkA receptor agonists protects monocytes from apoptosis induced by gliotoxin or UVB radiation by upregulation of the expression of the anti-apoptotic Bcl-2 family members, Bcl-2, Bcl-XL, and Bfl-1. In addition, it has been shown that TrkA stimulation does not change the antigen-presenting function of human monocytes from peripheral blood.

Hikawa et al. (2002) compared the Trk receptor expression in interstitial and alveolar macrophages in C57BL/6 mice. In contrast to interstitial macrophages, which express the truncated TrkB and the TrkC, no expression of neurotrophin receptors was detected in alveolar macrophages. In contrast to mice, small populations of neurotrophin receptor expressing alveolar macrophages were found in the human airways. Ricci et al. (2000) analyzed cells from human bronchoalveolar lavage fluid and demonstrated a positive immunoreactivity for TrkA in 3.5% of the alveolar macrophages. The full-length isoform for TrkB was expressed in 10%, and 2%

of the cells showed a positive staining for TrkC. No low-affinity p75[NTR] and TrkB-truncated isoform receptor-immunoreactive alveolar macrophages were found.

To date, the physiologic as well as the pathologic consequences of neurotrophin receptor expression in lung macrophages as well as in lymph nodes remains unclear.

The expression pattern of macrophages in the lymph nodes shows similarities to that of human alveolar macrophages. Shibayama and Koizumi, 1996 examined the cellular localization of all Trks in human nonneuronal tissues from adult individuals. Only a positive immunoreactivity for TrkB has been observed in monocytes/makrophages in lymph nodes (Shibayama and Koizumi, 1996) which are preferentially located in T cell zones (Garcia-Suarez et al., 1997).

Information about neurotrophin receptor expression on antigen-presenting cells other than macrophages, such as dendritic cells, is rare. As seen above, monocytes from peripheral blood express both functionally active receptors for NGF, the TrkA and p75[NTR]. During differentiation of human monocytes from peripheral blood to dendritic cells in vitro, a progressive loss of TrkA expression could be observed (la Sala et al., 2000), which is accompanied by the loss of survival-promoting characteristics of NGF. In lymph nodes, the expression pattern of neurotrophin receptors has changed. Human follicular dendritic cells isolated from secondary lymph follicles of tonsillar lymph nodes express both p75[NTR] and TrkA but no TrkB nor TrkC (Pezzati et al., 1992; Garcia-Suarez et al., 1997; Labouyrie et al., 1997). Results about the influence of neurotrophins on dendritic cells are missing completely.

In contrast, the impact of neurotrophins on lymphocyte function is relatively well characterized. Neurotrophin receptor expression in T-lymphocytes of bone marrow or their precursors, respectively, have not been examined. In peripheral blood, the described expression pattern is complex. TrkA expression in human T cells (Otten et al., 1989; Torcia et al., 1996) but no p75[NTR] (Torcia et al., 1996) was demonstrated. In contrast, Kittur et al. (1992) detected p75[NTR] in human lymphocytes and observed a PHA-dependent regulation. Furthermore,

TrkA expression was evaluated in Th1 and Th2 cell clones derived from human circulating mononuclear blood cells and has been detected in both examined subpopulations (Lambiase et al., 1997). TrkB is expressed in several subtypes of human T cells. Constitutive expression of the truncated form has been described in CD4 cells, whereas expression in CD8 cells requires anti-CD3 or PHA stimulation (Besser and Wank, 1999). Examination of the truncated TrkB isoform in T-cell clones reveals independence from cytokine profile and was observed to be constitutively expressed in Th1, Th2, TH0 as well as $\gamma\delta+$ T cell clones (Besser and Wank, 1999). The full-length form of TrkB, gp145TrkB, is also expressed in T cells from PBMC after adequate stimulation (PHA or anti-CD3). In contrast, Th1 T cell clones and the $\gamma\delta^+$ T cell clone expressed gp145TrkB constitutively, which could not be located to Th0 nor to Th2 clones. Detection of constitutively expressed TrkC in PBMCs failed. Stimulation with anti-CD3 caused an induction of this receptor in the Th0, Th1 and Th2 T cell clones, whereas no TrkC could be observed in the $\gamma\delta+$ T cell clone. Furthermore, a weak constitutive expression has been shown in two of the examined Th1 T cell clones (Besser and Wank, 1999). T cells derived from human lymph nodes show TrkA, but lack p75[NTR] (Torcia et al., 1996). Anyway, there is strong evidence for physiological importance of neurotrophin receptor expression on T cells as has been demonstrated in functional TrkB deficient mice (Garcia-Suarez et al., 2002). The thymus of homozygous functional TrkB-deficient animals showed structural and ultrastructural changes consistent with massive death of cortical lymphocytes, pointing out an important role for TrkB mediated survival of T lymphocytes. In addition, signaling of BDNF through TrkB, which is expressed in dependence on maturation level, is involved in the T cell differentiation process (Maroder et al., 1996).

Neurotrophin receptor expression in bone marrow B cells or lymphoid progenitors has not yet been described. In peripheral blood and lymphoid tissue, human B cells, which can be divided into resting and large B cells, both express TrkA as well as p75[NTR] on their surfaces (Otten et al., 1989; Torcia et al., 1996). Furthermore, expression of the truncated as well as the full-length form of TrkB has been demonstrated

in human peripheral blood B cells or a human B cell line, respectively, and is increased after stimulation with PHA (Besser and Wank, 1999). NGF acts as an autocrine survival factor in memory B cells but not in immature B lymphocytes, as seen in vitro B-cell preparations from human tonsils and in vivo experiments in the mouse. Incubation of human B-cells with neutralizing antibodies against NGF caused apoptotic cell death in B-cells expressing surface IgA/IgG, but not in 'virgin' B-lymphocytes bearing surface IgM/IgD. In vivo administration of neutralizing antibodies against NGF reduced the titer of specific IgG in mice immunized with tetanus toxoid, nitrophenol, or arsonate (Torcia et al., 1996). The possible mechanism for apoptosis induced by neutralization of endogenously produced NGF may be downregulation of Bcl-2 protein and prevention of Bcl-2 phosphorylation, inactivation of p38 MAPK, and cytochrome c release (Torcia et al., 1996, 2001). Besides promoting survival, NGF can influence B cell proliferation and stimulate murine and human B lymphocytes to produce IgM, IgA and IgG4 (Otten et al., 1989; Kimata et al., 1991; Torcia et al., 1996). Human IgG$_4$ production is Th2 cytokine-dependent like IgE synthesis. In a mouse model of asthma, Braun et al. (1998) have demonstrated that NGF also augments IgE and IgG$_1$ production in allergen-stimulated mononuclear cells from spleen. Whether this is a direct effect on B-cells or an indirect effect via modulation of Th2 cytokine release from T-cells remains unclear. However, immunoglobulin levels of p75[NTR] knockout mice as well as those of CCSP-NGF-tg mice are still normal even after allergic sensitization (Path et al., 2002; Kerzel et al., 2003).

Studies which elucidate the functional consequences of neurotrophin receptor expression are still absent and, in consideration of maturity-dependent and/or tissue dependent regulation of neurotrophin receptors, further studies are required to define the contribution of neurotrophins during sensitization.

Neurotrophins in early phase response

As indicated above, mast cells play a major role in the asthmatic early phase response. Although the NGF function in the immune system was first documented in these cells (reviewed by Aloe et al., 1997), little is known about the effects of other neurotrophins than NGF. In bone marrow, human mastocytes express TrkA, both isoforms of TrkB and TrkC, but no p75NTR (Labouyrie et al., 1999). Stimulation of murine bone marrow-derived cultured mast cells with NGF cause proliferation and differentiation after costimulation with IL-3 (Matsuda et al., 1991). NGF alone triggers these cells to produce eicosanoids (Marshall et al., 1999). In peripheral blood, NGF is responsible for comparable events and has been shown to induce differentiation of human basophil progenitor cells to mature cells (Matsuda et al., 1988b). NGF suppresses apoptosis (Kanbe et al., 2000), alters degranulation (Mazurek et al., 1986), and promotes histamine production of human umbilical cord-derived mast cells in costimulation with IL-3 (Richard et al., 1992). Because human cord blood-derived cultured mast cells constitutively express TrkA, which is upregulated during NGF-driven culture (Welker et al., 2000), but no p75NTR (Tam et al., 1997; Welker et al., 2000), all effects of NGF may be attributed to binding to this receptor (Kanbe et al., 2000). In addition, TrkC expression has been shown in these cells (Tam et al., 1997), but the influence of NT-3 on mast cell functions is not yet clear. In contrast to mast cells, it has been demonstrated that human basophils in peripheral blood express exclusively TrkA and not TrkB, TrkC and p75NTR (Burgi et al., 1996). Stimulation with NGF promotes survival in these cells (Miura et al., 2001), causes histamine release, primes basophils to produce lipid mediators in response to various cytokines or complement (Bischoff and Dahinden, 1992; Miura et al., 2001) and triggers synthesis and release of IL-13 (Sin et al., 2001). Interestingly, IL-13 production in response to NGF stimulation is increased in patients with allergic asthma but is not dependent on alterations of TrkA surface expression (Sin et al., 2001). Intrapulmonary human mature mast cells show a modified neurotrophin receptor expression pattern. TrkA, TrkB and TrkC expression has been observed in the lung tissue (Tam et al., 1997), but the functional consequences of neurotrophin stimulation remain unclear.

Recently published studies indicate that abrogation of the NGF signal by or intranasal application of neutralizing antibodies against NGF could inhibit allergen-induced early phase reaction in a mouse and guinea pigs model of allergic asthma (de Vries et al., 2002; Path et al., 2002). Furthermore, the early phase response to allergen was examined in sensitized CCSP-NGF-tg mice and compared to the reaction of wild type controls. In these animals, bronchoconstriction in response to aerosolized antigen was clearly increased, paralleled by a remarkable increase in serotonin levels in the airways, pointing out that NGF may be involved in mast cell degranulation in vivo which are located in the airways of asthmatics (Path et al., 2002).

Neurotrophins in late phase response

As described before, allergic late phase response is mainly induced by accumulation of eosinophils in the lung which become activated. Bronchoconstriction and tissue damage are consequences of mediator release from eosinophils. However, neurotrophins may be potent regulators of allergic late phase response.

Eosinophils are presumably recruited from bone marrow into the lung via peripheral blood. In bone marrow, mature eosinophils express neurotrophin receptors (Labouyrie et al., 1999) and therefore fulfill the prerequisite for mediating neurotrophin signals. In this study, expression of TrkB kinase, TrkC kinase and TrkC truncated was observed, but detection of TrkA as well as p75NTR failed. In human umbilical cord blood-derived eosinophil progenitors, NGF has no influence on eosinophil differentiation, underlining the absence of TrkA and p75NTR (Matsuda et al., 1988a). In contrast, peripheral blood eosinophils are responsible for NGF even though no receptor has been demonstrated yet. NGF stimulation causes eosinophil peroxidase (EPO) release both in human and murine cells (Solomon et al., 1998; Watanabe et al., 2001), and increases chemotactic activity and cytotoxicity as determined by larvicidal activity in human eosinophils (Hamada et al., 1996). The effect of NGF on eosinophil survival is controversely discussed (Hamada et al., 1996; Solomon et al., 1998). Although distribution of neurotophin receptors on human lung eosinophils has not yet been investigated, recently published studies indicate that blocking the p75NTR signal in vivo could prevent allergic late phase reaction as well as eosinophil accumulation (Tokuoka et al., 2001; Kerzel et al., 2003). Two groups of mice were examined, demonstrating that local intrapulmonary application of neutralizing p75^{NTR} antibodies could prevent the chronic inflammatory response as well as the general absence of this receptor as seen in p75^{NTR} knockout mice. Because local treatment led to similar changes as observed in *p75^{NTR}−/−* mice, it could be hypothesized that neurotrophin action on local lung cells plays the most important part in induction of inflammation.

First evidence of the particular role of NGF in this condition came from a study recently published by Path et al. (2002). These authors investigated the effects of intranasally applied NGF-neutralizing antibodies on inflammation of OVA-sensitized and challenged mice. In anti-NGF-treated mice, the number of eosinophils recruited into the airways was significantly reduced, whereas the numbers of macrophages, lymphocytes and neutrophils were not altered. Controversely to the effects of anti-NGF treatment, OVA-sensitized and challenged CCSP-NGF-tg mice, overexpressing NGF in the airways, demonstrated a higher influx of eosinophils, lymphocytes and neutrophils into the airways than corresponding wild type animals. It is not yet clear whether NGF directly alters eosinophil accumulation, e.g., by inhibiting apoptosis or increasing chemotactic activity, or whether this observation is due to indirect effects. Firstly, an altered local production of cytokines, which support survival, second, the facilitation or induction of mast cell degranulation (as described above) might contribute to eosinophil accumulation, and thirdly, neuronally mediated pathways may represent an additional important mechanism leading to airway inflammation.

Neuronal airway hyperresponsiveness

The neuronal changes in the airways play an important role in asthma. The function of nerves is not limited to modulation of airway inflammation, but may furthermore be responsible for the development and maintenance of AHR. The airway tone is regulated by cholinergic, adrenergic and nonadrenergic noncholinergic (NANC) nerves.

Sympathetic nerves are involved in the control of the airways. Adrenergic nerves arise from the upper six thoracic segments of the spinal cord and synapse in the sympathetic ganglia. The postganglionic fibers mainly innervate bronchial blood vessels and submucosal glands. Therefore, they do not directly control the airway smooth muscle tone but may influence bronchial obstruction by regulation of bronchial blood vessel tone (Belvisi, 2002).

Cholinergic excitatory nerves are thought to be the predominant bronchoconstrictors in the airways. The control of neural and nonneural target cells via nicotinic and muscarinic M3 receptors is mediated by acetylcholine (ACh). Interestingly, the most important control over ACh release from postganglionic cholinergic nerves is exerted by acetylcholine itself. The M2 autoreceptor is prejunctionally located on postganglionic nerves and stimulation limits the further release of ACh. A loss of function in the neuronal muscarinic M2 autoreceptor is observed after allergen exposure, ozone or virus infection (Fryer and Wills-Karp, 1991; Belmonte et al., 1998; Jacoby et al., 1998; Larsen et al., 2000). In consequence of neuronal M2-loss or altered function as induced by treatment with receptor antagonists (Evans et al., 2000), the vagally induced bronchoconstriction may become potentiated (Costello et al., 1999).

In human airways, sensory NANC neurons are involved in the control of respiratory functions. Inhibitory NANC (i-NANC) mechanisms are the only neural bronchodilatory mechanisms. The presumed neurotransmitters of the i-NANC system are vasoactive intestinal peptide (VIP) and nitric oxide (NO). Although one study indicates absence of VIP immunoreactivity in the lung of asthmatics (Ollerenshaw et al., 1989), no defect of the i-NANC system has been observed in further functional studies in humans using in vivo and in vitro methods (Michoud et al., 1988; Bai, 1990).

SP and neurokinin A (NKA) have been implicated as important neurotransmitters mediating the excitatory part of the NANC (e-NANC) nervous system. Three receptors, NK-1, NK-2 and NK-3 for tachykinins, predominantly located on blood vessels (NK-1), smooth muscle cells of both large and small airways (NK-2), and subpopulations of cholinergic neurons (NK-3), have been identified (Severini et al., 2002). They mediate part of the bronchoconstrictor effect of tachykinins. NK-1 shows highest affinity for SP, NK-2 for NKA, and NK-3 for NKB, respectively. NKA and SP are often colocalized in bronchial sensory nerves, mostly primary afferent chemosensory C-fibers (Lundberg et al., 1987; Uddman et al., 1997) innervating the airway mucosa. Polymodal nociceptor neurons (C-fibers and A-δ-fibers) could be stimulated by capsaicin, the principal pungent component of Capsicum peppers, via the vanilloid receptor 1 (VR1) (Szallasi and Blumberg, 1999), which is known to have a high specificity for neuropeptide-containing C-fibers (Fox et al., 1993). Reflex activity of these polymodal nociceptive terminals of sensory nerves is not restricted to capsaicin stimulation, and can be further induced by several physiological mediators, like bradykinin, or environmental irritants, like cold air, noxious agents including SO_2, NO_2 and ozone, changes in osmolarity, and smoke (Pedersen et al., 1998; Mutoh et al., 2000). Comprisingly, a wide range of stimuli causing allergen-independent bronchoconstriction, can affect C-fibers, which in turn secrete neuropeptides known to initiate bronchoconstriction directly by interaction with smooth muscle cells or indirectly by dysregulation of M2 receptors (Costello et al., 1998), cause vasodilatation and plasma extravasation, increase mucus secretion and alter immune cell function. Therefore, it has been suggested that these fibers may be involved in the pathogenesis of neurogenic inflammation and in the initiation of AHR (Hsiue et al., 1992; Joos et al., 2000). This suggestion is supported by demonstration of enhanced tachykinin content in the human airways of asthmatics as compared to healthy individuals (Nieber et al., 1992; Lilly et al., 1995; Tomaki et al., 1995). Allergen challenge both in humans as well as in guinea pigs increase neuropeptide levels in the airways, and, interestingly, increase the number of tachykinin-immunoreactive nodose ganglion neurons, projecting to the airways (Fischer et al., 1996). The sum of all quantitative and qualitative changes in neurons is summarized with the term 'neuronal plasticity', which is present in the lungs of asthmatics.

The role of neurotrophins, mainly NGF, in survival, differentiation, and maintenance of neurons is well-defined (Huang and Reichardt, 2001). In recent years, it became clear that neurotrophins play an important role in the function of nociceptive afferents during hyperalgesia (Lewin and Mendell, 1993a). NGF synthesis is upregulated during inflammatory pain (Weskamp and Otten, 1987), and local or systemic administration of NGF or the TrkB agonists, BDNF and NT-4 produces thermal and mechanical hyperalgesia (Lewin et al., 1993b; Rueff and Mendell, 1996; Shu et al., 1999). The necessity of NGF for induction of inflammatory pain has been elicited by blocking NGF itself or interrupting neurotrophin receptor signaling (Lewin et al., 1994; Woolf et al., 1994; McMahon et al., 1995). Under these conditions, the development of hyperalgesia was completely abolished, pointing out a pivotal role of NGF for inflammatory pain. Common hypotheses suggest that neurotrophins either influence neuronal innervation or/and modulate the sensitivity of neurons. Several studies indicate that NGF directly affects neuronal function by enhancing peptides such as SP and CGRP (Lindsay and Harmar, 1989; Woolf et al., 1994) and reducing the sensory threshold. In addition, there is evidence for functionally relevant NGF-dependent local sensory hyperinnervation in mice overexpressing NGF in different organs (Hassankhani et al., 1995 for heart; Hoyle et al., 1998 for lung).

Comparing the neuronal changes in inflammatory pain with those observed in the lungs of patients with asthma, it is remarkable that the pathophysiologic changes observed in hyperalgesia seem to be closely related to those elicited in airway hyperreactivity. Therefore, the role of neurotrophins for airway hyperresponsiveness has been investigated.

To address the question of whether neurotrophins are relevant in the pathogenesis of AHR, a sequence of experiments was performed. To respond to the first basic questions of whether neurotrophins are locally elevated in the lung in asthma, neurotrophin levels in BALF were measured in a well-characterized mouse model of asthma. Both NGF and BDNF levels were elevated after allergen treatment (Braun et al., 1998, 1999a). In parallel, segmental allergen provocation was performed in patients with mild allergic asthma to confirm the data in humans. Allergen challenge caused an increase of NGF, BDNF and NT-3 selectively in the allergen-provoked segment, indicating comparable allergen-dependent mechanisms (Virchow et al., 1998). To elucidate the effect of elevated pulmonary neurotrophin levels in asthma, mice were treated intranasally with neutralizing NGF antibodies and AHR assessed in vitro by electrical field stimulation of trachea segments. Application of antibodies to allergen-sensitized mice prevented the development of AHR (Braun et al., 1998). First evidence for the underlying mechanism of NGF-dependent AHR was given by Hoyle et al. (1998) and de Vries et al. (1999). NGF-transgenic mice over-expressing NGF locally in the lung showed an enhanced innervation of predominantly sensory neurons producing SP in the airways (Hoyle et al., 1998), pointing to the possibility of an indirect effect of NGF via enhanced tachykinin synthesis, comparable to that observed in hyperalgesia. Although hyperinnervation may be important, the study published by de Vries et al. indicates the presence of an additional rapid mechanism. Intravenously applied NGF in guinea pigs was sufficient to promote bronchoconstriction in response to inhaled histamine after 30 min. This effect was blocked by treatment with NK1 antagonists, indicating NGF-induced tachykinin release (de Vries et al., 2001). Although neurotrophins may induce AHR indirectly, it is not yet clear which neurotrophin receptor is responsible for mediating the NGF signal. The first study engaging in this problem was recently published by Kerzel et al. (2003). The contribution of p75^{NTR} to AHR was tested in sensitized and allergen-challenged $p75^{NTR}-/-$ mice as well as in wild type mice treated with antibodies against p75^{NTR}. It has been clearly demonstrated that this receptor plays a central role in the development of neuronal airway hyperreactivity. Either the absence of p75^{NTR} or the treatment with blocking agents almost completely abolished the development of capsaicin-induced sensory hyperreactivity. To understand the underlying mechanism, the hypothesis of a quantitatively changed sensory innervation was tested. Interestingly, the assessment of tachykinergic innervation revealed no alterations in genetically modified animals as compared to wild-type mice even after allergen treatment, indicating that qualitative neuronal changes due to the absence of the p75^{NTR} may play the more important role. This hypothesis is supported by results from the study of Graham et al. (2001) showing that the

amount of SP in the lungs of $p75^{NTR}-/-$ mice was reduced by approximately 50% as compared to wild type mice.

Although the study of Kerzel et al. (2003) has demonstrated the contribution of p75^{NTR}, it is not possible to exclude the functional relevance in the development of sensory hyperreactivity for the Trk receptors. An interaction between the low- and high-affinity neurotrophin receptors resulting in altered binding activities of neurotrophins and modified signal transduction has been demonstrated in numerous studies (Dobrowsky and Carter, 1998; Bibel et al., 1999).

Therefore, further studies are required to elucidate the role of the Trks and the signaling of the different neurotrophins in the development of neuronal hyperreactivity. New mouse models need to be generated to characterize the contribution of neurotrophins and neurotrophin receptors to the pathogenesis of asthma and to clarify whether inflammation is first required to initiate AHR, or alterations in neuronal functions are the first step in the development of airway inflammation.

Neurotrophins and neurogenic inflammation

The axon reflex describes a concept of how the excitatory NANC system may contribute to the pathophysiology of bronchial asthma (Barnes, 1986). The axon reflex is initiated by stimulation of vagal afferent nerve endings in the airway epithelium. Any damage to airway epithelium causes such stimulation. A broad range of trigger factors have been identified, including infectious agents such as viral and bacterial infections, inhaled environmental irritants such as NO_2, SO_2, ozone and others. Furthermore, various inflammatory mediators released from airway inflammatory cells also stimulate vagal afferent nerve endings. Stimulation of such afferent nerve fibers not only results in a central reflex activity, leading to enhanced efferent cholinergic activities with the result of increased acetylcholine release. In parallel, there exists a 'short-circuit' local reflex loop with the nonadrenergic noncholinergic afferent nerves. This peripheral reflex activity results in an increased release of neuropeptides and tachykinins including SP, NKA, NKB and CGRP. As a consequence, neuropeptide and tachykinin release causes airway smooth muscle restriction, mucus production, edema and release of inflammatory mediators from various inflammatory cells.

The inflammation-induced altered neuronal function can lead to an increased release of proinflammatory neuropeptides upon stimulation. Neuronally mediated inflammation is also subsumed by the term 'neurogenic inflammation' and describes the interaction between airway nerves and inflammation (Barnes, 1992). Sensory nerves may amplify inflammation in the airways through release of neurotransmitters such as neuropeptides. However, inflammatory mediators in turn may modulate cholinergic and sensory nerves in the airways by activating receptors on nerve terminals. Therefore, a close colocalization of immune cells and nerves in the airways is required and has been observed in several studies. First evidence for three-dimensional arrangement has been intensively studied in porcine and human airways. In this study, SP-containing sensory C-fibers have been detected in close relationship to bronchial epithelium (Lamb and Sparrow, 2002). An overview of the interaction of neurons and eosinophils is presented by Curran et al. (2002).

NANC system-associated neuropeptides with immunomodulatory functions are somatostatin (SOM), CGRP and the members of the tachykinin family, SP and NKA, which all act via G protein-coupled receptors and thus share several effector functions. Specific receptors for CGRP and SOM have been demonstrated on monocytes, B cells, and T cells (McGillis et al., 1991; van Hagen et al., 1994). Similarly to neuropeptide Y (NPY), CGRP and SOM have the capacity to induce T cell adhesion to fibronectin and to drive distinct Th1 and Th2 populations to an atypical expression pattern of Th2 or Th1 cytokines, respectively, and, therefore, break the commitment to a distinct T helper phenotype (Levite, 1998a; Levite et al., 1998b).

SOM exerts various inhibitory functions on immune responses via specific receptor activation (van Hagen et al., 1994). SOM affects the suppression of Ig production in B cells, including IgE (Kimata et al., 1992), modulation of lymphocyte proliferation (inhibitory effect at low concentrations or stimulatory effect at high concentrations) and reduction of (peritoneal) eosinophil infiltration in experimentally induced hypereosinophilia. Furthermore, SOM inhibits SP-induced mucus secretion in rats (Wagner et al., 1995).

CGRP inhibits SP-induced super oxide production in neutrophils and the proliferation as well as the antigen presentation by peripheral mononuclear cells (Tanabe et al., 1996). It also stimulates chemotaxis and adhesion of lymphocytes (Levite, 1998a) and causes eosinophilia in the rat lung (Bellibas, 1996). It has been demonstrated that a murine B cell line expresses CGRP receptors as well as B lymphocytes in bone marrow, spleen, and thymus (Nakamuta et al., 1986; McGillis et al., 1993; Mullins et al., 1993; Bulloch et al., 1998). CGRP itself, a 37-amino acid peptide, is synthesized by small sensory neurons which innervate the parenchyma of primary and secondary lymphoid organs (Popper et al., 1988; Elfvin et al., 1992a, 1992b), but also by human and rat T cells as well as in human by B lymphocytes (Xing et al., 2000; Bracci-Laudiero et al., 2002; Wang et al., 2002). NGF is a potent inductor for the release of sensory-nerve-derived CGRP (Lindsay and Harmar, 1989; Verge et al., 1995). Bracci-Laudiero et al. (2002) observed a homologous pathway for NGF-induced release of CGRP in tonsil-derived B cells; NGF also increases CGRP synthesis in B cells. CGRP mediates several proinflammatory but also anti-inflammatory effects. For example, it modifies antigen presentation in macrophages (Nong et al., 1989) and dendritic cells (Carucci et al., 2000) by alteration of expression patterns of costimulatory factors on their surfaces (Asahina et al., 1995; Fox et al., 1997), inhibits Th1 cytokine synthesis causing a disbalance of Th1/Th2 derived cytokines in favor of Th2 (Kawamura et al., 1998) and changes the cytokine profile in unstimulated Th0 clones (Levite, 1998a). Interestingly, it prevents edema induced by histamine and leukotrienes in rat paw as well as in human skin (Raud et al., 1991), and application of neutralizing antibodies against CGRP increases the severity of chronic inflammatory bowel diseases (Reinshagen et al., 1998). Furthermore, it also acts as an inhibitor of early B cell development (Fernandez et al., 2000).

The tachykinin family includes neuropeptides such as SP, NKA, and NKB, and they exhibit their function by binding to the three NK receptors, NK-1–3. NK-1 receptor expression has been identified on

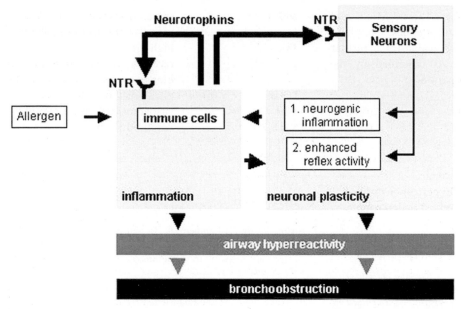

Fig. 1. Role of neurotrophins in the neuroimmune cross-talk in allergic asthma. NTR, neurotrophin receptors.

immune cells such as B and T cells (Braun et al., 1999b), monocytes and macrophages (Ho et al., 1997), and eosinophils and neutrophils (Iwamoto et al., 1993). Like VIP receptor, NK-1 expression by lung-infiltrating leukocytes in systemically and subsequently intratracheally allergen-challenged mice is strongly elevated (Kaltreider et al., 1997). Since SP binds to NK-1 with highest affinity, it is the predominant mediator of immunomodulatory effects among tachykinins. The actvities of SP on immune cells include a broad range of functional responses from neutrophils, eosinophils, mast cells, monocytes/macrophages and lymphocytes (see van der Velden and Hulsmann, 1999). SP stimulates a number of neutrophil functions, including chemotaxis, superoxide production and adherence to epithelium and endothelium. Most of these effects require high concentrations of SP, whereas at low doses SP primes the response to other stimuli that otherwise would be ineffective. SP has a degranulating effect on eosinophils and induces human eosinophil migration in vitro. In an in vivo study with allergic rhinitis patients, it was shown that SP administered after repeated allergen challenge enhanced the recruitment of eosinophils. It has been demonstrated that SP can cause histamine release from human lung mast cells

(Heaney et al., 1995). This is underlined by an in vitro model using trachea from the SP-hyperresponsive Fisher 344 rat, in which SP stimulation of mast cells represented a major factor leading to bronchoconstriction (Joos et al., 1997). Moreover, SP activates monocytes to release inflammatory cytokines, including TNF-α, IL-1, IL-6 and IL-10 (van der Velden and Hulsmann, 1999). In lymphocytes, SP inhibits glucocorticoid-induced thymocyte apoptosis (Dimri et al., 2000) and stimulates proliferation, cytokine production, and chemotaxis (van der Velden and Hulsmann, 1999) and also Th1/Th2 phenotype switch in T cells (Levite, 1998a; Levite et al., 1998b). In addition, SP induces differentiation and immunoglobulin switching in B cells (Braun et al., 1999b).

In summary, neurotrophins have two major functions in the pathogenesis of allergic asthma (Fig. 1). First, they act as potent proinflammatory cytokine-like factors on the different cellular levels of allergic inflammation. Almost all involved cells types, from antigen-presenting cells to T cells, B cells and effector cells such as eosinophils, are affected by the neurotrophin action. Second, they act as nerve growth factors, changing the function of mainly sensory neurons. These changes lead to a sensory hyperreactivity with the consequence of lung function changes as well as an increased production and release of proinflammtory neuropeptides. This neuropeptide release could be responsible for a further increase in inflammation and is described by the concept of neurogenic inflammation.

Abbreviations

ACh	acetylcholine
AHR	airway hyperresponsiveness
BDNF	brain-derived neurotrophic factor
[CCSP]-NGF-tg mice	[clara cell secretory protein promoter] NGF-transgenic mice
C-fibers	chemosensory nerve fibers
CGRP	calcitonin gene-related peptide
DC	dendritic cells
e-NANC system	excitatory nonadrenergic non-cholinergic system
EPO	eosinophil peroxidase
$Fc_{\varepsilon}RI$	high-affinity receptor for the Fc_{ε} part of the IgE molecule
i-NANC system	inhibitory nonadrenergic non-cholinergic system
JAK	janus family tyrosine kinase
LPS	lipopolysaccharide
MAPK	MAP kinase
NANC sytem	nonadrenergic noncholinergic system
NGF	nerve growth factor
NK	neurokinin
NO	nitric oxide
NO_2	nitrogen dioxide
NPY	neuropeptide Y
NT-3	neurotrophin-3
NT-4	neurotrophin-4
OVA	ovalbumin
p75NTR	low-affinity pan-neurotrophin receptor
PB	peripheral blood (derived)
SAC	staphylococcus aureus cowain strain
SO_2	sulphur dioxide
SOM	Somatostatin
SP	substance P
STAT	signal transducer and activator of transcription protein
Trk	high-affinity neurotrophin receptor, tyrosine protein kinase
UVB	ultraviolet B
VIP	vasoactive intestinal peptide
VR1	vanilloid receptor 1

References

Aloe, L., Bracci-Laudiero, L., Bonini, S. and Manni, L. (1997) The expanding role of nerve growth factor: from neurotrophic activity to immunologic diseases. Allergy, 52: 883–994.

Asahina, A., Moro, O., Hosoi, J., Lerner, E.A., Xu, S., Takashima, A. and Granstein, R.D. (1995) Specific induction of cAMP in Langerhans cells by calcitonin gene-related peptide: relevance to functional effects. Proc. Natl. Acad. Sci. USA, 92: 8323–8327.

Bacharier, L.B. and Geham, R.S. (2000) Molecular mechanisms of IgE regulation. J. Allergy Clin. Immunol., 105: S547–S558.

Bachert, C., Prohaska, P. and Pipkorn, U. (1990) IgE-positive mast cells on the human nasal mucosal surface in response to allergen exposure. Rhinology, 28: 149–158.

Bai, T.R. (1990) Abnormalities in airway smooth muscle in fatal asthma. Am. Rev. Respir. Dis., 141: 552–557.

Barnes, P.J. (1986) Asthma as an axon reflex. Lancet, 1: 242–245.

Barnes, P.J. (1992) Neurogenic inflammation and asthma. J. Asthma, 29: 165–180.

Bascom, R., Wachs, M., Naclerio, R.M., Pipkorn, U., Galli, S.J. and Lichtenstein, L.M. (1988) Basophil influx occurs after nasal antigen challenge: effects of topical corticosteroid pretreatment. J. Allergy Clin. Immunol., 81: 580–589.

Bellibas, S.E. (1996) The effect of human calcitonin gene-related peptide on eosinophil chemotaxis in the rat airway. Peptides, 17: 563–564.

Belmonte, K.E., Fryer, A.D. and Costello, R.W. (1998) Role of insulin in antigen-induced airway eosinophilia and neuronal M2 muscarinic receptor dysfunction. J. Appl. Physiol., 85: 1708–1718.

Belvisi, M.G. (2002) Overview of the innervation of the lung. Curr. Opin. Pharmacol., 2: 211–215.

Besser, M. and Wank, R. (1999) Cutting edge: Clonally restricted production of the neurotrophins brain-derived neurotrophic factor and neurotrophin-3 mRNA by human immune cells and Th1/Th2-polarized expression of their receptors. J. Immunol., 162: 6303–6306.

Bibel, M., Hoppe, E. and Barde, Y.A. (1999) Biochemical and functional interactions between the neurotrophin receptors trk and p75NTR. EMBO J., 18: 616–622.

Bingham 3rd, C.O. and Austen, K.F. (2000) Mast-cell responses in the development of asthma. J. Allergy Clin. Immunol., 105: S527–S534.

Bischoff, S.C. and Dahinden, C.A. (1992) Effect of nerve growth factor on the release of inflammatory mediators by mature human basophils. Blood, 79: 2662–2669.

Bonini, S., Lambiase, A., Bonini, S., Angelucci, F., Magrini, L., Manni, L. and Aloe, L. (1996) Circulating nerve growth factor levels are increased in humans with allergic diseases and asthma. Proc. Natl. Acad. Sci. USA, 93: 10955–10960.

Bousquet, J., Chanez, P., Lacoste, J.Y., Barneon, G., Ghavanian, N., Enander, I., Venge, P., Ahlstedt, S., Simony-Lafontaine, J. and Godard, P. (1990) Eosinophilic inflammation in asthma. N. Engl. J. Med., 323: 1033–1039.

Bowden, D.H. and Adamson, I.Y. (1972) The pulmonary interstitial cell as immediate precursor of the alveolar macrophage. Am. J. Pathol., 68: 521–537.

Bracci-Laudiero, L., Aloe, L., Buanne, P., Finn, A., Stenfors, C., Vigneti, E., Theodorsson, E. and Lundeberg, T. (2002) NGF modulates CGRP synthesis in human B-lymphocytes: a possible anti-inflammatory action of NGF? J. Neuroimmunol., 123: 58–65.

Braun, A., Appel, E., Baruch, R., Herz, U., Botchkarev, V., Paus, R., Brodie, C. and Renz, H. (1998) Role of nerve growth factor in a mouse model of allergic airway inflammation and asthma. Eur. J. Immunol., 28: 3240–3251.

Braun, A., Lommatzsch, M., Mannsfeldt, A., Neuhaus-Steinmetz, U., Fischer, A., Schnoy, N., Lewin, G.R. and Renz, H. (1999a) Cellular sources of enhanced brain-derived neurotrophic factor production in a mouse model of allergic inflammation. Am. J. Respir. Cell Mol. Biol., 21: 537–546.

Braun, A., Wiebe, P., Pfeufer, A., Gessner, R. and Renz, H. (1999b) Differential modulation of human immunoglobulin isotype production by the neuropeptides substance P, NKA and NKB. J. Neuroimmunol., 97: 43–50.

Braun, A., Quarcoo, D., Schulte-Herbruggen, O., Lommatzsch, M., Hoyle, G. and Renz, H. (2001) Nerve growth factor induces airway hyperresponsiveness in mice. Int. Arch. Allergy Immunol., 124: 205–207.

Bulloch, K., McEwen, B.S., Nordberg, J., Diwa, A. and Baird, S. (1998) Selective regulation of T-cell development and function by calcitonin gene-related peptide in thymus and spleen. An example of differential regional regulation of immunity by the neuroendocrine system. Ann. N.Y. Acad. Sci., 840: 551–562.

Burgi, B., Otten, U.H., Ochensberger, B., Rihs, S., Heese, K., Ehrhard, P.B., Ibanez, C.F. and Dahinden, C.A. (1996) Basophil priming by neurotrophic factors. Activation through the trk receptor. J. Immunol., 157: 5582–5588.

Calderón, M.A., Lozewicz, S., Prior, A., Jordan, S., Trigg, C.J. and Davies, R.J. (1994) Lymphocyte infiltration and thick-ness of the nasal mucous membrane in perennial and seasonal allergic rhinitis. J. Allergy Clin. Immunol., 93: 635–643.

Caroleo, M.C., Costa, N., Bracci-Laudiero, L. and Aloe, L. (2001) Human monocyte/macrophages activate by exposure to LPS overexpress NGF and NGF receptors. J. Neuroimmunol., 113: 193–201.

Carr, M.J., Hunter, D.D. and Undem, B.J. (2001) Neurotrophins and asthma. Curr. Opin. Pulm. Med., 7: 1–7.

Carucci, J.A., Ignatius, R., Wei, Y., Cypess, A.M., Schaer, D.A., Pope, M., Steinman, R.M. and Mojsov, S. (2000) Calcitonin gene-related peptide decreases expression of HLA-DR and CD86 by human dendritic cells and dampens dendritic cell-driven T cell-proliferative responses via the type I calcitonin gene-related peptide receptor. J. Immunol., 164: 3494–3499.

Cockcroft, D.W. (1988) Airway hyperresponsiveness and late asthmatic responses. Chest, 94: 178–180.

Constant, S.L., Brogdon, J.L., Piggott, D.A., Herrick, C.A., Visintin, I., Ruddle, N.H. and Bottomly, K. (2002) Resident lung antigen-presenting cells have the capacity to promote Th2 T cell differentiation in situ. J. Clin. Invest., 110: 1441–1448.

Costello, R.W., Evans, C.M., Yost, B.L., Belmonte, K.E., Gleich, G.J., Jacoby, D.B. and Fryer, A.D. (1999) Antigen-induced hyperreactivity to histamine: role of the vagus nerves and eosinophils. Am. J. Physiol., 276: L709–L714.

Costello, R.W., Fryer, A.D., Belmonte, K.E. and Jacoby, D.B. (1998) Effects of tachykinin NK1 receptor antagonists on vagal hyperreactivity and neuronal M2 muscarinic receptor function in antigen challenged guinea-pigs. Br. J. Pharmacol., 124: 267–276.

Curran, D.R., Walsh, M.T. and Costello, R.W. (2002) Interactions between inflammatory cells and nerves. Curr. Opin. Pharmacol., 2: 243–248.

de Monchy, J.G., Kauffman, H.F., Venge, P., Koeter, G.H., Jansen, H.M., Sluiter, H.J. and De Vries, K. (1985) Bronchoalveolar eosinophilia during allergen-induced late asthmatic reactions. Am. Rev. Respir. Dis., 131: 373–376.

de Vries, A., Dessing, M.C., Engels, F., Henricks, P.A. and Nijkamp, F.P. (1999) Nerve growth factor induces a neurokinin-1 receptor-mediated airway hyperresponsiveness in guinea pigs. Am. J. Respir. Crit. Care Med., 159: 1541–1544.

de Vries, A., van Rijnsoever, C., Engels, F., Henricks, P.A. and Nijkamp, F.P. (2001) The role of sensory nerve endings in nerve growth factor-induced airway hyperresponsiveness to histamine in guinea-pigs. Br. J. Pharmacol., 134: 771–776.

de Vries, A., Engels, F., Hendricks, P.A., Levsink-Muis, T., Fischer, A. and Nijkamp, F.P. (2002) Antibodies directed against nerve growth factor inhibit the acute bronchoconstriction due to allergen challenge in guinea pigs. Clin. Exp. Allergy, 32: 325–328.

Denburg, J.A., Sehmi, R., Saito, H., Pil-Seob, J., Inman, M.D. and O'Byrne, P.M. (2000) Systemic aspects of allergic disease: bone marrow responses. J. Allergy Clin. Immunol., 106: S242–S246.

Dimri, R., Sharabi, Y. and Shoham, J. (2000) Specific inhibition of glucocorticoid-induced thymocyte apoptosis by substance P. J. Immunol., 164: 2479–2486.

Dobrowsky, R.T. and Carter, B.D. (1998) Coupling of the p75 neurotrophin receptor to sphingolipid signaling. Ann. NY Acad. Sci., 845: 32–45.

Dormady, S.P., Bashayan, O., Dougherty, R., Zhang, X.M. and Basch, R.S. (2001) Immortalized multipotential mesenchymal cells and the hematopoietic microenvironment. J. Hematother. Stem Cell Res., 10: 125–140.

Ehrhard, P.B., Ganter, U., Stalder, A., Bauer, J. and Otten, U. (1993) Expression of functional trk protooncogene in human monocytes. Proc. Natl. Acad. Sci. USA, 90: 5423–5427.

Elfvin, L.G., Aldskogius, H. and Johansson, J. (1992a) Primary sensory afferents in the thymus of the guinea pig demonstrated with anterogradely transported horseradish peroxidase conjugates. Neurosci. Lett., 150: 35–38.

Elfvin, L.G., Aldskogius, H. and Johansson, J. (1992b) Splenic primary sensory afferents in the guinea pig demonstrated with anterogradely transported wheat-germ agglutinin conjugated to horseradish peroxidase. Cell Tissue Res., 269: 229–234.

Evans, C.M., Belmonte, K.E., Costello, R.W., Jacoby, D.B., Gleich, G.J. and Fryer, A.D. (2000) Substance P-induced airway hyperreactivity is mediated by neuronal M(2) receptor dysfunction. Am. J. Physiol. Lung Cell Mol. Physiol., 279: L477–L486.

Fernandez, S., Knopf, M.A. and McGillis, J.P. (2000) Calcitonin-gene related peptide (CGRP) inhibits interleukin-7-induced pre-B cell colony formation. J. Leukoc. Biol., 67: 669–676.

Fischer, A., McGregor, G.P., Saria, A., Philippin, B. and Kummer, W. (1996) Induction of tachykinin gene and peptide expression in guinea pig nodose primary afferent neurons by allergic airway inflammation. J. Clin. Invest., 98: 2284–2291.

Fox, A.J., Barnes, P.J., Urban, L. and Dray, A. (1993) An in vitro study of the properties of single vagal afferents innervating guinea-pig airways. J. Physiol., 469: 21–35.

Fox, F.E., Kubin, M., Cassin, M., Niu, Z., Hosoi, J., Torii, H., Granstein, R.D., Trinchieri, G. and Rook, A.H. (1997) Calcitonin gene-related peptide inhibits proliferation and antigen presentation by human peripheral blood mononuclear cells: effects on B7, interleukin 10, and interleukin 12. J. Invest. Dermatol., 108: 43–48.

Fryer, A.D. and Wills-Karp, M. (1991) Dysfunction of M2-muscarinic receptors in pulmonary parasympathetic nerves after antigen challenge. J. Appl. Physiol., 71: 2255–2261.

Fujimura, H., Altar, C.A., Chen, R., Nakamura, T., Nakahashi, T., Kambayashi, J., Sun, B. and Tandon, N.N. (2002) Brain-derived neurotrophic factor is stored in human platelets and released by agonist stimulation. Thromb. Haemost., 87: 728–734.

Garaci, E., Caroleo, M.C., Aloe, L., Aquaro, S., Piacentini, M., Costa, N., Amendola, A., Micera, A., Calio, R., Perno, C.F. and Levi-Montalcini, R. (1999) Nerve growth factor is an autocrine factor essential for the survival of macrophages infected with HIV. Proc. Natl. Acad. Sci. USA, 96: 14013–14018.

Garcia-Suarez, O., Blanco-Gelaz, M.A., Lopez, M.L., Germana, A., Cabo, R., Diaz-Esnal, B., Silos- Santiago, I., Ciriaco, E. and Vega, J.A. (2002) Massive lymphocyte apoptosis in the thymus of functionally deficient TrkB mice. J. Neuroimmunol., 129: 25–34.

Garcia-Suarez, O., Hannestad, J., Esteban, I., Martinez del Valle, M., Naves, F.J. and Vega, J.A. (1997) Neurotrophin receptor-like protein immunoreactivity in human lymph nodes. Anat. Rec., 249: 226–232.

Gergen, P.J. (2001) Understanding the economic burden of asthma. Allergy Clin. Immunol., 107: S445–S448.

Goedert, M., Stoeckel, K. and Otten, U. (1981) Biological importance of the retrograde axonal transport of nerve growth factor in sensory neurons. Proc. Natl. Acad. Sci. USA, 78: 5895–5898.

Graham, R.M., Friedman, M. and Hoyle, G.W. (2001) Sensory nerves promote ozone-induced lung inflammation in mice. Am. J. Respir. Crit. Care Med., 164: 307–313.

Hamada, A., Watanabe, N., Ohtomo, H. and Matsuda, H. (1996) Nerve growth factor enhances survival and cytotoxic activity of human eosinophils. Br. J. Haematol., 93: 299–302.

Hammarberg, H., Lidman, O., Lundberg, C., Eltayeb, S.Y., Gielen, A.W., Muhallab, S., Svenningsson, A., Linda, H., van Der Meide, P.H., Cullheim, S., Olsson, T. and Piehl, F. (2000) Neuroprotection by encephalomyelitis: rescue of mechanically injured neurons and neurotrophin production by CNS-infiltrating T and natural killer cells. J. Neurosci., 20: 5283–5291.

Hassankhani, A., Steinhelper, M.E., Soonpaa, M.H., Katz, E.B., Taylor, D.A., Andrade-Rozental, A., Factor, S.M., Steinberg, J.J., Field, L.J. and Federoff, H.J. (1995) Overexpression of NGF within the heart of transgenic mice causes hyperinnervation, cardiac enlargement, and hyperplasia of ectopic cells. Dev. Biol., 169: 309–321.

Heaney, L.G., Cross, L.J., Stanford, C.F. and Ennis, M. (1995) Substance P induces histamine release from human pulmonary mast cells. Clin. Exp. Allergy, 25: 179–186.

Hikawa, S., Kobayashi, H., Hikawa, N., Kusakabe, T., Hiruma, H., Takenaka, T., Tomita, T. and Kawakami, T. (2002) Expression of neurotrophins and their receptors in peripheral lung cells of mice. Histochem. Cell Biol., 118: 51–58.

Ho, W.Z., Lai, J.P., Zhu, X.H., Uvaydova, M. and Douglas, S.D. (1997) Human monocytes and macrophages express substance P and neurokinin-1 receptor. J. Immunol., 159: 5654–5660.

Hoey, T. and Grusby, M.J. (1999) STATs as mediators of cytokine-induced responses. Adv. Immunol., 71: 145–162.

Horwitz, R.J. and Busse, W.W. (1995) Inflammation and asthma. Clin. Chest Med., 16: 583–602.

Hoyle, G.W., Graham, R.M., Finkelstein, J.B., Nguyen, K.P., Gozal, D. and Friedman, M. (1998) Hyperinnervation of the airways in transgenic mice overexpressing nerve growth factor. Am. J. Respir. Cell Mol. Biol., 18: 149–157.

Hsiue, T.R., Garland, A., Ray, D.W., Hershenson, M.B., Leff, A.R. and Solway, J. (1992) Endogenous sensory neuropeptide release enhances nonspecific airway responsiveness in guinea pigs. Am. Rev. Respir. Dis., 146: 148–153.

Huang, E.J. and Reichardt, L.F. (2001) Neurotrophins: roles in neuronal development and function. Annu. Rev. Neurosci., 24: 677–736.

Hukkanen, M., Konttinen, Y.T., Rees, R.G., Gibson, S.J., Santavirta, S. and Polak, J.M. (1992) Innervation of bone from healthy and arthritic rats by substance P and calcitonin gene related peptide containing sensory fibers. J. Rheumatol., 19: 1252–1259.

Ip, N.Y., Ibanez, C.F., Nye, S.H., McClain, J., Jones, P.F., Gies, D.R., Belluscio, L., Le Beau, M.M., Espinosa 3d, R., Squinto, S.P., Persson, H. and Yancopoulos, G.D. (1992) Mammalian neurotrophin-4: structure, chromosomal localization, tissue distribution, and receptor specificity. Proc. Natl. cad. Sci. USA, 89: 3060–3064.

Study Report, ISAAC (1998) Worldwide variations in the prevalence of asthma symptoms: the International Study of Asthma and Allergies in Childhood (ISAAC). Eur. Respir. J., 12: 315–335.

Iwamoto, I., Nakagawa, N., Yamazaki, H., Kimura, A., Tomioka, H. and Yoshida, S. (1993) Mechanism for substance P-induced activation of human neutrophils and eosinophils. Regul. Pept., 46: 228–230.

Jacoby, D.B., Xiao, H.Q., Lee, N.H., Chan-Li, Y. and Fryer, A.D. (1998) Virus- and interferon-induced loss of inhibitory M2 muscarinic receptor function and gene expression in cultured airway parasympathetic neurons. J. Clin. Invest., 102: 242–248.

Janson, C., Chinn, S., Jarvis, D. and Burney, P. (1997) Physician-diagnosed asthma and drug utilization in the European Community Respiratory Health Survey. Eur Respir J., 10: 1795–1802.

Johansson, A., Lundborg, M., Skold, C.M., Lundahl, J., Tornling, G., Eklund, A. and Camner, P. (1997) Functional, morphological, and phenotypical differences between rat alveolar and interstitial macrophages. Am. J. Respir. Cell. Mol. Biol., 16: 582–588.

Johnson, D., Lanahan, A., Buck, C.R., Sehgal, A., Morgan, C., Mercer, E., Bothwell, M. and Chao, M. (1986) Expression and structure of the human NGF receptor. Cell, 47: 545–554.

Joos, G.F., Germonpre, P.R. and Pauwels, R.A. (2000) Neural mechanisms in asthma. Clin. Exp. Allergy, 30: 60–65.

Joos, G.F., Lefebvre, R.A., Bullock, G.R. and Pauwels, R.A. (1997) Role of 5-hydroxytryptamine and mast cells in the tachykinin-induced contraction of rat trachea in vitro. Eur. J. Pharmacol., 338: 259–268.

Kaltreider, H.B., Ichikawa, S. and Byrd, P.K. (1997) Upregulation of neuropeptides and neuropeptide receptors in a murine model of immune inflammation in lung parenchyma. Am. J. Respir. Cell. Mol. Biol., 16: 133–144.

Kanbe, N., Kurosawa, M., Miyachi, Y., Kanbe, M., Saitoh, H. and Matsuda, H. (2000) Nerve growth factor prevents apoptosis of cord blood-derived human cultured mast cells synergistically with stem cell factor. Clin. Exp. Allergy, 30: 1113–1120.

Kannan, Y., Matsuda, H., Ushio, H., Kawamoto, K. and Shimada, Y. (1993) Murine granulocyte-macrophage and mast cell colony formation promoted by nerve growth factor. Int. Arch. Allergy. Immunol., 102: 362–367.

Kaplan, D.R., Hempstead, B.L., Martin-Zanca, D., Chao, M.V. and Parada, L.F. (1991) The trk proto-oncogene product: a signal transducing receptor for nerve growth factor. Science, 252: 554–558.

Kassel, O., de Blay, F., Duvernelle, C., Olgart, C., Israel-Biet, D., Krieger, P., Moreau, L., Muller, C., Pauli, G. and Frossard, N. (2001) Local increase in the number of mast cells and expression of nerve growth factor in the bronchus of asthmatic patients after repeated inhalation of allergen at low-dose. Clin. Exp. Allergy, 31: 1432–1440.

Kawamura, N., Tamura, H., Obana, S., Wenner, M., Ishikawa, T., Nakata, A. and Yamamoto, H. (1998) Differential effects of neuropeptides on cytokine production by mouse helper T cell subsets. Neuroimmunomodulation, 5: 9–15.

Kerschensteiner, M., Gallmeier, E., Behrens, L., Leal, V.V., Misgeld, T., Klinkert, W.E., Kolbeck, R., Hoppe, E., Oropeza-Wekerle, R.L., Bartke, I., Stadelmann, C., Lassmann, H., Wekerle, H., Hohlfeld, R. (1999) Activated human T cells, B cells, and monocytes produce brain-derived neurotrophic factor in vitro and in inflammatory brain lesions: a neuroprotective role of inflammation? J. Exp. Med., 189: 865–870.

Kerzel, S., Path, G., Quarcoo, D., Raap, U., Kronenberg, D.A., Dinh, Q.T., Fischer, A., Nockher, A.W., Braun, A. and Renz, H. (2003) Pan-neurotrophin receptor p75 contibutes to airway hyperresponsiveness and airway inflammation in a murine model of allergic bronchial asthma. Am. J. Respir. Cell. Mol. Biol., 28: 170–178.

Kimata, H., Yoshida, A., Ishioka, C., Kusunoki, T., Hosoi, S. and Mikawa, H. (1991) Nerve growth factor specifically induces human IgG4 production. Eur. J. Immunol., 21: 137–141.

Kimata, H., Yoshida, A., Ishioka, C. and Mikawa, H. (1992) Differential effect of vasoactive intestinal peptide, somatostatin, and substance P on human IgE and IgG subclass production. Cell. Immunol., 144: 429–442.

Kittur, S.D., Song, L., Endo, H. and Adler, W.H. (1992) Nerve growth factor receptor gene expression in human peripheral blood lymphocytes in aging. J. Neurosci. Res., 32: 444–448.

Klein, R., Jing, S.Q., Nanduri, V., O'Rourke, E. and Barbacid, M. (1991a) The trk proto-oncogene encodes a receptor for nerve growth factor. Cell, 65: 189–197.

Klein, R., Nanduri, V., Jing, S. A., Lamballe, F., Tapley, P., Bryant, S., Cordon-Cardo, C., Jones, K. R., . Reichardt, L.F. and Barbacid, M. (1991b) The trkB tyrosine protein kinase is a receptor for brain-derived neurotrophic factor and neurotrophin-3. Cell, 66: 395–403.

Kobayashi, H., Gleich, G.J., Butterfield, J.H. and Kita, H. (2002) Human eosinophils produce neurotrophins and secrete nerve growth factor on immunologic stimuli. Blood, 99: 2214–2220.

la Sala, A., Corinti, S., Federici, M., Saragovi, H.U. and Girolomoni, G. (2000) Ligand activation of nerve growth factor receptor TrkA protects monocytes from apoptosis. J. Leukoc. Biol., 68: 104–110.

Labouyrie, E., Dubus, P., Groppi, A., Mahon, F.X., Ferrer, J., Parrens, M., Reiffers, J., de Mascarel, A. and Merlio, J.P. (1999) Expression of neurotrophins and their receptors in human bone marrow. Am. J. Pathol., 154: 405–415.

Labouyrie, E., Parrens, M., de Mascarel, A., Bloch, B. and Merlio, J.P. (1997) Distribution of NGF receptors in normal and pathologic human lymphoid tissues. J. Neuroimmunol., 77: 161–173.

Lamb, J.P. and Sparrow, M.P. (2002) Three-dimensional mapping of sensory innervation with substance p in porcine bronchial mucosa: comparison with human airways. Am. J. Respir. Crit. Care Med., 166: 1269–1281.

Lamballe, F., Tapley, P. and Barbacid, M. (1993) TrkC encodes multiple neurotrophin-3 receptors with distinct biological properties and substrate specificities. EMBO J., 12: 3083–3094.

Lambiase, A., Bracci-Laudiero, L., Bonini, S., Bonini, S., Starace, G., D'Elios, M.M., De Carli, M. and Aloe, L. (1997) Human CD4+ T cell clones produce and release nerve growth factor and express high-affinity nerve growth factor receptors. J. Allergy Clin. Immunol., 100: 408–414.

Larsen, G.L., White, C.W., Takeda, K., Loader, J.E., Nguyen, D.D., Joetham, A., Groner, Y. and Gelfand, E.W. (2000) Mice that overexpress Cu/Zn superoxide dismutase are resistant to allergen-induced changes in airway control. Am. J. Physiol. Lung Cell Mol. Physiol., 279: L350–L359.

Laurenzi, M.A., Beccari, T., Stenke, L., Sjolinder, M., Stinchi, S. and Lindgren, J.A. (1998) Expression of mRNA encoding neurotrophins and neurotrophin receptors in human granulocytes and bone marrow cells-enhanced neurotrophin-4 expression induced by LTB4. J. Leukoc. Biol., 64: 228–234.

Levine, J.D., Clark, R., Devor, M., Helms, C., Moskowitz, M.A. and Basbaum, A.I. (1984) Intraneuronal substance P contributes to the severity of experimental arthritis. Science, 226: 547–549.

Levite, M. (1998a) Neuropeptides, by direct interaction with T cells, induce cytokine secretion and break the commitment to a distinct T helper phenotype. Proc. Natl. Acad. Sci. USA, 95: 12544–12549.

Levite, M., Cahalon, L., Hershkoviz, R., Steinman, L. and Lider, O. (1998b) Neuropeptides, via specific receptors, regulate T cell adhesion to fibronectin. J. Immunol., 160: 993–1000.

Lewin, G.R. and Mendell, L.M. (1993a) Nerve growth factor and nociception. Trends Neurosci., 16: 353–359.

Lewin, G.R., Ritter, A.M. and Mendell, L.M. (1993b) Nerve growth factor-induced hyperalgesia in the neonatal and adult rat. J. Neurosci., 13: 2136–2148.

Lewin, G.R., Rueff, A. and Mendell, L.M. (1994) Peripheral and central mechanisms of NGF-induced hyperalgesia. Eur. J. Neurosci., 6: 1903–1912.

Lilly, C.M., Bai, T.R., Shore, S.A., Hall, A.E. and Drazen, J.M. (1995) Neuropeptide content of lungs from asthmatic and nonasthmatic patients. Am. J. Respir. Crit. Care Med., 151: 548–553.

Lindsay, R.M. and Harmar, A.J. (1989) Nerve growth factor regulates expression of neuropeptide genes in adult sensory neurones. Nature, 337: 362–364.

Lommatzsch, M., Braun, A., Mannsfeldt, A., Botchkarev, V.A., Botchkareva, N.V., Paus, R., Fischer, A., Lewin, G.R. and Renz, H. (1999) Abundant production of brain-derived neurotrophic factor by adult visceral epithelia. Implications for paracrine and target-derived neurotrophic functions. Am. J. Pathol., 155: 1183–1193.

Lundberg, J.M., Lundblad, L., Martling, C.R., Saria, A., Stjarne, P. and Anggard, A. (1987) Coexistence of multiple peptides and classic transmitters in airway neurons: functional and pathophysiologic aspects. Am. Rev. Respir. Dis., 136: S16–S22.

Maroder, M., Bellavia, D., Meco, D., Napolitano, M., Stigliano, A., Alesse, E., Vacca, A., Giannini, G., Frati, L., Gulino, A. and Screpanti, I. (1996) Expression of trKB neurotrophin receptor during T cell development. Role of brain derived neurotrophic factor in immature thymocyte survival. J. Immunol., 157: 2864–2872.

Marshall, J.S., Gomi, K., Blennerhassett, M.G. and Bienenstock, J. (1999) Nerve growth factor modifies the expression of inflammatory cytokines by mast cells via a prostanoid-dependent mechanism. J. Immunol., 162: 4271–4276.

Matsuda, H., Coughlin, M.D., Bienenstock, J. and Denburg, J.A. (1988a) Nerve growth factor promotes human hemopoietic colony growth and differentiation. Proc. Natl. Acad. Sci. USA, 85: 6508–6512.

Matsuda, H., Kannan, Y., Ushio, H., Kiso, Y., Kanemoto, T., Suzuki, H. and Kitamura, Y. (1991) Nerve growth factor induces development of connective tissue-type mast cells in vitro from murine bone marrow cells. J. Exp. Med., 174: 7–14.

Matsuda, H., Switzer, J., Coughlin, M.D., Bienenstock, J. and Denburg, J.A. (1988b) Human basophilic cell differentiation promoted by 2.5S nerve growth factor. Int. Arch. Allergy Appl. Immunol., 86: 453–457.

Mazurek, N., Weskamp, G., Erne, P. and Otten, U. (1986) Nerve growth factor induces mast cell degranulation without changing intracellular calcium levels. FEBS Lett., 198: 315–320.

McGillis, J.P., Humphreys, S., Rangnekar, V. and Ciallella, J. (1993) Modulation of B lymphocyte differentiation by calcitonin gene-related peptide (CGRP). I. Characterization of high-affinity CGRP receptors on murine 70Z/3 cells. Cell. Immunol., 150: 391–404.

McGillis, J.P., Humphreys, S. and Reid, S. (1991) Characterization of functional calcitonin gene-related peptide receptors on rat lymphocytes. J. Immunol., 147: 3482–3489.

McMahon, S.B., Bennett, D.L., Priestley, J.V. and Shelton, D.L. (1995) The biological effects of endogenous nerve growth factor on adult sensory neurons revealed by a trkA-IgG fusion molecule. Nat. Med., 1: 774–780.

Mekori, Y.A. and Metcalfe, D. (1999) Mast cell-T cell interactions. J. Allergy Clin. Immunol., 104: 517–523.

Metcalfe, D.D., Baram, D. and Mekori, Y.A. (1997) Mast cells. Physiol. Rev., 77: 1033–1079.

Michoud, M.C., Jeanneret-Grosjean, A., Cohen, A. and Amyot, R. (1988) Reflex decrease of histamine-induced bronchoconstriction after laryngeal stimulation in asthmatic patients. Am. Rev. Respir. Dis., 138: 1548–1552.

Miura, K., Saini, S.S., Gauvreau, G. and MacGlashan, D.W., Jr. (2001) Differences in functional consequences and signal transduction induced by IL-3, IL-5, and nerve growth factor in human basophils. J. Immunol., 167: 2282–2291.

Mullins, M.W., Ciallella, J., Rangnekar, V. and McGillis, J.P. (1993) Characterization of a calcitonin gene related peptide (CGRP) receptor on mouse bone marrow cells. Regul. Pept., 49: 65–72.

Mutoh, T., Joad, J.P. and Bonham, A.C. (2000) Chronic passive cigarette smoke exposure augments bronchopulmonary C-fibre inputs to nucleus tractus solitarii neurones and reflex output in young guinea-pigs. J. Physiol., 523: 223–233.

Nakamuta, H., Fukuda, Y., Koida, M., Fujii, N., Otaka, A., Funakoshi, S., Yajima, H., Mitsuyasu, N. and Orlowski, R.C. (1986) Binding sites of calcitonin gene-related peptide (CGRP): abundant occurrence in visceral organs. Jpn. J. Pharmacol., 42: 175–180.

Nieber, K., Baumgarten, C.R., Rathsack, R., Furkert, J., Oehme, P. and Kunkel, G. (1992) Substance P and beta-endorphin-like immunoreactivity in lavage fluids of subjects with and without allergic asthma. J. Allergy Clin. Immunol., 90: 646–652.

Nong, Y.H., Titus, R.G., Ribeiro, J.M. and Remold, HG. (1989) Peptides encoded by the calcitonin gene inhibit macrophage function. J. Immunol., 143: 45–49.

Ollerenshaw, S., Jarvis, D., Woolcock, A., Sullivan, C. and Scheibner, T. (1989) Absence of immunoreactive vasoactive intestinal polypeptide in tissue from the lungs of patients with asthma. N. Engl. J. Med., 320: 1244–1248.

Otten, U., Ehrhard, P. and Peck, R. (1989) Nerve growth factor induces growth and differentiation of human B lymphocytes. Proc. Natl. Acad. Sci. USA, 86: 10059–10063.

Path, G., Braun, A., Meents, N., Kerzel, S., Quarcoo, D., Raap, U., Hoyle, G.W., Nockher, W.A. and Renz, H. (2002) Augmentation of allergic early-phase reaction by nerve growth factor. Am. J. Respir. Crit. Care Med., 166: 818–826.

Pauwels, R. (1989) The relationship between airway inflammation and bronchial hyperresponsiveness. Clin. Exp. Allergy, 19: 395–398.

Payan, D.G. (1992) The role of neuropeptides in inflammation. In: J.I. Gallin, I.M. Goldstein and R. Snyderman (Eds.), Inflammation: Basic Principles and Clinical Correlates, Raven Press, New York, pp. 177–192.

Pearlman, D.S. (1999) Pathophysiology of the inflammatory response. J. Allergy Clin. Immunol., 104: S132–S137.

Pedersen, K.E., Meeker, S.N., Riccio, M.M. and Undem, B.J. (1998) Selective stimulation of jugular ganglion afferent neurons in guinea pig airways by hypertonic saline. J. Appl. Physiol., 84: 499–506.

Pezzati, P., Stanisz, A.M., Marshall, J.S., Bienenstock, J. and Stead, R.H. (1992) Expression of nerve growth factor receptor immunoreactivity on follicular dendritic cells from human mucosa associated lymphoid tissues. Immunology, 76: 485–490.

Popper, P., Mantyh, C.R., Vigna, S.R., Maggio, J.E. and Mantyh, P.W. (1988) The localization of sensory nerve fibers and receptor binding sites for sensory neuropeptides in canine mesenteric lymph nodes. Peptides, 9: 257–267.

Prokhorova, S., Lavnikova, N. and Laskin, D.L. (1994) Functional characterization of interstitial macrophages and subpopulations of alveolar macrophages from rat lung. J. Leukoc. Biol., 55: 141–146.

Radeke, M.J., Misko, T.P., Hsu, C., Herzenberg, L.A. and Shooter, E.M. (1987) Gene transfer and molecular cloning of the rat nerve growth factor receptor. Nature, 325: 593–597.

Raud, J., Lundeberg, T., Brodda-Jansen, G., Theodorsson, E. and Hedqvist, P. (1991) Potent anti-inflammatory action of calcitonin gene-related peptide. Biochem. Biophys. Res. Commun., 180: 1429–1435.

Reinshagen, M., Flamig, G., Ernst, S., Geerling, I., Wong, H., Walsh, J.H., Eysselein, V.E. and Adler, G. (1998) Calcitonin gene-related peptide mediates the protective effect of sensory nerves in a model of colonic injury. J. Pharmacol. Exp. Ther., 286: 657–661.

Ricci, A., Greco, S., Mariotta, S., Felici, L., Amenta, F. and Bronzetti, E. (2000) Neurotrophin and neurotrophin receptor

366

expression in alveolar macrophages: an immunocytochemical study. Growth Factors, 18: 193–202.

Richard, A., McColl, S.R. and Pelletier, G. (1992) Interleukin-4 and nerve growth factor can act as cofactors for interleukin-3-induced histamine production in human umbilical cord blood cells in serum-free culture. Br. J. Haematol., 81: 6–11.

Rueff, A. and Mendell, L.M. (1996) Nerve growth factor NT-5 induce increased thermal sensitivity of cutaneous nociceptors in vitro. J. Neurophysiol., 76: 3593–3596.

Schaffer, M., Beiter, T., Becker, H.D. and Hunt, T.K. (1998) Neuropeptides: mediators of inflammation and tissue repair? Arch. Surg., 133: 1107–1116.

Sebring, R.J. and Lehnert, B.E. (1992) Morphometric comparisons of rat alveolar macrophages, pulmonary interstitial macrophages, and blood monocytes. Exp. Lung Res., 18: 479–496.

Severini, C., Improta, G., Falconieri-Erspamer, G., Salvadori, S. and Erspamer, V. (2002) The tachykinin peptide family. Pharmacol. Rev., 54: 285–322.

Shibayama, E. and Koizumi, H. (1996) Cellular localization of the Trk neurotrophin receptor family in human non-neuronal tissues. Am. J. Pathol., 148: 1807–1818.

Shu, X.Q., Llinas, A. and Mendell, L.M. (1999) Effects of trkB and trkC neurotrophin receptor agonists on thermal nociception: a behavioral and electrophysiological study. Pain, 80: 463–470.

Sin, A.Z., Roche, E.M., Togias, A., Lichtenstein, L.M. and Schroeder, J.T. (2001) Nerve growth factor or IL-3 induces more IL-13 production from basophils of allergic subjects than from basophils of nonallergic subjects. J. Allergy Clin. Immunol., 108: 387–393.

Solomon, A., Aloe, L., Pe'er, J., Frucht-Pery, J., Bonini, S., Bonini, S. and Levi-Schaffer, F. (1998) Nerve growth factor is preformed in and activates human peripheral blood eosinophils. J. Allergy Clin. Immunol., 102: 454–460.

Sterk, P.J. (1995) The place of airway hyperresponsiveness in the asthma phenotype. Clin. Exp. Allergy, 25: 8–11.

Szallasi, A. and Blumberg, P.M. (1999) Vanilloid (Capsaicin) receptors and mechanisms. Pharmacol. Rev., 51: 159–212.

Tam, S.Y., Tsai, M., Yamaguchi, M., Yano, K., Butterfield, J.H. and Galli, S.J. (1997) Expression of functional TrkA receptor tyrosine kinase in the HMC-1 human mast cell line and in human mast cells. Blood, 90: 1807–1820.

Tanabe, T., Otani, H., Zeng, X.T., Mishima, K., Ogawa, R. and Inagaki, C. (1996) Inhibitory effects of calcitonin gene-related peptide on substance-P-induced superoxide production in human neutrophils. Eur. J. Pharmacol., 314: 175–183.

Tokuoka, S., Takahashi, Y., Masuda, T., Tanaka, H., Furukawa, S. and Nagai, H. (2001) Disruption of antigen-induced airway inflammation and airway hyper-responsiveness in low affinity neurotrophin receptor p75 gene deficient mice. Br. J. Pharmacol., 134: 1580–1586.

Tomaki, M., Ichinose, M., Miura, M., Hirayama, Y., Yamauchi, H., Nakajima, N. and Shirato, K. (1995) Elevated substance P content in induced sputum from patients with asthma and patients with chronic bronchitis. Am. J. Respir. Crit. Care Med., 151: 613–617.

Torcia, M., Bracci-Laudiero, L., Lucibello, M., Nencioni, L., Labardi, D., Rubartelli, A., Cozzolino, F., Aloe, L. and Garaci, E. (1996) Nerve growth factor is an autocrine survival factor for memory B lymphocytes. Cell, 85: 345–356.

Torcia, M., De Chiara, G., Nencioni, L., Ammendola, S., Labardi, D., Lucibello, M., Rosini, P., Marlier, L.N., Bonini, P., Dello Sbarba, P., Palamara, A.T., Zambrano, N., Russo, T., Garaci, E. and Cozzolino, F. (2001) Nerve growth factor inhibits apoptosis in memory B lymphocytes via inactivation of p38 MAPK, prevention of Bcl-2 phosphorylation, and cytochrome c release. J. Biol. Chem., 276: 39027–39036.

Tracey, K.J. (2002) The inflammatory reflex. Nature, 420: 853–859.

Turner, H. and Kinet, J.P. (1999) Signaling through the high-affinity IgE receptor Fc epsilonRI. Nature, 402: B24–B30.

Uddman, R., Hakanson, R., Luts, A. and Sundler, F. (1997) Distribution of neuropeptides in airways. In: P.J. Barnes (Ed.), Autonomic Innervation of the Respiratory Tract, Harwood Academic, London, pp. 21–37.

Undem, B.J., Hunter, D.D., Liu, M., Haak-Frendscho, M., Oakragly, A. and Fischer, A. (1999) Allergen-induced sensory neuroplasticity in airways. Int. Arch. Allergy Immunol., 118: 150–153.

van der Velden, V.H. and Hulsmann, A.R. (1999) Autonomic innervation of human airways: structure, function, and pathophysiology in asthma. Neuroimmunomodulation, 6: 145–159.

van Hagen, P.M., Krenning, E.P. and Kwekkeboom, D.J. (1994) Somatostatin and the immune and haematopoetic system; a review. Eur. J. Clin. Invest., 24: 91–99.

van Schoor, J., Joos, G.F. and Pauwels, R.A. (2000) Indirect bronchial hyperresponsiveness in asthma: mechanisms, pharmacology and implications for clinical research. Eu.r Respir. J., 16: 514–533.

Verge, V.M., Richardson, P.M., Wiesenfeld-Hallin, Z. and Hokfelt, T. (1995) Differential influence of nerve growth factor on neuropeptide expression in vivo: a novel role in peptide suppression in adult sensory neurons. J. Neurosci., 15: 2081–2096.

Virchow, J.C., Julius, P., Lommatzsch, M., Luttmann, W., Renz, H. and Braun, A. (1998) Neurotrophins are increased in bronchoalveolar lavage fluid after segmental allergen provocation. Am. J. Respir. Crit. Care Med., 158: 2002–2005.

Wagner, U., Fehmann, H.C., Bredenbroker, D., Yu, F., Barth, P.J. and von Wichert, P. (1995) Galanin and somatostatin inhibition of substance P-induced airway mucus secretion in the rat. Neuropeptides, 28: 59–64.

Wang, H., Xing, L., Li, W., Hou, L., Guo, J. and Wang, X. (2002) Production and secretion of calcitonin gene-related peptide from human lymphocytes. J. Neuroimmunol., 130: 155–162.

Wasserman, S.I. (1984) The human lung mast cell. Environ Health Perspect., 55: 259–269.

Watanabe, Y., Hashizume, M., Kataoka, S., Hamaguchi, E., Morimoto, N., Tsuru, S., Katoh, S., Miyake, K., Matsushima, K., Tominaga, M., Kurashige, T., Fujimoto, S., Kincade, P.W. and Tominaga, A. (2001) Differentiation stages of eosinophils characterized by hyaluronic acid binding via CD44 and responsiveness to stimuli. DNA Cell Biol., 20: 189–202.

Weiss, K.B., Sullivan, S.D. and Lyttle, C.S. (2000) Trends in the cost of illness for asthma in the United States, 1985–1994. J. Allergy Clin. Immunol., 106: 493–499.

Welker, P., Grabbe, J., Gibbs, B., Zuberbier, T. and Henz, B.M. (2000) Nerve growth factor-beta induces mast-cell marker expression during in vitro culture of human umbilical cord blood cells. Immunology, 99: 418–426.

Weskamp, G. and Otten, U. (1987) An enzyme-linked immunoassay for nerve growth factor (NGF): a tool for studying regulatory mechanisms involved in NGF production in brain and in peripheral tissues. J. Neurochem., 48: 1779–1786.

Williams, C.M. and Galli, S.J. (2000) The diverse potential effector and immunoregulatory roles of mast cells in allergic disease. J. Allergy Clin. Immunol., 105: 847–859.

Woolf, C.J., Safieh-Garabedian, B., Ma, Q.P., Crilly, P. and Winter, J. (1994) Nerve growth factor contributes to the generation of inflammatory sensory hypersensitivity. Neuroscience, 62: 327–331.

Xiang, Z. and Nilsson, G. (2000) IgE receptor-mediated release of nerve growth factor by mast cells. Clin. Exp. Allergy, 30: 1379–1386.

Xing, L., Guo, J. and Wang, X. (2000) Induction and expression of beta-calcitonin gene-related peptide in rat T lymphocytes and its significance. J. Immunol., 165: 4359–4366.

Progress in Brain Research, Vol. 146
ISSN 0079-6123

CHAPTER 23

Nerve growth factor and wound healing

Keiko Kawamoto and Hiroshi Matsuda*

*Laboratory of Clinical Immunology, Department of Veterinary Clinic, Faculty of Agriculture, Tokyo University of Agriculture
and Technology, Tokyo, Japan*

Abstract: The wound healing process following tissue injury consists of a highly regulated sequence of events. Besides many biological activities on both neuronal and nonneuronal cells, nerve growth factor (NGF) has been proposed as an important component of wound healing and tissue repair process in vivo and in vitro. For example, NGF accelerates the rate of wound healing both in normal mice and healing-impaired diabetic mice, and has a potent pharmacological effect in the treatment for ulcer of the skin and cornea in humans. This review summarizes the evidence for the role of NGF in wound healing and tissue repair, and introduces its clinical utility as a therapeutic agent for various diseases.

Keywords: NGF; wound healing; time repair; regeneration; mast cells; therapeutic potential

Introduction

Wound healing involves a complex process including induction of acute inflammation subsequent to injury, followed by parenchymal and mesenchymal cell proliferation, migration and activation with production and deposition of extracellular matrix. The healing process is highly regulated via sophisticated cytokine networks, and is characterized by three sequential events: inflammation, tissue formation and tissue remodeling (Clark, 1996; Martin, 1997). Earlier study (Huston et al., 1979) has shown that removal of the submandibular gland of mice retards the rate of contraction of experimentally-induced wounds. NGF abundantly exists in murine submandibular glands, and is secreted in high concentrations in murine saliva. Li et al. (1980) demonstrated the significant effect of topically-applied NGF upon the rate of wound contraction in sialoadenectomized animals, suggesting that licking wounds may introduce NGF as well as other

healing accelerators such as fibroblast growth factor (FGF) and epidermal cell growth factor (EGF) to promote cutaneous wound healing. Therefore, one of the physiological roles of NGF in saliva may be to promote wound healing in the licking behavior. Besides being a well-characterized neurotrophic factor, recent accumulating evidence has demonstrated that NGF possesses broad biological activities on nonneuronal cells including inflammatory cells (neutrophils, macrophages and mast cells) and other cell populations (keratinocytes, fibroblasts and endothelial cells) that are important cellular components in the wound healing process. The purpose of this review is to discuss roles of NGF and its mechanism of action in the would healing and tissue repair processes based on animal studies, and to comment on its potential for therapeutic agents to improve healing in various diseases.

Biology of wound healing

Time course of wound healing

The time course of wound healing of clean edges such as a surgical cut is summarized as follows. A fibrin clot

---

*Correspondence to: H. Matsuda, Laboratory of Clinical
Immunology, Department of Veterinary Clinic, Faculty of
Agriculture, Tokyo University of Agriculture and Technology,
Tokyo, Japan. E-mail: hiro@cc.tuat.ac.jp

DOI: 10.1016/S0079-6123(03)46023-8

that serves as a temporal shield closes the wound quickly. Neutrophil recruitment at the site peaks within the first 24 h after the wound, followed by epithelial cell migration along the wound edge within 48 h. Macrophages and granulation tissue predominate by day 3, with collagen fiber formation in the affected site. Granulation tissue and angiogenesis are evident by 3–5 days. Collagen bridges the wound by the end of the first week. Functional epidermis is formed, although dermal structures are permanently disrupted. After one week, wound strength is only 10% of normal, but is increased to 70–80% by 2–3 months. The rate of healing is influenced by many factors such as location of the injury, the type of tissue, infection, use of steroids and poor blood circulation.

Cellular events during wound healing process

Healing of wounds is a chain of processes necessary for not only the removal of invaded pathogens from lesions but also for complete or partial remodeling of injured tissues. In general, wound healing consists of three major interrelated and overlapping stages: inflammation, tissue formation (granulation, angiogenesis and reepithelization) and tissue remodeling (Fig. 1). The healing process is controlled by mediators derived from residential cells and inflammatory infiltrating cells at damaged sites.

Inflammation

Most injuries damage blood vessels, and coagulation is a rapid response leading to aggregation of circulating platelets within minutes and clot formation to protect from excessive blood loss. Activated platelets release growth factors including transforming growth factor-beta (TGF-β), platelet-derived growth factors BB homodimer (PDGF-BB) and vascular endothelial cell growth factor (VEGF), those of which cause vasodilation and increased permeability of blood vessels, and induce inflammatory cells to recruit at injured sites (Bennett and Schultz, 1993; Steed, 1997). For example, TGF-β1 elicits rapid chemotaxis of neutrophils and macrophages to the traumatic sites (Wahl, 1999). These cytokines and growth factors usually work through autocrine and/ or paracrine effects. Cytokines released from

damaged endothelial cells and parenchymal cells also stimulate inflammation and migration of leukocytes including neutrophils and macrophages. Macrophages are essential components in wound healing process because their depletion severely impairs the healing of wounds (Leibovich and Ross, 1975). Neutrophils and macrophages act in concert to phagocyte cellular debris and infectious invading microorganisms, and to produce cytokines, chemokines and growth factors that amplify and perpetuate the inflammatory responses.

Tissue formation

Fibroblasts, endothelial cells and epithelial cells migrate into the wound, and proliferate at damaged sites. The migration of residential cells depends upon various events such as cytokines and growth factors released from activated macrophages, and loss of contact in adjacent cells. Within several hours after injury, epithelial cells begin to proliferate, migrate and cover the exposed area. Reepithelization is an important event for optimal wound healing because of not only formation of a temporal barrier but also for its role in wound contraction. Various members of the matrix metalloproteinase (MMP) family are also upregulated by epithelial cells at the wound margin. The MMP not only degrade matrix components to allow the cell mobility but also function as regulatory molecules to process inactive forms of cytokines, adhesion molecules and matrix into biologically active forms by enzymatic cleavage (Ravanti and Kahari, 2000). The densely populated area with macrophages, fibroblasts and newly-formed capillaries (angiogenesis) within a loose extracellular matrix formed below the epithelium is called a granulation tissue. Fibroblasts begin to synthesize and deposit collagen that acts as a scaffold (framework) for migration and further fibroblast proliferation. Some fibroblasts modulate into myofibroblasts, which contract the wound and increase its tensile strength.

Tissue remodeling

During wound healing, there is, in turn, rapid synthesis and degradation of connective tissue components such as collagen and proteoglycans (Laurent, 1988).

Although more collagen is present in a wound at one week than the later stage, the later wound has greater tensile strength as remodeling of collagen proceeds. Degradation of these matrix molecules is driven by serine proteases and MMP secreted by neutrophils, injured epithelial cells and keratinocytes under the control of the cytokine network. Tissue inhibitor of MMP (TIMP) counteracts biological actions of MMP, and the regulation of these enzymes is important in the process of remodeling and wound healing.

From many studies using specific gene-targeted or -transferred mice, the cytokine network is believed to play a central role to orchestrate the healing process (Scheid et al., 2000). Consequently, cytokines and growth factors have become targets for therapeutic application to modulate the wound healing process.

Healing effects of NGF on various wounds and diseases

Peripheral nerve wound healing

The biology of nerve injury and repair is a complex process that depends on the interaction of the nerve cell body, the axonal fiber and the surrounding microenvironment (Ebadi et al., 1997). One of the most striking features of neurons in the mature peripheral nervous system (PNS) is their ability to survive and to regenerate their axons following axonal injury. Molecular changes in the distal stump involve in upregulation of neurotrophins and their corresponding receptors as well as upregulation of neural cell adhesion molecules and other cytokines. Since first described by Levi-Montalcini and Hamburger in 1953, NGF has been well known to promote the neural differentiation and survival of both basal forebrain cholinergic neurons and peripheral sensory neurons (Korsching, 1986; Levi-Montalcini, 1987). Responsiveness of target cells to NGF is governed by the expression of two classes of cell surface receptor, low-affinity neurotrophin receptor ($p75^{NTR}$) and high-affinity TrkA receptor (TrkA) (Yanker and Shooter, 1982; Sofroniew et al., 2001).

Within hours after axonal damage, mRNA levels of NGF and its receptors temporally increase (Sebert and Shooter, 1993), and show second peak of expression at 2–3 days after injury. Schwann cells

that play a key role in nerve regeneration in PNS produce NGF after nerve injury, which is induced by interleukin-1 (IL-1) released from distal end of transected nerves (Taniuchi et al., 1986; Clarke and Richardson, 1994; Liu et al., 1995; Frostick et al., 1998). Besides their phagocytic role, macrophages recruited to the damaged site release NGF, probably in response to local IL-1 and/or tumor necrosis factor-alpha (TNF-α). In addition, during the process of nerve regeneration, Schwann cells respond to loss of axons by demyelination, dedifferentiation, proliferation and finally aline in tubes to become a guide for axonal extension. NGF signaling via $p75^{NTR}$ induces differentiation and ceramide-mediated apoptosis in Schwann cells cultured from degenerating nerves (Hirata et al., 2001). NGF also induces sphingomyelin hydrolysis, which is correlated with the expression levels of $p75^{NTR}$ (Dobrowsky and Carter, 1998; Hirata et al., 2001). These findings suggest that NGF contributes to both phenotypic regulation and elimination of dedifferentiated Schwann cells, while supporting survival and regeneration of peripheral axons during nerve repair.

It is thought that VEGF, a most potent growth factor for angiogenesis that is a critical event in tissue formation of wound healing process, also plays an important role for regeneration of nervous tissue (Sondell et al., 1999; Hobson et al., 2000; Liekens et al., 2001; Meirer et al., 2001). Some stimuli induce neural VEGF expression, including growth factors, cytokines, hormones, hypoxia and nitric oxide (NO). Administration of NGF into newborn rats enlarges a size of dorsal root ganglia and increases a vessel density, which is associated with increased expression of VEGF and NO synthase. VEGF expression is also enhanced in DRG of NGF-treated healthy rats after sciatic nerve crush. In contrast, markedly decreased expression of VEGF is observed in NGF-treated animals with streptozotozin-induced type I diabetes, indicating NGF may stimulate VEGF production in normal controls (Samii et al., 1999). Impaired response to NGF in diabetic animals may be due to the abnormal expression of its receptors, $p75^{NTR}$ and TrkA (Tomlinson et al., 1997). In rodent models of diabetes, the expression levels of NGF receptors as well as NGF are decreased in peripheral nerves, which may reflect low peripheral NGF levels, which in consequence leads to the apparent deprivation of neuronal NGF in diabetic rats. To overcome the problem with unsuccessful NGF therapy for diabetic neuropathies, it may be important to normalize or increase the expression levels of NGF receptors leading to improvement in retrograde axonal transport of NGF and NGF-dependent support for peripheral sensory neurons.

Regeneration of neurons after traumatic injury by crush, axotomy, ischemia or inflammation is important not only in neuronal wound healing but also in the healing process of other tissues. Clinically, it is well known that areas which are denervated or poorly innervated heal insufficiently. For example, patients with paraplegic or quadriplegic disorders have a reduced capacity to heal wounds, as well as patients with peripheral neuropathies such as diabetes and leprosy. Experimental findings strongly support the speculation that an intact innervation is important for efficient healing of skin, epithelium, tooth pulp, bone and ligament.

Cutaneous wound healing

Although NGF is a neurotrophic factor essential for the development and function of central and peripheral neurons, recent findings have shown that NGF regulates immune and inflammatory responses in nonneuronal tissues. Levels of NGF are increased in the epidermis associated with skin innervation (Wyatt et al., 1990; Bothwell, 1997). The expression of NGF and its receptors is found in the skin (Shibayama and Koizumi, 1996; Bull et al., 1998). These findings suggest that NGF as well as other neurotrophins have an important function in the cutaneous biology (reviewed by Johansson and Liang, 1999). We have been investigating novel roles of NGF in the processes of inflammation and tissue repair. For example, NGF induces granulocyte differentiation from human and murine hematopoietic stem cells (Matsuda et al., 1988, 1991; Kannan et al., 1993). NGF suppresses apoptosis of both neutrophils and mast cells, and enchances functional properties of various inflammatory cells including neutrophils, macrophages, and mast cells (Kannan et al., 1991, 1992; Kawamoto et al., 1995; Susaki et al., 1996). In the skin, NGF is spontaneously produced by many types of residential cells such as fibroblasts, keratinocytes and mast cells, suggesting

possible role of NGF in cutaneous wound healing (Oger et al., 1974; Tron et al., 1990; Murase et al., 1992; Di Marco et al., 1993; Leon et al., 1994; Pincelli et al., 1994; Skaper et al., 2001).

We recently have demonstrated that topical application of NGF to cutaneous wounds accelerates the rate of wound healing in mice (Matsuda et al., 1998). Full-thickness wounds made by skin punching results in a rapid increase of serum NGF levels, while no NGF is detected in sera before wounds. The level of NGF in the wounded skin site was significantly increased as compared to the unwounded site, reached at maximal levels (7.8 ng/g wet volume) at 1 day after injury, and was gradually decreased to the baseline by 14 days. Sialoadenectomy before wounds completely inhibited the increase in serum NGF level but not that in the skin lesion, suggesting that NGF is released from salivary glands into the blood immediately after injury. The cellular source of NGF produced in the wounded skin was identified by in situ hybridization method. The signal of mRNA^{NGF} localized in epithelial cells, fibroblasts and granulation tissue at the wound site. We examined the effects of various inflammatory cytokines involved in wound healing process, including TGF-β1, EGF, FGF, PDGF-BB, TNF-α, IL-1β and interferon-gamma (IFN-γ), on NGF production by two cell lines of fibroblasts and keratinocytes. All cytokines tested except TNF-α enhanced NGF production in both the types of cells, suggesting that the local NGF synthesis after skin injury by residential cells is regulated by these cytokines, probably released from infiltrating cells including lymphocytes and macrophages.

Next, we examined a therapeutic effect of NGF on cutaneous wound healing in normal and healing-impaired diabetic KK/Ta mice. Decreased NGF levels have been reported in the submaximal gland of mutant *db/db* mice and experimentally-induced diabetic mice as well as in the serum and skin of patients with diabetes mellitus. The effect of NGF applied onto the wound of dorsal skins was evaluated by various parameters including the size of wounds, histological examination (the degree of cell infiltration, the degree of reepithelization, the thickness of granulation tissues and the density of extracellular matirix) and breaking strength of wounds. Topical application with NGF improved all of these healing parameters, and accelerated the rate of healing in cutaneous wounds in both normal and diabetic mice. The mechanism underlying NGF-mediated acceleration of wound healing will be discussed later. There is a possibility that NGF contributes to both inflammatory cell recruitment at the site of injury and cutaneous angiogenesis in wound healing process by direct and/or indirect mechanisms on endothelial cells. Promoting angiogenesis accelerates wound closure in diabetic animals. These findings propose NGF potentiality as a therapeutic tool for normal and healing-impaired individuals to improve their cutaneous wound healing.

Wound healing of other tissues

The eye

NGF has also been implicated as a potent stimulant of the wound healing process in the eye. Lambiase et al. (2002) have examined the levels of $mRNA^{NGF}$ and $mRNA^{TrkA}$ in both normal and experimentally-iridectomized rabbit ocular tissues. Both NGF and TrkA express in cornea, iris, ciliary body and lens. NGF is most abundant in iris (8.93 ± 3.97 ng/g), but little in the aqueous humor (0.023 ± 0.001 ng/g). After iridectomy, the concentrations of NGF in the iris, ciliary body and aqueous humor shows a rapid increase as compared to sham-operated groups, but not in cornea and lens. Experimental ocular hypertension in rabbits also induces retinal injury and enhances local production of NGF (Lambiase et al., 1997). NGF also exists constitutively in normal human and rat corneas (Lambiase et al., 1998b, 2000). Human and rat corneal epithelial cells produce, store and release NGF in vitro and also express TrkA. A transient increase of corneal NGF levels was observed after rat corneal epithelial injury, and inhibition of endogenous NGF activity by neutralizing anti-NGF antibodies aggravates the retinal damages as well as delays the corneal epithelial healing rate, whereas exogenous administration of NGF accelerates healing (Lambiase et al., 2000). These findings suggest a potential clinical usage of NGF in corneal epithelial diseases and an important role of NGF in the crosslink between the corneal epithelium.

The bone

When damaged, bone demonstrates a remarkable capacity of regeneration. The process of fracture healing is complex, but it can also be simplified into three phases as described above: inflammation, tissue formation and remodeling. During the tissue formation phase, a callus is formed by bone-making cells recruited to the site, osteoblasts, that produce cartilage-like bone matrix, which is mineralized in the end. In the final phase, the bone remodels through a process of resorption and deposition. The role of NGF in bone formation after experimental bone fracture has been investigated in a rodent model (Grills et al., 1997; Grills and Schuijers, 1998). NGF is constitutively expressed in bone tissues, but its expression, together with TrkA expression, is increased during the healing process after fracture (Grills and Schuijers, 1998). By immunohistochemical staining, periosteal osteoprogenitor cells, osteoblasts and immature osteocytes, some of chondrocytes, marrow stroma cells and endothelial cells in the fracture callus formed at damaged site are strongly positive for NGF immunohistochemically, but not in positioned osteoclasts and osteocytes. Bone-making cells of fractured rib expressed both NGF and its receptor, suggesting that NGF-mediated autocrine and/or paracrine mechanisms contribute to wound healing of the bone. To support this idea, topical application of NGF by using a miniosmotic pump to the fracture site for seven days significantly improves cartilage production, breaking strength and Young's modulus values of fractured bones (Grills et al., 1997).

The muscle

From a viewpoint of sports medicine, muscle injuries are the major problem in traumatotherapy. TrkA is constitutively detected in murine and human myoblast cell lines, suggesting that muscle cells may respond to NGF (Rende et al., 2000). In fact, addition of NGF increases the fusion rate of both primary myoblast cells and cell lines, as well as the proliferation of the slowly dividing primary myoblasts. In a mouse model for muscle contusion, large-scale muscle regeneration occurs within two weeks, and subsequently followed by the development of muscle fibrosis. Administration of NGF to injured sites enhances proliferation and differentiation of myoblasts and improves muscle healing (Kasemkijwattana et al., 2000). However, the same research group has reported that NGF is less effective than insulin-like growth factor (IGF) or FGF in wound healing of muscle injury (Menetrey et al., 2000). These data support the notion that NGF expression in skeletal muscle is not only associated with a classical target-derived neurotrophic function for PNS neurons, but also with an autocrine action which affects the proliferation, fusion into myotubes and cell morphology of developing myoblasts. The functional roles of NGF in muscle cells are still unclear and controversial regarding the responsiveness of muscle cells to NGF. To clarify the effect of NGF in skeletal muscle healing and repair, further detailed research should be done.

The tooth

NGF supports odontogenesis in a dose-dependent manner (Amano et al., 1999). Rat tooth pulpal cells induce neurite outgrowth from trigeminal neurons in vitro, suggesting secretion of neurotrophic factors such as NGF from the cells (Lillesaar et al., 2001).

Odontoblasts (Woodnutt et al., 2000), periodontal ligament cells and gingival fibroblasts and keratinocytes (Kurihara et al., 2003) express both NGF and TrkA receptors. Tooth pulp injury often results in extensive sprouting of sensory nerve fibers at the site of wounds associated with an increase in local levels of NGF produced by fibroblasts and dental pulp (Sullins et al., 2000). Localization of NGF and its receptor suggests that besides its apparent roles in the regulation of tooth innervation, NGF may serve nonneuronal, organogenetic functions during tooth formation and repair process. However, only fragmentary data are available at present.

Molecular mechanisms of NGF-mediated acceleration of wound healing

Molecular mechanisms underlying NGF-mediated acceleration of wound healing are not fully understood. However, NGF has various biological activities on all listed cell types in Fig. 1 that are involved in healing process, suggesting its broad role in wound healing and tissue repair. NGF exerts its biological activities via specific receptors p75[NTR] and TrkA expressed on the surface of target cells (Yanker and Shooter, 1982; Sofroniew et al., 2001). In skin,

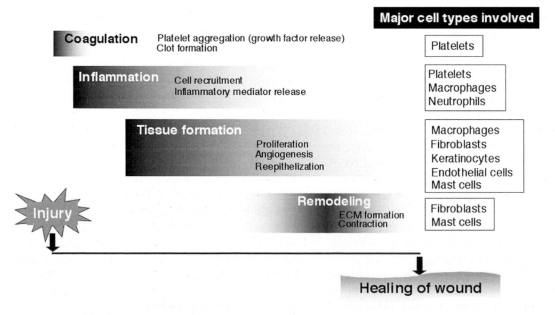

Fig. 1. A wound healing process.

there are various types of cells that express NGF receptors, including keratinocytes, melanocytes, fibroblasts and mast cells (Murase et al., 1992; Yaar et al., 1994; Nilsson et al., 1997; Pincelli et al., 1997; Bull et al., 1998; Johansson and Liang, 1999). In this chapter, we discuss about how NGF promotes a healing process with a focus on the events in cutaneous wound healing.

NGF acts as a powerful chemotactic factor to polymorphonuclear cells such as neutrophils and monocytes, whose recruitment is an important early event of wound healing process (Gee et al., 1983; Lawman et al., 1985; Kannan et al., 1991). In addition, NGF activates their functions: facilitating phagocytosis and superoxide production in neutrophils, and activating the functions of macrophages in parasite killing, and IL-1β production as well as phagocytosis (Kannan et al., 1991, 1992; Kawamoto et al., 1995, 2002; Susaki et al., 1996). Especially, macrophages are essential for the optimal healing process because the inhibition of their recruitment severely impairs healing.

Micera et al. (2001) have demonstrated that NGF exerted a chemotactic activity to skin fibroblasts and stimulated their migration in an in vitro wound model, but did not affect fibroblast proliferation, collagen production and MMP activities. NGF also significantly increased human fibroblast chemotaxis to fibronectin and PDGF-BB (Kohyama et al., 2002). After long-term culture with NGF, fibroblasts become to express p75^{NTR} which mediated apoptotic cell death signaling, suggesting that apoptosis induces by NGF in fibroblasts during the late stage such as a remodeling phase of the process of wound healing and tissue repair (Micera et al., 2001). Dermal myofibroblasts, a myodifferentiation form of fibroblasts characterized by the expression of alpha-smooth muscle actin (α-SMA), are a part of the granulation tissue and implicated in the generation of contractile forces during normal wound healing and pathological contractures. NGF not only increases gel collagen contraction but also induces α-SMA expression in skin fibroblasts, indicating that NGF changes their phenotype into myofibroblasts. In the wounded skin, the localization of α-SMA protein is identical with cells in which mRNA^{NGF} is detected (Hasan et al., 2000; Micera et al., 2001).

In contrast to fibroblasts, NGF not only stimulates proliferation of keratinocytes but also rescues from their apoptosis in both mice and humans through the surface-expressing TrkA receptor but not p75^{NTR} (Tron et al., 1990; Pincelli et al., 1994). Furthermore, keratinocytes are known to produce NGF in vivo and in vitro (Di Marco et al., 1993; Pincelli et al., 1994). NGF itself and other cytokines such as IL-1β stimulate NGF synthesis in keratinocytes. Thus, a marked upregulation of NGF at the edge of wounded epidermis indicates vigorous proliferation of keratinocytes, leading to smooth reepithelization through autocrine mechanisms associated with other inflammatory cytokines.

Extravascular migration of leukocytes into wounded sites is initially controlled by the interaction of cell surface adhesion molecules expressed on activated endothelial cells and infiltrating leukocytes. NGF has been reported to increase proliferation and ICAM-1 expression in human dermal microvascular endothelial cells (Raychaudhuri et al., 2001). RT-PCR analysis and immunological methods demonstrate that human umbilical vein endothelial cells express the TrkA receptor and p75^{NTR} (Cantarella et al., 2002). NGF stimulates a significant increase in human umbilical vein endothelial cell proliferation, and exerts a potent angiogenic activity in vivo as determined with chick embryo chorioallantoric membrane assay. These biological activities of NGF are abolished by K252a, a specific inhibitor for TrkA, suggesting that tyrosine phosphorylation signaling of TrkA is important in this response. In diabetic neuropathy as well, abnormalities of angiogenesis are responsible for the pathogenesis of impaired wound healing in diabetic patients. It has been previously reported that VEGF plays a central role in promoting angiogenesis during wound repair and that healing-impaired diabetic mice show decreased VEGF expression levels (Schratzberger et al., 2001). In addition, NGF has been demonstrated to stimulate angiogenesis and upregulate VEGF expression (Hammes et al., 1995; Samii et al., 1999; Sondell et al., 1999; Calza et al., 2001; Emanueli et al., 2002).

The positive involvement of mast cells in various phases during wound healing has been pointed out (Galli and Wershil, 1995; Artuc et al., 1999). Complex cytokine network play a central role in wound healing and tissue repair process via orchestrating and facilitating the functions of multiple cellular components, including the secretion of various cytokines and growth factors by mast cells. With the aid of a wide array of newly-formed or preformed mediators released by degranulation, the activated mast cell controls the key events of the healing phases: triggering and modulation of the inflammatory stage, proliferation of connective cellular elements and final remodeling of the newly-formed connective tissue matrix. The importance of the mast cell in regulating healing processes is also demonstrated by the fact that a surplus or deficit of degranulated biological mediators causes impaired repair, with the formation of exuberant granulation tissue (e.g., keloids and hypertrophic scars), delayed closure (dehiscence) and chronicity of the inflammatory stage.

Mast cells often locate near blood vessels, and generate and release a number of vasoactive mediators including histamine, serotonin and leukotrienes by crosslinking of $Fc\varepsilon R$-IgE with its specific Ag. In addition to the immunological stimulation, intradermal or subcutaneous administration of NGF to rats causes immediate vasodilatory responses characterized by the degranulation of local mast cells (Pearce and Thompson, 1986; Marshall et al., 1990). NGF has multiple biological activities on mast cells. For example, NGF not only induces the differentiation of mast cells but also supports their survival and functions via the TrkA receptor (Matsuda et al., 1991; Kawamoto et al., 1995, 2002). We have recently demonstrated that NGF collaboratively worked with membrane lysoPS of activated platelets to induce mast cell activation (Kawamoto et al., 2002). Thus, NGF released in response to injury may contribute to mast cell activation in collaboration with locally activated platelets that accumulate at damaged blood vessels in the process of inflammation and tissue repair. Being cultured with NGF, human mast cells upregulate mRNA expression for chymase (Tam et al., 1997; Welker et al., 2000). Chymase is activated immediately upon its release into the extracellular matrix in vascular tissues subsequent to activation of mast cells by stimuli such as vessel injury, and is thought to participate in the healing process as well as in fibrotic skin diseases (Nishikori et al., 1998; Artuc et al., 1999). Inhibition of chymase activity retards the rate of wound healing. Moreover,

there is some evidence that mast cells participate in angiogenesis, since heparin is able to stimulate endothelial cell migration and proliferation in vitro (Tharp, 1989; Norrby and Sorbo, 1992). Mast cells also produce and release VEGF upon various inflammatory stimuli (Boesiger et al., 1998).

All these findings have broad implications for understanding the role of NGF in natural wound healing responses. NGF may contribute aggressively to wound healing in various tissues throughout the process from the very initial stage to the final remodeling stage.

Clinical trials of NGF in wound healing and tissue repair

As mentioned above, the availability of NGF to promote wound healing and regeneration of nervous systems and peripheral tissues provides strong evidence for its potential therapeutic utilization. In this review, recent clinical studies of NGF are introduced and summarized in Table 1.

Skin vasculitic ulcers in rheumatoid arthritis

The study group of Aloe et al. (Tuveri et al., 2000) has done a clinical trial of NGF-mediated wound healing in rheumatoid arthritis (RA) patients with chronic cutaneous ulcer complication. The ulcers were treated with topical application of 50 µg NGF, which is purified from mouse submaxillary glands, daily for four weeks and twice a week for another month. Size and clinical appearance (inflammation, vascularization, detaching, necrosis, presence of fibrin, granulation, sanious secretion) of the lesions were monitored twice a week. During the first two weeks of NGF treatment, the ulcers showed improvement clinically: a significant reduction in the size, pain and inflammation; and an increase of granulation. By continuous NGF application, the size of the lesion is gradually decreased and the healing completed within eight weeks in all patients with RA, but not in patients with systemic sclerosis. Cutaneous ulcer, which is also called 'rheumatoid nodules', is a frequent complication of RA associated with cutaneous vasculitis that is characterized by inflammation and necrosis of small- and

TABLE 1
Clinical studies with NGF in various human diseases

Disease	Type of study	n	Application	Result	Side effects	Ref.
Alzheimer	Phase I	3	total 0.55–0.6 mg lateralcerebral infusion	Negative	Back pain, weight loss	Ericsdotter et al.
Diabetic neuropathy	Phase I–II Placebo control	250	0.3 μg/kg, s.c. 3 times/wk	Effective	Injected site pain	Apfel et al.
Diabetic neuropathy	Phase III Placebo control double blind	102	0.1 μg/kg, s.c. 3 times/wk	Negative	Injected site pain	Apfel et al.
HIV neuropathy	Phase II Placebo control	270	0.1–0.3 μg, s.c. twice a week	Negative	Injected site pain	McArthur et al.
HIV neuropathy	Phase II open label follow-up study	200	0.1–0.3 μg, s.c. twice a week	Negative Improvement in pain	Injected site pain	Schfitto et al.
Immune corneal ulcer	Preliminary uncontrolled	4	10 μg/drop 6 times/day	Complete cure No relapse (12 mo follow-up)	pain, photophobia (only initially)	Lambiase et al.
Neurotrophic keratitis	Preliminary	12	10 μg/drop 6–10 times/day	Complete cure No relapse (12 mo follow-up)	None	Lambiase et al.
Crush syndrome	Preliminary	1	10 μg s.c., 3 times/day 7 days	Improvement of the ischaemic area	None	Chiaretti et al.
Vasculitic ulcer	Preliminary	4	50 μg, topical applied < 8 wk	Complete cure (not in systemic sclerosis patients)	None	Tuveri et al.
Pressure ulcer	Preliminary		50 μg, topical applied	Complete cure		Bernabei et al.

medium-sized blood vessels. NGF is also effective to the patients with pressure ulcers (Bernabei et al., 1999). The healing effect of NGF on vasculitic ulcers may owe much to its angiogenic activity as well as the effects on keratinocytes and fibroblasts.

Corneal diseases

NGF has been reported to be effective in human ocular diseases such as neurotrophic keratitis and immune corneal ulcers (Lambiase et al., 1999). The cornea is the most densely innervated tissue in the mammalian organism. Axons from the trigeminal ganglion and sympathetic ganglion terminate in delicate endings among the epithelial cells of the cornea. Thus, the health of the corneal epithelium is largely dependent upon a normal innervation. Denervation of the cornea causes a degenerative change and inflammation in the corneal epithelium that is called neurotrophic (neuroparalytic) keratitis (NK). Bonini et al. (2000) performed clinical application of NGF in 45 eyes of NK patients unresponsive to other invasive therapies. The patients received murine NGF in eyedrops between four and twelve times a day until corneal ulcers were healed. During the treatment, size and depth of the lesion, corneal sensitivity and best corrected visual acuity were monitored. All the patients have been reported to show complete remission after 12 days to six weeks treatment with NGF. Visual functions such as corneal sensitivity and visual acuity were improved and the healing of corneal ulcers were promoted in NGF-treated eyes. Interestingly, except few cases, no relapse of the disease occurred during the follow-up period. The research group observed hyperemia and pain in the eye as a side effect at the first day of the NGF treatment.

Furthermore, Lambiase et al. (1998a) applied NGF eyedrops to four patients with corneal ulcer as a consequence of multiple immune disorders. Although patients did not show positive response to steroids and other immunosuppressive agents, NGF treatment healed corneal damage. And, no relapse was observed in any patients during 3–12 months follow-up. The adverse effect was pain at applied site and photophobia, which disappeared as the wound healing progressed. Although these tests were noncomparative studies, NGF eyedrops are expected to be a promising therapy for corneal diseases.

Diabetic neuropathy

Neuropathy is the most frequent complication of diabetes, and patients with diabetic neuropathy have a significant risk of chronic ulcers and impaired wound healing. The application of NGF aimed at diabetic neuropathy treatment has been assessed more intensively. After successful results of the Phase II study of recombinant human NGF for diabetic neuropathy, NGF therapy was conducted for further clinical study with enthusiastic expectations. However, no clinical benefit was observed in the Phase III study over the 12 months period (Apfel et al., 1998; Apfel, 2002). Several reasons for the discrepancy between the two clinical trials have been pointed out: inadequate dose, a robust placebo effect, the different criteria for clinical score, and changes to the formulation of recombinant human NGF for the Phase III trials. As a result, Genentech of San Francisco, California, decided not to proceed with further development of Neuleze, a trade name of recombinant human NGF. However, NGF is still expected to heal cutaneous ulcer in diabetic patients based on the animal studies. NGF accelerates healing rate of cutaneous wounds, and prevents both neuroretinal programmed cell death and capillary pathology in experimental diabetes. These data provide new insight into the mechanism of diabetic retinal vascular damage, and suggest that NGF may have potential as therapeutic agents for the prevention of human diabetic retinopathy.

Alzheimer's disease

The neurotrophins, particularly NGF, are currently under investigation as therapeutic agents for the treatment of neurodegenerative disorders such as Alzheimer's disease (AD), multiple sclerosis and Parkinson's disease (Hefti, 1997; Hefti and Weiner, 1986). Of the many neurotrophins now known to present in the central nervous system, NGF appears most important for the maintenance and function of cholinergic neurons of the basal forebrain that play a critical role in attention and other memory-related processes. In the brain that is damaged by AD, particularly basal forebrain cholinergic neurons are expected to respond to NGF. The Phase I study of NGF infusion into the lateral cerebral ventricle failed to show cognitive improvement of the disease (Eriksdotter Jonhagen et al., 1998). In animal studies, fibroblasts genetically modified to produce NGF prevent the death of certain nerve cells in both rat and monkey brains. Phase I clinical trial of gene therapy with NGF in AD patients have recently completed, and the results of this study will be reported in the near future.

Lameness in husbandry animals

Costa et al. (2002) recently have evaluated the effect of NGF in lame goats associated with cutaneous ulcers, abscesses and granulomas. The lesions were treated daily with 1 μg of purified murine NGF dissolved in sesame oil for three weeks. Before treatment, animals could not walk normally due to the severe ulcer of sole lesions. Although control group showed no clinical changes, healing of ulcers was promoted in all NGF-treated goats. The effect of NGF became significant within 3–4 days after the first application. As the ulcers were healed, animals show signs of improvement in lameness. These findings suggest that NGF is clinically useful for healing damaged soles in husbandry animals. Lameness by inflammation of soles such as sole ulcer, toe ulcer and sole abscess accounts for tremendous economic loss due to lower milk yields, reduced reproductive performance, higher involuntary culling rates, additional management, etc. Thus, treatment for lameness is clinically important in veterinary medicine, not only for the health of husbandry animals including goats, cows and horses, but also for the benefits of farmers.

Others

Chiaretti et al. (2002) have demonstrated the effect of NGF in a child with severe crush syndrome of the lower left limb. Subcutaneous administration of murine NGF (10 μg) every 8 h for 7 days improved the extensive ischemic skin lesion of the calcaneal area. No side effects such as injection site pain emerged.

Future directions

Chronic wounds are one of the major clinical problems with high morbidity. It is estimated that 1.25 million people are burned each year in the United States and that 6.5 million suffer form chronic skin ulcers caused by pressure, venous stasis or diabetes mellitus. Treatment is usually symptomatic and the management for a prolonged period brings down frustration, fair amount of medical expenses, and impaired quality of life in patients with chronic wounds. Biologically, the formation of chronic wounds is likely, in part, due to dysregulation of growth factors and infection. Some investigators have suggested that the nonhealing state is due to a disturbance of coordination of the temporal expression of these mediators. In some impaired wounds, such as diabetic foot ulcers and venous ulcers, a local deficiency of growth factors, or alteration in their receptors may lead to delayed wound healing.

The successful results of clinical application with NGF in wound repair raise hopes that NGF may have a beneficial effect especially on treating chronic wounds such as chronic ulcers. However, a variety of questions need to be answered prior to effective clinical application of NGF therapy. Much of the difficulty lies in the paucity of carefully-controlled clinical trials of wound healing. These include safety issues, treatment regimen including dosage and length of treatment, effective delivery systems, and adverse effects such as applied site pain. The site and extent of expression of NGF and its receptors in neuronal and nonneuronal cells determine the spectrum of biological actions essential not only for therapeutic use of NGF, but also for possible side effects. Furthermore, the majority of research has focused on the effect of NGF alone. It is likely that combinations with other cytokines and/or extracellular matrix, such as 3-D collagen gels and fibrin glue, will allow more clinical benefits in repair of both acute and chronic wounds by the NGF treatment. Refinement of therapeutic strategies with NGF will take place in the future.

Since wound healing is a complex process that can be influenced, both positively and negatively, by many factors, designing clinical trials has proved difficult. The exact mechanism by which exogenous NGF promotes wound healing is not completely understood. It is, however, suggested that the direct biological effect of NGF plays a key role not only in inflammation process by promoting the functions of neutrophils, macrophages, but also in granulation tissue formation and reepithelization by stimulating the mRNA or protein synthesis of residential cells such as fibroblasts, endothelial cells, kerationocytes and mast cells. Also, exogenous NGF may stimulate wound healing indirectly by increasing the production of other growth factors such as VEGF, a potent angiogenic factor. Ongoing research seeks the answers to the questions concerning what drives the healing process. The success of future therapies will be dependent on a growing understanding of the role of NGF in the pathophysiological processes that significantly influence wound healing. Hopefully, this will lead to strategies that allow better treatment and more rapid healing of acute and chronic wounds.

In addition, investigators have focused on the possibility of inserting genes encoding for growth factors into the cells participating in the wound healing response. This approach offers the potential of single-dose growth factor treatment of chronic wounds. There are several approaches for gene insertion, including viral vectors, gene guns and electroporation. A molecular genetic approach in which genetically modified cells synthesize and deliver the desired growth factor in a time-regulated manner is a powerful means to overcome the limitations associated with the topical application of recombinant growth factor proteins. Development of sophisticated delivery of NGF into the wounded sites has also stimulated its clinical application.

Conclusion

In summary, all these findings strongly suggest that NGF has wide-ranging biological activities that contribute to the acceleration of wound healing process by modulating its developmental stages: inflammation, migration, reepithelization, granulation, angiogenesis and tissue remodeling. Understanding the role of NGF in wound repair will contribute to the development of a new therapy to tissue injury and accelerate the improvement of patient care. With the collaboration of basic scientists and clinicians, it is hoped that these experimental strategies coupled with a greater understanding of the biology of NGF in wound healing will be translatable as a clinically effective formula in the near future.

Abbreviations

AD	alzheimer's disease
CAM	chorioallantoric membrane
EGF	epidermal cell growth factor
FGF	fibroblast growth factor
IFN-γ	interferon-gamma
IGF	insulin-like growth factor
IL	interleukin
MMP	matrix metalloproteinase
NGF	nerve growth factor
NK	neurotrophic keratitis
NO	nitric oxide
p75^{NTR}	low-affinity neurotrophin receptor
PDGF-BB	platelet-derived growth factor BB homodimer
PNS	peripheral nervous system
RA	rheumatoid arthritis
TGF-β	transforming growth factor-beta
TIMP	tissue inhibitor of MMP
TNF	tumor necrosis factor
TrkA	high-affinity tyrosine kinase A receptor
VEGF	vascular endothelial cell growth factor
α-SMA	alpha-smooth muscle actin.

References

Amano, O., Bringas, P., Takahashi, I., Takahashi, K., Yamane, A., Chai, Y., Nuckolls, G.H., Shum, L. and Slavkin, H.C. (1999) Nerve growth factor (NGF) supports tooth morphogenesis in mouse first branchial arch explants Dev. Dyn., 216: 299–310.

Apfel, S.C., Kessler, J.A., Adornato, B.T., Litchy, W.J., Sanders, C. and Rask, C.A. (1998) Recombinant human nerve growth factor in the treatment of diabetic polyneuropathy. NGF Study Group Neurology, 51: 695–702.

Apfel, S.C. (2002) Nerve growth factor for the treatment of diabetic neuropathy: what went wrong, what went right, and what does the future hold? Int. Rev. Neurobiol., 50: 393–413.

Artuc, M., Hermes, B., Steckelings, U.M., Grutzkau, A. and Henz, B.M. (1999) Mast cells and their mediators in cutaneous wound healing–active participants or innocent bystanders? Exp. Dermatol., 8: 1–16.

Boesiger, J., Tsai, M., Maurer, M., Yamaguchi, M., Brown, L.F., Claffey, K.P., Dvorak, H.F. and Galli, S.J.

(1998) Mast cells can secrete vascular permeability factor/vascular endothelial cell growth factor and exhibit enhanced release after immunoglobulin E-dependent upregulation of fc epsilon receptor I expression J. Exp. Med., 188: 1135–1145.

Bothwell, M. (1997) Neurotrophin function in skin J. Investig. Dermatol. Symp. Proc., 2: 27–30.

Bennett, N.T. and Schultz, G.S. (1993) Growth factors and wound healing: biochemical properties of growth factors and their receptors Am. J. Surg., 165: 728–737.

Bernabei, R., Landi, F., Bonini, S., Onder, G., Lambiase, A., Pola, R. and Aloe, L. (1999) Effect of topical application of nerve-growth factor on pressure ulcers Lancet, 24: 307.

Bonini, S., Lambiase, A., Rama, P., Caprioglio, G. and Aloe, L. (2000) Topical treatment with nerve growth factor for neurotrophic keratitis. Ophthalmol., 107: 1347–1351; discussion 1351–1352.

Bull, H.A., Leslie, T.A., Chopra, S. and Dowd, P.M. (1998) Expression of nerve growth factor receptors in cutaneous inflammation Br. J. Dermatol., 139: 776–783.

Calza, L., Giardino, L., Giuliani, A., Aloe, L. and Levi-Montalcini, R. (2001) Nerve growth factor control of neuronal expression of angiogenetic and vasoactive factors Proc. Natl. Acad. Sci. USA, 98: 4160–4165.

Cantarella, G., Lempereur, L., Presta, M., Ribatti, D., Lombardo, G., Lazarovici, P., Zappala, G., Pafumi, C. and Bernardini, R. (2002) Nerve growth factor-endothelial cell interaction leads to angiogenesis in vitro and in vivo FASEB J., 16: 1307–1309.

Chiaretti, A., Piastra, M., Caresta, E., Nanni, L. and Aloe, L. (2002) Improving ischaemic skin revascularisation by nerve growth factor in a child with crush syndrome Arch. Dis. Child, 87: 446–448.

Clarke, D. and Richardson, P. (1994) Peripheral nerve injury Curr. Opin. Neurol., 7: 415–421.

Clark, R.A.F. (1996). Wound repair: overview and general considerations. In: R.A.F. Clark (Ed.), The Molecular and Cellular Biology of Wound Repair. Plenum Press, New York, pp. 3–50.

Costa, N., Fiore, M. and Aloe, L. (2002) Healing action of nerve growth factor on lameness in adult goats Ann. Ist. Super. Sanita., 38: 187–194.

Di Marco, E., Mathor, M., Bondanza, S., Cutuli, N., Marchisio, P.C., Cancedda, R. and De Luca, M. (1993) Nerve growth factor binds to normal human keratinocytes through high and low affinity receptors and stimulates their growth by a novel autocrine loop J. Biol. Chem., 268: 22238–22846.

Dobrowsky, R.T. and Carter, B.D. (1998) Coupling of the p75 neurotrophin receptor to sphingolipid signaling Ann. N. Y. Acad. Sci., 845: 32–45.

Ebadi, M., Bashir, R.M., Heidrick, M.L., Hamada, F.M., Refaey, H.E., Hamed, A., Helal, G., Baxi, M.D., Cerutis, D.R. and Lassi, N.K. (1997) Neurotrophins and their receptors in nerve injury and repair Neurochem. Int., 30: 347–374.

Emanueli, C., Salis, M.B., Pinna, A., Graiani, G., Manni, L. and Madeddu, P. (2002) Nerve growth factor promotes angiogenesis and arteriogenesis in ischemic hindlimbs Circulation, 106: 2257–2262.

Eriksdotter Jonhagen, M., Nordberg, A., Amberla, K., Backman, L., Ebendal, T., Meyerson, B., Olson, L., Seiger, A., Shigeta, M., Theodorsson, E., Viitanen, M., Winblad, B. and Wahlund, L.O. (1998) Intracerebroventricular infusion of nerve growth factor in three patients with Alzheimer's disease, Dement. Geriatr. Cogn. Disord., 9: 246–257.

Frostick, S.P., Yin, Q. and Kemp, G.J. (1998) Schwann cells, neurotrophic factors, and peripheral nerve regeneration Microsurgery, 18: 397–405.

Grills, B.L., Schuijers, J.A. and Ward, A.R. (1997) Topical application of nerve growth factor improves fracture healing in rats J. Orthop. Res., 15: 235–242.

Galli, S.J. and Wershil, B.K. (1995) Mouse mast cell cytokine production: role in cutaneous inflammatory and immunological responses Exp. Dermatol., 4(4 Pt 2): 240–249.

Gee, A.P., Boyle, M.D., Munger, K.L., Lawman, M.J. and Young, M. (1983) Nerve growth factor: stimulation of polymorphonuclear leukocyte chemotaxis in vitro Proc. Natl. Acad. Sci. USA, 80: 7215–7218.

Grills, B.L. and Schuijers, J.A. (1998) Immunohistochemical localization of nerve growth factor in fractured and unfractured rat bone Acta. Orthop. Scand., 69: 415–419.

Hammes, H.P., Federoff, H.J. and Brownlee, M. (1995) Nerve growth factor prevents both neuroretinal programmed cell death and capillary pathology in experimental diabetes Mol. Med., 1: 527–534.

Hasan, W., Zhang, R., Liu, M., Warn, J.D. and Smith, P.G. (2000) Coordinate expression of NGF and alpha-smooth muscle actin mRNA and protein in cutaneous wound tissue of developing and adult rats Cell. Tissue Res., 300: 97–109.

Hefti, F. and Weiner, W.J. (1986) Nerve growth factor and Alzheimer's disease Ann. Neurol., 20: 275–281.

Hefti, F. (1997) Pharmacology of neurotrophic factors Annu. Rev. Pharmacol., 37: 239–267.

Hirata, H., Hibasami, H., Yoshida, T., Ogawa, M., Matsumoto, M., Morita, A. and Uchida, A. (2001) Nerve growth factor signaling of p75 induces differentiation and ceramide-mediated apoptosis in Schwann cells cultured from degenerating nerves Glia, 36: 245–258.

Hobson, M.I., Green, C.J. and Terenghi, G. (2000) VEGF enhances intraneural angiogenesis and improves nerve regeneration after axotomy J. Anat., 197(4): 591–605.

Huston, J.M., Nial, M., Evans, D. and Flowler, R. (1979) Effect of salivary glands on wound contraction in mice Nature, 279: 793–795.

Johansson, O. and Liang, Y. (1999) Neurotrophins and their receptors in the skin: a tribute to Rita Levi-Montalcini.

382

In: L. Aloe and G.N. Chaldakov (Eds.), Nerve Growth Factor in Health and Disease Biomedical Reviews, Vol. 10, Varna: The Bulgarian-American Center, pp. 15–23.

Kannan, Y., Ushio, H., Koyama, H., Okada, M., Oikawa, M., Yoshihara, T., Kaneko, M. and Matsuda, H. (1991) 2.5S nerve growth factor enhances survival, phagocytosis, and superoxide production of murine neutrophils Blood, 77: 1320–1325.

Kannan, Y., Matsuda, H., Ushio, H., Kawamoto, K. and Shimada, Y. (1993) Murine granulocyte-macrophage and mast cell colony formation promoted by nerve growth factor Int. Arch. Allergy Immunol., 102: 362–367.

Kannan, Y., Usami, K., Okada, M., Shimizu, S. and Matsuda, H. (1992) Nerve growth factor suppresses apoptosis of murine neutrophils Biochem. Biophys. Res. Commun., 186: 1050–1056.

Kasemkijwattana, C., Menetrey, J., Bosch, P., Somogyi, G., Moreland, M.S., Fu, F.H., Buranapanitkit, B., Watkins, S.S. and Huard, J. (2000) Use of growth factors to improve muscle healing after strain injury Clin. Orthop., 370: 272–285.

Kawamoto, K., Okada, T., Kannan, Y., Ushio, H., Matsumoto, M. and Matsuda, H. (1995) Nerve growth factor prevents apoptosis of rat peritoneal mast cells through the trk proto-oncogene receptor Blood, 86: 4638–4644.

Kawamoto, K., Aoki, J., Tanaka, A., Itakura, A., Hosono, H., Arai, H., Kiso, Y. and Matsuda, H. (2002) Nerve growth factor activates mast cells through the collaborative interaction with lysophosphatidylserine expressed on the membrane surface of activated platelets J. Immunol., 168: 6412–6419.

Kohyama, T., Liu, X., Wen, F.Q., Kobayashi, T., Abe, S., Ertl, R. and Rennard, S.I. (2002) Nerve growth factor stimulates fibronectin-induced fibroblast migration J. Lab. Clin. Med., 140: 329–335.

Korsching, S. (1986) The role of nerve growth factor in the CNS Trends Neurosci., 9: 570–577.

Kurihara, H., Shinohara, H., Yoshino, H., Takeda, K. and Shiba, H. (2003) Neurotrophins in cultured cells from periodontal tissues J. Periodontol., 74: 76–84.

Lambiase, A., Centofanti, M., Micera, A., Manni, G.L., Mattei, E., De Gregorio, A., de Feo, G., Bucci, M.G. and Aloe, L. (1997) Nerve growth factor (NGF) reduces and NGF antibody exacerbates retinal damage induced in rabbit by experimental ocular hypertension Graefes. Arch. Clin. Exp. Ophthalmol., 235: 780–785.

Lambiase, A., Rama, P., Bonini, S., Caprioglio, G. and Aloe, L. (1998a) Topical treatment with nerve growth factor for corneal neurotrophic ulcers N. Engl. J. Med., 338: 1174–1180.

Lambiase, A., Bonini, S., Micera, A., Rama, P., Bonini, S. and Aloe, L. (1998b) Expression of nerve growth factor receptors on the ocular surface in healthy subjects and during manifestation of inflammatory diseases Invest. Ophthalmol. Vis. Sci., 39: 1272–1275.

Lambiase, A., Rama, P., Aloe, L. and Bonini, S. (1999) Management of neurotrophic keratopathy Curr. Opin. Ophthalmol., 10: 270–276.

Lambiase, A., Manni, L., Bonini, S., Rama, P., Micera, A. and Aloe, L. (2000) Nerve growth factor promotes corneal healing: structural, biochemical, and molecular analyzes of rat and human corneas Invest. Ophthalmol. Vis. Sci., 41: 1063–1069.

Lambiase, A., Bonini, S., Manni, L., Ghinelli, E., Tirassa, P., Rama, P. and Aloe, L. (2002) Intraocular production and release of nerve growth factor after iridectomy Invest. Ophthalmol. Vis. Sci., 43: 2334–2340.

Laurent, G.J. (1988) The dynamic nature of collagen. The role of collagen degradation in the regulation of collagen mass Am. J. Physiol., 251: 1–9.

Lawman, M.J., Boyle, M.D., Gee, A.P. and Young, M. (1985) Nerve growth factor accelerates the early cellular events associated with wound healing Exp. Mol. Pathol., 43: 274–281.

Leibovich, S.J. and Ross, R. (1975) The role of the macrophage in wound repair. A study with hydrocortisone and anti-macrophage serum Am. J. Pathol., 78: 71–100.

Leon, A., Buriani, A., Dal Toso, R., Fabris, M., Romanello, S., Aloe, L. and Levi-Montalcini, R. (1994) Mast cells synthesize, store, and release nerve growth factor Proc. Natl. Acad. Sci. USA, 91: 3739–3743.

Levi-Montalcini, R. (1987) The nerve growth factor 35 years later Science, 237: 1154–1162.

Li, A.K., Koroly, M.J., Schattenkerk, M.E., Malt, R.A and Young, M. (1980) Nerve growth factor: acceleration of the rate of wound healing in mice Proc. Natl. Acad. Sci. USA, 77: 4379–4381.

Liekens, S., De Clercq, E. and Neyts, J. (2001) Angiogenesis: regulators and clinical applications Biochem. Pharmacol., 61: 253–270.

Lillesaar, C., Eriksson, C. and Fried, K. (2001) Rat tooth pulp cells elicit neurite growth from trigeminal neurones and express mRNAs for neurotrophic factors in vitro Neurosci. Lett., 308: 161–164.

Liu, H.M., Lei, H.Y. and Kao, K.P. (1995) Correlation between NGF levels in wound chamber fluid and cytological localization of NGF and NGF receptor in axotomized rat sciatic nerve Exp. Neurol., 132: 24–32.

Marshall, J.S., Stead, R.H., McSharry, C., Nielsen, L. and Bienenstock, J. (1990) The role of mast cell degranulation products in mast cell hyperplasia. I. Mechanism of action of nerve growth factor J. Immunol., 144: 1886–1892.

Martin, P. (1997) Wound healing-aiming for perfect skin regeneration Science, 276: 75–81.

Matsuda, H., Coughlin, M.D., Bienenstock, J. and Denburg, J.A. (1988) Nerve growth factor promotes human hemopoietic colony growth and differentiation Proc. Natl. Acad. Sci. USA, 85: 6508–6512.

Matsuda, H., Kannan, Y., Ushio, H., Kiso, Y., Kanemoto, T., Suzuki, H. and Kitamura, Y. (1991) Nerve growth factor induces development of connective tissue-type mast cells in vitro from murine bone marrow cells J. Exp. Med., 174: 7–14.

Matsuda, H., Koyama, H., Sato, H., Sawada, J., Itakura, A., Tanaka, A., Matsumoto, M., Konno, K., Ushio, H. and Matsuda, K. (1998) Role of nerve growth factor in cutaneous wound healing: accelerating effects in normal and healing-impaired diabetic mice J. Exp. Med., 187: 297–306.

Meirer, R., Gurunluoglu, R. and Siemionow, M. (2001) Neurogenic perspective on vascular endothelial growth factor: review of the literature J. Reconstr. Microsurg., 17: 625–630.

Menetrey, J., Kasemkijwattana, C., Day, C.S., Bosch, P., Vogt, M., Fu, F.H., Moreland, M.S. and Huard, J. (2000) Growth factors improve muscle healing in vivo J. Bone Joint Surg. Br., 82: 131–137.

Micera, A., Vigneti, E., Pickholtz, D., Reich, R., Pappo, O., Bonini, S., Maquart, F.X., Aloe, L. and Levi-Schaffer, F. (2001) Nerve growth factor displays stimulatory effects on human skin and lung fibroblasts, demonstrating a direct role for this factor in tissue repair Proc. Natl. Acad. Sci. USA, 98: 6162–6167.

Murase, K., Murakami, Y., Takayanagi, K., Furukawa, Y. and Hayashi, K. (1992) Human fibroblast cells synthesize and secrete nerve growth factor in culture Biochem. Biophys. Res. Commun., 184: 373–379.

Nilsson, G., Forsberg-Nilsson, K., Xiang, Z., Hallbook, F., Nilsson, K. and Metcalfe, D.D. (1997) Human mast cells express functional TrkA and are a source of nerve growth factor Eur. J. Immunol., 27: 2295–2301.

Nishikori, Y., Kakizoe, E., Kobayashi, Y., Shimoura, K., Okunishi, H. and Dekio, S. (1998) Skin mast cell promotion of matrix remodeling in burn wound healing in mice: relevance of chymase Arch. Dermatol. Res., 290: 553–560.

Norrby, K. and Sorbo, J. (1992) Heparin enhances angiogenesis by a systemic mode of action Int. J. Exp. Pathol., 73: 147–155.

Oger, J., Arnason, B.G., Pantazis, N., Lehrich, J. and Young, M. (1974) Synthesis of nerve growth factor by L and 3T3 cells in culture Proc. Natl. Acad. Sci. USA, 71: 1554–1558.

Pearce, F.L. and Thompson, H.L. (1986) Some characteristics of histamine secretion from rat peritoneal mast cells stimulated with nerve growth factor J. Physiol., 372: 379–393.

Pincelli, C., Sevignani, C., Manfredini, R., Grande, A., Fantini, F., Bracci-Laudiero, L., Aloe, L., Ferrari, S., Cossarizza, A. and Giannetti, A. (1994) Expression and function of nerve growth factor and nerve growth factor receptor on cultured keratinocytes J. Invest. Dermatol., 103: 13–18.

Pincelli, C., Haake, A.R., Benassi, L., Grassilli, E., Magnoni, C., Ottani, D., Polakowska, R., Franceschi, C. and Giannetti, A. (1997) Autocrine nerve growth factor protects human keratinocytes from apoptosis through its high affinity receptor (TRK): a role for BCL-2 J. Invest. Dermatol., 109: 757–764.

Raychaudhuri, S.K., Raychaudhuri, S.P., Weltman, H. and Farber, E.M. (2001) Effect of nerve growth factor on endothelial cell biology: proliferation and adherence molecule expression on human dermal microvascular endothelial cells Arch. Dermatol. Res., 293: 291–295.

Ravanti, L. and Kahari, V.M. (2000) Matrix metalloproteinases in wound repair Int. J. Mol. Med., 6: 391–407.

Rende, M., Brizi, E., Conner, J., Treves, S., Censier, K., Provenzano, C., Taglialatela, G., Sanna, P.P. and Donato, R. (2000) Nerve growth factor (NGF) influences differentiation and proliferation of myogenic cells in vitro via TrKA Int. J. Dev. Neurosci., 18: 869–885.

Samii, A., Unger, J. and Lange, W. (1999) Vascular endothelial growth factor expression in peripheral nerves and dorsal root ganglia in diabetic neuropathy in rats Neurosci. Lett., 262: 159–162.

Scheid, A., Meuli, M., Gassmann, M. and Wenger, R.H. (2000) Genetically modified mouse models in studies on cutaneous wound healing Exp. Physiol., 85: 687–704.

Schratzberger, P., Walter, D.H., Rittig, K., Bahlmann, F.H., Pola, R., Curry, C., Silver, M., Krainin, J.G., Weinberg, D.H., Ropper, A.H. and Isner, J.M. (2001) Reversal of experimental diabetic neuropathy by VEGF gene transfer J. Clin. Invest., 107: 1083–1092.

Sebert, M.E. and Shooter, E.M. (1993) Expression of mRNA for neurotrophic factors and their receptors in the rat dorsal root ganglion and sciatic nerve following nerve injury J. Neurosci. Res., 36: 357–367.

Shibayama, E. and Koizumi, H. (1996) Cellular localization of the Trk neurotrophin receptor family in human non-neuronal tissues Am. J. Pathol., 148: 1807–1818.

Skaper, S.D., Pollock, M. and Facci, L. (2001) Mast cells differentially express and release active high molecular weight neurotrophins Brain Res. Mol. Brain Res., 97: 177–185.

Steed, D.L. (1997) The role of growth factors in wound healing Surg. Clin. North Am., 77: 575–586.

Sofroniew, M.V., Howe, C.L. and Mobley, W.C. (2001) Nerve growth factor signaling, neuroprotection, and neural repair Annu. Rev. Neurosci., 24: 1217–1281.

Sondell, M., Lundborg, G. and Kanje, M. (1999) Vascular endothelial growth factor has neurotrophic activity and stimulates axonal outgrowth, enhancing cell survival and Schwann cell proliferation in the peripheral nervous system J. Neurosci., 19: 5731–5740.

Sullins, J.S., Carnes, D.L., Jr., Kaldestad, R.N. and Wheeler, E.F. (2000) Time course of the increase in trk A expression in trigeminal neurons after tooth injury J. Endod., 26: 88–91.

Susaki, Y., Shimizu, S., Katakura, K., Watanabe, N., Kawamoto, K., Matsumoto, M., Tsudzuki, M., Furusaka, T., Kitamura, Y. and Matsuda, H. (1996)

Functional properties of murine macrophages promoted by nerve growth factor Blood, 88: 4630–4687.

Tam, S.Y., Tsai, M., Yamaguchi, M., Yano, K., Butterfield, J.H. and Galli, S.J. (1997) Expression of functional TrkA receptor tyrosine kinase in the HMC-1 human mast cell line and in human mast cells Blood, 90: 1807–1820.

Taniuchi, M., Clark, H.B. and Johnson, E.M., Jr. (1986) Induction of nerve growth factor receptor in Schwann cells after axotomy Proc. Natl. Acad. Sci. USA, 83: 4094–4098.

Tharp, M.D. (1989) The interaction between mast cells and endothelial cells J. Invest. Dermatol., 93(Suppl. 2): 107S–112S.

Tomlinson, D.R., Fernyhough, P. and Diemel, L.T. (1997) Role of neurotrophins in diabetic neuropathy and treatment with nerve growth factors Diabetes, 46(Suppl. 2): S43–S49.

Tron, V.A., Coughlin, M.D., Jang, D.E., Stanisz, J. and Sauder, D.N. (1990) Expression and modulation of nerve growth factor in murine keratinocytes (PAM 212) J. Clin. Invest., 85: 1085–1089.

Tuveri, M., Generini, S., Matucci-Cerinic, M. and Aloe, L. (2000) NGF, a useful tool in the treatment of chronic vasculitic ulcers in rheumatoid arthritis Lancet, 356: 1739–1740.

Wahl, S.M. (1999). Transforming growth factor beta. In: J. Gallian and R. Snyderman (Eds.), Inflammation: Basic Principles and Clinical Correlates. Lippincott-Raven Publishers, Philadelphia, pp. 883–892.

Welker, P., Grabbe, J., Gibbs, B., Zuberbier, T. and Henz, B.M. (2000) Nerve growth factor-beta induces mast-cell marker expression during in vitro culture of human umbilical cord blood cells Immunology, 99: 418–426.

Woodnutt, D.A., Wager-Miller, J., O'Neill, P.C., Bothwell, M. and Byers, M.R. (2000) Neurotrophin receptors and nerve growth factor are differentially expressed in adjacent nonneuronal cells of normal and injured tooth pulp Cell Tissue Res., 299: 225–236.

Wyatt, S., Shooter, E.M. and Davies, A.M. (1990) Expression of the NGF receptor gene in sensory neurons and their cutaneous targets prior to and during innervation Neuron, 4: 421–427.

Yaar, M., Eller, M.S., DiBenedetto, P., Reenstra, W.R., Zhai, S., McQuaid, T., Archambault, M. and Gilchrest, B.A. (1994) The trk family of receptors mediates nerve growth factor and neurotrophin-3 effects in melanocytes J. Clin. Invest., 94: 1550–1562.

Yanker, B.A. and Shooter, E.M. (1982) The biology and mechanism of action of nerve growth factor Annu. Rev. Biochem., 51: 845–868.

Neurotrophins and neuro-inflammatory responses

Progress in Brain Research, Vol. 146
ISSN 0079-6123

CHAPTER 24

Interactions between the cells of the immune and nervous system: neurotrophins as neuroprotection mediators in CNS injury

Rinat Tabakman[1], Shimon Lecht, Stela Sephanova, Hadar Arien-Zakay and
Philip Lazarovici*

*Department of Pharmacology, School of Pharmacy, Faculty of Medicine, The Hebrew University of Jerusalem,
Jerusalem 91120, Israel*

Abstract: Inflammatory processes in the central nervous system (CNS) are considered neurotoxic, although recent studies suggest that they also can be beneficial and confer neuroprotection (neuroprotective autoimmunity). Cells from the immune system have been detected in CNS injury and found to produce and secrete a variety of neurotrophins such as NGF, BDNF, NT-3 and NT-4/5, and to express (similarly to neuronal cells), members of the tyrosine kinase (Trk) receptor family such as TrkA, TrkB and TrkC. Indeed, autocrine and paracrine interactions are observed at the site of CNS injury, resulting in a variety of homologic-heterologic modulations of immune and neuronal cell function. The end result of the inflammatory process, neurotoxicity and/or neuroprotection, is a function of the fine balance between the two cellular systems, i.e., of the complex signaling relationships between anti-inflammatory neuroprotective factors (neurotrophins and other chemical mediators) and proinflammatory neurotoxic factors (TNF, free radicals, certain cytokines, etc.). Autoimmune neuroprotection is a novel therapeutic approach aimed at shifting the balance between the immune and neuronal cells towards survival pathways in a variety of CNS injuries. This review focuses on data supporting this concept and its future therapeutical implications for optic nerve injury and multiple sclerosis.

Keywords: neuroprotection; neurotrophins; CNS injury; inflammation; lymphocytes; NGF; BDNF; Trk

Introduction

Neurotrophins are growth factors that play an essential role in the development and maintenance of the nervous system. This family includes nerve

[1]This review is part of a PhD thesis to be submitted to the Hebrew University of Jerusalem by RT.
*Correspondence to: Philip Lazarovici, Department of Pharmacology and Experimental Therapeutics, School of Pharmacy, Faculty of Medicine, The Hebrew University of Jerusalem, P.O. Box 12065, Jerusalem 91120, Israel. Tel.: +972-2-675-8729; Fax: +972-2-675-8741; E-mail: lazph@md.huji.ac.il

growth factor (NGF), brain-derived neurotrophic factor (BDNF), neurotrophin-3 (NT-3), neurotrophin-4/5 (NT-4/5) and neurotrophin-6 (NT-6) (Barbacid, 1995; Lazarovici et al., 1997). These are all synthesized as large precursor proteins, which are then cleaved to physiologically active neurotrophins by metalloproteinases (Lee et al., 2001; Chao and Bothwell, 2002). Neurotrophins, by binding and activating their respective receptors in different tissues, induce beneficial effects that may be exploited for treatment of neurological disorders and cancer (Apfel, 1997; Haase et al., 1998; Ochs et al., 2000). Two families of receptors that interact with

DOI: 10.1016/S0079-6123(03)46024-X

neurotrophins have been found so far: the high-affinity tyrosine kinase (Trk) receptor, and the low-affinity neurotrophin receptor p75 (p75^{NTR}), a member of the tumor necrosis factor receptor superfamily (Barbacid, 1995), which will not be addressed in this review. The trk family of receptors are similar to the receptor tyrosine kinases for other growth factors, but are selectively occupied and activated by neurotrophins: TrkA binds NGF; TrkB binds BDNF, NT-3 and NT-4/5, and TrkC binds NT-3 (Kaplan and Stephens, 1994). By interacting with their Trk receptors, the neurotrophins induce receptor autophosphorylation and recruitment of signaling protein substrates that bind to Trk. Activation of the Trk receptor triggers intracellular signaling cascades, resulting in the regulation of cell survival, proliferation, differentiation and synaptic plasticity (Chao et al., 1998). Intracellular signaling cascades activated by trk receptors of major interest include the Ras/extracellular signal regulated kinase (ERK) pathway, the phosphatidylinositol-3-kinase (PI3K)/Akt kinase (also known as protein kinase B) pathway, as well as phospholipase C-γ1 (PLC-γ1), which are all engaged in the promotion of cell survival (Lazarovici et al., 1997; Freidman and Greene, 1999; Patapoutian and Reichardt, 2001).

Diseases of the central nervous system (CNS) such as Alzheimer's, Parkinson's, Multiple Sclerosis, mechanical trauma and stroke, are accompanied by cell death, and are known as neurodegenerative diseases. Therefore, for therapeutic purposes we need to protect the CNS from neurodegeneration. Three different approaches may be considered: (i) drug therapy, to support the living neurons at the site of CNS injury (Weiss et al., 1990; Zuddas et al., 1992), (ii) cell therapy, such as cell transplantation, to replace damaged neuronal structures at the site of injury (Galvin and Jones, 2002; Okano et al., 2002; Storch and Schwarz, 2002), and (iii) gene therapy, e.g., expression of anti-appoptotic proteins, to prevent neuronal cell death at the site of CNS injury (Deigner et al., 2000; Hatmann et al., 2002; Ruth and Melamed, 2002). Clinical attempts to treat peripheral neurodegenerative diseases (neuropathies) using recombinant neurotrophins have proved promising so far, raising the hope that a similar strategy could be applied to the treatment of CNS neurodegenerative disorders (Haase et al., 1998; Skaper and Walsh, 1998; Ochs et al., 2000). A major problem encountered in all attempts to delivery therapeutic molecules to the CNS has been their inability to penetrate the blood-brain barrier (BBB). Hence, in order to develop effective therapeutic modalities based on neurotrophins, one must develop appropriate pharmacokinetic methods and formulations for their delivery to the CNS (Shoichet and Winn, 2000).

A challenging therapeutic modality for treating neurodegenerative diseases is exploitation of the ability of immune cells to secrete neurotrophins during inflamatory processes at the site of CNS injury. The immune system is considered a peripheral system that provides protection against foreign molecules. Its components can be divided into three main categories: cells of the innate immune system, cells of the adaptive immune system and those of the lymphoid tissues (Roitt et al., 2001). The term 'innate immune system' refers to the immediately responding system and comprises phagocytes (monocytes/macrophages and polymorphonuclear granulocytes) and natural killer cells. The term 'adaptive immune system' refers to cells particularly invoved in the specific immune response. This system contains antigen-presenting-cells (APCs) and lymphocytes such as T cells and B cells. The lymphoid system consists of lymphocytes, accessory cells (macrophages and APCs) and, in some tissues, epithelial cells. Lately, cumulative evidence points to a parallel structure-function in the immune and nervous systems. The basic unit is the neuron in the latter, and excitability is achieved by chemical or electrical connections (synapses) between different neurons. The precise connective pattern throughout the CNS is obtained by the generation of synapses between two different axonal projections. Several key rules are generally observed in synaptic assembly: (i) the neuronal endings participating in the synapse remain distinct (i.e., there is no fusion of plasma membranes at the synapse site), (ii) several adhesion molecules located pre and postsynaptically hold the two neurons together, (iii) the synaptic connection (cleft) is stable due to the presence of extracellular matrix, and (iv) neurotransmitter is secreted from the presynaptic site. Analogous to the neuronal synapse it was found that T cell–APC interactions fulfill these criteria, suggesting an organization of immunological synapses similar to that found in the CNS (Dustin and Colman, 2002). Under normal conditions, the CNS is considered an immune protected domain, mainly due to the BBB, which limits the access of systemic immune cells. However, in the diseased state, due to infection with exogenous antigens or autoimmunity, cells from the immune system are able to penetrate the CNS, enabling the generation of both immune and neuronal synapses. This pathological state may involve cross-talk between the neuronal and immunologic synapses. In addition to infiltration of the immune cells from the periphery to the injured CNS area, neuronal microglial cells can act as local potential APCs (Ulvestad et al., 1994; Carlson et al., 1998), suggesting that these and other neuronal cells may initiate, regulate and sustain an immune response (Becher et al., 2000). This cross-talk between the nervous and immune systems has implications not only for treatment, but also for the severity of the disease. Indeed, in many diseased states, an inflammatory process develops in the CNS. Although there are studies implying that the immune activity in the CNS is pathological (Lotan and Schwartz, 1994) and is even considered an important mediator of secondary damage (Carlson et al., 1998; Mautes et al., 2000), other investigations indicates that inflammation might beneficially affect the insult by promoting clearance of cell debris and secretion of neurotrophic factors and cytokines. In this chapter we will focus on the involvement of the neurotrophins as neuroprotective mediators in the cross-talk, during CNS injury, between the immune and nervous systems.

Neurotrophins in the immune system

Following the pioneer studies of Aloe and Levi-Montalcini (1977), Aloe (1988), Aloe and De Simone (1989), Aloe and Micera (1999) demonstrating the production of neurotrophins and the activity of NGF on mast cells, the last decade has seen a variety of direct and indirect evidence pointing to the production, expression and/or secretion of neurotrophins by different cell types of the immune system (Tables 1 and 2). Macrophages, T cells and B cells of murine and human origin, which are most abundant at the site of CNS injury, produce BDNF, NGF, NT-3 and NT-4/5 (Table 1). Other immune cells, such as mast cells, monocytes, neurotrophiles, granulocytes, basophiles, eosinophiles, splenocytes and peripheral blood mononuclear cells (PBMCs), as well as myeloid progenitor cells secrete one or more neurotrophins (Table 2). All these cells express different full length and/or truncated trk receptor isoforms (Tables 1 and 2). The lack of sequencing data on the different trk receptor forms present in the immune cells does not allow a clear conclusion to be drawn regarding their homology to the CNS trk receptors. However, treatment of the immune cells with recombinant human neurotrophin or native mouse NGF demonstrated that the cells are able to respond to neurotrophin stimulation (Tables 1 and 2). The major biological effects observed are increased proliferation, survival, immunoglobulin production, antigen presentation and migration, production of proinflammatory and anti-inflammatory chemical mediators, secretion of serotonin and histamine and differentiation (Tables 1 and 2). These findings strongly suggest that the immune system can produce most if not all members of the neurotrophin family. It is also conceivable that different types of immune cells are the targets of paracrine and/or autocrine activity, as they express trk receptors. Furthermore, these findings strongly suggest that neurotrophins can mediate heterologous interactions between the immune and nervous systems.

Dual role of the CNS cellular inflammatory process: neurotoxicity and neuroprotection

When a nerve in the peripheral nervous system (PNS) or a myelinated track in the CNS is sectioned surgically or damaged by mechanical trauma, the proximal segment often survives and degenerates. In the distal segment Wallerian degeneration takes place, with both axons and myelin disappearing rapidly. The debris is phagocytized by neural cells and macrophages. It has been established that the integrity of the myelin sheath depends on continuous contact with a viable axon. Any injury or disease causing general impairment of neuronal function will also result in axonal degeneration and the onset of myelin breakdown after the initial neuronal damage. Myelin breakdown and neuronal damage affect the

TABLE 1
Production, expression and/or secretion of neurotrophins, trk receptors presence and their biological role in cells of the immune system detected in CNS injury

Cell type	Origin	Trk receptor expression[a]	Neurotrophin expression or secretion	Biological effect	References
Macrophage	Mouse peritoneum	TrkA, TrkB, TrkC	BDNF, NT-3, NGF, NT-3	NO secretion, TNF-α, IL-1 production	Susaki et al., 1996, MacMicking et al., 1997; Brodie et al., 1999; Barouch et al., 2001a,b
	Mouse alveolar interstitial cells	Truncated TrkB, full length TrkC	NT-3, NT-4/5, BDNF	nf	Hikawa et al., 2002
	Human	TrkA, TrkB, TrkC	NGF	Differentiation, increase in oxidative burst	Otten and Gadient, 1995; Garcia-Suarez et al., 1997; Ricci et al., 2002
T-cell	Mouse	TrkA	NGF	nf	Ehrhard et al., 1993a
	Rat	TrkA[a]	NGF, BDNF, NT-3	Enhanced Ab synthesis, proliferation	Manning et al., 1985; Thorpe and Perez-Polo, 1987; Barouch and Schwartz, 2002; Muhallab et al., 2002
	Human	TrkA, no TrkB	BDNF, NGF, NT-3, NT-4/5	Proliferation, production of neuropeptide Y	Otten and Gadient, 1995; Bracci-Laudiero et al., 1996; Lambiase et al., 1997; Moalem et al., 2000; Stadelmann et al., 2002
Th1	Mouse	nf	no BDNF	nf	Ehrhard et al., 1993a, 1994; Ziemssen et al., 2002
Th2	Mouse	TrkB	NGF	nf	Besser and Wank, 1999; Bonini et al., 1999
	Human	TrkA, no TrkB	BDNF, NGF	nf	Besser and Wank, 1999; Ziemssen et al., 2002
B cells	Human	TrkB	NGF, BDNF, NT-3, NT-4	proliferation, differentiation, cytoskeletal effects	Otten et al., 1989; Otten and Gadient, 1995; Melamed et al., 1995, 1996; Schenone et al., 1996; Besser and Wank, 1999; Barouch et al., 2000

nf = no information found in the presented reference.
[a]The presence of trk receptors is based on mRNA and/or protein expression and/or neurotrophin autocrine activity.

TABLE 2
Production, expression and/or secretion of neurotrophins, trk receptors presence and their biological roles in selected cell types of the immune system

Cell type	Origin	Trk receptor expression[a]	Neurotrophin expression or secretion	Biological effect	References
Mast cell	Mouse	nf	NGF	nf	Xiang and Nilsson, 2000
	Rat peritoneum	Truncated TrkA	NGF, NT-3, NT-4	proliferation, upregulation of IL-3, survival, histamine serotonin and NGF release, chemotaxis,	Aloe and Levi-Montalcini, 1977; Bruni et al., 1982; Aloe, 1988; Aloe and De Simone, 1989; Horigome et al., 1993, 1994; Leon et al., 1994; Aloe and Micera, 1999; Skaper et al., 2001; Kawamoto et al., 1995; Tam et al., 1997
	Human, leukemic	TrkA, truncated and full length TrkB	NGF, BDNF, NT-3	expression of c-fos and NGF1-A; MAPK activation	Sawada et al., 2000; Xiang and Nilsson, 2000
	Lung	TrkC	nf	nf	
	umbilical cord	TrkA, TrkB, TrkC		increased expression of chymase	
Monocyte	Human	TrkA, TrkC	NGF, BDNF, NT-4	increased respiratory burst expression of Bcl-2, Bcl-XL and Bfl-1	Kannan et al., 1992; La Sala et al., 2000; Ehrhard et al., 1993b
		TrkA	NGF		
Neutrophil	Mouse, peritoneum	nf	NGF	survival, proliferation, chemotaxis, phagocytosis	Kannan et al., 1991; Brodie and Gelfand, 1992
Granulocyte	Human	Generally lacking	NGF, BDNF, NT-4	interaction with other effector cells	Laurenzi et al., 1998
Basophile	Human	TrkA	NGF	increased IL-13 secretion	Bischoff and Dahinden, 1992; Sin et al., 2001
	Human	TrkA[b]	nf	increased histamine release, differentiation	Matsuda et al., 1988b; Bischoff and Dahinden, 1992
Eosinophile	Human	TrkA, TrkB, TrkC	NGF, NT-3	increased peroxidase and IL-4 secretion	Solomon et al., 1998; Bonini et al., 1999; Kobayashi et al., 2002; Noga et al., 2002
Mixed splenocyte	Mouse	nf	NGF, BDNF, NT-3, NT-4	nf	Barouch et al., 2000
PBMC	Human	TrkB	nf	nf	Besser and Wank, 1999
Myeloid progenitor cell	Human	nf	NGF	colony growth and differentiation	Matsuda et al., 1988a

nf = no information found in the presented reference.

[a]The presence of trk receptors is based on mRNA and/or protein expression and/or neurotrophin autocrine activity.

neuron's electrical conductivity, resulting in its irreversible functional deficits (Kalb, 1995; Hauben et al., 2000). In addition, a destructive series of injury-induced secondary waves of events occur a phenomenon known as secondary degeneration. (Faden, 1993; Yoles and Schwartz, 1998; Hauben et al., 2000). The progression of CNS damage may be attributable to the fact that acute neuronal degeneration creates a hostile environment for the neighboring neuronal fibers that are still intact, thereby propagating the damage and eventually causing degeneration of the fibers (Kipnis et al., 2000). This process of 'secondary degeneration' is likely that seen after traumatic injury to other areas of the spinal cord and brain.

Several studies have shown that immune system intervention is necessary in order to minimize the damage and to activate healing-neuroregenerative processes (Bethea and Dietrich, 2002). The inflammatory process is one of the main host defense mechanisms that also fight/protect against CNS injury or infection. This is a biological phenomenon by which lymphocytes and soluble chemical mediators regulate tissue homeostasis (Bethea and Dietrich, 2002) and, by so doing, become an important 'house keeping-repair mechanism' in the peripheral tissues. The possibility that the CNS is dependent on a similar inflammatory activity for maintenance and repair, especially after injury, is an important issue that awaits investigation. Using the rat optic nerve or the spinal cord injury model, studies have yielded insights into novel mechanisms involved in degeneration and have led to the identification of specific molecules with neuroprotective properties (Schwartz et al., 1999). In the injured CNS, as in other injured tissues, activated T cells, B cells and macrophages are required for neuronal regeneration (Schwartz et al., 1999; Schwartz and Moalem, 2001).

Macrophage invasion of the damaged CNS is slower and less extensive than that observed after PNS injury (Schwartz et al., 1999). The macrophages infiltrating the damaged CNS are less efficient in clearing the myelin debris that inhibits the neuronal growth required for axonal regeneration and for the recovery of neuronal excitability (Schwartz et al., 1999). Also the microglia, CNS resident 'macrophages', are activated after injury, although their activity is not as high as that of the peripheral-blood macrophages infiltrating the CNS (Schwartz et al., 1999). Regrowth of the CNS axons was correlated with the rapid clearance of myelin from the treated axons and the abundant distribution of PNS-activated macrophages along the distal part of the damaged axons. The macrophages probably act in synchrony with other regenerating molecules reducing the toxic effect of the extracellular environment of the injured axons. Recruitment of macrophages to the injured site is an important physiological response and one of the earliest phases of host defense and of the wound healing process (Popovich and Jones, 2003). The generation and accumulation of CNS debris at the site of injury initiates the chemotaxis of neutrophils and macrophages from the periphery, resulting in phagocytosis and the release of neurotoxins at the CNS inflammatory site. Myelin debris and cytokines, eicosanoids, oxidative radicals and nitric oxide (Minagar et al., 2002), may be toxic to neurons and glia (Popovich and Jones, 2003), and might trigger the release of inflammatory mediators that can cause necrotic cell death and demyelination (Popovich and Jones, 2003).

The macrophages are not the only participants in the inflammatory process after CNS injury. The second arm of the immune system are the T cells which, in contrast to the macrophages, respond to specific antigens and 'remember' past experiences. When activated, the T cells can kill their target cells or produce signal molecules that activate or suppress the growth, movement or differentiation of other cells (Schwartz et al., 1999). The BBB, normally inaccessible to resting T cells, is permeable to activated T cells. The latter, however, do not accumulate in the healthy CNS unless they recognize and are able to react with their specific antigens (Schwartz et al., 1999). Some studies have revealed a greater accumulation of endogenous T cells in the injured PNS than in the damaged CNS (Moalem et al., 1999a). Moreover, it has been shown that elimination (by apoptosis) of T-cells from the CNS, is more effective than in the PNS, and that this phenomenon is virtually absent from tissues such as muscle and skin, again emphasizing the concept of the brain's immune privilege (Gold et al., 1997). The T cell response in the injured CNS is restricted and strictly regulated (Schwartz et al., 1999); the implication of this finding for regeneration of the injured CNS is not clear. It was shown that T cells specific for a myelin determinant can protect CNS neurons from secondary degeneration (Moalem et al., 1999b). Using the in vivo experimental model of partial lesion of the rat optic nerve, it was found that axonal injury was followed by the transient accumulation of endogenous T cells at the site of the lesion (Schwartz et al., 1999). Administration of in vitro activated syngeneic T cells specific for CNS self-antigen myelin basic protein (MBP) resulted in the local accumulation of T cells at the injured site in the optic nerve in vivo (Moalem et al., 1999b). Rats inoculated with the anti-MBP T cells (T_{MBP}) showed significantly less secondary degeneration of retinal neurons than animals that received phosphate-buffered saline or T cells specific for the foreign antigen ovalbumin (T_{OVA}) (Moalem et al., 1999b). The neuroprotective effect on retinal neurons was distinct despite the finding that the transferred anti-T_{MBP} cells induced a transient, monophasic paralytic disease known as experimental autoimmune encephalomyelitis (EAE)—a known animal model of multiple sclerosis (MS) (Schwartz et al., 1999).

The induction and activation of T lymphocytes require at least two signals from specific APCs. The first signal results from the binding of the T cell receptor to its antigen in the context of the major histocompatibility complex (MHC). The second is provided by various molecules expressed by the APCs (Friendl and Gunzer, 2001). There are at least two types of APCs that activate T cells: macrophages and B lymphocytes. The neuroprotective effect of T_{MBP} cells in the CNS might be the result of local cross-talk with APCs and/or a local effect on the CNS injury. In several CNS injury models it was found that the recovery of motor neurons after facial nerve transection in SCID mice, which lack functional T and B cells, is inferior to that in normal mice (Serpe et al., 1999, 2000; Barouch and Schwartz, 2002). Using the rat optic nerve injury model it was shown that administration of T_{MBP} induces a local immune response, which is manifested by an increase in the number of microglia/macrophages and B cells (Barouch and Schwartz, 2002).

The number of B cells at the site of injured optic nerve is very low. However, upon injection of T_{MBP} they accumulated at this site in far higher number than in the control rats, as evaluated 4 and 7 days after injury. No B cells were detected in the uninjured optic nerves of rats inoculated with T_{MBP}. The accumulation of B cells upon injection of T_{MBP} was transient and had decreased by 7 days after injury. It was also found that in addition to the augmented number of B cells, there was a transient increase in the expression of neurotrophins, particularly NGF. In this experimental model of injury the T cells did not express NGF but did express BDNF and NT-3; the B cells expressed NGF, BDNF and NT-3 and the macrophages-NGF and NT-3, but not BDNF. According to these findings, the type and quantity of neutrophins are dependent on the interplay between T cells, B cells and macrophages at the site of injury. It is conceivable that the accumulation of T cells at the injured site is a regulatory step in the recruitment of B cells and macrophages to the site of injury. It is also plausible that T cells secrete certain cytokines that mediate such accumulation, as well as the neuroprotective effect of T_{MBP}. The substantial expression of neurotrophins by B cells might have important implications for the survival of injured neurons. Recent studies demonstrated that also B cells can have a destructive effect after CNS insult (Schori et al., 2002). Therefore, it is possible that in CNS injury the beneficial B cell effect depends on proper T cell regulation and occurs in response to the accumulation of T cells at the site of CNS injury (Barouch and Schwartz, 2002).

Immune cell-derived neurotrophins as potential mediators of neuroprotection

As mentioned above, injection of activated T_{MBP} cells, but not T_{OVA} cells, reduces secondary neuronal degeneration after primary injury of CNS axons (Moalem et al., 1999b). Although, the mechanism underlying this autoimmune neuroprotection is not known, there are several findings suggesting that neurotrophins may serve as mediators between the immune and nervous systems. We will focus on the contribution of the neurotrophins to the autoimmune neuroprotective effect.

Based on the expression of different cytokine mRNAs, T cells accumulating at the site of the CNS injury may be classified as having a Th1, Th2 or Th3 phenotype (Moalem et al., 2000). Also, coexpression

TABLE 3

Differential production and secretion of neurotrophins by immune cells at the site of CNS injury

Neurotrophins secreted by immune system cells at the site of inflamation			
Injury model	Immune cell type detected at CNS site of injury	Produced and secreted neurotrophins	References
Optic nerve	macrophages	NGF, BDNF, NT3	Moalem et al., 2000; Barouch and Schwartz, 2002
	Th1	NT3	
	T cells: Th2	BDNF, NT4/5	
	Th3	BDNF	
	B cells	NGF, BDNF, NT3	
Multiple sclerosis	microglia	BDNF	Stadelmann et al., 2002; Ziemssen et al., 2002
	Th1	BDNF	
	T cells:		
	Th2	BDNF	
Nigrostriatal dopaminergic injury	microglia	GDNF, BDNF	Batchelor et al., 1999, 2002
EAE	T cells	BDNF, NT3	Hammarberg et al., 2000
	NK cells	BDNF, NT3	
Sustained cerebral embolism	T cells	NGF	Mizuma et al., 1999
HIV-1 encephalitis	microglia	BDNF, NT3	Soontornniyomkij et al., 1998

of specific cytokines: INFγ (Th1), IL-6 and IL-10 (Th2) and TGFβ (Th3) and of different neurotrophins enables the identification of neurotrophins according to T cell subtype. By evaluating mRNA levels, it was found that Th1 cells express NT-3, Th2 cells express NT-4/5 and Th3 cells express BDNF (Table 3). Neurotrophin detection by immunocytochemical techniques shows not only the presence of mRNA but also the production of neurotrophin proteins by the respective T cell populations. Using antibodies selective for individual neurotrophins and T cell receptor, flow cytometry analysis of blood-derived T cells or in vitro grown and activated T cells, confirmed the selective expression of BDNF, NGF, NT-3 by T cells. The use of ELISA has demonstrated that the secretion of neurotrophins by T cells is significantly augmented upon their exposure in vitro to MBP. Cumulatively, these findings suggest the expression, production and secretion of selective neurotrophins by T cells in the optic nerve injury model (Moalem et al., 2000).

Contribution of Trk receptors to the neuroprotective effect in CNS injury

Neuroprotection of the injured neuron at the site of inflammation can be afforded in two ways: (i) via an indirect effect through the autocrine activity of secreted neurotrophins on the respective Trk receptors in different T cell subpopulations and in other immune cells (e.g., B cells and macrophages); the possible physiological significance would be increased proliferation of the T lymphocytes, resulting in the production of an appropriate amount of neurotrophins; these would bind to the neuronal Trk receptor, leading to a significant neuroprotective effect at the injured target site; it is also conceivable that T lymphocytes, autocrinely activated by neurotrophins, produce additional mediators such as IL-6, which can affect neurons expressing IL-6 receptors (Carlson et al., 1999), and (ii) via a direct effect of secreted neurotrophins on the Trk receptors in the injured optic nerve. Indeed, expression of mRNA for TrkA and TrkB receptors, but not for TrkC receptors, was detected in the optic nerve, suggesting the presence of the respective receptors (Moalem et al., 2000). Therefore, it is very tempting to assume that the secreted neurotrophins activate the appropriate Trk receptors on the injured optic nerve and induce survival signal transduction pathways such as Ras/ERK and/or AKT/PI3 kinase. The final results would be the reduced cell death of retinal ganglion neurons, i.e., neuroprotection.

To substantiate the concept of Trk-induced neuroprotection, a pharmacological tool is available: K-252a, a relatively selective inhibitor of Trk family receptors, with a complex dual effect, which may be considered a mixed agonist/antagonist (Rasouly et al., 1992). In the presence of neurotrophins this compound behaves as an antagonist and blocks receptor activation by the neurotrophins. In their absence, K-252a acts as an agonist and stimulates Trk receptors (Knusel and Hefti, 1992). When applied locally as eye drops to animals with optic nerve injury, it inhibited Trk receptor activity. This, in turn, reduced the neuroprotective effect of the T_{MBP} cells (Moalem et al., 2000). In the same rat model, however, the time course and the severity of the EAE symptoms were not changed. The ineffectiveness may be explained by the low concentration of K-252a that entered into the systemic blood circulation, as opposed to its high local concentration in the eye. In SJL/J mice fed an oral formulation of K-252a, the appearance of EAE symptoms was delayed (neuroprotective effect), a phenomenon correlating with immunosuppression of the peripheral macrophages (Lazarovici et al., 1996). It is tempting to speculate that high systemic levels of K-252a inhibited the Trk receptors of the peripheral macrophages, resulting in less infiltration into the CNS, i.e., less destructive inflammation, as measured by a reduction in the severity of neurological symptoms (mean neurological score) of EAE model animals. It is also conceivable that K-252a inhibited the activated Trk receptors in the CNS neurons at the site of inflammation. Since overexpression of trk receptors or extensive trk receptor activity is oncogenic and/or neurotoxic (Nakagawara, 2001), such inhibition may impart neuroprotection. Assuming a high level of neurotrophins at the site of CNS injury resulting in hyperactivation, it is possible that under these circumstances the activity of K252a will predominate, promoting neuroprotection.

Multiple sclerosis as a potential target for neuroprotective autoimmunity

For many years it was considered that when the immune cells attack the nervous system, a neuroimmunological disease develops. Multiple sclerosis (MS) and its animal models provide the paradigm for such a deleterious interaction between cells of the

immune and nervous systems (Hohlfeld et al., 2000). Multiple sclerosis, a demyelinating disorder of the CNS of unknown etiology, it is thought to be an autoimmune disease mediated by T cells specific for the myelin antigen. MBP has been studied as a potential autoantigen in the disease because of its role as an encephalitogen in experimental autoimmune encephalomyelitis and postviral encephalomyelitis, and due to the presence in the blood of multiple sclerosis patients of in vivo activated T cells reactive to MBP (Ota et al., 1990). Immunological, clinical and pathological studies suggest that T lymphocytes directed against myelin antigens are involved in the pathogenesis of MS. Research on an experimental animal model of MS, EAE, demonstrated that MBP or proteolipid protein-(PLP)-specific T cells mediate the destruction of CNS myelin. In recent years, studies on EAE showed that encephalitogenic T cells recognize short MBP peptides (Martin, 1997). These findings suggest that injury-induced activation of myelin-reactive T cells may have physiological or pathological sequelae. Is it possible that the same inflammatory process is at the same time both neurodegenerative and neuroprotective in MS?

Recent studies strongly suggest that BDNF may be the key element for understanding the complex interactions between the immune and nervous system in MS patients. Neurons are considered to be one of the cellular sources of BDNF in the CNS (Stadelmann et al., 2002). Immune cells are capable of producing BDNF (Tables 1 and 2), which binds to TrkB receptors. Both are present in the CNS and in the immune cells. Immune cell-derived BDNF is detectable in human immune-mediated CNS inflammation (Stadelmann et al., 2002; Ziemssen et al., 2002, Table 3). Recent in vivo studies show that a higher percentage of inflammatory cells in actively demyelinating areas of MS lesions contain BDNF than in areas without ongoing myelin breakdown. In addition, a high percentage of BDNF$^+$ macrophages and lymphocytes are present in perivascular infiltrates (Stadelmann et al., 2002). These observations indicate that relatively high numbers of BDNF-containing immune cells are found in the early stages of CNS lesion development. BDNF also can be retrogradly transported from neurons outside the lesion area (Canals et al., 2001; Stadelmann et al.,

2002), such transport being upregulated after axotomy (Tonra et al., 1998; Stadelmann et al., 2002). It is possible that autocrine as well as paracrine interactions, through BDNF binding to the TrkB receptor, further enhance endogenous neurotrophic support in MS lesions. The neuroprotective capacity of immune cells is not restricted to BDNF. As mentioned above, a panel of other potentially neuroprotective and neurotrophic mediators is released at the site of the injured CNS. The complex neuronal-immune-neurotrophin network likely works in concert to preserve the axons in a microenviroment capable of producing significant neurotoxicity (Stadelmann et al., 2002). This concept may be extended to other models of CNS injury: nigrostriatal dopaminergic (Parkinson's), EAE (MS), cerebral embolism (stroke), and HIV encephalitis (CNS inflammation) (Table 3). In all these models of injury, like in the optic nerve and multiple sclerosis models, different members of the neurotrophin family are secreted by the immune cells at the site of inflammation, substantiating the notion that neuro-protective autoimmunity is provided by neurotro-phins released at the site of inflammation.

Conclusions

CNS injury is an inflammatory process involving heterologous interactions between immune cells and neural cells, brokered by chemical mediators inclu-ding cytokines and neurotrophins. For a better understanding of this process we need to identify all the molecules secreted at the site of CNS injury. It is clear that some of these molecules exert a neurotoxic effect while others yield neuroprotection. It is evident that immune cells produce and secrete neurotrophins such as NGF, BDNF, NT-3 and NT-4/5. This process also occurs at the site of CNS injury where it may even be amplified. Since both immune cells and neural cells express the neurotrophin trk receptors, it is reasonable to assume that the neurotrophins affect the immune and the neural cells both autocrinically and paracrinically. The end result will be neurotoxicity and/or neuroprotection, depending on the final balance between the released neurotoxic molecules and neuroprotective chemicals (neurotro-phins). CNS inflammatory responses occurring after injury contribute to both neuroconstructive and neurodestructive processes in the nervous system (Bethea and Dietrich, 2002). The key to successful neuronal regeneration is a balance between up-regulation of the repair potential while maintaining and/or reducing the destructive potential (Popovich and Jones, 2003).

The cellular therapeutic implications of the findings presented in this review may lead to novel manipulations of autoimmunite processes as a neuroprotective strategy for treatment of CNS injury (Schwartz and Kipnis, 2001). One approach would be to neutralize the neurotoxic molecules with anti-bodies. Another approach to be considered is inhibiting the gene expression of neurotoxic cytokines of infiltrating immune cells. Transcription factors of B and T cells, and macrophages, which regulate the expression of neurotoxic molecules, may be important targets for novel immunosupressive drugs. A cellular therapeutic approach could be the adoptive transfer of activated autoimmune cells specific for a CNS antigen. To this end, immune cells from the blood of CNS injured patients, would be isolated, propagated in vitro to secrete neurotrophins and reinjected. Presumably, the injected cells would reach the site of CNS injury where they could exert a neuroprotective effect as recently reported in the treatment of acute spinal cord injury with macrophages (http://www.proneuron.com).

Cell therapy approaches can not yet totally replace the use of low molecular weight drugs in the treatment of CNS injury. Therefore, novel drugs, such as Trk receptor agonists, need to be developed. The long-term pharmaceutical goal must be the development of agonist drugs with high selectivity towards individual Trks and the ability to distinguish between trk receptor subtypes belonging to immune cells and neuronal cells. The recent synthesis of CEPs (Roux et al., 2002), pure trk receptor antagonists, is an encouraging pharmacological development which may contribute to research on Trk receptor agonists. Drugs modulating the activity of Trk receptor activity may contribute to investiga-tions into the neuroprotective role of Trk receptors in CNS models of injury and cross-talk with the immune system and might prove useful in the clinical setting.

Abbreviations

NGF	nerve growth factor
BDNF	brain-derived neurotrophic factor
NT-3	neurotrophin 3
NT-4/5	neurotrophin 4/5
Trk	tropomyosin-related tyrosine kinase
PNS	peripheral nervous system
CNS	central nervous system
BBB	blood-brain barrier
APC	antigen presenting cell
MBP	myelin basic protein
EAE	experimental autoimmune encephalomy-elitis
T_{MBP}	T cells activated by MBP
T_{OVA}	T cells activated by ovalbumin
MHC	major histocompatibility complex
ELISA	enzyme-linked immunosorbent assay
IL-6	interleukin 6
mRNA	messenger RNA
$p75^{NTR}$	low-affinity neurotrophin receptor
K252a	a selective inhibitor of the Trk receptor
MS	multiple sclerosis.

References

Aloe, L. and Levi-Montalcini, R. (1977) Mast cells increase in tissues of neonatal rats injected with the nerve growth factor. Brain Res., 133: 358–366.

Aloe, L. (1988) The effect of nerve growth factor and its antibody on mast cell in vivo. J. Neuroimmunol., 18: 1–12.

Aloe, L. and De Simone, R. (1989) NGF primed spleen cells injected in brain of developing rats differentiate into mast cells. Int. J. Devl. Neurosci., 7: 565–573.

Aloe, L. and Micera, A. (1999) Nerve growth factor: basic findings and clinical trials. Biomed. Rev., 10: 3–14.

Apfel, S.C. (1997) Clinical Applications of Neurotrophic Factors. Lippincott-Raven, Philadelphia, pp. 1–209.

Barbacid, M. (1995) Neurotrophic factors and their receptors. Curr. Opin. Cell Biol., 7: 148–155.

Barouch, R., Appel, E., Kazimirsky, G., Braun, A., Renz, H. and Brodie, C. (2000) Differential regulation of neurotrophin expression by mitogens and neurotransmitters in mouse lymphocytes. J. Neuroimmunol., 103: 112–121.

Barouch, R., Appel, E., Kazimirsky, G. and Brodie, C. (2001a) Macrophages express neutrophins and neutrophin receptors. Regulation of nitric oxide production by NT-3. J. Neuroimmunol., 112: 72–77.

Barouch, R., Kazimirsky, G., Appel, E. and Brodie, C. (2001b) Nerve growth factor regulates TNF-α production in mouse macrophages via MAP kinase activation. J. Leukoc. Biol., 69: 1019–1026.

Barouch, R. and Schwartz, M. (2002) Autoreactive T cells induce neurotrophin production by immune and neural cells in injured rat optic nerve: implications for protective autoimmunity. FASEB J., 16: 1304–1306.

Batchelor, P.E., Liberatore, G.T., Wong, J.Y., Porritt, M.J., Frerichs, F., Donnan, G.A. and Howells, D.W. (1999) Activated macrophages and microglia induce dopaminergic sprouting in the injured striatum and express brain-derived neurotrophic factor and glial cell line-derived neurotrophic factor. J. Neurosci., 19: 1708–1716.

Batchelor, P.E., Porritt, M.J., Martinello, P., Parish, C.L., Liberatore, G.T., Donnan, G.A. and Howells, D.W. (2002) Macrophages and microglia produce local trophic gradients that stimulate axonal sprouting toward but not beyond the wound edge. Mol. Cell Neurosci., 21: 436–453.

Becher, B., Prat, A. and Antel, J.P. (2000) Brain-immune connection: immuno-regulatory properties of CNS-resident cells. Glia, 29: 293–304.

Besser, M. and Wank, R. (1999) Cutting edge: clonally restricted production of the neurotrophins brain-derived neurotrophic factor and neurotrophin-3 mRNA by human immune cells and Th1/Th2-polarized expression of their receptors. J. Immunol., 162: 6303–6306.

Bethea, J.R. and Dietrich, W.D. (2002) Targeting the host inflammatory response in traumatic spinal cord injury. Curr. Opin. Neurol., 15: 355–360.

Bischoff, S. and Dahinden, C. (1992) Effect of nerve growth factor on the release of inflammatory mediators by mature human basophils. Blood, 79: 2662–2669.

Bonini, S., Lambiase, A., Bonini, S., Levi-Schaffer, F. and Aloe, L. (1999) Nerve growth factor: an important molecule in allergic inflammation and tissue remodelling. Int. Arch. Allergy Immunol., 118: 159–162.

Bracci-Laudiero, L., Aloe, L., Stenfors, C., Tirassa, P., Theodorsson, E. and Lundberg, T. (1996) Nerve growth factor stimulates production of neuropeptide Y in human lymphocytes. Neuroreport, 7: 485–488.

Brodie, C. and Gelfand, E.W. (1992) Functional nerve growth factor receptors on human B lymphocytes: interaction with IL-2. J. Immunol., 148: 3492–3497.

Brodie, C., Barouch, R., Appel, E. and Kazimirsky, G. (1999) Expression of neutrophins and neutrophin receptors on macrophages. Induction of nitric oxide and activation of signal transduction pathways. Neurosci. Lett. Suppl., 54: S10–S11.

Bruni, A., Bigon, E., Boarato, E., Mietto, L., Leon, A. and Toffano, G. (1982) Interaction between nerve growth factor and lysophosphatidylserine on rat peritoneal mast cells. FEBS Lett., 138: 190–193.

Canals, J.M., Checa, N., Marco, S., Akerud, P., Michels, A., Perez-Navarro, E., Tolosa, E., Arenas, E. and Alberch, J.

398

(2001) Expression of brain-derived neurotrophic factor in cortical neurons is regulated by striatal target area. J. Neurosci., 21: 117–124.

Carlson, S.L., Parrish, M.E., Springer, J.E., Doty, K. and Dossett, L. (1998) Acute inflammatory response in spinal cord following impact injury. Exp. Neurol., 151: 77–88.

Carlson, N.G., Wieggel, W.A., Chen, J., Bacchi, A., Rogers, S.W. and Gahring, L.C. (1999) Inflammatory cytokines IL-1 alpha, IL-1 beta, IL-6, and TNF-alpha impart neuroprotection to an excitotoxin through distinct pathways. J. Immunol., 163: 3963–3968.

Chao, M., Casaccia-Bonnefil, P., Carter, B., Chittka, A., Kong, H. and Yoon, S.O. (1998) Neurotrophin receptors: mediators of life and death. Brain Res. Rev., 26: 295–301.

Chao, M.V. and Bothwell, M. (2002) Neurotrophins: to cleave or not to cleave. Neuron, 33: 9–12.

Deigner, H.P., Haberkorn, U. and Kinscherf, R. (2000) Apoptosis modulators in the therapy of neurodegenerative diseases. Expert Opin. Investig. Drugs, 9: 747–764.

Dustin, M.L. and Colman, D.R. (2002) Neural and immunological synaptic relations. Science, 298: 785–789.

Ehrhard, P.B., Erb, P., Graumann, U. and Otten, U. (1993a) Expression of nerve growth factor and nerve growth factor receptor tyrosine kinase Trk in activated CD4-positive T-cell clones. Proc. Natl. Acad. Sci. USA, 90: 10984–10988.

Ehrhard, P.B., Ganter, U., Bauer, J. and Otten, U. (1993b) Expression of functional trk protooncogene in human monocytes. Proc. Natl. Acad. Sci. USA, 90: 5423–5427.

Ehrhard, P.B., Erb, P., Graumann, U., Schmutz, B. and Otten, U. (1994) Expression of functional trk tyrosine kinase receptors after T cell activation. J. Immunol., 152: 2705–2709.

Faden, A.I. (1993) Experimental neurobiology of central nervous system trauma. Crit. Rev. Neurobiol., 7: 175–186.

Freidman, W.J. and Greene, L.A. (1999) Neurotrophin signaling via Trks and p75. Exp. Cell Res., 253: 131–142.

Friendl, P. and Gunzer, M. (2001) Interaction of T cells with APCs: the serial encounter model. Trends Immunol., 22: 187–191.

Galvin, K.A. and Jones, D.G. (2002) Adult human neural stem cells for cell-replacement therapies in the central nervous system. Med. J. Aust., 177: 316–318.

Garcia-Suarez, O., Hannestad, J., Esteban, I., Martinez del Valle, M., Naves, F.J. and Vega, J.A. (1997) Neurotrophin receptor-like protein immunoreactivity in human lymph nodes. Anat. Rec., 249: 226–232.

Gold, R., Hartung, H.P. and Lassmann, H. (1997) T-cell apoptosis in autoimmune diseases: termination of inflammation in the nervous system and other sites with specialized immune-defense mechanisms. Trends Neurosci., 20: 399–404.

Haase, G., Pettmann, B., Vigne, E., Castelnau-Ptakhine, L., Schmalbruch, H. and Kahn, A. (1998) Adenovirus-mediated transfer of the neurotrophin-3 gene into skeletal muscle of pmn mice: therapeutic effects and mechanisms of action. J. Neurol. Sci., 1: 97–105.

Hatmann, A., Mouatt-Prigent, A., Vila, M., Abbas, N., Perier, C., Faucheux, B.A., Vyas, S. and Hirsch, E.C. (2002) Increased expression and redistribution of the anti-apoptotic molecule Bcl-x$_L$ in Parkinson's disease. Neurobiol. Dis., 10: 28–32.

Hammarberg, H., Lidman, O., Lundberg, C., Eltayeb, S.Y., Gielen, A.W., Muhallab, S., Svenningsson, A., Linda, H., van der Meide, P.H., Cullheim, S., Olsson, T. and Piehl, F. (2000) Neuroprotection by encephalomyelitis: rescue of mechanically injured neurons and neurotrophin production by CNS-Infiltrating T and natural killer cells. J. Neurosci., 20: 5283–5291.

Hauben, E., Butovsky, O., Nevo, U., Yoles, E., Moalem, G., Agranov, E., Mor, F., Leibowitz-Amit, R., Pevsner, E., Akselrod, S., Neeman, M., Cohen, I.R. and Schwartz, M. (2000) Passive or active immunization with myelin basic protein promotes recovery from spinal cord contusion. J. Neurosci., 20: 6421–6430.

Hikawa, S., Kobayashi, H., Hikawa, N., Kusakabe, T., Hiruma, H., Takenaka, T., Tomita, T. and Kawakami, T. (2002) Expression of neutrophins and their receptors in peripheral lung cells of mice. Histochem. Cell Biol., 118: 51–58.

Hohlfeld, R., Kerschensteiner, M., Stadelmann, C., Lassmann, H. and Wekerle, H. (2000) The neuroprotective effect of inflammation: implications for the therapy of multiple sclerosis. J. Neuroimmunol., 107: 161–166.

Horigome, K., Pryor, J.C., Bullock, E.D. and Johnson, E.M. (1993) Mediator release from mast cells by nerve growth factor: neurotrophin specificity and receptor mediation. J. Biol. Chem., 268: 14881–14887.

Horigome, K., Bullock, E.D. and Johnson, E.M. (1994) Effect of nerve growth factor on rat peritoneal mast cells: survival promotion, and immediate-early gene induction. J. Biol. Chem., 269: 2695–2702.

Kalb, L.Y. (1995) Recovery from spinal cord injury: new approaches. Neuroscientist, 1: 321–327.

Kannan, Y., Ushio, H., Koyama, H., Okada, M., Oikawa, Y.T., Kaneko, M. and Matsuda, H. (1991) Nerve growth factor enhances survival, phagocytosis and superoxide production of murine neutrophils. Blood, 77: 1320–1325.

Kannan, Y., Usami, K., Okada, M., Shimizu, S. and Matsuda, H. (1992) Nerve growth factor suppresses apoptosis of murine neutrophils. Biochem. Biophys. Res. Commun., 186: 1050–1056.

Kaplan, D.R. and Stephens, R.M. (1994) Neurotrophin signal transduction by the Trk receptor. J. Neurobiol., 25: 1404–1417.

Kawamoto, K., Okada, T., Kannan, Y., Ushio, H., Matsumoto, M. and Matsuda, H. (1995) Nerve growth factor prevents apoptosis of rat peritoneal mast cells through the trk proto-oncogene receptor. Blood, 86: 4638–4644.

Kipnis, J., Yoles, E., Porat, Z., Cohen, A., Mor, F., Sela, M., Cohen, I.R. and Schwartz, M. (2000) T cell immunity to copolymer 1 confers neuroprotection on the damaged optic nerve: possible therapy for optic neuropathies. Proc. Natl. Acad. Sci. USA, 97: 7446–7451.

Knusel, B. and Hefti, F. (1992) K-252 compounds: modulators of neurotrophin signal transduction. J. Neurochem., 59: 1987–1996.

Kobayashi, H., Gleich, G.J., Butterfield, J.H. and Kita, H. (2002) Human eosinophils produce neurotrophins and secrete nerve growth factor on immunologic stimuli. Blood, 99: 2214–2220.

La Sala, A., Corinti, S., Federici, M., Saragovi, H.U. and Girolomoni, G. (2000) Ligand activation of nerve growth factor receptor trkA protects monocytes from apoptosis. J. Leukoc. Biol., 68: 104–110.

Lambiase, A., Bracci-Laudiero, L., Bonini, S., Bonini, S., Starace, G., D'Elios, M., De Carli, M. and Aloe, L. (1997) Human CD4 + T cell clones produce and release nerve growth factor and express high-affinity nerve growth factor receptors. J. Allergy. Clin. Immunoly., 100: 408–414.

Laurenzi, M.A., Beccari, T., Stenke, L., Sjolinder, M., Stinchi, S. and Lindgren, J.A. (1998) Expression of mRNA encoding neutrophins and neutrophin receptors in human granulocytes and bone marrow cells-enhanced neurotrophin-4 expression induced by LTB4. J. Leukoc. Biol., 64: 228–234.

Lazarovici, P., Rasouly, D., Friedman, L., Tabakman, R., Ovadia, H. and Matsuda, Y. (1996) K_{252a} and staurosporine microbial alkaloid toxins as prototype of neurotrophic drugs. In: B.R. Singh and A.T. Tu (Eds.), Natural Toxins II. Plenum Press, New York, pp. 367–377.

Lazarovici, P., Matsuda, Y., Kaplan, D. and Guroff, G. (1997) The protein kinase inhibitors K-252a and staurosporine as modifiers of neurotrophin receptor signal transduction. In: Y. Gutman and P. Lazarovici (Eds.), Cellular and Molecular Mechanisms of Toxin Action: Toxins and Signal Transduction, Vol. 1, Harwood Academic Publishers, Amsterdam, pp. 69–94.

Lee, R., Kermani, P., Teng, K.K. and Hempstead, B.L. (2001) Regulation of cell survival by secreted proneurotrophins. Science, 294: 1945–1948

Leon, A., Buriani, A., Dal Toso, R., Fabris, M., Romanello, S., Aloe, L. and Levi-Montalcini, R. (1994) Mast cells synthesize, store, and release nerve growth factor. Proc. Natl. Acad. Sci. USA, 91: 3739–3743.

Lotan, M. and Schwartz, M. (1994) Cross talk between the immune system and the nerve system in response to injury: implications for regeneration. FASEB J., 8: 1026–1033.

MacMicking, J., Xie, Q.W. and Nathan, C. (1997) Nitric oxide and macrophage function. Annu. Rev. Immunol., 15: 323–350.

Manning, P.T., Russell, J.H., Simmons, B. and Johnson, E.M. (1985) Protection from guanethidine-induced neuronal destruction by nerve growth factor: effect of NGF on immune function. Brain Res., 340: 61–69.

Martin, R. (1997) Immunological aspects of experimental allergic encephalomyelitis and multiple sclerosis and their application for new therapeutic strategies. J. Neural. Transm. Suppl., 49: 53–67.

Matsuda, H., Coughlin, M.D., Bienenstock, J. and Denburg, J.A. (1988a) Nerve growth factor promotes human hemopoietic colony growth and differentiation. Proc. Natl. Acad. Sci. USA, 85: 6508–6512.

Matsuda, H., Switzer, J., Coughlin, M.D., Bienenstock, J. and Denburg, J.A. (1988b) Human basophilic cell differentiation promoted by 2.5S nerve growth factor. Int. Arch. Allergy. Appl. Immunol., 86: 453–457.

Mautes, A.E., Weinzierl, M.R., Donovan, F. and Noble, L.J. (2000) Vascular events after spinal cord injury: contribution to secondary pathogenesis. Phys. Ther., 80: 673–687.

Melamed, I., Turner, C.E., Aktories, K., Kaplan, D.R. and Gelfand, E.W. (1995) Nerve growth factor trigges microfilament assembly and paxillin phosphorylation in human B lymphocytes. J. Exp. Med., 181: 1071–1079.

Melamed, I., Kelleher, C.A., Franklin, R.A., Brodie, C., Hempstead, B., Kaplan, D. and Gelfand, E.W. (1996) Nerve growth factor signal transduction in human B lymphocytes is mediated by $gp140^{trk}$. Eur. J. Immunol., 26: 1985–1992.

Minagar, A., Shapshak, P., Fujimura, R., Ownby, R., Heyes, M. and Eisdorfer, C. (2002) The role of macrophage/microglia and astrocytes in the pathogenesis of three neurologic disorders: HIV-associated dementia, Alzheimer disease, and multiple sclerosis. J. Neurol. Sci., 202: 13–23.

Mizuma, H., Takagi, K., Miyake, K., Takagi, N., Ishida, K., Takeo, S., Nitta, A., Nomoto, H., Furukawa, Y. and Furukawa, S. (1999) Microsphere embolism-induced elevation of nerve growth factor level and appearance of nerve growth factor immunoreactivity in activated T-lymphocytes in the rat brain. J. Neurosci. Res., 55: 749–761.

Moalem, G., Monsonego, A., Shani, Y., Cohen, I.R. and Schwartz, M. (1999a) Differential T cell response in central and peripheral nerve injury: connection with immune privilege. FASEB J., 13: 1207–1217.

Moalem, G., Leibowitz-Amit, R., Yoles, E., Mor, F., Cohen, I.R. and Schwartz, M. (1999b) Autoimmune T cells protect neurons from secondary degeneration after central nervous system axotomy. Nat. Med., 5: 49–55.

Moalem, G., Gdalyahu, A., Shani, Y., Otten, U., Lazarovici, P., Cohen, I.R. and Schwartz, M. (2000) Production of neurotrophins by activated T cells: implications for neuroprotective autoimmunity. J. Autoimmun., 15: 331–345.

Muhallab, S., Lundberg, C., Gielen, A.W., Lidman, O., Svenningsson, A., Piehl, F. and Olsson, T. (2002) Differential expression of neurotrophic factors and inflammatory cytokines by myelin basic protein-specific and other recruited T cells infiltrating the central nervous system during experimental autoimmune encephalomyelitis. Scand. J. Immunol., 55: 264–273.

Nakagawara, A. (2001) Trk receptor tyrosine kinases: a bridge between cancer and neural development. Cancer Lett., 169: 107–114.

Noga, O., Englmann, C., Hanf, G., Grutzkau, A., Guhl, S. and Kunkel, G. (2002) Activation of the specific neurotrophin receptors TrkA, TrkB and TrkC influences the function of eosinophils. Clin. Exp. Allergy, 32: 1348–1354.

Ochs, G., Penn, R.D., York, M., Giess, R., Beck, M., Tonn, J., Haigh, J., Malta, E., Traub, M., Sendtner, M., Toyka, K.V. (2000) A phase I/II trial of recombinant methionyl human brain derived neurotrophic factor administered by intrathecal infusion to patients with amyotrophic lateral sclerosis. Amyotroph. Lateral. Scler. Other Motor Neuron Disord., 1: 201–206.

Okano, H., Yoshizaki, T., Shimazaki, T. and Sawamoto, K. (2002) Isolation and transplantation of dopaminergic neurons and neural stem cells. Parkinsonism Relat. Disord., 9: 23–28.

Ota, K., Matsui, M., Milford, E.L., Mackin, G.A., Weiner, H.L. and Hafler, D.A. (1990) T-cell recognition of an immunodominant myelin basic protein epitope in multiple sclerosis. Nature, 346: 183–187.

Otten, U., Ehrhard, P. and Peck, R. (1989) Nerve growth factor induces growth and differentiation of human B lymphocytes. Proc. Natl. Acad. Sci. USA, 89: 10059–10063.

Otten, U. and Gadient, R.A. (1995) Neurotrophins and cytokines–intermediaries between the immune and nervous systems. Int. J. Dev. Neurosci., 13: 147–151.

Patapoutian, A. and Reichardt, L. (2001) Trk receptors: mediators of neurotrophin action. Curr. Opin. Neurobiol., 11: 272–280.

Popovich, P.G. and Jones, T.B. (2003) Manipulating neuroinflammatory reactions in the injured spinal cord: back to basics. Trends Pharmacol. Sci., 24: 13–17.

Rasouly, D., Rahamim, E., Lester, D., Matsuda, Y. and Lazarovici, P. (1992) Staurosporine-induced neurite outgrowth in PC12 cells is independent of protein kinase C inhibition. Mol. Pharmacol., 42: 35–43.

Ricci, A., Greco, S., Mariotta, S., Felici, L., Amenta, F. and Bronzetti, E. (2002) Neurotrophin and neurotrophin receptor expression in alveolar macrophages: an immunocytochemical study. Growth Factors, 18: 193–202.

Roitt, I., Brostoff, J. and Male, D. (2001) Immunology. Mosby, Philadelphia, pp. 15–64.

Roux, P.P., Dorval, G., Boudreau, M., Angers-Loustau, A., Morris, S.J., Makkerh, J. and Barker, P.A. (2002) K252a and CEP1347 are neuroprotective compounds that inhibit mixed-lineage kinase-3 and induce activation of Akt and ERK. J. Biol. Chem., 277: 49473–49480.

Ruth, D. and Melamed, E. (2002) New drugs in the future treatment of Parkinson's disease. J. Neurol., 249: 30–35.

Sawada, J., Itakura, A., Tanaka, A., Furusaka, T. and Matsuda, H. (2000) Nerve growth factor functions as a chemoattractant for mast cells through both mitogen-activated protein kinase and phosphatidylinositol 3-kinase signaling pathways. Blood, 95: 2052–2058.

Schenone, A., Gill, J.S., Zacharias, D.A. and Windebank, A.J. (1996) Expression of high- and low-affinity neutrophin receptors on human transformed B lymphocytes. J. Neuroimmunol., 64: 141–149.

Schori, H., Lantner, F., Shachar, I. and Schwartz, M. (2002) Severe immunodeficiency has opposite effects on neuronal survival in glutamate-susceptible and -resistant mice: adverse effect of B cells. J. Immunol., 169: 2861–2865.

Schwartz, M., Moalem, G., Leibowitz-Amit, R. and Cohen, I.R. (1999) Innate and adaptive immune responses can be beneficial for CNS repair. Trends Neurosci., 22: 295–299.

Schwartz, M. and Kipnis, J. (2001) Protective autoimmunity: regulation and prospects for vaccination after brain and spinal cord injuries. Trends Mol. Med., 7: 252–258.

Schwartz, M. and Moalem, G. (2001) Beneficial immune activity after CNS injury: prospects for vaccination. J. Neuroimmunol., 113: 185–192.

Serpe, C.J., Kohm, A.P., Huppenbauer, C.B., Sanders, V.M. and Jones, K.J. (1999) Exacerbation of facial motoneuron loss after facial nerve transection in severe combined immunodeficient (scid) mice. J. Neurosci., 19: RC7.

Serpe, C.J., Sanders, V.M. and Jones, K.J. (2000) Kinetics of facial motoneuron loss following facial nerve transection in severe combined immunodeficient mice. J. Neurosci. Res., 62: 273–278.

Shoichet, M.S. and Winn, S.R. (2000) Cell delivery to the central nervous system. Adv. Drug Deliv. Rev., 42: 81–102.

Sin, A.Z., Roche, E.M., Togias, A., Lichtenstein, L.M. and Schroeder, J.T. (2001) Nerve growth factor or IL-3 induces more IL-13 production from basophils of allergic subjects than from basophils of nonallergic subjects. J. Allergy Clin. Immunol., 108: 387–393.

Skaper, S.D. and Walsh, F.S. (1998) Neurotrophic molecules: strategies for designing effective therapeutic molecules in neurodegeneration. Mol. Cell Neurosci., 12: 179–193.

Skaper, S.D., Pollock, M. and Facci, L. (2001) Mast cells differentially express and release active high molecular weight neurotrophins. Brain Res. Mol. Brain Res., 97: 177–185.

Solomon, A., Aloe, L., Pe'er, J., Frucht-Pery, J., Bonini, S., Bonini, S. and Levi-Schaffer, F. (1998) Nerve growth factor is preformed in and activates human peripheral blood eosinophils. J. Allergy Clin. Immunol., 102: 454–460.

Soontornniyomkij, V., Wang, G., Pittman, C.A., Wiley, C.A. and Achim, C.L. (1998) Expression of brain-derived neurotrophic factor protein in activated microglia of human immunodeficiency virus type 1 encephalitis. Neurophatol. Appl. Neurobiol., 24: 453–460.

Stadelmann, C., Kerschensteiner, M., Misgeld, T., Bruck, W., Hohlfeld, R. and Lassmann, H. (2002) BDNF and gp145trkB in multiple sclerosis brain lesions: neuroprotective interactions between immune and neuronal cells? Brain, 125: 75–85.

Storch, A. and Schwarz, J. (2002) Neural stem cells and neurodegeneration. Curr. Opin. Investig. Drugs, 3: 774–781.

Susaki, Y., Shimizu, S., Katakura, K., Watanabe, N., Kawamoto, K., Matsumoto, M., Tsudzuki, M., Furusaka, T., Kitumara, Y. and Matsuda, H. (1996) Functional properties of murine macrophages promoted by nerve growth factor. Blood, 88: 4630–4637.

Tam, S.Y., Tsai, M., Yamaguchi, M., Yano, K., Butterfield, J.H. and Galli, S.J. (1997) Expression of functional trkA receptor tyrosine kinase in the HMC-1 human mast cell line and in human mast cells. Blood, 90: 1807–1820.

Tonra, J.R., Curtis, R., Wong, V., Cliffer, K.D., Park, J.S., Timmes, A., Nguyen, T., Lindsay, R.M., Acheson, A., DiStefano, P.S. (1998) Axotomy upregulates the anterograde transport and expression of brain-derived neurotrophic factor by sensory neurons. J. Neurosci., 18: 4374–4383.

Thorpe, L.W. and Perez-Polo, J.R. (1987) The influence of nerve growth factor on the in vitro proliferative response of rat spleen lymphocytes. J. Neurosci. Res., 18: 134–139.

Weiss, J.H., Hartley, D.M., Koh, J. and Choi, D.W. (1990) The calcium channel blocker nifedipine attenuates slow excitatory amino acid neurotoxicity. Science, 247: 1474–1477.

Ulvestad, E., Williams, K., Bjerkvig, R., Tiekotter, K., Antel, J. and Matre, R. (1994) Human microglial cells have phenotypic and functional characteristics common with both macrophages and dendritic antigen-presenting cells. J. Leukoc. Biol., 56: 732–740.

Xiang, Z. and Nilsson, G. (2000) IgE receptor-mediated release of nerve growth factor by mast cells. Clin. Exp. Allergy, 30: 1379–1386.

Yoles, E. and Schwartz, M. (1998) Degeneration of spared axons following partial white matter lesion: implications for optic nerve neuropathies. Exp. Neurol., 153: 1–7.

Ziemssen, T., Kumpfel, T., Klinkert, W.E.F., Neuhaus, O. and Hohlfeld, R. (2002) Glatiramer acetate-specific T-helper 1- and 2-type cell lines produce BDNF: implications for multiple sclerosis therapy. Brain, 125: 2381–2391.

Zuddas, A., Oberto, G., Vaglini, F., Fascetti, F., Fornai, F. and Corsini, G.U. (1992) MK-801 prevents 1-methyl-4-phenyl-1,2,3,6-tetrahydropyridine-induced parkinsonisms in primates. J. Neurochem., 59: 733–739.

Progress in Brain Research, Vol. 146
ISSN 0079-6123

CHAPTER 25

Role of nerve growth factor and other trophic factors in brain inflammation

Pablo Villoslada[1] and Claude P. Genain[2,*]

[1]*Neuroimmunology Laboratory, Department of Neurology, University of Navarra, Spain*
[2]*Department of Neurology, University of California, San Francisco, CA 94143-0435, USA*

Abstract: Inflammation in the brain is a double-edged process that may be beneficial in promoting homeostasis and repair, but can also result in tissue injury through the damaging potential of inflammatory mediators. Thus, control mechanisms that minimize the extent of the inflammatory reaction are necessary in order to help preserve brain architecture and restore function. The expression of neurotrophic factors such as nerve growth factor (NGF) is increased after brain injury, in part mediated by effects on astrocytes of pro-inflammatory mediators and cytokines produced by immune cells. Conversely, cells of the immune system express NGF receptors, and NGF signaling modulates immune function. Multiple sclerosis (MS) and the disease model experimental autoimmune encephalomyelitis are neurodegenerative disorders whereby chronic destruction of the brain parenchyma results from an autoaggressive, immune-mediated inflammatory process and insufficient tissue regeneration. Here, we review evidence indicating that the increased production of NGF and other trophic factors in central nervous system (CNS) during these diseases can suppress inflammation by switching the immune response to an anti-inflammatory, suppressive mode in a brain-specific environment. Thus, trophic factors networks in the adult CNS not only protects axons and myelin but appear to also actively contribute to the maintenance of the brain immune privilege. These agents may represent good targets for therapeutic intervention in MS and other chronic CNS inflammatory diseases.

Keywords: neural growth factors; autoimmune demyelination; multiple sclerosis; experimental autoimmune encephalomyelitis; cytokines; immunomodulation

Introduction

In addition to its the critical role during central nervous system (CNS) development, nerve growth factor (NGF) plays an important role in the maintenance of adult CNS homeostasis and in the response to brain tissue damage (Kernie and Parada, 2000). Due to the highly specialized architecture and function of the CNS, maintenance of brain tissue

*Correspondence to: C.P. Genain, Neuroimmunology Laboratory, Room C-440, University of California, 505 Parnassus Avenue, San Francisco, CA 94143-0435, USA. Tel.: + 1-415-502-5684; Fax: + 1-415-502-5899; E-mail: claudeg@itsa.ucsf.edu

integrity of it is fundamental for the survival of the individual. Bystander effects of inflammation, although well tolerated in most tissues such as epithelium or connective tissues, might be deleterious when occurring within the CNS. It has been proposed that the CNS is protected against such adverse consequences of inflammation through immune privilege (Streilein, 1995; Antel and Owens, 1999), which prevents invasion of brain parenchyma by inflammatory cells and/or partially restricts local immune responses by maintaining an immunosuppressive environment. In this short review we will focus on the roles of NGF during CNS inflammation and on newly discovered immunoregulatory properties of NGF and other neural growth factors that are

DOI: 10.1016/S0079-6123(03)46025-1

likely to contribute to the maintenance of CNS immune privilege.

During pathological situations such as CNS infections or autoimmune diseases, an inflammatory response is mounted against the CNS that may result in 'bystander' tissue destruction and loss of function. Multiple sclerosis (MS) is the prototypal inflammatory brain disease of probable autoimmune origin, in which infiltrating inflammatory immune cells and mediators contributes to CNS damage and lead to widespread destruction of the myelin sheaths, oligodendrocytes and axons, and gliosis (Steinman, 2001). The chronic accumulation of such lesions along the neuraxis profoundly disturbs nerve conduction and leads to significant disability, often in young adults. Autoimmune demyelination is a good example of disorders in which inflammation is detrimental to the CNS, which affords to the study of the complex interplay between inflammation, immune privilege, tissue destruction, and mechanisms of repair and neuroprotection.

NGF and immune system interactions

Several immune cell types express the NGF receptors TrkA and p75, and it has been demonstrated that NGF production is increased in inflammatory diseases. The role of NGF in the regulation of immune response is complex and not fully understood (Aloe et al., 1994; Levi-Montalcini et al., 1996). For example, NGF production is induced in brain cells by a variety of pro-inflammatory and anti-inflammatory cytokines such as interleukin-1 (IL-1), IL-4, IL-5, tumor necrosis factor-alpha (TNF-α), transforming growth factor-beta (TGF-β) and interferon-beta (IFN-β) mediated by NF$\kappa\beta$ signaling (Gadient et al., 1990; Awatsuji et al., 1993; Awatsuji et al., 1995; Friedman et al., 1996; Boutros et al., 1997). It is also of importance to keep in mind that the effects of NGF on inflammation may depend on the microenvironment, and immunomodulatory properties of NGF in the brain may be different from those in other tissues.

The TrkA receptor and p75 neurotrophin receptor (p75^{NTR}) are expressed both in cells of the innate immune system, especially in mast cells and macrophages, and in adaptive immune system cells, such as T cells (Ehrhard et al., 1993; Santambrogio et al., 1994) and B cells (Brodie and Gelfand, 1992; Melamed et al., 1996). NGF signaling may promote survival of memory B cells (Torcia et al., 1996), influences T cell and B cell proliferation, stimulates the synthesis of immunoglobulins (IgM, IgA and IgG4) (Brodie, 1996). The role of NGF in the innate immune system is also complex. Mast cells are most responsive to NGF, and themselves release NGF in an autocrine manner in response to inflammatory stimuli. Stimulation of mast cells by NGF promotes survival, induces the secretion of cytokines such as IL-3, IL-4, IL-10, TNF-α and granulocyte macrophage colony-stimulating factor (GMCSF), and modifies vascular permeability (Bullock and Johnson, 1996). NGF also influences differentiation of monocytes into macrophages, enhances phagocytosis and increases antimicrobial activity. Lipopolysaccharide (LPS), a pathogen product, induces the release of NGF and the expression of NGF receptors by macrophages (Caroleo et al., 2001). Together these findings provide evidence that NGF plays a complex role in the regulation of inflammatory responses, with specific effects that may vary depending on the organ or tissue involved.

Because of the existence of a blood–brain barrier (BBB) that ensures its immunologically privileged situation, immune responses in the CNS are mainly mediated by resident microglia, mast cells and astrocytes. However, in CNS inflammation the BBB is disrupted and immune mediators from the peripheral immune system may enter the parenchyma, which results in a fully developed adaptive immune response with loss of immune privilege. Perivascular mast cells and macrophages are highly sensitive to NGF, and it has been postulated that this interaction might contribute to the integrity of BBB and its restoration after CNS inflammation (Ransohoff and Trebst, 2000; Flugel et al., 2001). In adult CNS, NGF is mainly produced by astrocytes and stored into the extracellular matrix (ECM). The ECM represents a reservoir of trophic factors that interact with immune cells entering the CNS to modulate the ongoing immune response.

Role of NGF in CNS autoimmune demyelination

Experimental autoimmune encephalomyelitis (EAE) is a widely used disease model for MS of autoimmune pathogenesis (Bradl and Linington, 1996). In EAE, activated self-reacting CD4+ and CD8+ T cells, macrophages, and anti-myelin antibodies invade the CNS and mediate tissue damage either directly or through a variety of effector mechanisms such as production of nitric oxide, free radicals, and activation of complement or macrophage mediated cytotoxicity (Brosnan and Raine, 1996). Myelin sheaths and oligodendrocytes are the primary targets of the autoimmune attack, and axonal transection and gliosis are also present. During EAE, expression of NGF and NGF receptors is increased in astrocytes, oligodendrocytes and in cells from the subventricular zone (De Simone et al., 1996; Calza et al., 1998; Micera et al., 1998; Oderfeld-Nowak et al., 2001). The increased expression of NGF parallels that of a number of other neurotrophic growth factors, and may represent as in other forms of CNS tissue injury, a regulatory mechanisms aimed at promoting tissue repair. Another interesting possibility is that growth factors influence development and severity of autoimmune inflammatory demyelination, as suggested by reports of decreased severity after administration of insulin growth factor (IGF)-1, glial growth factor (GGF)-2, ciliary neurotrophic factor (CNTF) and NGF in rodent EAE models (de Webster, 1997; Cannella et al., 1998; Lovett-Racke et al., 1998; Arredondo et al., 2001; Linker et al., 2002), and aggravation upon administration of a neutralizing NGF antibody (Micera et al., 2000).

We recently evaluated the efficacy of recombinant human NGF using the marmoset model of EAE, which offers close phenotypic similarity with human MS including a relapsing remitting clinical course and pathologically, moderate perivascular inflammation accompanied by prominent demyelination at the acute stage (Genain and Hauser, 2001). Animals were treated intracerebroventricularly with 6 μg per day of recombinant human NGF or placebo, starting one week after immunization (typically one week before the onset of CNS infiltration) until two weeks after the onset of the clinical disease (first relapse). The original trial was designed in order to probe for

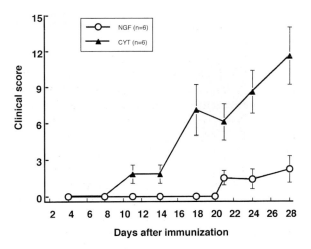

Fig. 1. Recombinant human NGF inhibits the development of clinical EAE in marmosets. EAE was induced by active immunization with myelin/oligodendrocyte glycoprotein in adjuvant, which in this species results in 100% incidence of disease. NGF (open circles) or placebo (cytochrome c, CYT, closed triangles) were administered intracerebroventricularly by continuous infusion (osmotic pump) beginning day 7 after immunization. EAE was graded using an expanded marmoset grading scale ranging from 0 to 45. Neuropathological findings throughout the CNS (optic nerves, brain and spinal cord) corroborated the observed clinical protection. *$p < 0.05$ student's t-test. For details please refer to Villoslada et al. (2000).

improved remyelination due to NGF therapy. However, clinical and neuropathological features of EAE were milder in NGF-treated animals than in placebo-treated controls (Fig. 1). Immunohistochemical studies demonstrated that CNS infiltrating mononuclear cells of the NGF-treated animals expressed lower amounts of interferon-gamma (IFN-γ), the prototypical Th1 cytokine in EAE, and higher amounts of IL-10, a major immunosuppressive cytokine, which indicated that the CNS-specific immune response had been induced to a switch from the proinflammatory Th1 phenotype to an anti-inflammatory Th2 phenotype. Perhaps more strikingly, glial cells within the unaffected white matter expressed higher amounts of IL-10 even in areas remote from inflammatory infiltrates (Figs. 2 and 3). The latter finding suggests that NGF had the ability to induce a CNS-wide immunosuppressive microenvironment that may have been the primary mechanism leading to decreased CNS infiltration, reduction in BBB breakdown and IFN-γ suppression

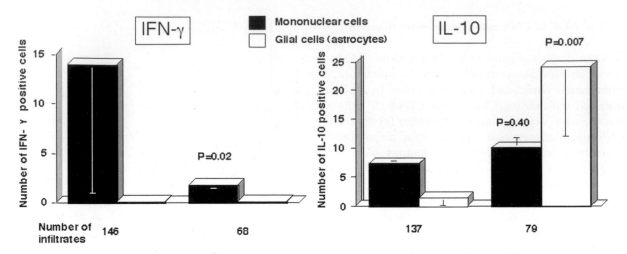

Fig. 2. Effects of recombinant human NGF on cytokine production (IFN-γ and IL-10) within CNS inflammatory infiltrates during marmoset EAE.

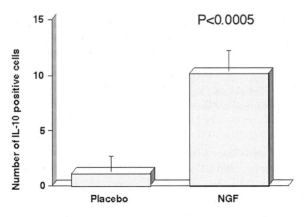

Fig. 3. Effects of recombinant human NGF upon IL-10 production by glial cells of normal, unaffected white matter during marmoset EAE. A striking upregulation of IL-10 is observed in areas at distance from the pathological infiltrates, suggesting that intraventricular administration of NGF induced a tissue-specific immune deviation that suppressed CNS inflammation mediated by IFN-γ-producing inflammatory infiltrating cells.

in treated animals (for details see Ransohoff and Trebst, 2000; Villoslada et al., 2000). These CNS-specific immunosuppressive properties of IL-10 may be specific for higher mammal species (human and nonhuman primates), since they have not been unequivocally demonstrated in rodents (Cannella et al., 1996; Bettelli et al., 1998; Cua et al., 1999).

The pronounced effects of intracranial administration of NGF on cytokine expression were specific to the CNS parenchyma, and were not observed in the periphery. T cell-mediated EAE in rodents can classically be inhibited by promoting an immune bias of myelin-reactive T cells towards the Th2 phenotype (immune deviation), or by administration of Th2 cytokines (Chen et al., 1994; Racke et al., 1994). This approach has yet to prove useful in human clinical trials, and immune deviation of the peripheral immune system may trigger lethal complications in CNS demyelinating disorders of complex pathophysiology because it has the potential to exacerbate pathogenic antibody responses (Genain et al., 1996). Peripheral administration NGF could also potentially exacerbate such responses, since NGF is a factor that promotes B cell survival and growth (Aloe et al., 1994; Levi-Montalcini et al., 1996; Torcia et al., 1996). However, intracranial administration of NGF in marmosets did no alter the expected development of T cell and humoral autoimmune responses against the immunizing antigen and did not induce immune deviation in the peripheral immune system, confirming that effector mechanisms for the beneficial effects were taking place within the CNS (Villoslada et al., 2000). To further confirm these findings, we evaluated the effect of administration of exogenous NGF on the production of Th1 (TNF-α, IFN-γ and IL-12) and Th2 (IL-4, IL-10, IL-6 and TGF-β)

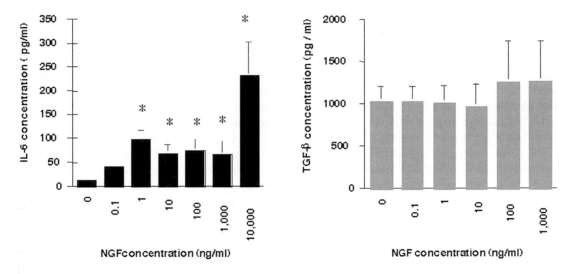

Fig. 4. Effects of exogenous recombinant human NGF on lectin-induced secretion of cytokines by PBMC. The only significant change detected was an increased secretion of IL-6. Please refer to the text for details of cytokines examined.

cytokines by cultured peripheral blood mononuclear cells (PBMC) and macrophages stimulated in vitro by antigen or lectins. In both human and marmoset systems, all cytokines examined remained unaffected by addition of NGF, except for IL-6 which was markedly increased with a dose-dependent response (Fig. 4). This additional effect of NGF on IL-6 production by peripheral blood mononuclear cells (PBMC) (which was not examined in the CNS of NGF-treated marmosets), is of interest since this cytokine has been shown to suppress disease both in EAE (Willenborg et al., 1995; Carr and Campbell, 1999) and in a viral experimental model of MS (Rodriguez et al., 1994). Nonetheless, the upregulation of IL-10 in CNS glial cells (and not PBMC) during treatment of marmosets with EAE highlights a unique property of NGF in inducing a form of tissue-specific immune deviation that is capable of modulating CNS autoaggressive responses and likely contributed to the maintenance of the CNS immune privilege.

The immunosuppressive properties of NGF in EAE were confirmed in a mouse model carrying a transgene encoding for the T cell receptor specific for an encephalitogenic epitope of myelin basic protein. NGF was expressed by T cells in response to exposure to a Th2 environment (IL-4), but in vitro administration of NGF did not change Th phenotypes or proliferation responses. Intraperitoneal administration of NGF during the induction phase of EAE ameliorated the disease without changing the encephalitogenic properties of the T cells. The interpretation of these results was that NGF may exert an indirect role in modulating the peripheral immune response, perhaps through enhanced sympathetic innervation of lymphoid tissue (Arredondo et al., 2001). A regulatory role for anti-inflammatory properties of NGF is also supported by the demonstration that antigen-specific Th1 cells engineered to deliver NGF in situ ameliorate disease and inhibit transmigration of inflammatory cells in several models of CNS or peripheral nervous system autoimmune demyelination (Kramer et al., 1995; Flugel et al., 2001). What mediates these anti-inflammatory properties in addition to, or via the secretion of suppressive cytokines such as IL-10 is not entirely understood, but may involve the down-regulation of the expression of major histocompatibility complex (MHC) molecules on the resident CNS antigen presenting cells, which in EAE function by reactivating invading antigen-specific T cells (Frei et al., 1994; Neumann et al., 1998).

In summary, data from EAE models provide enough evidence for a regulatory role of NGF in autoimmune demyelination, due to the overexpression of NGF and NGF receptors during CNS inflammation and in vivo experiments that demonstrate protection from disease. Studies conducted to date suggest that NGF is secreted both in paracrine and autocrine fashions, and possesses pleiotropic functions that cannot be regarded as classical immunosuppressive activities in lymphocytes such as for the suppressive cytokine IL-10. NGF may work by multiple mechanisms, including maintenance of BBB integrity, modulation of the overall immune system activity through sympathetic innervations, induction of an immunosuppressive microenvironment in situ in the CNS, and survival and restoration enhancement of oligodendrocytes and axons after immune-mediated damage. Finally, it is important to recognize that the relative importance of these different immunomodulatory pathways may vary between species.

Possible role of NGF in multiple sclerosis

In MS lesions, NGF receptors are expressed mainly by inflammatory cells such as microglia and monocytes and to a lesser extent by astrocytes but not by oligodendrocytes (Valdo et al., 2002). However, mRNA^{p75NTR} has been detected at increased levels in oligodendrocytes in MS plaques (Dowling et al., 1999). These findings and reports that oligodendrocytes expressing p75 may undergo apoptosis in vitro (Casaccia-Bonnefil et al., 1996; Frade et al., 1996) raises an interesting question, because oligodendrocyte death is a pathological hallmark in multiple sclerosis. By contrast, it has also been shown that p75^{NTR} signaling in stressed oligodendrocytes promotes survival (Lopez-Sanchez and Frade, 2002). The role of NGF during inflammatory demyelination in adult CNS may depend on the relative levels of expression of the two NGF receptors, and remains to be clarified taking into account the fact that the very same infiltrating lymphocytes that mediate parenchymal inflammation may express NGF as well as other neurotrophins such as brain-derived neurotrophic factor (BDNF) (Kerschensteiner et al., 1999; Moalem et al., 2000; Muhallab et al., 2002; Stadelmann et al., 2002). Thus, inflammatory effector cells themselves may, at least in part, support neuroprotective functions through the secretion of trophic factors, perhaps representing a self-limiting mechanism for inflammation that may be deficient in the case of CNS autoimmune damage.

During acute attacks, patients with MS have increased levels of NGF in the CSF compared with healthy controls or patients that are clinically stable, which may be interpreted as evidence that NGF is produced in an attempt to protect CNS tissue against inflammation (Laudiero et al., 1992). We have replicated these findings by assessing the levels of NGF in the CSF of patients with relapsing-remitting (RR) MS during an attack, patients with secondary-progressive (SP) MS, patients with a first manifestation suggestive of a demyelinating event (clinically isolated syndrome, CIS), healthy controls, and patients with other neurological disease than MS (P. Villoslada, unpublished results—Fig. 5). We found that NGF levels were higher in the CSF of patients with MS than in healthy controls or patients with other neurological diseases. Interestingly, NGF levels were on average higher in MS patients with RR-MS that typically display disseminated disease with evidence of active CNS inflammation and BBB breakdown (Gadolinium enhancement on magnetic resonance imaging), compared to patients with SP-MS in which inflammation is typically less prominent, and patients with a CIS who by definition lack evidence of disease dissemination. Together, our interpretation of these data is that NGF is produced during CNS damage as a means to recover brain homeostasis.

Role of other trophic factors in CNS immunosuppression and in neuroprotection

Other trophic factors than NGF clearly share similar immunosuppressive and neuroprotective properties. TGF-β is an immunosuppressive cytokine that is increased in tissues with immune privilege in order to prevent their damage by the immune system, and is able to induce immune deviation (Streilein et al., 1997). TGF-β KO mice develop spontaneous autoimmune diseases involving several organs including the brain (Schull et al., 1992), and treatment of animals induced for EAE with TGF-β prevents the disease (Racke et al., 1991). Insulin-like growth factor-1 (IGF-1) is another pleiotropic growth factor capable to suppress murine EAE when the treatment was started before disease onset, however in one study disease was exacerbated when treatment was

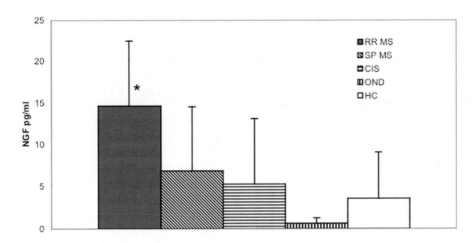

Fig. 5. Levels of NGF in the CSF from patients with relapsing-remitting MS (RR-MS) during a relapse ($n = 17$), secondary-progressive MS (SP-MS, $n = 16$), clinically isolated syndrome (CIS, $n = 16$), other neurological diseases (OND, $n = 16$) and healthy controls (HC, $n = 16$) from whom CSF was obtained in the differential diagnosis of acute headache. Levels of NGF in RR-MS patients during relapses were significantly higher than in the other groups ($p < 0.0001$, one factor ANOVA).

started after disease onset (de Webster, 1997; Lovett-Racke et al., 1998). Similarly, neuregulin or glial growth factor-2 (GGF-2) administered systemically prevented the development of murine EAE (Cannella et al., 1998). Disease attenuation appeared to be due to both increased remyelination and decreased CNS inflammatory activity which, reminiscent of our studies in marmosets may have been due to enhanced expression of the immunosuppressive cytokine IL-10.

Recently, two additional members of the neuro-cytokine family leukemia inhibitory factor (LIF) and ciliary neurotrophic factor (CNTF) were found to protect oligodendrocytes during murine EAE (Butzkueven et al., 2002; Linker et al., 2002). Oligodendrocytes from LIF-treated animals were preserved from inflammatory attack and stressed oligodendrocytes expressed increased levels of LIF receptor-β that might account for the protection against oligodendrocyte loss. However, an immuno-suppressive role was not demonstrated although immune cells expressed LIF receptors and treatment was administered systemically. CNTF deficient mice had more severe disease and recovery was poor with a 60% decrease in the number of proliferating oligodendrocyte precursors and myelin vacuolation. Thus, there may be differential contributions of trophic factors to maintenance of the immune privilege of the brain, to neuroprotection and CNS regeneration, and to immunosuppression.

In addition to NGF, we have recently analyzed CSF levels of LIF and glial cell line-derived neurotrophic factor (GDNF) in MS patients. We found CSF levels of LIF and GDNF are higher in patients with active disease than in those with chronic disease or healthy controls, which confirms that NGF is among a number of trophic factors released in response to brain damage (Table 1). These findings further support the role of the trophic factor network in promoting CNS regeneration after inflammatory damage and in maintaining an immunosuppressive microenvironment in the CNS.

Proposed model for trophic factors in CNS immune privilege and neuroprotection

The data reviewed here strongly argue that NGF (and other trophic factors) are capable of suppressing autoaggressive immune responses in a CNS-specific fashion which does not involve modulation of the peripheral immune system, likely indicating that this function can only be accomplished in the context of the CNS microenvironment. The immune privilege of the brain, although not fully understood, can be maintained by the additive effects of physical barriers such as the BBB and the lack of lymphatic vessels in CNS tissue, active downregulation of HLA antigen-presenting molecules (astrocytes and microglia), and

TABLE 1

Levels of trophic factors in serum and CSF from patients with relapsing-remitting MS (RR-MS), primary-progressive MS (PP-MS) and healthy controls (HC)

Trophic factor	Serum			CSF		
(pg/ml)	RR-MS	PP-MS	HC	RR-MS	PP-MS	HC
NGF	ND	ND	ND	14.6 ± 7	6.9 ± 7	3.62 ± 5
LIF	7.9 ± 13	ND	11.6 ± 21	11.7 ± 14	8.4 ± 19	1.7 ± 4
GDNF*	0	0	0	160 ± 10	70 ± 8	40 ± 7

*Total GDNF (after acidification treatment to measure both free and protein bind GDNF).

upregulation of immunosuppressive cytokines such as IL-10 or TGF-β. Another proposed mechanism is the expression of pro-apoptotic molecules such as the Fas-ligand that promote apoptosis of infiltrating activated lymphocytes, although a clear demonstration of such phenomena is still lacking (O'Connell et al., 2001).

We propose a model in which presence of trophic factors in the ECM, and/or their increased expression by astrocytes after brain damage might interact with specific receptors expressed by perivascular infiltrating mononuclear cells or macrophages inducing a partial suppression of the immune response (Fig. 6). This effect might be achieved by several mechanisms: (i) closing the BBB and diminishing the trafficking of activated inflammatory monocytes, via actions of NGF on brain perivascular mast cells or on expression of molecules responsible for lymphocyte adhesion and migration, (ii) inducing a switch of T cell phenotype from a Th1 phenotype to a Th2 phenotype, (iii) inducing the expression of immunosuppressive cytokines such as IL-10, and perhaps TGF-β, (iv) reducing activation of macrophages and microglial cells and downregulating expression of MHC class II molecules on these cells, (v) inducing the inactivation of toxic immune mediators by astrocytes, and (vi) modulating overall immune activity through sympathetic innervation. Several of these proposed mechanisms have been demonstrated, however comprehensive studies will be needed to understand the complex and dynamic nature of these processes. A fundamental point that arises from studies performed to date is that the effects of NGF must be seen as those of a pleiotropic cytokine in contrast to other specialized cytokines, and that overall effects will be the summary of NGF

signaling in the different systems expressing NGF receptors.

In addition to the immunosuppressive properties, trophic factors such as NGF most certainly contribute to the protection of injured neurons and oligodendrocytes, and promote survival and regeneration. This neuroprotective role may result from priming of anti-apoptotic or cell cycle progression pathways such as NF$\kappa\beta$ or Ras (Lopez-Sanchez and Frade, 2002). Although the effects of trophic factors in rescuing damaged neurons and oligodendrocytes are clearly demonstrated, little is known of their role in protecting demyelinated or damaged axons. This effect could be very important in MS, considering recent neuropathological analyses that show early axonal damage in this disease (Trapp et al., 1999). Thus, trophic factor networks might have more pleiotropic roles in maintaining homeostasis of the adult CNS than the very specific functions which they support during development.

Therapeutics opportunities for NGF in inflammatory brain diseases

The findings that NGF induces immunosuppression that is CNS-specific during autoimmune demyelination, in addition to its neuroprotective properties for oligodendrocytes and neurons, make this protein an attractive candidate molecule for treatment of CNS inflammatory diseases. Neurotrophic factors have been successfully used to treat animal models of central and peripheral nervous system disease with the rational of protecting neurons, axons and myelin (Yuen and Mobley, 1996; Kernie and Parada, 2000). Sadly, attempts to translate such therapies to human

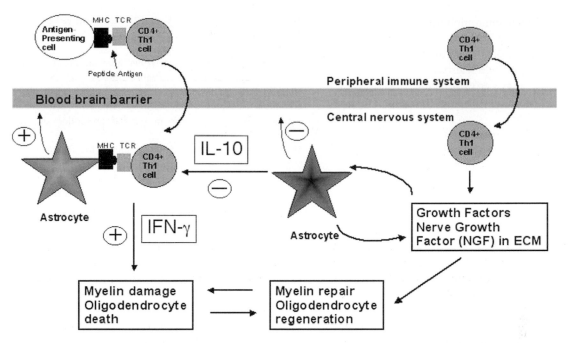

Fig. 6. Schematic model for the role of trophic factor networks in maintenance of CNS immune privilege. Right, simplified pathway of inflammation and myelin damage (demyelination) by myelin antigen-specific CD4 + Th1 cells that migrate into the CNS parenchyma and are reactivated by antigen presentation in context of MHC class II molecules expressed on astrocytes (represented in light gray). T cell, macrophage and glial cell activation further contributes to blood–brain barrier (BBB) permeability. Left, NGF and growth factors are secreted by both astrocytes and certain infiltrating T cells. In addition to their neuroprotective effects, these growth factors can stimulate astrocytes to produce IL-10, a suppressive cytokine that has the potential to downregulate inflammatory process and decrease BBB permeability (represented in dark gray).

diseases have failed to date, decreasing the current interest in these factors as therapeutics agents for neurological diseases (Apfel, 2002). Inherent weakness in most clinical trials are the use of systemic administration which results in failure to cross the BBB, and unacceptable side effects of treatment which are dose-limiting. A more comprehensive approach to disease pathogenesis that accounts for mechanisms of action of neurotrophins, and means of targeted delivery to specific organs might allow the development of promising NGF-based therapies for neurological diseases.

Acknowledgments

This work was supported by Roche Diagnostics, Penzberg, Germany, who provided recombinant human NGF and financial support for experiments, and the Spanish Ministry of Health (FIS 98/0961, 96/5122 and 97/5459 to P.V.). C.P.G. was a Harry Weaver Neuroscience Scholar of the National Multiple Sclerosis Society.

Abbreviations

BBB	blood–brain barrier
BDNF	brain derived neurotrophic factor
CIS	clinically isolated syndrome
CNS	central nervous system
CNTF	ciliary neurotrophic factor
EAE	experimental allergic encephalomyelitis
ECM	extracellular matrix
GDNF	glial derived neurotrophic factor
GGF	glial growth factor
GMCSF	granulocyte monocyte colony stimulating factor
IL-	interleukin-
IFN-	interferon-
IGF	insulin-like growth factor

LIF	leukemia inhibitory factor
LPS	lipopolysaccharide
MHC	major histocompatibility complex
MS	multiple sclerosis
NGF	nerve growth factor
PBMC	peripheral blood mononuclear cells
p75^{NTR} p75	neurotrophin receptor
RR-,SP,-PP-MS	relapsing remitting, secondary progressive, primary progressive multiple sclerosis
TGF	transforming growth factor
Th	T helper cell
TNF	tumor necrosis factor

References

Aloe, L., Skaper, S.D., Leon, A. and Levi-Montalcini, R. (1994) Nerve growth factor and autoimmune diseases. Autoimmunity, 19: 141–150.

Antel, J.P. and Owens, T. (1999) Immune regulation and CNS autoimmune disease. J. Neuroimmunol., 100: 181–189.

Apfel, S.C. (2002) Is the therapeutic application of neurotrophic factors dead? Ann. Neurol., 51: 8–11.

Arredondo, L.R., Deng, C., Ratts, R.B., Lovett-Racke, A.E., Holtzman, D.M. and Racke, M.K. (2001) Role of nerve growth factor in experimental autoimmune encephalomyelitis. Eur. J. Immunol., 31: 625–633.

Awatsuji, H., Furukawa, Y., Hirota, M., Furukawa, S. and Hayashi, K. (1995) Interferons suppress nerve growth factor synthesis as a result of interference with cell growth in astrocytes cultured from neonatal mouse brain. J. Neurochem., 64: 1476–1482.

Awatsuji, H., Furukawa, Y., Hirota, M., Murakami, Y., Nii, S., Furukawa, S. and Hayashi, K. (1993) Interleukin-4 and -5 as modulators of nerve growth factor synthesis/secretion in astrocytes. J. Neurosci. Res., 34: 539–545.

Bettelli, E., Das, M., Howard, E., Weiner, H., Sobel, R. and Kuchroo, V. (1998) IL-10 is critical in the regulation of autoimmune encephalomyelitis as demonstrated by studies of IL-10- and IL-4-deficient and transgenic mice. J. Immunol., 161: 3299–3306.

Boutros, T., Croze, E. and Yong, V.W. (1997) Interferon-beta is a potent promoter of nerve growth factor production by astrocytes. J. Neurochem., 69: 939–946.

Bradl, M. and Linington, C. (1996) Animal models of demyelination. Brain Pathol., 6: 303–311.

Brodie (1996) Differential effects of Th1 and Th2 derived cytokines on NGF synthesis by mouse astrocytes, FEBS Lett., 394: 117–120.

Brodie, C. and Gelfand, E.W. (1992) Functional nerve growth factor receptors on human B lymphocytes. Interaction with IL-2. J. Immunol., 148: 3492–3497.

Brosnan, C.F. and Raine, C.S. (1996) Mechanisms of immune injury in multiple sclerosis. Brain Pathol., 6: 243–257.

Bullock, E. and Johnson, E., Jr. (1996) Nerve growth factor induces the expression of certain cytokine genes and bcl-2 in mast cells. Potential role in survival promotion, J. Biol. Chem., 271.

Butzkueven, H., Zhang, J., Soilu-Hanninen, M., Hochrein, H., Chionh, F., Shipham, K., Emery, B., Turnley, A., Petratos, S., Ernst, M., Bartlett, P. and Kilpatrick, T. (2002) LIF receptor signaling limits immune-mediated demyelination by enhancing oligodendrocyte survival. Nat. Med., 8: 613–619.

Calza, L., Giardino, L., Pozza, M., Bettelli, C., Micera, A. and Aloe, L. (1998) Proliferation and phenotype regulation in the subventricular zone during experimental allergic encephalomyelitis: in vivo evidence of a role for nerve growth factor. Proc. Natl. Acad. Sci. USA, 95: 3209–3214.

Cannella, B., Gao, Y.L., Brosnan, C. and Raine, C.S. (1996) IL-10 fails to abrogate experimental autoimmune encephalomyelitis. J. Neurosci. Res., 45: 735–746.

Cannella, B., Hoban, C.J., Gao, Y.L., Garcia-Arenas, R., Lawson, D., Marchionni, M., Gwynne, D. and Raine, C.S. (1998) The neuregulin, glial growth factor 2, diminishes autoimmune demyelination and enhances remyelination in a chronic relapsing model for multiple sclerosis. Proc. Natl. Acad. Sci. USA, 95: 10100–10105.

Caroleo, M.C., Costa, N., Bracci-Laudiero, L. and Aloe, L. (2001) Human monocyte/macrophages activate by exposure to LPS overexpress NGF and NGF receptors. J. Neuroimmunol., 113: 193–201.

Carr, D. and Campbell, I. (1999) Transgenic expression of interleukin-6 in the central nervous system confers protection against acute herpes simplex virus type-1 infection. J. Neurovirol., 5: 449–457.

Casaccia-Bonnefil, P., Carter, B.D., Dobrowsky, R.T. and Chao, M.V. (1996) Death of oligodendrocytes mediated by the interaction of nerve growth factor with its receptor p75. Nature, 383: 716–719.

Chen, Y., Kuchroo, V.K., Inobe, J.-I., Hafler, D.A. and Weiner, H.L. (1994) Regulatory T cell clones induced by oral tolerance: suppression of autoimmune encephalomyelitis. Science, 265: 1237–1240.

Cua, D., Groux, H., Hinton, D., Stohlman, S. and Coffman, R. (1999) Transgenic interleukin 10 prevents induction of experimental autoimmune encephalomyelitis. J. Exp. Med., 189: 1005–1010.

De Simone, R., Micera, A., Tirassa, P. and Aloe, L. (1996) mRNA for NGF and p75 in the central nervous system of rats affected by experimental allergic encephalomyelitis. Neuropathol. Appl. Neurobiol., 22: 54–59.

de Webster, H. (1997) Growth factors and myelin regeneration in multiple sclerosis. Mult. Scler., 3: 113–120.

Dowling, P., Ming, X., Raval, S., Husar, W., Casaccia-Bonnefil, P., Chao, M., Cook, S. and Blumberg, B. (1999) Up-regulated p75NTR neurotrophin receptor on glial cells in MS plaques. Neurology, 53: 1676–1682.

Ehrhard, P.B., Erb, P., Graumann, U. and Otten, U. (1993) Expression of nerve growth factor and nerve growth factor receptor tyrosine kinase Trk in activated CD4-positive T-cell clones. Proc. Natl. Acad. Sci. USA, 90: 10984–10988.

Flugel, A., Matsumuro, K., Neumann, H., Klinkert, W.E., Birnbacher, R., Lassmann, H., Otten, U. and Wekerle, H. (2001) Anti-inflammatory activity of nerve growth factor in experimental autoimmune encephalomyelitis: inhibition of monocyte transendothelial migration. Eur. J. Immunol., 31: 11–22.

Frade, J.M., Rodriguez-Tebar, A. and Barde, Y.A. (1996) Induction of cell death by endogenous nerve growth factor through its p75 receptor. Nature, 383: 166–168.

Frei, K., Lins, H., Schwerdel, C. and Fontana, A. (1994) Antigen presentation in the central nervous system. The inhibitory effect of IL-10 on MHC class II expression and production of cytokines depends on the inducing signals and the type of cell analyzed. J. Immunol., 152: 2720–2728.

Friedman, W.J., Thakur, S., Seidman, L. and Rabson, A.B. (1996) Regulation of nerve growth factor mRNA by interleukin-1 in rat hippocampal astrocytes is mediated by NFkappaB. J. Biol. Chem., 271: 31115–31120.

Gadient, R.A., Cron, K.C. and Otten, U. (1990) Interleukin-1 beta and tumor necrosis factor-alpha synergistically stimulate nerve growth factor (NGF) release from cultured rat astrocytes. Neurosci. Lett., 117: 335–340.

Genain, C. and Hauser, S. (2001) Experimental allergic encephalomyelitis in the common marmoset C. jacchus. Immunol. Rev., 183: 159–172.

Genain, C.P., Abel, K., Belmar, N., Villinger, F., Rosenberg, D.P., Linington, C., Raine, C.S. and Hauser, S.L. (1996) Late complications of immune deviation therapy in a nonhuman primate. Science, 274: 2054–2057.

Kernie, S.G. and Parada, L.F. (2000) The molecular basis for understanding neurotrophins and their relevance to neurologic disease. Arch. Neurol., 57: 654–657.

Kerschensteiner, M., Gallmeier, E., Behrens, L., Leal, V., Misgeld, T., Klinkert, W., Kolbeck, R., Hoppe, E., Oropeza-Wekerle, R., Bartke, I., Stadelmann, C., Lassmann, H., Wekerle, H., Hohlfeld, R. (1999) Activated human T cells, B cells, and monocytes produce brain-derived neurotrophic factor in vitro and in inflammatory brain lesions: a neuroprotective role of inflammation. J. Exp. Med., 189: 865–870.

Kramer, R., Zhang, Y., Gehrmann, J., Gold, R., Thoenen, H. and Wekerle, H. (1995) Gene transfer through the blood-nerve barrier: NGF-engineered neuritogenic T lymphocytes attenuate experimental autoimmune neuritis. Nat. Med., 1: 1162–1166.

Laudiero, L.B., Aloe, L., Levi-Montalcini, R., Buttinelli, C., Schilter, D., Gillessen, S. and Otten, U. (1992) Multiple sclerosis patients express increased levels of beta-nerve growth factor in cerebrospinal fluid. Neurosci. Lett., 147: 9–12.

Levi-Montalcini, R., Skaper, S., Dal Toso, R., Petrelli, L. and Leon, A. (1996) Nerve growth factor: from neurotropin to neurokine. Trends Neurosci., 19: 514–520.

Linker, R.A., Maurer, M., Gaupp, S., Martini, R., Holtmann, B., Giess, R., Rieckmann, P., Lassmann, H., Toyka, K.V., Sendtner, M., Gold, R. (2002) CNTF is a major protective factor in demyelinating CNS disease: a neurotrophic cytokine as modulator in neuroinflammation. Nat. Med., 8: 620–624.

Lopez-Sanchez, N. and Frade, J.M. (2002) Control of the cell cycle by neurotrophins: lessons from the p75 neurotrophin receptor. Histol. Histopathol., 17: 1227–1237.

Lovett-Racke, A.E., Bittner, P., Cross, A.H., Carlino, J.A. and Racke, M.K. (1998) Regulation of experimental autoimmune encephalomyelitis with insulin-like growth factor (IGF-1) and IGF-1/IGF-binding protein-3 complex (IGF-1/IGFBP3). J. Clin. Invest., 101: 1797–1804.

Melamed, I., Kelleher, C.A., Franklin, R.A., Brodie, C., Hempstead, B., Kaplan, D. and Gelfand, E.W. (1996) Nerve growth factor signal transduction in human B lymphocytes is mediated by gp140trk. Eur. J. Immunol., 26: 1985–1992.

Micera, A., Properzi, F., Triaca, V. and Aloe, L. (2000) Nerve growth factor antibody exacerbates neuropathological signs of experimental allergic encephalomyelitis in adult lewis rats. J. Neuroimmunol., 104: 116–123.

Micera, A., Vigneti, E. and Aloe, L. (1998) Changes of NGF presence in non neuronal cells in response to experimental allergic encephalomyelitis in Lewis rats. Exp. Neurol., 154: 41–46.

Moalem, G., Gdalyahu, A., Shani, Y., Otten, U., Lazarovici, P., Cohen, I.R. and Schwartz, M. (2000) Production of neurotrophins by activated T cells: implications for neuroprotective autoimmunity. J. Autoimmun., 15: 331–345.

Muhallab, S., Lundberg, C., Gielen, A.W., Lidman, O., Svenningsson, A., Piehl, F. and Olsson, T. (2002) Differential expression of neurotrophic factors and inflammatory cytokines by myelin basic protein-specific and other recruited T cells infiltrating the central nervous system during experimental autoimmune encephalomyelitis. Scand. J. Immunol., 55: 264–273.

Neumann, H., Misgeld, T., Matsumuro, K. and Wekerle, H. (1998) Neurotrophins inhibit major histocompatibility class II inducibility of microglia: involvement of the p75 neurotrophin receptor. Proc. Natl. Acad. Sci. USA, 95: 5779–5784.

O'Connell, J., Houston, A., Bennett, M.W., O'Sullivan, G.C. and Shanahan, F. (2001) Immune privilege or inflammation? Insights into the Fas ligand enigma. Nat. Med., 7: 271–274.

414

Oderfeld-Nowak, B., Zaremba, M., Micera, A. and Aloe, L. (2001) The upregulation of nerve growth factor receptors in reactive astrocytes of rat spinal cord during experimental autoimmune encephalomyelitis. Neurosci. Lett., 308: 165–168.

Racke, M.K., Bonomo, A., Scott, D.E., Cannella, B., Levine, A., Raine, C.S., Shevach, E.M. and Rocken, M. (1994) Cytokine-induced immune deviation as a therapy for inflammatory autoimmune disease. J. Exp. Med., 180: 1961–1966.

Racke, M.K., Dhib-Jalbut, S., Cannella, B., Albert, P.S., Raine, C.S. and McFarlin, D.E. (1991) Prevention and treatment of chronic relapsing experimental allergic encephalomyelitis by transforming growth factor-beta 1. J. Immunol., 146: 3012–3017.

Ransohoff, R. and Trebst, C. (2000) Surprising pleiotropy of nerve growth factor in the treatment of experimental autoimmune encephalomyelitis. J. Exp. Med., 191: 1625–1629.

Rodriguez, M., Pavelko, K., McKinney, C. and Leibowitz, J. (1994) Recombinant human IL-6 suppresses demyelination in a viral model of multiple sclerosis. J. Immunol., 153: 3811–3821.

Santambrogio, L., Benedetti, M., Chao, M.V., Muzaffar, R., Kulig, K., Gabellini, N. and Hochwald, G. (1994) Nerve growth factor production by lymphocytes. J. Immunol., 153: 4488–4495.

Schull, M., Ormsby, I. and Kier, A. (1992) Targeted disruption of the mouse transforming growth factor b gene results in multifocal inflammatory disease. Nature, 359: 693–699.

Stadelmann, C., Kerschensteiner, M., Misgeld, T., Bruck, W., Hohlfeld, R. and Lassmann, H. (2002) BDNF and gp145trkB in multiple sclerosis brain lesions: neuroprotective interactions between immune and neuronal cells? Brain, 125: 75–85.

Steinman, L. (2001) Multiple sclerosis: a two-stage disease. Nat. Immunol., 2: 762–764.

Streilein, J.W. (1995) Unraveling immune privilege. Science, 270: 1158–1159.

Streilein, J.W., Ksander, B.R. and Taylor, A.W. (1997) Immune deviation in relation to ocular immune privilege. J. Immunol., 158: 3557–3560.

Torcia, M., Bracci-Laudiero, L., Lucibello, M., Nencioni, L., Labardi, D., Rubartelli, A., Cozzolino, F., Aloe, L. and Garaci, E. (1996) Nerve growth factor is an autocrine survival factor for memory B lymphocytes. Cell, 85: 345–356.

Trapp, B., Bo, L., Mork, S. and Chang, A. (1999) Pathogenesis of tissue injury in MS lesions. J. Neuroimmunol., 98: 49–56.

Valdo, P., Stegagno, C., Mazzucco, S., Zuliani, E., Zanusso, G., Moretto, G., Raine, C.S. and Bonetti, B. (2002) Enhanced expression of NGF receptors in multiple sclerosis lesions. J. Neuropathol. Exp. Neurol., 61: 91–98.

Villoslada, P., Hauser, S., Bartke, I., Unger, J., Heald, N., Rosenberg, D., Cheung, S., Mobley, W., Fisher, S. and Genain, C. (2000) Human nerve growth factor protects common marmosets against autoimmune encephalomyelitis by switching the balance of T helper type 1 and 2 cytokines within the central nervous system. J. Exp. Med., 191: 1799–1806.

Willenborg, D., Fordhan, S., Cowden, W. and Ramshaw, I. (1995) Cytokines and murine autoimmune encephalomyelitis. Inhibition or enhancement of disease with antibodies to select cytokines, or by delivery of exogenous cytokines using a recombinant vaccinia virus system. Scand. J. Immunol., 41: 31–41.

Yuen, E. and Mobley, W. (1996) Therapeutic potential of neurotrophic factors for neurological disorders. Ann. Neurol., 40: 346–354.

Progress in Brain Research, Vol. 146
ISSN 0079-6123

CHAPTER 26

Remyelination in multiple sclerosis: a new role for neurotrophins?

Hans H. Althaus*

Max-Planck-Institute for Experimental Medicine, RU Neural Regeneration, Göttingen, Germany

Abstract: Multiple sclerosis (MS) is a common neurological disease, which affects young adults. Its course is unpredictable and runs over decades. It is considered as an autoimmune disease, and is neuropathologically characterized by demyelination, variable loss of oligodendroglial cells, and axonal degeneration. Demyelination provides a permitting condition for axonal degeneration, which seems to be causative of permanent neurological deficits. Hence, the current treatment, which works preferentially immunmodulatory, should be complemented by therapeutics, which improves remyelination not only for restoring conduction velocity but also for preventing an irreversible axonal damage. One strategy to achieve this aim would be to promote remyelination by stimulating oligodendroglial cells remaining in MS lesions. While central nervous system neurons were already known to respond to neurotrophins (NT), interactions with glial cells became apparent more recently. In vitro and in vivo studies have shown that NT influence proliferation, differentiation, survival, and regeneration of mature oligodendrocytes and oligodendroglial precursors in favor of a myelin repair. Two in vivo models provided direct evidence that NT can improve remyelination. In addition, their neuroprotective and anti-inflammatory role would support a repair. Hence, a wealth of data point to NT as promising therapeutical candidates.

Keywords: neurotrophins; NGF; BNDF; remyelination; oligodendroglial cells; multiple sclerosis

Introduction

Multiple sclerosis (MS) is a neurological disease, which most commonly hits younger adults with an age between 20 and 30 years. The estimates, of how many people are afflicted, range from 650,000 for Europe and North-America (Dean, 1994) to 2,500,000 world wide (Compston and Coles, 2002). MS is considered as an autoimmune disease and is characterized by a variable loss of oligodendrocytes (OL) and oligodendroglial precursors (OLP), the myelin producing cells, within demyelinated plaques, in which signs of an inflammatory process can be

observed (Brück et al., 1994; Scolding et al., 1994; Allen and Kirk, 1997; Ewing and Bernard, 1998; Compston and Coles, 2002). Axons, which have lost their myelin sheath, at least for 2–3 subsequent internodes, are impaired in propagating action potentials. They could regain conduction ability to some extent due to reorganization of their membrane with expressing sodium channels in demyelinated axonal segments (Waxman, 1997). Attempts of myelin repair, present as so-called shadow plaques, have been described previously, and are now estimated to occur in about 40% of the plaques extending to more than 10% of the lesion area in acute lesions (Scolding et al., 1998) with a tendency to become less in chronic lesions. Concomitant with a less efficient remyelination was an oligodendroglial cell loss noted occurring more pronounced in later stages of the disease (Prineas et al., 1993a; Ozawa

*Correspondence to: H.H Althaus, Max-Planck-Institute for Experimental Medicine, RU Neural Regeneration, H.-Reinstr. 3, D-37075 Göttingen, Germany. Fax: +49-551-3899-317; E-mail: althaus@em.mpg.de

DOI: 10.1016/S0079-6123(03)46026-3

et al., 1994; Wolswijk, 2000). A combined morphological/immunohistochemical study examined lesions quantitatively and found that four principal patterns occur (Lucchinetti et al., 2000; Lassmann, 2002): in pattern I and II, demyelinated plaques, T-cell +/- antibody mediated, exhibit a variable loss of oligodendrocytes (OL) at the active lesional zone, reappearance of high numbers of OL in the inactive plaque center; remyelinated areas—shadow plaques—are present; in pattern III and IV, a higher incidence of oligodendroglial dystrophy and death can be detected compatible with the idea that OL and OLP may directly be targets of cytotoxic agents such as free radicals or glutamate (Noetzel and Brunstrom, 2001; Yamaya et al., 2002; Hsieh et al., 2003; Rosenberg et al., 2003); inactive lesions are almost devoid of OL, remyelination is sparse, if present at all. Lesions, where only a small number of eventual myelinating cells are present, indicate little migration of OLP towards the lesion (Perry, 1998), and/or a destruction of OL and OLP by recurrent inflammatory attacks. However, a recent study identified proliferating oligodendrocytes in active as well as in chronic inactive MS plaques in relatively high numbers, which again points to the variability of the course of the disease (Solanky et al., 2001).

Axonal degeneration in MS lesions, already described almost a 100 years ago (Marburg, 1936), was confirmed by recent immunocytochemical and morphological studies: Transected axons and a variable general reduction within the lesions were noticed (Mews et al., 1998; Trapp et al., 1998). Axonal injury and degeneration can occur in early acute lesions (Kuhlmann et al., 2002) as well as in chronic lesions (Kornek et al., 2000). Different mechanisms are discussed, of how damage of axons might come about (Bjartmar and Trapp, 2001). Axons in early lesions could directly be attacked by the inflammatory process, eventually independent of the demyelinating activity (Bitsch et al., 2000; Rieckmann and Smith, 2001; DeStefano et al., 2002), whereas in chronic lesions loss of OL and myelin are suggested as major cause for axonal damage. Results, which indicate the importance of myelin and myelin forming cells for the viability and organization of myelinated axons (Sanchez et al., 1996; Kaplan et al., 1997; Griffiths et al., 1998), support the view that axons require OL for a trophic supply and myelin as a protective

shield against noxious products, e.g., nitric oxide (NO) (Meyer-Franke et al., 1995; Bjartmar et al., 1999). The integrity of the axon-myelin-OL axis needs obviously to be intact to fulfill this task, since solely the presence of remaining OL and OLP in MS lesions does not seem to be sufficient to prevent axonal damage. The irreversible axonal damage is probably the major cause for permanent neurological deficits when the disease progresses (Bjartmar et al., 1999).

This long neglected fact of axonal damage has remyelination put forward for considering it more intensely as a therapeutical challenge (Scolding, 1999) than previously with restoring conduction velocity as the only demand.

Neurotrophins and their receptors in MS and EAE

Previous reports indicated that cytokines released by glial cells and lymphocytes will probably play a decisive role in damaging and repairing processes, hence the question was: are inflammatory cytokines harmful or helpful (Merrill and Benveniste, 1996; Aalto et al., 2002).

Neurotrophins (NT), a family branch of cytokines (Hopkins and Rothwell, 1995), constitute a group of 6 members, to which nerve growth factor (NGF), brain-derived neurotrophic factor (BDNF), neurotrophin-3 (NT-3), NT-4/5, NT-6 and NT-7 belong. They interact with two transmembrane receptors, namely p75 neurotrophin receptor (p75^{NTR}) and specific tyrosine kinase receptors, TrkA-C (Chao and Hempstead, 1995; Kaplan and Miller, 1997; Huang and Reichardt, 2001; Sofroniew et al., 2001). Not only neurons but also glial cells, microglial cells/macrophages, and lymphocytes produce NT (see below). These cells are also equipped with various NT receptors (Otten and Gadient, 1995; Levi-Montalcini et al., 1996; Nakajima et al., 1998; Althaus and Richter-Landsberg, 2000). Neurotrophins were considered in this context, since NT such as NGF, BDNF, NT-3, and NT-4/5 could potentially be involved in regeneration as well as in inflammatory actions as already pointed out by Aloe et al. (1994). Indeed, an increased level of NGF has been found in the central nervous system (CNS) of MS patients (Laudiero et al., 1992; Aloe et al., 1994; Aloe, 1998). BDNF has been shown to be produced by peripheral blood mononuclear cells (PBMC) of relapse-remitting (RRMS) MS patients at a significantly higher level during relapse and in the recovery phase in contrast to the lower BDNF values for PBMC of secondary progressing MS (SPMS); it was hypothesized that BDNF acts neuroprotective and favors remyelination in RRMS patients, while its reduced production correlates with the progression of the disease (Sarchielli et al., 2002). BDNF immunoreactivity was present in inflammatory infiltrates of MS produced by activated T and B cells, and monocytes. BDNF bioactivity was revealed by using sensory neurons (Stadelmann et al., 2002). A number of results support the view that, at least in experimental models, NT display a neuroprotective role and do not favor rather than suppress inflammatory activities at various sites thereby ameliorating EAE (Neumann et al., 1998; Villoslada et al., 2000; Arredondo et al., 2001; Flügel et al., 2001; Muhallab et al., 2002). The production of BDNF, NT-3, and glial cell line-derived neurotrophic factor (GDNF) by T and NK cells in EAE was suggested to be neuroprotective (Hammarberg et al., 2000). While NGF antibodies exacerbated neuropathological signs of EAE (Micera et al., 2000), was a reduction in the EAE disease score noticed when $p75^{NTR}$ antisense oligonucleotides were daily intraperitoneally injected. A $p75^{NTR}$ reduction was only seen at the blood-brain barrier (BBB) level, the mechanisms for EAE improvement remained speculative (Soilu-Hänninen et al., 2000).

Conflicting results exist concerning the expression of NT receptors in MS lesions. Findings in this respect could have important implications, among others, for the survival of OL, since previously it has been demonstrated that $p75^{NTR}$ expressing rat OL undergo apoptotic death when exposed to NGF (Casaccia-Bonnefil et al., 1996), and rescued when TrkA was expressed (Yoon et al., 1998). An upregulation of $p75^{NTR}$ was detected in OL and microglia/macrophages in MS lesions by Dowling et al. (1999), whereas Valdo et al. (2002) did not detect $p75^{NTR}$ expression on the majority of OL and OLP, but on microglia/macrophages; TrkA was never observed on glial cells. Chang et al. (2000), however, found $p75^{NTR}$ to be present on a subset of OL and OLP. In another report, both $mRNA^{TrkA}$ and $mRNA^{NGF}$ were shown to be expressed by OL in human optic nerves affected by MS (Micera et al., 1999).

Intrinsic sources of neurotrophins under normal and pathological conditions

Neuronal as well as nonneuronal cells can produce cytokines in the CNS (Merrill and Jonakait, 1995). The capability of glial cells to produce NT has recently been reviewed (Althaus and Richter-Landsberg, 2000), hence, it will here briefly be summarized.

Astrocytes express NGF (Furukawa et al., 1986) and $mRNA^{NGF}$, $mRNA^{NT-3}$, and $mRNA^{NT-4}$ (Condorelli et al., 1995) in culture. The level of BDNF and NGF increased significantly in reactive astrocytes (Goto and Furukawa, 1995; Rossner et al., 1997). Various substances have been shown to modulate NT synthesis in astrocytes. Interestingly, interleukin-1 (IL-1), a proinflammatory cytokine, enhanced NGF synthesis in reactive but not in normal astrocytes (Wu et al., 1998). Furthermore, a 40-fold increase of astroglial $mRNA^{NGF}$ was observed, when interferon-beta (IFN-β) was added to astroglial cultures. It was suggested that part of the beneficial IFN-β therapy for MS is related to this effect (Boutros et al., 1997).

Oligodendrocytes as well as their precursors might also be involved in NT synthesis. They seem to express NGF and BDNF in vitro and in vivo (Gonzalez et al., 1990; Byravan et al., 1994; Condorelli et al., 1995; Qu et al., 1995; Dai et al., 2001; Du and Dreyfus, 2002). In pig, however, OL NGF mRNA was not detected in Northern blots and only a faint band was seen in Western blots corresponding to comigrating NGF (unpublished data). In injured rat spinal cord, NGF has not been detected in OL (Krenz and Weaver, 2000).

Neuronal NGF synthesis has been found for cortical neurons in the forebrain, for neurons and interneurons of the hippocampus and striatum. NT immunoreactivity has also detected in retinal ganglion cells. Cerebral injuries upregulate neuronal NGF expression (reviewed by Raffioni et al., 1993; McAllister, 1999; Sofroniew et al., 2001).

Nonneural synthesis of NT was detected in macrophages/microglia (Elkabes et al., 1998; Heese et al., 1998; Nakajima et al., 2001), in cerebral endothelial cells (Bayas et al., 2002), in lymphocytes (Torcia et al., 1996; Kerschensteiner et al., 1999; Moalem et al., 2000; Barouch and Schwartz, 2002), and in choroid plexus cells, which contained high levels of mRNA^{NGF} and mRNA^{NT-4} and low levels of mRNA^{BDNF} and mRNA^{NT-3} (Timmusk et al., 1995). The latter finding is of particular interest, since the choroid plexus is bathed in the cerebrospinal fluid, which means that secreted NGF could reach subependymal cell layers on this way by diffusion.

Remyelination: which oligodendroglial cells could do the job?

Which oligodendroglial cells are capable to remyelinate? At least, four cellular sources are in question for their remyelinating capability: surviving mature OL (mOL) and oligodendroglial precursors (OLP) (Gensert and Goldman, 1997), immigrating OLP (Levine et al., 2001) and subventricular zone (SVZ) cells (Picard-Riera et al., 2002). The contribution of mOL has divided 'oligodendrogliologists' into two camps (see Norton, 1996). The one provided evidence that mOL can recapitulate the myelinating program, in particular, it has previously been shown that mOL furnish myelin in vitro, when carbon fibers, thereby replacing axons, were added to cultures of pig OL (Althaus et al., 1987). Furthermore, cuprizone induced demyelination was repaired by mOL when OL apoptosis was prevented by IGF-1; OLP failed to enter the lesion when mOL remained (Mason et al., 2000). The other debated their contribution to remyelination and suggested OLP as being the only source for remyelination (Keirstead and Blakemore, 1997; Carroll et al., 1998), however, a detailed discussion on the pros and cons should follow up elsewhere.

What immunocytochemical markers characterize mOL and adult OLP? The surface expression of galactosylcerebroside (GC), commonly accepted previously as marker for mOL, is obviously inadequate, since GC immunoreactivity was present on proliferative OLP (Shi et al., 1998). The final stage of oligodendroglial maturation is achieved by the acquisition of myelin proteins such as MBP, PLP, and MOG (Norton, 1996). But even a subpopulation of these cells can undergo proliferation at least for 3–4 cycles when stimulated via phorbol esters or NGF (Althaus et al., 1991, 1992). Similarly, recent results from aggregate cultures suggest that myelinating OL retain the capacity to reenter the cell cycle, when exposed to protein kinase C activating substances

(Pouly et al., 2001b). In this context, it is interesting to note that proliferating MOG$^+$ were observed in MS tissue (Schönrock et al., 1998). Eventually, these cells have entered a quiescent state during development prematurely, but followed their final differentiation program subsequently. Perinatal and adult OLP have markers in common such as NG2 and PDGF-α receptor (Scolding et al., 1998; Dawson et al., 2000), however, they differ in their expression of O4 with perinatal OLP as being O4$^-$ (Levine et al., 2001); perinatal OLP are A2B5$^+$, whereas only a fraction of adult OLP express this marker (Scolding et al., 1995). Interestingly, perinatal OLP seem to follow an intrinsic maturation program, when long-term cultured in a PDGF containing medium, and acquire characteristics of adult OLP with a reduced proliferation and migratory rate (Tang et al., 2000).

One disadvantage of mOL seems to be their restricted migratory capacity, which, however, only takes effect, when repopulation of lesions at a greater distance is concerned. A drastic reduction in their migratory ability was also noted for OLP, when they develop from O4$^-$ to O4$^+$ OLP (Warrington et al., 1993). Furthermore, perinatal and adult OLP differ substantially in their proliferating and migrating capacity, with less potency of adult OLP (Tang et al., 2000; Levine et al., 2001). A mixture of transplanted A2B5$^+$ (human, rat) and O4$^+$ OLP populated lesions, but did not migrate in normal white matter of the adult rat CNS (Targett et al., 1996; O'Leary and Blakemore, 1997), which indicates that only OLP surviving within or close to the lesion will probably remain as source for remyelinating axons (Windrem et al., 2002). An involvement of SVZ cells in myelin repair was recently suggested (Picard-Riera et al., 2002), taking up previous reports on the development of OL and astrocytes from SVZ cells (Levison and Goldman, 1993), and on their proliferation in experimental allergic encephalomyelitis (EAE) (Calza et al., 1998). The capacity of the adult rat brain to generate perinatal OLP by differentiating subependymal cells in vitro has been shown by Zhang et al. (1999). However, the intrinsic capabilities of SVZ cells seem in the human brain to be restricted, since close by periventricular lesions, pathognomonic for MS, are not privileged by populating and subsequently remyelinating cells. This might in part depend on the decreasing size of the SVZ with increasing age (Norton, 1996), but might also reflect a lack of appropriate growth factors (Chmielnicki and Goldman, 2002).

Whatever the percentage of OLP or mOL participating in myelin repair might be, the intrinsic capacity in MS lesions is relatively low and decreases further when the disease progresses despite the fact that a variable number of OL and adult OLP remain (Ozawa et al., 1994; Scolding et al., 1998; Wolswijk, 2000, 2002; Chang et al., 2002). Whether axons loose attractivity for OL (Chang et al., 2002) by, for example, expressing different subsets of Na^+ channels (Waxman, 2001) or PSA-NCAM (Durbec and Cremer, 2001; Charles et al., 2002), remains an open question. In contrast to the scarce remyelination in MS lesions are the results of those studies, where gliotoxin induced demyelinated areas are well remyelinated (Franklin, 2002). Hence, it was suggested that the growth factors present in MS lesions do only insufficiently support the remyelinating process. They might quantitatively or qualitatively not allow the regeneration of surviving OL or to drive proliferation and subsequent differentiation of OLP (for other reasons, see below).

Neurotrophins and oligodendroglial cells in vitro

Mature oligodendrocytes (mOL) as well as oligodendroglial precursors (OLP) were cultured to collect information about their properties, in particular, which factors direct their regeneration (major interest in OL) or their proliferation and differentiation (main focus in OLP). The species, from which the glial cells were isolated, range from mice and rats (OLP) to pigs (mOL) and human tissue (mOL and OLP).

The interest on mOL was based on the observation that a various number of OL survive a damaging hit within MS lesions (Wolswijk, 2000). The 'dying-back' of oligodendroglial processes could be one among other pathophysiological examples (Rodriguez et al., 1993). Indeed, a first morphological reaction of cultured mOL on a harmful but not killing event is a retraction of their processes. Surviving mOL would have to undergo the following sequential steps in terms of remyelination: Regeneration of their processes, finding and contacting axons, enwrapping of an axonal segment and compaction of the myelin sheath simultaneously with a production of myelin

components. Previous results have already indicated that the formation of new fibers is sensitive to environmental factors. For example, protein kinase C activating substances favor regeneration of longer (several times the body diameter) and branched oligodendroglial fibers, whereas protein kinase A activation via, for example, serum containing culture medium promotes relatively short flattened membranes, which would be less suitable for exploring the terrain at a distance. A third major pathway can be activated in pig OL by exposing the cells to NGF. The morphological effects were similar to those seen with PKC activation: enhanced oligodendroglial process regeneration and proliferation of a subset of these mature cells (Althaus et al., 1992). These effects could not be mimicked by BDNF or NT-3 suggesting the presence of NGF signaling competent receptors. Subsequently, TrkA as well as $p75^{NTR}$ were shown to be expressed, and major parts of the downstream signaling cascade revealed (Althaus et al., 1997). NGF transiently increased $[Ca^{2+}]_i$ in OL, which was dependent on a release of Ca^{2+} from internal stores (Engel et al., 1994); NGF stimulated the expression of several oligodendroglial proteins among which myelin proteins such as MBP and PLP were present (unpublished observation). Investigations on rat and mouse OL showed that NT receptors are present (Kumar et al., 1993), a survival effect of NGF and NT-3 was recognized (Cohen et al., 1996). Neurotrophins, in particular NGF, prevented OL from TNF-α mediated cell death by activating the Akt pathway (Takano et al., 2000), the signaling of which is necessary for OL survival (Vermuri and McMorris, 1996). An interrupted Akt pathway, induced by Wortmannin, a specific PI3-kinase inhibitor, interferes with the NGF mediated oligodendroglial process formation (Hempel, 1998) (Fig. 1).

Human mOL, obtained from temporal lobe resections of patients suffering from drug intractable epilepsy, seem to vary in the speed of regenerating their fibers, which eventually depend on the age of the brain, since OL from young epileptic patients regenerate their processes faster (Grenier et al., 1989) than those from elder patients (Althaus et al., 2001). Two groups have as yet reported about the presence of NGF receptors. Both agree upon the expression of $p75^{NTR}$ (Ladiwala et al., 1998), whereas TrkA was detected only by one (Fig. 2A, B). The good news

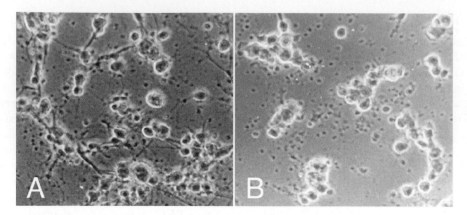

Fig. 1. Pig oligodendrocytes in culture: cells, which were exposed to NGF, have extended a number of processes (A); a concomitant treatment with the PI3-kinase inhibitor, Wortmannin, reduced the number of processes substantially (B) indicating that a retraction of processes is an early oligodendroglial response to an impairment of the 'survival pathway' PI3-kinase-Akt; removal of Wortmannin abolishes this effect and leads to a normal NGF response.

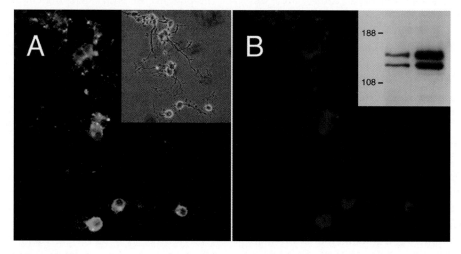

Fig. 2. Human oligodendrocytes, 4 weeks in vitro are identified by their expression of MBP (A); they also express TrkA as shown immunocytochemically and by Western blot (B); the inset presents TrkA, left lane human OL, right lane pig OL; note that the protein load for each lane was not adjusted.

said that exposure to NGF does not induce cell death (Ladiwala et al., 1998) even if only p75^{NTR} should be present. However, the regenerative response was not as pronounced as for pig OL, which might reflect age-dependent changes of NT signaling pathways (Althaus et al., 2001).

Studies on rat perinatal OLP showed that they express TrkC and respond to NT-3 with improved survival and proliferation (Barres et al., 1994; Cohen et al., 1996; Kumar and DeVellis, 1996; Bertollini et al., 1997). Under in vivo conditions it could also be demonstrated that NT-3 promotes survival and

proliferation (Kumar et al., 1998), mice lacking NT-3 and its receptor exhibited profound deficiencies in CNS glial cells (Kahn et al., 1999). Furthermore, it has been reported that NT-3, and to a lesser extent NGF, protect OLP against AMPA mediated excito-toxicity, to which OLP are more vulnerable than OL (Kavanaugh et al., 2000).

In vitro studies by Armstrong et al. (1992) revealed the presence of human OLP in the adult brain (13 to 51-year-old). This finding was confirmed by others. The number of slowly proliferating OLP in the human brain, which includes NG2$^+$, O4$^+$,

PDGF-α receptor[+] or A2B5[+] OLP, was estimated to be 1–4% (Gogate et al., 1994; Prabhakar et al., 1995; Scolding et al., 1998, 1999; Roy et al., 1999) compared to 5–8% in the rodent adult brain (Levine et al., 2001). In contrast, others have found NG2[+] OLP in the normal human brain as abundant as OL or astrocytes (Chang et al., 2000), however, expression of NG2 on astrocytes (Levine and Card 1987; Hirsch and Bähr, 1999), macrophages (Jones et al., 2002), and endothelial cells (Pouly et al., 2001a) should be taken as a caveat in terms of identification. In addition, sulfatide, which is detected by the O4 antibody, is also expressed by a subpopulation of astrocytes (Pernber et al., 2002). Attempts to manipulate adult OLP (mostly O4[+]) for stimulating their proliferation or for activating their differentiation program were up to now not very convincing at least with the set of growth factors, to which rat perinatal OLP respond. Human A2B5[+] OLP were also tested for their response to neurotrophins without a positive result (Scolding et al., 1995). No information exists concerning the expression of NT receptors.

In summary, mature OL and OLP express NT receptors. NGF and NT-3 induces a variety of responses, which are important for an eventual remyelination. The expression of full length TrkB remains to be shown (Althaus and Richter-Landsberg, 2000). Human OL from middle-aged brain tissue respond to NGF, however, less marked when compared to OL from young pig brains. It might be a common phenomenon that with increasing age of the brain OL as well as OLP are slowed down in their activity, indicating alterations in their intracellular signaling pathways.

Remyelination via neurotrophins

Transmission electron microscopy (TEM) and immunohistochemical studies had revealed two principal patterns of MS lesions, those where mature OL and adult OLP are present and those where a considerable loss of OL and OLP had occurred (Lucchinetti et al., 2001). To transplant myelin forming cells seems to be reasonable for the latter type of lesions, if a sufficient number of intact axons have remained (Paty and Arnold, 2002). In lesions, where a relevant number of OL and OLP survive, usually early MS cases and RRMS, some remyelination can be observed. It implies that it is worth to follow a strategy, which attempts to improve remyelination via a pharmacological intervention at an early stage of the disease.

Neurotrophins, in particular BDNF and NGF, could eventually act in MS lesions in two directions: anti-inflammatory (see above) and pro-repairing myelin. In fact, it has been suggested that elevated levels of NT contribute to the beneficial effect of IFN-β (Boutros et al., 1997). In addition, an anti-apoptotic, protective role for OL and OLP could support remyelination (Takano et al., 2000; Koda et al., 2002). Two in vivo studies, carried out in different animal species and models provide direct evidence that NT could be of benefit for remyelinating demyelinated areas.

The clinical background for one of the experimental approaches (McTigue et al., 1998) was the observation that part of the functional loss after spinal cord injury is due to a demyelination of trauma surviving axons. The study used a reproducible model of spinal contusion in the rat. The injury caused a central cavity, which was surrounded by an epicenter. Already 2 days after injury, a drastic reduction in the amount of axons occurred. At this time, the cavities were filled with sets of fibroblasts bioengineered to produce NGF, BDNF, NT-3, ciliary neurotrophic factor (CNTF), basic fibroblast growth factor (bFGF), and a control. At 10 weeks after injury, all transplants contained axons, but significant more in NT-3 and BDNF grafts. Nearly all of these axons were myelinated, whereas the others remained more or less unmyelinated. The lack of a detectable effect of CNTF was unexpected in the light of several studies, which report about enhanced survival and proliferation of oligodendroglial cells via CNTF (for references see below). It was shown that NT-3 and BDNF stimulated the proliferation of OLP with no effect on the number of astrocytes when compared with controls. It was concluded that the presence of bare axons alone is not sufficient to signal to myelinating cells.

The other model used young–adult minipigs for producing demyelinated lesions in the left or right part of the corpus callosum/centrum semiovale (ccs) (Althaus et al., 2001). Lysolecithin (0,01%, 10 μl) was injected via implanted cannules into the brain of anesthetized animals by using a stereotactic frame (Fig. 3). After 10 days, an almost complete demyelinated area had developed (Fig. 4A), in which a considerable number of axons remained morphologically intact (Fig. 4B). Mature OL were present within the lesion as revealed by mRNA^{PLP} in situ

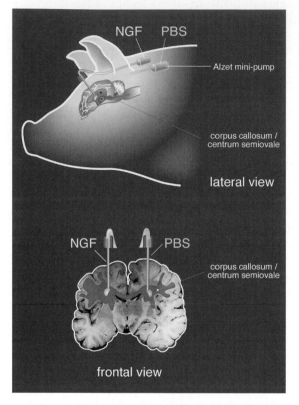

Fig. 3. Demyelination was induced in the white matter of mini pig brains; lysolecithin was injected into the corpus callosum/ centrum semiovale via a stereotactic frame; this area was superfused with PBS or PBS plus NGF 10 days thereafter.

hybridization and MOG$^+$ labeling; macrophages loaded with myelin debris were also visible. At this time, the lesions were superfused either with phosphate buffered saline (PBS) (control) or with PBS + NGF via Alzet minipumps connected to already implanted cannules (Fig. 3). The superfusion of NGF was performed with a concentration equivalent to the one used in vitro. It was continued for 10–12 days, after which the animals were sacrificed in deep anesthesia by perfusing them with paraformaldehyde via the common carotid arteries. PBS superfused areas remained largely demyelinated (Fig. 5A) with rare MOG$^+$ and MBP$^+$ islands presumably indicating a beginning myelin repair. In contrast, an intense remyelination occurred in NGF superfused areas (Fig. 5B). TEM showed that axons were surrounded by a thinner myelin sheath than normal indicating remyelination. The significance of the NGF effect was substantiated by an NIH image program. Although it is most likely that surviving mature OL were the cells mainly engaged in remyelination, a participation of OLP cannot be excluded. In any case, they were unable to furnish myelin within the time frame without NGF (Rohde, 1998; Althaus et al., 2001).

These two models directly indicated a benefit of NT in terms of remyelination. A third approach pointed to NGF as being involved in remyelination by acting protectively. The major impact of the study was to provide evidence that an exogenous administration of thyroid hormone (T4) can influence endogenous oligodendroglial precursors (Calza et al., 2002). It was shown that EAE rats upregulate the expression of markers for OLP and OL, when

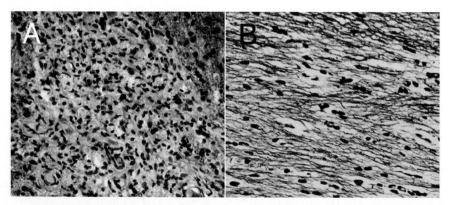

Fig. 4. Demyelinations. LFB staining (A), in which a reasonable number of axons remain; Bielschowsky staining (B).

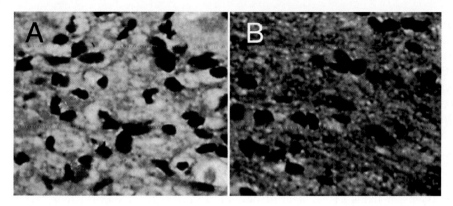

Fig. 5. Areas superfused with PBS show little, if any, remyelination (A), whereas superfusing with PBS plus NGF promotes remyelination considerably (B). LFB staining, myelin specific immunohistochemical stainings and TEM investigations showed that indeed remyelination has occurred.

receiving a subcutaneous dosage of T4. A coeffect was noticed in that T4 restored an NGF level to normal, which otherwise drops in the CNS during the acute phase of EAE.

Conclusions

Two major neural deficits characterize MS lesions: damage of the OL-myelin unit and degeneration of axons. The underlying inflammatory process can hit both structures almost at the same time, although demyelination seems to precede axonal injury (Kornek et al., 2000). Neurological disturbances result from both kind of impairments, and it might be idle speculation, which of the two contribute more to a permanent neurological disability as long as neither axonal destruction nor demyelination can effectively be repaired. However, improving remyelination, which could reduce axonal damage, might have a better chance at present. The predominant experimental route since a couple of years for achieving remyelination investigates the potential of transplanting myelin repairing cells. Indeed, a first phase I trial has recently commenced by using autologous Schwann cells (SC). The choice of SC has some apparent advantages, but one has to keep in mind that their survival in adult white matter could be limited (Iwashita et al., 2000), which might be improved when appropriate growth factors are simultaneously administered (Milward et al., 2000). Furthermore, in contrast to OL, SC are potentially antigen presenting cells. Hence, it will be interesting

to know not only whether SC survive or myelinate but also whether they influence the inflammatory process. The transplantation of SC or those of the oligodendroglial lineage seems to be of benefit for inherited demyelinating diseases and might be worth to be employed for certain 'key' lesions of chronic-progressive MS cases. Its use at the beginning of the disease seems to be problematic. An alternative option is prospectively more promising with a systemic transplantation of bioengineered hematopoietic stem cells (Akijama et al., 2002; Hintzen, 2002). However, its use at the beginning of the disease seems to be problematic for several reasons (Jeffery, 2002; Paty and Arnold, 2002).

The current pharmacologic treatment of MS is primarily focused on modulating the inflammatory immune response (Wiendl and Hohlfeld, 2002). The only exception is intravenous immunoglobulin G (IVIG), for which, in addition, a positive effect on remyelination was proposed. Clinical trials have so far given no convincing results (Wiendl and Hohlfeld, 2002). In addition, an in vitro study could not reveal an effect of IVIG on oligodendroglial proliferation, differentiation or migration, however, a subsequent study could show that IVIG protects OL and OLP from antibody-mediated complement injury (Stangel et al., 2000). The search for a remyelination promoting treatment became more challenging, when evidence was provided that myelin repair could be an essential factor to prevent irreversible axonal degeneration. Recently, a subfamily of cytokines, the growth factors, has been detected as

424

being of potential benefit for enhancing remyelination. Several studies in the past had already indicated that growth factors are important for oligodendroglial proliferation, differentiation, survival, and regeneration (reviewed by Norton, 1996; Ludwin, 1997; Althaus and Richter-Landsberg, 2000; Baumann and Pham-Dinh, 2001). The major pool of knowledge stems from investigations of rodents. Three growth factors, CNTF, insulin-like growth factor-1 (IGF-1), and neuregulin (GGF-2), gained particular interest. A number of reports support the view that they could be helpful agents in terms of remyelination (Barres et al., 1996; Butzkueven et al., 2002; Ransohoff et al., 2002; Stankoff et al., 2002 for CNTF; McMorris and Dubois-Dalcq, 1988; Liu et al., 1995; McMorris and McKinnon, 1996; Ye and D'Ercole, 1999 for IGF-1; Canoll et al., 1996; Vartanian et al., 1999; Fernandez et al., 2000; Calaora et al., 2001 for GGF/neuregulin). Clinical trials are on the way for all three. Preliminary data for IGF-1 showed that it is well tolerated, but yet without an effect on clinical measures of MS (Frank et al., 2002). On the basis of investigations on EAE animals, it was already doubted that IGF-1 is a good candidate for MS treatment, since it did not produce an enhancing effect on remyelination (Cannella et al., 2000).

Certainly, NT should also be nominated as promising candidates to ameliorate MS. Numerous studies in vitro and in vivo support this view as outlined above. In terms of remyelination, two different models in two different species have shown that NT, in particular NGF, BDNF and NT-3, have demonstrated its value to improve myelin repair by stimulating the proliferation of OLP or by enhancing OL regeneration, two decisive components of a repair program in MS lesions. In this context, it is important to note that mature human OL express TrkA as well as p75^{NTR}. The wide spectrum of NT effects on proliferation, differentiation, survival, and regeneration could provide an answer for a number of negative factors which are suggested to render remyelination in MS insufficient. For example, tumor necrosis factor-alpha (TNF-α) can perturb or even kill OL, whereas NGF can rescue OL (Takano et al., 2000). Oligodendroglial cells are vulnerable to AMPA mediated excitotoxicity, NT-3 and NGF protect them (Kavanaugh et al., 2000). Oxidative

stress could be deleterious to OL, NGF may probably prevent cell death (Althaus, unpublished observation) as already shown for PC12 cells (Satoh et al., 1999), and for granular cerebellar neurons (Skaper et al., 1998). A new aspect was recently provided, why OLP do not remyelinate MS lesions: OLP re-express the Notch pathway at the same time, where a transforming growth factor-beta 1 (TGF-β1) upregulation of Jagged-1 occurs in hypertrophic astrocytes. The close contact of OLP and astrocytes inhibits the process outgrowth from human OLP (John et al., 2002). During development, it is the axon, which expresses Jagged thereby inhibiting OLP differentiation; its downregulation allows OLP to enwrap axons (Wang et al., 1998). Hence, it would be interesting to know as to whether axons do also reexpress Jagged in MS lesions; furthermore, whether a similar scenario occurs in experimentally induced demyelinations, either promptly or as a matter of time. Since NT are known to influence the differentiation of embryonic stem cells and SVZ cells (Lachyankar et al., 1997; Schuldiner et al., 2000; Benoit et al., 2001; Kang et al., 2001), it would be interesting to prove, whether NT signaling could modulate or antagonize a reduced oligodendroglial process formation via Jagged-1/Notch.

Other reasons, why remyelination is reluctant in MS plaques are more complex and include axonal alterations, slowing of OLP proliferation, differentiation, and migration (Chari and Blakemore, 2002), which could be part of general aging effects on oligodendroglial cells. Several reports indicate that with increasing age, a reduction of oligodendroglial capabilities occurs (Wolf et al., 1986; Gilson and Blakemore, 1993; Targett et al., 1996; Althaus et al., 2001). This means that intrinsic myelin repairing mechanisms are delayed and less efficient, and that a pharmacologic intervention meets OL and OLP, which are less responsive. This problem becomes relevant, since MS is a disease, which usually runs several decades. Consequently, a myelin repairing treatment should start as soon as the disease is recognized, since in the initial stage of the disease, which becomes apparent usually between 20 and 30 years, a better potential for a repairing can be assumed (Jeffery, 2002; Paty and Arnold, 2002). An early improved remyelination will also minimize fatal axonal alterations; these could occur on two levels either because of an exposure of nude axons to noxious substances or due to endogenous membrane reorganizations over time. Once executed, it would oppose any remyelinating attempts. Early intervention and younger age of animals seems to contribute to a successful remyelination, where demyelination has experimentally been induced as a one time hit. A continuously ongoing deleterious inflammatory process would counteract remyelinating attempts. Hence, immunmodulatory agents should build a first frontline of defense and should provide a continuous downregulation of the inflammatory process, which otherwise could make the success of a remyelination worthless (Prineas et al., 1993b).The advantage of a combinatory treatment with NT would be that NT not only exert myelin promoting and neuronal protecting effects, but also contribute to a downregulation of the inflammatory process. Constructs of NT chimeric molecules, for example NGF/BDNF, may be of advantage by being synergistic in their actions (Ibanez, 1995).

A crucial point for making NT treatment effective is given by the fact that a local concentration between 1 and 10 ng/ml should be achieved. Endogenous levels of NT are too low (pg range) in the CNS (Lorigados et al., 2001), even though, for example, astroglial production of NGF can be enhanced several fold via IFN-β (Boutros et al., 1997) or via Clenbuterol (Zhu and Krieglstein, 1999). Using intravenous infusions for a delivery of NT would normally fail because of the impermeable blood-brain barrier (BBB). However, neurotrophins could cross the BBB and reach oligodendroglial cells, when it becomes permeable under pathological conditions such as MS (Loy et al., 1994; Hefti, 1997a; Butzkueven et al., 2002). In any case, various possibilities exist, to pass or to bypass the BBB (Tan and Aebischer, 1996; Hefti, 1997b; McGrath et al., 1997; Skaper and Walsh, 1998; Ankeny et al., 2001; Coumans et al., 2001; Martino et al., 2001; Thorne and Frey, 2001; Tuszynski et al., 2002).

Acknowledgments

Part of this work was supported by the Myelin Project. I wish to thank my former coworkers G. Rohde and R. Hempel, from whom part of their work is mentioned in this review. My thanks also go to N. Parvizi (Institute of Animal Breeding and Behavior, 31535, Mariensee) and W. Brück (Institute of Neuropathology, University of Göttingen), whose efforts contributed to the success of the minipig model used to demonstrate remyelination via NGF.

Abbreviations

BBB	blood–brain barrier
BDNF	brain-derived neurotrophic factor
bFGF	basic fibroblast growth factor
CNS	central nervous system
CNTF	ciliary neurotrophic factor
IFN-β	interferon-beta
mOL	mature oligodendrocytes
MS	multiple sclerosis
NGF	nerve growth factor
NT	neurotrophins
NT-3-4/5, -6, -7	neurotrophin-3, NT-4/5, NT-6, and NT-7
p75^{NTR}	p75 neurotrophin receptor
OL	oligodendrocytes
OLP	oligodendroglial precursors
RRMS	relapse-remitting MS
SPMS	secondary progressing MS
SVZ	subventricular zone
TGF-β1	transforming growth factor-beta 1
TNF-α	tumor necrosis factor-alpha
Trk	tyrosine kinase receptor

References

Aalto, K., Korhonen, L., Lahdenne, P., Pelkonen, P. and Lindholm, D. (2002) Nerve growth factor in serum of children with systemic lupus erythematosus is correlated with disease activity. Cytokine, 20: 136–139.

Akiyama, Y., Radtke, C., Honmou, O. and Kocsis, J.D. (2002) Remyelination of the spinal cord following intravenous delivery of bone marrow cells. Glia, 39: 229–236.

Allen, I.V. and Kirk, J. (1997). The anatomical and molecular pathology of multiple sclerosis. In: W.C. Russel (Ed.), Molecular Biology of Multiple Sclerosis. Wiley & Sons, Chichester, pp. 9–22.

Aloe, L. (1998) Nerve growth factor and autoimmune diseases: Role of tumor necrosis factor-alpha? Adv. Pharmacol., 42: 591–594.

Aloe, L., Skaper, S., Leon, A. and Levi-Montalcini, R. (1994) Nerve growth factor and autoimmune diseases. Autoimmunity, 19: 141–150.

426

Althaus, H.H. and Ritcher-Landsberg, C. (2000) Glial cells as targets and producers of neurotrophins. Int. Rev. Cytol., 197: 203–277.

Althaus, H.H., Bürgisser, P., Klöppner, S., Rohmann, A., Schröter, J., Schwartz, P., Siepl, C. and Neuhoff, V. (1987). Oligodendrocytes ensheath carbon fibers and produce myelin in vitro. In: H.H. Althaus and W. Seifert (Eds.), Glial-Neuronal Communication in Development and Regeneration. Springer Verlag, Berlin, pp. 780–798.

Althaus, H.H., Schröter, J., Spoerri, P., Schwartz, P., Klöppner, S., Rohmann, A. and Neuhoff, V. (1991) Protein kinase C stimulation enhances the process formation of adult oligodendrocytes and induces proliferation. J. Neurosci. Res., 29: 481–489.

Althaus, H.H., Klöppner, S., Schmidt-Schultz, T. and Schwartz, P. (1992) Nerve growth factor induces proliferation and enhances fiber regeneration in oligodendrocytes isolated from adult pig brain. Neurosci. Lett., 135: 219–223.

Althaus, H.H., Hempel, R., Klöppner, S., Engel, J., Schmidt-Schultz, T., Kruska, L. and Heumann, R. (1997) Nerve growth factor signal transduction in mature pig oligodendrocytes. J. Neurosci. Res., 50: 729–742.

Althaus, H.H., Rohde, G., Klöppner, S., Brück, W. and Parvizi, N. (2001a) Improved remyelination via nerve growth factor (NGF). J. Neurochem. (Suppl.), 78: 49.

Althaus, H.H., Mursch, K. and Klöppner, S. (2001b) Differential response of mature TrkA/p75NTR expressing human and pig oligodendrocytes: aging, does it matter? Microsc. Res. Tech., 52: 689–699.

Ankeny, D.P., McTigue, D.M., Guan, Z., Yan, Q., Kinstler, O., Stokes, B.T. and Jakeman, L.B. (2001) Pegylated brain-derived neurotrophic factor shows improved distribution into the spinal cord and stimulates locomotor activity and morphological changes after injury. Exp. Neurol., 170: 85–100.

Armstrong, R., Dorn, H.H., Friedman, E. and Dubois-Dalcq, M. (1992) Pre-oligodendrocytes from adult human CNS. J. Neurosci., 12: 1538–1547.

Arredondo, L.R., Deng, C.S., Ratts, R.B., Lovett-Racke, A.E., Holtzman, D.M. and Racke, M.K. (2001) Role of nerve growth factor in experimental autoimmune encephalomyelitis. Eur. J. Immunol., 31: 625–633.

Barres, B.A., Burne, J.F., Holtzmann, B., Thoenen, H., Sendtner, M. and Raff, M.C. (1996) Ciliary neurotrophic factor enhances the rate of oligodendrocyte generation. Mol. Cell. Neurosci., 8: 146–156.

Barres, B.A., Raff, M.C., Gaese, F., Bartke, I., Dechant, G. and Barde, Y. (1994) A crucial role for neurotrophin-3 in oligodendrocyte development. Nature, 367: 371–375.

Barouch, R. and Schwartz, M. (2002) Autoreactive T cells induce neurotrophin production by immune and neural cells in injured rat optic nerve: Implications for protective autoimmunity. FASEB J., 16: 5–27.

Baumann, N. and Pham-Dinh, D. (2001) Biology of oligodendrocyte and myelin in the mammalian central nervous system. Physiol. Rev., 81: 871–927.

Bayas, A., Hummel, V., Kallmann, B.A., Karch, C., Toyka, K.V. and Rieckmann, P. (2002) Human cerebral endothelial cells are a potential source for bioactive BDNF. Cytokine, 19: 55–58.

Benoit, B.O., Savarese, T., Joly, M., Engstroem, C.M., Pang, L.Z., Reilly, J., Recht, L.D. and Quesenberry, P.J. (2001) Neurotrophin channeling of neural progenitor cell differentiation. J. Neurobiol., 46: 265–280.

Bertollini, L., Ciotti, M.T., Cherubini, E. and Cattabeo, A. (1997) Neurotrophin-3 promotes the survival of oligodendrocyte precursors in embryonic hippocampal cultures under chemically defined conditions. Brain Res., 746: 19–24.

Bitsch, A., Schuchardt, J., Bunkowski, S., Kuhlmann, S. and Brück, W. (2000) Acute axonal injury in multiple sclerosis. Brain, 123: 1174–1183.

Bjartmar, C. and Trapp, B.D. (2001) Axonal and neuronal degeneration in multiple sclerosis: mechanisms and functional consequences. Curr. Opin. Neurol., 14: 271–278.

Bjartmar, C., Yin, X. and Trapp, B.D. (1999) Axonal pathology in myelin disorders. J. Neurocytol., 28: 383–395.

Boutros, T., Croze, E. and Yong, V.W. (1997) Interferon-beta is a potent promoter of nerve growth factor production by astrocytes. J. Neurochem., 69: 939–946.

Brück, W., Schmied, M., Suchanck, G., Brück, Y., Breitschopf, H., Poser, S., Piddleseden, S. and Lassmann, H. (1994) Oligodendrocytes in the early course of multiple sclerosis. Ann. Neurol., 35: 65–73.

Butzkueven, H., Zhang, J-G., Soilu-Hanninen, M.S., Hochrein, H., Chionh, F., Shipham, K.A., Emery, B., Turnley, A.M., Petratos, S., Ernst, M., Bartlett, P.F. and Kilpatrick, T.J. (2002) LIF receptor signaling limits immune-mediated demyelination by enhancing oligodendrocyte survival. Nat. Med., 8: 613–619.

Byravan, S., Foster, L.M., Phan, T., Verity, A.N. and Campagnoni, A.T. (1994) Murine oligodendroglial cells express nerve growth factor. Proc. Natl. Acad. Sci. USA, 91: 8812–8816.

Calaora, V., Rogister, B., Bismuth, K., Murray, K., Brandt, H., Leprince, P., Marchionni, M. and Dubois Dalcq, M. (2001) Neuregulin signaling regulates neural precursor growth and the generation of oligodendrocytes in vitro. J. Neurosci., 21: 4740–4751.

Calza, L., Giardino, L., Pzza, M., Bettelli, C., Micera, A. and Aloe, L. (1998) Proliferation and phenotype regulation in the subventricular zone during experimental allergic encephalomyelitis: in vivo evidence of a role for nerve growth factor. Proc. Natl. Acad. Sci. USA, 95: 3209–3214.

Calza, L., Fernandez, M., Giuliani, A., Aloe, L. and Giardino, L. (2002) Thyroid hormone activates oligodendrocyte precursors and increases a myelin-forming protein and NGF content in the spinal cord during experimental allergic encephalomyelitis. Proc. Natl. Acad. Sci. USA, 99: 3258–3263.

Cannella, B., Pitt, D., Capello, E. and Raine, C.S. (2000) Insulin-like growth factor-1 fails to enhance central nervous system myelin repair during autoimmune demyelination. Am. J. Pathol., 157: 933–943.

Canoll, P.D., Musacchio, J.M., Hardy, R., Reynolds, R., Marchionni, M.A. and Salzer, J.L. (1996) GGF/Neuregulin is a neuronal signal that promotes the proliferation and survival and inhibits the differentiation of oligodendrocyte progenitors. Neuron, 17: 229–243.

Carroll, W.M., Jennings, A.R. and Ironside, L.J. (1998) Identification of the adult resting progenitor cell by autoradiographic tracking of oligodendrocyte precursors in experimental CNS demyelination. Brain, 121: 293–302.

Casaccia-Bonnefil, P., Carter, B.D., Dobrowsky, R.T. and Chao, M.V. (1996b) Death of oligodendrocytes mediated by the interaction of nerve growth factor with its receptor p75. Nature, 383: 716–719.

Chang, A., Tourtellotte, W.W., Rudick, R. and Trapp, B.D. (2002) Premyelinating oligodendrocytes in chronic lesions of multiple sclerosis. N. Engl. J. Med., 346: 165–173.

Chang, A., Nishiyama, A., Peterson, J., Prineas, J. and Trapp, B.D. (2000) NG2-positive oligodendrocyte progenitor cells in adult human brain and multiple sclerosis lesions. J. Neurosci., 20: 6404–6412.

Chao, M.V. and Hempstead, B.L. (1995) p75 and TrK: a two-receptor system. Trends Neurosci., 18: 321–326.

Chari, D.M. and Blakemore, W.F. (2002) New insights into remyelination failure in multiple sclerosis: implications for glial cell transplantation. Mult. Scler., 8: 271–277.

Charles, P., Reynolds, R., Seilhan, D.G., Aigrot, S., Niezgoda, A., Zalc, B. and Lubetzki, C. (2002) Re-expression of PSA-NCAM by demyelinated axons: an inhibitor of remyelination in multiple sclerosis? Brain, 125: 1972–1979.

Chmielnicki, E. and Goldman, S.A. (2002) Induced neurogenesis by endogenous progenitor cells in the adult mammalian brain. Progr. Brain Res., 138: 451–464.

Cohen, R.I., Marmur, R., Norton, W.T., Mehler, M.F. and Kessler, J.A. (1996) Nerve growth factor and neurotrophin-3 differentially regulate the proliferation and survival of developing rat brain oligodendrocytes. J. Neurosci., 16: 6433–6442.

Compston, A. and Coles, A. (2002) Multiple sclerosis. Lancet, 359: 1221–1231.

Condorelli, D.F., Salin, T., Dell'Albani, P., Mudo, P., Corsaro, M., Timmusk, T., Metsis, M. and Belluardo, N. (1995) Neurotrophins and their trk receptors in cultured cells of the glial lineage and in white matter of the central nervous system. J. Mol. Neurosci., 6: 237–248.

Coumans, J.V., Lin, T.T.-S., Dai, H.D., MacArthur, L., McAtee, M., Nash, C. and Bregman, B.S. (2001) Axonal regeneration and functional recovery after complete spinal cord transaction in rats by delayed treatment with transplants and neurotrophins. J. Neurosci., 21: 9334–9344.

Dai, X., Qu, P. and Dreyfus, C.F. (2001) Neuronal signals regulate neurotrophin expression in oligodendrocytes of the basal forebrain. Glia, 34: 234–239.

Dawson, M.R.L., Levine, J.M. and Reynolds, R. (2000) NG2-expressing cells in the central nervous system: are they oligodendroglial progenitors? J. Neurosci. Res., 61: 471–479.

Dean, G. (1994) How many people in the world have multiple sclerosis? Neuroepidemiology, 13: 1–7.

DeStefano, N., Narayanan, S., Francis, S.J., Smith, S., Mortilla, M., Tartaglia, M.C., Bartolozzi, M.L., Guidi, L., Federico, A., Arnold, D.L. (2002) Diffuse axonal and tissue injury in patients with multiple sclerosis with low cerebral lesion load and no disability. Arch. Neurol., 59: 1565–1571.

Dowling, P., Ming, X., Raval, S., Husar, W., Casaccia-Bonnefil, P., Chao, M., Cook, S. and Blumberg, B. (1999) Up-regulated p75(NTR) neurotrophin receptor on glial cells in MS plaques. Neurology, 53: 1676–1682.

Du, Y.Z. and Dreyfus, C.F. (2002) Oligodendrocytes as providers of growth factors. J. Neurosci. Res., 68: 647–654.

Durbec, P. and Cremer, H. (2001) Revisiting the function of PSA-NCAM in the nervous system. Mol. Neurobiol., 24: 53–64.

Elkabes, S., Peng, L. and Black, I.B. (1998) Lipopolysaccharide differentially regulates microglial Trk receptor and neurotrophin expression. J. Neurosci. Res., 54: 117–122.

Engel, J., Althaus, H.H. and Kristjansson, G.I. (1994) NGF increases $[Ca^{2+}]i$ in regenerating mature oligodendroglial cells. NeuroReport, 5: 397–400.

Ewing, C. and Bernard, C.C. (1998) Insights into the aetiology and pathogenesis of multiple sclerosis. Immunol. Cell Biol., 76: 47–54.

Fernandez, P.-A., Tang, D.G., Cheng, L., Prochiantz, A., Mudge, A.W. and Raff, M.C. (2000) Evidence that axon-derived neuregulin promotes oligodendrocyte survival in the developing rat optic nerve. Neuron, 28: 81–90.

Flügel, A., Matsumoto, K., Neumann, H., Klinkert, W.E.F., Birnbacher, R., Lassmann, H., Otten, U. and Wekerle, H. (2001) Anti-inflammatory activity of nerve growth factor in experimental autoimmune encephalomyelitis: inhibition of monocyte transendothelial migration. Eur. J. Immunol., 31: 11–22.

Frank, J.A., Richert, N., Lewis, B., Bash, C., Howard, T., Civil, R., Stone, R., Eaton, J., McFarland, H., Leist, T. (2002) A pilot study of recombinant insulin-like growth factor-1 in seven multiple sclerosis patients. Mult. Scler., 8: 24–29.

Franklin, R.J.M. (2002) Why does remyelination fail in multiple sclerosis? Nat. Rev. Neurosci., 3: 705–714.

Furukawa, S., Furukawa, Y., Satayoshi, E. and Hayashi, K. (1986) Synthesis and secretion of nerve growth factors by mouse astroglial cells in culture. Biochem. Biophys. Res. Commun., 136: 57–63.

428

Gensert, J.M. and Goldman, J.E. (1997) Endogenous progenitors remyelinate demyelinated axons in the adult CNS. Neuron, 19: 197–203.

Gilson, J. and Blakemore, W.F. (1993) Failure of remyelination in areas of demyelination produced in the spinal cord of old rats. Neuropathol. Appl. Neurobiol., 19: 173–181.

Gogate, N., Verma, L., Zhou, J.M., Milward, E., Rusten, E., O'Connor, M., Kufta, C., Kim, J., Hudson, L., Dubois-Dalcq, M. (1994) Plasticity in the human adult oligodendrocyte lineage. J. Neurosci., 14: 4571–4587.

Gonzalez, D., Dees, W.L., Hiney, J.K., Ojeda, S.R. and Saneto, R.P. (1990) Expression of beta-nerve growth factor in cultured cells derived from the hypothalamus and cerebral cortex. Brain Res., 511: 249–258.

Grenier, Y., Ruijs, T.C.G., Robitaille, Y., Olivier, A. and Antel, J.P. (1989) Immunohistochemical studies of adult human glial cells. J. Neuroimmunol., 21: 103–115.

Goto, A. and Furukawa, S. (1995) Experimental changes in BDNF and NT-3-like immunoreactivities in the spinal cord following its transection. J. Jpn. Orthopaed. Assoc., 69: 506–516 (Engl. Abstr.).

Griffiths, I., Klugmann, M., Anderson, T., Yool, D., Thomson, C., Schwab, M.H., Schneider, A., Zimmermann, F., McCulloch, M., Nadon, N., Nave, K.A. (1998) Axonal swellings and degeneration in mice lacking the major proteolipid of myelin. Science, 280: 1610–1613.

Hammarberg, H., Lidman, O., Lundberg, C., Eltayeb, S.Y., Gielen, A.W., Muhallab, S., Svenningsson, A., Linda, H., van der Meide, P.H., Cullheim, S., Olsson, T., Piehl, F. (2000) Neuroprotection by encephalomyelitis: Rescue of mechanically injured neurons and neurotrophin production by CNS-infiltrating T and natural killer cells. J. Neurosci., 20: 5283–5291.

Heese, K., Hock, C. and Otten, U. (1998) Inflammatory signals induce neurotrophin expression in human microglial cells. J. Neurochem., 70: 699–707.

Hefti, F. (1997a) Pharmacology of neurotrophic factors. Annu. Rev. Pharmacol. Toxicol., 37: 239–267.

Hefti, F. (1997b) Neurotrophic factor therapy-keeping score. Nat. Med., 3: 497–498.

Hempel, R. (1998). Untersuchungen zur Nerve Growth Factor-induzierten Signalkaskade in Oligodendrozyten des Schweins. PhD Thesis, Universität Osnabrück, Germany.

Hintzen, R.Q. (2002) Stem cell transplantation in multiple sclerosis: multiple choices and multiple challenges. Mult. Scler., 8: 155–160.

Hirsch, S. and Bähr, M. (1999) Immunocytochemical characterization of reactive optic nerve Astrocytes and meningeal cells. Glia, 26: 36–46.

Hopkins, S.J. and Rothwell, N.J. (1995) Cytokines and the nervous system: expression and recognition. Trends Neurosci., 18: 83–88.

Hsieh, W.-Y., Hsieh, Y.-L., Liu, D.D., Yang, S.-N. and Wu, J.-N. (2003) Neural progenitor cells resist excitatory amino-acid-induced neurotoxicity. J. Neurosci. Res., 71: 272–278.

Huang, E. and Reichardt, L. (2001) Neurotrophins: Roles in neuronal development and function. Annu. Rev. Neurosci., 24: 677–736.

Ibanez, C.F. (1995) Neurotrophic factors: from structure-function studies to designing effective therapeutics. Trends Biochem. Sci., 13: 217–227.

Iwashita, Y., Fawcett, J.W., Crang, A.J., Franklin, R.J.M. and Blakemore, W.F. (2000) Schwann cells transplanted into normal and x-irradiated adult white matter do not migrate extensively and show poor long-term survival. Exp. Neurol., 164: 292–302.

Jeffery, D.R. (2002) Early intervention with immunomodulatory agents in the treatment of multiple sclerosis. J. Neurol. Sci., 197: 1–8.

John, G.R., Shankar, S.L., Shafit-Zagardo, B., Massimi, A., Lee, S.C., Raine, C.S. and Brosnan, C.F. (2002) Multiple sclerosis: Re-expression of a developmental pathway that restricts oligodendrocyte maturation. Nat. Med., 8: 1115–1121.

Jones, L.L., Yamaguchi, Y., Stallcup, W.B. and Tuszynski, M.H. (2002) NG2 is a major chonroitin sulfate proteoglycan produced after spinal cord injury and is expressed by macrophages and oligodendrocyte progenitors. J. Neurosci., 22: 2792–2803.

Kahn, M.A., Kumar, S., Liebe, D., Chang, R., Parada, L.F. and deVellis, J. (1999) Mice lacking NT-3, and its receptor TrkC, exhibit profound deficiencies in CNS glial cells. Glia, 26: 153–165.

Kang, S.K., Lee, R.H. and Jung, J.S. (2001) Effect of brain-derived neurotrophic factor on neural differentiation of mouse embryonic stem cells and neural precursor cells. Neurosci. Res. Commun., 29: 183–192.

Kaplan, D.R. and Miller, F.D. (1997) Signal transduction by the neurotrophin receptors. Curr. Opin. Cell Biol., 9: 213–221.

Kaplan, M.R., Meyer-Franke, A., Lambert, S., Bennett, V., Duncan, I.D., Levison, S.R. and Barres, B.A. (1997) Induction of sodium channel clustering by oligodendrocytes. Nature, 386: 724–728.

Kavanaugh, B., Beesley, J., Itoh, T., Itoh, A., Grinspan, J. and Pleasure, D. (2000) Neurotrophin-3 (NT-3) diminishes susceptibility of the oligodendroglial lineage to AMPA glutamate receptor-mediated excitotoxicity. J. Neurosci. Res., 60: 725–732.

Keirstead, H.S. and Blakemore, W.F. (1997) Identification of post-mitotic oligodendrocytes incapable of remyelination within the demyelinated adult spinal cord. J. Neuropathol. Exp. Neurol., 56: 1191–1201.

Kerschensteiner, M., Gallmeier, E., Behrens, L., Leal, V.V., Misgeld, T., Klinkert, W.E.F., Kolbeck, R., Hoppe, E., Oropeza-Wekerle, R.L., Bartke, I., Stadelmann, C., Lassmann, H., Wekerle, H., Hohlfeld, R. (1999) Activated human T cells, B cells, and monocytes produce brain-derived neurotrophic factor in vitro and in inflammatory brain lesions: A neuroprotective role of inflammation? J. Exp. Med., 189: 865–870.

Koda, M., Murakami, M., Ino, H., Yoshinaga, K., Ikeda, O., Hashimoto, M., Yamazaki, M., Nakayama, C. and Moriya, H. (2002) Brain-derived neurotrophic factor suppresses delayed apoptosis of oligodendrocytes after spinal cord injury in rats. J. Neurotrauma, 19: 777–785.

Kornek, B., Storch, M.K., Weissert, R., Wallstroem, E., Stefferi, A., Olsson, T., Linington, C., Schmidbauer, M. and Lassmann, H. (2000) Multiple sclerosis and chronic autoimmune encephalomyelitis. Am. J. Pathol., 157: 267–276.

Krenz, N. and Weaver, L.C. (2000) Nerve growth factor in glia and inflammatory cells of the injured rat spinal cord. J. Neurochem., 74: 730–739.

Kuhlmann, T., Lingfeld, G., Bitsch, A., Schuchardt, J. and Brück, W. (2002) Acute axonal damage in multiple sclerosis is most extensive in early disease stages and decreases over time. Brain, 125: 2202–2212.

Kumar, S. and DeVellis, J. (1996) Neurotrophin activates signal transduction in oligodendroglial cells: Expression of functional TrkC receptor isoforms. J. Neurosci. Res., 44: 490–498.

Kumar, S., Pena, L.A. and de Vellis, J. (1993) CNS glial cells express neurotrophin receptors whose levels are regulated by NGF. Mol. Brain Res., 7: 163–168.

Kumar, S., Kahn, M.A., Dirk, L. and deVellis, J. (1998) NT-3 mediated TrkC receptor activation promotes proliferation and cell survival of rodent progenitor oligodendrocyte cells in vitro and in vivo. J. Neurosci. Res., 54: 754–765.

Lachyankar, M.B., Condon, P.J., Quesenberry, P.J., Litofsky, N.S., Recht, L.D. and Ross, A.H. (1997) Embryonic precursor cells that express Trk receptors: Induction of different cell fates by NGF, BDNF, NT-3, and CNTF. Exp. Neurol., 144: 350–360.

Ladiwala, U., Lachance, C., Simoneau, S.J.J., Bhakar, A., Barker, P.A. and Antel, J.P. (1998) p75 neurotrophin receptor expression on adult human oligodendrocytes: signaling without cell death in response to NGF. J. Neurosci., 18: 1297–1304.

Lassmann, H. (2002) Mechanisms of demyelination and tissue destruction in multiple sclerosis. Clin. Neurol. Neurosurg., 104: 168–171.

Laudiero, L.B., Aloe, L., Levi-Montalcini, R., Buttinelli, C., Schilter, D., Gillesen, S. and Otten, U. (1992) Multiple sclerosis patients express increased levels of beta-nerve growth in cerebrospinal fluid. Neurosci. Lett., 147: 9–12.

Levi-Montalcini, R., Skaper, S.D., Toso, R.D., Petrelli, L. and Leon, A. (1996) Nerve growth factor: from neurotrophin to neurokine. Trends Neurosci., 19: 514–520.

Levine, J.M. and Card, J.P. (1987) Light and electron microscopic localization of a cell surface antigen (NG2) in the rat cerebellum: association with smooth protoplasmic astrocytes. J. Neurosci., 7: 2711–2720.

Levine, J.M., Reynolds, R. and Fawcett, J.W. (2001) The oligodendrocyte precursor cell in health and disease. Trends Neurosci., 24: 39–47.

Levison, S.W. and Goldman, J.E. (1993) Both oligodendrocytes and astrocytes develop from progenitors in the subventricular zone of postnatal rat forebrain. Neuron, 10: 201–212.

Liu, X., Yao, D.L. and deWebster, H. (1995) Insulin-like growth factor I reduces clinical deficits and lesion severity in acute demyelinating experimental autoimmune encephalomyelitis. Mult. Scler., 1: 2–9.

Lorigados, L., Pavon, N., Serrano, T., Robinson, M.A., Fernandez, C.I. and Alvarez, P. (2001) Cambios en los niveles de factor de crecimiento nervioso con el envejecimiento y el tratamiento neurotrofico en primates no humanos. Rev. Neurol., 33: 417–421.

Loy, R., Taglialatela, G., Angelucci, L., Heyer, D. and Perez-Polo, R. (1994) Regional CNS uptake of blood-borne nerve growth factor. J. Neurosci. Res., 39: 339–346.

Ludwin, S.K. (1997) The pathobiology of the oligodendrocyte. J. Neuropathol. Exp. Neurol., 56: 111–124.

Lucchinetti, C., Brück, W., Parisi, J., Scheithauer, B., Rodriguez, M. and Lassmann, H. (2000) Heterogeneity of multiple sclerosis lesions: Implications for the pathogenesis of demyelination. Ann. Neurol., 47: 707–717.

Lucchinetti, C., Brück, W. and Noseworthy, J. (2001) Multiple sclerosis: recent developments in neuropathology, pathogenesis, magnetic resonance imaging studies and treatment. Curr.Opin. Neurol., 14: 259–269.

Marburg, O. (1936). Multiple sklerose. In: O. Bumke and O. Foerster (Eds.), Handbuch der Neurologie. Band XIII, Springer Verlag, Berlin, pp. 546–693.

Martino, G., Furlan, R., Comi, G. and Adorini, L. (2001) The ependymal route to the CNS: an emerging gene-therapy approach for MS. Trends Immunol., 22: 483–490.

Mason, J.L., Ye, P., Suzuki, K., D'Ercole, A.J. and Matsushima, G.K. (2000) Insulin-like growth factor-1 inhibits mature oligodendrocyte apoptosis during primary demyelination. J. Neurosci., 20: 5703–5708.

McAllister, A.K. (1999) Neurotrophins and synaptic plasticity. Annu. Rev. Neurosci., 22: 295–318.

McGrath, J.P., Cao, X., Schutz, A., Lynch, P., Ebendal, T., Coloma, M.J., Morrison, S.L. and Putney, S.D. (1997) Bifunctional fusion between nerve growth factor and a transferrin receptor antibody. J. Neurosci. Res., 47: 123–133.

McMorris, F.A. and Dubois-Dalcq, M. (1988) Insulin-like growth factor 1 promotes cell proliferation and oligodendroglial commitment in rat glial progenitor cells developing in vitro. J. Neurosci. Res., 21: 199–209.

McMorris, F.A. and McKinnon, R.D. (1996) Regulation of oligodendrocyte development and CNS myelination by growth factors: prospects for therapy of demyelinating disease. Brain Pathol., 6: 313–329.

McTigue, D.M., Horner, P.J., Stokes, B.T. and Gage, F.H. (1998) Neurotrophin-3 and brain-derived neurotrophic factor induce oligodendrocyte proliferation and myelination of regenerating axons in the contused adult rat spinal cord. J. Neurosci., 18: 5354–5365.

Merrill, J.E. and Jonakait, G.M. (1995) Interactions of the nervous and immune system in development, normal brain homeostasis, and disease. FASEB J., 9: 611–618.

Merrill, J.E. and Benveniste, E.N. (1996) Cytokines in inflammatory brain lesions: helpful and harmful. Trends Neurosci., 19: 331–338.

Mews, I., Bergmann, M., Bunkowski, S., Gullotta, F. and Brück, W. (1998) Oligodendrocyte and axon pathology in clinically silent multiple sclerosis lesions. Mult. Scler., 4: 55–62.

Meyer-Franke, A., Kaplan, M.R., Pfrieger, F.W. and Barres, B.A. (1995) Characterization of the signaling interactions that promote the survival and growth of developing retinal ganglion cells in culture. Neuron, 15: 805–819.

Micera, A., Lambiase, A., Rama, P. and Aloe, L. (1999) Altered nerve growth factor level in the optic nerve of patients affected by multiple sclerosis. Mult. Scler., 5: 389–394.

Micera, A., Properzi, F., Triaca, V. and Aloe, L. (2000) Nerve growth factor antibody exacerbates neuropathological signs of experimental allergic encephalomyelitis in adult Lewis rats. J. Neuroimmunol., 104: 116–123.

Milward, E.A., Zhang, S.-C., Zhao, M., Lundberg, C., Ge, B., Goetz, B.D. and Duncan, I.D. (2000) Enhanced proliferation and directed migration of oligodendroglial progenitors co-grafted with growth factor secreting cells. Glia, 32: 264–270.

Moalem, G., Gdalyahu, A., Shani, Y., Otten, U., Lazarovici, P., Cohen, I.R. and Schwartz, M. (2000) Production of neurotrophins by activated T cells: Implications for neuroprotective autoimmunity. J. Autoimmun., 15: 331–345.

Muhallab, S., Lundberg, C., Gielen, A.W., Lidman, O., Svenningson, A., Piehl, F. and Olsson, T. (2002) Differential expression of neurotrophic factors and inflammatory cytokines by myelin basic protein-specific and other recruited T cells infiltrating the central nervous system during experimental autoimmune encephalomyelitis. Scand. J. Immunol., 55: 264–273.

Nakajima, K., Kikuchi, Y., Ikoma, E., Honda, S., Ishikawa, M., Liu, Y. and Kohsaka, S. (1998) Neurotrophins regulate the function of cultured microgilia. Glia, 24: 272–289.

Nakajima, K., Honda, S., Tohyama, Y., Imai, Y., Kohsaka, S. and Kurihara, T. (2001) Neurotrophin secretion from cultured microglia. J. Neurosci. Res., 65: 322–331.

Neumann, H., Misgeld, T., Matsumoto, K. and Wekerle, H. (1998) Neurotrophins inhibit major histocompatibility class II inducibility of microglia: involvement of the p75 neurotrophin receptor. Proc. Natl. Acad. Sci. USA, 95: 5779–5784.

Noetzel, M.J. and Brunstrom, J.E. (2001) The vulnerable oligodendrocyte. Neurology, 56: 1254–1255.

Norton, W.T. (1996) Do oligodendrocytes divide? Neurochem. Res., 21: 495–503.

O'Leary, M.T. and Blakemore, W.F. (1997) Oligodendrocyte precursors survive poorly and do not migrate following transplantation into the normal adult central nervous system. J. Neurosci. Res., 48: 1–9.

Otten, U. and Gadient, R.A. (1995) Neurotrophins and cytokines-intermediaries between the immuneand nervous systems. Int. J. Develop. Neurosci., 13: 147–151.

Ozawa, K., Suchanek, G., Breitschopf, H., Brück, W., Budka, H., Jellinger, K. and Lassmann, H. (1994) Patterns of oligodendroglia pathology in multiple sclerosis. Brain, 117: 1311–1322.

Paty, D.W. and Arnold, D.L. (2002) The lesions of multiple sclerosis. N. Engl. J. Med., 346: 199–200.

Pernber, Z., Molander-Melin, M., Berthold, C.-H., Hansson, E. and Fredman, P. (2002) Expression of the myelin and oligodendrocyteprogenitor marker sulfatide in neurons and astrocytes of adult rat brain. J. Neurosci. Res., 69: 86–93.

Perry, V.H. (1998) Reluctant remyelination: the missing precursors. Brain, 121: 2219–2220.

Picard-Riera, N., Decker, L., Delarasse, C., Goude, K., Nait-Oumesmar, B., Liblau, R., Pham-Dinh, D. and Baron-van Evercooren, A. (2002) Experimental autoimmune encephalomyelitis mobilizes neural progenitors from the subventricular zone to undergo oligodendrogenesis in adult mice. Proc. Natl. Acad. Sci. USA, 99: 13211–13216.

Pouly, S., Prat, A., Blain, M., Olivier, A. and Antel, J. (2001a) NG2 immunoreactivity on human brain endothelial cells. Acta Neuropathol., 102: 313–320.

Pouly, S., Matthieu, J.-M. and Honegger, P. (2001b) Remyelination in vitro following protein kinase C activator-induced demyelination. Neurochem. Res., 26: 619–627.

Prabhakar, S., D'Souza, S., Antel, J.P., McLaurin, J., Schipper, H.M. and Wang, E. (1995) Phenotypic and cell cycle properties of human oligodendrocytes in vitro. Brain Res., 672: 159–169.

Prineas, J.W., Barnard, R.O., Kwon, E.E., Sharer, L.R. and Cho, E.-S. (1993a) Multiple sclerosis: remyelination of nascent lesions. Ann. Neurol., 33: 137–151.

Prineas, J.W., Barnard, R.O., Revesz, T., Kwon, E.E., Sharer, L. and Cho, E.-S. (1993b) Multiple sclerosis. Brain, 116: 681–693.

Qu, P., Dai, X., Wu, H., Friedman, W.J., Black, I.B. and Dreyfus, C.F. (1995) Expression and regulation of neurotrophic factors in basal forebrain oligodendrocytes in vitro and in vivo. Soc. Neurosci. Abstr., 21: 2012.

Raffioni, S., Bradshaw, R.A. and Buxser, S.E. (1993) The receptors for nerve growth factor and other neurotrophins. Annu. Rev. Biochem., 62: 823–850.

Ransohoff, R.M., Howe, C.L. and Rodriguez, M. (2002) Growth factor treatment of demyelinating disease: at last, a leap into the light. Trends Immunol., 23: 512–516.

Rieckmann, P. and Smith, K.J. (2001) Multiple sclerosis: more than inflammation and demyelination. Trends Neurosci., 24: 435–437.

Rodriguez, M., Scheithauer, B.W., Forbes, G. and Kelly, P.J. (1993) Oligodendrocyte injury is an early event in lesions of multiple sclerosis. Mayo Clin. Proc., 68: 627–636.

Rohde, G. (1998). Effekt von NGF auf die Remyelinisation von experimentell demyelinisierten Arealen im Corpus callosum des Göttinger Miniaturschweines. MD Thesis, Tierärztliches Institut, Hannover, Germany.

Rosenberg, P.A., Dai, W., Gan, X.D., Ali, S., Fu, J., Back, S.A., Sanchez, R.M., Segal, M.M., Follett, P.L., Jensen, F.E., Volpe, J.J. (2003) Mature myelin basic protein-expressing oligodendrocytes are insensitive to kainate toxicity. J. Neurosci. Res., 71: 237–245.

Rossner, S., Schliebs, R., Hartig, W., Perez-Polo, J.R. and Bigl, V. (1997) Selective induction of c-Jun and NGF in reactive astrocytes after cholinergic degenerations in rat basal forebrain. NeuroReport, 8: 2199–2202.

Roy, N.S., Wang, S., Harrison-Restelli, C., Benraiss, A., Fraser, A.R., Gravel, M., Braun, P.E. and Goldman, S.A. (1999) Identification, isolation, and promoter-defined separation of mitotic oligodendrocyte progenitor cells from the adult human subcortical white matter. J. Neurosci., 19: 9986–9995.

Sanchez, I., Hassinger, L., Paskevich, P.A., Shine, H.D. and Nixon, R.A. (1996) Oligodendroglia regulate the regional expansion of axon caliber and local accumulation of neurofilaments during development independently of myelin. J. Neurosci., 16: 5095–5105.

Sarchielli, P., Greco, L., Stipa, A., Floridi, A. and Gallai, V. (2002) Brain-derived neurotrophic factor in patients with multiple sclerosis. J. Neuroimmunol., 132: 180–188.

Satoh, T., Yamagata, T., Ishiwaka, E., Yamada, M., Uchiyama, Y. and Hatanaka, H. (1999) Regulation of reactive oxygen species by nerve growth factor but not Bcl-2 as a novel mechanism of protection of PC12 cells from superoxide anion-induced death. J. Biochem., 125: 952–959.

Schönrock, L.M., Kuhlmann, T., Adler, S., Bitsch, A. and Brück, W. (1998) Identification of glial cell proliferation in early multiple sclerosis lesions. Neuropathol. Appl. Neurobiol., 24: 320–330.

Schuldiner, M., Yanuka, O., Itskovitz-Eldor, J., Melton, D.A. and Benvenisty, N. (2000) Effects of eight growth factors on the differentiation of cells derived from human emryonicstem cells. Proc. Natl. Acad. Sci. USA, 97: 11307–11312.

Scolding, N. (1999) Therapeutic strategies in multiple sclerosis. II. Long-term repair. Phil. Trans. R. Soc. Lond. B, 354: 1711–1720.

Scolding, N.J., Zajicek, J.P., Wood, N. and Compston, D.A.S. (1994) The pathogenesis of demyelinating disease. Progr. Neurobiol., 43: 143–173.

Scolding, N.J., Rayner, P.J., Sussman, J., Shaw, C. and Compston, D.A.S. (1995) A proliferative adult human oligodendrocyte progenitor. Neuroreport, 6: 441–445.

Scolding, N., Franklin, R., Stevens, S., Heldin, C.-H., Compston, A. and Newcombe, J. (1998) Oligodendrocyte progenitors are present in the normal adult CNS and in the lesions of multiple sclerosis. Brain, 121: 2212–2218.

Scolding, N., Rayner, P.J. and Compston, D.A.S. (1999) Identification of A2B5-positive putative oligodendrocyte progenitor cells and A2B5-positive astrocytes in adult human white matter. Neuroscience, 89: 1–4.

Shi, J., Marinovich, A. and Barres, B.A. (1998) Purification and characterization of adult oligodendrocyte precursor cells from the rat optic nerve. J. Neurosci., 18: 4627–4636.

Skaper, S.D. and Walsh, F.S. (1998) Neurotrophic molecules-strategies for designing effective therapeutic molecules in neurodegeneration. Mol. Cell Neurosci., 12: 179–193.

Skaper, S.D., Floreani, M., Negro, A., Facci, L. and Giusti, P. (1998) Neurotrophins rescue cerebellar granule neurons from oxidative stress-mediated apoptotic death-selective involvement of phosphatidylinositol 3-kinase and the mitogen-activated protein kinase pathway. J. Neurochem., 70: 1859–1868.

Sofroniew, M.V., Howe, C.L. and Mobley, W.C. (2001) Nerve growth factor signaling, neuroprotection, and neural repair. Annu. Rev. Neurosci., 24: 1217–1281.

Soilu-Hänninen, M., Epa, R., Shipham, K., Butzkueven, H., Bucci, T., Barrett, G., Bartlett, P.F. and Kilpatrick, T.J. (2000) Treatment of experimental autoimmune encephalomyelitis with antisense oligonucleotides against the low affinity neurotrophin receptor. J. Neurosci. Res., 59: 712–721.

Solanky, M., Maeda, Y., Ming, X., Husar, W., Li, W., Cook, S. and Dowling, P. (2001) Proliferating oligodendrocytes are present in both active and chronic inactive multiple sclerosis plaques. J. Neurosci. Res., 65: 308–317.

Stadelmann, C., Kerschensteiner, M., Misgeld, T., Brück, W., Hohlfeld, R. and Lassmann, H. (2002) BDNF and gp145trkB in multiple sclerosis brain lesions: neuroprotective interactions between immune and neuronal cells. Brain, 125: 75–85.

Stangel, M., Compston, A. and Scolding, N.J. (2000) Oligodendroglia are ptrotected from antibody-mediated complement injury by normal immunoglobulins ('IVIG'). J. Neuroimmunol., 103: 195–201.

Stankoff, B., Aigrot, M.-S., Noel, F., Wattilliaux, A., Zalc, B. and Lubetzki, C. (2002) Ciliary neurotrophic factor (CNTF) enhances myelin formation: A novel role for CNTF and CNTF-related molecules. J. Neurosci., 22: 9221–9227.

Takano, R., Hisahara, S., Namikawa, K., Kijama, H., Okano, H. and Miura, M. (2000) Nerve growth factor protects oligodendrocytes from TNF-alpha-induced injury through Akt-mediated signaling mechanisms. J. Biol. Chem., 275: 16360–16365.

432

Tan, S.A. and Aebischer, P. (1996). The problems of delivering neuroactive molecules to the CNS. In: Growth Factors as Drugs for Neurological and Sensory Disorders. Ciba Found. Symp. 196, Wiley, Chichester, pp. 211–239.

Tang, D.G., Tokumoto, Y.M. and Raff, M. (2000) Long-term culture of purified postnatal oligodendrocyte precursor cells: Evidence for an intrinsic maturation program that plays out over months. J. Cell Biol., 148: 971–984.

Targett, M.P., Sussman, J., Scolding, N., O'Leary, M.T., Compston, D.A.S. and Blakemore, W.F. (1996) Failure to achieve remyelination of demyelinated rat axons following transplantation of glial cells obtained from the adult human brain. J. Neuropathol. Appl. Neurobiol., 22: 199–206.

Thorne, R.G. and Frey, W.H. (2001) Delivery of neurotrophic factors to the central nervous system—Pharmacokinetic considerations. Clin. Pharmacokin., 40: 907–946.

Timmusk, T., Mudo, G., Metsis, M. and Belluardo, N. (1995) Expression of mRNAs for neurotrophins and their receptors in the rat choroid plexus and dura mater. NeuroReport, 6: 1997–2000.

Torcia, M., Bracci-Laudiero, L., Lucibello, M., Nencioni, L., Labardi, D., Rubatelli, A., Cozzolini, F., Aloe, L. and Garaci, E. (1996) Nerve growth factor is an autocrine survival factor for memory B lymphocytes. Cell, 85: 345–356.

Trapp, B.D., Peterson, J., Ransohoff, R.M., Rudick, R., Mörk, S. and Bö, L. (1998) Axonal transection in the lesions of multiple sclerosis. N. Engl. J. Med., 338: 278–285.

Tuszynski, M.H., Conner, J., Blesch, A., Smith, D., Merrill, D.A. and Lee Vahlsing, H. (2002) New strategies in neural repair. Prog. Brain Res., 138: 401–409.

Valdo, P., Stegagno, C., Mazzucco, S, . Zuliani, E., Zanusso, G., Moretto, G., Raine, C.S. and Bonetti, B. (2002) Enhanced expression of NGF receptors in multiple sclerosis lesions. J. Neuropathol. Exp. Neurol., 61: 91–98.

Vartanian, T., Fischbach, G.D. and Miller, R. (1999) Failure of spinal cord oligodendrocyte development in mice lacking neuregulin. Proc. Natl. Acad. Sci. USA, 96: 731–735.

Vermuri, G.S. and McMorris, F.A. (1996) Oligodendrocytes and their precursors require phosphatidylinositol 3-kinase signaling for survival. Development, 122: 2529–2537.

Villoslada, P., Hauser, S.L., Bartke, I., Unger, J., Heald, N., Rosenberg, D., Cheung, S.W., Mobley, W.C., Fisher, S. and Genain, C.P. (2000) Human nerve growth factor protects common marmorsets against autoimmune encephalomyelitis by switching the balance of T Helper cell type 1 and 2 cytokines within the central nervous system. J. Exp. Med., 191: 1799–1806.

Wang, S., Sdrulla, A.D., diSibio, G., Bush, G., Nofziger, D., Hicks, C., Weinmaster, G. and Barres, B.A. (1998) Notch receptor activation inhibits oligodendrocyte differentiation. Neuron, 21: 63–75.

Warrington, A.E., Barbarese, E. and Pfeiffer, S.E. (1993) Differential myelinogenic capacity of specific developmental stages of the oligodendrocyte lineage upon transplantation into hypomyelinating hosts. J. Neurosci., 32: 1–13.

Waxman, S.G. (1997) Molecular remodeling of neurons in multiple sclerosis: What we know, and what we must ask about brain plasticity in demyelinating diseases. Adv. Neurol., 73: 109–120.

Waxman, S.G. (2001) Transcriptional channelopathies: An emerging class of disorders. Nat. Rev. Neurosci., 2: 652–659.

Wiendl, H. and Hohlfeld, R. (2002) Therapeutic approaches in multiple sclerosis. Biodrugs, 16: 183–200.

Windrem, M.S., Roy, N.S., Wang, J., Nunes, M., Benraiss, A., Goodman, R., McKhann, G.M. and Goldman, S.A. (2002) Progenitor cells derived from the adult human subcortical white matter disperse and differentiate as oligodendrocytes within demyelinated lesions of the rat brain. J. Neurosci. Res., 69: 966–975.

Wolf, M.K., Brandenberg, M.C. and Billings-Gagliardi, S. (1986) Migration and myelination by adult glial cells: Reconstructive analysis of tissue culture experiments. J. Neurosci., 6: 3731–3738.

Wolswijk, G. (2000) Oligodendrocyte survival, loss and birth in lesions of chronic-stage multiple sclerosis. Brain, 123: 105–115.

Wolswijk, G. (2002) Oligodendrocyte precursor cells in the demyelinated multiple sclerosis spinal cord. Brain, 125: 338–349.

Wu, V.W., Nishiyama, N. and Schwartz, J.P. (1998) A culture model of reactive astrocytes: Increased nerve growth factor synthesis and reexpression of cytokine responsiveness. J. Neurochem., 71: 749–756.

Yamaya, Y., Yoshioka, A., Saiki, S., Yuki, N., Hirose, G. and Pleasure, D. (2002) Type-2 astrocyte-like cells are more resistant than oligodendrocyte-like cells against non-n-methyl-d-aspartate glutamate receptor-mediated excitotoxicity. J. Neurosci. Res., 70: 588–598.

Ye, P. and D'Ercole, A.J. (1999) Insulin-like growth factor I protects oligodendrocytes from tumor necrosis factor-alpha-induced injury. Endocrinology, 140: 3063–3972.

Yoon, S.O., Casaccia-Bonnefil, P., Carter, B. and Chao, M.V. (1998) Competitive signaling between TrkA and p75 nerve growth factor receptors determines cell survival. J. Neurosci., 18: 3273–3281.

Zhang, S.-C., Ge, B. and Duncan, I.D. (1999) Adult brain retains the potential to generate oligodendroglial progenitors with extensive myelination capacity. Proc. Natl. Acad. Sci. USA, 96: 4089–4094.

Zhu, Y. and Krieglstein, J. (1999) β2-Adrenoreceptor agonist Clenbuterol causes NGF expression and neuroprotection. CNS Drug Rev., 5: 347–364.

Progress in Brain Research, Vol. 146
ISSN 0079-6123

CHAPTER 27

Role of NGF and neurogenic inflammation in the pathogenesis of psoriasis

Siba P. Raychaudhuri* and Smriti K. Raychaudhuri

Psoriasis Research Institute and Stanford University School of Medicine, Palo Alto, CA 94306, USA

Abstract: A contributing role of neurogenic inflammation has provided a new dimension in understanding the pathogenesis of various cutaneous and systemic inflammatory diseases such as atopic dermatitis, urticaria, rheumatoid arthritis, ulcerative colitis and bronchial asthma. Several critical observations, such as (i) psoriasis resolves at sites of anaesthesia, (ii) neuropeptides are upregulated, and (iii) there is a marked proliferation of terminal cutaneous nerves in psoriatic plaques, encouraged us to search for a mechanism of neural influence in inflammation and inflammatory diseases. In immunohistochemical studies, we found that keratinocytes in lesional and nonlesional psoriatic tissue express high levels of nerve growth factor (NGF) and that there is a marked upregulation of NGF receptors, p75 neurotrophin receptor (p75^{NTR}) and tyrosine kinase A (TrkA), in the terminal cutaneous nerves of psoriatic lesions. As keratinocytes of psoriatic plaques express increased levels of NGF, it is likely that murine nerves will promptly proliferate into the transplanted plaques on a severe combined immunodeficient mouse. Indeed, we have noted marked proliferation of nerve fibers in transplanted psoriatic plaques compared with the few nerves in transplanted normal human skin. By double label immunofluorescence staining, we have further demonstrated that in these terminal cutaneous nerves there is a marked upregulation of neuropeptides, such as substance P and calcitonin gene-related protein. These observations, as well as recent findings about NGF-induced chemokine expression in keratinocytes, further substantiate a role of the NGF–p75^{NTR}–TrkA system in the inflammatory process of psoriasis. Currently, we are evaluating antagonists to selected neuropeptides and NGF/receptors, with the expectation of identifying pharmacological agents to counter neurogenic inflammation in psoriasis.

Keywords: psoriasis; neurogenic inflammation; NGF; TrkA; chemokines; NGF antagonists; therapy

Introduction

Psoriasis is a relatively common chronic inflammatory skin disease, affecting about 2% of populations worldwide (Farber and Peterson, 1961; Lomeholt, 1963). In most of the patients, the psoriatic plaques are distributed over the elbows, knees and scalp. Lesions of psoriasis are characterized by erythema, scaling and infiltration. Psoriasis is a nonfatal, lifelong disease, but on occasions psoriasis can be a source of significant morbidity. Generalized involvement of the body (erythroderma), extensive pustular lesions and an associated mutilating arthritis are severe forms of psoriasis.

Today, there is no cure for psoriasis, and its pathogenesis is unreveling. Cytokines, chemokines, growth factors, adhesion molecules, neuropeptides and T lymphocytes act in integrated ways to evolve in unique inflammatory and proliferative processes typical of psoriasis. The concept of neuroimmunology as it relates to psoriasis is relatively new. Farber

*Correspondence to: S.P. Raychaudhuri, Psoriasis Research Institute and Stanford University School of Medicine, 510 Ashton Avenue, Palo Alto, CA 94306, USA. Fax: + 1-650-725-8564; E-mail: raysiba@aol.com

DOI: 10.1016/S0079-6123(03)46027-5

et al. (1986) first proposed a possible role of neuropeptides in the pathogenesis of psoriasis. On the other hand, there is substantial evidence that activated T lymphocytes play a key role in the cellular mechanisms of this disease. It is therefore essential to understand the relationships between neurogenic factors and activation of T cells.

In this chapter, we will present the neuroimmune factors that influence the inflammatory and proliferative processes associated with psoriasis.

Neurogenic inflammation in psoriasis

Our research group has a special interest in cutaneous neurogenic inflammation (reviewed by Johansson and Liang, 1999; for other diseases, see Braun et al. and Renz et al. in this issue). Correlating the clinical observations that stress exacerbates psoriasis and that psoriasis induces symmetrically distributed lesions, we proposed a role for neuropeptides in the pathogenesis of psoriasis. Subsequently, many investigators, including us, have reported an upregulation of neuropeptides such as substance P (SP), vasoactive intestinal peptide (VIP) and calcitonin gene-related protein (CGRP) along with marked proliferation of terminal cutaneous nerves in psoriatic lesions (Wallengren et al., 1987; Naukkarinen et al., 1989; Al'Abadie et al., 1995; Farber and Raychaudhuri, 1999). Neuropeptides can play significant role in the inflammatory and proliferative processes of psoriasis. SP is chemotactic to neutrophils (Smith et al., 1993), activates T cells (Calvo et al., 1992), VIP is mitogenic to keratinocytes (Haegerstrand et al., 1989), CGRP acts synergistically with SP to stimulate keratinocyte proliferation (Wilkinson, 1989) and both VIP and CGRP are potent mitogenic for endothelial cells (Haegerstrand et al., 1990).

There is good clinical evidence that the absence of sensory nerve innervation equates to the absence of psoriasis (Raychaudhuri and Farber, 1993). A 68-year-old Caucasian male had chronic plaque psoriasis involving the elbows, forearms, knees and legs. The patient underwent a reconstructive surgery on the left knee for osteoarthritis. In 6–8 weeks following surgery, a large plaque on the lateral surface of the left leg resolved. On examination, the skin at the resolved site was found to be anesthetic probably due to nerve damage following surgery. A comparable plaque on the contralateral leg remained active. In another patient it was observed that psoriasis resolved at the anesthetic area over knee and with the return of sensation, psoriasis reappeared at the same site.

These unique observations that psoriasis resolves at sites of anesthesia, while the neuropeptides are upregulated and there is a marked proliferation of terminal cutaneous nerves in psoriatic plaques, encouraged us to look for the underlying cause of neural influence in the inflammatory processes of psoriasis. As nerve growth factor (NGF) and its receptors play a crucial role in regulating innervation (Levi-Montalcini, 1987) and upregulating neuropeptides (Lindsay and Harmar, 1989; Aloe et al., 1997) including in the skin (Wyatt et al., 1990; Johansson and Liang, 1999), we decided to investigate the expression of NGF and its receptors, p75 neurotrophin receptor (p75^{NTR}) and tyrosine kinase A (TrkA), in the lesional and nonlesional psoriatic skin, normal skin and other inflammatory skin diseases.

In an immunohistochemical study, we found that keratinocytes in lesional and nonlesional psoriatic tissue express high levels of NGF compared to the controls (Raychaudhuri et al., 1998). Fantini et al. (1995) have also demonstrated similar results. Several functions of NGF are relevant to the inflammatory and proliferative processes of psoriasis. NGF promotes keratinocyte proliferation and protects keratinocytes from apoptosis (Wilkinson et al., 1994; Pincelli et al., 1997). NGF degranulates mast cells and induces migration of these cells (Aloe and Levi-Montalcini, 1977; Pearce and Thompson, 1986; Aloe et al., 1997), both being early events in developing lesions of psoriasis. In addition, NGF activates T lymphocytes and recruits inflammatory cellular infiltrates (Thorpe et al., 1987; Bischoff and Dahinden, 1992; Lambiase et al., 1997). In one of our recent study, we have reported that NGF induces expression of a C–C chemokine, regulated upon activation, normal T cell expressed and secreted (RANTES), in the keratinocytes (Raychaudhuri et al., 2000). RANTES as well as C–X–C chemokines are chemotactic signals for resting CD4+ memory T cells, and activated naive and memory T cells (Schall, 1991), suggesting that a chemokine network plays an important role in the inflammatory process of psoriasis (Fukuoka et al., 1998; Goebeler et al., 1998; Raychaudhuri et al., 1999; Kanda and Watanabe, 2003). In this context, it is worth noting that a vitamin D3 analogue, tacalcitol, inhibits RANTES and interleukin-8 production in cultured epidermal keratinocytes, which may partly account for its antipsoriatic action (Fukuoka et al., 1998; also see Kanda and Watanabe, 2003 for inhibition of RANTES secretion by estradiol). Further, the level of circulating RANTES is elevated in patients with psoriatic arthritis compared to controls (Macchioni et al., 1998). It is therefore possible that in a developing psoriatic lesion, upregulation of NGF induces, via chemokines, the influx of mast cells and lymphocytes, which in turn may initiate an inflammatory reaction contributing to the pathogenesis of psoriasis.

Is psoriasis a neuroimmunologic disease?

Some investigators consider psoriasis as an autoimmune disease induced by an unidentified antigen (Bos et al., 1983; Valdimarsson et al., 1986; Chang et al., 1997). Up to now, the alleged role of an antigen in psoriasis is hypothetical; no antigen has yet been discovered for psoriasis. An antigen-induced T cell activation process alone fails to clarify various salient features of psoriasis. It does not explain the Koebner phenomenon, the symmetrical distribution of psoriasis lesions, proliferation of cutaneous nerves and the upregulation of neuropeptides in psoriatic tissue (Farber and Peterson, 1961; Bos et al., 1983; Bischoff and Dahinden, 1992; Calvo et al., 1992; Chang et al., 1997). It does not have an answer either for a striking clinical observation that psoriasis resolves at sites of anesthesia (Haegerstrand et al., 1990).

In last two decades extensive work has been done to explore the immunological mechanisms involved in psoriasis. An active role of T cells is strongly substantiated by the following observations: (i) immunotherapy targeted specifically against CD4 + T cells clears active plaques of psoriasis (Gottlieb et al., 2000), and (ii) in severe combined immunodeficient (SCID) mice, transplanted nonlesional psoriatic skin converts to a psoriatic plaque subsequent to intradermal administration of T cells activated with an antigen cocktail (Wilkinson et al., 1994). However, it is equally true that psoriasis treated with agents such as Calcipotriol (Dovonex) and Etritinate (Tegison) which affect on the differentiation process of keratinocytes are very effective in psoriasis. Neither Dovonex nor Tegison are effective in other T cell-mediated cutaneous diseases such as atopic dermatitis or contact dermatitis.

Though psoriasis has been claimed to be an autoimmune disease, the antigen or specific endogenous factors responsible for activation of T cells in psoriasis is still unknown. Regarding induction of psoriasis in transplanted nonlesional skin on the SCID mouse model, the T cells were activated with an antigen cocktail (Wrone-Smith and Nickoloff, 1996). As such artificial antigen cocktail does not exist in a lesioned or nonlesioned psoriatic skin, it is possible that local epidermal and dermal factors like NGF and SP may be responsible for T lymphocyte activation. Nickoloff and his group demonstrated that they could induce psoriasis by injecting lymphomononuclear cells activated with SP (personal communication). We have a novel observation showing that in the SCID mouse model of psoriasis, autologus immunocytes activated with NGF can convert a transplanted nonlesional skin to a psoriatic plaque in 3 weeks (Raychaudhuri et al., 2001).

All these observations suggest that NGF and neurogenic inflammation play a critical role in the pathogenesis of psoriasis. In a double-blinded, placebo controlled study, we addressed the potential role of NGF and its high-affinity receptor, tyrosine kinase A (TrkA), in an in vivo system, using the SCID mouse-human skin model of psoriasis. Transplanted psoriatic plaques on the SCID mice ($n = 12$) were treated with K252a, a blocker of TrkA. Psoriasis significantly improved following 2 weeks of therapy. The length of *rete pegs* changed from 308.57 ± 98.72 μm to 164.64 ± 46.78 μm ($p < 0.01$, Student t-test). This study further substantiates the role of NGF and TrkA receptor in the pathogenesis of psoriasis and provides insights to develop novel therapeutic modalities.

In the last decade, significant progress has been made to elucidate the pathogenesis of psoriasis. However, the molecular basis of the inflammatory process of psoriasis is not yet fully clarified. In this article, we have addressed certain key events responsible in the development of a psoriatic lesion. Significant are the proliferation of nerves and the upregulation of neuropeptides, chemokines, and NGF. Clearance of psoriatic lesions at sites of anesthesia following nerve injury suggests an indisputable role of neurogenic inflammation in the pathogenesis of psoriasis.

Abbreviations

CGRP	calcitonin gene-related protein
NGF	nerve growth factor
p75^{NTR}	p75 neurotrophin receptor
RANTES	regulated upon activation, normal T cell expressed and secreted
SCID	severe combined immunodeficient
SP	substance P
TrkA	receptor tyrosine kinase A
VIP	vasoactive intestinal peptide

References

Al'Abadie, M.S., Senior, H.J., Bleehen, S.S. and Gawkrodger, D.J. (1995) Neuropeptides and general neuronal marker in psoriasis-an immunohistochemical study. Clin. Exp. Dermatol., 20: 384–389.

Aloe, L. and Levi-Montalcini, R. (1977) Mast cells increase in tissues of neonatal rats injected with the nerve growth factor. Brain Res., 133: 358–366.

Aloe, L., Bracci, L., Bonini, S. and Manni, L. (1997) The expanding role of nerve growth factor: from the neurotrophic activity to immunological diseases. Allergy, 52: 883–904.

Bischoff, S.C. and Dahinden, C.A. (1992) Effect of nerve growth factor on the release of inflammatory mediators by mature human basophils. Blood, 79: 2662–2669.

Bos, J.D., Hulsebosch, H.J., Krieg, S.R., Bakker, P.M. and Cormane, R.H. (1983) Immunocompetent cells in psoriasis: in situ immunophenotyping with monoclonal antibodies. Arch. Dermatol. Res., 275: 181–189.

Calvo, C.F., Chavanel, G. and Senik, A. (1992) Substance P enhances interleukin-2 expression in activated human T cells. J. Immunol., 148: 3498–3504.

Chang, J.C., Smith, L.R., Froning, K.J., Kurland, H.H., Schwabe, B.J., Blumeyer, K.K., Karasek, M.A., Wilkinson, D.I., Farber, E.M., Carlo, D.J., Brostoff, S.W. (1997) Persistance of T-Cell Clones in psoriatic lesions. Arch. Dermatol., 133: 703–708.

Farber, E.M. and Peterson, J.B. (1961) Variations in the natural history of psoriasis. Calif. Med., 95: 6–11.

Farber, E.M., Nickoloff, B.J., Recht, B. and Fraki, J.E. (1986) Stress, symmetry, and psoriasis: possible role of neuropeptides. J. Am. Acad. Dermatol., 14: 305–311.

Farber, E.M. and Raychaudhuri, S.P. (1999) Is psoriasis a neuroimmunologic disease? Int. J. Dermatol., 38: 12–15.

Fantini, F., Magnoni, C., Brauci-Laudeis, L. and Pincelli, C. (1995) Nerve growth factor is increased in psoriatic skin. J. Invest. Dermatol., 105: 854–855.

Fukuoka, M., Ogino, Y., Sato, H., Ohta, T., Komoriya, K., Nishioka, K. and Katayama, I. (1998) RANTES expression in psoriatic skin, and regulation of RANTES and IL-8 production in cultured epidermal keratinocytes by active vitamin D3 (tacalcitol). Br. J. Dermatol., 138: 63–70.

Goebeler, M., Toksoy, A., Spandau, U., Engelhardt, E., Brocker, E.B. and Gillitzer, R. (1998) The C-X-C chemokine Mig is highly expressed in the papillae of psoriatic lesions. J. Pathol., 184: 89–95.

Gottlieb, A.B., Lebwohl, M., Shirin, S., Sherr, A., Gilleaudeau, P., Singer, G., Solodkina, G., Grossman, R., Gisoldi, E., Phillips, S., Neisler, H.M. and Krueger, J.G. (2000) Anti-CD4 monoclonal antibody treatment of moderate to severe psoriasis vulgaris: results of a pilot, multicenter, multiple-dose, placebo-controlled study. J. Am. Acad. Dermatol., 43: 595–604.

Haegerstrand, A., Jonzon, B., Dalsgaard, C.J. and Nilsson, J. (1989) Vasoactive intestinal polypeptide stimulates cell proliferation and adenylate cyclase activity of cultured human keratinocytes. Proc. Natl. Acad. Sci. USA, 86: 5993–5996.

Haegerstrand, A., Dalsgaard, C.J., Jonzon, B., Larsson, O. and Nilsson, J. (1990) Calcitonin gene-related peptide stimulates proliferation of human endothelial cells. Proc. Natl. Acad. Sci. USA, 87: 3299–3303.

Johansson, O. and Liang, Y. (1999) Neurotrophins and their receptors in the skin: a tribute to Rita Levi-Montalcini. Biomed. Rev., 10: 15–23.

Kanda, N. and Watanabe, S. (2003) 17beta-estradiol inhibits the production of RANTES in human keratinocytes. J. Invest. Dermatol., 120: 420–427.

Lambiase, A., Bracci-Laudiero, L., Bonini, S., Bonini, S., Starace, G., D'Elios, M.M., De Carli, M. and Aloe, L. (1997) Human CD4+ T cell clones produce and release nerve growth factor and express high-affinity nerve growth factor receptors. J. Allergy Clin. Immunol., 100: 408–414.

Levi-Montalcini, R. (1987) The nerve growth factor 35 years later. Science, 237: 1154–1162.

Lomeholt, G. (1963) Psoriasis: Prevalence, spontaneous course and genetics: a census study on the prevalance of skin disease in the Faroe Islands. In: G. Lomholt (Ed.), Prevalence, Spontaneous Course and Genetics: A Census Study on the Prevalence of Skin Disease in the Faroe Islands. GEC Gad, Copenhagen, pp. 5–50.

Lindsay, R.M. and Harmar, A.J. (1989) Nerve growth factor regulates expression of neuropeptides genes in adult sensory neurons. Nature, 337: 362–364.

Macchioni, P., Boiardi, L., Meliconi, R., Pulsatelli, L., Maldini, M.C., Ruggeri, R., Facchini, A. and Salvarani, C. (1998) Serum chemokines in patients with psoriatic arthritis treated with cyclosporin A. J. Rheumatol., 25: 320–325.

Naukkarinen, A., Nickoloff, B.J. and Farber, E.M. (1989) Quantification of cutaneous sensory nerves and their substance P content in psoriasis. J. Invest. Dermatol., 92: 126–129.

Pearce, F.L. and Thompson, H.L. (1986) Some characteristics of histamine secretion from rat peritoneal mast cells stimulated with nerve growth factor. J. Physiol., 372: 379–393.

Pincelli, C., Haake, A.R., Benassi, L., Grassilli, E., Magnoni, C., Ottani, D., Polakowska, R., Franceschi, C. and Giannetti, A. (1997) Autocrine nerve growth factor protects human keratinocytes from apoptosis through its high affinity receptor (TRK): a role for BCL-2. J. Invest. Dermatol., 109: 757–764.

Raychaudhuri, S.P. and Farber, E.M. (1993) Are sensory nerves essential for the development of psoriasis lesions? J. Am. Acad. Dermatol., 28: 488–489.

Raychaudhuri, S.P., Jiang, W-Y. and Farber, E.M. (1998) Psoriatic keratinocytes express high levels of nerve growth factor. Acta Dermatol. Venereol., 78: 84–86.

Raychaudhuri, S.P., Jiang, W-Y., Farber, E.M., Schall, T.J., Ruff, M.R. and Pert, C.B. (1999) Upregulation of RANTES in psoriatic keratinocytes: a possible pathogenic mechanism for psoriasis. Acta Dermatol. Venereol., 79: 9–11.

Raychaudhuri, S.P., Farber, E.M. and Raychaudhuri, S.K. (2000) Role of nerve growth factor in RANTES expression by keratinocytes. Acta Dermatol. Venereol., 80: 247–250.

Raychaudhuri, S.P., Dutt, S., Raychaudhuri, S.K., Sanyal, M. and Farber, E.M. (2001) Severe combined immunodeficiency mouse-human skin chimeras: a unique animal model for the study of psoriasis and cutaneous inflammation. Br. J. Dermatol., 144: 931–939.

Schall, T.J. (1991) Biology of the RANTES/SIS cytokine family. Cytokine, 3: 165–183.

Smith, C.H., Barker, J.N., Morris, R.W., MacDonald, D.M. and Lee, T.H. (1993) Neuropeptides induce rapid expression of endothelial cell adhesion molecules and elicit granulocytic infiltration in human skin. J. Immunol., 151: 3274–3282.

Thorpe, L.W., Werrbach-Perez, K. and Perez-Polo, J.R. (1987) Effects of nerve growth factor on the expression of IL-2 receptors on cultured human lymphocytes. Ann. N.Y. Acad. Sci., 496: 310–311.

Valdimarsson, H., Baker, B.S. and Jonsdottir, I. et al. (1986) Psoriasis: a disease of abnormal keratinocyte proliferation induced by T lymphocytes. Immunol. Today, 7: 256–259.

Wallengren, J., Ekman, R. and Sunder, F. (1987) Occurrence and distribution of neuropeptides in human skin. An immunocytochemical and immunohistochemical study on normal skin and blister fluid from inflamed skin. Acta Dermatol. Venereol. (Stockh), 67: 185–192.

Wilkinson, D.I., Theeuwes, M.I. and Farber, E.M. (1994) Nerve growth factor increases the mitogenicity of certain growth factors for cultured human keratinocytes: A comparison with epidermal growth factor. Exp. Dermatol., 3: 239–245.

Wilkinson, D.I. (1989) Mitogenic effect of substance P and CGRP on keratinocytes. J. Cell Biol., 107: 509A.

Wrone-Smith, T. and Nickoloff, B.J. (1996) Dermal injection of immunocytes induces psoriasis. J. Clin. Invest., 98: 1878–1887.

Wyatt, S., Shooeter, E.M. and Davies, A.M. (1990) Expression of the NGF receptor gene in sensory neurons and their cutaneous targets prior to and during innervation. Neuron, 2: 421–427.

Neurotrophic factors and potential therapeutic applications

Progress in Brain Research, Vol. 146
ISSN 0079-6123

CHAPTER 28

Nerve growth factor: from animal models of cholinergic neuronal degeneration to gene therapy in Alzheimer's disease

Mark H. Tuszynski[1,2,]* and Armin Blesch[1]

[1]*Department of Neurosciences-0626, University of California, San Diego, La Jolla, CA, USA*
[2]*Veterans Administration Medical Center, San Diego, CA 92093, USA*

Abstract: Over the last 20 years it has been recognized that neurotrophic factors profoundly influence the development of the nervous system and have the potential to modify disease processes in the adult nervous system. The ability of nervous system growth factors to prevent or reduce neuronal degeneration in animal models of neurodegenerative diseases has led to several clinical trials. One of the main obstacles to the success of these trials has been the method of growth factor delivery: sufficiently high doses of neurotrophic factors must be achieved in the target region of the brain to efficiently modify disease processes, but delivery must be restricted to specific brain regions to prevent adverse effects. Recent advances in molecular medicine have made gene therapy in the nervous system a potentially realistic approach for the delivery of therapeutic molecules such as growth factors. As an alternative to conventional drug delivery, several gene therapy trials for the treatment of central nervous system diseases have started or will start in the near future. This chapter reviews the development of neurotrophic factor gene therapy for neurodegenerative diseases focusing on the therapeutic potential of nerve growth factor in Alzheimer's disease, currently the subject of a phase I clinical trial.

Keywords: gene theraphy; neurological disease; growth factors; neurodegeneration; Alzheimer's disease; Parkinson's disease; Huntington's disease; amyotrophic lateral sclerosis

Introduction

Alzheimer's disease (AD) is the most common neurodegenerative disorder, affecting 4 million people in the United States alone. Its prevalence will double over the next 25 years, and therefore effective therapies that can prevent or slow the progressive neuronal degeneration in AD are urgently needed.

Although the neuropathological hallmarks of AD, including neurofillary tangles and amyloid plaques, have been identified for years, the exact neuropathological mechanisms leading to the widespread degeneration of synapses and neurons are not well understood. One of the neuronal populations severely affected in AD are cholinergic neurons of the basal forebrain (Perry et al., 1977; Whitehouse et al., 1981) and the loss of this neuronal population is thought to contribute to overall cognitive decline (Perry et al., 1978; Bartus et al., 1982; Whitehouse et al., 1982; Candy et al., 1983). Indeed, the only approved drugs for AD target the cholinergic system by inhibiting acetylcholine degradation, thereby partially compensating for the loss of cholinergic neurons in Alzheimer's disease.

The discovery of neurotrophic factors as naturally produced nervous system growth factors that support neuronal survival during development and neuronal

*Correspondence to: M.H. Tuszynski, Department of Neurosciences-0626, University of California, San Diego, La Jolla, CA 92093, USA. Tel.: +1-858-534-8857; Fax: +1-858-534-5220; E-mail: mtuszynski@ucsd.edu

DOI: 10.1016/S0079-6123(03)46028-7

function throughout adulthood generated a broad interest in the use of these factors to intervene in neurodegenerative diseases. The initial discovery that nerve growth factor (NGF), the first described neurotrophic factor, efficiently protects basal forebrain cholinergic neurons from lesion-induced degeneration in the adult nervous system stimulated research on the effects of NGF on cholinergic neurons and mnemonic deficits in rodents and primates over the last 20 years, leading to a current phase I clinical trial of ex vivo NGF gene delivery to the nucleus basalis in patients with AD.

In the following sections we will review some of the studies conducted in numerous laboratories demonstrating the ability of NGF to prevent basal forebrain cholinergic neuronal degeneration. These studies form the basis for an ongoing clinical evaluation of NGF gene therapy for AD.

Effects of NGF on cholinergic neuronal degeneration

In the 1980's it was shown that NGF not only influences neuronal survival and axon growth during development, but that NGF was also produced in the adult hippocampus and neocortex (Korsching et al., 1985). Subsequent studies indicated that infusions of NGF into the ventricles could prevent the lesion-induced degeneration of cholinergic neurons (Hefti, 1986) and reverse the atrophy of basal forebrain cholinergic neurons in the aged rat brain (Fischer et al., 1987). Moreover, NGF improved age related impairments in learning and memory (Fischer et al., 1987). Similar studies were extended to primate models of cholinergic neuronal degeneration and confirmed the previous findings in the rodent brain (Koliatsos et al., 1990; Tuszynski et al., 1990). Thus, NGF was the first neurotrophic factor that efficiently prevented neuronal degeneration in the adult brain. The potential of such effects in the treatment of neuronal degeneration in AD were immediately apparent. Cholinergic neurons in the basal forebrain undergo extensive degeneration in AD, and the deficiency of cholinergic function in AD correlates with cognitive impairment. Further, to date the partially effective therapy for AD consists of cholinesterase inhibition. Whereas cholinesterase inhibitors compensate for the loss of

cholinergic function, NGF has the potential to both *prevent* cholinergic degeneration and *stimulate* the production and release of acetylcholine from remaining neurons (Dekker et al., 1991).

In all of the studies described above, NGF was delivered into the brain by intracerebroventricular infusions because the size and polarity of NGF limit its distribution across the blood brain barrier. Although this technique is able to reach target neurons in the basal forebrain, other structures sensitive to NGF are also exposed to increased NGF levels, leading to adverse effects. Several studies in rodents indicated that NGF infusions can affect sensory nociceptive neurons, sympathetic neurons and Schwann cells, collectively resulting in pain, weight loss, and migration/proliferation of Schwann cells into the subpial space surrounding the brainstem and spinal cord (Crutcher, 1987; Williams, 1991; Winkler et al., 1997). Indeed, infusions of NGF into three patients with AD resulted in pain and weight loss, requiring termination of the study (Eriksdotter Jonhagen et al., 1998). Thus, an alternative delivery method was needed that would restrict NGF to the basal forebrain cholinergic neuron targets.

NGF gene therapy

In an effort to deliver NGF into the central nervous system in a spatially restricted, targeted manner, ex vivo gene therapy was explored beginning in the late 1980's. Gene therapy has several hypothetical advantages for the delivery of neurotrophic factors into the central nervous system (CNS). First, the delivery is restricted to a specific site in the brain, allowing for high growth factor concentrations at the target region without leading to broad NGF distribution and the consequent adverse effects. Thus, the safety of neurotrophic factor treatment is potentially enhanced. Second, gene therapy has the potential to deliver the growth factor for extended time periods without further manipulation. In contrast to chronic growth factor infusions, which would require refilling of a pump reservoir and are subject to infection and malfunction, a single gene treatment could last for years, potentially providing long-lasting neuronal protection.

In the first experiments testing ex vivo NGF gene delivery, retroviral vectors coding for NGF were used for the stable genetic modification of cells in vitro that were subsequently grafted adjacent to cell bodies of lesioned basal forebrain cholinergic neurons. These NGF cell 'factories' were able to prevent cholinergic neuronal cell death after fimbria-fornix lesion in adult rats (Rosenberg et al., 1988). Whereas initial studies used immortalized cell lines for genetic modifications and grafting, subsequent studies showed that primary cells that were obtained from a biopsy and were genetically modified to produce NGF, effectively reversed age-related cognitive decline in rats (Chen and Gage, 1995). Subsequent studies showed that NGF ex vivo gene therapy prevents lesion-induced degeneration of cholinergic neurons in the brains of adult primates (Kordower et al., 1994; Tuszynski et al., 1996). NGF secreting fibroblasts rescued 68% of cholinergic neurons from lesion-induced degeneration, whereas control monkeys continued to express cholinergic markers in only 25% of cholinergic cell bodies. The location of the NGF secreting graft had a significant impact on the degree of cholinergic rescue, reaching 92% in the primate with the most accurately placed graft (Tuszynski et al., 1996). The accuracy of graft placement plays an important role because NGF diffuses very short distances from the modified cell graft into surrounding host tissue. The placement of the cellular NGF source close to the cholinergic cell bodies achieved neuronal protection with doses that were 500-fold lower than earlier primate studies that used intraventricular infusions of NGF (Koliatsos et al., 1990; Tuszynski et al., 1990).

Additional studies investigated whether NGF delivered by genetically modified fibroblasts is also able to reverse age-related atrophy of cholinergic neurons in the primate brain (Smith et al., 1999). To accurately place the grafts, monkeys underwent preoperative MRI scans. Aged primates received grafts of primary fibroblasts expressing NGF, or, in control subjects, beta-glactosidase secreting cells, adjacent to the nucleus basalis of Meynert. Findings were compared to unoperated young monkeys. Stereological quantification of cholinergic neurons indicated that aged control animals showed a 40% decline in cholinergic neuronal markers compared to young subjects. In contrast, NGF secreting cell grafts reversed the age-related reduction in cholinergic markers, restoring the number of labeled cholinergic neurons in the nucleus basalis to levels equivalent to young monkeys. Aged monkeys also displayed a significant reduction in cortical innervation by cholinergic systems, and cellular delivery of NGF to cholinergic somata in the basal forebrain restored levels of cholinergic innervation in the cortex to levels of young monkeys (Conner et al., 2001). Thus, substantial evidence from rodent and primate studies in several laboratories indicated that NGF gene therapy might represent a means of slowing or preventing the degeneration of cholinergic neurons in Alzheimer's disease.

Preclinical safety studies of NGF gene therapy

Based on the efficacy data of NGF gene delivery in animal models, additional safety data were needed to determine if NGF gene delivery could restrict diffusion of the growth factor to the targeted regions and avoid the toxicity of general NGF ventricular infusions. Due to the irreversibility of gene therapy, broad evidence of safety and efficacy was required before considering the initiation of human trials. These studies focused on ex vivo gene delivery rather than in vivo gene therapy, because ex vivo gene therapy techniques were better characterized at the time that these studies were initiated in the early-to-mid 1990s. Current efforts are focused on in vivo gene delivery, and are described in more detail below.

Studies described above had shown that cellular delivery of NGF can prevent the lesion-induced degeneration of cholinergic neurons in the primate brain, reverse age-related atrophy of cholinergic neurons, and restore cholinergic innervation in the cortex to levels of young animals (Tuszynski et al., 1996; Smith et al., 1999; Conner et al., 2001). In these and many other studies, primary fibroblasts that were genetically modified with retroviral vectors had been used for the cellular delivery of NGF into the primate brain. Fibroblasts fulfill all requirements for ex vivo gene therapy: they can be easily obtained from a skin biopsy, cultivated in vitro to establish sufficient cell quantities, undergo transduction with retroviral vectors, survive for long periods in vivo, and sustain local delivery of the therapeutic protein for prolonged time periods in the CNS. Because the cells are derived from the same, individual host, cells are not immunogenic and problems with graft rejection are circumvented.

To establish a safety profile of ex vivo NGF delivery into the primate brain, eight adult rhesus monkeys received injections of escalating doses of fibroblasts expressing NGF into the cholinergic basal forebrain. Over a wide range of cell volumes, from the minimal effective dose to 50-fold higher volumes of NGF expressing cells, signs of toxicity were not observed. Primates showed no weight loss, no pathological response of Schwann cells, and no general indices of pain. NGF protein was not detectable by ELISA in samples of cerebrospinal fluid 6 months and 1 year postimplantation. Cellular grafts did not form tumors, and cells did not migrate from the injection site. Gene expression was persistent for at least 1 year in vivo, the longest time point examined, and hemorrhages did not occur during the introduction of cells into the primate brain. Taken together, substantial evidence indicated that NGF ex vivo gene delivery was an effective and safe means of protecting cholinergic neurons in the basal forebrain.

A phase I clinical study in Alzheimer's disease

Whereas animal studies indicated that NGF is effective in preventing cholinergic neuronal cell death and degeneration from injury, excitotoxicity, and aging, it cannot be predicted with certainty that cholinergic neurons in the Alzheimer's brain will respond to NGF delivery. The mechanism of cell death in AD is unknown and it is thus unclear that NGF will prevent the cell death of cholinergic neurons in the Alzheimer's brain. Only a clinical study will be able to answer this question because no other species develops the full set of pathological characteristics that occur in AD. It further remains to be determined whether NGF delivery can reduce clinical progression in AD. The cholinergic neurons that might respond to NGF comprise only a subpopulation of neurons undergoing degeneration in AD; other subsets of subcortical systems and cortical neurons also degenerate in AD. Nonetheless, cholinergic dysfunction has been most clearly correlated with synapse loss and amyloid plaque density in AD (Bartus et al., 1982; Coyle et al., 1983; Terry et al., 1991). Further, basal forebrain cholinergic neurons are thought to play an important role in

modulating normal neuronal function in the intact hippocampus and cortex. The disruption of the cholinergic system causes memory dysfunction in normal rats and humans (Bartus et al., 1982; Coyle et al., 1983; Everitt and Robbins, 1997) and impairs attention in primates (Voytko et al., 1994). Furthermore, cholinesterase inhibitors, approved for the treatment of AD, target the cholinergic system and provide some measure of clinical improvement. Thus, the stimulation of cholinergic function and the prevention of cholinergic cell death could have a significant impact on the clinical progression of AD.

Based on the evidence presented in the preceding paragraphs, a phase I clinical trial of NGF gene delivery for AD was initiated at the University of California, San Diego in 2001. Eight patients have been enrolled and treated in this study. The patients have received increasing doses of autologous fibroblasts (from skin biopsies) that have been genetically modified to secrete human NGF. The lowest dose was chosen based on the minimum effective dose that rescued cholinergic neurons in primate studies, whereas the highest dose is well below the maximum dose tested in nonhuman primates. NGF-secreting cells have been implanted in the region of the nucleus basalis of Meynert, to determine whether cholinergic neuronal loss can be reduced and whether the function of remaining neurons stimulated in the AD brain. Patients with early AD were the subjects of this first study. The rationale for the selection of early stage AD patients is 2-fold: first, for NGF to be effective, cholinergic neurons must be alive at the time of treatment, and second, patients must to be able to provide informed consent to undergo this experimental treatment, which hypothetically could result in adverse effects despite the range of preclinical safety data provided above. Potential adverse effects due to an inadvertent widespread distribution of NGF could include weight loss, Schwann cell responses and pain syndromes. Surgical complications such as hemorrhage or tumor formation are also potential risk factors. These adverse effects were not encountered in previous preclinical safety studies with more than 200 grafts in nonhuman primates. Nonetheless, should adverse events related to the distribution of NGF occur, they would be difficult to treat. Given the lack of effective neuroprotective therapies, the progressive nature of the disease and the extensive preclinical safety studies, this trial proceeded with approval of the appropriate regulatory authorities. The primary outcome measure of this Phase I trial is safety. Secondary outcome measures include assessment of cognitive function, serial MRI scans and PET scans. The study will be completed in November 2003, one year after enrollment of the last patient. A phase II trial will then begin to address the potential efficacy of NGF gene delivery in a blinded, placebo-controlled manner.

In vivo gene therapy

Whereas ex vivo gene therapy has proven to efficiently deliver genes for therapeutic, secreted proteins into the nervous system, the in vitro preparation, genetic modification and expansion of patient cells can be cumbersome and expensive. The direct delivery of gene therapy vectors to transduce brain cells in the region of interest (in vivo gene therapy) is simpler. The two most promising current candidate vectors for the long-term expression of therapeutic molecules in the CNS are lentiviral vectors and adenoassociated virus (AAV). Both vectors have the ability to transduce nondividing cells, a prerequisite for gene therapy in the adult nervous system. Unlike earlier vectors, current versions completely lack viral genes, thereby limiting immune responses and toxicity of gene delivery. AAV vectors appear to persist as extrachromosomal elements in the nucleus, whereas lentiviral vectors integrate into the genome. Depending on the site of integration, vector incorporation into the genome could hypothetically lead to the mutation, activation or inactivation of host genes, in turn leading to transformation of host cells. This could be a particular problem in cells that divide after genetic modification, as revealed in recent trials of genetically modified bone marrow stem cells in severe combined immunodeficiency (SCID). Because cells in the CNS are mostly quiescent, and cell transformation requires several mutations, the risk of tumor formation from viral integration might be substantially lower in the central nervous system.

Other hypothetical possibilities in the use of in vivo gene therapy in the CNS exist. For example, if adult neural cells are genetically modified to express high levels of growth factors intracellularly, it is possible that cells could modify their transmitter phenotype, or even receive new inputs from projecting neurons. The autocrine expression of neurotrophic factors by cells that normally receive trophic support from other cells could potentially lead to unforeseen consequences.

To date, studies in rodents demonstrate that in vivo NGF gene delivery is as effective as ex vivo NGF gene delivery in preventing cholinergic neuronal degeneration, using both AAV (Mandel et al., 1999) and lentiviral vectors (Blommer et al., 1998). Additional safety and efficacy studies are progress. Should in vivo gene therapy prove to be efficacious and safe, future clinical trials using in vivo NGF gene therapy for AD will be initiated.

In addition, the development of systems that allow for the regulation of gene product could substantially enhance the safety profile of gene therapy in the future. Regulatable gene expression would provide the ability to turn off gene expression should adverse effects occur, and even to optimize the dose of gene product in vivo to adjust the therapeutic dose if necessary. The best-studied system has been the tetracycline-regulated system (Gossen et al., 1994). Two variations of this system are available that either turn gene expression on in response to tetracycline (or its analog doxycycline), termed the 'tet-on system', or that repress gene expression in response to the regulator, termed the 'tet-off system'. Using the 'tet-off' system we have shown that NGF expression by genetically modified fibroblasts can be regulated in vivo using orally-administered doxycycline (Blesch et al., 2001), thereby regulating effects of NGF on cholinergic neurons in the rodent brain. However, additional studies are needed before this system can enter clinical trials. One recent study in primates indicated that immune reactions to the tetracycline activator could limit the duration of gene expression and regulation in vivo (Favre et al., 2002). Other experimental systems for the regulation of gene expression are under development, including systems regulated by ecdysone (Suhr et al., 1998) and rapamycin (Ye et al., 1999). All regulatable gene deliverysystems require further development, however, before entering clinical trials.

Neurotrophic factor gene therapy for other neurological disorders

Several other growth factors have been shown to prevent cell loss in animal models of neurodegenerative disorders. Parkinson's disease (PD) in particular is as a prime candidate for neurotrophic factor gene therapy because neuronal loss is primarily restricted to a single neuronal system, the nigrostriatal projection. The three most promising candidates for growth factor gene therapy in this circuitry are brain-derived neurotrophic factor (BDNF), glial cell line-derived neurotrophic factor (GDNF), and neuturin (NTN). Animal studies have shown that these factors can prevent the loss of neurons in the substantia nigra and improve behavioral outcomes (Spina et al., 1992; Frim et al., 1993; Lin et al., 1993; Hyman et al., 1994; Gash et al., 1996; Kotzbauer et al., 1996; Winkler et al., 1996; Connor et al., 1999; Bjorklund et al., 2000; Kordower et al., 2000). Intraventricular infusions of GDNF in patients with PD resulted in adverse effects due to the spread of growth factors to nontargeted structures, similar to problems encountered with intracerebroventricular NGF infusions (Nutt et al., 2003). More recently, studies using in vivo gene delivery of GDNF have shown morphological and functional recovery in rodent (Connor et al., 1999; Bjorklund et al., 2000) and primate models of PD (Kordower et al., 2000). Pending further safety and toxicity studies, delivery of these growth factors by gene therapy could be initiated in the near future in patients with PD. Direct intrastriatal infusion of GDNF protein has also been conducted in a Phase I trial, with preliminary indications that the physiology of the nigrostriatal system can be influenced by growth factors to modify certain clinical parameters of disease severity (Gill et al., 2003).

Huntington's disease (HD) and amyotrophic lateral sclerosis (ALS) are also potential candidates for growth factor therapy. In animal models of HD, ciliary neurotrophic factor (CNTF) has been shown to be neuroprotective and to promote functional recovery (Emerich et al., 1996; Emerich et al., 1997; also see Alberch et al. in this issue). The striatal region is a target that could be addressed with growth factor infusions or gene therapy. Studies in animal models of progressive motor neuron degeneration indicate the potential therapeutic effects of growth factors such as BDNF and CNTF. However, clinical studies using peripheral administration of these growth factors failed (Group, 1996; Group, 1999), presumably because the quantities of administered growth factors were insufficient to influence the targeted neurons.

In addition to the neuroprotective effects of growth factors, described above, axon growth-promoting effects of neurotrophic factors might be of use in CNS trauma, particularly spinal cord injury. BDNF and NT-3 have been shown to induce axonal regeneration and to improve functional outcomes in rodent models of spinal cord injury (Grill et al., 1997; Liu et al., 1999; Tuszynski et al., 2002). In contrast to neurodegenerative diseases that are likely to require long-lasting delivery of growth factors, spinal cord injury may require only transient expression of axon growth-promoting molecules at the site of injury, to prevent axons from becoming trapped at the source of the neurotrophic factor administration within the injury site. Effective systems for the regulation of growth factor expression by cellular grafts and in vivo gene therapy vectors may be needed before clinical trials can be considered in these disorders, however.

Summary

Gene therapy offers the unique opportunity to deliver therapeutic substances into specific regions of the brain. If growth factor gene therapy can prevent neuronal cell loss in early stages of neurodegenerative diseases, neuroprotective treatments could become available. The treatment of cholinergic neurons with NGF will be only one of many potential growth factor gene therapies to emerge over the next several years, as we reach a threshold for harnessing the potency of growth factors to impact the human condition.

Abbreviations

NGF	nerve growth factor
CNS	central nervous system
AD	Alzheimer's disease
MRI	magnetic resonance imaging

AAV	adeno-associated virus
SCID	severe combined immunodeficiency
PET	positron emission tomography
PD	Parkinson's disease
BDNF	brain-derived neurotrophic factor
GDNF	glial cell line-derived neurotrophic factor
NTN	neuturin
HD	Huntington's disease
ALS	amyotrophic lateral sclerosis
CNTF	ciliary neurotrophic factor

References

Bartus, R., Dean, R.L., Beer, C. and Lippa, A.S. (1982) The cholinergic hypothesis of geriatric memory dysfunction. Science, 217: 408–417.

Bjorklund, A., Kirik, D., Rosenblad, C., Georgievska, B., Lundberg, C. and Mandel, R.J. (2000) Towards a neuroprotective gene therapy for Parkinson's disease: use of adenovirus, AAV and lentivirus vectors for gene transfer of GDNF to the nigrostriatal system in the rat Parkinson model. Brain Res., 886: 82–98.

Blesch, A., Conner, J.M. and Tuszynski, M.H. (2001) Modulation of neuronal survival and axonal growth in vivo by tetracycline-regulated neurotrophin expression. Gene Ther., 8: 954–960.

Blommer, U., Kafri, T., Randolph-Moore, L., Verma, I.M. and Gage, F.H. (1998) Bcl-x1 protects adult septal cholinergic neurons from axotomized cell death. PNAS, 95: 2603–2608.

Candy, J.M., Perry, R.H., Perry, E.K., Irving, D., Blessed, G., Fairbairn, A.F. and Tomlinson, B.E. (1983) Pathological changes in the nucleus basalis of Meynert in Alzheimer's and Parkinson's diseases. J. Neurosci., 54: 277–289.

Chen, K.S. and Gage, F.H. (1995) Somatic gene transfer of NGF to the aged brain: Behavioral and morphological amelioration. J. Neurosci., 15: 2819–2825.

Conner, J.M., Darracq, M.A., Roberts, J. and Tuszynski, M.H. (2001) Non-tropic actions of neurotrophins: Subcortical NGF gene delivery reverses age-related degeneration of primate cortical cholinergic innervation. Proc. Nat. Acad. Sci. USA, 98: 1941–1946.

Connor, B., Kozlowski, D.A., Schallert, T., Tillerson, J.L., Davidson, B.L. and Bohn, M.C. (1999) Differential effects of glial cell line-derived neurotrophic factor (GDNF) in the striatum and substantia nigra of the aged Parkinsonian rat. Gene Ther., 6: 1936–1951.

Coyle, J.T., Price, P.H. and Delong, M.R. (1983) Alzheimer's disease: A disorder of cortical cholinergic innervation. Science, 219: 1184–1189.

Crutcher, K.A. (1987) Sympathetic sprouting in the central nervous system: A model for studies of axonal growth in the mature mammalian brain. Brain Res. Rev., 12: 203–233.

Dekker, A.J., Langdon, D.J., Gage, F.H. and Thal, L.J. (1991) NGF increases cortical acetylcholine release in rats with lesions of the nucleus basalis. Neuroreport, 2: 577–580.

Emerich, D.F., Lindner, M.D., Winn, S.R., Chen, E.Y., Frydel, B.R. and Kordower, J.H. (1996) Implants of encapsulated human CNTF-producing fibroblasts prevent behavioral deficits and striatal degeneration in a rodent model of Huntington's disease. J. Neurosci., 16: 5168–5181.

Emerich, D.F., Winn, S.R., Hantraye, P.M., Peschanski, M., Chen, E.Y., Chu, Y., McDermott, P., Baetge, E.E. and Kordower, J.H. (1997) Protective effect of encapsulated cells producing neurotrophic factor CNTF in a monkey model of Huntington's disease. Nature, 386: 395–399.

Eriksdotter Jonhagen, M., Nordberg, A., Amberla, K., Backman, L., Ebendal, T., Meyerson, B., Olson, L., Seiger, Shigeta, M., Theodorsson, E., Viitanen, M., Winblad, B. and Wahlund, L.O. (1998) Intracerebroventricular infusion of nerve growth factor in three patients with Alzheimer's disease, Dement. Geriatr. Cogn. Disord., 9: 246–257.

Everitt, B.J. and Robbins, T.W. (1997) Central cholinergic systems and cognition. Ann. Rev. Pscyhol., 48: 649–684.

Favre, D., Blouin, V., Provost, N., Spisek, R., Porrot, F., Bohl, D., Marme, F., Cherel, Y., Salvetti, A., Hurtrel, B., Heard, J.M., Riviere, Y. and Moullier, P. (2002) Lack of an immune response against the tetracycline-dependent transactivator correlates with long-term doxycycline-regulated transgene expression in nonhuman primates after intramuscular injection of recombinant adeno-associated virus. J. Virol., 76: 11605–11611.

Fischer, W., Wictorin, K., Bjorklund, A., Williams, L.R., Varon, S. and Gage, F.H. (1987) Amelioration of cholinergic neuron atrophy and spatial memory impairment in aged rats by nerve growth factor. Nature, 329: 65–68.

Frim, D.M., Uhler, T.A., Short, M.P., Ezzedine, Z.D., Klagsbrun, M., Breakefield, X.O. and Isacson, O. (1993) Effects of biologically delivered NGF, BDNF and bFGF on striatal excitotoxic lesions. Neuroreport, 4: 367–370.

Gash, D.M., Zhang, Z., Ovadia, A., Cass, W.A., Yi, A., Simmerman, L., Russell, D., Martin, D., Lapchak, P.A., Collins, F., Hoffer, B.J., Gerhardt, G.A. (1996) Functional recovery in parkinsonian monkeys treated with GDNF. Nature, 380: 252–255.

Gill, S.S., Patel, N.K., Hotton, G.R., O'Sullivan, K.O., McCarter, R., Bunnage, M., Brooks, D.J., Svendsen, C.N., Heywood, P. (2003) Direct infusion of glial cell line-derived neurotrophic factor in Parkinson disease, Nat. Med., 9: 589–595.

Gossen, M., Bonin, A.L., Freundlieb, S. and Bujard, H. (1994) Inducible gene expression systems for higher eukaryotic cells. Curr. Opin. Biotechnol., 5: 516–520.

Grill, R., Murai, K., Blesch, A., Gage, F.H. and Tuszynski, M.H. (1997) Cellular delivery of neurotrophin-3 promotes corticospinal axonal growth and partial functional recovery after spinal cord injury. J. Neurosci., 17: 5560–5572.

Group, A.-C.T.S. (1996) A double-blind placebo-controlled trial of subcutaneous recombinant human ciliary neurotrophic factor (rhCNTF) in amyotrophic lateral sclerosis. Neurology, 46: 1244–1249.

Group, T.B.S. (1999) A controlled trial of recombinant methionyl human BDNF in ALS: The BDNF Study Group (Phase III). Neurology, 52: 1427–1433.

Hefti, F. (1986) Nerve growth factor (NGF) promotes survival of septal cholinergic neurons after fimbrial transection. J. Neurosci., 6: 2155–2162.

Hyman, C., Hofer, M., Juhasz, M., Yancopoulos, G.P., Squinto, S.P. and Lindsay, R.M. (1994) Brain derived neurotrophic factor is a neurotrophic factor for dopaminergic neurons of the substantia nigra. Nature, 350: 230–232.

Koliatsos, V.E., Nauta, H.J., Clatterbuck, R.E., Holtzman, D.M., Mobley, W.C. and Price, D.L. (1990) Mouse nerve growth factor prevents degeneration of axotomized basal forebrain cholinergic neurons in the monkey. J. Neurosci., 10: 3801–3813.

Kordower, J.H., Winn, S.R., Liu, Y.-T., Mufson, E.J., Sladek, J.R., Hammang, J.P., Baetge, E.E. and Emerich, D.F. (1994) The aged monkey basal forebrain: Rescue and sprouting of axotomized basal forebrain neurons after grafts of encapsulated cells secreting human nerve growth factor. Proc. Nat. Acad. Sci. USA., 91: 10898–10902.

Kordower, J.H., Emborg, M.E., Bloch, J., Ma, S.Y., Chu, Y., Leventhal, L., McBride, J., Chen, E.Y., Palfi, S., Roitberg, B.Z., Brown, W.D., Holden, J.E., Pyzalski, R., Taylor, M.D., Carvey, P., Ling, Z., Trono, D., Hantraye, P., Déglon, N. and Aebischer, P. (2000) Neurodegeneration prevented by lentiviral vector delivery of GDNF in primate models of Parkinson's disease. Science, 290: 767–773.

Korsching, S., Auburger, G., Heumann, R., Scott, J. and Thoenen, H. (1985) Levels of nerve growth factor and its mRNA in the central nervous system of the rat correlate with cholinergic innervation. EMBO. J., 4: 1389–1393.

Kotzbauer, P.T., Lampe, P.A., Heuckeroth, R.O., Golden, J.P., Creedon, D.J., Johnson, E.M. and Milbrandt, J. (1996) Neurturin, a relative of glial-cell-line derived neurotrophic factor. Nature, 384: 467–470.

Lin, L.F., Doherty, D.H., Lile, J.D., Bektesh, S. and Collins, F. (1993) GDNF: A glial cell-line derived neurotrophic factor for midbrain dopaminergic neurons. Science, 260: 1130–1132.

Liu, Y., Kim, D., Himes, B.T., Chow, S.Y., Schallert, T., Murray, M., Tessler, A. and Fischer, I. (1999) Transplants of fibroblasts genetically modified to express BDNF promote regeneration of adult rat rubrospinal axons and recovery of forelimb function. J. Neurosci., 19: 4370–4387.

Mandel, R.J., Gage, F.H., Clevenger, S.K., Snyder, R.O. and Leff, S.E. (1999) Nerve growth factor expressed in the medial septum following in vivo gene delivery using a recombinant adeno-associated viral vector protects cholinergic neurons from fimbria-fornix lesion-induced degeneration. Exp. Neurol., 155: 59–64.

Nutt, J.G., Burchiel, K.J., Comella, C.L., Jankovic, J., et al. (2003) Randomized, double-blind trial of glial cell line-derived neurotorphic factor (GDNF) in PD. Neurology, 60: 69–73.

Perry, E.K., Perry, R.H., Blessed, G. and Tomlinson, B.E. (1977) Necropsy evidence of central cholinergic deficits in senile dementia. Lancet, 2: 143.

Perry, E.K., Tomlinson, B.E., Blessed, G., Bergmann, K., Gibson, P.H. and Perry, R.H. (1978) Correlation of cholinergic abnormalities with senile plaques and mental test scores in senile dementia. Brit. Med. J., 2: 1457–1459.

Rosenberg, M.B., Friedmann, T., Robertson, R.C., Tuszynski, M., Wolff, J.A., Breakefield, X.O. and Gage, F.H. (1988) Grafting genetically modified cells to the damaged brain: restorative effects of NGF expression. Science, 242: 1575–1578.

Smith, D.E., Roberts, J., Gage, F.H. and Tuszynski, M.H. (1999) Age-associated neuronal atrophy occurs in the primate brain and is reversible by growth factor gene therapy. Proc. Nat. Acad. Sci. USA, 96: 10893–10898.

Spina, M.B., Hyman, C., Squinto, S. and Lindsay, R.M. (1992) Brain-derived neurotrophic factor protects dopaminergic cells from 6-hydroxydopamine toxicity. Ann. N.Y. Acad. Sci., 648: 348–350.

Suhr, S.T., Gil, E.B., Senut, M.C. and Gage, F.H. (1998) High level transactivation by a modified Bombyx ecdysone receptor in mammalian cells without exogenous retinoid X receptor. Proc. Natl. Acad. Sci. USA, 95: 7999–8004.

Terry, R.D., Masliah, E., Salmon, D.P., Butters, N., DeTeresa, R., Hill, R., Hansen, L.A. and Katzman, R. (1991) Physical basis of cognitive alterations in Alzheimer's disease: synapse loss is the major correlate of cognitive impairment. Ann. Neurol., 30: 572–580.

Tuszynski, M.H., U, H.S., Amaral, D.G. and Gage, F.H. (1990) Nerve growth factor infusion in primate brain reduces lesion-induced cholinergic neuronal degeneration, J. Neurosci., 10: 3604–3614.

Tuszynski. M.H., Roberts, J., Senut, M.C., U, H.-S. and Gage, F.H. (1996) Gene therapy in the adult primate brain: intraparenchymal grafts of cells genetically modified to produce nerve growth factor prevent cholinergic neuronal degeneration. Gene Ther., 3: 305–314.

Tuszynski, M.H., Grill, R., Jones, L., McKay, H.M. and Belsch, A. (2002) Spontaneous and augmented growth of axons in the primate spinal cord: Effects of local injury and nerve growth factor-secreting cells. J. Comp. Neurol., 448: 88–101.

Voytko, M.L., Olton, D.S., Richardson, R.T., Gorman, L.K., Tobin, J.R. and Price, D.L. (1994) Basal forebrain lesions in monkeys disrupt attention but not learning and memory. J. Neurosci., 14: 167–186.

Whitehouse, P.J., Price, D.J., Clark, A., Coyle, J.T. and DeLong, M. (1981) Alzheimer's disease: Evidence for selective loss of cholinergic neurons in the nucleus basalis. Ann. Neurol., 10: 122–126.

Whitehouse, P.J., Price, D.L., Struble, R.G., Clar, A.W., Coyle, J.T. and Delong, M.R. (1982) Alzheimer's disease and senile dementia: Loss of neurons in the basal forebrain. Science, 215: 1237–1239.

Williams, L.R. (1991) Hypophagia is induced by intracerebroventricular administration of nerve growth factor. Exp. Neurol., 113: 31–37.

Winkler, C., Sauer, H., Lee, C.S. and Bjorklund, A. (1996) Short-term GDNF treatment provides long-term rescue of lesioned nigral dopaminergic neurons in a rat model of Parkinson's disease. J. Neurosci., 16: 7206–7215.

Winkler, J., Ramirez, G.A., Kuhn, H.G., Peterson, D.A., Day-Lollini, P.A., Stewart, G.R., Tuszynski, M.H., Gage, F.H. and Thal, L.J. (1997) Reversible Schwann cell hyperplasia and sprouting of sensory and sympathetic neurites after intraventricular administration of nerve growth factor. Ann. Neurol., 41: 82–93.

Ye, X., Rivera, V.M., Zoltick, P., Cerasoli, F., Schnell, M.A., Gao, G., Hughes, J.V., Gilman, M. and Wilson, J.M. (1999) Regulated delivery of therapeutic proteins after in vivo somatic cell gene transfer. Science, 283: 88–91.

Progress in Brain Research, Vol. 146
ISSN 0079-6123

CHAPTER 29

Viral vector-mediated gene transfer of neurotrophins to promote regeneration of the injured spinal cord

William T.J. Hendriks, Marc J. Ruitenberg, Bas Blits, Gerard J. Boer
and Joost Verhaagen*

Graduate School for Neurosciences Amsterdam, Department of Neuroregeneration, Netherlands Institute for Brain Research, Meibergdreef 33, 1105 AZ, Amsterdam, The Netherlands

Abstract: Injuries to the adult mammalian spinal cord often lead to severe damage to both ascending (sensory) pathways and descending (motor) nerve pathways without the perspective of complete functional recovery. Future spinal cord repair strategies should comprise a multi-factorial approach addressing several issues, including optimalization of survival and function of spared central nervous system neurons in partial lesions and the modulation of trophic and inhibitory influences to promote and guide axonal regrowth. Neurotrophins have emerged as promising molecules to augment neuroprotection and neuronal regeneration. Although intracerebroventricular, intrathecal and local protein delivery of neurotrophins to the injured spinal cord has resulted in enhanced survival and regeneration of injured neurons, there are a number of drawbacks to these methods. Viral vector-mediated transfer of neurotrophin genes to the injured spinal cord is emerging as a novel and effective strategy to express neurotrophins in the injured nervous system. Ex vivo transfer of neurotrophic factor genes is explored as a way to bridge lesions cavities for axonal regeneration. Several viral vector systems, based on herpes simplex virus, adenovirus, adeno-associated virus, lentivirus, and moloney leukaemia virus, have been employed. The genetic modification of fibroblasts, Schwann cells, olfactory ensheathing glia cells, and stem cells, prior to implantation to the injured spinal cord has resulted in improved cellular nerve guides. So far, neurotrophic factor gene transfer to the injured spinal cord has led to results comparable to those obtained with direct protein delivery, but has a number of advantages. The steady advances that have been made in combining new viral vector systems with a range of promising cellular platforms for ex vivo gene transfer (e.g., primary embryonic neurons, Schwann cells, olfactory ensheating glia cells and neural stem cells) holds promising perspectives for the development of new neurotrophic factor-based therapies to repair the injured nervous system.

Keywords: central nervous system; gene therapy; neurotrophins; regeneration; spinal cord injury; viral vectors

Introduction

Injuries to the adult mammalian spinal cord often lead to severe damage to both ascending (sensory) and descending (motor) nerve pathways. Although

still a matter of some controversy (Levi et al., 1996), the disruption of the main descending spinal motor pathways, including the rubrospinal tract (RST) and the corticospinal tract (CST), is responsible for the loss of voluntary motor control and organ function below the level of the lesion (Nathan, 1994). Despite the occurrence of some degree of spontaneous functional recovery, complete return of these functions does not occur due to the limited regenerative

*Correspondence to: J. Verhaagen, Netherlands Institute for Brain Research, Meibergdreef 33, 1105 AZ, Amsterdam, The Netherlands. Tel.: +31-20-566-5513; Fax: +31-20-696-1006; E-mail: j.verhaagen@nih.knaw.nl. Authors contributed equally.

DOI: 10.1016/S0079-6123(03)46029-9

potential of the adult central nervous system (CNS) (Schwab and Bartholdi, 1996; Raineteau and Schwab, 2001; Weidner et al., 2001; Blits, 2002; McDonald and Sadowsky, 2002).

The primary mechanical trauma, in most instances caused by fractures or dislocation of one or more vertebra, results in a local compression or laceration of the spinal cord at the impact site. This will cause disruption of axonal pathways and cellular membrane integrity as well as microvascular damage contributing to a state of ischemia at the injury site, shifts in electrolyte concentrations and excessive release of excitatory amino acids (e.g., glutamate). These processes initiate a destructive signaling cascade resulting in progressive cell death and demyelination, which induces the formation of a glial scar that impedes axonal regrowth (Schwab and Bartholdi, 1996; Carlson et al., 1998; Yong et al., 1998).

Ideally, future spinal cord repair strategies will comprise a multifactorial approach addressing the following issues: (i) minimalization of the deleterious effects of early trauma, inflammation and scar formation, (ii) optimalization of the survival and function of spared CNS fibers in partial lesions of the spinal cord, (iii) replacement of lost neural tissue at the impact site, (iv) integrated exploitation of trophic and inhibitory influences to promote and guide axonal regrowth, and (v) establishment of appropriate connections by regenerating axons to restore damaged neural circuits. At present, most repair strategies are therefore aimed at neuroprotection (e.g., preventing neuronal atrophy and cell death) or at the promotion of anatomical restoration (e.g., augmenting regrowth of injured axons), or a combination of both (Blits et al., 2002).

The paucity of regeneration in the mammalian CNS is widely believed to depend on a critical molecular balance between growth-promoting and growth-inhibiting factors (Liebl et al., 2001). Initially, within days after injury, severed axons begin to form sprouts but their growth usually stops at the lesion site. The expression of growth-inhibitory molecules by both hyperreactive astrocytes that surround the damaged area and cells that constitute the core of the neural scar, e.g., meningeal cells, is thought to be a major obstacle for regrowing axons (Dou and Levine, 1994; Mukhopadhyay et al., 1994; Chen et al., 2000; Pasterkamp and Verhaagen, 2001; De Winter et al.,

2002). In addition, the shortage of both extrinsic and intrinsic factors such as the growth-associated proteins B50/GAP43 (Fernandes et al., 1999) and the alterations in the expression of guidance molecules all contribute to the inability of axons to regenerate across the injury site (Fawcett, 1997). An imbalance thus arises between neuronal growth-promoting and neuronal growth-inhibiting molecules in favor of the latter. The delivery of neurotrophic factors, e.g., neurotrophins, has emerged as a promising strategy to manipulate this molecular balance towards a more stimulating, conducive environment for axonal regrowth (Emborg and Kordower, 2000; Blesch et al., 2002).

The term neurotrophins refers to a family of proteins that are highly conserved among mammals and essential for neuronal development and neuronal plasticity. Since the discovery of the prototype neurotrophin nerve growth factor (NGF) by Rita Levi-Montalcini and coworkers, three other mammalian neurotrophins have been identified. In the 1980s and early 1990s, brain-derived neurotrophic factor (BDNF), neurotrophin-3 (NT-3) and NT-4/5 were cloned and characterized (reviewed by Lewin and Barde, 1996). Neurotrophins exert their biological actions intracellularly through specific transmembrane receptors, i.e., tyrosine kinase receptors (Trk); NGF signals through TrkA, BDNF and NT-4/5 use TrkB as their preferred receptor and NT-3 mediates its effect mainly through TrkC. All neurotrophins additionally bind to the low-affinity p75 neurotrophin receptor (p75^{NTR}) (for reviews on neurotrophin receptors see elsewhere in this issue). Besides neurotrophins, two other neurotrophic factors, glial cell line-derived neurotrophic factor (GDNF) and ciliary neurotrophic factor (CNTF), have been widely studied in relation to CNS regeneration (Houenou et al., 1996; Kuzis and Eckenstein, 1996). GDNF signals through a multicomponent receptor composed of GDNFRα and Ret. The CNTF receptor is composed of the subunits CNTFRα, LIFRβ and gp130 (Frank and Greenberg, 1996; Ip and Yancopoulos, 1996; Airaksinen and Saarma, 2002). Besides being essential factors for neuronal development and plasticity, all of these neurotrophic factors appeared to be able to prevent cellular atrophy, enhance neuronal survival and facilitate axonal regeneration after injury of the adult CNS (reviewed by Thoenen and Sendtner, 2002).

Delivery of neurotrophic factors to the injured CNS

The application of neurotrophic factors is hampered by the lack of efficient means of delivery to the brain and spinal cord. Oral or systemic delivery of neurotrophins is not feasible due to the very short half-life of these proteins in vivo and inability to pass the blood–brain-barrier. These limitations of protein-based therapeutics have resulted in the development of direct application methods of neurotrophic factors to the nervous system by intrathecal, intracerebro-ventricular (ICV) or intraparenchymal infusion. More recently developed methods for neurotrophic factor delivery are direct gene transfer using (non)viral vectors or the implantation of ex vivo genetically engineered cells secreting neurotrophic factors. This review briefly discusses cerebrospinal fluid (CSF) and local delivery of neurotrophic factors to the injured spinal cord. Subsequently, gene and cell therapy as methods to administer neurotrophic factors to the nervous system will be reviewed and discussed.

Intracerebroventricular and intrathecal protein delivery

Delivery of neurotrophic factors to injured neurons via the CSF can be accomplished either via ICV or intrathecal infusion. Intrathecal delivery has been extensively used in rodent spinal cord models to asses the beneficial effects of these factors on the survival of neurons and regrowth of axons (Kishino et al., 1997; Jakeman et al., 1998; Ramer et al., 2000). Intrathecal infusion of BDNF into the subarachnoid cavity after spinal root avulsion prevented retrograde cell death of spinal motoneurons and counteracted the lesion-induced reduced synthesis of transmitter-related enzymes such as choline acetyltransferase (ChAT) and acetylcholine esterase (Kishino et al., 1997). When BDNF was administered for 4 weeks in a similar lesion model, significant axonal regrowth of damaged motoneurons was found at the avulsion site at 12 weeks after injury (Novikov et al., 1997). Besides enhancement of regeneration, prevention of retrograde cell death in descending pathways after

SCI has reportedly been observed after long-term intrathecal infusion of neurotrophins (Novikova et al., 2000). In adult rats with injured dorsal roots, treatment with NGF, NT-3 and GDNF resulted in selective regrowth of damaged sensory axons across the dorsal root entry zone and into the spinal cord (Ramer et al., 2000). After peripheral nerve stimulation in rats treated with NGF, NT-3 or GDNF, dorsal horn neurons were found to be electrophysiologically active, indicating functional reconnection. Behavioral analysis revealed that rats treated with NGF and GDNF also recovered sensitivity to noxious heat and pressure, thereby confirming the anatomical and electrophysiological observations.

Advantages of intrathecal and ICV neurotrophic factor delivery are that both the amount and duration of neurotrophic factor administration can be tightly controlled. A major disadvantage of ICV or intrathecal delivery of neurotrophic factors, however, is a limited diffusion into the CNS parenchyma (Yan et al., 1994; Anderson et al., 1995; Dittrich et al., 1996), which, in case of BDNF and NT-3, has been attributed to the abundant expression of truncated isoforms of the TrkB and TrkC receptor on the cell surface of ependymal cells (Biffo et al., 1995; Rubio, 1997). In contrast, for neurotrophic factors that do diffuse readily through the brain parenchyma, such as NGF, this type of application to the CNS has a widespread effect on multiple populations of NGF-sensitive neurons and glial cells. Continuous ICV administration of NGF reportedly led to vigorous sprouting of many sensory and sympathetic fibers in the spinal cord of rats and monkeys (Day-Lollini et al., 1997; Winkler et al., 1997). It has been suggested that this NGF-induced aberrant sprouting of nociceptive fibers in the spinal cord contributed to the development of chronic pain syndromes (Eriksdotter Jonhagen et al., 1998; Hao et al., 2000). For BDNF, adverse effects like weight loss and gait disturbances have been found in sheep after continuous intrathecal infusion (Dittrich et al., 1996). Another disadvantage of intrathecal delivery is the substantial damage to the spinal cord after pro-longed intrathecal infusion periods due to scar formation at the catheter tip and from the catheter itself (Jones and Tuszynski, 2001). Taken together, these drawbacks of neurotrophic factor delivery to the CSF, prompted the development of alternative delivery methods, e.g., more local intraparenchymal neurotrophic factor infusion and gene transfer technology.

Local protein delivery

The major advantage of local delivery is minimalization of the risk to induce side effects on neural systems that are not the target of the treatment as has been observed following CSF infusion (reviewed by Dijkhuizen and Verhaagen, 1999). Various methods have been developed to directly deliver neurotrophic factors to the CNS, which include: (i) a single injection at the lesion site (Schnell et al., 1994), (ii) stereotaxic placement of an intraparenchymal infusion cannula connected to an osmotic minipump (Giehl and Tetzlaff, 1996; Oudega and Hagg, 1996), or (iii) placement of slow release materials such as gelfoam, fibrin glue or polymerized collagen at the site of injury (Ye and Houle, 1997; Houweling et al., 1998a).

Direct injections of neurotrophins immediately rostral of the lesion at the lower thoracic level of the rat spinal cord revealed a highly significant effect of NT-3 on CST regenerative sprouting, whereas NGF had only little effect and no effect of BDNF was found (Schnell et al., 1994). These results are consistent with findings that TrkC, the functional receptor for NT-3, is highly expressed in the developing cortex, where the cells of origin of the CST are found. However, TrkB is expressed in the developing cortex as well (Ringstedt et al., 1993).

Intraparenchymal infusions of BDNF or NT-4/5 close to the cell bodies of axotomized rubrospinal neurons have been demonstrated to prevent lesion-induced cell death and were able to induce the upregulation of several regeneration-associated genes (RAGs), e.g., growth-associated protein B-50/GAP-43 (Kobayashi et al., 1997). When combined with a peripheral nerve implant, infusion of BDNF was found to increase the number of lesioned rubrospinal neurons that regenerated into the grafts following both acute and chronic injury (Kobayashi et al., 1997; Kwon et al., 2002). Intraparenchymal infusion of BDNF by means of an osmotic minipump with the tip of the cannula placed caudal to the level of

a midthoracic contusion injury or hemisection (vertebral T8), resulted in enhanced outgrowth of cholinergic fibers and improved hind limb stepping (Jakeman et al., 1998).

In a chronic spinal cord injury model gelfoam saturated with BDNF, NT-3 or CNTF was placed into the lesion cavity and replaced after 1 week by a piece of peripheral nerve inserted into the rostral wall of the lesion. This resulted in a significantly enhanced number of regrowing supraspinal neurons (Ye and Houle, 1997). A collagen matrix saturated with NT-3 and implanted at the site of hemisection in the spinal cord resulted in regenerative regrowth of the CST axons and in improved performance of hind limb function (Houweling et al., 1998b). Similarly, collagen-mediated delivery of BDNF following hemisection showed an improved hind limb function, which was suggested to be related to protection from secondary tissue damage, but not with anatomical regeneration of descending nerve tracts (Houweling et al., 1998c). Application of neurotrophic factors in a peripheral nerve injury has been reported to have an effect on CNS neurons as well. Local delivery of NGF or CNTF with a piece of gelfoam attached to the proximal stump of a transected rat sciatic nerve counteracted apoptosis of deafferented interneurons in the spinal cord (Oliveira et al., 2002).

The studies summarized above showed that the administration of neurotrophic factors can stimulate neuroregeneration after SCI. Although encouraging results have been achieved, these methods of delivery also have certain disadvantages. The use of saturated pieces of gelfoam or polymerized collagen does not guarantee a steady delivery of the neurotrophic factor because the source is usually rapidly depleted. In addition, invasive surgery is required for placement of these implants. Continuous intraparenchymal delivery by means of an osmotic minipump requires the chronic implantation of infusion cannulas, which after prolonged use leads to a tissue reaction around the cannula, with the risk that the cannula becomes clogged, thereby hampering continuous protein delivery. As discussed below, gene and cell therapeutic strategies to express neurotrophic factors in the injured CNS provide a more sophisticated solution to the delivery problem circumventing most of the problems outlined above (Fig. 1).

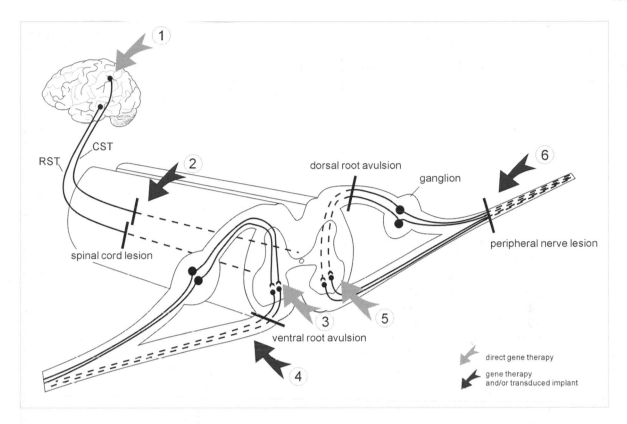

Fig. 1. Different gene transfer approaches for neurotrophins in spinal cord injuries to study regeneration in different lesion models. Both direct transduction of the tissue (gray arrows) and application of an ex vivo transduced implants are illustrated (black arrows). (1) To prevent neuronal atrophy and cell death of the long descending axonal projections after lesion, e.g., RST or CST, these neurons or their neighboring cells can be genetically modified by a single vector injection to generate neurotrophins. (2) Regenerative axonal growth of the RST and CST may be stimulated by overexpression of neurotrophins at the level of axotomy either via direct vector injection and/or via a cellular implant acting as a biological drug-delivery device, additionally acting as a matrix to bridge the lesion cavity. (3) After ventral root avulsion, resulting in axotomy of motorneuron efferents, direct application of a viral vector encoding a neurotrophin to the ventral motorneuron pool could provide neuroprotection. (4) Direct transduction of the distal root or an ex vivo transduced peripheral nerve implant that bridges the gap between spinal cord and avulsed root may be used to direct and stimulate regrowth of motorneuron efferents into the ventral root. Combinations of (1) and (2) may be possible as well as of (3) and (4). (5) After a dorsal root injury, sensory fibers can be rescued and stimulated to regenerate via direct application of a viral vector encoding a neurotrophin to the dorsal horn. (6) Regeneration of injured peripheral nerve fibers can be stimulated either via direct application of viral vectors distal from the transaction, or in case of the use of an implant, via transduced implants, or a combination of both.

Gene therapy

Experimental gene therapy is a strategy that transfers biologically relevant genetic material into somatic cells to treat disease. Transfer of a piece of DNA encoding neurotrophic factors into the CNS to enhance neuroregeneration, can be approached in two different ways: (i) via direct in vivo injection of (non)viral vectors (Palella et al., 1989; Yang et al., 1997), or (ii) via implantation of ex vivo genetically engineered cells (Gage et al., 1987).

In vivo or direct gene transfer can be achieved via viral or nonviral vector systems. Nonviral vector-mediated transfer entails the introduction of a plasmid containing the transgene under a relevant promoter into the cell by physical, receptor-mediated or cationic liposome-mediated methods (Yang et al., 1997). A main advantage of these nonviral vectors would be that they do not induce a host immune response. At present however, they have the disadvantage of poor in vivo delivery efficiency and a transient expression of the transgene (Emborg and

Kordower, 2000). So far, only viral vector-mediated gene transfer appeared feasible for application in the CNS in terms of efficiency and duration of transgene expression.

Ex vivo gene transfer entails the introduction of genes into (cultured) cells or tissue fragments that are subsequently implanted into the host organism either directly or following encapsulation in a microporous polymer tube. Such transgenic implants then act as a manufacturing reservoir, synthesizing and secreting therapeutic agents within the CNS. Cells used to express a transgene in the CNS should have specific properties. If neural cells from a donor or of the recipient of the cells itself (auto-implants) are employed, they should be relatively easy to obtain. If retrieved for ex vivo transduction of glial or nonneural cell transplants, the cells should divide and remain viable in tissue culture for prolonged periods of time to obtain enough cells for successful genetic manipulation and implantation. Finally, these cells should be well-tolerated by CNS tissue without eliciting a host immune response and without displaying uncontrolled growth that would destroy the normal architecture of CNS tissue (Dijkhuizen and Verhaagen, 1999; Lacroix and Tuszynski, 2000; Blesch et al., 2002).

Viral vector-mediated delivery of neurotrophic factor genes to the injured CNS

There are a number of requirements that would make a viral vector suitable and safe for gene transfer to the CNS. First, the vector should be replication-deficient to prevent cell damage and viral spread. Second, the vector has to be nontoxic in terms of direct toxicity of viral envelope and capsid proteins and in terms of intrinsic toxicity caused by viral expression of proteins from remaining wild type sequences in the recombinant genome. Third, viral vector stocks have to be free of wild type or helper virus. Finally, for CNS application, the viral vector should be able to transduce nondividing cells (Hermens and Verhaagen, 1998; Emborg and Kordower, 2000). Viral vectors capable of infecting CNS cells can provide a local source of neurotrophic factors either at the site of the cell soma to counteract neuronal atrophy or to stimulate neuronal survival or, at the site of injury, to promote restorative neurite outgrowth. An additional advantage of gene therapy is that the cellular production of a neurotrophic factor mimics its normal physiological situation. Viral vectors can be delivered to the CNS by means of a single (stereotactic) injection thereby minimizing neurosurgical intervention. Until now, four different viral vector systems have been used for direct gene transfer in the CNS, including herpes simplex viral (HSV), adenoviral (AdV), adeno-associated viral (AAV) and lentiviral (LV) vectors. The classical retroviral vectors, based on the Moloney murine leukemia virus (MLV), do not transduce nondividing cells such as neurons and they can therefore not be used for direct gene transfer in the nervous system (Miller et al., 1990). However, MLV vectors have been successfully used for the ex vivo genetic modification of cells used for transplantation purposes, such as fibroblasts and Schwann cells (Tuszynski et al., 1996; Grill et al., 1997a; Menei et al., 1998; Weidner et al., 1999; Blesch et al., 2002).

Direct viral vector-mediated gene transfer of neurotrophins

Herpes simplex viral vectors

The first viral vectors used for direct genetic manipulation of the nervous system were based on herpes simplex virus (HSV). This vector system was chosen because of its neurotropic nature and maintenance of a life-long latency in neurons after deletion of immediate-early (IE) genes needed for induction of a lytic cycle (Roizman and Sears, 1987; Ho and Mocarski, 1988; Palella et al., 1989). During latency, the viral genome is maintained in an episomal state, and expression of the vast majority of the genes in the HSV genome is silenced. However, part of the long repeat region of the viral genome is transcribed during latency, generating a population of RNA transcripts collectively known as the latency-associated transcripts (LATs). This has prompted the use of the LAT promoter to drive transgene expression (Lilley et al., 2001). An advantage of HSV vectors is their large cloning capacity (152 kb) allowing the insertion of large or multiple genes (reviewed by Hermens and Verhaagen, 1998).

Federoff and colleagues (1992) were the first to construct and apply HSV vectors encoding neurotrophins in the nervous system. They demonstrated the expression of biologically active NGF in the rat superior cervical ganglion following injection of an NGF-encoding HSV vector. Injection of this HSV vector after axotomy of a sympathetic ganglion prevented the decline in tyrosine hydroxylase expression that otherwise would have occurred in axotomized neurons (Federoff et al., 1992). In vitro, NGF under a human cytomegalovirus promotor (hCMV) or latency-active promotor (LAP2) element was able to protect neurons in cultured rat DRGs from the cytotoxic effects of hydrogen peroxide. In these studies, the hCMV promoter gave immediate expression which faded away with time, whereas the use of LAP2 resulted in a delayed expression of NGF that peaked at 2 weeks (Goins et al., 1999).

An HSV-1 vector encoding the anti-apoptotic protein Bcl-2 was used to protect motoneurons against lesion induced death after ventral root avulsion. Following injection of the vector into the spinal cord 1 week before lumbar root avulsion, increased survival of lesioned motoneurons was detected 14 days after injury as determined with fluorogold retrograde tracing. However, HSV vector-mediated Bcl-2 expression did not preserve ChAT expression (Yamada et al., 2001). In a subsequent study, the same group injected replication-defective HSV vectors encoding Bcl-2 or GDNF, or a combination of both, into the spinal cord 30 min after avulsion of L4, L5 and L6 spinal nerves. Transduction of motoneurons with either one of the vectors resulted in a substantial increase in the number of surviving motoneurons but again not in the preservation of ChAT expression. When a combination of HSV-Bcl-2 and HSV-GDNF was injected, however, the number of ChAT-positive cells in the lesioned ventral horn increased substantially (Natsume et al., 2002).

Inoculation of the gastrocnemius muscle with HSV-1 encoding LacZ has been reported to transduce motoneurons in the mouse spinal cord and transgene expression was detectable up to 1 month (Keir et al., 1995). In the rat, expression of β-galactosidase following direct injection of a replication-incompetent HSV-LacZ vector backbone into the spinal cord lasted for 1 month. Moreover, the vector was transported from the site of injection in the spinal cord to the cell bodies, i.e., dorsal root ganglia (DRG), brainstem and areas of the hind- and midbrain (Lilley et al., 2001).

Many studies using HSV vectors have focussed on the peripheral nervous system (PNS), since wild type HSV naturally targets with high efficiency to neuronal cell bodies in sensory ganglia (Leib and Olivo, 1993). This is due to the high level of its preferred high-affinity receptor nectin-1 on peripheral sensory axons (Mata et al., 2001). HSV vectors are retrogradely transported from the site of injection. This property makes HSV very well suited for gene therapeutic applications for conditions that affect the PNS including motoneuron disease and peripheral neuropathies. Although the above studies show the potential of HSV vectors in gene therapeutic approaches in the nervous system, the usefulness of the HSV vector for in vivo gene therapy in the CNS is limited due to the relatively poor infection efficiency of neuronal populations other than sensory neurons, the transient expression of inserted genes due to latency-associated shut-off of promoters and the virus-associated toxicity (Wood et al., 1994a; Wood et al., 1994b; Verhaage net al., 1995; Kennedy, 1997; Lilley et al., 2001). Considerable tissue damage and evidence of viral replication was also detected following application in the spinal cord (Keir et al., 1995).

Adenoviral vectors

In 1993, there were four independent reports on the generation of a first generation adenoviral (AdV) vector system for gene transfer to the nervous system (Akli et al., 1993; Bajocchi et al., 1993; Davidson et al., 1993; Le Gal La Salle et al., 1993). All of these studies showed efficient transduction of astrocytes, oligodendrocytes, ependymal cells and neurons with a replication-defective AdV vector encoding β-galactosidase (AdV-LacZ). These first generation AdV vectors were rendered replication defective through deletions in the E1-region, resulting in a significantly diminished albeit not complete deactivation of the expression of other early (E) and late (L) genes needed for adenoviral replication (Hermens and Verhaagen, 1998).

Adenoviral vectors encoding CNTF or NT-3 used to infect cultured primary astrocytes and fetal DRGs, respectively, yielded the production of sufficient amounts of functional protein to stimulate neurite outgrowth in vitro (Smith et al., 1996; Dijkhuizen et al., 1997). In addition, motoneurons plated onto astrocyte monolayers pretreated with AdV-CNTF showed a 2- to 4-fold increase in ChAT activity compared to motoneurons grown on astrocyte monolayers pretreated with AdV-LacZ (Smith et al., 1996).

Baumgartner and Shine (1998) tested AdV vectors for GDNF, BDNF, NGF, NT-3 and CNTF in a *sciatic nerve transection* model. The vectors were injected into the hind limb muscles of newborn rats, and taken up by the peripheral nerves for retrograde transport to the lumbar spinal cord thereby providing transgene expression in motoneurons. The use of AdV-GDNF protected a significant number of sciatic nerve motoneurons against axotomy-induced cell death at 1 and 3 weeks compared with AdV-LacZ treatment, though at 6 weeks, this neuroprotective effect did not persist (Baumgartner and Shine, 1998). This observation is in line with their earlier observations with AdV-GNDF, -CNTF and -BDNF injected into the facial nerve one week after unilateral axotomy of the facial nerve. Rescue of motoneurons was also possible following direct injection in the facial nerve (Gravel et al., 1997). In a study using a sciatic nerve crush, Dijkhuizen et al. (1998) also studied the possibility to transduce the Schwann cells near the injured area. Injection of Ad-LacZ into the uninjured sciatic nerve resulted in high levels of transgene expression lasting at least up to 12 days. The number of transduced Schwann cells gradually declined, however, until only a few transduced Schwann cells were present at 45 days. Infusion of the vector at the time of, or immediately after a crush of the nerve however failed to result in significant transduction (Dijkhuizen et al., 1998). When the vector was administered one day after the crush, this problem was overcome and the number of transduced Schwann cells became similar to that observed in the uncrushed nerve. Thus, AdV vectors can deliver transgenes for relatively short-lived, local production of neurotrophic factors in injured peripheral nerves.

Adenoviral vectors have been used in a number of spinal cord repair studies (Zhang et al., 1998; Huber et al., 2000; Romero et al., 2001). In a *dorsal root injury* model, in which the lumbar dorsal roots of adult rats were severed and reanastomosed Ad-LacZ or Ad-NT-3 were injected into the ventral horn of the lumbar spinal cord 2 weeks after the lesion. NT-3 attracted the injured sensory axons to regrow into the spinal cord (Zhang et al., 1998). Transgene expression was found in glial cells as well as motoneurons for up to 40 days. The regenerated sensory axons were mainly found in the gray matter, and had penetrated as deep as lamina V. Romero et al. (2001) applied AdV-NGF to the lumbar spinal cord 2 weeks after dorsal root avulsion. Robust axonal regeneration into normal as well as ectopic locations was observed within the dorsal spinal cord. This regeneration resulted in near-normal recovery of sensory function. In addition, the observed functional recovery was completely abolished after recutting the sensory (dorsal) roots. Injections of AdV-NGF also resulted in extensive sprouting of noninjured sensory axons, which may have led to the observed symptoms of hyperalgesia and chronic pain (Romero et al., 2000; Romero et al., 2001).

Gene transfer in the *dorsohemisected rat spinal cord* was investigated by injecting AdV-LacZ (Blits et al., 2002) or an AdV vector encoding green fluorescent protein (AdV-GFP) (Huber et al., 2000). Neurons, astrocytes, oligodendrocytes and peripheral cells infiltrating the lesion site were transduced by the adenoviral vector. In the study by Blits et al. (2002), transduction largely failed when the injection was given immediately following the lesion, but was prominent when injected a few hours prior to the lesion or 1 week later. In the study by Huber et al. (2000), vector injections were given 1 day prior lesion. AdV-GFP-transduced cells were then found within 3 mm on each side of the lesion. Both studies revealed that the high expression between 3 and 8 days practically had ceased by 21 days. Both virus-mediated inflammatory responses, direct toxicity, promotor downregulation and loss of the episomally present transgene has been put forward to explain the transient transgene expression with AdV vectors (see below).

Direct injections of AdV-BDNF and AdV-NT-3, placed 1 h prior and distally to a *lesion of the corticospinal tract* (CST) in the spinal cord, failed to induce significant regrowth of CST fibers, but retraction of CST end bulbs was less at 14 weeks compared to AdV-LacZ treatment (Blits, 2002; Blits, Boer and Verhaagen, unpublished data). It was postulated that the exposure of the injured CST axon tips to BDNF and NT-3 was not sufficient due to the distance between end bulbs and the site of transduction distal from the neural scar, or that adenoviral vector-mediated expression of neurotrophins lasted not long enough (Fig. 2). Nevertheless, amelioration of the number of toe drags was observed compared to the AdV-LacZ-treated animals. This indicates that the beneficial effects are not related to regrowth of the CST.

Recently, AdV-NT-3 was delivered to the spinal motoneurons through retrograde transport via the

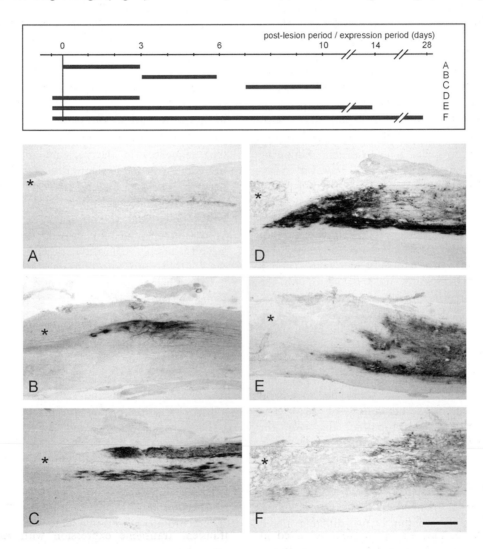

Fig. 2. In vivo adenoviral vector-mediated transgene expression in the dorsally hemisected rat spinal cord after various time intervals following the lesion. The scheme in the top panel shows the time points of AdV-LacZ injection (2×10^7 pfu in 1 µl) 5 mm distal to the center of the lesion site. Results are shown in panels A through F. The cords were sectioned longitudinally at 20 µm and representative results of the X-gal-staining for the expressed transgene β-galactosidase photographed. Three days following injection, transduction of cells was nearly absent when vector injection was placed at the time of lesioning (A). When injected at day 3 (B) or day 7 (C) postlesion transduction was prominent. A similar, or somewhat better transduction of cells was obtained when the AdV-LacZ injection was placed 1 h prior surgery (D). This expression however faded away in time with decreasing levels at 14 days (E) and nearly no expression at 28 days (F). Lesion, injection and staining procedures of this study have been published by Blits (2002). Bar indicates 500 µm. Asterisks denote lesion areas.

sciatic nerve 2 weeks after a contralateral CST lesion above the pyramidal decussation (Zhou et al., 2003). Expression of NT-3 was detected up to 3 weeks (longest time point studied) but the amount of NT-3 had decreased compared to the 1 week time point. No effect on sprouting of CST fibers was seen in unlesioned animals, but NT-3 production induced sprouting of CST fibers from the unlesioned to the contralateral lesioned side across the midline.

The studies summarized above, applying AdV vectors for the overexpression of neurotrophic factors in the sciatic nerve, spinal root and spinal cord injury models, can be regarded as proof of the principle that cell survival and nerve fiber regeneration can be stimulated by AdV vector-mediated neurotrophic gene transfer. Following direct application of AdV vectors, however, transgene expression is only transient and toxicity has now been documented extensively (Zhang and Schneider, 1994; Byrnes et al., 1995; Byrnes et al., 1996; Wood et al., 1996), in particular following injection of higher doses of vector particles (Hermens and Verhaagen, 1997; 1998; Huber et al., 2000). Though recombinant adenoviruses used as vector for the nervous system were initially rendered replication-deficient through deletions in the E1-region, at high multiplicities of infection (MOI) the E1 region becomes dispensable for replication, in particular in vivo (Jones and Shenk, 1979; Nevins, 1981). As a result E1-independent activation of the E2 promoter can lead to replication of wild-type adenoviral DNA, followed by accumulation of viral gene products that lead to cytopathic effects on the transduced cells (Zhang and Schneider, 1994) and activation of the host immune system (Byrnes et al., 1995). Also direct cytotoxic and immunogenic effects of AdV vector-mediated gene transfer have been shown in a number of studies. Injection of AdV-LacZ into the striatum of two different inbred rat strains led to inflammation in this brain area. Severity of the inflammation seemed to depend on the rat strain used (Byrnes et al., 1995). Although inflammation was not evident at 2 months after injection, transgene expression was greatly reduced at this time. Similar inflammatory responses were found after injection of the vector in the brain of adult mice (Kajiwara et al., 1997), whereas also reexposure to the AdV vector leads to severe inflammation and microglial activation resulting in local demyelination and further decrease in transgene expression (Byrnes et al., 1996). Transduction of the ventral root motoneurons with AdV vectors clearly results in neuronal death. In addition, the adenoviral genome exists as an episome in the infected host cell that leads to additional decline of transgene expression when the cells become mitotically active, e.g., activated Schwann cells following a peripheral nerve lesion.

Adeno-associated viral vectors

Problems of immunogenicity and direct neuronal toxicity of HSV and AdV vectors, mainly due to wild type sequences in the recombinant genome were overcome with the development of AAV vectors. In the current AAV vector system, the complete viral protein-encoding part of the genome has been deleted and is substituted with a transgene expression cassette. To provide the *rep* and *cap* genes for rescue, replication and encapsidation of the AAV DNA into infectious AAV vector particles initially a helper plasmid containing the AAV-coding sequence flanked by Ad type 5 terminal repeats and super-infection with AdV as a helper virus in a human cell-line was used (Kaplitt et al., 1994; McCown et al., 1996). AdV was removed from the resulting AAV vector stocks by heating the stocks at 56°C for 1 h or by density gradient centrifugation (reviewed by Hermens and Verhaagen, 1998). Todate, a helper virus-free production method is used with 2 helper plasmids (reviewed by Monahan and Samulski, 2000). AAV-LacZ injection in several areas of the rat brain resulted in efficient transduction of cells and LacZ expression lasted for months without cytopathic effects (Kaplitt et al., 1994). The efficiency of AAV vectors to transfer genes to the rat spinal cord has been tested as well. Following AAV-GFP injections in the rat cervical spinal cord, most GFP-positive cells were neurons which lasted for months (Peel et al., 1997; Dijkhuizen, 1999; Ruitenberg et al., 2002; Blits et al., 2002).

The efficient transduction of CNS neurons with AAV vectors encoding neurotrophic factors have been shown after injection into the septum, where overexpression of NGF and BDNF resulted in upregulation of ChAT and Trk expression

(Klein et al., 1999). In recent years several studies applied AAV vector-mediated overexpression of neurotrophic factors in spinal cord lesion models.

After a cervical *lesion of the RST*, rats received an injection with either AAV-BDNF or AAV-LacZ in the red nucleus. AAV-directed BDNF prevented lesion-induced atrophy of the perikarya in this nucleus (Ruitenberg et al., unpublished observations, 2003). The effect was not yet visible at 2 weeks but AAV-BDNF treatment fully reversed atrophy at 5 weeks postlesion. The lack of an effect at 2 weeks was thought to be caused by the fact that AAV-mediated expression of BDNF is starting slowly but substantially increased between 2 weeks and 1 month after AAV vector injection (Ruitenberg et al., unpublished observations, 2003). Also following chronic RST lesions, AAV-BDNF injections in the red nucleus 18 months following the lesion resulted in reversal of the neuronal atrophy compared to AAV-GFP injections. Thus, viral vector-mediated neurotrophin treatment can potentially also be effective in 'old' lesions.

In an attempt to lure the long descending fibers to cross Schwann cell implants bridging the cavity of a *completely transected rat spinal cord*, AAV-BDNF or -NT-3 have been injected distally from the implant (Blits et al., 2003). Up to four months postinjection numerous cells, predominantly neurons, expressed BDNF, NT-3 or control GFP transgene. The many (anterogradely traced) axons that were observed to regrow into the Schwann cell implant (Xu et al., 1995) failed however to cross the distal implant-host spinal cord interface (Blits et al., 2003). Retrograde tracing demonstrated however that twice as many neurons with processes extending towards the Schwann cell implant were present in the distal second lumbar segment compared to controls. Since assessment of functional locomotor recovery with the BBB locomotor test revealed that at 7 weeks after surgery the mean score of animals treated with AAV-BDNF and AAV-NT-3 was significantly higher than those for the control groups, it was postulated that BDNF and NT-3 caused modifications in the local axonal circuitry beneficial for recovery of locomotor function. Following the use of AdV-BDNF and -NT-3 in this experimental paradigm, short-term BDNF and NT-3 overexpression had no effect (Blits, 2002).

After unilateral chronic constriction injury (CCI) of the sciatic nerve, a model for *chronic neuropathic pain*, the effect of AAV-mediated overexpression of BDNF in the spinal cord was investigated (Eaton et al., 2002). When AAV-BDNF vector was injected into the dorsal horn of the rat thoracic spinal cord 1 week after injury, allodynia and hyperalgesia were permanently reversed as compared to control AAV-GFP-treated animals. Chronic neuropathic pain after peripheral nerve injury seems therefore beneficially sensitive to BDNF, as was also shown following intraspinal implantation of a rat raphe neuronal cell line genetically modified to produce BDNF (Cejas et al., 2000).

Above studies, although still limited in number, showed that the AAV vector system is an efficient, nontoxic tool for gene transfer in spinal cord repair strategies.

Lentiviral vectors

The first LV vector reported to stably transduce nondividing cells in vivo was based on the human immunodeficiency virus (HIV) (Naldini et al., 1996a; Naldini et al., 1996b). Recently primate and nonprimate lentiviral (LV) vectors have been developed from simian and feline immunodeficiency viruses and equine infectious anemia virus (Mitrophanous et al., 1999; Trono, 2000; Mazarakis et al., 2001). The LV and AAV vectors appear to fulfil most of the requirements for successful gene transfer into the CNS (Hermens and Verhaagen, 1998). The performance of LV vectors is superior to that of AdV vectors with regard to stability of gene expression and absence of an inflammatory response. The core particles of this viral vector system are usually pseudotyped with vesicular stomatitis virus glycoprotein G (VSV-G), which confers a broad tropism and allows to infect a large variety of different cell types (Naldini et al., 1996b; Watson et al., 2002). The vector has been shown to transduce both dividing and nondividing cells including neurons and to insert transgenes into the chromosomes of their target cells. Transduction with LV results in long-term expression of the transgene (Naldini et al., 1996a). LV-LacZ injected into the rat brain resulted in expression of β-galactosidase up to 6 months, without decrease in transgene expression and no detectable pathology (Blomer et al., 1997). In nonhuman primate models for Parkinson's disease, LV-GDNF was injected into the striatum and substantia nigra of nonlesioned aged rhesus monkeys or young adult rhesus monkeys treated with MPTP 1 week prior to the injection of the vector (Kordower et al., 2000). Extensive GDNF expression was seen in animals for up to 8 months. In aged monkeys, LV-GDNF augmented dopaminergic function. In MPTP-treated monkeys, LV-GDNF reversed functional deficits and prevented nigrostriatal degeneration.

No study has been reported yet on the application of LV vectors in spinal cord injury. Hottinger et al. (2000) delivered GDNF to motoneurons of the mouse facial nucleus that were axotomized 1 month later by a facial nerve transection. LV-directed expression of GDNF in the axotomized motoneurons completely protected against cell death observed 3 months later (95% of the motoneurons were present as demonstrated by both Nissl staining and ChAT immunoreactivity). In addition, GDNF prevented lesion-induced neural atrophy and maintained proximal motoneurons, despite the lack of target cell reinnervation. Only a mild inflammation consisting in minor perivascular cuffing and in aggregation of lymphocytes around the injection site was observed following LV vector application (Hottinger et al., 2000).

The LV vector system proofs to be a valuable tool in gene transfer to the CNS also because it has the advantage of a large cloning capacity. The vector integrates its genomic content into the host genome warranting expression of transgenes in mitotic active cells and has the ability to efficiently transduce postmitotic cells, e.g., neurons. There is, however, some controversy about the bio-safety of the LV vector system because belonging to the family of retroviruses, LV vector-mediated gene transfer might lead to insertional mutagenesis as seen with MLV-based ex vivo gene therapy (Check, 2002; Kaiser, 2003). Furthermore, it has been reported that HIV-1 inserts more often into genes than into noncoding sequences of the mammalian genome (Schroder et al., 2002).

Ex vivo gene transfer of neurotrophic factors to the injured CNS

Advances in neurotransplantation techniques and the refinement of gene delivery vector systems inspired the idea of utilizing ex vivo gene therapy: transplanting cells that are genetically modified to express exogenous therapeutic gene products. There are several criteria that characterize the ideal cell type for placement in the site of CNS/spinal cord injury. Such a cell would be (i) readily obtainable, (ii) survive grafting to the CNS for extended periods of time, (iii) produce a milieu or matrix that would support the growth of all classes of axons at a site of injury, (iv) be a suitable target cell for genetic modification, and (v) pose no ethical controversy following a possible clinical application (Lacroix and Tuszynski, 2000). At present, a number of cells have been identified that are promising candidates for genetic engineering and subsequent transplantation into the injured CNS to provide neuroprotection and stimulate neuroregeneration (see Table 1). These cells include primary fibroblasts, Schwann cells, olfactory ensheating glia cells and neural stem cells. The use of ex vivo transduced cells would be especially beneficial to improve experimental approaches using implantation of cellular matrices to bridge a nerve lesion gap or spinal lesion cavities. The use of engineered cells secreting the therapeutic protein may then serve as a better conduit for regrowth of damaged axons.

Primary fibroblasts

Most work to date has used primary fibroblasts as cellular vehicles for gene transfer in models of CNS injury. Fibroblasts are easily sustainable in vitro, they survive implantation to the nervous system, and they secrete molecules such as collagen and fibronectin that can reconstitute an extracellular matrix at the lesion site to which many types of neurons can attach and extend axons (Lacroix and Tuszynski, 2000). The therapeutic potential of genetically modified fibroblasts has been shown in several animal models for CNS disease. Cell line-derived fibroblasts, genetically modified with a retroviral vector (MLV) to secrete NGF and implanted into the brain of rats that had received surgical lesions of the fimbria-fornix, survived and produced sufficient NGF to prevent degeneration of cholinergic neurons. In addition, the treatment induced enhanced sprouting of axons in the direction of the cellular source of NGF (Rosenberg et al., 1988). Because immortalized fibroblasts implanted in the CNS can give rise to tumors (Horellou et al., 1990), follow up studies used implanted primary skin fibroblasts to circumvent the risk of tumorigenesis. The fibroblasts were genetically engineered to produce NGF and proved to be beneficial in a model that assesses the regeneration of cholinergic neurons of the rat septum. No transplant-derived tumors were observed (Kawaja et al., 1992).

Initially, NGF-secreting fibroblasts have been implanted to the uninjured adult rat spinal cord, survived for up to 1 year and induced a robust ingrowth of spinal neurites (Tuszynski et al., 1994). Control implants and basic fibroblast growth factor (bFGF)-producing implants promoted less neurite growth. Neurites penetrating the NGF implants were of sensory origin based on calcitonin gene-related peptide (CGRP) labeling. In order to see if NGF-secreting fibroblasts could induce sprouting from axotomized spinal neurons, these cells were implanted to sites of midthoracic *spinal cord dorsal hemisection* lesions (Tuszynski et al., 1996; Tuszynski et al., 1997). Recipients of NGF-secreting grafts showed deficits on locomotion over a wire mesh not different from control animals after 3 months. NGF-secreting implants elicited specific sprouting from spinal primary sensory afferent axons, local motor axons, and putative coerulospinal axons, but no specific responses from corticospinal axons were observed. Axons responding to NGF penetrated into the implant but did not exit into distal spinal cord tissue (Tuszynski et al., 1997). Fibroblasts secreting NGF have also been implanted into the chronically injured adult dorsohemisected rat spinal cord (Grill et al., 1997b). Animals with implants of syngenic fibroblasts secreting NGF placed 1 to 3 months after lesioning showed growth of coerulospinal and primary sensory axons of the dorsolateral fasciculus into the graft/lesion site indicating that axons retain responsiveness to neurotrophins in chronic stages of spinal cord injury (Grill et al., 1997b). In a study by Grill et al. (1997a), primary fibroblasts genetically modified to secrete NT-3 were grafted to acute dorsal hemisection lesion cavities. After 3 months, significant partial functional recovery occurred in NT-3 fibroblast grafted animals. In this study an increase in corticospinal axon growth at and for up to 8 mm distal to the injury site was measured. CST axons

extended through the spinal cord gray matter but not through the implant nor into white matter (Grill et al., 1997a).

In a *spinal cord contusion* model, BDNF and NGF secreting implants have been reported to accelerate functional recovery after implantation into the lesion cavity (Kim et al., 1996). Primary fibroblasts secreting BDNF were placed into the lesion cavity after a partial cervical hemisection lesion that completely interrupted one RST (Liu et al., 1999b). One to two months after lesion and implantation, anterograde and retrograde tracing showed RST axons regenerating through and around the transplants. These axons grew for long distances within white matter caudal to the transplant, and terminated in spinal cord gray matter regions that are the normal target areas of RST axons. Behavioral testing showed significant recovery of forelimb function in animals that received BDNF-producing implants and this recovery was abolished after a second lesion transecting the regenerated axons (Liu et al., 1999b; Kim et al., 2001; Liu et al., 2002). Besides inducing axonal regrowth of the RST, implantation of BDNF-secreting fibroblasts prevented neuronal atrophy and death in the red nucleus, indicating retrograde beneficial effects of the treatment (Liu et al., 2002).

Primary fibroblasts expressing the reporter gene LacZ, NT-3, BDNF, CNTF, NGF or bFGF, have been grafted to the *contused adult rat spinal cord* (McTigue et al., 1998). Ten weeks after injury, NT-3 and BDNF implants contained significantly more axons than control or other growth factor-producing implants. In addition, more myelin basic protein-positive profiles were detected in these implants, indicating enhanced myelination of ingrowing axons. To see if this myelination was associated with increased proliferation of oligodendrocyte lineage cells, BrdU labeling was applied and BDNF and NT-3-secreting implants were indeed found to have more BrdU-positive oligodendrocytes than control LacZ implants (McTigue et al., 1998).

Prior to implantation, primary fibroblasts have also been genetically modified by AdV vector-mediated gene transfer to express the reporter genes GFP and LacZ (Liu et al., 1998). Cell density and morphology of the implants were comparable with that of retrovirally transduced implants, suggesting

TABLE 1
Neurotrophins augment regeneration of spinal cord pathways[a]

Neurotrophin/ Preferred receptor	Experimental SCI	Delivery method	Additional intervention	Reported regenerative growth	Reference
NGF/TrkA				*sensory (DR), coerulospinal axons:*	
	dorsal HX (T7)	ex vivo engineered fibroblasts		into the graft	Tuszynski et al., 1996; Grill et al., 1997a
		ex vivo engineered Schwann cells		into the graft (partial myelination)	Weidner et al., 1999b
				sensory (DR) axons:	
	DR crush (L4-5)	in vivo gene transfer (AdV vector)		into the spinal cord	Romero et al., 2001
	DR crush (C5-C7)	intrathecal infusion		into the spinal cord	Ramer et al., 2000
	DR crush (L4)	intraparenchymal infusion		into the spinal cord	Oudega and Hagg, 1996
	DC lesion (T10)	intraparenchymal infusion	peripheral nerve graft	into the graft and beyond (DC)	Oudega and Hagg, 1996
BDNF/TrkB				*supraspinal axons:[b]*	
	Transection (T8)	ex vivo engineered Schwann cells		across the transection site[1,2,4,6]	Menei et al., 1998
	dorsal HX (C2/C3)	soaked gelfoam	peripheral nerve graft	into the graft[1,2,4]	Ye and Houle, 1997
	dorsal HX (C4/T6)	soaked gelfoam	fetal spinal cord graft	into the graft[1,2,5]	Bregman et al., 1997
	transection (T8)	distal infusion (icw NT-3)	Schwann cell graft	into the graft[1-4,6]	Xu et al., 1995
	transection (T6)	(intrathecal) infusion	fetal spinal cord graft	into the graft and beyond[1-6]	Coumans et al., 2001
				serotonergic (raphe-derived) axons:	
	dorsal HX (C3/C4)	ex vivo engineered fibroblasts		into the graft	Jin et al., 2000
				RST axons:	
		intraparenchymal infusion		into, around and beyond the graft	Liu et al., 1999b; Kobayashi et al., 1997
			peripheral nerve graft	into the graft	Kwon et al., 2002
				spinal axons (source unknown):	
	dorsal HX (T8)	intraparenchymal infusion (icw NT-3)	Schwann cell graft	into the graft and beyond	Bamber et al., 2001
				distal intraspinal axons:	
	transection (T9)	in vivo gene transfer (icw NT-3) (AAV vector)	Schwann cell graft	towords the graft	Blits et al., 2003
				sensory (DR) axons:	
	DC lesion (T9)	intraparenchymal infusion	peripheral nerve graft	into the graft and beyond (DC)	Oudega and Hagg, 1999
NT-3/TrkC				*supraspinal axons:[b]*	
	dorsal HX (C2/C3)	soaked gelfoam	peripheral nerve graft	into the graft[3,4]	Ye and Houle, 1997
	dorsal HX (C4/T6)	soaked gelfoam	fetal spinal cord graft	into the graft[1,2,5]	Bregman et al., 1997
	transection (T8)	distal infusion (icw BDNF)	Schwann cell graft	into the graft[1-4,6]	Xu et al., 1995
	transection (T6)	(intrathecal) infusion	fetal spinal cord graft	into the graft and beyond[1-6]	Coumans et al., 2001
				CST axons:	
	dorsal HX (T7)	ex vivo engineered fibroblasts		underneath and beyond the graft	Grill et al., 1997b
	dorsal HX (T12)	ex vivo engineered peripheral nerve		underneath and beyond the graft	Blits et al., 2000
	dorsal HX (T8)	NT-3 containing collagen matrix	collagen implant	into the collagen matrix	Houweling et al., 1998b
	dorsal HX (T8)	single injection/soaked gelfoam	IN-1 treatment	around the lesion,	Schnell et al., 1994

transection (T9)	intraparenchymal infusion (icw BDNF)	Schwann cell graft	limited growth beyond *spinal axons (unknown source):* into the graft and beyond	Von Meyenburg et al., 1998
DR lesion (L4-6)	in vivo gene transfer (icw BDNF) (AAV vector)	Schwann cell graft	*distal intraspinal axons:* towords the graft	Bamber et al., 2001 Blits et al., 2003
DC crush (T6)	in vivo gene transfer (AdV vector)		*sensory (DR) axons:* into the spinal cord beyond the DC lesion	Zhang et al., 1998
DC lesion (T9)	intrathecal infusion intraparenchymal infusion	peripheral nerve graft	into the graft and beyond (DC) *supraspinal axons:*	Bradbury et al., 1999 Oudega and Hagg, 1999
NT-4/TrkB ldorsal HX (C4/T6)	soaked gelfoam	fetal spinal cord graft	into the graft.[1,2]	Bregman et al., 1997

[a]Only those studies that investigated anatomical regeneration of lesioned CNS pathways in the spinal cord upon neurotrophin administration were included in this table. Abbreviations: AAV, adeno-associated virus; AdV, adenovirus; C, cervical; DC, dorsal column; DR, dorsal root; HX, hemisection; icw, in combination with; IN-1, monoclonal antibody against inhibitory myelin components NI-35/250; L, lumbar; T, thoracic.

[b]Supraspinal axons include spinal cord pathways that originate from the following brain regions: (1) locus coeruleus, (2) raphe nuclei, (3) red nucleus, (4) reticular formation, (5) sensorimotor cortex, (6) vestibular complex. If applicable, the observed regenerative responses of each of these pathways as described in the individual studies mentioned are indicated by their designated numbers.

similar long-term cell survival. However, the level of transgene expression rapidly decreased with time as described also for direct intraparenchymal application (see above). A mild immunological response to the grafted cells was reported, which together with episomal presence of the AdV genome during cell division, accounted for this loss in transgene expression (Liu et al., 1998). The AAV or LV vector systems have not been used for genetic modification of fibroblasts, though other cell types have been efficiently transduced with the LV vector system (Englund et al., 2000).

Schwann cells

The use of Schwann cells in CNS regeneration studies is likely to have an advantage not shared with primary fibroblasts, namely the fact that Schwann cells are myelinating cells of the peripheral nervous system. In order to regain function after spinal cord injury, the sole regeneration of injured nerve fibers is not enough. Real repair would also require remyelination of regenerated axons. Schwann cells play a crucial role both in supporting peripheral axon regeneration and remyelinating axons. They support regeneration by secreting trophic factors, extracellular matrix molecules like laminin and collagen, and they express cell adhesion molecules (Snyder and Senut, 1997). In addition, Schwann cells provide physical support to injured peripheral axons by their alignment into the so-called bands of Büngner that define pathways for axonal regrowth (Ide, 1996). These characteristics of Schwann cells make them promising candidates for transplantation to the injured CNS and several studies provide strong evidence for spinal regeneration following the use of cultured Schwann cell preparations or of Schwann cell-containing nerve pieces (Xu et al., 1995; Bamber et al., 2001). Increasing the trophic milieu in (or near) the implant might promote the extent of axon regrowth.

To test if Schwann cell-induced axonal growth in the CNS could be increased (Xu et al., 1995), cultured primary adult rat Schwann cells were genetically engineered to secrete NGF, by transducing them with a retroviral vector (MLV) (Tuszynski et al., 1998). These cells were then implanted to the midthoracic spinal cord of adult rats where they survived for up to 1 year. During the survival time some NGF secreting grafts slowly increased in size compared to nontransduced grafts. NGF-transduced transplants became more densely penetrated by primary sensory nociceptive axons originating from the dorsolateral fasciculus of the spinal cord than control implants. Ultrastructually, axons in both NGF and control implants became extensively myelinated by Schwann cells (Tuszynski et al., 1998). When applied in the lesion cavity of a dorsal hemisection of the rat spinal cord, a significant increase in growth of spinal cord axons into the transplant was observed (Weidner et al., 1999). More dense penetration of coerulospinal axons and of central processes of primary afferent sensory axons were found in 1 month compared to that seen in untransduced implants. In addition, these axons became ensheathed and in some instances remyelinated by Schwann cells. In this study, the implanted Schwann cells exhibited a phenotypic and temporal course of differentiation into a myelinating state while aligning spontaneously. At 3 days after grafting Schwann cells were still in a undifferentiated or nonmyelinating state, at 2 weeks they had upregulated the cell adhesion molecule L1, a marker for differentiated nonmyelinating Schwann cells. At 3 weeks, P0 glycoprotein, a major constituent of peripheral myelin, was detected, indicating that a number of Schwann cells had adopted a myelinating phenotype. As no differences for Schwann cell markers were found between NGF-secreting and control grafts, NGF itself did not appear to modulate the Schwann cell myelinating phenotype. The observed temporal course of Schwann cell differentiation after grafting to a CNS injury site recapitulated patterns seen after PNS injury, indicating that a physiological response of Schwann cells to injury is retained at an ectopic site, i.e., in the CNS (Weidner et al., 1999).

Schwann cells retrovirally modified to secrete BDNF and implanted as trails in and distal to the transection site of the adult rat spinal cord attracted more dorsal root ganglion and spinal and supraspinal fibers to sprout than in normal Schwann cell implants (Menei et al., 1998). The Schwann cell trail was maintained for at least 1 month and most fibers sprouted at the level of the transection. Many fibers that crossed the transection site and growing into the BDNF Schwann cell trail were CGRP-positive, and identified as dorsal root fibers. The BDNF-secreting Schwann cells, however, appeared not to myelinate the regenerated axons based on absence of P0 staining in contrast to normal Schwann cells. It was postulated that the Schwann cells in the presence of BDNF might be maintained in a less differentiated state, one that promotes fiber regeneration but not myelination (Menei et al., 1998). These studies demonstrate that supraphysiologic levels of neuro-trophic factor production from Schwann cells augment the regenerative potential of injured supra-spinal and other axons. In addition, NGF-transduced Schwann cells can appropriately remyelinate regen-erating axons and guide and provide permissive surfaces for growth of central axons.

Peripheral nerve bridges

The permissiveness of peripheral nerve for axon growth (David and Aguayo, 1981; Campbell et al., 1992; Anderson et al., 1998) is mainly attributed to the presence of Schwann cells. Nerve transplants were applied in the completely transected adult rat spinal cord (Cheng et al., 1996). Significant functional recovery as well as regrowth of CST fibers through the peripheral nerve graft into the distal cord was reported (Cheng et al., 1996). However, these results have never been reproduced thus far. Because ingrowth of CNS nerve fibers in such transplants is limited as not all neuronal populations are equally responsive to the molecular environment in the grafts (Anderson et al., 1998), genetic modification of peripheral nerve bridges to overexpress neurotrophic proteins should, in principle, improve the permissive properties of peripheral nerve transplants.

Direct AdV-mediated gene transfer of peripheral nerves has been achieved by injection procedures. The Schwann cells of cultured intercostals nerves displayed transgene expression up to 30 days (Blits et al., 1999). Ex vivo AdV-LacZ-transduced auto-logous intercostal nerve pieces were subsequently placed as bridge implants in the hemisected adult rat spinal cord (Blits et al., 1999). The implants survived and numerous β-galactosidase-expressing cells were observed at 1 week after implantation, although at 2 weeks the number of cells expressing the transgene

had declined. In a follow-up study, intercostal nerves were transduced with AdV-NT-3 and CST regenera-tion and motor recovery was studied (Blits et al., 2000). Neurofilament staining revealed ingrowth of many fibers from the host spinal cord into mock-, AdV-LacZ- or AdV-NT-3-transduced implants, but axons of the anterograde traced CST failed to regenerate into the intercostal nerve implants. Instead, CST axons grew underneath the implants in the spared gray matter and did so 3- to 4-fold better in the case of AdV-NT-3 implants. Regrowth of CST axons occurred over more than 8 mm distal to the lesion site. This effect was similar as described in a study using NT-3-producing fibroblasts (Grill et al., 1997a). Animals implanted with Ad-NT-3-trans-duced peripheral nerves also exhibited improved function of the hind limbs after 8 weeks. Thus, transient overexpression of NT-3 in peripheral nerve tissue bridges during the first 2 weeks is sufficient to stimulate regrowth of CST fibers and to promote recovery of hind limb function. Taken together, combining an established neurotransplantation approach with viral vector-mediated gene transfer promotes the regrowth of injured CST fibers through gray matter and improves the recovery of hind limb function.

Olfactory ensheathing glia cells

Recently, there has been increasing interest to use olfactory ensheathing glia (OEG) cells as cellular conduits to stimulate regeneration in the damaged CNS. Normally, OEG cells guide new primary olfac-tory axons growing from the olfactory epithelium in the periphery, towards their CNS target area, the olfactory bulb glomeruli. This interesting feature of OEG, i.e., the support of axonal growth within the CNS, has turned them into attractive candidates for neural transplantation purposes in damaged areas of the CNS. At present, a significant number of studies have demonstrated the potential of OEG cells to support regenerative growth of axons in the lesioned spinal cord (Ramon-Cueto and Nieto-Sampedro, 1994; Li et al., 1998; Ramon-Cueto et al., 1998), often accompanied with a certain degree of functional recovery (Li et al., 1997; Ramon-Cueto et al., 2000). Although the results obtained were encouraging, observed regenerative growth through these grafts was still far from optimal and did not result in the (complete) anatomical restoration of injured CNS pathways. The fact that OEG cells are CNS in origin, well-tolerated within the spinal cord and naturally express a variety of growth-promoting molecules (Boruch et al., 2001) makes these glial cells perhaps one of the most attractive cellular platforms for genetic engineering, with the aim to create a more conducive environment for CNS axon regeneration.

Recently, AdV and LV vectors have been successfully used to study the possibility of transducing OEG cells prior to implantation in the lesioned spinal cord (Ruitenberg et al., 2002). Reportedly, ex vivo transduction of primary OEG with AdV vectors resulted in a transient pattern of transgene expression following implantation in the lesioned spinal cord whereas persistent transgene expression was observed in LV vector-transduced OEG implants for at least up to 4 months.

Genetic modification of OEG cells to make these cells secrete neurotrophic factors could significantly enhance their regeneration-stimulating potential as was previously observed for engineered Schwann cell grafts (Menei et al., 1998). The main advantage of OEG implants over Schwann cell grafts is, however, that they do not upregulate inhibitory proteoglycans at the graft-host interface (Lakatos et al., 2000), nor seem to entrap regenerating axons at the implantation site (Ramon-Cueto et al., 1998). Recent results have indicated that the transgenic OEG, engineered with NT-3- and BDNF-encoding AdV vectors, in principle can stimulate regrowth of lesioned spinal axon populations that are normally refractory to regenerate (Ruitenberg et al., 2002; 2003). Behavioral analysis revealed that neurotrophin-transduced OEG implants improved hinbdlimb performance during horizontal rope walking. This locomotion performance moreover did directly correlate with lesion size, suggesting that the neuroprotective effects of the implants contributed to the level of functional recovery.

Neural stem cells

Stem cells are widely investigated for the development of cell therapies in a variety of diseases. Embryonic stem (ES) cells as well as (somatic) neural stem cells have been experimentally used to repair CNS and to promote regeneration. A neural stem cell has the potential to generate all cell types in the CNS, including neurons, astrocytes and oligodendrocytes (McKay, 1997). As stem cells are expanded and usually differentiated prior to implantation, the capacity of these cells can be genetically modified in vitro for dedicated and/or improved cell therapy. Only a limited number of studies are reported in relation to spinal cord repair and less so in relation with implantation plus gene therapy. Neural differentiated mouse embryonic stem cells have been transplanted into the rat spinal cord 9 days after *a contusive injury of the spinal cord* (McDonald et al., 1999). Histological analysis showed that transplant-derived cells survived and differentiated into astrocytes, oligodendrocytes and neurons, and migrated for as far as 8 mm from the lesion site. Functional analysis showed hind limb weight support and partial hind limb coordination not found in control animals. Although this study did not investigate the effect of the transplant on axonal regeneration, the authors postulated that the observed functional recovery was partly due to enhanced myelination as most stem cells differentiated into oligodendrocytes (McDonald et al., 1999). Functional recovery of weight-bearing and hind limb stepping is also reported for neural stem-containing scaffolds placed in hemisected spinal cord (Teng et al., 2002). Anterograde tracing showed CST fibers passing through the injury epicenter to the caudal cord, which was not seen in control animals. Using in vitro oligodendrocyte-differentiated ES cells in chemically demyelinated rat spinal cord or the spinal cord of myelin-deficient Shiverer mouse, large numbers of cells survived and remyelinated host axons indeed (Liu et al., 2000).

The fate of retrovirally infected neural stem C17 cell line following implantation in the spinal cord has been described by Liu et al. (1999a). The cells survived and expressed NT-3 for at least 2 months and the cells differentiated into both neurons and glia. Most C17.NT-3 cells incompletely differentiated or remained quiescent, based on positive nestin staining. In the C17.NT-3 grafts, axon ingrowth from the host spinal cord was reported based on neurofilament staining, whereas no axons were found in the control grafts (Liu et al., 1999a). In the first step to create new synaptic targets for regenerating spinal axons following injury, neural stem cells genetically engineered to overexpress TrkC neurotrophin receptor were transplanted in intact rat spinal cord (Castellanos et al., 2002). Combined with in vivo treatment with NT-3, nearly all cells survived. Astrocytic differentiation was reduced in the combined treatment, but most cells remained in an undifferentiated state.

No single study that combined ex vivo therapy of stem cells and implantation in injured spinal cord has yet been reported. Stem cells are good candidates to be used as cellular grafts in spinal cord injury, because they have the potential to differentiate into neurons and glial cells. ES cells have been used as transplants, but these might become tumorgenic if not completely predifferentiated (Nishimura et al., 2003). Neural stem cells isolated from embryonic or fetal brain are potent alternatives (Wu et al., 2002) and possible also nonneural somatic stem cells from various other sources such as bone marrow, skin, that have shown to give rise to neural phenotypic cells (Azizi et al., 1998; Sanchez-Ramos, 2002).

Concluding remarks

Gene transfer of neurotrophic factor genes to the injured spinal cord, either via direct viral vector injection or via genetically engineered cellular transplants, protects injured neurons from death, prevents atrophy, stimulates regeneration of severed axons and counteract the occurrence of secondary injury at the lesion site. The location of gene transfer (near the neuronal cell body or at the lesion site) determines whether the neurons are rescued or regeneration of injured neuritic processes is stimulated. Thus, the site of gene and cell therapeutic intervention should be carefully considered and is dependent on the type of injury that should be ameliorated.

So far the beneficial effects of neurotrophic factor gene therapy in spinal cord-related injury models are similar to the effects that have been achieved with various protein-infusion paradigms. Direct viral vector-mediated gene transfer is, however, much less invasive than the use of an osmotic minipump. The use of genetically engineered cell implants as biological minipumps has the additional advantage that the cells serve as a bridge or substrate for injured nerve fibers. Moreover, implants will fill up fluid filled cavities that often form after spinal injury. Local

production of neurotrophic factors in the cells surrounding the injured nerve fibers is a much more targeted approach than injection or infusion. Side effects following intracerebroventricular infusion of NGF could be overcome by implants of NGF-secreting cells in the basal forebrain. Not much is known, however, as yet about the absence of side effects of neurotrophic factor gene and cell therapy to repair the injured spinal cord.

Over the last 15 years, substantial improvements in viral vector-systems have been made. Gene transfer in the CNS with HSV and AdV vectors demonstrated that it is possible to directly and locally modify neurons and glia cells. These vectors systems induced an immune response, however, and cannot serve as vehicles for long term, safe expression of neurotrophic factor genes. With the development of AAV and LV vectors these problems have been largely overcome. Current efforts are focusing on constructing AAV and LV vectors that direct regulated gene expression by inserting specific promotors, e.g., tetracyclin-responsive promoters (Gossen and Bujard, 1992) and vectors that allow specific cell-targeting, e.g., through the use of different serotypes. Regulated transgene expression may be an essential requirement for any future gene therapy strategy to repair injured nerve tracts. It has been shown that continued high level expression of BDNF results in 'trapping' of regenerating axons (Blits, 2002). This prevents outgrowth and functional reconnection of regenerating axons to their target cells. This problem may well be solved by using vectors that harbor regulatable transgene expression cassettes (Blesch et al., 2001).

The steady advances that have been made in combining new viral vector systems with a range of promising cellular platforms for ex vivo gene transfer (e.g., primary embryonic neurons, Schwann cells, OEGs, and neural stem cells) holds promising perspectives for the development of new neurotrophic factor-based therapies to repair the injured nervous system.

Abbreviations

AAV	adeno-associated virus
AdV	adenovirus
BBB	Basso, Beattie and Bresnahan

BDNF	brain-derived neurotrophic factor
bFGF	basic fibroblast growth factor
CCI	chronic constriction injury
CGRP	calcitonin gene related peptide
ChAT	choline acetyl transferase
CNS	central nervous system
CNTF	ciliary neurotrophic factor
CSF	cerebrospinal fluid
CST	corticospinal tract
DRG	dorsal root ganglion
ES cell	embryonic stem cell
GDNF	glial cell line-derived neurotrophic factor
GFP	green fluorescent protein
hCMV	human cytomegalo virus
HIV	human immunodeficiency virus
HSV	herpes simplex virus
ICV	intracerebroventricular
IE gene	immediate early gene
LAP	latency active promotor
LAT	latency-associated transcript
LV	lentivirus
MLV	moloney leukemia virus
MOI	multiplicity of infection
MPTP	N-methyl-4-phenyl-1,2,3,6-tetrahydro-pyridine
NGF	nerve growth factor
NT-3	neurotrophin-3
NT-4/5	neurotrophin-4/5
OEG	olfactory ensheathing glia
PNS	peripheral nervous system
RAG	regeneration-associated gene
RST	rubrospinal tract
SCI	spinal cord injury
Trk	tyrosine kinase receptor
VSV-G	vesicular stomatitus virus-glycoprotein

References

Airaksinen, M.S. and Saarma, M. (2002) The GDNF family: signaling, biological functions and therapeutic value. Nat. Rev. Neurosci., 3: 383–394.

Akli, S., Caillaud, C., Vigne, E., Stratford-Perricaudet, L.D., Poenaru, L., Perricaudet, M., Kahn, A. and Peschanski, M.R. (1993) Transfer of a foreign gene into the brain using adenovirus vectors. Nat. Genet., 3: 224–228.

Anderson, K.D., Alderson, R.F., Altar, C.A., DiStefano, P.S., Corcoran, T.L., Lindsay, R.M. and Wiegand, S.J. (1995) Differential distribution of exogenous BDNF, NGF, and NT-3 in the brain corresponds to the relative abundance and distribution of high-affinity and low-affinity neurotrophin receptors. J. Comp. Neurol., 357: 296–317.

Anderson, P.N., Campbell, G., Zhang, Y. and Lieberman, A.R. (1998) Cellular and molecular correlates of the regeneration of adult mammalian CNS axons into peripheral nerve grafts. Prog. Brain Res., 117: 211–232.

Azizi, S.A., Stokes, D., Augelli, B.J., DiGirolamo, C. and Prockop, D.J. (1998) Engraftment and migration of human bone marrow stromal cells implanted in the brains of albino rats—similarities to astrocyte grafts. Proc. Natl. Acad. Sci. USA, 95: 3908–3913.

Bajocchi, G., Feldman, S.H., Crystal, R.G. and Mastrangeli, A. (1993) Direct in vivo gene transfer to ependymal cells in the central nervous system using recombinant adenovirus vectors. Nat. Genet., 3: 229–234.

Bamber, N.I., Li, H., Lu, X., Oudega, M., Aebischer, P. and Xu, X.M. (2001) Neurotrophins BDNF and NT-3 promote axonal re-entry into the distal host spinal cord through Schwann cell-seeded mini-channels. Eur. J. Neurosci., 13: 257–268.

Baumgartner, B.J. and Shine, H.D. (1998) Neuroprotection of spinal motoneurons following targeted transduction with an adenoviral vector carrying the gene for glial cell line-derived neurotrophic factor. Exp. Neurol., 153: 102–112.

Biffo, S., Offenhauser, N., Carter, B.D. and Barde, Y.A. (1995) Selective binding and internalisation by truncated receptors restrict the availability of BDNF during development. Development, 121: 2461–2470.

Blesch, A., Connor, J.M. and Tuszynski, M.H. (2001) Modulation of neuronal survival and axonal growth in vivo by tetracycline-regulated neurotrophin expression. Gene Ther., 8: 954–960.

Blesch, A., Lu, P. and Tuszynski, M.H. (2002) Neurotrophic factors, gene therapy, and neural stem cells for spinal cord repair. Brain Res. Bull., 57: 833–838.

Blits, B. (2002) Cell and gene therapy experimental strategies for spinal cord regeneration. Thesis, Vrije Universiteit Amsterdam.

Blits, B., Dijkhuizen, P.A., Boer, G.J. and Verhaagen, J. (2000) Intercostal nerve implants transduced with an adenoviral vector encoding neurotrophin-3 promote regrowth of injured rat corticospinal tract fibers and improve hindlimb function. Exp. Neurol., 164: 25–37.

Blits, B., Dijkhuizen, P.A., Carlstedt, T.P., Poldervaart, H., Schiemanck, S., Boer, G.J. and Verhaagen, J. (1999) Adenoviral vector-mediated expression of a foreign gene in peripheral nerve tissue bridges implanted in the injured peripheral and central nervous system. Exp. Neurol., 160: 256–267.

Blits, B., Boer, G.J. and Verhaagen, J. (2003) Pharmacological, cell, and gene therapy strategies to promote spinal cord regeneration. Cell Transplant., 11: 593–613.

Blits, B., Oudega, M., Boer, G.J., Bartlett Bunge, M. and Verhaagen, J. (2003) Adeno-associated viral vector-mediated neurotrophin gene transfer in the injured adult rat spinal cord improves hind-limb function. Neuroscience, 118: 271–281.

Blomer, U., Naldini, L., Kafri, T., Trono, D., Verma, I.M. and Gage, F.H. (1997) Highly efficient and sustained gene transfer in adult neurons with a lentivirus vector. J. Virol., 71: 6641–6649.

Boruch, A.V., Conners, J.J., Pipitone, M., Deadwyler, G., Storer, P.D., Devries, G.H. and Jones, K.J. (2001) Neurotrophic and migratory properties of an olfactory ensheathing cell line. Glia, 33: 225–229.

Bradbury, E.J., Khemani, S., Von, R., Priestly, J.V. and McMahon, S.B. (1999) NT-3 promotes growth of lesioned adult rat sensory axons ascending in the dorsal columns of the spinal cord. Eur. J. Neurosci., 11: 3873–3883.

Bregman, B.S., McAtee, M., Dai, H.N. and Kuhn, P.L. (1997) Neurotrophic factors increase axonal growth after spinal cord injury and transplantation in the adult rat. Exp. Neurol., 148: 475–494.

Byrnes, A.P., MacLaren, R.E. and Charlton, H.M. (1996) Immunological instability of persistent adenovirus vectors in the brain: peripheral exposure to vector leads to renewed inflammation, reduced gene expression, and demyelination. J. Neurosci., 16: 3045–3055.

Byrnes, A.P., Rusby, J.E., Wood, M.J. and Charlton, H.M. (1995) Adenovirus gene transfer causes inflammation in the brain. Neuroscience, 66: 1015–1024.

Campbell, G., Lieberman, A.R., Anderson, P.N. and Turmaine, M. (1992) Regeneration of adult rat CNS axons into peripheral nerve autografts: ultrastructural studies of the early stages of axonal sprouting and regenerative axonal growth. J. Neurocytol., 21: 755–787.

Carlson, S.L., Parrish, M.E., Springer, J.E., Doty, K. and Dossett, L. (1998) Acute inflammatory response in spinal cord following impact injury. Exp. Neurol., 151: 77–88.

Castellanos, D.A., Tsoulfas, P., Frydel, B.R., Gajavelli, S., Bes, J.C. and Sagen, J. (2002) TrkC overexpression enhances survival and migration of neural stem cell transplants in the rat spinal cord. Cell Transplant., 11: 297–307.

Cejas, P.J., Martinez, M., Karmally, S., McKillop, M., McKillop, J., Plunkett, J.A., Oudega, M. and Eaton, M.J. (2000) Lumbar transplant of neurons genetically modified to secrete brain-derived neurotrophic factor attenuates allodynia and hyperalgesia after sciatic nerve constriction. Pain, 86: 195–210.

Check, E. (2002) A tragic setback. Nature, 420: 116–118.

Chen, M.S., Huber, A.B., van der Haar, M.E., Frank, M., Schnell, L., Spillmann, A.A., Christ, F. and Schwab, M.E. (2000) Nogo-A is a myelin-associated neurite outgrowth inhibitor and an antigen for monoclonal antibody IN-1. Nature, 403: 434–439.

Cheng, H., Cao, Y. and Olson, L. (1996) Spinal cord repair in adult paraplegic rats: partial restoration of hind limb function. Science, 273: 510–513.

Coumans, J.V., Lin, T.T., Dai, H.N., MacArthur, L., McAtee, M., Nash, C. and Bregman, B.S. (2001) Axonal regeneration and functional recovery after complete spinal cord transection in rats by delayed treatment with transplants and neurotrophins. J. Neurosci., 21: 9334–9344.

David, S. and Aguayo, A.J. (1981) Axonal elongation into peripheral nervous system 'bridges' after central nervous system injury in adult rats. Science, 214: 931–933.

Davidson, B.L., Allen, E.D., Kozarsky, K.F., Wilson, J.M. and Roessler, B.J. (1993) A model system for in vivo gene transfer into the central nervous system using an adenoviral vector. Nat. Genet., 3: 219–223.

Day-Lollini, P.A., Stewart, G.R., Taylor, M.J., Johnson, R.M. and Chellman, G.J. (1997) Hyperplastic changes within the leptomeninges of the rat and monkey in response to chronic intracerebroventricular infusion of nerve growth factor. Exp. Neurol., 145: 24–37.

De Winter, F., Oudega, M., Lankhorst, A.J., Hamers, F.P., Blits, B., Ruitenberg, M.J., Pasterkamp, R.J., Gispen, W.H. and Verhaagen, J. (2002) Injury-induced class 3 semaphorin expression in the rat spinal cord. Exp. Neurol., 175: 61–75.

Dijkhuizen, P.A. (1999) Viral vectors as tools for neurotrophin gene delivery to the injured nervous system. Thesis, Vrije Universiteit Amsterdam, p. 160.

Dijkhuizen, P.A., Hermens, W.T., Teunis, M.A. and Verhaagen, J. (1997) Adenoviral vector-directed expression of neurotrophin-3 in rat dorsal root ganglion explants results in a robust neurite outgrowth response. J. Neurobiol., 33: 172–184.

Dijkhuizen, P.A., Pasterkamp, R.J., Hermens, W.T., de Winter, F., Giger, R.J. and Verhaagen, J. (1998) Adenoviral vector-mediated gene delivery to injured rat peripheral nerve. J. Neurotrauma, 15: 387–397.

Dijkhuizen, P.A. and Verhaagen, J. (1999) The use of neurotrophic factors to treat spinal cord injury: advantages and disadvantages of different delivery methods. Neurosci. Res. Comm., 24: 1–10.

Dittrich, F., Ochs, G., Grosse-Wilde, A., Berweiler, U., Yan, Q., Miller, J.A., Toyka, K.V. and Sendtner, M. (1996) Pharmacokinetics of intrathecally applied BDNF and effects on spinal motoneurons. Exp. Neurol., 141: 225–239.

Dou, C.L. and Levine, J.M. (1994) Inhibition of neurite growth by the NG2 chondroitin sulfate proteoglycan. J. Neurosci., 14: 7616–7628.

Eaton, M.J., Blits, B., Ruitenberg, M.J., Verhaagen, J. and Oudega, M. (2002) Amelioration of chronic neuropathic pain after partial nerve injury by adeno-associated viral (AAV) vector-mediated over-expression of BDNF in the rat spinal cord. Gene Ther., 9: 1387–1395.

Emborg, M.E. and Kordower, J.H. (2000) Delivery of therapeutic molecules into the CNS. Prog. Brain Res., 128: 323–332.

Englund, U., Ericson, C., Rosenblad, C., Mandel, R.J., Trono, D., Wictorin, K. and Lund, C. (2000) The use of a recombinant lentiviral vector for ex vivo gene transfer into the rat CNS. Neuroreport, 11: 3973–3977.

Eriksdotter Jonhagen, M., Nordberg, A., Amberla, K., Backman, L., Ebendal, T., Meyerson, B., Olson, L., Seiger, A., Shigeta, M., Theodorsson, E., Viitanen, M., Winblad, B. and Wahlund, L.O. (1998) Intracerebroventricular infusion of nerve growth factor in three patients with Alzheimer's disease. Dement. Geriatr. Cogn. Disord., 9: 246–257.

Fawcett, J.W. (1997) Astrocytic and neuronal factors affecting axon regeneration in the damaged central nervous system. Cell Tissue Res., 290: 371–377.

Federoff, H.J., Geschwind, M.D., Geller, A.I. and Kessler, J.A. (1992) Expression of nerve growth factor in vivo from a defective herpes simplex virus 1 vector prevents effects of axotomy on sympathetic ganglia. Proc. Natl. Acad. Sci. USA, 89: 1636–1640.

Fernandes, K.J., Fan, D.P., Tsui, B.J., Cassar, S.L. and Tetzlaff, W. (1999) Influence of the axotomy to cell body distance in rat rubrospinal and spinal motoneurons: differential regulation of GAP-43, tubulins, and neurofilament-M. J. Comp. Neurol., 414: 495–510.

Frank, D.A. and Greenberg, M.E. (1996) Signal transduction pathways activated by ciliary neurotrophic factor and related cytokines. Perspect. Dev. Neurobiol., 4: 3–18.

Gage, F.H., Wolff, J.A., Rosenberg, M.B., Xu, L., Yee, J.K., Schults, C. and Friedmann, T. (1987) Grafting genetically modified cells to the brain: possibilities for the future. Neuroscience, 23: 795–807.

Giehl, K.M. and Tetzlaff, W. (1996) BDNF and NT-3, but not NGF, prevent axotomy-induced death of rat corticospinal neurons in vivo. Eur. J. Neurosci., 8: 1167–1175.

Goins, W.F., Lee, K.A., Cavalcoli, J.D., O'Malley, M.E., DeKosky, S.T., Fink, D.J. and Glorioso, J.C. (1999) Herpes simplex virus type 1 vector-mediated expression of nerve growth factor protects dorsal root ganglion neurons from peroxide toxicity. J. Virol., 73: 519–532.

Gossen, M. and Bujard, H. (1992) Tight control of gene expression in mammalian cells by tetracycline-responsive promoters. Proc. Natl. Acad. Sci. USA, 89: 5547–5551.

Gravel, C., Gotz, R., Lorrain, A. and Sendtner, M. (1997) Adenoviral gene transfer of ciliary neurotrophic factor and brain-derived neurotrophic factor leads to long-term survival of axotomized motor neurons. Nat. Med., 3: 765–770.

Grill, R., Murai, K., Blesch, A., Gage, F.H. and Tuszynski, M.H. (1997a) Cellular delivery of neurotrophin-3 promotes corticospinal axonal growth and partial functional recovery after spinal cord injury. J. Neurosci., 17: 5560–5572.

Grill, R.J., Blesch, A. and Tuszynski, M.H. (1997b) Robust growth of chronically injured spinal cord axons induced by grafts of genetically modified NGF-secreting cells. Exp. Neurol., 148: 444–452.

Hao, J., Ebendal, T., Xu, X., Wiesenfeld-Hallin, Z. and Eriksdotter Jonhagen, M. (2000) Intracerebroventricular infusion of nerve growth factor induces pain-like response in rats. Neurosci. Lett., 286: 208–212.

Hermens, W.T.J.M.C. and Verhaagen, J. (1997) Adenoviral vector-mediated gene expression in the nervous system of immunocompetent Wistar and T-cell deficient nude rats: preferential survival of transduced astroglial cells in nude rats. Hum. Gene Ther., 8: 1049–1063.

Hermens, W.T. and Verhaagen, J. (1998) Viral vectors, tools for gene transfer in the nervous system. Prog. Neurobiol., 55: 399–432.

Ho, D.Y. and Mocarski, E.S. (1988) Beta-galactosidase as a marker in the peripheral and neural tissues of the herpes simplex virus-infected mouse. Virology, 167: 279–283.

Horellou, P., Brundin, P., Kalen, P., Mallet, J. and Bjorklund, A. (1990) In vivo release of dopa and dopamine from genetically engineered cells grafted to the denervated rat striatum. Neuron, 5: 393–402.

Hottinger, A.F., Azzouz, M., Deglon, N., Aebischer, P. and Zurn, A.D. (2000) Complete and long-term rescue of lesioned adult motoneurons by lentiviral-mediated expression of glial cell line-derived neurotrophic factor in the facial nucleus. J. Neurosci., 20: 5587–5593.

Houenou, L.J., Oppenheim, R.W., Li, L., Lo, A.C. and Prevette, D. (1996) Regulation of spinal motoneuron survival by GDNF during development and following injury. Cell, Tissue Res., 286: 219–223.

Houweling, D.A., Bar, P.R., Gispen, W.H. and Joosten, E.A. (1998a) Spinal cord injury: bridging the lesion and the role of neurotrophic factors in repair. Prog. Brain Res., 117: 455–471.

Houweling, D.A., Lankhorst, A.J., Gispen, W.H., Bar, P.R. and Joosten, E.A. (1998b) Collagen containing neurotrophin-3 (NT-3) attracts regrowing injured corticospinal axons in the adult rat spinal cord and promotes partial functional recovery. Exp. Neurol., 153: 49–59.

Houweling, D.A., van Asseldonk, J.T., Lankhorst, A.J., Hamers, F.P., Martin, D., Bar, P.R. and Joosten, E.A. (1998c) Local application of collagen containing brain-derived neurotrophic factor decreases the loss of function after spinal cord injury in the adult rat. Neurosci. Lett., 251: 193–196.

Huber, A.B., Ehrengruber, M.U., Schwab, M.E. and Brosamle, C. (2000) Adenoviral gene transfer to the injured spinal cord of the adult rat. Eur. J. Neurosci., 12: 3437–3442.

Ide, C. (1996) Peripheral nerve regeneration. Neurosci. Res., 25: 101–121.

Ip, N.Y. and Yancopoulos, G.D. (1996) The neurotrophins and CNTF: two families of collaborative neurotrophic factors. Annu. Rev. Neurosci., 19: 491–515.

Jakeman, L.B., Wei, P., Guan, Z. and Stokes, B.T. (1998) Brain-derived neurotrophic factor stimulates hindlimb stepping and sprouting of cholinergic fibers after spinal cord injury. Exp. Neurol., 154: 170–184.

Jin, Y., Tessler, A., Fischer, I. and Houle, J.D. (2000) Fibroblasts genetically modified to produce BDNF support regrowth of chronically injured serotonergic axons. Neurorehabil. Neural Repair, 14: 311–317.

Jones, L.L. and Tuszynski, M.H. (2001) Chronic intrathecal infusions after spinal cord injury cause scarring and compression. Microsc. Res. Tech., 54: 317–324.

Jones, N. and Shenk, T. (1979) An adenovirus type 5 early gene function regulates expression of other early viral genes. Proc. Natl. Acad. Sci. USA, 76: 3665–3669.

Kaiser, J. (2003) Gene therapy. Seeking the cause of induced leukemias in X-SCID trial. Science, 299: 495.

Kajiwara, K., Byrnes, A.P., Charlton, H.M., Wood, M.J. and Wood, K.J. (1997) Immune responses to adenoviral vectors during gene transfer in the brain. Hum. Gene. Ther., 8: 253–265.

Kaplitt, M.G., Leone, P., Samulski, R.J., Xiao, X., Pfaff, D.W., O'Malley, K.L. and During, M.J. (1994) Long-term gene expression and phenotypic correction using adeno-associated virus vectors in the mammalian brain. Nat. Genet., 8: 148–154.

Kawaja, M.D., Rosenberg, M.B., Yoshida, K. and Gage, F.H. (1992) Somatic gene transfer of nerve growth factor promotes the survival of axotomized septal neurons and the regeneration of their axons in adult rats. J. Neurosci., 12: 2849–2864.

Keir, S.D., Mitchell, W.J., Feldman, L.T. and Martin, J.R. (1995) Targeting and gene expression in spinal cord motor neurons following intramuscular inoculation of an HSV-1 vector. J. Neurovirol., 1: 259–267.

Kennedy, P.G. (1997) Potential use of herpes simplex virus (HSV) vectors for gene therapy of neurological disorders. Brain, 120(7): 1245–1259.

Kim, D., Schallert, T., Liu, Y., Browarak, T., Nayeri, N., Tessler, A., Fischer, I. and Murray, M. (2001) Transplantation of genetically modified fibroblasts expressing BDNF in adult rats with a subtotal hemisection improves specific motor and sensory functions. Neurorehabil. Neural Repair, 15: 141–150.

Kim, D.H., Gutin, P.H., Noble, L.J., Nathan, D., Yu, J.S. and Nockels, R.P. (1996) Treatment with genetically engineered fibroblasts producing NGF or BDNF can accelerate recovery from traumatic spinal cord injury in the adult rat. Neuroreport, 7: 2221–2225.

Kishino, A., Ishige, Y., Tatsuno, T., Nakayama, C. and Noguchi, H. (1997) BDNF prevents and reverses adult rat motor neuron degeneration and induces axonal outgrowth. Exp. Neurol., 144: 273–286.

Klein, R.L., Muir, D., King, M.A., Peel, A.L., Zolotukhin, S., Moller, J.C., Kruttgen, A., Heymach, J.V., Jr., Muzyczka, N. and Meyer, E.M. (1999) Long-term actions of vector-derived nerve growth factor or brain-derived neurotrophic factor on choline acetyltransferase and Trk receptor levels in the adult rat basal forebrain. Neuroscience, 90: 815–821.

Kobayashi, N.R., Fan, D.P., Giehl, K.M., Bedard, A.M., Wiegand, S.J. and Tetzlaff, W. (1997) BDNF and NT-4/5 prevent atrophy of rat rubrospinal neurons after cervical axotomy, stimulate GAP-43 and Talpha1-tubulin mRNA expression, and promote axonal regeneration. J. Neurosci., 17: 9583–9595.

Kordower, J.H., Emborg, M.E., Bloch, J., Ma, S.Y., Chu, Y., Leventhal, L., McBride, J., Chen, E.Y., Palfi, S., Roitberg, B.Z., Brown, W.D., Holden, J.E., Pyzalski, R., Taylor, M.D., Carvey, P., Ling, Z., Trono, D., Hantraye, P., Deglon, N. and Aebischer, P. (2000) Neurodegeneration prevented by lentiviral vector delivery of GDNF in primate models of Parkinson's disease. Science, 290: 767–773.

Kuzis, K. and Eckenstein, F.P. (1996) Ciliary neurotrophic factor as a motor neuron trophic factor. Perspect Dev. Neurobiol., 4: 65–74.

Kwon, B.K., Liu, J., Messerer, C., Kobayashi, N.R., McGraw, J., Oschipok, L. and Tetzlaff, W. (2002) Survival and regeneration of rubrospinal neurons 1 year after spinal cord injury. Proc. Natl. Acad. Sci. USA, 99: 3246–3251.

Lacroix, S. and Tuszynski, M.H. (2000) Neurotrophic factors and gene therapy in spinal cord injury. Neurorehabil. Neural Repair, 14: 265–275.

Lakatos, A., Franklin, R.J. and Barnett, S.C. (2000) Olfactory ensheathing cells and Schwann cells differ in their in vitro interactions with astrocytes. Glia, 32: 214–225.

Le Gal La Salle, G., Robert, J.J., Berrard, S., Ridoux, V., Stratford-Perricaudet, L.D., Perricaudet, M. and Mallet, J. (1993) An adenovirus vector for gene transfer into neurons and glia in the brain. Science, 259: 988–990.

Leib, D.A. and Olivo, P.D. (1993) Gene delivery to neurons: is herpes simplex virus the right tool for the job?. Bioessays, 15: 547–554.

Levi, A.D., Tator, C.H. and Bunge, R.P. (1996) Clinical syndromes associated with disproportionate weakness of the upper versus the lower extremities after cervical spinal cord injury. Neurosurgery, 38: 179–183.

Lewin, G.R. and Barde, Y.A. (1996) Physiology of the neurotrophins. Annu. Rev. Neurosci., 19: 289–317.

Li, Y., Field, P.M. and Raisman, G. (1997) Repair of adult rat corticospinal tract by transplants of olfactory ensheathing cells. Science, 277: 2000–2002.

Li, Y., Field, P.M. and Raisman, G. (1998) Regeneration of adult rat corticospinal axons induced by transplanted olfactory ensheathing cells. J. Neurosci., 18: 10,514–10,524.

Liebl, D.J., Huang, W., Young, W. and Parada, L.F. (2001) Regulation of Trk receptors following contusion of the rat spinal cord. Exp. Neurol., 167: 15–26.

Lilley, C.E., Groutsi, F., Han, Z., Palmer, J.A., Anderson, P.N., Latchman, D.S. and Coffin, R.S. (2001) Multiple immediate-early gene-deficient herpes simplex virus vectors allowing efficient gene delivery to neurons in culture and widespread gene delivery to the central nervous system in vivo. J. Virol., 75: 4343–4356.

474

Liu, Y., Himes, B.T., Murray, M., Tessler, A. and Fischer, I. (2002) Grafts of BDNF-producing fibroblasts rescue axotomized rubrospinal neurons and prevent their atrophy. Exp. Neurol., 178: 150–164.

Liu, Y., Himes, B.T., Solowska, J., Moul, J., Chow, S.Y., Park, K.I., Tessler, A., Murray, M., Snyder, E.Y. and Fischer, I. (1999a) Intraspinal delivery of neurotrophin-3 using neural stem cells genetically modified by recombinant retrovirus. Exp. Neurol., 158: 9–26.

Liu, Y., Himes, B.T., Tryon, B., Moul, J., Chow, S.Y., Jin, H., Murray, M., Tessler, A. and Fischer, I. (1998) Intraspinal grafting of fibroblasts genetically modified by recombinant adenoviruses. Neuroreport, 9: 1075–1079.

Liu, Y., Kim, D., Himes, B.T., Chow, S.Y., Schallert, T., Murray, M., Tessler, A. and Fischer, I. (1999b) Transplants of fibroblasts genetically modified to express BDNF promote regeneration of adult rat rubrospinal axons and recovery of forelimb function. J. Neurosci., 19: 4370–4387.

Liu, S., Qu, Y., Stewart, T.J., Howard, M.J., Chakrabortty, S., Holekamp, T.F. and McDonald, J.W. (2000) Embryonic stem cells differentiate into oligodendrocytes and myelinate in culture and after spinal cord transplantation. Proc. Natl. Acad. Sci. USA., 97: 6126–6131.

Mata, M., Zhang, M., Hu, X. and Fink, D.J. (2001) HveC (nectin-1) is expressed at high levels in sensory neurons, but not in motor neurons, of the rat peripheral nervous system. J. Neurovirol., 7: 476–480.

Mazarakis, N.D., Azzouz, M., Rohll, J.B., Ellard, F.M., Wilkes, F.J., Olsen, A.L., Carter, E.E., Barber, R.D., Baban, D.F., Kingsman, S.M., Kingsman, A.J., O'Malley, K. and Mitrophanous, K.A. (2001) Rabies virus glycoprotein pseudotyping of lentiviral vectors enables retrograde axonal transport and access to the nervous system after peripheral delivery. Hum. Mol. Genet., 10: 2109–2121.

McCown, T.J., Xiao, X., Li, J., Breese, G.R. and Samulski, R.J. (1996) Differential and persistent expression patterns of CNS gene transfer by an adeno-associated virus (AAV) vector. Brain Res., 713: 99–107.

McDonald, J.W., Liu, X.Z., Qu, Y., Liu, S., Mickey, S.K., Turetsky, D., Gottlieb, D.I. and Choi, D.W. (1999) Transplanted embryonic stem cells survive, differentiate and promote recovery in injured rat spinal cord. Nat. Med., 5: 1410–1412.

McDonald, J.W. and Sadowsky, C. (2002) Spinal-cord injury. Lancet, 359: 417–425.

McKay, R. (1997) Stem cells in the central nervous system. Science, 276: 66–71.

McTigue, D.M., Horner, P.J., Stokes, B.T. and Gage, F.H. (1998) Neurotrophin-3 and brain-derived neurotrophic factor induce oligodendrocyte proliferation and myelination of regenerating axons in the contused adult rat spinal cord. J. Neurosci., 18: 5354–5365.

Menei, P., Montero-Menei, C., Whittemore, S.R., Bunge, R.P. and Bunge, M.B. (1998) Schwann cells genetically modified to secrete human BDNF promote enhanced axonal regrowth across transected adult rat spinal cord. Eur. J. Neurosci., 10: 607–621.

Miller, D.G., Adam, M.A. and Miller, A.D. (1990) Gene transfer by retrovirus vectors occurs only in cells that are actively replicating at the time of infection. Mol. Cell Biol., 10: 4239–4342.

Mitrophanous, K., Yoon, S., Rohll, J., Patil, D., Wilkes, F., Kim, V., Kingsman, S.M., Kingsman, A.J. and Mazarakis, M. (1999) Stable gene transfer to the nervous system using a non-primate lentiviral vector. Gene Ther., 6: 1808–1818.

Monohan, P.E. and Samulski, R.J. (2000) Adeno-associated virus vectors for gene therapy: more pros than cons? Mol. Med. Today, 11: 433–440.

Mukhopadhyay, G., Doherty, P., Walsh, F.S., Crocker, P.R. and Filbin, M.T. (1994) A novel role for myelin-associated glycoprotein as an inhibitor of axonal regeneration. Neuron, 13: 757–767.

Naldini, L., Blomer, U., Gage, F.H., Trono, D. and Verma, I.M. (1996a) Efficient transfer, integration, and sustained long-term expression of the transgene in adult rat brains injected with a lentiviral vector. Proc. Natl. Acad. Sci. USA, 93: 11382–11388.

Naldini, L., Blomer, U., Gallay, P., Ory, D., Mulligan, R., Gage, F.H., Verma, I.M. and Trono, D. (1996b) In vivo gene delivery and stable transduction of nondividing cells by a lentiviral vector. Science, 272: 263–267.

Nathan, P.W. (1994) Effects on movement of surgical incisions into the human spinal cord. Brain, 117 (Pt 2): 337–346.

Natsume, A., Mata, M., Wolfe, D., Oligino, T., Goss, J., Huang, S., Glorioso, J. and Fink, J. (2002) Bcl-2 and GDNF delivered by HSV-mediated gene transfer after spinal root avulsion provide a synergistic effect. J. Neurotrauma, 19: 61–68.

Nevins, J.R. (1981) Mechanism of activation of early viral transcription by the adenovirus E1A gene product. Cell, 26: 213–220.

Nishimura, F., Yoshikawa, M., Kanda, S., Nonaka, M., Yokota, H., Shiroi, A., Nakase, H., Hirabayashi, H., Ouji, Y., Birumachi, J., Ishizaka, S. and Sakaki, T. (2003) Potential use of embryonic stem cells for the treatment of mouse parkinsonian models: improved behavior by transplantation of in vitro differentiated dopaminergic neurons from embryonic stem cells. Stem Cells, 21: 171–180.

Novikov, L., Novikova, L. and Kellerth, J.O. (1997) Brain-derived neurotrophic factor promotes axonal regeneration and long-term survival of adult rat spinal motoneurons in vivo. Neuroscience, 79: 765–774.

Novikova, L.N., Novikov, L.N. and Kellerth, J.O. (2000) Survival effects of BDNF and NT-3 on axotomized rubrospinal neurons depend on the temporal pattern of neurotrophin administration. Eur. J. Neurosci., 12: 776–780.

Oliveira, A.L., Risling, M., Negro, A., Langone, F. and Cullheim, S. (2002) Apoptosis of spinal interneurons induced by sciatic nerve axotomy in the neonatal rat is counteracted by nerve growth factor and ciliary neurotrophic factor. J. Comp. Neurol., 447: 381–393.

Oudega, M. and Hagg, T. (1996) Nerve growth factor promotes regeneration of sensory axons into adult rat spinal cord. Exp. Neurol., 140: 218–229.

Oudega, M. and Hagg, T. (1999) Neurotrophins promote regeneration of sensory axons in the adult rat spinal cord. Brain Res., 818: 431–438.

Palella, T.D., Hidaka, Y., Silverman, L.J., Levine, M., Glorioso, J. and Kelley, W.N. (1989) Expression of human HPRT mRNA in brains of mice infected with a recombinant herpes simplex virus-1 vector. Gene, 80: 137–144.

Pasterkamp, R.J. and Verhaagen, J. (2001) Emerging roles for semaphorins in neural regeneration. Brain Res. Brain Res. Rev., 35: 36–54.

Peel, A.L., Zolotukhin, S., Schrimsher, G.W., Muzyczka, N. and Reier, P.J. (1997) Efficient transduction of green fluorescent protein in spinal cord neurons using adeno-associated virus vectors containing cell type-specific promoters. Gene Ther., 4: 16–24.

Raineteau, O. and Schwab, M.E. (2001) Plasticity of motor systems after incomplete spinal cord injury. Nat. Rev. Neurosci., 2: 263–273.

Ramer, M.S., Priestley, J.V. and McMahon, S.B. (2000) Functional regeneration of sensory axons into the adult spinal cord. Nature, 403: 312–316.

Ramon-Cueto, A., Cordero, M.I., Santos-Benito, F.F. and Avila, J. (2000) Functional recovery of paraplegic rats and motor axon regeneration in their spinal cords by olfactory ensheathing glia. Neuron, 25: 425–435.

Ramon-Cueto, A. and Nieto-Sampedro, M. (1994) Regeneration into the spinal cord of transected dorsal root axons is promoted by ensheathing glia transplants. Exp. Neurol., 127: 232–244.

Ramon-Cueto, A., Plant, G.W., Avila, J. and Bunge, M.B. (1998) Long-distance axonal regeneration in the transected adult rat spinal cord is promoted by olfactory ensheathing glia transplants. J. Neurosci., 18: 3803–3815.

Ringstedt, T., Lagercrantz, H. and Persson, H. (1993) Expression of members of the trk family in the developing postnatal rat brain. Brain Res. Dev. Brain Res., 72: 119–131.

Roizman, B. and Sears, A.E. (1987) An inquiry into the mechanisms of herpes simplex virus latency. Annu. Rev. Microbiol., 41: 543–571.

Romero, M.I., Rangappa, N., Garry, M.G. and Smith, G.M. (2001) Functional regeneration of chronically injured sensory afferents into adult spinal cord after neurotrophin gene therapy. J Neurosci, 21: 8408–8416.

Romero, M.I., Rangappa, N., Li, L., Lightfoot, E., Garry, M.G. and Smith, G.M. (2000) Extensive sprouting of sensory afferents and hyperalgesia induced by conditional expression of nerve growth factor in the adult spinal cord. J. Neurosci., 20: 4435–4445.

Rosenberg, M.B., Friedmann, T., Robertson, R.C., Tuszynski, M., Wolff, J.A., Breakefield, X.O. and Gage, F.H. (1988) Grafting genetically modified cells to the damaged brain: restorative effects of NGF expression. Science, 242: 1575–1578.

Rubio, N. (1997) Mouse astrocytes store and deliver brain-derived neurotrophic factor using the non-catalytic gp95trkB receptor. Eur. J. Neurosci., 9: 1847–1853.

Ruitenberg, M.J., Plant, G.W., Christensen, C.L., Blits, B., Niclou, S.P., Harvey, A.R., Boer, G.J. and Verhaagen, J. (2002) Viral vector-mediated gene expression in olfactory ensheathing glia implants in the lesioned rat spinal cord. Gene Ther., 9: 135–146.

Ruitenberg, M.J., Plant, G.W., Hamers, F.P.T., Wortel, J., Blits, B., Dijkhuizen, P.A., Gispen, W.H., Boer, G.J. and Verhaagen, J. (2003) Ex vivo adenoviral vector-mediated neurotrophin gene transfer to olfactory ensheathing glia: effects on rubrospinal tract regeneration, lesion size, and functional recovery after implantation in the injured rat spinal cord. J. Neurosci., 23: 7045–7058.

Sanchez-Ramos, J.R. (2002) Neural cells derived from adult bone marrow and umbilical cord blood. J. Neurosci. Res., 69: 880–893.

Schnell, L., Schneider, R., Kolbeck, R., Barde, Y.A. and Schwab, M.E. (1994) Neurotrophin-3 enhances sprouting of corticospinal tract during development and after adult spinal cord lesion. Nature, 367: 170–173.

Schroder, A.R., Shinn, P., Chen, H., Berry, C., Ecker, J.R. and Bushman, F. (2002) HIV-1 integration in the human genome favors active genes and local hotspots. Cell, 110: 521–529.

Schwab, M.E. and Bartholdi, D. (1996) Degeneration and regeneration of axons in the lesioned spinal cord. Physiol. Rev., 76: 319–370.

Smith, G.M., Hale, J., Pasnikowski, E.M., Lindsay, R.M., Wong, V. and Rudge, J.S. (1996) Astrocytes infected with replication-defective adenovirus containing a secreted form of CNTF or NT3 show enhanced support of neuronal populations in vitro. Exp. Neurol., 139: 156–166.

Snyder, E.Y. and Senut, M.C. (1997) The use of nonneuronal cells for gene delivery. Neurobiol. Dis., 4: 69–102.

Teng, Y.D., Lavik, E.B., Xianlu, Q., Park, K.I., Ourednik, J., Zurakowski, D., Langer, R. and Snyder, E.Y. (2002) Functional recovery following traumatic spinal cord injury mediated by a unique polymer scaffold seeded with neural stem cells. Proc. Natl. Acad. Sci. USA., 99: 3024–3029.

Thoenen, H. and Sendtner, M. (2002) Neurotrophins: from enthusiastic expectations through sobering experiences to rational therapeutic approaches. Nat. Neurosci., 5 (Suppl.): 1046–1050.

Trono, D. (2000) Lentiviral vectors: turning a deadly foe into a therapeutic agent. Gene Ther., 7: 20–23.

Tuszynski, M.H., Gabriel, K., Gage, F.H., Suhr, S., Meyer, S. and Rosetti, A. (1996) Nerve growth factor delivery by gene transfer induces differential outgrowth of sensory, motor, and noradrenergic neurites after adult spinal cord injury. Exp. Neurol., 137: 157–173.

Tuszynski, M.H., Murai, K., Blesch, A., Grill, R. and Miller, I. (1997) Functional characterization of NGF-secreting cell grafts to the acutely injured spinal cord. Cell Transplant., 6: 361–368.

Tuszynski, M.H., Peterson, D.A., Ray, J., Baird, A., Nakahara, Y. and Gage, F.H. (1994) Fibroblasts genetically modified to produce nerve growth factor induce robust neuritic ingrowth after grafting to the spinal cord. Exp. Neurol., 126: 1–14.

Tuszynski, M.H., Weidner, N., McCormack, M., Miller, I., Powell, H. and Conner, J. (1998) Grafts of genetically modified Schwann cells to the spinal cord: survival, axon growth, and myelination. Cell Transplant., 7: 187–196.

Verhaagen, J., Hermens, W.T., Dijkhuizen, P.A., Holtmaat, A.J. and Gispen, W.H. (1995) Use of viral vectors to promote neuroregeneration. Clin. Neurosci., 3: 275–283.

Von Meyenburg, J., Brosamble, C., Metz, G.A. and Schwab, M.E. (1998) Regeneration and sprouting of chronically injured corticospinal tract fibres in adult rats promoted by NT-3 and the mAb IN-1, which neutralizes myelin-associated neurite growth inhibitors. Exp. Neurol., 154: 583–594.

Watson, D.J., Kobinger, G.P., Passini, M.A., Wilson, J.M. and Wolfe, J.H. (2002) Targeted transduction patterns in the mouse brain by lentivirus vectors pseudotyped with VSV, Ebola, Mokola, LCMV, or MuLV envelope proteins. Mol. Ther., 5: 528–537.

Weidner, N., Blesch, A., Grill, R.J. and Tuszynski, M.H. (1999) Nerve growth factor-hypersecreting Schwann cell grafts augment and guide spinal cord axonal growth and remyelinate central nervous system axons in a phenotypically appropriate manner that correlates with expression of L1. J. Comp. Neurol., 413: 495–506.

Weidner, N., Ner, A., Salimi, N. and Tuszynski, M.H. (2001) Spontaneous corticospinal axonal plasticity and functional recovery after adult central nervous system injury. Proc. Natl. Acad. Sci. USA, 98: 3513–3518.

Winkler, J., Ramirez, G.A., Kuhn, H.G., Peterson, D.A., Day-Lollini, P.A., Stewart, G.R., Tuszynski, M.H., Gage, F.H. and Thal, L.J. (1997) Reversible Schwann cell hyperplasia and sprouting of sensory and sympathetic neurites after intraventricular administration of nerve growth factor. Ann. Neurol., 41: 82–93.

Wood, M.J., Byrnes, A.P., Pfaff, D.W., Rabkin, S.D. and Charlton, H.M. (1994a) Inflammatory effects of gene transfer into the CNS with defective HSV-1 vectors. Gene Ther., 1: 283–291.

Wood, M.J., Byrnes, A.P., Rabkin, S.D., Pfaff, D.W. and Charlton, H.M. (1994b) Immunological consequences of HSV-1-mediated gene transfer into the CNS. Gene Ther., 1(Suppl. 1): S82.

Wood, M.J., Charlton, H.M., Wood, K.J., Kajiwara, K. and Byrnes, A.P. (1996) Immune responses to adenovirus vectors in the nervous system. Trends Neurosci., 19: 497–501.

Wu, S., Suzuki, Y., Kitada, M., Kataoka, K., Kitaura, M., Kataoka, K., Nishimura, Y. and Ide, C. (2002) New method for transplantation of neurosphere cells into injured spinal cord through cerebrospinal fluid in rat. Neurosci. Lett., 318: 81–84.

Xu, X.M., Guenard, V., Kleitman, N., Aebischer, P. and Bunge, M.B. (1995) A combination of BDNF and NT-3 promotes supraspinal axonal regeneration into Schwann cell grafts in adult rat thoracic spinal cord. Exp. Neurol., 134: 261–272.

Yamada, M., Natsume, A., Mata, M., Oligino, T., Goss, J., Glorioso, J. and Fink, D.J. (2001) Herpes simplex virus vector-mediated expression of Bcl-2 protects spinal motor neurons from degeneration following root avulsion. Exp. Neurol., 168: 225–230.

Yan, Q., Matheson, C., Sun, J., Radeke, M.J., Feinstein, S.C. and Miller, J.A. (1994) Distribution of intracerebral ventricularly administered neurotrophins in rat brain and its correlation with trk receptor expression. Exp. Neurol., 127: 23–36.

Yang, K., Clifton, G.L. and Hayes, R.L. (1997) Gene therapy for central nervous system injury: the use of cationic liposomes: an invited review. J. Neurotrauma, 14: 281–297.

Ye, J.H. and Houle, J.D. (1997) Treatment of the chronically injured spinal cord with neurotrophic factors can promote axonal regeneration from supraspinal neurons. Exp. Neurol., 143: 70–81.

Yong, C., Arnold, P.M., Zoubine, M.N., Citron, B.A., Watanabe, I., Berman, N.E. and Festoff, B.W. (1998) Apoptosis in cellular compartments of rat spinal cord after severe contusion injury. J. Neurotrauma, 15: 459–472.

Zhang, Y., Dijkhuizen, P.A., Anderson, P.N., Lieberman, A.R. and Verhaagen, J. (1998) NT-3 delivered by an adenoviral vector induces injured dorsal root axons to regenerate into the spinal cord of adult rats. J. Neurosci. Res., 54: 554–562.

Zhang, Y. and Schneider, R.J. (1994) Adenovirus inhibition of cell translation facilitates release of virus particles and enhances degradation of the cytokeratin network. J. Virol., 68: 2544–2555.

Zhou, L., Baumgartner, B.J., Hill-Felberg, S.J., McGowen, L.R. and Shine, H.D. (2003) Neurotrophin-3 expressed in situ induces axonal plasticity in the adult injured spinal cord. J. Neurosci., 23: 1424–1431.

Progress in Brain Research, Vol. 146
ISSN 0079-6123

CHAPTER 30

Neurotrophic factors and their receptors in human sensory neuropathies

Praveen Anand*

*Department of Neurology, Peripheral Neuropathy Unit, Imperial College London, Hammersmith Hospital,
Du Cane Road, London W12 ONN, UK*

Abstract: Neurotrophic factors may play key roles in pathophysiological mechanisms of human neuropathies. Nerve growth factor (NGF) is trophic to small-diameter sensory fibers and regulates nociception. This review focuses on sensory dysfunction and the potential of neurotrophic treatments. *Genetic neuropathy*. Mutations of the NGF high-affinity receptor tyrosine kinase A (Trk A) have been found in congenital insensitivity to pain and anhidrosis; these are likely to be partial loss-of-function mutations, as axon-reflex vasodilatation and sweating can be elicited albeit reduced, suggesting rhNGF could restore nociception in some patients. *Leprous neuropathy*. Decreased NGF in leprosy skin may explain cutaneous hypoalgesia even with inflammation and rhNGF may restore sensation, as spared nerve fibers show Trk A-staining. *Diabetic neuropathy*. NGF is depleted in early human diabetic neuropathy skin, in correlation with dysfunction of nociceptor fibers. We proposed rhNGF prophylaxis may prevent diabetic foot ulceration. Clinical trials have been disappointed, probably related to difficulty delivering adequate doses and need for multiple trophic factors. NGF and glial cell line-derived neurotrophic factor (GDNF) are both produced by basal keratinocytes and neurotrophin (NT-3) by suprabasal keratinocytes: relative mRNA expression was significantly lower in early diabetic neuropathy skin compared to controls, for NGF ($P < 0.02$), BDNF ($P < 0.05$), NT-3 ($P < 0.05$), GDNF (< 0.02), but not NT4/5, Trk A or p75 neurotrophin receptor (all $P > 0.05$). Posttranslational modifications of mature and pro-NGF may also affect bioactivity and immunoreactivity. A 53 kD band that could correspond to a prepro-NGF-like molecule was reduced in diabetic skin. *Traumatic neuropathy and pain*. While NGF levels are acutely reduced in injured nerve trunks, neuropathic patients with chronic skin hyperalgesia and allodynia show marked local increases of NGF levels; here anti-NGF agents may provide analgesia. Physiological combinations of NGF, NT-3 and GDNF, to mimic a 'surrogate target organ', may provide a novel 'homeostatic' approach to prevent the development and ameliorate intractable neuropathic pain (e.g., at painful amputation stumps).

Keywords: neuropathies; sensory dysfunction; NGF; rhNGF; GDNF; diabetes; leprosy; pain; therapy

Introduction

A neurotrophic factor may be defined as a substance that plays a role in the development, maintenance or regeneration of the nervous system. Trophic factors act via their high-affinity receptors on specific nerve cells to influence their survival and gene expression. The classical neurotrophic factor, beta nerve growth factor (βNGF), is a protein normally produced by cells in the target organ, such as skin; NGF is taken up by sympathetic and small sensory fibers via a high-affinity tyrosine kinase A (TrkA) receptor and retrogradely transported to the cell body. NGF plays a key role in the survival and properties of sympathetic and sensory neurones (Levi-Montalcini, 1987) and in the development of

*Correspondence to: P. Anand, Peripheral Neuropathy Unit, Imperial College London, Department of Neurology, Hammersmith Hospital, Du Cane Road, London W12 ONN, UK. Tel.: 020-8383-3309/3319; Fax: 020-8383-3363; E-mail: p.anand@imperial.ac.uk

DOI: 10.1016/S0079-6123(03)46030-5

human neuropathies (Anand et al., 1996). In addition, there is evidence that βNGF plays an important role in inflammatory processes (Levi-Montalcini, 1987; Otten et al., 1994; Levi-Montalcini et al., 1996), including inflammatory pain (McMahon et al., 1995; Woolf, 1996).

A number of factors trophic to sensory and autonomic fibers have been discovered, some of which belong to the NGF-superfamily, such as NT-3, while others do not, such as GDNF. There is a family of *Trk* genes and their products (TrkA-C) represent receptors for the NGF family—NGF, BDNF, NT-3 and NT-4/5. Subpopulations of primary afferent fibers have differential expression of the trks, which correspond to their dependence on different neuro-trophins. NGF is trophic to sympathetic and unmyelinated (including nociceptive) neurones (via TrkA), NT-3 (via TrkC) is trophic to large myelinated and small unmyelinated (proprioceptive, mechanosensitive) sensory fibers. The high affinity Trk receptors are coexpressed with the shared low-affinity neurotrophin receptor (p75^{NTR}). It is of interest that a target mutation of the gene coding for even the p75^{NTR} in the mouse leads to markedly decreased substance P (SP) and GRP innervation of footpad skin and to development of ulcers and mutilation of the feet (Lee et al., 1992).

Glial cell line-derived neurotrophic factor (GDNF) was isolated and cloned from the B49 glial cell line. It has been shown to have a broad range of neurotrophic actions both in vivo and in vitro. An age-dependent responsiveness to GDNF for sympathetic, parasympathetic, proprioceptive, small and large cutaneous sensory neurons and enteroceptive cells has been reported. The GDNF receptor is a multi-component receptor, consisting of a ligand-binding, cell surface glycosyl-phosphatidylinositol (GPI)-linked protein GDNFR-alpha and a trans-membrane orphan tyrosine kinase, c-ret (Trupp et al., 1996). Binding of GDNF at the cell surface leads to phosphorylation of c-ret and initiation of the signaling cascade. C-ret and GDNF-deficient mice have been shown to suffer defects in the development of their sympathetic and enteric nervous system. A subset of nociceptors switch their dependency from NGF to GDNF for postnatal survival in the rodents; both NGF and GDNF are produced by basal keratinocytes in the skin.

Animal models show that neurotrophic factor deficiency may play a role in the development of common acquired peripheral neuropathies and exogenously administered neurotrophic factors may provide new treatments (see Lewin and Mendell, 1993; Brewster et al., 1994). Crushing or cutting a peripheral nerve leads to loss of target-derived NGF; although there is induction of NGF synthesis in Schwann cells at the site of the lesion and in the distal stump, the amount of NGF available to the dorsal root ganglion is insufficient to maintain substance P (SP) levels and NGF receptors are downregulated in injured fibers. In diabetic rats, NGF levels are decreased in peripheral nerves and their target organs (e.g., skin) and there is a defect of axonal transport of NGF and of Trk A expression.

We have described the first human neuropathy attributed to deficiency of a neurotrophic factor, NGF (Anand et al., 1991) and studied endogenous NGF levels in patients with nerve trauma (Anand et al., 1997), diabetes mellitus (Anand et al., 1996) and leprosy (Anand et al., 1994), the most common causes of human peripheral neuropathy world-wide. We have also reported regional changes of neuro-trophic factor levels in patients with ALS/ motoneuron disease (Anand et al., 1995b). These studies aim to provide a rational basis for the clinical use of neurotrophic agents in peripheral neuropathy.

Hereditary and developmental neuropathies

Failure of trophic interactions between the target organ and its innervation may result in nerve dysfunction, degeneration and abnormal regeneration. Mutations of the NGF high-affinity receptor Trk A have been found in congenital insensitivity to pain and anhidrosis (CIPA), also termed hereditary sensory and autonomic neuropathy type IV (HSAN IV) (Indo et al., 1996); these are likely to be partial loss-of-function mutations, as axon-reflex vasodilatation and sweating can be elicited albeit reduced (our unpublished observations in CIPA patients), suggesting rhNGF treatment might restore nociception in some patients. To illustrate, in one patient with CIPA, intradermal capsaicin induced skin axon-reflex vasodilatation was measurable by laser Doppler fluxmetry but reduced to 50% of control values, while axon-reflex sweating induced by intradermal nicotine and measured by an evaporimeter was 40% of control values (for test details see Anand, 1992; Anand et al., 1996). CIPA is an autosomal recessive disorder, with failure to develop protective sensation which leads to nonpainful injury and trophic mutilation, loss of sweating which leads to fever and mental retardation. Thus the three sets of neurons which depend on NGF in development—sensory, sympathetic and cholinergic forebrain neurons—appear affected. In keeping with the view that the mutations produce only partial loss of function, larger sensory fibers and autonomic control of blood pressure appear unaffected, although all sensory and sympathetic fibers are considered to require NGF at some point in development, at least in rodents: these relatively preserved fibers may thus have a greater 'safety factor' with regard to NGF, or other factors may compensate. A further possibility that deserves consideration derives from animal experiments where NGF deprivation at different stages of development may lead to altered peripheral and central connectivity and sensory phenotype (e.g., in the absence of NGF at a stage of development, developing cutaneous A delta high-threshold mechanoreceptors do not die, but innervate novel targets in dermis and become hair afferents instead (see Lewin and Mendell, 1996)). In congenital insensitivity to pain *without* anhidrosis (HSAN V) there does not appear to be a Trk A mutation, at least in one study; other deficits in the NGF-Trk A signaling pathway may be responsible. The report of a Trk A mutation in a case of HSAN V is based on the finding of predominantly A delta fibers loss on sural nerve biopsy, though the patient had anhidrosis and thus phenotypically had HSAN IV. Varying degrees of receptor functional deficits and the mechanisms in the animal experiments cited above may explain the different phenotypes. In some patients with HSAN V there may be some improvement in nociception over the years—it may be postulated that in these cases the surviving nociceptors that become GDNF-dependent may sprout or compensate.

The role of endogenous BDNF and NT-3 and their receptors and the low-affinity NGF receptor, is not known in human developmental neuropathies. Germline mutations of the RET proto-oncogene, a component of the GDNF receptor complex, lead

either to multiple endocrine neoplasia (MEN) 2a, 2b, or to Hirschsprung's disease, depending on the location and nature of the base substitution. Gain of function mutations of RET exon 10 give rise to MEN 2a, where quantitative sensory and autonomic deficits are similar to early diabetic neuropathy (Knowles et al., 1998). C-ret mutations, which lead to multiple endocrine neoplasia Type 2b in humans, also involve sensory and autonomic deficits by quantitative tests (Dyck et al., 1979).

Leprous neuropathy

Leprosy affects between 10 and 15 million people: the earliest reported nerve lesions in human leprosy and animal models are in unmyelinated fibers and their Schwann cells (Shetty et al., 1988), in accord with early loss of pain sensation and trophic changes. Schwann cells of unmyelinated fibers serve as the host of *Mycobaterium leprae*. The skin lesions in the early indeterminate and tuberculoid forms of leprosy, which are superficial and relatively well circumscribed, provide a unique opportunity to study the role of NGF in neuropathy. These lesions show hypoalgesia and hypopigmentation, in addition to hypohidrosis. Although anti-bacterial drugs are effective, failure of nerve regeneration, especially nociceptor sprouting within skin, leads to trophic changes, which remain a major cause of disability.

NGF-related mechanisms were implicated in animal models of leprosy and in human leprous tissues. In a mouse foot-pad model of leprosy, SP levels were reduced in sciatic nerve and ipsilateral spinal cord (Anand et al., 1983a). SP-immunoreactive fibers were undetectable in skin biopsies from patients with leprosy, although markers for the presence of nerve fibers (PGP 9.5, neurofilaments) were seen in all leprosy cases of indeterminate type and a proportion of tuberculoid and lepromatous cases (Karanth et al., 1989). NGF levels were decreased in leprosy-affected human skin and nerve (Anand et al., 1994).

While sensory loss in leprosy skin is the consequence of invasion by *M. leprae* of Schwann cells related to unmyelinated fibers, early loss of cutaneous pain sensation, even in the presence of nerve fibers and inflammation, is a hallmark of leprosy and requires explanation. In normal skin, NGF is produced by basal keratinocytes and acts via its Trk A receptor on nociceptor nerve fibers to increase their sensitivity, particularly in inflammation. We therefore studied NGF- and Trk A-like immunoreactivity in affected skin and mirror-site clinically-unaffected skin from patients with leprosy and compared these with nonleprosy, control skin, following quantitative sensory testing at each site (Facer et al., 1998, 2000). Sensory tests were within normal limits in clinically-unaffected leprosy skin, but markedly abnormal in affected skin. Subepidermal PGP 9.5- and Trk A-positive nerve fibers were reduced only in affected leprosy skin, with fewer fibers contacting keratinocytes. However, NGF-immunoreactivity in basal keratinocytes and intra-epidermal PGP 9.5-positive nerve fibers, were reduced in both sites compared to nonleprosy controls, as were nerve fibers positive for the sensory neuron specific sodium channel SNS1/PN3, which is regulated by NGF and may mediate inflammation-induced hypersensitivity. Keratinocyte Trk A expression (which mediates an autocrine role for NGF) was increased in clinically affected and unaffected skin, suggesting a compensatory mechanism secondary to reduced NGF secretion at both sites. We concluded that decreased NGF- and SNS/PN3-immuno-reactivity and loss of intra-epidermal innervation, may be found without sensory loss on quantitative testing in clinically-unaffected skin in leprosy; this appears to be an early subclinical change and may explain the lack of cutaneous pain with inflammation. Sensory loss occurred with reduced subepidermal nerve fibers in affected skin, but these still showed Trk A-staining, suggesting NGF treatment may restore pain sensation.

A study in leprosy patients showed no difference in the number of melanocytes and amount of pigment in hypopigmented lesions when compared to adjacent normal skin: it was suggested that the hypopigmentation could be caused by defective transfer of melanin into keratinocytes (Shereef, 1992). Cell culture studies of melanocytes show that they express functional NGF receptors, that NGF is chemotactic to melanocytes and increases the dendricity of melanocytes (Yaar et al., 1991): pigment is transferred from melanocytes to keratinocytes to determine skin color. The evidence cited above has led to our hypothesis

that nerve fibers and melanocytes are deprived of NGF in leprosy.

A number of the clinical and neurochemical changes in leprosy skin described above thus appear to be the opposite of changes in models of sunburn . Ultraviolet irradiation of skin produces erythema, hyperpigmentation and pain: it induces $mRNA^{NGF}$ in cultured keratinocytes and long-term increase in CGRP levels in dorsal spinal cord. It may be speculated that sunburn is, in a sense, the 'opposite' of leprosy, with levels of NGF expressed by basal keratinocytes driving the changes.

Diabetic sensory polyneuropathy

Background

Diabetic neuropathy comprises of a number of clinical presentations that are likely to be caused by different mechanisms, which may coincide in the same patient (see Thomas and Tomlinson, 1993). The prevalence of neuropathy rises to about 50% after 25 years of diabetes (Pirart, 1978). It has been classified into symmetric and focal neuropathies. Neurotrophic mechanisms are more likely to involve symmetric polyneuropathy rather than focal neuropathy. Sensory and autonomic polyneuropathy is a common form of neuropathy in diabetic patients, for which no specific and effective treatment is available. Diabetic sensory and autonomic polyneuropathy affects both nerve fibers that are dependent on NGF and those that are not. It was hypothesized that NGF deprivation may determine its presentation, although metabolic or vascular abnormalities may be the cause of the neuropathy (Anand, 1992).

Role of NGF

Our studies in insulin-dependent young adult diabetics showed an early length-dependent dysfunction in small-diameter sensory but not sympathetic or larger sensory fibers, in accord with previous reports (Guy et al., 1985). NGF depletion correlated significantly with decreased capsaicin-evoked skin axon-reflex vasodilatation (Anand et al., 1996). Axon-reflex vasodilatation is a test of unmyelinated afferent fibers, the flare component of Lewis' triple response in skin. Intradermal capsaicin was used to induce the flare, as it selectively activates nociceptive fibers. The increased capillary flux, as a result of vasodilation, was measured using laser Doppler fluxmetry and may be a good surrogate for NGF activity (see Anand, 1992). NGF immunostaining was strongest in the basal keratinocyte layer in control skin and was decreased in diabetic skin. NGF deprivation, from target organ failure and decreased axonal transport, may thus reduce chemo- and warm/heat pain sensitivity in early diabetic neuropathy, even in asymptomatic diabetes. As loss of nociception and axon-reflex vasodilatation in the feet have been shown to contribute to foot ulceration (Parkhouse and Le Quesne, 1988), a major and serious complication, early and prolonged NGF treatment at an appropriate dose may provide prophylaxis.

More severe human diabetic neuropathy with denervation and coexistent ischemia or inflammation, may be associated with increased mRNA^{NGF} (Diemel et al., 1999) and Trk A/C mRNA (Terenghi et al., 1997) expression by keratinocytes, possibly as a compensatory mechanism. Upregulation of the receptors may result from decreased secretion of the respective proteins: an autocrine mechanism has been established for NGF in cultured human keratinocytes.

Relationship to risk factors

NGF-related mechanisms may at least partly explain the risk factors for diabetic neuropathy, which include duration of diabetes, age, male sex and height. The length-dependent effect may result from abnormalities of axonal transport, including retrograde transport of NGF, as observed in diabetic rats (Schmidt et al., 1985). Ageing appears to recapitulate these clinico-pathological changes and exacerbates them in diabetic subjects (our observations). NGF levels appear higher in female calf skin, as do effects of exogenous NGF administration: oestrogens/proestrus upregulate NGF receptor mRNA in sensory neurons (Sohrabji et al., 1994) and testosterone reduces mRNA^{NGF} in cultured fibroblasts (Siminowski et al., 1987).

Roles of BDNF, NT-3, GDNF and receptors

Clinical trials have been disappointing (see below), suggesting further understanding of underlying molecular mechanisms is necessary. We have therefore studied in early diabetic neuropathic skin the expression and posttranslational processing of NGF and other key trophic factors. NGF and GDNF are both produced by basal keratinocytes but are trophic to distinct subsets of unmyelinated afferents, BDNF is trophic to hair follicle afferents, and NT-3 is trophic to Merkel cell afferents and a subset of unmyelinated afferents. Relative mRNA expression was significantly lower in diabetic skin compared to controls for NGF ($P < 0.02$), BDNF ($P < 0.05$), NT-3 ($P < 0.05$), GDNF (< 0.02), but not NT4/5, Trk A or p75^{NTR} (all $P > 0.05$) (Sinicropi et al., 2001). Multiple trophic factor expression deficits thus precede clinically-detectable nerve dysfunction.

We have demonstrated the presence of NT-3 in the suprabasal epidermis, where many unmyelinated sensory fibers terminated (Kennedy et al., 1998). As these fibers are affected in early diabetic neuropathy and a clinical trial of recombinant human NT-3 in diabetic neuropathy is in progress, we have investigated the concentration of endogenous NT-3 in skin of 24 patients at different stages of diabetic polyneuropathy. NT-3 concentrations, measured with a specific immunoassay, were significantly higher in affected skin biopsies from patients with diabetic neuropathy than matched control skin, particularly in the later stages. The optical density of NT-3-immunostaining was also significantly greater in the epidermis in diabetics. No correlation was found between individual quantitative sensory tests and the increase of NT-3 concentration. The increase of NT-3 appears to reflect the degree of skin denervation in diabetic neuropathy and may represent a compensatory mechanism. The concentrations of NT-3 in other peripheral targets (e.g., muscle) deserve study in diabetic neuropathy.

Relationship to neuropeptides

Changes in neuropeptide expression may reflect the role of neurotrophic factors in peripheral neuropathy.

Decreased SP and increased VIP expression in sensory fibers has been found in leprosy, wound healing, nerve injury, diabetes and after capsaicin application to nerve trunks. In accord, clinical and experimental rat studies of skin axon reflex vasodilatation and neurogenic extravasation, which are at least partly mediated by SP and CGRP in unmyelinated afferent fibers, show abnormalities in the above conditions. The degenerative changes in injured and diabetic model sensory fibers, including downregulation of SP and neurofilament synthesis and axonal atrophy, can be reversed by administration of exogenous NGF (see Brewster et al., 1994).

NGF may regulate small sensory fiber sensitivity and function directly, or via changes in expression of their neuro-effector agents SP and CGRP (see Anand, 1995). In our studies, SP, but not CGRP, was significantly depleted in skin from diabetic patients with mild neuropathy. In accord with the functional studies, there may be sparing in mild diabetic neuropathy of larger sensory fibers that contain CGRP (but not substance P), which are not dependent on NGF. Alternatively, a subset of CGRP-containing sensory fibers innervating blood vessels, particularly related to sweat glands, may have better NGF availability than fibers that contain both SP and CGRP and take up NGF from keratinocytes. There is support for these explanations from previous immunocytochemical studies in calf skin from similar patients where variable and even increased numbers of CGRP-immunostaining fibers were reported in the dermis and related to sweat glands and from diabetic rats, where an early increase of CGRP-immunoreactive nerves was reported in skin. Endogenous NGF has been shown to regulate collateral sprouting sensory fibers in rat skin (Diamond et al., 1992).

Pain

NGF may regulate nociception in human neuropathies, with implications for clinical trials. Less than 10% of patients with diabetic neuropathy in most large series develop clinically significant persistent pain (see Pirart, 1978; Thomas and Tomlinson, 1993). Among these are unusual cases with distal hyperalgesia and allodynia, who have early or mild neuropathy, with preservation of large sensory fiber function. In some patients, pain may be precipitated, paradoxically, by treatment of the diabetes or improvement of its control and the pain is usually self-limiting. It has been hypothesized (Anand, 1995) that fewer (sprouting) fibers may be exposed to 'relative excess' of NGF in such cases of early painful neuropathy in humans and in rats. Most cases, including those in our study, may have different NGF-related and unrelated nociceptive mechanisms in operation and show no simple correlation of pain with morphological change in nerve biopsies (see Thomas and Tomlinson, 1993). In such cases, chronic systemic rhNGF treatment (e.g., at doses that just produce local hyperalgesia) may 'paradoxically' both improve sensation progressively from short to long fibers and ameliorate or prevent deafferentation pain. Sympathetic agents may increase NGF synthesis and secretion (Tuttle et al., 1993), with implications for NGF and its use in neuropathy with established sympathetically-driven pain.

Cellular mechanisms

A key question is why NGF and other trophic factors are reduced in early diabetic neuropathy skin. One simple explanation is nonenzymatic glycation, which may reduce NGF bioactivity and immunoreactivity (our unpublished observations, with D Sinicropi and colleagues). Although a number of potential agents, including cytokines and hormones, have been shown to affect NGF expression in cultured cells, it is not known whether these mechanisms operate in skin cells in human diabetic neuropathy, or after denervation. Reduced NGF retrograde transport alone, while accounting for the length-dependent effect, should not affect skin NGF levels, unless this results from a secondary failure of nerve-skin interactions. This is unlikely, as acute denervation is associated with increased NGF expression in rat skin (Mearow et al., 1993); in diabetic rats, $mRNA^{NGF}$ is reduced in skin (Fernyhough et al., 1994), while NGF protein but not the NGF receptor is reduced in nerve (Hellweg et al., 1991, 1994). Another potential simple explanation for the decrease of NGF in diabetic skin was that it results from decreased keratinocyte turnover: stratum corneum cell turnover is delayed in human diabetic and ageing skin and failure of an autocrine effect observed in vitro may further contribute to decreased NGF production (Di Marco et al., 1993). Cell culture studies show that NGF expression is highest in rapidly dividing keratinocytes (Di Marco et al., 1993; Anand et al., 1995a), in accord with in situ hybridization studies in rat skin (Mearow et al., 1993) and our immunostaining studies in human skin (Anand et al., 1996). However, our studies of markers of cell proliferation and apoptosis failed to show changes (unpublished observations). It has been suggested that corticosterone may decrease NGF expression in diabetes, as it reduces NGF expression in cultured firbroblasts and may be increased in streptozotocin-treated rats (Nevue et al., 1992). However, the relevance of this observation to human diabetics is unclear and there does not appear to be a significant neuropathy in diseases where corticosteroid levels are markedly elevated.

NGF may regulate small sensory fiber sensitivity and function directly, or via changes in expression of their neuro-effector agents SP and CGRP (see Anand, 1995). In our studies, SP, but not CGRP, was significantly depleted in skin from diabetic patients with mild neuropathy. In accord with the functional studies, there may be sparing in mild diabetic neuropathy of larger sensory fibers that contain CGRP (but not SP), which are not dependent on NGF. Alternatively, a subset of CGRP-containing sensory fibers The role of NGF production by Schwann cells and their expression of NGF receptors in Wallerian degeneration is not known. Diabetic nerves do not show a deficit of upregulation of p75^{NTR} NGF receptors by Bungner bands immunocytochemically (Bradley et al., 1995). There is little evidence for antibodies to NGF in diabetic neuropathy (Zanone et al., 1994). The expression of TrkA, TrkC or the GDNF-receptor complex in human diabetic sensory or sympathetic neurones and their role in pathogenesis, is not known and deserves study.

NGF precursor proteins

Potential functional roles of the different NGF precursor derived proteins deserve investigation, particularly in sensory neuropathy and inflammatory pain. The nucleotide sequence of the cDNA clone predicts that NGF is synthesized as part of a precursor protein, prepro-NGF. Both mouse and human genes are highly homologous but not identical (Ullrich et al., 1983). After removal of the hydrophobic signal peptide, the precursor is expected to generate pro-NGFs of either 34 kD or 27 kD depending upon the size of the NGF-transcript (Scott et al., 1983; Ullrich et al., 1983; Edwards et al., 1986). How these precursors are processed in different tissues is not yet clearly understood. The presence of three double basic residues and three N-linked multiple glycosylation sites within the precursor suggests the possibility that they can generate multiple molecular forms having molecular weights greater than that predicted from the precursor amino-acid sequence.

In male mouse salivary glands NGF is synthesized at exceptionally high levels, where it associates with two kallikrein subunits known as αNGF and γNGF to form a 7S complex. In these glands, immunoblot studies have shown the presence of an additional, larger, 53 kD molecular form of prepro-NGF, as well as the presence of the fully processed 13.5 kD form of βNGF (Lakshmanan et al., 1988). These authors suggested that one of the NGF precursors, or a product derived from it, may undergo posttranslational modification, including glycosylation, to yield the 53-kD prepro-NGF form. This high molecular weight NGF-like protein was shown to be biologically active, as it stimulated PC-12 cell flattening, neurite formation and augmented cell survival (Levi-Montalcini, 1987). The 53 kD molecular form is released in the circulation in aggressive adult male mice (Lakshmanan, 1987) and increased in inflamed colon in a rat model of colitis (Reinshagen et al., 1995). The presence of similar prohormone isoforms in commercial recombinant mouse and human NGF preparations has recently been described (Reinshagen et al., 1997, 2000).

We too have noted in several Western blot experiments that an affinity-purified antibody raised against recombinant human NGF consistently detects the 53 kD form in human tissue extracts (Yiangou et al., 2002). This may correspond to a novel neurotrophin, or more likely to the previously described larger modified precursor of NGF. We have detected, on Western blots of human and rat tissue extracts under reducing conditions, large molecular forms of NGF-like immunoreactivity, but not the 14-kD forms. The failure to detect the 14-kD forms may be expected, because of its low levels in tissue extracts. The level of NGF-immunoreactivity in skin by different enzyme linked-immunoassays (ELISA) for mature NGF is between 0.5 and 1.5 pg per mg-wet weight of tissue (Anand et al., 1996), well below the detection limit of our Western blots. When blots were performed with rhNGF under nonreducing conditions, a 28-kDa band, presumably representing the NGF dimer, was consistently detected. This demonstrated the high affinity of the two NGF monomers for themselves. Similar results have also been described for the neurotrophin GDNF where it was shown to run as a dimer of apparent molecular weight 32–42 kD under nonreducing SDS gels, compared with 18–22 kD on reducing gels (Lin et al., 1993).

The larger molecular forms, particularly the 53 kD band, may not be detected in ELISA even if they are more abundant, as their immunoreactive epitopes may be masked in nondenaturing conditions; electro-phoretic denaturing conditions may dissociate the larger NGF-like forms from associated molecules, exposing epitopes that are immunoreactive with anti-NGF antibodies. The Western blot results suggest that these larger NGF-like forms are much more abundant in tissue extracts than the mature form. The low levels of NGF-immunoreactivity measured in these extracts by ELISA therefore suggests that the ELISA does not detect the larger molecular forms significantly and certainly not fully, presumably because the proteins are not denatured in extracts for the ELISA.

It is therefore also unlikely that the reduced concentrations of NGF-immunoreactivity in human diabetic skin extracts measured by ELISA (Anand et al., 1996) are related to differences in concentra-tions of the larger molecular forms, even though these appeared markedly reduced in our Western blots. We need to study a larger sample of diabetic skin extracts, both from patients with early diabetic neuropathy and those with severe skin denervation and from patients with other types of neuropathy, to determine the significance of the apparent decrease of the 53 kD NGF form in diabetic extracts. The rate of

synthesis, processing, secretion, degradation and uptake by cells of NGF precursor-derived molecules will need to be studied, as these may all contribute to loss of NGF-like activity in diabetic skin.

Our results are consistent with the recent finding of large NGF precursor proteins and absence of mature NGF in human prostatic stromal cells (Delsite and Djakiew, 1999). Further studies are necessary to quantify and establish the identity of the 53-kD form, including sequencing its N-terminus. A 66-kD band was also seen and NGF-like bands of similar magnitude have been reported in human prostate-derived cells (Djakiew et al., 1991).

Our finding of high molecular weight NGF-like precursors with Western blot in intact nerves and dorsal root ganglia might indicated that the larger molecular weight form may itself also be secreted from skin and transported axonally. Antibodies to mature NGF or N-terminal sequences of pro-NGF showed the same immunostaining pattern in skin as previously published with anti-rhNGF antibody (Anand et al., 1996; Yiangou et al., 2002). Antibodies to both mature NGF and prepro-NGF N-terminal sequences immunostained a subpopulation of nerve fibers in normal rat sciatic nerve, although immu-nostaining with prepro-NGF N-terminal sequence antibodies was relatively weak. These fibers had an immunostaining pattern similar to antibodies to CGRP and both are likely to have immunostained sensory afferent fibers. After inflammation and 24 h of nerve ligation, fibers were easily detected distal to the ligation site with prepro-NGF-like antibodies, indicating a retrograde transport. While Schwann cells produce NGF when they have lost contact with axons, as in Wallerian degeneration, normal nerve is not expected to synthesize NGF locally. The classical proposal (Levi-Montalcini, 1987) that synthesized NGF is stored as a large modified unprocessed form in skin, which then gets processed to the mature NGF, secreted and transported by nerve, may therefore need to be readdressed. NGF precursor forms and processed or degraded fragments may be, potentially, taken up by nerves either by Trk A or p75^{NTR}, the high- and low-affinity receptors for NGF, respectively. It is unlikely that the immuno-reactivity in nerve fibers in ligated nerves represents NGF production in Schwann cells, as the duration of ligation is too short (24 h) to produce this effect several millimeters distal to ligation and a similar pattern was seen in intact nerves. The immunostaining of DRG cell bodies is also in accord with this view, as discussed below. Double-staining studies are in progress, to colocalize NGF-like immunoreactivity with CGRP and other markers in this model. Immunostaining with antibodies to CGRP and NT-3 confirmed the appropriate direction of transport of the neurotrophins and neuropeptide.

Surprisingly, immunostaining of rat L4 DRG showed the strongest staining for prepro-NGF N-terminal sequence antibodies in large/medium-sized neuronal cell bodies, whereas small-sized cells alone were immunostained with antibodies to mature rhNGF, as expected. High-affinity NGF receptors, Trk A, are known to be present in a few large sensory neurones, but their role here is not known; large diameter sensory cells commonly express $p75^{NTR}$, which may also take up and transport NGF molecular forms axonally. However, these results should be interpreted with caution, as the specificity of the immunostaining with the prepro-NGF N-terminal sequence antibodies has to be studied further with preabsorption—the appropriate peptides are not yet available. One possible explanation is that the limited quantities of mature endogenous NGF may compete favourably for $TrkA/p75^{NTR}$ in small sensory neurones and that the precursor protein or N-terminal products may be transported by $p75^{NTR}$ in some larger sensory fibers—the sensitivity of the detection method may have led to our result. Retrograde transport studies of radiolabeled prepro-NGF, its cleaved peptides and breakdown fragments are necessary to establish their translocation to different sized sensory cell bodies and physiological studies clearly are necessary to establish their biological significance.

In a recent study (Dicou et al., 1997), it was shown that the pro-region of the NGF precursor protein contains two bioactive peptides, corresponding to the sequences -71 to -43 (LIP1) and -40 to -3 (LIP2) respectively, which are flanked by pairs of basic residues, suggesting points of cleavage. ELISA and immunohistochemistry demonstrated these peptides in rat intestine. The comparative neurobiology of the different molecular forms derived from prepro-NGF, in particular their differential changes with inflammation and neuropathy, is likely to be of great interest in pain mechanisms and may lead to new clinical treatments.

Clinical trials with neurotrophins

Clinical trials have been performed with NGF (Apfel et al., 2000) and BDNF (Wellmer et al., 2001) in diabetic neuropathy and NGF in HIV neuropathy (Schifitto et al., 2001). While the Phase II trial of NGF in diabetic neuropathy showed some promise, the later substantive trial did not confirm this. It is probable that the dose of NGF was insufficient to produce an effect in sensory neurons, being limited by local tenderness at the injection site: it needs to be considered whether the latter, given some of its features, is not at least partly due to an inflammatory focus and not a pure neurotrophic effect, as Trk A is expressed by inflammatory cells which accumulate at the site of NGF injection in rats (Bennett et al., 1998). The doses required for efficacy in animal models exceeded the dose given in the clinical trial. Depletion of circulating neutrophils abolished the NGF-related thermal hyperalgesia in rat paw (Bennett et al., 1998).

We conducted a randomized double-blind placebo-controlled study of rhBDNF in diabetic neuropathy. There were no significant overall differences on analysis of variance between groups; however, when a subgroup of patients with abnormal cool detection threshold (CDT) at recruitment were analyzed with nonparametric tests, the rhBDNF treatment but not placebo groups were indistinguishable from matched normal subjects at all time-points after 43 days: 5 of 8 such treated subjects were now within the normal CDT range, which has not been reported previously in this condition for any treatment. Skin biopsies failed to show evidence of structural change, suggesting a possible functional effect of rhBDNF. Interestingly, assessment of innervation of hair follicles was not possible because of the marked loss of this end-organ in diabetic neuropathic skin. The only side-effects of rhBDNF were infrequent non-painful injection-site reactions and increased gut motility at the higher dose. rhBDNF may thus improve CDT and constipation in diabetics, although a larger and longer trial is warranted.

Traumatic neuropathy

Nerve injury, particularly of the brachial plexus in obstetric and road traffic accident cases, may result in lifelong disability and chronic pain, despite advances in reconstructive surgery (Berman et al., 1998); nerve fiber degeneration and poor regeneration account for the failure of the surgery. A number of different traumatic lesions may result in NGF deprivation of sensory neurones, including the removal of the target organ, cutting or crushing the nerve and blockade of axonal transport. Crushing a peripheral nerve results in the induction of NGF synthesis cells at the site of the lesion and the distal stump: the upregulation of NGF synthesis after injury has been attributed to factors released by invading macrophages, including interleukin-1 (Heumann et al., 1987). The amount of NGF available to the injured fibers is insufficient to restore SP levels; NGF receptor downregulation in injured fibers may contribute to this insufficiency. However, when regeneration is complete, the target tissue is able to supply sufficient NGF to restore SP levels and function in sensory fibers.

We have studied NGF expression in patients with peripheral nerve injury (Anand et al., 1997). In biopsies taken proximal and distal to the injured region from patient undergoing peripheral nerve repair, NGF levels were reduced when compared to intact nerve, but were generally higher acutely (less than 3 weeks) in distal when compared to proximal segments in the more complete nerve lesions. NGF staining was present in Schwann cells in distal segments, including pockets of high expression in neuromas, but not in proximal segments or control nerves. Our findings suggest that the proximal stump and cell bodies are deprieved of NGF; animal models indicate that adequate NGF treatment may prevent or reverse the reduced expression of neuropeptides, NGF receptors and cytoskeletal proteins in injured nerves (Lindsay and Harmar, 1989; Gold et al., 1991) and thereby ameliorate the degenerative changes which limit the success of surgical repair. Exogenous NGF treatment may also enhance nerve regeneration (Whitworth et al., 1996).

Pain

In patients with peripheral nerve injury, it has been hypothesized that fewer axonal sprouts with less competition for normal or even reduced NGF levels, but *relative* excess ('TrkA afferent—NGF disproportion'), either in nerve trunks or the target organ, could lead to hyperalgesia (Anand, 1995): all our adult plexus injury patients reported significant chronic pain and a relative excess of NGF in the target organ may also explain the borderzone and reinnervation hyperalgesia seen in some of these patients.

There is evidence that some processes involved in determining the course of pain following nerve injury may not be related solely to NGF in the periphery. In comparing the consequences of nerve injury in human and rat neonates and adults, on the adults develop intractable neurogenic pain or autonomy; as there does not appear to be a qualitative difference in NGF changes in neonatal and adult injured nerve, the lack of chronic pain in neonates may result from spinal cord plasticity and adaptation, which may fail in adults (Anand, 1995).

Exogenous NGF may reverse hypoalgesia and in excess may produce hyperalgesia by: (i) directly sensitizing nociceptors, (ii) increasing levels of SP and CGRP, which may play a role in central sensitization and neurogenic inflammation, and (iii) local effects, such as release from mast cells (see Lewin and Mendell, 1993; McMahon et al., 1995). NGF also regulates ion channels and receptors that may be increased in chronic pain states (Coward et al., 2000; Yiangou et al., 2001). Although largely speculative at this stage, human conditions may present with pain related to alterations of NGF activity: these may provide suitable models for the study of mechanisms of NGF and pain in man and new NGF-related prophylaxis and therapies (Anand, 1995). The 'hyperalgesic' conditions include arthritis, some small fiber neuropathies (including erythromelalgia), painful hypertrophic scars, sunburn, urinary bladder pain, migraine and vascular head pain. The 'hypoalgesic' group, with decreased NGF activity, includes leprosy neuropathy. Other conditions, including the major neuropathies that follow trauma and diabetes, are more complicated, as discussed above and may display different NGF-related features during their development.

NGF is increased in inflammatory conditions (Donnerer et al., 1992; Anand, 1995; Oddiah et al., 1998); reinnervation of pockets of relatively high NGF expression, disproportionately few regenerating fibers exposed to normal levels of NGF (with a relatively high quantity of NGF taken up by each fiber), or increased susceptibility of nerve sprouts to NGF, could also lead to hyperalgesia. This hypothesis unifies the role of NGF in inflammatory and neurogenic pain. In such conditions, anti-NGF treatments may reduce hyperalgesia.

However, NGF may play a different role in the development of chronic pain that results from cell death or atrophy and failure of adaptation in the spinal cord, for example in deafferentation pain. Changes of NGF expression may, at the early stages of nerve injury or disease, form part of an adaptive response. Failure of this response, or of secondary adaptation in the dorsal spinal cord, may contribute to the development of chronic pain. NGF may thus, administered appropriately, provide prophylaxis and treatment in these conditions.

While tissue engineering has been applied mainly to bridging large gaps at sites of nerve injury, recent advances in the neurobiology of injured nerves indicate that other strategies would be immediately feasible for common and important unmet clinical needs i.e., nerve regeneration across great distances in human limbs, functional recovery at the target organ and relief of chronic neuropathic pain. Our findings in injured human nerves (Anand et al., 1997; Bar et al., 1998), including gene-chip DNA microarray studies, demonstrate a cascade of events at different sites (e.g., for sensory neurons, in cell bodies and central and peripheral terminals); and at different times (i.e., acute and chronic changes, but also remarkable persistent 'switches' in sets of genes). These represent new therapeutic targets for nerve regeneration and pain. Thus, tissue engineering strategies for sensory nerve repair should be directed at each of the following:

(1) dorsal root ganglion, where trophic therapy may directly rescue sensory cells or maintain regenerative phenotypes. The advantages include: (i) circumvents known deficiency of axonal transport in injured nerves, with failure to transport factors from the injury site (ii) avoids systemic side-effects, or inflammatory cytokine-like effects at the injury site which may exacerbate pain, (iii) maintains neurons if immediate nerve repair is not possible e.g., infection at injury site, (iv) maintains neurons in regenerative mode without need to match trophic factor levels to variable location of sprouts and when length of nerve regeneration is great e.g., reinnervation of hand after brachial plexus injury, and (v) delivers treatment for intractable pain without systemic effects; (note that sensory soma develop ectopic impulse activity after spinal root or peripheral nerve injury, which includes many patients with spondylosis and disc protrusions);

(2) site of nerve injury; following injury the nerve may require a bridging graft. To advance the bridging of gaps new scaffolds and bioengineering methods to enhance neurone extension towards targets (Whitworth et al., 1996). Scaffolds will incorporate spatially organized Schwann cell guidance channels, neurotrophic factor gradients and specialized extra-cellular matrix components;

(3) peripheral terminals, where (i) collateral nerve sprouting, and (ii) phenotypic maturation, must occur for useful functional recovery and prevention of painful hypersensitivity (note: lack of recovery of sensation results from failure of collateral sprouting, e.g., after nerve injury, nerve or skin grafts and poor sensory recovery following skin grafting predisposes to cutaneous breakdown, whereas immature regenerative sprouts can be painfully hypersensitive even in the event of successful nerve regeneration—tissue engineering strategies may help maintain/create a suitable target to regulate sprouts and their sensitivity);

(4) central terminals—apart from targeting central sensory fiber regeneration to restore sensation, a new, more feasible tissue engineering strategy would focus on alleviation of severe intractable central pain, which is a major unmet clinical need. In patients with brachial plexus avulsion injury we find a strong correlation between reduction in pain and successful nerve

regeneration (see Berman et al., 1998); however, in many patients with nerve injury e.g., amputations, there is failure of regeneration and severe chronic pain.

Current strategies for pain relief after peripheral nerve damage are limited in efficacy and by side-effects. They fall into the empirical (e.g., gabapentin) or partly mechanistic (e.g., sympathetic blockade in causalgia) categories. Based on our studies in injured human sensory neurones of old and novel 'sensory neurone specific' targets for analgesics (Coward et al., 2000), we believe that it is unlikely that any single target will be sufficient, even for symptomatic relief of different phenomena in a single patient. We therefore propose a new strategy, based on the axiom that successful nerve regeneration and subsequent maturation (interactions with the target organ), is nature's method and the best way to prevent or relieve chronic neuropathic pain. We would term this approach 'Homeostatic/Restorative' and use the techniques of tissue engineering to create a milieu that would mimic the optimal regenerative conditions or target organs (e.g., a mini-surrogate target organ in the case of painful neuroma at an amputation stump). Recent advances in our understanding of trophic factors (NGF, GDNF, NT-3) that regulate the structure and function of the entire range of human sensory fiber subtypes and available delivery strategies (using human fibroblast/keratinocyte/skin stem cells as autografts/allografts) make this a feasible and exciting clinical application prospect. There are no significant regulatory, ethical or clinical obstacles for our proposed approach.

The rationale for the pain studies is as follows. There are two main sets of small sensory fibers in humans and rodents that respond to noxious stimuli—those dependent on Nerve growth factor (NGF) and those on Glial derived neurotrophic factor (GDNF). In normal skin, NGF and GDNF expression is virtually restricted to the basal layer, whereas there is dense selective NT-3 expression in suprabasal keratinocyte layers (Kennedy et al., 1998). There is increased suprabasal NGF in inflammation which sensitizes sensory fibers terminating in the superficial epithelium, while NT-3 is necessary only for the maintenance of intra-epidermal branches (Woolf, 1996; Lowe et al., 1997, unpublished observations

with C. Woolf). Injections of NGF but not NT-3 produce significant long-lasting allodynia/hyperalgesia in human and rat skin and both are taken up by the same receptor (Trk A) in nociceptors. Conversely, NGF and GDNF are decreased and NT-3 increased in early diabetic neuropathy skin, where there is hypoalgesia (Anand et al., 1996; Kennedy et al., 1998; Sinicropi et al., 2001). NT-3 treatment of rats has been shown to be anti-nociceptive (Malcangio et al., 1997). Thus, altering the ratio of NGF and NT-3 in favour of the latter in a tissue-engineered surrogate target, such as fibroblasts overexpressing NT-3, implanted at sites of irreversible nerve injury, such as a human amputation stump neuroma, may ameliorate pain and hypersensitivity. This may be the mechanism underlying the present surgical procedure to implant damaged painful nerves into muscle, which has reduced levels of NGF compared to NT-3; however, this helps only for a limited period, as regenerating sprouts soon recontact cutaneous tissue and blood vessels, with a return to hypersensitivity. The effect would be enhanced if maintained at high pH, thereby possibly reducing activation of acid sensing ion-channels (ASICs 1,2,3) and VR-1, which are expressed by small sensory neurons in rodents and humans and mediate pain that accompanies tissue acidosis in inflammation, ischemia, trauma, or bone metastases (see Caterina et al., 1997; Waldmann and Lazdunski, 1998).

The neurotrophic approach is thus most rational and needs to be revisited in the treatment of human peripheral neuropathies.

Abbreviations

BDNF	brain-dervied neurotrophic factor
GDNF	glial cell line-derived neurotrophic factor
NGF	nerve growth factor
NT-3	neurotrophin-3
$p75^{NTR}$	p75 neurotrophin receptor
SP	substance P
TrkA	receptor tyrosine kinase A

References

Anand, P., Rudge, P., Mathias, C.J., Springall, D.R., Ghatei, M.A., Naher-Noe, M., Sharief, M., Misra, V.P., Polak, J.M., Bloom, S.R., Thomas, P.K. (1991) A new autonomic and sensory neuropathy with loss of sympathetic adrenergic function and sensory neuropeptides. Lancet, 337: 1253–1254.

Anand, P. (1992). The skin axon-reflex vasodilatation-mechanisms, testing, and implications in autonomic disorders. In: R. Bannister and C. Mathias (Eds.), Autonomic Failure, 3rd Edition. Oxford University Press, Oxford, pp. 479–488.

Anand, P. (1995) Nerve growth factor regulates nociception in human health and disease. Br. J. Anaesth., 75: 201–208.

Anand, P., Foley, P., Navsaria, H.A., Sinicropi, D., Williams-Chestnut, R.E. and Leigh, I.M. (1995a) Nerve growth factor levels in cultured human skin cells: effect of gestation and viral transformation. Neurosci. Lett., 84: 157–160.

Anand, P., Parrett, A., Martin, J., Zeman, S., Foley, P., Swash, M., Leigh, P.N., Cedarbaum, J., Lindsay, R.M., Sinicropi, D.V. and Williams-Chestnut, R.E. (1995b) Regional changes of ciliary neurotrophic factor and nerve growth factor immunoreactivity in spinal cord and cerebral cortex in human motoneurone disease. Nat. Med., 2: 168–172.

Anand, P., Pandya, S., Ladiwala, U., Singhal, B., Sinicropi, D.V. and Williams-Chestnut, R.E. (1994) Depletion of nerve growth factor in leprosy. Lancet, 344: 129–130.

Anand, P., Terenghi, G., Warner, G., Kopelman, P., Williams-Chestnut, R. and Sinicropi, D.V. (1996) The role of endogenous nerve growth factor in human diabetic neuropathy. Nat. Med., 2: 703–707.

Anand, P. (1997) The role of neurotrophic factors in leprosy and other human peripheral neuropathies. In: N.H. Antia and V.P. Shetty (Eds.), The Peripheral Nerve in Leprosy and Other Neuropathies, Oxford University Press, Delhi, pp. 231–240.

Anand, P., Birch, R., Terenghi, G., Wellmer, A., Cedarbaum, J., Lindsay, R.M., Williams-Chestnut, R.E. and Sinicropi, D.V. (1997) Endogenous nerve growth factor and ciliary neurotrophic factor levels in human peripheral nerve injury. Neuroreport, 8: 1935–1938.

Apfel, S.C., Schwartz, S., Adornato, B.T., Freeman, R., Biton, V., Rendell, M., Vinik, A., Giuliani, M., Stevens, J.C., Barbano, R. and Dyck, P.J. (2000) Efficacy and safety of recombinant human nerve growth factor in patients with diabetic polyneuropathy: A randomized controlled trial. rhNGF Clinical Investigator Group. JAMA., 284: 2215–2221.

Bar, K., Saldanha, G., Carlstedt, T., Birch, R., Facer, P., Kennedy, A. and Anand, P. (1998) GDNF and its receptor component Ret in injured human nerves and dorsal root ganglia. Neuroreport, 9: 43–47.

Bennett, G., al-Rashed, S., Hoult, J.R. and Brain, S.D. (1998) NGF induced hyperalgesia in the rat hind paw is dependent on circulating neutrophils. Pain, 77: 315–322.

Berman, J., Anand, P. and Birch, R.B. (1998) Pain following human brachial plexus injury with spinal cord root avulsion and the effect of surgery. Pain, 75: 199–207.

Boettger, M.K., Till, S., Chen, M.X., Anand, U., Otto, W.R., Plumpton, C., Trezise, D.J., Tate, S.N., Bountra, C., Coward, K., Birch, R. and Anand, P. (2002) Calcium-activated potassium channel SK1- and IK1-like immunoreactivity in injured human sensory neurones and its regulation by neurotrophic factors. Brain, 125: 252–263.

Bradley, J.L., Thomas, P.K., King, R.H., Muddle, J.R., Ward, J.D., Tesfaye, S., Boulton, A.J., Tsigos, C. and Young, R.J. (1995) Myelinated fibre regeneration in diabetic sensory polyneuropathy: correlation with type of diabetes. Acta Neuropathol. (Berl.), 90: 403–410.

Brewster, W.J., Fernyhough, P., Diemel, L.T., Mohiuddin, L. and Tomlinson, D.R. (1994) Diabetic neuropathy, nerve growth factor and other neurotrophic factors. Trends Neurosci., 17: 321–325.

Caterina, M.J., Schumacher, M.A., Tominaga, M., Rosen, T.A., Levine, J.D. and Julius, D. (1997) The capsaicin receptor: a heat-activated ion channel in the pain pathway. Naute, 389: 816–824.

Chaudhry, V., Giuliani, M., Petty, B.G., Lee, D., Seyedsadr, M., Hilt, D. and Cornblath, D.R. (2000) Tolerability of recombinant-methionyl human neurotrophin-3 (r-metHuN T3) in healthy subjects. Muscle Nerve, 23: 189–192.

Coward, K., Plumpton, C., Facer, P., Birch, R., Carlstedt, T., Tate, S., Bountra, C. and Anand, P. (2000) Immunolocalisation of SNS/PN3 and NaN/SNS2 sodium channels in human pain states. Pain, 85: 41–50.

Delsite, R. and Djakiew, D. (1999) Characterization of nerve growth factor precursor protein expression by human prostate stromal cells: a role in selective neurotrophin stimulation of prostate epithelial cell growth. Prostate, 41: 39–48.

Diamond, J., Holmes, M. and Coughlin, M. (1992) Endogenous NGF and nerve impulses regulate the collateral sprouting of sensory axons in the skin of the adult rat. J. Neurosci., 12: 1454–1466.

Dicou, E., Pflug, B., Magazin, M., Lehy, T., Djakiew, D., Ferrara, P., Nerriere, V. and Harvie, D. (1997) Two peptides derived from the nerve growth factor precursor are biologically active. J. Cell Biol., 136: 389–398.

Diemel, L.T., Cai, F., Anand, P., Warner, G., Kopelman, P.G., Fernyhough, P. and Tomlinson, D.R. (1999) Increased nerve growth factor mRNA in human diabetic lateral calf skin biopsies. Diab. Med., 16: 113–118.

Di Marco, E., Mathor, E. and Bondanza, S. (1993) Nerve growth factor binds to normal human keratinocytes through high and low affinity receptors and stimulate their growth by a novel autocrine loop. J. Biol. Chem., 268: 22,838–22,846.

DiStephano, P.S., Friedman, B. and Radziejewski, C. (1992) The neurotrophins BDNF, NT-3, and NGF display distinct patterns of retrograde axonal transport in peripheral neurons. Neuron, 8: 983–993.

Djakiew, D., Delsite, R., Pflug, B., Wrathall, J., Lynch, J.H. and Onoda, M. (1991) Regulation of growth by a NGF-like protein which modulates paracrine interactions between a neoplastic epithelial cell line and stromal cells of the human prostate. Cancer Res., 51: 3304–3310.

Donnerer, J., Schuligoi, R. and Stein, C. (1992) Increased content and transport of substance P and CGRP in sensory nerves innervating inflamed tissue: evidence for regulatory function of nerve growth factor in vivo. Neuroscience, 49: 693–698.

Duberley, R., Johnson, I., Anand, P., Leigh, P.N. and Cairns, N.J. (1997) Neurotrophin-3-like immunoreactivity and Trk C expression in human spinal motoneurones in amyotrophic lateral sclerosis. J. Neurol. Sci., 148: 33–40.

Duberley, R., Johnson, I., Martin, J. and Anand, P. (1998) RET-like immunostaining of spinal motoneurones in amyotrophic lateral sclerosis. Brain Res., 789: 351–354.

Dyck, P.J., Carney, J.A., Sizemore, G., Okazaki, H., Brimijoin, W.S. and Lambert, E.H. (1979) Multiple endocrine neoplasia type 2b: phenotype recognition; Neurological features and their pathological basis. Ann. Neurol., 6: 302–314.

Dyck, P.J., Peroutka, S., Rask, C., Burton, E., Baker, M.K., Lehman, K.A., Gillen, D.A., Hokanson, J.L. and O'Brien, P.C. (1997) Intradermal recombinant human nerve growth factor induces pressure allodynia and lowered heat-pain threshold in humans. Neurology, 48: 501–505.

Edwards, R.H., Selby, M.J. and Rutter, W.J. (1986) Differential RNA splicing predicts two distinct nerve growth factor precursors. Nature, 319: 784–787.

Facer, P., Mathur, R., Pandya, S., Singhal, B. and Anand, P. (1998) Correlation of quantitative tests of nerve and target organ dysfunction with skin immunohistology in leprosy. Brain, 121: 2239–2247.

Facer, P., Mann, D., Terenghi, G. and Anand, P. (1999) Nerve Growth Factor (NGF) concentrations in cultured human keratinocytes exposed to mycobacterium leprae cell free extract. Lep. Rev., 70: 213–217.

Facer, P., Mann, D., Matahur, R., Pandya, S., Ladiwala, U., Singhal, B., Hongo, J.-A., Sinicropi, D.V., Terenghi, G. and Anand, P. (2000) Do NGF-related mechanisms contribute to loss of cutaneous nociception in leprosy? Pain, 85: 231–238.

Fernyhough, P., Diemel, L.T., Brewster, W.J. and Tomlinson, D.R. (1994) Deficits in sciatic nerve neuropeptide content coincide with a reduction in target tissue nerve growth factor messenger mRNA in streptozotocin-diabetic rats: effects of insulin treatment. Neuroscience, 62: 337–344.

Guy, R.J.C., Clark, C.A., Malcolm, P.N. and Watkins, P.J. (1985) Evaluation of thermal and vibration sensation in diabetic neuropathy. Diabetologia, 28: 131–137.

Gold, B.G., Mobley, W.C. and Matheson, S.F. (1991) Regulation of axonal calibre, neurofilament content, and nuclear localisation in mature sensory neurons by nerve growth factor. J. Neurosci., 11: 943–955.

Hellweg, R., Raivich, G., Hartung, H.D., Hock, C. and Kreutzberg, G.W. (1994) Axonal transport of endogenous nerve growth factor (NGF) and NGF receptor in experimental diabetic neuropathy. Exp. Neurol., 130: 24–30.

Hellweg, R., Wohrle, M., Hartung, H.D., Stracke, H., Hock, C. and Federlin, K. (1991) Diabetes mellitus-associated decrease in nerve growth factor levels is reversed by allogeneic pancreatic islet transplantation. Neurosci. Lett., 125: 1–4.

Heumann, R., Lorsching, S., Bandtlow, C. and Thoenen, H. (1987) Changes of nerve growth factory synthesis in non-neoruonal cells in response to sciatic nerve transection. J. Cell Biol., 104: 1623–1631.

Indo, Y., Tsuruta, M., Hayashida, Y., Karim, M.A., Ohta, K., Kawano, T., Mitsubuchi, H., Tonki, H., Awaya, Y. and Matsuda, I. (1996) Mutations in the TRKA/NGF receptor gene in patients with congenital insensitivity to pain with anhidrosis. Nat. Genet., 13(4): 485–488.

Jolliffe, V., Anand, P. and Kidd, B. (1995) Assessment of cutaneous sensory and autonomic reflexes in rheumatoid arthritis. Ann. Rheum. Dis., 54: 251–255.

Karanth, S.S., Spingall, D.R. and Lucas, S. (1989) Changes in nerves and neuropeptides in skin from 100 leprosy patients investigated by immunocytochemistry. J. Pathol., 157: 15–26.

Kennedy, A., Wellmer, A., Facer, P., Saldanha, G., Kopelman, P., Lindsay, R. and Anand, P. (1998) Neurotrophin-3 is increased in skin in human diabetic neuropathy. J. Neurol. Neurosurg. Psychiatry, 65: 393–395.

Knowles, C.H., Misra, V.P., Wellmer, A.C., Monson, J.P. and Anand, P. (1998) Peripheral sensory dysfunction in a kindred of multiple endocrine neoplasia type 2A with a codon 634 mutation of the RET proto-oncogene. J. Neurol. Neurosurg. Psychiatry, 65: 425.

Laemmli, U.K. (1970) Cleavage of structural proteins during the assembly of the head of bacteriophage T4. Nature, 227: 680–685.

Lakshmanan, J. (1987) Nerve growth factor levels in mouse serum: variations due to stress. Neurochem. Res., 12: 393–397.

Lakshmanan, J., Burns, C. and Smith, R.A. (1988) Molecular forms of nerve growth factor in mouse submaxillary glands. Biochem. Biophys. Res. Commun., 152: 1008–1014.

Lakshmanan, J., Beattie, G.M., Hayek, A., Burns, C. and Fisher, D.A. (1989) Biological actions of 53 kDa nerve growth factor as studied by a blot and culture technique. Neurosci. Lett., 99: 263–267.

Lee, K.F., Li, E., Huber, L.J., Landis, S.C., Sharpe, A.H., Chao, M.V. and Jaenishch, R. (1992) Targeted mutation of the gene encoding the low affinity NGF receptor p75 leads to deficits in the peripheral sensory nervous system. Cell, 69: 737–749.

Levi-Montalcini, R. (1987) The nerve growth factor 35 years later. Science, 237: 1154–1162.

Levi-Montalcini, R., Skaper, S.D., DalToso, R., Petrelli, L. and Leon, A. (1996) Nerve growth factor: from neurotrophin to neurokine. Trends Neurosci., 19: 514–520.

Lewin, G. and Mendell, L.M. (1993) Nerve growth factor and nociception. Trends Neurosci., 16: 353–359.

Lewin, G. and Mendell, L.M. (1996) Maintenance of modality-specific connections in the spinal cord after neonatal NGF deprivation. J. Neurosci., 8: 1677–1684.

Lin, L-F.H., Doherty, D.H., Lile, J.D., Bektesh, S. and Collins, F. (1993) GDNF: A glial cell line-derived neurotrophic factor for midbrain dopaminergic neurons. Science, 260: 1130–1132.

Lindsay, R.M. and Harmar, A.J. (1989) Nerve growth factor eguatles expession of neuropeptide geneses in adult sensory neurones. Nature, 337: 362–364.

Lowe, E., Anand, P., Terenghi, G., Williams-Chestnut, R.E., Sincicropi, D.V. and Osborne, J. (1997) Increased NGF levels in the urinary bladder of women with idiopathic sensory urgency and interstitial cystitis. Br. J. Urol., 79: 572–577.

Malcangio, M., Garrett, N.E., Cruwys, S. and Tomlinson, D.R. (1997) NGF and NT-3 induced changes in nociceptive threshold and the release of substance P from the rat isolated spinal cord. J. Neurosci., 17: 8459–8467.

Mearow, K.M., Krill, Y. and Diamond, J. (1993) Increased NGF mRNA expression in denervated rat skin. Neuroreport, 4: 351–354.

McMahon, S.B., Bennett, D.L.H., Priestley, J.V. and Shelton, D.L. (1995) The biological effects of endogenous Nerve Growth Factor on adult sensory neurons revealed by a TrKA-IGG fusion molecule. Nature Med., 1: 774–780.

Murray, H.J. and Boulton, A.J. (1995) The pathophysiology of diabetic foot ulceration (Review). Clin. Pediatr. Med. Surg., 12: 1–17.

Nevue, I., Jehan, F. and Wion, D. (1992) Alternations of levels of 1,25-(OH) 2D3 and corticosterone found in experimental diabetes reduces nerve growth factor expression in vitro. Life Sci., 50: 1769–1772.

Oddiah, D., Anand, P., McMahon, S.B. and Rattray, M. (1998) Rapid increase of NGF, BDNF and NT-3 mRNAs in inflamed bladder. Neuroreport, 9: 1455–1458.

Otten, U., Scally, J.L., Ehrrhard, P.B. and Gadient, R.A. (1994) Neurotrophins: signals between the nervous system and the immune system. Prog. Brain Res., 103: 293–305.

Parkhouse, N. and Le Quesne, P.M. (1988) Impaired neurogenic vascular response in patients with diabetes and neuropathic foot lesions. N. Engl. J. Med., 318: 1306–1309.

Pirart, J. (1978) Diabetes mellitus and its degenerative complications: a prospective study of 4400 patients observed between 1947 and 1973. Diab. Care, 1: 168–188.

Reinshagen, M., Geerling, I., Lakshmanan, J., Sonila, S., Lutz, M.P., Eysselein, V.E. and Alder, G. (1995) Localization and changes in the levels of NGF prohormones during the time course of colitis in the rat colon. Soc. Neurosci. Abstr., 21: 23.3.

Reinshagen, M., Geerling, I., Lakshmanan, J., Rohm, H., Lutz, M.P., Sonila, S., Eysselein, V.E. and Alder, G. (1997) Commercial mouse and human nerve growth factors contain nerve growth factor prohormone isoforms. J. Neurosci. Meth., 76: 75–81.

Reinshagen, M., Geerling, I., Eysselein, V.E., Alder, G., Huff, K.R., Moore, G.P. and Lakshmanan, J. (2000) Commercial Recombinant human B-nerve growth factor and adult rat dorsal root ganglia contain an identical molecular species of nerve growth factor prohormone. J. Neurochem., 74: 2127–2133.

Saldanha, G., Hongo, J., Plant, G., Acheson, J., Levy, I. and Anand, P. (1999) Decreased CGRP- but preserved Trk A-immunoreactivity in nerve fibres in inflamed human superficial temporal arteries. J. Neurol. Neurosurg. Psychiatry, 66: 390–392.

Schifitto, G., Yiannoutsos, C., Simpson, D.M., Adornato, B.T., Singer, E.J., Hollander, H., Marra, C.M., Rubin, M., Cohen, B.A., Tucker, T., Koralnik, I.J., Katzenstein, D., Haidich, B., Smith, M.E., Shriver, S., Millar, L., Clifford, D.B. and McArthur, J.C. (2001) Long-term treatment with recombinant nerve growth factor for HIV-associated sensory neuropathy. Neurology, 57: 1313–1316.

Schmidt, R.E., Plurad, S.B., Saffitz, J.E., Grabau, G.G. and Yip, H.K. (1985) Retrograde transport of 125I-nerve growth factor in rat ileal mesenteric nerves. Effect of streptozocin diabetes. Diabetes, 34: 1230–1240.

Scott, J., Selby, M., Urdea, M., Quiroga, M., Bell, G.I. and Rutter, W.J. (1983) Isolation and nucleotide sequence of a cDNA encoding the precursor of nerve growth factor. Nature, 302: 538–540.

Seidah, N.G., Benjannet, S., Pareek, S., Savaria, D., Hamelin, J., Goulet, B., Laliberte, J., Lazures, C., Chretien, M. and Murphy, R.A. (1996) Cellular processing of the nerve growth factor precursor by the mammalian pro-protein convertases. Biochem. J., 314: 951–960.

Shereef, P.H. (1992) Hypopigmented macules in leprosy—a histopathological and histochemical study of melanocytes. Indian J. Lepr., 64: 1989–2191.

Shetty, V.P., Anita, N.H. and Jacobs, J.M. (1988) The pathology of early leprosy. J. Neurol. Sci., 88: 115–131.

Shu, S., Ju, G. and Fan, L. (1988) The glucose oxidase-DAB-nickel method in peroxidase histochemistry of the nervous system. Neurosci. Lett., 85: 169–171.

Siminowski, K., Murphy, R.A., Rennert, P. and Heinrich, G. (1987) Cortisone, testosterone, and aldosterone reduce levels of nerve growth factor messenger ribonucleic acid in L929 fibroblasts. Endocrinology, 121: 1432–1437.

Sinicropi, D.V., McCaffrey, D.S., Williams, P.M., Kopelman, P., Bloom, S.R. and Anand, P. (2001) Decreased skin mRNAs for NGF, BDNF, NT-3, GDNF but not NT4/5 or trkA/p75 in sub-clinical human diabetic neuropathy. J. Neurol. Sci., 187: S367.

492

Sohrabji, F., Miranda, R.C. and Toran-Allerand, C.D. (1994) Estrogen differentially regulates estrogen and NGF receptor mRNAs in adult sensory neurons. J. Neurosci., 14: 459–471.

Suter, U., Heymach, J.V.Jr. and Shooter, E.M. (1991) Two conserved domains in the NGF propeptide are necessary and sufficient for the biosynthesis of correctly processed and biologically active NGF. EMBO J., 10: 2395–2400.

Terenghi, G., Mann, D., Kopelman, P. and Anand, P. (1997) TrkA and trkC expression is increased in human diabetic skin. Neurosci. Lett., 228: 1–4.

Thomas, P.K. and Tomlinson, D.R. (1993). Diabetic and hypoglycemic neuropathy. In: P.J. Dyck, P.K, Thomas, J.W. Griffin, P.A. Low and J.F. Poduslo (Eds.), Peripheral Neuropathy, 3rd Edition. WB Saunders & Co, Philadelphia/London, pp. 1219–1250.

Trupp, M., Arenas, E., Fainzilber, M., Nilsson, A.-S., Sieber, B.-A., Grigoriou, M., Kilkenny, C., Salazar-Grueso, E., Pachnis, V., Arumae, U., Sariola, H., Saarma, M. and Ibanez, C.F. (1996) Functional receptor for GDNF encoded by the c-ret proto-oncogene. Nature, 381: 785–789.

Tuttle, J.B., Etheridge, R. and Creedon, D.J. (1993) Receptor-mediated stimulation and inhibition of NGF secretion by vascular smooth muscle. Exp. Cell Res., 208: 350–361.

Ullrich, A., Gray, A., Berman, C. and Dull, T.J. (1983) Human βnerve growth factor gene sequence is highly homologous to that of mouse. Nature, 303: 821–825.

Waldmann, R. and Lazdunski, M. (1998) H (+)-gated cation channels: neuronal acid sensors in the NaC/DEG family of ion channels. Curr. Opin. Neurobiol., 8: 418–424.

Wellmer, A., Sharief, M.K., Kopelman, P.G. and Anand, P. (1998) A clinical trial of recombinant human brain derived neurotrophic factor (rhBDNF) in diabetic sensory polyneuropathy. J. Neurol. Neurosurg. Psychiatry, 65: 423.

Wellmer, A., Misra, V.P., Sharief, M.K., Kopelman, P.G. and Anand, P. (2001) A double-blind placebo-controlled clinical trial of recombinant human brain derived neurotrophic factor (rhBDNF) in diabetic polyneuropathy. J. Peripher. Nerv. Syst., 6: 204–210.

Whitworth, I.H., Brown, R.A., Dore, C., Anand, P., Green, C.J. and Terenghi, G. (1996) Nerve growth factor enhances nerve regeneration through fibronectin grafts. J. Hand Surgery, 21: 514–522.

Woolf, C.J. (1996) Phenotypic modification of primary sensory neurons: the role of nerve growth factor in the production of persistent pain. Phil. Trans. R. Soc. Lond., B, 351: 441–448.

Yaar, M., Grossman, K., Eller, M. and Gilchrest, B.A. (1991) Evidence for nerve growth factor-mediated paracrine effects in human epidermis. J. Cell Biol., 114: 821–828.

Yiangou, Y., Facer, P., Dyer, N.H.C., Baecker, P.A., Ford, A.P., Knowles, C. and Anand, P. (2001) Capsaicin receptor VR1 and ATP-gated ion channel P2X3 in human pelvic disorders. J. Neurol. Sci., 187: S106.

Yiangou, Y., Facer, P., Sinicropi, D.V., Boucher, T.J., Bennett, D.L., McMahon, S.B. and Anand, P. (2002) Molecular forms of NGF in human and rat neuropathic tissues: decreased NGF precursor-like immunoreactivity in human diabetic skin. J. Peripher. Nerv. Syst., 7: 190–197.

Zanone, M.M., Banga, J.P., Peakman, M., Edmonds, M. and Watkins, P.J. (1994) An investigation of antibodies to nerve growth factor in diabetic autonomic neuropathy. Diab. Med., 11: 378–383.

Progress in Brain Research, Vol. 146
ISSN 0079-6123

CHAPTER 31

Epithelial growth control by neurotrophins: leads and lessons from the hair follicle

Vladimir A. Botchkarev[1], Natalia V. Botchkareva[1,2], Eva M.J. Peters[2] and Ralf Paus[3,*]

[1]*Department of Dermatology, Boston University School of Medicine, Boston, MA, USA*
[2]*Biomedical Research Center, Charité, Campus Virchow-Hospital, Humboldt-University Berlin, Berlin, Germany*
[3]*Department of Dermatology, University Hospital Hamburg-Eppendorf, University of Hamburg,*
Martinistr. 52, D-20246, Hamburg, Germany

Abstract: Neurotrophins (NTs) exert many growth-regulatory functions beyond the nervous system. For example, murine hair follicles (HF) show developmentally and spatio-temporally stringently controlled expression of NTs, including nerve growth factor (NGF), brain-derived neurotrophic factor (BDNF), neurotrophin-3 (NT-3), and NT-4, and their cognate receptors, tyrosine kinase A-C (TrkA-C) and p75 neurotrophin receptor (p75^{NTR}). Follicular NT and NT receptor expression exhibit significant, hair cycle-dependent fluctuations on the gene and protein level, which are mirrored by changes in nerve fiber density and neurotransmitter/neuropeptide content in the perifollicular neural networks. NT-3/TrkC and NGF/TrkA signaling stimulate HF development, while NT-3, NT-4 and BDNF inhibit the growth (anagen) of mature HF by the induction of apoptosis-driven HF regression (catagen). p75^{NTR} stimulation inhibits HF development and stimulates catagen. Since the HF is thus both a prominent target and key peripheral source of NT, dissecting the role of NTs in the control of HF morphogenesis and cyclic remodeling provides a uniquely accessible, and easily manipulated, clinically relevant experimental model, which has many lessons to teach. Given that our most recent data also implicate NTs in human hair growth control, selective NT receptor agonists and antagonists may become innovative therapeutic tools for the management of hair growth disorders (alopecia, effluvium, hirsutism). Since, however, the same NT receptor agonists that inhibit hair growth (e.g., BDNF, NT-4) can actually stimulate epidermal keratinocyte proliferation, NT may exert differential effects on defined keratinocyte subpopulations. The studies reviewed here provide new clues to understanding the complex roles of NT in epithelial tissue biology and remodeling in vivo, and invite new applications for synthetic NT receptor ligands for the treatment of epithelial growth disorders, exploiting the HF as a lead model.

Keywords: morphogenesis; keratinocytes; hair cycle; catagen; NGF; BDNF; NT-3; NT-4; p75^{NTR}; TrkA; TrkB; TrkC; dermal papilla

Introduction

NGF functions beyond the neurobiological horizon

As summarized elegantly on the occasion of her Nobel Prize award (Levi-Montalcini, 1987), Rita

Levi-Montalcini and her coworkers, notably Luigi Aloe, long ago recognised that the effects of the prototypic neurotrophin (NT) nerve growth factor (NGF) extend far beyond the horizon of classical neurobiology. In fact, NGF is now appreciated to exert a bewildering array of modulatory functions outside the peripheral and central nervous systems. These include potent immunomodulatory and neuroendocrine activities (reviewed by Aloe et al., 1997; 1999; Aloe and Micera, 1999). More recently, it became evident that selected NT are also involved in

#Present address: The Gillette Company, Needham, MA, USA.
*Correspondence to: R. Paus, Department of Dermatology,
University Hospital Hamburg-Eppendorf, Martinistr. 52,
D-20246 Hamburg, Germany. E-mail: paus@uke.uni-hamburg.de

DOI: 10.1016/S0079-6123(03)46031-7

the control of kidney, tooth, muscle and heart development (reviewed by Sariola, 2001).

The common neuroectodermal origin of the cutaneous epithelium and the nervous system make it a reasonable hypothesis that the same growth factors that govern the development, maintenance, and axonal sprouting of neurones are also involved in skin morphogenesis and in the development of skin appendages. Certainly, it was not very surprising when the proliferation, survival and pigment production of the skin's neural crest-derived cells, the melanocytes, turned out to be deeply influenced by NTs (Yaar et al., 1994; Pincelli and Yaar, 1997; Yaar, 1999). This recognition complemented the previous appreciation of NGF as a potent mast cell secretagogue and as a factor that promotes mast cell growth and survival in the skin and elsewhere (Aloe et al., 1997, 1999; Aloe and Micera, 1999).

A series of studies established that NTs can be generated locally in the skin, e.g., by glia cells, epithelial cells, fibroblasts, and Merkel cells (see Seiber-Blum et al. in this issue) and that NGF and other NTs are critical for proper innervation of this crucial peripheral sensory organ (reviewed by Lewin and Barde, 1996). Not surprisingly, defects in NT signaling are associated with severe sensory skin defects that facilitate ulcer formation and inhibit wound healing (Lambiase et al., 2000). Adequate, NT-guided innervation of the skin vasculature is also an essential prerequisite for optimal skin perfusion, as is best illustrated by the immensely chronic 'neuropathic' foot ulcers of patients with diabetes mellitus: these show NT-controlled innervation abnormalities of the cutaneous vasculature, which result in deleterious tissue hypoxia up to the point of necrosis, especially after prolonged tissue compression or minor trauma (Bernabei et al., 1999; Christianson and Riekhof, 2003).

More than 15 years ago, the skin was identified as a rich source of NGF (Weskamp and Otten, 1987), and the epidermis was recognised as a site of NGF expression (Davies et al., 1987). Shortly thereafter, it became clear that epidermal KCs not only are important NGF sources, but are also NT targets expressing NT receptors, at least in cell culture (Di Marco et al., 1991; Yaar et al., 1991; Di Marco et al., 1993; Pincelli et al., 1994). This was followed by the first evidence (from murine skin organ culture) that NGF can actually modulate the proliferation of normal epidermal KCs in situ (Paus et al., 1994a).

Interest in the role of NGF and other NTs in skin biology and pathology has steadily grown over the past years. For example, NGF stimulates the proliferation and inhibits the apoptosis of cultured human epidermal KCs (Pincelli et al., 1997), may play an as yet ill-defined role in triggering epidermal lesions in psoriasis (reviewed by Johansson and Liang, 1999; Pincelli, 2000; Raychaudhuri et al., 2000), and is increased in the skin of patients with allergic contact eczema (Kinkelin et al., 2000). Our own data from murine skin suggest the concept that, during epidermal homeostasis, BDNF, NT-4 and NT-3 all act as 'epitheliotrophins', since they stimulate the proliferation, and inhibit the apoptosis, of epidermal KCs in situ (Botchkarev et al., 1999c). This basic concept is supported by a subsequent study in patients with atopic eczema, which showed a stimulatory effect of NT-4 on human epidermal KC proliferation in vitro (Grewe et al., 1997). The exploitation of NTs for modulating its own growth is not unique to the epidermis: not only numerous peripheral sites of NGF expression and synthesis have now been characterized (Sariola, 2001), but many adult visceral epithelia also show abundant BDNF production (Lommatzsch et al., 1999).

The hair follicle as a model for dissecting nonneuronal functions of neurotrophins

In the same skin organ culture study that had provided the first proof that NGF can modulate keratinocyte proliferation in situ, the hair follicle (HF) already entered the picture, since the response of epidermal KCs was peculiarly dependent on the activity status of the hair follicles in organ-cultured skin fragments: epidermal proliferation was stimulated in mouse skin with all hair follicles in telogen, the 'resting' stage of the hair cycle, but was inhibited when all the hair follicles were in the growth stage (anagen), characterized by massive epithelial proliferation in the epidermis and HF epithelium (Paus et al., 1994a).

Shortly thereafter, studies in our lab (then in Berlin), indicated that the HF—probably the densely and most intricately innervated extraneural organ in the mammalian body—cyclically remodels precisely defined segments of its innervation each time it runs through its life-long cycle of growth (anagen), apoptosis-driven regression (catagen) and relative resting (telogen) (Botchkarev et al., 1997a). This— previously denied—plasticity of the perifollicular innervation even around mature HF strongly suggested that NTs, including paracrine activities of intrafollicularly produced NGF, were at the basis of this intriguing phenomenon and that the HF offers an exquisite model for dissecting epithelial–neural interactions (Paus et al., 1997).

Around the same time, we noted that skin mast cells, another recognised target of NGF activities, are important, though nonessential, players in murine hair cycle control and show striking hair cycle-dependent changes in their number, activation status and nerve fiber-contacts (Paus et al., 1994b; Botchkarev et al., 1995; Maurer et al., 1995; Botchkarev et al., 1997b; Maurer et al., 1997). In addition, we observed that the relative percentage of the sensory and autonomic murine skin nerve fibers positive for defined neuropeptide or neurotransmitters also significantly changes in a hair cycle-dependent manner (Botchkarev et al., 1997b; Peters et al., 2001).

Taken together, all this suggested that the HF is placed at the centre of a complex and dynamic signaling network that joins mast cells, perifollicular axons and their glia cells, and the perifollicular vasculature into one, closely interdependent functional unit (Paus et al., 1997). These observations also made it rather intuitive that NGF and/or other NTs play a major, previously ignored role in this functional unit and that follicular NT synthesis and NT receptor expression is hair cycle-dependent. Since our initial studies were in line with this concept (Welker et al., 1996), we were encouraged to systematically explore the role of NGF, other members of the NGF family (BDNF, NT-3, NT-4), and of selected neurotrophic growth factors of the transforming growth factor-beta (TGF-β)/bone morphogenetic protein (BMP) family—glial cell line-derived neurotrophic factor (GDNF) and neurturin—in HF biology.

Specifically, we were interested in learning whether or not NTs are involved in the controls that govern HF morphogenesis and cycling, since we hoped that this would provide new insights into the role of NT in epithelial biology in general. The complex pilo-neural signaling networks established within and around the hair follicle make it particularly instructive to dissect whether and how NT receptor mediated signaling acts as a major molecular switch for controlling the HF's cyclic transformations from phases of relative quiescence (telogen), to rapid growth (anagen) and apoptosis-driven regression (catagen) (Paus and Cotsarelis, 1999; Stenn and Paus, 2001). Its embryonal origin, unrivalled accessibility and ease of experimental manipulation, the profound clinical importance of hair growth disorders, and the nature of the HF as a prototypic neuroectodermal–mesodermal tissue interaction system (Hardy, 1992; Paus et al., 1997; Paus and Cotsarelis, 1999; Stenn and Paus, 2001) all designate the HF an ideal model for NT research.

In this review, we summarize the findings we have made so far in this respect: intracutaneously generated NTs not only are required for appropriate skin innervation, mast cell and melanocyte functions in mammalian skin, but also play important roles during both HF development and cycling. We hope to convince researchers from outside the field of investigative dermatology, in particularly from the neurobiological research community, that the HF offers them an absolutely ideal, yet far too rarely exploited model for studying many of the key questions of current NT research and that systematic use of this new model is likely to offer substantial new insights into NT biology in general.

The neurotrophin family and its receptors

Before our experimental data on the role of NTs and their receptors in hair biology are delineated, it may be useful to summarize the basic tenants of NT biology that, at the time our experiments were designed, formed the basis of our NT studies in the HF model.

NTs represent a family of structurally and functionally related polypeptides that consists of four proteins. NGF, brain-derived neurotrophic factor (BDNF), NT-3 and NT-4. All four members of the NT family have a size of ~ 13 kD, share about 50% of amino acid sequence homology, and exert their biological effects as dimers interacting with specific receptors (for review see Ibanez, 1998; Bibel and Barde, 2000). High-affinity receptors for NT belong to the tyrosine kinase family: TrkA serves a high-affinity receptor for NGF, TrkB is a high-affinity receptor for BDNF and NT-4, and TrkC is a high-affinity receptor for NT-3 (see this issue). However, NT-3 may also bind—with low-affinity—to TrkA and TrkB receptors (reviewed by Bothwell, 1995; Klesse and Parada, 1999). All four NTs bind also to the p75 NT receptor (p75^{NTR}), which is a member of the TNF family of receptors containing the cytoplasmic death domain, involved in mediating apoptosis (reviewed by Dechant and Barde, 2002; Roux and Barker, 2002).

By interacting with Trk-receptors and p75^{NTR}, NTs induce a variety of biological reactions and control cell proliferation, differentiation, survival and apoptosis (Bibel and Barde, 2000). The signals promoting survival and differentiation are generated by NT interaction with Trk receptors and require receptor dimerization, autophosphorylation, and the subsequent involvement of a number of adaptor molecules coupling Trk receptors to the distinct signal transduction pathways. Recent data suggest that NT interaction with Trk receptors leads to the activation of at least three signalling cascades followed by the distinct cellular responses (for review see Bibel and Barde, 2000). Activation of PI3K/PKB/AKT pathway stimulates survival and is associated with phosphorylation and inactivation of the pro-apoptotic member of the Bcl-2 family BAD (Franke et al., 1997; Crowder and Freeman, 1998). Trk-dependent activation of Ras/Raf/MEK/MAPK cascade leads to the promotion of differentiation in the developing neurons (Qian et al., 1998; Kaplan and Miller, 2000). Furthermore, NTs modulate their own release and synaptic transmission via Trk-associated regulation of intracellular Ca^{2+} (Canossa et al., 1997; Schuman, 1999; Bibel and Barde, 2000).

p75^{NTR} performs diverse functions depending on whether it is coexpressed with Trk receptors or whether it is expressed alone. The coexpression of p75^{NTR} with Trk receptors increases high-affinity NT binding, enhances ability of Trk receptors to discriminate a preferred ligand from the other NTs, and promotes survival effects of NTs (reviewed by Friedman and Greene, 1999). NT interaction with

p75^{NTR} expressed alone leads to a variety of responses including apoptotic death, cell cycle regulation, and modulation of growth factor production by target cells.

It was recently shown that signaling through p75^{NTR} may be activated not only by neurotrophins, but also by nonneurotrophin ligands, such as beta-amyloid and fragment of the prion protein (Yaar et al., 1997; Della-Bianca et al., 2001). Activation of p75^{NTR} signaling recruits a variety of intracellular adaptor molecules (reviewed by Bibel and Barde, 2000; Miller and Kaplan, 2001; Roux and Barker, 2002). Particularly, it was demonstrated in several models that apoptotic signaling through p75^{NTR} require presence of the neurotrophin receptor interacting factors 1 and 2 (NRIF-1/2), neurotrophin receptor interacting MAGE homolog (NRAGE) and neurotrophin-associated death executor (NADE; reviewed by Roux and Barker, 2002). JNK-p53-Bax pathway involved in neurotrophin-induced apoptosis (Aloyz et al., 1998; Bamji et al., 1998) appears to be downstream of the adaptor molecules mentioned above. Activation of JNK pathway leads to upregulation of p53 and to increase the transcription of p53 target genes including Bax (Miller and Kaplan, 2001). This results in caspase activation, DNA fragmentation and cell death.

Signaling through p75^{NTR} may also be accompanied by cell survival. Intracellular adaptor molecules interacting with the C-terminus of p75^{NTR} (TRAF6, FAP-1 and RIP-2) link p75^{NTR} with the NF-kB pathway and promote survival (reviewed by Roux and Barker, 2002). p75^{NTR} signaling likely is also involved in cell cycle regulation. It was demonstrated that PC12 cells express NGF receptors at their surface in a cell cycle-specific manner: p75^{NTR} is expressed mainly in late G1, S and G2 phases, whereas TrkA is expressed most strongly in M and early G1 phases (Urdiales et al., 1998). p75^{NTR} expression is also markedly upregulated on proliferating neural crest-derived cells, compared to nondividing cells (Young, 2000). During muscle development, downregulation of p75^{NTR} is necessary for guiding myogenic cells from proliferation towards terminal differentiation (Seidl et al., 1998). Taken together, these data suggest that p75^{NTR} signaling may mediate either apoptosis or survival depending on the expression of Trk receptors and intracellular adaptor molecules (reviewed in Roux and Barker, 2002).

p75^NTR signaling also modulates the expression of growth factor/growth factor receptor expression by/on target cells. For example, in retinal bipolar cells, BDNF increases production or release of FGF-2 by binding to p75^{NTR} (Wexler et al., 1998), while in adult sensory neurons NGF upregulates bradykinin binding sites via interaction with p75^{NTR} (Petersen et al., 1998). Overexpression of p75^{NTR} in human melanoma cells augments invasiveness and increases heparanase production (Walch et al., 1999), and in PC12 cells NGF binding to p75^{NTR} stimulates amyloid precursor protein mRNA expression (Rossner et al., 1998). Most importantly, all those effects described above are Trk-independent (Petersen et al., 1998; Rossner et al., 1998; Wexler et al., 1998; Walch et al., 1999) suggesting the existence of p75^{NTR}-specific signaling pathways in realization of the trophic effects of NTs on their target cells. These growing, ever more complex insights into NT receptor-mediated signaling must be kept in mind when interpreting the data we have obtained so far on the role of NT in hair growth control. The role of NTs in HF pigmentation is not covered here; for introduction see Tobin and Paus (2001).

Cellular sources and nonneuronal targets for neurotrophins in skin

In mice, NTs are expressed very early during embryonic development in both the skin epithelium and the cutaneous mesenchyme. Onset of mRNA^{BDNF} and mRNA^{NT-3} expression in skin is found as early as E9.5 (Buchman and Davies, 1993), while NGF transcripts were first noted in embryonic skin at E10.5 (Davies et al., 1987). During skin development, expression of NT changes in a wave-like manner: NGF and BDNF transcripts are maximal at E12.5, mRNA^{NT-3} and mRNA^{NT-4} peak at E13.5, while levels of all four NT transcripts then decline progressively towards E16-E18 (Davies et al., 1987; Wheeler and Bothwell, 1992; Buchman and Davies, 1993; Ibanez et al., 1993; Ernfors et al., 1994).

Although, the developmentally regulated fluctuations of NGF skin levels coincide with the appearance of sensory nerves, it is interesting to note that the levels of NGF transcripts are identical in both intact and denervated embryonic skin (Rohrer et al., 1988). This indicates that the developmentally regulated changes of NGF synthesis in skin are innervation-independent. Indeed, the onset of NT expression in embryonic skin also coincides with the time-point of appearance of K5 and K14 in the epidermis (E9.5), while maximum of NT synthesis coincides with the beginning of vibrissa development in facial skin (E12.5), and with the initiation of tylotrich hair follicle induction in dorsal murine skin (E14.5) (Ernfors et al., 1994). This invites the hypothesis that NTs fulfill multiple nonneurotrophic functions during skin development.

In postnatal skin, NTs and their receptors are differentially distributed in distinct cell populations (reviewed by Johansson and Liang, 1999). Basal epidermal keratinocytes in humans and rodents express NGF and NT-4 (Di Marco et al., 1991; Di Marco et al., 1993; Pincelli et al., 1994; Pincelli et al., 1997; Botchkarev et al., 1999c). NT-3 and BDNF are produced by fibroblasts in vitro (Acheson et al., 1991; Cartwright et al., 1994; Yaar et al., 1994). In murine skin, BDNF and NT-3 are expressed in cutaneous nerve fibers and myocytes of the arrector pili and panniculus carnosus muscles (Botchkarev et al., 1999c). All four NT receptors (TrkA-C, p75^{NTR}) have by now been detected on human epidermal KCs (Vega et al., 1994; Bronzetti et al., 1996; Shibayama and Koizumi, 1996; Pincelli et al., 1997; Terenghi et al., 1997). In murine skin, only TrkA and TrkB isoforms are seen in epidermal KCs, while TrkC and p75^{NTR} are expressed in cutaneous nerves and in the HF (Botchkarev et al., 1999c).

It has by now become well-established that NTs are intimately involved in the control of epidermal homeostasis (Pincelli and Yaar, 1997; Yaar, 1999). NGF stimulates keratinocyte proliferation and suppresses apoptosis in keratinocytes and melanocytes in vitro (Di Marco et al., 1991; Di Marco et al., 1993; Pincelli et al., 1994; Zhai et al., 1996; Pincelli et al., 1997). NGF stimulates TrkA phosphorylation in KCs, while the alkaloid K252a, an inhibitor of Trk signaling, suppresses NGF-induced keratinocyte proliferation in vitro (Pincelli et al., 1997). As we have shown in transgenic (tg) and knockout (KO) models, BDNF, NT-3 and NT-4 also stimulate keratinocyte proliferation in murine epidermis in vivo (Botchkarev et al., 1999c). These effects were confirmed in skin organ culture, i.e., in the absence of functional skin nerves, suggesting that the action of NTs as 'epitheliotrophins' is not mediated by a modulation of neuropeptide/neurotransmitter release from skin nerve endings (Botchkarev et al., 1999c).

In summary, NTs can influence numerous cellular functions in skin, and NT effects on skin cell fate (survival, apoptosis, differentiation) are likely to strongly depend of the multiplicity of signaling pathways that are activated by NTs in different cells under different conditions. Also, many of the effects NTs exert on the HF may not be explained by direct ligand-receptor interactions, e.g., on HF KCs, but may also reflect more or less complex indirect modes of action (cf. Paus et al., 1997), e.g., by NT-induced mast cell activation or by alterations in the growth factor secretion of dermal papilla fibroblasts, i.e., the mesenchymal 'command centre' of the HF (cf. Paus and Cotsarelis, 1999; Stenn and Paus, 2001). This background deserves to be kept in mind when interpreting the data reported subsequently.

Neurotrophins in the control of hair follicle morphogenesis

HF morphogenesis occurs via an inductive signaling cascade between epidermal KCs committed to hair-specific differentiation, and a specialized population of dermal fibroblasts, which form a mesenchymal condensation beneath the hair placode and subsequently develop into the follicular dermal papilla (DP) (Hardy, 1992; Oro and Scott, 1998; Philpott and Paus, 1998; Paus et al., 1999; Millar, 2002). These bidirectional neuroectodermal-mesodermal interactions are stringently controlled by the balance between numerous growth stimulators and inhibitors and drive the developing HF through defined stages, ultimately resulting in the formation of a fiber- and pigment-producing mini-organ (Fig. 1).

Several indications suggest that NTs and their receptors are also expressed in the developing and postnatal HF (Bronzetti et al., 1996; Shibayama and Koizumi, 1996). Induction and early steps of vibrissa HF development in mice coincide with high levels of cutaneous NTs. At E11.5-E13.5, NGF, BDNF, NT-3 and NT-4 messages are maximal in skin of the maxillary process, while their expression progressively declines during later steps of embryonic development (Davies et al., 1987; Harper and Davies, 1990; Buchman and Davies, 1993; Schoring et al., 1993). In vibrissa HF, NT-3 and NT-4 transcripts are expressed in the epithelium, while mRNA^{BDNF} is seen in the mesenchyme (Ibanez et al., 1993; Ernfors et al., 1994).

Many components of vibrissa HF innervation are absent in knockout mice with deletion of defined NT genes, suggesting that vibrissa-derived NTs play critical role in the development of different innervation subsets of this HF type with specialized sensory functions (Fundin et al., 1997). However, the recognized expression patterns of NTs and NT

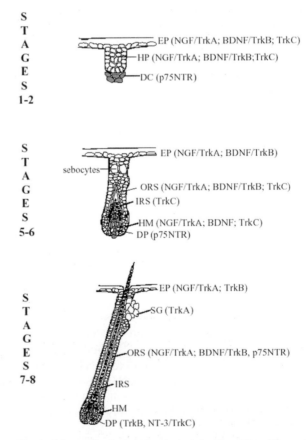

Fig. 1. Scheme illustrating the expression of neurotrophins and their receptors during hair follicle morphogenesis.

receptors during skin and hair development already suggest functions beyond the control of skin/hair innervation.

For example, synthesis of NGF in developing skin appears to be independent of innervation and coincides with the onset of cell differentiation programs in the epidermis (Rohrer et al., 1988; Schoring et al., 1993). Basal epidermal keratinocytes express TrkA, and mesenchymal cells around developing vibrissa HF show TrkB and TrkC receptors (Wheeler and Bothwell, 1992; Wheeler et al., 1998). During embryonic development of human skin, p75^{NTR} is expressed by nerve fibers and endothelial cells, as well as by the specialized population of dermal fibroblasts, which later forms the dermal papilla of the HF (Holbrook and Minami, 1991; Peters et al., 2002a).

This implicates that NTs not only control cutaneous innervation, but also in the regulation of HF development.

Expression of neurotrophins and their receptors during hair follicle morphogenesis

We characterized the expression of NTs and their receptors during all stages of HF development in the back skin of fetal and neonatal C57BL/6 mice (Botchkarev et al., 1998; Botchkareva et al., 1999a, 2000) (summarized in Fig. 1 and Table 1). During early stages of HF development (stages 0-2, E16.5-18.5), strong and fairly homogeneous expression of NGF, BDNF and their receptors TrkA and TrkB is found in the KCs of the hair plug as well

TABLE 1
NT and NT receptor distribution patterns in the developing and cycling hair follicle

Antigen	IR pattern in the during HF morphogenesis	IR pattern during HF cycling
NGF	• Vibrissae epithelium[rat] • Hair plug, distal outer root sheath[mouse]	• Distal outer root sheath (anagen)[mouse]
BDNF	• Vibrissae epithelium[rat] • Hair plug, distal outer root sheath[mouse]	• Proximal outer root sheath, matrix (anagen)[mouse, human] • Inner root sheath, dermal papilla (anagen)[human] • Outer root sheath, inner root sheath, proximal matrix, secondary hair germ (catagen)[mouse]
NT-3	• Vibrissae epithelium[rat] • Dermal papilla, epithelial hair bulb, proximal outer root sheath (late developmental stages)[mouse]	• Innermost outer root sheath (telogen)[mouse] • Innermost layer of the outer root sheath (isthmus region), dermal papilla (anagen)[mouse] • Proximal outer root sheath, matrix, epithelial strand, secondary hair germ (catagen)[mouse]
NT-4		• Outer root sheath (telogen)[mouse] • Proximal outer root sheath, inner root sheath, matrix (anagen)[mouse] • Outer root sheath, inner root sheath, proximal matrix, epithelial strand (catagen)[mouse]
P75	• Dermal papilla (early and mid developmental stages)[human, mouse]	• Proximal outer root sheath, keratogenous zone of the matrix (anagen)[human, mouse] • Outer root sheath, keratogenous zone, secondary hair germ (catagen)[mouse]
TrkA	• Hair plug, distal outer root sheath[mouse]	• Distal outer root sheath (anagen)[mouse] • Single cells in epithelial strand (catagen)[mouse]
TrkB	• Hair plug, distal outer root sheath, dermal papilla (upon completion)[mouse]	• Central and proximal outer root sheath, keratogenous zone of the matrix, dermal papilla (anagen)[mouse] • Inner root sheath, outer root sheath (anagen)[human] • Inner root sheath, matrix, epithelial strand, secondary hair germ, dermal papilla (catagen)[mouse]
TrkC	• Vibrissae epithelium[mouse] • Hair follicle epithelium[mouse] • Dermal papilla (late developmental stages)[mouse]	• Outer root sheath (telogen) • Central outer root sheath, matrix, dermal papilla (anagen)[mouse] • Proximal outer root sheath, inner root sheath, matrix, epithelial strand, secondary hair germ (catagen)[mouse]

as in the interfollicular basal epidermal KCs. TrkC was also expressed in the KCs of forming hair placode and epidermis, while NT-3 could be visible only in the single mesenchymal cells of the dermis. Interestingly, p75^{NTR} was expressed by population of fibroblasts, forming dermal condensation around hair placode and closely adjacent to placode epithelium (Fig. 1).

During advanced stages of HF morphogenesis (stages 4-5, E18.5-P2), we observed NGF in the distal and proximal compartments of the follicular outer root sheath (ORS), while TrkA was more prominent throughout the epithelium of developing HF (Fig. 1). Prominent expression of both BDNF and TrkB was observed in interfollicular epidermal KCs, and BDNF was broadly distributed in the follicular epithelium, while TrkB in the HF was restricted to the distal ORS. The expression of NT-3 was barely visible in isolated cells of the dermal papilla. However, its receptor (TrkC) was observed in the proximal ORS, hair matrix, and the developing inner root sheath (IRS). At the same period, p75^{NTR} was prominently expressed in the fibroblasts of the developing dermal papilla, and also in isolated dermal fibroblasts surrounding the HF, which likely go on to form the follicular connective tissue sheath.

During the final stages of HF morphogenesis (stages 6-8, P3-P6), expression of NGF and TrkA was restricted to the distal ORS, while other HF compartments were completely NGF- and TrkA-negative (Fig. 1). Basal epidermal KCs showed strong NGF- and TrkA-IR. Epidermal KCs became negative for BDNF, while they remained strongly positive for TrkB. In fully developed follicles, BDNF was found in the distal ORS. In addition, BDNF was seen in large round-shaped cells in the developing subcutis, morphologically resembling adipocytes. The expression of TrkB in the latest stages of HF development was present on the ORS KCs, and also was seen in the follicular papilla. NT-3 and TrkC were seen in KCs of the proximal hair bulb as well as in dermal papilla fibroblasts.

Intriguingly, appearance of TrkB and TrkC in the dermal papilla was accompanied by a substantial decline of p75^{NTR} expression, with subsequent complete disappearance in fully developed HF. At this time, a striking switch in the follicular expression patterns of p75^{NTR} occurred—i.e., from the HF mesenchyme to the HF epithelium- and p75^{NTR}-IR appeared in KCs of the central and proximal ORS (Fig. 1). In addition, p75^{NTR} and Trk receptors were expressed in skin nerves located around the developing HF. Thus, the expression of NTs and their receptors is spatio-temporally stringently regulated during HF morphogenesis, suggesting growth-modulatory functions for NTs (Table 1).

Gain or loss of neurotrophin signaling leads to alterations in hair follicle development

To define the functional roles of individual NTs and their receptors in HF development, we have compared the dynamics of HF morphogenesis between mutant mice with constitutive deletion of *TrkC* (−/−), *p75^{NTR}* (−/−), or partial deletion of *NT-3* (+/−) with the corresponding wild type mice (Botchkarev et al., 1998; Botchkarev et al., 1999a,b). This was complemented by studying the rate of HF development in transgenic mice over-expressing NGF or NT-3 (promoter: K14), or BDNF (promoter: alpha-myosin heavy chain) and the corresponding wild type controls (Botchkarev et al., 1998; Botchkarev et al., 2000). In addition, the effects of NGF and BDNF on fetal HF development was tested in skin organ culture (Botchkarev et al., 2000).

We found that constitutive overexpression of NGF in mice was associated with a discrete, but significant acceleration of HF morphogenesis, which was missing in BDNF transgenic mice (Botchkarev et al., 2000). Similar data were obtained in situ after culturing neonatal skin organ with either NGF or BDNF. In this assay, NGF significantly accelerated HF morphogenesis, while BDNF did not show any substantial effects (Botchkarev et al., 2000) (Fig. 2, Table 2). Newborn mice with overexpression of NT-3, showed significant acceleration of HF morphogenesis, whereas this was retarded in newborn heterozygous *NT-3* KO (+/−) mice, compared to the corresponding wild type animals (Botchkarev et al., 1998). These results are consistent with data obtained from *TrkC* KO mice: deletion of this high affinity receptor for NT-3 resulted in a

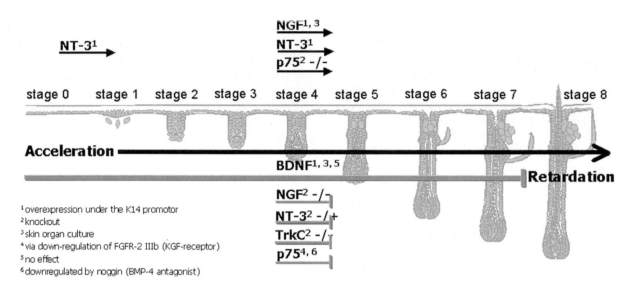

Fig. 2. NT differentially modulate hair follicle morphogenesis. Summary of our published data on NT effects on murine HF development in the back skin of neonatal mice (modified after Peters et al., 2003). Functional data were obtained by overexpression, gene knockout or organ culture. Note that NT-3, NGF and deletion of p75[NTR] all accelerate hair follicle morphogenesis and that NT-3 even induces premature hair follicle development, while p75[NTR] stimulation as well as deletion of NT-3, TrkC or NGF (see Crowley et al., 1994) all retard hair follicle development. BDNF, instead, has no significant modulatory effect on murine hair follicle development.

significant retardation of HF development, indicating that NT-3/TrkC signaling operates as a promoter of murine HF morphogenesis (Botchkareva et al., 1999b).

Growth-inhibitory functions of p75[NTR] during hair follicle morphogenesis are associated with regulation of the cell cycle and FGF7 signaling

Since p75[NTR] reportedly is one of the first growth factor receptors to become immuno-detectable during human foetal HF development (Holbrook and Minami, 1991), it was reasonable to assume that signaling via this NT receptor has a growth-promoting function during murine HF development. Unexpectedly, however, analysis of p75[NTR] KO (−/−) mice revealed an *inhibitory* function of this receptor on murine HF morphogenesis (Botchkareva et al., 1999a): Compared to age-matched wild-type animals, p75[NTR] KO (−/−) mice showed a significant acceleration of HF development.

Interestingly, this was not associated with an upregulation of apoptosis, as evident from the fact that DP fibroblasts in both p75[NTR] mutant and wild-type mice showed a high expression of Bcl-2 and apparently did not undergo apoptosis. These data are consistent with other reports that have shown a high resistance of dermal papilla fibroblasts against apoptosis, both during HF development (Polakowska et al., 1994; Magerl et al., 2001) and cycling (Seiberg et al., 1995; Lindner et al., 1997), even under chemotherapy (Lindner et al., 1997; Botchkarev et al., 2000b).

Instead, we found that deletion of p75[NTR] in mutants is associated with significant reduction of the proliferation rate in dermal papilla fibroblasts of stage 4–5 HF (Botchkareva et al., 1999a). As it shown in other models, downregulation of p75[NTR] is an important step in the cell transition from proliferation to differentiation (Erck et al., 1998; Urdiales et al., 1998). Downregulation of p75[NTR] in dermal papilla cells of stage 5 HF is followed by the onset of TrkB and TrkC expression in dermal papilla at stages 6–7 of HF development (Fig. 1) (Botchkarev et al., 1998; Botchkareva et al., 1999a, 2000). Thus, we speculate that p75[NTR] expression in dermal papilla fibroblasts during early stages of HF development plays a role in

TABLE 2
Neurotrophins and murine hair follicle morphogenesis

- **Spatiotemporally stringently controlled expression of NGF, BDNF, NT-3, TrkA, TrkB, TrkC and p75^{NTR}** in defined HF compartments during progressing HF development (e.g., early during HF morphogenesis, TrkA-C are expressed in epithelial hair placode, p75^{NTR} in underlying mesenchymal condensation that develops into follicular dermal papilla)

- NGF overexpression (K14-tg mice),
NGF administration (skin organ culture):
  - → **accelerated HF development**
  - → *NGF/TrkA signaling: promotes HF morphogenesis*

- **BDNF** overexpression,
BDNF administration (skin organ culture):
  - → **HF development unchanged**

- NT-3 or TrkC KO mice:
  - → **HF development retarded**

- NT-3 overexpression (K14-tg mice):
  - → **more HF induced at birth**
  - → **HF development accelerated**
  - → *NT-3/TrkC signaling: promotes HF morphogenesis*

- *p75^{NTR}* KO mice:
  - → **accelerated HF development**
    diminished inhibitor secretion by/ inhibitor activity on DP fibroblasts
  - → **upregulation of FGFR-2** (KGFR)
  - → KGF-neutralizing antibodies normalize accelerated HF development in *p75^{NTR}* KO mice
  - → noggin KO mice: severe retardation of HF development, upregulation of p75^{NTR} expression
  - → treament with noggin (= BMP-2/4 antagonist) inhibits p75^{NTR} expression
  - → *p75^{NTR}: molecular 'brake' on HF development, released by noggin?*
  - → *p75^{NTR}: negative control of HF development by downregulation of KGF/FGFR-2 signaling?*

Botchkarev et al., J. Invest. Dematol., 111: 279, 1998.
Botchkarev et al., J. Invest. Dermatol., 111:279, 1998.
Botchkareva et al., J. Invest. Derm., 113:425, 1999.
Botchkarev et al., Nat. Cell. Biol., 1:158, 1999.
Botchkareva et al., Dev. Biol., 216:135, 1999.
Botchkareva et al., J. Invest. Dermatol., 114:314, 2000.

mediating their transition from the proliferative mode into a more differentiated mode, associated with the onset of morphogen secretion by dermal papilla fibroblasts (Philpott and Paus, 1998; Millar, 2002).

Among a variety of growth factors and their receptors examined, only FGFR-2 (i.e., an isoform specific for the binding of FGF7/KGF) was upregulated in the dermal papilla and ORS of *p75^{NTR}* null HF (Botchkareva et al., 1999a). Administration of FGF7-neutralizing antibody significantly reduced the acceleration of HF morphogenesis seen in p75^{NTR} knockout mice, while this showed no effects in wild type mice (Botchkareva et al., 1999a). This suggests that, in developing murine HF, p75^{NTR} down-modulates FGF7-signaling through FGFR-2 and thus acts as a 'brake' on HF morphogenesis by limiting the effects of FGF7 on early steps of HF development (Fig. 2, Table 2).

Differential control of hair follicle development by Trk vs p75 signaling

Taken together, our data suggest that p75^{NTR} plays inhibitory roles during HF morphogenesis, while NGF and NT-3/TrkC induce opposite effects and accelerate HF development (Botchkarev et al., 1998; Botchkareva et al., 1999a,b, 2000). Stimulatory effects of NTs on HF development appear to be mediated via interactions with high affinity Trk receptors, which are expressed exclusively in the HF epithelium during early stages of HF development.

However, during early steps of HF development, NTs may not only promote KC proliferation via stimulation of Trk receptors expressed in the follicular epithelium, but NT interactions with p75^{NTR} may also control the proliferation/differentiation transition in dermal papilla fibroblasts, and may downregulate regulate FGFR-2 expression, thus limiting FGF7-stimulatory actions on developing HF. During late steps of HF development, when p75^{NTR} disappears from the dermal papilla and is expressed together with Trk-receptors in the follicular ORS, promoting effects of NTs on HF development apparently, are not limited anymore by their additional binding to p75^{NTR} on dermal papilla cells (Table 2). Therefore, Trk receptors and p75^{NTR} are important components of the molecular signaling network regulating HF development (Philpott and Paus, 1998; Millar, 2002).

One key challenge now is to elucidate the regulatory mechanisms that control expression of these NT receptors and their ligands during HF development. Our data suggest that BMPs are among the controls that regulate p75^{NTR} expression in the dermal papilla of developing HF, since p75^{NTR} is strongly upregulated in mutant mice deficient in the potent BMP antagonist noggin (Botchkarev et al., 1999a, Botchkarev, 2003). BMP signaling inhibits the early steps of HF development, and *noggin* KO mice show lack of induction of about 90% of HF in back skin (Botchkarev et al., 2002). This suggests that p75^{NTR} represents an important target for BMP regulation in developing HF, and that BMP receptor- and p75^{NTR}-mediated signaling operate in concert as two important pathways inhibiting HF development.

The hair follicle cycle as a unique model for neuroectodermal-mesodermal interactions during tissue remodeling

Immediately after the completion of its morphogenesis, as the only such organ in the mammalian body, the HF launches on a unique, life-long cycle of regression, growth and 'resting', during which the HF rhythmically undergoes dramatic, spontaneously

occurring transformation events which are controlled by a—still elusive—'hair cycle clock' that is located in the skin itself (Paus and Cotsarelis, 1999; Cotsarelis and Millar, 2001; Stenn and Paus, 2001). Since disturbances in HF cycling are at the basis of many hair growth disorders seen in clinical practise (Paus, 1996; Paus and Cotsarelis, 1999; Cotsarelis and Millar, 2001), the biological controls that drive the HF from periods of active growth and production of pigmented hair shafts (anagen) via by apoptosis-driven involution (catagen), followed by a period of relative 'resting' (telogen) back to anagen not only are biologically fascinating, but are also of profound clinical importance.

First cyclic changes of HF activity can be envisioned on days 14–17 of postnatal development in mice, when HF enters into catagen phase of the hair cycle. During catagen, proliferation and differentiation of hair matrix keratinocytes is dramatically reduced, pigment-producing activity of melanocytes is ceased, and hair shaft production is completed. This is accompanied by well coordinate apoptosis in the proximal HF epithelium and follicular melanocytes leading to the shortening of HF length up to 70% (Fig. 2) (Lindner et al., 1997; Paus and Cotsarelis, 1999; Cotsarelis and Millar, 2001; Stenn and Paus, 2001).

Catagen stage is followed by resting stage or telogen, characterized by minimal signaling exchange between dermal papilla fibroblasts and follicular keratinocytes and by absent or inconspicuous KC proliferation in the follicular germinative compartment (Fig. 2; Lindner et al. 1997; Paus and Cotsarelis, 1999; Cotsarelis and Millar, 2001; Stenn and Paus, 2001). Anagen induction and development is a process of programmed organ regeneration characterized by sudden activation of cell proliferation in the follicular germinative compartment located in most proximal part of the telogen hair follicle close to the dermal papilla (Paus and Cotsarelis, 1999; Cotsarelis and Millar, 2001, Panteleyev et al., 2001; Stenn and Paus, 2001). This leads to the downgrowth of the hair follicle into the subcutaneous mesenchyme, accompanied by the reactivation of a host of distinct differentiation programs in the HF epithelium. This is followed by the appearance of hair matrix and inner root sheath KCs, as well as by activation of melanogenesis in the HF melanocyte precursors (Fig. 2). Finally, full restoration of the hair fiber producing unit is associated with the formation of the so-called anagen hair bulb, which surrounds a morphogen-secreting dermal papilla and is located deep in the subcutis (Paus and Cotsarelis, 1999; Cotsarelis and Millar, 2001; Stenn and Paus, 2001).

The fact that HF cycling in mice is accompanied by a very significant reorganization of skin and HF innervation already suggests a prominent role for HF-derived NTs in the maintenance and remodeling of distinct subsets of skin nerves during pre- and postnatal ontogenesis and the striking neuronal plasticity associated with HF cycling in adult murine skin (Botchkarev et al., 1997a,b, 1999d; Peters et al., 2001, 2002a;). However, HF cycling becomes an even more intriguing (as well as a uniquely accessible and easily manipulated!) model for neuroectodermal-mesodermal interactions during tissue remodeling

if one considers the available indications that support a direct role for NT receptor-mediated signaling in the control of HF cycling itself.

Neurotrophins and hair follicle cycling: follicular expression of neurotrophins and their receptors is hair cycle-dependent

Indeed, in C57BL/6 mouse skin, the steady-state levels of NGF and NT-3 proteins significantly fluctuate in a hair cycle-dependent manner (Welker et al., 1996; Botchkarev et al., 1998). While NGF levels increased dramatically in early anagen skin, NT-3 protein was significantly upregulated during catagen. Also, the steady-state levels of NT-3, BDNF and NT-4 transcripts significantly rose prior to and during catagen onset (Botchkarev et al., 1998, 1999b) (Fig. 3).

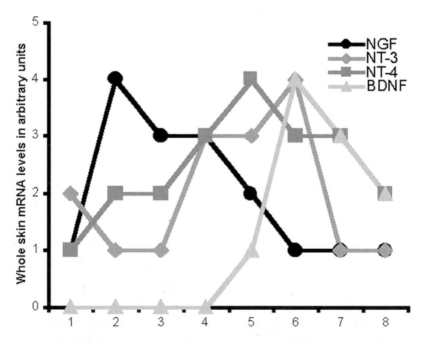

Fig. 3. NT gene expression in murine skin is hair cycle-dependent. Summary of our published data on hair cycle-dependent changes in NT steady-state transcript levels (modifed after Peters et al., 2003) as determined by semi-quantitative RT-PCR on RNA extracts from full thickness back skin of C57BL/6 mice before (1 = telogen) and after anagen induction by depilation (2 = early anagen one day after anagen induction, 3 = early-mid anagen, 4 = mid-late anagen, 5 = late anagen, 6 = catagen, 7 = catagen/telogen transition, 8 = telogen). Levels were determined by densitometry and expressed in arbitrary units on the basis of fold-increases. While NGF levels rise sharply immediately after anagen induction (possibly reflecting, at least in part, NGF upregulation as a component of the mild wound healing response associated anagen induction by depilation), NT-3, NT-4 and BDNF transcript levels all rise substantially only during the synchronized, spontaneously occurring anagen-catagen transformation of the hair follicle.

In unmanipulated murine back skin with all HF in telogen, NGF, NT-3, NT-4 and TrkA were found in the follicular ORS (Table 1, Fig. 4). In anagen II, increase of NGF-IR was observed in epidermis and growing HF, where NGF-IR together with TrkA-IR were prominently expressed in ORS and hair germ (Table 1, Fig. 4). During early and middle anagen, NT-4 was seen in single ORS KCs, while no expression of BDNF was found in the HF. In anagen IV HF, NT-3 becomes visible in single cells in the HF isthmus region, in the dermal papilla, and in perifollicular nerve fibers (Table 1, Fig. 4). At the same time, follicular TrkC-immunoreactivity became visible, yet only in the central outer root sheath.

With progressing anagen development (anagen IV–VI), HF-associated NGF-IR was restricted to central and distal ORS, while proximal ORS, IRS, hair matrix, and dermal papilla did not show NGF-IR (Table 1, Fig. 4). In addition to the distal and central ORS, TrkA-IR was found also in proximal ORS, IRS, and hair matrix of anagen VI HF. During progressing anagen development, mRNA^{BDNF} and BDNF-IR appeared in the proximal ORS and hair matrix of anagen VI HF. NT-4 was present not only in hair matrix KCs and in the proximal ORS, but also in the IRS of anagen VI HF. In anagen VI HF, expression of mRNA^{TrkB} and TrkB antigen was found in the central and proximal ORS. In anagen VI, NT-3 was widely expressed by proximal outer root sheath KCs (Table 1, Fig. 4), while many KCs in the hair matrix and outer root sheath were TrkC-IR. Prominent IR not only for NT-3, but also for TrkC appeared in dermal papilla of anagen VI HF. In anagen VI HF, p75^{NTR} was expressed by KC of the proximal and central outer root sheath (ORS), where its coexpression with TrkA-, TrkB- and TrkC-IR was

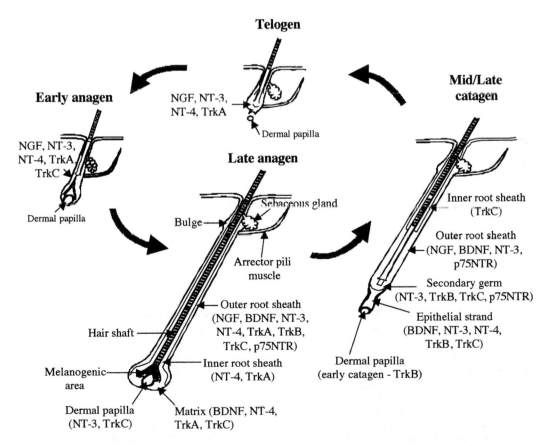

Fig. 4. Summary of the expression patterns of neurotrophins and their receptors during hair follicle cycling.

seen (Table 1, Fig. 4). No expression of $p75^{NTR}$-IR alone, without its coexpression with Trk-receptors, was observed in the anagen VI HF (Botchkarev et al., 1998, 1999b, 2000a).

In the beginning of HF regression (catagen II), a marked enhancement of $mRNA^{BDNF}$ expression, as well as of BDNF- and NT-4-IR was noted in the proximal hair matrix (Table 1, Fig. 4). During early catagen, $mRNA^{TrkB}$ and TrkB protein was observed in hair matrix KC and in dermal papilla fibroblasts (Table 1, Fig. 4). During mid/late catagen, NGF, BDNF and NT-3 were strongly expressed in the ORS, while all Trk receptors disappeared from this regressing HF compartment. ORS KCs of late catagen HF showed expression of the $p75^{NTR}$ (Table 1, Fig. 4). However, coexpression of $p75^{NTR}$- and TrkB-/TrkC-IR was found in the secondary hair germ of the catagen VI HF (Botchkarev et al., 1998, 1999b, 2000a).

In order to define, whether or not cells of the regressing HF compartments expressed $p75^{NTR}$ are colocalized with apoptotic markers during catagen, a double visualization of $p75^{NTR}$-IR and TUNEL, or $p75^{NTR}$ and Bcl-2 was performed. Interestingly, only cells in the regressing ORS, expressed $p75^{NTR}$ in catagen VI HF, showed colocalization of p75NTR and TUNEL, and no coexpression of $p75^{NTR}$- and Bcl-2-IR were seen here. In the secondary hair germ, where $p75^{NTR}$ was coexpressed with TrkB and TrkC, colocalization of $p75^{NTR}$- and Bcl-2-IR was found, and no coexpression of $p75^{NTR}$ and TUNEL was seen. Strong expression of Bcl-2 -IR was observed in the dermal papilla, while numerous TUNEL-positive cells were visible in the epithelial strand, a HF compartment showed no $p75^{NTR}$-IR and relatively weak Bcl-2-IR during catagen (Botchkarev et al., 1998, 1999b, 2000a).

In summary, NTs and their receptors show stringent spatio-temporal expression patterns in the hair follicle epithelium and mesenchyme during the hair cycle. While NGF is markedly upregulated in the follicular epithelial cells during early anagen, BDNF, NT-3 and NT-4 expressions are increased during late anagen–catagen. Given that $p75^{NTR}$ show coexpression with TUNEL in the regressing ORS, we, therefore hypothesized that NTs can promote catagen development and further explored the role of $p75^{NTR}$ in catagen control in functional assays.

Neurotrophins promote apoptosis-driven hair follicle involution via stimulating $p75^{NTR}$

Despite the prominent HF expression of all four NTs and their receptors, their central role in hair cycle control does not seem to be growth promoting. Functional assays in the murine system, using a variety of approaches such as organ culture of (denervated!) mouse, gene-knockout or overexpression revealed that NTs primarily *promote* catagen development in mature HF, i.e., operate as hair growth inhibitors once HF development has been completed (Table 3).

Using a transgenic approach, we showed that neonatal mice overexpressing NT-3 or BDNF are characterized by the acceleration of their first catagen development, and that BDNF overexpressing mice also display a significant shortening of their hair length compared to the corresponding age-matched wild type animals (Botchkarev et al., 1998, 1999b). *Vice versa*, neurotrophin (NT-3, BDNF, NT-4) null mutants show catagen retardation. Furthermore, NT-3, NT-4, and BDNF significantly stimulate catagen development in C57BL/6 murine skin organ culture (Botchkarev et al., 1998, 1999b).

Our data suggest that the catagen-stimulating effects of NTs are mediated by $p75^{NTR}$, since *$p75^{NTR}$* KO mice displayed significant catagen retardation and decrease of TUNEL-positive cells in the regressing ORS, compared to wild-type controls (Botchkarev et al., 2000a). In addition, a $p75^{NTR}$ antagonistic cyclic decapeptide shown to prevent apoptotic cell death in previous studies (Yaar et al., 1997; 2002), caused retardation of catagen by 33% in organ culture and prevented catagen-stimulatory effects of NTs. Finally, NTs failed to stimulate catagen in the organ-cultured skin of *$p75^{NTR}$* null mice, further suggesting the essential role of $p75^{NTR}$ in realization of the catagen-accelerating effects of neurotrophins (Botchkarev et al., 2003).

Neurotrophins also control human hair growth

Most recently, we have begun to evaluate how transferable these observations on the role of NTs and their receptors in murine hair biology are to the human system, employing organ culture of

TABLE 3
Neurotrophins and murine hair follicle cycling

- Spatio-temporally controlled, **strongly hair cycle-dependent intrafollicular expression of NGF, NT-3, BDNF, NT-4, TrkA, TrkB, TrkC and p75^{NTR}**

- **Upregulation** of intracutaneous expression for most of these NT and NTR **during late anagen VI and catagen** (s. Fig. 1), suggesting a role in apoptosis-driven hair follicle regression (catagen)

- **NT-3/TrkC or TrkB expression** apparently **colocalizing with apoptosis** of some HF keratinocyte during catagen

- Regressing HF epithelium: expression of p75^{NTR} in apoptotic 'hot spots' of catagen HF; **prominent p75^{NTR} expression in Bcl-2-negative, apoptotic HF keratinocytes that do** *not* **coexpress TrkB and/or TrkC with p75^{NTR}**

- *NT-3* KO mice($-/+$):	catagen retarded
- **NT-3 overexpression** (K14-tg mice)TG:	premature catagen
- *NT-4* KO mice:	catagen retarded
- *BDNF* KO mice:	catagen retarded
- BDNF overexpression (a-myosin-tg mice):	premature catagen
- NGF overexpression (K14-tg mice):	catagen accelerated
- BDNF, NT-4, NT-3, NGF (in anagen skin organ culture):	accelerate catagen in WT, but not in *p75^{NTR}* KO skin
- Synthetic p75 antagonist (anagen skin organ culture):	catagen retarded
- *p75^{NTR}* KO mice:	catagen retarded

Lindner et al., Am. J. Pathol. 151:1601, 1997.
Botchkarev et al., Am. J. Pathol. 153:758, 1998.
Botchkarev et al., FASEB J. 13:395, 1999.
Botchkarev et al, FASEB J. 14:1931, 2000.
Botchkarev et al., J. Invest. Dermatol. 120:168, 2003.

microdissected human scalp skin hair follicles in the anagen VI stage of the hair cycle (Peters et al., 2002b,c,d; Hansen et al., 2003a,b). Similar to the mouse hair follicle, the human hair follicle epithelium and dermal papilla express high steady state levels of NGF, BDNF, TrkA and TrkB (Peters et al., 2002b,c,d; Hansen et al., 2003a,b). Treatment of organ-cultured human scalp hair follicles (anagen VI) with NGF and BDNF results in decreased hair shaft elongation and HF KC proliferation, premature development of a catagen-like state and increased numbers of apoptotic cells in the hair bulb (Peters et al., 2002b,c,d; Hansen et al., 2003a,b). These data suggest that at least NGF, BDNF and p75^{NTR} signaling operate as catagen promoters in the anagen HF of human scalp skin, as well.

Neurotrophins regulate morphogen secretion by dermal papilla cells in vitro

Several indications suggest that indeed NTs may regulate the expression of TGF-β- and FGF-family growth factors, which are critically important for the initiation of HF anagen-catagen transition (Stenn and Paus, 2001): for instance, NGF increases TGF-β1 expression in PC12 cells, and BDNF is able to stimulate production and release of FGF-2 in the visual system via stimulation of p75^{NTR} (Cosgaya and Aranda, 1996; Wexler et al., 1998). We hypothesized that NTs may also accelerate HF anagen–catagen transition via altering the secretion of different growth factors by dermal papilla fibroblasts, which show the expression of TrkB and TrkC receptors prior and in the beginning of catagen (Table 1, Fig. 2).

To test this hypothesis, dermal papilla fibroblasts were isolated from normal human scalp HFs and cultured in the presence of different concentrations of NGF, BDNF, NT-3 or NT-4. Our data suggest that BDNF and NT-4 were capable of significantly reducing the amounts of stem cell factor and vascular endothelial growth factor secreted by dermal papilla cells, while the levels of TGF-β1 and hepatocyte growth factor remained unchanged (Tobin, 2001). However, in organ cultured human anagen HF, BDNF and NGF treatment upregulated TGF-β2 expression (Peters et al., 2002b,c,d; Hansen et al.,

2003a,b). These data suggest that NTs may regulate anagen-catagen transition at least in part via modulating the secretion of morphogens by dermal papilla cells.

Clinical and biological perspectives

That selected NTs, like NGF and BDNF, show highly comparable functional effects on mature anagen HF both in mice and man, suggests that selective NT receptor agonists and antagonists may become useful, novel therapeutic tools for the management of hair growth disorders. In principle, it is conceivable that, e.g., synthetic TrkB or p75^{NTR} antagonists (ideally applied topically via HF-targeting liposomes) might be exploited as hair growth-stimulatory agents in the management of various forms of hair loss (alopecia, effluvium), while corresponding agonists might be explored for the treatment of unwanted hair growth (hirsutism, hypertrichosis).

Also, p75^{NTR} antagonists and TrkA/TrkC agonists deserve be studied as candidate agents for aiding the *de novo* induction of HF wherever these have been permanently lost (i.e., in scarring alopecias). Instead, topical or intralesional p75^{NTR} agonists and TrkA/TrkC antagonists might be therapeutically useful in the treatment of multiple, surgically not manageable multiple hair follicle tumors.

However, our studies reveal that the same NTs which operate as inhibitors of hair growth can have opposite, growth-stimulatory effects on epidermal keratinocyte proliferation (Botchkarev et al., 1999c). This suggests that defined KC subpopulations in vivo are differentially regulated by selected NTs. Therefore, one key challenge in the clinical exploitation of the concepts extracted from our studies will be to selectively administer NT receptor ligands to the HF, e.g., by the use of HF-targeted liposomes, without altering epidermal growth patterns in a clinically undesired manner.

Dissecting the molecular and cellular mechanisms that underlie the hair growth-modulatory effects of NTs is an utterly daunting task, not only because of the extreme complexity of HF growth controls (Stenn and Paus, 2001; Millar, 2002), but also in view of the numerous signaling pathways and cross-talks that must considered in this respect (see discussion above). Besides direct signal transduction through surface NT receptors expressed by various different HF KC populations, multiple additional interactions must be taken into account, individually probed and verified or excluded. To name just a few, besides a modulation of neuropepetide/neurotransmitter release from skin nerve endings, these indirect hair growth-modulatory effects include by NT-induced changes in the expression of other cell surface receptors, and by the manipulation of perifollicular mast cell, fibroblast and macrophage as well as of intrafollicular melanocyte, Langerhans cell and lymphocyte activities (cf. Paus et al., 1997; Stenn and Paus, 2001).

While the use of intact HF (in vivo, in organ culture), as we have resorted to in our studies, clearly limits the analytical depth that one can expect to reach without very elaborate, costly and time-consuming assay designs, one can be relatively confident that most of the findings that are made in this—uniquely accessible—tissue interaction system are likely to be biologically and/or even clinically relevant. Also, the intriguing, previously unsuspected signaling cross-connections revealed in our murine HF studies (e.g., that between p75^{NTR} signaling on the one hand, and noggin/BMP receptor as well as FGF7/FGF-R2 signaling on the other) beautifully illustrate how a very complex tissue interaction system as the hair follicle, even on the in vivo level, can provide exciting new leads for basic NT biology that reach far beyond the boundaries of skin and hair research.

Acknowledgments

The authors' work summarized in this review was supported in part by grants from the Deutsche Forschungsgemeinschaft to RP (DFG Pa 345/6-1, 6-2, 6-4) and EMJP (DFG Pe 890/1-1), by grants from Wella AG, Darmstadt/Germany, and Cutech Srl, Padova/Italy to RP, and by grants from the National Alopecia Areata Foundation to VAB. We gratefully acknowledge the important contributions of all our collaborators in the cited studies, most notably Dr. K. Albers (University of Pittsburg, PA) and Dr. G.R. Lewin (Max-Dellbrück-Centrum for Molecular Medicine, Berlin-Buch).

Abbreviations

BDNF	brain-derived neurotrophic factor
BMP	bone morphogenetic protein
DP	dermal papilla of the hair follicle
HF	hair follicle(s)
IR	immunoreactive or immunoreactivity
IRS	inner root sheath
KC	keratinocyte(s)
KO	knockout
NGF	nerve growth factor
NT	neurotrophin(s)
NT-3	neurotrophin-3
NT-4	neurotrophin-4
ORS	outer root sheath
p75^{NTR}	p75 low-affinity neurotrophin receptor
TrkA-C	tyrosine kinase A-C (high-affinity neurotrophin receptors)

References

Acheson, A., Barker, P.A., Alderson, R.F., Miller, F.D. and Murphy, R.A. (1991) Detection of brain-derived neurotrophic factor-like activity in fibroblasts and Schwann cells: inhibition by antibodies to NGF. Neuron, 7: 265–275.

Aloe, L. and Micera, A. (1999) Nerve growth factor: basic findings and clinical trials. In: L. Aloe and G.N. Chaldakov (Eds.), Nerve Growth Factor in Health and Disease. Biomedical Reviews, Vol. 10, The Bulgarian-American Center, Varna, pp. 3–14.

Aloe, L., Bracci-Laudiero, L., Bonini, S. and Manni, L. (1997) The expanding role of nerve growth factor: from neurotrophic activity to immunologic diseases. Allergy, 52: 883–894.

Aloe, L., Simone, M.D. and Properzi, F. (1999) Nerve growth factor: a neurotrophin with activity on cells of the immune system. Microsc. Res. Techn., 45: 285–291.

Aloyz, R.S., Bamji, S.X., Pozniak, C.D., Toma, J.G., Atwal, J., Kaplan, D.R. and Miller, F.D. (1998) p53 is essential for developmental neuron death as regulated by the TrkA and p75 neurotrophin receptors. J. Cell Biol., 143: 1691–1703.

Bamji, S., Majdan, M., Pozniak, C.D., Belliveau, D., Aloyz, R., Kohn, J., Causing, C.G. and Miller, F.D. (1998) The p75 neurotrophin receptor mediates neuronal apoptosis and is essential for naturally occuring sympathetic neuron death. J. Cell Biol., 140: 911–923.

Bernabei, R., Landi, F.B.S., Onder, G., Lambiase, A., Pola, R. and Aloe, L. (1999) Effect of topical application of nerve-growth factor on pressure ulcers. Lancet, 354: 307.

Bibel, M. and Barde, Y.-A. (2000) Neurotrophins: key regulators of cell fate and cell shape in the vertebrate nervous system. Genes Dev., 14: 2919–2937.

Botchkarev, V.A. (2003) Bone morphogenetic proteins and their antagonists in skin and hair follicle biology. J. Invest. Dermatol., 120: 36–49.

Botchkarev, V.A., Yaar, M., Gilchrest, B. and Paus, R. (2003) p75 Neurotrophin receptor antagonist retards apoptosis-driven hair follicle involution (catagen). J. Invest. Dermatol., 120: 168–169.

Botchkarev, V.A., Botchkareva, N.V., Sharov, A.A., Funa, K., Huber, O. and Gilchrest, B.A. (2002) Modulation of BMP signaling by noggin is required for induction of the secondary (non-tylotrich) hair follicles. J. Invest. Dermatol., 118: 3–10.

Botchkarev, V.A., Botchkareva, N.V., Albers, K.M., Chen, L.-H., Welker, P. and Paus, R. (2000a) A role for p75 neurotrophin receptor in the control of apoptosis-driven hair follicle regression. FASEB J., 14: 1931–1942.

Botchkarev, V.A., Komarova, E.A., Siebenhaar, F., Botchkareva, N.V., Komarov, P.G., Maurer, M., Gilchrest, B.A. and Gudkov, A.V. (2000b) p53 is essential for chemotherapy-induced hair loss. Cancer Res., 60: 5002–5006.

Botchkarev, V.A., Botchkareva, N.V., Roth, W., Nakamura, M., Chen, L.-H., Herzog, W., Lindner, G., McMahon, J.A., Peters, C., Lauster, R., McMahon, A.P. and Paus, R. (1999a) Noggin is a mesenchymally-derived stimulator of hair follicle induction. Nat. Cell Biol., 1: 158–164.

Botchkarev, V.A., Botchkareva, N.V., Welker, P., Metz, M., Subramaniam, A., Lewin, G.R., Braun, A., Lommatzsch, M., Renz, H. and Paus, R. (1999b) A new role for neurotrophins: involvement of brain-derived neurotrophic factor and neurotrophin-4 in hair cycle control. FASEB J., 13: 395–410.

Botchkarev, V.A., Metz, M., Botchkareva, N.V., Welker, P., Lommatzsch, M., Renz, H. and Paus, R. (1999c) Brain-derived neurotrophic factor, neurotrophin-3, and neurotrophin-4 act as 'epitheliotrophins' in murine skin. Lab. Invest., 79: 557–572.

Botchkarev, V.A., Peters, E.M., Botchkareva, N.V., Maurer, M. and Paus, R. (1999d) Hair cycle-dependent changes in adrenergic skin innervation, and hair growth modulation by adrenergic drugs. J. Invest. Dermatol., 113: 878–887.

Botchkarev, V.A., Welker, P., Albers, K.M., Botchkareva, N.V., Metz, M., Lewin, G.R., Bulfone-Paus, S., Peters, E.M.J., Lindner, G. and Paus, R. (1998) A new role for neurotrophin-3: involvement in the control of hair follicle regression (catagen). Am. J. Pathol., 153: 785–799.

Botchkarev, V.A., Eichmüller, S., Johansson, O. and Paus, R. (1997a) Hair cycle-dependent plasticity of skin and hair

510

follicle innervation in normal murine skin. J. Comp. Neurol., 386: 379–395.

Botchkarev, V.A., Eichmuller, S., Peters, E.M., Pietsch, P., Johansson, O., Maurer, M. and Paus, R. (1997b) A simple immunofluorescence technique for simultaneous visualization of mast cells and nerve fibers reveals selectivity and hair cycle-dependent changes in mast cell–nerve fiber contacts in murine skin. Arch. Dermatol. Res., 289: 292–302.

Botchkarev, V.A., Paus, R., Czarnetzki, B.M., Kupriyanov, V.S., Gordon, D.S. and Johansson, O. (1995) Hair cycle-dependent changes in mast cell histochemistry in murine skin. Arch. Dermatol. Res., 287: 683–686.

Botchkareva, N.V., Botchkarev, V.A., Albers, K.M., Metz, M. and Paus, R. (2000) Distinct roles for NGF and BDNF in controlling the rate of hair follicle morphogenesis. J. Invest. Dermatol., 114: 314–320.

Botchkareva, N.V., Botchkarev, V.A., Chen, L.-H., Lindner, G. and Paus, R. (1999a) A role for p75 neurotrophin receptor in the control of hair follicle morphogenesis. Dev. Biol., 216: 135–153.

Botchkareva, N.V., Botchkarev, V.A., Metz, M., Silos-Santiago, I. and Paus, R. (1999b) Retardation of hair follicle development by the deletion of TrkC, high-affinity neurotrophin-3 receptor. J. Invest. Dermatol., 113: 425–427.

Bothwell, M. (1995) Functional interactions of neurotrophins and neurotrophin receptors. Annu. Rev. Neurosci., 18: 223 253.

Bronzetti, E., Ciraco, E., Germana, G. and Vega, J.A. (1996) Immunocytochemical localization of neurotrophin receptor proteins in human skin. Ital. J. Anat. Embryol., 100(Suppl.1): 565–571.

Buchman, V.L. and Davies, A.M. (1993) Different neurotrophins are expressed and act in a developmental sequence to promote the survival of embryonic sensory neurons. Development, 118: 989–1001.

Canossa, M., Griesbeck, O., Berninger, B., Campana, G., Kolbeck, R. and Thoenen, H. (1997) Neurotrophin release by neurotrophins: implications for activity-dependent neuronal plasticity. Proc. Natl. Acad. Sci. USA, 94: 13279–13286.

Cartwright, M., Mikheev, A.M. and Heinrich, G. (1994) Expression of neurotrophin genes in human fibroblasts: differential regulation of the brain-derived neurotrophic factor gene. Int. J. Dev. Neurosci., 12: 685–693.

Christianson, J.A., Riekhof, J.T. and Wright, D.E. (2003) Restorative effects of neurotrophin treatment on diabetes-induced cutaneous axon loss in mice, Exp. Neurol., 179: 188–199.

Cosgaya, J.M. and Aranda, A. (1996) Ras- and Raf-mediated regulation of transforming growth factor beta 1 gene expression by ligands of tyrosine kinase receptors in PC12 cells. Oncogene, 12: 2651–2660.

Cotsarelis, G. and Millar, S.E. (2001) Towards a molecular understanding of hair loss and its treatment. Trends Mol. Med., 7: 293–301.

Crowder, R.J. and Freeman, R.S. (1998) Phosphatidylinositol 3-kinase and Akt protein kinase are necessary and sufficient for the survival of nerve growth factor-dependent sympathetic neurons. J. Neurosci., 18: 2933–2943.

Davies, A.M., Bandtlow, C., Heumann, R., Korsching, S., Rohrer, H. and Thoenen, H. (1987) Timing and site of nerve growth factor synthesis in developing skin in relation to innervation and expression of the receptor. Nature, 326: 353–356.

Dechant, G. and Barde, Y.-A. (2002) The neurotrophin receptor p75NTR: novel functions and implications for diseases of the nervous system. Nature Neurosci., 5: 1131–1136.

Della-Bianca, V., Rossi, F.A.U., Dal-Pra, I., Costantini, C., Perini, G., Politi, V. and Della Valle, G. (2001) Neurotrophin p75 receptor is involved in neuronal damage by prion peptide-(106-126). J. Biol. Chem., 276: 38929–38933.

Di Marco, E., Marchisio, P.C., Bondanza, S., Franzi, A.T., Cancedda, R. and De Luca, M. (1991) Growth-regulated synthesis and secretion of biologically active nerve growth factor by human keratinocytes. J. Biol. Chem., 266: 21718–21722.

Di Marco, E., Mathor, M., Bondanza, S., Cutuli, N., Marchisio, P.C., Cancedda, R. and De Luca, M. (1993) Nerve growth factor binds to normal human keratinocytes through high and low affinity receptors and stimulates their growth by a novel autocrine loop. J. Biol. Chem., 268: 22838–22846.

Erck, C., Meisinger, C., Grothe, C. and Seidl, K. (1998) Regulation of nerve growth factor and its low-affinity receptor (p75NTR) during myogenic differentiation. J. Cell Physiol., 176: 22–31.

Ernfors, P., Lee, K.F. and Jaenisch, R. (1994) Target derived and putative local actions of neurotrophins in the peripheral nervous system. Prog. Brain Res., 103: 43–54.

Franke, T.F., Kaplan, D.R. and Cantley, L.C. (1997) PI3K: Downstream AKTion blocks apoptosis. Cell, 88: 435–437.

Friedman, W.J. and Greene, L.A. (1999) Neurotrophin signaling via Trks and p75. Exp. Cell Res., 253: 131–142.

Fundin, B.T., Silos-Santiago, I., Ernfors, P., Fagan, A.M., Aldskogius, H., DeChiara, T.M., Phillips, H., Barbacid, M., Yancopoulos, G.D. and Rice, F.L. (1997) Differential dependency of cutaneous mechanoreceptors on neurotrophins, trk receptors, and P75 LNGFR. Dev. Biol., 190: 94–116.

Grewe, M., Vogelsang, K., Ruzicka, T. and Krutmann, J. (1997) Human epidermal keratinocytes are a source for neurotrophic factors. J. Invest. Dermatol., 108: 574 [abstract].

Hansen, M.G., Overall, R., Arck, P.C., Paus, R. and Peters, E.M. (2003a) Nerve growth factor affects anagen-catagen transformation of cultured human anagen hair follicles, J. Invest. Dermatol., 121: 728.

Hansen, M.G., Overall, R., Schroeder, M., Pertile, P., Arck, P.C., Paus, R. and Peters, E.M.J. (2003b) Nerve Growth Factor and its receptor TrkA are expressed in the human pilosebaceous apparatus, and modulates catagen development of organ-cultured anagen scalp hair follicles. Arch. Dermatol. Res., 294: 463 [abstract].

Hardy, M.H. (1992) The secret life of the hair follicle. Trends Genet., 8: 55–61.

Harper, S. and Davies, A.M. (1990) NGF mRNA in developing cutaneous epithelium related to innervation density. Development, 110: 515–519.

Holbrook, K.A. and Minami, S.I. (1991) Hair follicle embryogenesis in the human. Characterization of events in vivo and in vitro. Ann. N.Y. Acad. Sci., 642: 167–196.

Johansson, O. and Liang, Y. (1999) Neurotrophins and their receptors in the skin: a tribute to Rita Levi-Montalcini. In: L. Aloe and G.N. Chaldakov (Eds.), Nerve Growth Factor in Health and Disease. Biomedical Reviews, Vol. 10, The Bulgarian-American Center, Varna, pp. 15–23.

Ibanez, C.F. (1998) Emerging themes in structural biology of neurotrophic factors. Trends Neurosci., 21: 438–444.

Ibanez, C.F., Ernfors, P., Timmusk, T., Ip, N.Y., Arenas, E., Yankopoulos, G.D. and Persson, H. (1993) Neurotrophin-4 is a target-derived neurotrophic factor for neurons of the trigeminal ganglion. Development, 117: 1345–1354.

Kaplan, D.R. and Miller, F.D. (2000) Neurorophin signal transduction in the nervous system. Curr. Opin. Neurobiol., 10: 381–391.

Kinkelin, I., Motzing, S.K.M. and Brocker, E.B. (2000) Increase in NGF content and nerve fiber sprouting in human allergic contact eczema. Cell Tissue Res., 302: 31–37.

Klesse, L.J. and Parada, L.F. (1999) Trks: signal transduction and intracellular pathways. Microsc. Res. Techn., 45: 210–216.

Lambiase, A., Manni, L., Bonini, S., Micera, A. and Aloe, L. (2000) Nerve growth factor promotes corneal healing: structural, biochemical, and molecular analyses of rat and human corneas. Invest. Ophthalmol. Vis. Sci., 41: 1063–1069.

Levi-Montalcini, R. (1987) The nerve growth factor 35 years later. Science, 237: 1154–1162.

Lewin, G.R. and Barde, Y.A. (1996) Physiology of the neurotrophins. Annu. Rev. Neurosci., 19: 289–317.

Lindner, G., Botchkarev, V.A., Botchkareva, N.V., Ling, G. and Paus, R. (1997) Analysis of apoptosis during hair follicle regression (catagen). Am. J. Pathol., 151: 1601–1617.

Lommatzsch, M., Braun, A., Botchkarev, V.A., Botchkareva, N.V., Paus, R., Fischer, A., Lewin, G.R. and Renz, H. (1999) Abundant production of brain-derived neurotrophic factor by adult visceral epithelia. Implications for paracrine and target-derived neurotrophic functions. Am. J. Pathol., 155: 1183–1193.

Magerl, M., Tobin, D.J., Muller-Rover, S., Hagen, E., Lindner, G., McKay, I.A. and Paus, R. (2001) Patterns of proliferation and apoptosis during murine hair follicle morphogenesis. J. Invest. Dermatol., 116: 947–955.

Maurer, M., Fische, E., Handjiski, B., Barandi, A., Meingasser, J. and Paus, R. (1997) Activated skin mast cells are involved in hair follicle regression (catagen). Lab. Invest., 319–332.

Maurer, M., Paus, R. and Czarnetzki, B.M. (1995) Mast cells as modulators of hair follicle cycling. Exp. Dermatol., 4: 266–271.

Millar, S.E. (2002) Molecular mechanisms regulating hair follicle development. J. Invest. Dermatol., 118: 216–225.

Miller, F.D. and Kaplan, D.R. (2001) Neurotrophin signaling pathways regulating neuronal apoptosis. Cell Mol. Life Sci., 58: 1045–1053.

Oro, A.E. and Scott, M.P. (1998) Splitting hairs: dissecting roles of signaling systems in epidermal development. Cell, 96: 575–578.

Paus, R. (1996) Control of the hair cycle and hair diseases as cycling disorders. Curr. Opin. Dermatol., 3: 248–258.

Paus, R. and Cotsarelis, G. (1999) The biology of hair follicles. New Engl. J. Med., 341: 491–498.

Paus, R., Luftl, M. and Czarnetzki, B.M. (1994a) Nerve growth factor modulates keratinocyte proliferation in murine skin organ culture. Br. J. Dermatol., 130: 174–180.

Paus, R., Maurer, M., Slominski, A. and Czarnetzki, B.M. (1994b) Mast cell involvement in murine hair growth. Dev. Biol., 163: 230–240.

Paus, R., Muller-Rover, S., Van Der Veen, C., Maurer, M., Eichmuller, S., Ling, G., Hofmann, U., Foitzik, K., Mecklenburg, L. and Handjiski, B. (1999) A comprehensive guide for the recognition and classification of distinct stages of hair follicle morphogenesis. J. Invest. Dermatol., 113: 523–532.

Paus, R., Peters, E.M.J., Eichmuller, S. and Botchkarev, V.A. (1997) Neural mechanisms of hair growth control. J. Invest. Dermatol. Symp. Proc., 2: 61–68.

Peters, E.M., Botchkarev, V.A., Botchkareva, N.V., Tobin, D.J. and Paus, R. (2001) Hair-cycle-associated remodelling of the peptidergic innervation of murine skin, and hair growth modulation by neuropeptides. J. Invest. Dermatol., 116: 236–245.

Peters, E.M.J., Botchkarev, V.A., Müller-Röver, S., Moll, I., Rice, F.L. and Paus, R. (2002a) Developmental timing of hair follicle and dorsal skin innervation in mice. J. Comp. Neurol., 448: 28–52.

Peters, E.M.J., Botchkarev, V.A. and Paus, R. (2003) Neurotrophins act as regulators of hair follicle morphogenesis and cycling. In: D. Van Neste (Ed.), Hair and Hair Technology, pp. 343–351.

Peters, E.M.J., Hansen, M.G., Arck, P., Pertile, P. and Paus, R. (2002b) Neurotrophins act as regulators of adult human hair growth. Arch. Dermatol. Res., 294: 18 [abstract].

Peters, E.M.J., Hansen, M.G., Overall, R.W., Nakamura, Arck, P.C., Klapp, B.F., Pertile, P. and Paus, R. (2002c)

Brain derived neurotrophic factor (BDNF) is expressed in human hair follicles and induces hair follicle regression (catagen). 9th Annual European Hair Research Society Conference, Bruxelles, Belgium: http://www.ehrs.org/conferenceabstracts/index.htm [abstract].

Peters, E.M.J., Hansen, M.G., Overall, R.W., Nakamura, M., Pertile, P., Klapp, B.F., Arck, P.C. and Paus, R. (2002d) Brain derived neurotrophic factor can induce human hair growth regression (catagen). J. Invest. Dermatol., 119: 277 [abstract].

Petersen, M., Segond von Banchet, G., Heppelmann, B. and Koltzenburg, M. (1998) Nerve growth factor regulates the expression of bradykinin binding sites on adult sensory neurons via the neurotrophin receptor p75. Neuroscience, 83: 161–168.

Philpott, M.P. and Paus, R. (1998) Principles of hair follicle morphogenesis. In: C.M. Chuong (Ed.), Molecular Basis of Epithelial Appendage Morphogenesis. Landes Bioscience Publ., Austin, TX, pp. 75–103.

Pincelli, C. (2000) Nerve growth factor and keratinocytes: a role in psoriasis. Eur. J. Dermatol., 10: 85–90.

Pincelli, C., Haake, A.R., Benassi, L., Grassilli, E., Magoni, C., Ottani, D., Polakowska, R., Franceschi, C. and Gianetti, A. (1997) Autocrine nerve growth factor protects human keratinocytes from apoptosis through its high affinity receptor (Trk): a role for Bcl-2. J. Invest. Dermatol., 109: 757–764.

Pincelli, C., Sevignani, C., Manfredini, R., Grande, Λ., Fantini, F., Bracci Laudiero, L., Aloe, L., Ferrari, S., Cossarizza, A. and Giannetti, A. (1994) Expression and function of nerve growth factor and nerve growth factor receptor on cultured keratinocytes. J. Invest. Dermatol., 103: 13–18.

Pincelli, C. and Yaar, M. (1997) Nerve growth factor: its significance in cutaneous biology. J. Invest. Dermatol. Symp. Proc., 2: 61–68.

Polakowska, R.R., Piacentini, M., Bartlett, R., Goldsmith, L.A. and Haake, A.R. (1994) Apoptosis in human skin development: morphogenesis, periderm, and stem cells. Dev. Dynam., 199: 176–188.

Qian, X., Riccio, A., Zhang, Y. and Ginty, D.D. (1998) Identification and characterization of novel substrates of Trk receptors in developing neurons. Neuron, 21: 1017–1029.

Raychaudhuri, S.P., Jiang, W.Y. and Farber, E.M. (2000) Nerve growth factor and its receptor system in psoriasis, Br. J. Dermatol., 143: 198–200.

Rohrer, H., Heumann, R. and Thoenen, H. (1988) The synthesis of nerve growth factor (NGF) in developing skin is independent of innervation. Dev. Biol., 128: 240–244.

Rossner, S., Ueberham, U., Schliebs, R., Perez-Polo, J.R. and Bigl, V. (1998) p75 and TrkA receptor signaling independently regulate amyloid precursor protein mRNA expression, isoform composition, and protein secretion in PC12 cells. J. Neurochem., 71: 757–766.

Roux, P.P. and Barker, P.A. (2002) Neurotrophin signaling through the p75 neurotrophin receptor. Progr. Neurobiol., 67: 203–233.

Sariola, H. (2001) The neurotrophic factors in non-neuronal tissues. Cell Mol. Life Sci., 58: 1061–1066.

Schoring, M., Heumann, R. and Rohrer, H. (1993) Synthesis of nerve growth factor mRNA in cultures of developing mouse whisker pad, a peripheral target tissue of sensory trigeminal neurons. J. Cell. Biol., 120: 1471–1479.

Schuman, E.M. (1999) Neurotrophin regulation of synaptic transmission. Curr. Opin. Neurobiol., 9: 105–109.

Seiberg, M., Marthinuss, J. and Stenn, K.S. (1995) Changes in expression of apoptosis-associated genes in skin mark early catagen. J. Invest. Dermatol., 104: 78–82.

Seidl, K., Erck, C. and Buchberger, A. (1998) Evidence for the participation of nerve growth factor and its low-affinity receptor (p75NTR) in the regulation of the myogenic program. J. Cell. Physiol., 176: 10–21.

Shibayama, E. and Koizumi, H. (1996) Cellular localization of the Trk neurotrophin receptor family in human non-neuronal tissues. Am. J. Pathol., 148: 1807–1818.

Stenn, K.S. and Paus, R. (2001) Control of hair follicle cycling. Physiol. Rev., 81: 449–494.

Terenghi, G., Mann, D., Kopelman, P.G. and Anand, P. (1997) TrkA and trkC expression is increased in human diabetic skin. Neurosci. Lett., 228: 33–36.

Tobin, D.J. and Paus, R. (2001) Graying: gerontobiology of the hair follicle pigmentary unit. Exp. Gerontol. 36: 29–54.

Urdiales, J.L., Becker, E., Andrieu, M., Thomas, A., Jullien, J., van Grunsven, L.A., Menut, S., Evan, G.I., Martín-Zanca, D. and Rudkin, B.B. (1998) Cell cycle phase-specific surface expression of nerve growth factor receptors TrkA and p75(NTR). J. Neurosci., 18: 6767–6775.

Vega, J.A., Vazquez, E., Naves, F.J., Del Valle, M.E., Calzada, B. and Represa, J.J. (1994) Immunohistochemical localization of the high-affinity NGF receptor (gp140-trkA) in the adult human dorsal root and sympathetic ganglia and in the nerves and sensory corpuscles supplying digital skin. Anat. Rec., 240: 579–588.

Walch, E.T., Albino, A.P. and Marchetti, D. (1999) Correlation of overexpression of the low-affinity p75 neurotrophin receptor with augmented invasion and heparanase production in human malignant melanoma cells. Int. J. Cancer, 82: 112–120.

Welker, P., Peters, E.M.J., Botchkarev, V.A., Petho-Schramm, A., Eichmuller, S. and Paus, R. (1996) Nerve growth factor and the murine hair cycle. J. Invest. Dermatol., 106: 910 [abstract].

Weskamp, G. and Otten, U. (1987) An enzyme-linked immunoassay for nerve growth factor (NGF): a tool for studying regulatory mechanisms involved in NGF production in brain and in peripheral tissues. J. Neurochem., 48: 1779–1786.

Wexler, E.M., Berkovich, O. and Nawy, S. (1998) Role of the low-affinity NGF receptor (p75) in survival of retinal bipolar cells. Visual Neurosci., 15: 211–218.

Wheeler, E.F. and Bothwell, M. (1992) Spatiotemporal patterns of expression of NGF and the low-affinity NGF receptor in rat embryos suggest functional roles in tissue morphogenesis and myogenesis. J. Neurosci., 12: 930–945.

Wheeler, E.F., Gong, H., Grimes, R., Benoit, D. and Vazquez, L. (1998) p75NTR and Trk receptors are expressed in reciprocal patterns in a wide variety of non-neural tissues during rat embryonic development, indicating independent receptor functions. J. Comp. Neurol., 391: 407–428.

Yaar, M. (1999) Neurotrophins in skin. In: M. Sieber-Blum (Ed.), Neurotrophins and the Neural Crest. CRC Press, Boston-London, pp. 117–140.

Yaar, M., Eller, M.S., DiBenedetto, P., Reenstra, W.R., Zhai, S., McQuaid, T., Archambault, M. and Gilchrest, B.A. (1994) The trk family of receptors mediates nerve growth factor and neurotrophin-3 effects in melanocytes. J. Clin. Invest., 94: 1550–1562.

Yaar, M., Grossman, K., Eller, M. and Gilchrest, B.A. (1991) Evidence for nerve growth factor-mediated paracrine effects in human epidermis. J. Cell. Biol., 115: 821–828.

Yaar, M., Zhai, S., Eisenhauer, P.B., Arble, B.L., Stewart, K.B. and Gilchrest, B.A. (2002) Amyloid beta binds trimers as well as monomers of the 75-kDa neurotrophin receptor and activates receptor signaling. J. Biol. Chem., 277: 7720–7725.

Yaar, M., Zhai, S., Pilch, P.F., Doyle, S.M., Eisenhauer, P.B., Fine, R.E. and Gilchrest, B.A. (1997) Binding of beta-amyloid to the p75 neurotrophin receptor induces apoptosis. A possible mechanism for Alzheimer's disease. J. Clin. Invest., 100: 2333–2340.

Young, H.M. (2000) Increased expression of p75NTR by neural crest-derived cells in vivo during mitosis. Neuroreport, 11: 725–728.

Zhai, S., Yaar, M., Doyle, S.M. and Gilchrest, B.A. (1996) Nerve growth factor rescues pigment cells from ultraviolet-induced apoptosis by upregulating BCL-2 levels. Exp. Cell Res., 224: 335–343.

Progress in Brain Research, Vol. 146
ISSN 0079-6123

CHAPTER 32

Nerve growth factor, human skin ulcers and vascularization. Our experience

Luigi Aloe*

Institute of Neurobiology and Molecular Medicine, National Research Council (CNR), Viale Marx 15, I-00137, Rome, Italy

Abstract: Cutaneous wound is known to elicit a series of typical cellular responses that include clotting, inflammatory infiltration, reepithelialization, the formation of granulation tissue, including new blood vessel, followed by tissue remodeling and wound contraction. The regulatory molecules implicated in these events are not well known. Neurotrophins and their receptors are trophic factors that are known to play important roles in cutaneous tissues, nerve development and reconstruction after injury. Among the neurotrophins, the nerve growth factor (NGF) was one of the earliest used for clinical studies. NGF has been tested for potential therapeutic application in neuropathies of the central and peripheral nervous system and more recently in human corneal and cutaneous ulcers. Here, I present and discuss data obtained in the last few years on the healing action of NGF in human and domestic animal skin ulcers.

Keywords: neurotrophins; NGF; skin ulcers; ocular inflammation; lameness; vasculitis; angiogenesis

Background

It has been estimated that over 4% of individuals all over the world experience suffering from burns and chronic skin ulcers. The latter is often associated with disorders such as diabetes, rheumatoid arthritis, pressure, and venous stasis. Cutaneous wound is known to elicit a series of typical cellular responses that include clotting, inflammatory infiltration, reepithelialization, the formation of granulation tissue including new blood vessels, followed by tissue remodeling and wound contraction. Growth factors are crucial cell biological mediators required for normal cell growth and cell differentiation during development and in adult life, but their success as therapeutic in wound treatment has, so far, been

limited. There is therefore a need to identify molecules with these properties.

Nerve growth factor (NGF) belongs to a family of neurotrophic factors, which include brain-derived neurotrophic factor and neurotrophin-3, -4/5, -6, -7, collectively called neurotrophins (reviewed by Sofroniew et al., 2001). It was first identified in malignant mouse sarcome 180 and 37, after their transplantation into the body wall of 3-day-old chick embryo and later in the mouse submaxillary gland, snake venom and guinea pig prostate gland (Levi-Montalcini, 1987). NGF is produced by a variety of cells of all vertebrates and released in the blood circulation in both physiological and pathological conditions (reviewed by Aloe and Micera, 1999; Aloe et al., 2001). NGF exerts a crucial role in survival, differentiation and function of peripheral sensory and sympathetic nerves and brain neurons of mammalians (Levi-Montalcini, 1987; Ebendal, 1992 Sofroniew et al., 2001). NGF is also involved in anti-inflammator responses (Banks et al., 1984; Aloe and Tuveri, 1987). These results lead to the

*Correspondence to: L. Aloe, Institute of Neurobiology and Molecular Medicine, National Research Council (CNR), Viale Marx 15, I-00137, Rome, Italy. Tel.: +39-06-8292-592; Fax: +39-06-8609-0370; E-mail: aloe@in.rm.cnr.it

DOI: 10.1016/S0079-6123(03)46032-9

hypothesis that NGF might have clinical potential application in neurodegenerative disorders (Anand, 1996; Anand et al., 1996; Apfel et al., 1998; Connor and Dragunow, 1998; Apfel, 2002). Indeed, animal and preclinical studies provided consistent evidence indicating that NGF administration may be useful to prevent the onset of neuropathy and to stimulate recovery of certain peripheral neuropathy (Donnerer et al., 1992; Anand, 1996; Apfel et al., 1998). Other studies carried out in our and other laboratories have recently demonstrated that NGF can also have therapeutic potentiality in human corneal and skin ulcers.

Aim of the present review is to report experimental and clinical data obtained in our laboratory in collaboration with clinicians demonstrating that topical application of NGF exerts healing action on human corneal ulcer, ocular inflammation, diabetes and rheumatoid arthritis-associated ulcers, pressure ulcer, and also ulcers of domestic animals.

NGF preparation and ethical issue

The NGF used for our studies was purified from submaxillary glands following Bocchini and Angeletti's methodology (1969). The purified molecule was dissolved, just before using in physiological solution (0.9% NaCl) at a concentration of 200 µg/ml and kept at 4°C, for no more than a week. The ulcers were treated with topical application of 50 µl of this solution 3–4 times a day or even more, depending on the size and the condition of the lesion. As for the ethical issue, all patients enrolled for these studies were on compassional basis, and only after all the available therapies failed to promote healing, patients were treated with NGF. All these studies were approved by the Ethical Committee of the hospital, where the patients were hospitalized; before treatment, written informed consent was obtained from each patient.

NGF treatment of human corneal ulcers

Human corneal neurotrophic ulcer is an ocular disorder which can be caused by a variety of endogenous and exogenous insults and can lead to blindness. The patients who come to our attention for possible application of NGF developed corneal ulcer as results of neurotrophic keratitis, chemical burns, topical anesthetic abuse, following surgery for orbital tumor, or acoustic neurinoma, after penetrating keratoplasty for unknown cause, and after lamellar keratoplasty for herpetic vascularized scar (Beuerman and Schimmelpfenning, 1980; Tervo et al., 1994; Mackie, 1995; Wilson and Kim, 1998). All ulcers were associated with impairment of corneal sensitivity. Before beginning the treatment with NGF, all patients had received conventional treatment, including artificial tears, antibiotics, eye patching, and soft contact lens bandaging, with little or no benefit. The criterion for enrollment of the patients in the study was: clinical evidence of corneal ulcer unresponsive to conventional therapy. The exclusion criteria were the presence of corneal infection and the presence of other ocular diseases.

A positive action of NGF was detectable few days after NGF application. The patients displayed a complete resolution of the corneal ulcer after 10 days to 6 weeks of NGF treatment. At this stage the dosage of the neurotrophin was reduced and 2 weeks later discontinued. Structurally, the first signs of healing were an advancement of epithelial cells from the margin toward the center of the ulcer and mild to moderate conjunctiva hyperemia. During epithelial proliferation, all patients had photophobia and burning of their eyes during slit lamp examinations, and most patients had improvement of corneal contact sensitivity that suggested functional recovery of corneal innervation. During the first week of treatment, some patients complained of a transient burning sensation. Superficial or deep corneal neovascularization in over 50% of NGF-treated patients was documented. All ocular discomfort disappeared when the corneal ulcer was completely healed (Fig. 1 A, B). The rate of healing was not associated with the severity of the ulcer or to its depth in the stroma, nor with the age of the patients or the cause of the ulcer. All the corneal ulcer treated so far ($n = 65$) with NGF healed completely.

The observation that exogenous NGF restored corneal integrity and sensitivity suggests that the progressive corneal damage that occurs in patients with corneal sensory nerve deficits was due to a deficit of endogenous NGF synthesis, release and/or utilization.

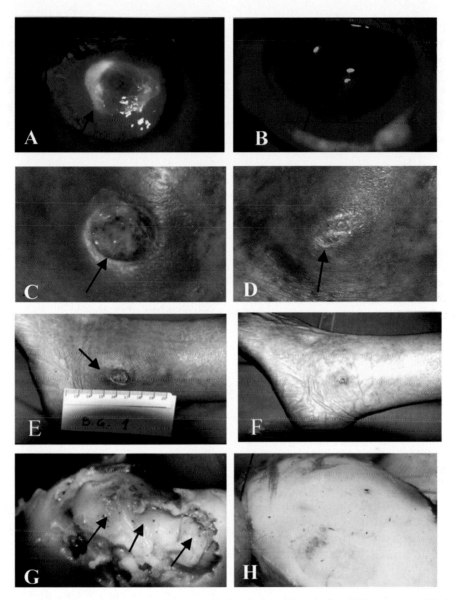

Fig. 1. An illustration of the effect of NGF on human corneal ulcer before (A) and after NGF treatment (B); human rheumatoid artritis-associated skin ulcer before (C) and after NGF treatmnet (D); human diabetic skin ulcer before (E) and after NGF treatment (F); goat sole (in lameness) before (G) and after NGF treatment (H). Arrows point to the damaged (A, C, E, G) and healed (B, D, F, H) tissues.

To further characterize the effect of NGF and understand the mechanism(s) implicated in corneal wound healing, we used animal models of corneal epithelial wound healing. Epithelial corneal debridement was performed in animals using n-heptanol and mechanical removal and the level of NGF measured at various time point postinjury and the effect of exogenous NGF or anti-NGF antibodies on corneal wound healing monitoring daily. Human and rodent corneas were also exposed in vitro to radiolabeled NGF, to assess the ability of damaged corneal cells to respond to the action of NGF. The results of these studies, reported in more details elsewhere (Lambiase et al., 1997,

1998a,b), clearly indicated that corneal cells expressed NGF receptors and that their interaction with NGF can promote tissue repair and complete healing.

Ocular inflammation

Immune corneal ulcers are rare ocular surface diseases with multiple etiologies (Donzis and Mondino, 1987). Immunosuppressive drugs and systemic or topical steroids are occasionally given to reduce and/or arrest the inflammatory process. This type of inflammation can lead to corneal melting and perforation. Four patients affected by ocular inflammation unresponsive to conventional therapy were selected for NGF treatment. All were affected with severe corneal melting, as consequences of immune related peripheral corneal ulcers. Topical treatment with steroids and systemic immunosuppressive therapy failed to heal these corneal ulcers. The first patient treated with NGF was a 45-year-old woman affected by lung tuberculosis, under treatment with ethambutol, rifampin and pyrazinamide, developed erythema, small leg ulcers and bilateral peripheral corneal ulcers associated with scleritis; the second patient was a 56-year-old male affected by rheumatoid arthritis with recurrent acute episodes of keratoconjunctivitis and blepharitis which were responsive to topical steroids, developed a peripheral corneal ulcer and scleritis in the left eye; the third patient was a 62-year-old male affected by recurrent scleritis responsive to topical steroids, developed a peripheral corneal ulcer associated with scleritis; and the fourth patient was a 63-year-old woman who underwent cataract extraction and after two months developed necrotizing scleritis with subsequent exposition of choroid. After two weeks of NGF treatment, all four immune-associated corneal ulcers healed completely and three weeks later the ocular inflammation disappeared. The only side effect observed during NGF treatment was transient local pain and photophobia which preceded the healing process but disappeared once complete ocular healing occurred (Lambiase et al., 1998b). No relapse of the disease was observed in any patients during a 7-year follow-up.

NGF was also applied in four dogs with dry eye, a chronic ocular surface disease due to impairment of tear film quantity and/or quality. Topical application of NGF at the ocular surface induced complete healing after three weeks. The NGF treatment enhanced corneal transparency and tear film production, leading to improvement of visual function (N. Costa et al., in submitted).

Rheumatoid arthritis- and systemic sclerosis-associated human skin ulcers

Vasculitis is a cutaneous disorder often associates with rheumatoid arthritis (RA) and systemic sclerosis (SSc). It is characterized by skin inflammation, damage of peripheral blood vessels and local tissue necrosis (Dahn, 1998; Tuveri et al., 2000). Four patients with RA-associated skin ulcers and four patients with SSc-associated skin ulcers, both being unresponsive to the available treatments, were selected for topical treatment with NGF. The vasculitis of the four RA patients showed a rapid improvement of their leg ulcers during the first 2–3 weeks of NGF application. The NGF treatment caused a significant size reduction of the ulcer as well as reduction of descriptive parameters such as pain, presence of granulation, absence of inflammation. Thereafter, the ulcer showed a slower but progressive improvement reaching complete healing after eight weeks of NGF treatment (Fig. 1 C, D).

The vasculitis of patients affected by SSc displayed a marked decrease of inflammation and pain during the first and second week of NGF treatment. Thereafter the healing process was blocked. Why the vasculitis induced by SSc responded differently is, at present, not known. One possible explanation is that the progressive microvascular fibrotic evolution, which characterizes the vasculitis in SSs, reduces the access of NGF to damaged cell layer (see also Tuveri et al., 2000).

Diabetes-associated human skin ulcers

It is known that skin ulcers associated primary or secondary to diabetes mellitus can become chronic and difficult to heal (Anand, 1996; Anand et al., 1996; Moulin et al., 1998; Adler et al., 1999). Early studies on animal models revealed that application of NGF into the wounds accelerated the rate of wound healing in normal mice and in healing-impaired diabetic mice (Matsuda et al., 1998). Based on this and our evidence with corneal ulcer, we treated a 61-year-old woman, affected by diabetes mellitus and progressive ulceration of her left lateral malleolus for over six years. Before the application of NGF, fibrin or scab were mechanically removed, the ulcer was cleaned with sterile saline solution and then wiped off with sterile gauze. Cleaning the ulcer by removing the dead tissues did cause neither bleeding nor pain to the patients, suggesting that in the proximity of the ulcer, nerve fibers and blood vessels were nearly absent. The ulcer was very deep and at the end the bone underneath was readily visible after cleaning. Most probably due to a particular chronic condition of the ulcer, the healing response to NGF of this type of ulcer was rather slow, as compared to other NGF-treated ulcers. Indeed, the first clear reduction of the size of the ulcer was noted after eight weeks of NGF treatment. After eight weeks, bleeding and pain sensation began to develop and the bone was no more visible, suggesting that NGF promoted the proliferation and/or differentiation of local immature cells as well, blood vessel formation and neurite outgrowth processes. Skin healing continued to improve and NGF application was stopped after 24 weeks, when the ulcer appeared completely healed. No relapse was observed after over two years of follow up.

Human pressure ulcers

Human pressure ulcer is a relevant cause of morbidity in older people and is one of the most important care problem in nursing home residents (Smith, 1995). Patients with pressure ulcer, characterized by a very slow response to local treatment with scarce tendency to heal, were selected for NGF treatment. Pressure ulcers localized at the lower portion of the foot are very frequent and more difficult to heal (Smith, 1995; Flour and Degreef, 1998; Kantor and Margolis, 1998). We treated two groups of these patients. One group of patients ($n = 8$) received NGF treatment while a second group ($n = 6$) received only conventional topical treatment. Daily observations revealed

that only the ulcers treated with topical application of NGF showed a significant acceleration of healing process. More specifically, after about two weeks of NGF treatment, advancement of epithelial tissue from the margin toward the center of the ulcer was clearly visible and four weeks after treatment, all the ulcers were reduced by nearly 50% in total area (Fig. 1 E, F). The lesions in the NGF-treated group displayed a marked peripheral scar indicating that the healing process was taking place. Moreover, it was found that the rate of recovering in the NGF-treated group was not related to the severity of the ulcer, the age of the patient, or the site of the ulcer. Only few ulcers (less than 10%) of patients who received the conventional treatment displayed a significant, but a slow recovery, while the majority of them failed to heal after three months of NGF treatment, which was the last time-point studied (Bernabei et al., 1999).

Effect of NGF on ischemic skin revascularization

About a year ago, we faced the uneasy decision to test on a young patient the effect of NGF in angiogenesis obtained in animal models. We had reported that NGF can influence angiogenic activity in a mouse model of foot ischemia (Emanueli et al., 2002; Turrini et al., 2002) and hypothesized that NGF might be able to stimulate growth of damaged blood vessels in humans. The decision was taken when a 5-year-old child was referred to the Pediatric Intensive Care Unit of the Catholic University Medical School, Rome, Italy, with crush syndrome following a road traffic accident. At the hospital, the child was diagnosed as having hypovolaemic shock as a result of splenic and left femoral fracture. The left lower limb appeared severely hypoperfused and marbled. Spontaneous mobility was absent even after application of painful stimuli and no arterial pulses were left. The left foot was generally edematous with peripheral ischemic zone. A Doppler ultrasound scan failed to detect any arterial flow at the level of the left popliteal artery. Flow reduction of the superficial femoral artery was shown with a good collateral circulation, while superficial femoral artery was interrupted. Limb perfusion did not improve and two days after fasciotomy, the lesion of the heel and toes worsened, with likely extensive foot necrosis with an impending risk of massive foot ischemia. At this stage, after obtaining approval of the University's Review Board and parents' informed consent, it was decided to treat the patient with local application of NGF to stimulate peripheral innervation and blood vessel formation.

We took the decision to treat the young patients with NGF after the failure of fasciotomy, and realize that there was a serious impending risk of a massive foot ischemia. The patient was treated daily with 50 µl of NGF at concentration of 200 µg/ml, in the calcaneal region, between healthy skin and ischemic area of the heal, because of a clearer evidence of a border to the ischemic area. The effect of NGF and the patient response to the treatment was followed closely. The daily observation revealed that the lesion close to the NGF injected site gradually improved, while in the untreated areas necrosis remain unchanged or worsened. No side effects were observed during NGF treatment. NGF infiltration was associated with an improvement of vascularization in a broad area of the heel, starting 4–5 days after the first NGF administration. After a week's treatment the necrotic evolution ceased. Moreover, NGF application, caused a marked reepithelization, without any scar formation. Three weeks later, the young patient was discharged in good clinical condition, after a femoral osteosynthesis (Glasgow Outcome Score 5).

Because this is the first known case of the NGF effect on human crush syndrome (skin ischemia), the question can be asked whether in the absence of NGF treatment, spontaneous recovery would have occurred. Though it is not possible to exclude this possibility, it should stress the point that the decision was taken after realizing that all other conventional therapies failed and a clinical evaluation to avoid the possibility of massive foot ischemia and the severe risk of foot amputation. It is also worth mentioning that previous findings obtained in animal models showed that NGF was able to promote, most probably acting on endothelial cells, neovessel formation (Samii et al., 1999; Calza et al., 2001; Emanueli et al., 2002; Turrini et al., 2002). Certainly, further cellular, biochemical and molecular studies need to be done to identify and characterize the effect and mechanism of NGF on the process of angiogen-

esis, before the clinical utilization of NGF on crush syndrome can be taken into consideration. The evidence that NGF can promote arteriolar formation and epithelial healing seems to suggest that this and/or other neurotrophins of the NGF family might be helpful to devise therapeutic strategies for burns and other ischemic skin lesions, if other therapies fail or are not available (see also Chiaretti et al., 2002).

Skin ulcers of domestic animals

The healing action of NGF on skin ulcer was also investigated on lameness, foot ulcers in farm animals. Lameness is a skin disorder affecting numerous farm animals, characterized by cutaneous ulcers, abscesses and granulomas in the interdigital and sole area of the anterior and posterior feet, causing foot pain and impairing locomotor activity (Dewes, 1978; Manson and Leaver, 1988; Hill et al., 1997). Several factors, including housing and flooring, nutritional, environmental, genetic and contagious factors are believed to be involved in the pathogenesis of lameness.

We have used 16 adult female goats with lameness, suffering with this skin ulcer for at least two months, unable to respond to the available therapies. We used the same protocol as used for human skin ulcers. It was found that NGF induced complete healing action of the skin ulcer in all goats treated with NGF, while no signs of recovery were observed in NGF untreated goats during the same period of treatment. The action of NGF was first observed about a week after NGF application, while complete ulcer healing occurred 3–4 weeks after the beginning of NGF treatment, see Fig. 1 G, H (Costa et al., 2002).

Concluding remarks

Skin ulcers are known to be associated with structural and functional damage of peripheral tissues, including altered sensory and sympathetic peripheral innervation, activation of inflammatory cells, also keratinocyte and fibroblast ability of releasing, in the wound space, various growth factors such as platelet-derived growth factor, epidermal growth factor, fibroblast growth factor, transforming growth factor-beta, and NGF (Li et al., 1980; Di Marco et al., 1991; Rothe and Falanga, 1992; Moulin, 1995; Waldorf and Fewkes, 1995; Matsuda et al., 1998; Moulin et al., 1998; Botchkareva et al., 2000). The results reported in this short review indicate that NGF plays an important role in healing of both ocular and skin ulcers in humans, as well as skin ulcers in domestic animals. Moreover, one relevant question raised by our findings regards the specific role of NGF in the mechanism(s) of wound healing. It has been shown that NGF can be secreted by fibroblasts/myofibroblasts, epithelial cells, keratinocytes and immune cells and that these cells, along with the endothelial cells and peripheral sensory and sympathetic nerves, have the ability, though in a different time and manner, to respond to NGF stimuli. Based on these and other available experimental and clinical evidences, one can reasonably hypothesise that NGF, via an autocrine and/or paracrine mechanism, acts on local undifferentiated cells (most likely stem cells). And, thus can stimulate and/or accelerate reepithelialization and tissue remodeling, most probably associated with blood vessel formation, leading ultimately to ulcer healing.

These findings and others, presented at the 7th International Conference on NGF and Related Molecules held in Modena in May 2002, raised a reasonable hope that in the near future, neurotrophic factors, including NGF, might find a therapeutic application not only for peripheral and brain neuropathies, but also for human ocular and skin ulcers of various etiologies.

Acknowledgments

These studies were in part supported by Progetto Strategico FISR/Neurobiotecnologie, MURST-CNR (grant nr. 1-2001-877), and Juvenile Diabetes Research Foundation International.

Abbreviations

NGF	nerve growth factor
RA	rheumatoid arthritis
SSc	systemic sclerosis

References

Adler, A.I., Bojko, E.J., Ahroni, J.H. and Smith, D.G. (1999) Lower-extremity amputation in diabetes. The independent effects of peripheral vascular disease, sensory neuropathy, and foot ulcer. Diabetes Care, 22: 1029–1035.

Aloe, L. and Tuveri, M.A. (1997) Nerve growth factor and autoimmune rheumatic diseases. Clin. Exp. Rheum., 15: 433–438.

Aloe, L. and Micera, A. (1999) Nerve growth factor: basic findings and clinical trials. In: Aloe L. and Chaldakov G.N. (Eds.), Nerve Growth Factor in Health and Disease, Biomedical Reviews, Vol. 10, The Bulgarian-American Center, Varna, pp. 3–14.

Aloe, L., Tirassa, P. and Bracci-Laudiero, L. (2001) Nerve growth factor in neurological and non-neurological diseases: basic findings and emerging pharmacological prospectives. Curr. Pharm. Des., 7: 113–123.

Anand, P. (1996) Neurotrophins and peripheral neuropathy. Philos. Trans. R. Soc. Lond. B. Biol. Sci., 351: 449–454.

Anand, P., Terenghi, G., Warner, G., Kopelman, P., Williams-Chestnut, R.E. and Sinicropi, D.V. (1996) The role of endogenous nerve growth factor in human diabetic neuropathy. Nat. Med., 2: 703–707.

Apfel, S.C. (2002) Nerve growth factor for the treatment of diabetic neuropathy: what went wrong, what went right, and what does the future holds. Int. Rev. Neurobiol., 50: 393–413.

Apfel, S.C., Kessler, J.A., Adornato, B.T., Litchy, W.Y., Sanders, C. and Rask, C.A. (1998) Recombinant human nerve growth factor in the treatment of diabetic polyneuropathy. Neurology, 51: 695–702.

Banks, B.E.C., Vernon, C.A. and Warner, J.A. (1984) Nerve growth factor has anti-inflammatory activity in the rat hindpaw oedema test. Neurosci. Lett., 47: 41–45.

Bernabei, R., Landi, F., Bonini, S., Onder, G., Lambiase, A., Pola, R. and Aloe, L. (1999) Effect of topical application of nerve-growth factor on pressure ulcers. Lancet, 24: 307.

Beuerman, R.W. and Schimmelpfenning, B. (1980) Sensory denervation of the cornea affects epithelial properties, Exp. Neurol., 69: 196–201.

Bocchini, V. and Angeletti, P.U. (1969) The nerve growth factor: purification as a 30,000-molecular-weight protein. Proc. Natl. Acad. Sci. USA, 64: 787–794.

Botchkareva, N.V., Botchkarev, V.A., Albers, K.M., Metz, M. and Paus, R. (2000) Distinct roles for nerve growth factor and brain-derived neurotrophic factor in controlling the rate of hair follicle morphogenesis. J. Invest. Dermatol., 114: 314–320.

Chiaretti, A., Piastra, M., Caresta, E., Nanni, L. and Aloe, L. (2002) Improving ischaemic skin revascularisation by nerve growth factor in a child with crush syndrome. Arch. Dis. Chil., 87: 446–448.

Calza, L., Giardino, L., Giuliani, A., Aloe, L. and Levi-Montalcini, R. (2001) Nerve growth factor control of neuronal expression of angiogenetic and vasoactive factors. Proc. Natl. Acad. Sci. USA, 98: 4160–4165.

Connor, B. and Dragunow, M. (1998) The role of neuronal growth factors in neurodegenerative disorders of the human brain. Brain Res. Brain Res. Rev., 27: 1–39.

Costa, N., Fiore, M. and Aloe, L. (2002) Healing action of nerve growth factor on lameness in adult goats. Ann. Ist. Super. Sanità, 38: 187–194.

Dahn, M.S. (1998) The role of growth factors in wound management of diabetic foot ulcers. Fed. Pract., 7: 14–19.

Dewes, H.F. (1978) Some aspects of lameness in dairy herds. N. Zeal. Vet. J., 26: 147–148.

Di Marco, E., Marchisio, P.C., Bondanza, S., Franzi, A.T., Cancedda, R. and De Luca, M. (1991) Growth regulated synthesis and secretion of biologically active nerve growth factor by human keratinocytes. J. Biol. Chem., 266: 21718–21722.

Donnerer, J., Schuligoi, R. and Stein, C. (1992) Increased content and transport of substance P and calcitonin gene-related peptide in sensory nerves innervating inflamed tissue: evidence for a regulatory function of nerve growth factor in vivo. Neuroscience, 49: 693–698.

Donzis, P.B. and Mondino, B.J. (1987) Management of noninfectious corneal ulcers. Surv. Ophthalmol., 32: 94–110.

Ebendal, T. (1992) Function and evolution in the NGF family and its receptors. J. Neurosci. Res., 32: 461–470.

Emanueli, C., Salis, M.B., Pinna, A., Graiani, G., Manni, L. and Madeddu, P. (2002) Nerve growth factor promotes angiogenesis and arteriogenesis in ischemic hindlimb. Circulation, 106: 2257–2262.

Flour, M. and Degreef, H. (1998) Pharmacological treatment of wounds. Sem. Cutan. Med. Surg., 17: 260–265.

Kantor, J. and Margolis, D. (1998) Efficacy and prognosis value of simple wound measurements. Arch. Dermatol., 134: 1571–1574.

Hill, N.P., Murphy, P.E., Nelson, A.J., Mouttotou, N., Green, L.E. and Morgan, K.L. (1997) Lameness and foot lesions in adult British dairy goats. Vet. Rec., 141: 412–416.

Lambiase, A., Centofanti, M., Micera, A., Manni, L., Mattei, E., De Gregorio, A., de Feo, G., Bucci, M.G. and Aloe, L. (1997) Nerve growth factor (NGF) reduces and NGF antibody exacerbates the retinal damage induced in rabbit by experimental ocular hypertension. Graefes Arch. Clin. Exp. Ophthalmol., 235: 780–785.

Lambiase, A., Rama, P., Bonini, S., Caprioglio, G. and Aloe, L. (1998a) Topical treatment with nerve growth factor for corneal neurotrophic ulcers. N. Engl. J. Med., 23: 1174–1180.

Lambiase, A., Bonini, S., Micera, A., Rama, P., Bonini, S. and Aloe, L. (1998b) Expression of nerve growth factor receptors on the ocular surface in healthy subjects and during manifestation of inflammatory diseases. Invest. Ophthalmol. Vis. Sci., 39: 1272–1275.

Levi-Montalcini, R. (1987) The nerve growth factor 35 years later. Science, 237: 1154–1162.

Li, A.C.K., Koroly, M.J., Schattenkerk, M.E., Malt, R.A. and Young, M. (1980) Nerve growth factor: acceleration of the rate of wound healing. Proc. Natl. Acad. Sci. USA, 77: 4379–4381.

Mackie, I.A. (1995) Neuroparalytic keratitis. In: F. Fraunfelder and F.H. Roy (Eds.), Current Ocular Therapy, 4th Ed., WB Saunders, Philadelphia, pp. 506–508.

Manson, F.J. and Leaver, J.D. (1988) The influence of concentrate amount on locomotion and clinical lameness in dairy cattle. Animal Prod., 47: 185–190.

Matsuda, H., Koyama, H., Sato, H., Sawada, J., Itakura, A., Tanaka, A., Matsumoto, M., Konno, K., Ushio, H. and Matsuda, K. (1998) Role of nerve growth factor in cutaneous wound healing: accelerating effects in normal and healing-impaired diabetic mice. J. Exp. Med., 187: 297–306.

Moulin, V., Lawny, F., Barritault, D. and Caruelle, J.P. (1998) Platelet releasate treatment improves skin healing in diabetic rats through endogenous growth factor secretion. Cell Mol. Biol., 44: 961–971.

Moulin, V. (1995) Growth factors in skin wound healing. Eur. J. Cell Biol., 68: 1–7.

Rothe, M. and Falanga, V. (1992) Growth factors and wound healing. Clin. Dermatol., 9: 553–555.

Samii, A, Unger, J. and Lange, W. (1999) Vascular endothelial growth factor expression in peripheral nerves and dorsal root ganglia in diabetic neuropathy in rats. Neurosci. Lett., 262: 159–162.

Smith, D.M. (1995) Pressure ulcers in the nursing home. Ann. Intern. Med., 123: 433–442.

Sofroniew, M.V., Howe, C.L. and Mobley, W.C. (2001) Nerve growth factor signaling, neuroprotection, and neural repair. Annu. Rev. Neurosci., 24: 1217–1281.

Tervo, K., Latvala, T.M. and Tervo, T.M. (1994) Recovery of corneal innervation following photorefractive keratoablation. Arch. Ophthalmol., 112: 1466–1470.

Turrini, P., Gaetano, C., Antonelli, A., Capogrossi, M.C. and Aloe, L. (2002) Nerve growth factor induces angiogenic activity in a mouse model of hindlimb ischaemia. Neurosci. Lett., 323: 109–112.

Tuveri, M., Generini, S., Matucci-Cerinic, M. and Aloe, L. (2000) NGF, a useful tool in the treatment of chronic vasculitic ulcers in rheumatoid arthritis. Lancet, 356: 1739–1740.

Waldorf, H. and Fewkes, J. (1995) Wound healing. Adv. Dermatol., 10: 77–96.

Wilson, S.E. and Kim, W.J. (1998) Keratocyte apoptosis: implications on corneal wound healing, tissue organization, and disease. Invest. Ophthalmol. Vis. Sci., 39: 220–226.

Overview and conclusion

Progress in Brain Research, Vol. 146
ISSN 0079-6123

CHAPTER 33

The nerve growth factor and the neuroscience chess board[1]

Rita Levi-Montalcini*

Institute of Neurobiology and Molecular Medicine, CNR, Viale Marx 15, I-00137, Rome, Italy

The date of 10 December 1986 marked the end of the errant life of the nerve growth factor (NGF) and its official recognition by the scientific community. The ceremony held in Stockholm on the occasion of its presentation to the Swedish royal family brought it to the attention of both biologists and nonscientists. Its solemn recognition as the firstborn of a constantly growing group of endogenous specific growth factors and as a precursor of the oncogens brought it to the attention also of researchers working in other scientific sectors, particularly in biochemistry, molecular biology, genetics and immunology. One constant feature of the long and tortuous path followed by NGF ever since, half a century earlier, had appeared on the biological scene, was the fairytale and adventurous nature of the path it followed. NGF was to open up increasingly broad vistas and horizons as we strove to find a place for it in the framework of neuroscience.

When, in the winter of 1951 the effects obtained by grafting a malignant tumor on to chick embryos were reported for the first time at a neuroembryology congress, the results aroused more perplexity than interest.

The discovery that nerve fibers could cross barriers, such as vein tunics, that normally barred their access, penetrate the circulatory system, and at the same time ramify chaotically in embryonic organs, was considered to be an abnormal phenomenon. This was because it reflected and extended to nervous tissue components a property typical of neoplastic cells from which were derived the hitherto unidentified substance synthesized and released by the cells themselves.

The use of in vitro culture techniques led to the discovery that the submaxillary gland of the adult male rate synthesizes and releases into circulation a protein molecule that is exactly the same as the one produced by tumors that is ten thousand times more active than that identified.

The discovery that NGF plays an essential role in promoting the differentiation of nerve cells receptive to its action suggested that other similar factors might exert a proliferative or differentiating action on both neuronal and nonneuronal populations.

The demonstration that NGF specific antibodies caused the death of the nerve cells receptive to its action represented undeniable proof of the essential role played by this molecule. Many years later its action mechanism was to be interpreted differently, that is, that the activation of the programmed death gene is normally inhibited by the action of the NGF gene.

Studies carried out in 1971 revealed the sequence of amino acids contained in NGF, i.e., its primary

---

[1]This text was previously published in the Archives Italiennes de Biologie, Volume 141(2–3): pp. 85–88.
*Correspondence to: R. Levi-Montalcini, Institute of Neurobiology and Molecular Medicine, CNR, Viale Marx 15, I-00137 Rome, Italy.
Tel.: +39-06-8609-0510; Fax: +39-06-8609-0370;
E-mail: r.levimontalcini@in.rm.cnr.it

DOI: 10.1016/S0079-6123(03)46033-0

structure. This demonstration then led to the discovery of the NGF gene.

The discovery that a neoplastic cell known as PC12, derived in vitro from a rat pheochromocytoma, responded to NGF by a mitotic arrest, taking on the phenotypic property of an adrenergic nerve cell, allowed the NGF action mechanism to be studied at the subcellular and molecular level.

In the 1970s and 1980s, it was discovered that the role played by NGF is likewise essential for primary or secondary cells of the immune system (mastocytes, T- and B-lymphocytes, macrophages and others) and endocrine system cells (hypophysis, thyroid and other endocrine glands).

At the same time it was discovered that the activity of the NGF molecule was not restricted to peripheral nervous system cells but extended also to cholinergic type central nervous system cells involved in cognitive (neocortical system) and emotional and affective (limbic system) activities. All the neuronal and nonneuronal cells receptive to NGF action in the above systems are subjected to programmed death on being deprived of the NGF molecule.

In the nineties, a low-affinity receptor, p75, and a high-affinity receptor tyrosine kinase, TrkA, for NGF were identified.. This opened up the possibility of identifying populations belonging to different cell lines in both normal and diseased conditions.

Research using animal models and clinical trials revealed the important role performed by the NGF and NGF-like molecules in certain forms of disorder of a neurodegenerative, inflammatory and autoimmune nature.

What position is occupied by the NGF molecule in the field of the neuroscience today?

In 1975, on the occasion of an MIT conference, "I said that despite all its extraordinary exploits, NGF had still not found its proper place on the neuroscience chess board" (Levi Montalcini, 1975). Today, nearly half a century after its discovery, the role of this molecule may be likened to that of a 'pawn' on a real chess board.

In the arrangement of the various pieces on the neuroscience chess board, genes are placed hierarchically. During early embryonic development, the genes known as homeotypic direct the formation and activity of individual body parts: among these of particular interest, from an anthropocentric point of view, are the activator genes in the neocortical and limbic system (like the King and Queen on a real chess board). Next on the hierarchical scale are the genes involved in the activity of the subcortical system (thalamus, hippo- thalamus, cerebellum and medulla oblongata) comparable to the Bishops, Rooks and Knights in the game of chess. Lastly, in the neuroscience chess board, the genes involved in functions which were considered in the past of second order such as that of the 'Pawns' of the real chess board, program the release of chemical substances (cytokines, endorphins, hormone factors, growth factors, neurotransmitters, etc.) which intervene in the complicated nervous and nonnervous functions of all vertebrates, from the lowest to the highest phylogenetic level.

One of these is the NGF which has not only found its rightful place on the neuroscience chess board but, at the same time, also led to the recognition of the role played by the other factors.

A deregulation of the role of 'pawns' in the chess board of the brain can be harmful. Their function can vary from defensive to offensive not on a topographical basis (that is, by proximity or remoteness) but according to the environmental conditions and to the complex of factors at work in the processes in which they are involved. When overactivated, they can have a harmful effect on the 'pieces' they are subject to or serve.

Another difference in the role displayed by the pawns in the real chess board and in the neuroscience chess board resides in the fact that 'cerebral pawns' differ in their activity from one another and that their strength depends on their interaction together continuously. Their activity has a strong modulating effect on the activity performed by all the 'pieces' on the board. One particularly important activity is that exerted on the nervous components having cognitive and emotive functions. The 'pawns' have the merit of having incorporated the nervous system in the body complex from which it had been previously excluded and of providing a more correct definition of the tasks assigned to them.

This reappraisal of Pawns in the real chess game was introduced more than two centuries ago, as Philidor claims, and was recently confirmed by the great Russian chess player Nimzowitsch (1989): '… the Pawn has another virtue compared with the pieces, and that is, that it is a born defender, we discover little by little that on all 64 squares a Pawn deserves all our respect... Who protects its own piece most securely? The Pawn. And who works at the lowest cost? Again the Pawn, since the pieces do not possess its capacity for work...'

Likewise on the neuroscience chess board, the discovery of the working capacity of 'pawns' opened up a new chapter in our knowledge of the mechanisms underlying not only homeodynamic functions but also the processes involving the whole organism.

The nervous system, the endocrine and the immune system, autonomously operating units, are no longer considered as elements of continuous interchange, but integrated through morphological and functional relations: they are parts of a single network regulated by mutual communication among the systems, and not through independent cell complexes as was believed in the past.

The evidence that NGF circulates without any limitation of organ or structure, nor of system boundaries, assigns to this molecule a key role in the body's global homeostasis.

The results achieved in recent decades have allowed great leaps forward in the process of our understanding of the physiological mechanisms underlying the health of man. These results were attained by combining the analytical method with a holistic approach in which individual parts are investigated in order to obtain an overall view. You cannot appreciate a mosaic by concentrating on individual tesserae; the most detailed analysis of the parts cannot in fact tell us anything about the overall view.

This basic philosophy gave rise to a new experimental approach designed to connect, combine and associate, just as the systems at work to maintain the body's equilibrium are connected, combined and associated.

These roughly outlined stages of development must not be considered as a 'list of independent facts' but as a logical sequence of events linked together by the thread of the need to elucidate properties or functions that gradually come to our knowledge.

The NGF provides confirmation of the difficulty of predicting at the outset of the research what the later developments will be and the contribution it will make to the field of investigation and neighboring fields. The studies that were to lead to the discovery of the NGF were originally aimed at investigating the role played by peripheral tissues and organs on the nervous centers innervated by them in the spinal cord. This was an emerging problem that was viewed at the time in the vast field of the neuroscience as 'peripheral' in both meanings of the term. First, because the active cells have a peripheral location and function. Second, because these cells have tasks that are relatively insignificant compared with those of the central nervous system, which are involved in cognitive functions: thinking, memory, creativity and in the complex psycho-emotional functions.

The research forming the subject of the present conference opens up new chapters having a huge potential for further development of the study of neuroendocrineimmune circuits involved in the homeostatic equilibrium of living beings.

The NGF saga, presented as a paradigm of the stages in the gradual development of scientific research, followed a winding and imperfect itinerary. This is a further evidence that it is imperfection and not perfection that lie on the basis of human endeavor.

References

Levi Montalcini, R. (1975) An uncharted route. In: F.G. Worden, J.P. Swazey and G. Adelman (Eds.), The Neurosciences Paths of Discovery, Cambridge, Mass., pp. 245–265.

Nimzowitsch, A. (1989) Il mio sistema, Collana "I giochi". Mursia, Milano.

Subject Index